DYNAMIC SYSTEMS BIOLOGY MODELING AND SIMULATION

DYNAMIC SYSTEMS BIOLOGY MODELING AND SIMULATION

Joseph DiStefano III

ELSEVIER

AMSTERDAM • BOSTON • HEIDELBERG • LONDON
NEW YORK • OXFORD • PARIS • SAN DIEGO
SAN FRANCISCO • SINGAPORE • SYDNEY • TOKYO

Academic Press is an imprint of Elsevier

AP

Academic Press is an imprint of Elsevier
32 Jamestown Road, London NW1 7BY, UK
225 Wyman Street, Waltham, MA 02451, USA
525 B Street, Suite 1800, San Diego, CA 92101-4495, USA

First edition 2014

Revised version, January 2015

Cover graphic credit to Allegra DiStefano

British Library Cataloguing-in-Publication Data
A catalogue record for this book is available from the British Library

Library of Congress Cataloging-in-Publication Data
A catalog record for this book is available from the Library of Congress

ISBN: 978-0-12-410411-2

For information on all Academic Press publications visit our website at elsevierdirect.com

Typeset by MPS Limited, Chennai, India
www.adi-mps.com

"For anyone interested in kinetic modeling of substances in physiological systems, Chapter 8 *is a must read. It so elegantly covers the three important areas the reader is likely to encounter in a wider reading of the subject, namely physiologically-based models of organs and whole-body systems, multicompartmental models applied to such systems, and application of noncompartmental models. It is rare to find in one chapter all three elements succinctly explored yet in sufficient depth to allow the reader meaningful insights into their application, pitfalls and limitations. Throughout the chapter the assumptions of system linearity and stationarity underpin much of the mathematical development, and the reader is rightfully cautioned that there are situations encountered, for example in toxicology and pharmacology, where one or more processes is saturable at the applied dose, rendering the system nonlinear, which can often be readily incorporated in the modeling process."*

Malcolm Rowland
Professor of Pharmacy,
University of Manchester, U.K.

*"I am just in awe of your ability to start with simple ideas and use them to explain sophisticated concepts and methodologies in modeling biochemical and cellular systems (*Chapters 6 and 7*). This is a great new contribution to the textbook offerings in systems biology."*

Alex Hoffmann
Director of the San Diego Center
for Systems Biology and the UCSD Graduate Program
in Bioinformatics and Systems Biology

"I found Chapter 1 *to be a marvel of heavy-lifting, done so smoothly there was no detectable sweat. Heavy-lifting because you laid out the big load of essential vocabulary and concepts a reader has to have to enter the world of biomodeling confidently. In that chapter you generously acknowledge some of us who tried to accomplish this earlier but, compared to your Chapter 1, we were clumsy and boring. For me, now, Chapter 1 was a "page-turner" to be enjoyed straight through. You have the gift of a master athlete who does impossible performances and makes them seem easy.*
"Your Chapter 9 *– on oscillations and stability – is a true jewel. I have a shelf full of books etc on nonlinear mechanics and system analyses and modeling, but nothing to match the clarity and deep understanding you offer the reader. You are a great explainer and teacher."*

F. Eugene Yates
Emeritus Professor of Medicine,
Chemical Engineering and Ralph and Marjorie Crump Professor
of Biomedical Engineering, UCLA

"Chapter 4 *covers many aspects of the notion of compartmentalization in the structural modeling of biomedical and biological models – both linear and nonlinear. Developments are biophysically motivated throughout; and compartments are taken to represent entities with the same dynamic characteristics (dynamic signatures). A very positive feature of this text is the numerous worked examples in the text, which greatly help readers follow the material. At the end of the chapter, there are further well thought out analytical and simulation exercises that will help readers check that they have understood what has been presented.*
"Chapter 5 *looks at many important aspects of multicompartmental modeling, examining in more detail how output data limit what can be learnt about model structure, even when such data are perfect. Among the many features explained are how to establish the size and complexity of a model; how to select between several candidate models; and whether it is possible to simplify a model. All of this is done with respect to the dynamic signatures in the model. As in Chapter 4, readers are helped to understand the often challenging material by means of numerous worked examples in the text, and there are further examples given at the end."*

Professor Keith Godfrey
University of Warwick,
Coventry, U.K.

Contents

Appendix E: Decomposition, Equivalence, Minimal & Canonical State Variable Models 809

Appendix F: More On Simulation Algorithms & Model Information Criteria 833

Preface to the First Edition

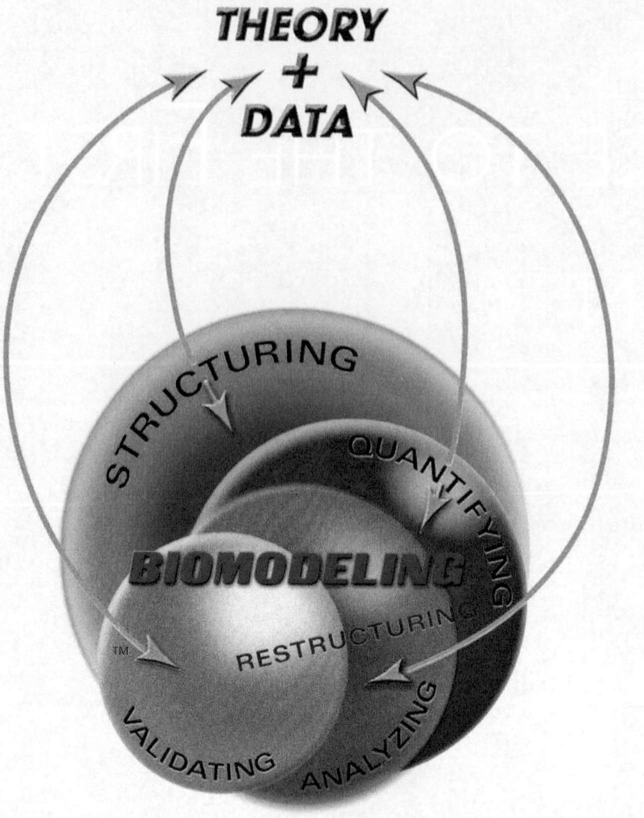

Pedagogical Struggles

I've been learning and teaching mathematical (abbrev: math) modeling and computer simulation of biological systems for more than 47 years. As a control systems engineering graduate student in the 1960s searching for a research area, I found myself quite attracted by biological control systems. This was an esoteric and lonely direction at the time; the primary alternatives for control systems engineering PhD students in southern California were well-funded military control system projects most of my peers were choosing. Worthy, but not my calling. Biological control system modeling otherwise fit neatly into the realm of my major field. For added value, it also provided living examples for another collaborative writing project (DiStefano III et al. 1967).

There was little integrated pedagogy or support at the time for the subject of *my* calling and I realized I faced a multidisciplinary learning task. I had to learn a great deal of new vocabulary (and jargon), and digest no small amount of new scientific knowledge – in biophysics and biochemistry as well as basic biology and physiology; all this before I could begin to develop models for addressing and helping solve real problems in life sciences. Credible math modeling – in any field – requires deep knowledge in the domain of the system being modeled.

Multicultural aspects of my quest also became evident, as did the varieties of different approaches to modeling science. The *cultures* in which each of these disciplines function are quite different, most different between life sciences and math. In the 1960s, biology was largely empirical – still firmly rooted in observation and experiment – with a highly disciplined "wet-laboratory" culture. Quite foreign to my math-systems-engineering "work-anywhere-anytime" culture. Biology remains much that way and, although this is changing, it's still very much reductionist. The other sciences and engineering rely on more theory, as well as empiricism, in the extreme often functioning successfully as "solo" (research) performances – no need for a culture!

I minored in physiology as a PhD student and followed an early path studying physiological systems and mathematical and systems engineering-inspired methods for best modeling them. Biomodeling in those days was done primarily at macroscopic and whole-organism levels. Not any more. As technological breakthroughs in measurement technologies have burgeoned in the last half-century, the spatial and temporal scales over which biological systems knowledge is unfolding has generated a need for deeper understanding of molecular and cellular biology, biochemistry and neurobiology at more granular levels. In lieu of specializing in all these areas, interdisciplinary scientists have typically "picked-up" the needed knowledge along the way, as I did. Modern biomedical engineering programs now include the bio-basics. This means courses with substantive content in molecular and cellular biology as well as physiology and biochemistry.

After earning my PhD in 1966, I began teaching modeling and simulation of dynamic biological systems via an *ad hoc* interdisciplinary major called "biocybernetics." This developed later into a formal PhD field, as well as the moniker of my laboratory at UCLA. I've been wrestling with optimizing my pedagogical path ever since, adjusting it continually along the way, with the goal of communicating it ever-better, and upgrading it as new approaches and discoveries have emerged.

Crystallizing and Focusing – My Way

MODELING, as such, can stand alone as a mature discipline, but it is done somewhat differently across the multidisciplinary spectrum of its practitioners, and has a history of being studied and developed on an "as-needed" basis, especially in the life sciences. There is, however, a substantial common core of methodologies for dynamic biosystem modeling widely disseminated in journals and books. Much of it is developed or described for different applications, or for single *scales* (e.g. molecular-cellular, organ-system, or population levels), using a variety of (and sometimes ambiguous) nomenclature. One of my goals as a teacher has been to help crystallize and unify the substance and language of this core of material – to make it more accessible to a larger audience – at the same time exposing and clarifying the ambiguities in some concepts and tools.

I've been drawing from and merging aspects of the several classical disciplines involved with math modeling in biology into a *unified subject matter*, maximally comprehensible to undergraduates (and graduates) in any of these sciences or engineering. This textbook codifies this process. It offers a basics-to-intermediate-level treatment of modeling and simulation of dynamical biological systems, focused on classical and contemporary multiscale methodologies, consolidated and unified for modeling from molecular/cellular, organ-system, on up to population levels. It will undoubtedly be of interest to individuals from a variety of disciplines, probably with widely varying degrees of mathematical as well as life science training. It is intended primarily as an upper division (advanced undergraduate), graduate level or summer program textbook in biomedical engineering (bioengineering), computational biology, biomathematics, pharmacology and related departments in colleges and universities.

The text material is written in a maximally tutorial style, also accessible to scientists and engineers in industry – anyone with an interest in math modeling and simulation (computational modeling) of biological systems. One or 2 years of college math are prerequisite and, for those with minimum preparation, the needed math (differential and difference equations, Laplace transforms, linear algebra, probability, statistics and stochastics topics, etc) is included either in methods development sections or in appendices.

My approach to biomodeling is drawn in large part from my dynamic systems engineering and control theory viewpoint and training (the theory). But my 30 years of "wet-lab" research (the data) – closely integrated with and guided by biomodeling – plays an equal role. The two have motivated each other, first serendipitously, and then by design. Much of the book pedagogy is a distillation and consolidation of my own and my students' modeling efforts and publications over half a century, including novel and previously unpublished features particularly relevant to modeling dynamic systems in biology. These include qualitative theory and methodologies for recognizing dynamical signatures in data, using structural (multicompartmental and network) models, and no small amount of algebra and graph theory for structuring models, discovering what they are capable of revealing about themselves – from data – and designing experiments for quantifying them from data.

Approaches to biomodel formulation by various practitioners – interdisciplinary scientists with basic training in a diversity of fields – have many common features, for example, as found in references like (Rashevsky 1938/1948; Jacquez 1972; Carson et al. 1983; Murray 1993; Edelstein 2005; Palsson 2006; Alon 2007; Klipp et al. 2009; Voit 2012). I've made every effort to maintain the best of these developing features as they morph into our communities' best traditions – toward developing a culture of its own. Following the practice of most expositions of modeling biological systems, the biology, biochemistry and biophysics needed to comprehend context, goals and biomodeling domain details are included within the chapters. Some of this supporting and complementary material is in the text proper, some is in footnotes – as much as needed and space considerations allow. Abundant citations to supplementary and advanced topics are included throughout.[1]

The chapters include exercises for students and solutions will be available for teachers on the book website. Ancillary material, including computer code and program files for many examples and exercises (*Matlab, Simulink, VisSim, SimBiology, Copasi, SBML, Amigo* model code, etc) also are included on the book website.

1 All citations in the text are formatted by name and year, instead of potentially distracting numbers, to facilitate reading without wondering about the origins of textual statements.

In Other Words... & Other Didactic Devices

The substance and style of my teaching and writing developed from my experience in the classroom – typically populated by students from mixed subject backgrounds. Indubitably,[2] interdisciplinary material typically needs *more* explanation, of one disciplinary sort or another. So, with major emphasis on being tutorial, exposition and development of biomodeling methods and applications in this text range from simple introductory material – understandable by any science student with high school math, some physics or chemistry, and maybe some biology – to fairly complex, requiring intermediate-level math skills. There may be more information than perhaps desired by one target group (e.g. the mathematicians) or the other, and I apologize for this necessity in advance. But that same target group should appreciate extra verbiage, examples or redundancy, when the extras are about what *they* know little about (e.g. cell biology). (So, maybe I should take back the apology?) In any case, such 'distractions' have been minimized, or isolated for purposes of ignoring them as desired – as I explain below.

For mathematical exposition (all *applied* math), I've aimed for a relatively low level of formality, or rigor, without sacrificing accuracy, and I provide very few proofs – only when it is instructive of the modeling point at hand. Most math beyond the basics is developed as needed in the chapters, with some developed more completely in appendices, detailed in item 3 below. The reference citations and bibliography provide additional theory. On the other hand, my expository style is methodical, with the purpose of providing foundations for comprehension and for developing these modeling methodologies further. I've used several devices in organizing the chapters and their content to simplify use of this book in different settings, and for students or readers having different backgrounds. These include:

1. Every chapter has a detailed table of contents of its own, an overview or introductory section describing its purpose and content, and a final section summarizing the main points of the chapter and providing readers with some guidance on the content of subsequent chapters relative to that in the present one.

2. This book has a "logo family" associated with it, depicted in its primary (parent) form at the beginning of this *Preface* and on the back cover. It depicts the rather circular nature of the overall process and steps in biomodeling, and their intimate connections with theory and data – a fundamental theme of the book – all as explained in Chapter 1. The logo[3] is repeated in slightly different forms (the children) as the chapters progress, each one depicting the focus of that chapter.

3. Several more traditional math or engineering topics pertinent to modeling dynamic biological systems are developed further or presented concisely in separate *appendices*. These topics have been studied elsewhere by many, and therefore might distract from the main exposition.

---

2 The famous comedic actor Jimmy Durante often used this word in his films and TV appearances – instead of the less efficient "without a doubt." It always made me smile, so I'm sharing this with you here.

3 Biomodeling/Theory + Data Images (The Logo: parent and children) are reproduced above and in the chapters and back cover of this book with permission by the author Joseph DiStefano III. It is a Trademark (™) and service mark of the author, Copyright © 2012.

Alternatively, these topics may be new or not so well-known – and are thus included for gaining depth of understanding, for completeness, or as a refresher or for separate study. They include pertinent extracts of subjects like Laplace transform and transfer function methods (***Appendix A***); linear algebra and matrices basics (***Appendix B***); some advanced systems and control theory – like model *equivalence* and model *reduction*; statistical methods and algorithms, and other topics useful for some biomodeling developments (***Appendices C – F***).

4. To highlight additional explanatory remarks following more difficult concepts, mathematical developments, or less well-known biology, I've included numerous *Footnotes, Remarks, Caveats* and also some special remarks called *In Other Words...* Hopefully, these will suffice to clarify denser or more complex material.

5. Numerous examples are included – from simple to moderately complex – many being published and unpublished biomodels developed from real biodata and used in real applications. These serve to put a practical face on the whole process. Some "simple" examples are carried forward to subsequent chapters, where they are used to build on and illustrate the additional methodologies and concepts therein.

6. ***Notation***: I use *italics* to highlight (and emphasize) words or phrases; **boldface** to designate definitions and their synonyms or variants; and ***boldface italic*** for strongest emphasis or for special headers. For the math, I've tried to stay as close as possible to the conventional and accepted terminology and symbology of applied math, for example in defining matrices and vectors as typically done in linear algebra – because it's standardized and imminently logical. This is invariably a little more difficult for specific modeling topics, e.g. compartmental modeling, because areas developed within different fields typically have their own pet jargon. But I've chosen the terminology I believe is most acceptable and consistent with applied math symbology.

7. Modeling is done in a variety of ways in the sciences, at the most obvious level with different nomenclature, jargon and the like. Pharmacology, physiology, molecular biology, biochemistry, physics and engineering all have their modicum of modeling distinctions and distinguishing features. These include some ambiguities and disagreements on nomenclature, definitions and other more or less important details. As a central theme throughout this book, I make every attempt to bridge these cultural differences, with additional explanations and information.

8. "Systems biology" software tools based in new biochemistry and cell biology languages are becoming widely available via the Internet, with user groups to support and further develop them. I illustrate the utility of some representative ones in modeling examples and do my best to clarify and bridge differences between older and newer nomenclature employed in these programs.

9. Not all published quantitative modeling methodologies applied to biological systems are included in this book. Instead, the focus is on primary ones I've found most useful in my research and teaching, those I believe are likely to persist as major sustainable approaches over the longer term. This should not be interpreted as any sort of judgement on the

importance of any not included. Both space and, in some cases, mathematical complexity, have been the main motivators for limiting coverage. There are many other fine textbooks out there that cover what I've left out.

How to Use this Book in the Classroom

The ordering of the material covered here is my best compromise toward systematically organizing the material in increasing complexity, for both teaching purposes and maximum comprehension. In part, it accounts for what I believe most university students in the sciences and engineering learn first, second and so on about the basic science and math that underlies modeling concepts. But it's only one way of presenting the subject matter. I've designed (and redesigned) the chapters with this in mind, so it can be used in multiple settings.

As such, the book can be taught in its entirety in a year of coursework – two semesters in most universities, or 3 quarters in those, like UCLA, on the quarter system. The chapters are laid out so this could be done sequentially. Ideally, in a year-long 2-semester sequence, chapters 1–9 are doable in the first semester and 10–17 in the second. In a 3-quarter modeling sequence of courses, chapters 1–5 can be covered more deeply in the first quarter, 6–11 in the second, and 12–17 in the third.

The even-numbered chapters (2, 4, 6, ...) include the basics of the different modeling topics covered throughout the book. With the exception of Chapter 3, on simulation methodology, the odd-numbered chapters provide extensions of the material in the even-numbered chapter that precedes it. Some teachers (or individual readers) might thus consider using the even-numbered chapters as the basis for initial course development of their own.

For a single course (or for a summer program), much of Chapters 1–9 can be covered selectively – with choices dependent on the student audience and their backgrounds; and it should include basic material on quantifying models, selected from Chapters 10–12. Advanced material – usually designated as such – with asterisks – can be skipped over in short courses.

Acknowledgements

So many former and current students, colleagues, friends and family have helped me with the book and deserve my heartfelt appreciation. They have assisted with producing it, critiquing it, contributing to it, motivating it, or have supported it in other important ways. Thank you: Celine Sin, Marisa Eisenberg, Robyn Javier, Thuvan Nguyen, Pamela Douglas, Greg Ferl, Sharon Hori, Nik Brown, Long Nguyen, Olivera Stajic, Rotem ben Shacher, Natalia Tchemodanov, Christine Kuo, Bill Greenwald, Patrick Mak, Gene Yates, Tom Chou, Van Savage, Elliot Landaw, Alan Garfinkle, Malcolm Rowland, Matteo Pellegrini, Marc Suchard, teD Iwasaki, Chris Anderson, Walter Karplus, Keith Godfrey, Nikki Meshkat, John Novembre, Mary Sehl, Fiona Chandra, Pep Charusanti, Eva Balsa-Canto, Alex Hoffmann, Paul Lee and Todd Millstein.

My daughter Allegra DiStefano contributed the beautiful original graphics. My wife Beth Rubin provided key criticism along the way – as well as infinite patience and loving support. My immediate progenitors, Joe and Angie, provided me *every* opportunity to learn. They put the pen in my hand early on and gently and firmly encouraged me to use it creatively and constructively.

I also thank the people of the State of California for supporting the great University of California educational system. For nearly half a century, it has granted me virtually complete academic freedom and support for my intellectual pursuits. It has promoted my interests in biomedical research and in teaching and honing great minds. Teaching and learning from students continues to be an honor for me – the reason I remain employed full-time at UCLA. I wrote this textbook for my students, past, present and future; and I dedicate it to them.

Los Angeles, California, July 4, 2013

References

Alon, U., 2007. An Introduction to Systems Biology. Chapman & Hall/CRC, Boca Raton, FL.
Carson, E., Cobelli, C., Finkelstein, L., 1983. The Mathematical Modeling of Metabolic and Endocrine Systems: Model Formulation, Identification, and Validation. J Wiley, New York.
DiStefano III, J., Stubberud, A.R., Williams, I.J., 1967. Feedback and Control Systems. McGraw-Hill, New York.
Edelstein, L., 2005. Mathematical Models in Biology. SIAM Books, Philadelphia.
Jacquez, J., 1972. Compartmental Analysis in Biology and Medicine: Kinetics of Distribution of Tracer-Labeled Materials. Elsevier Publishing, New York.
Klipp, E., Liebermeister, W., Wierling, C., Kowald, A., Lehrach, H., Herwig, R., 2009. Systems Biology: A Textbook. Wiley-VCH, Weinheim.
Murray, J., 1993. Mathematical Biology. Springer-Verlag, Berlin.
Palsson, B.O., 2006. Systems Biology: Properties of Reconstructed Networks. Cambridge University Press, Cambridge.
Rashevsky, N., 1938. Mathematical Biophysics: Physico-Mathematical Foundations of Biology, second ed. Univ. of Chicago Press, Chicago.
Voit, E., 2012. A First Course in Systems Biology. Garland Science.

Biosystem Modeling & Simulation: Nomenclature & Philosophy

Chapter 1

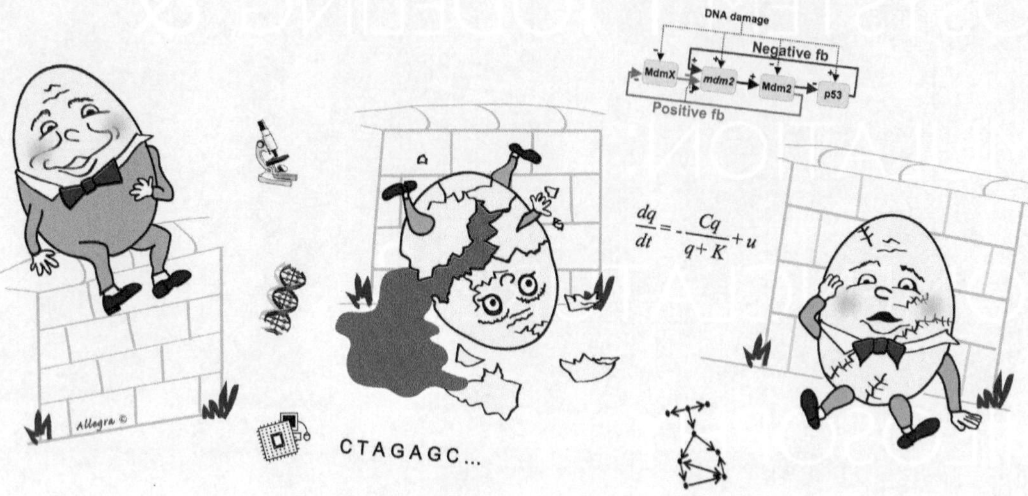

Cartoon design by Allegra DiStefano

> "Molecular biology took Humpty Dumpty apart; mathematical modeling is required to put him back together again."
>
> *Schnell et al. 2007*

Overview

Mathematical (abbrev: math) modeling is practiced in all sciences, each with its own culture and nomenclature developed within and for that culture. Biological and clinical sciences are somewhat of an exception. Math modeling has not been a priority in most life science curricula. Biomodeling methodologies have come largely from the outside, and thus the life sciences have inherited a cultural spectrum of nuanced differences across the languages of worlds that some consider alien or impenetrable. As with most things technical, much of the problem – inasmuch as it is a problem – is in large part due to naming conventions, with different nomenclature for the same entities across the spectrum of disciplines. That's not the whole story, of course, but it's a dominant thread. With this mindset, we attempt to help bridge some of the gaps, in part, by utilizing a common nomenclature and clarifying differences in meaning and intent.

We begin by defining and discussing some of the key terms – in plain words, minimizing the math. This is done with full realization that any given definition is unlikely to be agreed upon by all. Considerable effort has been made to choose from the most prominently acceptable ones in this field, and to explain pertinent differences, as needed. We integrate and apply this vocabulary and begin building a conceptual framework for modeling and simulation in biology – focusing on biological systems (abbrev: biosystems). A single, very simple dynamic system model, math included, illustrates the concepts as they unfold in this chapter. Along the way we also introduce several different paradigms used for *structuring* biomodels in various scientific communities. Structuring here means defining graphical or schematic topological schema from which model equations are typically derived. These include block diagrams, graphs, compartmental diagrams, reaction diagrams, and several others developed throughout the chapters. Chapter 1 concludes with a "top-down" model of the book – an overview of the remaining chapters and how they are systematically connected.

Modeling Definitions

Terms like *model*, *animal model*, *mathematical model*, *model system*, *system model*, *measurement model*, *data*, *data model*, *simulation model*, *stability*, *dynamics*, *kinetics*, *multiscale*, etc., are often confused or misunderstood, particularly across different disciplines. Other labels like bioinformatics, systems biology, integrative biology, computational biology, quantitative biology – and variants or combinations of these terms – can be equally vague. With the premise that "defining one's terms" is prerequisite to successful exposition, particularly of technical material infused with mathematics – and a great deal of jargon – we define our primary working vocabulary.

> *"When I use a word," Humpty-Dumpty said, "it means just what I choose it to mean – neither more nor less."*[1]

Model has several dictionary definitions. The most common is model as a *copy of an object*. Copies take many forms, often as recognizable physical objects, on a smaller scale, like toy

1 from Alice's Adventures in Wonderland, Lewis Caroll, 1865.

soldiers, or Barbie dolls. Models for our purposes are more abstract, ultimately mathematical. The terms *model* and *modeling* in the sciences are usually concept-driven, with different levels of abstraction. In the copy sense, *network diagrams* or *schematics*, system *block diagrams*, and *cartoons*, like the ones shown in **Fig. 1.1**, are useful model forms to begin with. These are (usually) qualitative representational forms for models in biology, often formulated as an early step toward developing quantitative models. We might call this the *organizational* step in modeling, i.e. gathering the major component parts (of a system) together into a whole, usually without specifying much or any quantitative detail.

For us, a **model** is a hypothetical description or representation of a (more-or-less) complex entity or process; in essence a *formal* representation of a hypothesis. We may even use the terms model and hypothesis interchangeably. In this sense, we model to overcome conceptual muddle. For example, the qualitative models in **Fig. 1.1** are hypotheses, the first about the *presumed* structure of a protein network, the second the *presumed* components and interconnections in a biocontrol system, and the third a cartoon of the *presumed* molecular components and signals in an intracellular hormone regulation pathway. Such cartoon models are common in molecular and cellular biology.

A **system** is simply a collection of objects, usually interconnected or interacting in some coordinated way, so a **system model** is a model of a collection of objects, or component parts, normally interconnected in some way (maybe only some parts are interconnected). Teleologically speaking, systems usually have a *purpose*, or *function*, and the component parts usually are connected in the particular way they are so as to satisfy that purpose. The diagram and cartoon models in **Fig. 1.1** exemplify use of biosystem models as statements of specific hypotheses about a biological system *structure* or *function*.

A **goal-oriented model** is a model developed for a particular purpose. This may seem obvious – maybe *too* wordy – but it's not, because two models of the same biosystem, possibly even built from the same data base, can have very different levels of complexity depending on a modeler's generally different goals.

Models of systems, or **system models**, also called **structured models** or **structural models**, are usually based on physical (e.g. biophysical or biochemical) principles (first-principles) and hypotheses, descriptive information about how a system is structured and possibly also how it functions. For our purposes, this includes models based on physical (which include chemical) laws and their consequences. For example, the law of mass action, product–precursor or other mass balance relationships, cell transduction processes, control theory, psychophysical concepts, noncompartmental and multicompartmental structures, Newton's second law, $F = Ma$ – the first math model discussed in Chapter 2 – and others. A system model has component parts based on such processes that interact in some organized way, with a topological, morphological, or mechanistic description. **Mechanistic models** are usually classified as structured models. **Physicochemical models**, those describing biomolecular transformations of all types, are structured models.

Remark: As with most manmade definitions, there is some ambiguity here, at least at the outset. Certain "structured" models might have no direct physical analog, if they have been generated

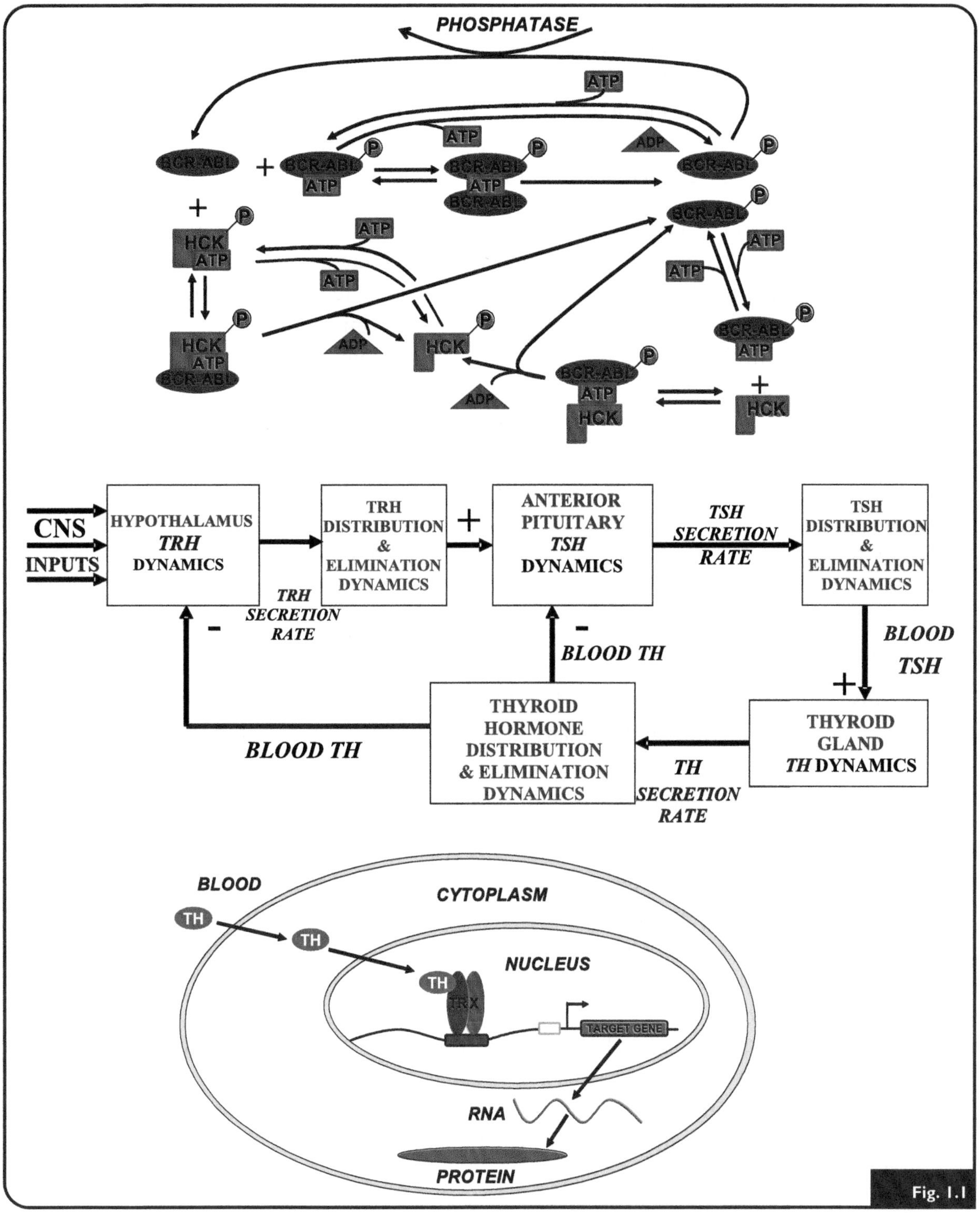

Fig. 1.1

Qualitative biomodel examples. (Top) Presumed structure of a protein control network and signaling pathway model of BCR-ABL, an oncoprotein that constitutively activates numerous cell signaling pathways (Charusanti 2006). (Middle) The components and interconnections in a biocontrol system block diagram model of thyroid hormone (TH) feedback regulation, adapted from (Eisenberg et al. 2006). (Bottom) A cartoon model of presumed signals and primary molecular components in an intracellular (growth) hormone pathway regulated by TH (Larsen et al. 2003). TR, thyroid hormone receptor; X, transcription factor. Cartoon models are common in molecular and cellular biology.

only from input–output time-series data. This includes some multicompartmental models (Chapter 4) and mathematically or structurally *equivalent* system models, or *canonical* models – developed in Chapter 5 and subsequent chapters with additional theory in Appendices D & E. These might, at first, be classified as models of data, rather than systems. But they might be transformable into system models, if structural information is available and utilized, as part of a multistep modeling process. (More on this in later chapters.)

Network models in biology and chemistry are structural models, e.g. models based on gene expression, or other genomic, proteomic or metabalomic interrelationships. Formally, a **network** is a connected (or partially connected) set of objects,[2] i.e. a *structured system*, or *structured model* – if the objects are symbolic. An organic molecule made up of carbon, hydrogen and oxygen (the objects) is a particular arrangement of atoms – connected by chemical bonds – and this is a network in the strict sense of the word. But it's not what is usually meant by networks in biology, where networks usually are more complex collections of biomolecules connected in a way that elucidates one or more specific pathways or functions of the set of biomolecules. Chemical reactions typically occur sequentially, in stages, together called **pathways** and, in biology, cellular chemical and biophysical reactions participate in *metabolic pathways* or *signaling pathways*, or both. Collections of such pathways are also called networks, i.e. networks of metabolic or signaling pathways, or both. The terminology here can be ambiguous, with pathway and network often used interchangeably. Context usually clarifies the meaning, if it's important.

Graphs or **graph models** (Chapter 4 and beyond) might be used to represent network and other structured models. Biology has many layers or levels, from the atomic to the organismic and beyond, and can thus be viewed – and modeled – as collections of networks, often represented graphically, using graph models with *nodes* as the objects and *edges* (directed lines) as the connections. For example, the schematic model in **Fig. 1.1** (top).

Remark: It's important to distinguish system model, as defined here, from the common expression *model-system* used by experimentalists. When an experimentalist cannot or does not do an experiment on the real object of interest, e.g. a real, particular animal or cell, they call the alternate system they work on the model-system (with or without the hyphen). For example, *Drosophila*, *C. elegans*, yeast, a different cell type, etc., may be the model-system for studying neural or metabolic pathways in related species. We'll hyphenate model-system, as above, when we have occasion to use it, to distinguish it from system model. The term *animal model* might also be used in this context, meaning a model-system, usually in reference to studies in an animal other than a human, for example, because it is not feasible or ethical to do in a human.

Modeling Science

Before moving on, it's useful here to frame the discussion and develop and integrate modeling definitions within a broader formalism, primarily because modeling is viewed and practiced

2 The Internet, or World Wide Web, is a network in the computer science domain. It is quite analogous to biological networks, with a major difference being that it appears to be ever-expanding, without bound.

somewhat differently in different disciplines. For pedagogical purposes, the science of mathematical modeling a physical process (minus the physics) can be rationally categorized into four subject areas, taught and studied in different disciplines in various forms and orders: *probability, statistics, signals* and *systems*. **Probability** is about modeling uncertainty, e.g. mathematically characterizing random events, like the tossing of a coin, or the timing of cell division in biology, using probability functions. **Statistics** is about quantification of uncertainty, e.g. establishing the numerical values of the probabilities associated with the coin tossing or cell division models. **Signals** are detectable physical quantities associated with a process that convey presumably meaningful information about that process, e.g. data collected in a laboratory experiment. Modeling signals is about characterizing them mathematically and analyzing them for properties of interest. Modeling **systems** means mathematically characterizing the whole structure or topology of a process as well as internal properties of its component parts, as described earlier.

From the viewpoint of modeling, these four subjects are strongly interrelated. Probability and statistics – not the *main* focus of this book – have a definite role in development of systems biology models at all levels and scales – in both structuring them and analyzing the signals driving them internally and measured from them in data. Measured signals typically have both deterministic (certain) and random (uncertain) components (further defined below). The random components – called **noise** – are usually contributed by the measurement process itself.[3] Noise, unless it's verifiably undetectable or negligible, can and should be modeled along with the more certain (deterministic) aspects of the signals and systems. As we shall see in subsequent chapters, *signal (data) analysis* can be used effectively to structure systems models – establishing data as a driving force for structural modeling, *with and without consideration of noise components of a signal*. Chapter 5 is particularly focused on establishing structure from noise-free signals.

Whereas the major focus of this book is *systems* modeling in biology, we nevertheless utilize all four of these paradigms in modeling methodology developments. Probability and statistics have a smaller role, used primarily for characterizing uncertainty (noise) in measurement models and for optimally quantifying models from measurements. They are also used in our introductory development of stochastic dynamic molecular systems biology models in Chapter 7.

Modeling Essential System Features

System modeling usually begins by specifying modeling goals and then isolating the *essential features* of the system, consistent with these goals. This is a critical and often most difficult step, effectively defining the boundaries, complexity, and thus the tractability of the modeling problem as well as the model. In brief, the **essential features** are that subset of specific component parts and properties of the system (system structure and function) necessary and sufficient to achieve the modeling goals. The principle of parsimony[4] is our guide here, with the added

3 Stochastic models, presented in Chapter 7, have inherently random component signals, independent of random measurement noise.

4 See (Thorburn 1918, pp. 345–353) for a discussion of the origins of "Occam's Razor," the principle that entities should not be multiplied beyond necessity, also known as the principle of parsimony.

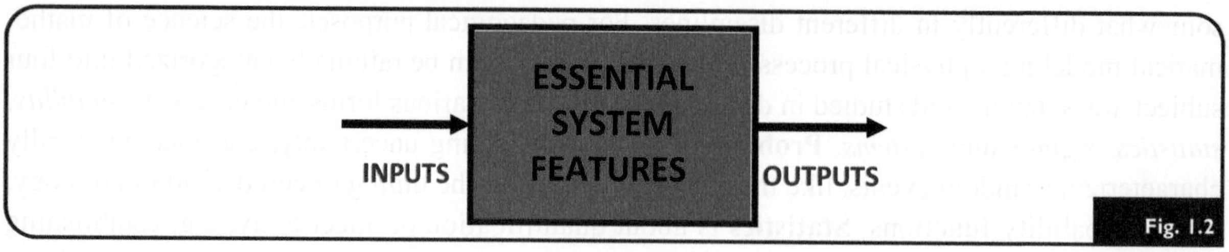

Characteristics of system models. They represent *essential features*, satisfy goals, include simplifications, and usually have inputs (stimuli) and outputs (responses).

caveat that "things should be made as simple as possible – but not simpler," wisdom attributed to Albert Einstein.[5]

To begin modeling essential features, draw a box around them, as in **Fig. 1.2**. Next, consider what's external to the box that can influence what's inside. External things, in this context, can be purposeful (e.g. "control" signals), or not (background "noise"). They might even be things excluded that perhaps should not have been. Let's circumvent these (albeit realistic) nuances until later. For the present, we consider what's most important at this early stage in system modeling pedagogy.

In the language of cybernetics and systems engineering, systems have **inputs** and **outputs**, as illustrated in **Fig. 1.2**. Inputs and outputs, which take on many forms and purposes in modeling, are usually system *stimuli* and *responses*. **Inputs** are stimuli generated external to the system proper; they enter it and influence it.[6] **Outputs** are system responses to input stimuli observed from outside the system proper. Because systems can generally have more than one output, modeled outputs of interest are typically selected by modeling goals. Overall, system models can be defined by the relationships describing object connections and interactions with these inputs and outputs, with (only) *essential features* of the system represented betweens inputs and outputs.

*Example 1.1: **Temperature Regulation***

The *block diagram* in **Fig. 1.3** is a model of a *system for controlling temperature in a room* (temperature control system), illustrating that it is a collection of individual components (simplified thermostat,[7] valves, louvers for adjusting warm or cold air flow, etc.) connected specifically as shown. The *input* is the temperature setting on the thermostat (constant *setpoint*) and the *output* is actual room temperature, which can be recorded periodically, generating a set of measurements modeled as a time-varying numerical sequence. This block diagram model captures the essential features of this control system, but of course there are many more. Features unessential in the

5 Perhaps hearsay, because the original source is lost. See (Calaprice 2000, p. 314). Whether it *was* in fact him, Einstein *would* say something like this.

6 Systems often have subsystems within them that can have inputs and outputs as well. These can still be interpreted as stimuli and responses – of the subsystem, but they then would likely not be exogenous (external), but rather interior signals within the system proper. More on this in the next section.

7 A more realistic room thermostat elicits no stimulus signal to the valves when room temperature is within an acceptable range, i.e. 1 or 2 degrees from the setpoint. It only activates when this range is exceeded.

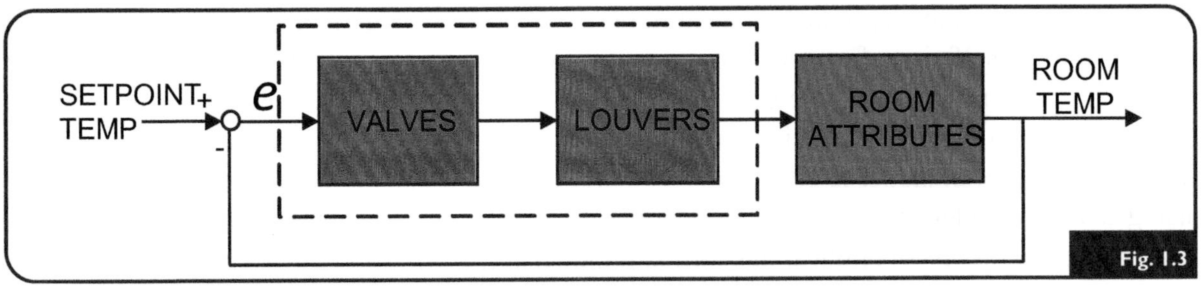

Fig. 1.3

Block diagram of a simplified room temperature control system model. An example of simple *negative feedback* control. The simplified *thermostat* is represented functionally by the circular component (called a "summer") at the input, which effectively subtracts the actual room temperature from the desired setpoint temperature. When the "error" e is positive, the valves and louvers are opened to warm air ducts, drawing warm air into the room; cool air flows when the error signal e is negative.

control system are not included, such as the heat/cold energy source, plumbing for carrying hot and cold air to the louvers, electric power for the thermostat and valves, color and price of the components, the ever-changing number of people in the room whose presence alters the room temperature, the opening and closing of windows and doors, wall and ceiling insulation, and so-on.

Remark: The biosystem regulating core body temperature in *homeothermic* (warm-blooded) animals is functionally similar to room temperature control, but it is far more complex, involving skeletal muscle control (shivering), blood vessel smooth muscle control (vasomotor activities for heat exchange), sweat gland control (evaporative cooling), thermoreceptors (peripheral "thermostats"), a region of the brain called the hypothalamus (central "thermostat"), and other regulatory components (Stolwijk 1980).

A mathematical model is a model represented symbolically by equations (or inequalities), whose terms are collections or arrangements of symbols joined by mathematical operations, e.g. addition, multiplication, etc. Mathematics serves to render models maximally abstract and unyieldingly specific, via use of unique symbols and logic defining the relationships among the symbols – and therefore among the real *system* components (elements) represented by symbols. This is a major advantage, as models go, because math models can be manipulated, simplified, solved, or otherwise transformed using established axioms, theorems, and other mathematical results. Math models permit investigation of properties of real objects or processes and, in many cases, prediction of future outcomes. They also facilitate *testing* of the hypotheses or assumptions upon which the model is built, against experimental data, and thereby pave the way for new hypotheses when the model does not fit the data.

Most of the models in this book are *differential equation* models, the traditional paradigm for dynamical systems in science. However, *algebra* and *geometry* – high-school level and beyond – play major roles in model development, e.g. algebra in model structuring and extracting essential features (Chapters 2–4 and beyond); and geometry, for distinguishing among candidate models and analyzing stability and other dynamical characteristics of a model (Chapters 5 and 9). Methods of algebraic and geometric *topology* also are useful in biomodeling at intermediate to advanced levels (Chapter 5 and beyond). In this category, graphs and graph theory are finding substantive utility in systems biology modeling, especially for modeling biomolecular pathways and networks, as noted earlier.

Simulation means reproduction or imitation of essential features of an object. The process of modeling, itself, is a form of simulation. However, in the context of math modeling, it usually means implementing the equations of the system model on a computer for the purpose of solving them, or for exercising the model to study its properties or its predictive value. The term *simulation model* is often used in reference to simulated biological experiments, done using a math model implemented on a computer rather than on the real biological system, a very important class of applications for math models.

***In numero* model** is another term for simulation model.[8] This term is often used in reference to (simulated) biological experiments done on a computer, i.e. *in numero* experimentation.

***In silico* model** is yet another, newer term for simulation model. It is popular among modelers of cellular and molecular level systems. This term also is often used in reference to (simulated) biological experiments done on a computer, i.e. *in numero* experiments.

A **computational model** is a simulation model. It usually refers to a complex mathematical model requiring a large amount of computational resources for studying the behavior of a complex system by computer simulation. But it can be a simple model too.

A **stochastic model** is governed by probabilistic mathematical laws for describing *inherent indeterminacy* in a system. Its variables and/or parameters are usually *random variables*, described by probability distributions of values rather than unique values. Stochastic models are used in molecular-level systems biology to precisely describe relationships among very small numbers of molecules.

A **deterministic model** has no indeterminacy. The components, mathematical variables and parameters are represented by symbols with unique (or multiple distinct) values, as opposed to random variables. Most models are deterministic. In systems biology deterministic models are used to describe relationships among large (large-enough) numbers of entities, e.g. molecules or species.

A **compartment model** is distinguished by discrete boundaries – often *abstract* boundaries – between components called **compartments.**[9] **Multicompartmental models** have more than one compartment. Further discussion of this important topic is reserved for Chapter 4 and beyond.

Data is information obtained from experiments, usually in the form of *numbers* or *facts.* System input (stimulus) or output (response, signal) data – or both – are usually expressed numerically, with values, or mathematical functions of values, typically as time-series data (signals) for dynamic systems and their models. System structural (topological or morphological) information is facts, or descriptors, capable of rendering the model a bio*system* model.[10]

8 I recall use of this term in publications of the 1960s, when digital computer simulation was becoming accessible to scientists via simulation languages. Works of D. Bartholomew may be the original source.

9 We include this rather abstract, minimalist definition here because it's likely one all can agree on. Otherwise, there are two different definitions of compartment being used in current systems biology literature, as discussed in Chapter 4.

10 Some modelers consider only quantitative numerical data as (experimental) "data." Biologists use the term more broadly – like I do – because many experiments are done to generate, for example, morphological information (data), or detailed topological data about networks or pathways, or other *biostructural* information (data).

Models developed from numerical or structural data (signals) are often called **data-driven (biosystem) models** or **data-inspired (biosystem) models**.

Pure **models of data** do not require, use, or portray any structural (mechanistic, topological, or morphological) hypotheses about the physical process (system) from which the data is obtained, except possibly the characteristics of the (external) measurement system itself. Most classical statistical models as such fall into this category. They often require only that the data are random samples drawn from a population with a certain probability distribution (*generative data*), or an approximation of it. An example is the *t*-test of significance, which depends on approximate normality (Gaussian distribution) of the data. So-called **empirical** or **phenomenological** models, integrated closely with experiment, are also usually models of data, but they also can be mixed models of data and systems, if they are based in part on first principles. Polynomial functions, and sums of exponential functions as such, are examples of these.

A major limitation of pure models of data is that, by themselves, they cannot be used with any degree of confidence for extrapolation or prediction purposes. In their purest form, they say nothing about mechanism. They usually represent **measurement models**, with components that reflect an *experiment*, with input, output and (typically noisy) measurement data. Measurement models are discussed further below.

Primary Focus: Dynamic (Dynamical) System Models

A **dynamic (or dynamical) system (DS) model** is, first, a model of a system, with inputs and outputs, and equations describing system *motion* in space and time.[11] What distinguishes it from nondynamic systems is that it has **memory**, i.e. information about its "whereabouts": where it "goes" or "is" depends on where it "was" (and when). This is also termed its **state**: DS have "states"; nondynamic systems do not. In other words, in order to know the *motions* of a DS at future times, one needs to know where the system is "at" now (the memory). In physics and engineering, this distinction translates into a DS having the property that it can *store energy* in some form. For example, an electrical capacitor stores charge when it is energized, and a pendulum (or arm or leg) stores kinetic energy when it is displaced. Mathematically, motions of dynamic systems are most often described by differential equations and, to solve them, i.e. to find future system motions, one needs to know where the system is now, called the *initial conditions*, or **initial state** – the memory.

Conceptually, understanding dynamic systems in these technical terms is important, not pedantic. Studying the basic properties of dynamic systems, using mathematics to build comprehension and intuition about them, is important because living systems *are* dynamic systems. And it is the "memory and motions" of such biosystems that occupy much of scientific investigation in the life sciences.

Remark: The word *dynamic* in systems theory and practice does not have the same "changing" meaning as in the English language. It does *not* imply that system motions *vary with time*,

11 *Motion in space and time* here means temporal changes in system variables and their derivatives.

as implied by nontechnical use of the term in everyday language, i.e. models that describe *longitudinal* behavior over time are not necessarily dynamical. The system-complement of dynamic system in system theory is *instantaneous*[12] (nondynamic) system, which is simply one with no memory. But it certainly can vary with time, as does the instantaneous system model described by Ohm's law, relating the voltage and current in an ideal resistor of resistance R ohms: $V(t) = Ri(t)$ Apply a voltage $V(t)$ across R and current $i(t)$ begins passing through it instantaneously. Remove the voltage source and, instantaneously, the current ceases to flow. Repeat the experiment beginning at another time $\tau \neq t$ and you get precisely the same result. There is no memory: no energy is stored in an ideal resistor. This situation fits a longitudinal model but not a dynamical model description.

Remark: The term **kinetics** is often used synonymously with dynamics. It usually means the same thing, as in physics (motion)[13] and chemistry (reaction kinetics).

Deterministic vs. Stochastic Dynamic System Models

System dynamics usually are represented with *deterministic* (not random) variables, typically as differential or difference equations[14] – uninfluenced by "indeterminacy." We focus mainly – but not exclusively – in this book on deterministic models for biosystem dynamics, coupled with measurement models that usually include indeterminacy, often in the form of errors with a probabilistic description. However, indeterminacy can be included in dynamic system model equations, using probabilistic notions, and this methodology can be important in some biosystem studies. States and outputs of **stochastic dynamic system models** are *stochastic processes* (also called *random* processes) evolving in time and governed by probability distributions, as well as by the basic system dynamics equations depicting (bio)system structure. They are especially useful for describing the evolution of biosystem dynamics when the numbers of objects represented by state variables are *very small*, e.g. for intracellular processes where numbers of molecules of order of magnitude 10 interact with a similar number of others. We introduce the methodologic basics for stochastic modeling and simulation of dynamic biomolecular systems in the latter part of Chapter 7, using the same mass action principles developed primarily for much larger numbers of molecules in most of Chapter 6 – using deterministic continuous-time (continuum) differential equations – but focusing that principle more microscopically on *small numbers of molecules* – using discrete-time stochastic event models.[15] In addition to describing detailed quantitative variability in such biosystems – by simulation and by statistical analysis of their random outputs – stochastic dynamic system models can be quite useful for studying their *qualitative* behavior. Stochastic models can behave fundamentally differently than their deterministic counterpart, e.g. when variabilities are large enough to drive the biosystem into an unstable regime of operation.

12 Nondynamic systems and models are sometimes called *stationary* rather than instantaneous.

13 Some people distinguish between *kinetics*, requiring causes (i.e. forces), and *kinematics*, relating to motions with no apparent forces – like Kepler's solution of planetary orbits, sweeping equal areas in equal times (Kepler 1609; Donahue and Kepler 1992).

14 These model-type distinctions are made clear in Chapter 2.

15 These model-type distinctions are also made clear in Chapter 2 and then more so in Chapters 7 and 9.

Markov Models

Markov models in statistics and computer science are direct analogs of dynamic and instantaneous system models, but developed stochastically over time instead of deterministically. A **Markov model** is a probabilistic (stochastic) process over a finite set of possible *states* evolving in time. Each state-transition generates a character from the "alphabet" of the process, i.e. from one of its possible states. Markov models are typically characterized as structured processes, often as *graphs*, depicting probabilities of given states coming up *next*, and this implies prior history, or memory.[16]

The *size of memory* in a Markov model determines its *order*. A first-order Markov model (also called Markov-1) has a memory of size 1, directly analogous to first-order difference or differential equations with one initial (or boundary) condition (initial state) to specify a solution uniquely. *n*th-order Markov (Markov-*n*) models have memory of size *n*, and *n*th-order difference or differential equations requiring *n* initial conditions to solve them going forward, and so on. Importantly, a **zero-order Markov model** has no memory, directly analogous to an *instantaneous* system or its model. Finally, a **hidden Markov model** is one in which the states of the model are unobservable in data (hidden, unmeasurable or unmeasured), a concept we develop further in Chapter 4 in the context of *hidden modes*, or *hidden compartments* of a compartmental model.

Dynamic system properties are developed as we begin discussing methods for dynamic biosystem modeling and analysis in Chapter 2, we continue their development throughout most of this book, and Appendices C-E provide a more advanced treatment.

What about *nondynamic* systems modeling? Dynamic biosystems do exist in approximately nondynamic states, when viewed on a much longer time scale, an important issue in model building. That is, they may appear to be nondynamic, or instantaneous, relative to other systems coupled to them and operating on a very different, longer time scale. Cell biology models, intermixing dynamical processes occurring at nuclear (millisecond to second) and cytoplasmic (minute to hour) levels (scales), are an important example. Nuclear *transcription* events, for example, appear approximately instantaneous relative to cytoplasmic *translational*[17] event dynamics. Michaelis-Menten theory of enzyme-substrate interactions is another, which also forms the mathematical foundations for receptor-ligand interactions (Michaelis and Menten 1913). The basic assumptions of Michaelis-Menten theory lead to simplifications, yielding some nondynamic *algebraic* systems of equations (constraint relationships) in their models, representing fast binding reactions for some variables, as described in detail in Chapter 6. These approximations are expressed with simpler mathematics as special cases, embedded within the framework of dynamic systems analysis, which rules the kingdom here.

---

16 Such processes are usually called Probabilistic Finite State Automata (PFSA) or Probabilistic Finite State Machines (PFSM) in computer science, because they are closely linked to finite state automata as used in formal language theory.

17 **Transcription** is the name given to the process of making a messenger for specific genetic information encoded on a DNA segment (gene), i.e. a gene transcript complement called **messenger RNA** (mRNA). Transcription is controlled by enzymes and other factors (RNA polymerases, **transcription factors**, etc). The message carried by mRNA is translated (decoded – **translation**) at the cellular site of protein biosynthesis, i.e. the production site (the **ribosome**) for making of a specific sequence of amino acids into a polypeptide chain (**protein**). In humans and other eukaryotic species, transcription occurs in the cell nucleus, translation in the cell cytoplasm. Prokaryotes (bacteria, etc) have no nucleus; both occur in the cytoplasm.

Example 1.2: ***A Simple Elimination Dynamics Model***

To keep things very simple mathematically – to assure comprehension – let's consider the following *in vitro* "thought experiment." Introduce substance X into a beaker of acid, an acid that presumably breaks down X in proportion to how much is in the beaker at time t. We say "presumably" because this may be a modeling assumption to keep in mind. Later we may discover that real data do not match this model as well as anticipated. Call the amount added $q(0)$, the zero here representing the time the experiment begins. Formally, this model can be stated as:

The elimination rate of X, $ER(t)$, is proportional to its amount $q(t)$ in the beaker. This amount falls over time as the acid does its job. Mathematically:

$$ER(t) = -kq(t) \tag{1.1}$$

Both variables are written explicitly as functions of time t here, because this is a time-varying process. $ER(t)$ has units of mass per unit time, and $q(t)$ has mass units. *Therefore the units of k are inverse time* (t^{-1}).

Now let's frame this as a dynamic system model and solve it for $q(t)$ for $t > 0$. Physically, the *rate* of elimination of X can be expressed as a derivative of $q(t)$ with respect to t, which, physically, is the elimination rate defined above:

$$\frac{dq(t)}{dt} = ER(t) = -kq(t) \tag{1.2}$$

Again, ER has a negative sign, because X is disappearing. Eq. (1.2) is a simple first-order differential equation model, and the intricacies of its solution are given in Chapters 2 and 4. For our purpose here, the general solution of Eq. (1.2) is $q(t) = Ae^{-kt}$, with A arbitrary. This is easily verified by substitution of $q(t) = Ae^{-kt}$ into both sides of Eq. (1.2). To get the *unique* solution to this model, we use the initial amount of X, $q(0)$, as an initializing value for the equation solution:

$$q(0) = Ae^{-k \cdot 0} = A \cdot 1 = A \tag{1.3}$$

Thus,

$$q(t) = q(0)e^{-kt} \tag{1.4}$$

The initial condition $q(0)$ is the *unique memory* (initial state) of this dynamic system model. This model and its decaying single exponential solution are ubiquitous in biology, with Eq. (1.1) being the most common model for constitutive elimination rates of proteins and other biomolecules. A graph of Eq. (1.4) is given in **Fig. 1.4** in Cartesian coordinates. On semi-log coordinates, i.e. ln $q(t)$ vs. t, the exponential curve

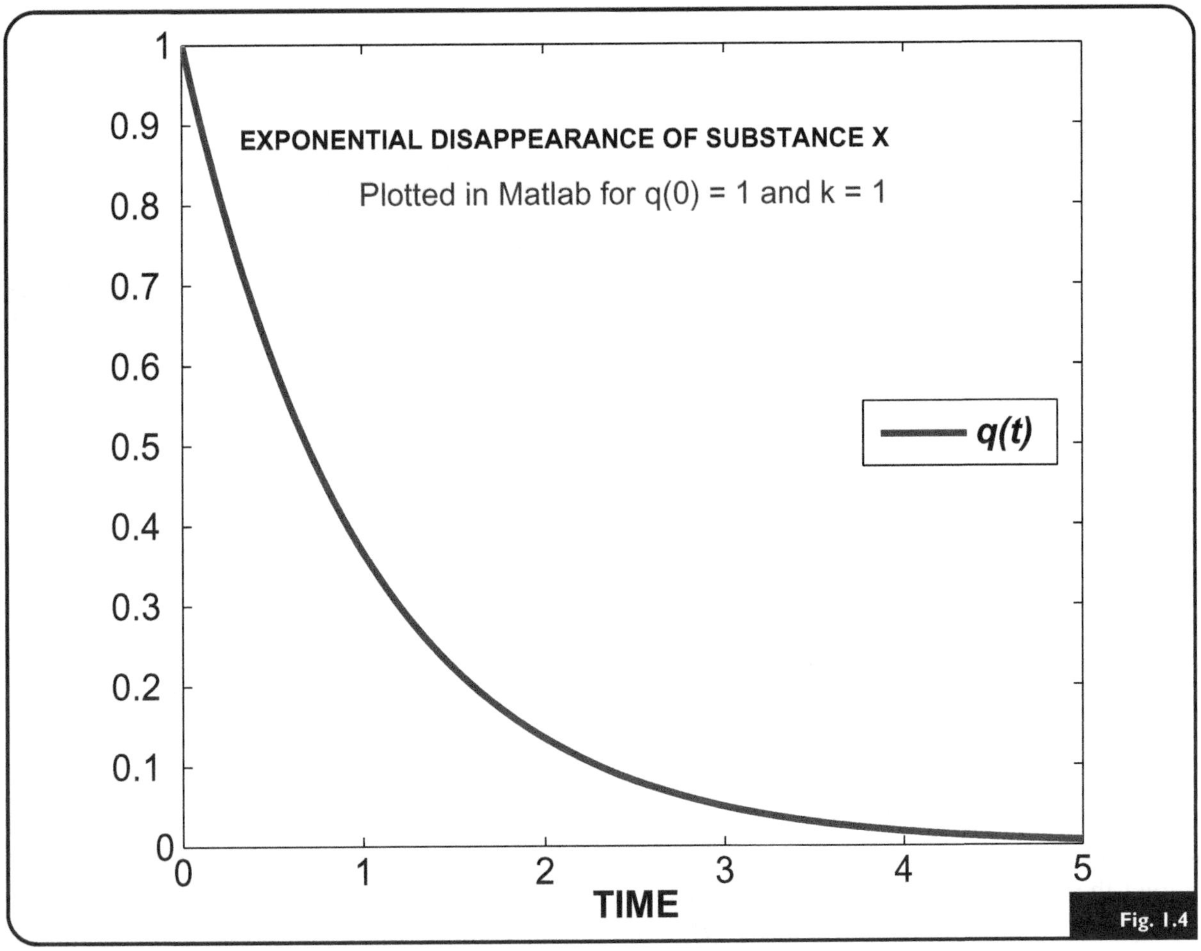

Fig. 1.4

Single-exponential elimination dynamics.

becomes a *straight line*, with slope $-k$. This exponent is often estimated from $q(t)$ data as the slope of the line fitted to this data on semi-log coordinates.

Measurement Models & Dynamic System Models Combined: Important!

As noted above, modeling only the dynamics of a biosystem, the part between outputs and inputs, does not finish the job. Laboratory data representative of the outputs and inputs also gets modeled, and the result usually is called the **measurement model**. Thus, completing the task means modeling both the biosystem and the biodata. We adopt the intuitively appealing system theory paradigm for dynamic system models (Zadeh and Desoer 1963), illustrated in **Fig. 1.5**, with separate blocks representing the dynamic system model and output measurement model distinctly. The two together generate a more complete dynamic system model. In Chapter 10 we present the **complete dynamic system model**, also called the **constrained structure** (DiStefano III and Cobelli 1980). This includes all additional information about the

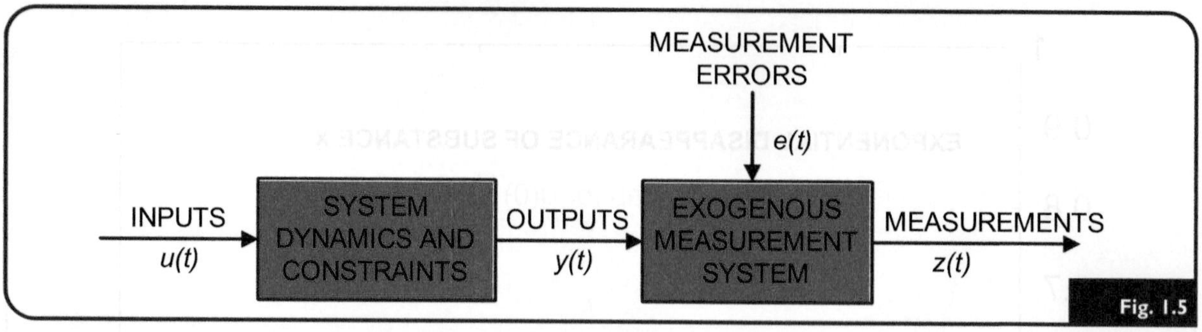

Fig. 1.5

A complete dynamic system model. The system dynamics block (left) embodies the model of the system as such, with inputs $u(t)$, outputs $y(t)$ and constraint relations among parameters or variables. The exogenous measurement system block embodies the model of the system generating noisy output data (the measurements), with measurement errors $e(t)$ corrupting the measured outputs, yielding the output measurement data $z(t)$. Statistics usually deals with the right-side, deterministic control theory with the left, and more general system theory with both acting together.

parameters and variables of the model, in the form of constraint relationships augmenting the system dynamics model.

For example, the system dynamics might be a set of differential equations, and these equations might represent temporal dynamic interactions among several proteins in living cells, following an exogenous input stimulus (e.g. constant external radiation, environmental toxins, etc.). Constraint relationships might include conservation equations for amounts of proteins that exist in two states, bound and unbound to receptors, the total amount being constrained as the sum of the two (for example). The output measurement model might then depict the concentration over time of one of these proteins, call it the *output* $y(t)$, measured say via QPCR,[18] and may take the form $z(t) = y(t) + e(t)$, where $z(t)$ are the "noisy" concentration data measured at times t. The **noise** or **measurement errors** $e(t)$ are the *irregularities* in the temporal measurements. They have an appropriate statistical description, e.g. time-dependent sample means and sample standard deviations.

Inputs (**input models**) are often expressed as mathematical functions (e.g. sinusoids, exponentials, etc.) and the usual convention is to include them in the dynamical model equations.[19] Strictly speaking, however, inputs are not part of system dynamics as such and it is important in many modeling applications to maintain input variables separate from system dynamical state variables, as depicted graphically in **Fig. 1.5**. For example, inputs are important "independent" experiment design variables for analysis of dynamic system models for quantification – the topic of Chapter 10, on *structural identifiability*. Alternatively, if the inputs are represented by measurement data, they can be included mathematically in the overall measurement model, if it is useful, by *augmenting* the measurement model with the input data model equations. In any case, input data or input functions must be provided to solve the model.

Remark: Given that noise in data is not deterministic, i.e. not numerically predictable, realistic measurements are essentially stochastic processes, and thus their models have implied

18 Quantitative polymerase chain reaction (see wikipedia.org/wiki/QPCR).

19 This somewhat unfortunate convention for inputs is convenient – because it is parsimonious – but can be confusing in modeling applications, especially where inputs are meant to be manipulatable experiment design variables rather than system variables constrained by internal system dynamics.

or explicit probabilistic (statistical) descriptions. However, because they are usually nondynamic – represented by algebraic equations – it is often unnecessary to deal with them formally as *stochastic process models* when developing or studying deterministic system dynamics models – our major focus in this book. Deterministic system dynamics models have deterministic states and outputs – assuming model *inputs* are deterministic, as in **Fig. 1.5**. However, if inputs $u(t)$ include stochastic (random) components (Papoulis 1965), then the entire model also is stochastic, and theory and analysis are more involved, as discussed in Chapter 7 in the section on *Stochastic Models and Stochastic Simulators.*

Example 1.3: ***Complete Model of Elimination Dynamics and its Measurements***

If we measure the changing concentration of X in the beaker of acid continuously as a function of time after time 0 in Example 1.2, and call this output measurement $y(t)$, then the noise-free measurement model is $y(t) = \frac{q(t)}{V} = \frac{q(0)}{V}e^{-kt}$, where $V > 0$ is the nonzero acid volume in the beaker and initial condition $q(0) > 0$ is another positive constraint on this model. The *complete dynamic system model* is:

$$\frac{dq(t)}{dt} = -kq(t), \quad y(t) = \frac{q(t)}{V}, \quad V > 0, \quad q(0) > 0 \tag{1.5}$$

If the measurements are taken at N discrete time instants instead of continuously, and the data are noisy, with noise $e_i(t)$, the complete model might have the form:

$$\begin{aligned} &\frac{dq(t)}{dt} = -kq(t) \\ &y(t_i) = \frac{q(t_i)}{V} + e(t_i) \;\text{ for }\; i = 1, 2, \ldots, N \\ &V > 0, \; q(0) > 0 \end{aligned} \tag{1.6}$$

Stability

The terms *stable, unstable, stability* and their variants mean different things in various scientific disciplines, some differences being subtle. It's useful to note some of these here, to distinguish them from notions of stability of dynamic systems and their models, which we begin introducing formally in Chapter 2.

Dictionary definitions of *stable* say, "steady, firm, not likely to move or change" – all implying some kind of constancy. In physics, stable means enduring, indefinitely long-lived, having no known mode of decay, as in stable (nonradioactive) isotopes of compounds. In chemistry, a stable compound is one that is not easily decomposed or otherwise modified chemically. Molecular biologists appear to generally adhere to this notion of molecule stability. Their focus is often on molecular *degradative* processes, as in controlled degradation of protein molecules, as part of protein or gene regulatory networks – an example being **proteolysis** by proteinase enzymes (**proteases**) or the **proteosome** in cells.

This is the focus of many *molecular systems biology* problems, i.e. proteins may interact – with one altering the *degradation rate* of another – as exemplified below. This kind of regulation is treated separately from what is called **constitutive** (also **basal**), or ever-present and balanced production and elimination processes, i.e. constitutive production equaling constitutive degradation. Balanced basal or constitutive activity of proteins in a cell usually occurs on a much longer time scale than other regulatory processes. We are emphasizing here that this kind of stability – of a protein (or similar molecule) – is not about the bio*system* participating in nonconstitutive regulation, but only about the degradative properties of the protein, i.e. it's about an *element* in the biosystem, not *system* stability.

Example 1.4: ***Mdm2 regulation of the tumor suppressor protein (transcription factor) p53***

The enzyme *Mdm2* serves to inhibit *p53* levels (abundance) in the nucleus of the cell by increasing the p53 degradation rate. This is what is meant by *Mdm2* decreasing the *stability* of *p53*. *Mdm2* also inhibits the *activity* of *p53* from doing its job as a transcription factor, by blocking it, without affecting its level, and therefore not its stability.

Overall regulation of the *p53* system includes another major component pathway: *p53* activation stimulates transcription of the *Mdm2 gene* – thereby eventually increasing production of *Mdm2* molecules in the cytoplasm, which then reenter the nucleus and interact with *p53* in the two ways described above. This "negative feedback loop" of *Mdm2* interaction with *p53* is a primary component of the essentially *stable system* controlling *p53* activities in the cell (Proctor and Gray 2008). The *p53* regulation system can exhibit (nonconstant) oscillatory behavior when confronted with DNA damage stress in the cell (Proctor and Gray 2008), but – as we shall see in Chapters 14 and 15, these are stable, bounded-magnitude oscillations. Thus, *system* stability here is not a property of the way *Mdm2* acts to "destabilize" the *p53* molecule, but a property of *all* the different interactions among these (and other unnamed or undiscovered) component molecules.

Notions of stability as they are commonly used in everyday discourse, and system stability to a large extent, carry forward to the technical world in a fairly straightforward way, as do instability and *system instability*, along with their negative connotation. However, as we have just illustrated, system stability is a property of all (essential feature) system components acting *together*. If the system also interacts with its environment, then that too must be included in stability analysis. The *parts* of the system can, separately, have the opposite stability properties.

A physically intuitive example of an unstable system is a rod (or pencil) balanced vertically at a point (**Fig. 1.6**); it readily falls when moved ever so slightly from this so-called **unstable equilibrium point**.[20] Importantly, the system here includes both the pencil and the

20 An equilibrium point of a system is a point where the system has no (net) motion, i.e. first derivative equals zero.

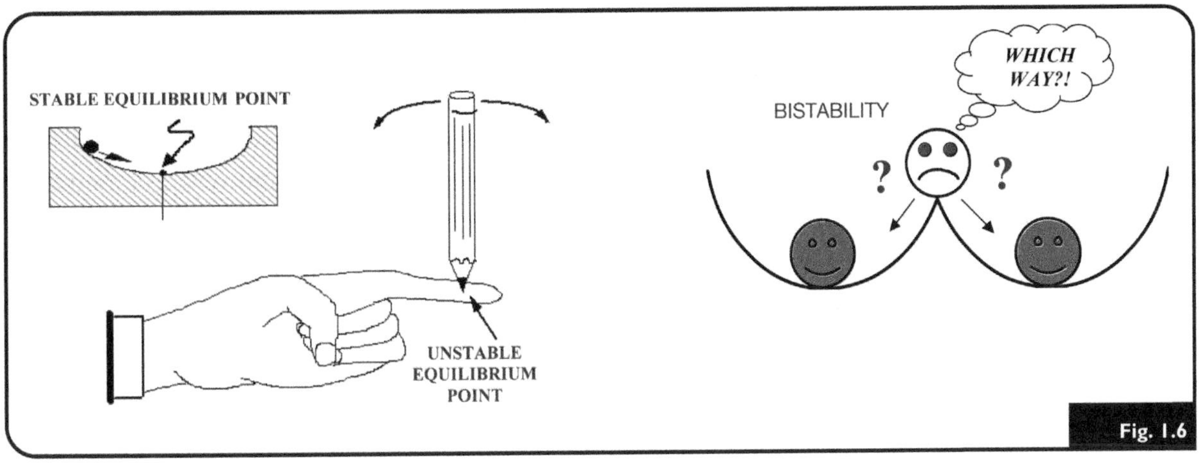

Fig. 1.6

Three kinds of equilbria.

hand attempting to balance it (the essential features). Similarly, a ball resting at the bottom of a hemispherical well (together, the system) is at its **stable equilibrium point (Fig. 1.6)**. Moving the ball to either side and releasing it yields damped oscillations of the ball about the **asymptotically stable** equilibrium point. Neither the pencil nor the ball are by themselves inherently stable or unstable; but the systems in these simple examples are unstable or stable.

A third example illustrates a much more complex phenomenon, called **bistability (Fig. 1.6)**, where the ball at the unstable equilibrium point has two ways to fall to one of two stable equilibrium points; like balancing the pencil, it appears that it can go either way depending on "the way the wind is blowing" (*sic*).

Robustness & Fragility

These important system stability notions were introduced in the control systems engineering literature more than half a century ago (Nyquist 1932), and have been developed and utilized for designing or analyzing control systems – biological or otherwise – for just as long. However, in the recent systems biology literature, they are being investigated more extensively because they relate to biosystem stability in a more complex *evolutionary dynamics* way. They are fundamental properties of evolving dynamic systems of any kind, in particular complex and evolvable biological systems.

Roughly speaking, *robustness* of a biosystem means – first – that it maintains overall "stable" behavior in response to (random or expected) perturbations or disturbances, akin to the usual notion of system stability.[21] Thus, a robust biological system should, first, handle short-term disturbances – at least well enough to survive them. Second, it should also adapt itself, as needed, so that it survives and thrives more effectively over longer evolutionary time.

Fragility is the opposite of robustness: fragile systems are highly sensitive to disturbances. Paradoxically, fragility can also be beneficial to the organism. Robust biological systems

21 Thiss is likely to bring to mind the classical notion of Claude Bernard, termed *homeostasis*, introduced briefly three sections below and later in several chapters.

evolved to handle most disturbances can be highly sensitive to perturbations rarely encountered, e.g. a new and deadly virus. In this sense, they are robust if their immune system responds quickly and handles the virus over the short term, or the patient dies but the species evolves to handle it (survival of the fittest). It has been argued that biological systems possess both properties – robustness and fragility – balancing survival, performance and resource demands (Carlson and Doyle 2002).

Stability, robustness and fragility analysis methods are developed in Chapters 9 and 11.

Top-Down & Bottom-Up Modeling

These two terms are used often in modeling, referring to the relative degree of detail in the model, top-down being more "macroscopic" and bottom-up being more "microscopic." Both, however, are usually involved – at least implicitly – in the modeling process, and deciding on a level of complexity is ultimately dependent on what the modeler considers *essential features* in the context of modeling goals *and* data availability. Thus, we often settle on a model that lies somewhere between the two extremes – **in-between** modeling.

Top-down modeling entails first modeling "holistically," i.e. capturing the overall dynamical features of the system, expressed (for example) as a collection of interacting subsystems, and then modeling the subsystems. **Bottom-up modeling** begins by modeling the subsystems. Then the overall system is modeled by interconnecting them, and so on. The block diagram models illustrated in **Fig.1.1** and **Fig.1.3** are models at the top level. If we developed detailed models of the subsystem blocks first, and then interconnected them, that would be bottom-up modeling. Bottom-up modeling is reductionist, the traditional scientific approach to studying biology.

Remark: A structural hierarchy of components, e.g. molecules, cells, tissues, organ systems, etc. is often used as a starting point for biosystem modeling. However, deciding on an impractical level of complexity at the outset can greatly hamper the success of a modeling project, maybe even kill it. Ultimately, model complexity is absolutely limited by data, whether available or anticipated. That is why a good starting philosophy is "let the data speak first." We have much more to say on this topic. Another dimension, a temporal instead of or in addition to a structural hierarchy, also may govern the modeling effort to some degree, in the context of multiscale modeling, described below.

Remark: Data is also sometimes distinguished as **top-down data** or **bottom-up data**. High-throughput data is usually top-down. For example, genome-scale mRNA expression profiling data, used to quantify the simultaneous expression of many or all genes under different conditions, is top-down data – often with some form of system modeling goal. In contrast, the relatively small numbers of data derived from classical genetics, biochemistry or physiology experimentation are bottom-up data – often used to validate, refine or otherwise complement high-throughput data.

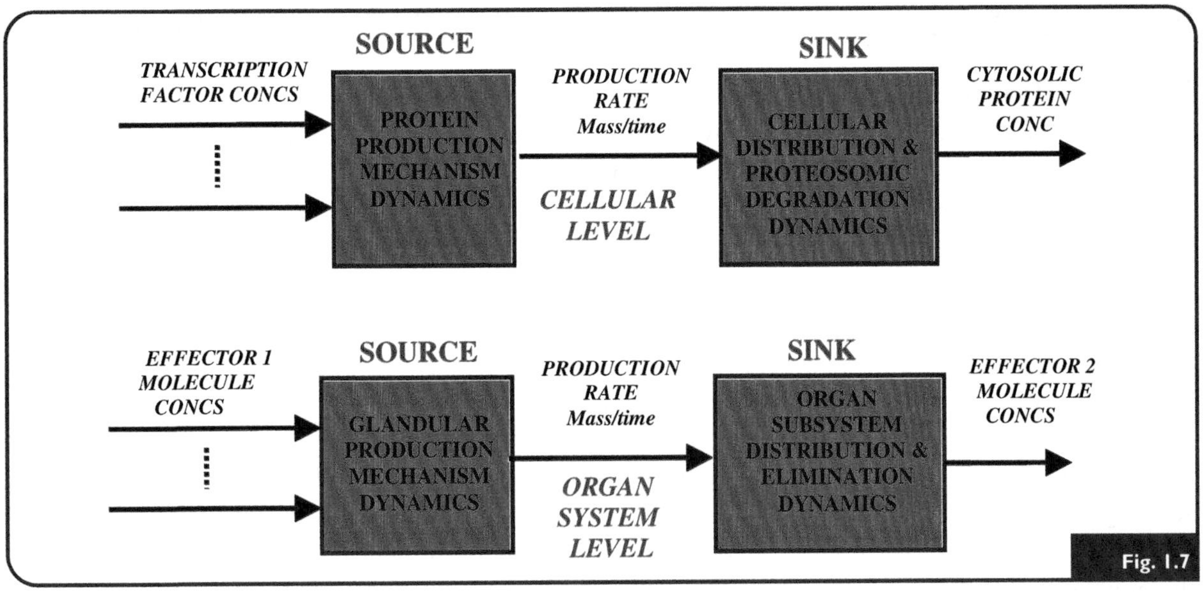

Fig. 1.7

Source and sink block paradigms for submodeling.

Source & Sink Submodels: One Paradigm for Biomodeling with Subsystem Components

It is often convenient and useful to separate system component models hierarchically into subsystems, or **submodels**. And it is particularly helpful to organize these components into dynamically and/or physically distinctive categories. Biosignal sources and sinks are two such categories. By **biosignals** here, we mean state variables representing masses, concentrations or other measures of chemical, electrical or other biophysical entities. **Sources** are locations/processes where/how these signals are produced (e.g. transcription of genes → translation of proteins, anabolism, *de novo* secretion, etc.). **Sinks** are where/how they are eliminated, and it is often convenient to include distribution with elimination processes in sinks (e.g. distribution in the organelle, cell or organism, plus disposition processes like degradation, proteosomic action, catabolism, excretion, etc.). Distribution and elimination also can be kept separate, if this is useful in satisfying the goals of the modeling.

Fig. 1.7 illustrates organization of source and sink processes using block diagram submodels at two distinctly different levels or scales. For this paradigm, *outputs of source blocks are fluxes or rates*, e.g. production rates in mass per time units. *Inputs are state variable signals* affecting these rate outputs, e.g. amounts, concentrations, volts, amps, lengths, etc. Sink inputs and outputs are the reverse, as shown. The figure uses cellular protein production, distribution and elimination (cellular level); and organ (e.g. glandular) production, distribution and elimination (organ system level), to illustrate the paradigm. Sources and sinks, however, being inherently physical concepts, are represented in different ways for different paradigms in different disciplines or modeling domains. In Chapter 4, for example, sources and sinks are represented by arrows into and out of "compartments" – from and to the "environment."

Systems, Integration, Computation & Scale in Biology

Systems Biology

Systems biology is an evolving field with various definitions, all involving biological *systems* and more or less inclusive of biology at different scales: hierarchical – from molecular to organismic; and temporal – from nanoseconds up to years and beyond. The goal of systems biology is to gain a better understanding of biological system function *as a whole*. Recent growth and interest in this field is primarily at the cellular and subcellular levels, involving protein network and metabolic pathway modeling (Alon 2007; Fiehn and Weckwerth 2002). However, systems biology has been with us (albeit with less fanfare) conceptually since the turn of the 19th century and mathematically since the middle of the 20th century.

A the end of the 19th century, French physiologist Claude Bernard introduced the notion of self-regulation in physiological systems – in terms of constancy of the "*millieu interieur;*" and Walter Cannon, at Harvard, later systematized this concept in what he termed "homeostasis" – the constancy of physiological system variables in the presence of disturbances (Adolph 1961). Biomathematics and biophysics – without a particular systems focus – began evolving decades earlier, spearheaded by pioneers like Otto Schmitt (biomimetrics) at the University of Minnesota (Schmitt and Schmitt 1931; Schmitt and Schmitt 1932) and Nicholas Rashevsky at the University of Chicago (Rashevsky 1938). Mathematical biosystems modeling began developing soon after, first in the guise of a new science called "cybernetics."

The focus on physiological systems – particularly neurophysiological systems – grew substantially in the 1940s. Norbert Weiner, Warren McCulloch, Arturo Rosenblueth, Julian Bigelow, Walter Pitts, W. Ross Ashby, W. Grey Walter and others together pioneered *cybernetics* – the study of feedback, control and communication processes in living organisms, machines and organizations (Ashby 1952; Ashby 1956; McCulloch and Pitts 1943; Rosenblueth et al. 1943; Walter 1953; Weiner 1948). Honing in more specifically on biology, advances in systems physiology accelerated in the 1950s and 1960s, motivated by rapid developments in systems engineering, control theory and computer simulation methodology, led by a growing population of bioengineers and biosystem scientists. Arthur Guyton[22] and H. Thomas Milhorn at the University of Mississippi, William Ganong at UCSF, Fred Grodins at USC, Gene Yates at Stanford, Larry Stark and Douglas Riggs at MIT and John Milsum at McGill were among the earliest proponents and practitioners of organ system modeling in physiology (Ganong 1991; Grodins 1963; Guyton et al. 1955; Milhorn 1966; Milsum 1966; Riggs 1970; Stark 1968; Yates and Urquhart 1962; Yates 1971; Yates 1973).

Systems Physiology & Pharmacology

Systems physiology developed with a focus on integrative biological function at the organ-system level, with emphasis on organization and regulation of organ and organ-system processes. Interest in math modeling in pharmacology also developed during this period – apparently independently of systems physiology – motivated by legislation requiring the pharmaceutical industry to

22 Arthur Guyton's computer model of the cardiovascular system was the first large-scale computer model integrating the many factors influencing the peripheral circulation – the heart, the endocrine systems, the autonomic nervous system, the kidneys and body fluids (Guyton et al. 1972). An analog computer program diagram of this "giant-of-a-model" can be found in Chapter 14.

establish standards for demonstrating that the drugs they produce and sell enter the bloodstream and have their intended effects. These were ideal questions to address with math models, and from this, the fields of **pharmacokinetics** (abbrev: PK) – roughly, what the body does to the drug, and **pharmacodynamics** (abbrev: PD) – what the drug does to the body, i.e. its effects, were born. More recently, physiology is playing a larger role in these models, and **physiologically-based** (abbrev: PB) **PK or PD (PBPK or PBPD) models** are somewhat the vogue (Chapter 8). A fairly new field emphasis in numerous pharmacology education departments is **systems pharmacology**, with emphasis on system- and organ-related aspects of pharmacology and with model-system studies as well as system models, with PK, PBPK, PD, and PBPD modeling playing supportive roles.[23] A new moniker, **pharmacometrics**, has been coined to encompass all of these quantitative math modeling methodologies in pharmacology, with a focus on patient populations and an emphasis on their variable responses to drugs. In chemistry and toxicology, physiologically-based **toxicokinetic** (PBTK) models[24] are becoming prominent in studies and policy decisions designed to protect the environment as well as the health of humans and other species (Villeneuve and Garcia-Reyero 2010; Nichols et al. 2011).

The granularity of dynamic systems biology is certainly growing sharper, especially at molecular levels. Recent interest in systems methods and math modeling in biology has been targeting more microscopic levels, and with great fanfare and support from research funding institutions. This new emphasis is due in large part to the explosion of *-omics* data emanating primarily from genomics level studies, which accelerated during the 1990s, and more recently from transcriptomics, proteinomics, lipidomics and metabolomics study data, generated primarily from newer high-throughput[25] experimental technologies. Elucidation of the genomes of many species, including human, has provided a very large catalog of materials and components. By the end of the 20th century, innovative molecular and cellular biologists were realizing that "Humpty Dumpty was getting divided into more parts than they knew what to do with," and they needed help. They discovered that the means for reassembly of these parts into physiological systems were at hand, via computational and systems modeling methods, and that it was time to begin that endeavor in earnest. Indeed, high-throughput methodologies essentially *forced* the systems viewpoint. Systems biology drew much greater interest from a larger group of biologists than ever before, entering a world with many more components than when dealing with macroscopic physiological systems. This subdivision of the field is appropriately and more often termed **molecular systems biology**, the primary components of which are at the molecular level. Molecular systems biology research is about how these molecular components are linked within networks, and the functional states associated with these dynamical networks. Participants at Harvard[26] and other venues have also called it the *new physiology*, or *modern physiology*, which do seem to require additional context for understanding.

---

23 See nigms.nih.gov/Training/IOSP.htm

24 A PBTK model example for Mn toxicokinetics is developed in Chapter 8 (Douglas et al. 2010).

25 **High-throughput** experimentation means *automation of experiments* such that large-scale repetition becomes feasible. For cell biology research, this technology permits rapid, highly parallel explorations of how cells function, interact with each other, and – in clinical applications – how pathogens exploit them in disease.

26 See http://focus.hms.harvard.edu/2004/Sept17_2004/systems_biology.html

"Science is facts; just as houses are made of stones, so is science made of facts; but a pile of stones is not a house and a collection of facts is not necessarily science."

Henri Poincare (1854–1912)

Earlier biologists – firmly rooted in experiment and observation, but without the great advantages of high-throughput experimental methodologies, such as DNA microarrays – paid little attention to mathematical modeling or systems biology methods. Fortuitously, in this instance, the new measurement technology had the effect of awakening much broader interest in quantitative modeling methods, methods that have worked well for many decades in physics, engineering and other areas of applied mathematics. Necessity did not need to "mother new invention"; extending old ones appears to be working effectively, at least for biomodeling at one or two scales, e.g. subcellular with intracellular – in gene network modeling (Smolen et al. 1998) and pharmacogenomic modeling (Jin et al. 2003). And classical dynamic systems modeling approaches readily accommodate many problems at the intersection of intracellular with organ levels in the biosystem hierarchy, e.g. in immune system (Ferl et al. 2005) and neuroendocrine system modeling (Eisenberg et al. 2006), to mention only a few.

High-throughput microarray technology is currently used to measure expressions (quantitative responses) of hundreds, thousands, or even millions of genes, proteins, metabolic products or other components of a measured sample, all simultaneously. Systems biologists are using systems modeling methods to make sense of this exponentially increasing data, integrating it into meaningful biosystem models, to hopefully explain how life works at multiple scales of space and time. The recent emergence of systems biology as an independent field of study in major universities (Harvard, UCLA, UCSD, UCI, UCSB, UCB, Stanford, Texas A&M, Pittsburg, MIT, Chicago, etc.) underscores the increasing awareness and need for systems biology and mathematical modeling methods. Furthermore, integration of systems-level modeling with experimentation is becoming the norm in systems-level biological studies, done more and more by the same study groups instead of by distant collaborators. Systems biology models provide an explicit structural focus for depicting coordinated interactions or couplings among individual biosystem elements. By their very nature, these models become a systematic guide for laboratory measurements, rendering the whole process more efficient and potentially fruitful.

Multiscale Modeling

The fact that dynamics at different structural levels are often occurring simultaneously over different time scales[27] appeared as an early challenge for molecular systems biology modeling. At the intracellular level, vastly different characteristic time scales exist. Transcription factor (TF) activation occurs in *milliseconds*, TF binding to its DNA site in *seconds*, followed by the dynamics of protein transcription and translation occurring in *minutes*, and so on. New questions and data spanning multiple boundaries like these have engendered new problems and discoveries in recent years. Regulation of the T-lymphocyte, the blood-borne white cell core of adaptive immunity, is one example spanning multiple time scales. Intracellular signaling

27 Multiscale dynamics can complicate the modeling substantially, as alluded to in our brief discussion of mixed dynamic–nondynamic modeling and repeated in many examples throughout this book, beginning with Example 1.5.

components play a role in T-cell signaling and responses to external stimuli, thereby playing a crucial role in regulating the whole-body (system level) immune response to specific pathogens (Kumar et al. 2009).

Dynamic systems modeling across widely divergent time scales is not straightforward, and not unique to biology. This technical modeling issue occurs often enough in physics and engineering – called "stiff" problems – and analytical and computational approaches exist for solving them. Beginning in Chapter 3, we shall see that structural and other data from multiple levels of information-signaling hierarchies can be integrated, thereby providing coupled dynamics of the parts, or among the parts, in both time and space. Formally describing such integrated dynamics is what is meant by *multiscale modeling*. **Multiscale biomodels** thus depict biosystem dynamics at a variety of spatial and temporal scales. A recent paradigm in systems biology is modeling so-called **scale-free networks** of regulated genes of many different (base-pair) sizes and receptor capacities, operating on many different time scales, none being dominant (Barbasi and Oltvai 2004). Problems like these motivated a multiscale modeling initiative in 2003 by several funding agencies,[28] as well as institutes dedicated to multiscale modeling, e.g. the Center for Integrative Multiscale Modeling and Simulation (CIMMS) at Caltech, and, in 2003, a new Society for Industrial and Applied Mathematics (SIAM) journal, *Multiscale Modeling and Simulation* (siam.org/journals/mms.php).

Example 1.5: ***A Simple Multiscale Model, Simplified and Solved***

As another variation of the simple elimination model in Example 1.2, we consider its solution, Eq. (1.4), over a *very large* time interval. In this case, k, the rate of fall of the exponential appears very large, relatively speaking, i.e. it appears to fall very rapidly and, for large enough t, it appears to be a "spike" of magnitude $q(0)$ – visualizing it over a very large time interval. The disappearing function becomes an algebraic constant[29] $q(0)$ at $t = 0$. The "dynamics" are, for all practical purposes, gone. The model is (approximately) nondynamic if we look at it from a great distance in time.

This demonstrates what we mean when we say (distinct) dynamic biosystems exist in approximately nondynamic states when viewed on a much longer time scale. Let's see now how the notion of *multiscale* dynamics, i.e. viewing dynamics across spatial and temporal scales, might enter the modeling mix.

Suppose we are interested in the dynamics of another molecule with mass $v(t)$ coupled to (influenced by) and thus integrated with this one, but with dynamics (or kinetics) changing 100 times slower, i.e. at rate $k/100$ per unit time. Let's assume the dynamics of this hypothetical molecule are governed by $dv(t)/dt = (k/100)q(t)$, i.e. growing at rate $k/100$ instead of disappearing at rate k like $q(t)$. Then our original elimination model $dq(t)/dt = -kq(t)$, Eq. (1.2), looks nondynamic – or relatively

28 Interagency Opportunities in Multi-Scale Modeling in Biomedical, Biological, and Behavioral Systems: Program Solicitation (http://www.nsf.gov/pubs/2004/nsf04607/nsf04607.htm), managed by the National Institute of Biomedical Imaging and Bioengineering (NIBIB), as well as institutes dedicated to multiscale modeling, e.g. the Center for Integrative Multiscale Modeling and Simulation (CIMMS).

29 Mathematically, the limit of $q(t)$ as $k \to \infty$ is the constant $q(0)$.

instantaneous, i.e. $q(t) = q(0)e^{-kt} \cong q(0)$ over the longer time scale of $v(t)$ dynamics. Then the *coupled dynamics model* can be approximated (simplified) as the single equation $dv(t)/dt \cong (k/100)q(0)$. The right-hand side of this equation is constant and therefore the solution $v(t) \cong (kq(0)/100)t$, is an increasing function of time. This one-dimensional solution approximately represents the essential dynamical features of the coupled system, i.e. it is influenced by q at $t = 0$, but appears to be uncoupled from $q(t)$ for all $t > 0$.

We have thus illustrated one way a multiscale model can be solved: by simplifying it to one that still reasonably accurately reflects its *essential dynamical features*. We shall see in Chapter 3 that "simplified" here is not only good for its own sake (Thorburn 1918, pp. 345–353). It also can be the best way to resolve difficult problems in simulating and solving multiscale biomodels numerically. Do keep in mind, however, that mathematical and computational methodology for coupling multiple scales effectively and seamlessly in biomodeling is still under development and we should be keeping an eye out for new methodologies, although fundamentally older approaches are still providing new results (Chu and Hahn 2008).

Summarizing, maximally effective systems biology research follows a back-and-forth path between theory (modeling) and experiment (data collection), each supporting and refining the other. Modeling the *biodata* along with the *biosystem* is essential. In this sense, systems biology integrates experimental data and computational modeling approaches to study and understand systems level interactions in biological processes at all levels – within and among cells, tissues and/or organisms. We subscribe here to this broadest, most inclusive view of systems biology, a primarily quantitative systems approach to all biology at multiple scales. The ultimate goal of systems biology "in a nutshell" is to gain a better understanding of function in a biological system holistically, the main focus on the whole, rather than the parts.

Bioinformatics

Bioinformatics is generally considered to be about computational, mathematical and statistical approaches for acquiring, organizing, archiving, visualizing and analyzing high-throughput biological *data*, and not about biological *systems* as such. Bioinformatics methods and algorithms establish, classify and correlate genetic and cellular information within and across species. They are based in statistics and computer science disciplines, with data-mining, data association, pattern recognition, machine-learning and visualization methods currently at the forefront. Examples of basic research areas currently being investigated with these approaches include DNA sequence and protein structure alignments, gene-finding and associating, modeling of evolution, and *geogenomics* – elucidation of the global planetary role of existing diversity of species genomes in their concerted interactions within local and regional ecosystems. Applied research areas include drug discovery and design, as well as personalized medicine based on patient sequence measurements.

Computational Systems Biology & Computational Biology

Computational systems biology is an integrative paradigm in computational biology, an aggregation of bioinformatics and its sister field, systems biology. The integrative goal here is creation of a finely detailed, mechanistic picture of cellular, molecular and organism level biology (the system), accomplishing this by combining the parts – genes, annotated genome objects, and established metabolic and other source and elimination pathways – with observations and data of transcriptional (via microarrays), translational (via proteomic tools) states of a cell and whole-organism kinetics (via tracer kinetics). System and data are merged (integrated) via algorithmic and other computational methods. At the time of writing (2013), these tools are still being developed primarily and independently by bioinformatics and systems biology researchers. Closer integration of efforts over the near term is anticipated, lest we fail in our goal of re-integrating the vital parts.

Computational biology is the "umbrella" field here, containing bioinformatics, systems biology and computational systems biology and the like as sub-areas. Consistent with this notion, the National Institutes of Health (NIH) web site defines computational biology as "*The development and application of data-analytical and theoretical methods, mathematical modeling, and computational simulation techniques to the study of biological, behavioral, and social systems*."

Overview of the Modeling Process & Biomodeling Goals

This book is about methods for goal-oriented modeling of dynamic biosystems – operating normally or abnormally – and application of these models to basic and clinical problems. Importantly, dynamic biosystem models include submodels for input–output measurements, as well as system dynamics and boundary conditions (constraints), as presented schematically in **Fig. 1.5**. Modeling usually begins with available theory – theory (hypotheses) about biomechanisms, biosystem components and their interconnectivity – and any other information about the biosystem (or its environment) relevant to formulating the model consistent with modeling goals.[30] Whereas modeling system dynamics and constraints on these dynamics is the focus, measurement models – the modeling of system or experiment *inputs* as well as dynamic system *outputs* – are included, along with their statistical or stochastic description, where appropriate. The measurement model is important at every stage of the modeling process, as discussed briefly below – and as a theme repeated throughout this book.

Goals range over many scales; from characterizing complex networks of presumably interacting proteins or other cellular components – together with their expression or concentration profile and input condition or perturbation data (for example, for the purpose of understanding their connectivity or function) – to characterizing sparse temporal biodata on a more macroscopic scale, by simple equations, like a sum of one or two exponential functions, for modeling of blood-borne kinetic data, perhaps for a therapeutic control system application.

Putting this all together, useful biomodeling science is a systematically integrated and multidisciplinary effort of experimental and mathematical modeling approaches. It is an

30 Other information often includes values or estimates of model parameters and/or boundary conditions, often called "literature data."

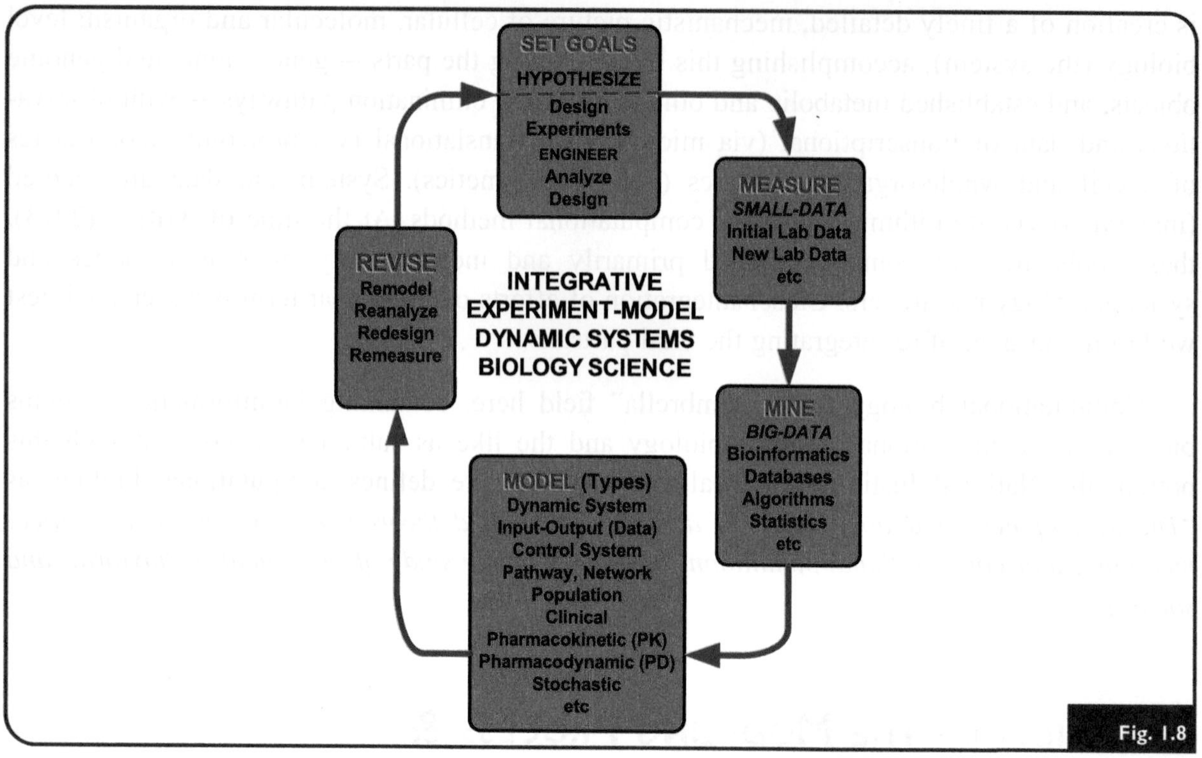

Integrative systems biology modeling paradigm.

iterative procedure that closely integrates theory, and experimental and mathematical approaches for the purpose of satisfying modeling goals. Whether these goals are testing mechanistic hypotheses in cellular or molecular biology, at one end of the spectrum of modeling domains, or developing clinically useful devices, or models, of populations at the other end, the procedure is essentially the same. **Fig. 1.8** illustrates this hypothesis/goal-driven paradigm in contemporary terms. The circular (repetitive) nature of the modeling process, integrating and supporting theory with data along the way, also is illustrated in **Fig. 1.8**.

To give biomodeling goals a biological flavor, I sometimes use the terms **diagnostic model** or **therapy model** or **synthesis model**, where appropriate, analogous to the terms modeling-for-*analysis* or modeling-for-*design* (or synthesis) in engineering. Modeling a cellular metabolic network, for example, would be modeling for "diagnosis" (analysis), if the primary goal is to understand the couplings, or their relative import, in affecting network behavior – or how the network affects entities outside it. On the other hand, if the modeling goal is to re-engineer a network model in a cell, so that it produces more of a desired substance (biofuel for energy), this might be called "synthesis modeling." Another example of "synthesis modeling" is re-engineering a process in a bacterium to consume undesirable entities (oil slick on the ocean). These are the realms of **synthetic biology**, which also usually requires a "diagnostic model" to accomplish the task. The exponential function model, if it represents (say) a pharmacokinetic response to oral dosing of a drug, would be a "therapy model," as it is usually intended for designing drug-dosing schedules for drug therapy.

As discussed in earlier sections, dynamic systems modeling methods are used for studying classes of living systems at all levels and scales, including: evolutionary dynamics (Nowak 2006), the dynamics[31] of socially interacting organisms and systems of organisms; pharmacological, physiological and cellular-level systems (Alon 2007; Chu and Hahn 2008; Gabrielsson and Weiner 2000; Khoo 2000); and for integration of physiological studies of biological systems at multiple scales (Ferl et al. 2005). This book includes developments or examples of all these. Whereas deterministic, ordinary differential equations (ODE) are the primary *modus operandi* for modeling biosystem dynamics, *stochastic system dynamics* modeling is also addressed – for systems involving small numbers of interacting molecules.

Modeling goals (of a diagnostic nature) may include better *understanding of organization and interactions* among system components in (for example) signaling networks, genetic networks and metabolic networks at mechanistic levels (modeling for *diagnosis*), as well as overall functional states of such biosystems. Biosystem models of dynamical cellular processes can be developed, not only for depicting transcriptional and translational regulation of proteins, as noted earlier, but also for cell cycle control, apoptosis, cell differentiation, division and migration; intracellular transport, membrane and organelle biogenesis; energy generation and utilization; host–pathogen interactions; or intercellular communications; or any combination of these. Biophysical and biochemical principles generally govern the modeling process at these levels.

A primary use of biosystem models is *testing competing hypotheses* about such processes, which we illustrate with examples throughout the book, and focus on specifically in Chapter 15. More common uses include *predicting behaviors* or dynamic responses to internal or external stimuli, or predicting interactions among organisms competing for environmental resources, or simulating *normal or abnormal biosystem dynamics* to aid in their repair when they are broken. Models intended to predict physiological and pathophysiological behavior in response to natural and artificial perturbations can contribute to the understanding and treatment of human diseases. Examples are included in the second half of the book.

As noted earlier, systems biology is playing a larger role in pharmacology and toxicology, in what were – until recently – primarily reductionist areas of investigation. Biosystem models depicting how genes, proteins, potential drugs and metabolites interact are being used early on to gain holistic understanding of drug dynamics – including drug absorption, distribution, metabolism, excretion, toxicity and interactions at target sites, from the cellular to whole-body levels (Jin et al. 2003; McMullin et al. 2007). The traditional approach in drug discovery has been based on bottom-up science, studying first how compounds interact with drug targets, followed by *in vitro* and laboratory animal testing, and then to testing in patients. It is often the case that previously unrecognized toxicity and effectiveness problems are discovered at this late endpoint. Effective systems biology modeling in drug discovery (pharmacometrics) can short-circuit some of these issues at earlier stages, with obvious benefits in clinical therapy and – in environmental toxicology – provide computational models to augment *in vitro* data for studying effects of toxins in various species and the environment (Villeneuve and Garcia-Reyero 2010). Illustrative examples of dynamical toxicokinetic modeling are provided in Chapter 8 and beyond.

31 See wikipedia.org/wiki/Social_dynamics for more information.

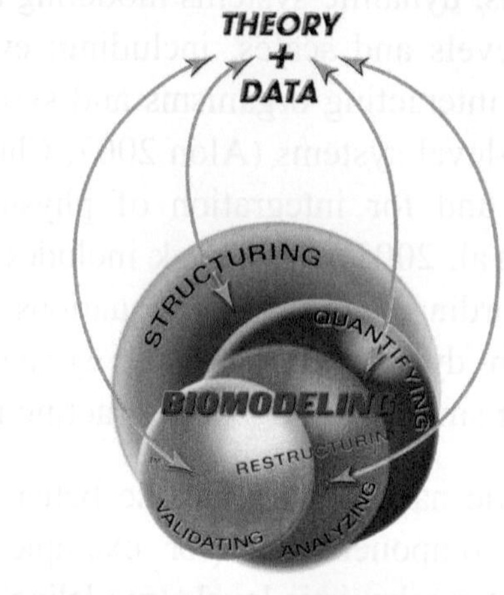

Fig. 1.9

Circularity and dynamics of data-driven biomodeling. (Biomodeling/Theory + Data Image (The Logo) is reproduced here and in other chapters and back cover of this book with permission by the author Joseph DiStefano III. It is a Trademark ™ and service mark of the author, Copyright © 2012.)

Data-Driven Biomodeling: Structuring, Quantifying, Analyzing and Restructuring

Data-driven biomodeling is a central theme here, because data usually drives theory (hypotheses), or "informs it," by driving adjustments or corrections to theory (**Fig. 1.9**). This then gets reflected in an evolving model as a new (or modified) hypothesis. Ultimately, only data can validate theory – and the model – when it satisfies the goals. What comes first – theory or data – is not necessarily easy to discern; and it is not important to dwell upon it here, because context and specifics usually make this clear. Typically, new data leads to adjustments, large or small (mostly small), in an existing theory; alternatively, they generate a new theory – along with genesis of a new model.[32]

The *form* or shape of this theoretical construct (or hypothesis) – the one that gets adjusted by or born from new data – is called the model **structure**. This usually begins with a diagram or schematic representation for the hypothesis, although (less likely) it could start with the equations of the model-hypothesis.[33] Structuring is about organizing distinct system or subsystem components, establishing their couplings (interactions), and expressing them – first schematically and then mathematically – as a mapping of the structure expressed by the schematic. Structuring is usually limited by essential system features – and is essential for satisfying modeling goals, as described in an earlier section.

An exception is when essential features have not yet been established. For example, a complex model might be structured first to include all known and hypothesized details about biosystem components and interconnections, in order to study what is essential to its dynamical behavior. Its complexity then can be reduced (with the aid of data) to essential features. The biochemical pathway schematic, the block diagram and the molecular biology cartoon

32 Most of Chapter 5 is specifically about generating (minimal) dynamic system models from biodata.

33 Or "law," if the hypothesis being modeled is a well-accepted one in the community, e.g. Ohm's law, Newton's laws etc.

schematic in **Fig. 1.1** are example structures. **Fig. 1.3** illustrates assembly of block diagram components into a feedback biosystem model structure.

The entire modeling process can be thought of as formulating and adjusting a model structure from old and new data, and using mathematical methods for expressing, quantifying, analyzing and validating evolving structures until they resolve into one (or more) that satisfies particular modeling goals. Pedagogically speaking, none of these modeling steps – structuring, expressing,[34] quantifying, analyzing or validating – can be fully separated from the others. This is not a semantic issue, but an operational one. They overlap and are interdependent to varying degrees, as illustrated in **Fig. 1.9**, with *restructuring* depending strongly on quantifying, analyzing and validating the model,[35] usually from new data. Importantly, existing data for quantifying and new data for validation or restructuring comes in the form of *input–output* data generated from dynamic biosystem experiments. This clearly establishes the importance of the measurement model for this data (see **Fig. 1.5**), the mathematical expression and modeler's handle on the empiricist's window into the biosystem. Without it, the model remains but a theoretical construct.

Biomodel structuring is done differently in different scientific disciplines, sometimes differently for different problem domains in the same discipline. Structuring basics are included in Chapters 2–9. Structuring based primarily on physics and engineering is included in Chapters 2, 8 and 14; on control systems engineering in 3, 9 and 14; compartmental and directed graph theory in 4 and 5; biochemistry and cell biology in 5, 6 and beyond; and physiology/anatomy and engineering paradigms in 7, 8 and 14. The different approaches and structural paradigms developed in these chapters have much in common, and all delineate a visual form for facilitating mathematization – mapping onto differential and algebraic equations representing biosystem dynamics and input–output experiments. Example paradigms are shown in **Fig. 1.10**. These include: an enzymatically controlled reaction diagram, with substrate S and enzyme E forming product P, in two different schematic representations – a classic enzymology diagram and a Systems Biology Graphic Notation (SBGN) form[36] (top); compartmental model diagrams – a simple one and a more complex physiological one (next two levels); a physio-anatomic diagram representing a typical physiologically based (PB) pharmacokinetic (PK and PBPK) structure (fourth level) and, next to it, a noncompartmental model diagram and a directed graph of a network pathway. Control system engineering diagrams, not included in **Fig. 1.10**, are illustrated in most other figures of this chapter. All of these structural representations (and others) are used for modeling throughout the remaining chapters.

Quantifying a model means quantifying its unknowns (e.g. parameters, boundary conditions), so that it matches data collected from a biosystem experiment, usually input–output data. Formal methods for quantification are developed beginning in Chapter 10.[37] In the context

34 Semantically, expressing the model in equation form can be included in either structuring or quantifying.

35 On the other hand, exposition of the mathematical methods for carrying out these steps, in whatever order needed, can be separated, as in the chapters of this book. The ordering is not fixed.

36 http://SBGN.org

37 Quantification methodology appears in the latter half (Chapters 10–13) rather than earlier in the book because the subject is more complex mathematically. This is my compromise to cater for differences in educational backgrounds of my own students studying from early versions of this book. Didactically, with sufficient applied math background, all or parts of them should be studied earlier – for deeper understanding of modeling *in toto*.

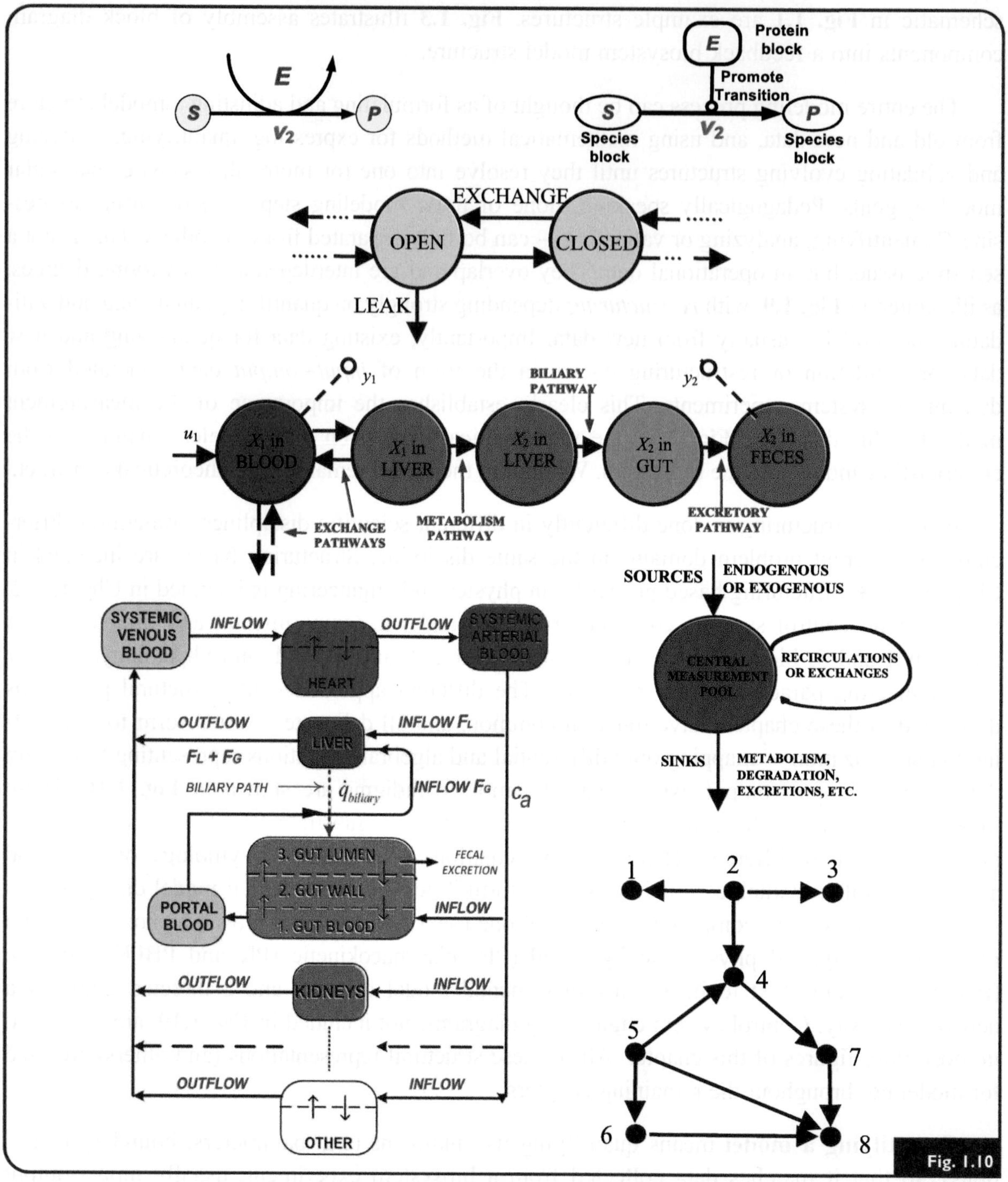

Fig. 1.10

Example modeling structures from different disciplines (see text). Engineering control system diagrams, illustrated earlier in the chapter, are not shown again here.

of quantification, **analyzing** means establishing the extent to which it does or does not match the data. This often then entails adjusting the model structure (**restructuring**), if the match is not (yet) good enough to satisfy modeling goals **(Fig. 1.9)**. **Analyzing** also means discovering dynamical characteristics of a model, e.g. its stability properties. Exercising (simulating, solving) the model to establish various dynamical characteristics, e.g. stability properties, means analysis as well. **Validating** usually means analyzing it by comparing measured model outputs to data not used in model construction. Failure of validation usually motivates or guides restructuring.

The deeper meaning of each of these steps is developed as the needed math modeling tools evolve in subsequent chapters.

The Biomodeling Process in Toto: *A Philosophical Recap*

Before moving on, we step back a bit for some perspective reinforcement of the nuances of the modeling process and its dynamical features described above and depicted in **Fig. 1.9**.

The figure graphically illustrates the integration of steps in the process of building biomodels and keeping them alive and pertinent. The fluid nature and oligarchical interdependencies of the biomodeling enterprise are prominent. *Theory* and *data* appear as overarching "gods," inseparable from the core process as they are from each other. The process depicted is intrinsically a dynamical system – dynamical in both the memory-building (state-determined) and changing sense defined in this chapter and framed in the exposition of this book.

In **Fig. 1.9**, the several modeling steps blend one into the other, in a Venn-like aggregate with *imprecise boundaries*. The figure reflects the difficulties in distinctly defining the borders of – or even ordering – this circularity of steps. Each is expressed overtly and grammatically as an action verb, meant for present and future,[38] reflecting that the biomodeling process is hardly ever complete, despite often well-defined goals pursued by those us with even the best of exteroceptive abilities. A faithful and complete (and goal satisfying) model is what we seek, but permanence is elusive – the nature of the beast. And for good reason!

Math models are always (or always should be) amenable to, and likely to be in need of, adjustment. Not just for fixing errors, or augmenting it with knowledge missed in earlier research. They require updating when new or more refined data comes along, following development of newer measurement schemes capable of probing deeper.[39] No matter how honest and how hard the effort, it is highly unlikely that any model with limited goals and based on current theory and data captures all essential features *in perpetuity*. How could it? Today's theory and data are surely incomplete in the larger frame of things. Biomodels either evolve, to remain faithful and useful, or should be abandoned and labeled as such.

In chapters to come, **Fig. 1.9** is used as a pedagogical guide, morphed and remodeled a bit, to reflect the goals and content of the particular chapter.

38 Grammatically speaking, the present and future continuous tense are used here – as in structuring, quantifying, validating, restructuring.

39 2013 Examples: High-throughput screening methodologies (ultra-fast robotic measurement) are greatly expanding model dimensionality and prediction abilities for the bioinformatics and pharmaceutical-discovery worlds. So is mass spectrometry and tracer-based isotopomer analysis for metabolomics. Functional magnetic resonance imaging (fMRI) is opening new windows for collecting detailed dynamic physiological data in living systems non-invasively. The list is ever-expanding, so model refinements – and the new knowledge they bring – must follow.

Looking Ahead: A Top-Down Model of the Chapters

We begin with fundamentals of *math modeling* in Chapter 2, including differential equations and transform methods, illustrated with classical examples from physics, population dynamics, biomathematics and control systems engineering. Then we cover programming and the basics, mechanics and limitations of dynamic biosystem *simulation* methods in Chapter 3, along with some model reformulation methods – like *scaling* of model variables – to facilitate analysis. Then we move on in Chapter 4 to structuring our first class of dynamic biosystem models – *multicompartmental* and related *structural biomodels*, both linear and nonlinear. This intuitively appealing model class is fundamental, and it gives us an opportunity to develop the conceptual framework for biomodeling in a pedagogically facile context. At the same time, it provides some important qualitative analysis methods for matching models to data, and ready-to-use tools for modeling what is commonly called biosystem "kinetics data," as well as many other biological systems – linear and nonlinear, in which compartmentalization is commonly used or useful. *Pool models*, another practical modeling class, are also introduced in Chapter 4; pool models capture inhomogeneities macroscopically, in a manner useful for mass-flux balance analysis (for example).

Chapter 5 is focused on developing algebraic, geometric and systems theory tools for sizing, distinguishing and simplifying models from input–output data. Graphs and computer science *graph theory* also play a role here. *Linearization* techniques, fundamental to nonlinear model stability analysis – among other dynamical properties – also are presented in this chapter.

The fundamentals of nonlinear dynamic system biomodeling based on *mass action* and mass-action-like principles are developed in Chapters 6 and 7. This is the basis for most mesoscale cellular-level, biomolecular, drug and *population dynamics* modeling. *Stochastic* dynamic biosystem modeling, for biosystems with very small numbers of interacting species, is also presented in Chapter 7.

In Chapter 8 we develop *physiologically-based*[40] (PB) modeling – including physiologically-based *pharmacokinetic* (PBPK), physiologically-based *pharmacodynamic* (PBPD) and physiologically-based *toxicokinetic* (PBTK) modeling, plus the related issue of *allometric scaling* – and also *whole-organism kinetics* and *noncompartmental* modeling and analysis methods. Dynamic biosystem model *oscillations and stability analysis* methods, for nonlinear and linear models, follow in Chapter 9.

Chapters 10–13 cover the methodologic essentials and examples of model *quantification* from data: identifiability, sensitivity and parameter estimation. In Chapter 10, we explore parameter *structural identifiability*, a particularly difficult technical issue in biomodeling. This subject reveals the intricacies of how biomodel topology and data input–output structure are related, and how they limit what can be established from a biomodel and its associated data – even when the data are perfect. In Chapter 11, we present and illustrate methods for model parameter *sensitivity* analysis, a key issue in model interpretation, usefulness, simplification

40 A one-page synopsis of vascular and tissue physiology is also presented in Chapter 8, to help with modeling approaches in that chapter.

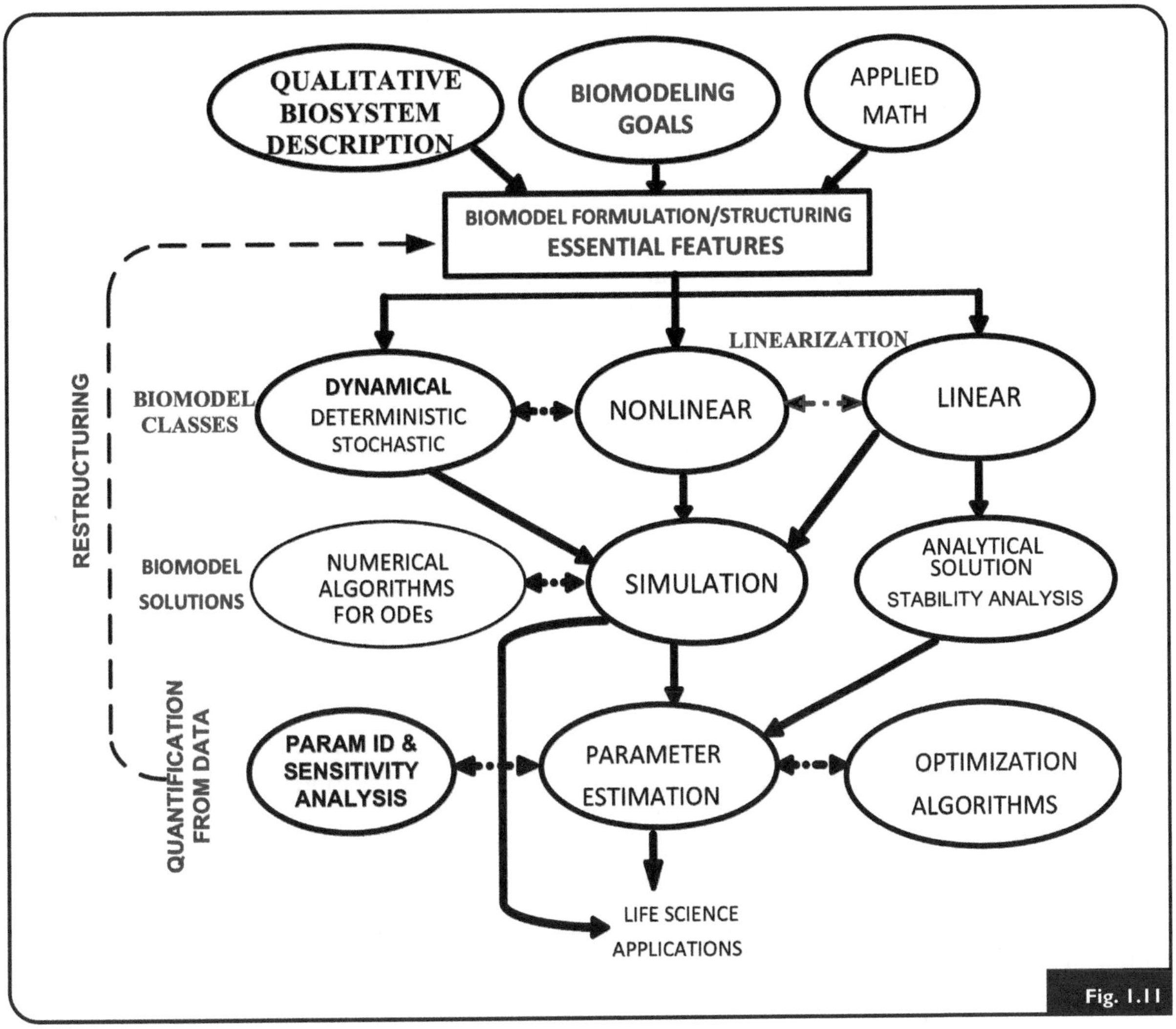

Fig. 1.11

Flow chart of the dynamic biosystem modeling and simulation process.

and followup experiment design. In Chapter 12 we present methods for model *parameter estimation* from real data, also called the *inverse problem*, and associated issues of *numerical identifiability* and estimation errors in parameter optimization, i.e. computing and interpreting parameter estimation variances and correlations. Identifiability, sensitivity and parameter estimation are intimately related problems in data-driven biomodeling. Chapter 13 is focused on methods for facilitating, simplfying and working with data for model parameter estimation.

Chapter 14 is about dynamic and steady-state feedback and feedforward *biocontrol system* modeling methods, with example applications and focus on human physiology, cell biology and metabolism modeling and *parameter estimation from sparse data*. The focus of Chapter 15 covers real physiological modeling problems more broadly, again using relatively sparse real data. Data-driven biomodeling for alternative *hypothesis-testing* is a focus.

Chapter 16 addresses use of identifiability methodology and graphs for kinetic *experiment design*, as well as information theory for *design of optimal sampling schedules* in kinetic experiments. *Model reduction* and *network inference* methods based in sensitivity analysis is the topic of Chapter 17.

Case studies and examples from real experimental studies at various scales and levels – from molecular to organismic, are developed throughout the chapters, to illustrate the concepts and methods presented. The several appendices cover Laplace transforms, pertinent linear algebra, more advanced control and systems theory modeling methods, and statistical methods for modeling and modeling research. Pseudo-code or real code (*Matlab, Simulink, VisSim*, etc.) is included with the solutions of many of the chapter examples, or can be found on the book website.

Fig. 1.11 is a conceptual organization of the overall biomodeling and simulation process, broken down into stages and steps. The ordering implied is somewhat flexible, as discussed in the last sections, but, for the most part, development of subject basics in this book follows this flow. The dashed "restructuring" line signifies the iterative nature of the process.

References

Adolph, E., 1961. Early concepts of physiological regulations. Physiol. Rev. 41, 737–770.
Alon, U., 2007. An Introduction to Systems Biology. Chapman & Hall/CRC, Boca Raton, FL.
Ashby, W., 1952. Design for a Brain. J Wiley, New York.
Ashby, W., 1956. An Introduction to Cybernetics. J Wiley, New York.
Barbasi, A., Oltvai, Z., 2004. Network biology: understanding the cell's functional organization. Nat. Rev. Genet. 5, 101–113.
Calaprice, A., 2000. The Expanded Quotable Einstein. Princeton University Press, Princeton, NJ.
Carlson, J.M., Doyle, J., 2002. Complexity and robustness. Proc. Natl. Acad. Sci. U. S. A. 99 (Suppl. 1), 2538–2545.
Charusanti, P., 2006. Math models of Gleevec effects on cellular-level and molecular-level signaling events in chronic myeloid leukemia. UCLA Dissertation.
Chu, Y., Hahn, J., 2008. Parameter set selection via clustering of parameters into pairwise indistinguishable groups of parameters. Ind. Eng. Chem. Res. 48 (13), 6000–6009.
DiStefano III, J., Cobelli, C., 1980. On parameter and structural identifiability: nonunique observability/reconstructability for identifiable systems, other ambiguities, and new definitions. IEEE Trans. Automat. Control. 25 (4), 830–833.
Donahue, W., Kepler, J. (Eds.), 1992. New Astronomy [Translation of Astronomia Nova]. Cambridge University Press, Cambridge, MA.
Douglas, P., Cohen, M.S., DiStefano III, J.J., 2010. Chronic exposure to Mn may have lasting effects: a physiologically based toxicokinetic model in rat. J. Toxicol. Environ. Chem. 92 (2), 279–299.
Eisenberg, M., Samuels, M., DiStefano III, J.J., 2006. L-T4 bioequivalence and hormone replacement studies via feedback control simulations. Thyroid. 16 (12), 1–14.
Ferl, G., Wu, A., DiStefano III, J.J., 2005. A predictive model of therapeutic monoclonal antibody dynamics and regulation by the neonatal Fc receptor (FcRn). Ann. Biomed. Eng. 33 (11), 1640–1652.
Fiehn, O., Weckwerth, W., 2002. Can we discover novel pathways using metabolomic analysis? Curr. Opin. Biotechnol. 13, 156–160.
Gabrielsson, J., Weiner, D., 2000. Pharmacokinetic and Pharmacodynamic Data Analysis: Concepts and Applications. Swedish Pharmaceutical Press, Stockholm.
Ganong, W., 1991. Cardiovascular regulatory mechanisms. Review of Medical Physiology. Appleton & Lange, Norwalk, CT.
Grodins, F., 1963. Control Theory and Biological Systems. Columbia University Press, New York.

Guyton, A., Lindsey, A.W., Kaufmann, B.N., 1955. Effect of mean circulatory filling pressure and other peripheral circulatory factors on cardiac output. Am. J. Physiol. 180, 463–468.
Guyton, A., Coleman, T.G., Granger, H.J., 1972. Circulation: overall regulation. Annu. Rev. Physiol. 34, 13–46.
Jin, J., Almon, R.R., DuBois, D.C., Jusko, W.J., 2003. Modeling of corticosteroid pharmacogenomics in rat liver using gene microarrays. J. Pharmacol. Exp. Ther. 207, 93–109.
Kepler, J., 1609. Astronomia Nova. Heidelberg, G. Voegelinus.
Khoo, M., 2000. Physiological Control Systems: Analysis, Simulation, and Estimation. IEEE Press, New York.
Kumar, H, Kawai, T., Akira, S., 2009. Pathogen recognition in the innate immune response. Biochem. J. 420, 1–16.
Larsen, P., Kronenberg, H.M., Melmed, S., Polonsky, K.S., 2003. Williams Textbook of Endocrinology. WB Saunders, Philadelphia.
McCulloch, W., Pitts, W., 1943. A logical calculus of ideas immanent in nervous activity. Bull. Math. Biophys. 5, 115–133.
McMullin, T., Hanneman, W.H., Cranmer, B.K., Tessari, J.D., Anderson, M.E., 2007. Oral absorption and oxidative metabolism of atrazine in rats evaluated by physiological moedling approaches. Toxicology. (In Press).
Michaelis, L., Menten, M., 1913. Die kinetik der invertinwirkung. Biochem. Z. 49, 333–369.
Milhorn, H., 1966. The Application of Control Theory to Physiological Systems. Saunders, Philadelphia.
Milsum, J., 1966. Biological Control System Analysis. McGraw-Hill, New York.
Nichols, J.W., Breen, M., Denver, R.J., DiStefano III, J.J., Edwards, J.S., Hoke, R.A., et al., 2011. Predicting chemical impacts on vertebrate endocrine systems. Environ. Toxicol. Chem. 30 (1), 39–51.
Nowak, M., 2006. Evolutionary Dynamics. Belnap Press of Harvard University, Cambridge, MA.
Nyquist, H., 1932. Regeneration theory. Bell Syst. Tech. J. 2, 126–147.
Papoulis, A., 1965. Probability, Random Variables, and Stochastic Processes. McGraw-Hill, New York.
Proctor, C.J., Gray, D.A., 2008. Explaining oscillations and variability in the p53-Mdm2 system. BMC Syst. Biol. 75, 1–20.
Rashevsky, N., 1938. Mathematical Biophysics: Physicomathematical Foundations of Biology. University of Chicago Press, Chicago.
Riggs, D., 1970. Control Theory and Physiological Feedback Mechanisms. Williams & Wilkins, Baltimore.
Rosenblueth, A., Wiener, N., Bigelow, J., 1943. Behavior, purpose, and teleology. Philos. Sci. 10, 18–24.
Schmitt, F., Schmitt, O., 1931. A vacuum tube method of temperature control. Science. 73, 289–290.
Schmitt, O., Schmitt, F., 1932. A universal precision stimulator. Science. 76, 328–330.
Schnell, S., Grima, R., Maini, P.K., 2007. Multiscale modeling in biology. Am. Sci. 95, 134–142.
Smolen, P., Baxter, D.A., Byrne, J.H., 1998. Frequency selectivity, multistability, and oscillations emerge from models of genetic regulatory systems. Am. J. Physiol., Cell Physiol. 274 (2), C531–C542.
Stark, L., 1968. Neurological Control Systems: Studies in Bioengineering. Plenum Press, New York.
Stolwijk, J., 1980. Mathematical models of thermal regulation. Ann. N. Y. Acad. Sci. 335 (1), 98–106.
Thorburn, W., 1918. The myth of Occam's razor. Mind. 27 (3), 345–353.
Villeneuve, D.L., Garcia-Reyero, N., 2010. A vision and strategy for predictive ecotoxicology testing in the 21st century. Environ. Toxicol. Chem. 30 (1), 1–8.
Walter, W., 1953. The Living Brain. Norton, New York.
Weiner, N., 1948. Cybernetics. MIT Press, Cambridge, MA.
Yates, F., 1971. Systems analysis in biology. In: Brown, J., Jacobs, J., Stark, L. (Eds.), Biomedical Engineering. F.A. Davis Company, Philadelphia, pp. 3–19.
Yates, F., 1973. Systems biology as a concept. In: Brown, J., Gann, D., Bronner, F. (Eds.), Engineering Principles in Physiology, 1. Academic Press, New York, pp. 3–12.
Yates, F., Urquhart, J., 1962. Control of the plasma concentrations of adrenal cortical hormones. Physiol. Rev. 42, 359–443.
Zadeh, L., Desoer, C., 1963. Linear System Theory: The State Space Approach. McGraw-Hill, New York.

Chapter 2

Math Models of Systems: Biomodeling 101

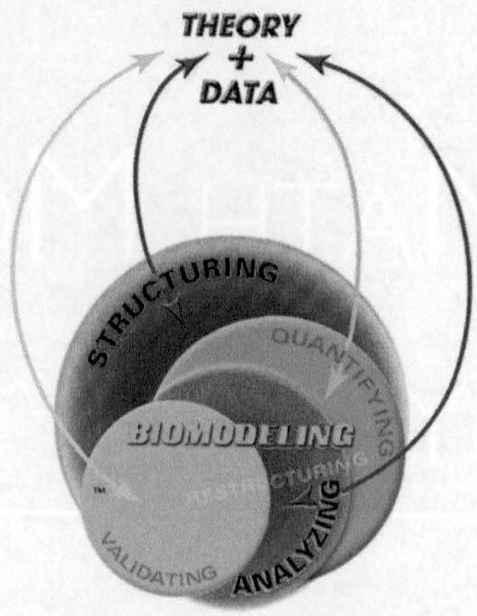

"How can a part know the whole? … and everything is both cause and effect, working and worked upon, mediate and immediate, all things mutually dependent. A bond that is both natural and imperceptible binds together things the most distant and the things most different. I hold it, therefore, impossible to know the parts without knowing the whole, any more than we can know the whole without knowing each particular part."

Pascal, B (1913) Pensees, Ed. de 1670 (French).

Some Basics & a Little Philosophy

In this chapter we present much of the language and some of the mathematical methods we need in this book, primarily based in algebra, geometry, differential and difference equations, plus some transformation methods. The math is at intermediate level and not very difficult. Little should be new for most science and engineering students. Most others should be more or less familiar with the terminology. In any case, after a little "walking," we should be up and running.

If needed, Appendix A is a short course on Laplace transforms and their application to solving differential equations and analyzing linear dynamic system models. Appendix B is a concise synopsis of linear algebra for biosystem modeling. More sophisticated readers can skim the chapter, as needed, at the very least for gaining familiarity with nomenclature and style, plus several likely unfamiliar topics.

"All science is either physics or stamp collecting."

This poignant and seemingly sarcastic quote is attributed to Ernest Rutherford (1871–1937) (Birks 1962). It succinctly summarizes the first half of our story.

The physics of biological systems, the *biophysics*, provides much of the mechanistic basis – or *theory* – for modeling biological systems. *Biochemistry* might be included in this discussion too, but we'll stay focused on biophysics for now. A major point here is that we can glean a great deal of intuition about modeling biological systems (abbrev: biomodeling) from physical analogies. The basic laws of physics, and biophysics, are represented symbolically by mathematical models, defining relationships between fundamental physical variables. We'll call these *math-models-based-on-theory*: **theory models**.

"It is a capital mistake to theorize before one has data."

These words, from Sherlock Holmes,[1] aptly summarize completion of the story and provide our first and foremost caveat to biomodeling.

Useful math models, especially mechanistic ones, are best developed from both theory and data.[2] As a practical matter, the two go hand-in-hand in biomodeling,[3] as was illustrated in **Fig. 1.9**. When we speak of *math-models-based-on-data,* **data-driven models** (introduced in Chapter 1), such models are usually not devoid of theory model components. Indeed, an ideal system model would be one developed only from well-validated theory component models.[4] And therein lies the caveat: theories are, by definition, hypotheses about the makeup or workings

1 The greatest detective of them all – penned by Arthur Conan Doyle (1859–1930) (Doyle n.d.).

2 Of course, most theory is (or was) based on real world data, as opposed to pure speculation. Thinking people would probably say that even the most speculative theories are likely to have been conceived from *some* data, no matter how obscure. When there is enough data to validate a theory – at least enough data to convince enough scientists that the theory is valid – then we usually say it's *established theory*, and call it a *law*, like Newton's and Ohm's laws in Example 2.1 & Example 2.2.

3 When considering data for modeling purposes, we first mean data for developing or structuring a model, e.g. informational data about biosystem morphology or topology, typically older data – expressed possibly as established theory – like Malthus' law in Example 2.3, or the *source-sink* models of **Fig. 1.7** in the last chapter. For an already structured model, new data – or recent data – might then be used for quantifying or validating (theoretical or speculative) components of a model, or the entire model.

4 A fully theoretical model would probably still require new data for validation.

of things – even when one is called a "law." Validation requires data. Practically speaking, the extent to which theory model components can be used in a biosystem model is limited by available data – data for both quantifying the model or validating use of this theory in the model. All this circumscribes our modeling approach.

ALGEBRAIC OR DIFFERENTIAL EQUATION MODELS

Equations describing a system model can take different forms. The first two examples below illustrate how two basic laws can be represented by either algebraic or differential equations. The same can be said for the third example, if we treat the algebraic form solution of the Malthusian differential equation model as a representation of the model as well, which it is!

Example 2.1: ***Newton's Second Law (Theory)***

Newton's second law states that if a force of magnitude F is applied to a mass of M units, the acceleration a of the mass is related to F by the equation $F = Ma$. If we classify this model by its mathematical properties, this equation is an *algebraic* model. Alternatively, this model can be written as a relationship between F and the rate of change of the velocity v of the mass with respect to time, $F = M\frac{dv}{dt}$ (a first-order *differential equation*), because, physically, $a \equiv dv/dt$. Another model is obtained if we express velocity as the rate of change of distance x traveled by the mass, i.e. $v = dx/dt$ which yields $F = M\frac{d^2x}{dt^2}$, a second-order differential equation. Newton's second law is illustrated later in the chapter, in Example 2.12, in a simple model.

Example 2.2: ***Ohm's Law (Theory)***

Ohm's law states that, if a voltage of magnitude V is applied across a resistor of R units, the electrical current i through the resistor is related to V by the algebraic equation $V = Ri$. Ohm's law can be written alternatively as a differential equation relationship between V, R and the time rate of passage of *charge* through the resistor, i.e. $V = \frac{dq}{dt}R$, because, physically, $i \equiv dq/dt$.

Remark: Functions of time t, such as $a(t)$, $v(t)$, $V(t)$, $x(t)$, $q(t)$ in the examples above, are sometimes written without the time t argument, i.e. $a(t)$ is written a and $v(t)$ as v, etc., to simplify the notation. This helps keep notation from masking meaning as the complexity of equations builds. Where the time-varying nature of the variable is important to maintain, we keep the argument t explicit, as in Eq. (2.1) below.

Remark: Newton's and Ohm's laws are used in several areas of quantitative biology, e.g. to model *biomechanical* systems – like skeletal muscle, walking or running dynamics, and electrical properties of neurons (in *neurophysiology)* and muscle (*myoelectric* phenomena). Newton's law is ubiquitous in modeling mechanical systems in motion.

Remark: Without loss of focus on the math here, these two "laws" surely are well validated, by virtually uncountable data – collected under appropriately controlled conditions. But they are not valid in some circumstances. For example, Newton's second law doesn't work well in very strong gravitational fields, for very small-scale phenomena, or for very high velocities. It was designed to work for macroscopic objects, under the earthly influence of constant gravitational fields. The "law" in the next example – albeit a major factor in motivating the science of population dynamics, is hardly ever valid in the real world.

Example 2.3: ***Malthusian Law: Simple Population Growth (Theory)***

The differential equation $\frac{dN}{dt} = cN$, is a representation of simple (oversimplified) *population growth* (Malthus 1798–1826), where N is the number of organisms in a population, e.g. the number of bacteria in a colony growing in a cell culture laboratory, and c is their **fractional growth (reproduction) rate** (with units of day^{-1}). The solution of this equation is: $N(t) = N(0)e^{ct}$ for $t > 0$. This is easily shown by substitution of $N(t) = N(0)e^{ct}$ and its first-derivative $cN(0)e^{ct}$ directly into the equation, i.e. $dN/dt = cN(0)e^{ct} = cN(0)e^{ct}$.

If the bacterial cell cycle takes 20 minutes, then $c = 1440$ min/20 min $= 72$ bacterial cell divisions per day, i.e. the fractional reproduction rate of the bacteria is 72 per day. After 2 days, then, one bacterium grows the colony to an impossible size: $e^{ct} = e^{144} = 3.45 \times 10^{62}$. Virtually unlimited resources – food and space – would be needed to make this happen.[5]

Remark: Let's face it: unlimited resources are impossible, so the simple exponential growth theory in the above example clearly needed "adjustment" to have practical significance – beyond the utopian dreams of philosophers of the day. This was clearly understood by Malthus (Malthus 1798–1826) and other theologians, scientists and philosophers influenced by his ideas and hopes during the 19th and 20th centuries (for example see Lennox 1993; Ariew 2007). Hard data and common sense data were the driving force for this during that same first half of the 19th century, generating two more realistic population demographic growth models: *logistic growth* and *Gompertz growth* "laws" (theories) – illustrated later in Example 2.10 and Example 2.11.

Differential & Difference Equation Models

The primary class of models we deal with are differential equation models, which model systems over continuous-time domains. We also need to appreciate difference equation (DE) models, which model systems over discrete-time domains, i.e. model equations defined only

5 If the cell cycle took twice the time, i.e. 40 minutes, then $c = 36$ and $N(t) = e^{72} = 1.86 \times 10^{31}$. Still too many!

for discretely separated instants of time. For example, dynamic systems *simulation* algorithms, the subject of Chapter 3, are based on DE models.

Differential equations are algebraic or transcendental (sine, cosine, etc.) equalities that include terms with either differentials (e.g. dx) or derivatives (e.g. dx/dt), defined continuously over one or more independent variables (e.g. t).

Remark: t usually represents *time* in this book, but space, location or other variables can be independent variables as well.

Difference equations (DE) are algebraic or transcendental equalities involving discrete sequences of values of the dependent variable(s) of the model, corresponding to discrete sequences of values of one or more of the independent variable(s) of the model.

Remark: Independent variables of DEs typically include time t, and possibly also space variables. Dependent variables of difference equations do not involve differentials or derivatives. However, if the latter are also present, the model is called a **differential-difference equation** model, which also has real world applications, including in systems biology.

Some nonphysical laws also can be represented by equations, a DE in the following example.

Example 2.4: ***Compound Interest[6] (Nonphysical Economics Theory)***

The *compound interest law* states that, if a principle amount $P(0)$ is deposited in an interest-bearing bank account for n equal periods of time, at an interest rate I for each time period, the balance will grow to $P(n) = P(0)(1+I)^n$ after n time periods, where n is any number in the sequence $n = 1, 2, 3, \ldots$. The compound interest law also can be written as a *difference equation relationship* between $P(k)$, the total amount of money after k periods of time (where sequence $k = 0, 1, 2, \ldots$), and $P(k+1)$ the total amount of money after $k+1$ periods of time. The difference equation is $P(k+1) = (1+I)P(k)$, also known as a **recursive** relationship. In this second form of the law, the integer k is variable and, in this manner, we use difference equations for relating the *evolution* of variables from one discrete instant of time to another.

Key Point: These several examples begin to illustrate how math modeling precisely formalizes concepts about real world processes, living or inanimate, into quantitative collections of symbols parsimoniously representing these conceptualizations. Modelers organize these symbols via this unique language of mathematics and, depending on their purpose, manipulate them into useful paradigms, using knowledge of worldly phenomena — like physics, biology and economics.

6 Einstein considered compound interest the most powerful force in the universe. It is surely a "law," as sure as $1+1=2$, given that it normally is used to compute interest precisely, not approximately, and needs no additional data to validate it.

Different Kinds of Differential & Difference Equation Models

A **partial differential equation (PDE)** is an equality involving one or more dependent and *two or more independent variables*, together with partial derivatives of the dependent variable(s) with respect to the independent variables. PDE models are sometimes called **distributed** models, or **distributed parameter** models.

Example 2.5: ***Diffusion Equation (Physical Theory)***

The one-dimensional **diffusion equation** $\frac{\partial c}{\partial t} = D\frac{\partial^2 c}{\partial x^2}$, also called the **heat equation**, describes the relationship between the time rate of change of a quantity in a medium (e.g. time-varying *density* of material, or *temperature*, along a narrow cylinder) and the positional rate of change of that quantity. D is called the **diffusivity** (or thermal conductivity) constant. This is a simple partial differential equation (PDE). The quantity $c = c(x, t)$ is its dependent variable (e.g. density or temperature); the position x (along the cylinder) and the time t are its *two independent* variables.

Remark: This class of PDEs, in 1-, 2-, and 3-dimensional spaces, is widely represented in biophysical modeling (Britton 1986).

Remark: PDE models are used for capturing essential features specifically and significantly dependent on space variables as well as time. Examples are found in intracellular process modeling, such as modeling electrodiffusion processes in subcellular systems (Keener and Sneyd 1998), or angiogenesis (new capillary formation) in tumor cells (Tee and DiStefano 2004). If the precise temperature in every region of a room is needed in designing room temperature control systems like that in Example 1.1 or **Fig. 1.3**, or in understanding fine details of body temperature control in homeothermic animals, PDEs are needed, because temperature depends on space variables as well as time in these systems.

An **ordinary differential equation (ODE)** is an equality involving one or more dependent variables, *only one independent variable*, and one or more derivatives of the dependent variables with respect to the independent variable. ODE models (our focus) are also called **lumped** models, or **lumped parameter** models.

Newton's second law, Ohm's law and the population growth models discussed above are ODEs. More general (and possibly familiar) ODEs of a particular class (linear) have forms like:

$$a_n(t)\frac{d^n y}{dt^t} + a_{n-1}(t)\frac{f^{n-1}y}{dt^{n-1}} + \ldots + a_1(t)\frac{dy}{dt} + a_0(t)y = u(t)$$

or more compactly,

$$\sum_{i=0}^{n} a_i(t)\frac{d^i y(t)}{dt^i} = u(t) \tag{2.1}$$

In this ODE, $a_0(t), a_1(t), \ldots, a_n(t)$ are generally time-varying coefficients (functions), $y(t)$ and $u(t)$ are dependent variables, and t is the single independent variable.

ODE (2.1) is called **time-invariant** (TI) (alternatively, **constant-coefficient**) if the $a_i(t) \equiv a_i$ are *constant*, exemplified by the Newton's or Ohm's law ODEs above. Otherwise, it is a **time-varying** (TV) ODE, and the coefficients usually retain their explicit time-dependent $a_i(t)$ designation. The $u(t)$ in Eq. (2.1) usually represents a forcing function or input. In general, it is time-varying, not constant. TI ODE models with $u(t) \equiv 0$ are sometimes called **autonomous**.

Difference equations also can be time-varying or time-invariant. For example, the difference equation $ky(k+1) + y(k) = u(k)$ where u and y are dependent variables, is time-varying because the term $ky(k+1)$ depends explicitly on the coefficient k, which represents the time **index** variable at the kth discrete time instant t_k, i.e. k varies with time.

Remark: For a time-invariant ODE model, the *waveform* of the dependent variable $y(t)$ (usually the *output*) does not depend on *when* the input $u(t)$ is applied.[7]

Remark: **Time-Varying Biomodels**. Time-varying parameters in biomodels are rare, an exception being the often-quoted "Steele equation" for glucose turnover, i.e. $\frac{dy}{dt} = -\frac{ky}{V(t)}$, where $y(t)$ represents glucose disappearance dynamics in a so-called "time-varying" (abstract) distribution volume $V(t)$ (Steele et al. 1976).[8]

Remark: **Time-Delay Differential Equations.** Nuclear transcription factor regulation of cytoplasmic protein synthesis, illustrated in Chapter 1 – **Fig. 1.1**, occurs at multiple cellular locations and PDEs might be used to accurately capture the coupled dynamics of individual component processes at the different locations. However, when dynamical properties are needed only at the whole-cell level, such *distributed* processes are often *lumped* together and represented as ODEs with time-delays, as in gene network modeling (Chen and Kazuyuki 2002; Klipp 2005), exemplified next.

*Example 2.6: **Gene Network Model (Mixed Theory–Multiscale Data Model)***

Suppose X_2 is a protein regulated by a specific gene that transcribes messenger-RNA (mRNA) X_1, with X_1 responsible for regulating production of protein X_2. Let lower case letters $x_1(t)$ and $x_2(t)$ be the time-varying masses (or numbers of molecules) of mRNA and protein. A typical **time-delay ODE** in a gene network model has the form:

$$\frac{dx_2(t)}{dt} = ax_1(t-\tau) + \ldots \tag{2.2}$$

The ODE represents the (net) rate of production of X_2 in the cell at time t, $\frac{dx_2(t)}{dt}$, in response to mRNA amount $x_1(t-\tau)$ transcribed earlier, at time $t-\tau$. The parameter a

7 If the input waveform is shifted in time by α, the output waveform will be shifted in time by α as well. For now, we reap the benefits of this property with greatly facilitated analyses. (See Appendix C for additional theory on time-invariance.)

8 This kind of oversimplification is not a recommended substitute for a more complex model that fits the data without rendering the biology so artificial. Improvements on the Steele model are presented in several references, e.g. Cobelli et al. 1987; Mari 1992.

is the fractional rate constant (t^{-1} units) for X_1 affecting X_2 production, i.e. it turns mass into a rate – dimensionally as well as quantitatively balancing the equation. The dynamics of the transcriptional and subsequent translocational and translational processes occurring at different locations (nucleus and cytoplasm) – at different times – are lumped together (approximated) here, their net effect thereby contracted as occurring τ time units later. This approximation is based on data indicating that translation evolves over a very much longer time span (>1000 times longer) than does gene transcription and translocation processes. These are *data-driven* parts of the otherwise simple dynamic system model, *simplified* as a pure time-delay.

Linear & Nonlinear Mathematical Models

Thorough understanding of the distinction between linear and nonlinear models is essential. The temporal and steady-state behavior (dynamical properties) of these two major model classes, in response to stimuli, can be quite different. Making the mathematical distinction, however – by inspection of the terms of the equations – is really quite simple. A **linear term** is one of *first degree* in the *dependent* variables and their derivatives. A **linear equation** is an equation consisting of a sum of linear terms. All others are **nonlinear** equations. That's it!

For example, $y(t) + u(t) = 0$ is linear, because it's first degree in both $y(t)$ and $u(t)$, while $y^2(t) + u(t) = 0$ is nonlinear, because it's second degree in $y(t)$. A linear model is usually a linear *equation* (e.g. $y + u = 0$), but sometimes a linear expression (e.g. $y + u$), not necessarily in equation form, is called a linear model.

For a differential equation model, if any term contains higher powers, products, or any transcendental functions of the *dependent* variables, it is nonlinear (abbrev: NL).

Example 2.7: ***Nonlinear Terms and Models***

NL terms include $\left(\frac{dy}{dt}\right)^4$, $y\frac{dy}{dt}$, and $\sin u$. Also, $\left(\frac{\sin 2t}{\cos t}\right)\left(\frac{d^2y}{dt^2}\right)$ is a term of first degree in the dependent variable y; $\frac{d^2y}{dt^2}$ is second *order*, not second degree. Similarly, $2uy^3\frac{dy}{dt}$ is a term of fifth degree in the dependent variables u and y. The ODEs $\left(\frac{dy}{dt}\right)^3 + y = 0$ and $\frac{d^2y}{dt^2} + \cos y = 0$ are both NL, because $\left(\frac{dy}{dt}\right)^3$ is third degree and $\cos y$ is *not* first degree because it is one of the transcendental functions.

Example 2.8: ***Second-Order NL Difference Equation Model***

$y(k+2) + u(k+1)y(k+1) + y(k) = u(k)$, in which u and y are dependent variables, is NL because $u(k+1)y(k+1)$ is second degree in u and y together. This type of nonlinear equation is sometimes called **bilinear** in u and y.

A linear ODE (LODE) generally has the form:

$$\sum_{i=0}^{n} a_i(t)\frac{d^i y}{dt^i} = \sum_{i=0}^{m} b_i(t)\frac{d^i u}{dt^i} \tag{2.3}$$

The coefficients $a_i(t)$ and $b_i(t)$ depend *only* upon the (continuous-time) *independent* variable t (or are *constant*), and not on y or u or their derivatives.

Remark: We noted earlier that the $u(t)$ on the right-hand side (RHS) of Eq. (2.1) typically represents a model input. It's also an input on the RHS of Eq. (2.3), but we have derivatives of input $u(t)$ as well. In this case, the inputs are not usually exogenous to the overall biosystem model. Rather, this is a form of ODE input–output relationship that often appears in internal (endogenous) *components* of dynamic system models, i.e. the subsystem *submodels*. These are typically found in *transfer function* submodel components of constant coefficient linear dynamic systems models, discussed further below.

Similarly, a **linear difference equation (LDE)** generally has the form:

$$a_0(k)y(k) + a_1(k)y(k+1) + \ldots + a_n(k)y(k+n) = b_0(k)u(k) + b_1(k)u(k+1) + \ldots + b_n(k)u(k+n)$$

or, more compactly,

$$\sum_{i=0}^{n} a_i(k)y(k+i) = \sum_{i=0}^{n} b_i(k)\ u(k+i) \tag{2.4}$$

The coefficients $a_i(k)$ and $b_i(k)$ depend only upon the (discrete-time) independent variable k (or are *constants*), and not on y.

A **homogeneous** ODE or DE has all input or forcing function terms on the RHS equal to or adding up to zero. This means $\sum_{i=0}^{m} b_i(t)\frac{d^i u}{dt^i} = 0$ in Eq. (2.3) or $\sum_{i=0}^{n} b_i(k)\ u(k+i) = 0$ in Eq. (2.4).

Example 2.9: ***Malthusian Growth Limited by Nutrient Supply: NL Logistic Population Growth – Continuous-Time Version (Mixed Theory–Data Model)***

To render the Malthusian population growth model in Example 2.3 realistic, Verhulst (1838) extended (augmented) it, to account for growth of organism number $N(t)$ in the presence of limited resources. The result is the most-common form of what is widely known as **sigmoidal growth**. We illustrate with a simple hypothesis about the relationship between organism reproduction (growth) rate $c(t)$ and nutrient availability in the population. Suppose c is *proportional* to the concentration η of nutrient in the cell culture medium (continuing with bacterial growth, as an example of population growth). Mathematically, this means $c(\eta) = k_1\eta$ (k_1 is a proportionality constant); and k_2 units of nutrient are consumed per cell cycle (population increment). This means the rate of consumption of nutrient is $d\eta/dt = -k_2 dN/dt = -k_1 k_2 \eta N$. The second part of this equation is nonlinear, because ηN is second degree. The negative sign on this ODE indicates that nutrient supply falls as it is consumed when it comes in contact with reproducing organisms. The equations need to be solved for both $\eta(t)$ and $N(t)$ to establish the longer term behavior of either or both bacterial reproduction dynamics and nutrient utilization. We do this analytically by first integrating both sides of the double-derivative nutrient consumption equation $d\eta/dt = -k_2 dN/dt$:

$$\int d\eta = -k_2 \int dN \rightarrow \eta(t) - \eta(0) = -k_2(N(t) - N(0)) \tag{2.5}$$

Rearranging terms and solving for $\eta(t)$ gives:

$$\eta(t) = -k_2 N(t) + \eta(0) + k_2 N(0) \tag{2.6}$$

This is an expression for the nutrient supply as a function of the number of organisms in the population at any time t and the initial number $N(0)$ and nutrient amount $\eta(0)$. To get an equation for the number of organisms remaining at any time t, we substitute this η relationship into the linear Malthus growth model ODE, with reproduction rate c varying linearly with nutrient concentration η, i.e. $c(\eta) = k_1 \eta$:

$$\begin{aligned} \frac{dN}{dt} &= cN \equiv c(\eta)N = k_1 \eta N = k_1 N([\eta(0) + k_2 N(0)] - k_2 N) \\ \frac{dN}{dt} &= k_1 N(A - k_2 N), \quad A \equiv \eta(0) - k_2 N(0) \end{aligned} \tag{2.7}$$

The final equation, in red, is the continuous-time version of the discrete-time (DE) logistic growth equation (Verhulst 1838) discussed in Example 2.10 below. Its solution can be obtained by direct integration (Exercise E2.7) resulting in:

$$N(t) = \frac{N(0)\dfrac{A}{k_2}}{N(0) + \left(\dfrac{A}{k_2} - N(0)\right)e^{-k_1 A t}} \equiv \frac{N(0)C}{N(0) + (C - N(0))e^{-Gt}} \tag{2.8}$$

where $C \equiv A/k_2 = [\eta(0) + k_2 N(0)]/k_2$ is called the **carrying capacity** and $G \equiv k_1 A = k_1[\eta(0) + k_2 N(0)]$ is called the **intrinsic growth rate** of the colony (reproduction rate in the presence of limited resources). The steady state solution $N(t)$ as $t \rightarrow \infty$ is readily evaluated by taking limits of the RHS of this solution. The exponential term goes to zero and this yields $N(\infty) = C$. **Fig. 2.1** illustrates this solution $N(t)$ for $t > 0$ for parameter values $C = 6$, $G = 0.07$, $N(0) = 1$ and $\eta(0) = 1000$.

Example 2.10: ***Gompertz Law of Mortality (More Theory than Data Model)***

Another continuous-time population growth model, due to Gompertz (1825), is based a little less directly on physics, and a little more on math. Like the Malthus model, population growth is exponential. But Gompertz assumes that the fractional growth

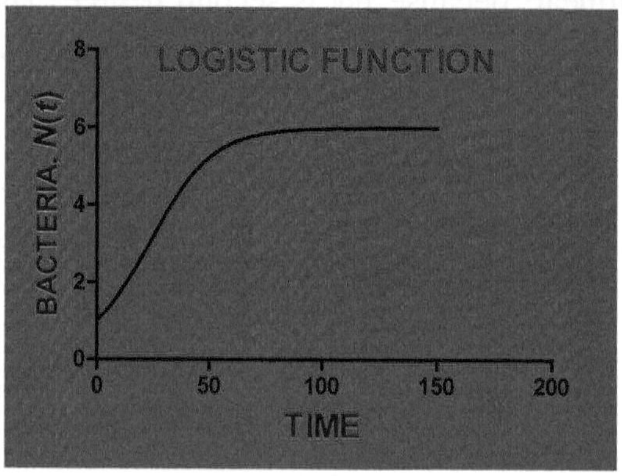

Fig. 2.1

Growth of organism number $N(t)$ in the presence of limited resources: sigmoidal growth. Shown here is this solution $N(t)$ for $t > 0$ for parameter values $C = 6$, $G = 0.07$, $N(0) = 1$ and $\eta(0) = 1000$.

rate exponent c in e^{ct} itself falls exponentially with time, instead of being constant or a given function of limited resources, i.e. $c(t) = e^{\alpha t}$, where α is the constant **fractional retardation rate** constant (governed presumably by limited resources).

The Gompertz model is the solution of two coupled ODEs, the first of which is *time-varying*:

$$\begin{aligned} \frac{dN(t)}{dt} &= c(t)N(t), \quad N(0) = N_0 \equiv \text{initial size} \\ \frac{dc(t)}{dt} &= \alpha c(t), \qquad c(0) \equiv c_0 \equiv \text{initial growth (or regression) rate} \end{aligned} \tag{2.9}$$

The Gompertz model solution[9] is then:

$$N(t) = N(0)e^{\frac{c_0}{\alpha}(1-e^{-\alpha t})} \tag{2.10}$$

The Gompertz model has found application in modeling *tumor growth*, where $N(t)$ is the size of the tumor. The sign of the initial growth rate c_0 determines whether tumor grows ($c_0 > 0$) or shrinks ($c_0 < 0$) (see for example, Laird 1964; d'Onofrio 2005). Its dynamics are similar to the logistic model, but with a slower initial increase.

Example 2.11: ***Logistic Population Growth – Discrete-Time Version (Extended Theory)***

The nonlinear **discrete-time logistic equation** (or **logistic map**) for population growth accounts for limited resources and deaths of organisms in a population – as does the continuous-time version above, but the dynamics are expressed as changes over distinct time intervals rather than over continuous time.[10] With $N(t_k) \equiv N(k)$ as the

9 Substitution of Eq. (2.10) into (2.9) demonstrates that this solution is correct.

10 Again, both models are approximate representations of the reality.

number or organisms at time t_k, as in the growth models above, it usually is written: $N(k+1) = N(k)[a - bN(k)] \equiv f(N(k))$. It is nonlinear because the RHS has a second-degree N^2 term. Intuitively, evolution of N as a function of k in this equation describes a process where the population $N(k)$ increases from one generation to the next when N is relatively small (resources for all), then $N(k)$ decreases from one generation to next as N gets larger (resources insufficient to sustain growth). Mathematically speaking, $f(N(k))$ increases monotonically over $0 < N(k) < a/2b$ from $f(0) = 0$ to $max f(N(k)) = f(a/2b) = a^2/4b$ at $N(k) = a/2b$ ($df/dx_k \equiv 0$ gives the max!). Then $f(N(k))$ decreases monotonically for $N(k) > a/2b$, from the max value back to zero at $N(k) = a/b$, generating a parabolic function of $N(k)$.

Further analysis of the dynamical properties of this famous equation (May 1976) is presented in Chapter 9, where we shall see that it's helpful to combine the two parameters a and b, by changing variables, which reduces the dimensionality of the parameter space and *normalizes* the state variable,[11] as: $x \equiv x(N(k)) \equiv bN(k)/a$. Effectively, this means x is measured in units of b/a, a *scaling factor*, i.e. N is scaled by it's maximum value, $x \equiv N/N_{max} = N/(a/b)$, with x ranging then over the unit interval:

$$x(k+1) = ax(k)[1 - x(k)] \quad \text{for} \quad 0 < x < 1 \tag{2.11}$$

Eq. (2.11) is simulated for several values of a in Chapter 3, illustrating the very different *qualitative behavior* of the response of this population dynamics model for different parameter values.

Remark: Equations like this are often written with a simpler notation for time-dependence, in subscripted form: $x_{k+1} = ax_k(1 - x_k)$.

Remark: Logistic (Verhulst) and Gompertz growth "laws" are each theoretical models approximating real population growth processes. Real-world population data would be typically used to quantify their parameters in applications (intrinsic growth, carrying capacity, etc.), and eventually establish their fidelity as well as utility, e.g. in actuarial work in the insurance industry, or in designing treatments in cancer. They (or variants of them) might each be considered potential hypotheses for understanding or treating the mechanisms of growth in a population, or in diseased tissue, and in this sense would be the work of theory and data integrated in the interest of prediction or scientific discovery.

Remark/Caveat: It's important to keep in mind when developing math models that the **dimensions** of the terms in equations – as well as the terms themselves – must all balance. This is typically a little more complicated for nonlinear than for linear equations. For example, if variable $x(t)$ has the dimension *mass*, then the k in linear ODE $dx/dt = -kx$ has the dimension $1/time$, whereas the k in nonlinear ODE $dx/dt = -kx^3$ has the dimension $1/time \cdot mass^2$. Dimensional analysis of equations is useful for checking correctness of the equation. The section on Normalization of ODEs in Chapter 3 further underscores and illustrates these issues.

11 Normalization of model state variables is presented in Chapter 3 in the context of simplifying simulation analyses.

Piecewise-Linearized[12] Models: Mild/Soft Nonlinearities

Probably no real dynamic system can be fully described *exactly* by a linear differential equation (ODE). However, many nonlinear dynamic systems can be represented over a *limited operating range,* or can be *approximated,* by such equations. These approximations can be useful in practice, because results for linear systems often can be used to analyze them. The next two examples illustrate linear and piecewise-linear approximations. They are presented as a purely physical system, but to give them a biological flavor, imagine they represent a simple model of "elastic" tissue.

Example 2.12: ***Nonlinear Spring Mass System Linearized***

Consider the spring-mass system illustrated in **Fig. 2.2** where the spring force $F_s(x)$ is a nonlinear function of the displacement x measured from the rest position, as shown in the graph of F_s vs. x in the left-hand side (LHS) of **Fig. 2.3**. From Newton's second law, $F = Ma$, the equation of motion of the mass is $M\dfrac{d^2x}{dt^2} + F_s(x) = 0$. First, suppose the abso-

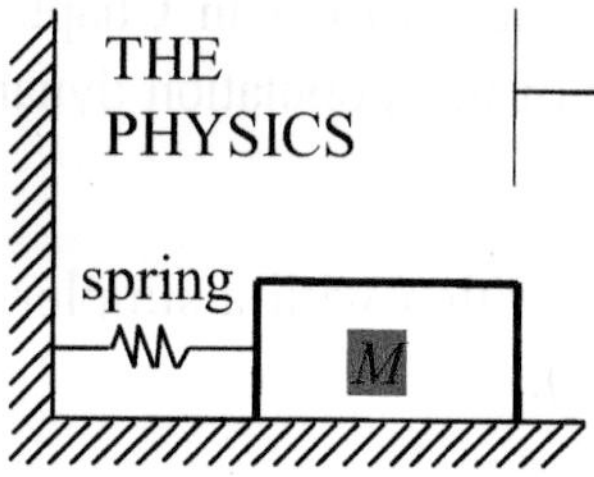

Fig. 2.2

Illustration of a spring-mass system.

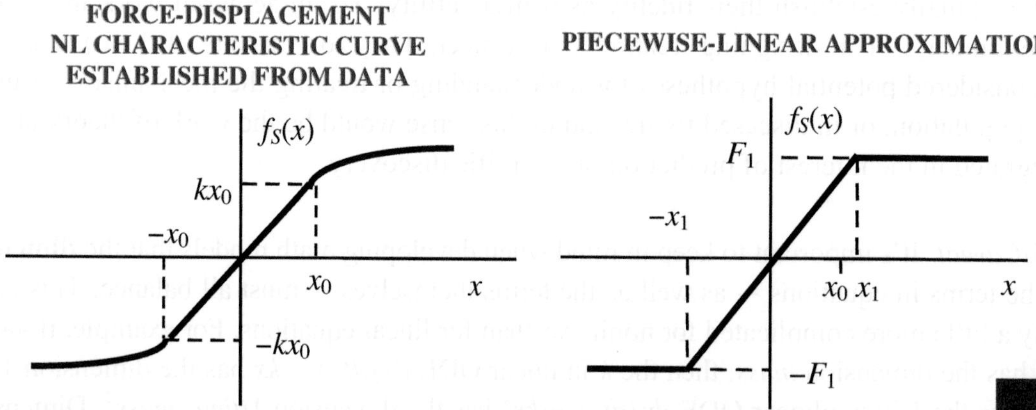

Fig. 2.3

Piecewise linearization.

12 These are usually called piecewise-linear and linearized. We invent the term *piecewise-linearized* here to distinguish the methodology in this section from *model linearization* via Taylor series approximation, introduced in the last section of Chapter 5 on Tracer Methodology and Linearization.

lute magnitude of the displacement does not exceed x_0. This means the mass moves only in the linear range of the spring, so $F_s(x) = kx$, where k is a proportionally constant. Thus, the equation of motion in this case is, approximately, a linear constant coefficient ODE: $M\frac{d^2x}{dt^2} + kx = 0$ which we note is valid only for $|x| \leq x_0$, the linear range of operation.

Example 2.13: ***Nonlinear Spring Mass System Piecewise-Linearized***

Now suppose the displacement x exceeds x_0. To model this problem, we let the spring force curve be approximated by the straight lines shown in the RHS of **Fig. 2.3**. In this manner, the system can be approximated by a piecewise-linear(ized) model, i.e. the linear equation $M\frac{d^2x}{dt^2} + kx = 0$ when $|x| \leq x_1$, and by the two other linear equations $M\frac{d^2x}{dt^2} \pm F_1 = 0$ when $|x| > x_1$. The plus sign applies if $x > x_1$ and the minus sign if $x < x_1$.

So, *approximate* linear(ized) models can sometimes be obtained directly, with "pencil and paper," for models with so-called *mild* nonlinearities. Nonlinear systems and models, however, are not always amenable to such treatments.

Solution of Ordinary Differential (ODE) & Difference Equation (DE) Models

Not many nonlinear or time-varying differential or difference equations can be solved analytically. We address numerical (computer) solution of ODEs by simulation in Chapter 3. Fortunately, many linear time-invariant (LTI) ODE and DE model classes can be solved analytically, and we study these first below.

Following is a brief synopsis of some basic results in classical linear differential equation theory, enough to begin understanding how to model with ODEs. Most of it is about constant-coefficient linear ODEs.

The Differential Operator *D* & Characteristic Equation: Characterizing Modes & Modal Dynamics of Linear ODE Models

For our purposes, the differential operator is a formalism for transforming a notationally complex ODE into a simpler algebraic relationship, making the ODE easier to manipulate, solve and use in developing modeling concepts. The **differential operator** is denoted $D \equiv \frac{d}{dt}$ and the ***n*th-order differential operator** is $D^n \equiv \frac{d^n}{dt^n}$. In these terms, Eq. (2.1) becomes:

$$D^n y + a_{n-1}D^{n-1}y + \ldots + a_1 Dy + a_0 y = (D^{n-1} + \ldots + a_1 D + a_0)y = u$$

The last expression has y factored out. The polynomial in D

$$D^n + a_{n-1}D^{n-1} + \ldots + a_1 D + a_0 \tag{2.12}$$

is called the **characteristic polynomial**, and the equation

$$D^n + a_{n-1}D^{n-1} + \ldots + a_1 D + a_0 \equiv \sum_{i=0}^{n} a_i D^i = 0 \tag{2.13}$$

is called the **characteristic equation**. This is an important equation in linear system modeling.

Example 2.14: ***Characteristic Polynomial & Equation***

The ODE $\frac{d^2y}{dt^2} + 3\frac{dy}{dt} + 2y = u$ has the characteristic polynomial $D^2 + 3D + 2$ and the characteristic equation $D^2 + 3D + 2 = 0$, which has two distinct solutions: $D = \{-1, -2\}$.

The *fundamental theorem of algebra* says that, in general, the characteristic equation has exactly n solutions, $D_1, D_2, \ldots, D_n$, not necessarily unique. These n solutions are components of the general ODE solution.

Why is this important? Because we use these notions for developing dynamic system model *structure* and establishing the *dimensionality of the model observable in data*, in terms of *modes* associated with the n characteristic equation solutions.[13] To begin to see this, we briefly introduce another concept from linear equation theory: *fundamental set*.

If the characteristic equation has *distinct* roots $D_1, D_2, \ldots, D_n$, then a **fundamental set** for the homogeneous equation $\sum_{i=0}^{n} a_i \frac{d^i y}{dt^i} = 0$ is the set of functions $y_1 = e^{D_1 t}, \ldots, y_n = e^{D_n t}$. If the characteristic equation has *repeated* roots, then for each root D_i of multiplicity n_i (i.e. n_i roots equal to D_i), the fundamental set has n_i elements: $e^{D_i t}, te^{D_i t}, \ldots, t^{n_i - 1}e^{D_i t}$. We can begin to write solutions to the ODE with these exponential functions, also called **modes** of the ODE solution.

Example 2.15: ***Fundamental Sets and Modes: Distinct and Repeated Roots***

$\frac{d^2y}{dt^2} + 3\frac{dy}{dt} + 2y = 0$ has characteristic equation $D^2 + 3D + 2 = 0$ and roots $D_1 = -1$ and $D_2 = -2$. A fundamental set for this equation is therefore the two modes $y_1 = e^{-t}$ and $y_2 = e^{-2t}$. A second ODE model: $\frac{d^2y}{dt^2} + 6\frac{dy}{dt} + 9y = 0$ has characteristic equation $D^2 + 6D + 9 = 0$. Factoring then yields repeated (double) roots at $D = -3$ and therefore a fundamental set consisting of the two modes e^{-3t} and te^{-3t}.

13 NL models also can have modes. Modes in NL dynamic system models are discussed in Chapter 9.

ODEs like Eqs. (2.1) and (2.3) typically have at least one set of n *linearly independent*[14] solutions (modes), and any such set of solutions is a fundamental set. However, *there is no unique fundamental set.* From any given fundamental set, other fundamental sets can be generated, in the form of linear combinations of linearly independent solutions, e.g. $e^{-3t} + te^{-3t}$ plus e^{-3t} or te^{-3t} in Example 2.15.

Solution of Linear Time-Invariant (TI) ODEs

We now use the concepts of characteristic equation and fundamental set to solve differential equations of the form of Eq. (2.1). Remember that a_i and b_i are constant. We assume $u = u(t)$ is a known (given) time function, e.g. a model *input,* and $y = y(t)$ is the unknown solution of the equation, e.g. the *output* of the model. The output is noise-free, also called **deterministic**. If this equation describes a causal physical system, e.g. a real biological system, then $m \le n$, and n is called the **order** of the ODE model.[15]

To completely specify the problem so that a *unique* solution $y(t)$ can be obtained over some desired time interval $t_0 \le t \le t_{end}$, a set of n **boundary conditions** are needed, i.e. n values of $y(t)$ or its derivatives at one or more fixed times. Typically, these boundary conditions are a set of n **initial conditions (ICs)** for $y(t)$ and its first $n - 1$ derivatives at time t_0:

$$y(t_0),\ \left.\frac{dy}{dt}\right|_{t=t_0} \equiv y^{(1)} \tag{2.14}$$

A problem defined over this interval and with these initial conditions is called an **initial value problem** (IVP). Computer simulations of ODE models are usually initial value problems.[16]

Remark: **Initial conditions codify the memory of the dynamic system**, compactly summarizing everything about its dynamics prior to time $t = t_0$.

One Way to View ODE Solutions: Classical Approach

The solution of an ODE of this class can be divided into two parts, a *free response* $y_{free}(t)$ and a *forced response* $y_{forced}(t)$.[17] The **total response** (solution) is $y(t) = y_{free}(t) + y_{forced}(t)$.

The **free response** $y_{free}(t)$, also called the **zero input response**, is the solution when input $u(t)$ is identically zero, i.e. when $\sum_{i=0}^{n} a_i \frac{d^i y}{dt^i} = 0$, and thus depends only on the n initial conditions. Note that it's the same as the solution of the homogenous equation $\sum_{i=0}^{n} a_i \frac{d^i y}{dt^i} = 0$.

14 A set of n functions of time $f_1(t), f_2(t), \ldots, f_n(t)$ is called **linearly independent** if the only constants $c_1, c_2, \ldots, c_n$ for which $c_1 f_1(t) + c_2 f_2(t) + \ldots + c_n f_n(t) = 0$ for all t are the constants $c_1 = c_2 = \ldots = c_n = 0$. A set of functions that is not linearly independent is called **linearly dependent**.

15 The $m \le n$ is a causality constraint: causes precede effects, i.e. the outputs cannot anticipate inputs. More on this later.

16 Final value problems, with boundary conditions specified at t_{end}, are encountered using models in optimal control (therapy) problems and other applications too.

17 This is not the only way to split the total response, as we see below, but it is basic to classical solution of ODEs.

Example 2.16: ***Free Response I***

The solution of $dy/dt + y = 0$ with IC $y(0) = c$ is $y(t) = ce^{-t}$, as can be verified by direct substitution (do it!). Thus, $y_{free}(t) = ce^{-t}$ is the free response of any differential equation of the form $dy/dt + y = u$ with initial condition $y(0) = c$.

One reason we introduce the free response concept is that it can always be written as a linear combination of the elements of a fundamental set, i.e. the modes of the ODE. That is, if $y_1(t), y_2(t), \ldots, y_n(t)$ is a fundamental set, then any free response $y_{free}(t)$ of the ODE can be represented as:

$$y_{free}(t) = \sum_{i=1}^{n} c_i y_i(t) \tag{2.15}$$

where each c_i can be determined from ICs (2.14) by direct substitution into the $y(t)$ and successive derivatives of $y(t)$ Eqs. (2.15)[18] evaluated at the ICs:

$$y(t_0) = \sum_{i=1}^{n} c_i y_i(t_0), \quad \left.\frac{dy_i}{dt}\right|_{t=t_0} = \sum_{i=1}^{n} c_i \dot{y}_i(t_0), \quad \ldots \quad \left.\frac{d^{n-1}y}{dt^{n-1}}\right|_{t=t_0} = \sum_{i=1}^{n} c_i y_i^{(n-1)}(t_0)$$

Example 2.17: ***Free Response II***

The free response $y_{free}(t)$ of $\frac{d^2y}{dt^2} + 3\frac{dy}{dt} + 2y = u$, with initial conditions $y(0) = 0$ and $\left.\frac{dy}{dt}\right|_{t=0} = 1$, is $y_{free}(t) = c_1 e^{-t} + c_2 e^{-2t}$. Coefficients c_1 and c_2 are unknown and e^{-t} and e^{-2t} are a fundamental set for the ODE (as shown in Example 2.15). Free response $y_{free}(t)$ must satisfy the initial conditions:

$$y_{free}(0) = c_1 + c_2 = y(0) = 0 \quad \text{and} \quad \left.\frac{dy_{free}}{dt}\right|_{t=0} = -c_1 - 2c_2 = 1$$

Solving this algebraic system of equations $c_1 + c_2 = 0$ and $-c_1 - 2c_2 = 1$, we get $c_1 = 1$ and $c_2 = -1$, and the free response is $y_{free}(t) = e^{-t} - e^{-2t}$.

The **forced response** $y_{forced}(t)$, also called the **zero state response**, is the solution of the ODE with all ICs identically zero. Thus, $y_{forced}(t)$ depends only on the input $u(t)$.

Convolution

An important result in signal and system theory (and practice) is that linearity of a system means its input–output model relationship (forced-response) can be written as what's called a

18 The linear independence of $y_i(t)$ guarantees a solution for $c_1, c_2, \ldots, c_n$.

convolution integral. For TI ODEs like Eq. (2.1), the forced response is the **convolution integral**:

$$y_{forced}(t) = \int_{t_0}^{t} w(t-\tau)u(\tau)d\tau \tag{2.16}$$

In this equation, $w(t-\tau)$ is called the **weighting function** (or **kernel** or **transition function**) of the ODE. For models of real systems, $w(t-\tau) \equiv 0$ for $t < \tau$. This is a property of all linear constant coefficient models of **causal** (real) systems, defined as those whose outputs cannot anticipate their inputs. *This special function w(t) embodies the intrinsic dynamical properties of the (linear) system and its model,* with the special property that it is itself a *free response* of the ODE. As a consequence, $w(t)$ can be written as a linear combination of n fundamental solutions (modes) $y_1(t), y_2(t), \ldots, y_n(t)$ for $t \geq 0$, and zero for $t < 0$:

$$w(t) = \begin{cases} \sum_{i=1}^{n} c_i y_i(t) & \text{for } t \geq 0 \\ 0 & \text{for } t < 0 \end{cases} \tag{2.17}$$

The coefficients $c_1, \ldots, c_n$ are constants and, as always, $y_1(t), y_2(t), \ldots, y_n(t)$ are a fundamental set of the ODE. To completely specify the free response $w(t)$ uniquely, we need n ICs, to evaluate $c_1, \ldots, c_n$. The special n initial conditions for accomplishing this computation are:

$$w(0) = 0, \quad \left.\frac{dw}{dt}\right|_{t=0} = 0, \quad \ldots \quad \left.\frac{d^{n-2}w}{dt^{n-2}}\right|_{t=0} = 0, \quad \left.\frac{d^{n-1}w}{dt^{n-1}}\right|_{t=0} = 1 \tag{2.18}$$

Example 2.18: ***Convolution Integral & Total Response***

The weighting function for $\frac{d^2y}{dt^2} + 3\frac{dy}{dt} + 2y = 1$ is a linear combination of e^{-t} and e^{-2t} (a fundamental set of the equation), i.e. $w(t) = c_1 e^{-t} + c_2 e^{-2t}$. We first determine c_1 and c_2 using the initial conditions $y(0) = 0$ and $\left.\frac{dy}{dt}\right|_{t=0} = 1$ and the two algebraic equations at $t = 0$: $w(0) = c_1 + c_2 = 0$ and $\left.\frac{dw}{dt}\right|_{t=0} = -c_1 - 2c_2 = 1$. This gives $c_1 = 1$ and $c_2 = -1$, and thus $w(t) = e^{-t} - e^{-2t}$. Then, from Eq. (2.16), the *forced response* is:

$$y_{forced}(t) = \int_0^t w(t-\tau)u(\tau)d\tau = \int_0^t [e^{-(t-\tau)} - e^{-2(t-\tau)}] \cdot 1 d\tau$$
$$= e^{-t}\int_0^t e^{\tau}d\tau - e^{-2t}\int_0^t e^{2\tau}d\tau = \frac{1 - 2e^{-t} + e^{-2t}}{2}$$

Finally, the **total response** $y(t)$ of this equation, with ICs $y(0) = 0$ and $\left.\frac{dy}{dt}\right|_{t=0} = 1$, is the sum of the free response, determined earlier, plus the forced response:

$$y_{total}(t) = y_{free}(t) + y_{forced}(t) = (e^{-t} - e^{-2t}) + \frac{1 - 2e^{-t} + e^{-2t}}{2} = \frac{1 - e^{-2t}}{2}$$

The Transient & Steady State Responses: The Modern Systems Approach

The *transient response* and the *steady state response* are another pair of response terms whose sum is equal to the total response.[19] These are the terms most commonly used in systems modeling applications and, for this reason, they take on added importance. However, as we see later, when we discuss *stability*, only *stable* (and marginally stable) systems and models have transient and steady state responses. In this sense, free and forced responses are more general.

The **transient response** is that part of the total response approaching zero as time approaches infinity. The **steady state response** y_{ss} is that part of the total response (of a stable system) *not* approaching zero as time approaches infinity.

Example 2.19: ***Transient and Steady State Responses***

The total response computed in Example 2.18, $y = \dfrac{1-e^{-2t}}{2}$, splits readily into the steady state response, $y_{ss} = \dfrac{1}{2}$, and the transient response is $-\dfrac{e^{-2t}}{2}$, because $\lim\limits_{t\to\infty}\left[-\dfrac{e^{-2t}}{2}\right] = 0$.

Remark: In this example, the steady state is constant, but this is not generally true. The total response $y(t) = \sin t + e^{-t}$, for example, has a transient response equal to e^{-t} and a steady state response equal to the periodic function $\sin t$. Thus, when a steady state *is* constant, we (should) call it a **constant steady state.**

Special Input Forcing Functions (Signals) & Their Model Responses: Steps & Impulses

Two special forcing functions in the study of system models are the *unit step* and the *unit impulse*. Each is related to the other by integration or differentiation. We henceforth use the term **input** for the forcing function $u(t)$ and **output** for the model solution $y(t)$, also called input or output *signals*, although not the only signals in a system or model. This is common terminology for *system* models, as illustrated in **Fig. 2.4**.

A **unit step function** $\mathit{1}(t-t_0)$, shown in **Fig. 2.5**, is defined by Eq. (2.19).

$$\mathit{1}(t-t_0) \equiv \begin{cases} 1 \; for \; t > t_0 \\ 0 \; for \; t \le t_0 \end{cases} \tag{2.19}$$

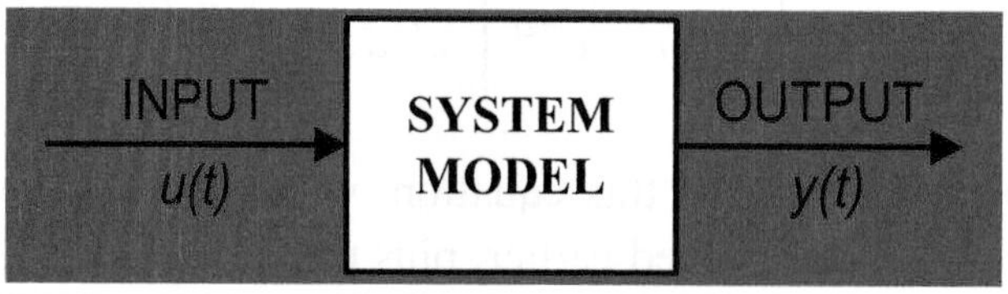

Fig. 2.4

Common terminology for system models.

19 This way of splitting the total response is quite common and practical in engineering and physics.

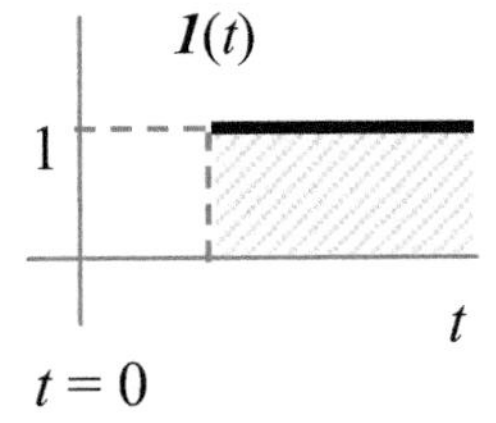

Fig. 2.5

A unit step function $I(t - t_0)$.

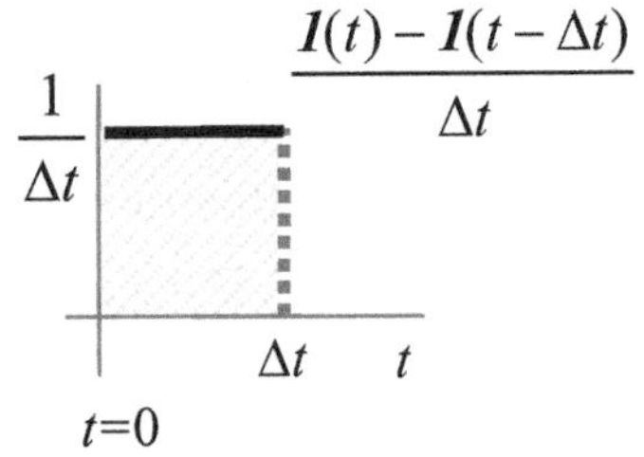

Fig. 2.6

The unit impulse function is often used in applications to approximate a brief pulse.

Remark: This function might be used to model constant endogenous production of material in an open biological system, or a constant delivery of input signal (e.g. substance) exogenously.[20] Step functions are also used to model switching functions in biological systems, e.g. protein *activation/deactivation* can be modeled as a step "turning on," from say zero to one, or off – from one to zero.

A **unit impulse function** $\delta(t)$, also called the Dirac delta function, is usually defined as shown in Eq. (2.20) and **Fig. 2.6**, where $\boldsymbol{1}(t)$ is the unit step function.

$$\delta(t) \equiv \lim_{\Delta t \to 0^+} \left[\frac{\boldsymbol{1}(t) - \boldsymbol{1}(t - \Delta t)}{\Delta t} \right] \tag{2.20}$$

Strictly speaking, this equation defines the *one-sided derivative* of the unit step function. $\Delta t \to 0^+$ means that Δt approaches zero *from the right*, to make it real, in the sense that this signal cannot really be applied at exactly $t = 0$, but a wee bit later, at $t = 0^+$. Nevertheless, an impulse is usually represented graphically as an arrow pointing up at $t = 0$, or at $t \equiv \tau \neq 0$, as in **Fig. 2.7**. The limiting process produces a function whose height approaches infinity and width approaches zero so that, by design, the area under the curve of the unit impulse is always equal to 1, for all values of Δt, i.e.

$$\int_{-\infty}^{\infty} \delta(t) dt = 1 \tag{2.21}$$

20 Exogenous means external to the system model dynamics, external to the box in **Fig. 2.4**, also implying the system is open to external inputs.

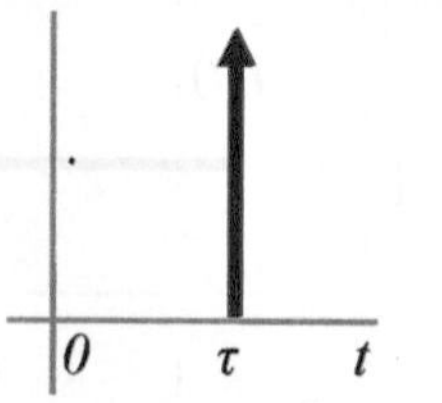

Fig. 2.7

Unit impulse.

Remark: The unit impulse function is often used in applications to *approximate* a brief pulse, e.g. a rapidly injected dose of a drug, often called a *pulse-dose* or *bolus-dose,* or just *bolus*. A pulse looks like the function in **Fig. 2.6**, for some finite width and height. The impulse function is the right-sided limit of this pulse, as defined in Eq. (2.20).

Remark: We explore bolus dose inputs and illustrate the quality or degree to which the unit impulse function approximates a finite pulse or bolus, in Chapter 3, in simulation Examples 3.2–3.4.

The unit impulse function satisfies the **screening** or **sifting property**, especially important in applications. The integral of the product of a unit impulse function $\delta(t-t_0)$ and a function $f(t)$, continuous at $t=t_0$ over an interval, which includes t_0, is equal to the function $f(t)$, evaluated at t_0:

$$\int_{-\infty}^{\infty} f(t)\delta(t-t_0)dt = f(t_0) \tag{2.22}$$

The **unit impulse response** of a model is the output $y(t)$ when $u(t)=\delta(t)$ and all initial conditions are zero. Similarly, the **unit step response** is the output $y(t)$ when the input $u(t)=\mathbf{1}(t)$ and all initial conditions are zero.

Example 2.20: ***Screening/Sifting Property***

If the input–output relationship of a linear model is given by the convolution integral, Eq. (2.16), then the unit impulse response $y_\delta(t)$ is:

$$y_\delta(t) = \int_0^t w(t-\tau)\delta(\tau)d\tau = \int_{-\infty}^t w(t-\tau)\delta(\tau)d\tau = w(t) \tag{2.23}$$

since $w(t-\tau)=0$ for $\tau>t$, $\delta(\tau)=0$ for $\tau<0$, and the screening (sifting) property of the unit impulse has been used to evaluate the integral. Thus, we have the important result:

The unit impulse response of a linear system model is identical to its weighting function.

The Unit Impulse Input: An Important Equivalence Property

When a model has an impulse function as input, this input often can be better represented as an initial condition (IC), rather than an input, especially for simulation, but also for other purposes. The basis for this is the following result, which we show for a simple example.

Equivalence Property: The response of a single-input–single-output (SISO) dynamic system model (linear or nonlinear) to an impulse input, with zero initial conditions (system relaxed or at rest), is equal to the zero-input response of the model with initial conditions set equal to the magnitude of the impulse input.

To illustrate, for a simple first-order ODE, the following two sets of equations are equivalent, yielding the same solution $x(t)$ for $t > 0$.

$$\left\{ \begin{array}{c} dx/dt = f(x) + x_0\delta(t) \\ x(0) = 0 \end{array} \right\} \Leftrightarrow \left\{ \begin{array}{c} dx/dt = f(x) \\ x(0) = x_0 \end{array} \right\} \tag{2.24}$$

To show this for $f(x) = -kx$, $w(t) = x_0 e^{-kt}$ is the fundamental solution of this simple inhomogeneous ODE (LHS), from Eq. (2.17). This is also the solution of the homogeneous ODE on the RHS of Eq. (2.24). (Q.E.D.)

An important consequence of this equivalence of impulse inputs with ICs is that *impulse inputs can be simply converted to initial conditions in computer simulation implementations of ODEs*. We illustrate and utilize this in the next chapter and beyond.

State Variable Models of Continuous-Time Systems

For most modeling problems, particularly dimensionally large or otherwise complex ones, it is convenient to describe a system by *n first-order* ODEs, called *state variable* form, rather than by a single nth-order ODE. Actually, system models are often built of interconnected submodels of such simpler subsystem components, based on mechanistic subsystem knowledge, e.g. the basic biophysics of the process being modeled. Sets of first-order ODEs, for example, are obtained *directly* for many model classes, such as multicompartmental and multipool models, introduced in Chapter 4, and models based on mass action principles, introduced in Chapter 6. Quite general and powerful results from the theory of linear algebra can then be readily applied in deriving solutions for and analyzing the dynamical properties of ODE models in state variable form. Many of these linear algebra results are given in Appendix B. We develop the math first for linear ODEs. The transformation applies to nonlinear ODEs as well, as we see in the next section.

Example 2.21: ***Newton's Second Law as Two ODEs***

From Example 2.1, we have shown Newton's second law, $F = Ma = M\dfrac{d^2x}{dt^2}$. It is clear from the meanings of velocity v and acceleration a that this second-order ODE can be replaced by two first-order ODEs, $v = \dfrac{dx}{dt}$ and $F = M\dfrac{dv}{dt}$.

There are numerous ways to transform an nth-order differential equation into n first-order equations. One is quite prevalent in the literature, and we introduce only this straightforward transformation here to illustrate the approach. Others are presented in Appendix E. Consider the *n*th-order, single-input linear ODE:

$$\sum_{i=0}^{n} a_i(t)\frac{d^i y}{dt^i} = u$$

This equation can always be replaced by the following *n* first-order ODEs:

$$\begin{aligned}
\frac{dx_1}{dt} &= x_2 \\
\frac{dx_2}{dt} &= x_3 \\
&\vdots \\
\frac{dx_{n-1}}{dt} &= x_n \\
\frac{dx_n}{dt} &= -\frac{1}{a_n(t)}\left[\sum_{i=0}^{n-1} a_i(t)x_{i+1}\right] + \frac{1}{a_n(t)}u
\end{aligned} \tag{2.25}$$

where we have chosen $x_1 \equiv y$. This choice is arbitrary, but $x_1 \equiv y$ is the most common. Using vector-matrix notation (see Appendix C), and the abbreviation $a_i(t) \equiv a_i$, this set of equations can be written as:

$$\begin{bmatrix} \frac{dx_1}{dt} \\ \frac{dx_2}{dt} \\ \vdots \\ \frac{dx_n}{dt} \end{bmatrix} = \begin{bmatrix} 0 & 1 & 0 & \cdots & 0 \\ 0 & 0 & 1 & \cdots & 0 \\ 0 & 0 & 0 & \ddots & \vdots \\ \vdots & \vdots & \vdots & \ddots & 1 \\ -\frac{a_0}{a_n} & -\frac{a_1}{a_n} & -\frac{a_2}{a_n} & \cdots & -\frac{a_{n-1}}{a_n} \end{bmatrix} \begin{bmatrix} x_1 \\ x_2 \\ \vdots \\ x_n \end{bmatrix} + \begin{bmatrix} 0 \\ 0 \\ \vdots \\ 0 \\ \frac{1}{a_n} \end{bmatrix} u \tag{2.26}$$

or more compactly as:

$$\frac{d\boldsymbol{x}}{dt} = A\boldsymbol{x} + \boldsymbol{b}u \tag{2.27}$$

In Eq. (2.27), $\boldsymbol{x} \equiv \boldsymbol{x}(t)$ is called the **state vector**, and its elements – the n time functions $x_1(t), x_2(t), \ldots, x_n(t)$ – are the **state variables** of the model. The scalar input of the model is $u(t)$. A is called the **system matrix**.[21]

21 *A* in this particular form is called a **companion matrix** (Appendix C).

More generally, multi-input (MI) continuous-time system models can be represented by a vector-matrix differential equation of the form:

$$\begin{bmatrix} \frac{dx_1}{dt} \\ \frac{dx_2}{dt} \\ \vdots \\ \frac{dx_n}{dt} \end{bmatrix} = \begin{bmatrix} a_{11} & a_{12} & \cdots & a_{1n} \\ a_{21} & a_{22} & \cdots & a_{2n} \\ \vdots & \vdots & \ddots & \vdots \\ -a_{n1} & a_{n2} & \cdots & a_{nn} \end{bmatrix} \begin{bmatrix} x_1 \\ x_2 \\ \vdots \\ x_n \end{bmatrix} + \begin{bmatrix} b_{11} & b_{12} & \cdots & b_{1r} \\ b_{21} & b_{22} & \cdots & b_{2r} \\ \vdots & \vdots & \ddots & \vdots \\ b_{n1} & b_{n2} & \cdots & b_{nr} \end{bmatrix} \begin{bmatrix} u_1 \\ u_2 \\ \vdots \\ u_r \end{bmatrix} \tag{2.28}$$

or more compactly as:

$$\frac{d\boldsymbol{x}}{dt} = A\boldsymbol{x} + B\boldsymbol{u} \tag{2.29}$$

In Eq. (2.29), $\boldsymbol{x}$ is defined as in Eq. (2.27), A (the system matrix) is the $n \times n$ matrix of coefficients a_{ij}, B (the input matrix) is the $n \times r$ matrix of coefficients b_{ij}, and u is the vector of r input functions, all as shown in Eq. (2.28). In Appendix C we show that, for constant coefficients a_i, the general solution of Eq. (2.29) for $t \geq t_0$ is:

$$\boldsymbol{x}(t) = e^{At}\boldsymbol{x}(t_0) + \int_{t_0}^{t} e^{A(t-\tau)}B\boldsymbol{u}(\tau)d\tau \tag{2.30}$$

The initial condition $\boldsymbol{x}(t_0)$ *is often referred to as the state of the system (model) at time* $t = t_0$. The first term in Eq. (2.30) is the free response and the second term is the forced response,[22] as in the previous section. Inspection of Eq. (2.30) indicates that knowledge of $\boldsymbol{x}(t_0)$ and the input $u(t)$ on the interval $t_0 \leq t$ gives the complete and unique solution for the state variables $\boldsymbol{x}$ (t) for all times $t \geq t_0$, once the model, i.e. the matrices A and B, have been specified.[23]

Importantly, the entire memory of this dynamical system, for all time passed, is embodied in the initial condition $\boldsymbol{x}(t_0)$.

This way of modeling is often called the **state space approach**, because the state vector $\boldsymbol{x}$ is a point in a linear vector space.[24] The output (noise-free measurement) model in state variable form is usually represented as:

$$\boldsymbol{y} = C\boldsymbol{x} \tag{2.31}$$

22 Note that the second term is the convolution of the weighting function (impulse response) with the input.

23 Actually, knowledge of the state at *any* time t, say $t' < t$, and knowledge of the input $\boldsymbol{u}(t)$ for $t \geq t'$ completely determine the state vector $\boldsymbol{x}(t)$ at all subsequent times $t \geq t'$.

24 Linear vector spaces are studied in linear algebra (Friedberg et al. 2003). A concise collection of results useful for state variable (state space) modeling is given in Appendix B.

where the **output matrix** C has dimension $m \times n$ and output $\boldsymbol{y}$ is $m \times 1$. For example, for $n = 2$, if the single output measure ($m = 1$) is $y = x_1 + x_2$, $C = [1\ 1]$, a 1 by 2 matrix.

Nonlinear (NL) State Variable Biomodels

The state space approach applies equally well to nonlinear (NL) ODEs. The usual vector-matrix form for NL ODEs is:

$$\dot{\boldsymbol{x}} = \boldsymbol{f}(\boldsymbol{x}, \boldsymbol{u}) \tag{2.32}$$

with outputs modeled as

$$\boldsymbol{y} = \boldsymbol{g}(\boldsymbol{x}, \boldsymbol{u}) \tag{2.33}$$

In these equations, f and g are NL n and m vector functions of x and u, $\boldsymbol{u}$ is an r-vector of inputs, and $\boldsymbol{y}$ is an m-vector of outputs.

Example 2.22: ***NL Vector ODEs***

The following NL model has $n = 2$, $r = 1$, $m = 1$:

$$\begin{aligned} \dot{x}_1 &= c_1 x_2 - c_2 x_1^2 + 6u \\ \dot{x}_2 &= \frac{c_3 x_1}{c_4 + x_1} - c_5 \\ y &= c_6 x_1^2 \end{aligned} \tag{2.34}$$

It is readily written in the form of Eqs. (2.32) and (2.33):

$$\begin{aligned} \dot{x}_1 &= c_1 x_2 - c_2 x_1^2 + 6u \equiv f_1(\boldsymbol{x}, u) \\ \dot{x}_2 &= \frac{c_3 x_1}{c_4 + x_1} - c_5 \equiv f_2(\boldsymbol{x}, u) \\ y &= c_6 x_1^2 \equiv g(\boldsymbol{x}) \end{aligned} \tag{2.35}$$

Expressed on column vector form, except for the scalar output:

$$\begin{aligned} \begin{bmatrix} \dot{x}_1 \\ \dot{x}_2 \end{bmatrix} &= \begin{bmatrix} c_1 x_2 - c_2 x_1^2 + 6u \\ \dfrac{c_3 x_1}{c_4 + x_1} - c_5 \end{bmatrix} \\ y &= c_6 x_1^2 \end{aligned} \tag{2.36}$$

Remark: The equivalence property and Eq. (2.24) are valid for NL Eq. (2.32). For example, if u were a unit impulse input in Eq. (2.35), the $6u$ term would be deleted and replaced by an IC of 6 on x_1, i.e. $x_1(0) = 6$.

Remark: We use this elegant and simple vector notation as often as possible, emphasizing that – in this form – scalar and vector equations are essentially the same, with boldface notation for vectors. The vector form permits use of the same equation for representing complex systems, of any finite dimensionality. And general results are then easier to apply for models of any order n. For specific cases, we often expand the vector equations completely.

Linear Time-Invariant (TI) Discrete-Time Difference Equations (DEs) & Their Solution

This development parallels that for continuous-time system models. Only a few basic results are presented here, for the simplest version of the time-invariant linear DE model given earlier in Eq. (2.4). This model is representative of a wide variety of physical and nonphysical system models. It has a special name, **autoregressive moving average** (ARMA) model, in some areas of engineering and economics (Ljung 1987). It has the general form:

$$\sum_{i=0}^{n} a_i y(k+i) = u(k) \tag{2.37}$$

In this model, $k = 0, 1, 2, \ldots$, is the integer-valued discrete-time variable (corresponding to the time t_k at the kth instant), the coefficients a_i and b_i are constant, and a_0 and a_n are nonzero. In the next section, we develop this model in state variable form.

The input $u(k)$, an abbreviation for $u(0)$, $u(1)$, $u(2)$, …, is a time sequence, and the output $y(k)$ is the sequence solution of the equation, i.e. $y(0), y(1), y(2), \ldots$. Since $y(k+n)$ is an explicit function of $y(k), y(k+1), \ldots, y(k+n-1)$, n is called the order of the difference equation.

To obtain a unique solution for $y(k)$ for $k = 0, 1, 2, \ldots$, we need n initial (or other boundary) conditions (ICs) for $y(k)$, i.e. $y(0), y(1), \ldots, y(n-1)$. Continuing the analogy with continuous system models, the free response of a difference equation is the solution when the input sequence is identically zero. Then, if $y_1(k), y_2(k), \ldots, y_n(k)$ is a fundamental set, any free response can be represented as

$$y_{free}(k) = \sum_{i=1}^{n} c_i y_i(k) \tag{2.38}$$

where the constants c_i are determined in terms of the initial conditions (ICs) $y_i(0)$ of the fundamental set of $y_i(k)$, from the set of n algebraic equations:[25]

$$y(0) = \sum_{i=1}^{n} c_i y_i(0), \quad y(1) = \sum_{i=1}^{n} c_i y_i(1), \ldots, \quad y(n-1) = \sum_{i=1}^{n} c_i y_i(n-1)$$

Linear independence of the $y_i(k)$ guarantees a solution for the constants $c_1, c_2, \ldots, c_n$.

25 Linear independence of the $y_i(k)$ guarantees a solution for the constants $c_1, c_2, \ldots, c_n$.

Example 2.23: ***Compound Interest as an ARMA Model***

The compound interest equation $P(k+1) = (1+I)P(k)$ presented earlier (Example 2.4) is an ARMA model, with input $u(k) \equiv 0$, $y(k) \equiv P(k)$, $a_0 = -1$ and $a_1 = \frac{1}{I+1}$, written in the form of Eq. (2.37). Recall that this homogeneous DE has IC $P(0) \neq 0$ and solution $P(k) = P(0)(1+I)^k$ for any $k = 1, 2, \ldots$.

The **forced response $y_{forced}(k)$ of a DE** is the solution of the difference equation when all initial conditions $y(0), y(1), \ldots, y(n-1)$ equal zero. The forced response of a linear constant coefficient DE can be written in terms of a ***convolution sum:***

$$y_{forced}(k) = \sum_{j=0}^{k-1} w(k-j)u(j) \quad \text{for } k = 0, 1, 2, \ldots \tag{2.39}$$

where $w(k-j)$ is called the **weighting sequence** of the DE. Note that $y_{forced}(0) = 0$ by definition of the forced response. This realistic form of the convolution sum assumes *causality* of the system, defined by the weighting sequence, $w(k-j) = 0$ for $k < j$.

The **total response** of a DE is the sum of the free and forced responses, just as it is for continuous systems. Similarly, the **transient response** of a difference equation is the part of the total response approaching zero as time approaches infinity; the part of the total response not approaching zero is called the **steady state response**, which, we repeat, is not necessarily constant.

Remark: Difference equation models (approximations) of ODE models are the basis of numerical simulation solutions of ODEs, which we develop in Chapter 3.

State Variable Models of Discrete-Time Systems

As with differential equations, it is often useful to model a discrete-time system by a set of first-order difference equations, rather than by one or more nth-order difference equations. We illustrate here how to make this conversion and solve the resulting vector-matrix equation.

Remark: The procedure and results closely resemble those for continuous-time system models.

The nth-order, single-input, linear constant coefficient difference Eq. (2.37) can always be replaced by the following set of n first-order difference equations:

$$\begin{aligned} x_1(k+1) &= x_2(k) \\ x_2(k+1) &= x_3(k) \\ &\vdots \\ x_{n-1}(k+1) &= x_n(k) \\ x_n(k+1) &= -\frac{1}{a_n}\left[\sum_{i=0}^{n-1} a_i x_{i+1}(k)\right] + \frac{1}{a_n}u(k) \end{aligned} \tag{2.40}$$

where we have chosen $x_1(k) \equiv y(k)$. Using vector-matrix notation, this set of equations can be written as the *vector-matrix* difference equation:

$$\begin{bmatrix} x_1(k+1) \\ x_2(k+1) \\ \vdots \\ x_n(k+1) \end{bmatrix} = \begin{bmatrix} 0 & 1 & 0 & \cdots & 0 \\ 0 & 0 & 1 & \cdots & 0 \\ 0 & 0 & 0 & \ddots & \vdots \\ \vdots & \vdots & \vdots & \ddots & 1 \\ -a_0/a_n & -a_1/a_n & -a_2/a_n & \cdots & -a_{n-1}/a_n \end{bmatrix} \begin{bmatrix} x_1(k) \\ x_2(k) \\ \vdots \\ x_n(k) \end{bmatrix} \begin{bmatrix} 0 \\ 0 \\ \vdots \\ 0 \\ 1/a_n \end{bmatrix} u(k) \tag{2.41}$$

or, more compactly, as

$$\boldsymbol{x}(k+1) = A\boldsymbol{x}(k) + \boldsymbol{b}u(k) \tag{2.42}$$

In these equations, $x(k)$ is an n-vector element of a time sequence called the **state vector**, made up of scalar elements $x_1(k), x_2(k), \ldots, x_n(k)$ called the **state variables** of the system at the *kth* time instant. Similarly, **multi-input (MI) discrete-time system models** can be represented by

$$\boldsymbol{x}(k+1) = A\boldsymbol{x}(k) + B\boldsymbol{u}(k) \tag{2.43}$$

where $x(k)$ is the state vector as above, A is an $n \times n$ matrix of coefficients a_{ij}, as above, but B is now an $n \times r$ *input matrix* of coefficients b_{ij}, each defined as in Eq. (2.28), and $u(k)$ is an r-vector element of a *multi-input* sequence. If the elements a_{ij} of A are constant and an IC vector $x(0)$ is given, the solution of Eq. (2.43) is:

$$\boldsymbol{x}(k) = A^k \boldsymbol{x}(0) + \sum_{j=0}^{k-1} A^{k-1-j} B\boldsymbol{u}(j) \tag{2.44}$$

Eq. (2.44) is analogous to Eq. (2.30) for continuous system models.

Remark: Eq. (2.43) can be used to solve *differential* equations without using ODE solvers (simulation algorithms), when the solutions are needed only at discrete instants of time, as illustrated next.

Example 2.24: ***Time-Discretization of Differential Equations (ODEs)***

Suppose we only need to characterize the state variables in continuous model Eq. (2.29) at periodic time instants $t = 0, T, 2T, \ldots, kT$, etc. In this case, we directly write the solution from Eq. (2.30) (by direct substitution of the time arguments) for the following *sequence* of state vectors:

$$\begin{aligned} \boldsymbol{x}(T) &= e^{AT}\boldsymbol{x}(0) + \int_0^T e^{A(T-\tau)}\boldsymbol{Bu}(\tau)d\tau \\ \boldsymbol{x}(2T) &= e^{AT}\boldsymbol{x}(T) + e^{AT}\int_T^{2T} e^{A(T-\tau)}\boldsymbol{Bu}(\tau)d\tau \\ &\vdots \\ \boldsymbol{x}(kT) &= e^{AT}\boldsymbol{x}((k-1)T) + e^{A(k-1)T}\int_{(k-1)T}^{kT} e^{A(T-\tau)}\boldsymbol{Bu}(\tau)d\tau \end{aligned} \tag{2.45}$$

If we suppress the parameter T, using the abbreviation $\boldsymbol{x}(k) \equiv \boldsymbol{x}(kT)$, and define a new *input sequence* by

$$\boldsymbol{u}'(k) \equiv e^{AkT}\int_{kT}^{(k+1)T} e^{A(T-\tau)}\boldsymbol{Bu}(\tau)d\tau \tag{2.46}$$

then the set of solutions above can be replaced by the single *vector matrix difference equation*:

$$\boldsymbol{x}(k+1) = e^{AT}\boldsymbol{x}(k) + \boldsymbol{u}'(k) \tag{2.47}$$

The solution of Eq. (2.47) is then

$$\boldsymbol{x}(k) = e^{AkT}\boldsymbol{x}(0) + \sum_{j=0}^{k-1} e^{(k-1-j)AT}\boldsymbol{Bu}'(j) \tag{2.48}$$

for $k = 1, 2, \ldots$. This equation can be solved *recursively*, given the initial state $\boldsymbol{x}(0)$ and the input sequence $\boldsymbol{u}'(0), \boldsymbol{u}'(1), \ldots$, determined from Eq. (2.46).

We often use a more abbreviated notation for discrete-time variables in discrete-time models, i.e. $x(k) \equiv x_k$, $y(k) \equiv y_k$, $u(k) \equiv u_k$, etc. For example, in these terms, Eqs. (2.43) and (2.44) become:

$$\boldsymbol{x}_{k+1} = \boldsymbol{A}\boldsymbol{x}_k + \boldsymbol{B}\boldsymbol{u}_k \tag{2.49}$$

$$\boldsymbol{x}_k = A^k\boldsymbol{x}_0 + \sum_{j=0}^{k-1} A^{k-1-j}\boldsymbol{B}\boldsymbol{u}_k \tag{2.50}$$

Remark: Discretization of an ODE was done here for equally spaced time intervals. It is also applicable for unequally spaced intervals, as illustrated in the ODE simulation algorithms presented in Chapter 3.

Remark: In applications, e.g. in imbedded (bio)control systems based on (bio)dynamic system models, these time variable state vectors $\boldsymbol{x}_k$ can be stored in memory (on a chip), along with the exponential matrix e^{AT}, for recomputing solutions for different inputs and ICs.

LINEARITY & SUPERPOSITION

The concept of linearity was presented earlier as a property of differential and difference equations. In this section, linearity is discussed as a property of *general systems* with one independent variable, time t. If all initial conditions (ICs) in the system are zero, i.e. if the system is *completely at rest* (*memory* zero or deleted), then the system is a **linear system** if it has the following property:

If an input $u_1(t)$ produces an output $y_1(t)$ and another input $u_2(t)$ produces a second output $y_2(t)$, then the input $c_1u_1(t) + c_2u_2(t)$ produces the output $c_1y_1(t) + c_2y_2(t)$, for all pairs of inputs $u_1(t)$ and $u_2(t)$ and all pairs of constants c_1 and c_2.

Remark: A *linear system* (or its model) need not obey the above property if the ICs are not zero.

Linear systems can often be represented by linear differential (ODE) or difference equation models, but these are not the only equations representing linear systems. For example, algebraic equations, like Newton's second law, $F = Ma$, and integral equations, like Eq. (2.45), are also linear.

The concept of linearity is also expressed by the **principle of superposition**. The response $y(t)$ of a linear system due to several inputs $u_1(t), u_2(t), \ldots, u_n(t)$ acting simultaneously is equal to the sum of the responses of each input acting alone, when all ICs in the system are zero. That is, if $y_i(t)$ is the response due to the input $u_i(t)$, then

$$y(t) = \sum_{i=1}^{n} y_i(t) \tag{2.51}$$

is the response to the several inputs acting simultaneously.

Example 2.25: ***Superposition***

Consider the linear system described by $y(t) = 2u_1(t) + u_2(t)$, where $u_1(t) = t$ and $u_2(t) = t^2$ are inputs, and $y(t)$ is the output. Assume all ICs are zero. When $u_1(t) = t$ and $u_2(t) = 0$, then $y_1(t) = 2t$. When $u_1(t) = 0$ and $u_2(t) = t^2$, then $y_2(t) = t^2$. The total output resulting from $u_1(t) = t$ and $u_2(t) = t^2$ is then equal to $y(t) = y_1(t) + y_2(t) = 2t + t^2$.

Key Point: Linearity and superposition are equivalent only when all ICs in the system are zero.

Example 2.26: ***Superposition of Forced but not Free Responses***

Consider the linear response in Eq. (2.30) with *nonzero* IC:

$$\mathbf{x}(t) = e^{At}\mathbf{x}(t_0) + \int_{t_0}^{t} e^{A(t-\tau)}B\mathbf{u}(\tau)d\tau$$

For inputs $\boldsymbol{u}_1$ and $\boldsymbol{u}_2$, solutions $\boldsymbol{x}_1$ and $\boldsymbol{x}_2$ are obtained by simply substituting these inputs into the integral term, assuming the same nonzero IC for both inputs. The solution $\boldsymbol{x}_3(t)$, when the input is the sum of these inputs, is then:

$$\boldsymbol{x}_3(t) = e^{At}\boldsymbol{x}(t_0) + \int_{t_0}^{t} e^{A(t-\tau)}B[\boldsymbol{u}_1(\tau) + \boldsymbol{u}_2(\tau)]d\tau$$
$$\neq x_1(t) + x_2(t) = 2e^{At}\boldsymbol{x}(t_0) + \int_{t_0}^{t} e^{A(t-\tau)}B[\boldsymbol{u}_1(\tau) + \boldsymbol{u}_2(\tau)]d\tau$$

Superposition applies to the inputs here, but not the nonzero ICs or the complete solution.[26]

Laplace Transform Solution of ODEs

We used a transformation based on the differential operator $D^n \equiv d/dt^n$ in an earlier section to simplify ODEs and their solution, facilitating – among other things – computation of dynamical *modes*[27] of ODEs. The *Laplace* transformation is more powerful, especially for building and analyzing linear time-invariant ODE models – along with their inputs, outputs and initial conditions. This is accomplished by first transforming it into a simpler algebraic frequency domain problem. Importantly, additional dynamical insights are gleaned by transformation into the frequency domain. This includes facile development of feature-rich submodel input–output *transfer functions* – presented in the next section. Laplace transformation also provides time-domain solutions of linear time-invariant ODE models, via *inverse* Laplace transformation, and we show how to do this next. This seemingly restricted modeling class remains of great value, surprisingly applicable in modeling practice, because many systems and their models satisfy these properties at least approximately or over some limited region of operation.

Following is a very abbreviated, minimalist version of Laplace transforms (LT) applied to ODE solution, just enough to get us into the next sections on transfer functions and their analysis. A more complete synopsis of Laplace transform methods is presented in Appendix A.

We formally define the LT and its inverse (abbrev: ILT) first, to illustrate their level of complexity and provide some insight, but hasten to add that one rarely – if ever – needs to evaluate them, because solutions for typical applications are published in tabular form, as LT–ILT pairs – exemplified in **Table 2.1**.

The **Laplace transform of $f(t)$** for $t > 0$ is $\mathcal{L}[f(t)] \equiv F(s) \equiv \int_{0^+}^{\infty} f(t)e^{-st}dt$, where $s = \sigma + j\omega$ is a complex variable, σ and ω are real variables, and $j = \sqrt{-1}$. The real part σ of s is often written Re(s) (the real part of s) and the imaginary part ω as Im(s) (the imaginary part of s). The lower limit on the integral $t = 0^+$ reflects the fact that real signals begin at $t > 0$; they don't turn on instantaneously.

26 We are reminded here of another use of the equivalence property, Eq. (2.24), noted earlier. If nonzero ICs are transformed into impulse inputs, so none remain explicitly in the equations of the model, and these are added to the corresponding ODE equations – or under the integral in their solutions – as in the second term of solution $\boldsymbol{x}(t)$ above, superposition then would apply.

27 The importance of modes in dynamic biosystem modeling, for transforming temporal biodata patterns into alternative biomodel structures, is clarified and developed in Chapter 5.

Table 2.1 Short table of LT-ILT pairs

TIME FUNCTION (ILT)		LAPLACE TRANSFORM (LT)
Unit impulse	$\delta(t)$	1
Unit step	$1(t)$	$\frac{1}{s}$
Unit ramp	t	$\frac{1}{s^2}$
Exponential	e^{-at}	$\frac{1}{s+a}$
Sine wave	$\sin \omega t$	$\frac{\omega}{s^2+\omega^2}$

The LT takes us from the t-domain to the s-domain. To go back, i.e. to obtain the time-domain solution from the s-domain solution, we use the *inverse Laplace transform*. If $F(s)$ is the LT of $f(t)$, for $t > 0$, then the **inverse Laplace transform of $F(s)$** is the *contour integral*: $f(t) \equiv \mathcal{L}^{-1}[F(s)] \equiv \frac{1}{2\pi j}\int_{c-j\infty}^{c+j\infty} F(s)e^{st}ds$ where $c = \text{Re}(s) = \sigma_0$ (some real number σ_0). This is admittedly not simple, but again, it is rarely if ever necessary to do the integration to find $F(s)$, or to perform the contour integration, to get $f(t)$. We get the solutions from a *table of transform pairs*, like abbreviated **Table 2.1**, or the larger one in Appendix A. We present several key properties of this pair of transforms, the ones most useful for obtaining ODE solutions. All are illustrated in Example 2.27 and Example 2.28 below.

Four Key Properties of LT & ILT

1. The LT and ILT are *linear transformations* between the t and s domains. This means they satisfy superposition principles: if $F_1(s)$ and $F_2(s)$ are the LT *of* $f_1(t)$ and $f_2(t)$ (and *vice versa* for the ILT), then $a_1F_1(s) + a_2F_2(s)$ is the LT of $a_1f_1(t) + a_2f_2(t)$ (and *vice versa* for the ILT), where a_1 and a_2 are arbitrary constants.

2. The ***LT of the derivative*** df/dt of $f(t)$ whose LT is $F(s)$ is

$$\mathcal{L}\left[\frac{df}{dt}\right] = sF(s) - f(0^+) \tag{2.52}$$

   where $f(0^+)$ is the initial value of $f(t)$, i.e. its initial condition (IC).

3. The ***LT of the integral*** $\int f(\tau)d\tau$ of $f(t)$ with whose LT is $F(s)$ is

$$\mathcal{L}\left[\int_0^t f(\tau)d\tau\right] = \frac{F(s)}{s} \tag{2.53}$$

4. The ***LT of the time-delayed function*** $f(t-T)$, where $T > 0$ and $f(t-T) = 0$ for $t \leq T$, is

$$\mathcal{L}[f(t-T)] = e^{-Ts}F(s) \tag{2.54}$$

Let's apply these properties to find the LT $Y(s)$ of the solution $y(t)$ of an example ODE. The result will motivate the *partial fraction expansion* method needed for inverting $Y(s)$ to obtain $y(t)$.

Example 2.27: ***LT of an ODE Solution***

To evaluate the LT of $\frac{d^2y}{dt^2} + 5\frac{dy}{dt} + 6y = u(t) \equiv \int_{0^+}^{t} \delta(t-1)dt$, with initial conditions (ICs) $y(0^+) = -1$ and $\left.\frac{dy}{dt}\right|_{t=0^+} = 2$, we need to first find the LT of the second derivative term, not in the list above. We get this by applying the first derivative property Eq. (2.52):

$$\mathcal{L}\left[\frac{d^2y}{dt^2}\right] = \mathcal{L}\left[\frac{d}{dt}\frac{dy}{dt}\right] = s(sY(s) - y(0^+)) - \left.\frac{dy}{dt}\right|_{t=0^+} = s^2Y(s) - sy(0^+) - \left.\frac{dy}{dt}\right|_{t=0^+} \tag{2.55}$$

Then we apply the linearity principle direction to the ODE, substituting Eq. (2.55) for the first term, using Eq. (2.53) on the second term, and simply transforming and adding the others:

$$\mathcal{L}\left[\frac{d^2y}{dt^2} + 5\frac{dy}{dt} + 6y\right] = \mathcal{L}\left[\frac{d^2y}{dt^2}\right] + 5\mathcal{L}\left[\frac{dy}{dt}\right] + 6\mathcal{L}[y] = (s^2+5s+6)Y(s) - s(-1) - 2 - (-1)$$
$$= \mathcal{L}[u(t)] = \frac{1}{s}\mathcal{L}[\delta(t-1)] = \frac{1}{s}\mathcal{L}[\delta(t)]e^{-s} = \frac{e^{-s}}{s}$$

The last term was evaluated first using the LT of $\delta(t)$ $(=1)$ given in **Table 2.1**, plus Property 4 above. Then the output transform $Y(s)$ is determined by rearranging the above equation:

$$Y(s) = \frac{-(s^2+s-e^{-s})}{s(s^2+3s+2)} = -\left(\frac{s+1}{(s+2)(s+3)} + \frac{e^{-s}}{s(s+2)(s+3)}\right) \tag{2.56}$$

We separated out the term with the time-delay on the RHS of Eq. (2.50), because it's defined only for $t > 1$, whereas the first term is valid for all $t > 0$. To obtain $y(t)$, the ILT of $Y(s)$, we need to expand $Y(s)$ into partial fractions, next.

Partial Fraction Expansions

Partial fraction expansion of rational algebraic functions like $Y(s)$ above greatly simplifies LT inversion. We present only enough of the methodology here to get started, with additional details and examples found in Appendix C and DiStefano III (1990). Consider any $F(s)$:

$$F(s) = \frac{\sum_{i=0}^{m} b_i s^i}{\sum_{i=0}^{n} a_i s^i} \tag{2.57}$$

where $a_n = 1$ and $n \geq m$ (because our model represents a *causal* system). By the fundamental theorem of algebra, the denominator polynomial equation $\sum_{i=0}^{n} a_i s^i = 0$ has n roots, some possibly repeated. Suppose there are n_1 roots equal to $-p_1$, n_2 roots equal to $-p_2, \ldots,$ and n_r roots equal to $-p_r$, where $\sum_{i=1}^{r} n_i = n$. Then $\sum_{i=0}^{n} a_i s^i = \prod_{i=1}^{r} (s+p_i)^{n_i}$ and $F(s)$ then becomes:

$$F(s) = \frac{\sum_{i=0}^{m} b_i s^i}{\prod_{i=1}^{r} (s+p_i)^{n_i}} \tag{2.58}$$

The partial fraction expansion (PFE) of $F(s)$ is

$$F(s) = b_n + \sum_{i=1}^{r} \sum_{k=1}^{n_i} \frac{c_{ik}}{(s+p_i)^k} \tag{2.59}$$

where $b_n = 0$ unless $m = n$. The coefficients c_{ik} $(c_{i1}, \ldots, c_{r1})$ – called the **residues** of $F(s)$ at $-p_{i1}, \ldots, -p_{r1}$, are given by

$$c_{ik} = \frac{1}{(n_i - k)!} \frac{d^{n_i - k}}{ds^{n_i - k}} [(s+p_i)^{n_i} F(s)] \bigg|_{s=-p_i} \tag{2.60}$$

If the roots are all distinct, then the PFE is simpler:

$$F(s) = b_n + \sum_{i=1}^{n} \frac{c_{i1}}{s + p_i} \tag{2.61}$$

with residues:

$$c_{i1} = (s+p_i)F(s)\big|_{s=-p_i} \tag{2.62}$$

Example 2.28: ***ODE Solution by PFE and ILT***

Consider the output transform Eq. (2.56) computed in the last example: $Y(s) = -\left(\frac{s+1}{(s+2)(s+3)} + \frac{e^{-s}}{s(s+2)(s+3)}\right) \equiv Y_1(s) + Y_2(s)e^{-s}$. From Eq. (2.61), the PFE of the first term is $Y_1(s) = -\left(b_2 + \frac{c_{11}}{s+2} + \frac{c_{21}}{s+3}\right)$. Since $m < n$, $b_2 = 0$. From Eq. (2.62), the residues are:

$$c_{11} = (s+2)Y_1(s)\big|_{s=-2} = \frac{-s-1}{s+3}\bigg|_{s=-2} = -1, \quad c_{21} = (s+3)Y_1(s)\big|_{s=-3} = \frac{-s-1}{s+2}\bigg|_{s=-3} = -1$$

This yields $Y_1(s) = \frac{1}{s+2} + \frac{1}{s+3}$. Similarly, the PFE parenthetical component of the second term is $Y_2(s) = -\left(b'_3 + \frac{c'_{11}}{s} + \frac{c'_{21}}{s+2} + \frac{c'_{31}}{s+3}\right)$, with $b'_3 = 0$, and the residues are:

$$c'_{11} = sY_2(s)\big|_{s=0} = \frac{-1}{(s+2)(s+3)}\bigg|_{s=0} = -1/6$$

$$c'_{21} = (s+2)Y_2(s)\big|_{s=-2} = \frac{-1}{s(s+3)}\bigg|_{s=-2} = 1$$

$$c'_{31} = (s+3)Y_2(s)\big|_{s=-3} = \frac{-1}{s(s+2)}\bigg|_{s=-3} = -1/3$$

Therefore, $Y_2(s) = \frac{1}{6}\left(\frac{1}{s} - \frac{6}{s+2} + \frac{2}{s+3}\right)$ and

$$Y(s) \equiv Y_1(s) + Y_2(s)e^{-s} = \frac{1}{s+2} + \frac{1}{s+3} + \frac{1}{6}\left(\frac{1}{s} - \frac{6}{s+2} + \frac{2}{s+3}\right)e^{-s}$$

The ILT of $Y(s)$ is obtained by simply inverting each term separately using **Table 2.1** and applying superposition to the result. We need only the ILT of the exponential, constant (step) and delay functions. The time-delay operator shifts the three terms it multiplies by 1 time unit, all yielding: $y(t) = e^{-2t} + e^{-3t} + \frac{1}{6}(\boldsymbol{1}(t-1) - 6e^{-2(t-1)} + 2e^{-3(t-1)})\boldsymbol{1}(t-1)$, for $t > 0$.

The $\boldsymbol{1}(t-1)$ delayed step function multiplying the terms in parentheses is needed to render the last three terms zero prior to $t = 1$. To further emphasize the domain over which each of the components of this solution are valid, we can use a step function $\boldsymbol{1}(t)$ to define the solution for values of $t > 0$ only, writing the total response as

$$y(t) = (e^{-2t} + e^{-3t})\boldsymbol{1}(t) + \frac{1}{6}(\boldsymbol{1}(t-1) - 6e^{-2(t-1)} + 2e^{-3(t-1)})\boldsymbol{1}(t-1)$$

The solution $y(t)$ is graphed in **Fig. 2.8**. We graphed this solution by simulating and then combining each of the functions in this analytically derived solution. Only constant and exponential solutions were needed, with the third term in the summation requiring all its components – and the whole term – delayed by one time unit. Most if not all simulation or math solving programs can do this.

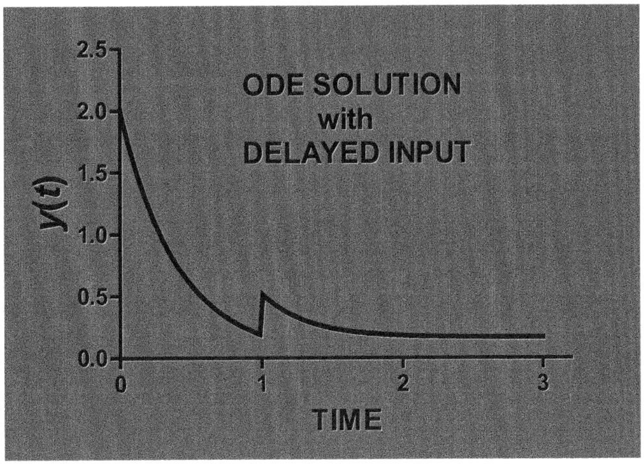

Fig 2.8

Transfer Functions of Linear TI ODE Models

Transfer functions (TFs)[28] are special Laplace domain representations of linear time-invariant system components, or the whole system – between input and output. Like the Laplace transform itself, TFs simplify solution and manipulation of ODE models. And they readily provide insights into dynamical properties, like stability.

In Appendix A, Eq. (A.16) or (A.22), a linear system *output* can be generally expressed in the time domain or Laplace domain as a sum of two terms, one due to the input and the other to ICs, like in Example 2.21 above. In the Laplace domain, this output has the form:

$$Y(s) = \left(\frac{\sum_{i=0}^{m} b_i s^i}{\sum_{i=0}^{n} a_i s^i} \right) U(s) + (\text{terms due to all initial conditions}) \tag{2.63}$$

The **transfer function (TF)** is defined as the *polynomial-ratio factor* in the equation for $Y(s)$ multiplying the input $U(s)$:

$$H(s) = \left(\frac{\sum_{i=0}^{m} b_i s^i}{\sum_{i=0}^{n} a_i s^i} \right) = \frac{b_m s^m + b_{m-1} s^{m-1} + \ldots + b_0}{a_n s^n + a_{n-1} s^{n-1} + \ldots + a_0} \tag{2.64}$$

The denominator is the *characteristic polynomial*, and $Y(s) = H(s)U(s) +$ (terms due to all initial conditions). If all ICs are 0, then:

$$\begin{aligned} &Y(s) = H(s)U(s), \quad y(t) = \mathcal{L}^{-1}[H(s)U(s)], \\ &\text{and the TF is:} \qquad H(s) = Y(s)/U(s) \end{aligned} \tag{2.65}$$

28 The transfer function concept can be applied to nonlinear systems as well, using generalized nonlinear system operators on inputs. This is a whole other subject (Isidori 1999).

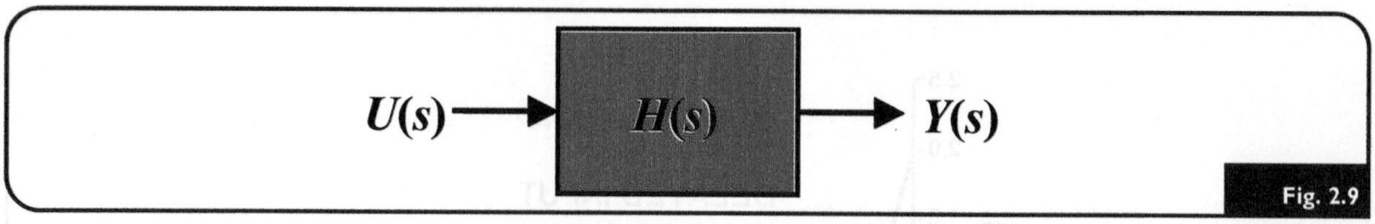

Fig. 2.9

Input–output transfer function block.

Remark: The LT of the unit impulse equals unity (**Table 2.1**). With $U(s) = 1$, we see that *H(s) is the Laplace transform of the impulse response,* an important result.

Transfer functions are typically expressed in block diagram form as in **Fig. 2.9**.

*Example 2.29: **Time-Delayed Functions***

The LT of $u(t)$ is $U(s)$ and the LT of $u(t\text{-}T)$ is $e^{-sT}U(s)$ (see Appendix A). Therefore the TF of a delayed function $y(t) = u(t - T)$ is $H(s) \equiv e^{-sT}$. (See also LT Property 4 above.)

*Example 2.30: **TFs from ODEs***

The input–output TF for the ODE $\dot{y} + 2y = \dot{u} + u$ is computed by taking the LT of the ODE, with ICs $\equiv 0$: $(s+2)Y = (s+1)U$. Then $H(s) \equiv \dfrac{Y(s)}{U(s)} = \dfrac{s+1}{s+2}$.

*Example 2.31: **ODEs from TFs***

To obtain an ODE from the TF, simply replace s by $D \equiv \dfrac{d}{dt}$. If $H(s) = \dfrac{2s+1}{s^2+s+1}$, substitute D for s. Then $y(t) = \left(\dfrac{2D+1}{D^2+D+1}\right)u(t)$, which simplifies to $D^2y + Dy + y = 2Du + u$ or $\dfrac{d^2y}{dt^2} + \dfrac{dy}{dt} + y = 2\dfrac{du}{dt} + u$.

Transfer Function (TF) Matrix of an ODE

This particular matrix representation of a linear constant-coefficient ODE model has many uses in construction and analysis of dynamic biosystem models. We'll encounter it in subsequent chapters. The math is straightforward for *single-input–single-output* (SISO) and *multi-input–multi-output* (MIMO) models in state variable form. For the **SISO case**, the ODE model is: $\dot{\boldsymbol{x}} = A\boldsymbol{x} + \boldsymbol{b}u$ and $y = \boldsymbol{c}^T\boldsymbol{x}$. To obtain the TF, we take the Laplace transform of the ODE, set $\boldsymbol{x}(0) \equiv \boldsymbol{0}$, then solve for the vector $\boldsymbol{X}(s)$ and then for the (scalar) output $Y(s)$:

$$s\boldsymbol{X}(s) = A\boldsymbol{X}(s) + \boldsymbol{b}U(s)$$
$$(sI - A)\boldsymbol{X}(s) = \boldsymbol{b}U(s)$$
$$\boldsymbol{X}(s) = (sI - A)^{-1}\boldsymbol{b}U(s)$$
$$Y(s) = \boldsymbol{c}^T(sI - A)^{-1}\boldsymbol{b}U(s)$$

From Eq. (2.65), the TF matrix is 1×1, a scalar:

$$H(s) = \frac{Y(s)}{U(s)} = \boldsymbol{c}^T(sI - A)^{-1}\boldsymbol{b} \tag{2.66}$$

This readily generalizes for multi-input–multi-output (**MIMO**) models, substituting the $n \times r$ input and $m \times n$ output matrices A and C for vectors $\boldsymbol{c}^T$ and $\boldsymbol{b}$:

$$H(s) = C(sI - A)^{-1}B \tag{2.67}$$

The characteristic equation of the model in both cases is $\det(sI - A) = 0$.

Example 2.32: ***TF Matrix of an ODE***

Consider the ODE model in state variable form:

$$\dot{\boldsymbol{x}} = \begin{bmatrix} 0 & 1 \\ -2 & -3 \end{bmatrix} x + \begin{bmatrix} 0 \\ 1 \end{bmatrix} u$$

$$y = \begin{bmatrix} 1 & 0 \end{bmatrix} \boldsymbol{x}$$

which corresponds to the second-order ODE $\ddot{y} + 3\dot{y} + 2y = u$. We use Eq. (2.68) to determine the TF. First, calculate[29] $(sI - A)^{-1}$:

$$sI - A = \begin{bmatrix} s & 0 \\ 0 & s \end{bmatrix} - \begin{bmatrix} 0 & 1 \\ -2 & -3 \end{bmatrix} = \begin{bmatrix} s & -1 \\ 2 & s+3 \end{bmatrix}$$

$$(sI - A)^{-1} = \frac{\mathrm{adj}(sI - A)}{|sI - A|} = \frac{\begin{bmatrix} s+3 & 1 \\ -2 & s \end{bmatrix}}{s^2 + 3s + 2}$$

$$H(s) = c^T(sI - A)^{-1}\boldsymbol{b} = \frac{\begin{bmatrix} 1 & 0 \end{bmatrix} \begin{bmatrix} s+3 & 1 \\ -2 & s \end{bmatrix} \begin{bmatrix} 0 \\ 1 \end{bmatrix}}{s^2 + 3s + 2} = \frac{1}{s^2 + 3s + 2}$$

Note that $|sI - A|$ is the characteristic polynomial of the original ODE.

The Complex Plane: Pole-Zero Maps of Transfer Functions

The "dynamics" of the ODE model in TF form can be visualized graphically by transformation into the complex plane, yet another intuition-builder for modeling dynamic biosystems.

29 See Appendix C if you need help with the matrix manipulations.

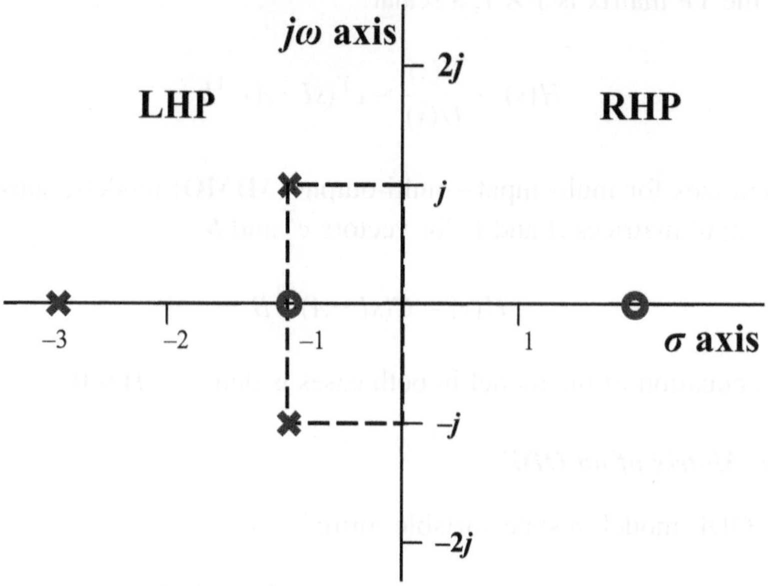

Fig. 2.10

Pole-zero map in the complex plane.

Consider Eq. (2.66) again, but in a different form, with the polynomials factored in the numerator and denominator:

$$H(s) = \frac{b_m \sum_{i=0}^{m} \frac{b_i}{b_m} s^i}{\sum_{i=0}^{n} a_i s^i} = \frac{b_m \prod_{i=1}^{m} (s + z_i)}{\prod_{i=1}^{n} (s + p_i)} \tag{2.68}$$

Factoring is usually readily accomplished by partial fraction expansion of $H(s)$, as in the previous section and in Appendix A. The z_i are called the **zeros** and the p_i the **poles** of the TF, these names chosen because $H(-z_i) = 0$ and $H(-p_i) = \infty$.

Poles and zeros are often graphed in the **complex plane**, illustrated in **Fig. 2.10**, poles shown with **x** and zeros with O symbols. The abscissa σ and ordinate $j\omega$ axes represent the real and imaginary parts of the complex number $s = \sigma + j\omega$ and thus $H(s)$ is represented graphically as a function of this complex number argument. The notations LHP and RHP mean the **left-hand plane** and the **right-hand plane** and these planes have significance in model dynamical stability analysis, as illustrated in the next section.

Example 2.33: ***TF and Pole-zero Map of a Drug PK Model***

Consider the simple model of drug X pharmacokinetics (PK) in blood, introduced intravenously (IV) into blood as a bolus dose (approximated by an impulse) in amount D: $\dot{x} + 2x = D\delta(t)$. The input–output TF of this ODE is easily established from the TF of the ODE: $sX + 2X = (s + 2)X = D$. We have: $H(s) = X(s)/U(s) = D/(s + 2)$. The pole-zero map of this $H(s)$ has a single pole at $s = -2$ in the LHP.

Example 2.34: ***Pole-Zero Map of a TF***

Fig. 2.10 includes a **pole-zero map** of the TF:

$$H(s) = \frac{2s^2 - 2s - 4}{s^3 + 5s^2 + 8s + 6} = \frac{2(s+1)(s-2)}{(s+3)(s+1+j)(s+1-j)} \begin{matrix} \leftarrow \text{Zeros} \\ \leftarrow \text{Poles} \end{matrix}$$

This TF has zeros at $s = -1$ and $s = 2$ and poles at $s = -3$, $s = -1 - j$, and $s = -1 + j$, as shown.[30]

More on System Stability

We introduced this topic briefly in Chapter 1. We extend it a bit here, having developed some of the math needed to appreciate this notion better. We develop the concept and stability analysis methodologies further, for nonlinear as well as linear system models, in Chapter 9.

The stability of a dynamic system is determined by its response to inputs or disturbances, as was illustrated graphically in **Fig. 1.6**. We begin here with what might be termed a robust, practical or semi-formal definition. A continuous or discrete-time dynamic system model with output $y(t)$ and driven by input $u(t)$ is **stable** if every *bounded* input $u(t)$ produces a *bounded* output $y(t)$. This is often called **input–output stable**. It is **input–output unstable** if a bounded input produces an unbounded output, one that increases without bound toward infinity. Many results in stability theory, however, are developed for unforced (zero-input) system models, focusing on the system and model response to brief disturbances, or equivalently, the response of the model to *initial conditions* as (input signal) models of physical disturbances. Looking back on the *equivalence property* of dynamic systems, Eq. (2.24), this seems reasonable enough, at least for SISO models with impulse inputs. It works for MIMO ones too; and impulses are very powerful inputs, so we can appreciate this generalization at least intuitively for now.

Suppose an unforced (zero-input) *stable* system is briefly disturbed away from a so-called *equilibrium point*[31] at time t_0, and the cause of the disturbance is immediately removed. This is equivalent to applying an impulse, or nonzero initial conditions to the model of the system at time t_0. Then the stable system should eventually return to the equilibrium point, or at least stay close to it. Returning to the equilibrium point is equivalent to saying the impulse response of a stable system approaches zero as time approaches infinity. This is easy to illustrate for linear time-invariant dynamic system models, by analyzing the model TF.

Stability of Linear TI Dynamic Systems

Continuous linear TI dynamic system models are stable if their impulse responses approach zero as $t \to \infty$, which is true if and only if the poles of the TF of the model are in the left-half-plane (LHP). **If the poles are in the RHP, the model is unstable.**[32] The models in the two previous examples are thus stable.

30 Complex roots always occur in complex conjugate pairs.

31 Where the first derivative of the state of the system is zero, i.e. initially at rest.

32 Proof of this can be found in DiStefano III 1990.

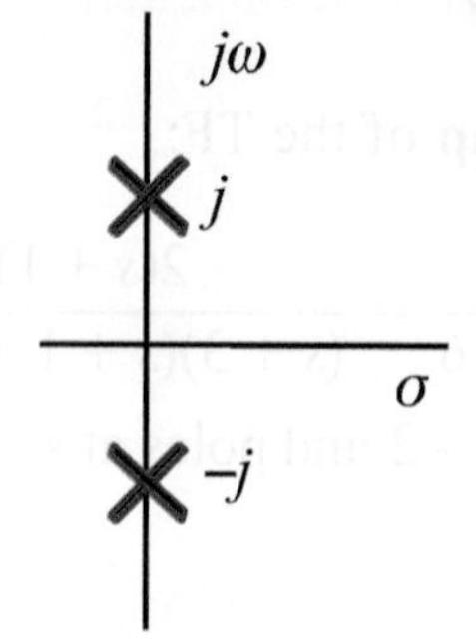

Fig 2.11

$s^2 + 1 = 0$. This equation has the two roots $\pm j$.

Example 2.35: ***Asymptotic Stability***

For $(s+1)Y(s) = U(s)$, the characteristic equation is $s + 1 = 0$, and $s = -1$, which is in the LHP. Then, for impulse input, $U(s) = 1$, the output transform is $Y(s) = \frac{1}{s+1}$, and its inverse is $y(t) = e^{-t}$. This response clearly approaches zero as time approaches infinity, i.e. it is asymptototically stable.

Example 2.36: ***Marginal Stability***

The system described by the Laplace-transformed ODE $(s^2 + 1)Y(s) = U(s)$ has the characteristic equation: $s^2 + 1 = 0$. This equation has the two roots $\pm j$ (**Fig. 2.11**). Since these roots have zero real parts, and are not in the LHP, the system is not stable. It is, however, not unstable, because the roots are not in the RHP. It is what is termed **marginally stable**. In response to most inputs or disturbances, the system oscillates with a bounded output.[33]

Remark: Again, as a general rule, **if the poles are in the LHP, the model is stable. Equivalently, if the roots of the characteristic equation (the poles) are in the LHP ⇔ stability for all ICs and inputs.**

The concept of system stability is quite broad and important in biomodeling, particularly rich and sometimes unpredictable for nonlinear dynamic systems. Chapter 9 is devoted to dynamic biosystem model stability and oscillations, mostly nonlinear. An important issue is assessing system behavior following *small* versus *big* (any) disturbances of the system state, the former type called *local* and the latter *global* stability. Dynamic systems and their models can be locally but not globally stable.

Looking Ahead

The basic language and applied math for modeling linear and nonlinear dynamical systems has been developed this chapter, drawn from classical mathematical biology as well as systems

33 Not true for some inputs. For input $u = \sin t$ the output contains a term of the form $y = t\cos t$, which is unbounded.

and control engineering. Examples illustrating these beginnings are drawn from both. On the biological side these include several well-known population biology models, plus a few biophysics and modern cell biology models. The engineering-flavored (applied math) examples are more illustrative of the mathematics of the methods themselves, applicable to any science in need of dynamic system models.

The analysis methodology developed in the last half of the chapter has focused largely on *linear* dynamic system models, for two reasons. First, analytical methods are well developed and fairly complete for linear – but not so well or complete for nonlinear dynamic system models. Second, there's a lot to learn about modeling in general by studying classical linear system methods, and they are widely applicable in biology as well as in other sciences. To build and evaluate nonlinear (NL) models, we touched upon linearization methods here – to be developed further later. The primary tool for nonlinear dynamic systems modeling and analysis is numerical, i.e. computer simulation, the subject of the next chapter.

Basic principles for structuring both linear and NL classes of dynamic *biosystem* models, then begins in earnest in Chapters 4 and 5. This is followed by NL modeling methods based on the principle of mass action – including molecular and cell biology modeling – in Chapters 6 and 7. Whole-organism, physiologically based and noncompartmental modeling follow in Chapter 8, and stability and oscillations of NL models in Chapter 9. The nitty-gritty of quantifying (completing and validating) biodynamic system models follows in Chapters 10 through 13. These basic steps are summarized in the block-flow representation of the overall process in **Fig. 2.12** – a simpler and more conventional way of viewing the modeling process – without the circularity and strong data-dependencies and overlappings expressed

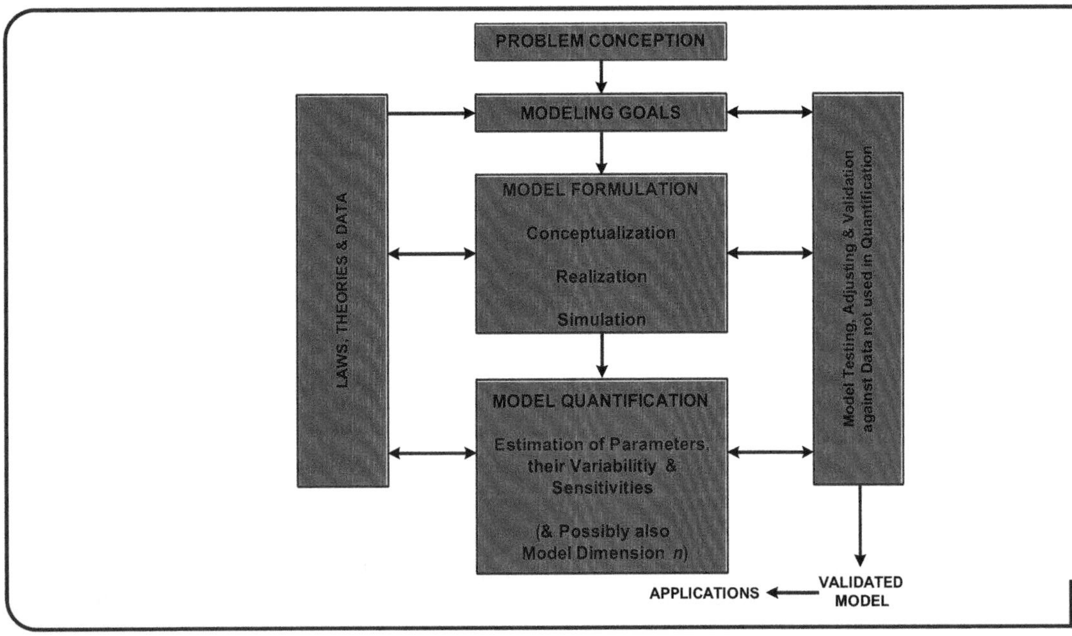

Fig. 2.12

Generic modeling approach.

in **Fig. 1.9**. (Some like to see it all laid out separately like this.) In the last several chapters we address biocontrol system modeling, biomodel-hypothesis testing, experiment design and network inference.

EXERCISES

E2.1 ***Theory* and *Data Discussion Question*:** Consider *data* to mean all information obtained empirically from experiments done on biosystems, not just "output data."

a) How is *theory* used in modeling dynamic biological systems?

b) How are *data* (all kinds of data) used in modeling dynamic biological systems?

c) Why are both data and theory needed to generate useful math models of dynamic biosystems?

d) Give at least one example of how both data and theory are used to generate useful math models of dynamic biosystems.

E2.2 Show that the homogeneous part of

$$\frac{d^4y}{dt^4} - \frac{d^3y}{dt^3} - 3\frac{d^2y}{dt^2} + 5\frac{dy}{dt} - 2y = u$$

has four independent solutions, two of which have the form $e^{\lambda_1 t}$ and $e^{\lambda_2 t}$. Then find λ_1 and λ_2.

E2.3 Transform the equation in Exercise E2.2 into four first-order ODEs and put the result in vector-matrix form, using the transformation: $y \equiv x_1, \frac{dy}{dt} \equiv x_2$, etc.

E2.4 Given the cascade system in **Fig. 2.13**, S_1 is modeled by:

$$y_1(k+2) + \frac{5y_1(k+1)}{6} + \frac{y_1(k)}{6} = u(k-1)$$

and S_2 by:

$$y_2(k+3) + \frac{y_1(k+2)}{2} + y_2(k+1) = 0$$

Transform this discrete-time model into first-order vector-matrix form, using the transformation Eq. (2.40). *Hint:* the final form should be $\boldsymbol{x}(k+1) = A\boldsymbol{x}(k) + \boldsymbol{f}(k)$, and the model must be fourth-order.

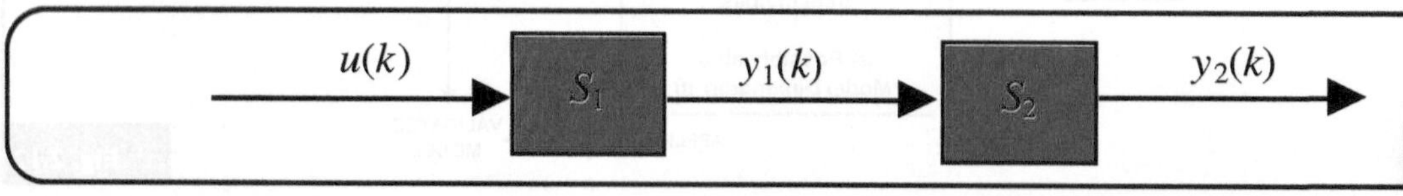

Fig. 2.13

Cascade system.

E2.5 **a)** Find the complete solution in the time domain for the systems with rational algebraic transfer function *H(s)* in Example 2.32.

b) Use the solution found in (a) to solve for the response to a step function of magnitude 5. Initial conditions are zero for parts (a) and (b).

E2.6 Find $y(t)$ for $Y(s) = \dfrac{1}{(s+1)^2(s+2)}$.

E2.7 Use the TF representation to simplify the following ODE model with a second-order approximation:

$$\frac{d^3y}{dt^3} + 12\frac{d^2y}{dt^2} + 22\frac{dy}{dt} + 20y = 20u$$

E2.8 Show that Eq. (2.7) $\dfrac{dx}{dt} = k_1x(A - k_2x)$ has the solution given by Eq. (2.8). Begin by letting $A/k_2 \equiv C$, $G \equiv k_1A$ and redefine the state variable x by $\xi \equiv x/C$, yielding the following simpler equation for solution: $\dot{\xi} = k_1k_2C\xi(1 - \xi)$. Direct integration of this equation, after multiplying both sides by dt and rearranging the terms, is a first step. Partial fraction expansion of the resulting quotient term can help, followed by consideration of the properties of logarithms.

E2.9 For a linear constant coefficient system model with zero ICs, the output $y_1(t)$ to input $u(t) = 1$ for $t > 0$ is $y_1(t) = 1 - e^{-t}$ for $t > 0$. What is **output** $y_2(t)$ for zero ICs and input equal to $10t$? (You can do this by Laplace transforms, but you don't have to. There's an easier way!)

References

Ariew, A., 2007. Under the influence of Malthus's law of population growth: Darwin eschews the statistical techniques of aldolphe quetelet. Studies in History and Philosophy of Science Part C. 38 (1), 1–19.

Birks, J.B., 1962. Rutherford at Manchester: British Chemist & Physicist. Heywood, London.

Britton, N., 1986. Reaction-Diffusion Equations and Their Applications To Biology. Academic Press, London.

Chen, L., Kazuyuki, A., 2002. Stability of genetic regulatory networks with time delay. IEEE Trans. Circuits Syst. 49, 602–608.

Cobelli, C., Mari, A., Ferrannini, E., 1987. Non-steady state: error analysis of Steeles model and developments for glucose kinetics. Am. J. Physiol. Endocrinol. Metab. 252 (5), E679–689.

DiStefano III, J., Stubberud, A.R., Williams, I.J., 1990. Feedback and Control Systems, 2nd Edition: Continuous (Analog) and Discrete (Digital). McGraw-Hill, New York.

dOnofrio, A., 2005. A general framework for modeling tumor-immune system competition and immunotherapy: mathematical analysis and biomedical inferences. Physica D. 208, 220–235.

Doyle, A.C., n.d. Arthur Conan Doyle – Sherlock Holmes. from <http://www.brainyquote.com/quotes/quotes/a/arthurcona131991.html>.

Friedberg, S., Insel, A., Spence, L.E., 2003. Linear Algebra. Pearson Education, Upper Saddle River, NJ.

Gompertz, B., 1825. On the nature of the function expressive of the law of human mortality, and on a new mode of determining the value of life contingencies. Philos. Trans. R Soc. Lond B Biol. Sci. 115, 513–585.

Isidori, A., 1999. Nonlinear control systems II. Communications and Control Engineering Series. Springer-Verlag, New York.

Keener, J., Sneyd, J., 1998. Mathematical Physiology. Springer, New York.

Klipp, E., 2005. Systems Biology in Practice: Concepts, Implementation and Application. Wiley-Interscience, Weinheim.

Laird, A.K., 1964. Dynamics of tumor growth. Br. J. Cancer. 18, 490–502.

Lennox, J.G., 1993. Darwin was a teleologist. Biol. Philos. 8, 4.

Ljung, L., 1987. System Identification: Theory for the User. Prentice-Hall, Englewood Cliffs, NJ.

Malthus, T.R., 1798–1826. An Essay on the Principle of Population. J. Johnson, in St. Paul's Churchyard, London.

Mari, A., 1992. Estimation of the rate of appearance in the non-steady state with a two-compartment model. Am. J. Physiol. Endocrinol. Metab. 263 (2), E400–415.

May, R.M., 1976. Simple mathematical models with very complicated dynamics. Nature. 261, 459–467.

Steele, R., Rostami, H., Altszuler, N., 1976. A 2-compartment calculation for the dot glucose compartment in the non-steady state. Fed. Proc. 33, 1869–1876.

Tee, D., DiStefano III, JJ, 2004. Simulation of tumor-induced angiogenesis and its response to anti-angiogenic drug treatment: mode of drug delivery and clearance rate dependencies. Cancer Res. Clin. Oncol. 130 (1), 15–24.

Verhulst, P.F., 1838. Notice sur la loi que la population poursuit dans son accroissement. Correspondance mathématique et physique. Nouveaux Mémoires de l'Académie Royale des Sciences et Belles-Lettres de Bruxelles. 10, 113–121.

Chapter 3

Computer Simulation Methods

"Done properly, computer simulation represents a kind of "telescope for the mind," multiplying human powers of analysis & insight just as a telescope does our powers of vision. With simulations we can discover relationships that the unaided human mind, or even the best mathematical analysis, would never grasp. Better market models alone will not prevent crises, but they may give regulators better ways for assessing market dynamics, and more important, techniques for detecting early signs of trouble. Economic tradition, of all things, shouldn't be allowed to inhibit economic progress."

From: This Economy Does Not Compute, by *Mark Buchanan, Nobel Laureate, Op-Ed Columnist,* New York Times, *Oct. 1, 2008.*

This essay was a timely critique of traditional economics theory dominating US economics policy – especially use of "equilibrium theory" models relying primarily on the balance of external forces (new information, external forcing functions, inputs) to predict market values – exclusive of internal (mechanistic) market dynamics. In other words, their models were criticized as being more "models of data" than (more reliably predictive) "models of dynamic systems."

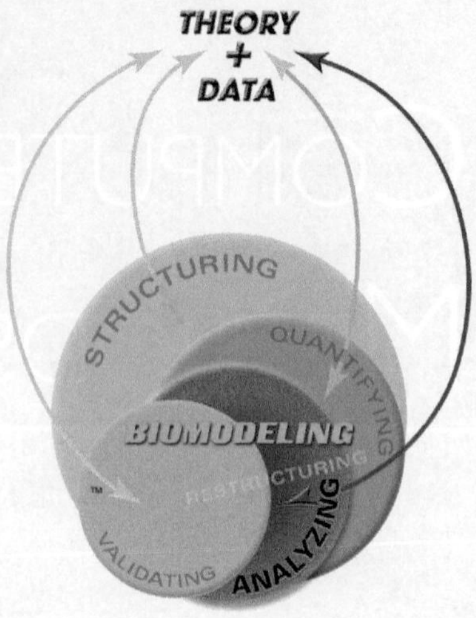

Overview

Simulation of a model has two aspects: *implementing or representing the model* in a particular form – in a simulation or modeling *program* (abbrev: **simulator**) – and solving the model, which usually is more about numerical solution methodology. The choice of simulator or modeling program is typically a matter of what's available or convenient; but in special areas like modeling dynamic biological systems, there are domain-specific choices, depending on a modeler's science and math training, and facility with various programming language or equation syntaxes. Both aspects are presented in this chapter, but the main focus is implementing and solving ordinary differential equation (ODE) models,[1] and the mathematical/computational basis for understanding and using the numerical methodology effectively.

This exposition is tailored for biomodeling users, employing transformation approaches and graphical language for ease of implementation and analysis, and (hopefully) enough numerical theory to maximize comprehension of numerical algorithms for ODE implementation and solution and minimize misuse and errors. Additional algorithms and mathematical details can be found in Appendix F and the references.

Procedures for solving models by simulation are essentially the same for both linear and nonlinear dynamic system models, so we make no distinction. Analytical solution approaches are preferred for linear time-invariant (LTI) models, but when LTI models are complex, we often resort to numerical methods for help with analysis or solution. Most time-varying linear and nonlinear ODE models require numerical solution.

Differential equations are solved using special numerical *integration* algorithms, usually packaged with other routines for numerical analysis of various types of equations. Procedurally, we introduce the model equations into an **ODE solver** (*simulation program, simulator*) and solve them using a built-in integration algorithm, with several such algorithms to choose from in most simulation programs. These include *graphical* simulation languages[2] like *Simulink* (a *Matlab Toolbox*) and *VisSim*. We introduce these first, in the form of a generic, control system block diagram graphical language, which applies to *Simulink* or *VisSim* and other languages based on control system diagrams. *SAAMII*[3] is another graph-based language, originally designed as a special purpose language for multicompartmental (MC) modeling, in which biomodels are programmed by assembling graphical MC model components. *SAAMII* also can handle more general ODE models, albeit somewhat awkwardly and with little additional effort, it also can "fit" MC or more general nonlinear ODE models to data – additional and important functionality for biomodeling. We introduce *SAAMII* and other programs (e.g. *Matlab*) for biomodeling in Chapters 4 and 5, and – for fitting models to data (quantifying them) – in Chapter 13. Discussion and/or illustration of other domain-specific languages and packages for modeling or simulation of biological or biochemistry-based systems, e.g. *SBML*, *SimBiology*, *COPASI*,

1 PDE models are not our focus, so their simulation – which is considerably more involved – is not presented here. Many other sources are available for PDE simulation methods, e.g. Asaithambi 1995.

2 *Simulink* (www.mathworks.com/products/simulink); *Matlab* (www.mathworks.com/products/matlabcentral); *VisSim* (www.vissim.com).

3 *SAAMII* (www.saamii.com).

BioSens and others, are postponed until later chapters. Some of these accomodate both math and chemistry language. *Matlab, Mathematica, Mathcad, Octave* (and others coming onto the horizon) are more comprehensive, general-purpose equation-based languages with ODE simulation functionality.[4]

Procedurally, simulation is relatively straightforward, but it is not without its inherent and less than transparent pitfalls. ODE simulation models, i.e. models represented in a simulator and solved using a numerical algorithm – are actually discrete-time models of continuous-time system models, i.e. approximations twice removed from reality. Simulated solutions must be scrutinized with this in mind. Simulated solutions invariably involve numerical errors, despite being computed with high precision, which can make even nonsense solutions look great or seem accurate. Some of these numerical errors are predictable and can be kept sufficiently small. Others can be unpredictable, unstable, or even *chaotic*, whether or not the original ODE model is unstable or chaotic (Stewart 1992). Understanding some of the basics of the numerical analysis underlying ODE solution algorithms, presented in abbreviated form here, should help in avoiding such problems – one of the goals of this chapter.[5]

Another goal is to expose troublesome numerical difficulties inherent in solving *multiscale biomodels* with numerical simulation tools. As noted in Chapter 1, **multiscale biomodels** are models that depict biosystem dynamics over a variety of spatial and/or temporal scales. They might have broadly separated temporal scales, producing dynamical responses with widely differing rates for different components of a response, e.g. widely spaced eigenvalues of a system matrix A in Eq. (2.29), or widely different *modes* – fundamental solutions[6] – of an ODE. The ODEs of such models are called *stiff*, and special methods designed for stiff models usually are needed for their solution. Crossing or mixing time-scales in biomodeling is common, and this typically generates ODEs that form a more-or-less stiff system of equations.

Any program that solves ODEs can be used for solving dynamic biosystem models, but – *from the viewpoint of learning how to model, graphical approaches are advantageous.* Simulation methodology is developed here by first presenting a generic programming language for *graphically* implementing ODE models for simulation. The goal is to develop graphical modeling skills, using what amounts to *feedback and feedforward control system diagrams.* We describe two ODE solver programs, each using minor variants of this language, namely *VisSim* (Visual Simulation Language) and the *Simulink* component of *Matlab,* along with examples of their use. Many more examples will surface in subsequent chapters. We then show how to *standardize*, *scale* and *normalize* ODEs, which can facilitate biomodel simulation analysis in applications, especially when the models are complex. Some of the special numerical analysis methodologies designed for ODE solution are then developed, beginning with the

4 *Systems Biology Markup Language* (*SBML*) (www.sbml.org); *SimBiology* (www.mathworks.com/products/simbiology); *Complex Pathway Simulator* (*COPASI*) (www.copasi.org); *BioSens* is a sensitivity and simulation tool for *BioSPICE* (www.biospicesourceforge.net), an open-source framework and software toolset for systems biology modeling and analysis; *Mathematica* (www.wolfram.com); *Mathcad* (www.ptc.com); *GNU Octave* (www.gnu.org/software/octave/).

5 In the last section of the chapter, we again clarify the distinction between deterministic and stochastic dynamic system models – and situations where the latter may be needed. Methodology for stochastic model simulation is deferred to Chapter 7, following development of mass action concepts upon which it depends.

6 See Chapter 2 or Appendix C for more on fundamental solutions.

Taylor series (TS) and TS-based methods, and ending with methods for solving *stiff* ODEs, with examples to motivate the math and illustrate numerical problems. We also show how to use an ODE solver to solve discrete-time ordinary *difference* equations.

Initial-Value Problems

ODE models are normally implemented for simulation as initial-value problems, i.e. with state variable initial conditions (ICs) given. **Final-value** problems are sometimes encountered, in optimal control systems, for example, where some function of the end-state or end-time is optimized. Such problems can be transformed into initial-value problems by redefining the time variable as $\tau = -t$ and thus $d\tau = -dt$. We treat only initial value problems in the remainder of this chapter, i.e. our goal is to solve: $\dot{\boldsymbol{x}} = \boldsymbol{f}(\boldsymbol{x}(t), \boldsymbol{u}(t), \boldsymbol{p}, t)$, with IC $\boldsymbol{x}(t_0) \equiv \boldsymbol{x}_0$ specified.

Graphical Programming of ODEs

Graphical language implementation is particularly advantageous for understanding and manipulating a biomodel. The big advantage is that it renders biosystem couplings and other model relationships among model variables and parameters *graphically explicit*, rather than representing them as sets of equations with variable and parameter symbols, connected by mathematical operators, using a symbolic syntax programming language. Overall, biomodel connectivity (topology, "wiring") is easier to build and assess graphically; and modeling intuition is built faster.

Interestingly, programming simulation models on digital computers using control system diagram language is directly analogous to programming 1950s-vintage analog computers. The basic idea is that an nth-order model, made up of n first-order ODEs, or one or more ODEs comprising up to nth-order derivatives, requires n integrations to solve it. This means n integrators must be included in the program. To complete the thought, i.e. to complete the program, these n integrators must be connected (wired) in a way that balances the equation.

Example 3.1: ***Using Graphical Programming Languages***

To solve $\dot{x} = kx + u$, we need $n = 1$ integrator to turn $\dot{x}$ into x; and then we have to feed $kx + u$ into this integrator to "wire" the correct solution at the output of the integrator. Graphically, the program starts out looking like the diagram in **Fig. 3.1**.

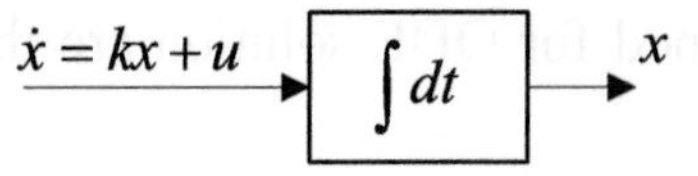

Fig. 3.1

To evaluate the input side, we adopt **control system block diagram language** (**Fig. 3.2**). We need to multiply x by k and then add u to it. First input x to a block with constant k in it. Then, assuming u is an exogenous model input, a summer circle does the addition, laid out in any direction, whichever is more convenient, as in **Fig. 3.2**:

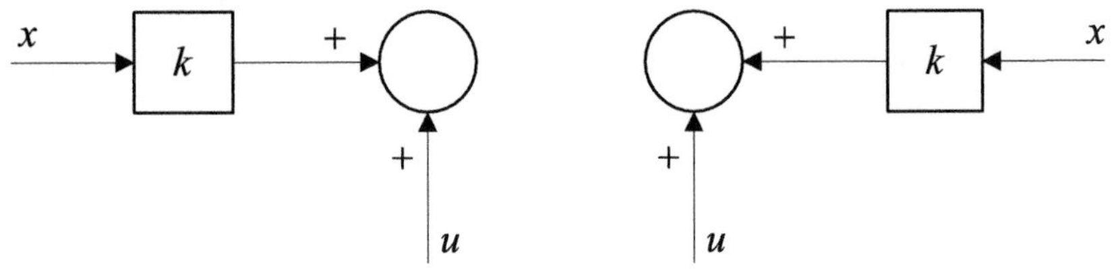

Fig. 3.2

Now we connect the two diagrams, as per the equation: $\dot{x} = kx + u$, as in **Fig. 3.3**. It's as simple as that. Just repeat this graphical procedure for all equations of the model, integrating and interconnecting (wiring) the state variables with their interconnecting parameters or functional relationships with appropriate blocks representing the needed mathematical operations.

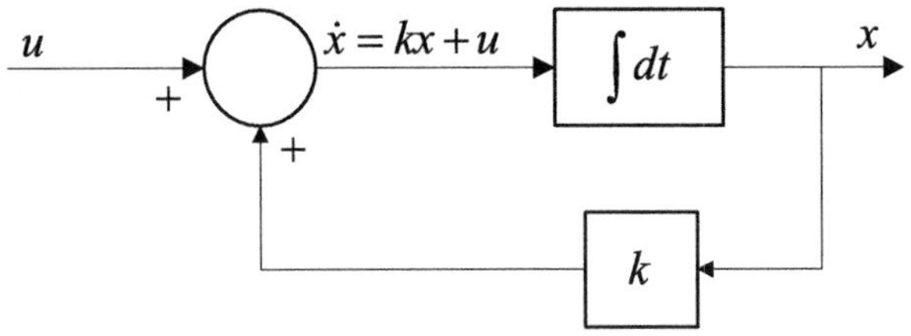

Fig. 3.3

Block Diagram Languages

The graphical objects used for mathematical operations and their interconnectivity in different programs differ somewhat, but they are usually fairly easy to decipher from the context. For example, *Simulink* and *VisSim* use different symbols for summers and integrators, different from the control system diagram language illustrated in Example 3.1. In both programs, integration is expressed in a block in Laplace transform notation, i.e. as $1/s$, instead of the integral operator shown in the figure above. It should also be noted that, to solve an ODE uniquely, initial conditions must be supplied for all integrators. This means that the initial values of each state variable (ICs) must be known and specified in the program. This is typically done by double-clicking on the integration block in the program diagram and filling in the requested info, which includes ICs, in the window that opens.[7]

Summers are expressed as circles with a $\sum$ symbol inside the circle in *VisSim*, and as a circle with the plus or minus signs *inside* the circle in *Simulink*, instead of outside, as in **Fig. 3.3** above. Constants or "gains" are expressed as numbers inside a triangular (*Simulink*) or five-sided polyhedron (*VisSim*). Other mathematical operations are typically represented inside a box.

Table 3.1 illustrates many of these. We use them interchangeably in examples to follow.

7 Nonzero ICs can also be set by adding a constant (via a constant block) to the output of the integrator, the constant value being the IC, i.e. the value $x(0)$ if the integrator has dx/dt at its input.

Table 3.1 Examples of Various Block Operations in *Simulink* and *VisSim*

	Simulink	*VisSim*
Summation (adjustable signs)	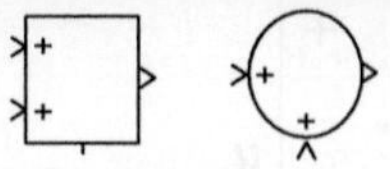	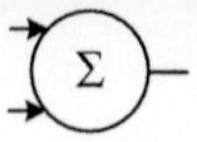
Absolute value	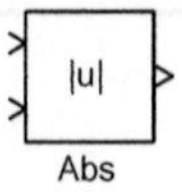	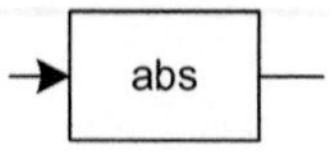
Negation	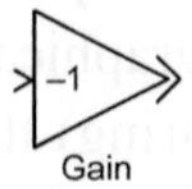	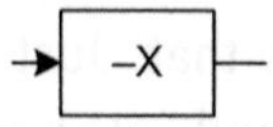
Multiply/divide (numerator on top)	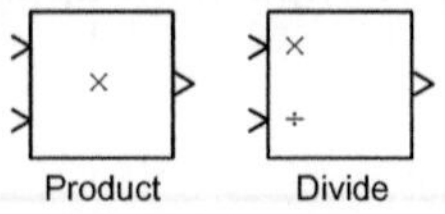	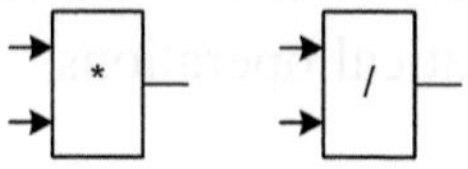
Dot product (vectors)	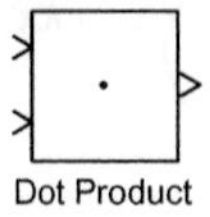	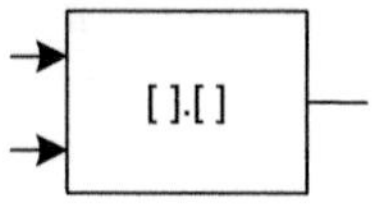
Inverse	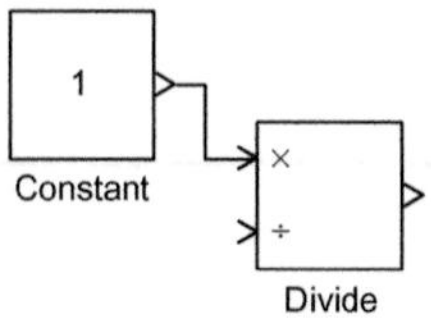	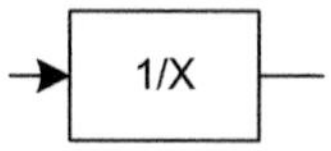
Gain (number is variable)	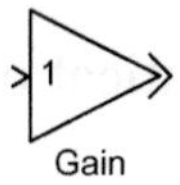	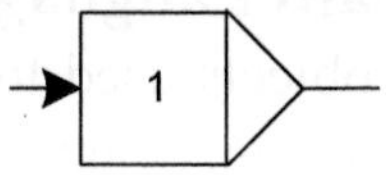
Derivative	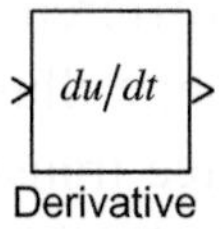	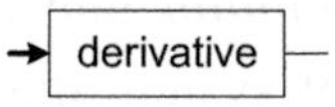
Integrator	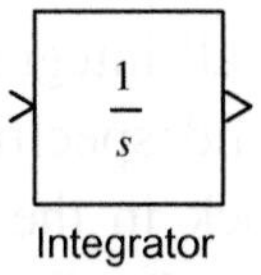	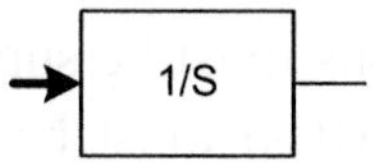
Constant input (number is variable)	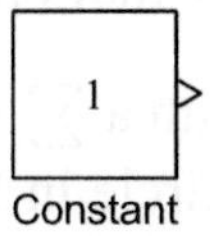	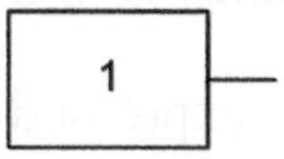
Ramp input	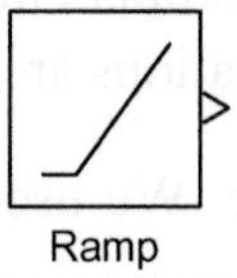	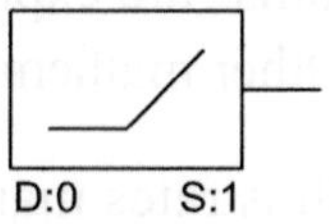

(Continued)

Table 3.1 (Continued)

	Simulink	*VisSim*
Pulse train input	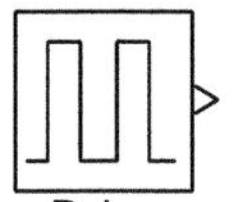 Pulse Generator	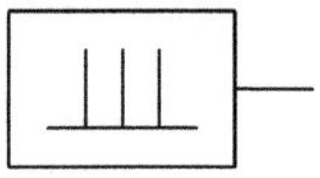
Unit step input	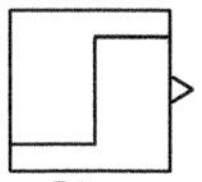 Step	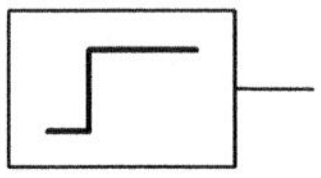
Output graph	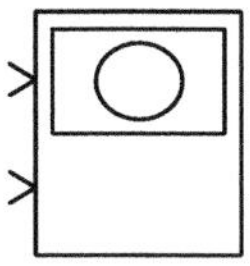 XY Graph	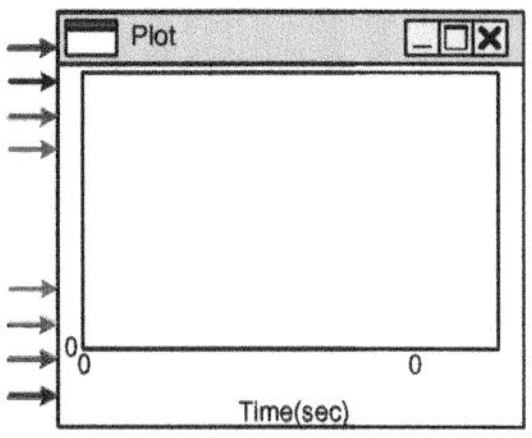

Special Input Function Simulations

In addition to mathematical operations, **Table 3.1** also includes several *signal functions*, typically used as inputs (constant, ramp, step, pulse-train inputs), or for generating other input waveforms in system simulation program diagrams (e.g. finite pulses, sawtooth waveforms, etc.).

Example 3.2: ***Implementing a Finite Pulse (Bolus) Input Waveform by Combining Blocks***

The bolus input is common in many physiological and pharmacological experiments – where the magnitude of the bolus is usually called a "dose," but it's really a dose *rate D*, and the dose (a mass) is the *area A* of the pulse, $D \times \tau$. It's a dose rate because inputs $u(t)$ in ODE models with mass-based state variables, e.g. $\dot{q} = -kq + u$, are mass rates of change with respect to time. A finite pulse input of magnitude D, width τ and area $D \times \tau = A$ is shown in the left-hand side (LHS) of **Fig. 3.4**. It can be readily programmed using the superposition principle, i.e. by adding a negative step input function of magnitude D starting at time τ, i.e. $-D \cdot \mathit{1}(t - \tau)$, to the positive step starting at time 0, i.e. $D \cdot \mathit{1}(t) - D \cdot \mathit{1}(t - \tau)$, as shown in right-hand side (RHS) of **Fig. 3.4**.

A finite pulse of width $\tau = 10$ and dose rate $D = A/\tau = 1/10$ is shown in the *VisSim* diagram and plot of **Fig. 3.5** ($D = 1/\mathrm{T1}$ in the figure), simulated by simply combining two input blocks with a summer (subtractor), each with step-input dose rate 1/10, as illustrated. The first step in the diagram is shown starting at $t = 0$ and the second (subtracted) at $\mathrm{T1} = \tau = 10$, equivalent to a total dose of magnitude 1, over 10 time units.

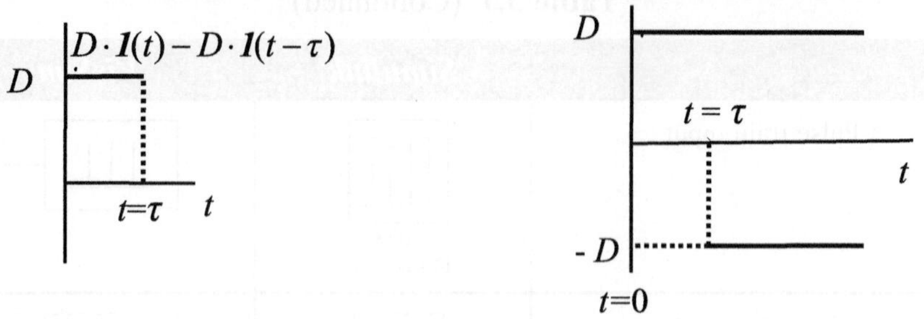

Fig. 3.4

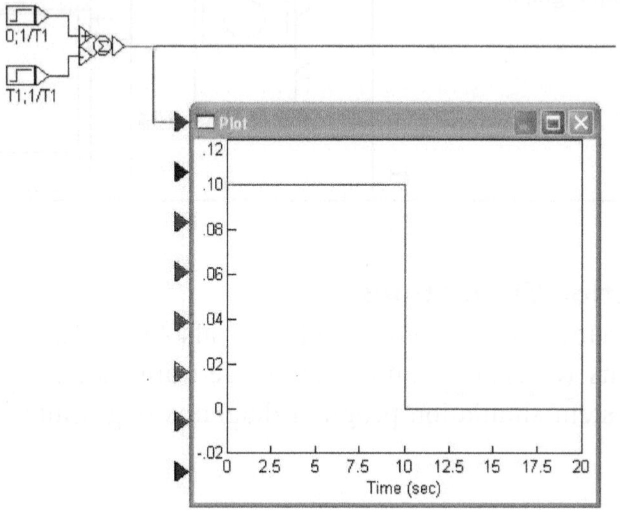

Fig. 3.5

Example 3.3: ***Dynamic Response Simulations of an ODE Model to Bolus Inputs***

Consider the simple ODE model $\dot{q} = -kq + u$, with $u(t)$ given by the pulse in **Fig. 3.5**, with $A = 1$, input dose rate $D/\tau = 1/10 = 0.1$, and the simulation run from $t = 0$ to $t_{end} = 20$. Four distinct solutions of this equation superimposed are given in the *VisSim* program and response plots of **Fig. 3.6**, illustrating the procedure with k set equal to 1. The program was simply copied four times to get the superposed $x(t)$ model output solutions, for four decreasing values of τ: 10, 1, 0.1 and 0.001. *VisSim* permits the abbreviated specification of parameter values using the gain blocks shown in the upper left of **Fig. 3.6**. The four *compound blocks in blue* contain the pulse input functions $u(t)$, for the four values of the parameter τ, all with $k = 1$.

A prominent and important feature in **Fig. 3.6** is that the pulse response looks more and more like an impulse response as the width of the pulse gets narrower and the pulse height gets taller, whereas the area under the "dose" response curve remains constant. We see that, in the limit as $\tau \to 0$, the initial value of the dose response goes to 1, i.e. $x(0) \to D \times \tau = 1$, the integral of the pulse input, which is also the integral of the impulse function.

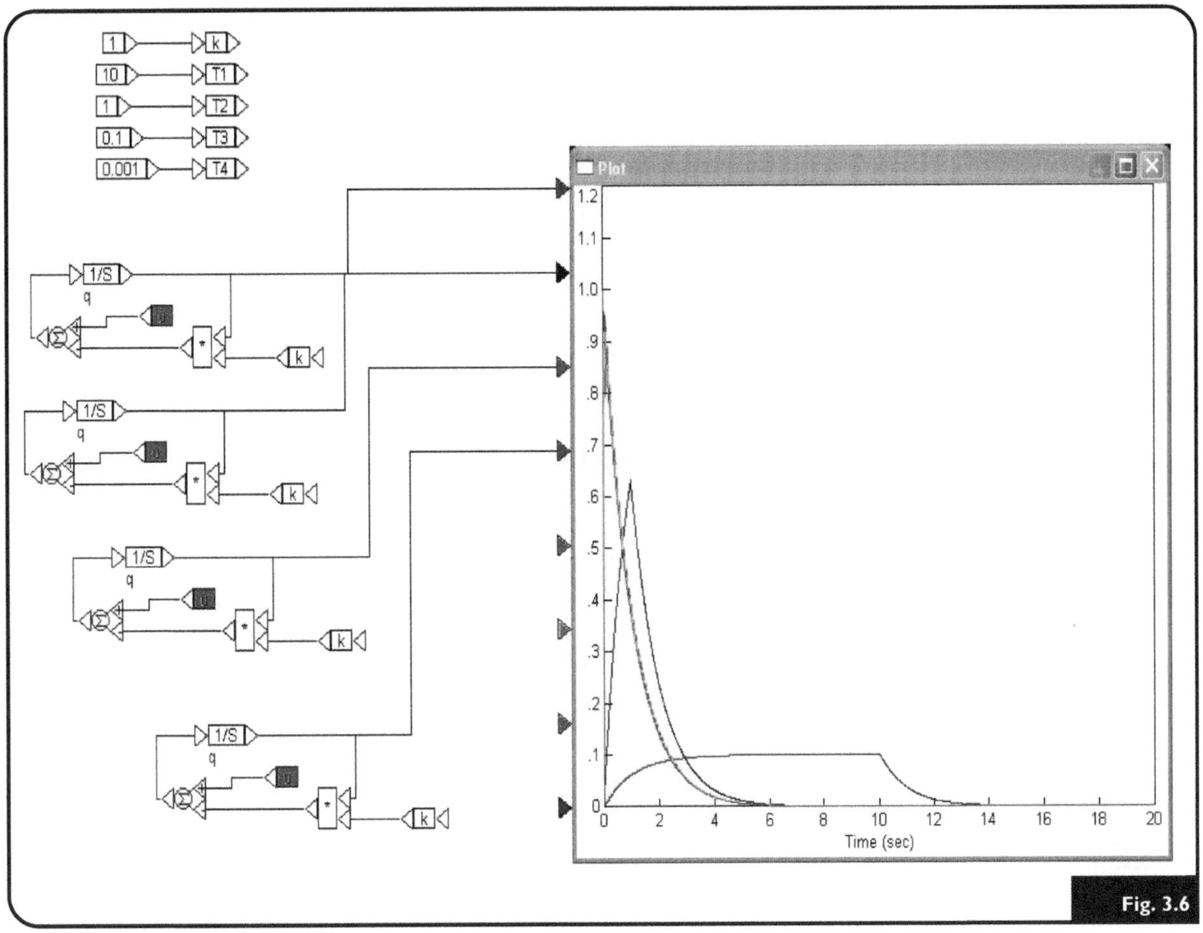

Fig. 3.6

Remark: Among other things, this example illustrates the equivalence property of the unit impulse input with ICs of ODEs, as presented in Chapter 2, Eq. (2.20). Practically speaking, it means you can implement impulse inputs in ODE simulations as ICs instead of forcing functions.

Remark: The value of $t_{end} = 20$ was chosen for the simulations in the above example so the transient responses for all values of τ can be fully visualized. The ratio $t_{end}/\tau = t_{end}/10$ is 2 in our simple example, with the dynamics uniquely characterized by the single k $(= 1/T,\ T =$ **time constant**). For smaller k values (longer time constants $k = 1/\tau$), t_{end} would need to be increased, because k determines the rate of rise when the pulse is turned on and the rate of fall when it's turned off. Generally, in more complex examples this would translate into dependence on the slowest mode or time constant, or what is commonly called the **dominant time constant** $T_{dominant}$ of the system. For linear system models, transient or homogeneous solutions typically fall to $<1\%$ of their original values in about 5 time constants, and $<3\%$ in about two time constants. So a rough rule of thumb for characterizing the response to the finite pulse of width τ might be $t_{end} \approx \tau + 3T_{dominant}$.

Impulse Inputs

The Dirac delta or impulse function $\delta(t)$ is often used to represent certain experimental inputs in ODE model applications, usually to *approximate boluses* or brief pulses, or repeated pulse inputs, called **pulse trains**. The quality or degree to which $\delta(t)$ can approximate a finite pulse input for a particular model depends primarily on the time-domain $[0, t_{end}]$ for the "experiment" conducted with this bolus input, but also on the particulars of the model dynamics. Intuitively, the pulse should "look approximately like" an impulse, viewed over the time interval from 0 to t_{end}, so t_{end} divided by the width or duration τ of the pulse presumably should be large. How large can be difficult to theorize about. It's easier to explore these limits using simulation tools.

Example 3.4: ***Impulse Approximation of Bolus (Finite Pulse) Inputs***

In Example 3.3 above, the green and pink response curves, corresponding to pulse widths of $\tau = 0.1$ and 0.001, are nearly identical to the response due to IC: $x(0) = 1$, the equivalent IC for impulse inputs and virtually superimposed on the plots of these responses. The blue curve, for $\tau = 1$, appears to be a poor approximation of the impulse response over the interval up to $t_{end} = 20$; and the red curve, for $\tau = 10$, is a very bad approximation of the impulse response – i.e. the response due to IC: $x(0) = 1$ with no other input applied. We could say that a ratio of $t_{end}/\tau \sim 100$ works well, at least in this example, for justifying use of $\delta(t)$ or equivalent ICs in place of a finite pulse input, and this two-orders of magnitude difference might be a good rule of thumb. Nevertheless, this is worth checking, by running the comparative simulations – in particular if doing so might save laboratory experimental resources, time, money, etc. (Isn't this what simulation analysis is for?)

Time-Delay Simulations

ODE models that include signal *time-delays* are easy to implement using block diagram languages, typically with a single block and block specifiers, as illustrated next in *VisSim* (Example 3.5) and then in *Simulink* (Example 3.6) below. They are more difficult to implement using code-based programs (e.g. *Matlab, Mathematica*) (Shampine et al. 2003),[8] as we see in the next example and in Exercise E3.6.

Example 3.5: ***Time-Delays in the Feedback Loop***

Two different feedback system simulations are shown in **Fig. 3.7**. Both are very simple, each with a unit-step input and unit-delay in the feedback path. They differ only in the component block in the feedforward path: the first with a simple unity gain, the second with a simple integrator, with IC = 0. Both results are interesting – and perhaps surprising. The upper diagram feedback system, without integration, turns the unit-step input into an oscillating but stable **pulse-train output**, with frequency 2 times the

8 "Solving Delay Differential Equations with DDE23" available at www.mathworks.com/dde_tutorial.

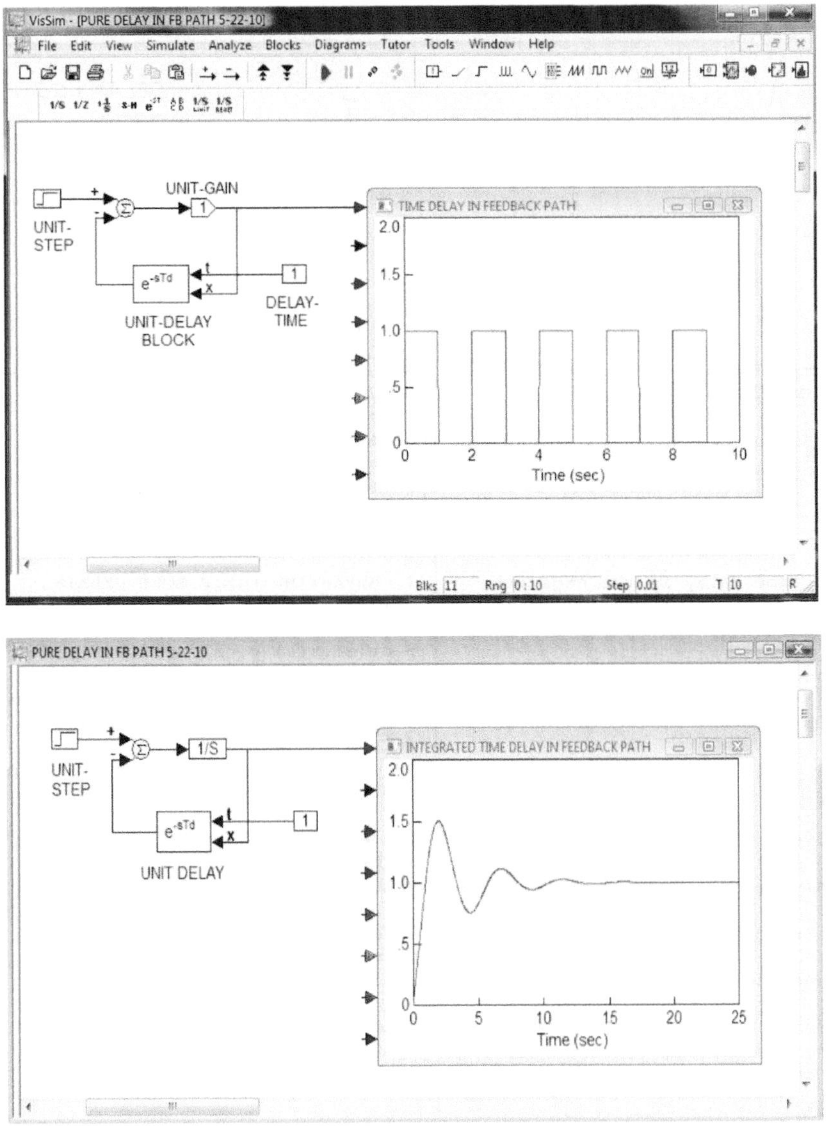

Fig. 3.7

delay-time. This configuration can be used to implement *pulse-train inputs* in more complex simulations. It is convenient in such applications to put the "pulse-train generator" depicted in **Fig. 3.7** into a single **compound block**, naming the new block "pulse-train generator." (This is done in *VisSim* by selecting the blocks to be encapsulated, and then choosing *Create Compound Block* in the Edit pulldown menu.)

The lower figure turns the unit-step into a smoothed, continuous asymptotically stable response, overshooting the value 1 at first and then settling down in a damped oscillatory fashion to a constant steady-state value of 1. The "smoothing" is due primarily to the integrator.[9] These are *fundamental properties of negative feedback systems*, which we revisit numerous times in subsequent chapters. However, the

9 Smoothing is one of the things integration does to a signal.

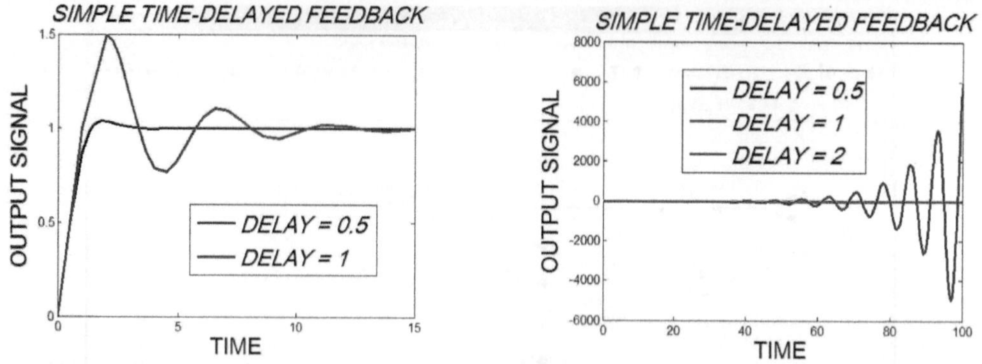


```
function ddeFbk

%   ddeFbk solutions run with the DDE23
solver
%
%   Simple example of a feedback loop,
%   with an integrator in the forward
path and
%   a unit time delay in the negative
feedback
%   path. Solutions are stable for small
delays
%   and unstable for large delays.
%
%   The ODE for this configuration is
simply:
%
%         y'(t) = 1 - y(t-1)
%
%   We solve it over [0, tfinal] with
history
%   (IC only) y(t) = 0 for t <= 0. The
pure
%   lag is specified as a 1-vector
(scalar)
%   here and the delay differential
equation
%   is coded in the subfunction DDFbkDE.
%   Because the history is constant,
%   it is supplied as a scalar IC.
%
```

```
%  SETTING THE RUN PARAMETERS...
IC = 0;
delay1 = 0.5;
tfinal = 100;
interval = [0, tfinal];
s1 = dde23(@ddeFbkDE,delay1,IC, interval);
plot(s1.x,s1.y, 'LineWidth',2);

%  REPEATING FOR 2 MORE DELAYS...
delay2 = 1;
s2 = dde23(@ddeFbkDE,delay2,IC, interval);
plot(s1.x,s1.y, s2.x,s2.y, 'LineWidth',2);

%   Note: suppress s3 and plot for running
%          only first 2 delays
delay3 = 2;
s3 = dde23(@ddeFbkDE,delay3,IC, interval);
plot(s1.x,s1.y, s2.x,s2.y,s3.x,s3.y,...
                'LineWidth',2);
title('SIMPLE TIME-DELAYED FEEDBACK');
xlabel('TIME');
ylabel('OUTPUT SIGNAL');
 ---------------------
% FUNCTION DEFINING THE ODE
function dydt = ddeFbkDE(t,y,Z)
% Differential equation function for
ddeFbk.
ylag = Z(:,1);
dydt = 1 - ylag;
```

Fig. 3.8

degree of "smoothing" in this solution is, not entirely accurate, given that we implicitly assumed a "pure" unit time delay, with a sharp discontinuity. However, what is implemented by the time-delay block chosen is a variable (continuous) transport delay, which smoothes away the sharper solution "corners" when the delay is reached each time the signal travels around the loop. Interestingly, this is likely the more realistic, real-world solution, for biological as well as other continuous physical system solutions. By definition, no physically realizable dynamic system state variable can have a jump discontinuity (change instantaneously).[10] **Fig. 3.8**, generated using native

10 See Appendix C for further insights into dynamic system properties.

Matlab code for solving delay *differential equations (dde)*, also shown in **Fig. 3.8**, also illustrates these points.

The ODE for the lower **Fig. 3.7** model is $\dot{y} = 1 - y(t - \tau)$, with IC $y(0) = 0$. It is also assumed that $y(t) = 0$ for all $t \leq 0$. This is important with *dde* solvers because state y has a history that must be defined prior to reaching the delay the first time, in this case for $-1 \leq t \leq 0$. This model is implemented, solved and solutions plotted using the *Matlab* code in **Fig. 3.8**, for three different delays, $\tau = 0.5$, 1 and 2. The plot for $\tau = 1$ is qualitatively the same as the corresponding one in **Fig. 3.7**, *but not as smooth*. It is continuous, but not smooth at the "corners," when the delayed feedback signal is subtracted from the step input.

Strictly speaking, the solution of $\dot{y} = 1 - y(t - \tau)$ in **Fig. 3.8** is more accurate than the *VisSim* solution[11] shown in **Fig. 3.7**[12] – more accurate at least with regard to what we thought the delay blocks represented mathematically. As noted above, the **Fig. 3.7** time-delay block is in fact a continuous and smooth *approximation* of a pure time delay; and the **Fig. 3.7** response is likely the more realistic solution of the real continuous system implemented in *VisSim* (or *Simulink*), using the variable transport delay block for more smoothly delaying the signal.

Finally, we note that, whereas the outputs for $\tau = 0.5$ and 1 are damped oscillatory and stable, for $\tau = 2$ the solution is *unstable* (becoming unbounded). Instabilities are characteristic of biological (and other) feedback control systems with time delays in the loop, as illustrated in simulation models of blood pressure regulation of heart rate (Ottesen 1997). Further analysis of this phenomenon is deferred to the discussion on stability in Chapter 9. For now we should recognize that simulations can generate surprise solutions – some very nuanced.

Multiscale Simulation and Time-Delays

The following relatively simple example illustrates several facets of simulation of dynamic systems biology models. It's our first biosystem simulation-modeling example, demonstrating first how to translate the descriptive biology of an intracellular process into mathematical equations that approximate the dynamics of the interactions described – in this example into a single equation approximation. More than one biochemical species is involved and this would normally demand more than one equation. However, the back-and-forth movements of molecules between the nucleus and cytoplasm of the cell cross temporal dynamical boundaries – or **scales** – as well. A simple and usually effective implementation of disparately mixed dynamics is to represent the (relatively) very fast dynamics by algebraic (rather than differential) functions that include *time-delays*.[13] Example 1.5 in Chapter 1 illustrated

11 The *Simulink* solution with the variable transport delay simulation is the same.

12 The mathematical explanation for this is that the integrator in the forward loop smoothes the derivative occurring at the discontinuity, keeping the signal continuous, but not differentiable at switching times.

13 This is one form of *time-scaling*, for the purpose of simplifying the ODE model. More on this in the next section.

the algebraic part of this concept, based on coupled equations spanning two very different time scales – and reducing them (approximately) to one. Realistically, of course, species cannot change instantaneously from one state to another. There is always some delay dynamics, albeit (relatively) very small in the problem context we are discussing. Nevertheless, pure time-delay functions, e.g. $f(t-\tau)$ representing $f(t)$ delayed by τ time units – with unspecified rapid dynamics among possibly several species – can be good approximations; and they invariably simplify the math, and simulation implementation of time delays is easy in *Simulink* and *VisSim*.

Example 3.6: ***Simplified Protein Transcription-Translation Dynamics – A Multiscale Simulation***

The cartoon model illustrated in **Fig. 3.9** and mathemetized in Eq. (3.1) below (adapted from Smolen et al. 1998), is a simplified version of autoregulation of the dynamics of a single nuclear protein *transcription factor* (TF) $P \equiv P(t)$ by a *dimer* of itself **(homodimer)**, formed in the cytoplasm of a cell following *ribosomal* production of the proteins.[14] The protein, produced at a basal rate b and degraded at a linear fractional rate k_{deg} (called **constitutive** synthesis and degradation – the first and third terms in Eq. (3.1) below), activates itself in a time-delayed, nonlinear saturating fashion leading to transcription of its messenger RNA (mRNA). Several steps in the regulatory process are aggregated into a single term in this model, rendering it one with only a single dependent variable – protein $P(t)$ – explicit in the model. Aggregated processes include the dynamics of mRNA in the transcription step, activation of the transcription factor homodimer – a phosphorylation step that enhances its action at the TF protein receptor element (TF-RE) in the DNA *promoter* region for this gene – and transport of macromolecules around the feedback path. These are all lumped together and modeled simply as a single aggregated and algebraic *time-delayed*

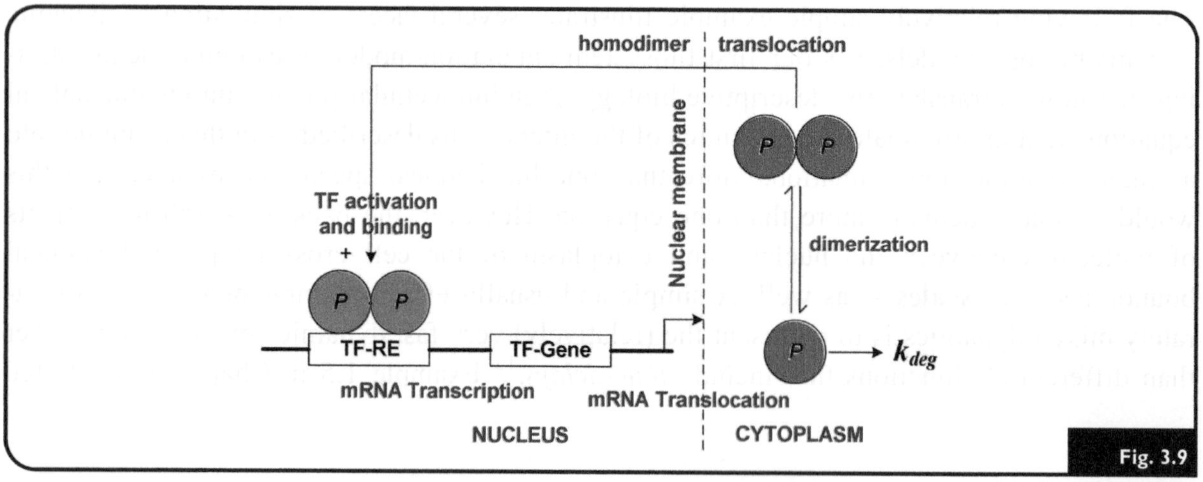

Fig. 3.9

14 **Transcription factors** are proteins that enhance or inhibit transcription of a gene in the nucleus of a cell. A **dimer** is a complex of two molecules. Proteins are produced in **ribosomes**, which reside in the cytoplasm, the cellular space external to the nucleus.

(τ) *regulatory term.* This is given by the second term on the RHS of this ODE – called a second-order Hill function:[15]

$$\frac{dP(t)}{dt} = b + \frac{K_1 P^2(t-\tau)}{P^2(t-\tau) + K_2} - k_{deg}P(t) \tag{3.1}$$

Implementation of Eq. (3.1) in *Simulink* or *VisSim* is straightforward. We need one integrator, two gain blocks – for multiplying signals by the parameters K_1 and k_{deg}, a power block – to square $P(t)$, a summer – to add K_2 to $P^2(t-\tau)$, a *time-delay* τ block – to delay the squared output, and a constant input block – for b, feeding a second summer, all as shown in the computer screen copies of *Simulink* diagrams in **Fig. 3.10**.

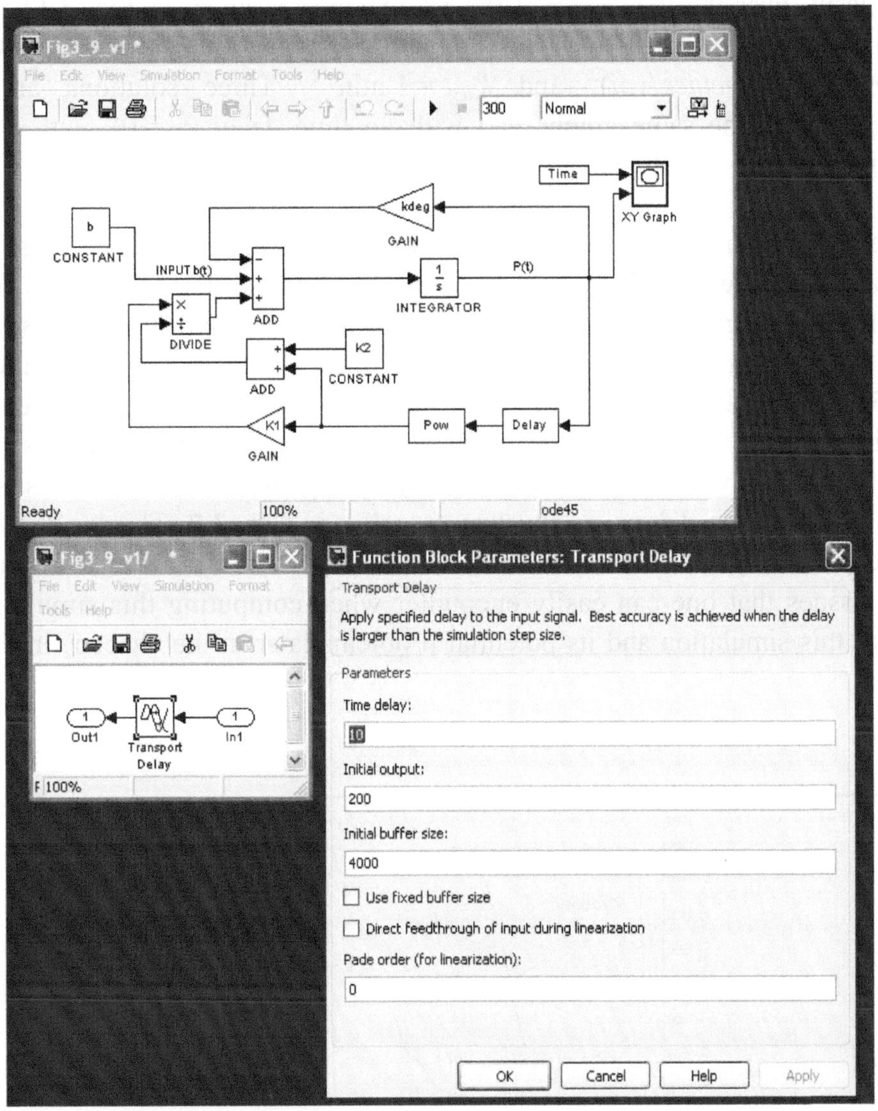

Fig. 3.10

15 Nonlinear models like this one are developed in greater detail in Chapter 6 and beyond.

Implementation of the time-delay function is shown in the lower part of the figure, for a 10-min delay. It's important to set the output ICs correctly, as shown, 200 in this case, otherwise the solution will be erroneous. Implied here is that the output has "historically" been 200 – at least for 10 minutes (the time-delay) prior to initiating the simulation.

The state variable $P(t)$ in this example might have *mass* or *concentration* units. Let's assume it's a *concentration*, which means b and K_1 have units of *concentration/time*, k_{deg} has units of $time^{-1}$ and K_2 has units of $concentration^2$ (squared), all to balance the units in the equation.

This simple model has complex dynamics. It is **bistable**, meaning it has two *stable steady state* solutions, which we explore further in Chapter 9. A plot of one solution of (3.1) for initial condition $P(0) = 200$ molecules of TF in some unit volume of cytoplasm, over the time interval $0 < t < 180$ min, is given in **Fig. 3.11** for basal synthesis rate $b = 20$ *(molecules/vol)/min* and parameter values $K_1 = 2,000\ \text{min}^{-1}$, $K_2 = 400,000\ (molecules/vol)^2$ and $k_{deg} = 1\ \text{min}^{-1}$. Three solutions are shown superimposed on the same graph: $P(t)$ with no time delay ($\tau = 0$), with $\tau = 5$ min delay, and with $\tau = 10$ min delay.

Remark: The simulation solution in this example is interesting for several reasons. First, the three responses shown predictably shift to the right as the delay increases from 0 to 10. With a little hindsight, this should be expected. Second, small oscillations, at frequency $1/\tau$, are apparent in the curve for $\tau = 10$, and they also appear with smaller amplitude for $\tau = 5$, but with insufficient resolution for visibility on this graph. Indeed, they exist for all time-delays $\tau > 0$. These oscillations are small when the signal is not changing much, and increase with the rate of change of the signal as well as τ. This is an inherent property of pure time-delays, as illustrated earlier in **Fig. 3.7**. Third, although the solutions shown in **Fig. 3.11** are reasonably accurate, there are ever-so-subtle numerical accuracy issues that one can easily encounter when computing this simulation solution. We revisit this simulation and its potential problems later in the chapter, in Example 3.17

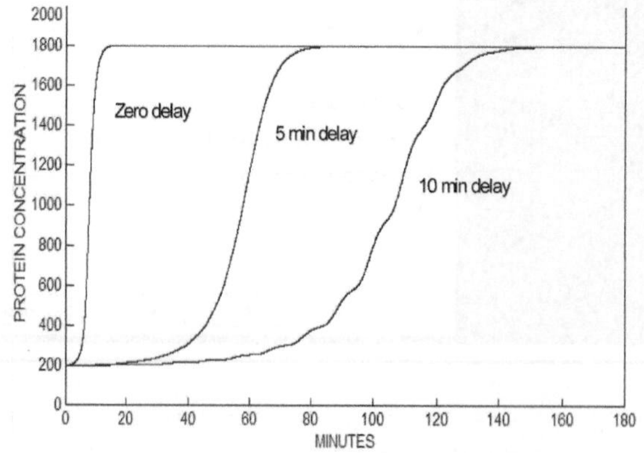

Fig. 3.11

and Example 3.24, where we illustrate that choosing the "right solver" for the job can be difficult.[16]

Remark: We belabor another point here. This model is an example of a multiscale model, rendered easier to solve by use of a time-delay ODE approximation. It effectively spans two different time scales –first the intranuclear, with millisecond range dynamics, second the cytoplasmic – with second to minute range dynamics – accounting for back-and-forth translocation and ribosomal protein production and proteosomic degradation process dynamics. The forward translocation of mRNA and return translocation of the homodimer back to the nucleus for transcription factor action takes substantial time, with dynamics approximated in this model as a minutes-range time delay τ.

Does Model Size Matter?

This could be a realistic concern when using graphical simulation languages: what to do if the dimensionality of the model, i.e. the number of equations, is very large? If the simulation language capacity can handle it,[17] then visual advantages of graphical modeling might be limited by how much of the model can be visualized on a computer screen or other output device. Fortunately, they don't have to be. One can readily use the *imbedding level functionality* of these programs: when a model gets "too big," all or part can be readily grouped into one or more *compound blocks* on the graphics user interface (GUI), as was done for combining several blocks that generated bolus inputs in **Fig. 3.6**.

Example 3.7: ***Compound Blocks Eliminate Clutter***

Fig. 3.12 illustrates the model of Example 3.6 imbedded in a *Simulink* implementation of a compound block. The compound block is created by left-clicking and highlighting

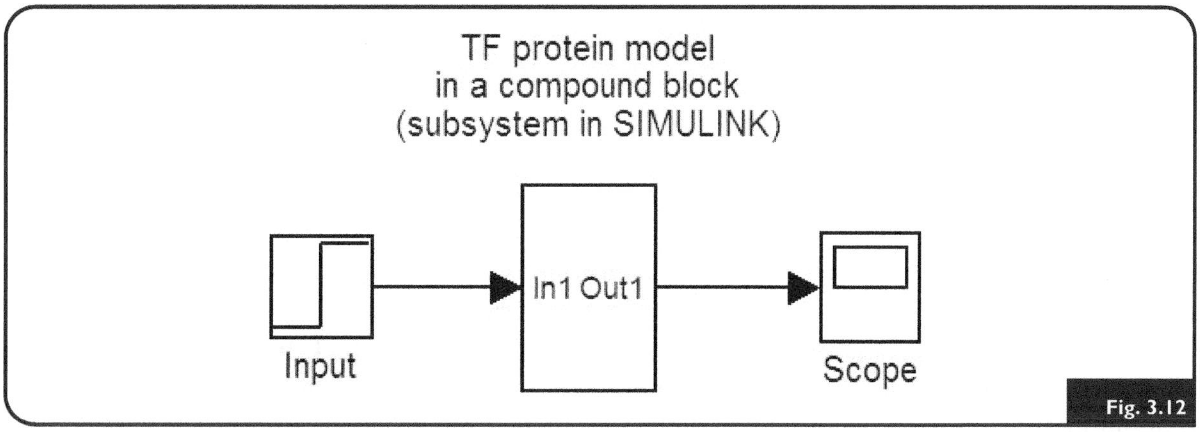

Fig. 3.12

16 Equation (3.1) is "stiff" and the solutions shown in **Fig. 3.11** were obtained and verified using several built-in *Matlab/Simulink solvers* appropriate for the task, as illustrated in the examples noted. We verified these using (different) *VisSim* solvers as well. Discussion of numerical methods for solving ODEs (solvers) begins in the next section, and so-called *stiff* problems and solvers in a later section.

17 *VisSim* and *Simulink*, for example, can handle very large models, with hundreds of equations and levels.

the blocks to be included and then right-clicking and choosing CREATE COMPOUND from the menu.

Double-clicking on the block created by the compounding operation – the "In1 Out1" block in the center –opens it up, revealing the details in earlier **Fig. 3.10** between the step-input and the $P(t)$ output graph shown on the Scope.

Normalization of ODEs: Magnitude- & Time-Scaling

Fractions and percentages are everyday ways of thinking relatively about numbers. Normalization is a variant of this, in which we treat physical variables relative to something fixed, thereby allowing us to think more easily about differences in or among values of these variables. For example, one can take the largest value of a set of numbers, divide all other values by this largest one, and then think about all of them as some fraction of the largest one. In this sense, they are all normalized to the largest number. Comparisons with *initial* – instead of maximum – values of variables are common in some areas of science. These are called **fold-differences**, e.g. in molecular biology, when describing changes in protein levels (amounts or concentrations) relative to an *initial* level in a cell signaling pathway. This idea is applied to time-dependent solutions $\boldsymbol{x}(t)$ of ODEs, by dividing $\boldsymbol{x}(t)$ by its maximum value attained over time, $x(t)/x_{\max}$, or by its initial value: $x(t)/x_0$. This is what is meant by normalizing the solution $\boldsymbol{x}(t)$ to its max or its initial value, or **magnitude scaling**. Normalization, or **scaling**, of the dependent state variables can also be applied to the independent variable $time = t$ in ODEs.[18] **Time-scaling** may be used to visualize the solution over a normalized unit interval (0,1), or over a time period (range) on which the dependent variable(s) *change significantly*.

Generally speaking, *scaled dimensionless variables* – whether dependent (state) or independent (time) variables – are obtained by dividing the variables by their scale factors, yielding *dimensionless variables with unity order of magnitude*. Also, different scales may be needed over different ranges of the time domain in some biomodel analysis problems, as in simulating enzyme–substrate, receptor–ligand and other biochemical interactions with vastly different dynamics over different time domains or scales.[19] Eqs. (3.1) above, and (3.4) below, are examples of **extreme time-scaling**, where very fast dynamics are replaced in the model by nondynamic (instantaneous) algebraic terms with pure time-delays.

Transformations rendering physical variables in an ODE dimensionless are also useful in studying ODE model solutions, in part because they usually reduce the number of interdependent physical variables, and thereby the complexity of the ODE. The **Buckingham Pi theorem** (Buckingham 1914) is the basis for this simplification. In brief, it says that for an equation (e.g. an ODE) involving n physical variables dependent on P independent physical quantities

18 Independent *space* variables in partial differential equations (PDEs) also can be conveniently scaled, thereby normalizing solutions over the domain – say $\{x,y,z,t\}$ – of the dependent variable – say temperature $T(x,y,z,t)$ – in a diffusion PDE, to say the "unit cube"; i.e. the scaled T over scaled x, y, z and t are computed over (0,0,0,0) to (1,1,1,1), instead of (a,b,c,d) to (e,f,g,h).

19 Derivation of the *quasi-steady state* solutions of the Michaelis-Menten enzyme-substrate equations in Chapter 6, detailed further in Chapter 6 Extensions, are classic and modern examples of using different scalings for two time domains over which solutions evolve.

(e.g. parameters or boundary conditions), we can construct an equivalent expression (ODE) involving $n - P$ *dimensionless* variables from the original set of n variables.[20]

To summarize, it is often useful to scale the ODEs, before solving them, and running and analyzing the solution of the resulting *normalized* or *scaled ODEs*. This helps in distinguishing effects of model parameters, inputs or ICs on solutions of coupled ODEs, effectively uncoupling some of these effects from their absolute values. Scaling also is important for simplifying solutions, as in *quasi-steady approximations* (QSSA) in biochemical reaction kinetics, which we study in Chapter 6. We present magnitude scaling first.

Magnitude-Scaling ODEs

Example 3.8: ***Simulating Normalized Protein Transcription-Translation Dynamics***

Let's normalize Eq. (3.1) by the initial value $P(0)$ of $P(t)$. The goal here is to simulate $P(t)/P(0)$ – the time-varying fold-difference – over some interval $0 < t < T$, which begins at unity at $t = 0$. We do this by dividing $P(t)$ appearing in all four terms of (3.1) by $P(0)$, balancing each term by an appropriate factor so as to satisfy the equality, i.e.

$$\frac{dP(t)/P(0)}{dt} = \frac{b}{P(0)} + \frac{(K_1/P(0))(P(t-\tau)/P(0))^2}{(P(t-\tau)/P(0))^2 + K_2/P^2(0)} - k_{deg}\frac{P(t)}{P(0)} \tag{3.2}$$

Now redefine $P(t)/P(0) \equiv P'(t)$ and (3.2) becomes:

$$\frac{dP'(t)}{dt} = \frac{b}{P(0)} + \frac{(K_1/P(0))(P'(t-\tau))^2}{(P'(t-\tau))^2 + K_2/P^2(0)} - k_{deg}P'(t) \tag{3.3}$$

Note that $P'(t)$ and the dissociation rate constant $K'_2 \equiv K_2/P^2(0)$ are now dimensionless in Eq. (3.3), whereas they were *concentration/time* and *concentration^2/time* units in (3.1). Actually, the state variable $P(t)$ and K_2 in this model were given as dimensionless in the original reference, because "effective concentrations of TFs and of DNA elements in the nucleus are generally not known." (Smolen, Baxter et al. 1998). We gave them arbitrary numerical values here to illustrate the normalization process.

A plot of one solution of (3.3) for normalized initial condition $P'(0) = 1$ over the interval $0 < t < 180$ min is given in **Fig. 3.13** for parameter values $\tau = 10$ min, $b' = 0.1$ min^{-1}, $K'_1 = 10$ min^{-1}, $K'_2 = 10$ (dimensionless), $k_{deg} = 1$ min^{-1}. Qualitatively, it's no different than the corresponding solution of the unnormalized solution in **Fig. 3.11**. Only the values on the protein concentration axis are different.

20 The choice of dimensionless variables or parameters in a set of equations is not unique and typically depends on goals of the modeling problem. We illustrate normalization to initial conditions in this section, common in molecular biology. But normalization to time or distance intervals also can be helpful in modeling and simulation, for example, from $[a,b]$ to $[0,1]$ for any a and b as in (Tee and DiStefano III 2004) for tumor growth models.

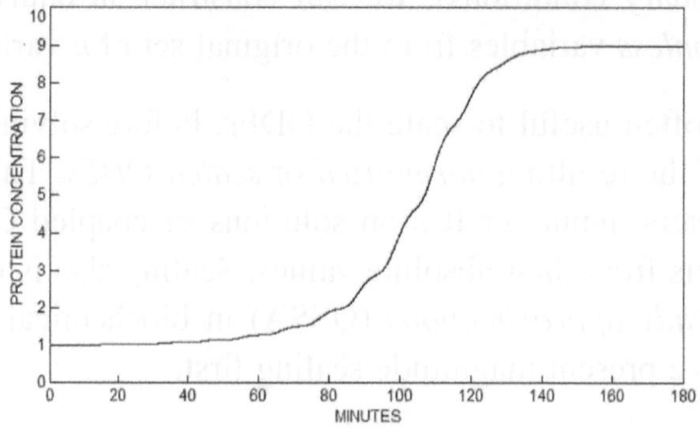

Fig. 3.13

*Example 3.9: **Simulating Two Variable Transcription–Translation Dynamics***

How does one normalize ODE models with more than one state variable? Consider a simpler protein production model – without autoregulation – with simplified, proportional *mRNA* dynamics included explicitly,[21] with the effect of the messenger on protein production delayed by τ, to account for (primarily) translocation to the site of translation in the cytoplasm, i.e.

$$\frac{dP(t)}{dt} = p \cdot mRNA(t-\tau) - k_{degP}P(t) + u(t) \tag{3.4}$$

This kind of time-delay ODE model, seen earlier in Example 2.6, might be justified if *mRNA*(t) and $P(t)$ dynamics are the focus of a problem, and may be roughly depicted as a fixed delay of the effect of *mRNA*(t) on $P(t)$ production, as expressed in Eq. (3.4). As in Eq. (3.1), this simplification is accomplished by *extreme time scaling*. The first term on the RHS of (3.4) represents production (synthesis) of protein P via the messenger *mRNA* (*mass* units) transcribing P, with rate constant p (*time*$^{-1}$ units), with time-delay τ approximating translocation dynamics. The second term on the RHS reflects linear degradation of P (*mass/time* units).[22] The initial protein state in this equation is $P(0)$. The $u(t)$ term might represent: (a) an experimental perturbation of $P(t)$ dynamics – perhaps an exogenous signal; or (b) an endogenous cellular signal not modeled as a dynamic ODE, but as a known time function.

Molecular biologists interested in protein fold-change would divide P by $P(0)$ in (3.4), balancing each term, as needed, with further divisions by $P(0)$, as in the second and fourth RHS terms:

$$\frac{dP(t)/P(0)}{dt} = \frac{P}{P(0)} \cdot mRNA(t-\tau) - k_{degP}\frac{P(t)}{P(0)} + \frac{u(t)}{P(0)} \tag{3.5}$$

21 Common when *mRNA* expression data are available.

22 Degradation of proteins (*P*) typically occurs intracellularly in **proteosomes**, very large protein complexes which degrade unneeded or damaged proteins by **proteolysis**, using **protease** enzymes that break peptide bonds.

Then setting $P(t)/P(0) \equiv P'(t)$ in (3.5), and replacing the constants and variables also by primed symbols, as needed, we have

$$\frac{dP'(t)}{dt} = p' \cdot mRNA(t-\tau) - k_{degP}P'(t) + u'(t) \tag{3.6}$$

where $u(t)/P(0) \equiv u'(t)$. To complete the process, so all model state variables are normalized and expressed as fold changes, (3.4) or (3.6) need normalization of the $mRNA$ variable, by its initial value $mRNA(0)$. In this case, we can maintain the balance of all terms in the equation by both multiplying and dividing the first term on the RHS of (3.6) by $mRNA(0)$ and also setting:

$$\begin{aligned}\frac{dP'(t)}{dt} &= p' \cdot mRNA(0)\frac{mRNA(t-\tau)}{mRNA(0)} - k_{degP}P'(t) + u'(t) \\ &= p''mRNA'(t-\tau) - k_{degP}P'(t) + u'(t)\end{aligned} \tag{3.7}$$

Both state variables in (3.7) are normalized, with values ranging from 1 at $t = 0$ to values that are *any-fold* above or below 1.

Note that parameter p'' and input $u'(t)$ in Eq. (3.7) are different than p and $u(t)$ in Eq. (3.4); but parameter k_{degP} associated with the linear degradation term remains the same when the equation is normalized. (Can you see why?)

Time-Scaling ODEs

Extreme time-scaling, to simplify models by replacing (relatively) extremely fast dynamics by pure time delayed algebraic functions – as in Eqs. (3.1) and (3.4) – was discussed in the previous section. Here we consider time-scaling for visualizing the solution over a normalized unit interval (0,1), and also time-scaling for speeding up or slowing down a solution. Time-scaling ODE models over two different time periods (ranges) – the first on which dependent variable(s) *change significantly* and the second over which they change very little – is developed in Chapter 6, beginning with the analysis of enzyme–substrate and other biochemical interaction models.

Remark: As a general rule, for models with several dependent variables $x_1, x_2, \ldots$ with different degrees of variation (slow, fast, faster, etc.), choose the time scale for the fastest varying dependent variable x_i first. Then the resulting scaled equations "will reflect the relative slowness of changes in the other dependent variables" (Segel and Slemrod 1989). This is particularly important in modeling biomolecular dynamical systems with (multiscale) subsystem components in quasi-steady state, a Chapter 6 topic we encounter explicitly when we study enzyme–substrate and receptor–ligand and other binding kinetics – with additional scaling discussion in the Extensions section of Chapter 6.

ODEs are readily time-scaled to make their solutions a function of *dimensionless time* τ instead of t – computed over the interval (0, 1) – instead of (t_0, t_{max}), where the units of (t_0, t_{max}) can be any time units – seconds, minutes, years. Dynamically correct ODE solutions are then computed over (0,1) – independent of the time units. All model equations are scaled, including their parametric components, and the dynamics of $\mathbf{x}(t)$ and its scaled form $\mathbf{x}'(\tau)$ look the same, with the exception that they fully evolve over (0,1) instead of (t_0, $t_{\max}$).

Time-Scale Formula: Derivation of the time-scaling transformation for mapping (t_0, $t_{\max}$) into (0,1) is relatively straightforward – and constructive – pedagogically speaking. We define the transformation as $\tau \equiv \alpha t + \beta$ of the t-domain onto the τ-domain, because we are in part stretching (or shrinking) the t-domain time-interval by α, and also shifting it by β. The result must satisfy the endpoint constraints (boundary conditions): $\tau = 0$ when $t = t_0$, and $\tau = 1$ when $t = t_{max}$. Substituting these values into $\tau \equiv \alpha t + \beta$ and solving the two resulting equations for the two unknowns α and β yields the desired result:

$$\tau = \frac{t - t_0}{t_{\max} - t_0}, \quad dt = (t_{\max} - t_0)d\tau, \quad t = (t_{\max} - t_0)\tau + t_0 \tag{3.8}$$

The procedure then is to substitute τ and $d\tau$ into the model equations and simulate the transformed model over (0,1).

Example 3.10: ***Time-Scaling the Observation Interval I: Time-Invariant Model***

We transform *time-invariant* NL ODE $dx(t)/dt = f[x(t), u(t), p]$, defined over ($t_0$, $t_{\max}$), into $dx(\tau)/d\tau = f'[x(\tau), u(\tau), p]$, so we can solve it over (0,1). Rewriting the ODE and substituting (3.8) gives $dx = f[x, u, p]dt = f[x, u, p](t_{\max} - t_0)d\tau$. The transformed ODE then becomes:

$$\frac{dx(\tau)}{d\tau} = f'[x(\tau), u(\tau), p] = (t_{\max} - t_0)f[x(\tau), u(\tau), p] \tag{3.9}$$

The ODE $dx(t)/dt = kx(t)$ on the interval $1 < t < 10$ is thus transformed into $dx(\tau)/d\tau = 9kx(\tau)$ on the interval $0 < \tau < 1$.

If f were explicitly a function of time $t, f \equiv f(x, u, p, t)$ – rendering the ODE *time-varying* – then the substitution $t = (t_{\max} - t_0)\tau + t_0$ also would be needed in f.

Example 3.11: ***Time-Scaling the Observation Interval II: Time-Varying Model***

Time-varying ODE $dx(t)/dt = -tx(t)$ over the interval (1,3) is transformed into

$$\frac{dx(\tau)}{d\tau} = f[x(\tau), u(\tau), p, t] = (t_{\max} - t_0)f[x(\tau), u(\tau), p, t] = 2[-tx(\tau)]$$

$$= 2[-(t_{\max} - t_0)\tau - t_0] = -2(2\tau + 1)x(\tau) = -(4\tau + 2)x(\tau)$$

on the interval $0 < \tau < 1$.

Multiplication of the RHS of the original ODE by the factor $(t_{\max} - t_0) \equiv \Delta T$ in Eq. (3.9) suggests another application of this transformation. The dynamic response of a model can

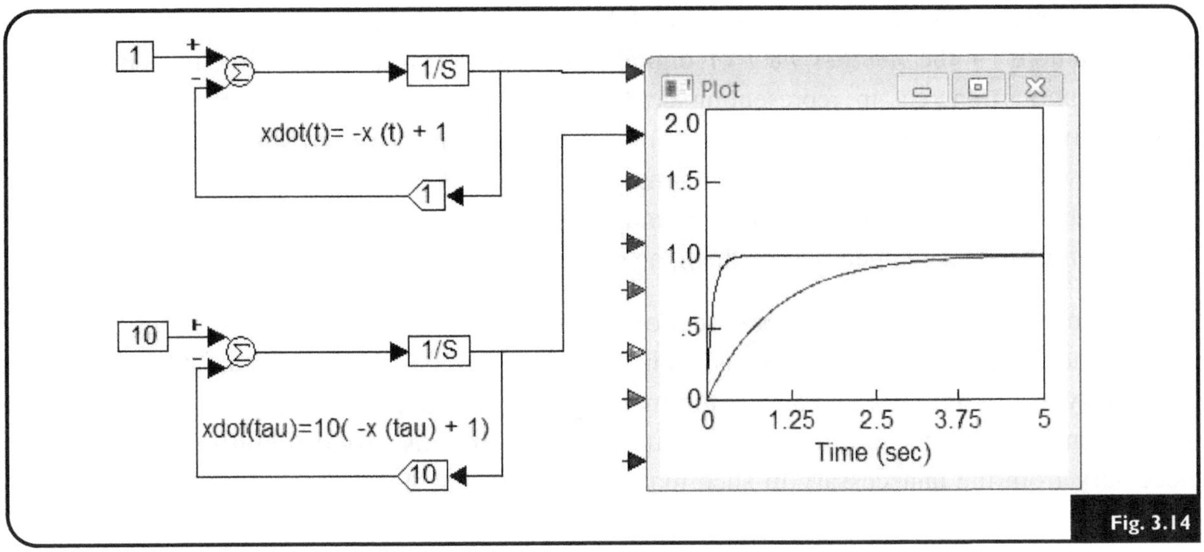

Fig. 3.14

be adjusted, to make it faster or slower, by time-scaling its ODEs by this factor – without regard to the interval of observation (t_a, t_b). If $\Delta T > 1$, this speeds up the solution, and *vice versa*. This might be needed – for example – in designing the dynamics of control system components in a prosthetic device.

Example 3.12: ***Speeding up System Dynamics by Time-Scaling the ODE***

This simple example illustrates the point. *VisSim* **Fig. 3.14** shows a ODE with 10 times faster dynamics in the bottom panel, the simulated output of the ODE scaled by a factor of $\Delta T = 10$, using Eq. (3.9).

Numerical Integration Algorithms: Overview

In the old days – about half a century ago, ODEs were solved using analog computers. Analog computers use collections of continuous-time, physical components – electronic or mechanical integrators, multipliers, continuous function generators and the like – for computing mathematical operations needed for ODE solutions. In this sense, as direct physical analogs – analog computers need no special approximations for ODE math. Precision of solutions depends primarily on the dynamic ranges and quality of the components, not on any fundamental mathematical limitations.

ODEs are now solved almost exclusively on digital computers, fundamentally capable of high computational precision – controlled by user software choices and limited by time allotted for solution; and one can do this for very large systems of ODEs,[23] at a tiny fraction of the cost

23 In an analog computer, solving – for example – twice as many equations takes the same time, but requires about twice as many components and a machine twice the size, compared with digital computers, which can solve larger systems of ODEs in a single size machine – at a cost of longer computation times.

for the hardware. However, the fundamental limitations of numerical (digital) computation of ODE solutions – the *Achilles heel* of digital simulators, if you will – are imbedded in the approximations inherent in representing continuous-time functions sufficiently precisely by a finite number of discretely sampled functions. For this reason, we need some basic understanding of how the numerical algorithms in simulation programs are constructed. Most of all, users of simulation tools need to anticipate the kinds of errors they generate – so they can recognize, avoid, minimize and compensate for them, as needed in applications.

The material in the next several sections is necessarily mathematical – about algorithm development, rather than modeling as such. I've done my best to keep it as abbreviated and expository as possible – maintaining the major goal – to expose the utility and limitations of simulation as a primary tool in modeling methodology. It is usually difficult to distinguish the necessary from the unnecessary in such matters, so most has been retained within the chapter. The Exercises include some additional details and extensions. Appendix F goes further.

THE TAYLOR SERIES

This infinite series expansion of solution $\boldsymbol{x}(t)$ of ODE $\dot{\boldsymbol{x}} = \boldsymbol{f}$ over a time interval – say t_0 to t_{end} – is the basis for most numerical algorithms for solving ODEs on a digital computer.[24] It also stands on its own as one such method, particularly its simplest form – Euler's Method, detailed in the next section.

The notation in the next several equations is mathematically "wordy," to keep the meaning of each operation and term clear. The "evaluation points" for each term are important in understanding and using the Taylor series. Notation is simplified as we move along. Appendix B can help with the vector-matrix algebra appearing in some equations, the goal being to simplify the notation and better expose both its meaning and generality.

For a scalar-valued function x of t, i.e. $x(t)$, the value of x at a small distance Δt from t, i.e. at $t + \Delta t$ can be written:

$$\begin{aligned} x(t+\Delta t) &= \sum_{k=0}^{\infty} \frac{(\Delta t)^k}{k!} \frac{d^k x}{dt^k}\bigg|_t \\ &= x(t) + \frac{dx}{dt}\bigg|_t \Delta t + \frac{1}{2!}\frac{d^2x}{dt^2}\bigg|_t \Delta t^2 + \frac{1}{3!}\frac{d^3x}{dt^3}\bigg|_t \Delta t^3 \cdots \\ &= x(t) + \frac{dx}{dt}\bigg|_t \Delta t + \frac{1}{2!}\frac{d^2x}{dt^2}\bigg|_t \Delta t^2 + O(\Delta t^3) \end{aligned} \tag{3.10}$$

In Eq. (3.10), $O(\Delta t^3)$ means that the terms remaining in the series are (at worst) of "order Δt^3." This is a result of Taylor's theorem (Kline 1998). The derivatives in each term of Eq. (3.10) are "evaluated" at t, whereas x is evaluated at $t + \Delta t$. Equivalently, the value of x

24 The Taylor series is also a basis for *linearizing* nonlinear ODE *models*, not just computing their solutions, as we see in Chapter 5. Developments here will be useful there as well.

at $t + \Delta t$ is written as a series in terms of x and its derivatives evaluated at time t. This is the interpretation of Eq. (3.10) we need for solving initial-value ODE problems. To solve an ODE like $\dot{x}(t) = f[x(t), u(t), t]$ over a time interval t_0 to t_{end}, we thus need to evaluate x at many future times, $t + \Delta t$, $t + 2\Delta t, \ldots$, etc.

Example 3.13: ***Taylor Series (TS) Expansion Δt Ahead***

Using Eq. (3.10), the TS expansion of $x(t) = t^2$, at $t + \Delta t$, a small distance Δt from t, is:

$$x(t + \Delta t) = t^2 + (2t)\Delta t + \frac{1}{2}(2)\Delta t^2 + \ldots = t^2 + 2t\Delta t + \Delta t^2 + O(\Delta t^3) \tag{3.11}$$

The term $O(\Delta t^3)$ contains the infinite number of terms remaining after the second-order term shown and, by Taylor's theorem, all of these terms together can be represented by this single term, of the order of Δt^3. The first two terms in Eq. (3.10) have a special name; they are called the **zeroth-order** and **first-order** terms, and together they make up the *linear approximation* of $x(t + \Delta t)$, with $x(t) = t^2$ for this example, namely: $x(t + \Delta t) \cong t^2 + 2t\Delta t$.

Generalization to an n-vector of state variable solutions, i.e. $\boldsymbol{x} = \begin{bmatrix} x_1 & x_2 & \cdots & x_n \end{bmatrix}^T$, is easy if we use vector-matrix notation. Eq. (3.10) becomes:

$$\begin{aligned} \boldsymbol{x}(t + \Delta t) &= \boldsymbol{x}(t) + \left.\frac{d\boldsymbol{x}}{dt}\right|_t \Delta t + \frac{1}{2!}\left.\frac{d^2\boldsymbol{x}}{dt^2}\right|_t \Delta t^2 + \frac{1}{3!}\left.\frac{d^3\boldsymbol{x}}{dt^3}\right|_t \Delta t^3 + \cdots \\ &= \boldsymbol{x}(t) + \dot{\boldsymbol{x}}(t)\Delta t + \frac{1}{2!}\ddot{\boldsymbol{x}}(t)\Delta t^2 + \frac{1}{3!}\dddot{\boldsymbol{x}}(t)\Delta t^3 + \boldsymbol{O}(\Delta t^4) \end{aligned} \tag{3.12}$$

In the last form of (3.12) we have used the **overdot notation** for derivatives, and expressed the evaluation points $|_t$ as the argument of $\boldsymbol{x}$ and its derivatives, i.e. as $\boldsymbol{x}(t)$ etc, to simplify the notation – now that we understand what it means. The remainder term $\boldsymbol{O}(\Delta t^{k+1})$ (with $k = 3$) in (3.12) is given a special name and more pneumonic symbol: **truncation error** E_T^{k+1} (sometimes just E_T), i.e.

$$O(\Delta t^{k+1}) \equiv \text{truncation error} \equiv E_T^{k+1} \equiv \text{h.o.t.} \tag{3.13}$$

where k is called the **order** of the truncated series algorithm. For example, in (3.12), the $\boldsymbol{O}(\Delta t^4) \equiv \boldsymbol{O}(\Delta t^{3+1})$ is the remainder of the series truncated after third-order terms (third-order algorithm). It's the error in approximating the series by the k terms up to and excluding these **higher-order terms**, (abbrev: h.o.t.s).

The importance of the Taylor series is based on two properties. First, under reasonable differentiability conditions on $\boldsymbol{x}$, the series converges. Second, for small enough Δt, higher-order terms in Δt^2, Δt^3, etc., depending on where one chooses to terminate the approximation, may be small enough to be neglected, as we see below. This means $\boldsymbol{x}$ may be approximated by a linear function for small enough Δt.

Taylor Series Algorithms for Solving Ordinary Differential Equations

Consider now the general nonlinear or linear ODE:

$$dx/dt \equiv \dot{x}(t) = f[x(t), u(t), t] \tag{3.14}$$

The goal in this section is to understand the basics of how algorithms are developed for computing numerical solutions $x(t)$ of Eq. (3.14) for values of t greater than the initial $t \equiv t_0$. Without loss of generality, we subsume given inputs $u(t)$ into the solution $x(t)$. They are normally given as specific functions of time within the ODE – in $f[x(t), u(t), t]$ on the RHS of (3.14). Recasting our problem with $u(t)$ included implicitly in the function f, we seek a unique numerical solution to:

$$\dot{x}(t) = f[x(t), t] \quad \text{with} \quad x(t_0) \equiv x^0 \tag{3.15}$$

This initial-value problem has a unique solution if f satisfies a so-called Lipschitz condition, which we assume is met for all problems we deal with here. The existence and uniqueness theory can be found in most numerical analysis books, e.g. (Burden and Faires 2005).

To obtain the first step in the solution of Eq. (3.15), i.e. x evaluated at $t + \Delta t$, expand the *solution function* $x(t + \Delta t)$ of Eq. (3.15) in a Taylor series (a small distance Δt from t), using Eq. (3.12). We recognize that the first derivative of x in (3.12) is known from the ODE $\dot{x} = f$ (nice!). So, to simplify it a bit, we substitute Eq. (3.15) into (3.12):

$$x(t + \Delta t) = x(t) + f[x(t), t]\Delta t + \frac{\dot{f}[x(t), t]}{2!}\Delta t^2 + \frac{\ddot{f}[x(t), t]}{3!}\Delta t^3 + \cdots \tag{3.16}$$

Two more changes in notation below further simplifies it. First, let $t \equiv t_m$, the mth arbitrary time point. Then $t + \Delta t$ is the next time point, $t_m + 1$. Now convert Eq. (3.16) into *algorithmic* form, with Δt called the **step-size** of the algorithm. Again, the goal here is to develop numerical algorithms based on the Taylor series to solve our ODE. In these terms, Eq. (3.16) becomes:

$$x(t_{m+1}) = x(t_m) + f[x(t_m), t_m]\Delta t + \frac{\dot{f}[x(t_m), t_m]}{2!}\Delta t^2 + \frac{\ddot{f}[x(t_m), t_m]}{3!}\Delta t^3 + \cdots \tag{3.17}$$

Eq. (3.17) is an infinite series solution of the ODE in Eq. (3.14) at an arbitrary time point t_{m+1} in terms of x and function f and its derivatives at the previous time point t_m. Given that t_m is arbitrary, i.e. any point in time, the same relationship would give the solution x at t_m in terms of x at t_{m-1}, and so on. These solution points are illustrated in **Fig. 3.15**, where we recognize the *recursive* nature of algorithmic solution of the ODE, from one time-step to the next, and so on.

For brevity, we make two additional notational simplifications. First, let $x(t_{m+1}) \equiv x^{m+1}$. In these terms, Eq. (3.17) becomes:

$$x^{m+1} = x^m + f[x^m, t_m]\Delta t + \frac{\dot{f}[x^m, t_m]}{2!}\Delta t^2 + \frac{\ddot{f}[x^m, t_m]}{3!}\Delta t^3 + \cdots \tag{3.18}$$

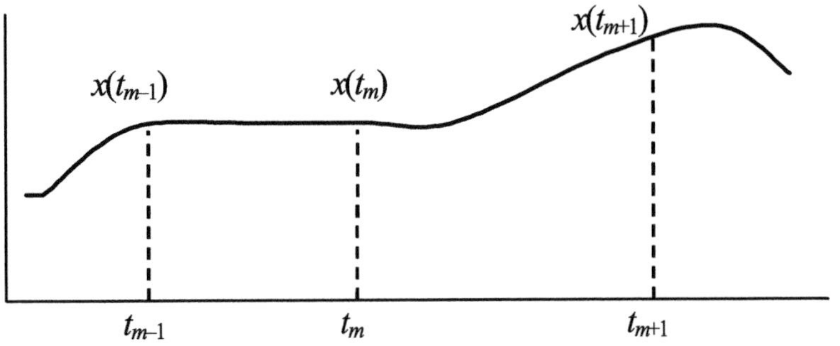

Fig. 3.15

Then, with $f[x^m, t_m] \equiv f^m$, Eq. (3.18) simplifies to:

$$x^{m+1} = x^m + f^m \Delta t + \frac{\dot{f}^m}{2!}\Delta t^2 + \frac{\ddot{f}^m}{3!}\Delta t^3 + \text{h.o.t.} \tag{3.19}$$

Red-colored Eq. (3.19) is the most basic form of the algorithm for **Taylor methods**. The terms shown are the algorithm *of order* $n = 3$, if we eliminate higher-order terms (h.o.t.s) in Δt^4 and beyond.

For (3.19) to be useful, higher-order *derivatives* of f are needed. With some effort, they can be determined by differentiating the ODE using the chain-rule, e.g.

$$\dot{x}^m = f(x^m, t_m) \equiv f^m$$

$$\ddot{x}^m = \dot{f}^m = \left[\frac{\partial f}{\partial x}\frac{dx}{dt} + \frac{\partial f}{\partial t}\frac{dt}{dt}\right]\Bigg|_{t=t_m} = \frac{\partial f^m}{\partial x} f^m + \frac{\partial f^m}{\partial t}\cdot 1 \equiv f^m f_x^m + f_t^m \tag{3.20}$$

Abbreviation subscripts x and t here designate partial differentiation with respect to (w.r.t.) x and t. Substitution of the RHS of Eq. (3.20) into Eq. (3.19) then yields:

$$x^{m+1} = x^m + \Delta t\left[f^m + \frac{\Delta t}{2}(f_t^m + f^m f_x^m)\right] + \frac{\ddot{f}^m}{3!}\Delta t^3 + \cdots \tag{3.21}$$

The third-order term also can be written in terms of f^m and its partial derivatives, and so on. If we terminate the series in (3.21) after second-order terms, with truncation error E_T^3, we get a modified form of Eq. (3.19) – a **second-order Taylor formula**, requiring (partial) derivates only of f and not $\dot{f}$:

$$x^{m+1} = x^m + \Delta t\left[f^m + \frac{\Delta t}{2}(f_t^m + f^m f_x^m)\right] + E_T^3 \tag{3.22}$$

This is a lot simpler looking, but its solution potentially requires additional computations or approximations – of derivatives, as we illustrate next – in part motivating this level of detailed development.

Computing Derivatives

Like Eq. (3.19), Eq. (3.22) is a Taylor series expansion of $\boldsymbol{x}$, in the form of a first-order difference equation solution for $\boldsymbol{x}(t)$ at discrete instants of time t_1, t_2, etc. Importantly, it is expressed only in terms of $\boldsymbol{x}$, $\boldsymbol{f}$ and partial derivatives of $\boldsymbol{f}$, which we assume exist and, in principle, can be evaluated. Thus, Eq. (3.22) is a **recursive algorithmic solution** to the original ODE, Eq. (3.14), for $m = 1, 2, \ldots,$ yielding $\boldsymbol{x}^1$, $\boldsymbol{x}^2$, etc. at successive, discrete time points, beginning at the known IC: $\boldsymbol{x}^0$.

To compute $\boldsymbol{x}^{m+1}$ from Eq. (3.22), even though higher-order derivatives of $\boldsymbol{f}$ are not needed, we still need first-order *partial derivatives* of the function $\boldsymbol{f}$ at $t = t_m$. There are three ways to compute these partial derivatives: (1) *analytically* – by direct differentiation – for specific problems with (usually) differentiable $\boldsymbol{f}$; (2) *symbolically* – using the symbolic math functionality of programs like *Matlab*, *Mathematica*, *Maple* (and others), which yield mathematical expressions for the derivatives, for substitution into Eq. (3.22); and (3) using numerical derivative *approximations*. Derivatives in ODE algorithms that cannot be evaluated analytically can be approximated numerically.

Derivative Approximation Formulas

Differentiation is an inherently *continuous* operation, defined as a limit across both sides of $t = t_m$, i.e. before and after t_m, e.g.

$$\frac{df}{dt} \equiv \lim_{t_m^- \to t_m^+} \frac{f(t_m^+) - f(t_m^-)}{t_m^+ - t_m^-}, \; t_m^- < t < t_m^+ \tag{3.23}$$

For example, when $t = t_0$, the initial time for solution, the derivative is undefined because t_m^- is always less than t. The most common *numerical* approximations of derivatives are **backward-difference** (BD) formulas when only one "past" value of $\boldsymbol{f}$ is available:

$$\dot{\boldsymbol{f}}^m = \frac{\boldsymbol{f}^m - \boldsymbol{f}^{m-1}}{t_m - t_{m-1}} + E_{BD} \cong \frac{\boldsymbol{f}(t_m) - \boldsymbol{f}(t_{m-1})}{t_m - t_{m-1}} \tag{3.24}$$

or **central-difference** (CD) formulas when two "past" values of $\boldsymbol{f}$ are available:

$$\dot{\boldsymbol{f}}^m = \frac{\boldsymbol{f}^{m+1} - \boldsymbol{f}^{m-1}}{t_{m+1} - t_{m-1}} + E_{CD} \cong \frac{\boldsymbol{f}(t_{m+1}) - \boldsymbol{f}(t_{m-1})}{t_{m+1} - t_{m-1}} \tag{3.25}$$

E_{BD} and E_{CD} in (3.24) and (3.25) are the truncation errors for backward and central difference approximation formulas, respectively. For the partial derivative terms in Eq. (3.22), we can use backward-difference approximation formulas:

$$\boldsymbol{f}_x^m \equiv \left.\frac{\partial \boldsymbol{f}}{\partial \boldsymbol{x}}\right|_{t=t_m} \cong \frac{\boldsymbol{f}^m - \boldsymbol{f}^{m-1}}{x_m - x_{m-1}} \quad \text{and} \quad \boldsymbol{f}_t^m \equiv \left.\frac{\partial \boldsymbol{f}}{\partial t}\right|_{t=t_m} \cong \frac{\boldsymbol{f}^m - \boldsymbol{f}^{m-1}}{t_m - t_{m-1}} \tag{3.26}$$

yielding

$$\boldsymbol{x}^{m+1} = \boldsymbol{x}^m + \Delta t \left[\boldsymbol{f}^m + \frac{\Delta t}{2} \left(\frac{\boldsymbol{f}^m - \boldsymbol{f}^{m-1}}{t_m - t_{m-1}} + \boldsymbol{f}^m \frac{\boldsymbol{f}^m - \boldsymbol{f}^{m-1}}{\boldsymbol{x}^m - \boldsymbol{x}^{m-1}} \right) \right] + E_T^3 + \text{BD errors} \tag{3.27}$$

A Taylor series algorithm for approximating $\boldsymbol{x}$ based on Eq. (3.27) is thus:

$$\boldsymbol{x}^{m+1} \cong \boldsymbol{x}^m + \Delta t\left[\boldsymbol{f}^m + \frac{\Delta t}{2}\left(\frac{\boldsymbol{f}^m - \boldsymbol{f}^{m-1}}{t_m - t_{m-1}} + \boldsymbol{f}^m\frac{\boldsymbol{f}^m - \boldsymbol{f}^{m-1}}{\boldsymbol{x}^m - \boldsymbol{x}^{m-1}}\right)\right] \tag{3.28}$$

Example 3.14: ***More Data Needed***

Let's try to solve the (nonlinear) ODE $\frac{dx}{dt} = -x^2$ at $t = 1$, with $x(0) = 1$ given, using Eq. (3.28). Writing everything explicitly:

$$\begin{aligned} x^1 &\cong x^0 + \Delta t\left[f^0 + \frac{\Delta t}{2}\left(\frac{f^0 - f^{-1}}{t_0 - t_{-1}} + f^0\frac{f^0 - f^{-1}}{x^0 - x^{-1}}\right)\right] \\ &= 1 + 1\left[(-1)^2 + \frac{1}{2}\left(\frac{-1^2 - f^{-1}}{0 - t_{-1}} + (-1)^2\frac{-1^2 - f^{-1}}{1 - x^{-1}}\right)\right] \end{aligned} \tag{3.29}$$

We see that f and x and t are needed at t_{-1}, which is undefined, so we can't compute $x(1) \equiv x^1$. If we try to compute $x(2)$ at $t = 2$, we face a slightly different dilemma: whereas $x(0) = 1$ and $f(0) = 0$ at $t = 0$ and $f(1) = (-1)^2$ at $t = 1$, we still need x^1 to complete the calculation, and this is not available. We say that Eq. (3.28) is not *self-starting*; it is computable only if *two* past values of $\boldsymbol{x}$ are available.

In Other Words: **Higher-order derivatives** (H.O.D.) in Taylor formulas require more information to start a solution, even when H.O.D. are reduced to first-order partial derivatives. ICs are not enough. Taylor's formulas of order >1 are not self-starting, whether or not the H.O.D. are transformed to first-order.

Local vs. Global Truncation Errors

The **truncation error** E_T^{k+1} in Eq. (3.12) above was defined as the Taylor theorem remainder term $\boldsymbol{O}(\Delta t^{k+1})$ for kth-order Taylor-based algorithms. Strictly speaking, this is the **local truncation error** at each step Δt in the solution interval. **Global truncation error** is the cumulative total truncation error at the end of the solution interval. Roughly speaking, with $N = \boldsymbol{O}(1/\Delta t)$ steps of size Δt having local truncation error $\boldsymbol{O}(\Delta t^{k+1})$, then one can imagine that global truncation error will be N times the local error,[25] which gives $\boldsymbol{O}(\Delta t^{k+1}/\Delta t) \approx \boldsymbol{O}(\Delta t^k)$. So, kth-order Taylor-based algorithms have global truncation errors of about the same order.

Roundoff Errors

Nearly all numerical computations performed with digital computers involve number *rounding*. ODE solution computations thus include rounding errors in addition to algorithm truncation errors. Rounding occurs because numbers must fit within finite word lengths on digital computers. For example, ⅓ = 0.333333...3...3..., and so-on. If there's only room for five characters, say, then ⅓ is represented by the number 0.333, which is slightly smaller than ⅓. Similarly,

25 We forego the complex theoretical calculations needed to show this more precisely (Ascher and Petzold 1998).

⅔ = 0.66666. . ..6. . . would be roundly represented by the five characters 0.667, which is slightly larger than ⅔. The net sum of all these negative and positive errors at the end of a series of calculations is called the **roundoff error** for that computation.

In the early days of digital computation, when machine word lengths were typically 16-bits, roundoff errors were a far more important factor in ODE solution errors than they are now. Most CPUs have at least 32-bit word lengths, with 64-bits becoming more common at the time of this writing. So, be happy, but not ecstatic. Roundoff errors can still generate unacceptable and difficult-to-detect errors in ODE simulations (Kollár and Widrow 2008).

Computational/Numerical Stability

The previous section revealed the primary numerical errors in ODE methods, namely truncation errors, with roundoff errors a less important error-source in modern day general-purpose computers. It should come as no surprise that these errors can build up over time when using ODE solution algorithms, possibly making a solution worthless. Perhaps worse, the virtually limitless precision of digital computations can easily mask such worthlessness!

Surely, an ODE solution algorithm gives "satisfactory" results for a particular problem if it generates a solution sufficiently close to the actual solution of the ODE. We'd like a way to quantify this. Given that we don't know the actual solution, and are seeking one close enough to it "for our purposes," we have little recourse but to use as much theory as possible to circumscribe the desirable and undesirable properties of the algorithms available to us.[26] We continue this quest with a few definitions related to algorithm performance measures, realizing there is no universal answer for all model simulation problems – but we usually have choices and can benefit from understanding the knowledge boundaries better.

An algorithm is called **consistent** with the ODE if its truncation error E_T approaches zero as the step size Δt approaches zero. It's **convergent** if the approximate solution approaches the true solution, denoted $\boldsymbol{x}^*(t)$, as Δt approaches zero. Consistency and convergence, however, do not guarantee a satisfactory solution, even for small-enough step-sizes. Other factors, like small errors in ICs for starting computations, round-off errors at any stage in computation, etc., need to be factored in. We want our results to depend predictably on such errors, so that small changes in ICs, or roundoff errors, for example, produce similarly small changes in subsequent approximations in the algorithms, i.e. generating results that depend *continuously* on ICs, or computed values at any step. The following robust definition includes these considerations, for both self-starting and *multistep* methods presented later. For this we need to define **relative error:**

$$\text{Relative error} \equiv \frac{\text{total computational error}}{\text{actual (theoretical) solution}} = \frac{E_T + E_R}{x^*(t)} \tag{3.30}$$

In this equation total computational error is equal to *all* truncation errors plus *all* roundoff errors. This means that all errors in approximations used in the computational algorithm are

26 Analysis of stability and convergence of numerical algorithms can be fairly complex. Abbreviated results are presented here, to give the flavor of the problem and summarize the main results for our applications. Additional details can be found in most numerical analysis texts, e.g. Burden and Faires 2005.

lumped in E_T, not just the Taylor series truncation error. For example, $E_T \equiv E_T^3 + \text{BD}$ errors in Eq. (3.27) – if one uses up to second-order terms of the Taylor series. Similarly, all roundoff errors are included in E_R in Eq. (3.30).

In these terms, an algorithm is said to be **computationally stable** or **numerically stable** if *relative errors* remain bounded as the solution evolves. Note that this is not the same as *system* stability; however, it is related, because difference equations also have characteristic polynomials with stability properties (Burden and Faires 2005). Numerical instability is a major issue for "stiff" equations and their special solution algorithms discussed later in the chapter.

To summarize, questions of numerical or computational stability of simulation solutions do not have a universal answer. They depend on the specifics of both the problem and the method chosen to solve it. ODE dimensionality and complexity, and other particulars – like the size of the time-interval of solution are important determinants of numerical stability, as are the properties of the algorithms chosen for solution. This issue is central in simulation and is discussed further as we move along in the chapter, along with suggestions for making good choices and maximizing the reliability of solutions.

Self-Starting ODE Solution Methods

A **self-starting** algorithm is one that requires only a single value, usually the single IC of the ODE, to proceed with computation of subsequent values. Eq. (3.28) is not self-starting, because it requires two "past" values, $\boldsymbol{x}_m$ and $\boldsymbol{x}_{m-1}$, to compute $\boldsymbol{x}_{m+1}$, which motivates an even simpler version of Eq. (3.28), with only zero and first-order terms retained and truncation error E_T^2, called Euler's method, developed next. Runge–Kutta methods, a more practical and widely used set of self-starting algorithms, are developed in the section that follows.

Euler Method

This algorithm includes only the first two (the zeroth- and first-order) terms of the Taylor series expansion in Eq. (3.19) or (3.28) to predict the solution at t_{m+1}:

$$x^{m+1} \cong x^m + f^m \Delta t \tag{3.31}$$

Unfortunately, the truncation error for Euler's method E_T^2 can be large, even for relatively small step sizes Δt, so it's less commonly used for solving ODEs. But its simplicity renders it a very useful pedagogical tool for understanding more complex algorithms and their inherent errors.

The approximation in Eq. (3.31) is illustrated in **Fig. 3.16**, where the slope (derivative) of x at x^m is represented as the tangent line at x^m projected to x^{m+1}, a *linear* extrapolation. The predicted error E_T, also shown in **Fig. 3.16**, is not "small."

Example 3.15: ***Euler Algorithm: Simple NL ODE***

Using the same ODE as in Example 3.14:

$$\frac{dx}{dt} = -x^2, \quad \text{for } x(0) = 1 \tag{3.32}$$

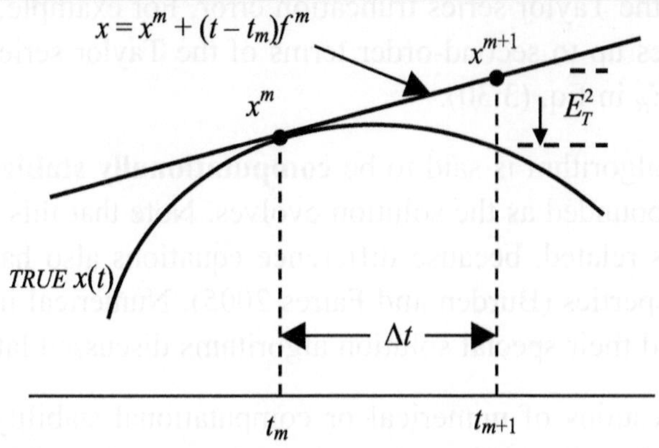

Fig. 3.16

The exact solution is: $x = \dfrac{1}{1+t}$, readily verified by substitution into Eq. (3.32). Applying Euler's algorithm, $x^m = x^m + (-x^m)^2\Delta t$, we evaluate x^1, x^2, etc. with step-size $\Delta t = 0.1$: $x^1 = 1 + (0.1)(-1^2) = 0.9$. The exact solution at this point is evaluated as $x(0.1) = \dfrac{1}{1+0.1} = 0.909$. Next, at $t = 0.2$, $x^2 = 0.9 + (0.1)(-(0.9)^2) = 0.819$. The exact solution is: $x(0.2) = \dfrac{1}{1+0.2} = 0.833$. After 10 steps with $\Delta t = 0.1$, $x^{10} = 0.463$, with exact solution: $x(1) = \dfrac{1}{1+1} = 0.5$. We see that E_T is building up fairly rapidly.

Intuition suggests that truncation error E_T would be much smaller in the example above if we use a much smaller step-size Δt. This is true, but with negative consequences if Δt is made *too small*, because roundoff errors generally build-up as the number of steps increases.[27] Theoretically, the Euler method is consistent and convergent, but it isn't particularly efficient, and more sophisticated methods are available.

Runge–Kutta Methods

For most problems, Taylor's self-starting methods have the disadvantage of requiring complicated and time-consuming computation of high-order derivatives to obtain small truncation errors. In contrast, Runge–Kutta (R–K) methods, (e.g. Forsythe et al. 1977), can achieve similarly small truncation errors without the need for computing high-order derivatives. The trade-off is in additional computations of ODE function f in R–K methods, usually easier than computing derivatives.

Main Features of R–K Methods

1. They are *one-step, self-starting*: to find x^{m+1}, we only need information available at the preceding point, x^m.

27 Albeit relatively less important with modern 32-bit and 64-bit machines, as noted earlier.

2. They agree with the Taylor series through terms in Δt^k, with k called the **order** of the method. Thus, they have the same precision as Taylor methods of the same order.

3. They *do not* require evaluation of any derivatives of f, only f itself. To accomplish this trade-off, $f(x, t)$ is evaluated not only at (x^m, t_m) but also at one or more closely adjacent points between (x^m, t_m) and (x^{m+1}, t_{m+1}). Then the solution x_{m+1} at t_{m+1} is obtained by means of a *weighted sum* of these expressions for $f(x, t)$.

Derivation of the different R–K methods requires Taylor series expansion of f in two variables, x and t, instead of just t, because derivatives of f in the expansion are replaced by evaluations of f at intermediate values of x between (x^m, t_m) and (x^{m+1}, t_{m+1}). This expansion of $f(x, t)$ evaluated a small distance Δx from x and Δt from t, i.e. in a "small neighborhood" of the point (x, t), is:

$$\begin{aligned} f(x+\Delta x, t+\Delta t) = f(x,t) + \left.\frac{\partial f}{\partial x}\right|_{(x,t)} \Delta x + \left.\frac{\partial f}{\partial t}\right|_{(x,t)} \Delta t + \\ \frac{1}{2!}\left[\left.\frac{\partial^2 f}{\partial x^2}\right|_{(x,t)} \Delta x^2 + 2\left.\frac{\partial^2 f}{\partial x \partial t}\right|_{(x,t)} \Delta x \Delta t + \left.\frac{\partial^2 f}{\partial t^2}\right|_{(x,t)} \Delta t^2\right] + \cdots \\ \equiv f(x,t) + f_x(x,t)\Delta x + f_t(x,t)\Delta t + \frac{1}{2!}(f_{xx}(x,t)\Delta x^2 \\ + 2f_{xt}(x,t)\Delta x \Delta t + f_{tt}(x,t)\Delta t^2) + \cdots \end{aligned} \tag{3.33}$$

The f_{xx}, f_{xt}, and f_{tt} in Eq. (3.33) are the second-order partial derivatives of f evaluated at x and t. Further derivation details for different order R–K methods can be found in various references, (e.g. Butcher 1987). We present and discuss two of the most popular ones here.

Fourth-Order Runge–Kutta Method

This most popular of the R–K methods, with truncation error $E_T \sim O(\Delta t^5)$, is:

$$x^{m+1} = x^m + \frac{\Delta t}{6}(k_1 + 2k_2 + 2k_3 + k_4) \tag{3.34}$$

where

$$\begin{aligned} k_1 &= f(x^m, t_m) \\ k_2 &= f\left(x^m + \frac{\Delta t}{2}k_1, t_m + \frac{\Delta t}{2}\right) \\ k_3 &= f\left(x^m + \frac{\Delta t}{2}k_2, t_m + \frac{\Delta t}{2}\right) \\ k_4 &= f(x^m + \Delta t k_3, t_m + \Delta t) \end{aligned} \tag{3.35}$$

The *ks* in (3.35) are the f evaluated at intermediate values of $\boldsymbol{x}$ between (x^m, t_m) and (x^{m+1}, t_{m+1}). The algorithm begins with computation of the *ks*, followed by Eq. (3.34), for $m = 1, 2, \ldots$ up to the final time t_f of the computation interval (t_0, t_f) (Butcher 1987).

Example 3.16: ***Fourth-Order Runge–Kutta solution of a NLTV ODE***

In the following, function f is evaluated four times using R–K fourth-order formulas, thereby replacing four derivative evaluations. Starting from $x(0) = x^o = x^m = 3$ $(m = 0)$, compute the first predicted value $x(0.1) = x^{m+1} = x^{-1}$ for the nonlinear time-varying (NLTV) ODE:

$$\frac{dx}{dt} = \frac{4t}{x} - tx \equiv f(x, t) \quad \text{with } x(0) = 3 \text{ and step-size } \Delta t = 0.1$$

For $m = 0$ and $t_0 = 0$:

$$k_1 \equiv f(x^o, 0) = \frac{4(0)}{3} - (0)3 = 0$$

$$t^{1/2} = t_0 + \frac{\Delta t}{2} = 0 + 0.05 = 0.05$$

$$x^{1/2^*} = x^o + \frac{\Delta t}{2} k_1 = x^o + \frac{\Delta t}{2} f(x^o, 0) = 3 + 0.05(0) = 3$$

$$k_2 \equiv f(x^{1/2^*}, t_{1/2}) = \frac{4}{3} 0.05 - 0.05(3) = -0.083333$$

$$x^{1/2^{**}} = x^o + \frac{\Delta t}{2} k_2 = x^o + \frac{\Delta t}{2} f(x^{1/2^*}, t_{1/2}) = 3 + 0.05(-0.083333) = 2.995833$$

$$k_3 \equiv f(x^{1/2**}, t_{1/2}) = \frac{4(0.05)}{2.995833} - 0.05(2.995833) = -0.083033$$

$$x^{1^*} = x^o + \Delta t f(x^{1/2^{**}}, t_{1/2}) = 3 + 0.1(-0.083033) = 2.991697$$

$$k_4 \equiv f(x^{1^*}, t_1) = \frac{4(0.1)}{2.991697} - 0.1(2.991697) = -0.165466$$

$$x^1 = x^o + \frac{\Delta t}{6}(k_1 + 2k_2 + 2k_3 + k_4)$$

$$= 3 + \frac{0.1}{6}[0 + 2(-0.083333) + 2(-0.083033) - 0.165466] = 2.991697$$

Example 3.17: ***Fourth-Order Runge–Kutta Solution of Multiscale Eq. (3.1)***

Solution of Eq. (3.1) for $\tau = 10$ only – shown earlier in **Fig. 3.11** – was reevaluated for several fixed stepsizes ($\Delta t = 0.05$, 0.1, 0.5, 1 and 2 min), using the fixed-step R–K fourth-order formulas (R-K4 solver – ODE4 in *Simulink*). Results are shown superimposed in **Fig. 3.17**.

None of the solutions are the same, the red one – for step-size = 2 mins, is shifted substantially to the left of the others. These four other solutions fall closely in a cluster at the resolution shown and they are likely reasonably accurate – within a few percent, based on reevaluations with several independent solvers (not shown). Visually, the red solution differs from the others by as much as 15–20%. On the other hand, 180/2 = 90 steps in model time (not computer runtime) were needed to solve the problem with $\Delta t = 2$ and at the other extreme it took 180/0.05 = 3600 steps to get a solution for $\Delta t = 0.05$ mins. Actual runtimes on a modern computer are very short.

Stepwise Errors, Tolerances & Error Control

It's no surprise that solution accuracy falls with increasing step-size. The "right step-size" is problem-dependent and often requires trial-and-error simulations to establish what works best when using any fixed step solver, as illustrated above. Fixed step-sizes capable of providing solutions with a desired accuracy for all equations and all solution time-intervals do not exist.

ODE solvers more sophisticated than fixed-step algorithms like Euler's and R-K4 methods have provision for automatic control of step-size during computational solution, thereby

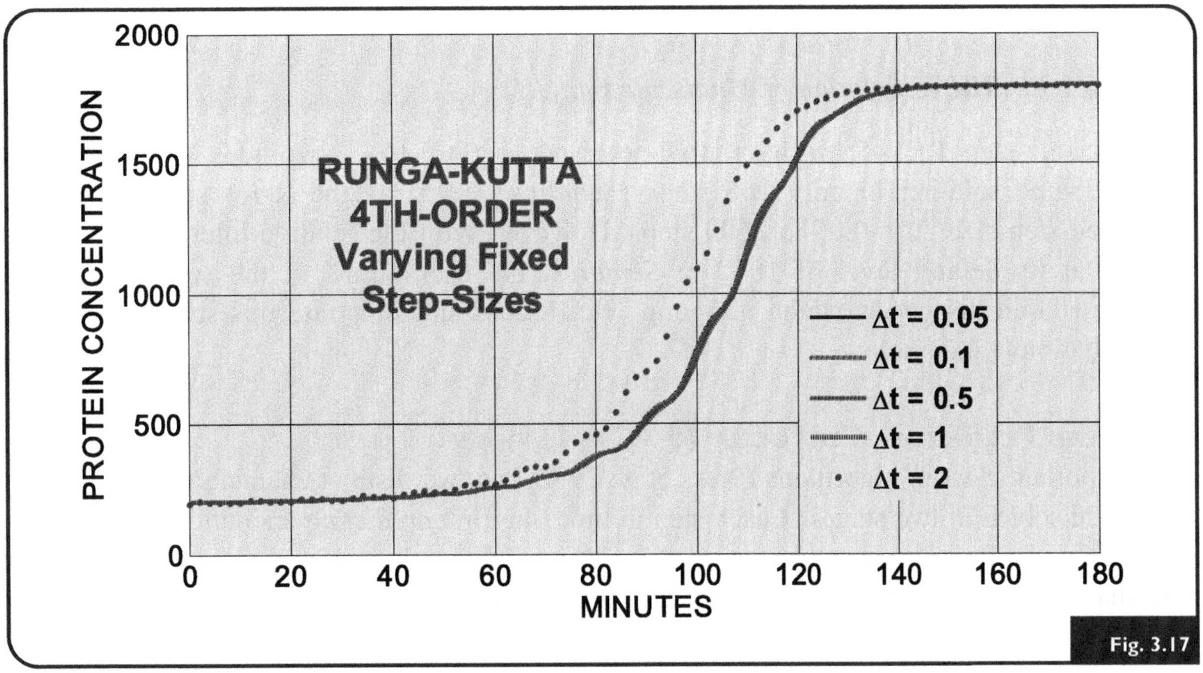

Fig. 3.17

circumventing user intervention or trial-and-error runs (as illustrated above) to establish the "best" step-size to use in a run. They provide for computation of *truncation error estimates* at each step, often called the **error-per-step (EPS)**, or **absolute (local) error**. **Relative error** is the relative size of the absolute local error compared with the size of the variable (say $x(\tau)$) being computed at time τ. For example, if EPS = 0.1 and $x(\tau) = 42.35$, then *relative error* = 0.1/42.35 = 2.6313×10^{-3}. The idea is to keep the EPS smaller than an acceptable tolerance level.

Importantly, error tolerances are user-selectable in sophisticated program packages for doing math (e.g. *Matlab*), typically as *absolute* or *relative tolerances*, or both. **Absolute tolerance** is an upper bound on the absolute error. **Relative tolerance** is an upper bound on relative error. These are called **error-per-step (EPS) criteria**. Again, the idea is keep this local error estimate below the absolute and/or relative tolerance. If the estimated EPS in an ODE solver does not pass this test, the step is rejected and repeated with a smaller step-size.

Algorithms for Estimating and Controlling Stepwise Precision

Runge–Kutta–Fehlberg (R–K–F)

These algorithms are made up of different order R–K methods modified and combined to provide a means of estimating and controlling truncation error at each step (Forsythe et al. 1977). They are Taylor series formulas with slopes k_i at t_i combined in two ways: one to give an estimate x^{m+1} of order Δt^m and another to give an estimate z^{m+1} of order Δt^{m+1}. Then $z^{m+1} - x^{m+1}$ is considered a reasonable estimate of the error in x^{m+1}, used to control the precision of the computation. One can also think of the higher-order method as giving the "exact" answer and use this to determine the error in the lower-order method. In the **RKF45** formulas used in most simulation packages, z^{m+1} is a fifth-order formula using six stages (the minimum possible) and x^{m+1} is fourth-order, using five stages.[28]

Example 3.18: ***RKF45 Solution of Multiscale Eq. (3.1)***

Variable-step RKF45, with relative tolerance set to 10^{-5}, provides essentially the same solution in only 166 steps (*Simulink* RK45) as the R-K4 solver with fixed-step-size $\Delta t = 0.05$ in 3600 steps. However, with the relative tolerance level set at its default level of 10^{-3} in *Simulink*, the oscillations in the solution for $\tau = 10$ min disappear, which is wrong. We discuss and illustrate this subtle result in Example 3.24 below.

Multistep Predictor–Corrector (P–C) Methods

This important class of algorithms (Press et al. 1992) has two distinct characteristics. First, the ODEs are solved in two stages at each step in time, the first *predicting* a solution x^{m+1} at each step, with one formula, and then *correcting* the predicted solution, with a second, corrector formula – in one of two variations. They can be *explicit* or *implicit*. Explicit corrector

28 RKF45 solvers are available in both *Simulink* and *VisSim*. *Simulink* RK45 is a Dormand–Prince algorirthm.

formulas generate solutions in one iteration (one stage of the algorithm), sometimes by adding an estimate of the truncation error E_T to the corrector formula, determined using Taylor's theorem with remainder. Implicit ones require multiple iterations of the algorithm to obtain a solution, i.e. they *recorrect* it any number of times by iteration, instead of adding a fixed correction. The second distinct characteristic is that, unlike Runge–Kutta methods, predictor-corrector algorithms use values of the solution x determined at two or more preceding points, rather than a single initial condition (IC) point, hence the designation *multistep*. Therefore, they are *not* self-starting.

Example 3.19: ***Adams–Bashforth Predictor***

The following predictor formula requires two values of x – present and past, x^m and x^{m-1}, to compute (predict) x^{m+1}:

$$x^{m+1} = x^m + \frac{\Delta t}{2}(3f^m - f^{m-1}) \tag{3.36}$$

Example 3.20: ***Second-Order Adams–Moulton Corrector***

The following corrector formula requires only one value of x in the past, x^m, but the current value of x^{m+1} is needed to compute f^{m+1} on the RHS. x^{m+1} is thus implicit and requires an iterative algorithm to compute x^{m+1} on both sides of the formula simultaneously:

$$x^{m+1} = x^m + \frac{\Delta t}{2}(f^{m+1} + f^m) \tag{3.37}$$

Example 3.21: ***Fourth-Order Adams Predictor***

The following predictor formula requires *four* previous values of x, i.e. x^{m-3}, x^{m-2}, x^{m-1}, and x^m to predict x^{m+1}:

$$x^{m+1} = x^m + \frac{\Delta t}{24}\left[55f^m - 59f^{m-1} + 37f^{m-2} - 9f^{m-3}\right] \tag{3.38}$$

where

$$f^{m-i} = f(x^{m-i}, t_{m-i})$$

$$f^{m+1} = f(x^{m+1}, t_{m+1})$$

for $i = 1, 2, 3$.

This is a fourth-order algorithm, hence errors are of the order of Δt^5, similar to fourth-order Runge–Kutta algorithm.

Simple Multistep Formulas Based on Taylor Series

As with self-starting algorithms, predictor–corrector formulas are derived using the Taylor series as a base:

$$x^{m+1} = x^m + \Delta t\left[f^m + \frac{\Delta t}{2!}\dot{f}^m + \frac{\Delta t^2}{3!}\ddot{f}^m + \cdots\right] \tag{3.39}$$

Typically, derivatives $\dot{f}, \ddot{f}$, etc. are replaced by backward-difference approximations in the series, e.g. $\dot{f}^m \cong \frac{f^m - f^{m-1}}{\Delta t}$. For example, the second-order Adams–Moulton formula in Eq. (3.36) is derived as:

$$\begin{aligned} x^{m+1} &= x^m + \Delta t\left[f^m + \frac{\Delta t}{2}\left(\frac{f^m - f^{m-1}}{\Delta t}\right) + O(\Delta t^3)\right] \\ &= x^m + \Delta t\left[\frac{3}{2}f^m - \frac{1}{2}f^{m-1}\right] + O(\Delta t^3) \end{aligned} \tag{3.40}$$

A third-order Adams formula is generated by also substituting a two-point backward difference approximation for the second-derivative in the Taylor series of Eq. (3.39):

$$\ddot{f}^m \cong \frac{f^m - 2f^{m-1} + f^{m-2}}{(\Delta t)^2} \tag{3.41}$$

This yields:

$$x^{m+1} = x^m + \Delta t\left[\frac{23}{12}f^m - \frac{16}{12}f^{m-1} + \frac{5}{12}f^{m-2}\right] + O(\Delta t^4) \tag{3.42}$$

Hopefully, use of the Taylor series in developing ODE solution algorithms – along with the error components – has been successfully communicated with the material presented thus far. Several additional and important algorithms based on the Taylor series are presented in Appendix F. These include explicit Euler P–C formulas modified to give higher-order (smaller) truncations errors; implicit P–C algorithms, that employ iterations on the corrector formula that in principal converges to a better approximation of x at each step; and non-iterative predictor–modifier–corrector algorithms that employ estimates of the predictor and corrector formula truncation errors to generate better approximations of x at each step.

Taylor Series-Based Method Comparisons

Runge–Kutta (R–K) *methods* have several positive attributes. They are self-starting, generally stable, non-iterative, and permit easy change in Δt, but not truncation error estimation in their basic form. In place of derivative evaluations, needed in direct Taylor series methods, they substitute several evaluations of $f(x,y)$, the number equal to the order of the method.

Generally speaking, *predictor–corrector* (P–C) or *predictor–modifier–corrector* (P–M–C) methods are more efficient than R–K methods, because they substitute information about prior points for repeated evaluations of $f(x, y)$. They also allow easy estimation of truncation errors and are quite good for efficiently getting highly accurate solutions for non-stiff problems (see below). But they are not self-starting, and changes in step size are difficult, requiring self-starting methods – like Runge–Kutta – for this purpose. They also have more stringent step-size restrictions, due to smaller regions of absolute stability, and thus can become computationally unstable. Fortunately, Δt can be adjusted to compensate, providing greater numerical stability.

Example 3.22: ***Taylor Series-Based Method Approximations***

This informative example illustrates how three methods developed in this chapter approximate the Taylor series expansion of x, using a very simple ODE to enhance the comparisons.

$$\dot{x} = f(x) = x \tag{3.43}$$

Note that all derivatives in this problem are equal, i.e.

$$f(x) = \dot{x} = x = \frac{df(x)}{dt} = \frac{d^2f(x)}{dt^2} = \cdots \tag{3.44}$$

The Taylor series (TS) expansion of this function is:

$$\begin{aligned} x^{m+1} &= x^m + \frac{df^m}{dt}\Delta t + \frac{d^2f^m}{dt^2}\bigg|\frac{\Delta t^2}{2!} + \frac{d^3f^m}{dt^3}\bigg|\frac{\Delta t^3}{3!} + \frac{d^4f^m}{dt^4}\bigg|\frac{\Delta t^4}{4!} + \cdots \\ &= x_m\left[1 + \Delta t + \frac{\Delta t^2}{2} + \frac{\Delta t^3}{6} + \frac{\Delta t^4}{24} + \cdots\right] \end{aligned} \tag{3.45}$$

The **Euler formula** is the same as the TS to first-order:

$$x^{m+1} = x^m + f(x^m)\Delta t = x^m + x^m\Delta t = x^m[1 + \Delta t] \tag{3.46}$$

Multistep Euler Corrector Formula:

$$\begin{aligned} x_C^{m+1} &= x^{m+1}\frac{\Delta t}{2}\left[f^{m+1} + f^m\right] = x^m + \frac{\Delta t}{2}\left[x_P^{m+1} + x^m\right] \\ &= x^m + \frac{\Delta t}{2}[x^m(1 + \Delta t) + x^m] = x^m[x^m(1 + \Delta t) + x^m] \end{aligned} \tag{3.47}$$

Fourth-order Rung–Kutta algorithm – same as TS to fourth order:

$$\begin{aligned} k_1 &= x_m \\ k_2 &= x^m + k_1\frac{\Delta t}{2} = x^m\left(1 + \frac{\Delta t}{2}\right) \\ k_3 &= x^m + k_2\frac{\Delta t}{2} = x^m\left(1 + \frac{\Delta t}{2} + \frac{\Delta t^2}{4}\right) \\ k_4 &= x^m + k_3\Delta t = x^m\left(1 + \Delta t + \frac{\Delta t^2}{2} + \frac{\Delta t^3}{4}\right) \end{aligned} \tag{3.48}$$

$$x^{m+1} = x^{m+1} + \frac{\Delta t}{6}[k_1 + 2k_2 + 2k_3 + k_4] = x^{m+1}\left[1 + \Delta t + \frac{\Delta t^2}{2} + \frac{\Delta t^3}{6} + \frac{\Delta t^4}{24}\right] \tag{3.49}$$

Stiff ODE Problems

Stiff equation problems are ubiquitous in biology, chemical kinetics, control theory, weather prediction, physics, electronics and many other fields of science utilizing ODE models. Intuitively, a *stiff* ODE has some "modes" that change very rapidly compared with others, or it has solutions that sometimes change very rapidly, and slowly at other times, e.g. very rapid transient behavior and a constant steady state. If such a system is perturbed, it typically "springs back" toward a constant steady state very quickly. Or a change might define a new and different steady state, e.g. when a *gene* is switched on and moves rapidly to a new steady state. **Fig. 3.18** illustrates the stiffness problem very simply, with the two terms of the solution $x(t) = e^{-100t} - e^{-t}$ plotted separately: one disappears almost instantaneously relative to the other, which remains relatively constant. The first clearly requires very small and the latter much larger step-sizes to be computed precisely.

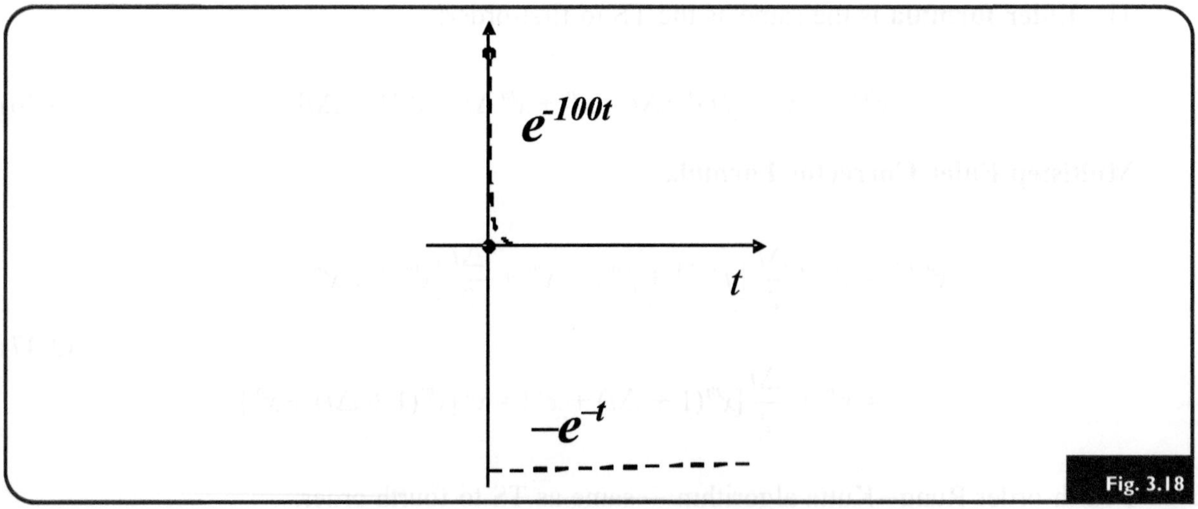

Fig. 3.18

Stiff ODEs are common in *multiscale* models, with coupled dynamics represented over different time scales or levels in a biosystem, e.g. as in gene regulated protein production with couplings between cytoplasmic and nuclear components acting over different time scales, and biosystems in general, with mixed biophysics and biochemistry submodels. We encountered the approximation of translation and transport dynamics of a transcription factor by a simple nondynamic time-delay in Example 3.6 and Example 3.9. This circumvented at least part of the problem by substituting a simple time-delay for the relatively very slowly changing ODEs that would be needed to represent "lagging" translation and transport dynamics. This pure time-delay (*extreme time-scaling*) simplification is an important approach to modeling across scales in a multiscale biomodel. In some cases, however, it may not solve all stiffness problems in a particular solution scenario, as we illustrate in Example 3.23 below.

In general, stiff ODEs are in need of *time-scaling*, but – except for very simple models where the need may be apparent – stiffness is not obvious and special simulation methods are needed that effectively do time-scaling automatically as the solution proceeds. Recognizing stiffness in ODEs is generally difficult, and you can't tell by looking at only one solution (Ascher and Petzold 1998). Standard algorithms can readily yield numerically unstable results, which might be obvious if the solutions blow up at some point in the computation. But if the instability is more subtle, more-or-less poor solutions to the ODEs are obtained. If standard methods are used, like step-size controlled (variable-step-size dependent on truncation error estimates at each step) R–K or P–C algorithms, results may not appear unusual, despite typically requiring a *very large* number of steps to achieve even modest accuracy. The problem is that the number of steps taken is not usually transparent to the user, and most likely goes unnoticed for problems run on fast computers. For moderately stiff problems, the program may seem to be running okay, with no hint of problems. When a *really* stiff problem is encountered, the program *may* seem like it isn't working right, but maybe not; this depends on the specifics of the ODE model, and the solver chosen, as exemplified later.

The following formalization of stiffness can be helpful in understanding the computational difficulties. Mathematically, a nonlinear ODE is **locally stable** at (x, t) if the eigenvalues $\lambda_i(x,t)$ of the local **Jacobian matrix** $\frac{\partial f}{\partial x}$ have negative real parts. Fortunately, we usually work with ODE models that are inherently stable, at least locally – in the region of interest of the system dynamic response. An ODE is **stiff** if the relative range of the inverse of the magnitude of these local eigenvalues, i.e. the min to max of $|1/\lambda_i|$, is large, say larger than 100, for any point (x, t). We would like to not have to check these properties each time we simulate an ODE. We should use methods appropriate to the problem, of course, but we need to understand what underlies potential stiffness problems. This is easier with linear than with NL models.

Example 3.23: ***A Simple Stiff ODE Model***

The linear ODE:

$$\frac{d^2x}{dt^2} + 1001\frac{dx}{dt} + 1000x = 0 \tag{3.50}$$

is stiff. A little analysis demonstrates this. The roots (eigenvalues) of the quadratic characteristic equation for this model are easily calculated analytically as -1000 and -1. The solution therefore has a very "fast" mode, a_1e^{-1000t} and a "slow" mode, a_2e^{-t}, 1000 times slower, similar to the modes in **Fig. 3.18**, only 10-fold more separated.

How would you choose a single step-size for computing the solution using a fixed step-size method like Runge–Kutta fourth-order? Not easily, given that roundoff error *might* generate instability in the solution with very small step-sizes at the outset – to capture the fast transient. But it doesn't for this example. Truncation errors appear to dominate the solution space, discovered by trial-and-error with the simulation tools.

Fig. 3.19 illustrates two solutions, using the R–K fourth-order solver in *Simulink*. The first, on the left side of the figure – for both $x(t)$ (yellow) and $dx(t)/dt$ (red) is over the time interval [0,1], with step-size = 0.0025 (400 steps). It appears stable, illustrating the solutions as they should appear. Interestingly, the solutions appear to be stable even over the interval [0,10,000] (4 million steps!) at this step-size = 0.0025.

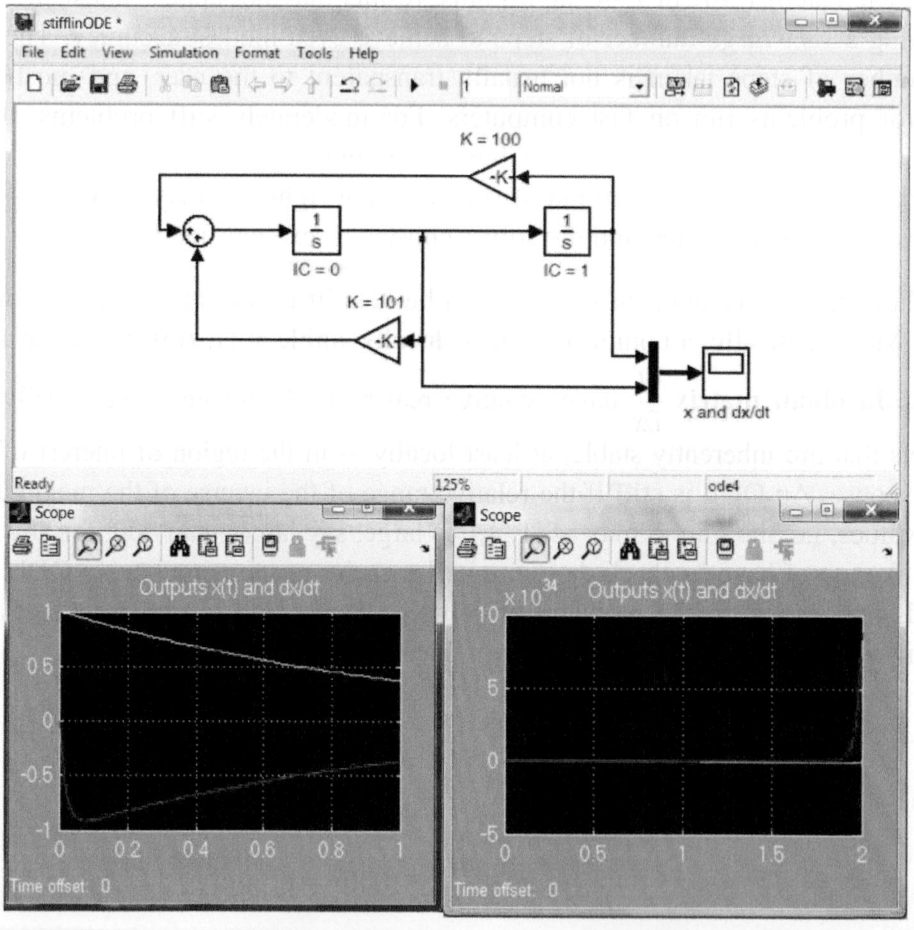

Fig. 3.19

In contrast, with an ever so slightly larger step-size, 0.003, the solution blows up almost immediately, at t values as small as 0.01. The accuracy of the solution may be degraded by roundoff errors for the 4 million-step solution, but *truncation error* seems to dominate at step-sizes in this range, with a critical value between 0.0025 and 0.003. One way to verify this and – at the same time – compute the computational errors with this algorithm, is to evaluate the analytical solution of this linear equation, and subtract it from the solver solution. Because we already know the two solution modes, the analytic solution is simply their sum. For ICs $x(0) = 1$ and $\dot{x}(0) = 0$, the analytic solution is easily computed as $x(t) = (1000/999)e^{-t} - (1/999)e^{-1000t}$. (This is left as an exercise, E3.5.)

Remark: This example is very simple and intuitive. More complex examples are not likely to be so friendly.

Stiff Solvers

Before 1970, users had little choice: either use a great deal of computer time with very small steps, and risk numerical instabilities, or mathematically manipulate the model to eliminate or reduce stiffness, as in Example 3.6 noted above. More "manipulations" like this are presented in Chapter 6 – where we study inherently multiscale enzyme and receptor–ligand dynamics. One of the first stiff-equation solvers was Gear's method (Press et al. 1992). Solvers that can work reasonably well for both stiff and nonstiff problems include: Bogacki–Shampine (ODE23 in *Matlab/Simulink*), Rosenbrock (ODE23s), trapezoidal (ODE23t), multistep NDF (ODE15s), Backward-Euler and Bulirsch–Stoer (in *VisSim*) solvers (Bulirsch and Stoer 1966; Calahan 1968). Bulirsch and Stoer (1966) is a *Richardson method* – a class not based on the Taylor series, with the advantage of being robust, able to solve a large spectrum of ODEs, including moderately stiff ones reasonably accurately (Frank 1960). The multistep implementation of TR-BDF2 (ODE23tb in *Matlab/Simulink*) is an implicit Runge–Kutta formula with a trapezoidal rule first stage and a second stage consisting of a backward differentiation formula of order two, particularly well suited for very stiff models (Hosea and Shampine 1996). For stiff systems, *Adaptive BDF* in *VisSim* has the advantageous property that discontinuities are detected and the step-size is automatically shrunk around them.

How to Choose a Solver?

A very good question, indeed! No single program or method is uniformly best for all problems, so one should always experiment with several, certainly for large, but even for small numbers of equations.

You don't choose one solver. *You should try at least two different methods – several step-sizes and solution tolerances, and compare the results.* If they concur, i.e. give the same solution, they are likely to – but are not guaranteed to be "correct." If a fixed-step-size solver, like R–K fourth-order, for example, yields a stable solution – and not too many steps have been taken to get a solution, the solution should be uniformly stable – meaning that making the step-size a little smaller or larger – should give approximately the same solution. When the model is stiff, error tolerances (usually under user-control in the program) may need adjustments to obtain accurate stable solutions, as illustrated in the next example.

Example 3.24: ***Different Solver Solutions of Stiff Multiscale Eq. (3.1)***

Eq. (3.1) was solved in Example 3.6 and solutions for three different time-delays were presented in **Fig. 3.11**. In deriving Eq. (3.1), major stiffness was eliminated beforehand from this multiscale model by approximating fast dynamics (transcription, activation, etc.) as occurring instantaneously relative to the remaining regulatory dynamics, with a pure time-delay attached in the second-order Hill function regulatory term – the delay representing aggregated intracellular translocation and cytoplasmic protein translation delays. However, Eq. (3.1) is still moderately stiff, because of the very large coefficients (order 10^3 to 10^5), further squared in the second-order Hill function term – relative to the protein degradation rate equal to 1.

In Example 3.17, we used *Matlab/Simulink* ODE solver RK45 with the relative tolerance control parameter set to a value 100 times smaller than the default value (10^{-3}) for this solver, namely 10^{-5}, and we got the solutions shown for $\tau = 10$ min in the left panel of **Fig. 3.20**, in 166 steps. In this run, the step-size was automatically varied to keep the relative truncation errors (tolerance) $<10^{-5}$. In contrast, when we reset the relative tolerance to 10^{-3}, which is 100 times larger – but nevertheless the *default value* for this solver – the value that would likely be automatically specified unless the user changes it – we got the wrong solution, in 73 steps. This shown in the right panel of **Fig. 3.20**. This right-panel solution is a *smoothed* version of the correct solution, with oscillations for time-delay $\tau = 10$ min not visible at all. For relative tolerance level 10^{-4} (10 times larger than the default), the solution (not shown) is less smoothed, but still *wrong*.

Stiff solvers faired a bit better. Moderately stiff solver Rosenbrock (ODE23s in *Simulink*) yielded reasonably accurate solutions in 123 steps, with relative tolerance

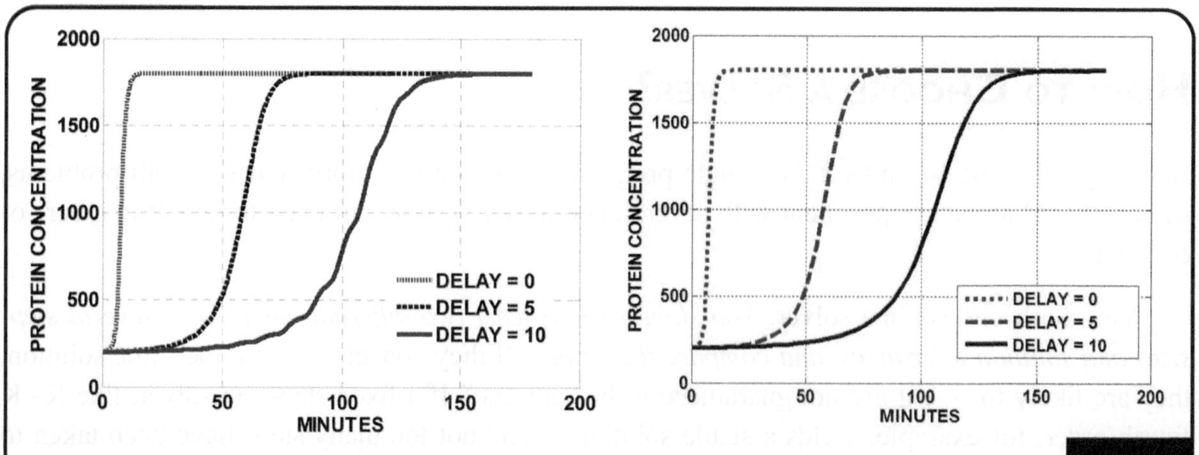

Fig. 3.20

level set to the default value of 10^{-3}. Stiff solvers NDF (ODE15s in *Simulink*), in 220 steps, *Simulink* ODE23t, in 228 steps, and very stiff solver ODE 23tb, in 193 steps, all required a relative tolerance of 10^{-4} or smaller to get a reasonably correct solution.

This example problem and its various computed solutions reveals some unanticipated subtleties in numerical solution of ODEs, emphasized further by the simplicity of the model, Eq. (3.1). The smoothed and incorrect solutions were "too easy" to get with control parameters set to their default values. Only Rosenbrock worked well at default settings. The other stiff solvers generated inappropriately smoothed results when the precision tolerance control parameters were not set finely enough (min 10^{-4}); but the nonstiff solver ODE45 (and others not shown) required even more stringent relative tolerances (10^{-5}) for correct solutions.

Another Caveat: It is important to note that some of the numerical details (e.g. tolerances, number of steps) of this example may not be reproducible precisely on a computer different than a Gateway 32-bit machine, on which they were evaluated. In fact, the numbers were slightly different on a newer Dell 64-bit machine. Different central processing units (CPUs) round numbers differently – even if they have the same bit-length, as some of us have experienced over the years we've been computing ODE solutions. Expect that most simulated ODE solutions – for any example, will be different in some details when computed on different machines. Bottom line? Get at least two methods to agree, as discussed earlier, on at least one machine.

Solving Difference Equations (DEs) Using an ODE Solver

First-order difference equations (DE)

$$\boldsymbol{x}^{m+1} = \boldsymbol{g}(\boldsymbol{x}^m, \boldsymbol{u}^m, \boldsymbol{t}_m) \equiv \boldsymbol{g}^m \tag{3.51}$$

where $\boldsymbol{g}$ is generally a nonlinear vector function of the indicated arguments, can be solved *directly*, beginning with the initial value, assumed known, by solving *recursively* for the solution at each succeeding discrete time-step.

Example 3.25: ***Recursive Solution of a DE***

If $x(0) = 1$, the solution of $x(k+1) = -7 \times {}^2(k)$ for $k = 1, 2, \ldots$ is:

$$x(1) = -7(1)^2 = -7,\ x(2) = -7(-7)^2 = -7^3,\ x(3) = -7(-7^3)^2 = +7^7,\ \ldots,\ x(k) = (-1)^k 7^{2k+1}$$

This procedure can be more or less difficult, depending on the complexity of the equation.

Another approach is to use an ODE solver that includes the simple Euler method (EM) for ODEs:

$$\mathbf{x}^{m+1} = \mathbf{x}^m + \dot{\mathbf{x}}^m \Delta t = \mathbf{x}^m + \mathbf{f}^m \Delta t \tag{3.52}$$

where $\Delta t = t_{m+1} - t_m$ and the ODE is $\dot{\mathbf{x}} = \mathbf{f}(\mathbf{x}, \mathbf{u}, t)$. For $\Delta t = 1$,

$$\mathbf{x}^{m+1}\big|_{\Delta t=1} = \mathbf{f}^m + \mathbf{x}^m \tag{3.53}$$

Comparing Eq. (3.53) with Eq. (3.51), the only difference is the addition of the term x^m in Eq. (3.53). We therefore get the solution of the difference equation (3.51) by using EM with a step-size $\Delta t = 1$ to solve the ODE:

$$\dot{\mathbf{x}}(t) = \mathbf{g}(\mathbf{x}, \mathbf{u}, t) - \mathbf{x}(t) \tag{3.54}$$

Example 3.26: ***Solving a DE Using Euler's Method***[29]

To solve the difference equation $x^{m+1} = g^m \equiv -x^m + 1$ with IC x^0 using Eq. (3.53) and EM, we need to solve $\dot{x} = -x + 1 - x = -2x + 1$ with IC x^0 and compute $x(t_m) = x^m$ at the discrete instants required for the difference equation solution.

Example 3.27: ***Solving the Logistic DE Using Euler's Method***

The discrete-time logistic population growth model was presented in Example 2.11. It accounts for limited resources and death of organisms – as well as growth – changing over distinct time intervals. It has the (normalized) form: $x_{k+1} = ax_k(1 - x_k)$, with constraints $0 < x < 1$ and $0 < a < 4$. This simple model can exhibit very complex dynamic responses, depending on the value of constant a. Full stability analysis of this DE is involved, and is reserved for Chapter 9, but we use simple simulations here to illustrate several interesting and qualitatively very different solutions to this equation. As above, we solve the DE using the Euler method with step-size 1 for the ODE $\dot{x} = ax(1 - x) - x$.

The dynamic responses to an initial value of $x(0) = 0.01$ (and no other input) is shown for various values of the single parameter a in **Fig. 3.21**, implementing the equivalent ODE $\dot{x} = ax(1 - x) - x$. Plots of x_{k+1} vs. x_k, called the **phase plane** solutions, are also shown, with line-plots outlining the parabolas limited by the built-in plotter resolution. For $a = 0.9$ in the upper left panel, the response is a simple monotonically decreasing function of time. For larger values of a, e.g. $a = 1.2$ in the upper right panel, the response becomes constant after an initial transient. For $a = 3.3$, the dynamic response is sustained, constant range oscillations – termed a *limit cycle* – after an initial transient, as shown in the lower left panel. And, for $a = 3.65$ (any value of $a > 3.57$),

29 I thank my colleague Alan Garfinkel in *Integrative Biology and Physiology* at UCLA for showing me this neat trick.

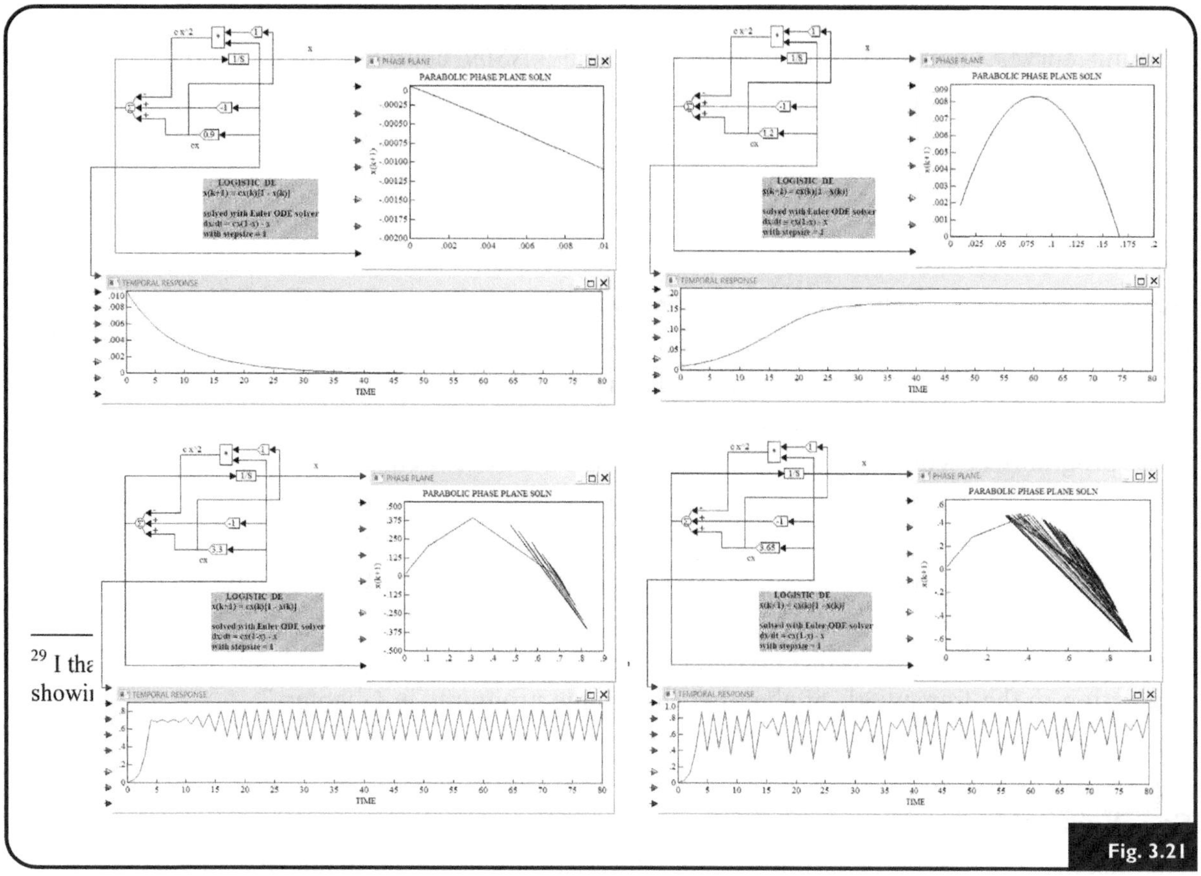

Fig. 3.21

[29] I tha
showin

the response is *chaotic*, as in the lower right panel. Chaos, limit cycles, phase planes and further explanation of these results is reserved for Chapter 9, where it is also shown that a is physically bounded by $0 < a < 4$.

Remark: Besides illustrating the use of ODE solvers for computing solutions to DEs here, the main purpose of this example is to illustrate how simulation can be used to discover surprising dynamical responses from simple nonlinear models. The shark–tuna problem below is another simulation example of a continuous-time (continuum) population dynamics (predator–prey) model with a periodic solution.

Other Simulation Languages & Software Packages

Math-Based Simulation Languages & Software

Comprehensive, general-purpose math programs like *Mathematica* (Research 2008), *Mathcad* (www.ptc.com/products/mathcad/) and *Matlab* (www.mathworks.com/matlabcentral/) incorporate sophisticated ODE simulation capabilities, along with numerous other math modeling tools.[30] They are primarily direct equation-entry based, with *Mathcad* syntax being the most

30 Another is *Advanced Continuous Simulation Language* (ACSL), designed for simulating ODE models directly from equations or block diagrams. *ACCSLXtreme* is a sophisticated, commercial version of ACSL (www.acslsim.com/products).

mathematically natural. All have a GUI for displaying simulation results. They all have steeper learning curves compared with the *Simulink* and *VisSim* control system diagram simulation packages presented in this chapter. The simple linear transcription/translation problem below illustrates use of *Matlab* scripting language to simulate, analyze and do a "what-if" experiment using another simple model of gene regulation of protein production.

Special Purpose Simulation Packages in Biochemistry & Cell Biology

Program packages are available from commercial and non-commercial sources for specifically simulating models in cell and molecular biology and biochemistry. Some notable ones were mentioned in the Introduction of this chapter. With some exceptions, entry of ODE models in these programs is typically done primarily in terms of chemical reactions or reaction network schematics, rather than math equations or block diagrams, with either special graphical or command line syntax. *SBML* is a programming language and syntax for representing models in biology, is used directly in some packages, and is supported in numerous computational modeling and analysis programs with other primary modes of model entry. We begin modeling biochemical reactions and molecular and cell biology interactions in Chapter 7, so we'll revisit simulation in the framework of chemistry-language programs in Chapter 7.

Stochastic Model Simulators

The distinction between deterministic and stochastic models was made first in Chapter 1. The modeling methodology developed since then has been primarily deterministic, with attention to "indeterminacy" only in measurement models associated with dynamic system experiments. Stochastic modeling methods are used to model indeterminacy in the whole of dynamic systems. This methodology can be important in biosystem dynamics studies when the number of objects represented by state variables is *very small*, e.g. for intracellular processes where interacting molecules are no more than double-digit in number. Gene expression, for example, is a natural candidate for stochastic modeling. Genes are activated and inactivated by *random* association and dissociation of *transcription factors* or *repressors* – very few in number – to DNA, with gene transcription often occurring only a few times per cell cycle. Stochastic simulation algorithms, based in probability theory, handle these situations.

Introduction of the methodologic basics for stochastic modeling and simulation of dynamic biomolecular systems is deferred to the latter part of Chapter 7, where we develop the Gillespie algorithm (Gillespie 1976), the most widely used one for stochastic simulation. The theory is based on the same mass action principle used for characterizing the dynamics of much larger numbers of molecules in most of Chapter 6 – using deterministic ODEs. For stochastic modeling, mass action is focused more microscopically on small numbers of molecules – using discrete-time event models. Stochastic simulation algorithms are included in some program packages, such as *Dizzy*,[31] *Copasi*,[32] *SimBiology*[33] and others, and we develop

31 *Dizzy* (www.magnet.systemsbiology.net/software/Dizzy/).

32 *Copasi* (www.copasi.org/).

33 *SimBiology* (www.mathworks.com/products/simbiology).

several examples using software packages, including a stochastic variation of the simple linear transcription/translation problem solved deterministically below.

Two Population Interaction Dynamics Simulation Model Examples

Simpler & Linear Transcription/Translation Dynamics

Adapted from (Thattai and van Oudenaarden 2001).

This simple and linear "birth and death" protein production model captures – in the simplest possible way – the primary dynamical interactions among DNA, RNA and protein, the major players in this basic cell biology system. It also provides an opportunity to introduce some additional cell biology terminology and some quantitative analysis – including a "what-if" experiment – in simpler language. It's also a starting point for including several real system complexities in later chapters, for analyzing their contributions to overall dynamics.

The model begun in Example 2.6 of Chapter 2 and extended in Example 3.6 of this chapter is fully simplified by ignoring the time-delay, and by assuming linear production and elimination of mRNA ($\equiv M$) and protein (P). All other factors (transcription factors, promoters, repressors, etc.) that may affect mRNA or protein dynamics are also neglected. Another important (over)simplifying assumption here is that mRNA regulates production of a single protein. In reality, a single mRNA is capable of regulating production of more than one protein, by **alternative splicing**, a process by which exon fragments of the transcribed RNA – primary gene transcripts called **pre-mRNA** – are reconnected in multiple ways during RNA splicing. The resulting different mRNAs can be translated into different protein isoforms, thus generating multiple proteins.

DNA is included here as a constant source for mRNA during transcription (constitutive gene expression). DNA dynamics evolve over a much longer time-scale *relative* to changes in M and P. The goal here is to simulate simple mRNA regulation of protein production, and analyze and interpret the results biologically. The model ODEs are:

$$\begin{aligned} \dot{M}(t) &= k_1 DNA - k_2 M(t) \\ \dot{P}(t) &= k_3 M(t) - k_4 P(t) \end{aligned} \tag{3.55}$$

The units of the ks are $\min^{-1}$ and the units of DNA, M and P are molecule number (integers). The ICs are [1 0 0], i.e. the process begins with 1 constant DNA and no mRNA or protein molecules, i.e. the gene is transcribed *constitutively* at constant rate $k_1 DNA$, a constant input (step) in the simulation. The steady state solutions of Eq. (3.55) (with $\dot{P} = \dot{M} = 0$ in Eq. (3.55)) are $M_{ss} = k_1 DNA/k_2$, $P_{ss} = k_3 M/k_4$ numbers of M and P. An important molecular biology parameter of this process is the average number of proteins produced per transcript in the life cycle of its messenger mRNA. This is called the **burst factor** $b \equiv k_3/k_2$ $(= k_4 P_{ss}/k_1)$.

The model is coded in the *Matlab* script in **Fig. 3.22**, with b rendered explicit in the code by writing $k_3 = k_2 b$, also recognizable as a *scale factor*. The model solutions are also shown in the figure. The upper graphs illustrate the solution for two different values of b, $b = 2$ and $b = 20$,

```
% SIMPLE TRANSCRIPTION/TRANSLATION MODEL
%          RUN FILE FOR function TVHJ.m
% code adapted from: Hayot &
Jayaprakash(2008)
  % Time in minutes, Vol = 1
  % k1*DNA is mRNA production rate with
D=1(#/t)
  % k3*M is protein production rate (#/t)
  % k2*mRNA degradation rate (#/t)
  % k4*P is Protein degradation rate (#/t)
global b  % pass burst factor b btwn the
files
tspan=[0 667]; % run and plot to  ~10 hrs
yzero=[0;0];       % zero ICs on mRNA and P
b = 2; % set burst factor
[t,y]=ode45(@TVHJ,tspan,yzero);
plot(t,y,'linewidth', 2); hold;
title('PROTEIN NUMBER vs. TIME')
xlabel('MINUTES')
ylabel('PROTEIN NUMBER')
b = 20; % adjust burst factor
[t,y] = ode45(@TVHJ,tspan,yzero);
plot(t,y(:,2),'linewidth', 2, 'color', 'r');
legend('mRNA', 'Protein with b = 2',...
   'Protein with b = 20'); hold;
Pss = ya(length(ta),2) % steady state values
Mss = ya(length(ta),1)          % printed
%.............................................
%          FUNCTION FILE FOR TVhj.m
function dydt = TVHJ(t,y)
global b  % pass variable b btwn the files
k1=0.6; % set mRNA fractional prod rate
(1/MIN).
% half-life of mRNA = 2 MIN, protein = 60 MIN
   k2=log(2)/2; % rate const from half-life
   k4=log(2)/60; % rate const from half-life
% b = BURST FACTOR = avg # proteins produced
in mRNA lifetime b*k2 = k3 in the eqn direct:
dydt = [k1-k2.*y(1); b*k2.*y(1)-k4.*y(2)];
```

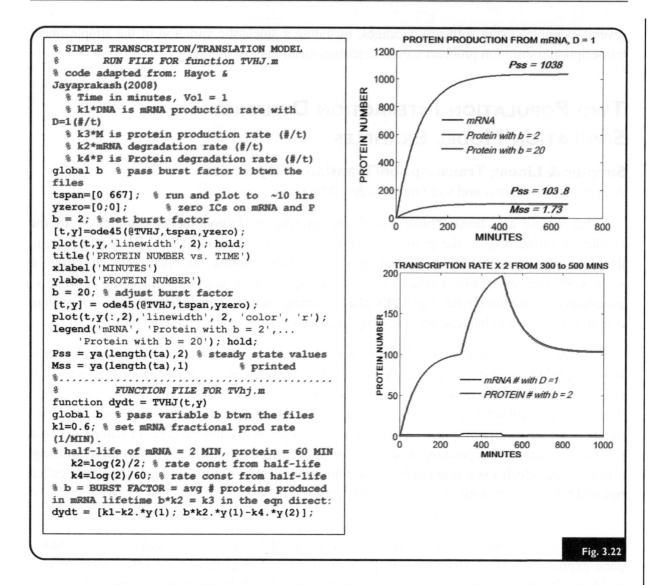

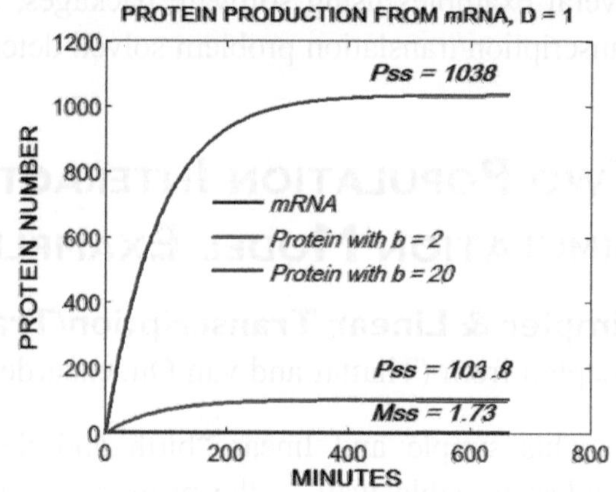

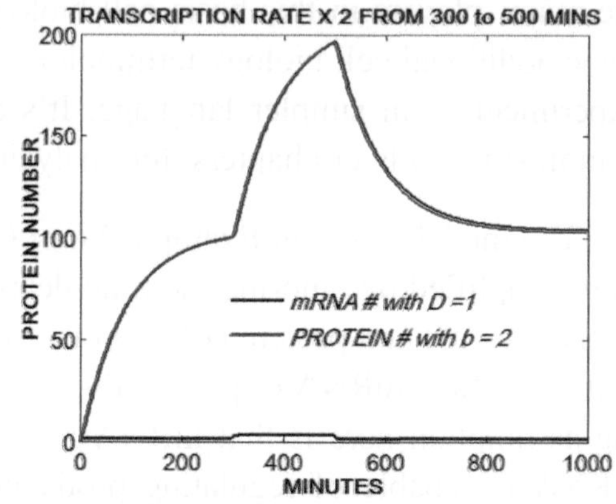

Fig. 3.22

and for $k_1 = 0.6$/min, $k_2 = 0.346$/min ($= \ln 2/2$) and $k_4 = 0.0116$/min ($\ln 2/60$). Parameters k_2 and k_4 are determined from measured *half-lives*[34] of the mRNA (2 min) and protein (60 min).

What-If: In a second simulation result, shown in the lower part of **Fig. 3.22** we do an *in silico experiment* on this model (a growing-more-common synonym for computer-simulated experiments in life sciences). Parameter $b = 2$ is fixed and k_2 and k_4 are the same. Only k_1 is changed here, to approximately simulate an unspecified *regulatory factor* effect on the mRNA *transcription rate* k_1, by artificially adjusting k_1, as if there were an *extrinsic* factor changing its value. This is accomplished in **Fig. 3.22** (lower) by doubling k_1, from 0.6 to 1.2/min. A more complete model would explicitly include this effect using a third ODE, representing the specific effect of the factor (e.g. a transcription factor) altering mRNA transcription on DNA explicitly, as in Exercise E3.8 below.

34 Biologists usually measure disappearance rates in terms of half-lives ($T_{1/2}$) of linearly ($\sim kx$) disappearing substances. A typical experiment is to introduce the substance exogenously into the mix (e.g. cell culture, blood, etc.), measure its concentration as a function of time, and estimate the time at which the exponentially falling curve reaches half its initial value. More on this very common kinetic parameter in Chapter 8.

Biological Results: With $k_1 = 0.6/\text{min}$, the predicted steady state mRNA values are: $M_{ss} = k_1 DNA/k_2 = k_1/k_2 = 1.73$ – the same for both $b = 2$ and $b = 20$ – over a tenfold increase in burst factor b. For the protein, steady state values change by a factor of ten: $P_{ss} = k_3 M/k_4 = 1038$ for $b = 20$ and 103.8 for $b = 2$. This is clear, because protein production is directly proportional to burst factor, and mRNA production is independent of b. These values are accurately demonstrated on the plots, and they are likely large enough to be reasonable estimates of actual mean values. This might justify using this deterministic model to predict average protein levels. However, the average mRNA level of 1.73 molecules is way too small to make predictions about numbers of mRNA molecules transcribed to make this many proteins, putting the entire deterministic formalism used here likely inadequate to the task.

What-If Results, shown in the lower graph of **Fig. 3.22**, illustrate a smooth transition to a doubling of the protein number produced when the transcription rate k_1 is temporarily doubled over the period $300 < t < 500$ minutes, and a smooth transition back when it is returned to its previous rate. This rather crude **parameter forcing function** approach to otherwise regulating transcription (more appropriately) with another state variable in the mix presumably mimics the effects on protein production generated by a transcription factor stimulating the process. This example is continued in stochastic form and with greater mechanistic complexity in Chapter 7.

Simple Predator–Prey Dynamics

This example is based on a simplification of the Lotka–Volterra (L–V) model, classical in ecology (Berryman 1992). It describes the dynamics of two interacting species, a predator species feeding on a prey species. The following assumptions about the behavior of the two populations and their environment are the basis for the model. The prey have an unlimited food supply. Predators eat only the prey. Birth and death rates of both are proportional to their size (number). The generations of predator and prey are continually overlapping. The environment does not favor one species over the other, and there is no genetic adaptation.

Following is a simple L–V model of the time-varying numbers (dynamics) of predator sharks $S(t)$ and tuna prey $T(t)$ in ocean populations. When a shark encounters a tuna close up, it makes a meal of the tuna at rate k_1ST. The product ST is the average probability of an encounter and k_1 is the average probability per unit time the encounter results in a meal. Together they represent the rate at which the shark population *grows*. Disappearance of the shark population by natural causes (death rate) is k_2S. We assume the tuna have no other predators and their net growth (birth–death rates) is proportional to their net population number, k_3S, and (of course) they also disappear in proportion to the product of their encounters with predator sharks, k_4ST, not necessarily equal to growth rate of the sharks. The ODEs for this model are:

$$\begin{aligned} \dot{S} &= k_1 ST - k_2 S \\ \dot{T} &= k_3 T - k_4 ST \end{aligned} \tag{3.56}$$

Fig. 3.23, produced using *VisSim*, illustrates cycling behavior of the shark and tuna populations, after a brief transient response from the ICs. The shark population increases until the supply

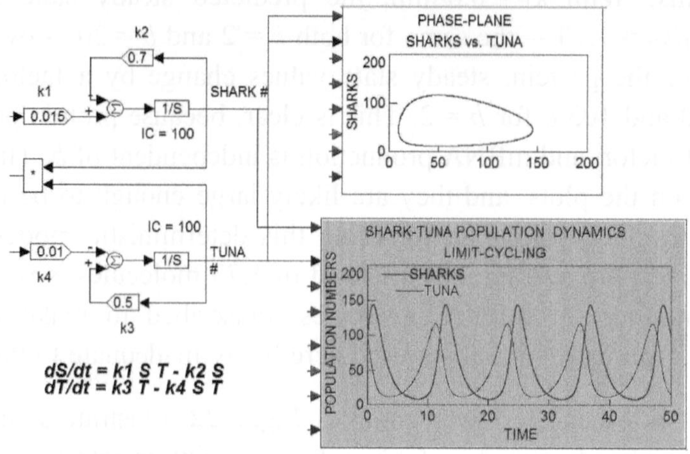

Fig. 3.23

of tuna falls enough to make that meal inadequate to sustain the increase. The tuna population rises and falls as well, falling when the sharks have consumed enough to stem the tuna number increase, and so on. This process would increase *ad infinitum* if nothing in the mix changed.

Parameter values for the run shown are (written as column *vector*) $\boldsymbol{k} = [0.015\ 0.7\ 0.5\ 0.01]^T$, and initial conditions are: $S(0) = T(0) = 100$. The ODE solver used for this run was Adaptive Bulirsch–Stoer.[35] Runge–Kutta fourth-order (RK4) worked just as well. The second (upper) plot, of $S(t)$ vs. $T(t)$, is called a *phase plane* plot, discussed in detail in Chapter 9. Phase plane plots of oscillating systems are always closed curves. Can you see why?

Taking Stock & Looking Ahead

The primary purpose of this chapter was to develop the conceptual and mathematical basics of simulation methodology – hopefully enough to intelligently use these powerful tools for implementing and solving any linear or nonlinear ODE, given the model parameters, initial conditions and inputs. Classical biosystem examples were used to illustrate the simulation process, along with some of its cautionary limitations, caveats and subtleties, and also to further exemplify math modeling applications in the biosciences.

We introduced several subsidiary modeling tools and methods along the way, for augmenting and facilitating the pedagogy and simulation experience. Graphical, control system block diagram representations of ODEs were introduced, for programming them and – at the same time – "visualizing" their structure and dynamical interactions. Normalization, magnitude- and time-scaling of ODE variables were introduced to facilitate model implementation and analysis and introduce *multiscale modeling*. And we detailed and exemplified ways for implementing common inputs, time-delays and other mathematical operations typically encountered in the biomodeling and simulation process.

35 I use this as default, because it works for a wide variety of models.

The next several chapters are specifically about the process of *building* and *structuring* model topologies from biophysical, biochemical and other physical principles. The biomodeling process introduced graphically in **Fig. 1.9** of Chapter 1 will be morphed accordingly for these chapters, highlighting the structuring process. Simulation methodology will help with the building as well as with model analysis and, later, it will play an important subsidiary role in quantifying models from data, validating it and ultimately using it for satisfying modeling goals. Thus, **Fig. 1.9** is illustrated here, *in toto*, with all of its several components, because simulation is a major tool in all aspects of the modeling enterprise.

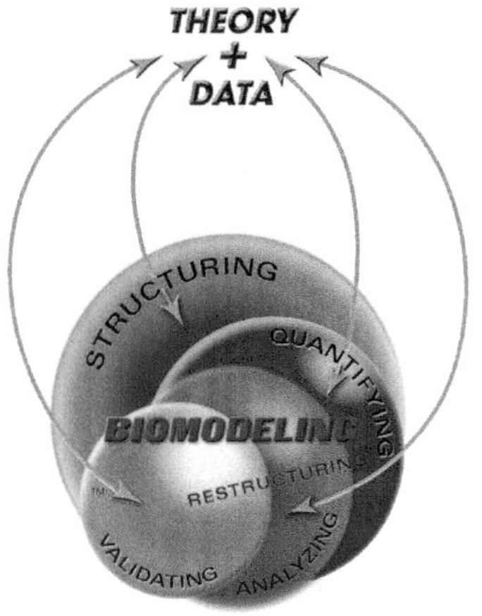

EXERCISES

E3.1 ***Feedback Control System Simulation of Simple Linear Constituent Production and Elimination Processes***. Consider the model in **Fig. 3.24**. It's the simplest model of constituent (basal) production and elimination of a biomolecular species, to be encountered in subsequent chapters. The equation is $\dot{y} = u - ky$, where u is the production rate and ky is the elimination rate. Using any modeling software, e.g. *Matlab, Simulink, VisSim, Mathematica,* etc., simulate this model for $k = 1$ and then $k = 5$ over the interval $0 \leq t \leq 2$, for zero initial conditions, with the following inputs:

a) A unit-impulse function (Dirac delta) input.

b) A unit-step function input.

c) A finite pulse input of unit magnitude and unit width ($\Delta t = 1$).
Plot the input and output on the same axes as the input, using program functions, for each part and each k value. How does the value of k affect the responses?

d) Repeat part (c) after changing the model by fixing $k = 1$ and changing the gain G of the forward loop path from $G = 1$ (implied in the figure) to $G = 2$ and then 5. How do these changes affect the response?

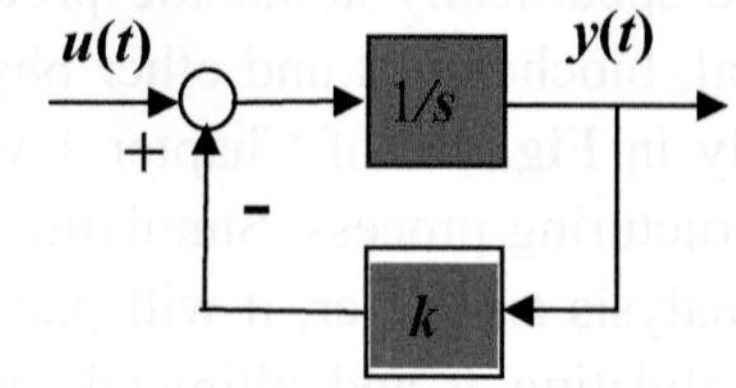

Fig. 3.24

E3.2 ***Coding a Solver.*** In this exercise, use any programming language familiar to you (e.g. *Matlab* script, C++, etc.). The idea here is to first get a deeper understanding of what these solvers do, step-by-step, what they need to do it, how the program handles "input and output" commands – and then see how errors develop as a function of the step-size and method. You also might want to compare your recursive (sequential) solutions with the easier to program block diagram solutions generated by *Simulink* or *VisSim*, which are equally computationally recursive in the background, but appear to be computing "in parallel."

The rate of emission of radioactivity of a radioactively labeled tracer substance – measured in microcuries (μCi) – is proportional to the amount y remaining, i.e. $\dot{y} = -ky$ Suppose $k = 0.01$ (t^{-1} units), and $y(0) = 100$ μCi. The exact solution is $y = 100e^{-t}$ and $y(100) = 36.788$ μCi.

a) Draw two separate flow charts (programs) for numerical solution of this equation using Euler and fourth-order Runge–Kutta formulas. Specify all variables that need to be specified by the user, e.g. Δt, solution interval [0,T].

b) Code and run your program with $\Delta t = 1$, 10, 25, 50, and 100, using Runge–Kutta fourth-order and Euler methods.

c) Compare the computational errors for each method and each Δt by computing the simulated minus exact solutions for each.

E3.3 ***Normalization.***

a) Normalize the state variables in the following two ODEs *to their nonzero ICs* and draw a simulation block diagram for solving them over the time interval a to b, with a and $b > 0$. Parameters k_{ij} and A and $C > 0$

$$\dot{q}_1 = -\left(k_{21} + \frac{C}{A + q_1}\right)q_1 + k_{12}q_2 + u,$$

$$\dot{q}_2 = k_{21}q_1 + k_{22}q_2$$

$$q_1(0) \equiv q_{10}, \;\; q_2(0) \equiv q_{20}, \;\; y(t) = q_1(t)$$

b) Normalize the *time variable* so the computation proceeds from normalized time $\tau = 0$ to normalized time $\tau = 1$. How does this change simulation results?

E3.4 ***Stiffness I.*** Solve the model $\dot{x} + x + \sin 1000t = 0$ (for any ICs) using different stiff and nonstiff ODE solvers. Compare the results. What method works best, worst? Why?

E3.5 ***Stiffness II.*** Show analytically that $x(t) = (1000/999)e^{-t} - (1/999)e^{-1000t}$ is the unique solution to Eq. (3.50) for ICs $x(0) = 1$ and $\dot{x}(0) = 0$.

E3.6 ***Math Programming Language Tools.*** Repeat Example 3.6 using *Matlab* or other math programming language tools for solution of delay differential equations (DDEs). See the tutorial at: www.mathworks.com/dde_tutorial.

E3.7 A model of the blood-pressure controlled baroreflex-feedback mechanism, with time-delay in the feedback loop, is given in a research article by J. T. Ottesen (1997). This relatively simple model has been coded as *Matlab* demo example *ddex2* and can be found at the *Matlab* website. This demo contains the complete code for this example and can be found using the *Matlab search* function www.mathworks.com/help/matlab/ref/ddesd.html.

As an exercise in simulation and observing complex effects of time-delays in the loop, download and run the code (too long to include here), first as is. To download the code in an editor, click the example name, or type *edit ddex2* at the command line. To run the example type *ddex2* at the command line.

Vary the time-delay parameter tau (default value 4) and plot the results for tau values 0.1, 1, 4, 5, 8, 30, 40 and 50. Note the very different qualitative behavior, in particular minimal overshoot and damped oscillations for small delays, wild oscillations from 8 until about 30 and clustered oscillatory bursts between 40 and 50.

E3.8 Extend the simple transcription/translation model in the last section of this chapter to one positively regulated by a *transcription factor F*, similar to but simpler than that in Example 3.6. F is bound reversibly to the TF-RE site on the DNA site where mRNA is transcribed. The site thereby becomes an activated site, with enhanced but saturable transcription described by a nonlinear term $k_1^* DNA^*(F(t)) = k_1^* F(t)/(K + F(t))$ replacing D in (3.56), i.e. $\dot{M}(t) = k_1^* DNA^*(F(t)) - k_2 M(t)$, with $k_1^* = 2k_1$. The protein equation and other k parameters remain the same, and the dynamics of $F(t)$ are described by $\dot{F} \cong 0$, with $F(0) = 1$ and $K = 1$. Simulate and plot the evolution of all state variables over 10 hours. What can you say about these results compared with those of the simpler, no-F model of the last section?

E3.9 Program the *VisSim* (or *Simulink* version) of the Shark–Tuna simulator. Run it with ICs $= [1\ 0]^T$ and also $[1\ 1]^T$. With one shark and no tuna, the model predicts a falloff to zero for the sharks. Why? For one shark and one tuna, the shark apparently eats a tuna, but the tuna population nevertheless survives to cycle many times over, with a zero nadir occurring with each cycle. Is this possible? Why? If not, fix the model to make it "right" for these ICs.

References

Asaithambi, N.S., 1995. Numerical Analysis. Saunders College Publishing, Fort Worth, TX.

Ascher, U.M., Petzold, L.R., 1998. Computer Methods for Ordinary Differential Equations and Differential-Algebraic Equations. SIAM, Philadelphia.

Berryman, A.A., 1992. The origins and evolution of predator–prey theory. Ecology. 73, 1530–1535.

Buckingham, E., 1914. On physically similar systems, illustrations of the use of dimensional equations. Physical Review. 4 (4), 345–376.

Bulirsch, R., Stoer, J., 1966. Numerical treatment of ordinary differential equations by extrapolation methods. Numerishe Mathematic. 8 (1), 1–13.

Burden, R., Faires, J., 2005. Numerical Analysis. Thomson Brooks/Cole, Belmont, CA.

Butcher, J.C., 1987. The Numerical Analysis of Ordinary Differential Equations: Runge–Kutta and General Linear Methods. Wiley Interscience, New York.

Calahan, D.A., 1968. A stable, accurate method of numerical integration for nonlinear systems. Proceedings of the IEEE. 56 (4), 744.

Forsythe, G.E., Malcolm, M.A., Moler, C.B., 1977. Computer Methods for Mathematical Computations (Chapter 6). Prentice-Hall, Englewood Cliffs, NJ.

Frank, W.L., 1960. Solution of linear systems by Richardson's method. J. ACM. 7 (3), 274–286.

Gillespie, D.T., 1976. A general method for numerically simulating the stochastic time evolution of coupled chemical reactions. Journal of Computational Physics. 22, 403–434.

Hosea, M.E., Shampine, L.F., 1996. Analysis and Implementation of TR-BDF2. Applied Numerical Mathematics. 20, 21–37.

Kline, M., 1998. Calculus: An Intuitive and Physical Approach. Dover, London.

Kollár, I., Widrow, B., 2008. Quantization Noise: Roundoff Error in Digital Computation, Signal Processing, Control, and Communications. Cambridge University Press, Cambridge.

Ottesen, J.T., 1997. Modelling of the baroreflex-feedback mechanism with time-delay. Journal of Mathematical Biology. 36, 41–63.

Press, W.H.F., B. P., Teukolsky, S.A., Vetterling, W.T., 1992. Multistep, multivalue, and predictor-corrector methods. Numerical Recipes in FORTRAN: The Art of Scientific Computing. Cambridge University Press, Cambridge, 740–744.

Research, W., 2008. Mathematica, Version 7.0. Wolfram Research, Inc., Champaign, IL.

Segel, L.A., Slemrod, M., 1989. The quasi-steady-state assumption: a case study in perturbation. SIAM Review. 31 (3), 446–477.

Shampine, L.F., Gladwell, I., Thompson, T., 2003. Solving ODEs in MATLAB, Applied Numerical Mathematics. Cambridge University Press, Cambridge.

Smolen, P., Baxter, D.A., et al., 1998. Frequency selectivity, multistability, and oscillations emerge from models of genetic regulatory systems. American Journal of Physiology–Cell Physiology. 274 (2), C531–C542.

Stewart, I., 1992. Warning - handle with care!. Nature. 335, 303–304.

Tee, D., DiStefano III, JJ, 2004. Simulation of tumor-induced angiogenesis and its response to anti-angiogenic drug treatment: mode of drug delivery and clearance rate dependencies. Cancer Research and Clinical Oncology. 130 (1), 15–24.

Thattai, M., van Oudenaarden, A., 2001. Intrinsic noise in gene regulatory networks. Proceedings of the National Academy of Sciences. 98 (15), 8614–8619.

Structural Biomodeling from Theory & Data: Compartmentalizations

Chapter 4

Toe bone connected foot bone
Foot bone connected leg bone
Leg bone connected knee bone
Don't you hear the word of the Lord?

Leg bone connected knee bone
Knee bone connected thighbone
Thighbone connected hipbone
Don't you hear the word of the Lord?

Hipbone connected backbone
Backbone connected shoulder bone
Shoulder bone connected neck bone
Don't you hear the word of the Lord?

Dem bones, dem bones, dem dry bones
Dem bones, dem bones, dem dry bones
Dem bones, dem bones, dem dry bones
Don't you hear the word of the Lord?

Dry Bones *(Original version)*
http://www.negrospirituals.com/news-song/dry_bones.htm

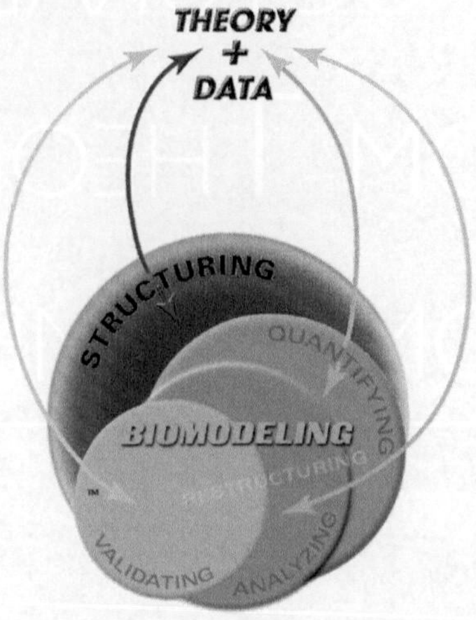

INTRODUCTION

Conscious of it or not, we are all likely compartmentalizing things while participating in the activities of daily life. We organize (compartmentalize) our clothing in closets and drawers, we organize food and provisions in our (compartmentalized) cabinets, storage areas and refrigerators. We are being guided by "compartmental models" developed in part by those who designed the architectures of the storage areas, and in part by ourselves, in deciding what goes where. Countries are compartmentalized into states, states into counties, and so on. Indeed, virtually everything around us (and in us!) is compartmentalized. There seems to be something innate in the process of compartmentalization, that "compartmental modeling" of one sort or another is deeply ingrained in people and organizations at every level.

The notion of compartmentalization is fundamental to structural modeling of biological systems. Its utility as a simple organizational tool is immediately apparent. The popular notion of a compartment as a bounded space (container) for locating well-defined biological entities (molecules, chemical "species," etc.) has long been utilized for conveniently conceptualizing the organization of biological systems. In this chapter and the next we develop compartmentalization in a deeper way, leveraging the comprehensive literature on multicompartmental modeling in life and other physical sciences, and algebraic and geometric innovations that serve to connect models with structural and transient dynamical and steady state kinetic data. For the most part, we maintain a completely physical interpretation of the concept and its mathematics – based on the dynamical characteristics of homogeneous entitities. We begin by showing how to build such biomodels from structural (system topology) and other mechanistic data. In the process, we show how structural and kinetic data complement each other, like an antigen and an antibody or a receptor and a ligand, but somewhat more abstractly. We address initiation of model construction from kinetic time-series data in the next chapter.

Development of the mathematics of multicompartmental methods then provides a solid, biophysically motivated and well-developed methodology for modeling a variety of linear and nonlinear biosystems and biosystem experiments. Multicompartmental theory is based on deriving the dynamics of mass-based state variables from conservation principles – and also the dynamics of concentration-based and rate-based variables – with a bit more effort. Mathematically, homogeneity of dynamical compartments means that ordinary differential equations (ODEs) rather than partial differential equations (PDEs) define the model and its rate components. This transcends the more macroscopic compartment-as-container approach. It reaches conceptually and physically into deeper compartments – as deep as the data permits – to represent different entities with *common dynamical characteristics*, as a compartment $\equiv$ state variable isomorphism. This dynamical definition facilitates analysis of biodata and their "dynamical signatures" – a feature of the next chapter – at the same time it provides the *basic compartment-unit* for formulation of *candidate* multicompartmental model structures and analyzing their responses to input stimuli. Input and output measurement (data) models are also included in the developments, as are data-driven models of indirect and time-delayed inputs. Nonlinear compartmental disease dynamics models are included as a special class in a separate section.

Inhomogeneous entities and spaces are also accommodated, without the need for partial differential equations, in a separate section on *pools* and *pool models*. The math models in this section are algebraic and limited to systems in a constant steady state. Nevertheless, this is a useful paradigm for steady state analysis of biosystems and experiments, e.g. for flux balance analysis (FBA) in metabolism studies.

Much of the material in this chapter may be new (and hopefully enlightening) to readers following developments of software tools in biochemistry and molecular systems biology over the past decade or two. In particular, the lexicon of many of these specialized tools employ a definition of compartment different than the dynamical one[1]. Fortunately, the distinctions and ambiguities are not a barrier to using old or new tools and we make every effort here to clarify them and help with making choices.

Compartmentalization: A First-Level Formalism for Structural Biomodeling

The goal here is to present some pertinent history, biological context and the basic elements of formulating compartmental models – with the aid of several tutorial examples. Formal definitions are deferred until the next section, to temporarily circumvent the different lexicons noted above. They are easier to clarify and distinguish following presentation of these several examples of compartmentalization and structural aspects of multicompartmental modeling. We do this first with minimal math, and note that the structural models can be linear or nonlinear. Math models are developed in the sections that follow, mostly in the form of ODE model structures and all (except *pool models*) with homogeneous compartments.

Basic Multicompartmental (MC) Model Formulation from Structural & Kinetic Data

Multicompartmental (abbrev: *compartmental* or MC) *models* were invented for qualitative and quantitative analysis of physical systems and data, based on the familiar idea that "compartmentalizing" variables into (approximately) "homogeneously distinct" entities in complex systems can make them easier to understand, explain and analyze. For dynamic biosystems, both linear and nonlinear, these variables represent *kinetically* (or dynamically) distinct entities in these homogeneously distributed spaces, i.e. they are *state variables* – dependent variable functions of the single independent variable time t, e.g. $q_1(t)$, $q_2(t)$, ... (Chapter 2).

Compartmental modeling theory has its origins in applied physics, with models representing exchange of matter or mass among quantized physical spaces. In MC models, compartments interact by means of *material exchange and flow* governed by conservation laws. Material implies physical matter, because the primary applications involve exchange (or interconversion) of mass. Examples include exchange of energy, nutrients, money, electrons, insects, and people – almost anything exchangeable via flows from one approximately homogeneous entity to another. More generally, material is interpreted more abstractly, with compartments exchanging "information" via intercompartmental "signaling." In cell-signaling systems for example, some molecular

1 The alternate definition – compartment as "a bounded space in which *species* are located" (i.e. a container of state variables).

entities can exist in an *active* form – where they transmit particular information – and an *inactive* form – where they don't (Chapters 6 and 7). These forms comprise two compartments of the biosystem model, because their dynamics are different. Schematically, we could express this as *Inactive X→Active X.* And if the active form can revert back to inactivity, we could put a head on both ends of the arrow: *Inactive X↔Active X.* We'll do this more formally below, so we can express the components of this biological process more precisely, and thereby analyze or otherwise use the model more systematically in generating useful results.

Structural & Kinetic Data Complement Each Other

MC models are, first of all, structured from mechanistic and/or other topological information about real systems *S* (structural data). These structures are validated and/or quantified from input-output experiments (kinetic time-series data) done on *S*. We might know a great deal about system structure (e.g. an established or hypothesized protein network or metabolic pathway), and our first-cut model topology might then be quite complex. In this situation we'll likely have to depend more on available kinetic input–output data to guide simplifications. As we shall see, kinetic data helps in establishing the complexity of the simplest model representing the data, which we call the **minimal model**. On the other hand, we might know very little about the structure, so we'll have to depend more on the kinetic input-output data to help configure the model structure – major topics of the next chapter. In either case, model state variables and parameters, wherever feasible, should have direct analogies in the biological system and thereby be *located similarly* in the MC (or other structured) model. We discuss here how to do this, facilitated by examples that illustrate how to translate the biology into structured models/graphs, and a bit about how kinetic data are used in the process.

The first steps in formulating compartmental models are deceptively simple. We begin by drawing a diagram corresponding to what we see or know about a real-world system or problem. This is followed by additions, subtractions and other adjustments of the diagram to render it maximally realistic and consistent with available data as the modeling process evolves. Model structuring in this fashion takes maximum advantage of what we see and what we already know about a system, and this goes far towards achieving a successful model. For example, if the output of the block diagram model of **Fig. 2.5** of Chapter 2 has a sinusoidal or other periodic waveform, and the input to that block is not periodic, then we anticipate the existence of components within the block capable of generating that sinusoidal output, i.e. there's an *oscillator* in the block. We'll express this more formally with diagrams and mathematics in due course. On to the process first.

Input and Output Compartments First

To begin with, let's assume we're modeling the biodynamics of hypothetical substance *X*. A model of *X* dynamics can be formulated without inputs and outputs, but it's more natural – and ultimately more effective – to start structuring the model by delineating the compartments where experimental inputs are introduced and from where output variables are measured. These are designated the input and output **ports**.[2] Importantly, the combination of input

2 This is the common practice in engineering but not in much of biochemical and cell biology modeling.

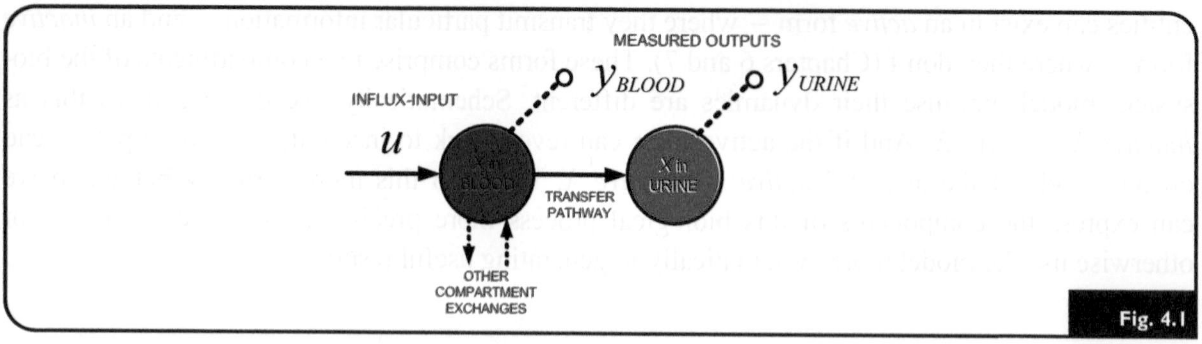

Fig. 4.1

and output *signals* at these ports of entry and observation together reflect fundamental system characteristics – essential dynamical features – that help define the physical boundaries and dimensionality of the model (as we see below[3]). Therefore, model compartments should first be assigned within the probed and/or directly measured physical spaces. Simple examples include X in blood and X in urine in a whole-body physiological system model, or X in cell medium in a cell culture model, or X in whole-cell homogenate in an *in vitro* kinetic study, etc.[4] Blood and urinary spaces are typical for introducing material as *input* (into blood) and measuring it as *output* (in blood and/or urine) in *in vivo* whole organism studies, which introduces our first example of compartmental modeling.

Example 4.1: ***Blood/Urine Measurement MC Submodel***

Blood and urine are singled out as candidate physical spaces for compartments in this model because X is measurable in both blood and urine (**Fig. 4.1**). We assume input (influx) of X is into blood only,[5] from where it eventually finds its way unidirectionally into urine – depicted in the figure by the unidirectional *transfer pathway* arrow. If X is homogenously distributed in each, X in blood and X in urine are each valid compartments for depicting substance X dynamics in this MC model, shown as circles in the figure. The blood compartment for X in this context usually means the entire vascular pool of blood and, if mixing of X in blood is fast relative to transfer and exchange dynamics with other compartments, the assumption of homogeneity of X in blood is usually OK.

X in urine is represented as a **dead-end compartment** (also called a **trap**), i.e. one with no **leak** arrow directed from it, because – in live animal studies – X is typically measured in urine collected into a container as it accumulates. We assume X is distributed homogenously in collected urine. Possible exchanges of substance X between other compartments and blood are represented by the dashed arrows entering and leaving the blood X compartment in **Fig. 4.1**. Blood and urine samples then can be

3 We begin exploring this data-driven component of the model below and then more so in Chapter 7. We also shall see in Chapter 10 that inclusion of inputs and outputs is essential for exploring what's quantifiable in the model (called *identifiability analysis*).

4 Note that the physical spaces were X resides are *not* the compartments. Other materials, e.g., Y, Z, etc., also can reside in the same spaces.

5 Many *in vivo* experiments entail input entry via direct intravenous injection. Inputs entering blood indirectly, e.g. via absorption from the gut from an oral dose input, or via subcutaneous (SC), intramuscular (IM) or intraperitoneal (IP) inputs, are developed in a separate section below, and modeled as in **Fig. 4.28**. Oral inputs – as with the other indirect inputs – require a gut submodel at the input end, as in Exercise E4.13.

assayed (measured via the dashed lines with circles) for output $y(t)$ concentrations or amounts of X – designated generally here as $y_{BLOOD}(t)$ and $y_{URINE}(t)$ – over some (implied) time interval $[0,t_{END}]$, or at discrete times $t_0, t_1, \ldots, t_{END}$.

For simplicity, time t arguments are omitted in the figure output symbols. And the specifics of additional compartments and exchanges (dashed arrows) are also not shown, to emphasize in this example that modeling begins with the compartments measured.

The temporal dynamical characteristics (kinetic patterns) revealed in the collected blood and urinary data will contain "dynamical signatures" for characterizing, validating and further developing the structure and (later) quantifying this system-experiment model. We develop this along with the mathematical theory in this and the next chapter. But, without any additional modeling details (e.g. ODEs), we can think of a hypothetical experiment, do a little algebra and likely glean some practical quantitative results about model parameters from the structural modeling so far. For example, suppose the input into blood is a mass dose $u(t) = Q_0\delta(t)$ of X given as an approximate impulse (brief pulse). Assuming instantaneous mixing in the blood compartment, the concentration (*mass/volume*) at time $t = 0^+$ measured as $y_{BLOOD}(0^+)$ right after dose Q_0 is administered into blood is Q_0/V_{BLOOD}. Therefore, if this first measurement $\hat{y}_{BLOOD}(0^+)$is taken sufficiently close to $t = 0$, which means before $y_{BLOOD}(t)$ can change very much, blood volume can be estimated as: $V_{BLOOD} \cong Q_0/\hat{y}_{BLOOD}(0^+)$.

Example 4.2: ***Two-Compartment Metabolic Interconversion Submodel***

Fig. 4.2 is structured the same as **Fig. 4.1** and represents a common intracellular metabolic subsystem process in which substance X is converted irreversibly into another substance in the same cell. In this example, hormone thyroxine (T_4) is metabolized into the more active thyroid hormone metabolic product (metabolite), triidothyronine (T_3) (DiStefano III and Feng 1988).[6] In this case, unlike the **Fig. 4.1** model, *both compartments are in the same biospace* – the cell.[7] For these two metabolic processes, the

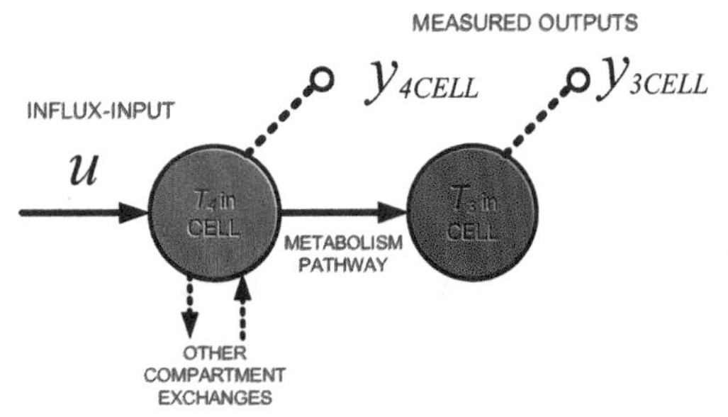

Fig. 4.2

6 Another example is metabolic conversion of glucose into glucose-6-phosphate.

7 Enzymes are involved in the conversion process, but these are neither shown nor important in representing the approximate and dominant dynamics shown, as clarified in Chapters 6 and 7.

input $u(t)$ is T_4 influx into the cell from T_4 in blood (blood T_4 source not shown); and T_{4CELL} concentration ($y_{T4CELL}(t)$) is measured along with metabolic product T_{3CELL} concentration ($y_{3CELL}(t)$) in that same cell (dashed lines with circles).

This example illustrates that different molecules in the same biospace can have the same dynamical structure as a single molecular entity measured in two different biospaces (Example 4.1), discounting any dynamics contributed by possible compartments connected to the dashed arrows and not shown in either figure.

More Compartments in More Organs

Let's extend the ideas in the previous two examples to a more complex biosystem, one with additional physiological processes structurally modeled with homogeneous compartments.

In a biochemical or physiological system, the knowledge base typically designates X and its precursors or metabolites within organs, cells, organelles, etc. Dynamically, these are sites of production, distribution, action (regulation, control), or elimination processes. We might begin by modeling these by at least one compartment each. Related biological/biochemical system information specifies multicompartmental (network) interconnectivity, which we represent by directed arrows among compartments, and "leaks" directed from them, all as illustrated in the next example. This is a good starting point for developing the model further.

Example 4.3: ***Multiorgan Compartment Model with Exchange, Metabolism and Excretion***

In the physiological model illustrated in **Fig. 4.3**, X_1 first enters blood via (influx) input u_1 and exchanges reversibly with X_1 in liver. Some X_1 is metabolized irreversibly in liver, to metabolite X_2, and some is exchanged back with X_1 in blood. X_2 leaves the liver only via the (special) biliary pathway directly into gut, from where it is excreted in feces. Accumulation of X_2 in feces is modeled with a dead-end compartment (no leak, only accumulation), similar to the urinary leak in Example 4.1. Exchanges of X_1 are possible

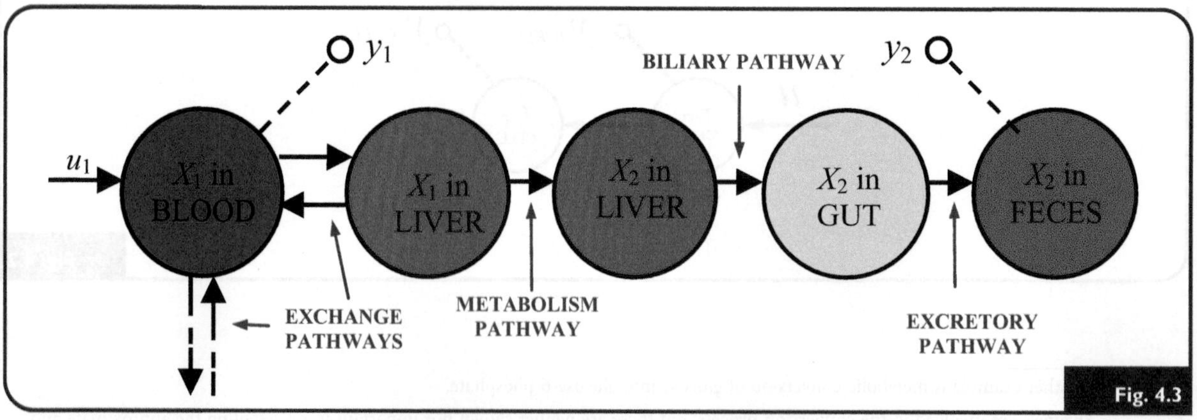

Fig. 4.3

MC model with metabolism of X_1 in liver, etc.

between blood and other biosystem compartments, e.g. X_1 in other organs, as shown by the dashed arrows.[8]

Remark: In this hypothetical but physiologically plausible example, it is noteworthy that two biomolecules are represented as five compartments interchanging and metabolizing via three different organs, blood, liver and gut, and then one of the molecules (X_2) exits into feces.

The specific interconnectivity of the multicompartmental diagram was structured from this basic physio-anatomic and biochemical knowledge. However, to be useful as a *dynamic system model* of the processes depicted, it must satisfy the fundamental assumption of compartmental modeling methodology, i.e. that all compartments are distinctly (approximately) homogeneous (well-mixed). This is what renders each compartment dynamically distinct and thereby associatable with a single state variable (when we do the math). *Kinetic output data* can be used to test this assumption, thereby validating the structure in as far as that data permit. For this model, kinetic data are modeled as outputs $y_1(t)$ and $y_2(t)$ collected from the probed compartments in **Fig. 4.3** over some time interval $[0,t_{END}]$. More on the validation phase later.

Handling Structural & Kinetic Data Mismatches: Caveat #1

The next example is a bit more complicated. Its a simple illustration of how theory and data sometimes appear to clash – a common modeling issue that can readily cloud or otherwise complicate the modeling process. It also justifies the presence of so-called *central compartments* in some kinetic models, e.g. in modeling drug dynamics. We simply introduce the problem here – descriptively, with words and graphs – to further emphasize how theory and data work hand-in-hand in the modeling process. We follow up with more details and generalizations in the next chapter.

Context: Suppose the structure of a multicompartment model is formulated from biochemical, physiological or other biological data about the biosystem structure, as illustrated in the example above. A model structure formulated in this fashion is called a **candidate model structure**. How do we validate that this candidate is a "correct"[9] one?

The other kind of data, *kinetic output data (output signal data)*, hold the key to distinguishing or validating the separation of compartments in such a model. This means the kinetic data has to be sufficiently revealing in this regard, otherwise compartments in the candidate model structure can lose their distinction from each other. Since kinetic data usually take the form of a sequence or continuum of output measurements over time, this means the data have to be collected over time periods that capture these distinctions. A common problem is that data collection is not always begun soon enough to capture fast-changing early transients. We illustrate this explicitly – using simulations – and the structure established in **Fig. 4.3** and show how to partially "model our way out" of the problem. The first output $y_1(t)$, from blood in **Fig. 4.3** is particularly susceptible, because liver blood flow, the biophysical "force" for X_1 exchange between blood and liver, is typically relatively rapid relative to other material fluxes in the

8 Exercise E4.20 extends this model with additional enterohepatic (gut–liver interchange) pathways for X_1, in a way that further illustrates the meaning of dynamical compartments and the organs that contain them.

9 "Correct" here means that the structure is a good approximation of the biosystem, consistent with modeling goals and available data.

system, i.e. metabolism and excretion in **Fig. 4.3** We discuss the physiology and necessary assumptions here. The math and additional data-model interpretation issues come later.[10]

Example 4.4: ***X_1 in Blood & X_1 in Liver as One Lumped "Central" Compartment***

Consider substance X_1 in blood, call it X_{1B}, exchanging reversibly with X_1 in liver (call it X_{1L}). X_1 is measured in blood only, as $y_1(t) = y_{1B}(t)$. In liver, X_{1L} is partially metabolized irreversibly (arrow directed outward), with some X_{1L} but no metabolite returning to blood, via the exchange pathway. We normally would model the structure of this system with two distinct compartments, one each for X_1 in blood and liver, as shown in the left-hand side (LHS) of **Fig. 4.4**. Note that this is identical to the first two compartments of **Fig. 4.3**.

In practice, however, if measurements begin only after a significant degree of exchange of X_1 has occurred with liver, because exchange and therefore mixing is very rapid, it can be difficult to distinguish these two compartments in the data. The output $y_1(t)$ would likely appear to be from a single "lumped" or "composite" compartment, with the relatively rapid exchange of X_1 between blood and liver obscured by delayed sampling of X_1 in blood (see **Fig. 4.5**). The "dynamic signature" in the measurement would not be readily interpretable as two compartments, physiologically, because of the rapid mix of blood X_1 and liver X_1 compartments in the measurement. Nevertheless, some simple analysis can help with this ambiguity, which we continue more fully and generally in the next chapter. We're developing an appreciation for reality here, in the simplest possible way.

Let's first assume blood-liver compartment mixing is rapid enough to consider the pair as approximately homogeneous. Then, to maintain model fidelity with the data, we can lump the two compartments into a single blood–liver "compartment" with $X \equiv X_{1B} + X_{1L}$, chosen as a new, *aggregated* state variable, all as illustrated in the right-hand side (RHS) of **Fig. 4.4**. In this case, extrapolating the response data back

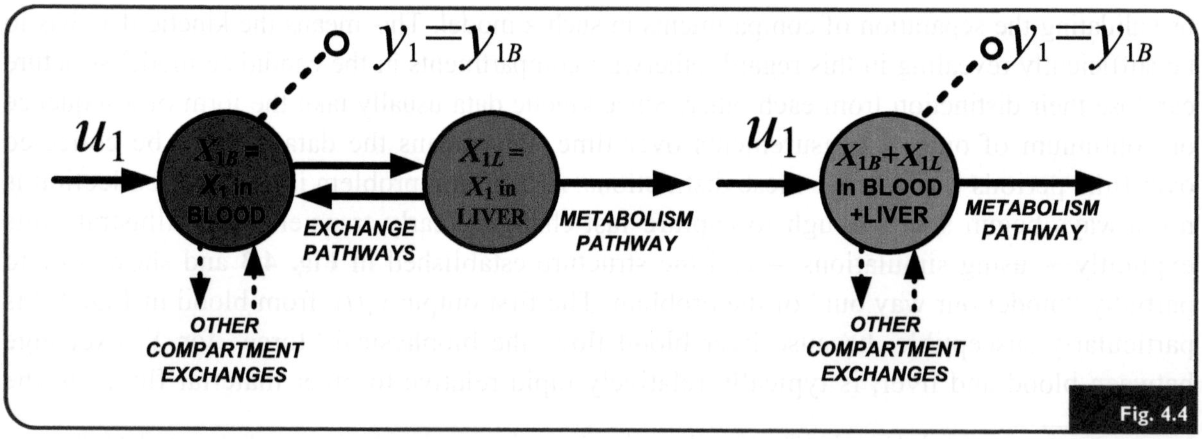

Fig. 4.4

10 Does this suggest *stiff* ODEs (Chapter 3)?

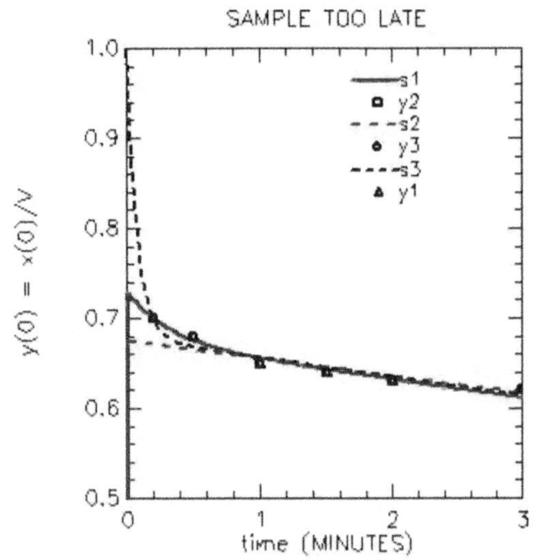

Fig. 4.5

SAAMII simulation responses of the blood–liver two-compartment model of **Fig. 4.4**. The dashed blue line (s3) is the ideal simulated response in blood to a unit impulse input, with blood volume $V_B = 1$. The solid green line (s1) is the model fitted to all the data points shown, the earliest one at $t = 0.2$ min. It crosses the y-axis at $0.725 = y_{composite}(0) = 1/V_{componsite} = 1/1.38$. The dashed red line (s2) was fitted to the data beginning at $t = 1$ min. It crosses the y-axis at $0.675 = y_{BLOOD+LIVER}(0) = 1/V_{BLOOD+LIVER} = 1/1.48$. Note that $1 = V_{BLOOD} < V_{composite} < V_{BLOOD+LIVER} = 1.48$.

to zero time will correspond to a zero time output value originating in the aggregated compartment, with a size equal to the sum of the volumes of blood and liver, as shown in **Fig. 4.5**. Volume is in the denominator of the concentration measurement, which explains the lower value of:

$$y_{BLOOD+LIVER}(0) = x(0)/(V_{BLOOD} + V_{LIVER}) < y_{BLOOD}(0) = x(0)/V_{BLOOD}.$$

However, if mixing is not rapid enough to support the homogeneous assumption, the aggregated compartment volume $V_{compsite}$ will be larger than V_{BLOOD} but smaller than the sum of blood and liver, so $y_{1B}(0) \equiv y_{BLOOD}(0) > y_{composite}(0) > y_{BLOOD+LIVER}(0)$, also as illustrated in **Fig. 4.5**. This means that the **initial volume of distribution** – the **volume of the central compartment** – is overestimated, more or less, depending on when the earliest data point is taken. It also means blood and liver compartments cannot be well differentiated from each other. (More on this later.)

Remark: *Modeling Practice*. As a practical matter, the lumped blood–liver compartment is a useful approximation if its identity remains clear. It's the amount of X_1 in blood and liver combined, not either one. In principle, *then*, $y_1(0^+) \approx Q_0/(V_{BLOOD} + V_{LIVER})$ for a dose input of Q_0. As a corollary, it is common in whole-body compartmental modeling – especially in *in vivo* drug pharmacokinetics or toxicokinetics studies – to define a single approximately well-mixed ***central compartment*** consisting of blood plus the ***highly perfused organs***: heart, brain, liver and kidney. These organs receive a large fraction of blood flow per unit mass and are hence relatively well mixed with the intravascular blood compartment. Together they are a good candidate for being a homogeneous central compartment for substance X dynamics, under the assumption that these dynamics are relatively much slower in their

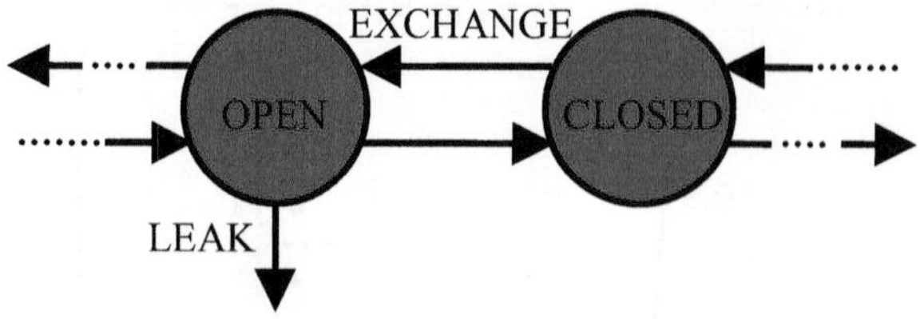

Fig. 4.6

extravascular, tissue distribution phase. Again, the identity of the "container-spaces" containing central compartment state variables must be made explicit, as above.

Remark: *More Modeling Practice*. For most drugs, toxins (xenobiotics) and many endogenous substances, the enzyme systems responsible for their metabolism (Chapters 6 and 7) reside in the liver or kidney, or they are excreted more directly in urine, approximately justifying "leak" arrows directed from the central compartment – again assuming relatively rapid exchange of the drug between blood and these organs, as in **Fig. 4.1** and **Fig. 4.4**.

Remark: *Experiment Design Recommendation*. Generalizing from this example, one way of assuring a probed compartment retains its distinct physical identity is to sample it sufficiently soon and often enough after applying a test-input. Otherwise it will be difficult to distinguish it in output measurement data collected after it has begun to be transferred to (taken up in significant amount by) unprobed compartments with which it exchanges relatively rapidly. If sampling is begun too late in the above sense, model connectivity as well as order may change, usually becoming more obscure. In this case, otherwise distinct compartments are merged, or lumped with each other, as illustrated above.

Following these motivational examples, we formalize multicompartmental modeling nomenclature and schematic representations and move on to the mathematics of multicompartmental modeling.

Compartments, State Variables, Graphs, Pools & Chemical Species Defined

We formalize and further clarify our primary nomenclature, paraphrasing the classical definitions collectively.[11]

A **compartment** is an amount of material X that acts kinetically (within a dynamic system) in a *homogeneously distinct* manner. Homogeneously distinct are the key words here. This means entities within a compartment are indistinguishable from each other, they are "mixed completely." If they are not, they comprise more than one compartment.

A compartment is **open** if it "leaks" to the environment; it is **closed** if it does not (**Fig. 4.6**). Most (not all) compartmental models of real systems have at least one open compartment.

11 Following is a representative sampling of the numerous references following the classical compartment definitions, used in physics, biophysics, various areas of biology and ecology, since the early 20th century and to the present: (Hevesey 1923; Atkins 1969; Jacquez 1972; Rubinow 1973; Dickson et al. 1982; Godfrey 1983; Gydesen 1984; Tardiff and Goldstein 1991; Jacquez 1996; Cobelli, Sparcino et al. 2006; Schramski, Patten et al. 2009; Rowland and Tozer 2011).

Every (physical) space containing compartments has a presumed volume in which the material substance (chemical "species," etc.) is homogeneously distributed. This volume usually is the **compartment volume** and might be measurably real, as in Example 4.1 above, or it might be virtual – called an *equivalent volume.* The composite compartment Example 4.4 above has an equivalent volume – or *apparent volume* of distribution for substance *X.* More on volumes later.

In Other Words: A compartment is an idealized store of a substance, characterized by its environment (Example 4.1), physiochemical properties (Example 4.2), or both. *Other examples*: all the substance in a *given form* qualifies as a compartment, e.g. *active* or *inactive*, or *receptor-bound* or *unbound* form – one compartment for each – because they generally have different dynamics. Alternatively, all of the substance in a homogeneously distributed (well-mixed) location, or all of it in a given form and location, each qualify as compartments, e.g. the mass of unbound drug D_F in blood, or the mass of receptor-bound drug D_R in blood, or either in liver (or other organ). Similarly, the total drug mass $D_F + D_R$ in blood or liver, are each valid compartments in principle.

A **multicompartmental (MC) model** consists of a graph of two or more compartments interconnected so there can be exchange of material, or change of state, among some or all of them, and possible **leaks** from any of them (i.e. **Fig. 4.6**). **Exchange** may occur by material transgressing some physical barrier (e.g. by diffusion or membrane transport), or by undergoing some physical or chemical transformation (metabolism, inactive to active state, etc.). Each **compartment** is represented by a single **state variable** in a multicompartmental (MC) model. Changes in location of material generate different compartments (different state variables) for the same material, typically with different volumes. Changes in material form (precursor to product, hormone to metabolite, etc.) generate different compartments (different state variables) – possibly in the same "space" or "container" and with the same volume, and so on.

Circles are used to represent compartments, **arrows** represent unidirectional transfers – which include inputs, influxes and effluxes (leaks) – and **dashed lines**, with a small open-circle endpoint, denote **measured outputs (Fig. 4.7)**.

Directed graphs (digraphs) also are used to represent multicompartmental models, with **nodes** (vertices) representing compartments and **edges** as the arrows interconnecting compartments (nodes) or the environment, as in **Fig. 4.7**.

Remark: Some authors use rectangles for compartments, as well as for input–output (e.g. transfer function) *subsystem* blocks. We maintain a distinction, because they are basically different modeling paradigms. One of the important differences is that outputs generally have

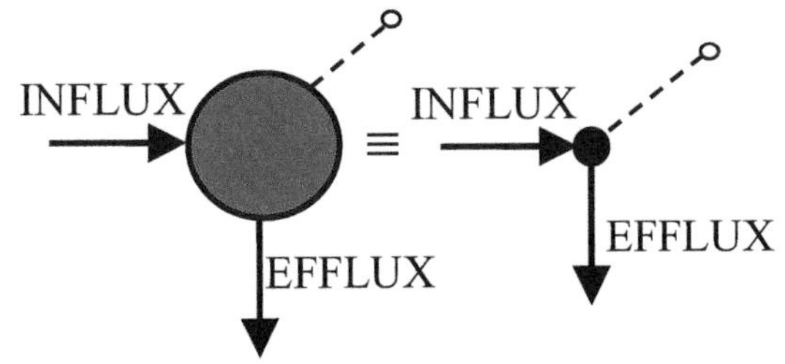

Fig. 4.7

different physical units than effluxes, because they are typically amount (abundance) or concentration variables. Inputs may have the same units as influxes, because they are usually rates of change of amounts, etc.

A **pool** is the total amount of a substance in a system or subsystem (Bergner and Lushbaugh 1967, p. 18). This term was defined purposely loosely to suit the variety of ways different investigators were using it. We are a bit more precise here, defining **pool** – similarly, as a mass or volume size measure for a real or abstract space or material, emphasizing that it is not necessarily uniformly (homogeneously) distributed.[12] A compartment may be a pool, but not necessarily *vice versa*. On the other hand, we can speak of the "pool of *X* in compartments *i* and *j*" or the "pool of *X* in the whole organism" (or "whole compartmental model or system") because the term pool circumvents the homogeneity requirement on any of these entities. The section on Pool Models below illustrates this concept in quasi-compartmentalized multi-pool models useful for steady state *flux balance analysis* (FBA).

Species is a term commonly used in chemistry and the molecular systems biology literature, with nomenclature embodied in the latest version (Level 3) of the *Systems Biology Markup Language SBML*, a "*bridge* language (*lingua franca*) for communication or exchange of mathematical models across programming tools" (*www.sbml.org*). They define species – short for *chemical species* – differently than in biology.[13] **Chemical species** are defined as distinct molecular entities sharing the same chemical properties. In our terms, same chemical properties effectively means homogeneous. In a biochemical kinetic model, distinct species are thus represented as distinct state variables – the classical definition of (homogeneous) compartment in a dynamic system model.

Remarks: All well and good. There would be no incompatibility problem if the differences stopped there, because species = state variable, or classical compartment = state variable, would amount to simply renaming things. Chemical species and compartment appear to be one and the same. But the SBML documention, amplifying on the definition, adds that species refers to "a *pool* of entities considered indistinguishable from each other for the purposes of the model, participate in reactions, and are located in a specific compartment (*classical "space" or container*)." Their *pool* then is our classical (homogeneous) compartment; and their (differently defined) "compartment" is a "bounded space in which species are located" (i.e. container of state variables). As noted above, in classical terms, *pool* is a superset of *compartment*, meant to represent an entity possibly distributed inhomogeneously in a distinct space (e.g. a biomolecular species in a poorly mixed container). They've redefined pool as well as compartment, with potential for confusion. The definitions of biostructural components chosen by molecular systems biologists are thus irreconcilable with the classical definitions in current use in other areas of life sciences. The meanings of *species*, *compartment* and *pool* – taken together – generates potential ambiguities for modeling. So, we'll be especially cautious with their use.

12 This definition of *pool* is reaffirmed in another publication on establishing "compartmental" nomenclature (Rescigno et al. 1990).

13 As widely understood, in biology **species** is a basic unit of biological classification and taxonomic rank, e.g. the species *Homo sapiens* (humans) or species *Canis lupus* (grey wolf).

MATHEMATICS OF MULTICOMPARTMENTAL MODELING FROM THE BIOPHYSICS

Fig. 4.8 and **Fig. 4.9** illustrate and **Table 4.1** summarizes compartment model nomenclature used in much of this chapter and remainder of the book. Table entries are expressed in terms of a generic species type X (substance, molecule, etc.). For the symbols and equations, t is time, lower case $x(t)$ is any function of time, $\dot{x} \equiv dx/dt$ is its derivative, $\delta(t) \equiv$ unit impulse function (t^{-1}), and $\boldsymbol{1}(t) \equiv$ unit step function (dimensionless).

In other Words: To keep things clear (albeit redundant), the following notation is worth emphasizing: X symbolizes a substance or entity (species type) modeled by a particular compartment, with $q(t)$ designating the mass or amount state variable for X, or $c(t)$ the concentration state variable for X. V is the volume of the compartment vessel. In a multi-compartment model, lower case $q_i(t) \equiv q_i$ is the amount or abundance of X homogenously distributed (well mixed) in a particular location i, i.e. it's *the* compartment, with location volume V_i. We informally often say that q_i is the compartment, thereby associating the state variable q_i with the compartment. We might also say that the concentration variable c_i is the compartment, if the equations are written in terms of concentrations (typical in biochemistry). When there is more than one compartment in a model, different subscripts designate different compartments (and state variables). Lower case letters other than $c(t)$ are also used to represent time varying concentrations in other chapters, e.g. $w(t)$ for species named W; $q(t)$ is always an amount or mass variable.

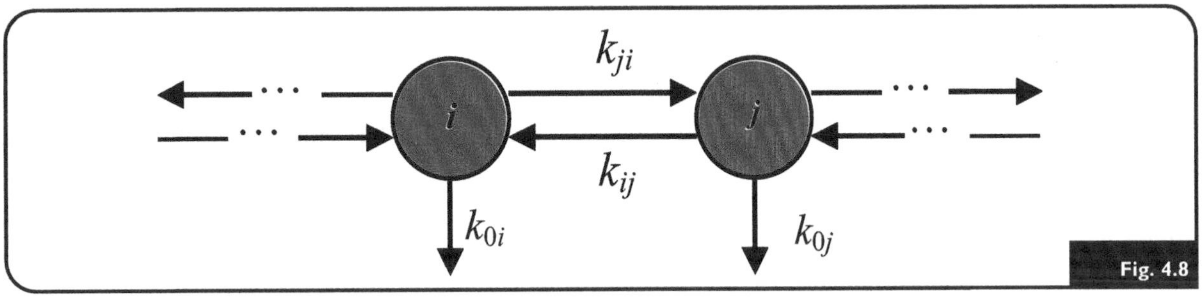

Fig. 4.8

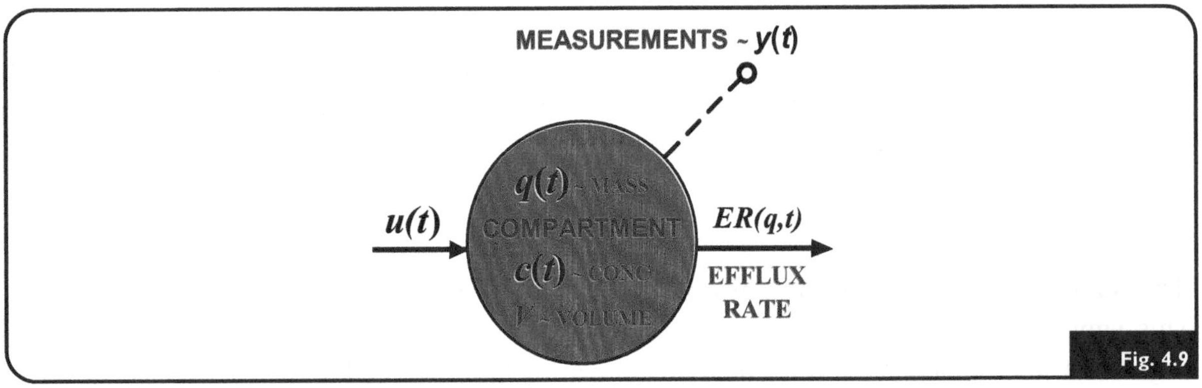

Fig. 4.9

Table 4.1 Basic Compartmental Nomenclature, Nonlinear & Linear

$q_i(t) \equiv$ amount of X (or # of X, if X is a count of molecules) in compartment i at time t (mass or #).
$c_i(t) \equiv$ concentration of X in compartment i at time t (mass/vol or #/vol or m/sample mass).
$y_i(t) \equiv$ measured output of compartment i at time t, usually $c_i(t)$ or $q_i(t)$.
$ER_i(t) \equiv ER_i =$ elimination rate (removal/disposal/catabolism/efflux rate) of X from compartment i at time t (m/t). It can be linear or nonlinear.
$k_{oi} \equiv$ fractional elimination rate (leak) of X from compartment i to the environment, also called *elimination rate constant* (t^{-1}) if the model is linear. It is a function of $q_i(t)$ if the rate is nonlinear.
$TR_{ij}(t) \equiv TR_{ij} =$ transfer rate of X from compartment j to compartment i at time t (mass/t). It can be linear or nonlinear.
$k_{ij} \equiv$ fractional rate of transfer of X from compartment j to compartment i, also called *exchange rate constant* (t^{-1}) if the model is linear. It is a function of $q_j(t)$ if the rate is nonlinear.
$PR_i(t) \equiv PR_i =$ endogenous production rate of X in compartment i at time t (mass/t).
$IR_i(t) \equiv$ exogenous influx rate into compartment i at time t (mass/t).
$Q_0 \equiv$ impulsive (pulse-dose) input magnitude (mass).
$q_i(0) \equiv q_{0i} =$ initial value of q (at time $t = 0$).
$q_i(\infty) \equiv q_{i\infty} \equiv q_{iss} =$ steady state value of $q_i(t)$ (at $t \to \infty$).
$V_i \equiv$ distribution volume of the compartment $= q_{i\infty}/c_{i\infty}$ (*Note*: V_i may be real or abstract, i.e. *equivalent*, as discussed later.)

Fig. 4.8 illustrates two compartments, designated i and j in a MC model. In this case, mass-transfer between compartments, or from compartments to the environment (called **leaks**), are represented by **fractional transfer rate parameters** k_{ij} (time^{-1} units). This means that parameter k_{ij} is the *fraction* of compartment j transferred to compartment i or the environment, compartment "0," per unit time. They may be constant – and thus called **rate constants** – or may be a function of the compartment mass q_j. In the latter case, the compartment model is nonlinear (NL). The k_{ij} are also called microparameters in the pharmacological modeling community.

Remark: Some authors, particularly in pharmaceutical sciences, use a somewhat different nomenclature. For example, they define the k_{ij} ordering for $i \neq j$ the other way around, i.e. k_{ji} instead of k_{ij}. The distinctions and rationale for our choice in **Fig. 4.8** are clarified later in the chapter, when we generalize MC models to n-compartments and adopt matrix notation.

A major advantage of compartmental models is the k_{ij} can be assigned and thus have useful biophysical, biochemical, physiological or other bioscience interpretations. But the structure, in particular the presence or absence of these rate parameters at various locations, must be accurate if such interpretations are to have biophysical or biological meaning, a message repeated and exemplified often below.

The One-Compartment Model and Its Solution

This simplest model of a system representable with a compartmental structure is shown in **Fig. 4.9**. The basic assumption for the one-compartment model is that the "substance" X under study is distributed homogeneously in the entire physical system in which it resides, e.g. a drug X in the whole body, or a protein in a whole cell, or just in the cell nucleus. As simple as it is, the one-compartment model has found wide application as a first-approximation (coarse-grained oversimplification), because additional complexity is sometimes unnecessary; in such cases the one-compartment model of the dynamics is good enough. For now, it represents the first pedagogical step in the development of this class of models.

The influx rate $u(t)$ in **Fig. 4.9** may be endogenous, say $PR(t)$, with P implying production, or exogenous influx $IR(t)$. The efflux rate ER is typically dependent on the mass of material in the compartment, as well as time, i.e. $ER(q,t)$, and ER may be nonlinear, if it is not first degree in q. By conservation of mass flow, the rate of change of mass,[14] $\dot{q} \equiv dq/dt$, equals the influx mass rate minus the efflux mass rate:

$$\dot{q}(t) \equiv \frac{dq}{dt} = u(t) - ER(q,t) \tag{4.1}$$

Remark: As noted earlier, the output y(t) in **Fig. 4.9** – designated graphically by the dashed line with the circle on the end – is how we represent outputs from any compartment in a multicompartment model. Other ways of depicting outputs are found in the literature, including a dashed or solid line with the circle on the end filled in, or an "eye" caricature, looking into the compartment, or similar such representation of observing state variables. They all mean the same thing.

Impulse-(Bolus)-Input Response: One-Compartment Model

If $ER(q,t)$ in (4.1) is nonlinear (NL), solution for $q(t)$ over some time interval $[t_0,T]$ usually requires numerical simulation. We defer further discussion of the NL one-compartment model to a later section. Under a linear elimination rate assumption, $ER \equiv kq$, and Eq. (4.1) becomes:

$$\dot{q} = u - kq \tag{4.2}$$

This equation is linear and time-invariant and we can solve it analytically for most inputs $u(t)$.[15] Consider a kinetic experiment in which the exogenous test input into a linear one-compartment model is a single (approximately) impulsive input $u \equiv Q_0\delta(t)$ of magnitude Q_0. If the system and model are initially relaxed, then Eq. (4.2) becomes:

$$\dot{q} = -kq + Q_0\delta(t) \tag{4.3}$$

and the equivalence property, Eq. (2.16), is applicable:

$$\left\{ \begin{array}{c} \dot{q} = -kq + Q_0\delta(t) \\ q(0) = 0 \end{array} \right\} \Leftrightarrow \left\{ \begin{array}{c} \dot{q} = -kq \\ q(0) = Q_0 \end{array} \right\} \tag{4.4}$$

Thus, the solution of inhomogeneous Eq. (4.3) is simply the solution of the homogeneous RHS of Eq. (4.4) with intial condition (IC) equal to Q_0, as shown in **Fig. 4.10**: $q(t) = Q_0e^{-kt}$.

14 or number (#) of X, if q is a count of X, e.g. # of molecules.

15 Note that we have omitted the argument t in Eq. (4.2) and shall continue this practice when it promotes clarity rather than ambiguity.

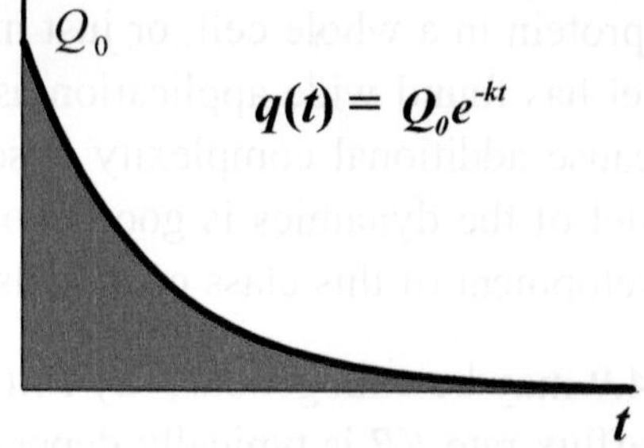

Fig. 4.10

Impulse-response of the one-compartment linear model.

Step-Input Response: One-Compartment Model

If the input is a step function, for example a *constant influx rate*, $u = IR \cdot \mathbf{1}(t) \equiv IR$ (endogenous or exogenous), with *arbitrary* initial condition q_0, then the model is:

$$\dot{q} = -kq + IR, \quad q(0) = q_0 \tag{4.5}$$

Using Laplace transforms (Chapter 2 and Appendix A), with the simplified notation $\mathcal{L}[q(t)] \equiv Q(s) \equiv Q$, we have:

$$sQ - q(0) = -kQ + \frac{IR}{s}, \quad (s+k)Q = \frac{IR}{s} + q_0$$

$$Q = \frac{IR}{s(s+k)} + \frac{q_0}{s+k} = \frac{A}{s} + \frac{B}{s+k} + \frac{q_0}{s+k}$$

The last expression is written as a partial fraction expansion. From Chapter 2 results,

$$A = s\frac{IR}{s(s+k)}\bigg|_{s=0} = \frac{IR}{k}, \quad B = (s+k)\frac{IR}{s(s+k)}\bigg|_{s=-k} = -\frac{IR}{k}$$

Thus,

$$Q = \frac{IR}{k}\left(\frac{1}{s} - \frac{1}{s+k}\right) + \frac{q_0}{s+k} \tag{4.6}$$

Finally, inversion of Eq. (4.6) gives:

$$\mathcal{L}^{-1}[Q(s)] \equiv q(t) \tag{4.7}$$

Rewriting (4.7) as steady state (∞) and transient responses:

$$q(t) \equiv q_\infty + q_{transient}(t) = \frac{IR}{k} + \left(q_0 - \frac{IR}{k}\right)e^{-kt} \tag{4.8}$$

This is graphed in **Fig. 4.11** for both $q(0) < IR/k$ and $q(0) > IR/k$, to illustrate how ICs can alter the transient but not the steady state response. The *Matlab* script code for making the figure is also shown.

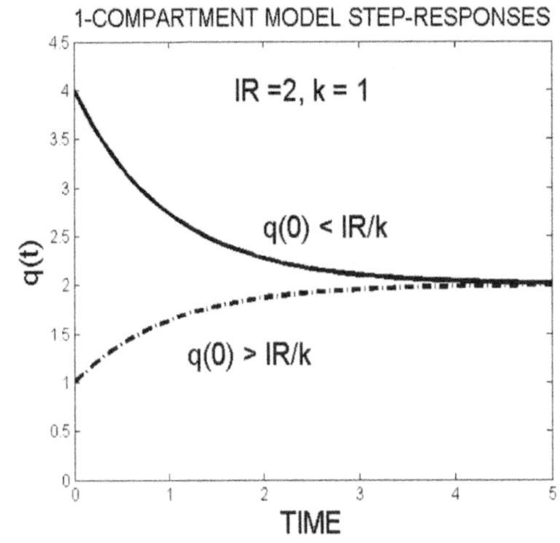


```
%1-Comp Model Step-Responses from 2 ICs
% MATLAB CODE
t=0:0.1:5;
qss = 2; q0 = 4;
qa = qss + (q0-ss)*exp(-t);
qss = 2; q0 = 1;
qb = qss + (q0-ss)*exp(-t);
plot(t,qa,t,qb,'Linewidth',3);
axis([0 5 0 4.5]);
title('1-COMPARTMENT MODEL STEP-RESPONSES',...
      'FontSize',15);
xlabel('TIME','FontSize',18);
ylabel('q(t)','FontSize',18);
text(2,2.6,'q(0) < IR/k','FontSize',18);
text(1.2,1.3,'q(0) > IR/k','FontSize',18);
text(1.7,4,'IR =2, k = 1','FontSize',18);
```

Fig. 4.11

The Two-Compartment Model, its Forms, Properties & Solutions

The most general two-compartment model structure is shown in **Fig. 4.12**, with the smaller red diagram without symbols in *node-edge* **directed graph** form. This model has six nonnegative parameters: k_{01}, k_{02}, k_{12}, k_{21}, V_1, and V_2; a maximum of two independent input variables $u_1(t)$ and $u_2(t)$; and a maximum of two possible independent measurement (output) variables $y_1(t)$ and $y_2(t)$ in compartments 1 and 2, denoted by broken lines and open circles.

Remark: Linear combinations (e.g. sums) of inputs or outputs to two compartments are possible; and additional edges could be included that parallel the ones shown. But this structure is the most general in the sense that it has its full complement of independent parameters (edges) and two *independent* inputs and outputs.

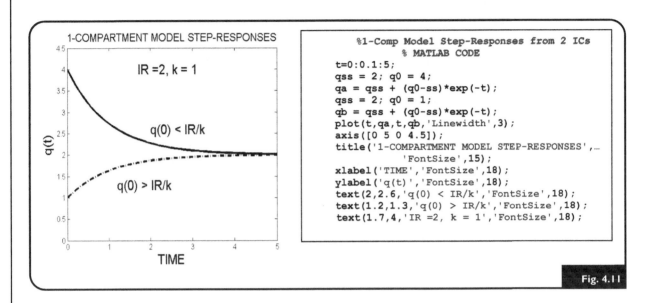

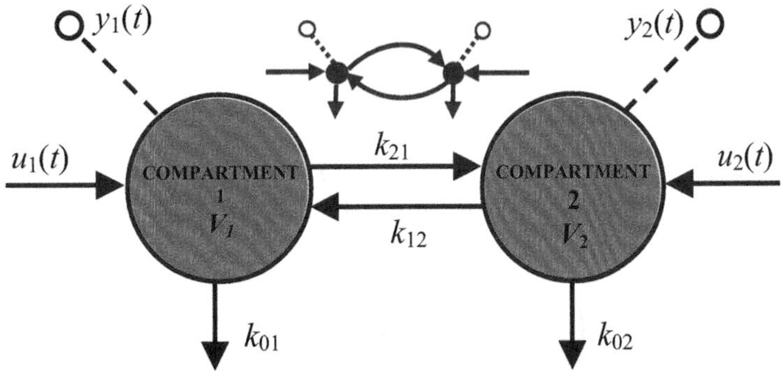

Fig. 4.12

We develop the math for the linear two-compartment model first here, but this figure and model are easily extended for NL systems, by redefining the parameters k_{ij}. We develop this later in the chapter. For the present, the k_{ij} are (fractional) rate constants, with dimensions of time^{-1}; the V_i are volumes; the u_i are mass flux rates (mass/time); and the outputs y_i may be concentrations (mass/volume) or amounts (mass). The differential equations (dynamics) of the model can be written by choosing the two state variables as either amounts in compartments 1 and 2, $q_1(t)$ and $q_2(t)$, or as concentrations $c_1(t) = q_1(t)/V_1$ and $c_2(t) = q_2(t)/V_2$. As we see below, mass equations are mathematically simpler, directly translatable from the biophysics and arguably easier to manipulate and interpret. Both are common in the literature. We begin by presenting the full equations, written out, then in the simpler vector-matrix form. We'll do this both ways for awhile, to get accustomed to the more concise notation. The vector-matrix nomenclature is eminently systematic and it is easier to generate and extend useful results. The full equations can be written out as needed.

Model Dynamics with Mass State Variables $q_i(t)$

$$\begin{aligned} \dot{q}_1 &= -(k_{01} + k_{21})q_1 + k_{12}q_2 + u_1 \\ \dot{q}_2 &= k_{21}q_1 - (k_{02} + k_{12})q_2 + u_2 \end{aligned} \tag{4.9}$$

or

$$\dot{\boldsymbol{q}} = \begin{bmatrix} -(k_{01} + k_{21}) & k_{12} \\ k_{21} & -(k_{02} + k_{12}) \end{bmatrix} \boldsymbol{q} + \begin{bmatrix} 1 & 0 \\ 0 & 1 \end{bmatrix} \boldsymbol{u} \equiv K\boldsymbol{q} + \begin{bmatrix} u_1 \\ u_2 \end{bmatrix} \tag{4.10}$$

or, more compactly,

$$\dot{\boldsymbol{q}} = K\boldsymbol{q} + \boldsymbol{u} \tag{4.11}$$

As shown later, the form of (4.11) is completely general, for compartment models with any number of compartments n (K is n by n).

Model Dynamics with Concentration State Variables $c_i(t) \equiv q_i(t)/V_i$

Mass flux must remain balanced, so – after dividing both equations in (4.9) by V_1 or V_2, the model becomes:

$$\begin{aligned} \dot{c}_1 &= -(k_{01} + k_{21})c_1 + \left(\frac{k_{12}V_2}{V_1}\right)c_2 + \frac{u_1}{V_1} \\ \dot{c}_2 &= \left(\frac{k_{21}V_1}{V_2}\right)c_1 - (k_{02} + k_{12})c_2 + \frac{u_2}{V_2} \end{aligned} \tag{4.12}$$

or

$$\dot{c} = \begin{bmatrix} -(k_{01}+k_{21}) & \dfrac{k_{12}V_2}{V_1} \\ \dfrac{k_{21}V_1}{V_2} & -(k_{02}+k_{12}) \end{bmatrix} c + \begin{bmatrix} 1/V_1 & 0 \\ 0 & 1/V_2 \end{bmatrix} u \tag{4.13}$$

Defining a diagonal **distribution volume matrix** V as:

$$V = \begin{bmatrix} V_1 & 0 \\ 0 & V_2 \end{bmatrix} \tag{4.14}$$

gives:

$$q(t) \equiv Vc(t) \tag{4.15}$$

Substituting (4.15) into (4.11) and rearranging terms then gives the vector-matrix form:

$$\dot{c} = V^{-1}KVc + V^{-1}u \tag{4.16}$$

Remark: This $q \rightarrow c$ transformation amounts to a change of basis in linear vector space. (Appendices B and E).

Output Measurement Models

Outputs usually are measured as either amounts, q_1 and/or q_2, or concentrations, c_1 and/or c_2.

Model Outputs with Mass Measurements

$$y_1 = q_1 \text{ and } y_2 = q_2 \tag{4.17}$$

$$y = \begin{bmatrix} 1 & 0 \\ 0 & 1 \end{bmatrix} q \tag{4.18}$$

Model Outputs with Concentration Measurements

$$y_1 = c_1 = \frac{q_1}{V_1} \text{ and } y_2 = c_2 = \frac{q_2}{V_2} \tag{4.19}$$

$$y = \begin{bmatrix} 1 & 0 \\ 0 & 1 \end{bmatrix} c = \begin{bmatrix} 1/V_1 & 0 \\ 0 & 1/V_2 \end{bmatrix} q = V^{-1}q \tag{4.20}$$

Remark: Outputs also may be measured as integrals of mass or concentrations, e.g. in urinary accumulation experiments, which entails using additional *dead-end* compartments, or *traps* as they are also called. A dead-end (measurement) compartment representing accumulation of substance X in urine was shown in **Fig. 4.1**.

Biosystem & Experiment Model Combined

In the context of an input–output experiment, *the biomodel has two parts: the dynamical equations (ODEs) of the biosystem and the output measurement model.*[16] For the linear two-compartment model, the choice of Eqs. (4.11) or (4.13) and (4.18) or (4.20) depends on the specific problem, or system and experiment being modeled. If convenient, it is preferable to choose Eq. (4.11) for the dynamics, because of its simplicity. Eq. (4.20) is designated for the outputs, if measurements are concentrations rather than amounts. Concentration measurements are more common, e.g. when any fluid is sampled for assessment of one of its molecular components. We choose this combination for the remaining analysis, keeping in mind that the procedure is essentially the same for the other choice of state and measurement variables. Thus, with mass-based dynamics and concentration measurements, we have the model:

$$\dot{\boldsymbol{q}} = \begin{bmatrix} \dot{q}_1 \\ \dot{q}_2 \end{bmatrix} = \begin{bmatrix} -(k_{01}+k_{21}) & k_{12} \\ k_{21} & -(k_{02}+k_{12}) \end{bmatrix} \begin{bmatrix} q_1 \\ q_2 \end{bmatrix} + \begin{bmatrix} u_1 \\ u_2 \end{bmatrix} = K\boldsymbol{q} + \boldsymbol{u} \tag{4.21}$$

$$\boldsymbol{y} = \begin{bmatrix} y_1 \\ y_2 \end{bmatrix} = \begin{bmatrix} 1/V_1 & 0 \\ 0 & 1/V_2 \end{bmatrix} \begin{bmatrix} q_1 \\ q_2 \end{bmatrix} = \begin{bmatrix} q_1/V_1 \\ q_2/V_2 \end{bmatrix} = V^{-1}\boldsymbol{q} \tag{4.22}$$

Remark: It is often the case that only compartment 1 can be presented with an input u_1 and sampled for concentration y_1, e.g. blood, or cell culture medium. In this case, $u_2 \equiv 0$ in Eq. (4.21) and Eq. (4.22) becomes

$$\boldsymbol{y} = y_1 = \begin{bmatrix} \frac{1}{V_1} & 0 \end{bmatrix} \boldsymbol{q} = \frac{q_1}{V_1} \tag{4.23}$$

Remark: Measurement Eq. (4.22) is a continuous function of time t, over $0 \le t \le T$. In most cases for molecular (mass-based) signals, continuous measurements are infeasible and observations are made at N discrete times: $t_1, t_2, \ldots, t_N$. In this case, it is more appropriate to write Eq. (4.22) as:

$$\boldsymbol{y}(t_i) = V^{-1}\boldsymbol{q}(t_i) \tag{4.24}$$

or more compactly, as

$$\boldsymbol{y}_i = V^{-1}\boldsymbol{q}_i \tag{4.25}$$

16 See **Fig. 1.5**. In Chapter 6 and beyond, we begin to add constraint equations, to *complete* the model.

for time-samples $\boldsymbol{y}(t_i)$ for all components of the vector of outputs at $t_1, t_2, \ldots, t_N$ (not to be confused with output y_j of compartment j). To be more realistic, we should include the fact that measurements are usually made with some error (they are "noisy"). For **additive noise**:

$$z_i = \boldsymbol{y}_i + \boldsymbol{e}_i, \quad \text{for } i = 1, 2, \ldots, N \tag{4.26}$$

where $\boldsymbol{e}_i$ represents the noise in the measurement at time t_i, and it is often given a probabilistic or stochastic description (e.g. Gaussian distributed errors, with mean zero and variance σ_i^2). We ignore these fine points for the present and introduce them again later, when needed.

Transfer Function Matrix for the Combined Biosystem & Experiment Model

In Chapter 2, we developed the transfer function (TF) model representation and the transfer function matrix H for a linear system model in state variable form. We do the same here, specifically for the two-compartment model. The TF matrix yields all the fundamental input–output (I–O) characteristics of the model, for all I–O pairs, and is derived directly from the vector-matrix forms of Eqs. (4.21) and (4.22). This is the simplest approach. Then we show how to obtain the model solution for any Laplace transformable inputs $u_1(t)$ and $u_2(t)$ directly from this matrix. The Laplace transforms of Eqs. (4.21) and (4.22),[17] with all initial conditions zero, are:

$$s\boldsymbol{Q} = K\boldsymbol{Q} + \boldsymbol{U} \tag{4.27}$$

$$y = V^{-1}\boldsymbol{Q} \tag{4.28}$$

Then

$$\begin{gathered} s\boldsymbol{Q} - K\boldsymbol{Q} = (sI - K)\boldsymbol{Q} = \boldsymbol{U} \\ \boldsymbol{Q} = (sI - K)^{-1}\boldsymbol{U} \end{gathered} \tag{4.29}$$

and

$$\boldsymbol{Y} = V^{-1}(sI - K)^{-1}\boldsymbol{U} \tag{4.30}$$

i.e.

$$\begin{aligned} H = V^{-1}(sI-K)^{-1} &= \begin{bmatrix} 1/V_1 & 0 \\ 0 & 1/V_2 \end{bmatrix} \begin{bmatrix} s+k_{01}+k_{21} & -k_{12} \\ -k_{21} & s+k_{02}+k_{12} \end{bmatrix}^{-1} \\ &= \begin{bmatrix} 1/V_1 & 0 \\ 0 & 1/V_2 \end{bmatrix} \begin{bmatrix} \dfrac{s+k_{02}+k_{12}}{D} & \dfrac{k_{12}}{D} \\ \dfrac{k_{21}}{D} & \dfrac{s+k_{01}+k_{21}}{D} \end{bmatrix} \end{aligned} \tag{4.31}$$

17 All the steps are here, but review Chapter 2 or Appendix B if you need help with this.

where we have substituted D for the characteristic polynomial:

$$D = s^2 + (k_{01} + k_{02} + k_{12} + k_{21})s + k_{01}k_{02} + k_{01}k_{12} + k_{02}k_{21} \equiv (s - \lambda_1)(s - \lambda_2) \tag{4.32}$$

The "eigenvalues" or roots of the polynomial $D = 0$ are readily determined algebraically as:

$$\lambda_{1,2} = \frac{-(k_{01} + k_{02} + k_{12} + k_{21}) \pm \sqrt{(k_{01}+k_{02}+k_{12}+k_{21})^2 - 4(k_{01}k_{02} + k_{01}k_{12} + k_{02}k_{21})}}{2} \tag{4.33}$$

Remark: $\lambda_{1,2} \leq 0$ (real and nonpositive) for all $k_{ij} \geq 0$ in Eq. (4.33). Also, $\lambda_{1,2} < 0$ for all $k_{ij} > 0$. Showing this is left as an exercise.

Finally, multiplying the matrices gives:

$$H = \frac{1}{D}\begin{bmatrix} \dfrac{s + k_{02} + k_{12}}{V_1} & \dfrac{k_{12}}{V_2} \\ \dfrac{k_{21}}{V_2} & \dfrac{s + k_{01} + k_{21}}{V_2} \end{bmatrix} \equiv \begin{bmatrix} H_{11} & H_{12} \\ H_{21} & H_{22} \end{bmatrix} \tag{4.34}$$

Remark: TF matrix H is the **impulse response matrix**. Element H_{ij} is the impulse response at the ith output "port" (measurement in compartment i) to a unit impulse applied at the jth input "port" (into compartment j). Let's interpret this as **experiment** H_{ij}, as it describes the input/output characteristics between input port j and output port i.

Unit-Impulse Response from H(s)

The model in Eq. (4.21), and alternatively in Eq. (4.22), can be solved by classical ODE methods, as in Chapter 2 and Appendix A. Instead, we find the impulse responses $y_1(t)$ and $y_2(t)$ to a single unit impulse input $u_1(t) = \delta(t)$ into compartment 1 (with zero initial conditions) here using the TF matrix derived above. This requires inversion of $H_{11}(s)$ and $H_{21}(s)$, i.e. if we let $\mathcal{L}$ represent the **Laplace transform operator**, $y_1(t) = \mathcal{L}^{-1}[H_{11}(s)]$ and $y_2(t) = \mathcal{L}^{-1}[H_{21}(s)]$, and

$$y_1(t) = \mathcal{L}^{-1}\left[\frac{s + k_{01} + k_{12}}{V_1(s - \lambda_1)(s - \lambda_2)}\right] \equiv \mathcal{L}^{-1}\left[\frac{A}{s - \lambda_1} + \frac{B}{s - \lambda_2}\right] \tag{4.35}$$

where A and B are determined by partial fraction expansion of Eq. (4.35), i.e.

$$A = (s - \lambda_1)H_{11}\big|_{s=\lambda_1} = \left.\frac{s+k_{01}+k_{12}}{V_1(s-\lambda_2)}\right|_{s=\lambda_1} = \frac{k_{02} + k_{12} + \lambda_1}{V_1(\lambda_1 - \lambda_2)} \tag{4.36}$$

$$B = (s - \lambda_2)H_{11}\big|_{s=\lambda_2} = \frac{k_{02} + k_{12} + \lambda_2}{V_1(\lambda_2 - \lambda_1)} \tag{4.37}$$

Thus, (for $\lambda_1 > \lambda_2$),

$$y_1(t) = \frac{1/V_1}{\lambda_1 - \lambda_2}\mathcal{L}^{-1}\left[\frac{k_{02} + k_{12} + \lambda_1}{s - \lambda_1} - \frac{k_{02} + k_{12} + \lambda_2}{s - \lambda_2}\right] = \frac{(k_{02} + k_{12} + \lambda_1)e^{\lambda_1 t} - (k_{02} + k_{12} + \lambda_2)e^{\lambda_2 t}}{V_1(\lambda_1 - \lambda_2)} \tag{4.38}$$

Because $(k_{02} + k_{12} + \lambda_1) > 0$ and $(k_{02} + k_{12} + \lambda_2) < 0$, both coefficients of $e^{\lambda_i t}$ in Eq. (4.38) are greater than 0. Similarly,

$$y_2(t) = \mathcal{L}^{-1}\left[\frac{k_{21}/V_2}{(s - \lambda_1)(s - \lambda_2)}\right] = \frac{k_{21}(e^{\lambda_1 t} - e^{\lambda_2 t})}{V_2(\lambda_1 - \lambda_2)} \tag{4.39}$$

These solutions are shown in **Fig. 4.13**. The maximum value of $y_2(t)$, $\boldsymbol{ymax \equiv y_2(t_M)}$, and the time $\boldsymbol{tmax \equiv t_M}$ it occurs, are needed, for example, in developing pharmaceuticals, where $y(t)$ represents the **pharmacokinetics** (PK) of a drug.[18] These two important **PK parameters** are determined as follows. Setting $\dot{y}_2(t) \equiv 0$ in Eq. (4.39) to determine the maximum, $\lambda_1 e^{\lambda_1 t_M} = \lambda_2 e^{\lambda_2 t_M}$, $e^{(\lambda_1 - \lambda_2)t_M} = \lambda_2/\lambda_1$ and $t_M = \ln(\lambda_2/\lambda_1)/(\lambda_1/\lambda_2)$. Thus,

$$y_2(t_M) = \frac{k_{21}e^{\lambda_2 t_M}[e^{(\lambda_1 - \lambda_2)t_M} - 1]}{V_2(\lambda_1 - \lambda_2)} = \frac{-k_{21}e^{\lambda_2 t_M}}{V_2\lambda_1} = \frac{-k_{21}}{V_2\lambda_1}\left[\frac{\lambda_2}{\lambda_1}\right]^{\frac{\lambda_2}{(\lambda_1 - \lambda_2)}} \equiv ymax \tag{4.40}$$

The **area under the curve**, the **AUC** (third important PK parameter) of responses like $y_2(t)$ is also of interest in pharmacokinetic and other biological system kinetic studies. Computation of the AUC for $y_2(t)$, from $t = 0$ to $t = \infty$, in Eq. (4.39) is an exercise for the reader, Exercise E4.14.

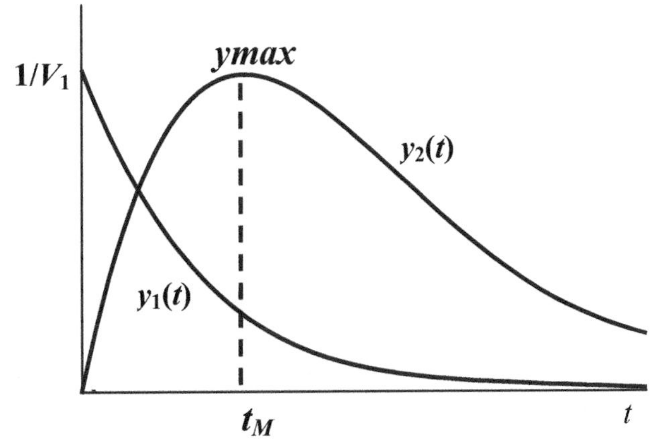

Fig. 4.13

18 More on structuring PK models in subsequent chapters.

Unit-Step Response: LT Solution

From Eq. (4.30), $Y = HU$, we have $Y_1 = H_{11}U_1 + H_{12}U_2 = H_{11}U_1$ (since $U_2 \equiv 0$) and $Y_2 = H_{21}U_1$. Now, for a step input, $U_1 \equiv U_1(s) = 1/s$ and zero ICs $q_1(0) \equiv q_2(0) \equiv 0$, $y_1(t) = \mathcal{L}^{-1}[H_{11}/s]$ and $y_2(t) = \mathcal{L}^{-1}[H_{21}/s]$. By partial fraction expansion:

$$\begin{aligned}\frac{H_{11}}{s} &= \frac{(s + k_{02} + k_{12})/V_1}{s(s-\lambda_1)(s+\lambda_2)} \\ &= \frac{1}{V_1\lambda_1\lambda_2}\left[\frac{k_{02}+k_{12}}{s} - \frac{\lambda_1(k_{02}+k_{12}+\lambda_2)}{(\lambda_1-\lambda_2)(s-\lambda_2)} + \frac{\lambda_2(k_{02}+k_{12}+\lambda_1)}{(\lambda_1-\lambda_2)(s-\lambda_1)}\right]\end{aligned} \tag{4.41}$$

and output 1 is the inverse transform of Eq. (4.41):

$$y_1(t) = \frac{1}{V_1\lambda_1\lambda_2}\left[k_{02}+k_{12}+\frac{\lambda_2(k_{02}+k_{12}+\lambda_1)e^{\lambda_1 t} - \lambda_1(k_{02}+k_{12}+\lambda_2)e^{\lambda_2 t}}{\lambda_1-\lambda_2}\right] \tag{4.42}$$

Similarly,

$$\frac{H_{21}}{s} = \left[\frac{k_{21}/V_2}{s(s-\lambda_1)(s-\lambda_2)}\right] = \frac{k_{21}}{V_2\lambda_1\lambda_2}\left[\frac{1}{s} + \left(\frac{1}{\lambda_1-\lambda_2}\right)\left(\frac{\lambda_2}{s-\lambda_1} - \frac{\lambda_1}{s-\lambda_2}\right)\right] \tag{4.43}$$

and output 2 is

$$y_2(t) = \frac{k_{21}}{V_2\lambda_1\lambda_2}\left(1 + \frac{\lambda_2 e^{\lambda_1 t} - \lambda_1 e^{\lambda_2 t}}{\lambda_1 - \lambda_2}\right) \tag{4.44}$$

These solutions are shown in **Fig. 4.14**.

Remark: Solution of the two-compartment model also can be achieved via simulation, and also by directly solving the ODEs analytically in the time-domain, as illustrated in Exercise E4.27.

Steady State Solutions[19]

We recall that the steady state (SS) solutions of a system of equations, linear or nonlinear, are not necessarily constant. The SS solution is that part of the complete solution that *does not* go to zero as t grows without bound and, for example, this may be a "steadily" oscillating sinusoidal function. This is easy to visualize if the input is oscillatory, but what if the input is constant for all $t \geq 0$? Does this mean the output also must be constant in the steady state?

The answer is no, in general. For example, the differential equation $\ddot{x} + x = u = 0$ (constant zero input), with initial condition $x(0) = 1$, has the solution $x(t) = \cos t$ for all $t \geq 0$. On the other hand, if the SS solution is constant for all state variables, then all derivatives of state variables are zero in the steady state.

19 This subject is developed more fully in Chapter 9, on stability.

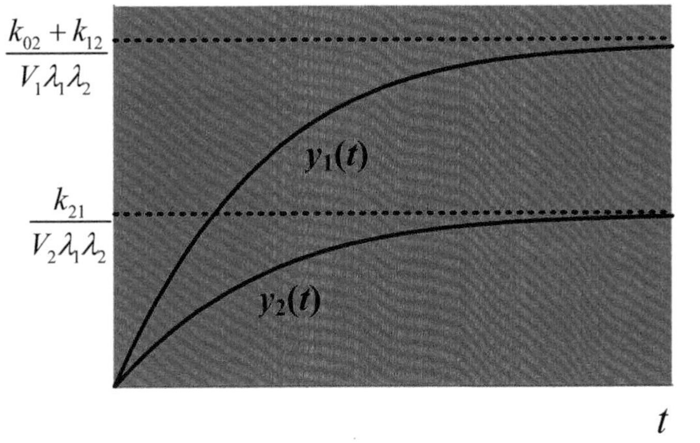

Fig. 4.14

The **zero-derivative property** can almost always[20] be used to test for constancy of the steady state, Set all derivatives of all state variables equal to zero and attempt to solve for the resulting state variables of the system from the resulting equations. In general, there is more than one steady state solution and these are called the **equilibrium points** or **fixed points** of the system in classical thermodynamics. If one or more nonzero and constant solutions are feasible, the steady state is constant. More than one feasible solution means more than one steady state. If no such steady state solution can be found for all state variables, the steady state is not constant. This approach always works for asymptotically stable systems, like the two-compartment model discussed next.

Steady State Solutions for the Linear Two-Compartment Model

The steady state (SS) solutions $q_1(\infty)\equiv q_{1\infty}$ and $q_2(\infty)\equiv q_{2\infty}$ are constant when the inputs $u_1(t)\equiv u_{1\infty}$ and $u_2(t)=u_{2\infty}$ are constant. To see this, we set $\dot{q}_1\equiv 0$ and $\dot{q}_2\equiv 0$ in Eq. (4.9) and solve for $q_{1\infty}$ and $q_{2\infty}$. First, let $k_{11}=-(k_{01}+k_{21})$ and $k_{22}=-(k_{02}+k_{12})$, for brevity. Then:

$$0=k_{11}q_{1\infty}+k_{12}q_{2\infty}+u_{1\infty} \tag{4.45}$$

$$0=k_{21}q_{1\infty}+k_{22}q_{2\infty}+u_{2\infty} \tag{4.46}$$

Solving Eq. (4.45) for $q_{1\infty}$, we have:

$$q_{1\infty}=\frac{u_{1\infty}-k_{12}q_{2\infty}}{k_{11}} \tag{4.47}$$

Substituting Eq. (4.47) into (4.46) then gives:

$$q_{2\infty}=\frac{k_{21}u_{1\infty}-k_{11}u_{2\infty}}{k_{11}k_{22}-k_{12}k_{21}} \tag{4.48}$$

20 In certain special cases, unlikely in applications, this generalization is not valid, e.g. for for $\ddot{x}+x=u=C=$ constant, with $x(0)\equiv C,\ddot{x}\equiv 0$ provides the constant and unique steady state solution $x_\infty=C$. But if $x(0)\neq C$, then $x(t)=[x(0)-C]\cos t+C\equiv x_{ss}$, which is clearly not constant.

and substitution of Eq. (4.48) into (4.47) finally gives:

$$q_{1\infty} = \frac{k_{12}u_{2\infty} - k_{22}u_{1\infty}}{k_{11}k_{22} - k_{12}k_{21}} \tag{4.49}$$

These constant solutions clearly exist and are unique.

Special Case: The most common use of this model is for one input, $u_{1\infty} > 0$, with $u_{2\infty} \equiv 0$. In this case Eqs. (4.48) and (4.49) become:

$$q_{1\infty} = \frac{-k_{22}u_{1\infty}}{k_{11}k_{22} - k_{12}k_{21}} \tag{4.50}$$

$$q_{2\infty} = \frac{k_{21}u_{1\infty}}{k_{11}k_{22} - k_{12}k_{21}} \tag{4.51}$$

and we also can solve for the input $u_{1\infty}$ in terms of $q_{1\infty}$, from Eq. (4.50):

$$u_{1\infty} = \left(\frac{k_{11}k_{22} - k_{12}k_{21}}{k_{22}}\right)q_{1\infty} \tag{4.52}$$

Thus, the steady state masses $q_{1\infty}$ and $q_{2\infty}$ can be determined from Eqs. (4.50) and (4.51) if the k_{ij} and input $u_{1\infty}$ are known, and an unknown input $u_{1\infty}$ can be determined from Eq. (4.52) if the k_{ij} are known and $q_{1\infty}$ is measured. Finally, dividing Eq. (4.50) by Eq. (4.51) gives:

$$\frac{q_{1\infty}}{q_{2\infty}} = -\frac{k_{22}}{k_{21}} \tag{4.53}$$

Remark: The ratio of steady state masses in the two compartments is dependent only on the ratio of rate constants, $-k_{22}/k_{21}$, and is independent of k_{01}, the leak from compartment 1, as well as the magnitude of the input $u_{1\infty}$. We'll return to this ratio following introduction of equivalent distribution volumes in the section after next.

Nonlinear Multicompartmental Biomodels: Special Properties & Solutions

Most *linear* compartmental model parameters are fractional rate *constants* k_{ij}, with units of time^{-1}. Nonlinearities are introduced in various ways, often parametrically – *by representing fractional transfer rate or leak rate parameters as functions of the compartment state variable associated with the parameter*. This makes sense physically when such parameters represent nonlinear *fluxes*, i.e. influx, exchange or efflux rates in compartmental systems, e.g. as in the defining equation of the one-compartment model, Eq. (4.1).

The first-order *Hill function* – or *Michaelis–Menten function*, both developed in detail in Chapters 6 and 7, are common representations for NL saturating efflux or influx rates in

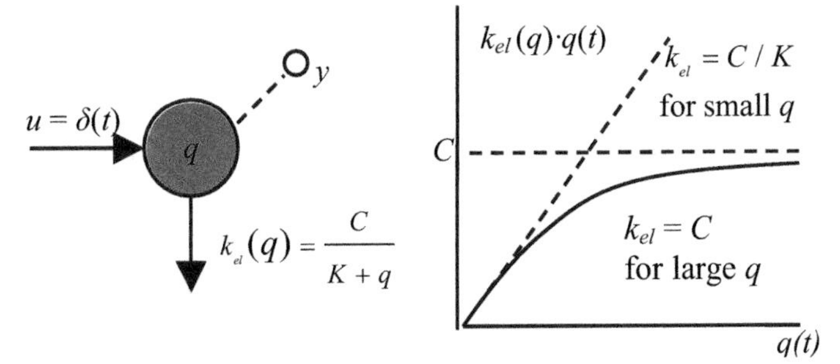

Fig. 4.15

compartmental models.[21] The **fractional rate parameter** associated with this function can be written:

$$k(q(t)) = \frac{C}{K+q(t)} \equiv k(q) \equiv \frac{C}{K+q} \quad \text{time}^{-1} \tag{4.54}$$

where C (mass/time units here) and K (mass units) are constants and $q(t) \equiv q$ is the compartment mass state variable associated with the influx or efflux. This function falls in value (reduced *fractional* transfer) as q increases. The following two examples illustrates use of $k(q)$, first for NL efflux – common in pharmacokinetic and biochemical kinetics models – and then for NL influx or "effects"– common in pharmacodynamic and protein network modeling (Chapter 7 and beyond).

Example 4.5: ***One-Compartment Model with Nonlinear Elimination***

In this model, $q(t)$ is the amount of substance in the single compartment, and $k_{el}q(t)$ is its mass efflux rate (amount per unit time), which depends nonlinearly on $q(t)$, as shown in the graph of $k_{el}q(t)$ versus $q(t)$ in **Fig. 4.15**. We note that the mass efflux rate $k_{el}q(t)$ is approximately linear for low values of $q(t)$, saturating for high $q(t)$. To see this more formally, we take limits:

$$\begin{aligned} &\lim_{q\to\infty} \frac{Cq}{K+q} = \lim_{q\to\infty} \frac{C}{K/q+1} = C, \text{ and } \cong C \text{ for } q \gg K \\ &\lim_{q\to 0} \frac{Cq}{K+q} = 0, \text{ and } \cong \frac{C}{K} q \text{ for } q \ll K \end{aligned} \tag{4.55}$$

The mass balance ODE model is simply:

$$\dot{q} = u - \frac{Cq}{K+q} \tag{4.56}$$

This model can be appropriate for substances extensively metabolized or eliminated in a single large organ, such as ethanol (ethyl alcohol) in the liver (Holford 1987). At low concentrations alcohol elimination is approximately linear, but at high concentrations, the metabolic process exhibits saturable inhibition and behaves like a zero-order

21 This is an example of a **kinetic law** in programming languages for chemical kinetics, e.g. *Systems Biology Markup Language (SBML)*.

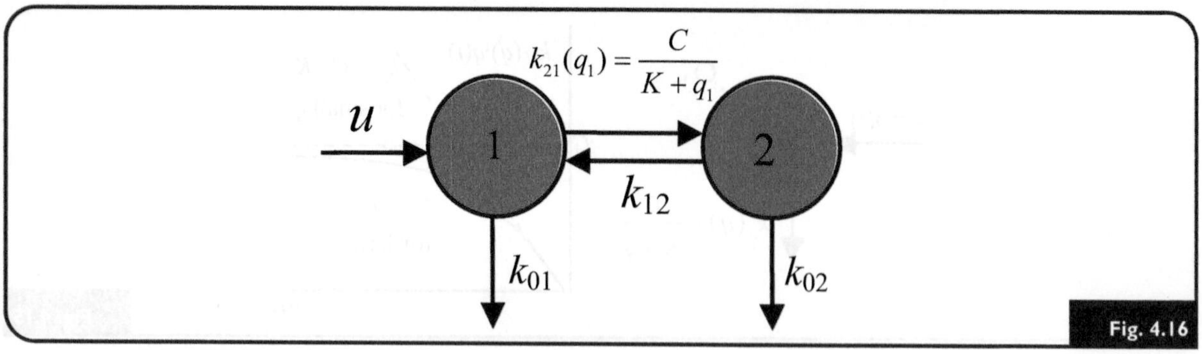

Fig. 4.16

NL influx to compartment 2 from 1.

elimination (constant) process. Thus, alcohol is metabolized faster when its blood concentration is lower, a scientific justification for keeping oral influx of alcohol low!

Example 4.6: ***Two-Compartment Model with Nonlinear Transfer Flux***

The model in **Fig. 4.16** has nonlinear influx of material into compartment 2 from 1, of Hill function form – as in Eq. (4.55), and we also have linear elimination components from both compartments. The nonlinearity here might represent receptor-mediated action of a biomolecule in compartment 1 on cell receptors in compartment 2 – for example a drug effect on a target cell (elaborated upon in Chapter 6). The ODE model is readily written from the diagram:

$$\begin{aligned} \dot{q}_1 &= -\left(k_{01} + \frac{C}{K+q_1}\right)q_1 + k_{12}q_2 + u \\ \dot{q}_2 &= \frac{Cq_1}{K+q_1} - (k_{02} + k_{12})q_2 \end{aligned} \tag{4.57}$$

Nonlinear (NL) Compartment Model Building & Analysis

The basic assumption of homogeneity of compartments remains for nonlinear ODE compartmental models, each compartment remains associated with a different state variable, and the procedures for structuring them is generally the same as that described for linear models in much of this chapter. However, solutions for the state variables of NL models (with some exceptions) are not computable analytically – only numerically. Furthermore, their dynamic responses and other properties are generally more complex and in some cases can be surprisingly unpredictable, as we see in Chapter 9. A major difference between linear and NL models is that dynamic responses of NL models can be quite different – qualitatively as well as quantitatively – depending on the *magnitude* as well as the shape[22] of the inputs $\boldsymbol{u}(t)$. *NL models do not obey the linear superposition principle for input–output relationships* elucidated in Chapter 2. This is illustrated indirectly in the two examples above. The output responses $q(t)$ are "manipulated" in

22 In contrast, linear system models responses (zero-state responses, i.e. zero initial conditions) are qualitatively the same, and quantitatively satisfy the superposition principle for inputs.

a NL way by their own instantaneous magnitude, which in turn are dependent on the magnitude of the inputs. Very small inputs generate approximate linearity of responses, and larger inputs drive NL responses towards "saturation," or no change in the qs. The next example clearly demonstrates this characteristic phenomenon quantitatively, by comparative simulations.

Example 4.7: ***MATLAB Solution for the Two-Compartment Model with NL Transfer Flux***

A simple *MATLAB* program for solving Eqs. (4.57) is given in **Fig. 4.17**. The program has two files, one representing the *function NL2COMP.m*, for solving $\dot{\boldsymbol{q}} = \boldsymbol{f}(\boldsymbol{q}, u, \boldsymbol{p})$, and the other the "run" file with program specifications for the inputs, initial conditions, time interval and choice of solver. We chose the adaptive step-size Dormand–Prince solver *ODE45* in *MATLAB*, a variant of the Runge–Kutta–Fehlberg algorithm discussed in Chapter 3. Solutions for inputs of different step-input magnitudes D are explored, to illustrate simply the dynamic (nonlinear) complexity of NL model responses $q_2(t)$ in compartment 2 to different inputs into compartment 1.

The ODEs (4.57) represent the model of **Fig. 4.16**. For simplicity, we set the values of all five parameters = 1 and explore three different ranges of the NL function $\frac{q_1}{1+q_1}$: the linear range – when the input magnitude is very small (up to about $D = 0.1$), the middle NL range ($\sim D = 0.1$ to 5), and the saturated range – when D is large (>10). These ranges are consistent with the limits of the NL function (4.54):

Defining the ODE and params in a function routine (filename: NL2COMP.m)

```
function dqdt = NL2COMP(t, q)
global k01 k02 k12 C K D;
u = D;
q1_dot = -(k01 + C/(K+q(1)))*q(1) + k12*q(2)+ u;
q2_dot = C*q(1)/(K+q(1)) - (k02+k12)*q(2);
dqdt = [q1_dot; q2_dot];
```

Running the pgm (filename: NL2COMP_NLRange.m)

```
global k01 k02 k12 C K D;
k01= 1; k02 = 1; k12 = 1; C = 1; K = 1;
time = [0, 10]; ICs = [0, 0];
D = 1;
[t1, q1] = ode45(@NL2COMP, time, ICs);

D = 2.5;
[t2, q2] = ode45(@NL2COMP, time, ICs);

D =5;
[t3, q3] = ode45(@NL2COMP, time, ICs);

plot(t1,q1(:,2), t2,q2(:,2), t3,q3(:,2),...
'linewidth', 3)
```

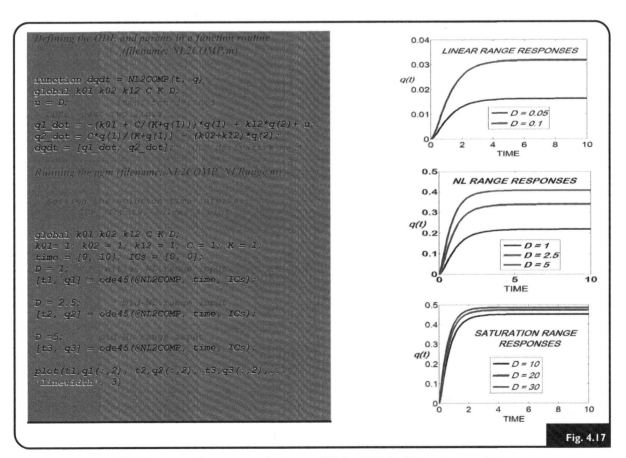

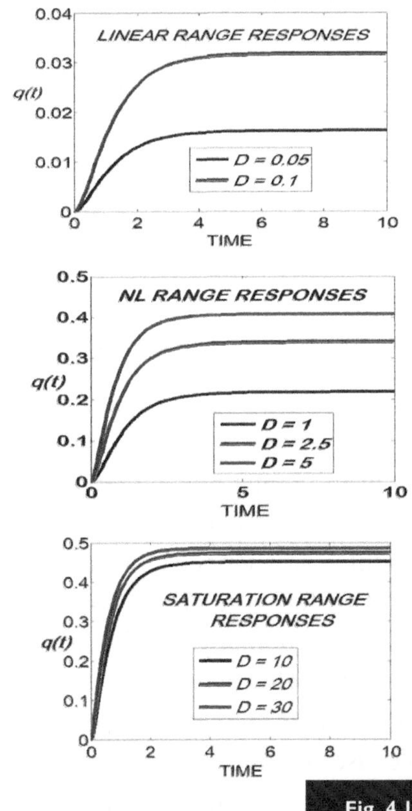

Fig. 4.17

MATLAB program and compartment 2 solutions of NL Eqs. (4.57) for different input magnitudes.

$\dfrac{q_1}{1+q_1} \rightarrow q_1$ (linear) for $q_1 \ll 1$, and $\dfrac{q_1}{1+q_1} \rightarrow 1$ (constant) for $q_1 \gg 1$. Constant inputs are chosen, which generate responses that go to approximate steady state over the time interval for simulation [0,10], thereby facilitating solution comparisons.

The three sets of solutions for $q_2(t)$ in *compartment 2* are shown in **Fig. 4.17**. Compartment 2 is the *receiving* compartment here, where the NL effects are most pronounced (e.g at a hormone or drug receptor site on a cell membrane). Compartment 1 dynamics (not shown) are affected, but not nearly as much.

The linear range plots are for $D = 0.05$ and 0.1, verifying that the steady state response for $q_2(t)$ is doubled for a doubling of the input magnitude. The NL range plots are for $D = 1$, 2.5 and 5 and evidently no linearity exists among these responses. The saturated range plots are for $D = 10$, 20 and 40, twice doubling the input magnitude. The responses have nearly equal (saturated) steady state values.

Dynamic System Nonlinear Epidemiological Models

Nonlinear (NL) compartmental models are convenient for studying the spread of disease epidemics. We introduce three classical formulations of disease process dynamics here, exemplifying how they are formulated and solved.

In this modeling domain, an at-risk human (or other animal) population of total size N typically is divided into two or three groups (simplest models): the number S – **susceptible** to the disease; the number with the disease – denoted I for **infectives**; and in some the number R –**recovered** from the disease. Each group is assumed to be homogeneous with respect to these states of health and disease, i.e. all of S are equally susceptible, etc., providing the primary basis for multicompartmental modeling.

Special Nomenclature: S, I and R are the typical notation used in modeling epidemics. To maintain a consistent notation in this chapter and most of the rest of the book, we redefine S, I and R here as q_S, q_I and q_R, the time-varying numbers of susceptible, infective and recovered individuals in a population of size N. The assumption that these groups are homogeneous means that each can be modeled as a single compartment.

Flows between compartments are represented less typically, because one includes a nonlinear product term – similar to the shark–tuna predator–prey example in Chapter 3 – where we did not use a compartmental representation. In Example 4.5 and Example 4.6 above, we encountered a similar representational difficulty for nonlinear models in compartmental form, in those cases with a nonlinear quotient representing fractional flows from a compartment. There is no established standard for representing nonlinear flows in compartmental models and the choice in those examples is one of many possible ways to represent flows in those one- and two-compartment NL model structures. However, a currently developing standard in the molecular systems biology community for biochemical reaction flow rates – one that reappears in Chapter 6 and beyond – is useful here. We use a lowercase v_{ij} to represent total flow from compartment j to compartment i, in this case flow in number of individuals in the population per unit time units. For example, in the models below, $v_{IS} = k_{IS}q_Iq_S$, and k_{IS} is the fraction of the product q_Iq_S of susceptible and infective individuals themselves that become infective, given that only a fraction of interactions among individuals leads to infection. The similarity here with predator–prey models is no coincidence. Epidemic models are clearly in the same class. The remaining interaction terms in our three models below are linear, but for consistency, we'll represent these in the form of whole flow rates instead of fractional flow rates in their respective compartmental diagrams.

SIS, SIR and SIRS Models

All three of these models are shown in **Fig. 4.18** without inputs or outputs. Any of the compartments can receive inputs or be measured as outputs. Input–output model details are needed for model quantification and analysis, as illustrated below and in the exercises as the end of the chapter.

The simplest model – mathematically speaking – includes infectives and temporarily recovered individuals who become (reversibly) susceptible again. This model has two compartments and is called an **SIS model**, with $N = S + I \equiv q_S + q_I$, as shown in **Fig. 4.18**. Either of these two state variables can be measured, as $y_S(t)$ or $y_I(t)$, and either can receive inputs, $u_S(t)$ or $u_I(t)$. For example, the number of susceptible individuals might be increased by an exogenous increase $u_S(t) = \Delta S \equiv \Delta q_S(t)$ in healthy, susceptible individuals, e.g. a traveling group unaware of the presence of the disease at their destination. This would increase the total population, i.e. $N(t) = \Delta q_S(t) + q_S(t) + q_I(t)$. From the figure, the SIS model equations, without exogenous inputs or measured outputs specified, are:

$$\dot{q}_I = -\dot{q}_S = k_{IS}q_Iq_S - k_{SI}q_I = k_{IS}q_I(N - q_I) - k_{SI}q_I = (k_{IS}N - k_{SI})q_I - k_{IS}q_I^2 =$$
$$\equiv a(1 - q_Ik_{IS}k_{SI})q_I = a(1 - q_Ik_{SI}/r)q_I = a(1 - bq_I)q_I \tag{4.58}$$

with $a \equiv k_{IS}N - k_{SI}$ and $b \equiv k_{SI}/r$

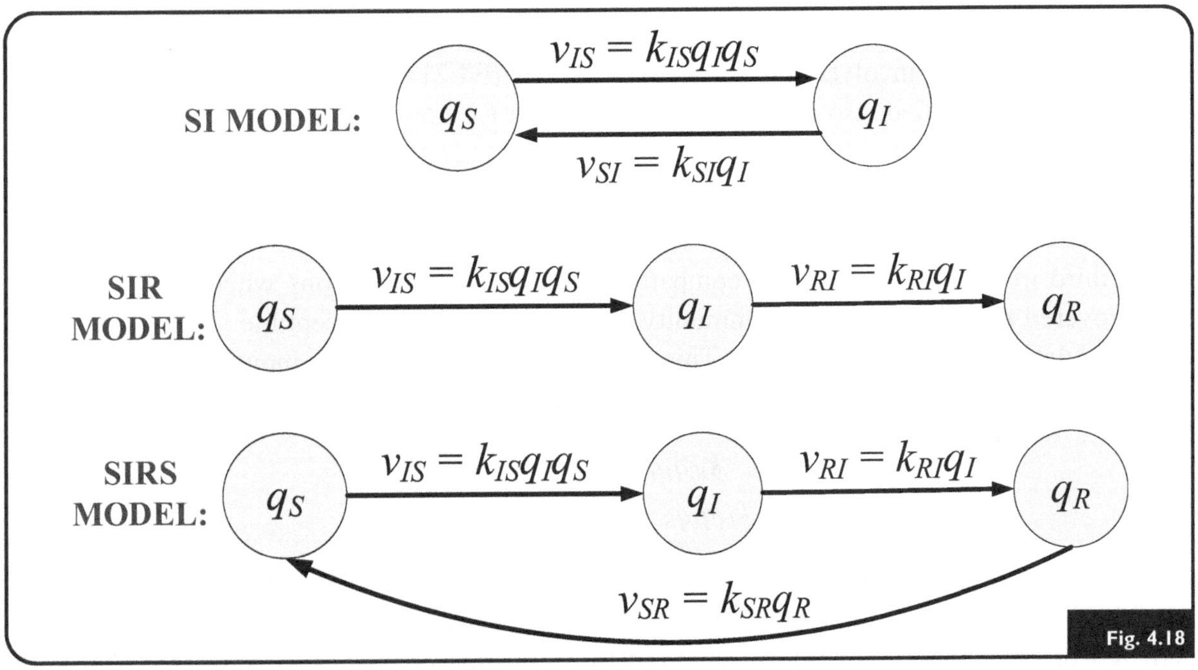

Fig. 4.18

Compartmental models of infections.

Note that we substituted the conservation constraint $N = q_S + q_I$ in the second equality. Eq. (4.58) is the familiar **logistic equation**, with **intrinsic growth rate** a and **carrying capacity** $1/b$ (see Example 2.9 in Chapter 2). The solution of (4.58), by direct integration – as in Example 2.9 – is:

$$q_I(t) - 1/b(1 + ce^{-at}) \text{ with } c \text{ an arbitrary constant of integration} \tag{4.59}$$

Let's see how the epidemic progresses with this model. If a fixed population of size N has one individual infected at the start, $q_I(0) = 1$ (equivalent to a unit-impulse input), then from (4.59) $c = (1 - b)/b$ and the steady state (eventual) number of infected individuals becomes $q_{I\infty} = 1/b = N - k_{SI}/k_{IS}$ as t gets large enough. This means the epidemic will not affect the entire population N, and the actual number depends on the relative ratio of susceptible to infected encounters, k_{SI}/k_{IS}. **Fig. 2.1** in Chapter 2 illustrates these dynamics.

A slightly more complex three-compartment model is for a presumably less serious disease (e.g. measles) in which recovered individuals are (irreversibly) no longer susceptible, due to acquired immunity. The model for this is called an **SIR model**, with $N = S + I + R$, also as shown in **Fig. 4.18**. This model also can potentially have three inputs and/or measured outputs. To develop the model, we recognize that the recovery rate is donor controlled, because the fraction k_{RI} of infectives do recover per unit time (fractional **recovery rate**). From the figure, the SIR model equations – again without inputs or outputs specified – are:

$$\begin{aligned} \dot{q}_S &= -k_{IS}q_Iq_S \\ \dot{q}_I &= k_{IS}q_Iq_S - k_{RI}q_I \\ \dot{q}_R &= k_{RI}q_I = \dot{q}_S - \dot{q}_I \end{aligned} \tag{4.60}$$

The constraint $q_R \equiv N - q_I - q_S$ renders the last ODE superfluous, i.e. dependent, because $\dot{q}_S + \dot{q}_I + \dot{q}_R = \dot{N} = 0$ for a constant population N.

Solution and analysis of (4.60) is, as above, in response to fixed ICs. Analytic solution is mathematically more involved and left as an exercise (E4.21). Simulation solution with parameters and ICs specified also is illustrated, in Exercise E4.22.

Remark: An epidemic can be avoided completely if $q_S(0)$ is sufficiently small. This is in part the basis for vaccinating a population, thereby keeping initial susceptibles $q_S(0)$ small.

A third model, also with three compartments, represents situations where fraction k_{SR} of the recovered population loses its immunity and again becomes susceptible. This is called the **SIRS model**, as shown in **Fig. 4.18**. This model also can have three inputs and/or measured outputs. From the figure, the SIRS model equations, without inputs or outputs specified, are:

$$\begin{aligned} \dot{q}_S &= -k_{IS}q_Iq_S + k_{SR}(N - q_I - q_S) \\ \dot{q}_I &= k_{IS}q_Iq_S - k_{RI}q_I \end{aligned} \tag{4.61}$$

In (4.61), we made the substitution $q_R \equiv N - q_I - q_S$ in the first ODE, thereby eliminating the need for the equation for $\dot{q}_R$. Solution of (4.61) cannot be obtained in closed form, with either non-zero ICs or other inputs, but can be analyzed by simulation or by *linearization* (see Chapter 5, Exercise E5.10).

Remark: Implicit in these models is the assumption that infection and recovery have much faster dynamics than the time scale of births and deaths from all causes in the population. Therefore birth and death dynamics are ignored in the simplest model forms above, because – in accordance with this assumption – there is no net change in births and deaths. On the other hand, if disease dynamics are relatively long-term, birth and death dynamics may be important and can be readily included, without the need for additional compartments; they change only individual compartment component interactions and ODEs, with total population N remaining the same: $N = q_S + q_I$ (see Exercise E4.24).

Remark: The presumably non-fatal disease models above are often used, with some simple adjustments, to model deadly diseases (e.g. malaria). For the SIR and SIRS models, R then represents a mix of recovered and removed (dead) individuals, with deaths occurring from the disease – on the same time scale of infection and recovery. R must be measured to make this distinction, the number of compartments remains the same, but the structures and dynamic responses are different (Exercise E4.25).

Remark: More complex disease models have additional compartments, e.g. a fourth compartment for **exposed** (E) individuals who may or may not come down with the disease, or for pathogens (W) in the water supply (Eisenberg et al. 2012). Also see (Daley and Gani 2005; Vynnycky and White 2010).

Epidemics & Reproduction Number R_0

A very important application of infectious disease models is to establish whether or not a disease can spread through a population. A useful metric for this purpose is the **reproduction number** R_0, i.e. the average number of cases an infected individual generates over the course of their infectious period. R_0 signifies whether an epidemic occurs or a disease simply dies out. This metric can be computed from a disease model. For example, if the entire population is susceptible, i.e. $q_S(0) \approx N$, then an infective individual makes $k_{IS}N$ contacts per unit time, producing new infections with a mean (fractional) recovery rate k_{RI} (mean infectious period of $1/k_{RI}$). Then: $R_0 = k_{IS}N/k_{RI}$ (also see Exercise E4.26). When $R_0 < 1$, the infection will eventually die out. But, if $R_0 > 1$, it will spread – an **epidemic**. If $R_0 = 1$, the disease is **endemic**, meaning it remains in the population (van den Driessche and Watmough 2002; Keeling and Rohani 2007). More general models utilizing this metric are given in many references, e.g. in Blower et al. (1995) for tuberculosis and in Eisenberg et al. (2012) for cholera.

Compartment Sizes, Concentrations & the Concept of Equivalent Distribution Volumes

The *size* of a compartment in a multicompartmental model (or pool in a pool model) is expressed in one of two ways: size in mass terms or size in volume terms. Mass means amounts, numbers of molecules, abundance, etc. Concentration means mass per unit volume, mass densities (mass per unit mass), number of molecules per unit mass or volume, etc. For systems with molecular signals, the mass interpretation is natural. It appeals to intuition, as well as satisfying physical conservation laws easily transformed into and manipulated in mathematical equations. In this context compartment volumes can have more than one interpretation. The two most common are: real (true) volumes, associated with well-defined physical spaces, and *equivalent distribution* volumes – abstract and equivalent in an operational sense. Real volumes are often difficult or impossible to measure, so an "equivalent" V is used often to denote the hypothetical volume that a substance would occupy if its concentration in a region of interest were the same as that in some measurable reference region, for example, plasma or whole blood in organ-level models. This concept is important because, in steady state, *plasma* (or other space) *equivalent distribution volumes*, denoted V_i for a single compartment i and V_D for the whole organism, can provide true total quantities of substance Q_i or Q_{tot}, in regions of interest.

Equivalent distribution volumes[23] provide computable measures of relative compartment sizes for situations where true volumes are either undefined or experimentally indeterminate. The one-compartment model provides a first, physically appealing step in defining the concept, because it represents a single homogeneously distributed physical mass. The **equivalent distribution volume** V_D for the one-compartment model is given by the (true) mass of substance in the system at any time t, $q(t)$, divided by the (uniform) concentration $c(t)$ of that substance, at the same time t, i.e. $V_D = q(t)/c(t)$. By introducing the concentration variable $c(t)$, we are implicitly restating the fundamental assumption of the one-compartment model: that the mass $q(t)$ is *uniformly distributed* throughout the entire system, which means that its concentration $c(t)$ is assumed to be the same at every point in the space the system occupies.

Remark: It is usually assumed that compartment volumes, real or equivalent, are constant in multi-compartmental models, although this is not always strictly valid. For example, total blood volume in humans varies, at least diurnally, due to gravitational effects of standing versus lying down. We bypass this issue for now, and assume V is constant, and thus $V = q(t)/c(t)$ at any time t.

Let's focus on the uniformity of concentration of material in a one-compartment model and apply this idea to the two-compartment model. By definition of a compartment, it must be that the concentration of material in compartment 2 is uniform in compartment 2 and the concentration of material in compartment 1 is uniform in compartment 1. Now we make what at first appears to be an unjustified assumption: that the concentrations of material in each compartment are equal in steady state, i.e. $c_{1\infty} \equiv c_{2\infty}$. In other words,

$$c_{1\infty} \equiv \frac{q_{1\infty}}{V_1} \equiv \frac{q_{2\infty}}{V_2} \equiv c_{2\infty} \tag{4.62}$$

If this assumption were realistic, then V_1 and V_2 would be *real* or *true* volumes, with each compartment possibly representing a well-defined physical space. But, if $c_{1\infty} \neq c_{2\infty}$, then either V_1 or V_2, or both, are (nonphysical) *equivalent* volumes, based on the assumption of equality of concentration throughout. These V_i are useful in applications, as we see below. This is the essence of the concept. We are now ready to state the definitions more generally and formally.

23 In pharmacokinetics, equivalent distribution volumes are also called *apparent* distribution volumes. More on this in Chapter 8.

The **equivalent distribution volume V_i of single compartment i** in a multicompartment model is the equivalent volume the substance would occupy in steady state if it were distributed uniformly and with the same concentration in compartment i as the **reference concentration** of the substance c_{ref} in some (other) **reference compartment** (compartment j, $j \neq i$) in steady state, i.e. $c_{i\infty} \equiv c_{j\infty} \equiv c_{ref}$.

The **total system/model equivalent distribution volume V_D** is the equivalent volume the substance would occupy in the steady state if it were distributed uniformly and with the same concentration in *every* compartment in steady state, i.e. $c_{1\infty} \equiv c_{2\infty} \equiv \ldots \equiv c_{n\infty} \equiv c_{ref}$. Therefore, for an n-compartment model,

$$V_D = \sum_{i=1}^{n} V_i \tag{4.63}$$

Remark: Usually, c_{ref} is a real, measured concentration, but it doesn't have to be.

The reason for defining this concept for the steady state, rather than at any time t, given that V_i and V_D are assumed to be constant, is illustrated by rearranging the simple relationship for the two-compartment model in Eq. (4.62):

$$\frac{q_{1\infty}}{q_{2\infty}} = \frac{V_1}{V_2} \tag{4.64}$$

This says that the ratio of the true steady state masses is equal to the ratio of the equivalent distribution volumes.[24] This result is important because equivalent distribution volumes provide a measure of relative compartment sizes, in mass or equivalent volume terms, when no such measures are available or convenient from direct observations.

If all compartments in a system can be probed, e.g. if both true compartment concentrations, $c_1(t) \equiv y_1(t)$ and $c_2(t) \equiv y_2(t)$ in the two-compartment model can be *measured* directly over some finite time interval $0 \leq t \leq T$ or in steady state, these concentration measurements can provide measures of the *true* distribution volumes of both compartments 1 and 2, V_1^{true} and V_2^{true}. In this case $V_1^{true} \neq V_1$ or $V_2^{true} \neq V_2$, and the real $c_{1\infty} \neq c_{2\infty}$. However, Eqs. (4.62) to (4.64) are still valid if V_1 and V_2 retain their not necessarily real, equivalent volume interpretations.

Remark: Total plasma equivalent distribution volumes V_D are often of primary interest in a kinetic study, especially comparative clinical or pharmacological system studies.[25] If the reference concentration c_{ref} of material is measured (e.g. in blood or plasma), and the total body equivalent distribution volume V_D can be determined from the kinetic study, then the *total body mass* of material is readily computed as: $Q_{tot} = V_D c_{ref}$.

Example 4.8: ***Total Body Essential-Metal Pool Sizes***

Metals like zinc and manganese are essential for proper nutrition, but only in trace amounts. Too much can be harmful (Chowdhury and Chandra 1987; Douglas et al.

24 This is not true for finite times, only in steady state; for finite t, $q_1(t)/q_2(t) \neq V_1/V_2$.

25 These are developed in detail in Chapter 8.

2010). Blood plasma concentrations $c_{p\infty}$ (presumably measured in steady state) indicate their presence and level in plasma. But their total body amount Q_{tot} is what's important, i.e. whether too much or too little is present. Q_{tot}, in turn, depends on numerous physiological factors governing the sizes of tissue pools of the metals, differing in different individuals.[26] These are aggregated in the total body equivalent distribution volume V_D for the metal, and V_D often can be computed (estimated) from the data collected in a kinetic study using a tracable form of the metal introduced exogenously (Chowdhury and Chandra 1987). For example, if the system can be approximated by a one-compartment model – a common assumption in many pharmacokinetic studies – then V_D can be estimated by extrapolation of the impulse response data, i.e. the single exponential output $y(t) = Q_0 e^{kt}$, backward to zero-time. That is, with drug dose Q_0 and $u(t) = Q_0\delta(t)$, concentration measurement $y(0) = Q_0/V_D$, $V_D = Q_0/y(0)$ and $Q_{tot} \equiv V_D c_{p\infty} \cong Q_0 c_{p\infty}/y(0)$.

General *n*-Compartment Models with Multiple Inputs & Outputs

We now consider n compartments with r inputs and m outputs, and use both scalar and vector-matrix notation introduced earlier to represent these, as needed. *The development here is for linear n-compartment models.* Extension to NL systems usually entails generalizing the interpretation of the fractional rate parameters k_{ij}, allowing them to be functions of the state variables, e.g. $k_{ij} \equiv k_{ij}(q_j)$, as discussed earlier in the chapter, or as NL mass fluxes $v_{ij}(\boldsymbol{q})$ as in the section on epidemiological models above. Exceptions are noted as needed.

Remark: Vector-matrix notation is more than just "elegant," it is maximally parsimonious and enhances understanding of concepts as well as facilitating computation.

Model Dynamics

Assume we are modeling a mass-based system and thus the (vector) inputs $\boldsymbol{u}(t) \geq \boldsymbol{0}$ and states $\boldsymbol{q}(t) \geq \boldsymbol{0}$ for all times $t > 0$. Actually $\boldsymbol{q}(t) \geq \boldsymbol{0}$ is a *consequence* of the nonnegativity of inputs and the physical constraints $k_{ij} \geq 0$ for $i \neq j$ (Bellman 1970). In general, the ith, jth, and sth compartments of an n-compartment model can be fully interconnected as shown in **Fig. 4.19**.

By conservation of mass flux (the "derivative" of conservation of matter), we have for compartment j:

$$\begin{aligned} \dot{q}_j &= \text{influx} - \text{efflux} = k_{ji}q_i + k_{js}q_s + \ldots - (k_{0j} + k_{ij} + k_{sj} + \cdots)q_j + u_j \\ &= \sum_{\substack{l=1 \\ l\neq j}}^{n} k_{jl}q_l - \sum_{\substack{l=0 \\ l\neq j}}^{n} k_{lj}q_j + u_j \end{aligned} \tag{4.65}$$

26 It would be better if one could compute the total amount in usually inaccessible organs, like brain, where toxins have their effects, like Mn in brain (Douglas et al. 2010). PK studies, requiring only blood measurements, are the next best thing.

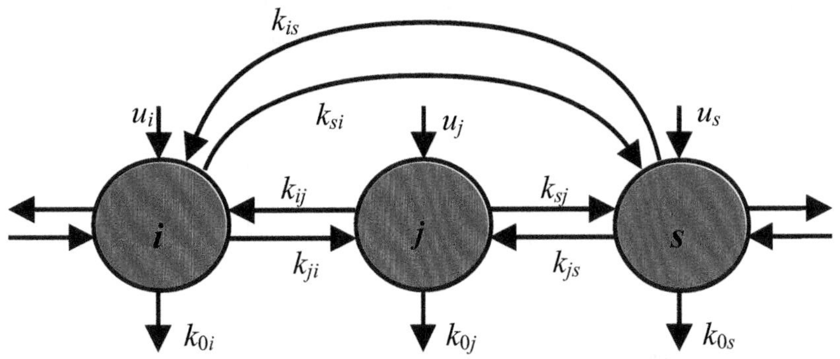

Fig. 4.19

Three compartments of a general n-compartment model.

To simplify the equations, we define the **turnover rate** of the jth compartment:

$$\text{turnover rate of compartment } j \equiv -k_{jj} \equiv \sum_{\substack{l=0 \\ l \neq j}}^{n} k_{lj} = k_{0j} + k_{1j} + \cdots + k_{nj} \tag{4.66}$$

Remark: The turnover rate $-k_{jj}$ is the sum of the arrows leaving the compartment. The inverse of $-k_{jj}$ is called the **mean transit time** through the compartment, $\tau_j \equiv -1/k_{jj}$, the mean time spent by the material in the compartment each time it passes in and out of the compartment.

Combining Eq. (4.65) and (4.66):

$$\dot{q}_j = k_{j1}q_1 + \cdots + k_{jn}q_n + u_j = \sum_{l=1}^{n} k_{jl}q_l + u_j \tag{4.67}$$

and, in vector-matrix form, the ***n*-compartment model** becomes:

$$\dot{\boldsymbol{q}} = \begin{bmatrix} \dot{q}_1 \\ \dot{q}_2 \\ \vdots \\ \dot{q}_n \end{bmatrix} = \begin{bmatrix} k_{11} & k_{12} & \cdots & k_{1n} \\ k_{21} & k_{22} & \cdots & k_{2n} \\ \vdots & \vdots & \ddots & \vdots \\ k_{n1} & k_{n2} & \cdots & k_{nn} \end{bmatrix} \begin{bmatrix} q_1 \\ q_2 \\ \vdots \\ q_n \end{bmatrix} + \begin{bmatrix} u_1 \\ u_2 \\ \vdots \\ u_n \end{bmatrix} + K\boldsymbol{q} + \boldsymbol{u} \tag{4.68}$$

where we simplified the notation with $K \equiv [k_{ij}]$, and our equations are constrained by $k_{ij} \geq 0$ for $i \neq j$, $\boldsymbol{u}(t) \geq \boldsymbol{0}$ and $\boldsymbol{q}(t) \geq \boldsymbol{0}$ for all $t \geq 0$.

Remark: If we use this standard vector-matrix notation, the equations for the most-general n-compartment model reduce to this very simple vector-matrix, constant coefficient linear ODE: $\dot{\boldsymbol{q}} = K\boldsymbol{q} + \boldsymbol{u}$, with the constraints noted. This form is much easier to remember, as well as to algebraically manipulate.[27]

27 Nonlinear (NL) biochemical reaction kinetics models are introduced in Chapter 6 using a similar notation. The **stoichiometric matrix** N is a matrix of stoichiometric coefficients of biochemical reactions and linear-looking ODEs $\dot{\boldsymbol{q}} = N\boldsymbol{v} + \boldsymbol{u}$. The difference from $\dot{\boldsymbol{q}} = K\boldsymbol{q} + \boldsymbol{u}$ is that the v are NL mass fluxes $v_i(\boldsymbol{q})$. More on this in Chapter 6.

Remark: The **compartment matrix** K is expressed here in standard matrix notation, the notation that predominates in fields that utilize compartmental methods. Specifically,

$$K = \begin{bmatrix} k_{11} & k_{12} & \cdots & k_{1n} \\ k_{21} & k_{22} & \cdots & k_{2n} \\ \vdots & \vdots & \ddots & \vdots \\ k_{n1} & k_{n2} & \cdots & k_{nn} \end{bmatrix} \tag{4.69}$$

Some authors, particularly in pharmaceutical sciences, define turnover rates $-k_{11}$, $-k_{22}$, etc. with the opposite sign; and they define the k_{ij} notation for $i \neq j$ the other way around as well, i.e. k_{ji} instead of k_{ij}, e.g. (Wagner 1993; Rescigno 2003). There is nothing intrinsically wrong in using this or any other notation. It does, however, make analysis of more than the simplest models more difficult when matrix manipulations are needed. Nonstandard notation is especially problematic when using matrix manipulation software, which usually follows standard conventions, as in K in Eq. (4.69).

Output Measurements Model

If all state variables (masses in this case) are measured directly, $\boldsymbol{y} = \boldsymbol{q}$ $(m = n)$. If measured indirectly, or in linear combinations, $\boldsymbol{y} = C\boldsymbol{q}$, where C is the necessary linear transformation. If the outputs instead are concentrations, and if all compartment concentrations are measured,

$$\boldsymbol{y} = \begin{bmatrix} y_1 \\ y_2 \\ \vdots \\ y_n \end{bmatrix} = \begin{bmatrix} 1/V_1 & & 0 \\ & \ddots & \\ 0 & & 1/V_n \end{bmatrix} \begin{bmatrix} q_1 \\ q_2 \\ \vdots \\ q_n \end{bmatrix} = V^{-1}\boldsymbol{q} \tag{4.70}$$

where $V \equiv [V_{ii}]$ is the diagonal **volume matrix**, because each concentration $y_i = q_i/V_i$. In this case, all $V_{ii} \equiv V_i^{true}$ are real volumes, not equivalent ones.

Remark: Only rarely are all concentrations (or masses) measured. Typically, for practical reasons, only one is, e.g.

$$\boldsymbol{y} = \begin{bmatrix} \frac{1}{V_1} & 0 & \cdots & 0 \end{bmatrix} \boldsymbol{q} = \frac{q_1}{V_1} = y_1$$

Remark: Eqs. (4.68) and (4.70) represent linear systems. But they also typically represent linearized versions of a perturbation experiment performed on a nonlinear system, e.g. a tracer experiment on a biological system – the subject of the last section of the next chapter. For this mass-based form of the model, the biological system is represented by a compartment model, with the state variables chosen as the amounts of tracer in each compartment, $q_i(t)$, and the outputs (measurements) are concentration variables.

Remark: If the model is nonlinear (NL), the k_{ij} in (4.65) and (4.68) are functions of the q_i, so the same vector matrix form can be used for NL models.

The Constituent Compartmental Equations

For all k_{ij} with $j \neq i$, we always have $k_{ij} \geq 0$. Then, from Eq. (4.66), the turnover rate of compartment j can be written: $k_{jj} \equiv -\sum_{\substack{i=0 \\ i\neq j}}^{n} k_{ij} = -\sum_{\substack{i=1 \\ i\neq j}}^{n} k_{ij} - k_{0j} \leq 0$, a nonpositive sum of rate constants.

Then, solving for k_{0j} yields $k_{0j} = -\sum_{i=1}^{n} k_{ij}$ for all $j = 1, 2, \ldots, n$. This, incidentally, says that the sum of the terms in column j of matrix K equals $-k_{0j}$. Transposing terms finally yields:

$$\sum_{i=0}^{n} k_{ij} = 0 \tag{4.71}$$

In summary, Eqs. (4.68) and (4.71) completely define the dynamics of a compartmental model, exclusive of its outputs, i.e.

$$\dot{\boldsymbol{q}} = K\boldsymbol{q} + \boldsymbol{u}, \text{ constrained by } \boldsymbol{u}(t) \geq \boldsymbol{0},\ k_{ij} \geq 0 \text{ for } i \neq j, \text{ and } \sum_{i=0}^{n} k_{ij} = 0 \text{ for all } j \tag{4.72}$$

Remark: Relations (4.72) are the **constituent relations of a compartmental model**. In addition, because the state variables $q_i(t)$ represent masses, we also have the constraint $q(t) \geq \boldsymbol{0}$ for all t. This also renders all terms $\varphi_{ij}(t-\tau) \geq 0$ for the **state transition matrix**[28] of a compartmental model (Bellman 1970), a useful result for some model analyses.

In the next several examples we see how algebraically complex different three-compartment model configurations can be. First, the most general:

Example 4.9: ***The Most General Three-Compartment Model***

In **Fig. 4.20** we assume for generality that all inputs and outputs and all parameters are present. We write the model equations and derive the transfer function matrix of the model from them. This model has nine nonnegative and intrinsic rate constants: k_{01}, k_{02}, k_{03}, k_{12}, k_{13}, k_{21}, k_{23}, k_{31}, k_{32}, and three nonnegative volumes. With all compartments directly measured, the volumes are real: V_1^{true}, V_2^{true}, and V_3^{true} instead of equivalent. We also have a maximum of three possible independent inputs, $u_1(t)$, $u_2(t)$, and $u_3(t)$, and a maximum of three independent measurements or outputs, $y_1(t)$, $y_2(t)$ and $y_3(t)$, in compartments 1, 2, and 3, respectively. The ks are (fractional) rate constants (with dimensions of time^{-1}), the Vs are volumes, the us are mass flux rates (mass/time), and the ys are measured as concentrations (mass/volume) or

28 State transition matrices $\Phi(t,\tau)$ are discussed in Appendix C. For a linear compartment model, $\Phi(t,\tau)$ is the solution of $\dot{\Phi} = K\Phi$.

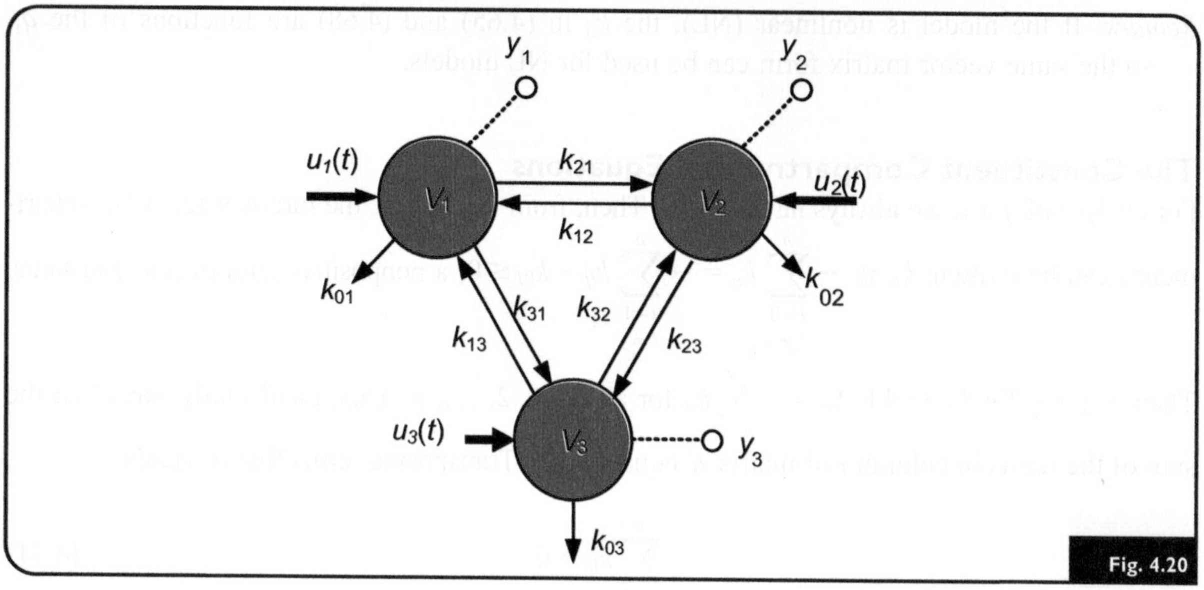

Fig. 4.20

amounts (mass). The three state variable and concentration measurement equations are written directly from **Fig. 4.20**:

$$\begin{aligned} \dot{q}_1(t) &= (k_{01} + k_{21} + k_{31})q_1 + k_{12}q_2 + k_{13}q_3 + u_1 \\ \dot{q}_2(t) &= k_{21}q_1 - (k_{02} + k_{12} + k_{32})q_2 + k_{23}q_3 + u_2 \\ \dot{q}_3(t) &= k_{31}q_1 + k_{32}q_2 - (k_{03} + k_{13} + k_{23})q_3 + u_3 \end{aligned} \tag{4.73}$$

$$y_1(t) = \frac{q_1(t)}{V_1^{true}}, \quad y_2(t) = \frac{q_2(t)}{V_2^{true}}, \quad y_3(t) = \frac{q_3(t)}{V_3^{true}} \tag{4.74}$$

In state variable form these become:

$$\begin{aligned} \dot{\boldsymbol{q}} = K\boldsymbol{q} + \boldsymbol{u} &= \begin{bmatrix} k_{11} & k_{12} & k_{13} \\ k_{21} & k_{22} & k_{23} \\ k_{31} & k_{32} & k_{33} \end{bmatrix} \boldsymbol{q} + \boldsymbol{u} \\ \dot{\boldsymbol{q}} &= \begin{bmatrix} -(k_{02} + k_{21} + k_{31}) & k_{12} & k_{13} \\ k_{21} & -(k_{02} + k_{12} + k_{32}) & k_{23} \\ k_{31} & k_{32} & -(k_{03} + k_{13} + k_{23}) \end{bmatrix} \boldsymbol{q} + \boldsymbol{u} \end{aligned} \tag{4.75}$$

$$\boldsymbol{y} = V^{-1}\boldsymbol{q} = \begin{bmatrix} 1/V_1^{\text{true}} & 0 & 0 \\ 0 & 1/V_2^{\text{true}} & 0 \\ 0 & 0 & 1/V_3^{\text{true}} \end{bmatrix} \boldsymbol{q} \tag{4.76}$$

Then, from Eq. (2.66), the *transfer function (TF) matrix* for this model is:

$$H(s) = V^{-1}[sI-K]^{-1}$$

$$= \begin{bmatrix} 1/V_1^{true} & 0 & 0 \\ 0 & 1/V_2^{true} & 0 \\ 0 & 0 & 1/V_3^{true} \end{bmatrix} \begin{bmatrix} s+k_{01}+k_{21}+k_{31}) & -k_{12} & -k_{13} \\ -k_{21} & s+k_{02}+k_{12}+k_{32}) & -k_{23} \\ -k_{31} & -k_{32} & s+k_{03}+k_{13}+k_{23}) \end{bmatrix}^{-1} \tag{4.77}$$

Inversion requires the *determinant* of *K*, which we recognize as the *characteristic polynomial:*

$$\begin{aligned} D \equiv s^3 - (k_{11} + k_{22} + k_{33})s^2 + (k_{11}k_{22} + k_{11}k_{33} + k_{22}k_{33} - k_{12}k_{21} - k_{13}k_{31} - k_{23}k_{32})s \\ + k_{11}k_{23}k_{32} + k_{22}k_{13}k_{31} + k_{33}k_{12}k_{21} - k_{11}k_{22}k_{33} - k_{12}k_{23}k_{31} - k_{13}k_{21}k_{32} \end{aligned} \tag{4.78}$$

The remaining steps of symbolically inverting the matrix can be done the old-fashioned way, manually dividing the matrix adjoint by *D* (see Appendix B), or using a program like *Matlab, Mathematica, Maple*, or *Reduce*.[29] The result is:

$$H(s) = \begin{bmatrix} \dfrac{s^2-(k_{22}+k_{33})s+k_{22}k_{33}-k_{23}k_{32}}{V_1^{true}D} & \dfrac{k_{12}s-k_{12}k_{33}+k_{13}k_{32}}{V_1^{true}D} & \dfrac{k_{13}s+k_{12}k_{23}-k_{13}k_{22}}{V_1^{true}D} \\ \dfrac{k_{21}s-k_{21}k_{33}+k_{23}k_{31}}{V_2^{true}D} & \dfrac{s^2-(k_{11}+k_{33})s+k_{11}k_{33}-k_{13}k_{31}}{V_2^{true}D} & \dfrac{k_{23}s-k_{11}k_{23}+k_{13}k_{21}}{V_2^{true}D} \\ \dfrac{k_{31}s+k_{21}k_{32}-k_{31}k_{22}}{V_3^{true}D} & \dfrac{k_{32}s-k_{11}k_{32}+k_{12}k_{31}}{V_3^{true}D} & \dfrac{s^2-(k_{11}+k_{22})s+k_{11}k_{22}-k_{12}k_{21}}{V_3^{true}D} \end{bmatrix} \tag{4.79}$$

We can write the nine individual transfer functions between each of the three inputs and three outputs directly from the TF matrix. These are useful in studying the dynamics of specific experiments using these inputs and outputs (Chapter 16), and for *identifiability analysis* of such experiments, the subject of Chapter 10. For example, the first two TFs are:

$$\begin{aligned} H_{11}(s) &= \frac{s^2-(k_{22}+k_{33})s+k_{22}k_{33}-k_{23}k_{32}}{V_1^{true}D} \\ H_{12}(s) &= \frac{k_{12}s-k_{12}k_{33}+k_{13}k_{32}}{V_1^{true}D} \end{aligned} \tag{4.80}$$

29 Program websites: *Matlab:* www.mathworks.com/products/matlab/; *Mathematica:* www.wolfram.com/products/mathematica/index.html; *Maple*; www.maplesoft.com/products/maple; *Reduce*; www.zib.de/Symbolik/reduce/.

As noted earlier, and in Chapter 2, these two TFs can be interpreted as representing two different I/O experiments: the first an experiment with impulse input in compartment 1 and output from compartment 1, and the second with impulse input in compartment 2 and output from compartment 1. Information about the different sets of parameters in each of these TFs can be gleaned from these experiments – a subject in Chapter 16.

Remark: By definition, all multicompartment model rate constants are constrained by $k_{ij} \geq 0$ for $i \neq j$. Otherwise, they are not compartmental, e.g. if $k_{12} < 0$, the model is not compartmental. Also, physical constraints require $V_i > 0$.

Example 4.10: ***A Closed (Leakproof) Cyclic System Model***

This is our first special configuration of the three-compartment model, with one input and three outputs measured (**Fig. 4.21**). It's called **cyclic** because a **cycle** is present, i.e. *a closed unidirectional path of connections among three or more compartments and back to the same compartment.* The model and equations are:

$$\dot{\boldsymbol{q}} = \begin{bmatrix} -k_{21} & 0 & k_{13} \\ k_{21} & -k_{32} & 0 \\ 0 & k_{32} & -k_{13} \end{bmatrix} \boldsymbol{q} + \begin{bmatrix} 1 \\ 0 \\ 0 \end{bmatrix} u_1$$

$$\boldsymbol{q}(0) = \boldsymbol{0}, \quad y = q$$

The characteristic equation is:

$$D \equiv |\lambda I - K| = \lambda[\lambda^2 + (k_{13} + k_{21} + k_{32})\lambda + k_{13}k_{21} + k_{13}k_{32} + k_{21}k_{32}] = 0$$

The three roots (eigenvalues) of this equation are: $\lambda_1 = 0$ and

$$\lambda_{2,3} = \frac{-(k_{13} + k_{21} + k_{32}) \pm \sqrt{(k_{13} + k_{21} + k_{32})^2 - 4(k_{13}k_{21} + k_{13}k_{32} + k_{21}k_{32})}}{2}$$

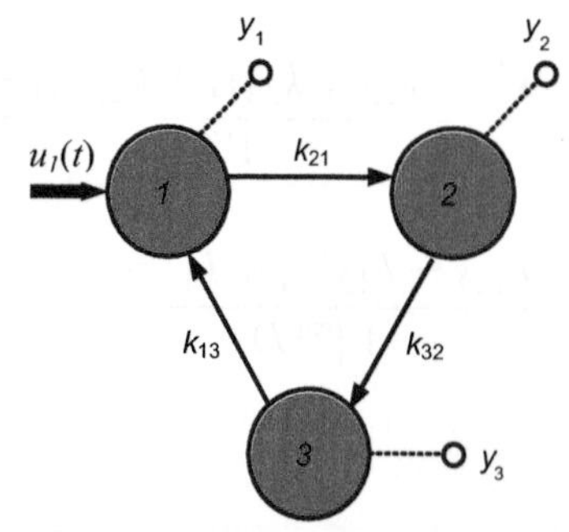

Fig. 4.21

Cyclic model equations and diagram.

An obvious question is: can the quantity under the radical sign be negative, making λ_2 and λ_3 complex? The answer greatly affects the dynamic responses of this model, because complex roots generate oscillatory responses and real roots do not. Rather than analyze the algebra symbolically, we try some numbers. Let all rate constants = 1. Then:

$$\lambda_{2,3} = \frac{-3 \pm \sqrt{9-12}}{2} = \frac{-3 \pm j\sqrt{3}}{2}$$

This demonstrates that *a cyclic system can have complex modes.* All solutions for $\boldsymbol{q}(t)$ and the given $\boldsymbol{q}(0)$ can then be obtained from the TF matrix:

$$H(s) = I[sI-K]^{-1}B = [sI-K]^{-1}\begin{bmatrix}1\\0\\0\end{bmatrix} = \frac{1}{D}\begin{bmatrix}\alpha_{11} & \alpha_{12} & \alpha_{13}\\ \alpha_{21} & \alpha_{22} & \alpha_{23}\\ \alpha_{31} & \alpha_{32} & \alpha_{33}\end{bmatrix}\begin{bmatrix}1\\0\\0\end{bmatrix} = \frac{1}{D}\begin{bmatrix}\alpha_{11}\\ \alpha_{21}\\ \alpha_{31}\end{bmatrix}$$

$$= \frac{1}{D}\begin{bmatrix}(s+k_{13})(s+k_{32})\\ k_{21}(s+k_{13})\\ k_{21}k_{32}\end{bmatrix} = \frac{1}{s(s^2+3s+3)}\begin{bmatrix}(s+1)^2\\ s+1\\ 1\end{bmatrix}$$

Then, for an impulse input $u_1(t) = \delta(t)$, $U_1(s) = 1$, and $Q(s) = H(s)$:

$$Q_1(s) = \frac{(s+1)^2}{s(s^2+3s+3)} = \frac{1}{3}\left[\frac{1}{s} + \frac{2s+3}{s^2+3s+3}\right]$$

$$Q_2(s) = \frac{s+1}{s(s^2+3s+3)} = \frac{1}{s^2+3s+3}Q_3(s)$$

$$Q_3(s) = \frac{1}{s(s^2+3s+3)}$$

Inverting these, we get the three state variable responses:

$$q_1(t) = \frac{1}{3}\left(1 + 2e^{-3t/2}\cos\frac{\sqrt{3}}{2}t\right)$$

$$q_2(t) = \frac{1}{3}\left(1 - e^{-3t/2}\cos\frac{\sqrt{3}}{2}t - \sqrt{3}\sin\frac{\sqrt{3}}{2}t\right)$$

$$q_3(t) = \frac{1}{3}\left(1 - e^{-3t/2}\cos\frac{\sqrt{3}}{2}t + \sqrt{3}\sin\frac{\sqrt{3}}{2}t\right)$$

Remark: For this leakproof model, the impulse responses $q_1(t)$, $q_2(t)$, and $q_3(t)$ have both a constant steady state component and a decaying oscillatory transient component (complex poles in the LHP). The constant steady state is due to the pole at the origin (no model leaks). The *Matlab* solution for $q_1(t)$ is shown in **Fig. 4.22**. Note that damping of the oscillations (slight undershoot at $t \sim 2$, slight overshoot at ~ 5) is nearly complete in one or two cycles, a property of cyclic models. With more compartments, oscillations are more pronounced.

Mammillary & Catenary Compartment Models

The two most common classes of multicompartmental models are *mammillary* and *catenary* model structures. **Mammillary models** consist of a central compartment surrounded by and connected with peripheral (noncentral) compartments, none of which are connected to each other (**Fig. 4.23**).

Catenary models have all compartments arranged in a chain, with each connected (in series) only to its nearest neighbors (**Fig. 4.24**). Note that the two-compartment model is both mammillary and catenary.

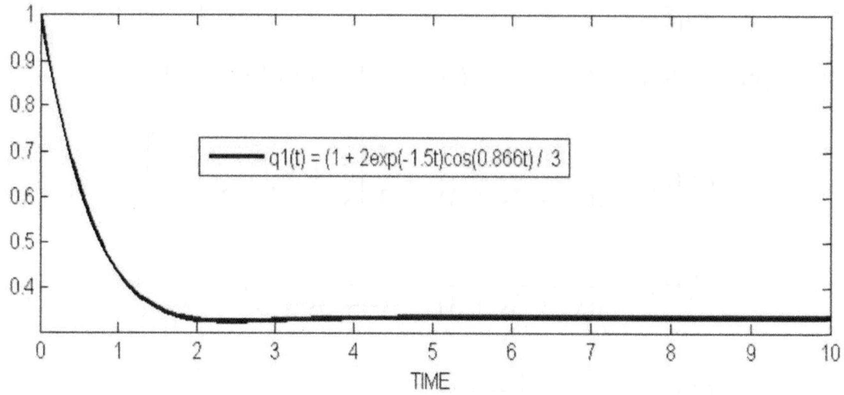

Fig. 4.22

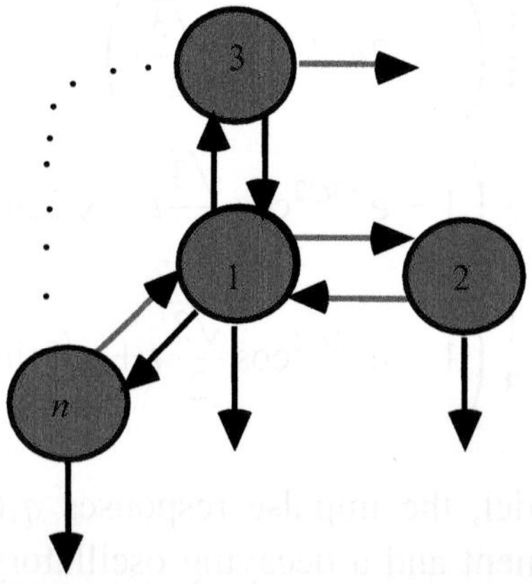

Fig. 4.23

Mammillary.

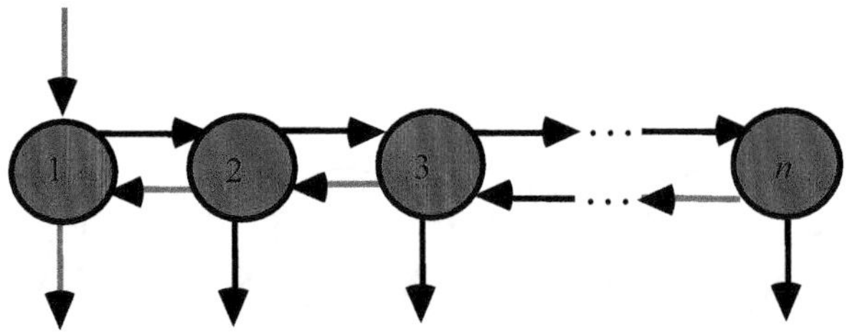

Fig. 4.24

Catenary.

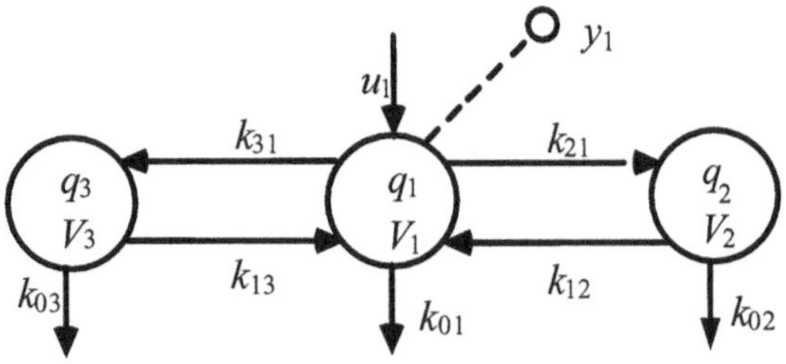

Fig. 4.25

Three-compartment mammillary model.

Example 4.11: ***Three-Compartment Mammillary Model***

The q_i represent masses in each compartment, and y is a concentration measurement in compartment 1 only. The model equations are determined from the diagram (**Fig. 4.25**) by application of mass flux balance to each compartment, as in Eq. (4.66). For $k_{ij} \geq 0$, $i \neq j$, and $V_i > 0$ (nonnegative, by definition), these are given by:

$$\dot{\boldsymbol{q}} = \begin{bmatrix} -(k_{02}+k_{21}+k_{31}) & k_{12} & k_{13} \\ k_{21} & -(k_{02}+k_{12}) & 0 \\ k_{31} & 0 & -(k_{03}+k_{13}) \end{bmatrix} \boldsymbol{q} + \begin{bmatrix} 1 \\ 0 \\ 0 \end{bmatrix} u_1 = \begin{bmatrix} k_{11} & k_{12} & k_{13} \\ k_{21} & k_{22} & 0 \\ k_{31} & 0 & k_{33} \end{bmatrix} \boldsymbol{q} + \begin{bmatrix} 1 \\ 0 \\ 0 \end{bmatrix} u_1$$

$$y = \begin{bmatrix} \frac{1}{V_1} & 0 & 0 \end{bmatrix} \boldsymbol{q} \tag{4.81}$$

where $-k_{11} \equiv k_{01} + k_{21} + k_{31}$, $-k_{22} \equiv k_{02} + k_{12}$ and $-k_{33} \equiv k_{03} + k_{13}$ define the **turnover rates** of the three compartments. The parameter vector has seven elements:

$$\boldsymbol{p} \equiv [k_{02} \quad k_{03} \quad k_{12} \quad k_{13} \quad k_{21} \quad k_{31} \quad V_1]^T$$

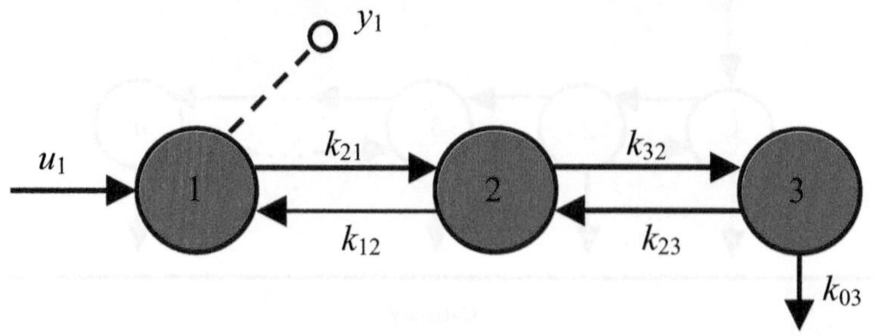

Fig. 4.26

SISO three-compartment catenary model.

Note that V_2 and V_3 do not appear in the equations (why not?) and therefore they are not included in $\boldsymbol{p}$. The *transfer function matrix* of this single-input–single-output (SISO) model is:

$$H(s) = C[sI - K]^{-1}B = \begin{bmatrix} \frac{1}{V_1} & 0 & 0 \end{bmatrix} \begin{bmatrix} s-k_{11} & k_{12} & k_{13} \\ k_{21} & s-k_{22} & 0 \\ k_{31} & 0 & s-k_{33} \end{bmatrix}^{-1} \begin{bmatrix} 1 \\ 0 \\ 0 \end{bmatrix}$$

$$= \begin{bmatrix} \frac{1}{V_1} & 0 & 0 \end{bmatrix} \begin{bmatrix} y_{11} & y_{12} & y_{13} \\ y_{21} & y_{22} & y_{23} \\ y_{31} & y_{32} & y_{33} \end{bmatrix} \begin{bmatrix} 1 \\ 0 \\ 0 \end{bmatrix} = y_{11}/V_1 = \frac{(s-k_{22})(s-k_{33})/V_1}{s^3 + \alpha_3 s^2 + \alpha_2 s + \alpha_1} \equiv H_{11}(s) \tag{4.82}$$

where

$$\alpha_3 = -(k_{11} + k_{22} + k_{33})$$

$$\alpha_2 = k_{11}k_{22} + k_{11}k_{33} + k_{22}k_{33} - k_{12}k_{21} - k_{13}k_{31}$$

$$\alpha_1 = k_{12}k_{21}k_{33} + k_{13}k_{22}k_{31} - k_{11}k_{22}k_{33}$$

Remark: Comparing this model (**Fig. 4.25**) to the most general three-compartment model in **Fig. 4.19** we see that the zero entries in the K matrix are where connections in **Fig. 4.25** are missing, e.g. there are no direct connections between compartments 2 and 3 in **Fig. 4.25**.

*Example 4.12: **SISO Three-Compartment Catenary Model***

This model is illustrated in **Fig. 4.26**. If $k_{01} \equiv k_{02} \equiv 0$ and only y_1 (H_{11}) is measured, the resulting model is still catenary and the TF matrix has only one element, because the model is SISO. Writing this TF is left as an exercise.

Data-Driven Modeling of Indirect & Time-Delayed Inputs

In the last sections, compartmental model dynamical equations and output measurement equations were developed for linear systems. For nonlinear systems, the ODEs have nonlinear terms, but the input–output structure is similar, in particular with inputs entering compartments, any number of them, directly. A typical input can enter compartment 1 directly, for example, by direct intravenous (IV) injection into the blood (or "central") compartment in a clinical or experimental animal study. However, exogenous inputs of material also can be administered less directly, via other routes, all or some of which eventually arrives in compartment 1, e.g. in blood. These situations usually require additional compartments or other structural components between the administration port and the compartment of entry. We address this model-augmentation issue here, in this familiar clinical experiment context, emphasizing that the same data-driven modeling approach is applicable in non-clinical experiments, e.g. in intracellular or population studies. **Fig. 4.27** serves to illustrate the data-modeling issues.

An IV bolus dose (D_{ose}) of drug administered directly into blood has a measured response in blood typical of that shown in green in **Fig. 4.27** (for $D_{ose} = 1$). There is a rapid rise, slowed somewhat by mixing in blood,[30] before the drug concentration falls in accordance with overall system dynamics, often termed the **disposition phase** in pharmacokinetic studies. Now let's idealize these realistic conditions and assume the drug mixes instantaneously in an already and continuously well-mixed (homogeneous) blood compartment. Also assume the bolus can be approximated as an impulse of magnitude D_{ose}. Then the ideal response begins at time 0^+ – with magnitude scaled by blood volume, D_{ose}/V_B – falling for $t > 0$ in accordance with overall system dynamics – as illustrated by the blue dashed-line curve in **Fig. 4.27** (for $D_{ose}/V_B = 1$).

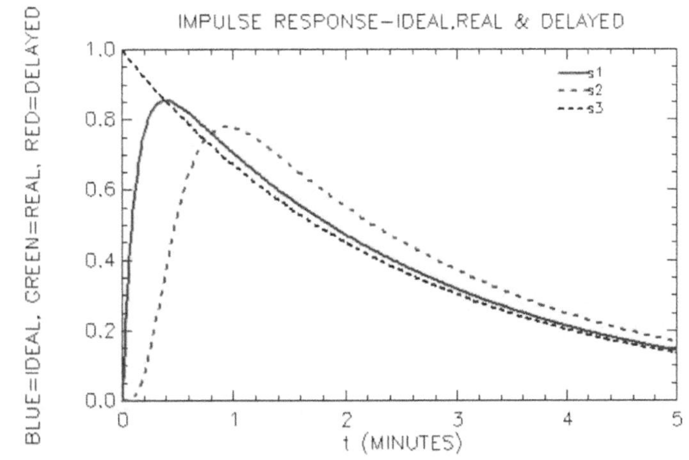

Fig. 4.27

30 A real IV pulse dose takes a finite time to administer. In reality, it is a constant infusion for very short period of time, the time it takes to empty a syringe, for example. So, finite mixing times are only part of the approximation problem near time zero.

Now suppose the same dose D_{ose} is given **subcutaneously** (SC) – under the skin – or **intramuscularly** (IM) – into a muscle – or into another organ, instead of IV. In this case, entry of drug into blood is delayed, because it must "traverse," or is "slowed down" by some tissue space before entering capillaries and the general circulation. This is also illustrated in **Fig. 4.27**, by the superimposed red dashed-line curve. How do we model this somewhat more encompassing system, which includes additional tissue space?

A classical compartmental model approximation for the delayed response to input $u(t)$ in a central compartment, e.g. blood, is the one-parameter, catenary **time-delay model** (**mixing-cell model, transit-time model**) shown in **Fig. 4.28**. The nth-compartment here is the one the central compartment "sees" entering it instead of $u(t)$ directly. This model is an often-used one in place of pure lag-time delays, primarily because it is often more physiological.

Let's write the ODEs for this n-compartment delay model:

$$\begin{aligned} \dot{q}_1 &= u - kq_1 \\ \dot{q}_2 &= kq_1 - kq_2 = k(q_1 - q_2) \\ &\vdots \\ \dot{q}_n &= k(q_{n-1} - q_n) \end{aligned} \tag{4.83}$$

Taking Laplace transforms, we get

$$Q_1 = \frac{U}{s+k}, \quad Q_2 = \frac{kU}{(s+k)^2}, \quad \cdots \quad Q_n = \frac{k^{n-1}U}{(s+k)^n} \tag{4.84}$$

The transfer function is therefore:

$$H_{n-DELAY} = Q_n/U = \frac{k^{n-1}}{(s+k)^n} \tag{4.85}$$

The temporal response in the nth compartment to input $u(t)$ is the inverse Laplace transform of Q_n. For impulsive input dose of magnitude D_{ose} , $U(s) = D_{ose} \times 1$. The inverse transform of Q_n is then readily obtained from the table of Laplace transform pairs (for $1/t^n$), combined with the complex translation property, all in Appendix A. The result is:

$$q_n(t) = \frac{D_{ose}(kt)^{n-1}}{(n-1)!}e^{-kt} \tag{4.86}$$

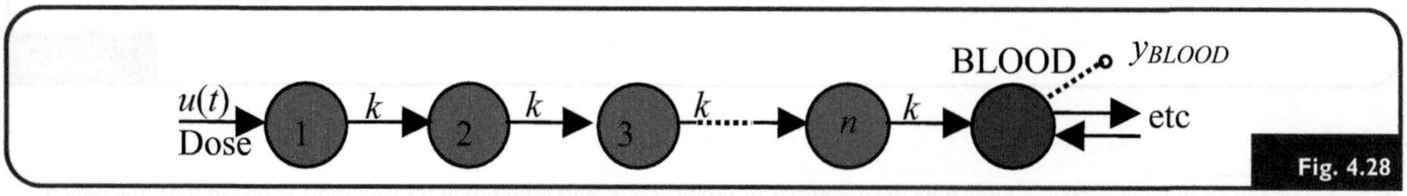

Fig. 4.28

Time delay (transit time) model.

The input function into blood (what the blood *sees*) is then:

$$u_{BLOOD}(t) = kq_n(t) = D_{ose}k^n t^{n-1}e^{-kt}/(n-1)! \tag{4.87}$$

The big question is: *how does this model provide an appropriate time delay in moving the input signal through the catenary chain of compartments into the central compartment?* A second question is: *how do we establish the magnitude of this time delay?* In other words, how do we find k and n so we can use this submodel to represent indirect inputs or other delays in a system?

The answers – and the design of this simple submodel – are based on the fact that each compartment has a dynamical *lag effect* on the signal moving through the model. This lag effect is reinforced and reshaped – sharpened – as the signal passes from compartment to compartment. And, as might be anticipated, it is a function of the value of k, as well as the number of compartments n, which together govern the **mean transit time (MTT)** of the signal going from compartment to compartment. In the previous section, following Eq. (4.66), we defined what we need here: the average time a substance spends in transit through a compartment each time it passes in or out. It's the inverse of the compartment turnover rate $-k_{jj}$ for any compartment j. The MTT to reach compartment 2 from 1 is thus $-1/k_{11} = 1/k$. By simple induction, the effect is cumulative, so the **overall mean transit time (OMTT)** – for the whole chain of compartments – is simply: OMTT $= n/k$.

Remark: As n gets larger, the time delay becomes sharper and sharper, approaching a pure time lag function.

Signal Delay with Attenuation

Depending on the site of entry of an input, the signal may also be attenuated, for example by local degradation at the entry site, or by elimination of input material somewhere along the way, prior to entry into blood (or other entry compartment). Either circumstance can be accommodated in the delay model by including a leak from one of more compartments. The simplest addition would be a single leak from an additional compartment, say compartment $n + 1$, as shown in **Fig. 4.29**. The $n + 1$st-compartment here is commonly called the **absorption compartment**, which becomes the dynamical (compartment) the blood compartment "sees" entering it, as illustrated in **Fig. 4.29**. The elimination or leak parameter is called k_{elim} here, and

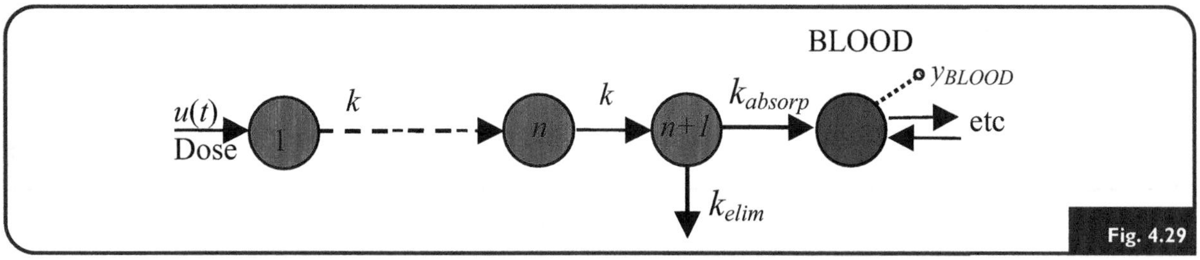

Fig. 4.29

Signal delay with attenuation.

the absorption parameter k_{absorp}. With one more "leaky" compartment, it is easy to show that the delayed input into blood is now:

$$u_{BLOOD}(t) = \frac{D_{ose}K_{absorp}}{k_{absorp} + k_{elim}} q_{n+1}(t) = \frac{D_{ose}K_{absorp}}{k_{absorp} + k_{elim}} k^n t^n e^{-kt}/n! \tag{4.88}$$

Eq. (4.88) can be simplified by noting that the leading rate constant ratio would be the same if k_{absorp} is set equal to k. We also then redefine the elimination rate constant with a prime, because it is different. With this simplification, (4.88) becomes:

$$u'_{BLOOD}(t) = \frac{D_{ose}k}{k + k'_{elim}} q_{n+1}(t) = \frac{D_{ose}}{k + k'_{elim}} k^{n+1} t^n e^{-kt}/n! \tag{4.89}$$

Example 4.13: ***Oral Dosing Losses & Delays in the Gastrointestinal (GI) Tract***

Suppose entry of an unit oral dose of drug ($D_{ose} = 1$) is delayed an average time of ½ hour before absorption begins, and 20% of it is lost by degradation and (fecal) excretory processes within the GI tract.[31] If we use $n + 1 = 6$ compartments to generate the mean delay, then MTT = ½ = $n + 1/k = 6/k$, so $k = 12$. To reflect 80% absorption in the model, k'_{elim} is computed from: $0.8 = 12/(12 + k'_{elim})$, i.e. $k'_{elim} = 3$.

A *SAAMII* simulation of this model is shown in **Fig. 4.30** for a unit dose input. All program windows are shown. The *Equations* window demonstrates how we make

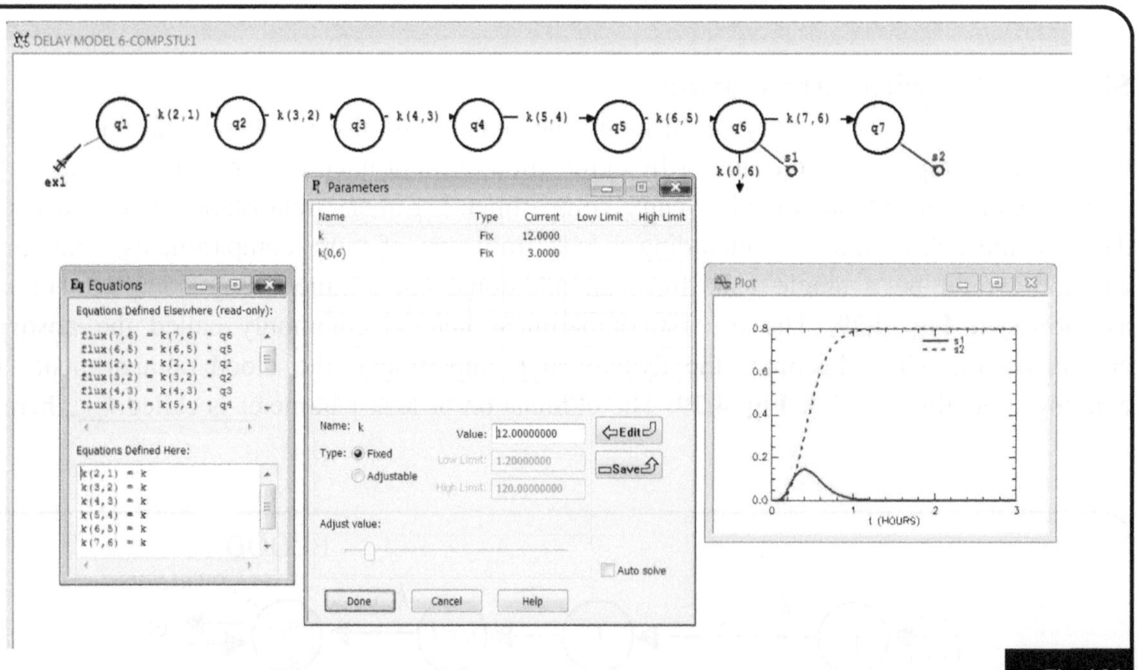

Fig. 4.30

Oral dosing losses and delays in the GI tract

31 In Chapter 14, this will translate into 100 − 20 = 80% of the drug being *bioavailable* for therapeutic effects.

all *SAAMII* (or other simulator) model parameters equal, so only k and the leak parameter k(0,6) need to be specified in the *Parameters* window.

The output of compartment 6, $s1 = q_6$ (green curve) illustrates the impulse dose input delayed by ½ hour and reshaped along the way. The output of compartment 7, $s2 = q_7 = q_{BLOOD}(t)$ representing drug in blood (red dashed curve) is the integrated version of the signal received from compartment 6 (same as AUC), because (unrealistically) there is no leak from the blood compartment in this model. Note that the signal has been attenuated by 20%, with a steady-state value of 0.8 instead of 1, due to the leak in compartment 6. The AUC of the green impulse response is 0.8, for the same reason.

Remark: a single time delay block operator is available in *SAAMII* for performing time delay operations, with and without degradation functionality. It could have been used instead of the implementation in this example.

Remark: Delay of a signal caused by physical transport from one distant location to another, as in material flow through a pipe from one end to the other, is often modeled as a pure time delay, e.g. output $y(t) = u(t - \tau)$, for $t > 0$ and delay $\tau > 0$. Flow system (e.g. vascular) models, for example, discussed in Chapter 8 employ these kinds of delays.

Remark: Time delay operations are handled in different ways in different programming packages. *MATLAB*, for example, has special integration algorithms for solving delay-differential equations, i.e. ODEs with one or more pure time delays.

Pools & Pool Models: Accommodating Inhomogeneities

As noted earlier, *pools* and *pool models* are different than compartments and compartmental models, and are particularly useful for modeling biosystems with nonhomogeneous "nodes." Compartmental models have homogeneous "nodes." **Pool** was defined earlier as a mass or volume size measure for a real or abstract space or material, emphasizing that it is not necessarily uniformly (homogeneously) distributed.[32] A compartment may be a pool, but not necessarily *vice versa*. On the other hand, we can speak of the "pool of X in compartments i and j" or the "pool of X in the whole organism" (or "... whole compartmental model or system") because the term pool circumvents the homogeneity requirement on any of these entities.

Pool Models

Pool models are a superset of multicompartmental models. For pool models, the requirement or assumption of homogeneity of compartments is relaxed, as noted above. A *pool* does not necessarily represent a homogeneously distinct entity, such as a uniform concentration of material in an (idealized) physical space, but it does typically represent a distinct entity, such as the total *amount* of material (in mass terms) or the *size* of a physical space (in mass or

32 (Bergner and Lushbaugh 1967; Rescigno et al. 1990).

volume terms). Thus, pool models can be used to represent the transfer of mass or volume among pools arranged graphically, as do compartmental models.

The caveat is that the lack of homogeneity in a pool restricts its applicability to lumped parameter (ODE) systems in constant steady state, represented by algebraic equations. There are other ways to formalize inhomogeneous distribution. Partial differential equations (PDEs) are often used to describe their transient dynamics, because they are dependent on space variables as well as time t (more than one independent variable). Another approach is to quantize each state variable of the inhomogeneous space model into several approximately homogeneous compartments for representation of transient or periodic (nonconstant) steady state responses.

Pool vs. Compartmental Model Applications: Flux-Balance & Cut-Set Analysis

Pool models can facilitate biosystem steady state analysis, e.g. flux balance analysis (FBA). For example, if the total *steady state efflux* of material from both compartments of the two-compartment model in **Fig. 4.12** is of interest, i.e. $k_{01}q_{1\infty} + k_{02}q_{2\infty}$, the simpler *one-pool model* of **Fig. 4.31** (LHS) can be used. In this case, we assume the steady state is constant. If it is periodic, we might still use this approach to obtain an approximate *average* efflux. The total mass of material in both must be the same, i.e. $q_{1'}(t) \equiv q_1(t) + q_2(t)$ and, in steady state, equivalence between the two models requires $u_{1'\infty} \equiv u_{1\infty} + u_{2\infty}$ and $k_{01'}q_{1'\infty} = k_{01}q_{1\infty} + k_{02}q_{2\infty}$. This is an explicit statement of *mass flux balance* for the overall two-compartment model, illustrated by drawing a big, single-pool ellipse – called a **cut-set** (Feng and DiStefano III 1991) – that encloses both compartments (**Fig. 4.31**, RHS), and then equating influxes and effluxes across the cut-set.[33] Remember, however, that the one-pool model cannot represent transient responses for the same material represented by the two-compartment model.

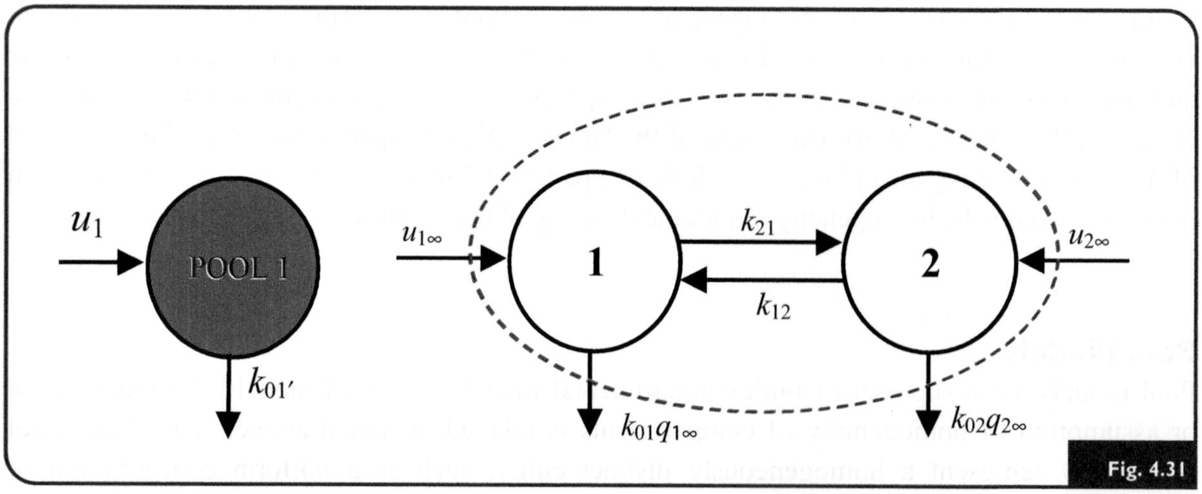

Fig. 4.31

One-pool equivalent model for flux-balance analysis of a two-compartment model (RHS). Fluxes in and out of the cut-set balance in steady state.

33 We develop flux balance analysis (FBA) – a standard method in the systems biology of metabolism – beginning in Chapter 7.

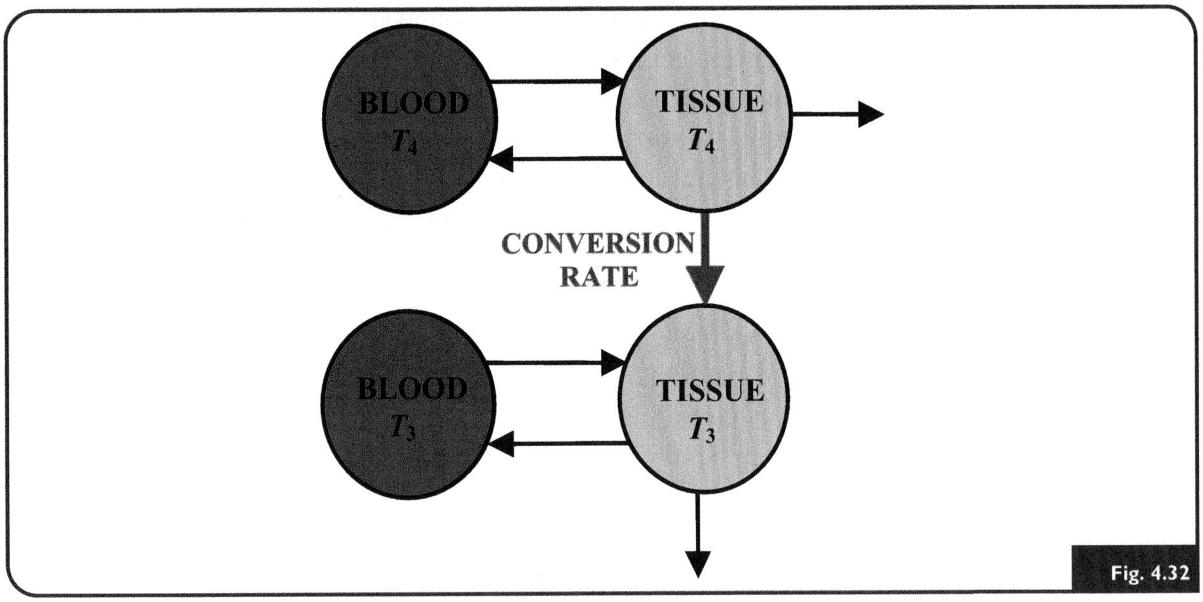

Fig. 4.32

Example 4.14: ***Thyroid Hormone Steady State Pool Model***

Fig. 4.32 is a simple steady state pool model of thyroid hormone (T_4 and T_3) distribution in blood and lumped tissue pools, none of which are homogeneously distributed pools for these substances. T_4 is converted to T_3 at different rates in virtually all cells of the body, but not in blood. The goal of this pool model is to provide a means for evaluating the whole-body hormone conversion rate, in steady state only – shown in red, which we pursue in Exercise E4.17. The steady state for these substances is mildly periodic, so the solution we get using this model is an averaged one. The arrangement of pools isolates the desired steady state variable to be quantified. The pools are physio-anatomically well-defined spaces, realistically connected. The exchange and "leak" arrows represent total material transfer between blood and all organs, or from all organs, and the bold red arrow labeled CONVERSION RATE represents total body conversion of T_4 to T_3 (mass/time units), known to occur only outside the vascular pool. No conversion and no significant leaks occur in or from blood pools of the animals and conditions for which this model is designed. This model is used again in Chapter 16 to design steady state experiments to evaluate CONVERSION RATE, using *cut-set analysis* (Feng and DiStefano III 1991), a facile graph theory method for experiment design. More complex versions of this model and its analysis are given in the reference and in (Feng and DiStefano III 1992).

Recap & Looking Ahead

We've developed the notion of compartmentalization as fundamental to structural modeling of biological systems, both linear and nonlinear, and shown how to build multicompartmental biomodels from structural and other mechanistic data. We've elaborated compartmentalization in a deeper way than the popular notion of compartment as a bounded space (container) for locating well-defined biological entities (e.g. molecules, chemical "species," etc.). Compartments go as deep as the data permits – representing different entities with *common dynamical characteristics*, as a **compartment ≡ state variable isomorphism**. This dynamical definition, based on the characteristic dynamics of homogeneous entitities, facilitates analysis of biodata and their "dynamical signatures" – featured in the next chapter, where the focus is on model structuring from transient kinetic data. It also provides the *basic compartment-unit* for formulation of **candidate** multicompartmental model structures and analyzing their responses to input stimuli. Input and output measurement (data) models are included in all modeling developments. And a few special modeling classes of particular practical utility, namely data-driven models of indirect and time-delayed inputs, and nonlinear compartmental disease dynamics models, are also included – to facilitate practical applications – before moving on to other domain based modeling methods from biochemistry, cellular, pharmacological and other systems biology areas in the next four chapters.

Exercises

E4.1 ***One-Compartment Model with Finite Pulse Input***. Analytically find and graph the solution of a single-compartment model with a finite pulse input of magnitude P and pulse-width T, with $q(0) = 0$. Repeat using any simulation package and compare your results.

E4.2 ***Step Responses from Impulse Responses***. Derive the solutions, Eqs. (4.42) and (4.44), from the impulse responses, Eqs. (4.38) and (4.39).

E4.3 ***Step Responses for Nonzero Initial Conditions***. Derive both step responses of the two-compartment model for a nonzero initial condition $q_1(0) = a$.

E4.4 ***Equivalent Input Modeling***. Transform the two-compartment model equations for a system with nonzero initial state $q(0) \neq 0$ to one with zero initial state and additional impulsive inputs.

E4.5 ***Two-Compartment Model with Finite Pulse Input***. Solve the two-compartment model with a finite pulse input into compartment 2 (only) using Eqs. (4.42) and (4.44).

E4.6 ***Analytic and Simulated Solutions***. Find the complete solution for the output $y(t)$ of the following model, both analytically and by simulation:

$$\dot{x} = \begin{bmatrix} -1 & 3 \\ 0 & -2 \end{bmatrix} x + \begin{bmatrix} 0 \\ u \end{bmatrix}$$

E4.7 ***Steady State Solutions***. Find the constant steady state solution(s) of the general compartmental model in Eq. (4.11) in terms of K and whatever else you need.

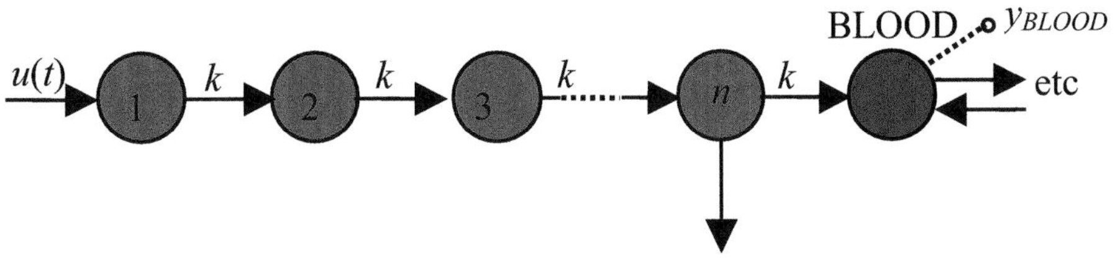

Fig. 4.33

E4.8 ***Indirect Input Modeling***. Find k, n and k_{elim} in **Fig. 4.33** to delay the input $u(t)$ by 3 minutes for an administered substance absorbed 95% into blood.

E4.9 Determine the solutions $\boldsymbol{q}(t)$ for the closed and cyclic model of Example 4.10 analytically and/or by simulation, but with:

a) Compartment 1 (only) open and $k_{01} = 1$.

b) Compartment 3 (only) open and $k_{03} = 1$.

E4.10 ***Theory***. Show that the matrix K for a *closed* compartmental model is *singular*.

E4.11 ***Cyclic Modeling***. Why do (linear) *cyclic* models require additional care during compartmental analysis and when fitting them to real data?

E4.12 ***Pharmacokinetics Model Analysis***. An impulsive dose of one unit strength of drug X is given to a subject intravenously, and its disappearance from blood is biexponential. The drug in blood exchanges reversibly with drug in tissues and is catabolized in tissues. It is not eliminated in or from the body in any other manner. A dose of ½ unit strength produces a blood response curve of half the magnitude of a one unit strength dose, and a dose of two unit strength produces a blood response curve 1.5 times the magnitude of a one unit strength dose.

a) Draw a diagram of a compartmental model of the disposition of this drug in blood and tissues and *state a major assumption or restriction on its applicability*.

b) How many modes are there in the response measured in the blood?

c) Is it possible to obtain the rate of catabolism of this drug from blood-borne measurements of the drug concentration? Explain your answer and, again, state any assumptions or restrictions concerning such results.

d) Describe how you would estimate the total body equivalent distribution volume V_D of the drug from the same intravenous impulse response experiment described above. (Remember that V_D is a steady state *concept*, so you might want to consider a steady state "thought experiment" with a constant input to derive some useful relationships.) There is more than one way to solve this problem.

E4.13 ***Pharmacokinetics of an Oral Dose Model***. Write the ODEs for the five-compartment human thyroxine kinetics model shown in **Fig. 4.34**, where the first two compartments

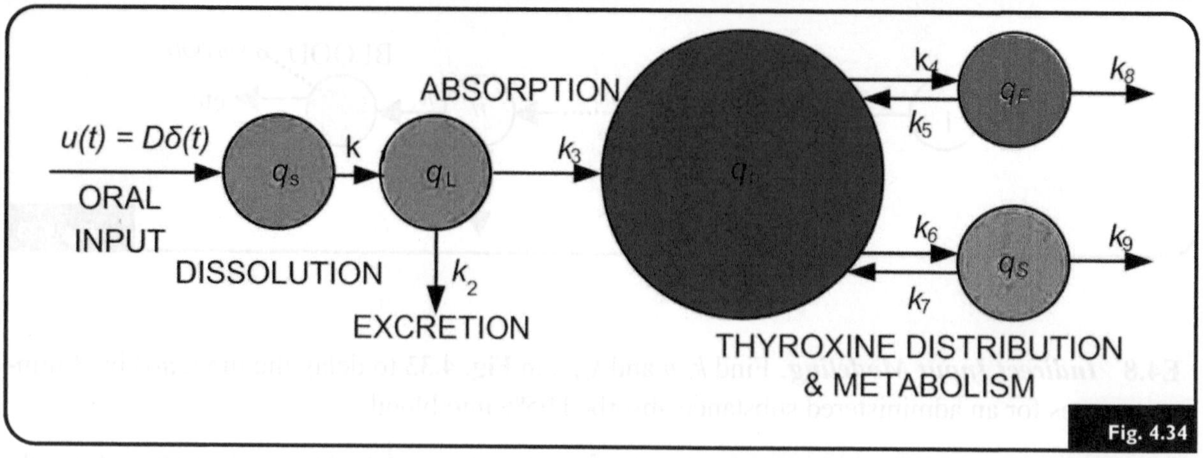

Fig. 4.34

Thyroxine distribution model following an oral dose input.

represent dissolution, absorption and fecal excretion of an oral dose input of thyroxine, and the other compartments represent the distribution and elimination of thyroxine. Parameter values $k_4, \ldots, k_9 = 1.35, 1.03, 0.14, 0.0548, 0.0046, 0.00208$ (DiStefano III, Wilson et al. 1975; DiStefano III and Mak 1979). Simulate the whole model for both an oral dose D of T_4. Use the following oral T_4-dose parameters: $k_1 = 1.3$, $k_2 = 0.62$, and $k_3 = 0.62$. Units for all ks are $1/h$.

Note: If dissolution of the drug is very fast relative to absorption and excretion from gut, a single compartment is often used to represent an oral input pathway to blood.

E4.14 ***Area under the Curve (AUC)***. Find the AUC for compartment 2 of the two-compartment model with input to compartment 1 (see **Fig. 4.13**).

E4.15 ***VisSim or Simulink Experiences***. Implement the three-compartment mammillary model of **Fig. 4.25** in Example 4.11 using a block diagram simulation language, e.g. *VisSim* or *Simulink*. For parameter values, use:

$$V_1 = 20, k_{21} = 1.0, k_{12} = 0.2, k_{31} = 0.1,$$
$$k_{13} = 0.009, k_{01} = 0, k_{02} = 0.02, k_{03} = 0.001$$

a) Run the simulation using the following initial conditions, inputs and simulation algorithm parameters: $q(0) = 0, u_1 = \delta(t)$, over the time interval [0, 600], using all integration algorithms in your package. Use several step-sizes – small to large – for fixed-step methods. Plot all state variables on the same axes, for comparison purposes. Which methods provide consistent results?

b) Repeat (a) for $u_1 = \mathit{1}(t)$, $t \geq 0$ (a unit step), and then again for $u_1 = \mathit{1}(t)$, for $0 < t \leq 20$ and $u_1 = 0$ for $t > 20$ (a unit pulse).

E4.16 ***SAAMII Simulations***. Repeat Exercise E4.15 using the compartmental modeling program *SAAMII* instead of a block diagram simulator. What are the attributes and deficiencies of these two simulation approaches, demonstrated for this model?

E4.17 ***Pool Model Application***. Assume for the model in Example 4.14 that all hormone (T_3 and T_4) masses can be measured in all four pools in response to two different constant infusion inputs: the first is a constant influx u_A of exogenous T_4 into the BLOOD T_4 pool; and the second is a constant influx u_B of exogenous T_3 into the BLOOD T_3 pool. The system is in approximately constant steady state when the measurements are made. Find a solution for the constant CONVERSION RATE mass flux arrow in terms of the measurements. You can do this by algebraically solving the steady equations for the system. Or you can do it more simply using cut-set analysis.

E4.18 ***More Simulation Experiences***. Solve for the state variables in each of the NL models in Example 4.5 and Example 4.6 by simulation, for parameter values $C = 1$, $K = 1$ and $t_{max} = 10$. Overlay the plots for step input magnitudes 0.1, 1 and 10, for each model.

E4.19 ***Some Theory***. Show that the eigenvalues (characteristic roots) are negative in Eq. (4.33) if all $k_{ij} > 0$, $i \neq j$.

E4.20 ***Physiological Variant of the Example 4.3 Model – More Compartments in Same Organs***. This problem begins with reference to Example 4.3 in which X_1 in blood exchanges reversibly with X_1 in liver, where it is metabolized to X_2, and possibly exchanges with other organs – via the dashed arrows. Suppose unmetabolized X_1 in blood also exchanges reversibly with cells of the gut, one of those other organs, via the mesenteric vasculature of the gut. In addition, X_1 in liver can reach the gut via the biliary pathway from liver to gut – as does X_2. Together, these are called the entero-hepatic circulation. Some X_1 in gut can then be returned to blood via mesenteric veins and some of it can also be excreted as X_1 directly in feces, where it can be measured. Draw a modified version of the Example 4.3 model that includes all of this additional physio-anatomic information.

E4.21 ***SIR Model Solution and Analysis***. Solve and analyze Eq. (4.60) for its dynamical properties.

E4.22 ***SIR Model Simulation Analysis***. Simulate and analyze Eq. (4.60) for its dynamical properties, for parameter and IC values: $k_{RI} = 0.01$, $k_{IS} = 0.001$, $q_I(0) = 1$ and $N = 100$. Will this model lead to an epidemic?

E4.23 ***Reduced Disease Models from Constraints***. Rearrange the first ODE in each of Eqs. (4.60) and (4.61), as was done for (4.58), and redraw the NL compartmental diagrams for these two reduced two-compartment SIR and SIRS models. The reduced models account for the conservation of the total population N constraints. Are these useful representations? For what?

E4.24 ***Disease Models with Births and Deaths Included***. Derive a new SIR model structure and ODEs with birth and death processes included, assuming they occur on the same time scale as the disease. How would SIS and SIRS models be adjusted for such birth and death processes?

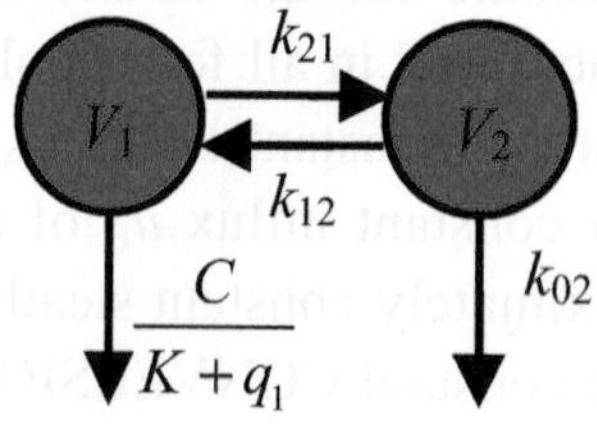

Fig. 4.35

E4.25 ***Models for Deadly Diseases***. Derive new SIR and SIRS model structures with R measured and only some infectives recovering.

E4.26 ***Reproductive Number in SIR Models***. Starting from the idea that an epidemic will spread if $\dot{q}_I > 0$ from the start, show that the reproduction number$R_0 \equiv k_{IS}N/k_{RI} > 1$ for an SIR model.[34] Begin with a single infective individual.

E4.27 ***Two-Compartment Model Time-Domain Solutions***. Analytically generate time-domain solutions the two-compartment model of Eqs. (4.9), with $u_2 = 0$ and u_1 arbitrary, two ways. First, convert the two ODEs to a single second-order ODE and find the free and forced responses. Second, write the general solution for vector q(t) in terms of the zero-input and zero state (convolution integral) responses (see Chapter 2).

E4.28 ***Modeling Different Input Routes***. Consider the adjacent two-compartment model of the pharmacokinetics of a drug X with NL leak from compartment 1 (**Fig. 4.35**). No input is shown. *Augment* the model, as needed, to render it one with an: (a) intravenous (IV) input; (b) an oral input; (c) an intramuscular (IM) input; and (d) a cutaneous input of X via a skin patch that includes a slow release drug. Compartment 1 is X in *blood*, and 2 is X in *kidney tissue*. You only need to augment this diagram (add to it) with what's minimally needed for each of the input types. No math, only pictures! Keep it as simple as possible.

REFERENCES

Atkins, G.L., 1969. Multicompartment Models for Biological Systems. Methuen, London.
Bellman, R., 1970. Introduction to Matrix Analysis. McGraw-Hill, New York.
Bergner, P., Lushbaugh, C. (Eds.), 1967. Compartments, Pools, and Spaces in Medical Physiology. US Atomic Energy Commission, Division of Technical Information, Washington, DC.
Blower, S.M., Mclean, A.R., Porco, T.C., Small, P.M., Hopewell, P.C., Sanchez, M.A., 1995. The intrinsic transmission dynamics of tuberculosis epidemics. Nat. Med. 1, 815–821.
Chowdhury, B.A., Chandra, R.K., 1987. Biological and health implications of toxic heavy metal and essential trace element interactions (Review). Prog. Food Nutr. Sci. 11 (1), 55–113.
Cobelli, C., Sparcino, G., et al., 2006. Compartmental models of physiological systems. In: Bronzino, J. (Ed.), Biomedical Engineering Fundamentals. CRC Press, Boca Raton, pp. 9–14.
Daley, D.J., Gani, J., 2005. Epidemic Modeling: An Introduction. Cambridge University Press, NY.

34 Kindly provided by my friend Marisa Eisenberg.

Dickson, K.L., Maki, A.W., Cairns Jr., J., 1982. Modeling the Fate of Chemicals in the Aquatic Environment. Ann Arbor Science Publishers, Ann Arbor.
DiStefano III, J., Feng, D., 1988. Comparative aspects of the distribution, metabolism, and excretion of six iodothyronines in the rat. Endocrinology. 123, 2514–2525.
DiStefano III, J., Mak, P., 1979. On model and data requirements for determining the bioavailability of oral therapeutic agents: application to gut absorption of thyroid hormones. Am. J. Physiol. Regul. Integr. Comp. Physiol. 236, R137–R141.
DiStefano III, J., Wilson, K., et al., 1975. Identification of the dynamics of thyroid hormone metabolism. Automatica. 11, 149–159.
Douglas, P., Cohen, M.S., DiStefano III, J.J., 2010. Chronic Exposure to Mn May have Lasting Effects: A Physiologically Based Toxicokinetic Model in Rat. J. Toxicol and Environ. Chem. 92 (2), 279–299.
Eisenberg, M.C., Robertson, S.I., Tien, J.H., 2012. Identifiability and estimation of multiple transmission pathways in cholera. J. Theor. Biol. 324, 1–17.
Feng, D., DiStefano III, J., 1991. Cut set analysis of compartmental models with applications to experiment design. Am. J. Physiol. Endocrinol. Metab. 261, E269–E284.
Feng, D., DiStefano III, J., 1992. Decomposition-based qualitative experiment design algorithms for compartmental models. Math. Biosci. 110, 27–43.
Godfrey, K., 1983. Compartmental Models and their Application. Academic Press, New York.
Gydesen, H., 1984. Mathematical models of the transport of pollutants in ecosystems. Ecol. Bull. 36, 17–25.
Hevesey, G., 1923. The absorption and translocation of lead by plants. Biochem. J. 17, 439–445.
Holford, N., 1987. Clinical pharmacokinetics of ethanol. Clin. Pharmacokinet. 13, 273–292.
Jacquez, J., 1972. Compartmental Analysis in Biology and Medicine: Kinetics of Distribution of Tracer-Labeled Materials. Elsevier Publishing, New York.
Jacquez, J.A., 1996. Compartmental Analysis in Biology and Medicine. Biomedware, Ann Arbor.
Keeling, M., Rohani, P., 2007. Modeling Infectious Diseases: In Humans and Animals. Princeton University Press, Princeton.
Rescigno, A., 2003. *Foundations of Pharmacokinetics*. Springer, New York.
Rescigno, A., Thakur, A., Bertrand, K., Brill, A., Mariani, G., 1990. Tracer kinetics: a proposal for unified symbols and nomenclature. Phys. Med. Biol. 35 (3), 449–465.
Rowland, M., Tozer, T.N., 2011. Clinical Pharmacokinetics and Pharmacodynamics: Concepts and Applications. Lippincott Williams & Wilkins, Baltimore.
Rubinow, S., 1973. Mathematical Problems in the Biological Sciences. Society for Industrial and Applied Mathematics, Philadelphia.
Schramski, J.R., Patten, B.C., et al., 2009. The reynolds transport theorem: application to ecological compartment modeling and case study of ecosystem energetics. Ecol. Modell. 220 (22), 3225–3232.
Tardiff, R.G., Goldstein, B.D., 1992. Methods for assessing exposure of human and non-human biota SCOPA 46, IPCS joint symposia, SGOMSEC 5, Chichester, 1991. J. Appl. Toxicol. 12(3): 233.
van den Driessche, P., Watmough, J., 2002. Reproduction numbers and sub-threshold endemic equilibria for compartmental models of disease transmission. Math. Biosci. 180 (1–2), 29–48.
Vynnycky, E., White, R.G. (Eds.), 2010. An Introduction to Infectious Disease Modelling. Oxford University Press, Oxford.
Wagner, J.G., 1993. Pharmacokinetics for the Pharmaceutical Scientist. Technomic Publishing, Lancaster, PA.

Chapter 5

Structural Biomodeling from Theory & Data: Sizing, Distinguishing & Simplifying Multicompartmental Models

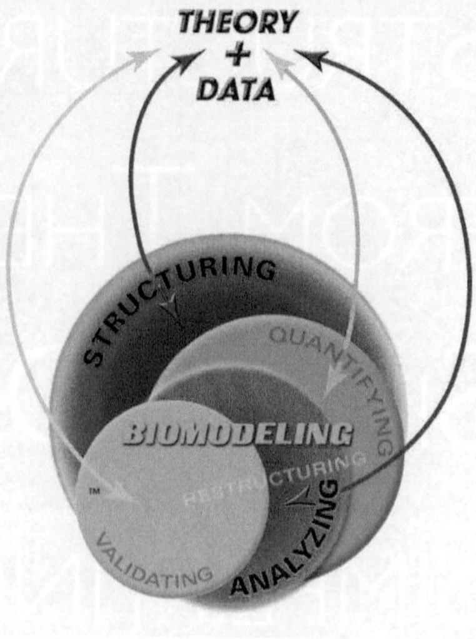

"Models should be as simple as possible. *But not too simple.*"

An adage like this has been attributed to Einstein. Whether or not the ascription is correct, the deep meaning is especially pertinent for this chapter.

INTRODUCTION

In this chapter, we take a closer look at how kinetic output data circumscribe and limit what we can learn and validate about model structure. The blood–liver–central compartment Example 4.4 (and **Fig. 4.4**) of the last chapter introduced this complexity issue. In that example, lack of early enough output data prevented blood and liver compartments from being clearly distinguishable from each other, depending on the earliest data collection point t_1. It didn't matter whether the data were otherwise perfect (error-free, noise-free), or not. Key dynamics were missing on the interval $0 < t < t_1$. More generally, detecting, distinguishing and validating the presence of hypothesized or known compartments in a model from specific input–output data is a much larger problem. This is the case even when data are collected early enough to distinguish the input entry compartment from any others and whether the data are perfect or not. The problem is inherently structural[1] and, fortunately, amenable to not-very-complex mathematical analysis, mostly algebraic, some geometric; there is no need for probability or statistics in this chapter. At this pedagogical stage we're interested is in what's possible under ideal conditions and imperfect data are an unnecessary complicating factor. We thus assume noise-free data for developments here. We introduce noise and *output error models* – with statistics as needed – in later chapters, when we study the numerical aspects of model quantification and analysis – a higher level of modeling complexity. This begins in earnest in Chapters 11 and 12.

A central theme of this chapter is analysis of *dynamical signatures* in hypothetical (perfect) input–output data associated with a model. The goals are: to establish model **dimensionality** (size, complexity); to select among multiple candidate models, called model **distinguishability;** and model **reduction,** a form of model **simplification** – all with respect to dynamical signatures in the data. **Modes**, *visible* and *hidden* modes, play an important role in understanding dynamical signatures in data.[2] We develop notions of visible and hidden compartments and modes and their dynamical properties. We include model **linearization** here too, another form of model simplification – over a limited dynamical range. Tracer perturbation methodology and Taylor series approximations for linearizing NL ODE system and experiment models are developed in the last section and applied to nonlinear biosystem problems beginning in the next chapter.

OUTPUT DATA (DYNAMICAL SIGNATURES) REVEAL DYNAMICAL STRUCTURE

We motivate the problem first with some simple examples of how dynamical data – non-steady state, temporal data – inform about systems structure. All were presented in earlier chapters, but the context here is different.

1 Whole model structure, i.e. specific input–output structure as well as biosystem dynamics structure.

2 For those familiar with large-scale data analysis in, say, bioinformatics or computational statistics, finding the *modes* in dynamic system data is akin to finding: the *principal components* (principal component analysis – PCA), or *singular-value decomposition* (SVD), of a large data matrix. This means visualizing a multivariate dataset – typically expressed in a high-dimensional data space, from its "most informative" viewpoint – its principal components – in a lower-dimensional space. The distinct *modes of chords and scales* in music are another example, with each mode sounding different but having a different arrangement of the same notes played in particular orders. The basics of SVD and PCA are presented in Appendix C.

What's in the Box?

Three sets of idealized output data in response to an impulse (brief pulse, bolus) input $u(t) = Q_0\delta(t - t_0)$ applied at $t = t_0$ are shown in **Fig. 5.1**. At the top, the response $Ae^{-k(t-t_0)}$ is graphed as $y(t) = Ae^{-kt}$, a single exponential decay, graphed for $A = 1$ and $t_0 = 0$ to simplify notation in the figure. We've encountered two subtly different structures that might "fit" this data, shown in the two adjoining input–output blocks: (1) a linear one-compartment model, e.g. of a material degrading in a well-mixed container of solvent; (2) another linear one-compartment model, represented as a *central* compartment consisting of an assumed homogeneous aggregate of more than one homogeneous compartment, like the blood + liver compartment in **Fig. 4.4** (right-hand side; RHS) of the last chapter. Recall that with material

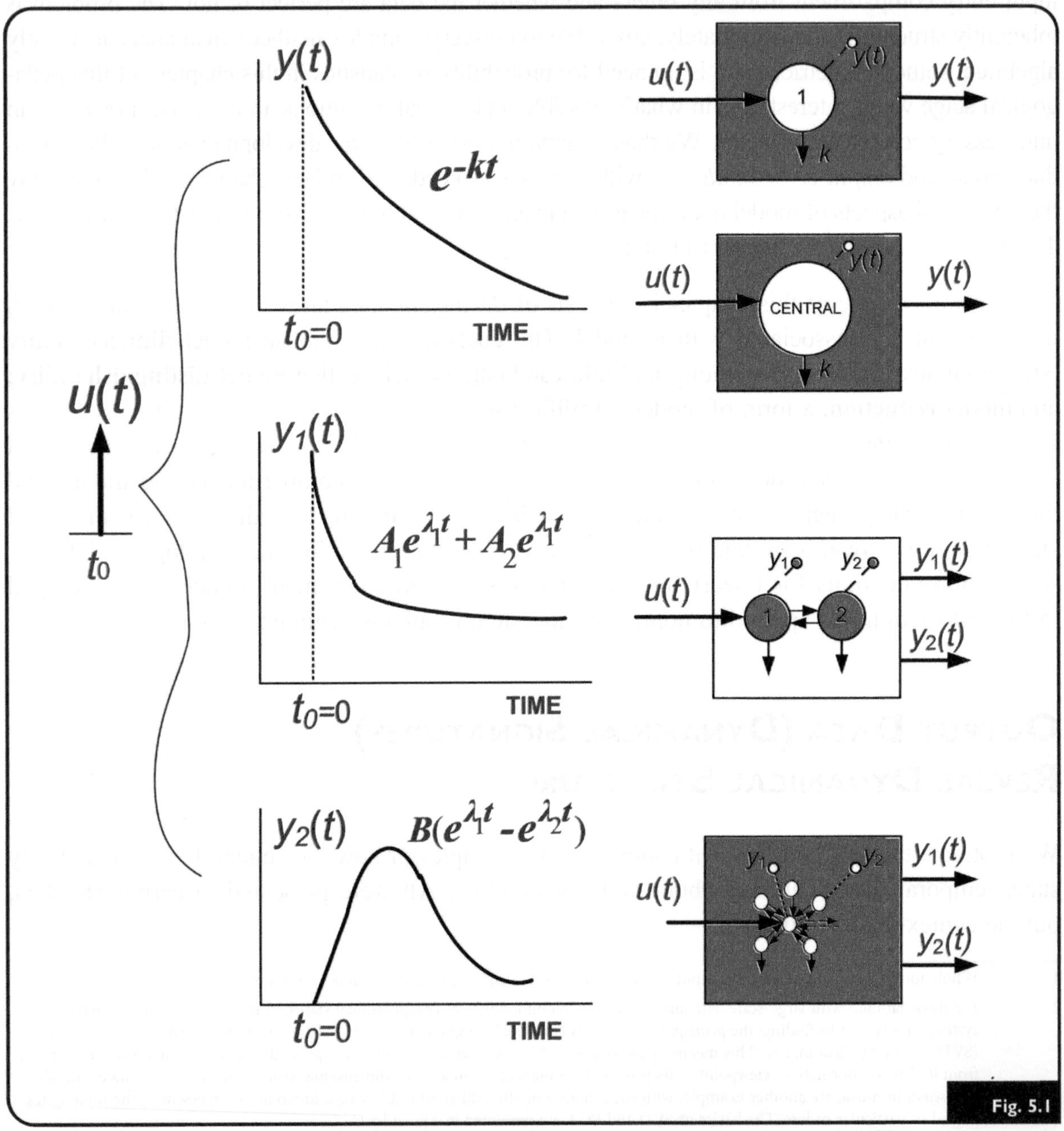

Fig. 5.1

flow between liver and the peripheral blood compartment being very rapid, blood sampling begun "too late" rendered the two essentially indistinguishable. Thus, its hard to know from the data what's precisely in the box, but we can begin to know its dynamical complexity and – from this, coupled with context – narrow down the possibilities.

The second set of graphs, of $y_1(t)$ and $y_2(t)$, are (noise-free) output data measured from two different compartments in response to the same impulse input. $y_1(t)$ is a monotonically decreasing function, measured in the same compartment where the input is given; and $y_2(t)$ – from a different compartment – increases from zero, peaks, then falls asymptotically to zero as time increases. These responses are meant to represent sums of exponential functions, as shown in the figure, in this case just two exponentials. The adjoining input–output blocks are two different models that "fit" this data well. The first is no surprise – a two-compartment model, like that in **Fig. 4.12** of the last chapter. The second, which should surprise the novice modeler, is a linear mammillary model with n compartments – *with n as large as we like.* Nevertheless, both sets of output data correspond to a sum of only two exponential functions! We show why in a later section.

Sinusoidal or other **periodic waveforms** are also common in output data, as illustrated in **Fig. 5.2** for two kinds: oscillating signals that dissipate over time and constant amplitude, sustained oscillating signals. In both cases the input is constant (step function) and thus *not* periodic. We therefore anticipate components or processes within each input–output block capable of generating periodic waveforms, i.e. there must be *oscillators* in the blocks. For example, the oscillating dissipative system might be a complex linear or nonlinear (NL) multicompartmental (MC) model-containing cycles, the oscillator might be a NL predator – prey model like in Chapter 3. We'll express all these more formally with diagrams and mathematics in due course.

It's hard to know from the data what's precisely "in the box," but we can begin to know it's dynamical complexity and, from this – coupled with context – narrow down the possibilities.

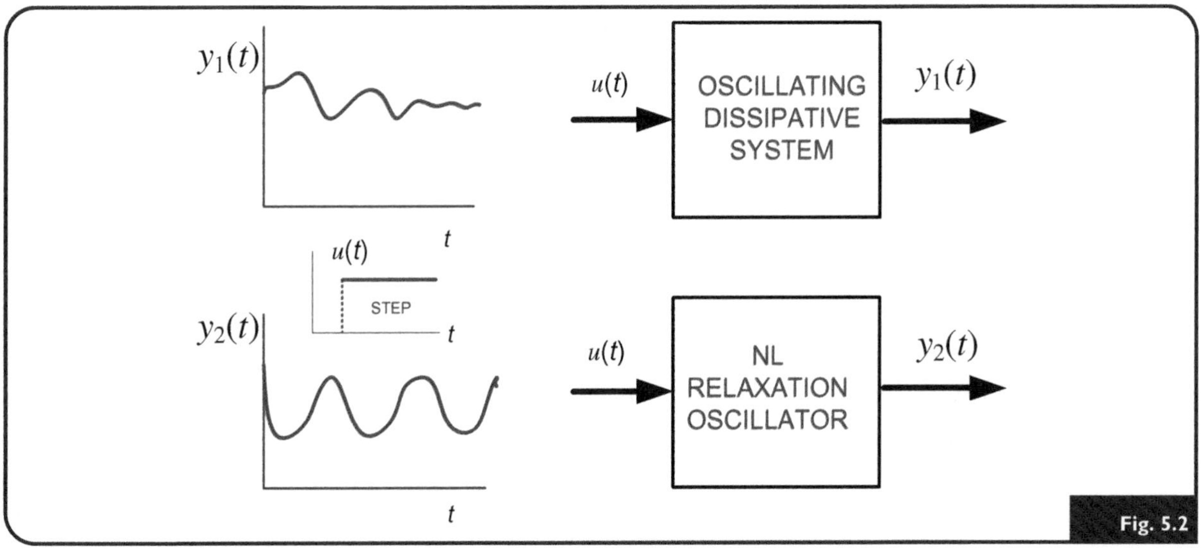

Fig. 5.2

Multicompartmental (MC) Model Dimensionality, Modal Analysis & Dynamical Signatures

We begin by addressing the granularity – or extent – of multicompartmentalization, i.e. the dimensionality of the model – as limited by the locations of inputs and outputs and (noise-free) data. A MC model can have any number of compartments. We show in this section that the minimum dimension – the smallest number of possible compartments – is equal to the number of its distinct *modes* visible in its output response, for any particular specification of input and output locations.

Multiexponential Impulse-Responses, Modes & Mode Parameters

We saw in Chapter 2 that *multiexponential* output response functions are fundamental solutions of (zero-input) homogeneous linear ODEs. This was illustrated in Example 2.15. The n linearly independent exponential terms of these responses are the response **modes** – a **fundamental set** of solutions of the ODE. We just caught a glimpse of the relationships between modes, compartment number and input/output locations in the **Fig. 5.1** illustrations.

Single-input–single-output (SISO) linear MC models of the form: $\dot{\boldsymbol{q}} = K\boldsymbol{q} + \boldsymbol{u}$, $y = \boldsymbol{c}^T\boldsymbol{q}$, are common in kinetic analysis applications. We show here that they generally have a multiexponential output when the input is an impulse or, equivalently, a nonzero IC and no input. Step or ramp or any integral of impulse inputs also yield multiexponential outputs. This general result has broad consequences.

Let the input be an impulse (or sufficiently brief pulse), which we convert to the equivalent homogeneous system, with IC vector $\boldsymbol{q}(0) \neq \boldsymbol{0}$, only one nonzero element, and $1 \times n$ output matrix $\boldsymbol{c}^T$ with one nonzero element, i.e.

$$\dot{\boldsymbol{q}} = K\boldsymbol{q} \quad \text{with ICs:} \quad \boldsymbol{q}(0) = \begin{bmatrix} q(0) \\ \vdots \\ 0 \end{bmatrix} \quad \text{and output: } y(t) = q_1(t) = \boldsymbol{c}^T\boldsymbol{q}(t) = [\,1 \quad \cdots \quad 0\,]\boldsymbol{q}(t) \tag{5.1}$$

The exponential matrix solution of this model is given by Eq. (2.22) in Chapter 2: $y(t) = \boldsymbol{q}(0)e^{Kt}$. All the modes of the response of model (5.1) are specified by this solution function.

Deeper insight into this problem is evident from a more general solution of the homogeneous equation. By direct substitution, it is easy to show that $\boldsymbol{q} = e^{\lambda t}\boldsymbol{v}$ is a general solution of Eq. (5.1), where λ is a (scalar) constant and $\boldsymbol{v}$ is a vector of n constants (see Appendix B if you need help with this). Differentiating $\boldsymbol{q}$ and substituting the result in Eq. (5.1), we get $\lambda e^{\lambda t}\boldsymbol{v} = Ke^{\lambda t}\boldsymbol{v}$. Dividing both sides by $e^{\lambda t}$ then yields:

$$\lambda\boldsymbol{v} = K\boldsymbol{v} \tag{5.2}$$

This is a classic result in linear algebra, defining the n **eigenvalues** λ and n, **eigenvectors** $\boldsymbol{v}$ of the matrix K. Thus, for our MC model output, we can say $e^{\lambda t}\boldsymbol{v}$ is a solution of Eq. (5.1) if λ is an eigenvalue of K and $\boldsymbol{v}$ is the corresponding eigenvector.

This completely general result provides the fundamental connection between linear MC and multiexponential models, in terms of multiexponential *modes*. This modal solution concept is extended in Chapters 6 and 9 to a similar pair of fundamental solution components (modes) in nonlinear MC systems biology models.

Simplest Case: Distinct Eigenvalues (Distinct Exponents)

The dynamical shape of the solution modes, for recognition purposes as well as math analysis, depends in part on relationships among the n eigenvalues $\lambda_1, \lambda_2, \ldots, \lambda_n$. If they are *distinct* (none are repeated) and the corresponding eigenvectors are $v_1, v_2, \ldots, v_n$, the general solution of our SISO ODE in Eq. (5.1) can be written:

$$\boldsymbol{q}(t) = \sum_{i=1}^{n} c_i \boldsymbol{v}_i e^{\lambda_i t} \tag{5.3}$$

The c_i are chosen to match the initial conditions on $\boldsymbol{q}(t)$, i.e. $c_1 v_1 = q(0)$, and all others zero, i.e.

$$\boldsymbol{q}(0) = \begin{bmatrix} q(0) \\ \vdots \\ 0 \end{bmatrix} = \sum_{i=1}^{n} c_i \boldsymbol{v}_i = \begin{bmatrix} v_1 \\ \vdots \\ 0 \end{bmatrix} \tag{5.4}$$

We have thus shown that, for impulsive (e.g. single trace-dose, small perturbation) test-inputs, a very common experimental scheme, the state vector – and hence the output $y(t)$, is a sum of n exponential terms (modes). That is, multiexponential functions are the natural responses to real systems if Eqs. (5.1) are an appropriate model for the system. Then, by linearity, the system response to a step, ramp, etc., will have essentially the same form as Eq. (5.3), i.e. a sum of exponential modes. Furthermore, the intrinsic parameters of this solution are the eigenvalues and eigenvectors of K, i.e. a set of **structural invariants**[3] of K. Each mode is a two-dimensional function, parameterized by two components, a "frequency" (eigenvalue) and a "shape" (eigenvector) component.

Remark: If the eigenvalues are not distinct, the solution in Eq. (5.3) or (5.4) is more complex (and looks different) but still multiexponential, and each repeated eigenvalue has its own mode, e.g. $c_1 e^{\lambda t}, c_2 t e^{\lambda t}$ are two different modes for repeated eigenvalue λ, developed below in Eqs. (5.6) and (5.7) for MC models with repeated and complex eigenvalues in their responses.

Following are some primary considerations for formulating, applying and interpreting MC models, plus some important properties based on consequences of these notions.

3 **Structural invariants** are one of many smallest sets of parameters completely determining or characterizing model dynamics. More on this in Chapter 10.

Dynamical Dimensionality: Establishing the Number of Compartments

We began exposition of the modeling process in the last chapter by transforming the structural knowledge base into biological state variables (e.g. molecule amounts) resident in well-defined biological (population, organ, intracellular, subcellular, etc.) spaces, with measured variables as primary identifiers of some of these state variable biospecies and spaces. We also need to establish the dynamical complexity of the model required to represent biosystem dynamics faithfully – including the compartments we don't know about or cannot "see" – all consistent with modeling goals. This is governed in large part by the dimensionality (size) of the state space (number of ODEs) needed for this purpose, and thus the minimum number "n" of compartments in the MC model. For this, we turn to *input–output* (I–O) *data*.

We begin by assuming the model has a single-input into one compartment and a single-output is measured from the same or other compartment (**SISO model**). This assumption is relaxed and multiple inputs and outputs (**MIMO models**) are allowed as the methodology develops.

Modes in Data ≡ Minimum Compartment Number

An overarching goal here is to know how to characterize system dynamics from temporal kinetic input–output (I–O) data collected from accessible I–O ports, which are delimiting. The number of compartments observable in dynamical data, and thus system complexity, depends on where (what compartments) *input* stimuli are applied and from what ports (what compartments) are *output* measurements collected. These probed compartment(s) are likely included in the count, but – at least in principle – the dynamical data embody a "hidden" number of additional unprobed compartments needed to establish minimum *dynamical* dimensionality n. In the next several subsections, this notion is developed into a set of useful tools based on classical and novel *modal* analyses of the model.

The basic idea is based on the linear system theory developed in Chapter 2: the n modes of linear dynamic system (including multicompartmental) models are **fundamental solutions** of the ODEs representing the model, one for each of the n state variables, and *any zero-input response* – i.e. response due only to ICs – *is a linear combination of these n fundamental solutions (modes)*. Importantly, these modes are exponential functions of the general form $t^j e^{\lambda t}$, where $j = 0, 1, \ldots$, etc.[4] Thus, for linear multicompartmental models, these n fundamental solutions (*exponential modes) correspond to the n compartments – one for each state variable (compartment) in the model*. This means **mode-count ≡ minimum compartment-count**.

Remark: The actual number of system compartments – identified as such or not – is typically greater than the minimum needed to represent the dynamics, in many cases very much larger than this minimum number. This can be challenging for the modeler, no matter what the goals are. But this apparent dilemma also can have positive as well as negative consequences, as we see later.

The following example illustrates these issues, at this juncture without detailed analysis. That comes later, after developing more theory.

4 In fact, model output solutions for any combination of nonzero inputs and ICs are functions of these exponential modes (see e.g. Eqs. (2.15) and (2.30) and Example 2.18). Our analysis, however, is based only on zero-input responses.

Example 5.1: ***Two-Mode Response in a Six-Dimensional Model***

The six-compartment graph in **Fig. 5.3** has six nodes, corresponding to six state variables, the dimensionality of this model. Assume all initial conditions are zero and an input is applied to compartment 1. If outputs are measured from compartments 1 or 2, only two of the six possible modes of the response would be present in either output (shown later). This would define the minimal dimensionality as two for this model with these inputs and outputs. As we see later, if the output were measured from compartment 6, six modes would be detected – three times as many found by simply choosing a different place to measure from.

This is an example of how doing the "right experiment" can lead to knowing "what's in the box." This clearly can be quite helpful in experiment design. On the other hand, we also glean from this example that the dominant dynamical response of this network – when viewed from compartments 1 or 2 – has only two modes. Then, if signals from nodes 1 or 2 are needed to control something (e.g. drug dynamics), for example, the remainder of the model between nodes 2 and 6 might be replaceable by a single leak from node 2. For this application, a much simpler two-compartment model with a single leak can replace **Fig. 5.3**. (More on this later.)

This notion of addressing the question of the number of modes in the data facilitates deep understanding of structural model complexity in relation to kinetic data, concepts that carry over into understanding nonlinear dynamic biosystem models, which we address in Chapter 6.

SISO Multiexponential Minimal Modeling: Some Answers & Insights

Consider a linear or appropriately linearized[5] experiment model under single-input–single-output (SISO) conditions. The exponential nature of modes suggests that the minimum compartment number corresponds to the number n of exponential terms in a generic multiexponential impulse response model $y(t)$. Indeed, given that multiexponential functions are intrinsic, fundamental solutions to linear compartmental model equations, fitting them to data[6] emanating from compartmental systems seems like a reasonable thing to do to quantify the minimum dimension n

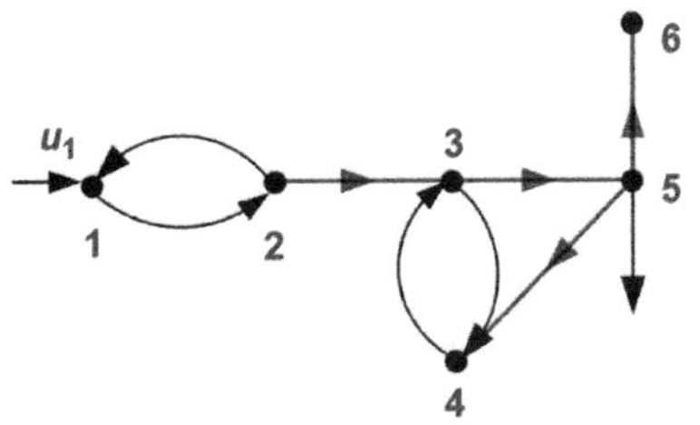

Fig. 5.3

5 For a reasonably large class of nonlinear biosystems, a *linearizing experiment* can render it sufficiently linear to make this kind of analysis practical, for a corresponding *linearized model* of these systems – the subject of the last section of this chapter.

6 Fitting models to data is a major topic in itself and is deferred to Chapters 12 and 13.

(as well as A_i and λ_i)[7] and – at the same time – gain other insights about the real system. For example, a time-series of output kinetic data can be fitted with a sequence of multiexponential (sum of n exponential terms) data-models, for increasing values of n. For impulse (or step, or ramp, or any such integral) response measurements, n is the number of *modes* or *exponential* terms inherent in the output data (and thus the model). Thus, in principle, the best-fit model yields n. As a practical matter, this can even be done – sometimes should be done – before the question of how compartment (or network) connectivity is resolved.

For multiexponential models, the usual assumption is that the n exponents (eigenvalues) are real and distinct, i.e. the noise-free SISO response is the familiar:

$$y(t) = \sum_{k=1}^{n} A_k e^{\lambda_k t} \tag{5.5}$$

Finding n by fitting sums of exponential functions (modes) of increasing order to impulse- or step-response data is a long established approach (Riggs 1970), and there are interactive internet application programs[8] for accomplishing this task efficiently (e.g. Harless and DiStefano III 2005), one of these – web application W^3DIMSUM – is available online at www.biocyb.cs.ucla.edu.

Remark: For several reasons, including the effects of *lumping,* described below, the number of exponential terms n that can be fitted accurately and meaningfully to a set of N time-series data $z(t_1)$, $z(t_2)$, ..., $z(t_N)$ is small, *no matter how large N is.* Because each exponential mode is characterized by two independent parameters – its coefficient and exponent – we have $n \leq N/2$ as a first limitation, but there are others. In practice, when fitting SISO models to real noisy data, n is rarely greater than 3 or 4, even for large N (DiStefano III and Landaw 1984).

The next topic, composite compartments, illustrates how establishing the real compartment number – rather than the minimum n – can be elusive, even if the data are not noisy, because very complex systems typically behave dynamically like they have relatively small dimension n, i.e. the minimum n. This takes us back to *lumping*. This usually means that compartments not directly measured are likely to be present in the system, albeit hidden, and these are typically *composite* or lumped compartments with more or less similar kinetic characteristics (e.g. approximately proportional concentrations) when measured from a given set of I–O ports. The blood–liver Example 4.4 of the last chapter illustrated lumping. Data from additional I–O ports is usually needed to expose them.

Composite Compartments: More Ambiguity & Insights

Material in a compartment, by definition, should be homogeneously distributed, i.e. it should have the same concentration throughout the compartment at all times. This means that, in applications, a multicompartmental model can be a good approximation if the rate of mixing

7 The coefficients A_i and exponents λ_i of a multiexponential function are also called **macroparameters** in pharmacology literature (Gabrielsson and Weiner 2000).

8 See www.biocyb.cs.ucla.edu/dimsum.

within each compartment is rapid relative to rates of transfer among the compartments, and it can be poor if this is not the case; but not necessarily. This may seem like a contradiction, but it isn't.[9]

"Homogeneous" compartments may include a subset of hidden compartments, and thus be a composite of several (homogeneous) compartments. For example, compartments with concentration dynamics in fixed proportion to each other at all times, and not directly measured, are indistinguishable in the outputs of other measured compartments, e.g. cells in the same organ, organs with similar turnover rates, each connected and exchanging with a central compartment – as in a mammillary model; these are indistinguishable in measurements in blood, the central compartment. When developing an MC model, a practical approximation is to "lump" distinct physical compartments with proportional state variables into a single one for analysis or quantification purposes. And such lumped compartments can have useful physical interpretations, e.g. the overall size of lumped tissue or cellular compartments, or fluxes between blood and (lumped) tissue or cellular *milieu*. Of course, one doesn't know if one or two compartments are hidden in this fashion – if the kinetic data reveals only one mode – then only one is dynamically present when viewed from the particular input–output probes specified by the model.

To help resolve – or avoid – this dilemma, the notion of *homogeneous distribution* for compartments is best interpreted dynamically, rather than physically, with homogeneity designating *uniformity of dynamical properties* of distinct objects lumped together. A lumped compartment consisting of several compartments with proportional concentrations is a *dynamically* homogeneous compartment. More often than not, however, lumping is not a matter of choice. Individual, unmeasured compartments are often hidden in kinetic data, as we show below.

Thus, we see one way in which compartmental model complexity is limited by available output measurement data. But this, in itself, does not restrict the utility of models with lumped compartments, if they are otherwise structured and interpreted appropriately.

Example 5.2: ***Composite (Lumped) Compartments***

Suppose physical knowledge suggests the three-compartment *mammillary* model shown in **Fig. 4.25** of Chapter 4, which we designate as a *structural hypothesis*. However, in an experiment to validate this structural hypothesis, one finds that a sum of two exponentials ($n = 2$) – not three – clearly fits the kinetic data "best." Then, as we shall show in Example 5.8, one possibility is that the two noncentral compartments can be lumped into one, either because they have similar turnover rates ($k_{22} \approx k_{33}$), or one compartment's size is much less than the other two (e.g. $V_2 << V_3$ and V_1). In either case, one can use the model to discuss material transfer to and from the *composite of both compartments*, e.g. two organs such as liver and kidney considered together (DiStefano III, Malone et al. 1982; DiStefano III, Jang et al. 1985; Jang and DiStefano III 1985). And we mean dynamical transfers of material, at any time t, not just in steady state.

9 Alas, things are often more complex than they initially appear to be.

If two exponentials fit the data best ($n = 2$), but physical knowledge is nonspecific about the model structure – only that three compartments are involved, not necessarily mammillary connected, then there are other lumping possibilities. Two adjacent compartments may be lumped, e.g. X in liver and blood (Example 4.4 of Chapter 4) and the three-compartment structure then may be catenary (e.g. **Fig. 4.26**), or the most general three-compartment structure (**Fig. 4.20**).

Remark: Measured compartments can also be a composite of more than one. Example 4.4 in Chapter 4 illustrated that sampling compartments are typically lumped with others when data sampling is begun too late (or too infrequently) to capture very fast dynamics. Again, peripheral compartment 2, X in liver, was indistinguishable from compartment 1, X in blood, in blood measurements, because the sampling "missed" the rapid transfer of test input substance from blood to liver (and back), with only one "mixed" mode visible in the output.

Remark: The examples illustrate how models with lumped or composite compartments can still have useful physiological interpretations, but again assuming they are otherwise structured correctly, or that their structures are interpreted correctly.

Remark: **MC models as models of data**. Compartmental models are also used as abstract models of data, having little or no specific physical analogies. This occurs when there is little or no natural partitioning of the system to guide compartmentalization, e.g. as a discretized approximation (quantization) of a continuous model or system,[10] or when no analogy of model to system states or parameters is desired in a particular application, e.g. when only an input–output equivalent model[11] is needed. Bottom line: *whether a MC model is used as a model of a system or as a model of data, the ultimate adequacy of the modeling depends on how well it satisfies the objectives, i.e. the model application, and the assumptions upon which it is based.*

The Bigger Picture – Visible & Hidden Modes: Distinguishable & Indistinguishable Compartments (State Variables or Species)

Lumped compartments, composite compartments and other notions of ambiguities in compartmental boundaries are merged here and in the next several sections with model *distinguishability*, a fundamental concept likely to have been born with the first notions of formal modeling in science. We introduce the notions *visible* (or *distinguishable*) and *hidden* (*indistinguishable*), along with a few complexities about multiexponential response functions first. These notions are then extended directly to compartmental diagrams, with multiexponential responses (modes) playing an important role.

Caveat #2 – Too Dynamically Simple?

Strictly speaking, this caveat applies to multiexponential functions fitted to output data assumed to have the form of Eq. (5.5). This multiexponential form may be wrong; the output

10 See Appendix C.

11 See Appendix D.

response function may have modes with *repeated* real eigenvalues or *complex* eigenvalues, because MC models can have these more complex dynamical characteristics. If the exponents (eigenvalues) λ_k are real, with some possibly *repeated*, then the correct modal output form is:

$$y(t) = A_{10}e^{\lambda_1 t} + A_{11}te^{\lambda_1 t} + \cdots + A_{20}e^{\lambda_2 t} + \cdots + A_{m,v_{k-1}}t^{v_k - 1}e^{\lambda_m t} = \sum_{k=1}^{m}\sum_{j=0}^{v_k - 1} A_{kj}t^j e^{\lambda_k t} \tag{5.6}$$

Distinguishing response data fitted to Eq. (5.6) versus Eq. (5.5) can be difficult, as illustrated in Example 5.5 below. If some λ_i are *complex*, the most general case, then the output includes sinusoidal terms:

$$y(t) = \sum_{k=1}^{m}\sum_{j=0}^{v_k - 1} A_{kj}t^j e^{\sigma_k t}\cos(\omega_k t - \theta_{kj}) \tag{5.7}$$

where $\lambda_k = \sigma_k + i\omega_k\left(i \equiv \sqrt{-1}\right)$ and $m \le n$ is the number of distinct eigenvalues. Deciding whether to fit Eq. (5.7) to data should be easier, because oscillating data are easier to recognize.

Remark: The specifics here are for linear models, but the complexity notions carry over for nonlinear models, unfortunately less predictively and without the benefit of the numerous results available for linear systems.

Example 5.3: ***Cyclic Model with Three Possible Dynamic Response Patterns***

The model shown in **Fig. 5.4** is **cyclic**, i.e. it has at least three compartments that include a closed unidirectional path among them. In Example 4.10 of Chapter 4 we saw that this model can have complex eigenvalues and thus solutions with periodic components. We show here that the compartmental matrix K of this model can have real distinct, real repeated, as well as complex eigenvalues – depending on the values of parameters p_1 and p_2. K is easily written directly from the diagram (if unclear, redefine the parameters as k_{ij}s). Then:

$$K = \begin{bmatrix} k_{11} & 0 & k_{13} \\ k_{21} & k_{22} & 0 \\ 0 & k_{32} & k_{33} \end{bmatrix} = \begin{bmatrix} -1 & 0 & p_2 \\ 1 & -p_1 & 0 \\ 0 & p_1 & -p_2 \end{bmatrix}$$

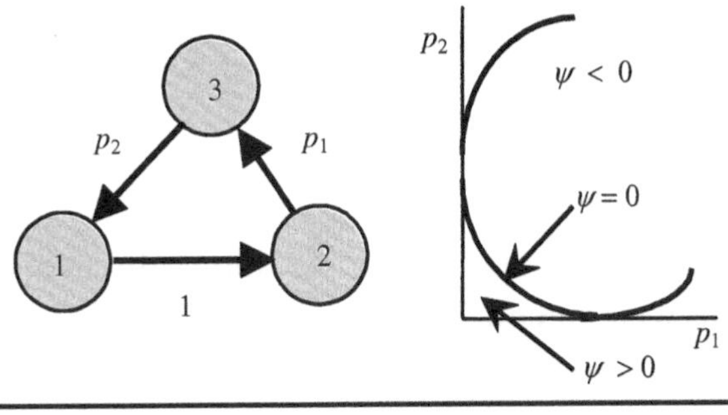

Fig. 5.4

The characteristic polynomial is:

$$\Delta(s) = |sI - K| = s[s^2 + (p_1 + p_2 + 1)s + p_1 p_2 + p_1 + p_2]$$

with discriminant:

$$\psi \equiv (p_1 + p_2 + 1)^2 - 4(p_1 p_2 + p_1 + p_2) = (p_1 - p_2)^2 - 2(p_1 + p_2) + 1$$

If $\Psi > 0$, the λ_k are real and distinct (try $p_1 = p_2 = 0.1$); if $\Psi = 0$, they are real and repeated; if $\Psi < 0$, they are complex. These three regions are illustrated on the graph of the p_2 vs. p_1 parameter space in **Fig. 5.4**. Interestingly, for a single input in any compartment of **Fig. 5.4**, measured output from any one of the three compartments has three visible modes, i.e. no modes (or λ) are hidden.

Multicompartmental models also can have ***hidden oscillations*** (hidden oscillatory modes) in some compartments – with complex eigenvalues (modes) hidden from outputs measured in a central compartment. This interesting phenomenon can occur in potentially physiologically realizable systems, modeled with ***generalized mammillary models*** (as developed in Fagarasan and DiStefano III 1986).

Caveat #3 – Hidden Compartments?[12]

Model dynamical dimensionality n is the minimum number of compartments, because one or more are often hidden or not detectable in output data. If one considers the transfer function (TF) as representing I–O data, a useful result is that a linear SISO multicompartmental model has hidden compartments if and only if its TF has at least one pole-zero cancellation. Conversely, it has no hidden compartments if and only if there are no pole-zero cancellations (Fagarasan and DiStefano III 1986).

We've been belaboring the point here and in the last few sections that not all modes and compartments (state variables) they represent are detectible in data, no matter how perfect, because some modes may not be visible in a particular output $y(t)$, in response to a particular input $u(t)$. The problem clearly depends on the particular input–output pair specified by the model. This then complicates model selection, with potentially profound consequences on our ability to fully comprehend biological system structures from data. Hidden compartments with hidden modes may (and often do) exist. Thus some portions of a system or system model (subsystems, submodels) may not be visible (detectable) in the specific data. But they may be visible in other data. We demonstrate below how mode and compartment visibility depends on input–output configurations.

Example 5.4: ***Mode Visibility***

The model shown in **Fig. 5.5** has the same input directed simultaneously into two separate and otherwise uncoupled compartments, with leak rate constants k_{0i}

12 Hidden compartments are directly analogous to hidden states in a Markov model, called a *hidden Markov model*.

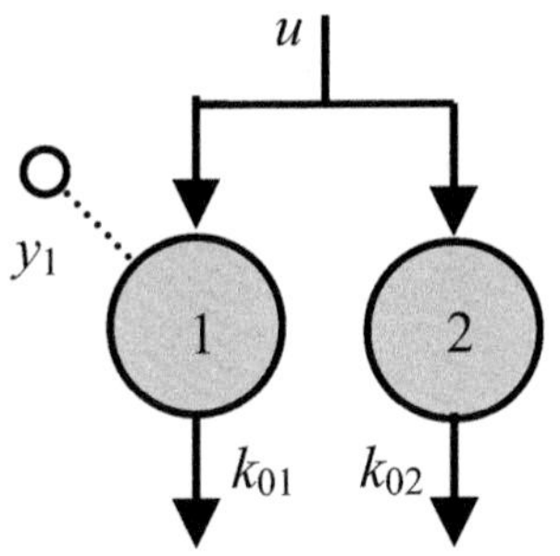

Fig. 5.5

A state unobservable model with all modes visible.

equal to 1 for both, and the output measured only in compartment 1. The model equations are:

$$\dot{\boldsymbol{q}} = \begin{bmatrix} -1 & 0 \\ 0 & -1 \end{bmatrix} \boldsymbol{q} + \begin{bmatrix} 1 \\ 1 \end{bmatrix} u$$

$$y = [1 \quad 0]\boldsymbol{q} = q_1$$

The two same diagonal terms in the diagonal K matrix clearly indicate this two-compartment model has repeated eigenvalues $\lambda_{1,2} = -1$. The impulse response of this model is obtained by solving for $q_1(t) = y(t)$, which is simply e^{-t}, or $h(t) = [1 \quad 0]e^{Kt}B = e^{-t}$ if we use matrix manipulations (Appendix B). This output is a *single* mode. Actually, this mode is visible in the zero-input (free) response for any choice of ICs in the two compartments: $\boldsymbol{q}_0 = [q_1(0)\ q_2(0)]^T$, with $q_1(0) \neq 0$, because the output is always

$$y(t) = Ce^{Kt}q_0 = [1 \quad 0]\begin{bmatrix} e^{-t} & 0 \\ 0 & e^{-t} \end{bmatrix}\begin{bmatrix} q_1(0) \\ q_2(0) \end{bmatrix} = q_1(0)e^{-t}$$

Furthermore, even if the leak parameter in compartment 2 is different from 1, rendering the two intrinsic modes different from each other, the output is still the same and thus never includes *any* hint of the second, distinct and hidden mode in compartment 2, because it is obviously disconnected from the only output. In the language of control theory, this model is called **state unobservable** (Appendix D).

Remark: This example may seem contrived, but it makes the point that a two-compartment model can "look like" a one-compartment model in the output data, one compartment being hidden and not showing up in the output. The next one is more realistic.

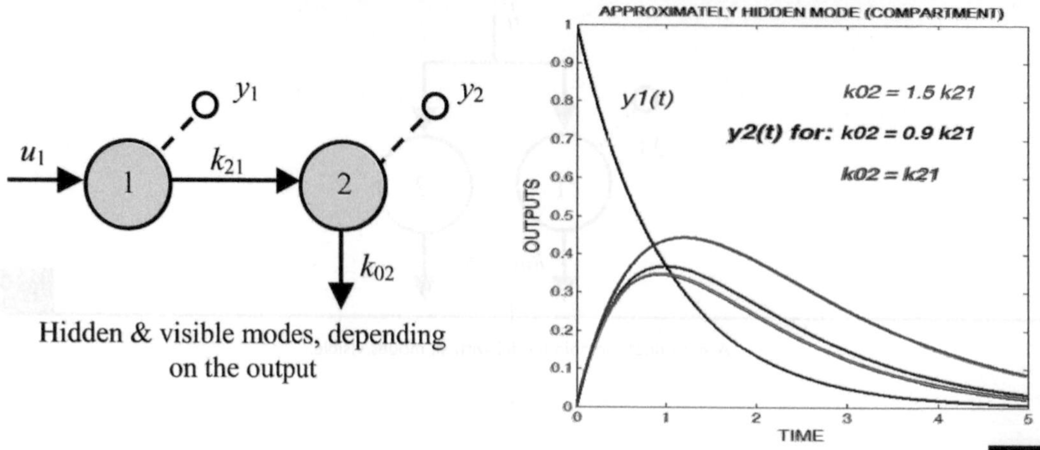

Fig. 5.6

Example 5.5: ***Hidden Eigenvalues: Totally Hidden and Approximately Hidden Compartments***

For $k_{21} \neq k_{02}$, the two-compartment model in **Fig. 5.6** has only one visible eigenvalue $\lambda_1 \equiv k_{21}$ in the first (visible) output mode: $y_1(t) = e^{-k_{21}t}$. But both modes and their corresponding eigenvalues are visible in the second output:

$$y_2(t) = \frac{k_{21}}{k_{21} - k_{02}}(e^{-k_{02}t} - e^{-k_{21}t})$$

Now suppose that the two rate constants are equal: $k_{02} = k_{21}$. In this case, the single eigenvalue is *repeated*. Then $y_1(t)$ is still the same, the single mode $y_1(t) = e^{-k_{21}t}$. But (by l'Hôpital's rule) $y_2(t) = k_{02}te^{-k_{02}t} \equiv k_{21}te^{-k_{21}t}$, a *single* mode, albeit different, rather than two modes when the *k*s were not equal.

This situation is difficult to detect because the overall shape of the response $y_2(t)$ is approximately the same, whether or not k_{21} is the same or different than k_{02}, as shown in the simulated results for equal *k*s and for $k_{21} = 0.9k_{02}$ or $1.5k_{02}$ in the graph in **Fig. 5.6**. For $k_{21} = 0.9k_{02}$ the plots are virtually indistinguishable and for a 50% change in k_{21}, at $k_{21} = 1.5k_{02}$, they differ little, which would be difficult to distinguish in real data.

Remark: Note that the two rate constants are the turnover rates of their respective compartments, i.e. $k_{21} = -k_{11}$ and $k_{02} = -k_{22}$, because $k_{01} \equiv 0$ and $k_{12} \equiv 0$. This means that the two modes visible from compartment 2 are these turnover rates, which is not true in general. Modes usually have exponents (eigenvalues) that are more complicated functions of the rate constants, e.g. as in Example 5.3 above, or in the next example below.

Finding Modes (or Compartments) Visible in Output Data by Graphical Inspection

The number of modes – and therefore compartments detectable in particular outputs of multicompartmental models – can be established by inspection of the compartmental model graph. This approach sharpens intuition by more graphically illustrating why full dimensionality of a model and the system it represents is difficult to establish. It also provides a more facile tool for discovering hidden information about the underlying system.

The following examples illustrate the step-by-step procedure illustrating further how the information content in dynamical responses is limited by the sites of input–output probes and also initial conditions, i.e. system *memory*, as well as input forms other than impulses.

Example 5.6: ***Visible Modes by Inspection***

In the simple two-compartment model in **Fig. 5.7**, we assume zero ICs and an impulse input. We also relax these assumptions, later, to discover how ICs and other inputs affect results. We are interested in the number of modes visible in outputs 1 and 2 under various conditions. We evaluate this number first mathematically, then more simply using our developing graphical-inspection intuition.

The Math: We saw earlier in Eqs. (4.38) and (4.39) of Chapter 4, repeated here, that both outputs are a sum of two distinct exponential modes:

$$y_1(t) = \frac{(k_{02} + k_{12} + \lambda_1)e^{\lambda_1 t} - (k_{02} + k_{12} + \lambda_2)e^{\lambda_2 t}}{V_1(\lambda_1 - \lambda_2)}$$

$$y_2(t) = \frac{k_{21}(e^{\lambda_1 t} - e^{\lambda_2 t})}{V_2(\lambda_1 - \lambda_2)}$$

and from Eq. (4.33) of Chapter 4, the two exponents (eigenvalues) are:

$$\lambda_{1,2} = \frac{-(k_{02} + k_{12} + k_{21}) \pm \sqrt{(k_{02} + k_{12} + k_{21})^2 - 4k_{02}k_{21}}}{2}$$

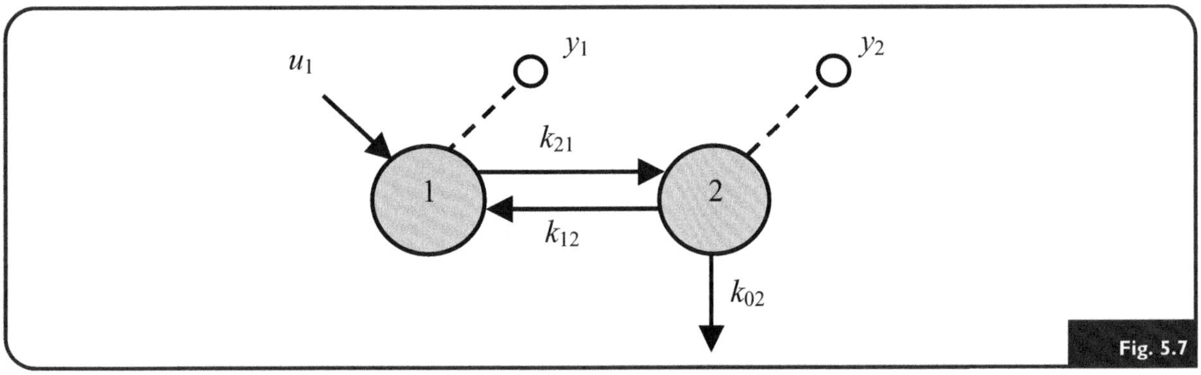

Fig. 5.7

In other words, there are exactly *two* (*exponential*) *modes* measurable in each compartment for an impulse input. Now suppose $k_{21} = 0$ (and ICs remain zero), then $\lambda_2 = k_{22}$, which nulls the second coefficient and the second exponential term in the equation for y_1, i.e. the solution for y_1 reduces to only one mode, and that for y_2 reduces to zero modes because the coefficient goes to zero. Thus, the model with no forward connection from compartment 1 to 2 will exhibit one mode in 1 and none in 2. That's what the math tells us.

The Simpler Graphical Solution: Now let's redo the problem by reasoning physically about each case, k_{21} nonzero and k_{21} zero, again, developing our intuition along the way. *The principle here is that each compartment, when it is stimulated by a signal entering it, elicits a distinct mode associated with that compartment in its dynamic response.* Whether this effect is visible in an output depends on the locations of the input–output probes, and the interconnections among compartments.

An input signal entering compartment 1 (an impulse input in this example) will first stimulate the first mode represented by the state variable – dynamic system compartment 1 itself. Then, if k_{21} is nonzero, the signal travels to 2 via k_{21}, where it stimulates the second distinct mode, represented by the state variable – dynamic system compartment 2. Finally, material travels back to 1 via k_{12}, with a dynamic reflecting both the stimulus from 1 along k_{21} *and* the corresponding dynamic due to stimulation of the distinct mode in 2. So we see *two* modes in both compartments.

This is how we get to see both modes of the model in the outputs of both compartments. They go back and forth, feeding their signals forward and back, because they are connected reversibly.[13]

Now, using the same reasoning and inspection of the graph, we see that if $k_{12} = 0$ and k_{21} is nonzero, we find only one mode in compartment 1 and two in compartment 2, for the same input and ICs, because the mode associated with compartment 2 is not visible in 1, because of the lack of connection back to 1 ($k_{12} = 0$). We discovered this using mathematics, earlier, but this graphical solution is easier.

Nonzero ICs: Going one step further, if we started with *zero* input, $k_{12} \neq 0$, $k_{21} \neq 0$, and nonzero ICs in both compartments,[14] both would have *two* modes visible in them, because the ICs are equivalent to impulse inputs (Chapter 2). And, unlike the zero IC case above, when $k_{21} = 0$, compartment 1 would still show two modes, because mode 2 would be stimulated by the nonzero IC in 2 and travel to 1 along k_{12}.

So we see again that ICs are as important as inputs in establishing model complexity, because – in essence – they are equivalent to impulse inputs.

13 *Music Analogy:* My teaching assistant (TA) this Fall quarter 2012 at UCLA, PhD student Long Nguyen, made a useful analogy for modes in compartmental models, by associating them with notes in a musical chord sounding simultaneously. The analogy is not perfect, but it was helpful to the students learning the concept. Two compartments interconnected interchangably have two fundamental modes, one for each compartment, both of which are visible (heard) simultaneously by observing (listening to the output of) any one.

14 We recall that nonzero ICs are equivalent to impulse inputs.

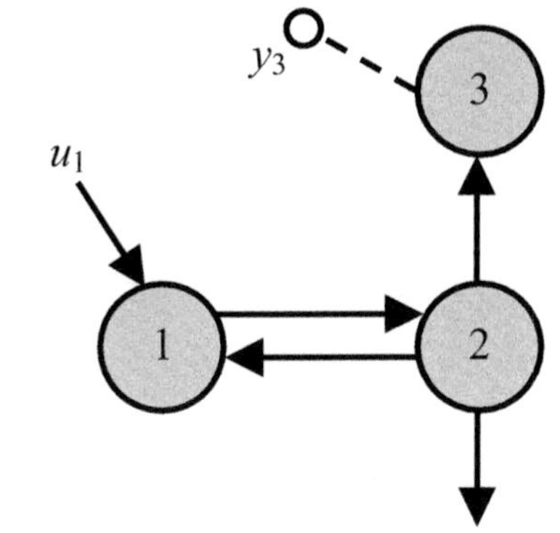

Fig. 5.8

Example 5.7: ***Number of Modes Visible in Different Compartments***

We establish the number of modes visible in the output of compartment 3 in **Fig. 5.8** as follows, using only graphical reasoning. We again assume zero ICs everywhere and an impulse input. In the previous example we established that two modes would be visible in compartment 2, if it were measured. Despite the fact that compartment 2 is not measured in this problem, there are still two "excited" modes in 2, and they travel to compartment 3 along nonzero path k_{32}. So, at least two modes are visible in compartment 3 measurements. But, compartment 3 is itself a dynamic state variable compartment, and when it is stimulated it generates its own mode. Stimulated by two modes coming from 2, it therefore exhibits *three* modes in its output. No math needed!

Different Inputs: Going one step further, if the input u_1 were an *exponential function* distinct from the fundamental modes of the model, e.g. a single exponential input, then *four* modes would be visible at y_3, because the input "mode" would also be visible in the output. Inputs are external to the model; therefore exponential inputs can be thought of as the output of a compartment external to the model, another mode! And so on....

The Exercises at the end of the chapter include several more complex examples. Completing these should greatly facilitate use of this graphical methodology on real modeling problems.

Remark: In principle, the results above do not depend on parameter values, only structure, assuming all $k_{ij} > 0$ in the models. In reality, the relative magnitudes of the parameter values matter, because – depending on parameter values, some modes may be very small and difficult to detect in real data. Chapters 11 and 17 address this very important issue in the context of model parameter sensitivity analysis.

Automated Mode Detection Using Graph Theory Algorithms

The manual mode-finding algorithm of the last section is readily automated using computer science graph theory tools. Either **breadth-first search (BFS) or depth-first search (DFS)**

```
Require: Graph G(V,E), node (compartment) i receives input
Ensure: Modes[j] mode-count for all compartments, for all j ϵ V
// Determine compartments reachable from the input

 while v ϵ V
 if v is reachable from i then // Using DFS or BFS:
  TrueMode[v] ← 1
  end if
  end while
 // Count modes based on connectivity:
 while v ϵ V do
  while v' ϵ V do
   if v is reachable from v' then // Using DFS or BFS:
     if TrueMode[v] == 1 then
     Modes[v] ← Modes[v] + 1
     end if
  end if
 end while
end while
```

Fig. 5.9

Pseudo Code for Calculating Mode Number for each Measurable Compartment.

(Cormen et al. 2009) can be used to count compartments with a directed path between any measurable compartments, i.e. those **reachable** from perturbed compartments, i.e. those receiving inputs or having nonzero initial conditions (ICs are the same as equivalent impulse inputs). Directed connectedness and the permutations of paths between inputs to different compartments and follow-up compartments is key to counting modes. **Pseudo code**[15] for the imbedded mode-counting algorithm of the last section is given in **Fig. 5.9**. Notation ***G*(*V*,*E*)** means a graph with *V* vertices (nodes, compartments) and *E* edges (rate constant parameters).

This algorithm actually finds the maximum number of modes in each compartment, because it is completely symbolic – with no regard for numerical values of model parameters. *Hidden* modes can be missed, e.g. those due to model parameters coinciding, like the $k_{22} = k_{33}$ in the model-reduction example of the next section.

Model Simplification: Hidden Modes & Additional Insights

Hidden modes and compartments were the "bad news" limitations in the last sections. There's also an upside to hidden modes, in the context of model simplification. For state variable models, model **reduction**, or **aggregation** as it is also called, is the process of finding a simpler representation of a given structure appropriate or useful for some purpose. The given structure

15 This algorithm was originally written by Farhad Hormozdiari, and augmented by teaching assistant Long Nguyen, both PhD students at UCLA at the time of this writing.

may or may not be in some canonical form, and *simple* in this context means fewer state variables, fewer parameters, or both.

We are interested in this subject for two reasons: first for its intrinsic value in finding simpler representations for specific applications, and second because it provides additional insights into real limitations on the model building process imposed by the dynamical characteristics of the system and the quality of input/output data. Again, this is why specific input/output probes retrieve only limited information about real system structure and connectivity, even in ideal circumstances with noise-free data. In the process, we gain insights into how to design or redesign input/output experiments so that system attributes "muddied" by measurement noise in real data might be clarified.

In Other Words: how does one distinguish between intrinsically unavailable and thereby hidden system attributes, and those not intrinsically hidden, but undetectable – because they are "buried" in high level-noise? Also, what does "simple model," or simple (low-order) model dynamics, imply about the seemingly complex biomechanisms they presumably represent? These are the big questions. The exposition here should help in finding answers.

State Variable Aggregation (Reduction): Transfer Functions Tell All

The underlying principle is straightforward for linear time-invariant SISO models: *nth-order SISO structures are reducible to lower-order if one or more pole-zero cancellations occur in the model transfer function, as the following example illustrates.*

Example 5.8: ***Pole-Zero Cancellations, Hidden Modes & Proportional States***

Suppose we know from biophysical considerations that the structure of a real biosystem is well represented by the three-compartment SISO mammillary model given in **Fig. 5.10**. This is the same model as in Example 4.12 of Chapter 4, but with $k_{01} = 0$.

Using the results of that example, the transfer function H_{11} is:

$$H_{11} = \frac{(s - k_{22})(s - k_{33})}{(s - k_{11})(s - k_{22})(s - k_{33}) - k_{12}k_{21}(s - k_{33}) - k_{13}k_{31}(s - k_{22})} \tag{5.8}$$

If the denominator were in factored form, we could easily determine the conditions for pole-zero cancellations. Analytically, this requires solution of a third-order polynomial. Instead, we establish the conditions by closer inspection of Eq. (5.8). Only a

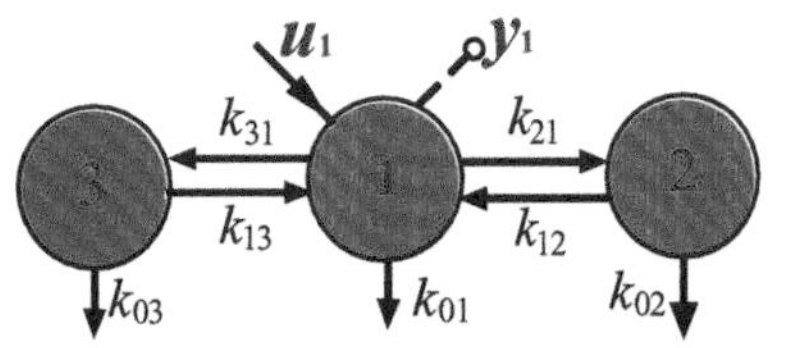

Fig. 5.10

factor of $s-k_{22}$ or $s-k_{33}$ in the denominator can cancel a zero in Eq. (5.8). We see that if $k_{22}=k_{33}$, we have such a factor, i.e.

$$
\begin{aligned}
H_{11}\big|_{k_{22}=k_{33}} &= \frac{(s-k_{22})^2}{(s-k_{11})(s-k_{22})^2-(s-k_{22})(k_{12}k_{21}+k_{13}k_{31})} \\
&= \frac{s-k_{22}}{(s-k_{11})(s-k_{22})+k_{12}k_{21}-k_{13}k_{31}}
\end{aligned}
\tag{5.9}
$$

Thus, $k_{22}=k_{33}$ is a sufficient condition, but we have not shown it to be necessary for a second-order reduced model. Now, if $k_{22}\approx k_{33}$, which means that the turnover rates of both compartments external to the central compartment are approximately the same, we have an approximate pole-zero cancellation, and the model will likely appear to be no more than second-order in real, noisy data.[16] This condition also can be given an intuitively appealing time domain interpretation. If the dynamic responses in compartments 2 and 3 to an impulse input in compartment 1, i.e. $h_{21}(t)$ and $h_{31}(t)$, are proportional, they will be impossible to distinguish from each other from measurements in any compartment. To see this, using results from Example 4.3, we evaluate the ratio of Laplace transforms of the masses in these compartments:

$$
\frac{Q_3(s)}{Q_2(s)}=\frac{k_{31}}{k_{21}}\left(\frac{s-k_{22}}{s-k_{33}}\right)
\tag{5.10}
$$

This ratio is constant if and only if $k_{22}=k_{33}$, and we have thus shown that this yields a second-order model for H_{11}. Furthermore, we have:

$$
H_{21}=\frac{k_{21}(s-k_{33})}{|sI-K|}
\tag{5.11}
$$

and

$$
H_{31}=\frac{k_{31}(s-k_{22})}{|sI-K|}
\tag{5.12}
$$

where $|sI-K|$ is the denominator of Eq. (5.8). Therefore, a pole-zero cancellation occurs in compartment 2 and compartment 3 responses when $k_{22}=k_{33}$. *This means that outputs measured in compartment 1, in particular, will exhibit only two modes, because the compartments 2 and 3 are aggregated, dynamically speaking.*

Remark: With all other model parameters nonzero, we get the reduced model if and only if $k_{22}=k_{33}$, which cannot be established simply by inspection. The symmetry in mammillary model structures strongly suggests looking for this possibility by examining turnover rate values. We discuss input–output equivalence and model distinguishability further in the next sections, and more deeply in Appendix E.

16 H_{11} is the input-output (i.e. data) transfer function.

Remark: In the model building process, approximate pole-zero cancellations can be used to generate an approximate simplified model. If the system parameter values are such that pole-zero cancellations occur *approximately*, the model may *appear* to be of smaller dimension, depending on the quantity and quality of input/output data. This is one way in which system attributes can be "muddied" by noise in data.

Remark: Using this same example 3-compartment model, Exercises E5.7 and E5.8 illustrate how initial conditions in unmeasured compartments also can "muddy the water" – so to speak – and affect mode visibility and thus apparent model dimensionality and structure.

Remark: Nonlinear (NL) ODE models also have "transfer functions" (TFs), i.e. input–output solution response functions. Unlike linear models, however, their dynamic responses are dependent on inputs – input shapes or magnitudes. So we can't write them only in terms of model parameters, like we do for linear model TFs. Nevertheless, the principle result above carries over to NL models, i.e. "modes" in NL systems can cancel each other – at least approximately – thereby rendering complex structure less visible in data characterized by specific inputs and outputs. Similarly simplified NL models, also with input-dependent dynamics, are characteristic of enzyme kinetic models studied in the next chapter. NL modes are specifically addressed in Chapter 9.

Remark: In Chapter 17, model reduction is readdressed – for NL as well as linear ODE models – in part as a problem in model *parameter sensitivitity analysis*, with important implications in systems biology. Complex cell signaling network hypotheses can be assessed for validity, and simplified, by eliminating network edges and nodes (parameters and associated state variables) with negligible effects on output data.

Biomodel Structure Ambiguities: Model Discrimination, Distinguishability & Input–Output Equivalence

The ambiguity problem addressed here is more nuanced than the model reduction problem discussed above: different structural models *with the same dimension n* can be fitted to the same data, in principle even perfectly to the same perfect (noise-free) data. We discuss and illustrate this problem first in the context of fitting compartmental models to real data. Then we treat the theory.

Compartment sizes (e.g. V_i) and intercompartmental transfers (k_{ij}) and irreversible losses (leaks) to the environment (k_{0j}) are depicted explicitly in MC models, whether linear or nonlinear – one of their major features. Quantification of a MC model from data usually means determining the values of these k_{ij} and V_i that render the model fit to real experimental data best in some sense, e.g. the k_{ij} and V_i that minimize the sum of the squared errors between the data and its corresponding *output* variable in the model.[17]

17 Model quantification methodology is presented in Chapters 10–13.

In a typical experiment, the system is perturbed with a known test-input in order to infer (compute) the values of model parameters, state variables, or combinations of these, from the measured output (an ***input–output*** or ***stimulus–response*** experiment). In some cases, some k_{ij} or V_i can be established from independent experiments, e.g. by studying isolated subsystems, as in organ perfusion studies, or from published values for the same or analogous systems, scaled for the system under study. Parameters often can be scaled allometrically[18] to humans, from values estimated for other species (for example, see Chapter 8). Searching the published literature for parameter values, even if from other species – should be high on the priority list when quantifying a biomodel. This can save time and experimental resources.

If the model structure reflects the true system structure in a sense consistent with modeling goals, these parameters can be given a biological interpretation. One difficulty, however, is distinguishing among a finite but typically small number of alternative and feasible compartmental structures that may fit the biostructural and experimental data bases equally well. Such models are called **input–output equivalent**, or **input–output indistinguishable**. In other words, more than one model can fit data equally well and the result is an ambiguity – fortunately often resolvable with some additional analysis and/or experimentation. The first step, though, is detecting the ambiguity.

Caveat #4: In addition to appropriately locating input–output ports, biomodel topology must faithfully reflect known system structure and connectivity, at least reproducing the connectivity of essential features. This means that leaks to the environment and intercompartmental transfers and their associated parameters *must be located accurately within the structure* if they, or functions of them, are to be given a physical interpretation.

Example 5.9: ***Leak Ambiguities in a Two-Compartment Model***

The two models shown in **Fig. 5.11** have the same transfer function,[19] i.e. they are *input–output equivalent* and exhibit the same modes in $y(t)$. But they are not acceptable as equivalent models for many applications, because their leaks are in different compartments. If it is known that only compartment 2 "leaks," the model on the left – with the wrong leak location – cannot be used, for example, to compute any of the k_{ij} or

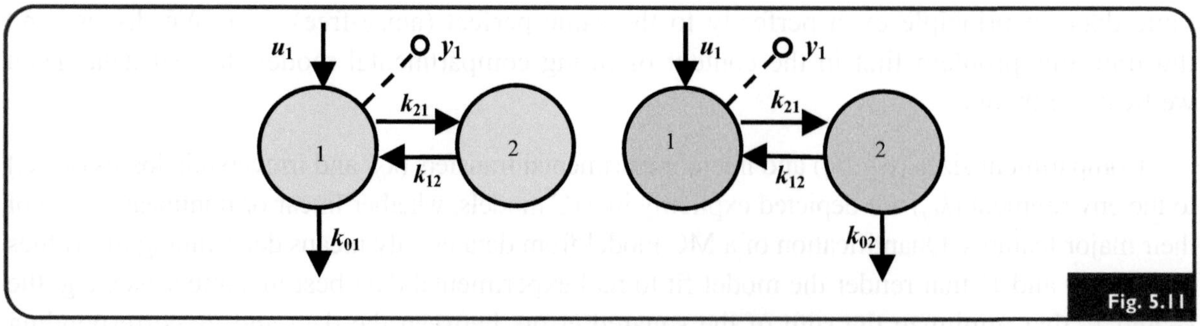

Fig. 5.11

Two different models with the same TF.

18 Relationship of body size to shape, anatomy or physiology of organisms.

19 Exercise E5.16.

V_i, because they don't correspond to the known biology. Input–output equivalence issues are very common in modeling and we address them as they arise in the remainder of this book. The general topic of equivalent models is treated in some depth in Appendix E.

We assume in the following paragraphs that the number of model compartments n has already been determined. Several examples below illustrate the problem and some approaches to its solution. The first two are different models of single-input–single-output (SISO) small-perturbation tracer[20] hormone distribution and disposal dynamics in different species. In each, multiexponential models of increasing order were first fitted to impulse response kinetic data, generating a biexponential model for data from trout and salmon, and a triexponential model for rat, human, sheep, and tilapia fish data. The problem is how to transform these "data models" into two- or three-compartment models, the primary problem being assignment of their specific connectivity and disposal pathways. We also see how data from additional experiments are also helpful. The third example illustrates the role of explicit multi-input/ multi-output experiment design and analysis in structuring a biomodel.

Example 5.10: ***SISO Two-Compartment Models of Thyroid Hormone (TH) Dynamics – Another Variation on Leak Ambiguities & Alternative Models***

The experiments here were done *in vivo*, i.e. in live fishes, by injecting radioactively labeled thyroxine (T_4) intravenously (IV), as a pulse dose of tracer[21] T_4 (the test-input), and then measuring the dynamical tracer kinetic response in blood plasma (the output). This was done by sampling blood at sufficiently frequent intervals, with the data fitting biexponenential functions well (Eales et al. 1982; Specker et al. 1984).

The most general two-compartment model can be structurally associated with this solution, as illustrated in **Fig. 5.12**. Assuming compartment 1 is plasma T_4 and compartment 2 represents all aggregated (lumped) tissue T_4, then both k_{12} and k_{21} are nonzero – because T_4 surely exchanges between blood and tissue compartments –

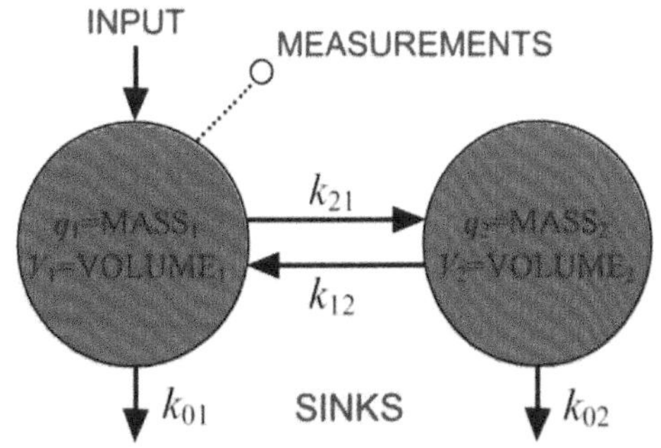

Fig. 5.12

20 These systems are nonlinear, but the small-perturbation tracer inputs stimulate their dynamical responses in linear ranges of operation, so the system-experiment models are linear. Linearizing experiments are discussed in the last section of this chapter.

21 A small distinguishable dose for tracking T_4 kinetics. Tracer theory is developed in the last section.

and three alternative structures are generally possible: (a) leaks exist from both plasma and tissue; (b) there is no leak from plasma ($k_{01} = 0$); or (c) there is no leak from tissue ($k_{02} = 0$). Now, if the experiment were planned so that blood sampling began sufficiently soon after IV injection, which means before significant T_4 has had a chance to be taken up by tissue (via k_{21}), and if subsequent samples were drawn often enough, compartment 1 could be associated with plasma only, as this is indeed the initial distribution space under such circumstances. In this case, only choice (b) is feasible, because it is known that T_4 is degraded or otherwise eliminated directly in extravascular tissues, but not in blood. With sampling beginning too late to designate compartment 1 as plasma only – and plasma then containing some extravascular T_4 spaces within various other organs – then choice (a) must also be considered. Urinary excretion is often reasonably well modeled as a direct leak from the plasma pool (Gabrielsson and Weiner 2000). So, even if sampling is begun early enough, if any T_4 leaks in urine, then leaks in both compartments – choice (a) – must be considered.

We used biophysical and physiological information in the example above to structure the model, as we do again in the next example. But model structure was ambiguated by experiment particulars, in this case the blood-sampling design. Data-blood sampling schedules typically are critical experiment design variables when modeling from transient kinetic response data.[22]

Example 5.11: ***Candidate Three-Compartment Models of TH Dynamics: Several Alternative Structures***

Linearizing tracer triiodothyronine (T_3) and thyroxine (T_4) kinetic data are each fitted well by tri-exponential functions in rats, humans and sheep (Chen 1970; Wilson et al. 1977; DiStefano III 1985) and also in trout (Sefkow et al. 1996). The model structure discrimination problem here is more complex than in the previous example, because there are more possibilities for structuring a three-compartment model. We reduce these by first assuming the experiment design is consistent with a correct interpretation of compartment 1 as plasma only. The alternative three-compartment model structures in this case are: (a) pure mammillary (**Fig. 5.13**); (b) pure catenary (**Fig. 5.14**); or (c) a combination of both, the most general case shown in **Fig. 5.15**.

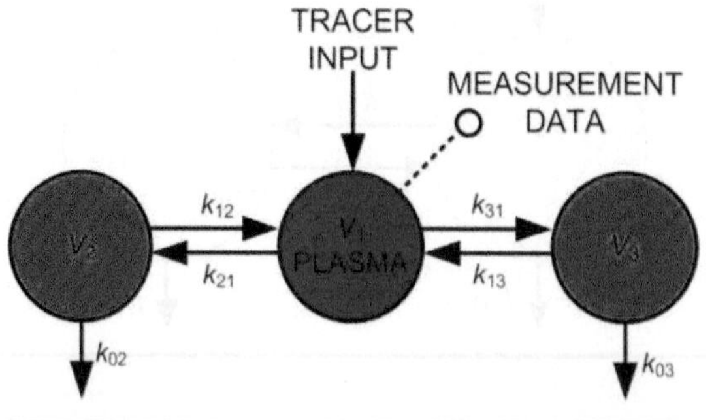

Fig. 5.13

22 Chapter 16 addresses experiment design issues.

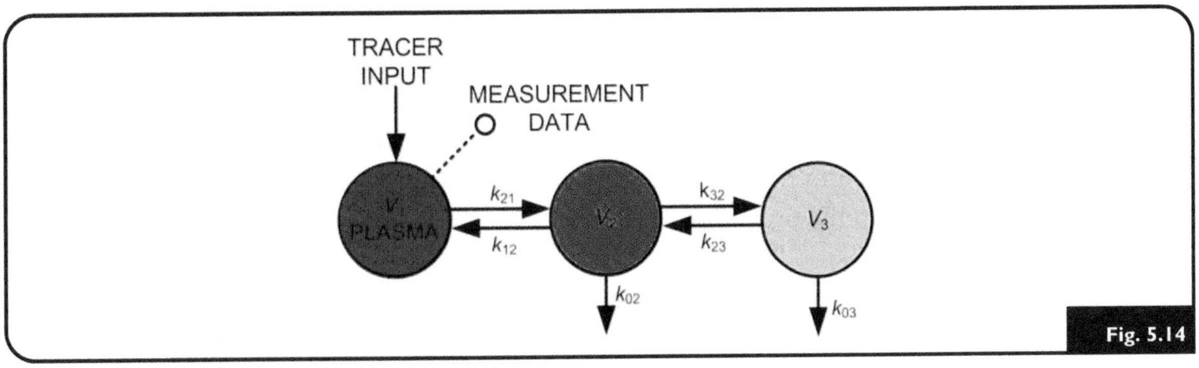

Fig. 5.14

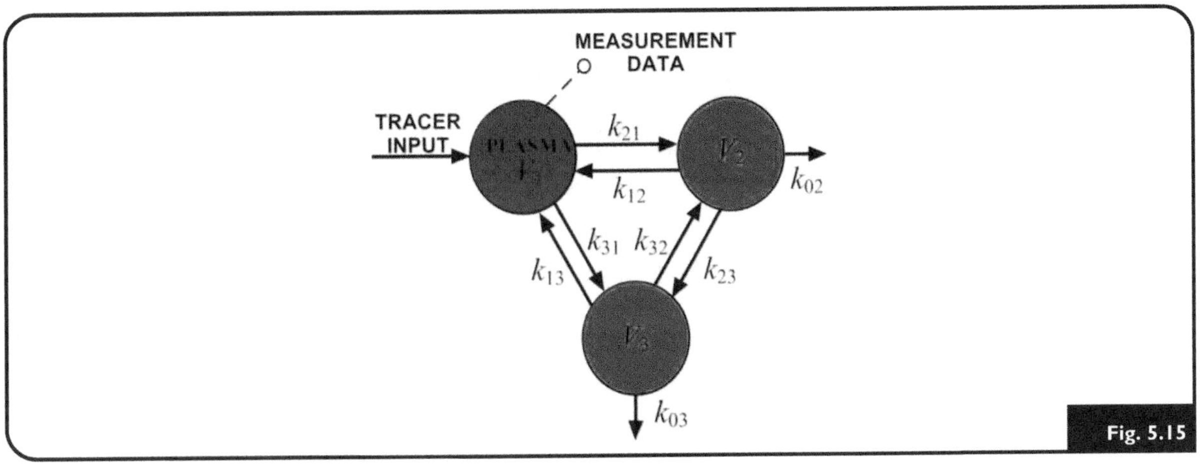

Fig. 5.15

Using Additional Independent Data: The three models can be distinguished using certain additional information (data) about peripheral tissue compartments, such as information about compartment size. If the total amount of hormone in one or more organs matches the amount predicted (by the model equations) to be in compartments 2 and/or 3 of a particular model structure, this would favor that structure over the others. For example, the steady state amounts of endogenous T_3 and T_4 (separately) predicted to be in compartment 2 of the mammillary model are about the same as the T_3 and T_4 reported as measured in whole rat liver and kidneys combined. Independently derived information about kinetic exchange rates also can be useful. For example, kinetic exchange rates for T_3 or T_4 are quite similar between plasma and liver and plasma and kidney (Hasen et al. 1968). Together, this quantitative information (data) suggests liver and kidneys compartments can be lumped and these results strongly suggest the mammillary model, with compartment 2 consisting chiefly of liver and kidney and compartment 3 relegated to all other tissues combined. The catenary model is difficult to rationalize, given the above additional organ data, chiefly because all tissue must be contained in compartments 2 and 3 in the sequential manner depicted.

The final, general model might be rationalized physiologically if, for example, compartment 2 includes liver, compartment 3 rather than compartment 2 includes gut, and significant hormone is transferred in bile from liver to gut and from gut to liver in portal blood (DiStefano III et al. 1992). Clearly there are quantitative as well as qualitative questions embedded in this alternate hypothesis, resolution of which requires additional kinetic data from the organs involved.

This example illustrates the sequential nature of the modeling process, again coupled with that of experiment design, but in a different way than in the previous example – where the sampling schedule design was an issue.

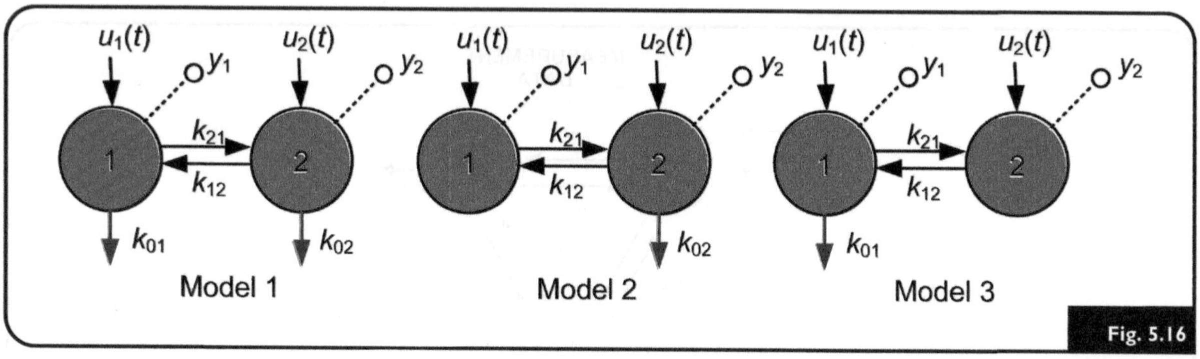

Fig. 5.16

Three alternative 2-compartment models.

Example 5.12: ***Discriminating Among Three Two-Compartment MIMO Models***

As just illustrated, data from additional inputs and outputs can help to distinguish between alternative structures. In this example, adapted from (Wagner et al. 1981), we discriminate among the three different two-compartment MIMO model configurations (structures) shown in **Fig. 5.16**, using idealized input–output data from experiments done in both compartments, with two inputs and two outputs as shown.

Both compartments 1 and 2 are accessible to test-input probes, $u_1(t)$ and $u_2(t)$, and measurements, $y_1(t)$ and $y_2(t)$, and we know there is interchange between compartments 1 and 2. But we do not know where the "leaks" are.

To solve this problem, we model the two impulse–response (pulse–dose) experiments. In the first, denoted Study A, $u_1^A = D^A\delta(t)$ and $u_2^A = 0$, and concentrations $y_1^A(t)$ and $y_2^A(t)$ are measured. From Eqs. (4.38) and (4.39) of Chapter 4, we have as impulse responses *for all three models*:

$$y_1^A(t) = \frac{D^A}{V_1(\lambda_1 - \lambda_2)}[(\lambda_1 - k_{22})e^{\lambda_1 t} - (\lambda_2 - k_{22})e^{\lambda_2 t}]$$
$$y_2^A(t) = \frac{D^A k_{21}}{V_2(\lambda_1 - \lambda_2)}(e^{\lambda_1 t} - e^{\lambda_2 t}) \tag{5.13}$$

The second experiment, Study B, has $u_1^B = D^B\delta(t)$ and $u_2^B = 0$, and the two measured impulse responses are easily determined from Eq. (5.13) by simply reversing all compartment 1 and 2 notations:

$$y_1^B(t) = \frac{D^B k_{12}}{V_1(\lambda_1 - \lambda_2)}(e^{\lambda_1 t} - e^{\lambda_2 t})$$
$$y_2^B(t) = \frac{D^B}{V_2(\lambda_1 - \lambda_2)}[(\lambda_1 - k_{11})e^{\lambda_1 t} - (\lambda_2 - k_{11})e^{\lambda_2 t}] \quad (5.14)$$

In principle, Eqs. (5.13) and (5.14) provide *all* the quantitative information obtainable from Studies A and B. Our task is to develop model discrimination conditions from these results.

Distinguishing features among Models 1, 2, and 3 in Eqs. (5.13) and (5.14) are embodied in the parameters k_{11}, k_{22}, and the eigenvalues λ_1 and λ_2 characterizing each of the four model outputs. These four intrinsic output parameters – the *structural invariants* of the model, can be quantified from these experiments, which we show in Chapter 10. Each is a function of k_{01} and k_{02}, both present in Model 1, and one each is zero in Models 2 and 3, respectively. We show here by *algebraic inequality analysis* of the model that knowledge of k_{11}, k_{22}, λ_1 and λ_2 provide enough information to discriminate among the three models. This is done without resorting to parameter estimation or otherwise knowing any parameter values – results, in principle, discernable from structure only.

We use a consequence of the linearity of these models to simplify calculations. The *definite integrals* from time zero to infinity of each of the responses in Eqs. (5.13) and (5.14) are easy to calculate analytically, from the results of Studies A and B. Equivalently, these integrals can be obtained as the responses to step functions or, equivalently, *infusion rate inputs* IR^A and IR^B (units of mass/time), the integrals of the impulse inputs in studies A and B. The desired integral $\int_0^\infty y(t)dt$ is the ***AUC***, the **area under the curve**. From Eqs. (5.13) and (5.14) we easily obtain:

$$AUC_1^A = \int_0^\infty y_1^A(t)dt = \frac{-k_{22}IR^A}{V_1\lambda_1\lambda_2}$$
$$AUC_2^A = \int_0^\infty y_2^A(t)dt = \frac{k_{21}IR^A}{V_2\lambda_1\lambda_2}$$
$$AUC_1^B = \int_0^\infty y_1^B(t)dt = \frac{k_{12}IR^B}{V_1\lambda_1\lambda_2} \quad (5.15)$$
$$AUC_2^B = \int_0^\infty y_2^B(t)dt = \frac{-k_{11}IR^B}{V_2\lambda_1\lambda_2}$$

It's convenient here to set $IR^A \equiv IR^B \equiv IR$ in Studies A and B. If this is not practical or desirable, the results below need only be altered by the factor IR^A/IR^B. To obtain the required relationships, we solve for k_{11}, k_{22}, and $\lambda_1\lambda_2$ for Models 1–3, in terms of the V_i and other k_{ij} $(i \neq j)$, with the help of the equation for the general two-compartment model presented in Chapter 4. These results are given in **Table 5.1**, along with corresponding values for the AUCs, ratios of AUCs, and final conditions based on the latter.

It is these six relationships that must be tested to discriminate among the three possible models fusing this approach. For example, if it is found that $AUC_2^A = AUC_2^B$ and

Table 5.1 Two-Compartment Model Discrimination Conditions

	Models		
	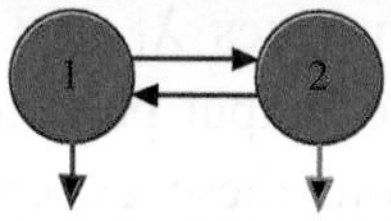	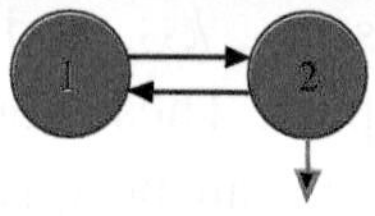	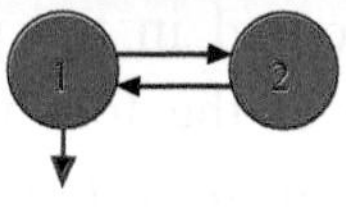
$-k_{11}$	$k_{01}+k_{21}$	k_{21}	$k_{01}+k_{21}$
$-k_{22}$	$k_{02}+k_{12}$	$k_{02}+k_{12}$	k_{12}
$\lambda_1\lambda_2$	$k_{01}k_{02}+k_{01}k_{12}+k_{02}k_{21}$	$k_{02}+k_{21}$	$k_{01}+k_{12}$
AUC_1^A/D	$\dfrac{k_{02}+k_{12}}{V_1(k_{01}k_{02}+k_{01}k_{12}+k_{02}k_{21})}$	$\dfrac{k_{02}+k_{12}}{V_1k_{02}k_{21}}$	$\dfrac{1}{V_1k_{01}}$
AUC_2^A/D	$\dfrac{k_{21}}{V_2(k_{01}k_{02}+k_{01}k_{12}+k_{02}k_{21})}$	$\dfrac{1}{V_2k_{02}}$	$\dfrac{k_{21}}{V_2k_{01}k_{12}}$
AUC_1^B/D	$\dfrac{k_{12}}{V_1(k_{01}k_{02}+k_{01}k_{12}+k_{02}k_{21})}$	$\dfrac{1}{V_1k_{02}k_{21}}$	$\dfrac{1}{V_1k_{01}}$
AUC_2^B/D	$\dfrac{k_{21}+k_{01}}{V_2(k_{01}k_{02}+k_{01}k_{12}+k_{02}k_{21})}$	$\dfrac{1}{V_2k_{02}}$	$\dfrac{k_{01}+k_{21}}{V_2k_{01}k_{12}}$
$\dfrac{AUC_1^A}{AUC_2^A}$	$\dfrac{V_2}{V_1}\left(\dfrac{k_{02}+k_{12}}{k_{21}}\right)$	$\dfrac{V_2}{V_1}\left(\dfrac{k_{02}+k_{12}}{k_{21}}\right)$	$\dfrac{V_2k_{12}}{V_1k_{21}}$
$\dfrac{AUC_2^A}{AUC_2^B}$	$\dfrac{k_{21}}{k_{01}+k_{21}}$	1	$\dfrac{k_{21}}{k_{01}+k_{21}}$
$\dfrac{AUC_1^A}{AUC_1^B}$	$\dfrac{k_{02}+k_{12}}{k_{12}}$	$\dfrac{k_{02}+k_{12}}{k_{12}}$	1
$\dfrac{AUC_2^A}{AUC_1^B}$	$\dfrac{V_1k_{21}}{V_2k_{12}}$	$\dfrac{V_1k_{21}}{V_2k_{12}}$	$\dfrac{V_1k_{21}}{V_2k_{12}}$
	Resulting Conditions		
	$AUC_2^A < AUC_2^B$	$AUC_2^A = AUC_2^B$	$AUC_2^A < AUC_2^B$
	$AUC_1^A > AUC_1^B$	$AUC_1^A > AUC_1^B$	$AUC_1^A = AUC_1^B$

$AUC_1^A > AUC_2^B$, Model 2 is the correct model, and so on. Of course, *when dealing with data obtained in real experiments, these AUC relationships must be examined with appropriate statistical tests of significance.*

*Algebra and Geometry of MC Model Distinguishability

The distinguishability problem is central to modeling biological systems. As discussed and exemplified above, two MC model structures are **indistinguishable** if they have the same input–output (I–O) dynamics, i.e. if they are **input–output equivalent**. This means that the same I–O data can be fitted by more than one model, typically many more than one. For example, the most general three-compartment model in **Fig. 5.15** has a total of nine rate constant parameters (graph edges), three of which – the turnover rates – are linearly dependent (see Eq. (4.75)), leaving six independent k_{ij}. This means there can be as many as $2^6 = 64$ different I–O equivalent three-compartment (candidate) models, because any k_{ij} can equal zero or non-zero. Fortunately, this is an upper bound. Only 13 of the 64 structures are feasible.

In this section, we develop a formal approach to reducing the number of candidate equivalent models that fit a set of I–O data. Attention is restricted to linear time-invariant ODE models, for which the theory is reasonably well developed (Godfrey and Chapman 1990; Zhang et al. 1991). The distinguishability problem for NL models is much more complex, with results available for some simple NL compartmental biomodels (Godfrey et al. 1994; Evans, Chappell et al. 2004). Conditions for distinguishability of the four different NL two-compartment models shown in **Fig. 5.17** are developed in (Godfrey et al. 1994). Distinguishability relationships also are developed in Evans et al. (2004) for epidemiological SIR (susceptible → infected → recovered) and SIRS (susceptible → infected → recovered → susceptible) models (Chapter 4, **Fig. 4.18**) and for a few simple biochemical reactions, noted in Chapter 6.

Whereas distinguishability analysis of linear systems is not directly applicable to NL models, it does provide substantial insight into the overall problem and, hopefully, its solution. Appendices C–E provide some additional linear system theory underlying this development. The following definition, from Godfrey and Chapman (1990), is abbreviated for simplicity.

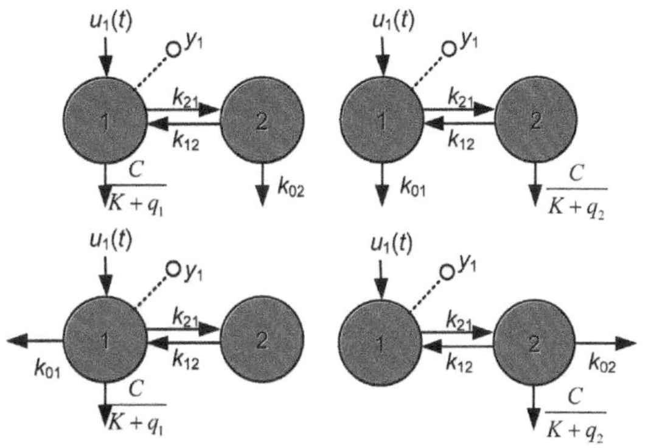

Fig. 5.17

Consider two different linear constant coefficient ODE models.

$$\begin{aligned} \dot{q} &= K(p)q + B(p)u \quad & \dot{q}' &= K'(p')q' + B'(p')u \\ y &= C(p)q & y &= C'(p')q' \end{aligned} \tag{5.16}$$

Inputs and outputs are the same in both, and we assume any model constraints $h(q,p) \geq 0$ are already included in (5.16). The presumption here is that the primed model on the right is **I−O equivalent to** and therefore **indistinguishable** from the original model on the left. This means they have the same transfer function, i.e.

$$H(s,p) = C(p)[sI - K(p)]^{-1}B(p) \equiv C'(p')[sI - K'(p')]^{-1}B'(p') = H(s,p') \tag{5.17}$$

In Eq. (5.17), both p and p' must be feasible and compatible members of parameter space for this set of models. In these terms, **indistinguishability analysis** is about finding all the candidate (primed) models with different internal structures from the original model, with the same dimension n and the same TF form as the original model − over some feasible parameter regions − and all candidates must fit the I−O data equally well. This all requires that no model with a smaller number of compartments can be found that fit the input−output data.[23]

Procedurally, for any pair of candidate structural models, say $H(s,p)$ and $H(s,p')$, checking for indistinguishability is then accomplished by equating the coefficients of the powers of s (called **structural invariants**) in the numerator and denominator polynomials of $H(s,p)$ and $H(s,p')$. In general, this leads to as many as $2nmr$ equations in the p and p' parameters (n states, m outputs and r inputs). If the p' can be solved in terms of the p, the models are indistinguishable. If not, they are distinguishable. *The difficulty here is attempting to solve* $2 \times n \times m \times r$ *polynomial equations for P different p and p'*. The following section describes how to minimize this number.

Graph Properties & Geometric Rules

We need additional notions of *connectedness*, plus the notion of a *trap*. From these, plus some geometric rules, we can again use graph theory based **breadth first search** (BFS) and **depth first search** (DFS) (Cormen et al. 2009) to help solve the problem.

A MC model is **input connectable** if every compartment has a path leading to it from the input compartment. It is **output connectable** if every compartment has a path leading from it to a measured compartment. It is **strongly connected** if there exist paths from any one compartment to any other. A **trap** is a strongly connected subset of compartments from which no path exists to any compartment outside the trap, or to the environment. We've encountered traps as single compartments with no leaks and no returns to any other compartment (dead-end compartments), e.g. the urine measurement trap in **Fig. 4.1**. But traps can include more than one compartment.

The following (necessary but not sufficient) rules eliminate many if not most candidate models (Godfrey and Chapman 1990), thereby reducing the symbolic algebra problem in a BFS or DFS search quite substantially.

23 Technically, this means that candidate indistinguishable models must be both **structurally controllable** and **structurally observable**, concepts developed in Appendices D−E.

Rule 1: The length of the shortest path from any input compartment to any output compartment must be preserved.

Rule 2: The number j of compartments with a path to any given output compartment i ($j \neq i$) must be preserved.

Rule 3: The number of compartments i that can be reached from any given input compartment j must be preserved.

Rule 4: The number of traps must be preserved.

In practice, the rules are applied first – prior to equating moment invariants of the TFs. As noted, this minimizes the resulting computational algebra problem. The set of all indistinguishable mammillary and catenary MC models was found using this approach (Kuhn de Chizelle and DiStefano III 1994). The next example, adapted and summarized from Godfrey and Chapman (1990), illustrates the procedure.

Example 5.13: ***Indistinguishable 3-Compartment Models***

We first apply geometric rules to the general three-compartment model in **Fig. 5.15**, with input in compartment 1 and output in compartment 3. *Rule 1*: The shortest path from 1 to 3 must be two, so $k'_{31} = 0$, $k'_{21} \neq 0$ and $k'_{32} \neq 0$. *Rule 2*: compartments 1 and 2 both have a path to output compartment 3. This gives no new information. *Rule 3*: Compartments 2 and 3 can both be reached from input compartment 1. This also gives no new information. *Rule 4*: There must be one trap, in any of three possible configurations: (1) compartment 3 is a trap, with $k'_{13} = k'_{03} = k'_{23} = 0$; (2) compartments 2 and 3 together form a trap, with $k'_{13} = k'_{12} = k'_{02} = k'_{03} = 0$ and $k'_{23} \neq 0$; (3) the whole model is a trap, with $k'_{01} = k'_{03} = k'_{02} = 0$ and either $k'_{13} \neq 0$ or $k'_{12}k'_{23} \neq 0$. For any of these three possibilities $k'_{03} = 0$. This constraint plus the three possibilities reduce the total number of candidates for this three-compartment model from 64 to 13. All 13 candidates for indistinguishability form a subset of all submodels of **Fig. 5.15** that have $k'_{03} = k'_{31} = 0$, so the arrows (edges) for these two parameters can be eliminated from the diagram. These 13 submodels need to have their TFs tested for indistinguishability.

The complete TF matrix for the general three-compartment model was given in Eq. (4.79) of Chapter 4, from which $H_{31}(s,\boldsymbol{k})$ and $H'_{31}(s,\boldsymbol{k}')$ can be extracted and equated (not repeated here – very complex), as in Eq. (5.17). The moment invariant equations thus obtained for the much-reduced set of 13 remaining candidates must satisfy all three Rule 4 possibilities. Equating these moment invariants for the different candidates yields fairly complicated, but solvable sets of algebraic equations for the remaining k'_{ij}. The details – resolved by symbolic computation – can be found in Godfrey and Chapman (1990). They result in the two five-parameter structures shown in **Fig. 5.18**, seven other four-parameter models and four other three-parameter models, for a total of 13 such submodels (11 not shown – see Godfrey and Chapman 1990).

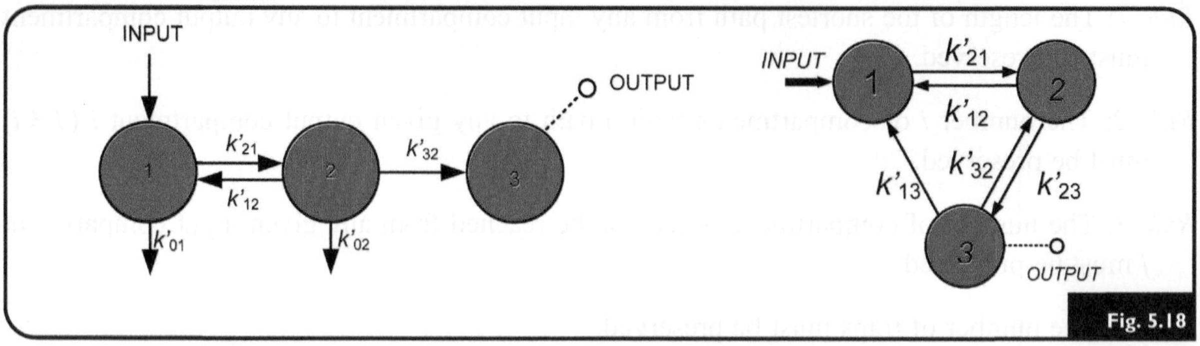

Fig. 5.18

Two indistinguishable candidates with 5 parameters, each I-O equivalent to the general 3-compartment model with input into 1 and output from 3.

Automation: DISTING

As noted above, the four geometric rules can be automated, as described by Zhang et al. (1991). A BFS search is readily adapted to check Rules 1–3. Rule 1 – the shortest path from input to output compartments; Rule 2 – finding the number of compartments with a path to an output compartment; Rule 3 – finding the number of compartments reachable from an input compartment; and output connectivity – finding the compartments with paths leading to an output compartment. A BFS search also can be used to check input connectivity – finding every compartment with a path leading to it originating from an input compartment. A depth first search (DFS) is more efficient for checking Rule 4 – finding the number of traps.

The algorithms for making these computations were described by Zhang et al. (1991), but no software or code was published or provided elsewhere by the authors. The algorithms were recently implemented and optimized using *Python* tools and verified against the published results by UCLA Computer Science graduate student Natalie Davidson. The package, called DISTING, will be made available as a web application at www.biocyb.cs.ucla.edu.

Reducible, Cyclic & Other MC Model Properties

The following additional properties of MC models can be useful in model building. **Reducible** models are those that include a **trap**, i.e. a subset of compartments from which no path exists to any compartment outside the trap, or to the environment. **Irreducible** models contain no traps and are **strongly connected**, i.e. a particle in any compartment can reach any other compartment in the model. A **cycle** was defined earlier (see Example 5.3) as a path between at least three compartments that returns to the original one. **Cyclic** models have at least one cycle, e.g. as in **Fig. 5.19**. Following are additional properties of compartmental matrices *K*.

1. *K* is **invertible** ($|K| \neq 0$) if the model is both **open**, i.e. it has at least one compartment with loss to the external environment, and contains no **traps**, i.e. no closed subsystem of compartments which can only receive material from other compartments, and have no losses either to the external environment or compartments outside the subsystem. This means all particles entering the system via any compartment will eventually leave the system.

2. For **reducible** K, compartment numbering and indices ij can be adjusted so that K can be written in block triangular form, with **irreducible** K_{11} and K_{22}, each of dim $<n$.

$$K \equiv \begin{bmatrix} K_{11} & 0 \\ K_{21} & K_{22} \end{bmatrix} \tag{5.18}$$

If irreducible submodels K_{11} and K_{22} are connected to sinks, i.e. have at least one open compartment, then **reducible** K have $\mathrm{Re}(\lambda_i) < 0$ for all i. Also, if $\mathrm{Re}(\lambda_i) = 0$, then $\lambda_i = 0$. Multiple eigenvalues at the origin are associated with different submodels.

3. **Open and irreducible** models have all eigenvalues λ_i of K in the left half of the complex plane (LHP), i.e. $\mathrm{Re}(\lambda_i) < 0$ for all i, and thus they are asymptotically stable (Chapter 2).

4. **Closed and irreducible** models have one $\lambda_j = 0$ and all other λ_i are in the LHP, $i \neq j$.

5. All **irreducible** models are **stable** (a consequence of 3 and 4).

6. **Cycle-free** models, e.g. *mammillary* and *catenary* models, have all λ_i real.

7. For a **cyclic** model, some λ_i can be *complex* numbers (but not necessarily, as we have seen in Example 5.3). Models with complex λ_i must have cycles.

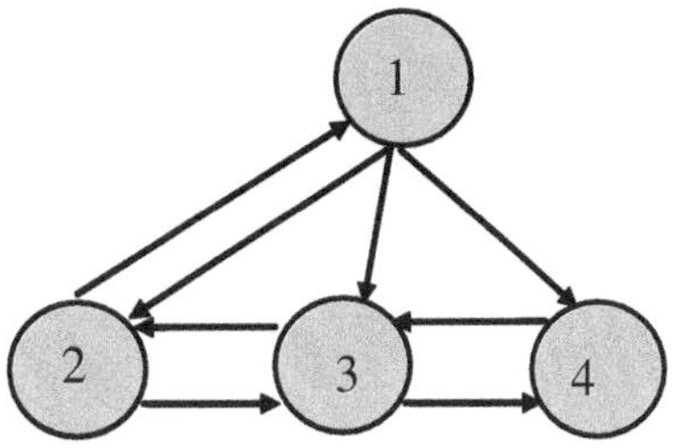

Fig. 5.19

A cyclic model with two cycles (1-4-3-2-1 and 1-3-2-1).

Tracers, Tracees & Linearizing Perturbation Experiments

Linearization approaches can be helpful in modeling and analyzing nonlinear (NL) biosystems. If experiments on a NL biosystem utilize small enough inputs – so-called "perturbation experiments" – linear methods can be appropriate.[24] Such experiments are usually performed using trace (small) amounts of discernable substances called **tracers**, and are valid when these experimental probes generate dynamic responses (for the tracer) in an approximately linear range. The major math tool for this is the Taylor series – introduced in Chapter 3, but used differently here – because solutions are sought for perturbations, or variations, in the state variable $\boldsymbol{x}$ instead of time t.

24 If they are not, and the experiment can be redesigned to make them so, this approach can have substantial value.

Suppose we introduce a small amount of a substance X into a biological system that normally contains the same substance, but in much larger amount. For example, if we inject X intravenously into an experimental animal *in vivo*, or directly into a cell culture *in vitro*. What we are doing (among other things) is "perturbing" the endogenous system and its dynamic state with the exogenous input. In the context of the problem at hand, the exogenous substance is normally called a **tracer** of the endogenous material, if it is measurably distinguishable from it in the measurement process. The endogenous material is called the **tracee.**[25] Usually, the tracer dynamic response can be and is separately measured in kinetic stimulus–response studies, possibly along with tracee-based data. For example, the tracer is often a radioactively labeled form of the tracee, so it can be measured with a radiation-counter; and tracee concentrations also may be measured separately, in the same samples. Or it may be a substance labeled with a *stable* (nonradioactive) isotope of an atom like carbon – ^{13}C replacing ^{12}C in an organic molecule – called an **isotopomer** (Lee et al. 2010). Thus, it is natural to seek a model of the tracer dynamics, and it is important that we know how this model reflects the dynamics of the endogenous substance it is tracing. What does the tracer dynamics reveal about the dynamics of the tracee, say under physiological or pathological conditions? Two simple examples illustrate the approach – a first step in answering this question. In the first example the endogenous system-model is linear and, in the second, it is NL.

Example 5.14: ***Simplifed Glucose Perturbation Control System Model***

The oversimplified but elegant and didactically useful Bolie model (Bolie 1961) for glucose–insulin interactions can be described by the control system block diagram in **Fig. 5.20**. In the figure, $G \equiv$ total glucose mass in plasma, $I \equiv$ total insulin mass in plasma, SR_G and SR_I are glucose and insulin secretion rates into plasma from their respective *source* organs of storage and secretion – for glucose – and production and secretion organ – for insulin. The distribution and elimination (**D&E**) blocks represent *sinks* for excretion and metabolism of glucose and insulin. As in the original Bolie model, glucose and insulin elimination rate terms are assumed to be first-order (linear), with fractional rate constants a and d (time^{-1}); glucose also is "removed" from the circulation (by tissue uptake) in proportion to the insulin amount present – the reason for the negative sign in the block diagram – with fractional uptake rate constant b, i.e. $ER_G = aG(t) + bI(t)$.

Insulin secretion is stimulated by glucose – assumedly proportional to glucose concentration, with fractional stimulation rate constant c, i.e. $SR_I = cG(t)$ – the " + " sign in the block diagram. Then, with ER_G and ER_I denoting overall elimination rates of glucose and insulin, the mass flux balance equations are:

$$\dot{G} = SR_G - ER_G \equiv SR_G - (aG + bI) \tag{5.19}$$

$$\dot{I} = SR_I - ER_I \equiv cG - dI \tag{5.20}$$

25 If there is no tracee, i.e. if the input-perturbation material is an exogenous drug not present endogenously, demonstration of linearity – for validation of this approach – might be difficult, but can be a good first-approximation.

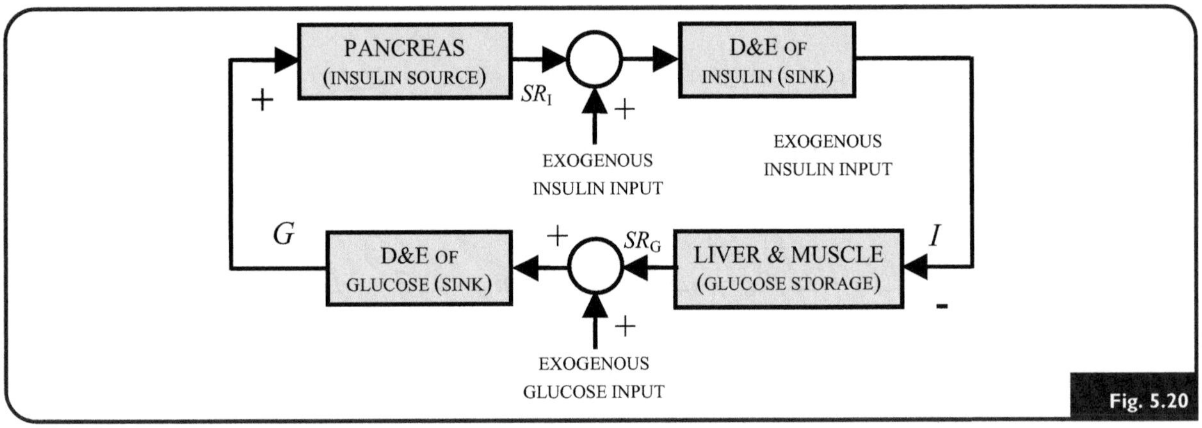

Fig. 5.20

A block diagram adaptation of the Bolie Model.

The steady state solutions G_∞ and I_∞ are easily found by setting $\dot{G}$ and $\dot{I}$ equal to zero:

$$0 = SR_{G\infty} - (aG_\infty + bI_\infty) \tag{5.21}$$

$$0 = cG_\infty - dI_\infty \tag{5.22}$$

i.e.

$$G_\infty = \frac{d}{c} I_\infty \tag{5.23}$$

$$I_\infty = \frac{c}{ad + bc} SR_{G\infty} \tag{5.24}$$

Both solutions are constant if $SR_{G\infty}$ is constant.

Perturbation Model: We derive this by defining $\Delta G(t)$ and $\Delta I(t)$ as the **perturbation solutions** about the steady state solution. Assuming an exogenous glucose rate input perturbation of only labeled tracer glucose into the model, i.e. $u \equiv \Delta SR_G(t)$ and $\Delta SR_I \equiv 0$, we define:

$$SR_G(t) \equiv SR_{G\infty} + \Delta SR_G(t) \tag{5.25}$$

$$G(t) \equiv G_\infty + \Delta G(t) \tag{5.26}$$

$$I(t) \equiv I_\infty + \Delta I(t) \tag{5.27}$$

Then, substituting Eqs. (5.25) to (5.27) into Eqs. (5.19) and (5.20) gives:

$$\dot{G} \equiv \dot{G}_\infty + \Delta\dot{G} = SR_{G\infty} + \Delta SR_G - (aG_\infty + bI_\infty) - (a\Delta G + b\Delta I)$$

$$\dot{I} = \dot{I}_\infty + \Delta\dot{I} = cG_\infty + c\Delta G - dI_\infty - d\Delta I$$

Solving for ΔG and ΔI using Eqs. (5.21) and (5.22) yields the desired **perturbation model**:

$$\Delta\dot{G} = \Delta SR_G - a\Delta G - b\Delta I \equiv u - a\Delta G - b\Delta I \tag{5.28}$$

$$\Delta\dot{I} = c\Delta G - d\Delta I \tag{5.29}$$

Remark: The Bolie model is not a compartmental model; it does not satisfy the constituent equation or positivity requirements of a compartmental model, Eq. (4.72) in Chapter 4. It's a simple feedback control system model.

Remark: Perturbation Eqs. (5.19) and (5.20) have precisely the same form as Eqs. (5.28) and (5.29), a consequence of the linearity of the Bolie model. This result is completely generalizable. If the (tracee) model is linear for the endogenous substance, we can immediately write the perturbation model for the exogenous tracer substance, by inspection, because they have exactly the same form.

Remark: If we measure labeled glucose in plasma, the measurement model is then $y_1(t) \equiv \Delta G(t)$. If we also measure the unlabeled, endogenous insulin response to the tracer glucose input perturbation, then $y_2(t) = I(t) = I_\infty + \Delta I$. In this case, total unlabeled insulin must be measured before as well as after $u(t)$ is introduced, to obtain the perturbation solution $\Delta I(t)$.

Example 5.15: ***Nonlinear Dynamic System Perturbation Model: Not the Same Form***

Consider

$$\dot{x} = -ax^2 + u \tag{5.30}$$

As in the previous example, we perturb the input and develop the perturbation solution to this NL ODE. Let $u \equiv u_\infty + \Delta u$ (the perturbed input) and $x \equiv x_\infty + \Delta x$ becomes the perturbed state variable in response to the perturbation. Substituting for x and u in (5.30), we get

$$\begin{aligned}\dot{x}_\infty + \Delta\dot{x} &= -a(x_\infty + \Delta x)^2 + u_\infty + \Delta u \\ &= -a[x_\infty^2 + (\Delta x)^2 + 2x_\infty \Delta x] + u_\infty + \Delta u \\ &= -ax_\infty^2 + u_\infty - a(\Delta x)^2 \Delta u - 2ax_\infty \Delta x\end{aligned}$$

Setting $\dot{x} \equiv \dot{x}_\infty \equiv 0$ in Eq. (5.30) gives $u_\infty = ax_\infty^2$, so the steady state solution of Eq. (5.30) is constant if u_∞ is constant. The resulting perturbation model for this system is:

$$\Delta\dot{x} = -a(\Delta x)^2 + \Delta u - 2ax_\infty \Delta x \tag{5.31}$$

In this example, unlike the previous (linear) one, Eqs. (5.30) and (5.31) do not have the same form. Eq. (5.31) has an extra term.

This dissimilarity is generally true for NL systems. Thus, the perturbation solution for NL systems is not simply a "translation" of the solution for the endogenous substance. Consequently, we cannot draw the same conclusions about the endogenous system from the perturbation equation when it is NL, as we did for the linear case. We can, however, employ perturbation solutions to effectively study NL systems – again – if the perturbations are small enough. Furthermore, we can employ all the powers of linear analysis in this case, as we see below. First, we review the pertinent Taylor series expansion, introduced in Chapter 3 for variations Δt in the time argument t, but different in the present application, where the argument is the state variable $x(t)$ and perturbations are $\Delta x(t)$.

Linearization of NL ODE Models

In Chapter 3, Taylor series (TS) expansions were defined and used to developed simulation algorithms for solving ODEs. Solutions $\boldsymbol{x}(t+\Delta t)$ at time $t+\Delta t$ were expressed as a TS expansion about $\boldsymbol{x}(t)$ at time t. We use the same theory here, in much abbreviated form, to linearize NL ODE models using only the zero- and first-order terms of the Taylor series. The difference is that now the derivative function $\boldsymbol{f}[\boldsymbol{x}(t)]=\dot{\boldsymbol{x}}(t)$ is expressed a small distance $\Delta \boldsymbol{x}(t)$ about state variable $\boldsymbol{x}(t)$, instead of Δt about the time t, i.e. we are varying $\boldsymbol{x}(t)$ in state space by $\Delta \boldsymbol{x}(t)$ about $\boldsymbol{x}(t)$ at the same time t. We apply this theory to the general NL ODE:

$$\dot{\boldsymbol{x}}(t)=\boldsymbol{f}[\boldsymbol{x}(t),\boldsymbol{u}(t),t], \quad t_0<t<T \tag{5.32}$$

Our interest is the behavior of the biosystem described by Eq. (5.32) for *small perturbations* $\Delta \boldsymbol{x}$ about a generally time-varying solution $\boldsymbol{x}_N(t)$, termed the **nominal solution**. That is, we define $\boldsymbol{x}(t)\equiv \boldsymbol{x}_N(t)+\Delta \boldsymbol{x}(t)$, and emphasize that $\boldsymbol{x}_N(t)$ is *not necessarily in steady state, constant or otherwise*. The input $\boldsymbol{u}(t)$ is assumed perturbed by $\Delta \boldsymbol{u}$, i.e. $\boldsymbol{u}(t)\equiv \boldsymbol{u}_N(t)+\Delta \boldsymbol{u}(t)$. This is natural, because changes, $\Delta \boldsymbol{x}$, in $\boldsymbol{x}$ usually occur as a result of changes $\Delta \boldsymbol{u}$ in the input. The time argument t here is treated as a "fixed parameter" in the TS expansion. So we consider $\boldsymbol{f}(\boldsymbol{x},\boldsymbol{u})$, a (column) vector-valued function[26] $\boldsymbol{f}\equiv[f_1 \;\; f_2 \;\; \ldots \;\; f_n]^T$ of two vector arguments, state vector $\boldsymbol{x}\equiv[x_1 \;\; \ldots \;\; x_n]^T$ and input vector $\boldsymbol{u}\equiv[u_1 \;\; \ldots \;\; u_r]^T$.

The value of $\boldsymbol{f}$ a small distance $\Delta \boldsymbol{x}$ from $\boldsymbol{x}$, i.e. at $\boldsymbol{x}+\Delta \boldsymbol{x}$, can be written:

$$\dot{\boldsymbol{x}}(t)\equiv\dot{\boldsymbol{x}}_N(t)+\Delta\dot{\boldsymbol{x}}(t)=\boldsymbol{f}[\boldsymbol{x}_N(t)+\Delta \boldsymbol{x}(t),\boldsymbol{u}_N(t)+\Delta \boldsymbol{u}(t),t] \tag{5.33}$$

26 Appendix B can help with the vector-matrix algebra.

Expanding the RHS in a TS (with two arguments) and retaining only zero- and first-order terms gives:

$$\boldsymbol{f}[\boldsymbol{x}_N(t) + \Delta\boldsymbol{x}(t), \boldsymbol{u}_N(t) + \Delta\boldsymbol{u}(t), t] \cong \boldsymbol{f}[\boldsymbol{x}_N(t), \boldsymbol{u}_N(t), t] + \left.\frac{\partial \boldsymbol{f}}{\partial \boldsymbol{x}}\right|_{(\boldsymbol{x}_N, \boldsymbol{u}_N)} \Delta\boldsymbol{x} + \left.\frac{\partial \boldsymbol{f}}{\partial \boldsymbol{u}}\right|_{(\boldsymbol{x}_N, \boldsymbol{u}_N)} \Delta\boldsymbol{u} \quad (5.34)$$

Under quite reasonable differentiability conditions on f, the Taylor series converges; and for small-enough Δx_i, higher order terms of the series, in Δx_i^2, ,. . ., etc, are small enough to be neglected in many applications. This means that, for "small-enough" Δx_i, (5.35) is a "good-enough" approximation for the entire series. Subtracting $\boldsymbol{f}(\boldsymbol{x}_N, \boldsymbol{u}_N, t) \equiv \dot{\boldsymbol{x}}_N$ from both sides of (5.33) and (5.34) then gives the desired **perturbation ODE**:

$$\Delta\dot{\boldsymbol{x}}_N(t) \cong \left.\frac{\partial \boldsymbol{f}}{\partial \boldsymbol{x}}\right|_{(\boldsymbol{x}_N, \boldsymbol{u}_N)} \Delta\boldsymbol{x} + \left.\frac{\partial \boldsymbol{f}}{\partial \boldsymbol{u}}\right|_{(\boldsymbol{x}_N, \boldsymbol{u}_N)} \Delta\boldsymbol{u} \quad (5.35)$$

The ***Jacobian matrices*** $\frac{\partial \boldsymbol{f}}{\partial \boldsymbol{x}}$ and $\frac{\partial \boldsymbol{f}}{\partial \boldsymbol{u}}$ in Eq. (5.35) are evaluated at $\boldsymbol{x}_N \equiv \boldsymbol{x}_N(t)$ and $\boldsymbol{u}_N \equiv \boldsymbol{u}_N(t)$, which means they are functions of t in general and we can simplify (5.35) notationally, as an approximately ***linear time-varying ODE***:

$$\Delta\dot{\boldsymbol{x}} \cong A(t)\Delta\boldsymbol{x} + B(t)\Delta\boldsymbol{u} \quad (5.36)$$

where

$$A(t) \equiv \left.\frac{\partial \boldsymbol{f}}{\partial \boldsymbol{x}}\right|_{(\boldsymbol{x}_N, \boldsymbol{u}_N)} \equiv \left[\frac{\partial f_{iN}}{\partial x_j}\right] \quad (5.37)$$

$$B(t) \equiv \left.\frac{\partial \boldsymbol{f}}{\partial \boldsymbol{u}}\right|_{(\boldsymbol{x}_N, \boldsymbol{u}_N)} \equiv \left[\frac{\partial f_{iN}}{\partial u_j}\right] \quad (5.38)$$

If x_N and u_N are constant, e.g. a constant steady state solution and constant input, then $A(t) \equiv A$ and $B(t) \equiv B$ are constant and we have a constant coefficient or time-invariant ODE for the perturbation model:

$$\Delta\dot{\boldsymbol{x}} \cong A\Delta\boldsymbol{x} + B\Delta\boldsymbol{u} \quad (5.39)$$

This result, Eq. (5.39), is the theoretical basis for using small-perturbation tracer experiments to study the kinetics of biological systems, linear or NL, but only if two conditions on this approximation are met: (1) the perturbations $\Delta\boldsymbol{u}$ and $\Delta\boldsymbol{x}$ must be "small enough" if any linear analysis is to be applied, using (5.36) or (5.39); and (2) the system must be in a constant endogenous steady state if the power of linear analysis for time-invariant systems is to be useful, using

Eq. (5.39). If either condition is not strictly met, all is not lost, but data resulting from such an experiment on a NL biosystem usually are more difficult to interpret quantitatively.

If a nominal solution $x_N(t)$ is *time-varying*, perturbation solutions $\Delta x(t)$ about $x_N(t)$ still can be good linearized solutions, if they are small-enough perturbations. For example, if $x_N(t)$ is time varying but constant plus a periodic oscillation, e.g. $A + B\sin t$, and the amplitude of the oscillation – B in this example, is small relative to constant A, a time-invariant model of the form (5.39) – based on small-perturbation experiments done in this periodic steady state, might still be a good approximation.

Linearized Output Measurement Model

In general, the output model is also nonlinear and, in the absence of measurement noise, it is usually of the form:

$$\boldsymbol{y}(t) = \boldsymbol{g}[\boldsymbol{x}(t)] \tag{5.40}$$

where $\boldsymbol{g} \equiv [g_1 \quad g_2 \quad \cdots \quad g_m]^T$ is a NL vector-valued function of the state vector $\boldsymbol{x}$, with m different output variables assumed. Defining $\boldsymbol{y}(t) \equiv \boldsymbol{y}_N(t) + \Delta\boldsymbol{y}(t)$, linearization yields

$$\Delta\boldsymbol{y}(t) \cong \left.\frac{\partial \boldsymbol{g}}{\partial \boldsymbol{x}}\right|_{\boldsymbol{x}_N} \Delta\boldsymbol{x} \equiv C(t)\Delta\boldsymbol{x} \tag{5.41}$$

This first-order approximation, has $m \times n$ matrix $C \equiv C(t)$ defined as the indicated Jacobian. If the nominal $\boldsymbol{x}_N(t)$ is constant, then C is constant.

Remark: Eq. (5.41) might also include inputs $\boldsymbol{u}(t)$, if they are also measured instead of being just “known”, i.e. $\boldsymbol{y}(t) = \boldsymbol{g}[\boldsymbol{x}(t), \boldsymbol{u}(t)]$.

Remark: Taylor series expansion of the ODE solution Eq. (5.40) is widely applicable. It is the basis for an analytical method for identifiability analysis of NL models (Pohjanpalo 1978), presented and exemplified in Chapter 10, for *stability* analysis in Chapter 9, as well as many more applications.

The Complete Linearized Model

The linearized model, Eqs. (5.36) and (5.41), may be written in the cumbersome but more directly usable form:

$$\begin{aligned} \Delta\dot{x}_i &\cong \sum_{k=1}^{n}\left[\frac{\partial f_{iN}}{\partial x_k}\right]\Delta x_k + \sum_{k=1}^{r}\left[\frac{\partial f_{iN}}{\partial u_k}\right]\Delta u_k \quad \text{for } i = 1, 2, \ldots, n \\ \Delta y_j &\cong \sum_{k=1}^{n}\left[\frac{\partial g_{jN}}{\partial x_k}\right]\Delta x_{ki} \quad \text{for } j = 1, 2, \ldots, m \end{aligned} \tag{5.42}$$

where $\dfrac{\partial f_{jN}}{\partial x_k} \equiv \left.\dfrac{\partial f_j}{\partial x_k}\right|_{\boldsymbol{x}_N}$ etc., i.e. each partial derivative is evaluated over the nominal solution $\boldsymbol{x}_N(t)$.

Example 5.16: ***Linearized Bilinear Model***

We linearize:

$$\begin{aligned} \dot{x} &= f(x, u) = -c_1 ux - c_2 x \\ y &= g(x) = c_3 x \end{aligned} \tag{5.43}$$

The bilinear term $-c_1 ux$ in Eq. (5.43) may represent (for example) an inhibitory multiplicative reaction between a control substance (u) and a target substance (x); the second term $-c_2 x$ is typical of linear constitutive elimination of x. Applying Eqs. (5.42) to Eqs. (5.43) yields:

$$\begin{aligned} \Delta\dot{x} &\cong \left[\frac{\partial f_N}{\partial x}\right]\Delta x + \left[\frac{\partial f_N}{\partial u}\right]\Delta u = -c_1 u_N \Delta x - c_2 \Delta x - c_1 x_N \Delta u = (-c_1 u_N - c_2)\Delta x - c_1 x_N \Delta u \\ \Delta y &= [\partial g/\partial x]\Delta x = c_3 \Delta x \end{aligned} \tag{5.44}$$

The nominal input u_N and nominal solution x_N can be constant *or* time-varying, respectively yielding a constant coefficient or time-vary linear perturbation model (5.44). The ODE perturbation model Eq. (5.44) does not have the same form as the original bilinear model (5.43) – because the bilinear model is nonlinear (NL), and (5.44) could be time-varying, whereas the bilinear model is time-invariant. The output equations, however, are the same – because the original output Eq. in (5.43) is linear.

Recap and Looking Ahead

In this chapter we more closely examined how kinetic output data circumscribe and limit what we can learn and validate about model structure, even when data are perfect. A central theme was analysis of dynamical signatures in hypothetical (perfect) input–output data associated with a structural model. The goals were to establish model dimensionality (size, complexity); to select among multiple candidate models – model distinguishability; and model simplification or reduction – all with respect to dynamical signatures in the data. As always, data rules. Modes, visible and hidden, played an important role in understanding dynamical signatures in data. Some automation of these highly technical aspects of biomodeling also was introduced – with more to come in subsequent chapters.

Taylor series approximations for linearizing NL ODE system and experiment models were developed at the end. These will be applied to NL biosystem problems beginning in the next two chapters (6 and 7), where we begin formal modeling of biochemical and cellular dynamical systems and then – importantly – when we study biosystem stability and oscillations in Chapter 9.

The encumbrances of noisy data and development of *output error* models to handle these additional realities – with statistics as needed – begins in Chapter 11–13, where we study the numerical aspects of model quantification and analysis from real data.

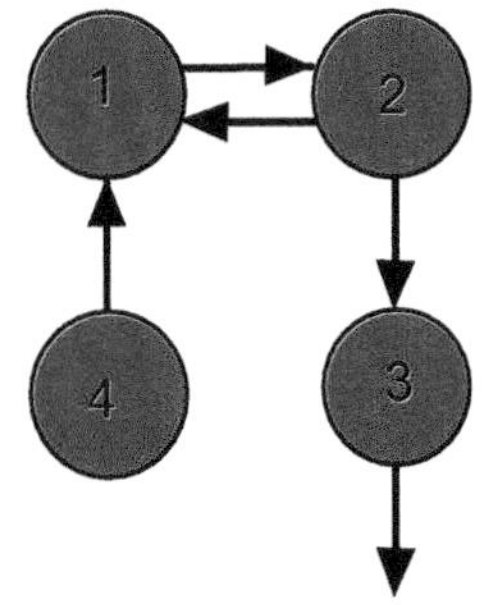

Fig. 5.21

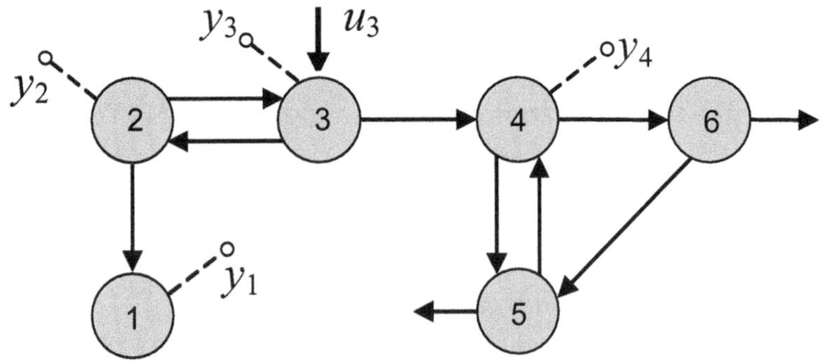

Fig. 5.22

Exercises

E5.1 Given the zero-input model in **Fig. 5.21** suppose the initial condition vector is $q(0) = [1\ \ 0\ \ 0\ \ 0]^T$. How many different exponential *modes* would you expect in the output responses in each of compartments 1, 2, 3, and 4, if all λ_i are distinct?

E5.2 Consider the six-compartmental model in **Fig. 5.22**:

If $u_3(t) = \delta(t)$, an impulse:

a) What are the expected responses in compartments 1, 2, and 3 (i.e. find the forms of $y_1(t)$, $y_2(t)$, and $y_3(t)$ for $t > 0$)?

b) How many *modes* would you expect in the responses of each of the compartments 1, 2, 3, and 4? Explain your answers.

c) If compartment 1 also had a *leak* k_{01}, does the number of modes in 1 change? Why?

d) If compartment 1 has an IC not equal zero, how does that change your answers to (a), (b) or (c)?

e) If compartment 3 has an IC not equal zero, how does that change your answers to (a), (b) or (c)?

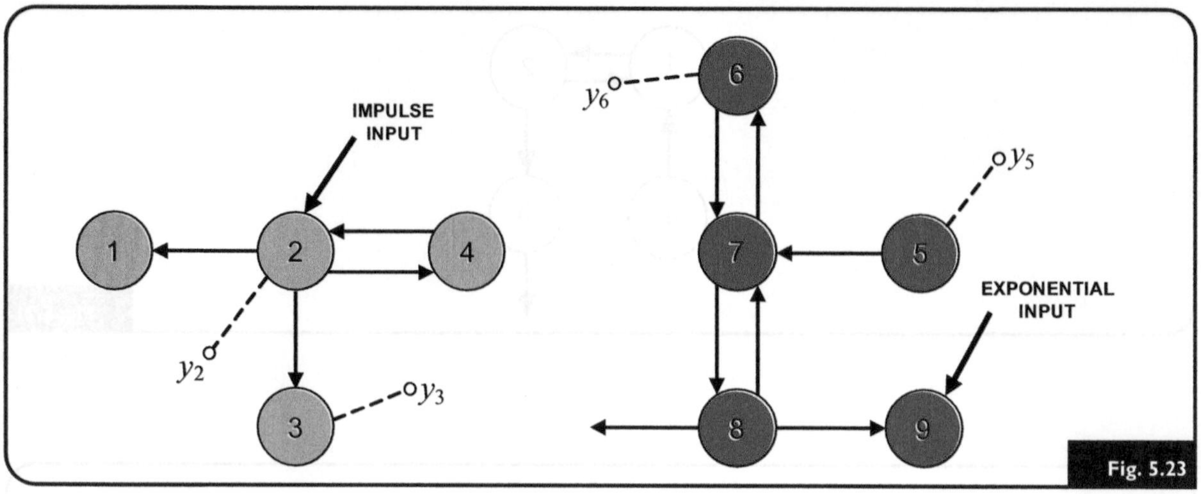

Fig. 5.23

E5.3 For the following two multicompartment models (**Fig. 5.23**), initial conditions (ICs) are zero everywhere and all k_{ij} arrows (edges) shown are nonzero.

a) Determine how many *distinct, nonconstant* modes can be measured at each of the outputs y_2, y_3, y_5 and y_6, given the inputs at the input arrows shown, *impulse* on the left and a *single-exponential* input on the right.
State any assumptions needed about the eigenvalues of the solutions, and possibly also the single-exponential input.

b) Repeat part (a), but with two nonzero ICs, namely: $y_1(0) = 1$ and $y_6(0) = 1$. The rest are zero.

c) Repeat part (a) again, with all ICs = 0, but *step* inputs in both places.

d) In Part (a), suppose the exponential input has an exponent identical to one of the poles of the transfer function of the compartment model structure. How does this change the answers to (a)? Use transfer function analysis to answer this.

e) Are there any *cycles* in these models? Where?

E5.4 In **Fig. 5.24**, how many modes can be measured at each of the named outputs, given simultaneous impulse inputs into compartments 1 and 9?

E5.5 You are given lab instrumentation for generating any input function with which to probe any node (compartment) in the network model schematized in **Fig. 5.25**.

You also can measure the dynamic response from any node. Answer and explain your answer to the following questions. Some questions have more than one possible answer, or maybe none at all. Unless otherwise specified, all initial conditions are zero in all 8 nodes (compartments).

a) You measure a single exponential mode from the output port of node 3. What is the only possible *input function* that could generate this output from 3, and where must it be applied (to what node)?

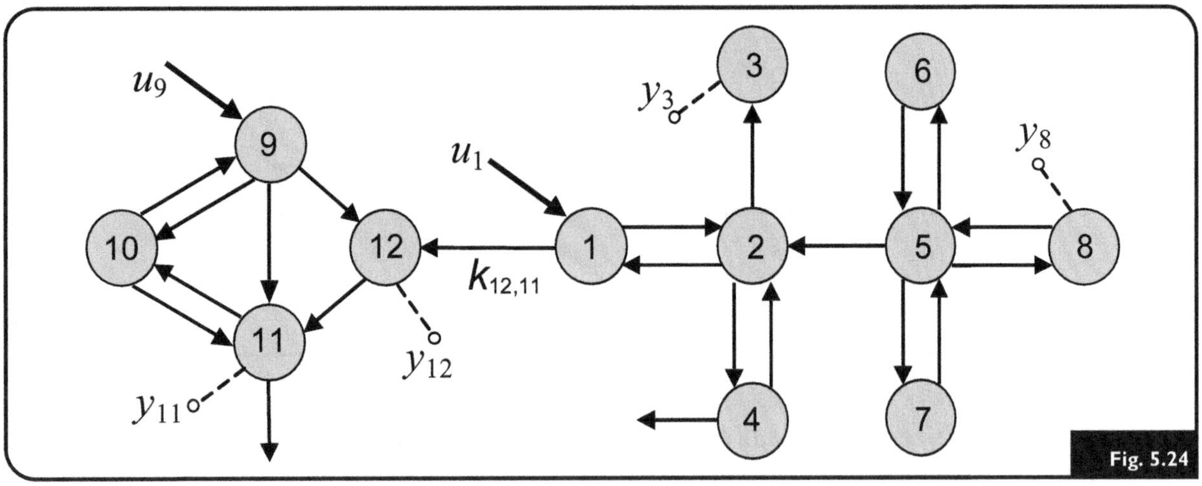

Fig. 5.24

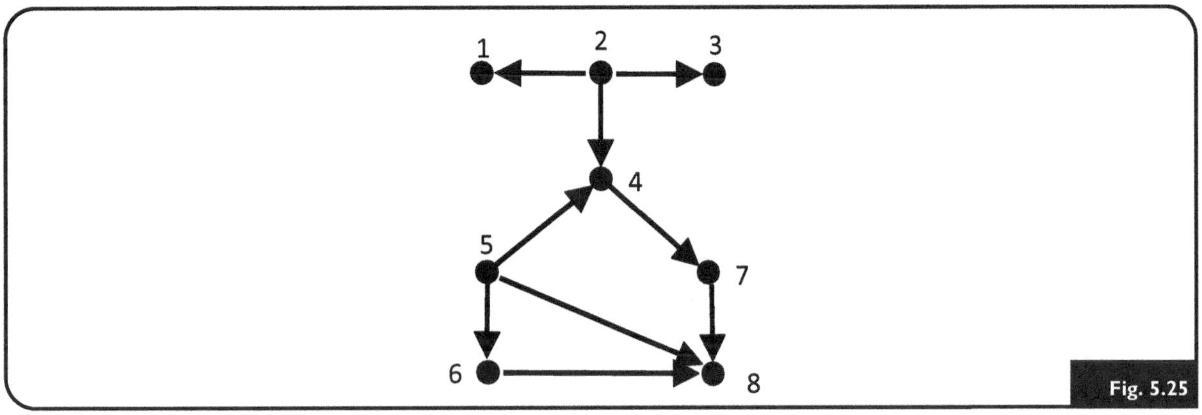

Fig. 5.25

b) What *input functions* will yield biexponential responses (two modes) from node 1 and where should these inputs be applied (to what nodes)? (Two separate answers.)

c) What input can you apply, and where (which node), to get five exponential modes from node 8?

d) What set of *initial conditions* on all eight nodes will yield a single mode response at node 2?

e) What set of *initial conditions* (only) on all eight nodes will yield a two-mode (biexponential) response at node 1?

E5.6 Consider M and M' in **Fig. 5.26**. In terms of compartment masses, distribution volumes, k_{ij}, and inputs and outputs, what ways are M and M': (a) equivalent, or (b) not equivalent (with all initial conditions zero), if both models have the same transfer function?

E5.7 ***Hidden Modes without Pole-Zero Cancellations.*** This exercise is an extension of Examples 5.2 and 5.8. Suppose $k_{22} \neq k_{33}$, so compartments 2 and 3 are not aggregated.

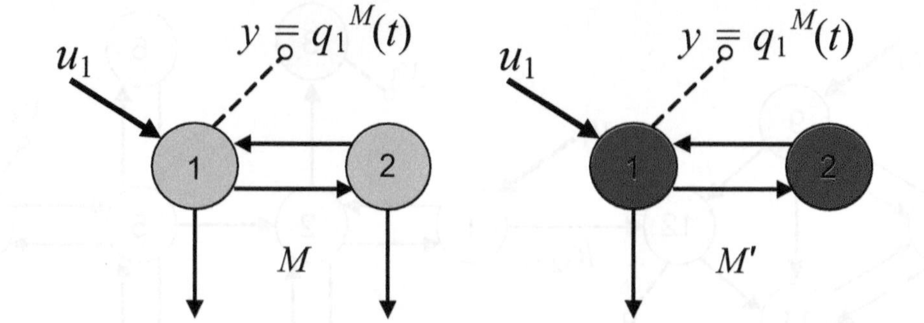

Fig. 5.26

Also assume zero ICs in all compartments. Explore the consequences of one or more rate constants vanishing (rather than approximately equal) to see other ways modes can be hidden for this three-compartment model.

E5.8 In ***the model of*** Example 5.8, input is only into compartment 1; and the simplifications (model reductions) shown are based on there being zero ICs in compartments 2 and 3. Alternatively, suppose the system was in endogenous steady state – with nonzero amounts in all compartments prior to applying the input u_1, and this model represented an exogenous perturbation of the endogenous system via the single input into 1, with measured output remaining in compartment 1. Two circumstances need to be considered.

1) the exogenous input material (call it X) is indistinguishable from the endogenous X after it enters the system and in measurements $y_1(t)$; and

2) the X material introduced exogenously is labeled (call it X^*), and X^* is distinguishable from X in the system and in measurements of the output of compartment 1, i.e. $y_1(t) = y_1^*(t)$.

What do the measurements indicate about the dimensionality of the model-system in each of these circumstances, for the two conditions illustrated in **Fig. 5.27a** and **b**?

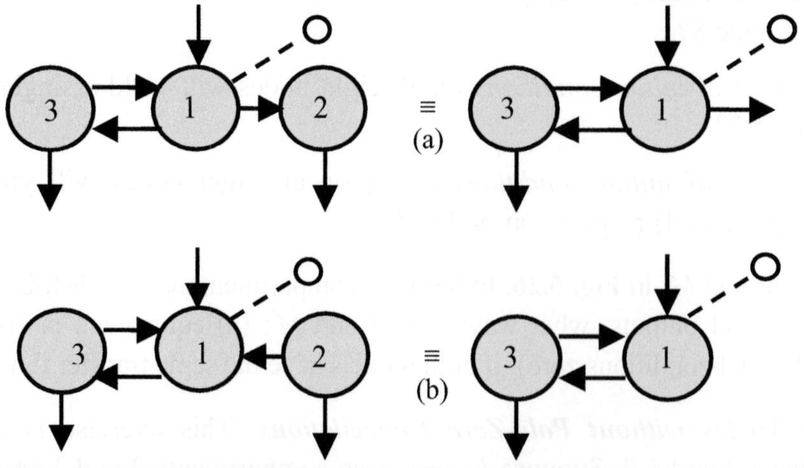

Fig. 5.27

Some Reducible Models.

E5.9 Why do (linear) *cyclic* models require additional care during compartmental analysis and when fitting them to real data?

E5.10 ***SIRS Model Analysis*** – Solve and analyze Eq. (4.61) of Chapter 4 for its dynamical properties, using linearization methods.

E5.11 ***Reducibility Conditions*** – Some models of biological systems are based on coupled biophysical processes with vastly different dynamics, e.g. very rapid drug–receptor, binding–unbinding reactions, mixed with relatively slow elimination rates, e.g. by metabolic degradation or by excretion. In many cases, the coupled differential equations for the combined processes can be reduced to lower order (or fewer number), thereby simplifying the overall biomodel. Show that the following three-dimensional coupled model can be approximated by a two-dimensional model, i.e. two first-order differential equations and any necessary constraints, over the interval $0 < t \le 10$. (Let $\boldsymbol{x}(0) = 0$ for convenience.)

$$\begin{aligned} \dot{x}_1 &= -x_1 + x_2 + x_3 \\ \dot{x}_2 &= -1000(x_2 + 1) \\ \dot{x}_3 &= -2x_1 \end{aligned}$$

Begin by solving for $x_2(t)$, and then think "approximation" for $x_i(t)$ on $0 < \mathrm{t} \le 10$.What is the approximate model?

E5.12 ***Reducibility Conditions*** – Use long division to divide the denominator polynomial in the H_{11} term of the mammillary three-compartment model in Eq. (5.8) by $\mathrm{s} - k_{22}$ or $s - k_{33}$ and determine any possible relationships among $k_{ij} \neq 0$ that might yield a reducible model. Are these conditions necessary?

E5.13 ***Reducibility Conditions*** – Show by induction that an n-compartment mammillary model is reducible if $k_{ii} = k_{jj}, i \neq j \neq 1$.

E5.14 ***Reducibility Conditions*** – Assume two inputs, u_1 and u_2, into the general two-compartment model, with $V_2 << V_1$. Using relationships among the k_{ij}, find conditions for reducibility of the model.

E5.15 ***Reducibility Conditions*** – Derive similar reducibility conditions on the k_{ij} of the n-compartment SISO (in compartment 1) mammillary model.

E5.16 ***I–O Equivalence*** –Show that the two models in **Fig. 5.11** have the same TF, i.e. are I–O equivalent.

References

Bolie, V., 1961. Coefficients of normal blood glucose regulation. J. Appl. Physiol. 16, 783–788.
Chen, C., 1970. Introduction to Linear System Theory. Holt, Rinehart, & Winston, New York.

Cormen, T.H., Leiserson, C., Rivest, R.L., Stein, C., 2009. Introduction to Algorithms. MIT Press, Cambridge.

DiStefano III, J., 1985. Modeling approaches and models of the distribution and disposal of thyroid hormones. In: Hennemann, G. (Ed.), Thyroid Hormone Metabolism. Marcel Dekker, New York, pp. 39–76.

DiStefano III, J., Jang, M., et al., 1985. Optimized kinetics of reverse-triiodothyronine (rT3) distribution and metabolism in the rat: dominance of large, slowly exchanging tissue pools for iodothyronines. Endocrinology. 116 (1), 446–456.

DiStefano III, J., Landaw, E., 1984. Multiexponential, multicompartmental, and noncompartmental modeling: physiological data analysis and statistical considerations, part I: methodological and physiological interpretations. Am. J. Physiol. Regul. Integr. Comp. Physiol. 246, R651–R664.

DiStefano III, J., Malone, T., et al., 1982. Comprehensive kinetics of thyroxine (T4) distribution and metabolism in blood and tissue pools of the rat from only six blood samples: dominance of large, slowly exchanging tissue pools. Endocrinology. 111 (1), 108–117.

DiStefano III, J., Nguyen, T., et al., 1992. Sites and patterns of absorption of 3,5,3-triiodothyronine (T3) and thyroxine (T4) along rat small and large intestines. Endocrinology. 131, 275–280.

Eales, J., Chang, J., et al., 1982. Effects of temperature on plasma thyroxine and iodide kinetics in rainbow trout. Gen. Comp. Endocrinol. 47, 295.

Evans, N.D., Chappell, M.J., et al., 2004. Structural indistinguishability between uncontrolled (autonomous) nonlinear analytic systems. Automatica. 40 (11), 1947–1953.

Fagarasan, J., DiStefano III, J., 1986. Hidden pools, hidden modes, and visible repeated eigenvalues in compartmental models. Math. Biosci. 82, 87–113.

Gabrielsson, J., Weiner, D., 2000. Pharmacokinetic and Pharmacodynamic Data Analysis: Concepts and Applications. Swedish Pharmaceutical Press, Stockholm.

Godfrey, K.R., Chapman, M.J., 1990. Identifiability and indistinguishability of linear compartmental models. Math. Comput. Simul. 32 (3), 273–295.

Godfrey, K.R., Chapman, M.J., et al., 1994. Identifiability and indistinguishability of nonlinear pharmacokinetic models. J. Pharmacokinet. Biopharm. 22, 229–251.

Harless, C., DiStefano III, J., 2005. Automated expert multiexponential biomodeling interactively over the internet. Comput. Methods Programs Biomed. 79 (2), 169–178.

Hasen, J., Bernstein, G., Volpert, E., Oppenheimer, J.H., 1968. Analysis of the rapid interchange of thyroxine between plasma and liver and plasma and kidney in the intact rat. Endocrinology. 82 (1), 37–46.

Jang, M., DiStefano III, J., 1985. Some quantitative changes in iodothyronine distribution and metabolism in mild obesity and aging. Endocrinology. 116 (1), 457–468.

Kuhn de Chizelle, A., DiStefano III, J., 1994. MAMCAT: an expert system for distinguishing between mammillary and catenary compartmental models. Comput. Biol. Med. 24, 189–204.

Lee, W., Wahjudi, P.N., Xu, J., Go, V.L., 2010. Tracer-based metabolomics: concepts and practices. Clin. Biochem. 43, 1269–1277.

Pohjanpalo, H., 1978. System identifiability based on the power series expansion of the solution. Math. Biosci. 41, 21–34.

Riggs, D., 1970. The Mathematical Approach to Physiological Problems: A Critical Primer. MIT Press, Cambridge, MA.

Sefkow, A., DiStefano III, J., et al., 1996. Thyroid hormone secretion and interconversion in 5-day fasted rainbow trout, oncorhynchus mykiss. Gen. Comp. Endocrinol. 101, 123–138.

Specker, J., DiStefano III, J., et al., 1984. Development-associated changes in thyroxine kinetics in juvenile salmon. Endocrinology. 115 (1), 399–406.

Wagner, J., DiSanto, A., et al., 1981. Reversible metabolism and pharmacokinetics: application to prednisone-prednisolone. Res. Commun. Chem. Pathol. Pharmacol. 32, 387–405.

Wilson, K., Fisher, D.A., Sack, J., DiStefano III, J.J., 1977. System analysis and estimation of key parameters of thyroid hormone metabolism in sheep. Ann. Biomed. Eng. 5 (1), 70–84.

Zhang, L.-Q., Collins, J.C., et al., 1991. Indistinguishability and identifiability analysis of linear compartmental models. Math. Biosci. 103 (1), 77–95.

Nonlinear Mass Action & Biochemical Kinetic Interaction Modeling

Chapter 6

* All or parts of the sections denoted with asterisks are more advanced topics – mathematically speaking – and might be skipped over on first pass, or in a more introductory course of study. This does not mean they are unimportant, or less important, for understanding the practical modeling topics of this chapter.

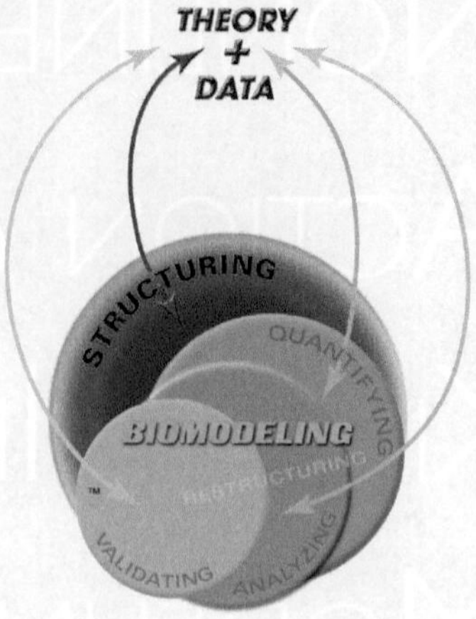

"It was obvious – to me at any rate – that the answer was to why an enzyme is able to speed up a chemical reaction by as much as 10 million times. It had to do this by lowering the energy of activation – the energy of forming the activated complex. It could do this by forming strong bonds with the activated complex, but only weak bonds with the reactants or products."

Linus Pauling,[†]
Nobel Laureate in both Chemistry (1954) and Peace (1962).

† Quoted in Thomas Hager, Force of Nature: The Life of Linus Pauling (1995).

OVERVIEW

At micro- and macroscopic levels, the *law of mass action* is the basic principle governing the rate of chemical reactions among one or more molecules. For most practical purposes, it says that the rate of a reaction is proportional to the product of the *molar concentrations* of its reactants. It's one of the most important biomodeling principles, with application beyond chemical and biochemical reactions.

Mass action and mass action-like interactions are ubiquitous in nature. At the intracellular level, the examples of transcription–translation of DNA–RNA–protein dynamics in Chapter 3 are based on mass-action molecular interactions. Epidemiological and predator–prey interaction models (e.g. virus–host) are similarly based and include the population growth models of Chapter 2, the shark–tuna predator–prey system model in Chapter 3 and the compartmental disease dynamics models of Chapter 4. Ordinary differential equation (ODE) models for these systems all involve interacting species – organismic or molecular – each represented as model state variables. Their primary distinction, mathematically speaking, is that these state variables are included as *products* in one or more terms of the ODEs and the models are thus nonlinear. Linear terms and models, however, do play an important role – as good approximations of micro- and macroscopic biological process dynamics, as well as linearizing experiments performed on nonlinear (NL) systems – as described at the end of the last chapter.

The goal here is to introduce the formalities of dynamic systems biology modeling with mass action or mass action-like components, with a focus on molecular systems biology processes. The biological domains and underlying theories for these models are primarily biochemical, so some biochemistry, physical chemistry and chemical kinetics are needed. These are very large and complex areas of physical science, with university departments among the largest, and textbooks among the most numerous. Only a fraction of the basics of these subjects are presented here, in a somewhat unconventional way – to maintain a focus on biosystem dynamics modeling. Hopefully, it's enough to appreciate the modeling methodology specifics associated with this fundamentally important systems biology domain. Because the mass action principle in chemistry and physics is about multiplicative *concentration* relationships,[1] the literature on mass action and mass action-like mechanisms is dominated by models developed in terms of products of continuum[2] *concentrations* of species, chemical or otherwise. Nonetheless, as we emphasized in Chapter 4, there are advantages to writing differential equation models in terms of *mass*-based state variables, primarily because mass rates (fluxes) are typically balanced in modeling. For example, dynamical interchanges among biosystem compartments with different volumes usually need volume adjustments when concentration state variables are used for ODE modeling. For these reasons, we pay due attention to terminology and dimensional analysis as we develop chapter topics and their associated deterministic ODE models here.

The exposition in the first few sections uses simple probability notions to describe some basic results about chemical reaction kinetics. These are integrated and developed into a systems

1 The probability of two or more independent interactions occurring together equals the product of the probabilities of each occurring separately.

2 Continuum means that the model independent variable *time* is continuous-time, as opposed to discrete-time.

modeling paradigm – from simple to more complex dynamical biosystem models – primarily deterministic continuum approximation[3] ODE models of biodynamic systems involving large numbers (thousands to millions) of molecules or species number. Energetics (thermodynamics) plays a role and therefore the dynamics of the biochemical processes described are dependent on physical properties like temperature and volume, which we briefly integrate into the modeling process. Substantial attention is given to modeling the basics of enzyme kinetic interactions, which dominate cellular processes. Further application of these modeling paradigms in biochemistry, molecular, cellular and organ level systems biology, pharmacology and toxicology is picked up again in the next and in subsequent chapters. In Chapter 7, we also extend the probability notions of chemical reaction kinetics to stochastic dynamic system models of cellular systems, those with very small numbers of interacting components.

Kinetic Interaction Models

Kinetic interactions in biology typically involve multiple factors. They may be subject to inhibition or enhancement signals, feedback or feedforward regulation, etc., and can be reversible or irreversible. Biochemical reactions at the cellular level, for example, are catalyzed by enzymes, accelerating their rates by up to a trillion-fold (10^{12}). Enzymes and other physicochemical species also provide fine control of metabolism or other cellular functions, by interacting with other molecules – via cellular signal transduction pathways. We present some of the simplest such interactions in this section – along with their ODE-based models, for enzymes, proteins and other biomolecules or biospecies. In the sections that follow, we introduce some of the prominent methodologies for extending them to more advanced biodynamic systems – including dynamic systems biology models of metabolism and cellular regulation processes – which we begin to develop more fully in the next chapter.

We restrict attention here to processes where the numbers of individual molecules or species are *large*, so the resulting equations – which begin as probability statements – result in deterministic (nonprobabilistic) ODE models of *average* kinetic behavior.[4] Event probabilities are inherent in kinetic theory and their inclusion in the pedagogy renders it more realistic and, hopefully, more rather than less intuitive.[5] For the present, spatial homogeneity of the molecular interaction environment is also assumed. More realistic conditions, such as cellular structures (membranes, channels, pores, etc.) inhibiting free movement of molecules throughout cells (for example), also can be modeled with ODEs,[6] and these two are introduced in the next chapter. We begin here with linear interactions that are not necessarily about reacting chemicals.

3 ODE continuum models are approximations because chemical reaction events are inherently probabilistic, and their evolution over time is stochastic in nature, more precisely described at the molecular interaction level by stochastic (random process) models. Stochastic models and applications are developed in the next chapter.

4 It is assumed that for large enough numbers of molecules, or species number, real dynamics are well represented by average behavior. Continuum (large-scale) ODE models are the most prevalent variety.

5 This probability-based approach also provides an introduction to stochastic modeling in the next chapter.

6 Spatial inhomogeneities are typically modeled using partial differential equations (PDEs). But one can use ODE models as well, by dividing up the spatial domains into separate (approximately) homogeneous compartments.

Linear & Monomolecular Interactions

These are the simplest: random contact resulting in growth or disappearance of individual units (species) in a large group of such units, numbering $\mathcal{N}(t)$ at time t. The detailed microscopic biophysics or chemistry may be complex, but at the macro- and organism levels, a simple model often suffices. The Malthusian growth model in Chapter 2, Example 2.3, illustrated this phenomenon for growth, albeit unrealistic growth – adjusted for limited resources in subsequent examples, to make it more real. We focus on disappearance (elimination) after contact[7] here, disappearance of atoms, molecules, particles, viruses, fishes, etc., present at time t. Assume each unit is subject to random elimination at time t, with uniform (constant) *probability P*. The goal is to develop an ODE model for this process.

Let $\Delta\mathcal{N}(t)$ be the number of units of $\mathcal{N}(t)$ eliminated in interval $(t,\ t+\Delta t)$. The *relative frequency definition of probability* – valid for large enough $\mathcal{N}(t)$ – then defines the ratio $\dfrac{\Delta\mathcal{N}(t)}{\mathcal{N}(t)} \cong P$, or $\Delta\mathcal{N}(t) \cong P\mathcal{N}(t)$. Dividing both sides by Δt and including a minus sign, because units are disappearing with time, then gives $-\dfrac{\Delta\mathcal{N}(t)}{\Delta t} \approx \dfrac{P}{\Delta t}\mathcal{N}(t)$. To get the desired ODE, we take limits of both sides as $\Delta t \to 0$, yielding:

$$\frac{d\mathcal{N}}{dt} \equiv \dot{\mathcal{N}} = -(P/dt)\mathcal{N} \equiv -k\mathcal{N} \quad \text{(number/time)} \tag{6.1}$$

For brevity, we henceforth use this **overdot notation for derivatives**, i.e. $dx/dt \equiv \dot{x}$ for any variable x.

We assume $\mathcal{N}_0$ units present at time 0 – the initial condition (IC). The ratio $P/dt \equiv k$ is the **rate constant**[8] (time^{-1} units) for this process, i.e. the (average) probability that any unit will be eliminated in infinitesimal interval dt. This simple model has the familiar single-exponential disappearance solution: $\mathcal{N}(t) = \mathcal{N}_0 e^{-kt}$.

This kind of linear elimination is a common model for *constitutive* elimination or degradation for numerous biomolecular species. **Constitutive** here means under basal or normal, or unperturbed conditions, e.g. **proteolysis**, or protein degradation in a basal steady state.

Example 6.1: ***Isotopic Decay Model: Probability of Atomic Disintegrations***

Radionuclide decay (spontaneous disintegration of one atom into another) is modeled well by Eq. (6.1) (Rutherford, 1906). One of the radioactive isotopes of iodine, ^{131}I, has a physical **half-life** of about 8 days – the time it takes to reduce the original number of atoms of ^{131}I, say $\mathcal{N}_0$ atoms, to $\mathcal{N}_0/2$. Substitution of these values into $\mathcal{N}(t) = \mathcal{N}_0 e^{-kt}$ then gives: $\mathcal{N}_0/2 = \mathcal{N}_0 e^{-8k}$ or $1/2 = e^{-8k}$. Taking logarithms of both sides, we get $-8k = \ln(1/2) = -\ln 2$, or $k = \ln 2/8 = 0.693/8 = 0.08663\ \text{day}^{-1} \cong 10^{-6}\ \text{sec}^{-1}$.

This is the probability that any atom of ^{131}I will disintegrate in a second.

7 Random, equally probable contact.

8 *In other words*, the rate constant k is the average number of individual species "disintegrations" per unit time that result in elimination. More on this in the section on biomolecular reactions.

Remark: Rutherford (Rutherford 1906) developed the theory of successive radioactive transformations in 1906, in which consecutive isotopic disintegrations – one into another – like the single one above, were each modeled as one-compartment models – each with single exponential modes, and all arranged as a unidirectional catenary model. This is said to have been the birth of compartmental modeling (Rescigno 2003).

We continue with a more general linear reaction model, based in classical chemistry, illustrating both the possibility of multiple reactant products and reactant reversibility.

Monomolecular Chemical Reactions

A **monomolecular reaction** (also called **unimolecular**) is one in which *single* molecules of a **reactant** change "spontaneously" into one or more **reactant products**, possibly reversibly.

Monomolecular Irreversible

Radioisotopic decay above is a simple example of "spontaneous" change of a single reactant into a single product. Let's now consider one reactant becoming two products: one molecule of W spontaneously and irreversibly becoming two molecules of Y and one of Z, expressed quantitatively by the balanced chemical equation:

$$W \xrightarrow{k_w} 2Y + Z \tag{6.2}$$

The parameter k_w is the *rate constant* (time^{-1} units) for this reaction. In probabilistic terms, k_w is the average probability per unit time that W will decompose into Y and Z during the infinitesimal interval dt. Following the same logic leading to (6.1) above, we have

$$\dot{\mathcal{N}}_w = -k_w \mathcal{N}_w \tag{6.3}$$

The convention is to express such equations using *concentration* state variables, typically *molar* concentrations. Dividing both sides by **Avogadro's number** $\equiv A_V \approx 6.022 \times 10^{23}$ mole^{-1}, the *number of molecules in one mole of a substance*, we have: $\dot{\mathcal{N}}_w / A_V = -k_w \mathcal{N}_w / A_V \equiv -k_w m_w$, or

$$\dot{m}_w = -k_w m_w \tag{6.4}$$

where m_w is the (still mass-based) **number of moles** of W, and the sign is negative because W is *lost* in the reaction. Since one mole of W produces 2 mols of Y in (6.2), a *gain* in molecules of Y,

$$\dot{m}_y = -2\dot{m}_w = 2k_w m_w \tag{6.5}$$

To convert moles (a normalized mass number) into a concentration, we divide both sides by volume V (L, liters) of the reaction vessel. Defining $m_y/V = y \equiv$ **molar concentration** of Y etc. (often abbreviated $\boldsymbol{M}$ (=mol/L, meaning molar concentration), also called **molarity**) then gives the desired concentration rate equations for each of the species in the reaction vessel. *Using lower case letters to represent concentrations*:

$$\dot{w} = -k_w w = -(1/2)\dot{y} = -\dot{z} \equiv -v \tag{6.6}$$

Eq. (6.6) states that the rate of a monomolecular reaction is directly proportional to the concentration of the reactant, etc., a single reactant in this simple case.

The 1/2 coefficient of $\dot{y}$ appears in Eq. (6.6) because – again – *Y appears* at 2 times the rate of *W*. And *Z* appears at the same rate *W disappears*, hence the minus sign. Note that rates are balanced in these ODEs; *W* is lost at half the rate *Y* is gained, etc.

The integer coefficients in (6.2) and balanced chemical equations like it have the special name **stoichiometric coefficients** – the (signed) number of molecules of each component that balances the equation, with (6.2) a balanced **stoichiometric equation**. The signs chosen are a matter of convention. Losses may get minus and gains positive signs (or *vice versa*). With this convention, *W* has **stoichiometry** -1. Stoichiometry for *Y* is $+2$ and for *Z* it's $+1$.

The rightmost symbol v in Eq. (6.6) is called the **reaction velocity** or **reaction rate**. Rewriting (6.6) in terms of v, we get $\dot{w} = -v$, $\dot{y} = 2v$, $\dot{z} \equiv v$ and in vector-matrix form:

$$\begin{bmatrix} \dot{w} \\ \dot{y} \\ \dot{z} \end{bmatrix} = \begin{bmatrix} -1 \\ +2 \\ +1 \end{bmatrix} v \equiv Nv \quad \text{(molar concentration/time)} \tag{6.7}$$

The matrix *N* of stoichiometric coefficients[9] in (6.7) is called the **stoichiometric matrix** for this reaction. Note that the number of rows in *N* is equal to the number of state variables (molecular species) in the equation, 3 in this example; and the number of columns in *N* is equal to the number of reaction rates, 1 in this example.

First Remark on Units Conventions (& Terminology): As noted in the Introduction, state variables in ODEs for chemical and biochemical reactions like those above and many others in the remainder of this chapter are, by convention, concentration rather than mass functions of time, which follows naturally from the mass action law (discussed below), where reaction rates are proportional to concentrations and their products. We used, favored and continue to favor mass-based state variables in compartmental and other structured system modeling in Chapter 4. Mass and concentration rate balances do not differ when all reactions take place in the same vessel volume, but not when exchanges occur between vessels of different volumes – the general case for real biosystems. Concentration based ODEs generally need adjustments for different volumes when modeling biosystems consisting of more than one vessel or compartment.

As always, each distinct species will have one state variable associated with it, generally located in different well-mixed (homogeneous) "spaces" – and thus with different volumes. If you subscribe to the classical definition of compartment as state variable, then each concentration state variable will have a different volume associated with it. If these volumes are all the same, the alternate definition of compartment-as-container overlaps – quantitatively speaking. If not, some care must be taken in interpreting model variables and volume parameters. This is exemplified in Example 6.5 below.

9 Use of *N* for this matrix is common in the literature. The letter *S*, which would be more natural, is reserved as a designation for the *relative sensitivity matrix* introduced in Chapter 11.

Monomolecular Reversible Conversion: Protein Folding and Unfolding

A **reversible reaction** can have exchange in both forward and backward directions, i.e. reactants to products and products to reactants. Suppose a single reactant molecule exists simply in two different states,[10] X and X'. For example, the spontaneous fluctuations of the three-dimensional (3-D) structure of a protein molecule from one **conformational state** X – a random unfolded coil state – to a folded native state X', as illustrated in **Fig. 6.1**. The reversible chemical equation for this reaction is

$$X \underset{k_{x'}}{\overset{k_x}{\rightleftharpoons}} X' \tag{6.8}$$

Two linear ODEs, in terms of the *concentrations* x and x' of X and X' represent this reaction.[11] Both *k*s have units of time^{-1}. The forward and reverse reactions are summed to get the total changes in X and X' concentrations:

$$\dot{x} = -k_x x + k_{x'} x' = -\dot{x}'$$

or (concentration per time) (6.9)

$$\begin{pmatrix} \dot{x} \\ \dot{x}' \end{pmatrix} = \begin{pmatrix} -k_x & k_{x'} \\ k_x & -k_{x'} \end{pmatrix} \begin{pmatrix} x \\ x' \end{pmatrix}$$

Since they are negatives of each other, we have $\dot{x} + \dot{x}' = 0$. Integration of this equation yields $x + x' = \text{constant} = x_0$, which means total protein (concentration), denoted x_0 here, is *conserved*. Solution of (6.9) is left as an exercise, E6.7. (Recognize that it's exponential?)

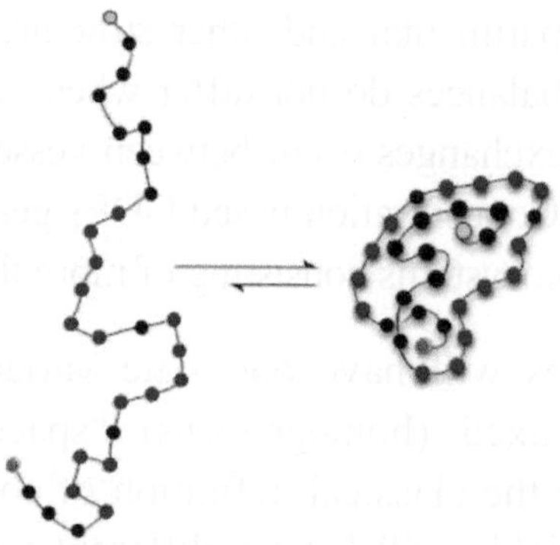

Fig. 6.1

Protein-folding. Circles are amino acids. Left: primary 2-D structure. Right: folded 3-D structure.

10 These two different states might be the *inactive state X* and the *active state X** of an enzyme or gene network transcription factor; or **isomerization** of a molecule, i.e. the reversible transformation from one isomeric form into another; or they might represent the **gating** of an ensemble of *intracellular ion channels*, each with an open and a closed state (Ullah et al. 2006).

11 In classical compartmental theory, this model has two compartments, and the resulting model is linear. This designation is not particularly useful in this simple example, but it can be in more complex reaction pathways.

Remark: Conservation equations like $x + x' = \text{constant} = x_0$ prevail as constraints in virtually all reactions and help with solution of the ODEs. However, if the reaction vessel (e.g. cell) is open to its environment, and/or X is produced or eliminated in the vessel, x_0 will be variable ($=x_0(t)$), with dynamics of its own governed by these processes. We expand on this notion in Example 6.5 below, illustrating the effects of such "openness."

If we rewrite Eq. (6.9) in terms of reaction rates, i.e. concentration reaction rates (velocities) $v_1 \equiv k_x x$ and $v_2 \equiv k_{x'} x'$ (conc/time units) in each direction on the right-hand side (RHS), similar to Eq. (6.7), but **bidirectionally**, we get

$$\begin{pmatrix} \dot{x} \\ \dot{x}' \end{pmatrix} = \begin{pmatrix} -1 & 1 \\ 1 & -1 \end{pmatrix} \begin{pmatrix} v_1 \\ v_2 \end{pmatrix} \equiv N\mathbf{v} \quad (\text{concentration/time}) \tag{6.10}$$

N in (6.10) is the **stoichiometric matrix** for this bidirectional reaction; it numerically depicts the stoichiometry of the reaction. Writing the ODEs in N-matrix form is quite useful in quantitatively studying metabolism and its regulation[12] and dynamic cell signaling (Chapter 14), and also for developing stochastic biosystem models (next chapter).

Second Remark on Units Conventions: The rates v_1 and v_2 have units of concentration per unit time. We'll continue to call these rates, or the more traditional velocities, and reserve the terms **mass flux**, or just **flux**, to mean mass rate, or mass per unit time. It is also common practice to express reaction equations like $X \underset{k_{x'}}{\overset{k_x}{\rightleftharpoons}} X'$ in different symbolic forms. Examples with different reaction rate arrow symbols include: $X \underset{v_2}{\overset{v_1}{\rightleftharpoons}} X'$ or $X \underset{v_2}{\overset{v_1}{\rightleftarrows}} X'$, or $X \underset{v_2}{\overset{v_1}{\longleftrightarrow}} X'$ or $X \overset{v}{\longrightarrow} X'$, where v above the single arrow in the last equation designates **net reaction rate** $v \equiv v_1 - v_2$.

Nonlinear Bimolecular Reactions & Other Species Interactions

Molecular & Other Collisions: Combinatorial Product Laws

The ODE models above – Eqs. (6.1), (6.4), (6.6) and (6.9) – are *linear* ODEs. This is not the case in general for mass action or mass action-like interactions. Interactions among different molecules or species follow product probability and combinatorics laws and hence have products of state variables in their ODE rate law models. For example, to model the interaction of two species or molecules – say W and X: $W + X \xrightarrow{k_{W+X}} Y + Z + \ldots$ reacting in some medium – interactions depend first on the chance of encountering each other. This depends on details like concentration of the reactants in the medium, the effective collision cross-section (geometry), the speed of motion of the molecules in the medium – which is dependent on the temperature of the medium, etc. These factors influence their *probability of collision* per unit time, i.e. *number*

12 Bernard Palsson's texts (Palsson 2006, 2009) are extensively devoted to developing the mathematical properties of the N-matrix and their application in metabolic network analysis.

of opportunities for collision per unit time, denoted P_{coll}/dt $(\equiv h \equiv (\mathcal{N}_w, \mathcal{N}_x))$. Let's see how many combinatorial opportunities there are for collision.

If there is one molecule of each, then there is one opportunity for collision: 1×1. For two molecules of one colliding with one of the other, there are two opportunities for collision: 2×1. Similarly, two and two give: $\begin{matrix} W_1 & \rightarrow & X_1 \\ & \nearrow\searrow & \\ W_2 & \rightarrow & X_2 \end{matrix}$ $\Rightarrow$ four opportunities for collision: 2×2, etc. So *the number of distinct pairs of molecules with opportunities for collision*[13] *is proportional to* $\mathcal{N}_w\mathcal{N}_x$. Then the probability of collision in the interval $(t,\ t+\Delta t)$ is $(P_{coll}/dt)\Delta t \simeq \Delta\mathcal{N}_{coll}(t)/\mathcal{N}_w(t)\mathcal{N}_x(t)$.

Dividing both sides by Δt, $\Delta\mathcal{N}_{coll}(t)/\Delta t = (P_{coll}/dt)\mathcal{N}_w(t)\mathcal{N}_x(t)$. Taking limits as $\Delta t \to 0$, we get the *nonlinear* ODE for the expected (mean) rate of collisions of distinct W and X per unit time: $d\mathcal{N}_{coll}(t)/dt = (P_{coll}/dt)\mathcal{N}_w(t)\mathcal{N}_x(t)$.

Simple collisions, however, are not enough to cause an actual *reaction* between W and X. The reaction needs a "spark of energy" (so to speak) to activate it, as in Eq. (6.1), and thus make it happen.[14] We model the interaction (with activation) of two molecules – say W and X – dissolved in a solvent with volume V_0 and temperature T_0 (Kelvin).

Energetics, Reaction Activation & Transformation

A **bimolecular reaction** is one in which two molecules (same or different **reactants** on the left-hand side (LHS)) combine to form one or more **products** (on the RHS), either irreversibly: $W + X \xrightarrow{k_{W+X}} Y + \ldots$ or reversibly:

$$W + X \underset{k_{Y+\ldots}}{\overset{k_{W+X}}{\rightleftharpoons}} Y + \ldots \tag{6.11}$$

In reality, they first form an active complex C that then decomposes into the products. For example, for an irreversible reaction producing only one product, the complete reaction might look like: $W + X \underset{k_{-C}}{\overset{k_{W+X}}{\rightleftharpoons}} C \xrightarrow{k_C} Y$. Usually, either the second reaction – from C to Y – is much faster than the forward reaction to C, or the back reaction from C is much faster than the forward one. This usually generates an active complex that exists for an extremely short time (order of picoseconds), achieving negligible complex concentration relative to the reactant and product species. So the active complexing step is usually ignored in writing equations, e.g. we consider only $W + X \xrightarrow{k_{W+X}} Y$ for this irreversible one-product illustration.

There are important exceptions to this simplification. The sections below, beginning with enzyme–substrate kinetics, the quasi-steady state assumptions, and also ligand–receptor

13 In a bimolecular reaction consisting of $\mathcal{N}$ same molecules ($W = X$) combining in distinct pairs to form a single molecule (dimer), the total number of molecules that can possibly react combinatorially is precisely given by the *binomial coefficient* $\binom{\mathcal{N}}{2} \equiv \mathcal{N}(\mathcal{N}-1)/2!$, the number of *distinct* pairs of W.

14 If you are swimming in the ocean and bump into a shark, the shark *might not* harm you.

kinetics, include the complex explicitly. Different boundary conditions and parameter values (the ks) in enzyme–substrate, ligand–receptor and protein–protein interactions often require explicit inclusion of the complex in the stoichiometric and dynamical equations. Dynamical responses of biochemical reactions in cells can depend strongly on ICs and parameter values and the kinds of assumptions made about the complexing reaction.

Remark: Model-Reduction-Hidden-Compartment Interpretation: Valid simplifications of reactions like $W + X \underset{k_{-C}}{\overset{k_{W+X}}{\rightleftharpoons}} C \xrightarrow{k_C} Y$ to $W + X \xrightarrow{k_{W+X}} Y$ can be interpreted as a reduction of a four-compartment (four-species) model to a three-compartment (three-species) reduced model. In this paradigm, the "complex" compartment ($c(t)$) – being of relatively negligible size – as well as relatively instantaneous kinetic response – is eliminated from the model, because it is hidden,[15] yielding an approximate model with three ODEs. Validity requires that $c << w$ and x for $t > 0$. If c is more comparable to w or x, the approximation is degraded accordingly. More on this later.

Remark: It is convenient in some modeling developments to consider only irreversible reactions. In this case, one writes reversible reactions as two irreversible reactions, one among the reactants, the second among the products, with the names reversed accordingly. More on this later.

For simplicity, consider the irreversible bimolecular reaction forming one product: $W + X \xrightarrow{k_{W+X}} Y$. The **rate constant** k_{w+x} is the average probability that $\mathcal{N}_w$ molecules of W and $\mathcal{N}_x$ molecules of X *collide* in *dt and* have between them the minimum **energy of activation** E_a needed to form the active complex. Thus, k_{w+x} is generally a product of two terms: $h \equiv$ the total number of opportunities for collision per unit time – which may or may not produce a reaction – multiplied by $b \equiv$ the average probability that any collision results in a reaction.[16] Thus, the *nonlinear* ODE for the reaction is:

$$\dot{\mathcal{N}}_{w+x} = b_{w+x}h_{w+x} \equiv k_{w+x}\mathcal{N}_w(t)\mathcal{N}_x(t) \tag{6.12}$$

The rate "constant" k_{w+x} therefore depends on the likelihood of molecular collision and the proportion of collisions that occur with energy exceeding some threshold activation for the reaction. Importantly, this depends on the physical attributes of the reaction, e.g. volume, temperature, medium, pH, etc., all of which can differ in different biological milieu and therefore affect reaction rate constants.

Again, even if the two molecules meet, reaction between them is not guaranteed: they must form an activated complex for interaction to occur; and activation of the coupling requires minimum activation energy E_a. Fortunately, biosystem modelers don't normally have to address these details, because rate constants are known for most well known reactions in cellular media. *The*

15 See Chapter 5, section on Model Simplification: Hidden Modes and Additional Insights. This is an example of a hidden *nonlinear mode*, the latter described in Chapter 9.

16 In these terms $k \equiv bh$ be can expressed in chemical thermodynamics terms, using the **Arrhenius equation** $k = e^{-E_aRT}h$ (Eisenberg and Donald 1979), where $b \equiv e^{-E_a/RT}$, with E_a the (Arrhenius) activation energy for the reaction (the energy needed to make it "go"), T temperature (degrees Kelvin) and B the Boltzmann constant.

main message here is that interaction is not enough, and a brief explanation of the complexity of reaction kinetics in chemistry and biochemistry helps to solidify this message. For *interactions* in more general biology, the situation is analogous, and needs little or no explanation. In animal populations, for example, males and females capable of reproducing interact closely in many ways, not necessarily leading to reproduction.

The **kinetic model** for all three molecules W, X, Y is established by noting the directions of the changes and writing the rates of turnover of each. In molar concentration terms, the ODEs for this simplest irreversible bimolecular reaction $W + X \xrightarrow{k_{W+X}} Y$ is the *nonlinear* set:

$$\begin{aligned} \dot{w} &= -k_{w+x}wx \\ \dot{x} &= -k_{w+x}wx = \dot{w} \quad \text{(at } V_0, T_0) \quad \text{(concentration per time)} \\ \dot{y} &= k_{w+x}wx \equiv v \end{aligned} \qquad (6.13)$$

where the **reaction rate** is denoted v, as in the last section, with $v = k_{w+x}wx$. The stoichiometric coefficients in this bimolecular irreversible association reaction differ only in sign and the stoichiometric ODE form is: $\begin{bmatrix} \dot{w} \\ \dot{x} \\ \dot{y} \end{bmatrix} = \begin{bmatrix} -1 \\ -1 \\ +1 \end{bmatrix} v \equiv Nv.$

Third Remark on Units Conventions: To balance Eq. (6.13), with wx having (concentration)^2 (concentration squared) units, the units of the rate constant k_{w+x} must be (time × concentration)$^{-1}$, i.e. the inverse of time × concentration.

Example 6.2: ***ODEs for Same Molecules Combining Depend on Reaction Rate Conventions***

If $W \equiv X$, then $2X \xrightarrow{k_{x+x}} Y$. Getting the ODEs right here can be tricky, because it depends on the convention chosen to represent the reaction. In most chemistry books, the convention is that *reaction rate* or *velocity* $v = k_{x+x}x \cdot x = k_{x+x}x^2 = \dot{y}$ (concentration/time) is the variable to be balanced in the ODEs. For this reaction, one Y is gained at this rate. And, each time the reaction occurs (at rate $v = k_{x+x}x^2$) two Xs are lost, so we have $\dot{x} = -2k_{x+x}x^2 = -2v$ (one for each molecule reacting).

We might also follow a different logic for adjusting the ODEs, balancing (conserving) mass (or concentration) in the equations. For our example, this would be "the mass of X is lost at minus twice the rate the mass of Y is gained." The conservation equation would then be: $\dot{x} = -2\dot{y}$. Then, (6.13) would yield: $\dot{x} = -k_{x+x}x \cdot x = -k_{x+x}x^2$, and $\dot{y} = (1/2)k_{x+x}x^2$.

We use the **reaction rate balance convention** in this book and the stoichiometric ODE for $2X \xrightarrow{k_{x+x}} Y$ is $\begin{bmatrix} \dot{x} \\ \dot{y} \end{bmatrix} \equiv \begin{bmatrix} -2 \\ +1 \end{bmatrix} v = Nv.$

Fourth Remark on Units Conventions: Different **subscripting conventions** for rate constants (or velocities v) illustrate different ways they can be more readily recognized when separated from their equations. The k_{w+x} (W and X going forward) and k_{y+z} (Y and Z returning) in $W + X \underset{k_{Y+Z}}{\overset{k_{W+X}}{\rightleftharpoons}} Y + Z$ might be found numbered and signed, i.e. $k_{w+x} \equiv k_1$ for forward reactions and $k_{y+z} \equiv k_{-1}$ for reverse reactions; or just signed, i.e. $k_{w+x} \equiv k_+$ and $k_{y+z} \equiv k_-$; or without any subscript if there is only one rate constant. Numbering is helpful for identifying individual rate constants when more than one reaction equation is involved.

Example 6.3: ***Bimolecular Reaction with Different Stoichiometry & Ambiguous Order***

Consider the reaction: $nW + mX \xrightarrow{k} Z$. The theory above then gives

$$\dot{w} = -nkw^n x^m, \quad \dot{x} = -mkw^n x^m \quad \text{and} \quad \dot{z} = kw^n x^m \text{ (concentration/time)} \tag{6.14}$$

In this equation $\boldsymbol{n + m}$ is called the **order of the reaction**. Now suppose $w >> x$ in this example reaction. Then, relatively speaking, $w \sim$ constant and, using our rate balance convention, $\dot{x} = -(mkw^n)x^m \cong -k'x^m$. In this case, if $x(t)$ is *measured* in an experiment at a sufficient number of times, and the equation is appropriately fitted to this data, the order of the reaction will be best estimated as parameter m in the last equality, which is also called the ***empirical order*** of the reaction. Note that, if $m = 1$, the reaction *appears* to be linear and, more important – behaves linearly – although it is based on a nonlinear ODE, one that is behaving approximately linearly. (Also see Exercise E6.3.)

Volume & Temperature Effects

Suppose the *volume* of the reaction medium (vessel) gets bigger (or smaller). This, naturally, presents fewer (or more) opportunities for collision. Let $V = nV_0$ (bigger if $n > 1$). Then V_0 contains $(1/n)^{\text{th}}$ of the concentration each of W and X, and since $V = nV_0$, the rate equation for W in bigger volume V is: $\dot{w}(t, V) = nk_{w+x}\frac{w}{n}\frac{x}{n} = k_{w+x}w\frac{x}{n} = \frac{V_0}{V}k_{w+x}wx$ (in V at T_0), a proportional – or *linear* – change. This is exemplified below for open systems, where a molecular species is transferred between different compartments with different volumes.

What about *temperature* changes? They definitely affect chemical reaction rates, which is especially important for enzymatic reactions in living cells. In gases, the average speed of a molecule, and hence the number of collisions per unit time, is proportional to $\sqrt{T}$ (Atkins and Jones 2005). Unfortunately, this square root law does not apply well to biological fluids; k is a function of temperature T in a more complicated way than a square root ratio, due to strong intermolecular interactions present in the liquid state (Eisenberg and Donald 1979; Lescovac 2003). For most situations, temperature has a stronger effect. Arrhenius' law (Eisenberg 1979) $k = Ae^{-E_a/RT}$ is generally considered an accurate measure of temperature effects on reaction rates in gases, where T = absolute temperature (degrees K), R = ideal gas law constant, and E_a is the minimum energy of activation. It is also used to approximate temperature effects in biological fluids and may be appropriate under particular conditions (Fujita and Arimitsu 1991; Peterson et al. 2007).

Many chemical reactions among one or more molecules in biological (e.g. cellular) systems have special names. **Ligands** are molecules that bind specifically to proteins, usually proteins acting as **receptors**. **Monomer receptors** have only one binding site, and **oligomer** receptors have more than one such site. We use these terms freely below.

Law of Mass Action

Monomolecular and bimolecular reactions are generalized, by induction, into the law of mass action for any number of reactants: *the rate of a one-step (elementary) chemical reaction is directly proportional to the product of the molar concentrations*[17] *of each of the molecular reactants, whether they are different or the same substances* (Atkins and Jones 2005).

Reaction Rate Equations

Reaction rate equations specify the rates at which individual species involved in the reaction change over time. Usually they occur in a sequence of coupled or uncoupled reaction steps called **elementary reactions**. The *rate v of each elementary step* is derived from the law of mass action; it's the rate at which the reactant species *disappear, i.e.* $k \times$ product of reactants $\equiv -v$.

Example 6.4: ***Sequence of Coupled Elementary Reactions***

Reactants Y and Z form compound (complex) YZ by elementary step 1:

$$Y + Z \xrightarrow{k_1} YZ \tag{6.15}$$

This could be (for example) the way protein Y is sequestered (captured) in a cell by receptor Z. YZ might then react (complex) with another species X, in elementary step 2, producing XY – and freeing Z to capture another Y:

$$X + YZ \xrightarrow{k_2} XY + Z \tag{6.16}$$

We use lower case letters x, y, z to represent concentrations of X, Y, Z and c_{yz} and c_{xy} for concentrations of *complexes* YZ and XY. The elementary reaction rates for each step, 1 and 2 *individually*, are: $v_1 \equiv k_1yz = -\dot{y} = -\dot{z} = +\dot{c}_{yz}$ and $v_2 \equiv k_2xc_{yz} = -\dot{x} = -\dot{c}_{yz} = +\dot{c}_{xy} = +\dot{z}$.

However, because the *sequence* of steps 1 and 2 are inherently coupled, and both Z and YZ participate in both reactions, the correct rate equations for Z and YZ are:

$$\begin{aligned} \dot{z} &= k_2xc_{yz} - k_1yz \\ \dot{c}_{yz} &= k_1yz - k_2xc_{yz}(= -\dot{z}) \end{aligned} \tag{6.17}$$

Combining (6.17) with the (uncoupled) reactions:

$$\dot{y} = -k_1yz, \quad \dot{x} = -k_2xc_{yz}, \quad \dot{c}_{xy} = -k_2xc_{yz} \tag{6.18}$$

17 Strictly speaking, the rate is proportional to the product of the *thermodynamic activities*, but this is proportional to the product of the molar concentrations in dilute solutions.

this set of five reaction rate equations are called the **overall (coupled) elementary rate equations**.

Elimination (removal, negative) terms in these elementary reaction rate equations involve the concentrations of the species being removed. This means that chemical concentrations can never be negative, otherwise they would contradict the physical reality – that removal rates are *ipso facto* negative, by definition. This is true, in general, for removal processes involving biomolecular species in open or closed biosystems.

Most real kinetic mechanisms of interest involve elementary steps coupled in complex ways, difficult to discern by theory or experiment. In these cases, overall process dynamics are pertinent and these are typically established experimentally as empirical rate laws (e.g. Example 6.3 above), i.e. **phenomenological models**. *Autocatalysis* and so-called *Brusselator* models (Field et al. 1974) are two such examples with complex dynamics, explored in *Stability* Chapter 9.

Reversible Reactions & Dynamic Equilibrium

Consider the reversible reaction:

$$W + X \underset{k_{-1}}{\overset{k_1}{\rightleftharpoons}} Y + Z \tag{6.19}$$

It has the nonlinear ODE model (four equations):

$$\dot{w} = -k_1 wx + k_{-1} yz = \dot{x} = -\dot{y} = -\dot{z} \quad \text{(concentration/time)} \tag{6.20}$$

At equilibrium (also steady state here), $\dot{w} = \dot{x} = \dot{y} = \dot{z} \equiv 0$ and the **equilibrium association constant** for this reaction is:

$$K_a \equiv \frac{k_1}{k_{-1}} = \frac{y_\infty z_\infty}{w_\infty x_\infty} \quad \text{(unitless)} \tag{6.21}$$

The infinity subscripts designate steady state values. The inverse of K_a: $1/K_a \equiv K_d$, is called the **equilibrium dissociation constant**.

Reversible Eq. (6.19) *can be interpreted as* ***two unidirectional elementary reactions***. The forward unidirectional reaction is W and X interacting to make Y and Z. The reverse unidirectional reaction is the combination of Y and Z to make W and X.

Equilibrium and equilibrium constants have fundamental importance in chemical and biochemical kinetics. Much of the modeling and laboratory experimentation in these fields involves estimating steady state system parameters like K_d, a measure of the tendency of the reactants to "come apart"; its reciprocal, the association constant K_a, is a measure of them "coming together." Their units – not necessarily unitless – as in Eq. (6.21), depend on how many concentration terms appear on each side of reaction equation, in this case the same number, 2 on each side, so the dimensions cancel each other.

Reaction Dynamics in Open Biosystems

In the reactions discussed so far, and in similar textbook treatments of this subject, it is implicitly assumed that the reaction is occurring in a closed system, e.g. with only fixed amounts or concentrations present at initiation of the reaction, i.e. w_0 and x_0 (or y_0 and z_0) in (6.19), with none added (or subtracted) afterwards – i.e. total reactant and reaction products remain constant. In biological systems, when reactant is added or subtracted from a reaction medium from *outside* the medium, additional terms are needed in equations like Eq. (6.20), and possibly also additional ODEs, to account for these additional kinetic processes.[18] For example, if the reaction is occurring in a cell and a reactant exits and enters the cell to and from the fluid surrounding it, the mass transfer kinetics (flow) back and forth must be included, typically as at least two additional terms in equations like Eq. (6.20). A reactant also might be degraded (a sink) or produced (a source), and additional terms may be needed for these dynamic processes, with due attention to units and volume changes in going from one to the other.

Example 6.5: ***Reversible Reaction Plus Degradation in an Open Biosystem***

Consider a group of cultured cells of known volume V_{Cell} in a homogeneous fluid culture medium (*Med*) of constant volume V_{Med}. The cells initially contain reactant *X*, but not reactant *W*, which is initially outside the system, entering the culture medium exogenously, as shown in **Fig. 6.2**. *W* enters the cells from the culture medium. When in the cells, *W* can react 1:1 with *X* molecules, as in Eq. (6.19). The (egg-shaped) cellular membrane inside the culture medium is assumed semi-permeable and *W* can move back and forth across it, presumably by simple linear diffusion. The components making up the open portions of this system are shown in **RED** in **Fig. 6.2**. In addition, *W* can be degraded (a *sink*) in the cell by a distinctly independent cellular process (not via *X*) at a rate proportional to its concentration in the cells.

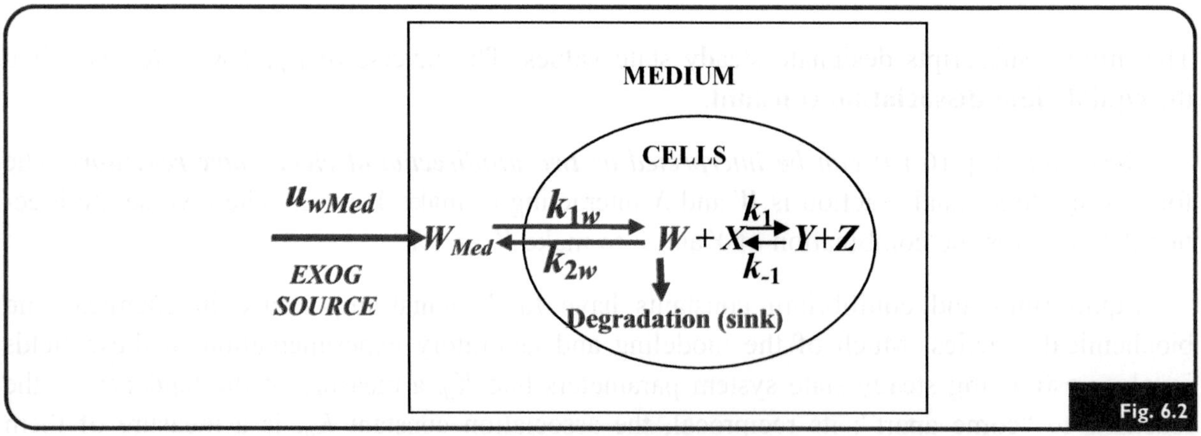

Fig. 6.2

Open biosystem: cell-culture dynamics.

18 As a result, open biochemical and biological systems have a much richer variety of dynamical behaviors than closed ones.

To model the overall process dynamics in this *open* biosystem, we need to augment the ODE for W (not X, Y and Z) in Eq. (6.20) with additional terms. We emphasize here that this cannot be done without due consideration for volume changes, and therefore concentration changes due to volume changes, in going from one compartment to the other. *Mass fluxes* need to be balanced.

In concentration rate terms, the two linear (diffusion) *exchange* rates can be expressed as $+k_{1w}w_{Med}$ and $-k_{2w}w$, where w_{Med} is the concentration of W in the medium, w (with no subscript) is the W concentration in the cells. The third added term, for *degradation* rate, is $-k_{\deg,w}w$. Given that the two terms $-k_{2w}w$ and $-k_{\deg,w}w$ are both in the same cellular volume V_{Cell}, they can be simply added to Eq. (6.20) for the rate of change of concentration w in the cells. But the W entering from the medium with different volume V_{Med} needs more care in adjusting Eq. (6.20).

The change in volumes must be included in establishing the rate parameter k_{1w} (time^{-1} units). In q_{Med} mass terms, let's call the *mass rate* of W leaving the medium and entering cells kq_{Med}. In w_{Med} concentration terms, since $w_{Med} \equiv q_{Med}/V_{Med}$, this mass rate equals $kq_{Med} \equiv kw_{Med}V_{Med}$, still the mass rate of W leaving. When it arrives in cells it is immediately diluted (or concentrated) due to the different volume of the cells V_{Cell}. So the mass rate $kV_{Med}w_{Med}$ leaving the medium – upon entering cells (and getting diluted by V_{cell}) – becomes: $k(V_{Med}/V_{cell})w_{Med} \equiv k_{1w}w_{Med}$, i.e. the adjusted and now *concentration* rate $k_{1w}w_{Med}$. Adding all three terms to Eq. (6.20), we get the open system model:

$$\dot{w} = -k_1 wx + k_{-1}yz - k_{2w}w - k_{\deg,w}w + k_{1w}w_{Med} \tag{6.22}$$

The overall system is also open to an exogenous source of W beyond the culture medium (**Fig. 6.2**). An additional ODE is needed to model W dynamics in the culture medium, taking into account (for generality) the time-varying exogenous mass input rate (mass/time) $u_{wMed}(t)$ of W into the medium. This is balanced by the bidirectional transfers of W into and out of the cells, modeled in this example as simple first-order diffusion into and out of the cells – and with no degradation of W within the medium (assuming, for example, degradation enzymes only within the cell).

The concentration rate equation for w_{Med} inside the medium is then:

$$\dot{w}_{Med} = \frac{u_{wMed}}{V_{Med}} - k_{1w}w_{Med} + k_{2w}w \tag{6.23}$$

If the transfer or degradation processes in the red-colored terms of (6.22) or (6.23) are, instead, nonlinear, they should be adjusted accordingly.

Given that (6.22) and (6.23) have terms in common, they are coupled, and these two are also coupled with the ODEs for X, Y and Z in Eq. (6.20). These five ODEs, based on the reactions and material exchanges illustrated in **Fig. 6.2**, make up the open biosystem model. *The input, five state variable ICs, five k and two volume parameters need to be specified to simulate the five ODEs of the model.*

Modelers must additionally pay attention to the dynamical time-scales mixed in the resulting equations, dependent of course on the dynamics of the process represented inside and outside the cells in the medium. Cell entry and exit were diffusion-limited in this example, so the rates of entry and exit of reactant W might be more or less slower (or faster) than the intracellular dynamics process under study. This may make the resulting ODEs more or less stiff and may require special attention in simulation (Chapter 3), or conversion of the fast dynamics equations to approximate algebraic relations, as in the developments below, if *quasi-steady state assumptions* can be invoked.

Closed chemical systems can exhibit damped oscillatory behavior, but not sustained oscillations about their unique equilibrium state (Scott 1994). In contrast, open chemical (biochemical, biological) systems – those that exchange mass with their surrounding – or have some implicit constant input – are capable of a greater diversity of characteristic behaviors, including sustained oscillations (e.g. relaxation oscillations) and chaos. We examine these topics in *Stability* Chapter 9.

Enzymes & Enzyme Kinetics

An **enzyme** E is a complex organic catalyst[19] with the primary function of *very greatly increasing the rate* of highly specific chemical reactions. It does this by *vastly decreasing the energy of activation* E_a a molecule of **substrate** $\mathcal{S}$ must acquire before it can participate in the reaction. Enzymes[20] are highly efficient in this regard, in particular under the rather mild conditions of the normal physiological state, i.e. low temperature, standard pressure and an aqueous medium. They also act as regulators of biological processes, like metabolism and cell signaling, and the principal way to understand the behavior of enzymes and characterize their properties is by studying enzyme kinetics, the temporal dependence and dynamical interactions among enzymes and substrates.

Most of the standard nomenclature for enzyme systems is retained in this section. The exceptions are minor, to maintain consistency with nomenclature developed earlier in the chapter, and generalized somewhat for eventual application in modeling open and multicompartmental biological systems (e.g. as in Example 6.5) and signal transduction networks, with multiple enzymes, other proteins and other component species.

19 Most enzymes, but not all, are proteins; e.g. *ribozyme* is a nonprotein catalyst made of RNA. They are quite specific in their action, have what is called a *catalytic center*, and often function complexed with cofactors (e.g. vitamins) capable of activating or inhibiting their action.

20 Enzyme classes include: *oxidoreductases*, which catalyze oxidation/reduction reactions; *transferases* transfer a functional group (e.g. a phosphate or methyl group); *hydrolases* catalyze hydrolysis bonds; *lyases* cleave bonds by other means; *isomerases* catalyze isomerization adjustments within a single molecule; and *ligases* join two molecules with covalent bonds. *Kinase* enzymes are important cellular transferases that modify other proteins by chemically phosphorylating them, which usually changes their functional state by changing enzyme activity, location or association with other proteins.

Substrate–Enzyme Interactions: Michaelis–Menten Theory

The classic model for the simplest interaction is attributed to Leonor Michaelis and Maud Menten (Michaelis and Menten 1913), for a single enzyme interacting in a closed system with a single substrate, producing one or more products:

$$S + E \underset{k_{-1}}{\overset{k_1}{\rightleftharpoons}} ES \xrightarrow{k_2} P_1 + P_2 + E \tag{6.24}$$

In the Michaelis–Menten (abbrev: M–M) theory, the enzyme E and substrate S initially react to reversibly form the complex $ES \equiv C$, and this complexing reaction is treated explicitly in developing model equation dynamics.[21] The complex then breaks down to form the free enzyme plus one or more products; P_1 and P_2 are two possible products. A back reaction: $SE \xleftarrow{k_{-2}} P_1 + P_2 + E$, which must exist on thermodynamic grounds, is considered to be relatively negligible, i.e. $k_{-2} \approx 0$, and is usually neglected in classical developments.

With lower case letters representing concentrations of the five species (state variables) in (6.24), the law of mass action applied to the three reactions in (6.24) leads directly to the following set of nonlinear ODEs:

$$\dot{s} = -k_1 es + k_{-1} c \tag{6.25}$$

$$\dot{e} = -k_1 es + k_{-1} c + k_2 c \tag{6.26}$$

$$\dot{c} = k_1 es - (k_{-1} + k_2) c \tag{6.27}$$

$$\dot{p}_1 = \dot{p}_2 = k_2 c \tag{6.28}$$

The k_i, k_{-i} are positive rate constants,[22] some notably with different units. Dimensional analysis indicates that k_{-1} and k_2 have inverse time t^{-1} units, whereas k_1 has inverse mass × time units, $m^{-1}t^{-1}$, as in the *Third Remark on Units Conventions* noted earlier. Initially, only substrate and enzyme are present, in amounts s_0 and e_0, *usually much more substrate than enzyme.* The initial condition vector for this system of equations is:

$$\begin{bmatrix} s(0) \\ e(0) \\ c(0) \\ p_1(0) \\ p_2(0) \end{bmatrix} = \begin{bmatrix} s_0 \\ e_0 \\ 0 \\ 0 \\ 0 \end{bmatrix} \tag{6.29}$$

21 This provides a more general model for bimolecular reactions than in the previous section, with a much wider range of applications – beyond enzyme–substrate kinetics.

22 Rate constants k_1 and k_{-1} are often called the **complexing** and **decomplexing** and k_2 is the **catalysis** rate constant.

A **conservation equation**[23] follows by adding and then integrating Eqs. (6.26) and (6.27) with ICs (6.29), or simply realizing that enzymes always exist in either free or bound (complexed) form c and must always have a sum equal to the initial e_0:

$$e(t) + c(t) = e_0 \tag{6.30}$$

This relation is used to eliminate $e \equiv e(t)$ from Eqs. (6.25) and (6.27), so they become

$$\dot{s} = -k_1 e_0 s + (k_1 s + k_{-1})c \tag{6.31}$$

$$\dot{c} = k_1 e_0 s - (k_1 s + k_{-1} + k_2)c \tag{6.32}$$

Strictly speaking, to find a solution for the product concentrations p_1 or p_2 from (6.28), three differential equations, Eqs. (6.28), (6.31) and (6.32), need to be solved simultaneously, in particular to obtain $c(t)$, and then p_1 or p_2 can be obtained by integration. However, $c(t)$ can be obtained without solving *any* ODEs, if we invoke the **basic assumption of M–M theory**, that: *equilibrium is very rapidly established among E, S and C in accordance with the first (the reversible) reaction, virtually independent of the second reaction*, i.e. $\dot{c}(t) \cong 0$ for all $t > 0$. This was originally called the **pseudo-steady state** condition (Briggs and Haldane 1925), these days more commonly called the **quasi-steady state assumption (QSSA)**.

The Quasi-Steady State Assumption (QSSA) & the M–M Approximate Equations

In early developments of the theory, the quasi-steady state assumption was validated by the strict assumption that, initially, $s_0 >> e_0$. Physically, this means there is always enough substrate for the enzyme released from the second reaction to continuously and rapidly recomplex with substrate, thereby maintaining an approximately constant complex concentration. In other words, the first reaction, forming the bound transition state, is in "approximate equilibrium" *at all times* relative to the second reaction. Actually, an excess of substrate is sufficient, but not necessary, and the QSSA is valid for such reactions even – for example – if $s_0 \approx e_0$ (Frenzen and Maini 1988). As discussed further below, validity of the QSSA[24] depends on all the rate constants of the reaction, k_1, k_2 and k_{-1}.

To see this, it is useful to argue differently, but equivalently, about this set of M–M reactions (Segel and Slemrod 1989). Let the time it takes to reach quasi-steady state (the initial transient period) be denoted t_C. Similarly, denote by t_S the time the overall system remains in quasi-steady state, when $t > t_C$ and substrate and product dynamics unfold. During the initial

23 This is a simple example of the age-old practice in science of using *additional information* about the environment or *boundaries* of the system under study for the purpose of reducing the number of unknowns in a model. This classical concept has been adapted and formalized into an approach called *constraint-based modeling*, for helping to solve complex systems biology models (Schilling and Palsson 1998).

24 The two Extensions at the end of the chapter provide additional details and derivations of the quasi-steady state assumption. The first is based on the classical literature and the second on the *differently scaled equations* development in (Segel and Slemrod 1989), motivated by more than half a century of excellent but apparently incomplete contributions to this not-so-simple theory.

transient period, up to t_C, there must only be a negligible fractional depletion of substrate, i.e. $s(t) \approx s_0$. (This is critical.) Then what is actually needed is that t_C be very short relative to t_S, which means $t_S >> t_C$ is required, rather than (as we shall see) the stricter $s_0 >> e_0$. The short time t_C is called the **pre-steady state** or **initial transient period** – or **fast time-scale** t_C – and the quasi-steady state period that follows – the period we are interested in most – is called the **slow time-scale**[25] t_S. Segel and Slemrod (1989) show these time scales[26] to be:

$$t_C = 1/k_1(s_0 + K_m) \quad \text{and} \quad t_S = (s_0 + K_m)/k_2 e_0 \\ \text{with } K_m \equiv (k_{-1} + k_2)/k_1 \tag{6.33}$$

Importantly, they also showed that the *decisive error measure* for validating the QSSA is the *fractional change in substrate:* $|\Delta s/s_0| \cong e_0/(K_m + s_0) \equiv \varepsilon$ during the initial fast transient. This error ε must be as small as possible, requiring $s(t) \approx s_0$ up to t_C. This translates to:

$$|\Delta s/s_0| \cong \frac{e_0}{K_m + s_0} \equiv \varepsilon \ll 1 \tag{6.34}$$

Thus the primary necessary condition for validity of the QSSA is $K_m + s_0 >> e_0$ (compared with $s_0 >> e_0$) and this guarantees that $t_S \gg t_C$ is valid. In other words, if $K_m + s_0 >> e_0$, then the fast time scale t_C is small enough, relatively speaking, and the slow time scale t_S is big enough for application of the simplifying QSSA to Eqs. (6.29)–(6.32). Clearly, **the magnitude of the constant K_m has a lot to do with validity of the QSSA.**

Summarizing, the **quasi-steady state assumption** (QSSA)[27] is the approximation:

$$\dot{c}(t) \cong 0 \quad \text{for all } t > 0 \tag{6.35}$$

And it is valid when:

$$K_m + s_0 \gg e_0 \quad \rightarrow \quad \varepsilon \equiv \frac{e_0}{\frac{k_{-1} + k_2}{k_1} + s_0} \ll 1 \tag{6.36}$$

Importantly, relations (6.36) mean that any combination of parameters and ICs that satisfy this condition render the QSSA (more-or-less) valid. For example, $s_0 >> e_0$, as in classical developments; or the opposite: $s_0 < e_0$ with $k_1 << k_{-1}$ or $k_1 << k_2$ and $e_0 << K_m$. And we see that $s_0 \approx e_0$ is ok too, if $K_m >> e_0 - s_0$. The latter is illustrated further in Example 6.6 below.

25 Note the similarity here with two "modes" of a linear dynamic system response being very far apart, i.e. a "stiff" ODE response.

26 Derived in the Extensions section, second approach, at the end of the chapter.

27 By itself the $\dot{c}$ equation is not null during the fast time-scale period, up to t_C. It has a normalized time-varying solution $c(t) = [1 - e^{-k_1(K_m + s_0)t}]e_0 s_0/(K_m + s_0)$, which reaches the normalized value 1 very quickly relative to t_S (Segel and Slemrod 1989).

The QSSA approximation (6.35) renders Eq. (6.32) effectively algebraic instead of differential, i.e. (6.32) becomes:

$$0 = k_1 e_0 s_{QSSA}(t) - (k_1 s + k_{-1} + k_2) c_{QSSA}(t) \quad \text{for all } t > 0 \tag{6.37}$$

where we designate $c_{QSSA}(t)$ and $s_{QSSA}(t)$ as the QSSA approximations of $c(t)$ and $s(t)$, to maintain the distinction from the exact complex and substrate concentrations. $s_{QSSA}(t)$ is different from $s(t)$ because Eq. (6.31), the exact ODE for $s(t)$, is solved with approximation c_{QSSA} substituted for c. It is also important to note that time t is the explicit argument in (6.35) and (6.37), despite this being a "steady state" solution of the ODE for $c(t)$. This means the complex equation is in an approximate (quasi) steady state that's changing (approximately instantaneously) with time, *relative to the coupled rates of loss of substrate and rate of product formation.*

Collecting terms and solving Eq. (6.37) for $c_{QSSA}(t)$, then gives

$$c_{QSSA}(t) = \frac{e_0 s_{QSSA}(t)}{\dfrac{k_{-1} + k_2}{k_1} + s_{QSSA}(t)} \cong c(t) \quad \text{for all } t > 0 \tag{6.38}$$

Then, using Eq. (6.28) – the exact ODE for $\dot{p}_1 = \dot{p}_2 \equiv \dot{p} = k_2 c$, the approximate (QSSA) rate of production of product, designated $\dot{p}_{QSSA}(t) = k_2 c_{QSSA}(t)$, is:

$$\dot{p}_{QSSA}(t) = \frac{k_2 e_0 s_{QSSA}(t)}{K_m + s_{QSSA}(t)} \cong k_2 c(t) \quad \text{for all } t > 0 \tag{6.39}$$

where we have substituted $(k_{-1} + k_2)/k_1 \equiv K_m$. Finally, we use (6.39) to devise alternative ODE approximations for $\dot{p}$ and $\dot{s}$. From Eqs. (6.25)–(6.28), we have the exact relationships $\dot{c} = -\dot{s} - k_2 c = -\dot{s} - \dot{p}$. If $\dot{c} \cong 0$, then $-\dot{s} = \dot{p} \equiv v$ and – in biochemistry notation – product is formed and substrate is lost at rate (velocity $v(t)$):

$$\dot{p}_{QSSA}(t) = \frac{v_{max} s_{QSSA}(t)}{K_m + s_{QSSA}(t)} = -\dot{s}_{QSSA}(t) \equiv v(t) \quad \text{for all } t > 0 \tag{6.40}$$

The two Eqs. (6.40) are the QSSA equations for the M–M reaction. They are also called *the* **Michaelis–Menten (M–M) equations**, where:

$v(t)$ represents the **reaction velocity** (rate)

$v_{max} \equiv k_2 e_0$ **is the maximum value of** v and

$K_m \equiv (k_{-1} + k_2)/k_1$ is the Michaelis constant

Remark: It is common practice to *not* use the QSSA subscripts in the above equations and we drop them in subsequent developments below, to simplify the notation. We have used them here to pedagogically emphasize that the variables c_{QSSA}, s_{QSSA} and p_{QSSA} are *approximations* of c, s and p in the exact M–M reaction equations.

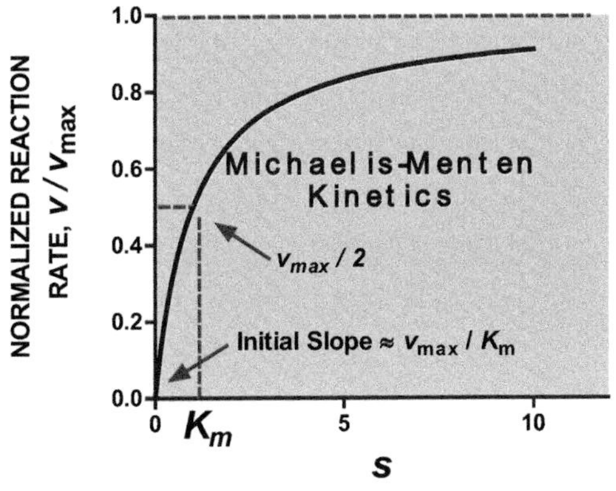

Fig. 6.3

Graphically, v as a function of s is a rectangular hyperbola,[28] drawn normalized to v_{max} in **Fig. 6.3**, making it a **reaction fraction**, which we designate as $F \equiv v/v_{max}$. Note that $s = K_m$ when $v = v_{max}/2$. The fact that v_{max} is the maximum velocity of the reaction means that $S \rightarrow E + P$ is typically the *rate-limiting step* of the reaction,[29] with P being produced (at the expense of S) at fractional rate k_2 (t^{-1} units). Again, the first, complexing reaction in Eq. (6.24), is – by the quasi-steady state assumption – virtually instantaneous relative to this much slower generation of product and release of enzyme over the time-domain of the reaction – assuming the QSSA condition $K_m + s_0 >> e_0$ is valid.

M–M QSSA Equation Extremes

It is important to recognize that nonlinear M–M regulation of enzyme activity is approximately linear (**first-order**) for very low values of substrate, and saturated (**zero-order**) for high enough values. This can be gleaned from the graph in **Fig. 6.3**. More formally, if we write the function as: $v = \dfrac{v_{max}s}{s + K_m} = \dfrac{v_{max}}{1 + K_m/s}$, we see that it becomes v_{max}, the (zero-order) saturation value, as $s \rightarrow \infty$, and is approximately v_{max}/K_m for $K_m \gg s$, i.e. for relatively very small s – the linear range. The linear range $K_m \gg s$ is thus favored by QSSA condition $K_m + s_0 >> e_0$. The nonlinearity prevails between the half-max value of v_{max} and somewhat beyond.

M–M QSSA Equation Representations, Transformations & Quantification

The four original M–M equations, (6.25) to (6.28), can be represented by four compartments (with one product) and, by analogy with linear system modes (Chapter 5), we could say that each compartment has one "NL mode" associated with it. This means that, in principle, there can be four distinct "dynamic signatures" in the set of state variables $s(t)$, $e(t)$, $c(t)$, $p(t)$. With linear dependency in the conservation equation (6.30) invoked, the ODE for $e(t)$ is eliminated and three distinct dynamical signatures ("NL modes") remain – for $s(t)$, $c(t)$ and $p(t)$.

28 This hyperbola is, in effect, a **dose–response plot** (v vs. s) *at any time t*, with the complexing reaction in quasi-steady state.

29 Note that it is not *necessary* for $k_{-1} >> k_2$ for this to happen, as noted in the remarks above, about the weaker requirement for QSSA validity, expressed in (6.36). The reverse can be true, or the rate constants can be comparable (Nelson and Cox 2000). The second Extension at the end of the chapter develops this further.

If the QSSA is valid, the model is further reduced to two ODEs, Eqs. (6.40), meaning two modes – for $e(t)$ and $c(t)$ – are "hidden." This means their dynamical signatures are not visible distinctly, because they are combined with others. Thus the overall M–M model can be said to be approximated by a *closed two-compartment model*, in graph form: $s(t)\bullet\xrightarrow{v}\bullet p(t)$, with NL velocity $v = v_{max}/(s(t) + K_m)$, or as:

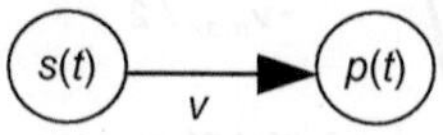

The M–M function in Eq. (6.40) also can be represented as a NL block in a control system block diagram. It mathematically transforms substrate "input" $s(t)$ into reaction velocity "output" $v(t)$, without explicitly involving the other variables in the reaction (circumvented by a semblance of "pole-zero cancellation" in a linear transfer function). In this sense it is a kind of time-domain *nonlinear transfer function*, loosely speaking – a black-box mesoscale[30] submodel for the M–M reaction:

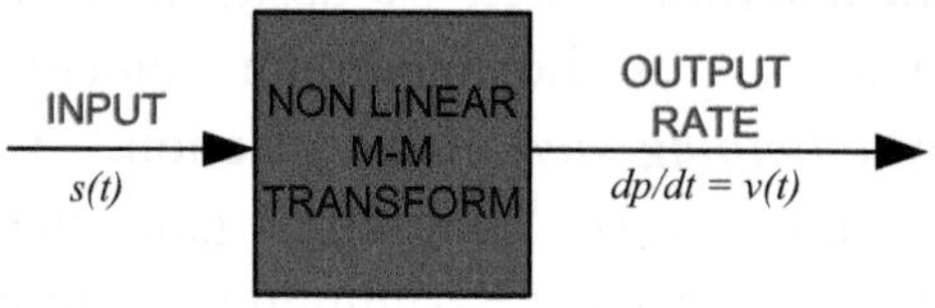

Several *linearizing transformations of the Michaelis–Menten relation* are useful for visualizing and *graphically quantifying* v_{max} and K_m. One is the **Lineweaver–Burk** plot (Lineweaver and Burk 1934), which displays $1/v$ as a linear function of $1/s$, illustrated in **Fig. 6.4**. Others include the Eadie–Hofstee, Hanes–Woolf and Eisenthal and Cornish-Bowden equations, each with advantages and disadvantages (Cornish-Bowden 1975). Computation of v_{max} and K_m by nonlinear least-squares regression (Chapter 12), directly from NL Eq. (6.40), is the preferred method if one of the many programs with functionality for facile enzyme kinetics analysis (e.g. *Curvefit* in *Graphpad*[31]) is available – or general program packages like *Matlab* or *Mathematica*.

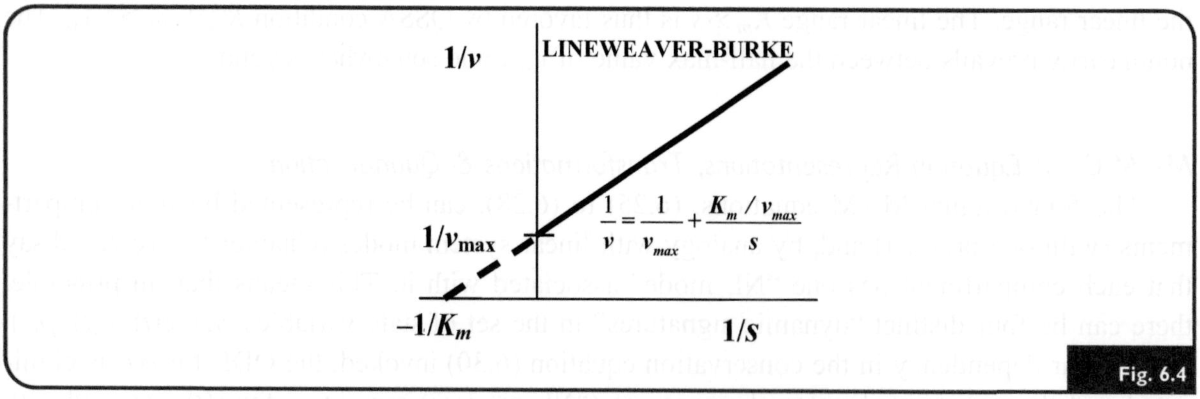

Fig. 6.4

Lineweaver–Burk plot.

30 The scale between micro- and macroscopic.

31 http://www.graphpad.com/curvefit/find_vmax_km.htm

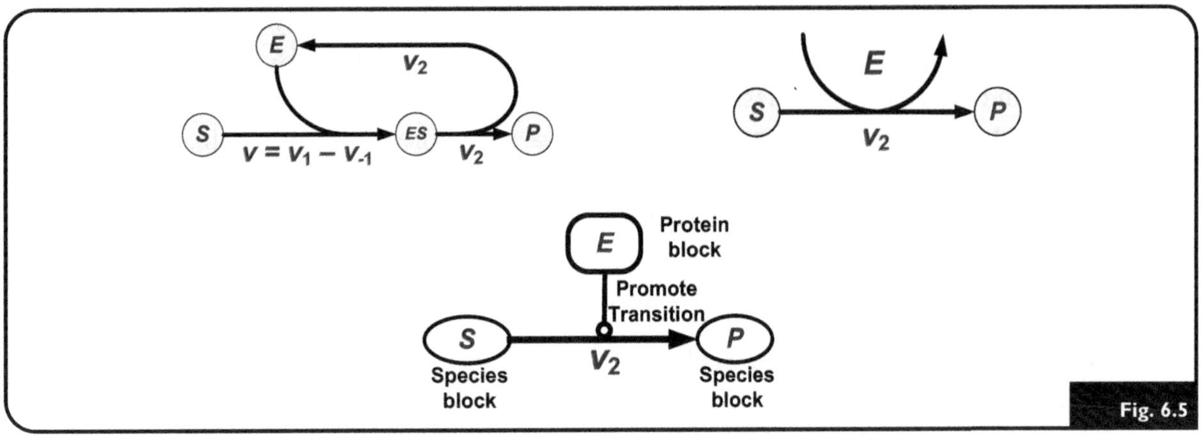

Fig. 6.5

Three example graph representations of the Michaelis–Menten reaction. Top left one is complete, the others simplified. Reaction velocities shown instead of rate constants, to illustrate that both are commonly in use. The lower graph is in Systems Biology Graphical Notation (SBGN), with block and line types noted.

M–M reaction equation (6.24) – and others like it – are also expressed graphically, in different ways by different authors, with no apparent unique convention. Two of these are illustrated in **Fig. 6.5**, with reaction velocities v shown instead of rate constants k. In the top leftmost, more complete one, summation of reactants is expressed by the enzyme E arc converging (summing) with substrate reactant S, to form complex ES at net rate v. ES then becomes product P and released-enzyme E, both at rate v_2. This type of diagram, showing all species involved in an enzymatically controlled reaction, is called **unpacked** by some authors. In the simplified (**packed**) graph on the top right, the complex ES is suppressed and the enzyme arc sums with S on the downslope and is released on the upslope, balancing the reaction represented by the graph. This alternate graph illustrates the catalytic function of the enzyme, for facilitating the production of P from substrate S. The enzymatic path might often be only implied and left out of the graph altogether (not shown). The third graph illustrates the same model in Systems Biology Graphical Notation (SBGN).[32]

Remark: The full dynamical set of M–M ODEs, Eqs. (6.25) to (6.28), are *stiff*. The QSSA, Eq. (6.40), when valid, eliminates this stiffness.

Total Quasi-Steady State Assumption (tQSSA): When the QSSA is not Valid

Several alternative approximations for M–M-like reactions are available that extend the domain of feasible parameter values considerably beyond that circumscribed by Eq. (6.36) (Bhorghans et al. 1996; Tzafriri and Edelman 2004, 2007). Each such approximation, called a total quasi-steady state assumption (tQSSA), is derived by treating substrate variables as **total intact substrate** $\hat{s} \equiv s + c$, i.e. free s plus complexed substrate c. This aggregation of substrate variables yields new equations and parameter constraints that produce approximations more robust and accurate than the QSSA. This is particularly important for cases when there is excess enzyme and K_m is not big enough to compensate, so that QSSA condition $K_m + s_0 >> e_0$ is *not* satisfied.

32 The *SBGN* group (www.sbgn.org/) is attempting to standardize nomenclature used in systems biology.

Recall that when the QSSA is valid, $s_0 + K_m \gg e_0$ and there must be only negligible loss of substrate during the initial transient period, so that $s(t_C) \approx s_0$, When the QSSA condition is *not* met, $s_0 \leq e_0 - K_m$ and s disappears too fast, which means the fast time scale "t_C" needs to be shorter. In contrast, the **total aggregate substrate** variable, $\hat{s} \equiv s + c$ (free + bound) unlike s, cannot be depleted as complex c is formed, i.e. it cannot disappear as fast as s alone, i.e. $\varepsilon_{tot} \equiv |\Delta\hat{s}/\hat{s}_0| < |\Delta s/s_0| = \varepsilon$, so it's a good candidate for computing new quasi-steady state conditions. The following tQSSA conditions were developed in (Bhorghans et al. 1996), with results and some extensions summarized below. Detailed derivations can be found in the references.

New **(total) fast and slow time scales** (t_C^{total} and t_S^{total}) for examining the validity of the tQSSA, are:

$$t_C^{total} = \frac{1}{k_1(e_0 + s_0 + K_m)} \quad \text{and} \quad t_{\hat{S}}^{total} = \frac{e_0 + s_0 + K_m}{k_2 e_0} \tag{6.41}$$

Both time scales are for the *aggregated* substrate variable, $\hat{s} \equiv s + c = s_0 - p$. They differ from the time scales for the QSSA in (6.33) by addition of e_0 to $s_0 + K_m$ in both equations in (6.41). A *necessary condition for validity of the tQSSA* is $t_C^{total} \ll t_{\hat{S}}^{total}$. By rearrangement of (6.41), this gives:

$$k_2 e_0 \ll k_1 (e_0 + s_0 + K_m)^2 \tag{6.42}$$

It is useful to rewrite (6.42) in an equivalent form:

$$\left(1 + \frac{e_0 + s_0}{k_2/k_1} + \frac{k_{-1}}{k_2}\right)\left(1 + \frac{s_0 + K_m}{e_0}\right) \gg 1 \tag{6.43}$$

This makes it clear that inequality (6.42) is valid if any of the following three inequalities are satisfied:

$$e_0 + s_0 \gg k_2/k_1, \quad k_{-1} \gg k_2, \quad s_0 + K_m \gg e_0 \tag{6.44}$$

The first inequality in (6.44) is favored by slow product formation (k_2) relative to complexing (k_1). The last inequality, $s_0 + K_m \gg e_0$, is precisely the condition for validity of the standard QSSA. This means if the QSSA is valid then so is the tQSSA. The second inequality, $k_{-1} >> k_2$, is particularly important – first, because it is a sufficient condition for the validity of (6.43), and second, because it is intuition-building. Kinetically speaking, it says that the tQSSA is valid if decomplexing is much faster than productive complexing. Tzafriri and Edelman also have shown that the tQSSA can be a reasonable approximation even for an *irreversible* forward first-reaction, i.e. $k_{-1} = 0$ (Tzafriri 2003; Tzafriri and Edelman 2004).

Application of the tQSSA conditions is similar to the standard QSSA, i.e. $t_C^{total} \ll t_{\hat{S}}^{total}$ yields the approximation: $\dot{c}(t) \cong 0$ for all $t > 0$. For standard M–M reactions of the form $S + E \leftrightarrow C \rightarrow E + P$, the simplest form of the tQSSA is an approximate first-order solution for the complex (Bhorghans et al. 1996):

$$c_{tQSSA}(t) = \frac{e_0 \hat{s}_{tQSSA}(t)}{e_0 + K_m + \hat{s}_{tQSSA}(t)} \tag{6.45}$$

The **first-order tQSSA ODEs** for $p_{tQSSA}(t)$ and $\hat{s}_{tQSSA}(t)$, based on using this linearized **Padé approximant**, are:

$$\dot{p}_{tQSSA}(t) = k_2 c_{tQSSA}(t) = \frac{v_{max}\hat{s}_{tQSSA}(t)}{e_0 + K_m + \hat{s}_{tQSSA}(t)} = -\dot{\hat{s}}_{tQSSA}(t)$$
$$\text{with } \hat{s}_{tQSSA} \equiv s_{tQSSA} + c_{tQSSA} \text{ and } v_{max} \equiv k_2 e_0 \tag{6.46}$$

Remark: As with Eqs. (6.40) for the QSSA, we use the "tQSSA" subscript in Eqs. (6.45) and (6.46) to emphasize the approximate nature of this set of equations.

Remark: The validity domain of the first-order approximation of the tQSSA, given by (6.44), is significantly larger than that of the QSSA.

The simple substitution $\hat{s} \equiv s + c$ and the tQSSA lay a foundation for assessing validity of QSSA-like assumptions for a much larger class of enzyme, protein and other substrate interactions in molecular and cell biology (Ciliberto et al. 2007), and also in predator−prey interaction modeling (Bhorghans et al., 1996). This is particularly important because modelers sometimes avoid the hard work of separating timescales and simply use the QSSA to model enzyme-catalyzed reactions in intracellular dynamical systems. *For protein interaction networks* (PINs), for example, the QSSA is often unjustified because many enzymes have multiple substrates, or substrates are acted upon by multiple enzymes, or enzymes and substrates swap roles, like two kinases phosphorylating each other. Application of the tQSSA in cell biology is introduced in the next chapter.

Alternative simplifications and more complex and complete solutions for the tQSSA rate law and associated validity domains, including systems with $e_0 >> s_0$ and competitive enzyme reactions, are developed in (Tzafriri and Edelman 2004, 2007; Pedersen et al. 2008; Pedersen and Bersani 2010).[33]

Summary Equations & Conditions for the QSSA and First-Order tQSSA

$$\text{QSSA ODEs:} \quad \dot{p}(t) \cong \frac{v_{max}s(t)}{K_m + s(t)} \cong -\dot{s}(t) \equiv v(t) \quad \text{for all } t > 0 \tag{6.47}$$

Validity: $e_0 \ll K_m + s_0$

$$\text{total tQSSA ODEs:} \quad \dot{p}(t) \cong \frac{v_{max}\,\hat{s}(t)}{e_0 + K_m + \hat{s}(t)} \cong -\dot{\hat{s}}(t)$$
$$\text{with } \hat{s} \equiv s + c \text{ and for all } t > 0 \tag{6.48}$$

Validity: $e_0 + s_0 \gg k_2/k_1$ or $k_{-1} \gg k_2$ or $s_0 + K_m \gg e_0$

33 Further refinements of tQSSA methodology are likely in the near future, so stay tuned to the literature on the subject.

Remark: For simplicity, we've dropped the "QSSA" and "tQSSA" subscripts in (6.47) and (6.48), using "$\cong$" signs to emphasize the approximate nature of these relationships. These subscripts also are not used below, unless they convey a particular point about their approximate nature – as in the next example.

Example 6.6: ***QSSA and tQSSA M–M Approximations Tested Analytically & by Simulation***

We examine three M–M models, each with different rate constants, and all for which the "classical" QSSA criterion $s_0 >> e_0$ is not met, because $s_0 = e_0 = 1$. This provides an opportunity to demonstrate the relative errors of the QSSA and the efficacy of the first-order tQSSA for "difficult" M–M-like models. Exact M–M Eqs. (6.25)–(6.28), QSSA Eqs. (6.47) and tQSSA Eqs. (6.48) are implemented for simulation and analysis, following analytical exploration of the conditions for meeting QSSA and tQSSA conditions.

Assume only one product $P \equiv P_1$ is formed ($P_2 = 0$) and ICs from Eq. (6.29) are $[1\ 1\ 0\ 0]^T$. Parameters $k_1 = 10/\text{sec} \times nM$ and $k_2 = 0.02/\text{sec}$ are the same for all three models. Only k_{-1} differs among the models, the second two models with 1 and then 2 orders of magnitude greater decomplexing rates $k_{-1} = \{2, 20, 200\}/\text{sec}$. These M–M equation parameters yield $K_m = \{0.202, 2.002, 20.002\}\ nM$. QSSA time scale parameters: $t_c = \{0.083, 0.0333, 0.0048\}$ sec and $t_s = \sim\{1, 2.3, 18\}\ min$ are from (6.33). Clearly, $t_s >> t_c$ for all values of k_{-1}. However, the necessary condition for the QSSA, namely $e_0 = 1 << K_m + s_0 = K_m + 1$, from Eq. (6.36), is not met well for $k_{-1} = 2$ and $k_{-1} = 20$. How much does this degrade QSSA responses?

The validity of the QSSA approximations at early times depends on the fractional error estimate for $s_{QSSA}(t)$ at $t = t_C$. This initial s_0 error estimate is: $|\Delta s/s_0| \cong \varepsilon = e_0/(K_m + s_0) = \{0.83,\ 0.33\ 0.048\}$ for the three models. Thus, for $k_{-1} = 2$ and 20, $\varepsilon = 0.33$ (33% initial s_0 error) and 0.83 (83% initial s_0 error) suggest poor compliance with the QSSA, *at least at early times*. For $k_{-1} = 200$, $\varepsilon = 0.048$ ($\sim$5% initial s_0 error), suggesting good compliance with the QSSA, also at least at early times.

For the tQSSA, t_C^{total} values from (6.41) are even smaller and $t_{\hat{S}}^{total} = \{1.83, 3.3, 18.3\}$ min. Again, $t_C^{total} \ll t_S^{total}$ for all values of k_{-1}, even more so than for the QSSA. The tQSSA also does a better job at early times, for all three models, because the total substrate variable $\hat{s} \equiv s + c$, unlike s, cannot disappear as fast as s alone: it cannot be depleted as complex c is formed, so $\varepsilon_{tot} \equiv |\Delta\hat{s}/\hat{s}_0| < |\Delta s/s_0| = \varepsilon$. More explicitly, from (6.48), $k_{-1} = \{2, 20, 200\} >> k_2 \equiv 0.02$ for all three models, so the tQSSA should yield good approximations for all three. Let's see how well these analytical results are borne out by simulations, which visually portray the precision of the QSSA and tQSSA approximations.

Exact model, QSSA and tQSSA model solutions are shown in **Fig. 6.6**. Seconds are converted to minutes, to visualize these intervals on the graphs, plotted for times beyond the slow time scale, so that final steady state concentrations are visualized. Exact solutions for the four species $s(t)$, $e(t)$, $c(t)$ and $p(t)$ are shown superimposed with QSSA solutions

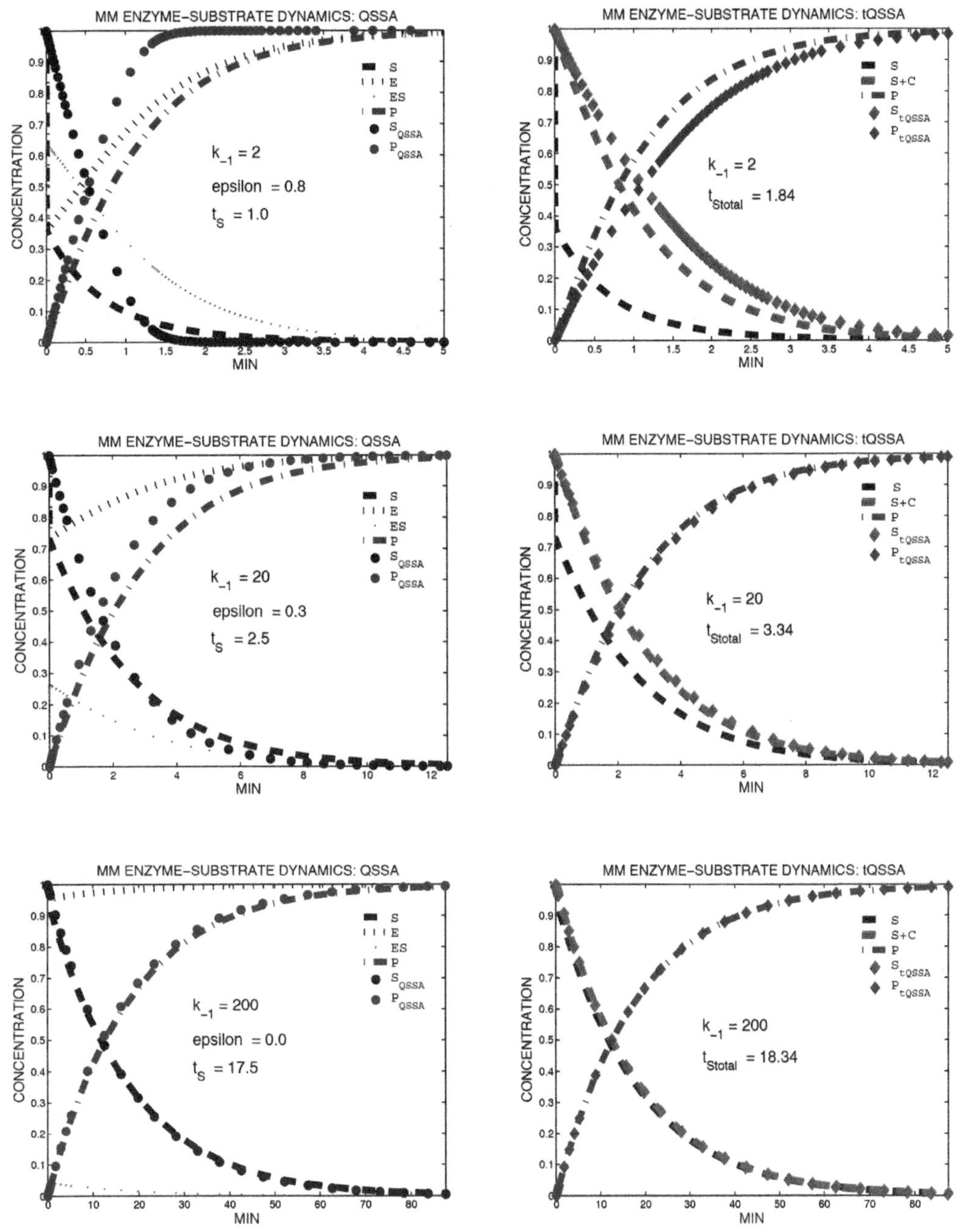

Fig. 6.6

M–M exact, QSSA (LHS) and first-order tQSSA (RHS) dynamic responses for three models, all with $k_1 = 10$/sec × nM and $k_2 = 0.02$/sec, but three different k_{-1} values 2, 20 and 200/sec. ICs are $s_0 = e_0 = 1$, $c_0 = p_0 = 0$. The QSSA is a poor approximation when the initial errors ε for $s(t)$ at $t = t_C$ are large (LHS top & middle). The tQSSA solutions are significantly better for all three k_{-1} values, but – for $k_{-1} = 2$ (RHS top) – still have ~10% errors in product formation and total substrate utilization over times up to ~$2t_S$ min, with errors falling to negligible values with increasing time. Note that $p_{tQSSA}(t)$ tracks $p(t)$ and $s_{tQSSA}(t)$ tracks $s(t) + c(t)$, but the latter overlap when the tQSSA approximation is very good (LHS bottom).

$s_{QSSA}(t)$ and $p_{QSSA}(t)$ on the LHS graphs. Total tQSSA solutions for $s(t)$, $\hat{s}(t) \equiv s(t) + c(t), \hat{s}_{tQSSA}(t)$, $p(t)$ and $p_{tQSSA}(t)$ are superimposed on the RHS graphs.

All models go through the fast time scales (0 to t_C) relatively instantaneously on the scales of the graphs. As anticipated analytically, the solutions for $k_{-1} = 200$ (bottom graphs) support validity of both the QSSA and the tQSSA. On the LHS, the exact and QSSA solutions for product $p(t)$ and substrate $s(t)$ are approximately superimposed on the graphs (asterisks and dashed lines); and the tQSSA solutions on the RHS are even more accurate for $k_{-1} = 200$.

In contrast, for $k_{-1} = 2$ and 20, the QSSA solutions start too low – as shown in the top and middle graphs on the LHS – as predicted by the 33% and 83% initial errors ε for $s(t)$ at $t = t_C$, i.e. $s(t_C) \approx 0.33s_0$ and $0.83s_0$, respectively. And they continue with relatively large $\pm$ residual errors when compared with exact solutions for product and substrate, well beyond the slow time scales, asymptotically approaching their final steady state solutions with diminishing errors – as anticipated.

For $k_{-1} = 2$ and 20, the tQSSA solutions on the RHS are nearly superimposed for $k_{-1} = 20$, but not nearly as well for $k_{-1} = 2$. The first-order tQSSA solutions for both product and total substrate are $\sim$10% higher than the exact $p(t)$ and $\hat{s}(t)$ over the intermediate time scale for $k_{-1} = 2$. This top-RHS graph also reinforces the fact that first-order tQSSA solutions – which approximate $p(t)$ and total substrate $\hat{s}(t) \equiv s(t) + c(t)$ generally do not reflect the original system in the same way as do the QSSA solutions, which approximate $p(t)$ and $s(t)$. But they do coincide when the approximation is very good, as in the $k_{-1} = 200$ solutions (bottom graphs).

Note: Greater accuracy of the QSSA and first-order tQSSA would be achieved in this example if initial substrate were greater than initial enzyme, rather than $s_0 = e_0$. Simulations (not shown) of the model with $k_{-1} = 2$, $e_0 = 1$ and $s_0 = 10$ (instead of 1) or more generate superimposed QSSA and exact solutions, with tQSSA *nearly* superimposed. This anomaly is likely due to the more approximate nature of the first-order tQSSA equations (6.48). The primary code for the *Matlab* script files for all simulations is shown in **Fig. 6.7**.

Enzymatic Regulation of Biochemical Pathways

Some broad generalizations can be made about enzymatic reactions. First, virtually every biochemical reaction in the transformation of one compound or molecular state into another is catalyzed by an enzyme. Second, complicated changes take place by a sequence of simple reactions, each catalyzed by a separate and specific enzyme, with a small change in substrate. A single enzyme, however, can catalyze more than one reaction.[34]

Remark: Pathways, networks and enzyme-controlled reactions are graphed (schematized) in a variety of ways. Those in **Fig. 6.8** are simple examples of variations in biochemical pathways. With the exception of the example coenzyme template, they should be fairly easy to comprehend without further explanation. All are shown with irreversible paths between nodes, but any could be reversible. The coenzyme graph has the arc characteristics of two of the enzyme reaction graphs in **Fig. 6.5**. Some detailed mechanisms and their

34 Analogously, a single transcription factor can regulate more than one gene, and one gene can regulate many proteins, via *alternative splicing* (Black 2003).

```
% FULL M-M ENZYME + QSSA + tQSSA
DYNAMICS
% Reactions---R1: S + E -- k1 --> C;
% R2: C--kminus1-->E + S; R3: C--k2--
>E + P
% Parameters(60 converts k's to per
min)
global K1 KMINUS1 K2 e0 s0 Km;
% Set kminus1 = 2/sec or 20 or 200/sec
K1 = 60*10; K2 = 60*0.02; KMINUS1 =
60*2;
Km = (KMINUS1+K2)/K1;
% Initial Conditions
x0 = [s0; e0; c0; p0; s0; p0; s0+c0;
p0];
s0 = 100; e0 = 1; c0 = 0; p0 = 0;
% Calculate QSSA & tQSSA metrics
tc = 1/(K1*(s0+Km));
ts = (s0+Km)/(e0*K2);
epsilon = e0/(Km+s0);
tstotal = (s0+e0+Km)/(e0*K2);
% Solve ODEs
tspan = [0 10*ts];
[t,x]=ode15s(@MM_QSSA_tQSSA, tspan,
x0);
% % Plot
fig1=figure('PaperPosition',[2 2 8
6.3]);
ax =
```

```
function [dxdt] =
MM_QSSA_tQSSA(t,x)
%rate eqns for exact M-M solver
global K1 KMINUS1 K2 Km e0;
% Complete equations
% sdot=-k1es+(kminus1)c
dxdt(1)= -K1*x(2)*x(1)
+KMINUS1*x(3);
% edot=-k1es+(kminus1+k2)c
dxdt(2)= -
K1*x(2)*x(1)+(KMINUS1+K2)*x(3);
% cdot=k1es-(kminus1+k2)c
dxdt(3)= K1*x(2)*x(1) -
(KMINUS1+K2)*x(3);
%pdot=k2c
dxdt(4)= K2*x(3);
% QSSA approximations
dxdt(5)= -K2*e0*x(5)/(x(5) +
Km);
dxdt(6)= K2*e0*x(5)/(x(5) + Km);
% tQSSA approximations
dxdt(7)= -K2*e0*x(7)/(e0 + Km +
x(7));
dxdt(8)= K2*e0*x(7)/(e0 + Km +
x(7));

dxdt=dxdt';
return
```

Fig. 6.7

Matlab script (two files) for simulating and plotting full M–M, QSSA and tQSSA equations.

math models follow. We develop several more pathways in other programming languages and other notations in the next chapter.

Enzymes & Introduction to Metabolic and Cellular Regulation

As noted earlier, enzymes also function as *cellular signals* in the regulation of metabolism and other cellular functions. Regulatory enzymes do this by participating in reactions specific to these processes, where they can be acted upon by *effectors.* **Effectors** are small proteins or other molecules that can act as *inhibitors* or *activators* of an enzyme, thereby effecting the reaction rate of a cellular process.[35] In this manner, they regulate or *tune* dynamical processes like cellular metabolism, DNA repair, etc. We focus on the basic mechanisms here, in simplified form, to illustrate the modeling process.

35 More generally, effectors also serve to regulate non-enzymatic cellular processes. Even particular cell-types can act as effectors, e.g. T-cell or B-cell regulation of immune system function.

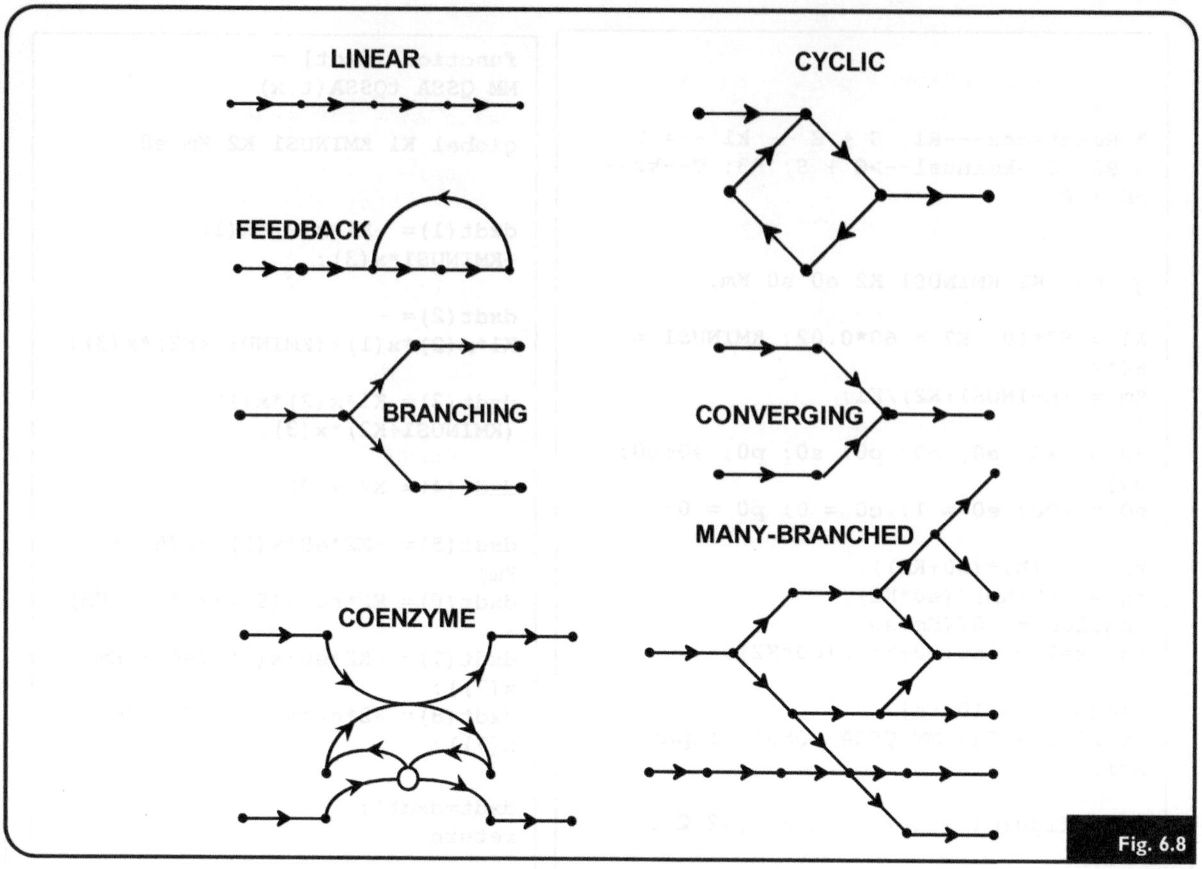

Fig. 6.8

Simple examples of variations in biochemical pathways.

Inhibition is a primary effector mechanism, with well-described schema, based on alterations of M−M reactions in several possible ways. Four of these are called *competitive*, *noncompetitive*, *uncompetitive* and *mixed inhibition*, differing by the step in the M−M reaction effected. We examine the first of these in detail here. The others are derived similarly and are left as an Exercise E6.5.[36]

A Simple Competitive Reaction: One Enzyme & Two Substrates

Consider one enzyme E and two substrates, S_1 and S_2, *both of which compete for the same, single reaction site* (**catalytic center**) on the enzyme. This is called **competitive inhibition**. Either substrate can be arbitrarily called "inhibitor" I in this simple schema, i.e. their roles are interchangeable. Two pathways are modeled:

$$\begin{aligned} &S + E \underset{k_{-1}}{\overset{k_{+1}}{\rightleftharpoons}} C_1 \xrightarrow{k_{+2}} E + P \quad \text{productive pathway} \\ &I + E \underset{k_{-3}}{\overset{k_{+3}}{\rightleftharpoons}} C_2 \quad \text{unproductive pathway} \end{aligned} \tag{6.49}$$

36 Do the exercise if you want to further develop the skills needed for translating verbal descriptions (reaction logic) into math models.

This pair of reactions also can be expressed in a more coupled manner as:

$$
\begin{gathered}
S + E \underset{k_{-1}}{\overset{k_1}{\rightleftharpoons}} C_1 \xrightarrow{k_2} E + P \\
+ \\
I \underset{k_{-3}}{\overset{k_3}{\rightleftharpoons}} C_2
\end{gathered} \tag{6.50}
$$

Applying the mass action principle, the kinetic ODEs corresponding to these reactions are:

$$
\begin{aligned}
&\dot{c}_1 = k_1 es - (k_{-1} + k_2)c_1 && \dot{i} = -k_3 ei + k_{-3}c_2 \\
&\dot{c}_2 = k_3 ei - k_{-3}c_2 && \dot{p} = k_2 c_1 \\
&\dot{e} = k_1 es - k_3 ei + (k_{-1} + k_2)c_1 + k_{-3}c_2 && \dot{s} = -k_1 es + k_{-1}c_1
\end{aligned} \tag{6.51}
$$

We assume the QSSA assumption is valid for both substrates (substrate and inhibitor), i.e. $K_{ms} + s_0 \gg e_0$ and $K_{mi} + i_0 \gg e_0$. If it is not, then the tQSSA would be applicable, and the equations below would involve $\hat{s} = s + e$ and $\hat{i} = i + e$ instead of s and i, as in Eq. (6.46).

To solve these ODEs, we simplify them first as follows, using conservation and QSSA constraints:

1. Recognize the conservation law $e_0 = e(t) + c_1(t) + c_2(t)$.
2. Eliminate $e(t)$ from the ODEs using this law.
3. Compute the QSSAs by setting $\dot{c}_1 = \dot{c}_2 \cong 0$.
4. Solve the resulting equations for approximate reaction velocities $v \equiv -\dot{s} = \dot{p}$ (or its inverse) as a function of s, i, and e_0.

This gives

$$
v = -\dot{s} = \dot{p} \simeq \frac{v_{max\,s}\, sK_{mi}/K_{ms}}{i + K_{mi} + sK_{mi}/K_{ms}} = \frac{v_{max\,s}\, s}{s + K_{ms}i} \equiv \frac{v_{max\,s}\, s}{s + \alpha(i)K_{ms}} \tag{6.52}
$$

where the subscripts s and i denote substrate and inhibitor, and $\alpha(i) \equiv \left(1 + \dfrac{i}{K_{mi}}\right)$ generates an **apparent Michaelis constant**, due to presence of inhibitor in the denominator term of v. In this form, we see how i *inhibits* the reaction: an increase in i decreases v, the forward rate of the rate limiting reaction, a kind of nonlinear feedback effect. It also illustrates how the value of K_{mi} influences the inhibitory term and thus v.[37] Equivalently,

37 A real-world example of the use of Eq. (6.52) in a physiologically based pharmacokinetic model is given in Example 8.1 of Chapter 8. That example involves competitive inhibition of iodine uptake into the thyroid gland by perchlorate contamination of drinking water. Iodine is needed to make thyroid hormone in the gland.

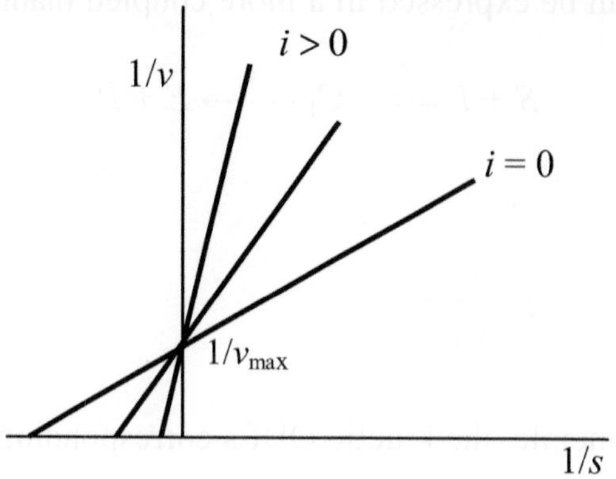

Fig 6.9

Lineweaver–Burk plot using inverse form.

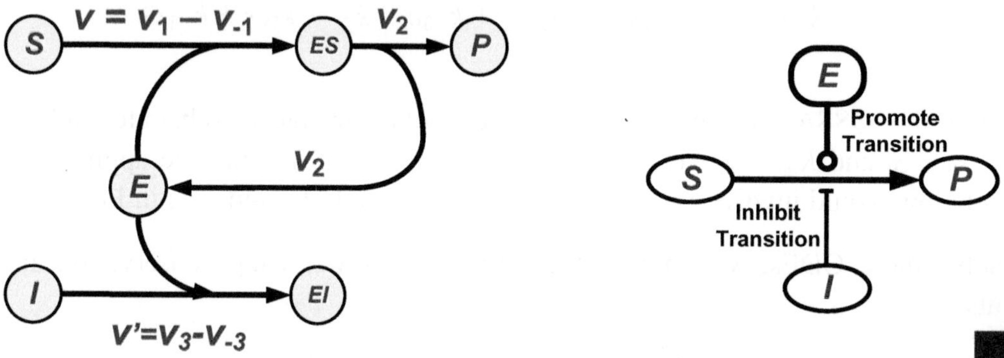

Fig. 6.10

Two reaction graph forms for competitive inhibition.

$$\frac{1}{\dot{s}} = \frac{1}{v} = \frac{1}{v_{maxs}}\left[1 + \frac{K_{ms}}{s}\left(1 + \frac{i}{K_{mi}}\right)\right] \tag{6.53}$$

Using this inverse form, we note that Lineweaver–Burk plots of v^{-1} versus s^{-1} are again straight lines, with slopes depending on the particular value of i (**Fig. 6.9**). Importantly, *fully competitive reactions are recognized by such experimentally determined curves*. They all cross at the same point, $1/v_{maxs}$, characteristic of this type of inhibition, another important result about mechanism for your "modeler's toolbox."

Remark: Models for uncompetitive, noncompetitive, mixed and partial inhibition also produce reaction rate equations with inhibitor i in their denominators – in different quantitative configurations – generating different degrees of NL inhibition. Model details and results can be found in standard references (Cleland 1963; Voit 2000; etc.) or derived as an exercise – as noted earlier (Exercise E6.5).

The "unpacked" competitive inhibition reaction depicted in Eq. (6.50) is illustrated graphically in **Fig. 6.10**, using the same notation as the M–M reaction in **Fig. 6.5**, and also with reaction

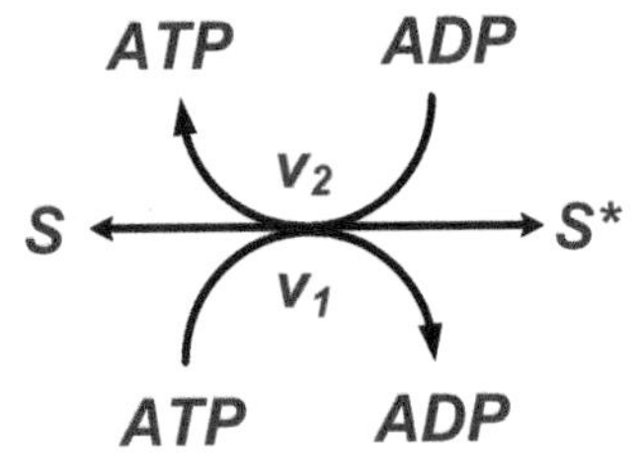

Fig. 6.11

Reversible release of phosphate from ATP (active form) – to provide energy in a reaction, leaving ADP (the inactive form) behind – ready to be reactivated to ATP.

rates v instead of rate constants k shown, again for illustration. The simplified graph on the right is in Systems Biology Graphical Notation (SBGN), with the reaction inhibitor shown with the horizontal line end and the enzyme line with the circle end shown promoting the reaction.

Local Regulation by Enzymatic Activation–Deactivation

An important set of elementary enzyme–substrate reactions regulate *activation* and *deactivation* of receptors and other proteins. **Activation** means conversion of an inactive form to one with metabolic activity, and *vice versa* for deactivation. **Phosphorylation** and **dephosphorylation** (adding and subtracting a phosphate group) are major activation–deactivation regulatory subprocesses in cell signaling transduction pathways. **Protein kinases** (abbrev: **kinases**) are enzymes for substrate activation by phosphorylation. **Phosphatases** remove a phosphate group from its parent molecule. An important example is the reversible *release of phosphate from adenosine triphosphate (ATP)* (active form) – to provide energy in a reaction, leaving adenosine diphosphate (ADP) (the inactive form) behind – ready to be reactivated to ATP, as illustrated in **Fig. 6.11**.

The basic scheme is:

$$\begin{aligned} &Activation: S + E_1 \rightleftharpoons C_1 \longrightarrow S^* + E_1 \\ &Deactivation: S^* + E_2 \rightleftharpoons C_2 \longrightarrow S + E_2 \end{aligned} \tag{6.54}$$

The asterisk denotes the activated form. The following generalized scheme is useful as a **subprocess model** in larger pathways and network models, where E_{j1} is a kinase and E_{j2} is a phosphatase:

$$\begin{aligned} &j^{\text{th}}\text{-activation: } S_j + E_{j1} \underset{d_{j1}}{\overset{a_{j1}}{\rightleftharpoons}} C_{j1} \xrightarrow{k_{j1}} S_j^* + E_{j1} \\ &j^{\text{th}}\text{-deactivation: } S_j^* + E_{j2} \underset{d_{j2}}{\overset{a_{j2}}{\rightleftharpoons}} C_{j2} \xrightarrow{k_{j2}} S_j + E_{j2} \end{aligned} \tag{6.55}$$

Invoking the QSSA, i.e. assuming $e_{j0} << s_{j0} + K_{jm}$ for both reactions, we can write the approximate reaction velocities (from Eq. (6.40)) as:

$$v_{phos} \equiv v_{j1} \cong \frac{k_{j1}e_{j10}s_j}{K_{j1} + s_j} \quad v_{dephos} \equiv v_{j2} \cong \frac{k_{j2}e_{j20}s_j}{K_{j2} + s_j} \tag{6.56}$$

In (6.56), the K_{ji} are Michaelis constants, the k_{ji} are the v_{max} values, and e_{j10}, e_{j20} are initial values of e_{j1}, e_{j2}. The NL rate equation for this reaction, with other possible coupled reactions included, is:

$$\dot{s}_j^* = v_{j1} - v_{j2} + \text{terms coupling other interactions with this one} \tag{6.57}$$

For any substrate–enzyme pair with comparable concentrations, i.e. with $s_{j0} \leq e_{j0} - K_{jm}$, the tQSSA is a better approximation and (6.56) becomes:

$$v_{phos} \equiv v_{j1} \cong \frac{k_{j1} e_{j10} \hat{s}_j}{K_{j1} + e_{j10} + \hat{s}_j} \qquad v_{dephos} \equiv v_{j2} \cong \frac{k_{j2} e_{j20} \hat{s}_j}{K_{j2} + e_{j20} + \hat{s}_j} \tag{6.58}$$

with $\hat{s}_j \equiv s_j + c_j$ and (6.57) becomes $\dot{\hat{s}}_j^* = v_{j1} - v_{j2} + \ldots$.

*More Complex Regulation: Cooperativity and Allosterism[38]

Regulatory enzymes involved in the control of metabolism typically respond with extraordinary sensitivity to changes in metabolite concentrations. Most of these responses do not obey simple M–M kinetics, but are regulated more finely by *effector* molecules. We examined this in the previous section, where an inhibitor affected reaction rate, by competing for the same (monomeric) site as the substrate (ligand). More generally, regulation of cellular processes is governed by *cooperative* or *allosteric* mechanisms (or both), in enzymes or proteins with *multiple* (**multimeric**) binding sites for substrates or ligands, and possible interactions among these sites (Lescovac 2003). These kinds of interactions are facilitated by the flexible macromolecular 3-D structure of enzymes and receptors, as illustrated in **Fig. 6.1** which permit their participation in a variety of regulatory activities, utilizing different kinds of chemical bonds in various conformational arrangements, as illustrated in the figure.[39]

Cooperative effects (cooperativity) among different binding sites on the same molecule (dimers, tetramers, etc.) either enhance or diminish reaction rates relative to simple single-site (monomeric) M–M kinetics. For cooperativity, all of these binding sites are or become **active centers** on binding and contact between the enzyme or receptor and its substrate (ligand) is limited to the active center. When a substrate binds to an enzyme or receptor subunit, other subunits are stimulated and become active centers as well. The complex 3-D structure of these enzymes and receptors is critical in these interactions. The mathematics is flatter, in 2-D. Reaction rate v versus substrate concentration s is a **sigmoidal** curve in cooperative enzyme reactions – characterized by an inflection point (**Fig. 6.12**) – instead of being a rectangular hyperbola as would be predicted by simple Michaelis–Menten theory (another mechanistic item for your modeler's toolbox).

38 This section is undoubtedly more complex both biochemically and mathematically. It might be skimmed or skipped over on first reading, returning to it when the need arises for understanding the nuances of cellular regulation processes in greater detail.

39 Protein folding can be quite complex in three dimensions. Only primary (linear amino acid sequence) and simplified tertiary (3-D) structure are illustrated in **Fig. 6.1**.

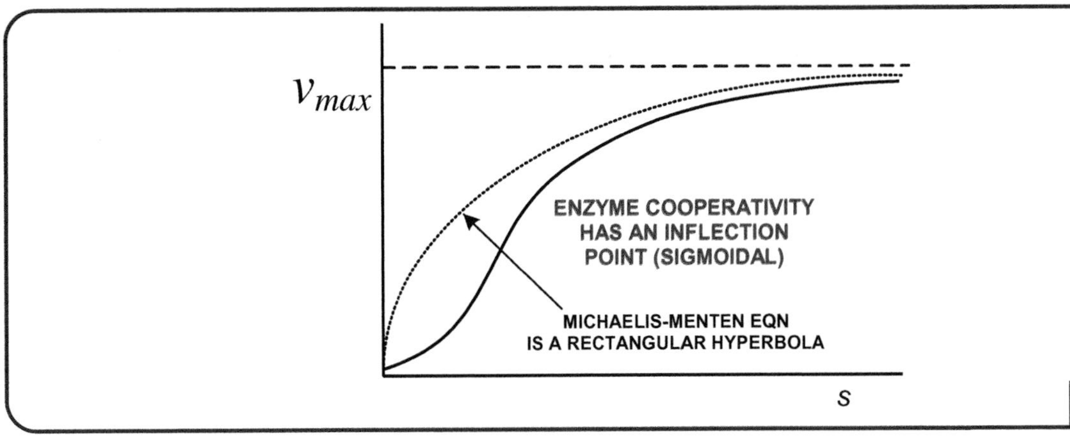

Fig. 6.12

Reaction rate v vs. substrate concentration s is a sigmoidal curve in cooperative enzyme reactions – characterized by an inflection point.

One possible explanation is that a single enzyme molecule, after having bound a substrate molecule at one active site, then binds another at another site. This is called a **cooperative system**, and this kind of enzyme reaction is called a **cooperative phenomenon**. Following is the theory of such a system for the simple case of an enzyme containing two active sites (Rubinow 1973). The results are qualitatively similar for any number of active sites.

Remark: The phenomena under discussion here are more ubiquitous in systems biology, i.e. they are not confined to just enzymatically controlled reactions. For example, "enzyme" in this development can be replaced by a "DNA" (nonprotein nucleotide) segment with several sites ("active centers") where protein effector molecules can bind and form a cooperative system involved in gene regulation.

For illustrating the modeling, we consider **positive homotropic cooperativity** here, where two bound substrate molecules, denoted *S*, *enhance* the rate of reaction. The enzyme has three possible states: (i) it is a free molecule *E*; (ii) one center is free and the other is occupied (bound, complexed) by a substrate molecule (C_1); (iii) both centers are occupied by substrate molecules (C_2). Schematically, this is represented by:

$$\begin{aligned} S + E &\underset{k_{-1}}{\overset{k_1}{\rightleftharpoons}} C_1 \xrightarrow{k_2} E + P \\ S + C_1 &\underset{k_{-3}}{\overset{k_3}{\rightleftharpoons}} C_2 \xrightarrow{k_4} C_1 + P \end{aligned} \tag{6.59}$$

where *P* is a product. The ODEs for the concentrations, denoted e, c_1 and c_2, are:

$$\dot{e} = -k_1 se + (k_{-1} + k_2)c_1 \tag{6.60}$$

$$\dot{c}_1 = k_1 se - (k_{-1} + k_2 + k_3 s)c_1 + (k_{-3} + k_4)c_2, \tag{6.61}$$

$$\dot{c}_2 = k_3 sc_1 - (k_{-3} + k_4)c_2, \tag{6.62}$$

In addition, conservation of enzyme molecules requires: $e_0 = e + c_1 + c_2$, with $e_0 = e(0)$ the initial enzyme concentration. Assuming QSSA conditions, i.e. $e_0 << s_0 + K_m$,[40] we set $\dot{c}_1 = \dot{c}_2 \cong 0$ in Eqs. (6.61) and (6.62) and solve the resulting algebraic equations, using constraint $e_0 = e + c_1 + c_2$, for the unknowns c_1 and c_2 in terms of s and e_0. The initial reaction velocity v from the ODE for p is then:

$$v = \dot{p} = k_2 c_1 + k_4 c_2 \tag{6.63}$$

Algebraic substitution of the solutions for c_1 and c_2 then yields:

$$v \equiv v(s) \cong \frac{s(v_{max} K_m + v'_{max} s)}{K_m K'_m + K'_m s + s_2} \tag{6.64}$$

a ratio of second-order *polynomials*, with coefficients

$$K_m = \frac{k_{-1} + k_2}{k_1}, \quad K'_m = \frac{k_{-3} + k_4}{k_3}, \quad v_{max} = k_2 e_0, \quad \text{and} \quad v'_{max} = k_4 e_0 \tag{6.65}$$

It can be shown that $v(s)$ has an inflection point[41] for particular values of v'_{max}/v_{max} and K'_m/K_m, i.e. a graph of v vs. s is sigmoidal rather than hyperbolic (**Fig. 6.12**). *The existence of such inflection points, observable in data, suggests that a cooperative phenomenon is at work, with the enzyme containing two or more interacting active centers.*

Remark: Eq. (6.64) would be third-order in s in the numerator and denominator if there were three binding sites for the substrate. In general, cooperative reaction velocities are nth-order polynomial quotients, with order number equal to the number of binding sites.

Remark: Cooperativity can be negative as well as positive, where binding of one ligand decreases the affinity of the enzyme (or other protein) to binding another ligand – where the distinction is made in comparison with simpler M–M kinetics, with only one binding site. If substrate binding is greater (less) than expected with the M–M model, it is positive (negative) cooperativity. These are further classified as **homotropic** and **heterotropic cooperativity**, where binding to one ligand affects the affinity of the enzyme or protein to accepting another ligand of the same or another type.

Example 6.7: ***Complete Positive Homomorphic Cooperativity Gives the Hill Function***

Consider the protein P with n identical binding sites, where binding of the first ligand substrate greatly facilitates binding of all the others. Assume this happens quickly and completely, so all the proteins in the medium are (approximately) either fully bound or empty. Then the first and then all other intermediate reactions in Eq. (6.59) (assuming n instead of two reactions) are rapidly bypassed, so we have protein P

40 Again, as in the previous section, the tQSSA gives a better approximation if enzyme–substrate pairs have comparable concentrations.

41 Second-derivative is zero at the inflection point and changes sign as it passes through it.

binding with n ligand substrate sites directly, i.e. $S+S+S+\ldots+P\underset{k_{-3}}{\overset{k_3}{\rightleftharpoons}}C_n$.
The ODE for c_n is: $\dot{c}_n = k_3ps^n - k_{-3}c_n$. In quasi-steady state $\dot{c}_n \cong 0$ and solving for c_n gives the concentration of bound substrate: $c_n = pK_Bs^n = p_0K_Bs^n/(1+K_Bs^n)$, where $K_B \equiv k_3/k_{-3}$ and we have eliminated p using the conservation equation $p_0 \equiv p + c_n$. The bound fraction as a function of substrate concentration s is then $F \equiv c_n/p_0 = K_Bs^n/(1+K_Bs^n)$ and we recognize this fraction also as the forward reaction rate normalized by v_{max}. Therefore,

$$\dot{p} = v \equiv v_{max}\,F = \frac{v_{max}K_Bs^n}{1+K_Bs^n} = \frac{v_{max}s^n}{K_B^{-1}+s^n} \tag{6.66}$$

This is the general (activating-stimulatory-promoting) nth-order **Hill function**. As n increases, this sigmoidal function becomes increasingly steep. We elaborate on Hill functions and their properties in the next chapter.

*Allosterism Modeling

An **allosteric effect (allosterism)** exists when binding of a molecule at one enzyme or protein site, an **allosteric site**, affects activity at a *different* site, sites other than active centers – the mechanism for cooperativity.[42] Thus the name **allosteric enzyme**[43] **or protein**. As with cooperativity, this is due to a 3-D conformational change in the protein (as in **Fig. 6.1**): a change in the relative position of the atomic nuclei making up the protein, but no change in the nature of their bonds. As noted earlier, a **ligand** is a small molecule bound to a receptor protein or enzyme. An **effector** or **modifier** ligand affects the activity of the receptor or enzyme, but is not itself affected. The following quantitative theory of allosteric enzymes and, more generally, allosteric proteins (or nucleotides), is due to Monod and coworkers (Monod et al. 1965).

An allosteric protein is assumed to be an **oligomer** made up of n identical subunit **protomers**, each with only one active site able to combine with a given ligand, as shown in **Fig. 6.13**. The active sites are assumed to combine independently with the ligand molecules.[44] However, the oligomer is assumed to exist in two active states, called **tensed** and **relaxed**, which differ in their affinity for the substrates – the relaxed ones binding more readily. The ability of the protein to exist in these two active states is the essential allosteric effect built into the theory. The transition from one state to another presumably is due to a conformational change in the protein, possibly requiring a modifier (effector) molecule to affect it, but the mathematics is the same, independent of such inferences (Monod et al. 1965).

The two states of the free (unoccupied) protein are denoted P_0^* and P_0, and the ligand molecule is S. The protein can exist in either of two conformational states with i ligand

42 This is how these subtly distinct mechanisms more finely control enzyme or protein reaction rates.

43 The term allosteric enzyme is used ambiguously in the literature, sometimes designating an enzyme with more than one active site (e.g. cooperative), other times an enzyme displaying a sigmoidal curve for $v(s)$.

44 In this sense, the protomer is not allosteric.

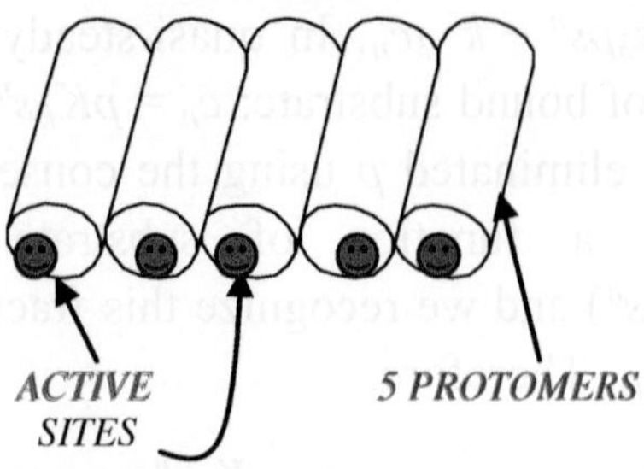

Fig. 6.13

An allosteric protein is assumed to be an oligomer made up of *n* identical subunit protomers, each with only one active site able to combine with a given ligand.

molecules attached, denoted P_i^* – the **tensed state**, or P_i – the **relaxed state**.[45] Assume n available sites, $i = 1, 2, \ldots, n$. Also, the protein complexes are assumed to be in equilibrium with a large number of ligand molecules S, so that a quasi-steady state exists. We ignore the possibility of transformation of the ligand molecules into product molecules. Further, the states P_0^* and P_0 are assumed to interconvert in a quasi-steady state manner, and the equilibrium constant for this interconversion is denoted $L \equiv P_0/P_0^*$, the **allosteric constant**. Schematically, these processes are represented as

$$\left.\begin{aligned} P_0^* &\underset{k_{-0}}{\overset{k_0}{\rightleftharpoons}} P_0 \\ S + P_i^* &\underset{k-1}{\overset{k_1}{\rightleftharpoons}} P_{i+1}^* \\ S + P_i &\underset{k_{-2}}{\overset{k_2}{\rightleftharpoons}} P_{i+1} \end{aligned}\right\} i = 0, 1, 2, \ldots, n-1 \tag{6.67}$$

With ligand concentration s treated as fixed (constant, relatively very large), the *rate equations governing these reactions are linear, and we can use linear compartmental analysis* to solve them. The linear compartmental graph shown in **Fig. 6.14** is drawn directly from these reaction equations (Volkenshtein 1966). The states are represented by nodes, with **tensed states** in **red**, and the coefficients in the rate equations by directed line segments or branches. Assuming quasi-steady state conditions, i.e. $\dot{p}_{i+1}^* \cong 0$ and $\dot{p}_{i+1} \cong 0$, yielding $k_1 s p_i^* = k_{-1} p_{i+1}^*$ and $k_2 s p_i = k_{-2} p_{i+1}$ for all i. These algebraic equations are readily solved for the concentrations of P_i^* and P_i in terms of P_0^* and P_0 for $i = 1, 2, \ldots, n$. From these results we compute the fraction of total sites actually bound by the ligand (normalized reaction velocity), called the saturation function or **fraction bound** F_{bound}:

45 For example, in *gene regulation*, P^* might be the activated transcription factor ($P^* \equiv TF^*$) and $P = TF$ its inactive form. The ligand S then might be the small protein signal that initiates activation or suppression when it binds to the promoter site on the gene (on the DNA).

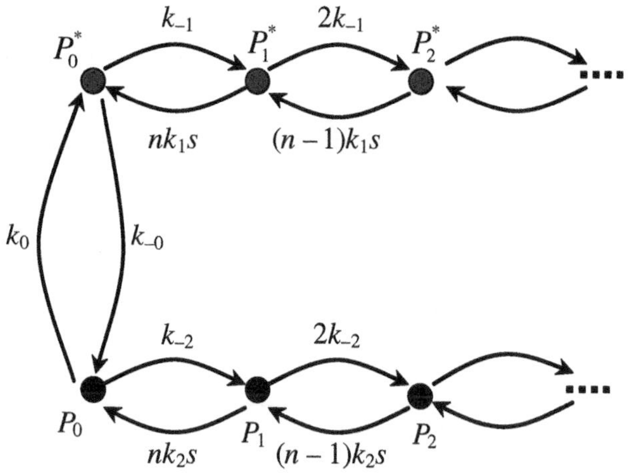

Fig. 6.14

Linear compartmental graph model for protein P combining allosterically with excess substrate S at different sites, either tensed sites shown in red, or untensed sites, in black (see text).

$$F_{bound} \equiv \frac{v}{v_{max}} = \frac{LC\alpha(1+C\alpha)^{n-1} + \alpha(1+\alpha)^{n-1}}{L(1+C\alpha)^n + (1+\alpha)^n} \tag{6.68}$$

where $\alpha = s/K_{P^*}$, $C \equiv K_{P^*}/K_P$, $K_{P^*} = k_{-1}/k_1$, $K_P = k_{-2}/k_2$ and $L = k_{-0}/k_0 = P_0/P_0^*$. We can also write F_{bound} in Eq. (6.68) in terms of the *fraction* F_{tensed} of ligand S in the P^* (tensed) state: $F_{tensed} = \sum_{i=0}^{n} p_i^* / \left(\sum_{i=0}^{n} p_i^* + \sum_{i=0}^{n} p_i\right)$. This gives:

$$F_{bound} = \frac{C\alpha(1 - F_{tensed})}{1 + C\alpha} + \frac{\alpha F_{tensed}}{1 + \alpha} \tag{6.69}$$

Several theoretical curves for the saturation function F_{bound} for various values of the **nonexclusive binding coefficient** C, with **allometric constant** $L = 1000$ and $n = 4$ (a tetrameric protein), are shown in **Fig. 6.15** (Monod et al. 1965). Note that, with $C = 0$ and $n = 4$, the sigmoidal character of the curve becomes less (more) pronounced as L becomes smaller (larger). ***Hemoglobin (Hb)***, one of the most extensively investigated proteins, is a tetramer. It displays such a sigmoidal curve when combining with oxygen, as shown for $C = 0$. In the forward direction, each succeeding O_2 molecule in contact with *Hb* in the lungs is bound more strongly – the cooperativity effect, and a rapid dissociation constant k_{-2} provides the mechanism for releasing O_2 molecules more freely in tissues.

In the allosterism model above, if *substrate can only bind to the tensed form* P^*, i.e. the relaxed conformational state binds no new substrate, then k_{-2} and therefore K_P are zero in Eq. (6.67) and (6.68) and the reaction rate v computed for this model becomes:

$$v = \left(\frac{v_{max}s}{s + 1/K_{P^*}}\right)\left(\frac{1}{1 + (L/(1+sK_{P^*})^n)}\right) \tag{6.70}$$

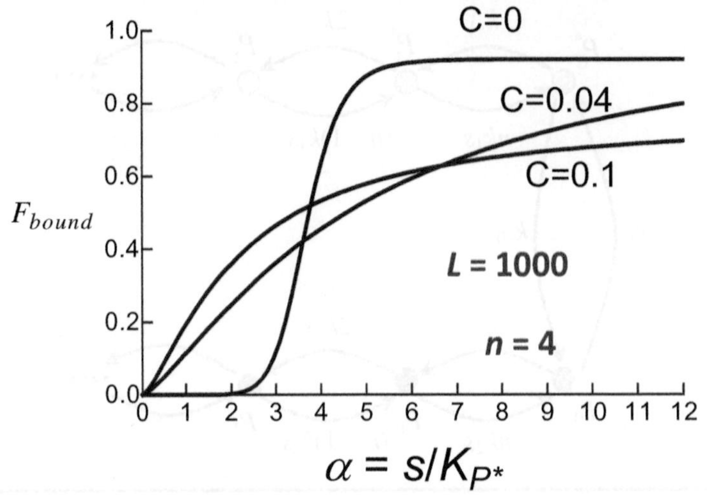

Fig. 6.15

Fraction bound versus scaled substrate for tetrameric allosteric protein P for three different binding coefficients. Hemoglobin binding to oxygen in the lung is represented by the $C = 0$ curve (see text).

In the limiting case, for allosteric constant $L = P_0/P_0^* = 0$, v is the simpler hyperbolic M–M reaction, the first term on the RHS of (6.70). For $L > 0$, the curve becomes sigmoidal, according the "regulatory" second term in (6.70), illustrating how allosterism can affect binding equilibria and reaction rates.

Other Allosterism Models & Summary Remarks

Eq. (6.68) is the principal result of the Monod ("MWC, concerted") model theory. We note that it is a ratio of *higher-order polynomials in substrate concentration s*. This is different than the simpler nth-order Hill function form, $As^n/(B^n + s^n)$, which has only a single-order term, s^n, in s. There are other theories (candidate models) for allosteric proteins, e.g. the "sequential KNF" model (Koshland et al. 1966) and other variations of allosteric or partial allosteric reactions, with and without effectors (inhibitors or stimulators). Chapter 14 in the reference (Lescovac 2003) provides many of these details.

The primary graphic distinction between monomeric, multimeric and these other variations of regulatory functions is evidenced by the relative absence or presence of inflection points (sigmoidal functions) in plots of binding data like that in **Fig. 6.15**. *Model distinguishability* (choosing among the available candidate models, see Chapter 5) is thus a practical issue. It is difficult to make a selection among them on the basis of only one experimentally determined curve, such as any F_{bound} in **Fig. 6.15** (maybe two or three curves), because these models are distinguished by several parameters whose values can be difficult to establish experimentally.[46] The model distinguishability problem can be further complicated if quasi-steady state (QSSA) conditions are applicable, even for otherwise perfect (noise-free) data. Cooperative reactions with models otherwise distinguishable from non-cooperative ones can be rendered indistinguishable under QSSA conditions (Evans et al. 2004). Nevertheless, recognizing and quantifying sigmoidal data are first steps toward solving these kinds of biocontrol problems.

46 They are often *numerically unidentifiable*, not quantifiable in practice (Chapter 12).

EXERCISES

E6.1 Solve Eq. (6.1) by the *separation of variables* method.

E6.2 Show that the stoichiometry of the (a) monomolecular reaction X→Y requires that $y(t) = x(0) - x(t)$ (concentrations); and (b) bimolecular reaction X + W→Z requires $x(t) = x(0) - z(t)$ and $w(t) = w(0) - z(t) = w(0) - x(0) - x(t)$.

E6.3 Show first that: (a) the solution $z(t)$ of the *biomolecular* reaction in the previous exercise (E6.2) is:

$$z(t) = x(0)w(0)(1 - e^{[x(0)-w(0)]kt})/(w(0) - x(0)e^{[x(0)-w(0)]kt}) \tag{6.71}$$

Then

a) Solve for $x(t)$, $w(t)$, $z(t)$ for equal initial concentrations of the reactants X, Y.

b) Show that if $w(0) >> x(0)$ the bimolecular reaction can be approximated by a monomolecular reaction. *Hint*: the rate constants of this nonlinear reaction are dependent on the ICs.

E6.4 Following the logic in Example 6.1, another radioactive isotope of iodine, ^{125}I, has a physical half-life of about 60 days. What is the probability that any atom of ^{125}I will disintegrate in a second?

E6.5 Equations for *competitive inhibition* were developed in (6.49) and (6.50). Two other forms of enzyme inhibition are called *uncompetitive* and *mixed-noncompetitive* inhibition. **Uncompetitive inhibition** occurs when an inhibitor combines with the enzyme substrate complex, but not with the enzyme. In mixed and noncompetitive inhibition, the inhibitor can combine with both the enzyme, i.e. both *EI* and *EIS* complexes exist. If the inhibitor has different affinities for the enzyme and the enzyme–substrate complex, then inhibition is **mixed**. If the affinities are the same, inhibition is **noncompetitive**. All three types of inhibition (including competitive) are included in reactions (6.72) below.

$$\begin{array}{ccl} E + S & \rightleftharpoons & ES \longrightarrow E + P \\ + & & + \\ I & & I \\ \Updownarrow & & \Updownarrow \ \text{(reversible reactions)} \\ EI + S & \rightleftharpoons & EIC \end{array} \tag{6.72}$$

a) Derive the ODEs for uncompetitive and mixed-noncompetitive inhibition. One or more of the reactions are missing in (6.72) for each case.

b) Derive the QSSA equations for the rate of product formation in each case.

E6.6 Based on Arhenius' law, the rates of many common chemical reactions double when the temperature is changed by 10°C from room temperature, i.e. $Q_{10} = 2$. What is the activation energy E_a for these reactions?

E6.7 Solve Eq. (6.9).

E6.8 *Effects of Temperature on Reaction Rate: Square-Root vs. Arrhenius Laws*
Consider a 5° rise in temperature 298°K (25°C)→303°K, for a gas-phase reaction with activation energy E_a of 25 kcal/mol. Both E_a and A are assumed independent of temperature. Compute the percentage change in the reaction rate constant using both the Arrhenius and the square root law formulas. Big or small difference?

E6.9 Write Eq. (6.13) in stoichiometric matrix form.

E6.10 *Some Acid-Base Biochemistry Modeling: The Henderson–Hasselbach (H–H) Equation*
An **acid** is defined as a **proton donor**, denoted **AH**, i.e. any compound which, in solution, is able to give up (donate) a hydrogen ion, H^+. A **base** is a **proton acceptor,** A^-, i.e. any compound which, in solution, is able to accept (associate with) an H^+. The dissociation of an acid in solution is often written as: $AH \underset{k_{-1}}{\overset{k_1}{\rightleftharpoons}} A^- + H^+$. Assume chemical equilibrium and let $pH \equiv -\log_{10} c_H$ and $pK_a \equiv -\log_{10} K_a$, where $c_H \equiv \dfrac{K_a c_{AH}}{c_A}$ and $\dfrac{k_1}{k_{-1}} \equiv K_a = \dfrac{c_A c_H}{c_{AH}} \equiv$ the acid dissociation constant (equilibrium association constant). Derive the H–H equation for this reaction:

$$\log_{10}\left(\frac{c_A}{c_{AH}}\right) = pH - pK_a \tag{6.73}$$

E6.11 *Gut Absorption and Acid-Base Buffering Absorption.* The permeability of cell membranes to most ions is severely restricted. What about aspirin in the gut? How much is actually absorbed? Use the H–H equation (6.73) to determine how much of it is absorbed by the gut into the bloodstream. That is, find the percent of acetylsalicylic acid (aspirin) in unionized form in gastric juice, which has pH = 1.5, given $pK_a = 3.5$ for aspirin.

Extensions: Quasi-Steady State Assumption Theory

It's instructive to consider the two main developments of the quasi-steady state assumption (QSSA), first from classic Michaelis–Menten theory, and second from a recent and more geometric point of view – which extends the theory substantially for modern applications in dynamic systems biology. Both utilize **singular perturbation analysis**, introduced pedagogically below, and the second approach introduces a more systematic way of choosing scaled variables and yields more robust results.

*Classic Michaelis–Menten Theory

Following is a detailed version of the classic development of the QSSA, i.e. rapid achievement of an approximate steady state for the complex of enzyme and substrate, i.e. $\dot{c} \approx 0$, following an initial very fast transient forming the complex. It is based on singular perturbation analysis repeated in similar forms in many references (e.g. Bowen et al. 1963; Rubinow 1975). The common approach introduces nonnegative dimensionless variables normalized by parameters or initial conditions, and is time-scaled as well: $s/s_0 \equiv s'$, $c/e_0 \equiv c'$, $e_0/s_0 \equiv \varepsilon$, $K \equiv (k_{-1} + k_2)/k_1 s_0 > 0$, $\lambda \equiv k_2/k_1 s_0 > 0$ and $k_1 e_0 t \equiv \tau$. Note that ε here represents the initial amount of enzyme relative to initial substrate and τ is the new time-scale variable. Eqs. (6.31) and (6.32) normalized with this scaling simplify to:

$$ds'/d\tau = -s' + (s' + K - \lambda)c' \tag{6.74}$$

$$\varepsilon dc'/d\tau = s' - (s' + K)c' \tag{6.75}$$

The initial conditions are $s'_0 \equiv s'(\tau = 0) = 1$ and $c'_0 \equiv c'(\tau = 0) = 0$. Now *assume much more substrate than enzyme*,[47] i.e. $\varepsilon << 1$. Typical values for ε are 10^{-7} to 10^{-2}. These equations are solved formally by seeking a solution in the form of the singular perturbation expansion,[48] shown here for terms up to first-order for singular perturbation ε:

$$\begin{aligned} s'(\tau, \varepsilon) &= s'_0(\tau) + \varepsilon s'_1(\tau) + O(\varepsilon^2) \\ c'(\tau, \varepsilon) &= c'_0(\tau) + \varepsilon c'_1(\tau) + O(\varepsilon^2) \end{aligned} \tag{6.76}$$

The goal is to establish the *quasi-steady state* (QSS) solutions as $\varepsilon \to 0$. Substituting Eq. (6.76) into Eqs. (6.74) and (6.75) and equating only the zero-order terms yields the zero-order equations:

$$ds_0/d\tau = -s'_0 + (s'_0 + K - \lambda)c'_0 \tag{6.77}$$

$$0 = s'_0 - (s'_0 + K)c'_0 \tag{6.78}$$

The zero-order approximation[49] to the solutions $s'_0(\tau)$ and $c'_0(\tau)$ is consistent with the QSSA because it nulls $dc'_0/d\tau$ (with $\varepsilon \equiv 0$) but not $ds'_0/d\tau$. Solving Eq. (6.78) for the complex yields

$$c'_0(\tau) = \frac{s'_0(\tau)}{s'_0(\tau) + K} \tag{6.79}$$

47 The classic quasi-steady state theory depends on this assumption.

48 Similar to a Taylor series expansion for (singular) perturbation ε, shown here to first-order + remainder.

49 The first-order term $O(\varepsilon)$ can be used to estimate a first-order correction for the zero-order approximation (Segel and Slemrod 1989).

Note that this expression for the normalized complex concentration $c'_0(\tau)$ does not satisfy the initial condition, because ε is not exactly zero; it is asymptotic, but not uniformly valid for all τ. Expression (6.76) represents what is called the **outer expansion**, valid for values of τ away from the origin as it gets bigger. Substituting (6.79) into (6.77) gives an equation for $ds'_0/d\tau$ alone,

$$ds'_0/d\tau = -\lambda c'_0 = -\frac{\lambda s'_0(\tau)}{s'_0(\tau) + K} \tag{6.80}$$

Integrating then yields the (substrate) solution:

$$s'_0(\tau) + K\log s'_0(\tau) + A = -\lambda\tau, \tag{6.81}$$

where A is the constant of integration. However, we cannot apply the initial condition to determine A, because the solution is not valid at $\tau = 0$. To obtain the asymptotic solution valid for small τ, we need a solution that matches this one at the boundary of this outer solution with the solution of the system near $\tau = 0$.

To accomplish this, introduce the **inner time-scaled variable**: $T \equiv \tau/\varepsilon$ $(dT = d\tau/\varepsilon)$. Then, to match solutions at the boundaries of the outer expansion, define the compatible inner variables $s''(T, \varepsilon) \equiv s'(\tau, \varepsilon)$, $c''(T, \varepsilon) \equiv c'(\tau, \varepsilon)$. Then Eqs. (6.74) and (6.75) become:

$$\begin{aligned} ds''/dT &\equiv \varepsilon ds''/d\tau = \varepsilon[-s'' + (s'' + K - \lambda)c''] \\ dc''/dT &\equiv dc''/d\tau = s'' - (s'' + K)c'' \end{aligned} \tag{6.82}$$

with

$$\begin{bmatrix} s''(0, \varepsilon) \\ c''(0, \varepsilon) \end{bmatrix} = \begin{bmatrix} 1 \\ 0 \end{bmatrix} \tag{6.83}$$

Now define a singular perturbation **inner expansion**:

$$\begin{aligned} s''(T, \varepsilon) &= s''_0(T) + \varepsilon s''_1(T) + O(\varepsilon^2) \\ c''(T, \varepsilon) &= c''_0(T) + \varepsilon c''_1(T) + O(\varepsilon^2) \end{aligned} \tag{6.84}$$

As above, substituting Eq. (6.82) into Eqs. (6.84) and equating only the zero-order terms (with $\varepsilon \equiv 0$) yields the zero-order equations:

$$\begin{aligned} ds''_0/dT &= 0 \\ dc''_0/dT &= s''_0 - (s''_0 + K)c''_0 \end{aligned} \tag{6.85}$$

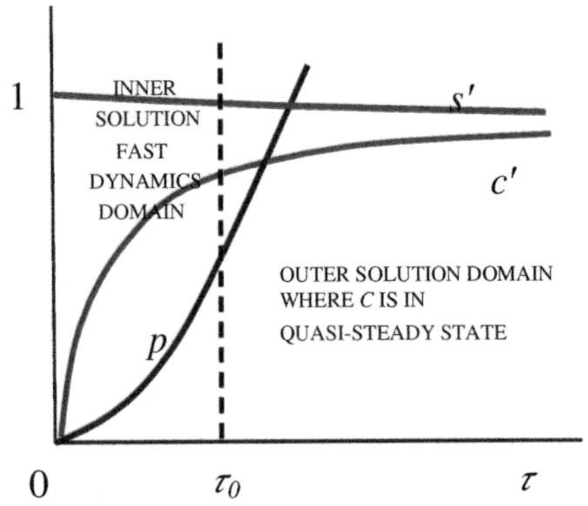

Fig. 6.16

The initial very rapid adjustment of the complex and enzyme concentrations to steady state on a time scale of order ε, followed by much slower conversion of substrate into product on a time scale of order 1.

with ICs $s_0''(0,\varepsilon)=1$ and $c_0''(0,\varepsilon)=0$, and solutions:

$$s_0''(\tau)=1, \quad c_0''(\tau)=\frac{1-e^{-(1+K)\tau}}{1+K} \tag{6.86}$$

This **inner expansion** solution satisfies the initial condition.

Matching Solutions: We assume now that the two expansion solutions (6.79) and (6.86), the former valid for large t and the latter for small t, have a common domain of validity in which the expansions are asymptotically equal, i.e. the inner and outer expansions are both valid for intermediate times $\varepsilon << \tau << 1$. The common domain (matched asymptotic expansion) is assumed to be small in normal time t and large in scaled time τ. The match comes as $T \to \infty$ and $\tau \to 0$ as $\varepsilon \to 0$ in the limit. By *reversion of series*, we expand $s_0'(\tau)$ in (6.81) as a power series in τ, and similarly for $s_0''(\tau)$ in Eq. (6.86). By comparison,[50] it can be shown that the inner and outer solutions agree when $A = -1$.

Biochemists typically use only the first term in the outer expansion. They measure substrate $s(t)$ over a short period of time (about 5 to 30 minutes), and then determine $\left.\frac{ds}{dt}\right|_{t=0}$ or, equivalently, $\left.\frac{dp}{dt}\right|_{t=0}$. To zero-order in ε, these quantities are equal in magnitude. The magnitude is the (initial) velocity of the reaction, denoted $\mathbf{v_0}$ at τ_0. Setting $t=0$ in Eq. (6.80), we find that, to zero-order in ε and with $s_0'(0)=1$,

$$v_0=\left|\frac{ds_0(0)}{dt}\right|=\frac{\lambda}{1+K} \tag{6.87}$$

This is the same as unscaled Eq. (6.40), with $K \equiv K_m/s_0$. **Fig. 6.16** illustrates the initial very rapid adjustment of the complex and enzyme concentrations to steady state on a time-scale of order ε, followed by much slower conversion of substrate into product on a time-scale of order 1.

50 Reversion of series means computation of the coefficients of the inverse function of a series given those of the forward series function. In our case, these are the coefficients that generate a match for the inner and outer solutions.

*Modern Quasi-Steady State Assumption (QSSA) Analysis & a Scaling Lesson

The QSSA was approached and validated differently by Segel and Slemrod (1989). Their solution is based on what they recognized as a general characteristic of ODE models like the M−M equations with an initial fast transient. In their derivations, they revisit the dimensional analysis part of the problem first, because the choice of scaling parameters is critical and key to their results. They use specific M−M model dynamics to normalize the time scales for the two independent variables, substrate $s(t)$ and complex $c(t)$, and use the same dimensionless scales for both. This decisive dimensional analysis step is based on the idea that all dimensionless variables should be chosen to be of *unit order magnitude*, rather than what appears convenient.

In the classical approach, a time scale for the outer solution is pre-defined beforehand, a scale which Segel and Slemrod (1989) ultimately conclude is too restrictive. Instead, they compute the two time domain intervals, first for the fast "inner solution" and then for the slow "outer solution." This (different) ordering is based on their conclusion that the *fastest varying dependent variable should be scaled first*, because the remaining dependent variables scaled the same way will then automatically reflect the relative slowness of their dynamics.

The **fast time scale** is defined as the short duration t_C of the "pre-steady-state" period. By assumption, substrate concentration does not change appreciably during this period, so s equals s_0 throughout and thus NL Eq. (6.32) is approximately linear, with solution:

$$c(t) \cong \frac{e_0 s_0}{K_m + s_0}\left(1 - \exp^{-k_1(K_m+s_0)t}\right) \tag{6.88}$$

The authors choose the *time constant* of this rising exponential solution (when $\exp(-x) = \exp(-1) = 0.37$) as t_C:

$$t_C = 1/k_1(s_0 + K_m) \tag{6.89}$$

The **slow time scale** t_S is estimated as the "linear approximation:"

$$t_S = s_{max}/|ds/dt|_{max} = s_0/|k_2 e_0 s_0/(K_m + s_0)| = (s_0 + K_m)/d_2 e_0 \tag{6.90}$$

where $|ds/dt|_{max}$ is the same as the max of Eq. (6.40). *In these terms,* ***the modern QSSA is simply***: $\boldsymbol{t_C \ll t_S}$. Making the substitutions from (6.89) and (6.90) and defining *new dimensionless parameters:*

$$\sigma \equiv s_0/K_m, \quad \eta \equiv e_0/K_m, \quad \kappa \equiv k_{-1}/k_2 \tag{6.91}$$

the inequality $t_C \ll t_S$ becomes:

$$k_2 e_0 \ll k_1(s_0 + K_m)^2 \tag{6.92}$$

For the two time scales to match up at the boundary where they meet, the value of $s(t)$ must not change appreciably from its initial value, i.e. $s(0) = s_0$. Equivalently, the *relative change* $|\Delta s/s_0|$

must be negligible (i.e. $<<1$ in dimensionless space) during the initial pre-steady state transient. The authors estimate $|\Delta s/s_0|$ using the linear approximation:

$$|\Delta s/s_0| \cong t_C |ds/dt|_{max}/s_0 \tag{6.93}$$

They use the original ODE for $s(t)$ with $c(t) = 0$ to get $|ds/dt|_{max} = -k_1 e_0 s_0$. Then (6.93) gives the additional necessary condition for the QSSA:

$$|\Delta s/s_0| \cong \frac{e_0}{K_m + s_0} \ll 1 \tag{6.94}$$

In terms of the dimensionless parameters (6.91), this means $\eta - \sigma \ll 1$, which provides a good candidate for the **new dimensionless small parameter** $\varepsilon(\ll 1)$ for further singular perturbation analysis:

$$\varepsilon \equiv \eta/(1+\sigma) = e_0/(K_m + s_0) \tag{6.95}$$

When this $\varepsilon << 1$, we get an even stronger condition than $t_C \ll t_S$, which is itself implied by (6.95). Furthermore, this ε is a more liberal small parameter than e_0/s_0 used in the original derivations of the QSSA, summarized in the first Extension above.

To get the scaled equations for the pre-steady state inner solution, note that $s \approx s_0$ and $c \approx \dfrac{e_0 s_0}{K_m + s_0} \equiv c_{ss}$, the steady state value of $c(t)$, during that brief interval (the coefficient of Eq. (6.88)). Thus, the **new scaled dimensionless variables** are immediately evident:

$$s'' \equiv s/s_0, \quad c'' \equiv c/c_{ss}, \quad T \equiv t/t_C \tag{6.96}$$

In these terms, the original M–M model equations (6.31) and (6.32) become:

$$\begin{aligned} \frac{ds''}{dT} &= \varepsilon\left(-s'' + \frac{\sigma c'' s''}{\sigma+1} + \frac{\kappa c''}{(\kappa+1)(\sigma+1)}\right) \\ \frac{dc''}{dT} &= s'' - \frac{\sigma c'' s''}{\sigma+1} - \frac{c''}{\sigma+1} \\ s''(0) &= 1, \qquad c''(0) = 0 \end{aligned} \tag{6.97}$$

The small parameter ε $(<<1)$ in (6.97) is defined by (6.95) above. Applying the singular perturbation expansion for Eqs. (6.97) for terms up to first-order:

$$\begin{aligned} s''(T,\varepsilon) &= s''_0(T) + \varepsilon s''_1(T) + O(\varepsilon^2) \\ c''(T,\varepsilon) &= c''_0(T) + \varepsilon c''_1(T) + O(\varepsilon^2) \end{aligned} \tag{6.98}$$

The zero-order terms are the constant 1 as $\varepsilon \to 0$, and $c(T)$ is the solution of a linearized ODE, i.e.:

$$s_0''(T) \equiv 1, \quad c_0''(T) = 1 - e^{-T} \tag{6.99}$$

To get the scaled equations for the **outer solution** – after the pre-steady state, only the time variable needs rescaling, as:

$$\tau \equiv \frac{t}{t_S} = \frac{k_2 e_0 t}{K_m + s_0} \tag{6.100}$$

Then, for the dependent variables $s''(\tau)$ and $c''(\tau)$ scaled the same as in (6.96), the equations governing the inner solution are:

$$\begin{aligned} \frac{ds''}{d\tau} &= (\kappa+1)(\sigma+1)\left(-s'' + \frac{\sigma c'' s''}{\sigma+1} + \frac{\kappa c''}{(\kappa+1)(\sigma+1)}\right) \\ \varepsilon\frac{dc''}{d\tau} &= (\kappa+1)(\sigma+1)\left(s'' - \frac{\sigma c'' s''}{\sigma+1} - \frac{c''}{\sigma+1}\right) \end{aligned} \tag{6.101}$$

with the same initial conditions $s_0'' \equiv s''(\tau = 0) = 1$ and $c_0'' \equiv c''(\tau = 0) = 0$, supplying matching conditions at the boundary between inner and outer solutions. Again, $0 < \varepsilon << 1$, and singular perturbation theory is applied to Eqs. (6.97) and (6.101):

$$\begin{aligned} s''(\tau,\varepsilon) &= s_0''(\tau) + \varepsilon s_1''(\tau) + O(\varepsilon^2) \\ c''(\tau,\varepsilon) &= c_0''(\tau) + \varepsilon c_1''(\tau) + O(\varepsilon^2) \end{aligned} \tag{6.102}$$

Substituting Eq. (6.76) into (6.101) and equating only the zero-order terms yields the zero-order equations:

$$\begin{aligned} \frac{ds_0''(\tau)}{d\tau} &= -\frac{\sigma+1}{\sigma s_0''(\tau)+1}s_0''(\tau) \\ c_0'' &= \frac{\sigma+1}{\sigma s_0''(\tau)+1}s_0''(\tau) \end{aligned} \tag{6.103}$$

The zero-order approximation to the solutions $s_0''(\tau)$ and $c_0''(\tau)$ is consistent with the QSSA because it nulls $dc_0''/d\tau$ (with $\varepsilon \equiv 0$) but not $ds''/d\tau$.

Error Analysis – Bounds on the QSSA: The first-order correction terms $s_1''(\tau)$ and $c_1''(\tau)$ in (6.76) were used to estimate a correction for the zero-order approximation, which provided bounds for the validity of the QSSA (Segel and Slemrod 1989). The author's comprehensive error analysis – which included convincing simulations – shows that the **decisive error measure**, denoted δ_S, is the *fractional change in substrate* during the initial fast transient:

$$\delta_s \equiv |\Delta s/s_0| \cong \frac{e_0}{K_m + s_0} \equiv \varepsilon \tag{6.104}$$

The approximation in (6.104) is from Eq. (6.94) and the last equality in (6.104) indicates that this approximation of the error bound *equals* the "small parameter" ε ($<< 1$) in the singular perturbation analysis above.

In practical terms, the QSSA, which amounts to, $\delta_s \cong \varepsilon \ll 1$, can be used if a fractional error equal to δ_s can be tolerated. For example, if a 1% error $\delta_s = 0.01$ (or less) can be tolerated, the authors show that the QSSA is very well satisfied. And, even for a 10% error, i.e. $\delta_s = 0.1$, the approximation still remains quite reasonable – a result of their simulation analysis. The details are lengthy and can be found in the reference (Segel and Slemrod 1989). Importantly, the error bound is computable from the model parameters and ICs of the particular model in question, as $\delta_s \cong e_0/(K_m + s_0)$. The error in using the QSSA is very small when $s_0/K_m >> 1$ and we confirmed these results and illustrated them earlier in this chapter, in Example 6.6 and associated simulation results in **Fig. 6.6**.

The QSSA for Arbitrary ICs: $c(0)$ and $p(0)$ are zero in all of the above. For nonzero ICs $c(0) = c_0 \neq 0$ and $p(0) = p_0 \neq 0$, the QSSA remains valid if the following substitutions are made in the appropriate equations above:

$$e_0 \rightarrow e_T \equiv e_0 + c_0 \quad \text{and} \quad \varepsilon \rightarrow e_T/(K_m + s_0) \tag{6.105}$$

The Reverse QSSA: Segel and Slemrod (1989) also show the interesting result that the QSSA is valid when the substrate $s(t)$ is in QSS w.r.t. complex $c(t)$, instead of the (usual) other way around.

Bottom Line: Strictly speaking, the QSSA is valid if $|\Delta s/s_0| \cong \frac{e_0}{K_m + s_0} \equiv \varepsilon \ll 1$. If this fractional combination of model parameters and ICs is small enough – even as great as ~ 0.1, the QSSA can be a reasonable approximation. The more classical and stringent condition $e_0/s_0 \ll 1$ is not necessary, because $K_m > 0$ is included in ε.

The total tQSSA: QSSA assumptions were developed and extended considerably further in (Bhorghans et al. 1996). The basic results for the tQSSA are presented fairly completely earlier this chapter and are not repeated here.

References

Atkins, P., Jones, L., 2005. Chemical Principles: The Quest for Insight. W. H. Freeman and Company, New York.
Bhorghans, J.A.M., DeBoer, R.J., Segel, L.A., 1996. Extending the quasi-steady state approximation by changing variables. Bull. Math. Biol. 58, 43–63.

Black, D.L., 2003. Mechanisms of alternative pre-messenger RNA splicing. Annu. Rev. Biochem. 72, 291–336.

Bowen, J.R., Acrivos, A., Oppenheim, A.K., 1963. Singular perturbation refinement to quasi-steady state approximation in chemical kinetics. Chem. Eng. Sci. 18 (3), 177–188.

Briggs, G., Haldane, J.B., 1925. A note on the kinetics of enzyme action. Biochem. J. 19 (2), 338–339.

Ciliberto, A., Capuani, F., Tyson, J., 2007. Modeling networks of coupled enzymatic reactions using the total quasi-steady state approximation. PLoS Comput. Biol. 3 (3), e45.

Cleland, W., 1963. The kinetics of enzyme-catalyzed reactions with two or more substrates or products. I. Nomenclature and rate equations. Biochim. Biophys. Acta. 67, 104–137.

Cornish-Bowden, A., 1975. The use of the direct linear plot for determining initial velocities. Biochem. J. 149, 305–312.

Eisenberg, D.S.C., Donald, M., 1979. Physical Chemistry with Applications to the Life Sciences. Addison Wesley.

Evans, N.D., Chappell, M.J., Chapman, M.J., Godfrey, K.R., 2004. Structural indistinguishability between uncontrolled (autonomous) nonlinear analytic systems. Automatica. 40 (11), 1947–1953.

Field, R.J., N., R.M., 1974. Oscillations in chemical systems. IV. Limit cycle behavior in a model of a real chemical reaction. J. Chem. Phys. 60 (5), 1877–1884.

Frenzen, C.L., Maini, P.K., 1988. Enzyme kinetics for a two-step enzymic reaction with comparable initial enzyme-substrate ratios. J. Math. Biol. 26, 689–703.

Fujita, S., Arimitsu, T., 1991. Statistical mechanical theory of bimolecular reaction. Derivation of the arrhenius law. Characterization of a catalyst. Fortschritte der Physik/Progress of Physics. 39 (1), 21–40.

Koshland, D.E., Nemethy, G., Filmer, D., 1966. Comparison of experimental binding data and theoretical models in proteins containing subunits. Biochemistry. 5 (1), 365–385.

Lescovac, V., 2003. (14) cooperative and allosteric effects (15) effects of temperature on enzyme reactions. Comprehensive Enzyme Kinetics. Springer, 243–282.

Lineweaver, H., Burk, D., 1934. The determination of enzyme dissociation constants. J. Am. Chem. Soc. 56, 658–666.

Michaelis, L., Menten, M., 1913. Die kinetik der invertinwirkung. Biochem. Z. 49, 333–369.

Monod, J., Wyman, J., Changeux, J.P., 1965. On the nature of allosteric transitions: a plausible model. J. Mol. Biol. 12, 88–118.

Nelson, D., Cox, M.M., 2000. Lehninger Principles of Biochemistry. third ed. Worth Publishers, New York, USA.

Palsson, B., 2009. Metabolic systems biology. FEBS Lett. 583 (24), 3900–3904.

Palsson, B.O., 2006. Systems Biology: Properties of Reconstructed Networks. Cambridge University Press, Cambridge.

Pedersen, M., Bersani, A., 2010. Introducing total substrates simplifies theoretical analysis at non-negligible enzyme concentrations: pseudo first-order kinetics and the loss of zero-order ultrasensitivity. J. Math. Biol. 60 (2), 267–283.

Pedersen, M., Bersani, A., Bersani, E., Cortese, G., 2008. The total quasi-steady-state approximation for complex enzyme reactions. Math. Comput. Simul. 79 (4), 1010–1019.

Peterson, M.E., Daniel, R.M., Danson, M.J., Eisenthal, R., 2007. The dependence of enzyme activity on temperature: determination and validation of parameters. Biochem. J. 402, 331–337.

Rescigno, A., 2003. Foundations of Pharmacokinetics. Springer, New York.

Rubinow, S., 1973. Mathematical Problems in the Biological Sciences. Society for Industrial and Applied Mathematics, Philadelphia.

Rubinow, S., 1975. Introduction to Mathematical Biology. John Wiley, New York.

Rutherford, E., 1906. Radioactive Transformations. The Yale University Press, Cambridge, USA.

Scott, S.K., 1994. Oscillations, Waves, and Chaos in Chemical Kinetics. Oxford University Press, Oxford, New York Toronto.

Segel, L.A., Slemrod, M., 1989. The quasi-steady-state assumption: a case study in perturbation. SIAM Rev. 31 (3), 446–477.

Tzafriri, A.R., 2003. Michaelis–Menten enzyme kinetics at high enzyme concentrations. Bull. Math. Biol. 65, 111–1129.

Tzafriri, A.R., Edelman, E.R., 2004. The total quasi-steady-state approximation is valid for reversible enzyme kinetics. J. Theor. Biol. 226 (3), 303–313.

Tzafriri, A.R., Edelman, E.R., 2007. Quasi-steady-state kinetics at enzyme and substrate concentration in excess of the Michaelis–Menten constant. J. Theor. Biol. 245, 737–748.

Ullah, M., Schmidt, H., Cho, K.-H., Wolkenhauer, O., 2006. Deterministic modelling and stochastic simulation of biochemical pathways using MATLAB. IEE Proc. – Syst. Biol. 153 (2), 53–60.

Voit, E.O., 2000. Computational Analysis of Biochemical Systems: A Practical Guide for Biochemists and Molecular Biologists. Cambridge University Press, Cambridge.

Volkenshtein, M.V., Goldstein, B.N., 1966. Allosteric enzyme models and their analysis by the theory of graphs. Biochim. Biophys. Acta. 115, 478–485.

Cellular Systems Biology Modeling: Deterministic & Stochastic

Chapter 7

* All or parts of the sections denoted with asterisks are more advanced topics – mathematically speaking – and might be skipped over on first pass, or in a more introductory course of study. This does *not* mean they are unimportant, or less important, for understanding the practical modeling topics of this chapter.

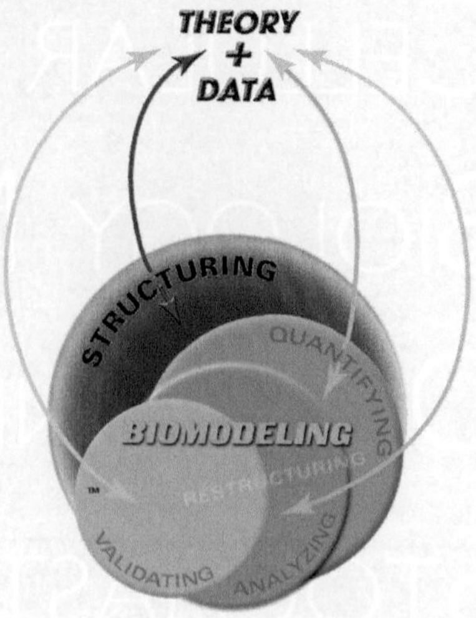

"Mathematics compares the most diverse phenomena and discovers the secret analogies that unite them."

Joseph Fourier, ~1800

"The first entirely vital action, so termed because it is not effected outside the influence of life, consists in the creation of the glycogenic material in the living hepatic tissue. The second entirely chemical action, which can be effected outside the influence of life, consists in the transformation of the glycogenic material into sugar by means of a ferment."

Claude Bernard, 1857[1]

1 *Sur le Méchanisme de la Fonction du Sucre dans le Foie* (1857). Translated in Joseph S. Fruton, *Proteins, Enzymes, Genes: The Interplay of Chemistry and Biology* (1999).

Overview

In Chapter 6 the mass action principle guided development of mechanistic models for interactions among chemical and other species, in particular, biochemical species and enzyme–substrate interactions, with some emphasis on Michaelis–Menten (M–M) kinetics and quasi-steady state assumptions (QSSA). Development of models of other basic and somewhat more complex enzymatically regulated intracellular processes followed from these principles. These included competitive, cooperative and allometric ones, and a ubiquitous cellular system model of a metabolic energy-regulating process – activation and deactivation of receptors and other proteins by adenosine triphosphate–adenosine diphosphate (ATP–ADP) phosphorylation–dephosphorylation enzymatic reactions. This served to illustrate use of the total tQSSA. We continue development of intracellular regulation models in this chapter, building on these fundamentals. Some attention is given to pharmacodynamic (PD) effects modeling. We do this first deterministically, using continuum approximation[2] ODE models of biodynamic systems involving large numbers (thousands to millions) of species or molecules. Then we turn our attention to *stochastic modeling* methodology, for situations where the number of interacting molecules is very small and "stochasticity" cannot be ignored.

The very simple and linear transcription/translation dynamics model developed and simulated at the end of Chapter 3 (**Fig. 3.23**) is a good place to begin study of modeling cellular systems. Review of that example is suggested prior to proceeding with the material in this chapter. Notably, Example 7.15 at the end of this chapter extends that same deterministic example into a somewhat more realistic stochastic model. Application of these modeling paradigms in biochemistry, molecular, cellular and organ level systems biology, pharmacology and toxicology are introduced in the remaining sections and developed in much of the rest of this book.

We also return to software tools and introduce several special purpose biochemistry/cell biology modeling packages. We illustrate how two of them – *SimBiology* and *Copasi* – can be programmed for solution of the simple two-compartment model previously solved in Chapter 4 using basic *Matlab* scripts. Another, software package *AMIGO*,[3] is discussed and used to implement and analyze dynamic systems biology models in later chapters.

Enzyme-Kinetics Submodels Extrapolated to Other Biomolecular Systems

Simple Models of Production Rate Regulation

We begin with some simple enzyme-specific submodels here and generalize them to other monomeric biomolecules or species. These are **phenomenological** (black-box, transfer function) submodels approximating regulation mechanisms[4] for enzyme synthesis (and production of other molecules), patterned on the mathematics of M–M theory. As submodels, they typically

2 ODE continuum models are approximations because chemical reaction events are inherently probabilistic, and their evolution over time is stochastic in nature, more precisely described at the molecular interaction level by stochastic (random process) models.

3 Balsa-Canto and Banga (2011): http://www.iim.csic.es/~amigo/papers.html

4 Macroscopically rather than microscopically mechanistic.

represent modular components of a biosystem or, equivalently, production rate terms in an ODE model of a biosystem.

Remark: Unless otherwise noted, substrate s in this and sections to follow means *free* s if reaction kinetics follow the QSSA. Alternatively, substrate means *total* $s \equiv s + c$ if the tQSSA is needed to better represent fast time scale dynamics. Both approximate slow time scale (longer-term) dynamics equally well.

Induction-Stimulation Submodel: We assume specific monomeric induction (stimulation) of enzyme E formation (production) by compound S (concentration s). Stimulation by S is normally limited, or saturating, as with a simple M–M function:

$$PR_e = \frac{\gamma s}{s + K_s} \tag{7.1}$$

The constant γ here is the maximum production rate of E. The units of PR_e are concentration, or mass, per unit time, depending on the units of the model ODE terms in which this function is used. If more than one molecule or more than one binding site is involved, as with cooperative or allosteric enzymes (for example), the relationship is more complex – as in the last sections of Chapter 6 – with inclusion of higher-order rational polynomial terms, as presented in the next section.

Repression-Inhibition Submodel: By analogy with Eq. (7.1), monomeric repression (inhibition) of enzyme formation by a specific compound, possibly a reaction product X, can inhibit enzyme production as follows:

$$PR_e = \frac{\gamma}{x + K_x} \tag{7.2}$$

Activity vs. Production Regulation: X can stimulate or inhibit the *activity* rather than the formation (production) of the enzyme. The submodels would be like (7.1) or (7.2) above, with e^* (meaning activated e) substituted for e.

Remark: Enzyme E in these submodels can be replaced by any biomolecule or population species being "produced," e.g. proteins, hormones, insects, etc.

Hill Function Regulation Models

All the mono- or multimeric enzyme/protein regulation mechanisms and models above and in the last chapter can be generalized to represent phenomena similarly based on mass action or mass action-like interactions, i.e. not necessarily enzymatically regulated processes. By extrapolation, they can be used for production or activity of proteins, nucleotides or other substances, and as more general network signaling stimulation or inhibition functions, and for modeling higher level, e.g. population, dynamics as well.

Let's generalize the meaning and nomenclature a bit. Consider production of any species, denoted Y – molecular or otherwise – regulated by X. In systems language, we are modeling the *rate* $dy_i(t)/dt$ of a subsystem S_i output $y_i(t)$ in response to (regulatory) *input* $x_i(t)$, as illustrated

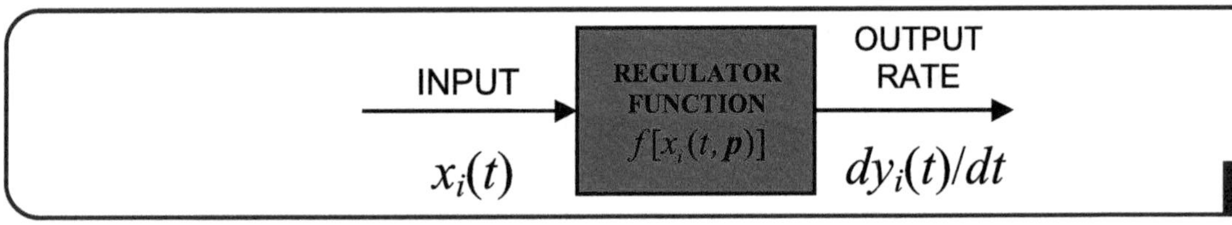

Fig. 7.1

in **Fig. 7.1**. Any dynamic system regulatory function $dy_i(t)/dt \equiv f[x_i(t,\boldsymbol{p})]$ can be represented inside the submodel box in the figure, e.g. the enzyme production regulation functions in (7.1) and (7.2) above. The $\boldsymbol{p}$ in the argument of f is a vector of **control parameters** governing the dynamics of X. Alon (2007) calls functions like this the **input function for a gene** in describing models of regulated transcription rates, involving interactions at the DNA level among repressor and promoter molecules with transcription factors. But they have broader application.

The **Hill function** (Hill 1910) is a commonly used approximation for mesoscale regulatory "transfer" functions in biosystem modeling,[5] in place of $f[x_i(t,\boldsymbol{p})]$ in **Fig. 7.1**. It was derived in the last chapter for a special circumstance, a protein with n identical binding sites, yielding Eq. (6.66). We generalize Hill functions here and begin by dropping the subscript i in the following definitions to simplify the notation.[6]

A Hill function of order $\boldsymbol{n}$, with **control parameters** $\boldsymbol{p} \equiv \{n, \gamma, K\}$, has one of two forms, depending on whether it represents *stimulation* (*activation, promotion*) or *inhibition* (*repression*):[7]

$$dy_S(t)/dt \equiv f_S[x(t,\boldsymbol{p})] = \frac{\gamma x^n}{x^n + K^n} \equiv \textbf{\textit{Hill stimulation fcn}} \tag{7.3}$$

$$dy_I(t)/dt \equiv f_I[x(t,\boldsymbol{p})] = \frac{\gamma}{\dfrac{x^n}{K^n} + 1} = \frac{\gamma K^n}{x^n + K^n} \equiv \textbf{\textit{Hill inhibition fcn}} \tag{7.4}$$

Each function is parameterized by n – the **order** of the Hill function, γ – the **maximum regulation rate** (usually concentration/time units), and K – the **stimulation (or inhibition) coefficient** (concentration units). *For $n = 1$, the Hill Activation function is identical to the Michaelis–Menten (M–M) function*; and it's also the same as phenomenological model (7.1); and inhibition function (7.4) has the same form as (7.2). For integers $n > 1$, the Hill Function in (7.3) represents a *cooperative-like* regulation of Y by X when n molecules (or other species) of X bind or combine and form a **multimer**. With this model, *multimers with fewer than n are assumed to have no effect* (having n interactive or bound is effective, less than n: zero effect). More complex (non-Hill) regulatory equations are needed if fewer than n also are bound.

5 Weiss (1997) reminds us that the Hill function is not a precise representation of most oligomeric ligand–receptor binding reactions in producing (and modeling) functional effects of multimeric receptors. Rational polynomial expressions with terms in s^{n-1}, s^{n-2} etc would be more appropriate, similar to those provided in the reference.

6 We can reintroduce the subscript index i, as needed, to designate different Hill functions in different subsystem model equations.

7 We prefer the more generic terms stimulate and inhibit here (and elsewhere), because activation, promotion, repression have very specific meanings in chemistry.

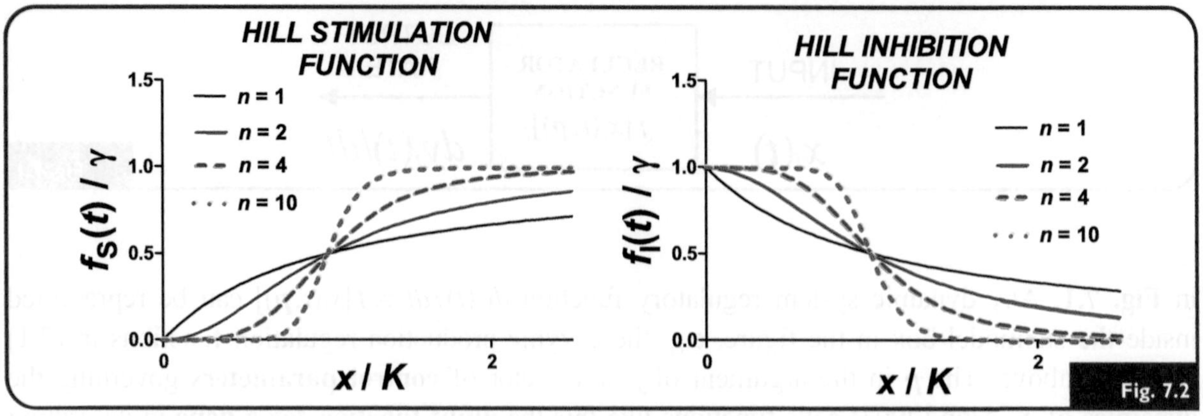

Fig. 7.2

Hill functions.

Both stimulation and inhibition Hill functions are graphed in **Fig. 7.2** for several values of n. For $n = 1$, f is monotonic (the hyperbolic M–M function). For increasing $n > 1$, they are *sigmoidal* negative images of each other about the x-axis. As n becomes larger, Hill functions become more "step-like," approaching binary **switching (step) functions** in the limit. As x becomes larger, the activator Hill function saturates (maximum possible activation). For the repressor, there is no repression when $x = 0$, half-repression when $x = K$, and repression plateaus as x increases further, with $f \to 0$ as $x \to \infty$.

Hill functions are common in gene regulation models, because transcription factors are typically composed of dimers or tetramers, i.e. repeated protein (X) subunits, each possibly binding gene inducer molecules and thereby initiating transcriptional regulation. Note that sigmoidal functions approximate a smoothed stepwise change in f_S or f_I, i.e. "switching" smoothly by distinctly transferring from one level to another.

Example 7.1: ***DNA-Repair Regulation***

The transcription factor protein p53 is in part regulated by the ligase enzyme Mdm2, and *vice versa*, in a negative feedback loop (Lev Bar-Or et al. 2000). It is shown in simplified cartoon schematic form in **Fig. 7.3** – in one of the ways biochemists commonly show it, and also the way systems engineers show it, with + and – signs to indicate the direction of the effect. Transcription of the mRNA for Mdm2 production is believed to be controlled in part by tetrameric subunits of p53 (Chène 2001), which can be modeled (approximately) as a Hill function:

$$f_S\left[p53(t, K_{transcription}, \gamma)\right] = \frac{\gamma[p53(t)]^4}{[p53(t)]^4 + K_{transcription}^4} \tag{7.5}$$

We write $p53(t)$ here explicitly as a function of time t and the two constants parameterizing f, to emphasize the distinction between time-varying **control variable** $p53(t)$ and **control parameters** K and γ in this regulatory equation.

Hill functions with integer coefficients $n = 1, 2, \ldots$ have a sound theoretical base, recently reinforced in (Sorribas et al. 2007). However, they are also used as phenomenological models

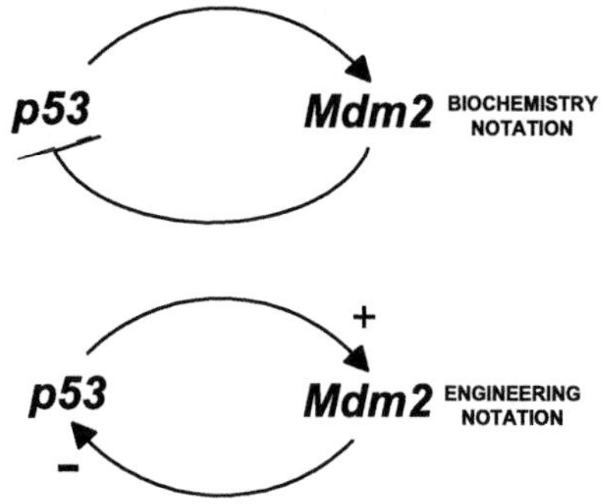

Fig. 7.3

Cartoon representations of negative feedback between p53 and Mdm2.

of molecular interactions that fit real data, more generally allowing non-integer Hill coefficients (Weiss 1997).

If a binding reaction is affected by more than one control variable X_1, X_2, ..., then the function f in (7.3) or (7.4) is multivariable: $dy(t)/dt \equiv f[x_1, x_2, \ldots, \mathbf{p}]$. The result for noninteracting X_i has been described by (Alon 2007) as a ratio of polynomials of the form:

$$dy(t)/dt \equiv f[x_1, x_2, \ldots] = \frac{\sum_{i=1}^{n_i} \gamma_i (x_i/K_i)^{n_i}}{1 + \sum_{i=1}^{m_i} (x_i/K_i)^{m_i}} \tag{7.6}$$

In this model, the Hill-like coefficients are: $n_i = 0$ and $m_i > 0$ for inhibitors; and $n_i = m_i$ for stimulators; all with binding coefficients K_i and maximum rates γ_i.

Ligand–Receptor Interactions & Drug Dynamics

In cellular systems, the two reactants on the left-hand side (LHS) of reversible bimolecular reactions might be a ligand L and a receptor R, to codify a process in which a ligand molecule binds to a single receptor site on another molecule – or surface made of many molecules, forming a complex – a **receptor-bound ligand** LR (also written $L \cdot R$) complex. For example, hormones that exist in unbound and protein-bound forms in blood are endogenous ligands; and drugs introduced exogenously also can be ligands. Receptors for drugs, hormones or other molecules may be "carrier" proteins – like albumin in blood, or protein sites on a cell membrane, among other possibilities. Receptors on a cell membrane are often **transmembrane proteins** – "gateways" receiving a ligand signal on one side (e.g. ligand in blood), and transmitting it to the other side (e.g. into the cytoplasm of the cell). When a transmembrane protein (e.g. "inactive" insulin receptor) senses a signal (e.g. insulin), it is often transformed (activated) into a form ("active" insulin receptor) able to initiate intracellular processes. Uptake of glucose into the cell – activated by the action of insulin at the cell membrane – is a ligand–receptor interaction, modeled – for example – as in (Hori et al. 2006). Movement through a transmembrane protein receptor can be in the other direction as well, e.g. when waste products are removed from the cell into blood.

Monomer receptors have only one binding site, as exemplified below, and oligomer proteins have more than one such site – which may interact and generate more complex dynamics, e.g. as described in the section on cooperativity in the last chapter.

Example 7.2: ***Reversible Drug/Receptor Binding & Steady State Bound Fractions*** F_{bound}

We make this example somewhat more realistic, with a description of an *ex-vivo/in-vitro* experiment, and thus the experiment model as well as the process model, and define some additional receptor dynamics terms useful later. Drugs typically bind to receptors to initiate their effects. The simplest (monomeric) such reaction has the form:

$$D + R \underset{k_{-1}}{\overset{k_1}{\rightleftharpoons}} DR \tag{7.7}$$

where DR (or $D \cdot R$) denotes the *receptor-bound ligand* – the *complex of drug and receptor*. Let d(t) represent free drug[8] concentration. Suppose the medium is an appropriate solvent of volume V in an *in-vitro* experimental vessel. Initial amounts of drug D_0 and receptors R_0 are added to it. Then $\mathrm{d}_0 = D_0/V$ in concentration terms and $r_0 = R_0/V$. Concentration units in *moles/liter* ≡ M are appropriate here, because we are combining whole numbers of molecules in (7.7). Nomenclature for the reacting concentrations (lower case letters) in this example are:

d ≡ unbound (free) drug concentration
r ≡ concentration of unoccupied (free) binding sites on R
≡ molar concentration of receptor × number of **free binding sites per molecule**
c ≡ concentration of **occupied binding sites** ≡ concentration of **complexed** receptor-bound drug, DR

The law of mass action applied to (7.7) gives:

$$\dot{c} = k_1 \mathrm{d}r - k_{-1}c = -\dot{\mathrm{d}} = -\dot{r} \tag{7.8}$$

At equilibrium (steady state) in the vessel, $\dot{c} \equiv 0$ and the *equilibrium dissociation constant* K_d is: $\frac{k_{-1}}{k_1} = \frac{r_\infty \mathrm{d}_\infty}{c_\infty} \equiv K_d$ (M ≡ mol/L units), as defined in the last chapter.

Now apply the constraining *conservation relations*: $\mathrm{d} + c \equiv \mathrm{d}_0$ and $r + c \equiv r_0$, i.e. total drug and receptor must be conserved at all times, including at steady state (as $t \to \infty$). We use the first of these here to solve for the steady state **fraction of bound drug**[9] $\equiv F_{bound}^{drug} \equiv \frac{\text{bound drug}}{\text{total drug}} = \frac{c_\infty}{\mathrm{d}_0}$, by successive substitution:

$$K_d = \frac{r_\infty(\mathrm{d}_0 - c_\infty)}{c_\infty} = r_\infty\left(\frac{\mathrm{d}_0}{c_\infty} - 1\right) = r_\infty\left(\frac{1}{F_{bound}^{drug}} - 1\right) \tag{7.9}$$

8 Drug *concentration* is denoted lower case d (no italic) so as to not confuse it with *derivative* notation *d*.

9 An important clinical/pharmaceutical parameter, a measure of the effectiveness of the drug.

Therefore,

$$F_{bound}^{drug} = \frac{r_\infty}{r_\infty + K_d} \tag{7.10}$$

Similarly, using the second conservation relation $r + c \equiv r_0$, we compute the steady state **fraction of bound receptors**:

$$F_{bound}^{receptors} \equiv \frac{\text{bound receptors}}{\text{total receptors}} = \frac{c_\infty}{r_0} = \frac{\mathrm{d}_\infty}{\mathrm{d}_\infty + K_d} \tag{7.11}$$

Note that, if $K_d \ll r_\infty$, or d_∞, $F_{bound} \approx 1$, all of the drug or receptors are bound. It is thus easy to see why receptor K_ds are widely quoted parameters when discussing kinetics. Receptor K_d comparisons with drug (or hormone, or ...) concentrations clearly inform about binding effectiveness. *Food for Thought:* Does this example suggest anything about effective design of therapeutic drugs?

Example 7.3: ***A Simple Pharmacodynamic Effects Model Based on the QSSA***

In living systems, the complex of drug and receptor (as above) typically produces a drug effect E (activation of drug action), i.e. $D + R \underset{k_{-1}}{\overset{k_1}{\rightleftharpoons}} DR \xrightarrow[k_e]{} E$. This looks like the M–M reaction, but the analogy ends after the first complexing reaction. The unidirectional arrow leading to E from DR, shown here in red, is not a chemical reaction step, with an associated ODE representing mass action kinetics.[10] The simplest effect equation is a *proportional* one, i.e. $e(t) = k_e c(t)$, rather than a mass-action based one, meaning $de(t)/dt \neq k_e c(t)$ (Sheiner and Steimer 2000). This is based on the simplest assumption that one drug molecule interacts with one receptor molecule – e.g. on the cell surface, or inside it – and, when the receptor is occupied by the drug, the receptor becomes (instantaneously) activated and produces an effect inside the cell proportional to the amount or concentration of bound and activated drug, denoted $c^*(t)$ instead of $c(t)$ below, to emphasize that the complex is the activated form of drug.[11]

The system here is assumed to be the human or other animal body, exposed periodically to exogenous drug input to circulating blood.[12] Our goal to find a useful relationship for drug effect $e(t)$ inside the cells targeted by the drug as a function of the concentration $\mathrm{d}(t)$ of drug in blood, the measurable response to exogenous drug input. Drug–receptor interactions and subsequent effects E are assumed to be occurring on and in target cells of the body. For sake of simplicity, we assume homogeneity of all cells affected, i.e. a one-compartment target for the drug.

10 This representation follows the convention in pharmacology literature. The complex of drug and receptor are considered the activated drug and thereby produce the effect.

11 We explicitly maintain the * superscript notation for activated molecular species – activated drug receptors in this example.

12 Modeling the dynamics of how and from where the drug gets into blood is circumvented here, to keep the main point of this example focused on finding the drug blood concentration → effect relationship. A model of *oral* drug absorption into blood was given in Exercise E4.14 of Chapter 4, and Exercise E4.8 and associated Chapter 4 material provides a more general model for indirect inputs into blood.

We also assume drug–receptor complexing dynamics are very fast – or in approximate equilibrium at all times – relative to effect dynamics (the QSSA). This means $\dot{c}^* \cong 0$ for all $t > 0$. Consequently, fraction-bound equation (7.11) applies at all times $t > 0$, not just $t = \infty$. We assume total receptors r_0 remain constant and drug concentration in blood $\mathrm{d}(t)$ is time-varying, governed by its own pharmacokinetics (PK).[13] Solving (7.11) for $c^*(t)$:

$$c^*(t) = \frac{\mathrm{d}(t) r_0}{K_d + \mathrm{d}(t)} \quad \text{for } t > 0 \tag{7.12}$$

Now, when total target receptors r_0 for the drug are fully occupied, i.e. $K_d << \mathrm{d}(t)$, $F_{bound} \approx 1$ and $c^*_{max}(t) = r_0$. Then drug effect $e(t) = k_e c^*(t)$ is maximum: $e_{\max}(t) = k_e r_0$, or $k_e = e_{max}(t)/r_0$. Substituting this k_e in (7.12) then yields:

$$e(t) = k_e c^*(t) \equiv \left(\frac{e_{max}(t)}{r_0}\right) c^*(t) = \frac{e_{max}(t)\mathrm{d}(t)}{K_d + \mathrm{d}(t)} \tag{7.13}$$

This is the desired equation for drug effect $e(t)$ in terms of blood drug concentration $\mathrm{d}(t)$. This time-varying, algebraic (nondifferential), saturating function is a first-order Hill function.

Variables $\mathrm{d}(t)$ (and thus $c^*(t)$ and $e(t)$) are functions of time in this open pharmacodynamic system example, to emphasize the biosystem is open to entry of drug, so effects vary with time, as in real dynamic biosystems. Eq. (7.13) is used widely in modeling overall drug pharmacokinetics (PK) and pharmacodynamics (PD) (Rowland and Tozer 2011), as illustrated in Example 7.5 below.

Example 7.4: ***PD Model for Lethal Dose or 50% Effects***

The K_d in Eq. (7.13) is often given different interpretations in applications. It might be the concentration of drug d_{50} that yields a **50% effect**, $e_{max}/2$. It might also be the **LD50**, i.e. **lethal dose of the drug (or toxin) for half the population**. In either case,

$$e(t) = \frac{e_{max}\mathrm{d}(t)}{\mathrm{d}_{50} + \mathrm{d}(t)} = \frac{e_{max}}{\mathrm{d}_{50}/d(t) + 1} = \frac{e_{max}}{d_{50}/d_{50} + 1} \equiv \frac{e_{max}}{2} \tag{7.14}$$

Example 7.5: ***A Simple, Combined Pharmacokinetic–Pharmacodynamic (PK–PD) Model***

The simplest linear PK model for a drug D with concentration d in blood volume V_{blood}, introduced exogenously at mass rate $u(t)$ is:

$$\dot{\mathrm{d}}(t) = u(t)/V_{blood} - k_{el}\mathrm{d}(t) \quad \text{OR} \quad V_{blood}\dot{\mathrm{d}}(t) = u(t) - k_{el}V_{blood}\mathrm{d}(t) \tag{7.15}$$

13 Drug may enter and may be consumed within the cell, or just act on the cell surface in this example. The net effect of its PK in the whole body generate the drug concentration in blood $\mathrm{d}(t)$ seen by the cell (Example 7.5).

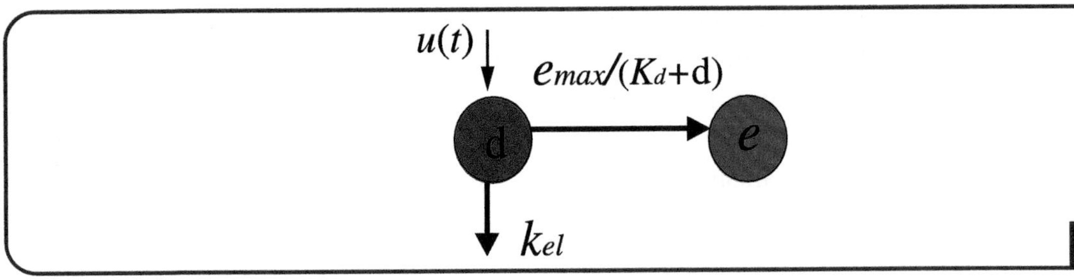

Fig. 7.4

Simple PK–PD model.

In the first concentration rate based form of Eq. (7.15), k_{el} is the drug **elimination rate** (time^{-1}). In the second mass (flux) rate based form, the product $k_{el}V_{blood}$ is called the drug **clearance rate** (volume/time). Both have units balanced by these different parameter definitions. The red compartment in **Fig. 7.4** depicts this PK part of the PK–PD model.

The PD effects part of this combined PK–PD model is given by the green compartment in **Fig. 7.4**. The effect $e(t)$ here is proportional to bound drug concentration $c^*(t)$ at an receptor–effector site directly accessible to blood, i.e. $e(t) = k_e c^*(t)$, as in Eq. (7.13) of Example 7.3. This representation maintains compartmental conventions, i.e. compartment e receives drug from compartment d at mass flux rate equal to concentration d(t) multiplied by the NL rate constant associated with the forward arrow, or $e(t) = \dfrac{e_{max}\mathrm{d}(t)}{K_d + \mathrm{d}(t)}$.

Pharmacodynamic Stimulus & Inhibitory Approximation Functions

The first-order Hill function is commonly used in pharmacodynamic modeling to approximate nonsigmoidal saturating *effects* (pharmacodynamics) of a drug (ligand) acting at a *single* receptor site on a cell. This is usually considered modeling at a macroscopic level, although it can be close to microscopic, for example when the dynamic behavior of individual molecules are well approximated by such functions. The simple nonlinear two-compartment model (**Fig. 4.17**) in Example 4.7 of Chapter 4 is a simple example, with the M–M function representing the effect of q_1 on compartment 2 dynamics. More generally, the first order Hill function can be used to macroscopically represent *stimulatory* (S) or *inhibitory* (I) effects of a substance X with concentration x, using one of the following terms appropriately placed in the ODE representing the interactions (Sharma and Jusko 1996):

$$S = \left(1 + \frac{\gamma_S x}{x + K_S}\right) \qquad I = \left(1 - \frac{\gamma_I x}{x + K_I}\right) \tag{7.16}$$

Here the γs represent the maximal degree of (unitless) stimulation or inhibition, and the constant K is the concentration or amount of X at which half this value is attained. For inhibition, γ is usually chosen as unity, to assure that the overall form remains nonnegative (Jin et al. 2003). This use of Hill functions is often applied in a more macroscopic way, lumping multiple processes together into one of these terms, to describe the overall effects of X

rather than reflecting a specific mechanism such as receptor–ligand binding or one enzymatic reaction. Note the similarity of the nonlinear (NL) terms in Eq. (7.16) to Eqs. (7.1)–(7.4) (for $n = 1$). Whether terms containing a Hill function can be considered more macroscopic or microscopic depends on their specific use within a biomodel, i.e. whether the components involved follow the precise mechanisms represented or simply have (approximately) the same input–output behavior.

Coupled-Enzymatic Reactions & Protein Interaction Network (PIN) Models

When the tQSSA is More Appropriate than the QSSA

The quasi-steady state assumption (QSSA: $K_m + s_0 >> e_0$) and resulting Michaelis–Menten (M–M) kinetics ($v(t) \cong v_{max}s(t)/(s(t) + K_m)$) work very well for representing and analyzing the dynamics of networks of carbohydrates and most other metabolic systems. Metabolites – as substrates – usually are present in great excess over the enzymes that catalyze their interconversion, i.e. $s_0 >> e_0$. An example is cellular conversion of glucose to glucose-6-phosphate by a phosphorylation reaction involving hexokinase with ADP and ATP, as in the model developed in Eqs. (6.54)–(6.56) and **Fig. 6.11** of the last chapter.

Protein interaction networks (PINs) are another matter. In cells, enzymes typically have multiple substrates; and substrates are generally acted upon by multiple enzymes. In PINs, the main issue here is that substrates and enzymes are all proteins with comparable concentrations ($s_0 \approx e_0$). Complexing reactions ($S + E \leftrightarrow C$) thus require more explicit attention, in part because $K_m \equiv (k_{-1} + k_2)/k_1$ is often not large enough to render $K_m + s_0 >> e_0$ when $s_0 \approx e_0$; and it is therefore likely then that the standard QSSA is not valid. Extra care is needed for appropriate application of QSSA for PINs and other dynamical interactions that mimic M–M reactions. Inadvertent use, or misuse, of the QSSA under conditions that lack validity can be not only quantitatively imprecise, but even qualitatively wrong (Bhorghans et al. 1996; Tzafriri and Edelman 2004; Flach and Schnell 2006; Ciliberto et al. 2007).

For PINS and similarly structured systems biology models that do not satisfy QSSA conditions, the *total* QSSA (tQSSA) comes to the rescue. The tQSSA is an excellent kinetic formalism for formulating and analyzing PINs.[14]

Brief theory, tQSSA conditions and equations were presented in Chapter 6. Their application is illustrated in the following example. We compare full model to both QSSA and tQSSA simulated solutions for an **intercellular** model – a cellular receptor–cellular substrate interaction mimicking a M–M enzyme–substrate reaction. Additional PIN model example applications of the tQSSA can be found in (Tzafriri and Edelman 2004; Ciliberto et al. 2007). This is a fertile topic and we expect to see many more examples in the near future.

14 For the **Fig. 6.11** model development of the last chapter, tQSSA Eqs. (6.57) and (6.58) also were developed, for phosphorylation–dephosphorylation reactions in systems with comparable reactant concentrations.

Example 7.6: ***T-Cell Proliferation in Response to Antigen***

This model, adapted from (Bhorghans et al. 1996), has M–M form, with the exception that "substrate" is a replicating immune system T-cell, denoted T, "enzyme" is (instead) a receptor site on an antigen-presenting cell (APC$\equiv A$), "complex" (C) is the T-cell complexed with the receptor, and product P is the T-cell and its replicate, denoted $2T$.[15] All together, $T + A \underset{k_{-1}}{\overset{k_1}{\rightleftharpoons}} C \xrightarrow{k_2} A + 2T$, with ODEs:

$$\dot{T} = -k_1(a_0 - c)T + (k_{-1} + 2k_2)c \quad \text{and} \quad \dot{c} = k_1(a_0 - c)T - K_m c \tag{7.17}$$

Breaking with our usual convention, we use *cap T* instead of t for T-cell (T) concentration, to distinguish it from the time variable.

The standard QSSA ODE (only one, for T-cell substrate and replicate product) is then derived from $\dot{c} \simeq 0$ as:

$$\dot{T} = k_2 c \cong \frac{k_2 a_0 T}{K_m + T} \tag{7.18}$$

The sign is plus here, unlike in the standard M–M reaction, because T is both the "substrate" and the product and increases by replication. Typically, APC cells ("enzyme") are initially large in number (large a_0) relative to initial T-cell number ("substrate", small T_0). This suggests that the corresponding QSSA condition, $T_0 + K_m \gg a_0$, is hard to satisfy, at least initially, near time $t = 0$, when cell number T is small. So conditions on K_m are relatively "tight."

To get the tQSSA ODEs, we define $\hat{T} \equiv T + c$ and get:

$$\dot{\hat{T}} = k_2 c, \quad \dot{c} = k_1[(a_0 - c)(\hat{T} - c) - K_m c] \cong 0 \tag{7.19}$$

with solutions obtained directly from Eq. (7.19):

$$c(t) \cong \frac{a_0 \hat{T}(t)}{a_0 + K_m + \hat{T}(t)}, \quad \dot{\hat{T}}(t) \cong \frac{k_2 a_0 \hat{T}(t)}{a_0 + K_m + \hat{T}(t)} \tag{7.20}$$

The corresponding tQSSA conditions different from $T_0 + K_m \gg a_0$ are, from Eq. (6.48) of Chapter 6: $a_0 + T_0 \gg k_2/k_1$ and $k_{-1} \gg k_2$, based on $\dot{c} \cong 0$ in Eq. (7.19). At first glance, these appear to be less dependent on ICs, except as a sum $a_0 + T_0$.

Fig. 7.5 illustrates "substrate" T-cell solutions for full – Eqs. (7.17), QSSA – Eq. (7.18) and tQSSA – Eq. (7.20) models, based on simulations of their corresponding ODEs over early and longer time intervals. Parameter values are $k_1 = 10$, $k_{-1} = 1$ and $k_2 = 0.1$ ($K_m = 0.11$). Initial conditions are $c(0) = 0$, $T(0) = \hat{T}(0) = 0.1$ and $a(0) = 100$, i.e. **100 × more "enzyme" than "substrate"**. This yields an invalid QSSA condition: $T(0) + K_m = 0.1 + 0.11 = 0.21$ is **not** $\gg a(0) = 100$, an invalid second tQSSA condition: $k_{-1} = 1 \ll k_2 = 10k_{-1}$(**not** $k_{-1} \gg k_2$), but a **strongly valid** first tQSSA condition: $a(0) + T(0) = 100.1 \gg k_2/k_1 = 0.1/10 = 0.01$.

15 Same as the M–M reaction, with product P being $2S$.

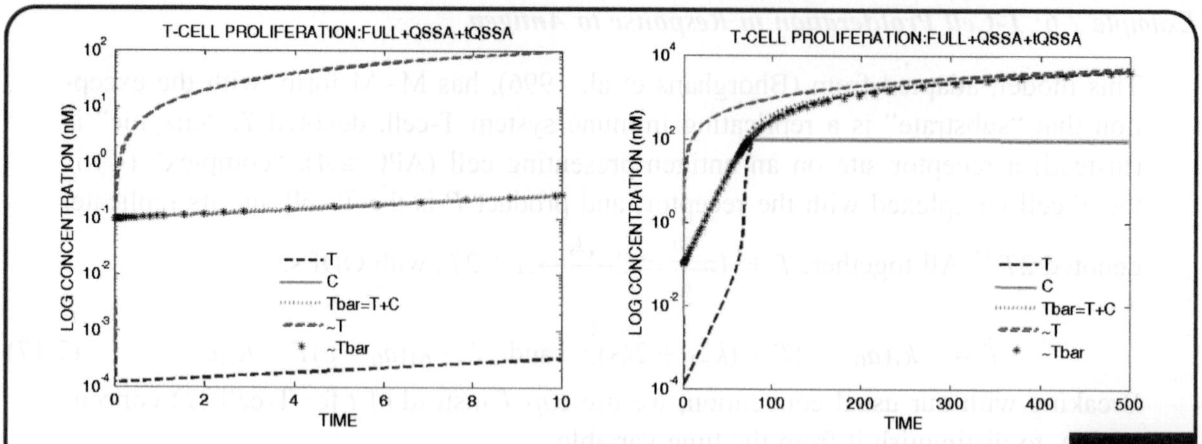

Fig. 7.5

Short and long term T-cell replication dynamics. LHS is over 10 and the RHS over 500 time-units. Free T, complex C and total (Tbar = free + bound) T-cell concentrations using: full ODE and approximate QSSA (~T) and tQSSA (~Tbar) models. Parameter values are $k1 = 10$, $k_{-1} = 1$ and $k2 = 0.1$ ($Km = 0.11$). Initial conditions are $c(0) = 0$, $T(0) = \bar{T}(0) = 0.1$ and $a(0) = 100$, i.e. 100 × more "enzyme" than "substrate". This yields invalid QSSA for free T-cells (green vs. blue curves), but valid tQSSA conditions for total T-cells (red vs. black).

These analytical results are borne out by the plots in **Fig. 7.5**, the LHS over only 10 time units, the right-hand side (RHS) over 500 time units. That the two time scales ($t_C^{total} \cong 0.001$, $t_S^{total} \cong 100$) for total "substrate" are very well separated is also evidenced by the virtually instantaneous jump at time zero – from fast-to-slow scales (LHS) – from 0 to 0.1 in complex concentration (yellow curves), followed by the slower rise in free and total substrate. Note that both approximations are equally good during the latter part of the slow time scale, the green curve being close to the blue and * curve close to the red curve (RHS). This is characteristic of these two approximate solutions. They tend to be most inaccurate near time zero, both incorporating the **hidden nonlinear mode** for complex *c*, the tQSSA more accurately than the QSSA.

Production, Elimination & Regulation Combined: Modeling Source, Sink & Control Components

Systematic modeling of some biological control or regulatory systems can be facilitated by distinguishing different operating states of the system and designating or developing terms in the model to distinctly characterize these states.

We loosely define the **basal** state as "undisturbed, normal or steady-state operating conditions," made more precise as we develop this approach. In physiology, basal conditions are traditionally called **constitutive**, meaning the (usually) balanced dynamical conditions that maintain constancy,[16] or **homeostatic** conditions – in the absence of disturbances. We illustrate the modeling procedure with three examples, continuing first with the tumor-suppressor regulation system in Example 7.1 above. We model production dynamics of the messenger *mRNA* for enzyme

16 Constancy in the context of homeostasis does not necessarily mean that state variables are constant over time. They can be steadily oscillatory, for example, as in "oscillatory steady state."

Mdm2 in the nucleus first, and then the enzyme in the cytoplasm of the cell – including terms in the model representing basal production and degradation, and terms separately representing regulation or control.

Example 7.7: ***Dynamics of Basal Production & Degradation of Mdm2-mRNA Production***

In the basal (quiescent) state, we assume a *constant basal Mdm2-mRNA production rate* $= p_{basal.\ mRNA2} > 0$, where *mRNA2* is an abbreviation for *Mdm2-mRNA*. This is traditionally designated **constitutive production**, normally balanced in quiescent steady state by a similar and linear **constitutive degradation** rate $= k_{\deg\ mRNA2} mRNA2$ in the absence of a DNA damage signal acting on the p53 tumor-suppressor control system (when $dmRNA2/dt = 0$). The regulated model can be written:

$$\frac{dmRNA2}{dt} = p_{basal.mRNA2} - k_{\deg\ mRNA2} mRNA2 + \textbf{regulatory terms} \tag{7.21}$$

Basal Production & Degradation

Example 7.8: ***Regulation of Mdm2-mRNA (mRNA2) Production by Tetrameric p53***

Enzyme *Mdm2* is a key player in tumor suppressor protein *p53* regulation. *Mdm2* production (protein translation) is governed first by transcription of its messenger *mRNA2* in the cell nucleus. *p53*, acting as a transcription factor, affects *mRNA2* production only as a tetramer, with negligible affect on *Mdm2* by *p53* monomers, dimers, etc. (Chène 2001), suggesting simple Hill function regulation, with $n = 4$, as shown in Example 7.1. From the perspective of signaling in a cell, *p53* signaling effects occur within milliseconds when the multimer receptor is fully occupied by ligands of *p53*.

If we assume no other factors significantly involved in regulation,[17] the messenger equation (7.21) becomes:

$$\frac{dmRNA2}{dt} = p_{basal.mRNA2} - k_{\deg\ mRNA2} mRNA2 + \left(\frac{\gamma_{mRNA2} p53^4}{p53^4 + K^4_{transcription}} \right) \tag{7.22}$$

Basal Production & Degradation **Regulation**

We also develop a corresponding ODE for *Mdm2*, produced in cytoplasm by translation of messenger *mRNA2* translocated from the nucleus, with *overall dynamics on a much longer time scale* (~20–60 minutes). Recalling Example 2.6 in Chapter 2, and Example 3.6 in Chapter 3, time-scale disparities as large as this can be approximated by pure time-delay terms in the model. In our developing model, the term representing the effect of *mRNA2*(t) on *Mdm2*(t) production is thus assumed to be a (generally nonlinear) function f of *mRNA2* with a fixed delay τ, $f[mRNA2(t\text{-}\tau)]$, designated as the first term in Eq. (7.23) below.

17 Other factors are surely involved, but probably not as strongly as these, so further complication is not called for here.

Degradation of *Mdm2* is more complex than that of *mRNA2* in (7.22). *Mdm2* is *autodegraded* by enzymatic *ubiquitination* of itself (Wahl et al. 2005), which can be represented by the Hill-like function in the second term of (7.23). Autodegration could be represented more simply as the numerator product, but the denominator allows for nonlinear degradation dynamics in a seemingly realistic and simplest possible way.

$$\underbrace{\frac{dMdm2(t)}{dt} = f[mRNA2(t-\tau)] - k_{\deg\ Hdm2}}_{\textbf{Production \& Degradation}} \underbrace{\left(\frac{Mdm2(t) \bullet Mdm2(t)}{Mdm2(t) + K_{mdm2-Ub}}\right) + \cdots}_{\textbf{Regulation}} \tag{7.23}$$

Additional terms are needed to represent transient regulation of this enzyme in response of the overall system to stress (double-stranded DNA breaks). To the extent that they are complete, the model ODEs (7.22) and (7.23) generally have validity only as a coupled set, not individually. We follow up with overall biocontrol system model for cellular *p53* regulation in Chapter 15.

In some circumstances, the model can be simplified (approximated), with basal production and degradation terms omitted from the equations, especially if they are operating on a much longer time scale. For example, if these processes are in quasi-steady state relative to the dynamics of the Regulation terms, Regulation terms dominate the dynamics. They also dominate if basal production and degradation terms contribute negligibly to the response effects of regulation terms. These can be complicating or confounding factors, but better to have them explicit so they might be dealt with appropriately.

Remark: In principle, constant basal production rates like p in Eq. (7.22) can be determined by measuring the degradation rate k (or half-life, $\ln 2/k$) in a pulse–dose kinetic experiment with the species (*mRNA2* here) introduced exogenously, with the biosystem in basal steady state. Measurement of steady state species concentration ($mRNA2_\infty$ here) is also required ($p_{basal\ mRNA2} = k_{\deg\ mRNA2} mRNA2_\infty$).

The Stoichiometric Matrix *N*

The stoichiometric matrix[18] N and its associated nomenclature were introduced and exemplified in the sections on mono- and bimolecular reactions of the last chapter. A simple plant cell growth model example, adapted from (Rios-Estepa and Lange 2007), is presented here first, to illustrate its use in studying metabolism, and summarize the concept for use in subsequent chapters.

Fifth Remark on Units Conventions: If we follow the usual convention in chemistry, reaction velocities v are expressed in concentration/time units, as when we first introduced stoichiometric matrix equations in the last chapter. It may, however, be useful and easier in some applications, e.g. metabolism, to express them in mass/time units.

18 Use of N for this matrix is common in the literature. The letter S, which would be more natural, is reserved as a designation for the *relative sensitivity matrix* introduced in Chapter 11.

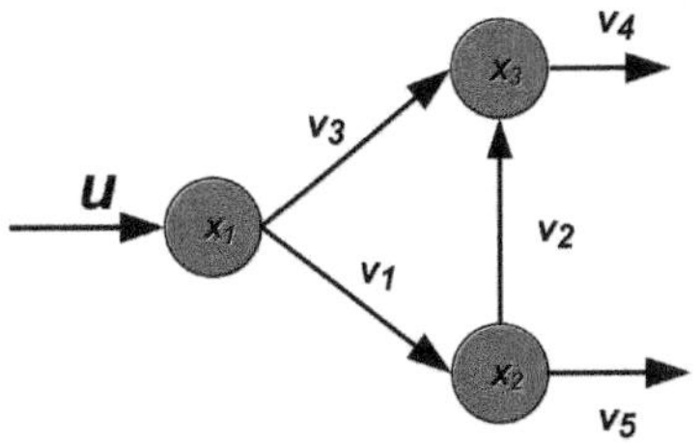

Fig. 7.6

The reactions in Eqs. (7.24) represent the metabolic network in **Fig. 7.6**. They depict import of metabolite X_1 into a plant cell at exogenous rate u (mass/time units), followed by formation of metabolites X_2 and X_3 from X_1 at mass flux rates (mass/time) v_1 and v_3 in the cell cytosol and conversion of X_2 into X_3 at mass flux rate v_3; X_2 and X_3 are then transported out of the cell to **plastids**,[19] denoted X_{2plast} and X_{3plast}, at mass rates v_4 and v_5, where they are acted upon by light. Only rates into, inside and out of the cell are part of this **flux rate balance model** for the cell, described first by the reaction Eqs. (7.24).

$$
\begin{aligned}
&X_{1exog} \xrightarrow{u} X_1 \xrightarrow{v_1} X_2 \xrightarrow{v_2} X_3 \\
&X_1 \xrightarrow{v_3} X_3 \xrightarrow{v_4} X_{3\ plast} \\
&X_2 \xrightarrow{v_5} X_{2\ plast}
\end{aligned}
\tag{7.24}
$$

For these reactions, the rate arrows between compounds are shown represented by their mass fluxes (mass/time), instead of their fractional rate constants k (time^{-1}). Think of this as a multicompartment model, with the compartment state variables x_i representing masses[20] in each compartment of the cell. The mass flux balance ODEs for (7.24) are then:

$$
\begin{aligned}
\dot{x}_1 &= u - v_1 - v_3 \\
\dot{x}_2 &= v_1 - v_2 - v_5 \\
\dot{x}_3 &= v_2 + v_3 - v_4
\end{aligned}
\tag{7.25}
$$

In vector-matrix form, Eq. (7.25) becomes:

$$
\dot{\boldsymbol{x}} = \begin{bmatrix} -1 & 0 & -1 & 0 & 0 & 1 \\ 1 & -1 & 0 & 0 & -1 & 0 \\ 0 & 1 & 1 & -1 & 0 & 0 \end{bmatrix} \boldsymbol{v} \equiv N\boldsymbol{v} \tag{7.26}
$$

where, as usual, $\boldsymbol{x} = [x_1\ x_2\ x_3]^{\mathrm{T}}$, and – *in this specialized notation – we also absorb the input term into the matrix*, as the last row of $\boldsymbol{v} = [v_1\ v_2\ v_3\ v_4\ v_5\ u]^{\mathrm{T}}$. The entire matrix is the **stoichiometric matrix** $\boldsymbol{N}$ – a matrix of signed ones (in this case) and zeros, with the three rows

19 Plastids are major organelles (cellular components) found in plant and algae cells, where chemical compounds used by the cell are manufactured and stored.

20 It is also common practice to choose the state variables as molecule concentrations, instead of masses, with adjustments for vessel volumes. Masses are easier.

representing the mass state variables x_1, x_2 and x_3, and six columns the mass fluxes $v_1,\ldots,v_5$ and u.

Equation (7.26) appears to be linear and time-invariant, at least in $\dot{\mathbf{x}}$ and $\mathbf{v}$, but not in the state vector $\mathbf{x}$. The vector of mass fluxes $\mathbf{v}$ is actually a nonlinear function of $\mathbf{x}$.Implementing the model (7.26) in a simulation language normally requires that these functions be expressed explicitly. For example, the linear or nonlinear functions of $\mathbf{x}$ components comprising the components of $\mathbf{v}$might be proportional degradations: $v_r = k_{ji}x_i$ and Hill functions: $v_i = \gamma_i x_i^n/(K_i^n + x_i^n)$.

The stoichiometric matrix N is a useful paradigm for systems biology modeling, with broad application to biosystems of any complexity. It is widely used to solve important problems in network analysis. In general, it is a **global connectivity matrix**, defining the couplings among the state variables (compartments, species) $\mathbf{x}(t)$ of $\dot{\mathbf{x}} = N\mathbf{v}$. Specification of the particular rate laws embedded in velocity vector $\mathbf{v}(t)$ then seamlessly integrates the multiple biological processes modeled by these ODEs. Nice. The architectures of systems biology modeling programs like *SimBiology* and *Copasi,* and others are based on this logical framework, as exemplified later.

The most widely established use is in **steady state metabolic flux balance analysis (FBA)**, where $\dot{\mathbf{x}} = N\mathbf{v} = 0$ is solved for the $v_{i\infty}$ in terms of the $x_{i\infty}$ components, providing metabolic fluxes from presumed measurable metabolite steady state masses or concentrations. Practical results of FBA in agriculture (and biofuel production) includes designing ways to increase plant growth (or inhibit it, if it's a weed), by manipulating component molecular species controlling specific v_i identified with growth (optimizing plant growth) (Rios-Estepa and Lange 2007). These applications are developed further in Chapter 14.

Example 7.9: ***Changing from Concentration to Mass Fluxes in Biomolecule ODE Models***

This example illustrates how to adjust terms in species concentration-based chemical/biochemical/cellular reaction equations into mass-based equations, by balancing fluxes and/or their units.

Simple flux terms like *concentration flux* = ks, where s is a (generally time-varying) species *concentration* (mass/volume) and k (time^{-1} units) is a fractional rate, are easily converted to mass fluxes. Simply multiply by volume V, i.e. *mass flux* = $ksV = kq$, with mass $q = sV$.

The procedure is somewhat more complex for nonlinear terms. A Michaelis–Menten (M–M) flux, for example, is conventionally expressed as a *concentration* flux, i.e. for substrate (species) concentration s in $v \equiv v_{max}s/(s + K_m)$. If species are expressed instead in masses q, mass flux for the M–M reaction is determined first by dividing both sides by volume V. But we also adjust the unit definitions of parameters v_{max} (concentration/time) and K_m (concentration) to yield an equation of the same form, with redefined parameter units. We do this by also multiplying numerator and denominator by V, i.e.

$$v' \equiv \frac{v}{V} = \frac{\frac{v_{max}}{V}s}{s + K_m} = \frac{\frac{v_{max}}{V}Vs}{Vs + VK_m} \equiv \frac{v'_{max}q}{q + K'_m} \text{ mass/time units} \tag{7.27}$$

This equation has the same M−M form, with mass instead of concentration per time units for species mass q and scaled Michaelis constant K'_m; the scaled v_{max}, i.e. v'_{max}, retains the same inverse time units.

Remark: In some applications (see Chapter 8), concentration-rate-balanced equations are expressed in terms of volume *clearance rates* or *flows*, e.g. $CR \equiv kVs \equiv Fs$, with F having units of volume/time.

Special Purpose Modeling Packages in Biochemistry, Cell Biology & Related Fields

A myriad of software tools and program packages are available from commercial and non-commercial sources for specifically simulating models in cell biology and biochemistry. Some of these have designated facilities for pharmacological modeling as well (e.g. in *SimBiology*,[21] an optional Toolbox *in Matlab*). Typically, they have been developed by different groups, often with particular domain applications, so they can be quite different from each other, with learning curves that can be steep, and nomenclature that can be inconsistent in some respects. Attempts are continually being made to standardize the platforms, frameworks, interfaces and nomenclature for these disparate tools, with some success. The *Systems Biology Workbench* (SBW)[22] is an open source framework connecting scores of heterogeneous software "modules" – applications for biomodeling users – for model representation and editing, simulation and analysis. The SBW website is a good source for locating these tools, as well as downloading the open-source SBW framework. The main SBW framework (broker-module) allows users to work relatively seamlessly with application modules.

A second standardization success in the making is the *Systems Biology Markup Language* (SBML, SBML.org), an XML-based programming language and syntax for representing (encoding) models in biology. SBML language syntax is being developed specifically as an exchange language for math models of biochemical reaction networks. It is not meant to be as mathematically transparent as, for example, *Mathematica* or *Matlab*. Instead, it is made up of "minimum information in the annotation of biochemical models" (abbreviated MIRIAM) statements (Novere et al. 2005). SBML is used directly in some packages, and is otherwise supported in an increasing number of programs with other primary modes of model entry. SBML code can be input into many of these programs, and some can generate SBML code as output to other programs, making it an ideal exchange language for centralized modeling frameworks like SBW.

Another open-source toolbox, *AMIGO*,[23] is illustrated in later chapters. *AMIGO* is a highly sophisticated package designed for dynamic system model quantification, optimization and experiment design. It supports nonlinear dynamic system models in mathematical formats,

21 www.Mathworks.com

22 www.SBW.org – the site has a long list of applications that can be run from the Workbench platform.

23 http://www.iim.csic.es/Bamigo/papers.html

using a simple syntax, C, FORTRAN or *Matlab*, and provides for import of SBML and black-box user-defined models as well.

With some exceptions, typical (non-SBML) entry of ODE models in most software tools for specifically simulating models in cell biology and biochemistry is done in terms of chemical reactions and/or pathway diagrams – with either special graphical or command line syntax – rather than math equations or block diagrams. We introduce and compare several of these here, by redoing simple *Matlab, Simulink* or *VisSim* modeling and simulation examples completed in earlier chapters. More complex models are developed or solved with these tools in subsequent chapters.

Example 7.10: ***Two-Compartment Nonlinear Modeling with*** **SimBiology (Matlab *Toolbox)***

We reproduce the model of Example 4.7 in Chapter 4 here, implemented and simulated using *SimBiology*. Recall that the NL influx of material into compartment 2 from 1 might represent receptor-mediated action of a biomolecule in compartment 1 on cell receptors in 2, e.g. a drug effect on a target cell. The nonlinearlity is characterized by a first-order Hill function and all other fluxes are linear.

SimBiology has *Desktop* and *Command-Line* input modalities. Both are illustrated here. In *Simbiology* syntax, *reactions* encode the model structure and *reaction rates* encode the relevant math. Models are built in these terms.

SimBiology Desktop: *Reactions* and *reaction rates* are entered in a *Simbiology table template* for creating the model (abbreviated in **Table 7.1**) Every flux in a model table needs a source and a sink. For example, reaction "q1→q2" is a textual representation of the flux from compartment 1 to 2, the NL function in this case. Similarly, for the reverse flux, q2→q1. SimBiology (and other chemistry based languages) uses "null" to indicate an exogenous flux source or sink, as is common in the language of chemical reactions, where "null" indicates that a reactant or product is not in the model. Following this format and ordering, the software automatically combines the fluxes when constructing the rate of change of q1 and q2.

A *SimBiology diagram view* of the model is generated automatically. It's available as illustrated in **Fig. 7.7**, although it doesn't usually begin in this neatly organized form. The connections are otherwise correct and the diagram is easily edited to generate a desired form, in this case arranged to mirror the original "classical" two-compartment diagram. SimBiology diagram symbols follow a different standard – in two ways shown here. First, instead of just arrows, the rate constants symbols include yellow circles (called *nodes*). A node allows a flux to connect (affect) more than two species. (One or two arrows would accomplish this in a "classic" multicompartment model.) Another is that the state variables (species) are represented by blue rectangles with rounded corners, instead of circles. Chemistry-based software packages typically use different shapes to represent different species types, not yet standardized in all. Step-input simulation results for both species are shown in **Fig. 7.8**, with the plot edited within *Matlab* for clarity.

Table 7.1 *SimBiology* Model Specification (Simplified)

#	Reaction	Reaction Rate
1	null→q1	30
2	q1→q2	C*q1/(K + q1)
3	q2→q1	k12*q2
4	q1→null	k01*q1
5	q2→null	k02*q2

#	Name	Value	Value Units
1	u	1.0	mole/second
2	C	1.0	mole/second
3	K	1.0	mole
4	k12	1.0	1/second
5	k01	1.0	1/second
6	k02	1.0	1/second

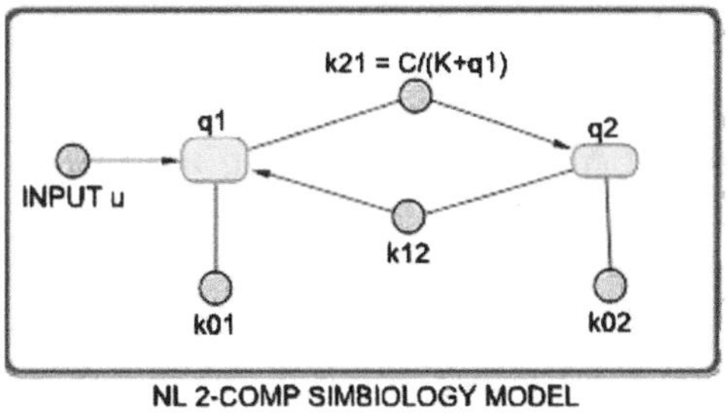

Fig. 7.7

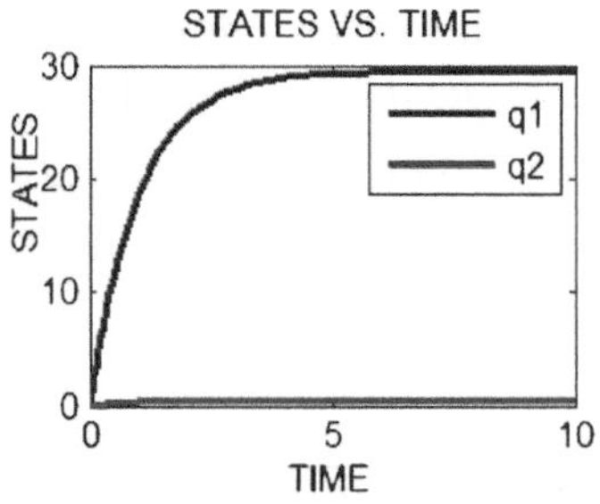

Fig. 7.8

Command-line SimBiology: In this modality, the model is created and simulated from the *Matlab* command-line, using only the "textual" language of *SimBiology*, as follows, where *species* means state variable in biochemistry language.

```
% Specify the model
model = NL2Comp('2-Compartment Model with Nonlinear Transfer Flux');
% Specify the species
model.addspecies('q1');
model.addspecies('q2');
% Specify the parameters
model.addparameter('k01', 1);
model.addparameter('k02', 1);
model.addparameter('C', 1);
model.addparameter('K', 1);
model.addparameter('k12', 1);
% Specify the reactions for the model
model.addreaction('null→q1', 'ReactionRate', 'A');
model.addreaction('q1→null', 'ReactionRate', 'k01*q1');
model.addreaction('q1→q2', 'ReactionRate', 'C/(K + q1)*q1');
model.addreaction('q2→q1', 'ReactionRate', 'k12*q2');
model.addreaction('q2→null', 'ReactionRate', 'k02*q2');
% Simulate the model
[t, q] = sbiosimulate(model);
```

Example 7.11: ***Two-Compartment Nonlinear Modeling with Copasi (Open-Source)***

The *Copasi* interface is also based primarily on the language of biochemical kinetics, a little less easy to follow than *SimBiology* model entry, e.g. users have to carry volumes throughout problem entry and solution, even for mass-based ODEs. *Copasi* does permit direct entry of ODEs, via an option in the Biochemical entry section, after defining a **global compartment** (the model "container"), which we call compartment "0" for this problem. **Fig. 7.9** illustrates three windows for this model.

At least one "compartment" (a "container") must be defined in *Copasi*. We set compartment 0 volume equal 1 at the outset and defined our species compartment symbols in molar mass terms. The "exogenous" input forcing function $u(t)$ was implemented similar to that in *SimBiology*, using Reaction mode model entry, represented as reaction 1: null→q1, as shown in the summary entry tables in **Fig. 7.9** But it took some experimentation to discover how to do this first time around. Plotted results for the same step input of 30 (not shown) are essentially the same as in **Fig. 7.8**.

Copasi, like the other packages exemplified above, also provides additional analysis tools once the model is programmed. The left column of the figure lists the various

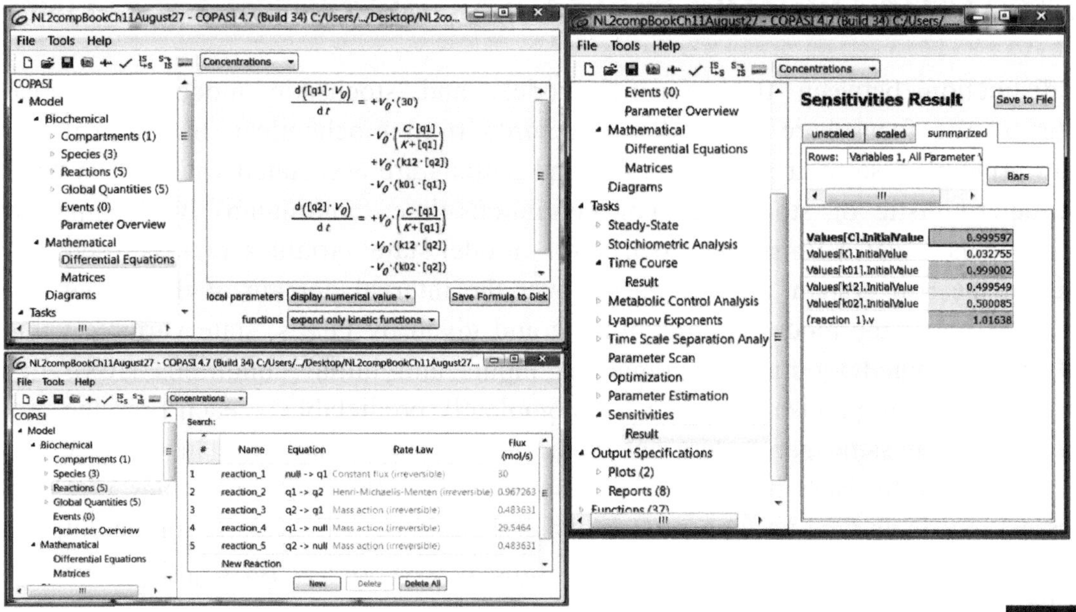

Fig. 7.9

Copasi windows for ODE & reaction rate specifications (left) & model results available from *Copasi* (right).

analyses that can be done – many potentially useful modeling fitting and analysis tools. Discussion of "Sensitivities," for example, shown in the third panel, awaits presentation of sensitivity theory in Chapter 11.

Some Afterthoughts: Each of the programs used to solve this simple NL example model problem have somewhat steep learning curves, primarily because they are designed to handle large problems, with sophisticated tools for specifically representing and simulating complex biochemical network models – in the special language of biochemical reactions. If you are used to earlier standards in defining compartment terminology, as discussed in the first sections of Chapter 4, there are some nomenclature issues to work around, as noted above. These software tools are worth learning if your problems are in this domain; and *SBW* modules, *SimBiology* and *Copasi* (and others) can provide various analyses of these models, e.g. sensitivity analysis and parameter estimation, albeit with language that may take some getting used to if you were educated in modeling in other domains. Simple problems like the one used in the two examples above – as well as many medium to large size systems biology problems – might be implemented, simulated and analyzed more readily in *Simulink, VisSim, SAAMII, Matlab, AMIGO* or *Mathematica*, directly from the ODEs. These more "mathematical" software system tools do provide all or most of the same analytical and numerical resources for solving dynamic systems biology problems, and *AMIGO* is particularly effective for fitting models to data and optimally designing experiments.

There is likely no "best" tool to use in general. Consider it fortunate that there are choices. And there will be more and better ones tomorrow.

*Stochastic Dynamic Molecular Biosystem Modeling

The distinction between deterministic models and stochastic models was made in Chapter 1. We focus here on models structured from biochemical, biophysical or other domain principles, so their state variables and outputs are associated with real system variables, deterministic or stochastic. For continuous-time (continuum) biodynamic system (BDS) models – deterministic or stochastic – model state variables $x_i(t)$ represent physical entities in the real system, for all times t over the interval domain of the model $[0, t_{end}]$, i.e. $0 \le t \le t_{end}$. If the model is deterministic and given by ODEs, state variable values are uniquely determined by model parameters, inputs and initial conditions (ICs). In other words, solutions are precisely the same – completely predictable – each time the model is solved, given the same parameters and deterministic inputs and ICs. In contrast, physical entities in stochastic BDS models are represented by *random state variables* $x_i(t)$ **(random processes)** and their time-varying probability distributions, which assign probabilities $P(x_i,t)$ to each possible value of the states $x_i(t)$ at time t, over all $t \in [0, t_{\mathrm{end}}]$. Solutions evolve according to their random variable equation solutions over time, resulting in different, "fluctuating" responses each time the model is solved. Sounds and is more complicated, mathematically speaking, so when do we want or need a stochastic model?

We address the purpose, basic theory and methods for stochastic modeling here, building on the simple probabilistic notions of the early sections of Chapter 6 on mass action modeling of first and second order chemical reactions.

When a Stochastic Model is Preferred

It is often reasonable to start modeling with deterministic, continuum approximation ODE models – as in earlier sections and chapters – and follow up with stochastic representations – as needed. Under what circumstances, then, is a stochastic model needed or to be preferred?

An underlying assumption of deterministic ODE modeling is that the physical entities represented by model state variables are *large* in number, so their responses – on average – represent the variables well. Models depicting interactions among large numbers of molecules – for example, in blood, whole solid organs or in cellular aggregates (cell cultures or homogenates) – usually satisfy this assumption. Large populations of insects or other animals in ecological systems also usually satisfy this assumption. However, in single cells and at subcellular levels, dynamic processes can involve *very small numbers* of molecules, e.g. signaling proteins and transcription factors in gene regulation. Stochastic process[24] models are needed to adequately describe dynamical interactions among very small numbers (tens to hundreds) of entities (state variables) and their less certain evolution.

* This section is undoubtedly more complex both biochemically and mathematically. It might be skimmed or skipped over on first reading, returning to it when the need arises for understanding the nuances of cellular regulation processes in greater detail.

24 Formally, stochastic process models are collections of random variables $x_i(t)$ indexed by time t, either at discrete-time instants t_0, t_1, t_2, . . ., t_{final} or continuous intervals $[t_0, t_{\mathrm{final}}]$.

Example 7.12: ***Stochastic Fluctuations Quantified***

We illustrate the magnitude of the variability that might be expected in mechanisms modeled more precisely – and possibly more appropriately – by stochastic than by deterministic models. Random fluctuations in chemical species concentrations or reaction rates are known to scale with a factor $\sqrt{N}$ (Gardiner 2004). For some intracellular events, e.g. some cell signaling processes, numbers of molecules of order of magnitude $N = 10$ may interact with a similar number of others. In this case, fluctuations of about $100\sqrt{10}/10 \simeq 32\%$ would be expected, and this may call for a stochastic model to characterize these rather large fluctuations in model state variables over time. Importantly, fluctuations *can* matter, for example, if they drive the system – usually a nonlinear one – into unstable regimes from stable ones in state space. This is an example of a stochastic model being more appropriate. In contrast, for $N = 100$, fluctuations are about $100\sqrt{100}/100 \simeq 10\%$ and, for $N = 1000$, only about $100\sqrt{1000}/1000 \simeq 3\%$. This is a basis for continuum ODE modeling being a reasonable approximation for molecular interactions numbering in at least the hundreds, becoming much better approximations for N in the thousands and greater.

*Stochastic Process Models & the Gillespie Algorithm

In the first few sections of Chapter 6 we began developing reaction kinetics with probability notions and integer numbers of molecules (or other entities) possibly reacting when they randomly collide. Deterministic ODE models were then developed for biosystems with large enough numbers of interacting molecules to warrant using the deterministic paradigm. The basics of stochastic dynamic system modeling are introduced here in that same integer number context, for smaller numbers of interacting biomolecules. In Chapter 6, interactions were occurring in a constant volume, spatially homogeneous (well stirred) vessel at constant temperature. These conditions are applied here as well. Reaction kinetic rates can be adjusted for different volumes and temperatures – if needed – as in sections on *Volume and Temperature Effects* and on *Open Biosystems* in the previous chapter.

In brief, a **dynamical stochastic process model** is defined by three things: its *states* (state variables); its *state transitions* (model topology and dynamics); and its *propensities* (random transition rates). The **propensities** are the probabilities per unit time of each of the state transitions (time^{-1} units), and they are generally nonlinear functions of the state variables. Equations are written from these properties based on known or postulated biochemistry or biophysics, e.g. by mass action interactions (kinetic theory), which also provide the propensities.

The Gillespie algorithm – also called a *direct* algorithm (Gillespie 1976, 1977) is widely used for stochastic simulation. It is a particular implementation of dynamic stochastic process models in chemistry, equivalent to solution of the chemical *Master equation*. But it "models" chemically reacting species collisions more directly, as a *discrete-time Markov stochastic process*, with stochastic events evolving over time at nonuniformly spaced intervals.

More efficient variants of the Gillespie algorithm also are available (e.g. Wilkinson 2006; Sehl et al. 2009; Ramaswamy and Sbalzarini 2011). We maintain our pedagogical focus here on the Gillespie algorithm. Most program packages that accommodate stochastic modeling and simulation of biodynamic systems include the Gillespie algorithm and/or variants of it (e.g. *Copasi, SimBiology* and others), with detailed differences available in the references.

Nomenclature & Equations

The nomenclature is modified slightly from the original here, for consistency with that in the last chapter and beyond. The Gillespie algorithm is developed for *n interacting species*: X_1, $X_2, \ldots, X_n$, possibly involved in *M reactions*: $R_1, R_2, \ldots, R_M$, with corresponding reaction *propensities* (fractional rates, time^{-1}) $a_1, a_2, \ldots, a_M$, where each a_j is generally a function of the state variables involved in reaction R_j, i.e. vector $\boldsymbol{a} \equiv \boldsymbol{a}(\boldsymbol{x})$. Lower case $x_1, x_2, \ldots, x_n$, with each x_i designating a time varying state variable $x_i(t)$, represent the integer *number of molecules* (units) of X_i involved in a reaction at any time *t*. The $x_i(t)$ are distributed with probability $P(x_i,t)$ at each *t*. Changes in x_i occur only by discrete amounts, which can become zero (extinct). The x_i also can be expressed as number per unit volume (concentration) units, with scaling usually applied to the propensities to maintain their units as time^{-1}, as illustrated below.

Propensities: Each a_j is the "propensity" for reaction R_j to occur in the next moment, i.e. probability per unit time of reaction R_j occurring. The vector of propensities $\boldsymbol{a} \equiv \boldsymbol{a}(\boldsymbol{x})$ is established by mass action principles, by separating it into a product of terms, as follows. **Propensity** $a_j(x_1, x_2, \ldots, x_n)$ is defined as:

$$a_j(\boldsymbol{x}) = c_j h_j(\boldsymbol{x}) \tag{7.28}$$

In (7.28), c_j is the specific *probability rate constant* (or **stochastic rate constant**) for reaction R_j. That is, $c_j dt$ is the average probability that a randomly chosen pair of R_j reactant molecules will actually react in the next infinitesimal time interval *dt*. It depends only on the physical properties of the reacting molecules and the temperature of the well-mixed system.[25] It is therefore directly related to but is not necessarily the same as the kinetic rate constant *k* for the reaction, because scaling may be needed, as illustrated in the examples below.

The second term on the RHS of (7.28), $h_j(\boldsymbol{x})$, is the number of distinct combinations of R_j reactant molecules available in state $\boldsymbol{x}$, i.e. the *number of opportunities for collision*. $h_j(\boldsymbol{x})$ depends on the number of reacting molecules of each species, and is resolved by combinatorial analysis – but not as simply as in the section on bimolecular reactions in the previous chapter, as illustrated below.

Remark: At most two species can react in any reaction R_j. Simultaneous collision of three species is highly unlikely and therefore considered negligible. Because reactions in stochastic theory are defined in this way as *elementary*, and also *unidirectional*, reversible reactions are represented as two reactions, one forward and one back, each with its unique propensity. We introduced elementary reactions in the previous chapter, in part, for this reason.

25 See the section on bimolecular reactions in the previous chapter.

Each a_j in reaction R_j (forward or backward) equals the product of the stochastic rate constant c_j times the number $h_j(x)$ of opportunities for collision of possible reactants in the given unidirection, with c_j and $h_j(x)$ generally different in each direction.

In the next two examples, we examine the deterministic ODE model forms first, to illustrate the relationship between deterministic rate constant k_j with stochastic reaction constant c_j, as well as evaluation of $h_j(\boldsymbol{x})$ in two simple situations.

Example 7.13: ***Propensity for a Stochastic Replacement Reaction***

Consider reaction R_1 in which reactant species X_1 combines with X_2 to form X_3, called a **replacement reaction** in a well-mixed system with volume V:

$$R_1 \colon X_1 + X_2 \xrightarrow[v_1]{k_1} X_3 \tag{7.29}$$

In continuum model development of the mass action principle in the previous chapter, Eq. (6.13) provided the ODEs and reaction velocity v in terms of rate constant k for this bimolecular reaction, namely: $-\dot{x}_1 = -\dot{x}_2 = \dot{x}_3 = v_1 = k_1 x_1 x_2$. Note that k_1 here has units of (time • concentration)$^{-1}$ if the x_i are concentrations; or (time • number)$^{-1}$ if the x_i are molecule *number*. The stoichiometric equation equivalent is: $\dot{\boldsymbol{x}} = \begin{bmatrix} -1 & -1 & 1 \end{bmatrix}^T v_1$.

To develop the stochastic algorithm, we need the *propensity* for R_1 to occur, i.e. the probability P_1 that R_1 occurs in the next time interval dt. This can be obtained from the reaction rate equation $v_1 = k_1 x_1 x_2$. However, the units need to be volume adjusted to get the right units for $a(\boldsymbol{x})$. Emulating Eq. (5.12) for a bimolecular reaction, with the stochastic rate constant c_1 adjusted by V, and the number of opportunities for collision $h_1(\boldsymbol{x}) = x_1 x_2$, we get:

$$P_1 = \frac{k_1}{V} x_1 x_2 dt \equiv a_1(\boldsymbol{x}) dt \Rightarrow a_1(\boldsymbol{x}) \equiv c_1 h_1(\boldsymbol{x}) = \frac{k_1}{V} x_1 x_2 \tag{7.30}$$

Example 7.14: ***Propensity for a Stochastic Dimerization Reaction***

If $X_1 = X_2$ in (7.29), $2X_1 \xrightarrow[v_2]{k_2} X_3$, with X_3 the *dimer*, and the reaction kinetic equations are $\dot{x}_3 = k_2 x_1^2 / V = v_2 = -\dot{x}_1/2$, $\dot{\boldsymbol{x}} = \begin{bmatrix} -2 & 1 \end{bmatrix}^T v_2$. The stochastic propensity for this reaction is *not* simply $k_2 x_1^2 / V$, despite scaling for volume. The number of distinct pairs of X_1 (distinct subsets with two elements), possibly combining to form one X_3 is $h_2(\boldsymbol{x}) = x_1(x_1 - 1)/2$, obtained by *combinatorial analysis*, using the **binomial coefficient** $\binom{x_1}{2} \equiv x_1(x_1 - 1)/2$! This is the number of pairs of distinct X_i that can possibly react combinatorially. Then, from the definition of propensity, the probability that reaction R_2 will occur *is* $a_2(\boldsymbol{x}) \equiv c_2 h_2(\boldsymbol{x}) = c_2 x_1(x_1 - 1)/2$, which should approach $k_2 x_1^2 / V$ in the limit as x_1 gets large enough. So the relationship between k_2 and c_2 is $2k_2/V = c_2$ and $a_2(\boldsymbol{x}) \equiv k_2 x_1 (x_1 - 1)/V$.

Remark: We see that the deterministic rate constant k and the stochastic reaction probability constant c generally differ by possibly two constant factors: the volume $V^{(1-M)}$, and the integer M, where M is the number of reactants. For one reactant, no volume or other adjustment is needed and $c = k$. For two reactants, the volume adjustment is V^{-1} and[26] $c = 2k/V$.

The "linear" *stoichiometric ODE model* is the most convenient form for implementing the Gillespie algorithm, with propensity $\boldsymbol{a}$ replacing rate $\boldsymbol{v}$:

$$\dot{\boldsymbol{x}}(t) \equiv \dot{\boldsymbol{x}} = N\boldsymbol{a} = N\boldsymbol{a}[\boldsymbol{x}(t)], \quad t \in [0, t_{\text{end}}] \tag{7.31}$$

N is the constant n by $(M+1)$ *stoichiometric matrix*, with (vector) input $\boldsymbol{u}$ included as the last column of N, thereby providing for possible exogenous inputs from the reaction environment. The elements η_{ij} of N are the *stoichiometric coefficients* of the reactions $R_1, R_2, \ldots, R_M$. Collectively they specify the algorithmic changes in the numbers of molecules in each x_i at each state transition, e.g. $x_i \rightarrow x_i + \eta_{ij}$, if reaction M_j with propensity a_j occurs.

Strictly speaking, the dynamics of this system should be written in terms of probabilities $P(x_i,t)$ instead of state variables x_i. Eq. (7.31) is formally correct with x_i treated as an individual **realization** (an experimental outcome) of the stochastic process over $[0,t_{end}]$. The statistical properties of the **ensemble** collection of these ***a posteriori*** realizations (histories) is of particular interest in model analysis, as discussed and illustrated below.

*The Chemical Master Equation

The continuum stochastic process defined above is governed by continuous-time state transition dynamics with integer-valued states (particle numbers). The *differential-difference equations* describing this process are called the **chemical master equations** for the dynamics of the probabilities $P(x_i,t)$ of state x_i transitions at times $t \in [t_0, t_{end}]$, one for each state variable $x_1, x_2, \ldots, x_n$. The n *linear* ODEs have the form:

$$\dot{P}(x_i, t) = \sum_{j=1}^{M} a_j(x_i - s_{ij})P(x_i - s_{ij}, t) - \sum_{j=1}^{M} a_j(x_i)P(x_i, t), \quad i = 1, \ldots, n \tag{7.32}$$

over $t \in [0, t_{\text{end}}]$. Eq. (7.32) is very difficult to solve, analytically or numerically, for any but the simplest models. But it does provide insight into how state transitions can be computed using

26 Although we have assumed it is highly unlikely, it is instructive to consider the effects on stochastic reaction parameters of more than two molecules colliding at the same time. Combinatorial analysis indicates that, if three molecules *did* combine, $c = 3!k/V_2 = 6k/V_2$, etc. In general, for elementary *forward or backward* reaction j, if more than two species did combine, i.e. if $\eta_{1j}X_1 + \eta_{2j}X_2 + \ldots + \eta_{nj}X_n \xrightarrow[v_j]{k_j} \ldots$, the *stochastic reaction constant* c_j for elementary forward or backward reaction j is: $c_j = \frac{k_j}{(VA_v)^{R_j-1}} \prod_{i=1}^{n} \eta_{ij}!$, with $R_j = \sum_i \eta_{ij}$. Avogadro's constant $A_v = 6.022 \times 10^{23}\ \text{mol}^{-1}$ is also included in this formula, to convert propensities to *molar* units. Model parameters in biochemical reaction models usually are expressed in molar units, e.g. the Michaelis constant K_m in mols per liter. Similarly, the h component of $a_j(\boldsymbol{x}) = c_j h_j(\boldsymbol{x})$ generally equals the product of binomial coefficients: $h_j(\boldsymbol{x}) = \binom{x_1}{\eta_{1j}}\binom{x_2}{\eta_{2j}}\cdots\binom{x_n}{\eta_{nj}}$ with $\binom{x}{\alpha} \equiv \frac{x(x-1)\ldots(x-(\alpha-1))}{\alpha!}$.

other approaches. The Gillespie algorithm provides a direct, unit-process driven realization (solution) for the evolution of state variables $\boldsymbol{x}(t)$ in Eq. (7.32), in discrete "jumps" (a random-walk) at random times, as in a continuous-time Markov jump process, described briefly in Chapter 1.

For each i, the RHS of (7.32) describes the $2M$ possible transitions that can change $P(x_i, t)$, and thus $x_i(t)$, each time resulting in a gain or loss of s_{ij} molecules of x_i. The first term in probability flow equation (7.32) accounts for all positive transitions from other states to x_i. The second accounts for all negative transitions from x_i to other states.

*The Gillespie Algorithm & Variants

This algorithm codifies the answers to two questions about reactions among the entities $X_i(t)$ of a dynamical biosystem model: *when* reactions R_j take place; and *which* reaction R_k among the M possible ones it will be.

Two random numbers are drawn to make these adjustments at each iteration step. Random and variable time-steps τ and next-reactions R_j are determined from these random numbers, by *Monte Carlo simulation* at each step. The τ are drawn from an *exponential* distribution of probabilities (the physics of variable time-steps). Reaction number j is drawn from a *uniform* probability distribution (because it is completely random). Model responses, i.e. the stepwise evolution of the transitions resulting from these choices, are then computed. The molecule count is then updated for all x_i based on the reaction that occurred, using the stoichiometric model equations (7.31), and the procedure is repeated until t_{end} is reached, or the reactants are exhausted. A step-by-step program *flowchart* of the algorithm is given in **Fig. 7.10**. Additional explanatory details in the flowchart are given below.

Monte Carlo Steps: The stepping *time variable* (time stepsize) τ must be drawn randomly from an *exponential* distribution of probabilities. It is computationally easier and equivalent to instead draw a random number r_1 from a *uniform* distribution on [0, 1] (Gillespie 1976) and compute random stepsize τ from r_1 using the logarithmic transformation:

$$\tau = \ln(1/r_1)/a_0 \quad \text{with} \quad a_0 \equiv \sum_{j=1}^{M} a_j \tag{7.33}$$

The *random reaction* R_j is chosen directly from a *uniform* distribution, by selecting a second random number r_2 from the unit interval [0, 1]. To make the choice, the propensities $a_i(\boldsymbol{x})$ are used to establish relative reaction weights. These weights are: $P(\text{transition } M_i) = a_i/a_0 = a_i/\sum_j a_j$. Then transition via reaction R_i at time $t + \tau$ is chosen if i satisfies:

$$SUM_{i-1} \equiv a_1 + a_2 + \cdots + a_{i-1} \leq a_0 r_2 \leq a_1 + a_2 + \cdots + a_i \equiv SUM_i \tag{7.34}$$

For example, if random number r_2 falls between 0 and a_1/a_0, R_1 is the reaction randomly chosen at time $t + \tau$. If, instead, r_2 falls between $(a_1 + a_2)/a_0$, then R_2 is the reaction chosen for time $t + \tau$. And so on.

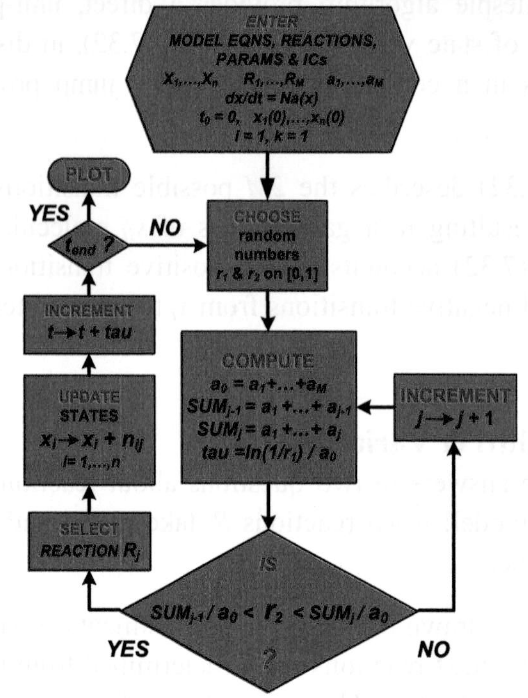

Fig. 7.10

Gillespie algorithm flowchart.

Once the reaction takes place at $t+\tau$, the number of molecules x_j changes, so these numbers must be updated each cycle. This is how the state variables evolve from one random time-step to the next, in accordance with stoichiometric Eq. (7.31). For example, for R_1 in (7.30), state variable x_1 is replaced by $x_1 - 1$, x_2 by $x_2 - 1$ and z by $z + 1$, because the stoichiometric coefficients η are -1, -1 and $+1$, representing the number of molecules subtracted and added to the two species each cycle.

The random process begins again with these newly updated state variable numbers, as in **Fig. 7.10**.

Algorithm Properties

The algorithm is typically run through many cycles, collecting an ensemble of histories, all different – because they began with different random r_1 and r_2 – but together providing inherent statistical information needed to evaluate results. For example, these statistics inform about the magnitude of fluctuations expected from real biosystem state variables.

The Gillespie algorithm can be very computationally intensive, especially for "stiff" problems, where certain reactions are much faster than others and thus occur much more frequently. In these cases, other algorithms (e.g. hybrid, tau leaping; Gillespie 2001; Sehl et al. 2009; Mélykúti et al. 2010), mentioned earlier, can be more efficient.

The Gillespie algorithm or its variants are the most widely used, but other approaches to stochastic simulation are available. One is to run deterministic ODE models with stochastic (random) input forcing functions added, e.g. as in Geva-Zatorsky et al. (2010). This approach may follow the chemical Langevin equation or Fokker Plank equations (Gillespie 2000), with stochastic Weiner process models (Sen and Bell 2006; Mélykúti et al. 2010), which typically have Markov-1 Gaussian white noise inputs that model the statistics (means and covariance parameters) of the input stochastic process. Statistical properties are typically either known from data or estimated from data along with ODE model parameters (Sung and Lee 2004). Some math and simulation software packages (*Matlab, Simulink, VisSim, Mathematica*, etc.) provide for such functionality, thereby permitting computation of noisy or fluctuating model state variable and output responses to noisy inputs.

*Stochastic Model Analysis: Ensemble Statistics

Ensembles (collections of individual runs) of say Q realizations collected by running the algorithm many (Q) times over the interval $[0, t_{end}]$ are typically analyzed by computing mean-value functions, covariances or time-correlations among the x_i, and possibly also probabilities for particular kinds of behavior. These statistics are useful, for example, to estimate the dynamical characteristics of the same process in Q different cells. They are obtained from the **first and second moment functions** by averaging the first or second powers of the numbers recorded for $x_i(t)$ at times t in these runs, i.e.

$$x_i^{(1)}(t) = (1/Q)\sum_{l=1}^{Q} x_{il}(t) \text{ and } x_i^{(2)}(t) = (1/Q)\sum_{l=1}^{Q} x_{il}^2(t) \tag{7.35}$$

for all $x_i(t)$ computed on the interval $[0, t_{end}]$. Then **sample mean** $\bar{x}_i(t) \equiv x_i^{(1)}(t)$ and **sample variance** $\bar{\sigma}_i^2(t) = x_i^{(2)}(t) - \bar{x}_i(t)$ functions measured over time interval $[0, t_{end}]$ (and other functions of these) are computed from these moments.

Following are several different measures of the *reliability* with which random state variable x_i might be estimated from a time window $[0, t_{end}]$ that on average contains several random events.

Root-mean-square (RMS) fluctuations (sample standard error functions) about average responses are typical of statistical results characterizing a stochastic model:

$$RMS_i(t) \equiv \sqrt{x_i^{(2)}(t) - \bar{x}_i^2(t)} = \sqrt{\bar{\sigma}_i^2(x_i(t))}, \quad 0 \le t \le t_{end} \tag{7.36}$$

The **coefficient of variation CV** is a common statistical measure of the *relative variability* (relative noise-to-signal ratio, dispersion) in an experimental outcome, namely the sample standard deviation $SD = \bar{\sigma}$ divided by the mean. For this stochastic process,

$$CV_i(t) = \bar{\sigma}(x_i(t))/\bar{x}_i(t), \quad 0 \le t \le t_{end} \tag{7.37}$$

The quite similar **Fano factor**[27] $F(x_i)$ is another relative measure, in this case of the relative *variance* dispersion of a distribution measured over a time interval $[0, t_{end}]$ (Fano 1947):

$$F(x_i) = \frac{\bar{\sigma}_i^2(x_i(t))}{\bar{x}_i(t)}, \quad 0 \le t \le t_{end} \tag{7.38}$$

27 $F(x_i)$ is 1 in the limit for Poisson processes.

The **sample covariance matrix** function $\overline{COV}(x(t))$ for the entire n-dimensional random x-vector is also available from ensembles of Q realizations:

$$\overline{COV}(x(t)) \equiv \frac{1}{Q-1}\sum_{j=1}^{Q}(x(t)-\overline{x}(t))(x(t)-\overline{x}(t))^{\mathrm{T}} \tag{7.39}$$

State vector $x(t)$ in Eq. (7.39) is the unaveraged values of the Q data sets. The $Q-1$ in the denominator of Eq. (7.39), instead of Q, eliminates bias in sample estimation of covariances (Cramér 1946). The sample variances $\overline{\sigma}_i^2$ are on the diagonal of $\overline{COV}(x)$.

Example 7.15: ***Stochastic mRNA–Protein Dynamics via the Gillespie Algorithm***

This example is based on the same simple, single gene model of protein (P) molecule production from mRNA simulated and analyzed in Chapter 3 (**Fig. 3.23**), using a deterministic ODE model. The Gillespie algorithm is applied here to $Q=200$ different cells (200 independent runs begun with different random numbers), each with the same parameter values, burst factor $b=2$, and beginning again with one DNA (D) molecule, and no mRNA (M) or protein P molecules present.

A *Matlab* script for this set of 200 runs is shown in **Fig. 7.11**. It is a contracted and modified version of code for this model published by Hayot and Jayaprakash (2008). It essentially follows the algorithm outlined in the flowchart of **Fig. 7.10**. About half of this code is the algorithm, the remaining half is setting up the run variables and parameters and plotting results.

Results are illustrated in **Fig. 7.12**. From the top, the first panel is the sample mean of the protein number over all 200 cells. It's roughly the same as the deterministic

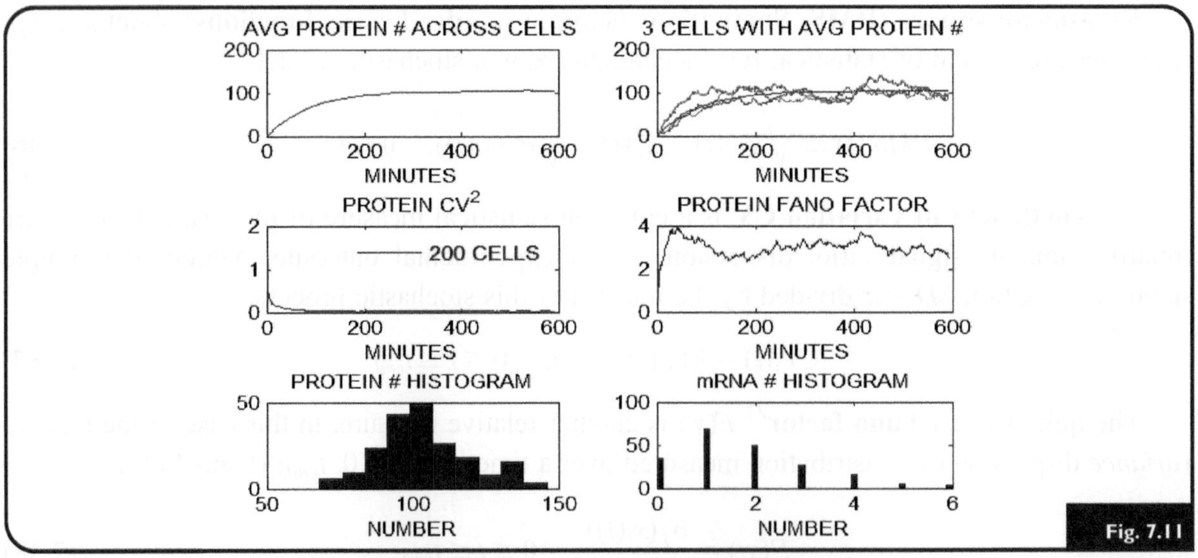

Fig. 7.11

Stochastic single-gene model of mRNA-protein dynamics and statistics, run for Q5200 individual cells, each beginning with 1 DNA and no mRNA or protein molecules. Highlights: Individual patterns (upper right) follow the mean curve fairly well, but histogram dispersions about average protein and mRNA numbers (103 and 1.73) are quite large (bottom).

```
%SINGLE GENE MODEL from Thattai-van Oudenaarden 2001
% Matlab script adapted from: Hayot & Jayaprakash(2008)
% time in minutes, Vol = 1, r = mRNA, d = DNA, p = protein
%  run parameters
 itime=1000; % file size for protein values & averages
 nens=200;   % # of runs (or individual cells)
 timelimit=600;  % tend = 600 min (10 hrs)
 b=2;c1=0.6;c2=log(2)/2;c3=b*c2;c4=log(2)/60;
 avp=0; avp2=0; tim=zeros(itime-1,1);npt=zeros(itime-1,1);
 si1=zeros(itime-1,1);si2=zeros(itime-1,1);
 si3=zeros(itime-1,1);npt2=zeros(itime-1,1);
 np1=zeros(itime-1,1);np12=zeros(itime-1,1); for is=1:nens
% intialization
 d=1; r=0; p=0; time=0; izt=1;
   while time < timelimit
 sum(1)=0;sum(2)=0;sum(3)=0;sum(4)=0;sum(5)=0;
 a1=c1*d;a2=c2*r;a3=c3*r;a4=c4*p;a0=a1+a2+a3+a4;
 r1=rand(1,1);r2=rand(1,1);tau=-(1./a0)*log(r1);yr2=r2*a0;
 sum(2)=sum(1)+a1;sum(3)=sum(2)+a2;sum(4)=sum(3)+a3;
 sum(5)=sum(4)+a4;  for k=2:5;
   if (sum(k) >= yr2) && (sum(k-1) < yr2) mu=k-1;end; end;
   if (mu == 1) r=r+1; end;
   if (mu == 2) r=r-1; end;
   if (mu == 3) p=p+1; end;
   if (mu == 4) p=p-1; end;
       ztime=timelimit./itime;
   if time > izt*ztime time;izt;izt*ztime;
        np1(izt)=np1(izt)+p;np12(izt)=np12(izt)+p^2;
   if (is ==23)  si1(izt)=p;  end;
   if (is ==43)  si2(izt)=p;  end;
   if (is ==63)  si3(izt)=p;  end;
        izt=izt+1;  end; time=time + tau;  end
%   x vector contains steady state values for all cells
   np1;x(is)=p;xr(is)=r;avp=avp+p;avp2=avp2+p.^2; end
%  steady state statistics, proteins avgd over all cells
   avp=avp./nens;avp2=avp2./nens;var=avp2-avp.^2;
  standarddev=sqrt(var);fano=var./avp;cvar=sqrt(var./avp^2);
%  time behavior of proteins avgd over all cells
  for jt=1:itime-1; tim(jt)=ztime*jt;  end
  np1=np1./nens;np12=np12./nens;coeff=(np12-np1.^2);
  fao=coeff./np1; cvars=fao./np1;
  tim1=tim(1:itime-10,1);fao1=fao(1:itime-10,1);
  cvars1=cvars(1:itime-10,1);npt1=np1(1:itime-10,1);
  si11=si1(1:itime-10,1);si21=si2(1:itime-10,1);
  si31=si3(1:itime-10,1);
  subplot(3,2,1), plot(tim1,npt1);
  title('AVG PROTEIN # ACROSS CELLS');xlabel('MINUTES');
  subplot(3,2,2), plot(tim1,npt1);
  title('3 CELLS WITH AVG PROTEIN #');xlabel('MINUTES');
  hold on; subplot(3,2,2), plot(tim1,si11,'r');hold on
  subplot(3,2,2), plot(tim1,si21,'g');hold on
  subplot(3,2,2), plot(tim1,si31,'m')
  subplot(3,2,3),  plot(tim1,cvars1)
  title('PROTEIN CV^2');xlabel('MINUTES');
  legend('200 CELLS',1);legend('boxoff');
  subplot(3,2,4),plot(tim1,fao1)
  title('PROTEIN FANO FACTOR');
  xlabel('MINUTES');
%  histogram of steady state protein numbers
  subplot(3,2,5);hist(x,10);title('PROTEIN # HISTOGRAM');
  xlabel('NUMBER');hold off;subplot(3,2,6);hist(xr,50);
  title('mRNA # HISTOGRAM');xlabel('NUMBER');  hold off
```

Fig. 7.12

Matlab script for running the Gillespie algorithm on the single-gene, mRNAprotein dynamics stochastic model. Evaluating primary statistics and plotting results are included.

solution simulated in **Fig. 3.23**. The top right panel is the same, but with three individual cell responses superimposed, illustrating the stochastic response variability for those three cells. Protein CV^2 falls to ≈ 0.17 (about 17% "error"), which is relatively small, and the Fano factor illustrates an approximately constant dispersion of about 2.8 for the whole period, because the sample variance is about 285, SD about 17. The two histograms in the bottom panel indicate a fairly wide dispersion among individual cells, clearly demonstrating that neither the average protein number (103) nor average mRNA number (1.73) predict their actual number in any single cell.

Remark: Although the Gillespie algorithm is relatively simple, the code for setting it up, running it and printing results is fairly involved if programmed in a general math programming system like *Matlab*, or *Mathematica*. One or more of the special purpose packages for systems biology modeling, one that includes provision for stochastic modeling (e.g. *SimBiology*), may provide more facile solutions.

Exercises

E7.1 ***Opening a Closed Model***. Consider the drug-receptor model of Example 7.3. Given the parameters, we can solve the ODEs, which have no input terms, uniquely using the given ICs. Suppose only receptors are in the medium at time zero (no drug) and drug is introduced exogenously.

a) Adjust the equations of the model accordingly.

b) Repeat (a) assuming drug is also metabolized linearly in the medium.

E7.2 ***Hill Function Applications***. Under what biologically mechanistic circumstances can one use one or more of the Hill functions shown in **Fig. 7.2** (for some n) in: (a) a gene network model; (b) an enzyme kinetics model; and (c) a receptor binding model. Give examples.

E7.3 ***Linearized Mass-Action-Based Kinetics:*** Linearize the following model about nominal solutions x_N and u_N.

$$\dot{x}_1 \equiv f_1(\mathbf{x}, u) = c_1 u x_2 - c_2 x_1^2 \tag{7.40}$$

$$\dot{x}_2 \equiv f_2(\mathbf{x}, u) = \frac{c_3 x_1}{c_4 + x_1} - c_5 \tag{7.41}$$

$$y \equiv g(\mathbf{x}) = c_6 x_1 \tag{7.42}$$

The term $c_1 u x_2$ in Eq. (7.40) may represent (for example) a mass action reaction between a protein (u) and a substrate (x_2); the term $-c_2 x_1^2$ may represent a second-order elimination process; the first term in Eq. (7.41) may represent a reaction with relatively very rapid Michaelis–Menten kinetics; and the term $-c_5$ may represent a zero-order detoxification or elimination process.

E7.4 ***Cofactor-Coupled Reactions*** are those in which one compound (e.g. ATP) donates a moiety (e.g. a phosphate group P) to another compound (e.g. a substrate S), yielding a new compound (SP = phosphorylated S) containing the moiety, plus a transformed version of the first compound (e.g. ADP from ATP): $S + ATP \underset{v_{-1}}{\overset{v_1}{\rightleftharpoons}} SP + ADP$. Derive the *stoichiometric matrix N* for this reaction.

E7.5 ***Equilibrium Association Constant***. Find the K_a for the hypothetical reversible reaction:

$$n_1X_1 + n_2X_2 + \cdots + n_rX_r \underset{k_{-1}}{\overset{k_1}{\rightleftharpoons}} m_1Y_1 + m_2Y_2 + \cdots + m_sY_s \tag{7.43}$$

E7.6 Find the **stoichiometric matrix** N for reaction (7.43).

E7.7 ***Reversible Dimerization of 2 Proteins***. In some intracellular systems, two proteins P_1 and P_2, or two of the same proteins, combine reversibly to form a dimer prior to having a specific regulatory action (Example 3.6 of Chapter 3 illustrated dimerization in a simple gene network). A simple bimolecular association reaction might take the form:

$$P_1 + P_2 \underset{k_{-1}}{\overset{k_1}{\rightleftharpoons}} P_1 \bullet P_2 \equiv D \quad \text{(dimer)} \tag{7.44}$$

Write the ODEs, equilibrium dissociation constant K_d and stoichiometric matrix N for this reaction.

E7.8 ***NL Modeling from Multiple Experiments***. A constant unit-step input of drug X is given to a subject intravenously for two weeks, and its transient response appearance in blood is initially approximately biexponential, followed by an approximately constant steady state – achieved after about a week. There was no drug in the system to begin with. Kinetically, the drug in blood exchanges reversibly with drug in tissues over time, and is catabolized in tissues. It's not eliminated in or from the body in any other manner.

a) Draw a diagram of a compartmental model of the pharmacokinetics of this drug in blood and tissues, one that might be compatible with these experimental results.
After a few weeks another experiment is done on the same individual. With zero initial conditions again, an exogenous step-input of half the magnitude of the earlier study, producing a steady state value of about half the magnitude of the unit-step response given earlier.
The whole experiment was repeated again a month later, this time infusing a dose five times greater than in the first (unit-step) input study. This study yielded a steady state response about three times the magnitude of the first (unit-step) input study.

b) Based on these two additional studies, do you need to modify your original model developed in part (a)? Why?

c) Suggest a few alternative "mechanistic" modifications you might make to the original model diagram to make it more consistent with all the data.

d) How might you distinguish these alternative models? (Brief answer, not an involved one.)

E7.9 ***NL Modeling***. The equations for a two-compartment model with nonlinear influx of material into compartment 2 from 1 are given below. The nonlinearity represents receptor-mediated action of a biomolecule in compartment 1 on cell receptors in compartment 2, e.g. a drug effect on a target cell. The nonlinearity is characterized by a second-order Hill function and all other fluxes are linear.

$$\dot{q}_1 = -k_{01}q_1 - \frac{Aq_1^2}{B + q_1^2} + k_{12}q_2 + u$$

$$\dot{q}_2 = \frac{Aq_1^2}{B + q_1^2} - (k_{02} + k_{12})q_2$$

Draw the compartmental model, labeling its components (with symbols or words).

E7.10 ***NL Modeling***. Consider a *saturable metabolic reaction pathway* in which species 1 (with mass q_1) is converted to species 2 (with mass q_2) within a well-mixed vessel, e.g. in (idealized) liver cells. A fraction α of the total mass flux $v_{tot} = \frac{Aq_1}{B + q_1}$ (mass/time units) from species 1 goes to species 2. Additionally, the remaining fraction $1 - \alpha$ of v_{tot} is irreversibly lost to a sink for species 1, e.g. the biliary pathway from liver to gut. Species 2 is, in turn, converted back to species 1 (in liver cells) at mass flux $k_{12}q_2$ by a different metabolic mechanism, linear under the conditions of the reactions, and is also lost irreversibly at a linear mass flux rate $k_{02}q_2$. Concentration is measured in compartment 1. Draw the compartmental diagram and write the equations for this model.

E7.11 ***Simple Stochastic Modeling & Analysis using the Master Equation***. The simple irreversible isomerization reaction $X \xrightarrow{c} Z$ has the deterministic kinetic model $\dot{x} = -cx$ with solution $x(t) = x_0 e^{-ct}$.

a) Write the Master equation (7.32) for this process.

b) Show that the solution of the Master equation is the binomial probability function:

$$P(x,t) = \frac{x_0!}{x!(x_0 - x)!} e^{-cxt}[1 - e^{-ct}]^{x_0 - x} \qquad x = 0, 1, \ldots, x_0 \tag{7.45}$$

Show that the mean value and RMS deviation functions, Eq. (7.36), for this stochastic process are:

$$x^{(1)}(t) = x_0 e^{-ct}$$
$$RMS(t) \equiv \sigma(t) = \sqrt{x^{(2)}(t) - [x^{(1)}(t)]^2} = \sqrt{x_0 e^{-ct}(1 - e^{-ct})} \quad (7.46)$$

where the mean value function is the first moment $x^{(1)}(t)$ and $x^{(2)}(t)$ is the second moment.

E7.12 ***Simple Stochastic Modeling & Analysis using the Gillespie Algorithm.*** Repeat the previous exercise using Gillespie's direct algorithm, with $c = 0.5$ and $x_0 = 10$, 100 and 1000 (three different starting numbers of X molecules). Run the algorithm enough times for each x_0 to collect statistics and plot the results of a typical run, overlayed with sample mean value functions $\bar{x}(t) \pm 2\bar{\sigma}(t)$, which provides the 95% confidence limits of the simulated stochastic process. Compare the results obtained with both methods. In what sense are they the same (different)? To accomplish these tasks, coding and running your own code for this simplest problem can go far toward solidifying understanding of the algorithm.

E7.13 ***Transcription-Factor–Transcription–Translation Stochastic Simulation.*** Repeat the gene transcription problem in Exercise E3.8 (with transcription factor included) by stochastic simulation using the Gillespie algorithm. You can do this by modifying the *Matlab* script in **Fig. 7.11**. Or, if you have access, use another software package, e.g. *SimBiology* or *Copasi*.

E7.14 Redo Example 7.15 using an available program package, e.g. *Simbiology*, *Copasi*.

REFERENCES

Alon, U., 2007. An Introduction to Systems Biology. Chapman & Hall/CRC, Boca Raton, FL.

Balsa-Canto, E., Banga, J.R., 2011. AMIGO, a toolbox for advanced model identification in systems biology using global optimization. Bioinformatics. 27 (16), 2311–2313.

Bhorghans, J.A.M., DeBoer, R.J., Segel, L.A., 1996. Extending the quasi-steady state approximation by changing variables. Bull. Math. Biol. 58, 43–63.

Chène, P., 2001. The role of tetramerization in p53 function. Oncogene. 20, 2611–2617.

Ciliberto, A., Capuani, F., Tyson, J., 2007. Modeling networks of coupled enzymatic reactions using the total quasi-steady state approximation. PLoS Comput. Biol. 3 (3), e45.

Cramér, H., 1946. Mathematical Methods of Statistics. Princeton University Press, Princeton, NJ.

Fano, U., 1947. Ionization yield of radiations. II. The fluctuations of the number of ions. Phys. Rev. 72, 26–29.

Flach, E.H., Schnell, S., 2006. Use and abuse of the quasi-steady-state approximation. IEE Proc. – Syst. Biol. 153 (4), 187–191.

Gardiner, C.W., 2004. Handbook of Stochastic Methods for Physics, Chemistry and the Natural Sciences. Springer-Verlag, Berlin.

Geva-Zatorsky, N., Dekel, E., Batchelor, E., Lahav, G., Alon, U., 2010. Fourier analysis and systems identification of the p53 feedback loop. Proceedings of the National Academy of Sciences. 107 (30), 13550–13555.

Gillespie, D.T., 1976. A general method for numerically simulating the stochastic time evolution of coupled chemical species. J. Comput. Phys. 22, 403–434.

Gillespie, D.T., 1977. Exact stochastic simulation of coupled chemical reactions. J. Phys. Chem. 81 (25), 2340–2361.

Gillespie, D.T., 2000. The chemical Langevin equation. J. Chem. Phys. 113, 297.

Gillespie, D.T., 2001. Approximate acceleerated stochastic simulation of chemically reacting systems. J. Chem. Phys. 115, 1716–1733.

Hayot, F., Jayaprakash, C., 2008. A tutorial on cellular stochasticity and Gillespie's algorithm. PRIME Technical Report 2008. <http://www.cs.princeton.edu/picasso/mats/matlab/princeton_spring06.pdf>.

Hill, A.V., 1910. The combination of haemoglobin with oxygen and with carbon monoxide. Int. J. Physiol. 40, iv–vii.

Hori, S.S., Kurland, I.J., DiStefano III, J.J., 2006. Role of endosomal trafficking dynamics on the regulation of hepatic insulin receptor activity: models for fao cells. Ann. Biomed. Eng. 34, 879–892.

Jin, J., Almon, R.R., DuBois, D.C., Jusko, W.J., 2003. Modeling of corticosteroid pharmacogenomics in rat liver using gene microarrays. J. Pharmacol. Exp. Ther. 207, 93–109.

Lev Bar-Or, R., Maya, R., Segel, L.A., Alon, U., Levine, A.J., Oren, M., 2000. Generation of oscillations by the p53-Mdm2 feedback loop: a theoretical and experimental study. Proc. Natl. Acad. Sci. U S A (PNAS). 97 (21), 11250–11255.

Mélykúti, B., Burrage, K., et al., 2010. Fast stochastic simulation of biochemical reaction systems by alternative formulations of the chemical Langevin equation. J. Chem. Phys. 132(16), 164109. Available from: http://dx.doi.org/10.1063/1.3380661.

Novere, N., Finney, A., Hucka, M., Bhalla, U.S., Campagne, F., Collado-Vides, J., et al., 2005. Minimum information requested in the annotation of biochemical models (MIRIAM). Nat. Biotechnol. 23 (12), 1509–1515.

Ramaswamy, R., Sbalzarini, I.F., 2011. A partial-propensity formulation of the stochastic simulation algorithm for chemical reaction networks with delays. J. Chem. Phys. 131(1), 014106. Available from: http://dx.doi.org/10.1063/1.3521496.

Rios-Estepa, R, Lange, B.M., 2007. Experimental and mathematical approaches to modeling plant metabolic networks. Phytochemistry. 68 (16–18), 2351–2374.

Rowland, M., Tozer, T.N., 2011. Clinical Pharmacokinetics and Pharmacodynamics: Concepts and Applications. Lippincott Williams & Wilkins, Baltimore.

Sehl, M., Alekseyenko, AV., Lange, KL., 2009. Accurate stochastic simulation via the step anticipation τ-leaping (SAL) algorithm. J. Comput. Biol. 16 (9), 1195–1208.

Sen, P., Bell, D., 2006. A model for the interaction of two chemicals. J. Theor. Biol. 238 (3), 652–656.

Sharma, A., Jusko, W., 1996. Characterization of four basic models of indirect pharmacodynamic responses. J. Pharmacokinet. Pharmacodyn. 24 (6), 611–635.

Sheiner, L.B., Steimer, J.L., 2000. Pharmacokinetic/pharmacodynamic modeling in drug development. Ann. Rev. of Pharm. Toxicol. 40, 67–95.

Sorribas, A., Hernández-Bermejo, B,E, Vilaprinyo, E, Alves, R, 2007. Cooperativity and saturation in biochemical networks: a saturable formalism using Taylor series approximations. Biotechnol. Bioeng. 97 (5), 1259–1277.

Sung, S.W., Lee, J., 2004. Modeling and control of Wiener-type processes. Chem. Eng. Sci. 59 (7), 1515–1521.

Tzafriri, A.R., Edelman, E.R., 2004. The total quasi-steady-state approximation is valid for reversible enzyme kinetics. J. Theor. Biol. 226 (3), 303–313.

Wahl, G.M., Stommel, J.M., Krummel, K.A., Wade, M., 2005. Gatekeepers of the guardian: p53 regulation by post-translational modification, Mdm2 and Mdmx. 25 Years of p53 Research. K. W. a. P. Hainaut, Springer.

Weiss, J.N., 1997. The Hill equation revisted: uses and misuses. FASEB J. 11, 835–841.

Wilkinson, D.J., 2006. *Stochastic Modelling for SystemsBiology*. CRC Press, Boca Raton.

Chapter 8

Physiologically Based, Whole-Organism Kinetics & Noncompartmental Modeling

"It is these boundary regions of science which offer the richest opportunities to the qualified investigator. They are at the same time the most refractory to the accepted techniques of mass attack and division of labor. If the difficulty of a physiological problem is mathematical in essence, ten physiologists ignorant of mathematics will get precisely as far as one physiologist ignorant of mathematics and no further. If a physiologist, who knows no mathematics, works together with a mathematician who knows no physiology, the one will be unable to state his problem in terms that the other can manipulate, and the second will be unable to put the answers in any form that the first can understand. ... a proper exploration of these blank spaces on the map of science could only be made by a team of scientists, each a specialist in his own field, but each possessing a thoroughly sound and trained acquaintance with the fields of his neighbors; all in the habit of working together, of knowing one another's intellectual customs, and of recognizing the significance of a colleague's new suggestion before it has taken on a formal expression. The mathematician need not have the skill to conduct a physiological experiment but he must have the skill to understand one, to criticize one, and to suggest one. The physiologist may not be able to prove a mathematical theorem, but he must be able to grasp the physiological significance, and to tell the mathematician for what he should look."

Norbert Weiner[1]

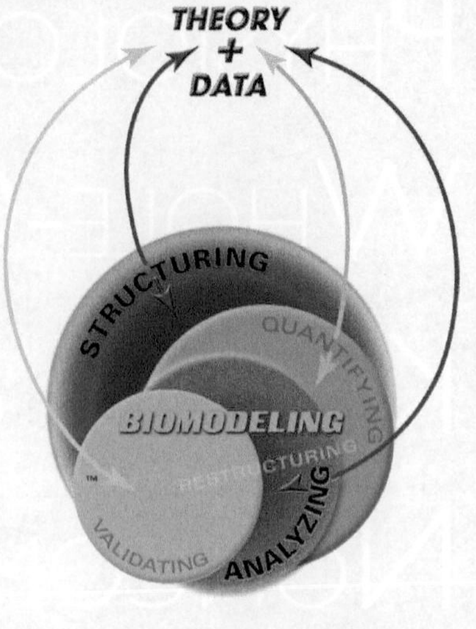

1 *Cybernetics*, MIT Press, 1948.

Overview

We develop three specialized biomodeling methodologies here – *physiologically based* (PB), *whole-organism* (WO) and *noncompartmental* (NC) methods – for primarily addressing quantitative clinical, environmental and health-related problems, for the most part at whole-organism and organ-system levels. Together they reflect an amalgam of paradigms for quantifying model attributes in physiological, engineering, pharmaceutical, toxological and biophysical sciences, and they often are used – all or in part – in combination with each other. Modern engineering-inspired *systems physiology* and *systems pharmacology* have a large role in model structuring and quantification in PB modeling. This approach was introduced in the 1960s using more vivid and highly recognizable depictions of realistic mammalian physiology and anatomy,[2] with the vascular system playing a more explicit role. This focus on organ-level/vascular dynamics has distinct advantages in model quantification – especially for developing population-level models from mixed experimental animal and human data. For this reason, PB modeling is being used increasingly by industry and regulatory agencies to more directly estimate dosages of toxic agents and their metabolites reaching explicitly modeled target tissues, with increased application in health risk assessment of consumer products, drugs and chemicals.

We begin discussion of *PB modeling* with a concise summary of *vascular and tissue physiology* – providing an overview of organ-system function and nomenclature for modeling blood and lymphatic vessels components and flow of substances into and through organs: a "short-course" in vascular and tissue anatomy and physiology.[3] For the non-specialist, this is needed for understanding much of the modeling in this chapter.

We then develop the mathematics of PB modeling methodology, which we call *modern PB* modeling, if there's potential for ambiguity. The reason for the distinction is that most modeling of dynamical processes in biological organisms is patterned on physiology, biochemistry or both. To a large extent, compartmental and pool modeling, and pathway and protein network modeling, are physiologically based; but they can be somewhat abstract and not necessarily consistent in all respects with anatomy or morphology. Compartments in MC models, for example, generally represent dynamical states and may or may not represent single organs or organ subcompartments or organelles or chemical species. PB modeling methodology was initially developed as a more explicity physiological and anatomical approach than earlier methodologies for studying pharmacokinetics (PK) and pharmacodynamics (PD). Multicompartmental (MC), whole-organism (WO) and noncompartmental (NC) analysis methods are among these earlier approaches and modern PB modeling (PBPK and PBPD) has characteristics of all of them. Interestingly, if MC models are structured with primarily physiological and anatomical components and intercompartmental exchanges – which they often are – they are likely interconvertible into modern PB models – and *vice versa*. Bridging commonalities and clarifying differences in these structural variants are among the goals of the chapter. A section is included on *inter-species scaling* (*allometry*), the basis for scaling up models based on PB laboratory animal studies to models primarily for the human. Facility for physiological

2 Not unlike the return to "realism" in 19th century French painting.

3 Highly abbreviated, of course.

parameter scaling is a major attribute of PB modeling. It's important when physical values are not known in the species of interest, but are known in others.

Next, we discuss two important *experiment design conditions* on applicability of the modeling methods that follow – common dependence on *linearity* of system or experiment and *stationarity* of the system during the experiment. For the most part, these apply to PB modeling as well, but they arise more explicitly in WO and NC modeling. Two other constraining conditions (more or less) limit applicability of all modeling methods developed in this chapter, namely: *steady state* conditions for the endogenous system during an experiment, and biosystem *structural* assumptions. These are addressed as they emerge during exposition.

An older and less organ-specific methodology – *whole-organism kinetic modeling* (WO) – follows. Analysis of kinetic (dynamical) properties of biological systems at the whole-organism level is usually directed at quantifying particular whole-organism parameters or *kinetic indices.* These "global" measures of *size* or *rate* are commonly used in metabolic and other physiological and pharmacological system kinetic studies, particularly in clinical research. Compartmental, pool model and other biophysical modeling approaches are natural choices for evaluating biosystem kinetic indices at the whole-organism level. Steady state limitations of these methods are prominent, and linearity and stationarity conditions also govern their applicability to varying degrees.

Noncompartmental[4] (NC) modeling, the least organ-specific, is presented last. This is another often-exercised paradigm for analyzing biodata and computing kinetic indices – especially in pharmacology and toxicology, also with theoretical roots in biophysics and engineering. It has some of the most stringent structural requirements or assumptions for strict applicability, as discussed in the NC analysis sections.

Not discussed in this chapter are physiologically based feedback control system (fbcs) models. These are reserved for Chapter 14 – Biocontrol System Modeling, Simulation and Analysis. Their overall feedback structures are typically organ-system based, and have other characteristics common with modern PB models, but they represent *control signal* interactions among components rather than *material flow dynamics through organs*, as in PB models.

Physiologically Based (PB) Modeling

This methodology was developed in the 1960s and '70s, primarily by Kenneth Bischoff and Robert Dedrick, as reviewed in (Bischoff 1975), in the context of physiologically based **pharmacokinetic (PBPK)** modeling. In this approach, physiology and anatomy are paramount, with model structure developed along these **physio-anatomic** lines, usually on a whole-body basis. The earliest such structural representation was called the **minimammal** model (Bischoff 1966). **Fig. 8.1** illustrates the concept using several recognizable organs. Additional organs and tissue spaces are readily included – mostly in parallel with the ones shown, with other material transfer in blood, lymph or other signal pathway flow connections added as needed, e.g. as shown for the two pathways to liver – the hepatic artery in red and the portal vein in blue.

4 The designation *noncompartmental* model is somewhat ambiguous. It is a special class of models, with particular properties. It does not mean the set of all modeling paradigms *not* compartmental.

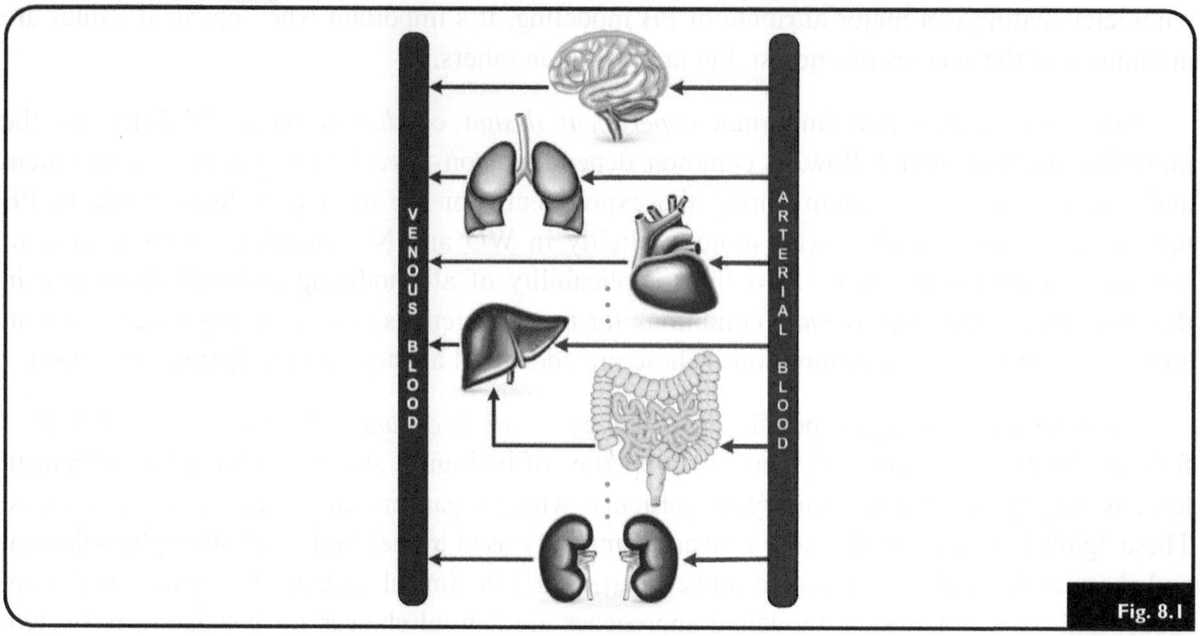

Fig. 8.1

The physiologically based (PB) model paradigm.

This modeling approach uses physiological and anatomical information advantageously – not only for model structure development – but also for model quantification, from human and other animal data sources. When human data is not available, PB models provide a bridge for parameter estimation from experimental laboratory animal data – typically more accessible and invasive than from human data. Scaling-up from lab animal to human parameters is then accomplished using *allometric models*, as described last in this section.

Nomenclature

Most of the terminology used in pharmacological modeling is standardized for that community and can be found in recent books on pharmacokinetics and pharmacodynamics (e.g. Rowland and Tozer 2011). A large fraction of PBPK/PBPD nomenclature is shared in the physiological, toxicological, engineering, systems and other biological modeling communities. All more or less utilize aspects of one or more of the quantitative methods developed here and in earlier chapters, for modeling systems with similar dynamical characteristics across the different problem domains. Inconsistencies in some terminology, however, have been problematic and differences eliciting confusion and misunderstanding in pertinent terminology have been addressed by several groups, resulting in proposed international standards for PB modeling nomenclature (Tofts et al. 1999; Innis et al. 2007). Most of the choices in this chapter, or close enough resemblances, are consistent with most of this nomenclature and in common use in biomodeling areas, particularly in physiology, and – in particular – with nomenclature in earlier chapters. I've done my best to clarify them when they are not.

Simplified Vascular & Tissue Physiology & Anatomy for PB Modeling

The distinguishing and explicit anatomical structuring and physiological features of PB models include key anatomical and physiological descriptors of the organism, like organ volumes, volumetric flows, material perfusion, breathing rates and tissue permeabilities – thus motivating our brief review of this anatomy and physiology here. Many of the parameters described can be estimated from published data, usually published tables of values for different populations or species (Brown et al. 1997). As needed, they also can be *allometrically scaled* to humans, minimizing the amount of human data needed for model quantification and validation. Alternatively, as introduced below – and detailed in Chapters 10–15 – parameters can be estimated by fitting model outputs to independently measured time series data.

The dynamics of either *endogenous* or *exogenous* material (or signal) X can be modeled using this approach, e.g. hormones, metabolites, drugs, chemicals, environmental toxins, or labeled tracers of any of these. Endogenous X is usually found in blood and exogenous X usually finds it way into blood after ingestion or exposure. The blood and lymphatic vascular systems then normally carry X as a constituent in blood (or lymph) to and from body organs. Let's review how this happens, in the context of modeling the biodynamics of any constituent X in flowing blood (or lymph), for any physiologically based (PB) model. We shall see that not every tissue/organ is needed in such models.

Blood & Lymph Vessels, their Constituents & Tissue Distributions

Refer to **Fig. 8.2** and **Fig. 8.3** to visualize the physio-anatomy described here.[5]

Briefly and simply, the **blood circulatory system** consists of **arteries, arterioles, capillaries, venules** and **veins**, which together carry blood and its constituents throughout the organism

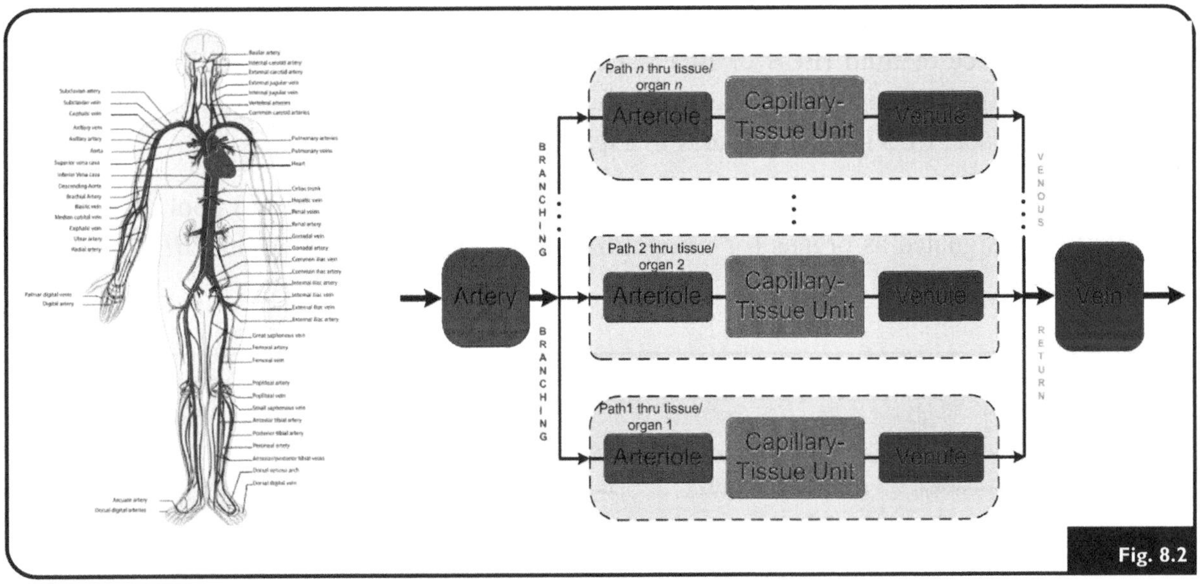

Fig. 8.2

Simplified blood circulation in the human (LHS), from right heart into **arteries** (red), through tissues/organs and then returning through **veins** (blue) to the left heart for recirculation. Detailed branching (RHS) from arteries to **arterioles**, then through tissue via **capillaries** – where molecular exchanges occur with tissue cells. Returning blood is via **venules** to veins. Not shown explicitly – and included within the capillary-tissue unit – is the **interstitial space** (ISS), with drainage of blood to venules (included in **Fig. 8.3**).

5 The drawings on the LHS of **Fig. 8.2** and RHS of **Fig. 8.3** are from the Wikipedia Commons (http://commons.wikimedia.org/wiki/Main_Page) and are public domain for all purposes worldwide.

(**Fig. 8.2** left-hand side (LHS)). In the human, resting blood flow through the heart (cardiac output) is 5 to 6 liters per minute. Blood leaves the left heart and flows through arteries, which gradually divide into arterioles and then the smallest vessels – **capillaries** in **tissues** (aggregates of similar cells) (**Fig. 8.2** right-hand side (RHS)). Capillaries are the vessels from which biomolecules and other constituents of blood can be exchanged between the circulating blood and the tissues. Exchange is limited by blood flow and membrane permeability (described below). The blood then travels from tissues first in venules, then in veins, back to the right heart, where it is pumped directly to the lungs. The primary function of the lungs is to effect bidirectional carbon dioxide exchange for oxygen; and oxygenated blood flows back to the left heart, from where it is again pumped into arteries and recirculated. As with any organ, material or signal X in blood also may have interactions with lung tissue, affecting O_2:CO_2 exchange – or other pulmonary processes. **Blood** (B) consists of the watery fluid **plasma** (p) containing suspended red and white cells, proteins, hormones, nutrients, dissolved gases, waste products and other compounds (blood **constituents**) transported into and from one location or organ to another in the body.

An **organ** is a group of tissues performing a specific function or group of functions for the organism – the heart, liver, kidneys, glands, etc. (**Fig. 8.2** RHS). Organ tissue contains well-differentiated, specific cells/cell-types bathed in a watery fluid, the **interstitial** space or fluid, (ISS or ISF) (**Fig. 8.3** LHS). The blood supply to the organ enters from arterioles, then – inside the organ – into miniscule capillaries passing in multiple directions through the ISS. Blood constituents may or may not pass to or from cells in the organ via the capillaries (**Fig. 8.3** LHS). The space occupied by the interior of all the cells inside a whole organ is called the **intracellular** space (ICS) or fluid (ICF). The space exterior to the ICS is called the **extracellular** space or fluid (ECS or ECF). The ECS, then, is the sum of the ISS plus blood volume in capillaries inside the organ. For this reason, it is also called the **extravascular-extracellular** space, EES (top two spaces in **Fig. 8.3** LHS).

Remark: The **blood** pool itself, or its **plasma** pool or **red blood cell (RBC)** pool constituents, might be designated as organs (or compartments) in some models (e.g. in *PBPK* models – see below).

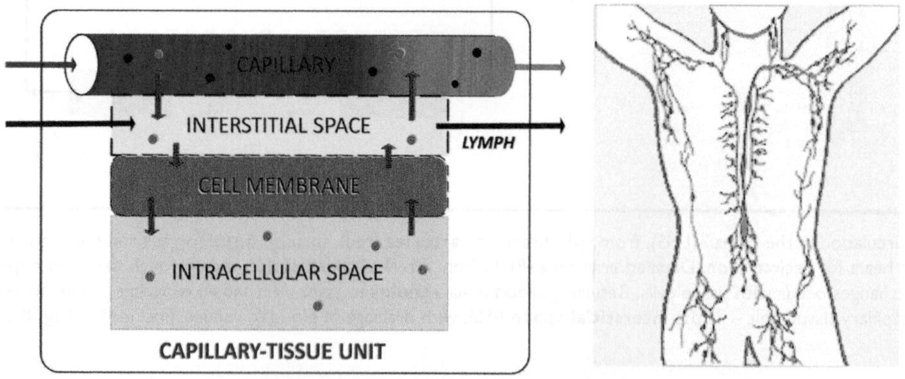

Fig. 8.3

LHS: Capillary-Tissue Unit: Transmembrane transfer of X reversibly between capillary and ISS, ISS and cell membrane, and cell membrane and ICS. X flows in blood (red and blue arrows) and lymph (black arrows). RHS: The lymphatic system. All flow is toward the heart and enters blood via the thoracic duct located just above the heart. (See text for abbreviations.)

The **lymphatic system** consists of organs, **ducts** and **nodes** containing and transporting a clear watery fluid called **lymph** (L) and its constituents unidirectionally towards the heart from tissues, via **lymphatic ducts**, and eventually to blood via the **thoracic duct** located just above the heart **(Fig. 8.3** RHS). The major function of the lymphatic system is to drain lymph and its constituents from the ISS in tissues, parallel to blood drainage in venules from capillaries. As such, it's an accessory route for returning ~3 L/day of **plasma** (blood minus blood cells) and its constituents to the blood circulation **(Fig. 8.3)**. The lymphatics absorb and transports fatty acids and fats from ISS in tissues to the circulatory system; and it has an important immune system function – containing and transporting immune cells – such as monocytes, lymphocytes and antibody-producing plasma cells, which protect the body against microbial infection. Lymphatic flow is very much slower than blood flow, with widely differing flow rates in various tissues (Modi et al. 2007) and, in contrast to the blood circulation, it does not recirculate on itself.

Structuring & Parameterizing a PB Model

Diagram Notation

Let X be our *generic* state variable for PB modeling developments. In the pharmacology modeling community (and some others), organs in which X dynamics are involved and being modeled are represented schematically by *boxes*. We reserve boxes for input signal–output signal (e.g. transfer function) diagrams, where the input and output arrows represent input and output signals (stimuli and responses), with any units. For PB models, arrows going in and coming out of organs (**Fig. 8.1**) represent influxes and effluxes of X, as in compartmental-style models. We emphasize the distinctions among these modeling classes (and communities), by representing organs in PB models as *boxes with rounded corners*, beginning with **Fig. 8.2**.

PBPK (PBPD, PBTK) Model Organs & Compartments

Physiologically based (PB) models are, in part, a hybridization of compartmental and tank and flow system concepts in chemical engineering and fluid dynamics. State variables are size (mass or concentration) based, as in earlier sections and chapters, but flow dynamics is based largely on physiologic *volumetric* flow instead of mass transfers.

In deciding on compartmentalizations, organs/tissues involved in X dynamics are often classified or grouped on the basis of how much blood flow they receive relative to others. One such *blood-uptake-kinetics* classification is: **rapidly perfused (richly perfused)** organs, like lungs and kidneys; **moderately perfused** tissue, like muscle; and **slowly perfused** tissue, like fat. Organs might be included separately or lumped together in these categories, if X material uptake into these organs is kinetically similar. Functional classifications are also made: **metabolizing tissue** (organs), often include liver and kidneys, where many metabolic and excretory processes occur.

Remark: Importantly, as with other modeling paradigms, PB modeling language tends to blur the distinction between the *state variables* (material "X" objects) within organs – whose structural connectivity is being modeled – with their (anatomical) organ "containers." This sometimes is unavoidable, but it can create ambiguity if the context is not ever-present. We try to maintain the distinctions as much as possible here.

Organ/tissue components in PB models usually represent more or less well-defined anatomical regions involved in *X* dynamics, typically called **compartments**, possibly with two or three (or more) physiological **subcompartments** in which material *X* is distributed within the organ, as shown for example in **Fig. 8.6** (see page 361). Compartments as such usually correspond to *X* in specific organs (with or without subcompartment subspaces of *X* in organs) – or to grouped organs/tissues having common characteristics, as described above. Groupings can be in accordance with the relative proportion of the cardiac output they receive, or with similar blood flow values per gram of tissue, or in accordance with the affinity they have for material *X*. Organs/tissues that have the largest effects on the dynamics of *X* substances/materials being modeled, e.g. lungs, liver, gut, kidneys, might be chosen as separate model *X* compartments. Those with lesser individual effects are usually grouped together into tissue *X* compartments having common characteristics, e.g. rapidly perfused and slowly perfused tissue *X* compartments. Transfers of *X* material in these localized tissue spaces (compartments and subcompartments) are based on the biophysics and chemistry of diffusion through and properties of membranes.

Remark: Strictly speaking, we should call these submodel organ spaces **pools** (of *X*) and subspaces **subpools** (of *X*) in the organ, instead of compartments (of *X*) and subcompartments (of *X*), because – in general – they may not have uniform distributions of material *X*. This is especially the case for the one-compartment organ model described below, represented as an approximately uniform (well-mixed) pool (compartment) of *X*. It is common for material *X* in anatomical regions such as the vascular, extracellular or cellular spaces to be each considered homogeneously distributed, with uniform concentrations in each. If the assumption is a good one, these are valid compartments. They can include nonlinear (e.g. saturable uptake, binding, etc) as well as linear interactions.

For modeling whole-body dynamics of **endogenous** *X, essential features of the model structure* should include compartments for all body organs (or organ/tissue groups) significantly affecting overall dynamics of *X*. This includes organs of production and elimination of *X* – in the form of synthesis or secretion, accumulation, metabolism and excretion – where there is some quantitatively significant interaction or other effect, alteration or processing of *X*. Organs in which *X* affects other processes may also be included, e.g. hormonal action or drug effects on target organs. Organs that do not produce, accumulate, metabolize or eliminate *X* (or significant amounts of *X*) are excluded. For **exogenous** substances *X* (e.g. drugs, chemicals, toxins), organs included are those that store or respond in some way to *X*, e.g. by accumulating significant amounts of *X* – enough to either alter its dynamics, produce a desired (therapeutic) or detrimental (toxic) effect, or both – in the case of therapeutic drugs with unintended side-effects, for example.

Remark: Good practice in multicompartmental (MC) modeling (Chapters 4 and 5) follows similar rules for choosing model compartments.

Remark: In modeling biosystems with more than one interacting state variable, more than one PB model may be needed – as for iodide and perchlorate in a human model of their interacting dynamics in Example 8.1 below.

Plasma vs. Blood Concentrations & Flow Rates

Many models have plasma (subscript p) instead of blood (subscript B) concentrations or masses in model equations, primarily because most substances are measured in plasma rather than whole blood. Typically, whole blood samples are centrifuged and the plasma or serum (plasma minus clotted material) is separated from the red blood cells (RBC) and assessed for X concentrations. If X is naturally excluded from RBC, nothing need be done in the equations except adjust the volume parameters, from blood to plasma volumes and flow rates. However, some X may associate with RBC, e.g. certain hormones with RBC (DiStefano III et al. 1992). Plasma instead of blood concentrations or masses in equations still can be a good approximation if X in blood equilibrates relatively instantaneously with X in RBC,[6] i.e. the fraction associated with RBC remains constant. Nevertheless, some substances, like glucose, are usually measured in whole blood. Care must be taken in modeling with either one; the appropriate volume measures must be included in the equations, with plasma volumes being smaller than blood volumes – thereby generating smaller (volumetric) plasma than blood flow rates. Specifically, **plasma volume** V_p is related to **blood volume** V_B by the hematocrit (Hct), the measured fraction of RBC in blood, i.e. $V_p = (1 - Hct)V_B$. Therefore **plasma flow rate** F_p is related to **blood flow rate** F_B as $F_p = F_B(1 - Hct)$.

Distribution Dynamics of Blood Constituents in Tissues

For modeling, we are interested primarily in the dynamics of constituents in blood, rather than blood flow as such. Blood constituent X in capillary blood can penetrate into (get taken up by) particular tissues, a process called **extravasation**. If it penetrates (gets wholly or partially **extracted** or **sequestered**), it must cross one or more cell membrane structures (**transmembrane transfer or transport**). The rate of uptake (penetration) of X from blood into tissue is limited by blood flow (perfusion – a *delivery* limitation) or membrane permeability (resistance to crossing membranes), or both. The LHS of **Fig. 8.3** illustrates these transmembrane fluxes, via both vascular and lymphatic flows through the tissue space. The relative rates of transfer of X among and through these spaces – their *kinetic heterogeneity* – is the primary factor determining the number of compartments/subcompartments needed to accurately represent X dynamics for this organ within the PB model.

If movement of X through the tissue space is only **flow-limited** (same as **perfusion-limited**), X diffuses freely and rapidly through the capillary membrane into the interstitial space (ISS), limited primarily by the regional blood flow rate F (vol/time) in the capillary of the organ. This means that membranes in flow-limited tissues offer no appreciable resistance to molecules that flow into the tissue space; they move freely within it, the entire tissue approximated as a *well-stirred (homogeneous) tank*. Kinetically, the basic assumption of such flow-limited transfer is that X in entering blood rapidly reaches equilibrium with that in the whole tissue space, i.e. diffusion[7] of X across cell membranes into tissue is very fast in comparison with its transport by the blood. *This rapid equilibrium is equivalent to a quasi-steady state assumption* (Chapter 6), i.e. after a very fast transient, the time-varying concentration $c(t)$ of X in tissue must follow the dynamic

6 X in blood is in quasi-steady state with X in RBC. Care must be taken in these circumstances to assure that material flux through the compartment or subcompartment is conserved.

7 Diffusion is transport/transfer by random thermal motion of molecules down a concentration gradient.

response pattern characteristic of the entering and leaving blood concentration of X. Under this assumption, the tissue concentration $c(t)$ of X is (approximately) proportional to the outgoing venous concentration $c_v(t)$ for all times t, i.e. $c(t) \equiv Pc_v(t)$, with P an important proportionality constant, the equilibrium or well-stirred **tissue:blood partition coefficient**, as described further below. In these terms, the *net* mass rate of **extravasation** is: $V\dot{c} = F\Delta c \equiv F(c_a - c_v) = F(c_a - c/P)$. Substances X may exist in bound as well as unbound (free) forms in blood. In flow-limited organs, it is the free form that normally diffuses very rapidly into tissue and concentrations of *free X* in blood and tissue are proportional.

Permeability-limited (membrane-limited) tissues may have different extravasation dynamics and different partitioning coefficients for X. They resist substance penetration and partitioning among compartments to widely varying degrees, depending primarily on cell membrane properties.[8] These include substance solubility in the membrane as well as other biophysically or biochemically mediated transmembrane transport properties. Substance X transfer between subcompartments can be modeled in different ways in permeability-limited organs, depending on available information about membrane properties and partitioning of material between blood and tissue and between subcompartments, represented by **biophysical partition coefficients** relating X concentrations between subcompartments (Thompson et al. 2012). This is described below and illustrated in the examples.

PBPK Model Parameters

Four main categories of parameters characterize PBPK model dynamics for substance X: system-specific properties like physiological and anatomical *sizes and flow/flux rates*, common across a species and independent of X; *source, sink and metabolic transformation* parameters; *partition coefficients* for quantitatively partitioning X between various media (compartments, subcompartments) modeled; and *membrane transport characteristics*, for permeability-limited uptake of X. Importantly, many of these physiological, biophysical and biochemical parameters are available from published and tabulated values for humans and other species e.g. (Brown et al. 1997) and can be used to partially quantify a PB model – at least approximately.

1. **Sizes & Flow/Flux Rates**: These include organ masses (q_{org}) and (real) volumes (V_{org}), regional (organ) blood flows F_{org} (vol/time) and ventilation rates. *Note:* Easier-to-measure organ masses are readily converted into real organ volumes via organ specific gravities. Many of these are close to 1 and can be approximated as such, because 1 kg $\approx$ 1 liter of tissue – composed primarily of water.
2. **Source, Sink or Metabolic Transformation Parameters**: These represent biochemical reaction kinetics (*ks* or *vs*) or other transformation (e.g. production/secretion rate *PR, SR*, activation) properties for X within organ compartments or subcompartments. Examples include: metabolic clearances, saturated uptake, plasma-protein binding affinity, tissue affinities, membrane permeabilities, enzymatic stability, and transporter activities.

8 X also can enter tissue via gaps between endothelial cells, which might be lumped with membrane properties.

3. **Partition Coefficients**: These specialized PBPK model parameters, introduced above, are denoted P (or K, or Z, or other symbol), with super- or subscript designating organ compartment or subcompartment locations, as needed. For *flow-limited tissues*, P is the ratio of total tissue concentration $c(t)$ of X to its outgoing total venous concentration $c_v(t)$ for all times t, and X in tissue and venous blood act kinetically as one compartment in equilibrium (*equilibrium P* or *well-stirred P*), with proportional (homogeneous) concentrations of X, i.e. $c(t) \equiv Pc_v(t)$ and $\dot{q} = F(c_a - c_v) = F(c_a - c/P)$. For *permeability-limited tissues*, somewhat different partition coefficients are used to quantify rates of transfer of X into and out of tissue subcompartments, designated *biophysical P* (Thompson et al. 2012), to distinguish them from equilibrium P.

   Ps also provide useful relationships among whole organ pool masses q_B of X and whole tissue volume V and blood B (or plasma) volume V_B (in equilibrium or steady state), namely:

$$P \equiv c/c_B \equiv \frac{q/V}{q_B/V_B} = \frac{V_B}{V}\frac{q}{q_B} \quad \text{or} \quad P\frac{V}{V_B} = \frac{q}{q_B} \tag{8.1}$$

   Thus, P scales real tissue and blood volumes into real tissue pool masses of X relative to blood (or plasma) pool masses of X, quite similar to the way *equivalent distribution volumes* scale real tissue compartment masses.[9] Ps are tabulated with other physiological parameters like organ blood flow rates and organ volumes.

4. **Membrane Transport Parameters**: For material transport in *permeability-limited* organs, these represent biophysical kinetic properties of X transport across membranes in tissue compartments or subcompartments. The net rate (mass/time) of diffusion across a membrane depends on the product of: the concentration difference Δc (mass/vol) across the membrane, membrane permeability k (distance/time) and membrane surface area SA (distance^2). Permeability k of substance X depends on its molecular size, lipophilicity and degree of ionization (charge) of X, as well as physical properties of the membrane.[10] Transfer is often assumed to be the same in both directions, with bidirectional permeability expressed as a **permeability area cross-product** (or **diffusion permeability coefficient**) $k \cdot SA \equiv k^{PA}(\text{distance}^3/\text{time} \equiv \text{vol}/\text{time units})$ and thus $\dot{q} = k \cdot SA \cdot \Delta c = k^{PA}\Delta c$. If permeability is expressed on a (normalized) per unit mass of tissue basis, as k^{PA} (vol/time · mass), then $\dot{q} = k^{PA}Q\Delta c$, where q is the tissue compartment mass.

   All four parameter types are illustrated in the methodology and examples below.

Dynamical Equations of PB Models

We derive ODEs for X dynamics in flow-limited and permeability-limited tissue compartments and the whole-body PB models in which they are included. Several well-established model types are developed first. In the process, we illustrate how they are augmented with needed source, sink

9 Eq. (8.1) is revisited later in this chapter in terms of both real and equivalent (apparent) distribution volumes (Chapter 4), for scaling real tissue compartment masses in steady state, using partition coefficients.

10 The biophysics of transmembrane transport can be complicated. Besides passive diffusion, it can be governed by physical properties that enhance (or impede) transport, e.g. **facilitated** or **active** transport by carrier molecules or energetic processes. Perturbation in these modes of transport of X can impact the value of P. Chapter 4 of (Rowland and Tozer 2011), on membranes and distribution of drugs, provides an excellent summary in the context of pharmacokinetics.

or metabolic transformation terms, and included in a whole-body PB model in accordance with their physiological functionality. A recent variation on the standard models – based on a quasi-steady state (QSSA) analysis – is also presented. In practice, these novel paradigms can provide improved modeling precision without additional model complexity, and can even be simpler.

Simplest Single-Organ Model

Single-organ compartment dynamics for material X is represented here by mass flow balance through an assumed (approximately) homogeneous organ/tissue – **Fig. 8.4**, also called the **well-stirred tank (WST) model** (Thompson and Beard 2011). Blood containing X at concentration c_a (mass/vol units)[11] flows at regional organ flow rate F (vol/time units) in blood of the systemic arterial circulation into a one-compartment organ of constant volume V. The simple single-compartment organ here does not distinguish between extracellular and cellular spaces, and does not explicitly include a lymph effluent. X flows out in blood at the same regional organ blood flow rate F but at a venous concentration c_v assumed to be proportional to the "total" for X (average) concentration c of X in this one-compartment organ. Inside the organ, metabolic transformations or interactions or elimination of X may occur and we lump these all into rate R (mass/time units) – designated *intraorgan source or sink processes* in **Fig. 8.4**. R can be positive or negative, linear or nonlinear. For example, $R = -kq$ if mass q of X is eliminated (or metabolized, or cleared) at a linear fractional rate k from the organ. R can be positive if X is produced in the organ. For *flow-limited tissue*, under the assumption of rapid equilibrium of X between blood and tissue, with single-compartment kinetics, concentrations c and c_v are related by the *partition coefficient* $P = c/c_v$ and mass flow balance through the organ gives:

$$\dot{q} = V\dot{c} = F(c_B - c_v) + R = F(c_B - c/P) + R \qquad (8.2)$$

In practice, an **empirical equilibrium partition coefficient** for X, $P^{equil} \equiv c^{equil}/c_v^{equil}$, can be used to approximate P and, in principle, it can be measured. *Importantly, all of the processes that alter tissue extraction – membrane effects, heterogeneities, binding, lymph efflux, etc. – are effectively lumped into this single (approximately) constant* P^{equil}. In other words, all tissue processes are lumped together and are thus accounted for in this single-compartment organ approximation, which means it is – strictly speaking – not restricted to only flow-limited tissues. Eq. (8.2) is thus approximated as:

$$\dot{q} \cong F\left(c_a - \frac{c}{P^{equil}}\right) + R \qquad (8.3)$$

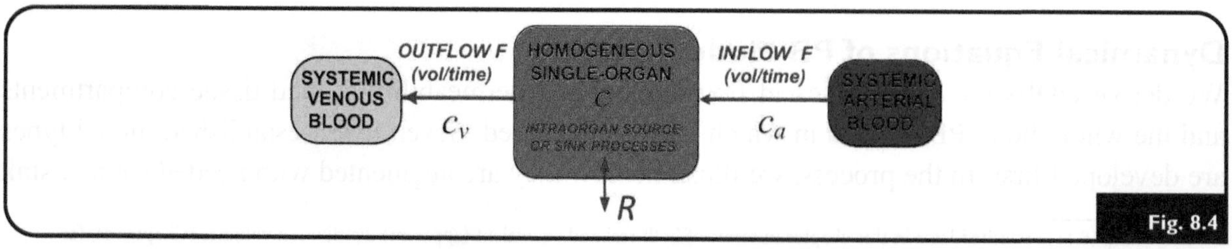

Fig. 8.4

11 Substitute plasma concentrate c_p throughout if the measurements are in plasma.

P^{equil} is typically estimated experimentally,[12] e.g. from tissue biopsy data, and extrapolated to the whole organ, assuming uniform concentration throughout.

Remark: If no source of X exists in the tissue – typical of exogenous drugs – only a fraction, say $(1 - P)$ of X $(P \leq 1)$ entering in capillaries is typically returned in the venous or lymphatic effluent. Fraction P is "removed" from blood by the tissue into cells. This partitioning depends on a variety of factors, including: steric exclusion of X from cells by membranes; other membrane transport effects – in either direction; organ-space heterogeneities – including association of X with different tissue structures; and the specifics of lymphatic drainage in the organ.

Permeability-Limited Two-Subcompartment (PLT) Organ Model

Fig. 8.5 has a single organ split dynamically (not necessarily anatomically) into two subcompartments, to explicitly accommodate the simplest model of membrane-limited intra-organ dynamics. Separate *vascular space* (compartment 1 with volume V_1) and *extravascular space* (compartment 2 with volume V_2) subcompartments for membrane-limited (permeability-limited) uptake and return are common. Extracellular and intracellular subcompartments are an alternative.

The whole-organ PLT model requires two ODEs, one for each subcompartment. It is assumed that the total concentration of X in vascular compartment 1 is homogeneously distributed and the same as that for X in blood (or plasma). X in the extravascular compartment, with total concentration c_2, is also assumed homogeneously distributed. Defining the permeability-limited mass exchange rates of X between the subcompartments as $\dot{q}_{21}$ and $\dot{q}_{12}$ (mass/time units[13]), and R_1 and R_2 as source or sink processes, mass rate balance gives:

$$\begin{aligned} \dot{q}_1 &= V_1\dot{c}_1 = F(c_a - c_1) - \dot{q}_{21} + \dot{q}_{12} + R_1 \\ \dot{q}_2 &= V_2\dot{c}_2 = \dot{q}_{21} - \dot{q}_{12} + R_2 \end{aligned} \tag{8.4}$$

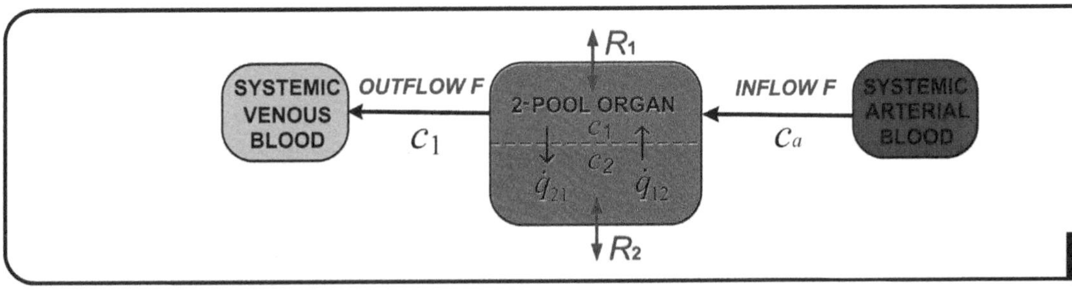

Fig. 8.5

12 See references (Parham et al. 1997; Patrick and Frank-Peter 2000) for details on how such tissue : blood partition coefficients are measured and used to quantify P_{equil} parameters in PB models.

13 The subscript notation for these inter-compartment exchange rates is the same as that defined earlier for multicompartmental models (Chapter 4).

If exchange between subcompartments 1 and 2 is linear and equally permeability-limited, then we can write the simplest model for compartment 2 dynamics as:

$$\dot{q}_{21} - \dot{q}_{12} \equiv k^{PA}\Delta c = k^{PA}\left(c_1 - \frac{c_2}{P}\right) \tag{8.5}$$

where k^{PA} is the *transmembrane permeability area cross product* (permeability coefficient) described in the previous subsection, and P is the extravascular tissue–blood biophysical partition coefficient, partitioning X between the two subcompartments. The simplest PLT model is then:

$$\begin{aligned} \dot{q}_1 &= V_1\dot{c}_1 = F(c_a - c_1) - k^{PA}\left(c_1 - \frac{c_2}{P}\right) + R_1 \\ \dot{q}_2 &= V_2\dot{c}_2 = k^{PA}\left(c_1 - \frac{c_2}{P}\right) + R_2 \end{aligned} \tag{8.6}$$

P in Eq. (8.6) is often approximated by the empirical equilibrium partition coefficient P^{equil}, as in Eq. (8.3). Lymph efflux is also likely lumped into the empirical partition parameter P^{equil} in this model, as in the one-compartment organ model.

Three-Compartment Organ in a Whole-Body Model

Organs/tissues in a PBPK model can have any number of subcompartments. **Fig. 8.6** has several multicompartment organs in a more general PB configuration, illustrating the more complex vascular interconnectivity of the liver and gut, plus some details for a three-compartment gut – to illustrate writing equations for a three-compartment organ. HEART and OTHER organs are shown with only two subcompartments, nominally vascular and extravascular subcompartments – as in **Fig. 8.5**.

Liver Equations: We assume a one-compartment liver here, for simplicity. Blood enters liver via the *hepatic artery* at concentration c_a and rate F_L and also from gut via the *portal vein* at concentration $c_v^{portal} \equiv c_G/P_G$ at rate F_G. Liver has only one venous outflow, at concentration and flow rate $F_L + F_G$. It also has an efflux pathway via the *bile duct* to gut, which we call $\dot{q}_{biliary}$, and other liver *source* and *sink* terms are lumped as mass flux rate R_L. Mass flux balance thus yields:

$$\begin{aligned} \dot{q}_L &= F_L(c_a - c_L/P_L) + F_G(c_G/P_G - c_L/P_L) - \dot{q}_{biliary} + R_L \\ &= F_L c_a + F_G c_G/P_G - (F_L + F_G)c_L/P_L - \dot{q}_{biliary} + R_L \end{aligned} \tag{8.7}$$

Saturable Biliary Excretion Model: Substances can be excreted from liver into gut via the bile duct in a saturable kinetic fashion, e.g. the drug methotrexate (Bischoff and Brown 1966). The simplest model for this transfer is:

$$\dot{q}_{biliary} = \frac{k_L c_L/P_L}{K_L + c_L/P_L} \tag{8.8}$$

where k_L and K_L are constants analogous to v_{max} and K_m in a M–M model.

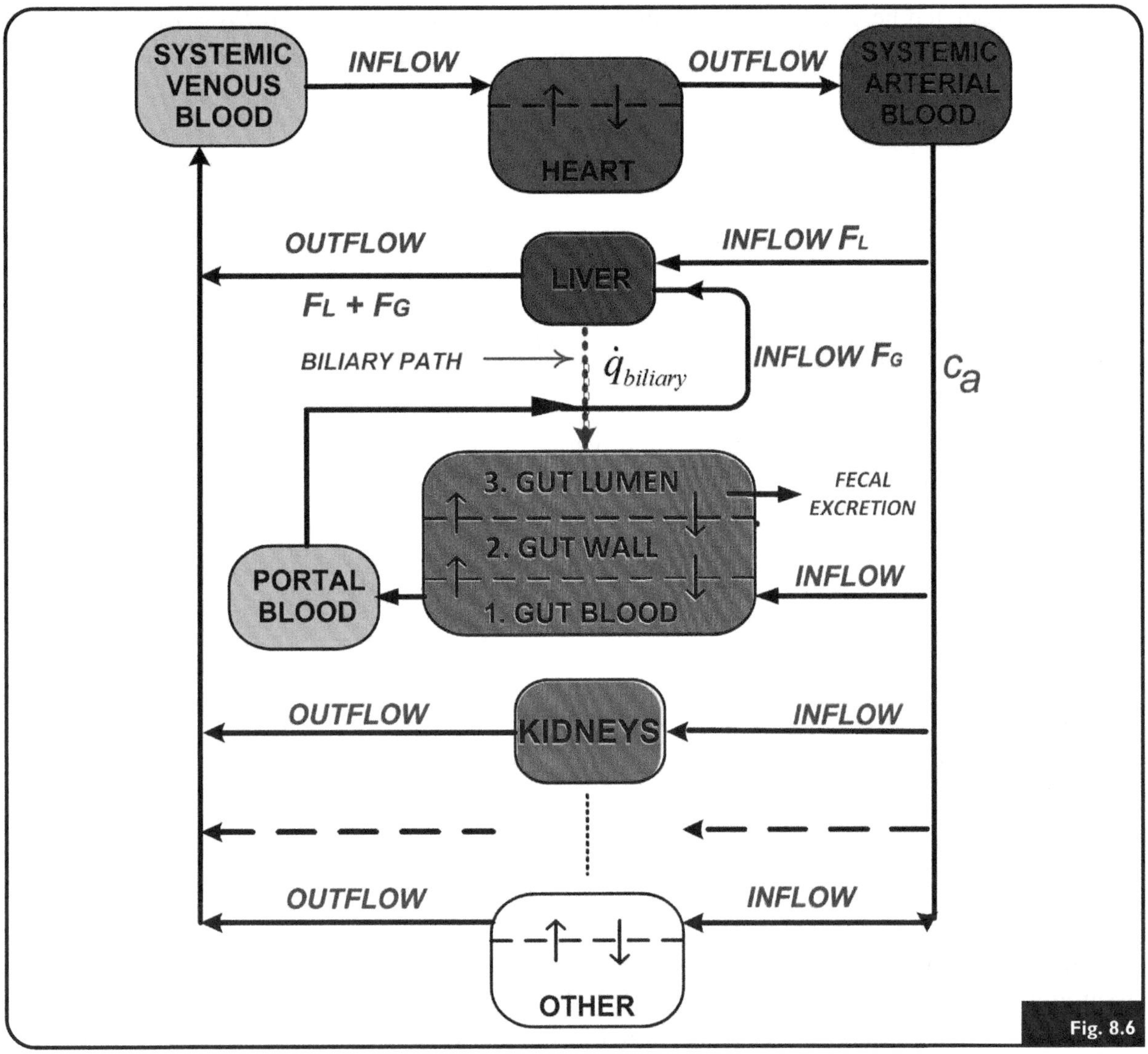

Fig. 8.6

One- two- and three-compartment organ submodels arranged in an example PBPK model, illustrating the special connectivity of liver and gut organs. The three-compartment gut submodel includes bidirectional exchanges of X among gut blood, wall and lumen and entry of material directly from liver via the bile duct. Some X is typically excreted in feces, directly from the gut lumen subcompartment, and some may be reabsorbed via portal blood. X may also enter the gut lumen via an exogenous oral dose input, not shown here, but shown in **Fig. 8.7**.

Three-Compartment Gut Equations: X in mesenteric arterial blood first enters the capillary space (compartment 1) of the gut at concentration c_a and regional gut blood flow rate F_G. It may pass right through, into portal venous blood, or some X may enter gut wall (compartment 2) and/or lumen spaces (compartment 3) and return or not, at (generally nonlinear) exchange rates $\dot{q}_{21}$, $\dot{q}_{12}$, $\dot{q}_{32}$, $\dot{q}_{23}$, within the expanded gut block – for illustration – in **Fig. 8.6**. X may enter gut via the biliary pathway from liver into the gut lumen, as shown, at rate $\dot{q}_{biliary} = \frac{k_L c_L/P_L}{K_L + c_L/P_L}$. It may also enter gut via an oral exogenous dose input, not shown in the figure, or the equations. X may undergo metabolic source or sink transformations in compartments 1 or 2 or 3 at (generally nonlinear) rates R_1, R_2 or R_3; and R_3 often has a fecal excretory rate component. We assume only $R_3 \equiv -k_{cl}c_3$ (linear fecal excretion of X) is present in the equations below. The tissue-blood equilibrium concentration ratio, defined above for the whole organ, is associated here with the gut wall and venous blood and denoted $P_G \equiv c_2/c_v^{portal}$. With the foregoing assumptions, the ODEs for these three gut compartments are:

$$\dot{q}_1 = V_1\dot{c}_1 = F_G(c_a - c_1) + \dot{q}_{12} - \dot{q}_{21} = F_G(c_a - c_1) - k_2^{PA}\left(c_1 - \frac{c_2}{P_G}\right)$$

$$\dot{q}_2 = V_2\dot{c}_2 = -\dot{q}_{12} + \dot{q}_{21} - k_3^{PA}\left(c_2 - \frac{c_3}{P_G}\right) = k_2^{PA}\left(c_1 - \frac{c_2}{P_G}\right) - k_3^{PA}\left(c_2 - \frac{c_3}{P_G}\right) \tag{8.9}$$

$$\dot{q}_3 = V_3\dot{c}_3 = -\dot{q}_{23} + \dot{q}_{32} + \dot{q}_{biliary} + R_3 = k_3^{PA}\left(c_2 - \frac{c_3}{P_G}\right) + \frac{k_L c_L/P_L}{K_L + c_L/P_L} - k_{el}c_3$$

It is useful to also have an equation for average concentration of X in the organ $\bar{c}$, obtained by adding the three pool masses and then the three equations in (8.9):

$$\begin{aligned} \bar{c} &\equiv (V_1c_1 + V_2c_2 + V_3c_3)/V \\ V &\equiv V_1 + V_2 + V_3 \text{ and } R \equiv R_2 + R_3 \\ V\frac{d\bar{c}}{dt} &= F_G\left(c_a - \frac{c_2}{P}\right) + R \end{aligned} \tag{8.10}$$

Eq. (8.10) is both convenient and practical, better appreciated by the now familiar approximation:

$$V\frac{d\bar{c}}{dt} \cong F_G\left(c_a - \frac{\bar{c}}{P^{equil}}\right) + R \tag{8.11}$$

Approximation (8.11) clarifies aspects of the simpler models above. It is obtained by assuming: first, that the equilibrium concentration ratio P^{equil} is constant;[14] second, that the total tissue concentration of X is proportional to the ISS concentration of X, $\bar{c} = kc_2$; third, that $c_v = k'c_a$; and, finally, k and k' are incorporated into P, denoted P^{equil} in (8.11), as in the simplest single organ model above. This means approximate Eq. (8.11) is the same as Eq. (8.3) for the one-pool organ! This can be very useful for model quantification, because P^{equil} can be measured.

Remark: When important enough in applications, additional experimental verification would be one way of testing approximations of constant P^{equil} for each organ involved in PB models. An alternative approach might be to fit the more complex model, without these approximations to experimental kinetic data – thus involving additional unknown parameters. This might require additional experiments (Parham et al. 1997; Poulin and Theil 2000).

Remark: If X passes right through the organ, with no entry into ISS (or ICS), then $c_v^{portal} = c_a$ and this organ can be neglected in an overall PB model.

14 Strictly speaking, the steady state (equilibrium) concentration ratio P^{equil} is not constant, but a nonlinear function of total tissue concentration (Bischoff 1975). The degree of nonlinearity depends on the type of binding or association X has inside the organ. If binding dynamics are in the linear range, for example, and uptake of X is flow-limited, P^{equil} is closer to being constant.

Remark: *Lungs* may be included explicitly, as in **Fig. 8.1**, as an additional organ situated above the heart in models assuming the structure of **Fig. 8.6**. Lungs oxygenate the returning venous blood, remove CO_2, and may also take part in metabolic exchanges of molecules other than oxygen and CO_2. The latter is illustrated in Example 8.2 below.

Remark: *Blood* or *Plasma* compartments may be included explicitly in a PB model structure, as in **Fig. 8.7** of the first example below, when dynamical processes involving subcompartments of X in blood need to be modeled explicitly, or when blood receives exogenous inputs of X. Both are included in the model of **Fig. 8.7**. An IV input of X into Plasma is illustrated in modeling practice Exercise E8.5. Other examples are given in (Pang and Durk 2010), where the Blood compartment is called a **Reservoir** for X in the system. The *influx* (concentration x flow rate) into a Blood or Plasma compartment equals the *sum of the venous effluxes* from all other organ compartments and the *efflux* from Blood or Plasma is $c_A F_{Cardiac}$.

Remark: Physicochemical processes are prominent in the pharmaceutical arena, particularly related to gastrointestinal (*GI*) *absorption of orally administered solid systems*, in the forms of tablets, capsules, suspensions, emulsions, etc. The dynamics of their dissolution and absorption, in particular, are modeled with ODEs in GI and gut submodels of PBPK structures. **Fig. 4.34** of Exercise E4.13 (Chapter 4) is an example of an oral input dose of a hormone in tablet form dissolving in the gut and being partially absorbed into blood.

Two-Region Asymptotically Reduced (TAR) Models

The spatial and temporal assumptions upon which the one-compartment (WLT) and two-compartment (PLT) organ model approximations are based – as expressed in Eqs. (8.2) to (8.6) – were recently addressed using a quasi-steady state analysis (QSSA) approach (Thompson and Beard 2011), similar to that developed in (Segel and Slemrod 1989) for mass action reactions. These authors develop two novel model forms: a permeability-limited two-region asymptotically reduced (P-TAR) model and a two region flow-limited (F-TAR) model that approximate organ dynamics more realistically. The P-TAR model formulation enables modeling of X concentrations in the vascular and extravascular spaces – as in the PLT (two-compartment) tissue model – with only one ODE instead of two (due to QSSA approximation $\dot{c}_1 \cong 0$). With no eliminations or sources in the organ, the **P-TAR model** equations are simply:

$$\begin{aligned} c_1 &= \frac{c_a + (k^{PA}/F)(c_2/P)}{1 + k^{PA}/F} \\ \dot{q}_2 &= V_2\dot{c}_2 = F\left(\frac{k^{PA}/F}{1 + k^{PA}/F}\right)(c_a - c_2/P) \end{aligned} \tag{8.12}$$

The equation for concentration $c_1(t)$ in (8.12) is algebraic, instead of an ODE; and the ODE for $\dot{q}_2$ in (8.12) differs from the PLT model (8.6) by the simple factor $1/(F + k^{PA})$ on the RHS. These variations are advantageous for modeling permeability-limited organs, because (8.12) requires solution of only one ODE, i.e. it's computationally simpler and, in practice, it has been shown to have nearly identical response dynamics when compared with the

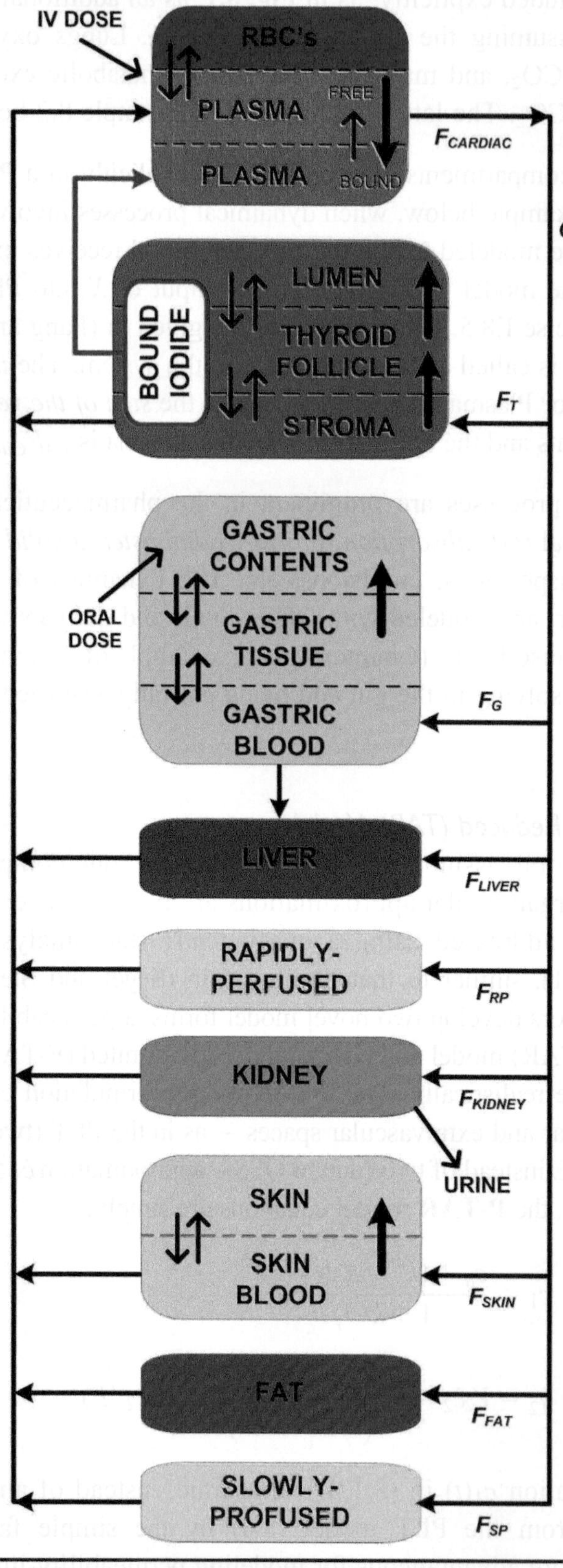

Fig. 8.7

PBPK model of competitive dynamics of iodide and contaminant perchlorate in the human. Bold arrows indicate active uptake at NIS sites (with exception of plasma binding, also indicated with a bold arrow). Small arrows indicate passive diffusion. Figure adapted from (Merrill et al. 2005).

standard PLT model, especially when X inflow kinetics (of $c_a(t)$) operate on a slower time scale (e.g. hours) than transit time through the organ (Thompson and Beard 2011).

For flow-limited organs, the new F-TAR model is biophysically more realistic than the WST model, because it retains the anatomical two region distinctions between vascular and extravascular spaces, a first-order improvement over the WLT model – all without introducing additional parametric or computational complexity. The **F-TAR model** is given by:

$$
\begin{aligned}
c_1 &= c_2/P \\
\dot{q}_2 &= V_2\dot{c}_2 = F(c_a - c_2/P)
\end{aligned}
\tag{8.13}
$$

Remark: The F-TAR model thus maintains the distinction between the vascular and extravascular spaces in an organ and thereby more accurately captures the behavior of a more realistic two-region organ under flow-limited, high permeability conditions. This is achieved with no additional parametric complexity, and only one ODE, like the classic WLT model.

PB Examples in Toxicology

We apply these concepts in the next two examples of published physiologically based models applied to problems in toxicology (Merrill et al. 2005; Douglas et al. 2010). We focus on the model-building *process* utilized in these publications, primarily to illustrate how to apply the methodology developed above.

The biology, motivation-for-modeling and model structure development are described for both examples. The first closely follows the modern PB methodology described above and provides additional depth about parameterization of PBPK models and derivation of ODEs for subcompartment exchange dynamics from the biophysics/physiology. The second example begins the same way and then deviates from strict PB modeling conventions, to illustrate how both PB and compartmental modeling can be applied in an either/or – or a hybrid – fashion, yielding the same representation. The key is maintaining maximum physiological fidelity in structuring intercompartmental influxes and effluxes. Further equation development, simplifications and quantification are postponed to Chapter 15 for the second example, following presentation of identifiability, sensitivity and parameter estimation methods in Chapters 10–13.

Example 8.1: ***A PBPK Model of Iodide & Perchlorate Distribution Kinetics in Humans***

Motivation & Biology: Chronic low-level exposure of humans to environmental perchlorate (ClO_4^-) in drinking water can affect normal endocrine function. It competes with and thus effectively inhibits normal uptake of dietary and recycled iodide (I^-) to the thyroid gland. I^- is needed for synthesis and secretion of thyroid hormone by the gland. The Merrill group (Merrill et al. 2005) developed a comprehensive PBPK model of this system, with the goal of quantitatively evaluating chronic effects of ClO_4^- exposure on healthy adult humans. They integrated human and animal data from multiple published sources with data from their own laboratories to develop and quantify the model. This included studies of the kinetics of distribution of trace doses of radioactively labeled I^- (I^*) and/or ClO_4^- (p^*) in human subjects or rats. In this example, we outline development of the PBPK model structure and its partial parameterization and quantification from available physiological, biochemical and biophysical data, to illustrate practical application of the PB modeling methodology developed above. Some of this quantification required *allometric scaling* of rat parameters not available from human data. Allometric scaling methodology is presented in the next section below. Discussion of model quantification by parameter estimation from experimental study data, and validation and predictive simulation studies, are given in the reference (Merrill et al. 2005).

Structuring the Model: An extracted overview of the structural modeling process is presented here. Detailed development and validation of the PBPK model structure, shown for purposes of this example in adapted and modified form in **Fig. 8.7**, is more fully described by Merrill et al. (2003, 2005). In brief, iodide (I^-) and perchlorate (ClO_4^-) anions are distributed similarly, yielding nearly identical structures for the two published models (one for each), differing in only a few details. They are shown superimposed in one PBPK structure in our **Fig. 8.7** adaptation. Thyroid gland and stomach organs have I^- or ClO_4^- uptake into one of three subcompartments, and – for skin – into one of two subcompartments, each with nonlinear (NL) saturable uptake kinetics. In each organ, entry of I^- into cells is reduced by presence of ClO_4^- competing for sites on an **active symporter** (NIS) uptake mechanism. This is modeled as **competitive inhibition**, as in Eq. (6.52) of Chapter 6 (described below). The thyroid has a fourth iodide subcompartment, as shown in white (modeled separately in the original). This **bound iodide** compartment in the gland represents primarily iodide "organification" into thyroid hormone –assumed to obey first-order linear kinetics – which is then secreted into venous blood, also linearly – as shown by the **red** arrow directed to a plasma subcompartment in blood. In contrast, peroxidase is unreactive and diffuses through thyroid subcompartments unchanged. Anion transfers between subcompartments in these three organs are by either passive linear membrane-limited diffusion (small arrows) or NL active transport (bold arrows). Organ data supported the number of required subcompartments – two, three or four in this model – for both anions.

The other organ and grouped-organ compartments (liver, kidneys, fat, slowly perfused and rapidly perfused) are assumed singularly homogeneous, with linear **flow-limited** uptake. Both anions are excreted linearly in urine, as shown, and fecal excretion is negligible. Because significant deiodination of hormone occurs in kidney cells, urinary excretion is modeled as a sink from the kidneys, instead of the more common pathway directly from plasma via kidney tubules.

In empirical PK studies, stomach also receives oral dosing inputs, and blood intravenous (IV) inputs, both as shown.

Model Parameters: Twenty-six physiological parameters were obtained from the literature: tissue volumes – mostly from average body and organ weights – and cardiac and regional blood flow rates. These are summarized in Table 1 of the reference (Merrill et al. 2005). Some were allometrically scaled from nonhuman animal data. Thirty-two additional biochemical and other kinetic parameters include tissue:blood *equilibrium partition coefficients* (Ps), *permeability area cross-products* (diffusion permeability coefficients) (k^{PA}s), NL Hill function kinetics (V_{max}s and K_ms) and clearance rates of anions (CLs) from or between subcompartments. These too were obtained from the literature or by fitting model outputs to empirical data. Again, some parameters were obtained by allometric scaling from experimental animal data. All are summarized in Table 2 of the reference (Merrill et al. 2005).

Model Equations: For ODE model development, kinetic coefficients for transfer of anions in flow-limited compartments were each described using partition coefficients, combined with regional blood flow rates (Fs) as e.g. in **Fig. 8.5** and **Fig. 8.6**. Inside the organs, permeability area cross-products (k^{PA}s) and Ps were used to describe anion transfer rates between the capillary bed, tissue and between inner compartments (small arrows in **Fig. 8.7**). These kinetic transfers result from inherent electrochemical gradients within the tissues, attenuated by the permeability coefficient geometries (porosities) of subcompartment membranes.

Several of the 32 ODEs representing iodide (i) and peroxidase (p) dynamics depicted in **Fig. 8.7** are detailed in (Merrill et al. 2005). For modeling, we illustrate four representative ones here, for i in the thyroid STROMA (S) – the structural encasement and extracellular fluid of the gland; the FOLLICLE (F) – cells; the LUMEN (L) – storage pool; and the BOUND IODIDE (hormone) compartments. Equations are given for uptake of i and p into the cellular compartment – represented as competitive inhibition (Eq. (6.48)) – and for i and p transfers between S, F and L, and hormonal (bound iodide) secretion from F into blood plasma.

Eq. (8.14) below is for i turnover in the STROMA. The first term is the flow-limited diffusion component, flux of i from arterial to venous blood at regional (thyroid) blood flow rate F_T. The second term is the linear transfer in and out of the FOLLICLE via the small arrows – with transfer to F scaled by the *biophysical partition coefficient* $P_F^i(=0.15)$ for *thyroid tissue : thyroid blood* (the c_F^i/P_F^i term), and exchange both ways limited by the *permeability area cross product* (permeability coefficient) $k_F^{PAi}(=6\times 10^{-4})$ for fluid flow through porous membranes separating F and S. The third term is the active uptake of iodide i into the FOLLICLE cells from STROMA, inhibited competitively by perchlorate p, and represented by Eq. (6.48):

$$\dot{c}_S^i = F_T(c_B^i - c_S^i) + k_F^{PAi}\left(\frac{c_F^i}{P_F^i} - c_S^i\right) - \left(\frac{V_{\max F}^i c_S^i}{c_F^i + K_{mF}^i(1 + c_S^p/K_{mF}^p)}\right) \tag{8.14}$$

For the FOLLICLE cells, there is additional exchange with the LUMEN storage pool (fourth term) – with exchange both ways limited by the *permeability area cross product* (permeability coefficient) $k_L^{PAi}(=10^{-4})$ for fluid flow through porous membranes separating L and F, and a *biophysical partition coefficient* $P_F^i(=7)$ *for thyroid lumen : thyroid tissue* (the c_L^i/P_L^i term); and competition with perchlorate (p) actively transported to F (third term), and linear clearance (CL_H) to the (hormonal) BOUND IODIDE compartment (fifth term):

$$\dot{c}_F^i = \left(\frac{V^i_{\max F} c^i_S}{c^i_F + K^i_{mF}(1 + c^p_S/K^p_{mF})}\right) - k_F^{PAi}\left(\frac{c^i_F}{P^i_F} - c^i_S\right) - \frac{V^i_{\max L} c^i_F}{c^i_F + K^i_{mL}(1 + c^p_F/K^p_{mL})}$$
$$+ k_L^{PAi}\left(\frac{c^i_L}{P^i_L} - c^i_F\right) - CL_H c^i_F \tag{8.15}$$

For the LUMEN, there is linear bidirectional exchange with the FOLLICLE (second term) and competition with p transported to L (first term):

$$\dot{c}_L^i = \frac{V^i_{\max L} c^i_F}{c^i_F + K^i_{mL}(1 + c^p_F/K^p_{mL})} - k_L^{PAi}\left(\frac{c^i_L}{P^i_L} - c^i_F\right) \tag{8.16}$$

For BOUND IODIDE (bdi), there is clearance (entry) from F (first term) and linear secretion of hormonal iodide out of the thyroid into the venous plasma compartment (second term):

$$\dot{c}_{bdi}^i = CL_H c^i_F - CL^{bdi}_{secretion} c^i_{bdi} \tag{8.17}$$

Example 8.2: ***A Hybrid PBTK Model for Manganese Distribution Dynamics in the Rat***

The Biology: In humans, inhaled manganese (Mn) can initiate neurodegeneration in the striatum – a tiny midbrain structure – and produces **manganism**, a disorder phenotypically mimetic of Parkinson's disease. It is not completely clear how Mn has this effect, nor how much inhaled Mn gets to the striatum and by what routes. The two routes inhaled Mn can get there are: direct transneuronal transport, via the olfactory system, and indirect transfer to brain across the blood–brain barrier – via lungs-to-blood. Olfactory transport of Mn means that Mn is transported transneuronally into brain, without passing into blood or other fluid first. Prior to this modeling study, it was not clear whether one or the other pathway dominated. The main modeling goal was to assess the relative contributions of each of these pathways quantitatively, using the rat as an experimental animal model of this process.

Structuring the Model: As in Example 8.1, an extracted overview of the structural modeling process is provided here. Additional details are provided in the reference (Douglas et al. 2009). This 11-state variable PBTK model is a hybrid of the physiologically based (PB) modeling methodology described above, and physiologically based multicompartmental (MC) modeling described in Chapters 4 and 5. We show that it can be visualized in either form. The published compartmentalized version is shown in **Fig. 8.8** with rounded-corner boxes rather than circle compartments – to emphasize the hybrid PB–MC nature of the model. The rightmost diagram in the figure has *mass fluxes* associated with arrows between and from compartments, rather

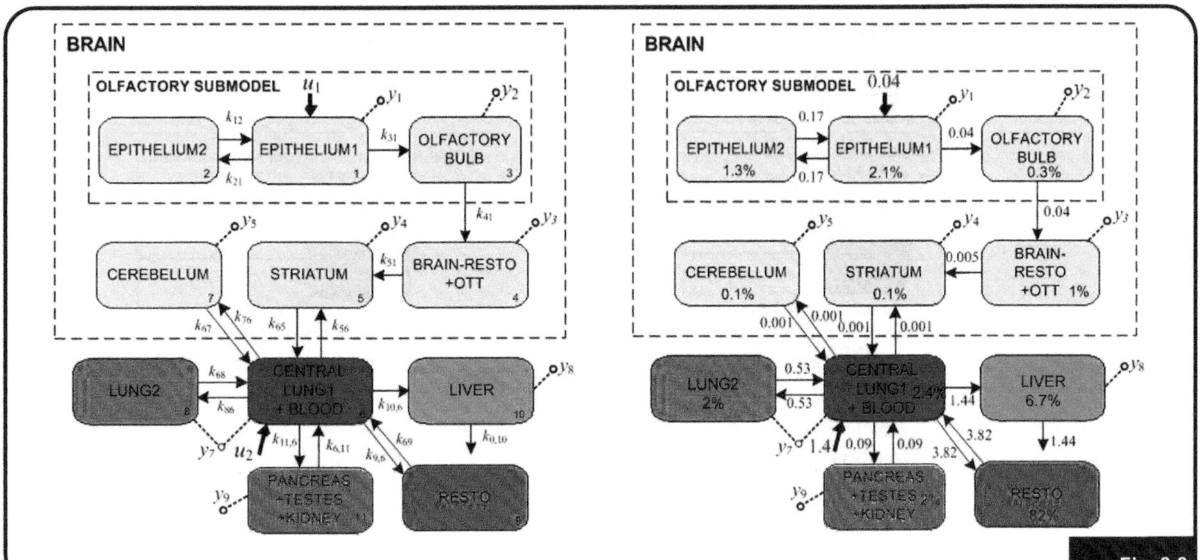

Fig. 8.8

Whole body PBTK model structure. Left: PBTK model with rate constants (kij day^{-1}) shown on arrows. Right: Same model structure, but with whole-body Mn distribution (% of total) and Mn compartment interchange rates as *mass fluxes* (μg Mn/day) shown on arrows. The **olfactory submodel** nodes: OLFACTORY EPITHELIUM 1 (EP1), OLFACTORY EPITHELIUM 2 (EP2) and OLFACTORY BULB (BULB) (boxed in); **brain compartments**: OLFACTORY submodel components + STRIATUM, CEREBELLUM and brain-resto/olfactory tract and tubercle (B-RESTO/OTT) (boxed in). **Whole body components** (nodes) include (lumped) CENTRAL + LUNG1 + BLOOD, LUNG2, (lumped) Rest-of-body (RESTO), and (lumped) PANCREA/TESTES/KIDNEY (P–T–K), and LIVER, (gray). EP2 and LUNG2 were needed to accommodate best 2-exponential fits to epithelium and lung data in preliminary analyses. Exogenous inputs are u_1 into EP1 and u_2 into CENTRAL + LUNG1 + BLOOD. Measurements are from 8 of the 11 compartments, as shown.

than multicompartmental rate constants as in the leftmost diagram, making it easier to visualize exchanges of mass among the various compartments of the model.

Equivalent PB Structure: Several organs/tissue compartments in **Fig. 8.8** are *lumped*, e.g. pancreas + testes + kidneys, *on the basis of each having similar transient dynamical responses in the data*. This is similar to grouping rapidly perfused or slowly perfused tissues in modern PB models like that in **Fig. 8.7**. To underscore the physiological basis for **Fig. 8.8**, which appears more MC than a traditional PB model, we recognize that the arrows representing exchange of material between the central composite compartment (lumped peripheral blood and part of the lungs compartment) and each organ compartment are exchanging along arterial and venous pathways; so we can reorganize the organ compartments outside the brain as simple one-pool organs, as shown in **Fig. 8.9**. The VENOUS pathways are shown separated in the leftmost version, again to emphasize the hybrid nature of the model. In the version on the right they are merged into one venous return pathway and the distinctions between the modern PB and classical MC forms are further blurred. The dynamical equations are written from either diagram representation (in Chapter 15).

The BRAIN pools are divided into two subsystems – the olfactory and the cerebellum-striatum-OTT-RESTO), each containing three subcompartments, as in **Fig. 8.8**. The two A-V pathways involved with brain subsystems are combined in the **Fig. 8.8** representation, and there is local unidirectional inter-transfer of material between olfactory and the other subcompartments, as in **Fig. 8.8**. In the experimental studies used to

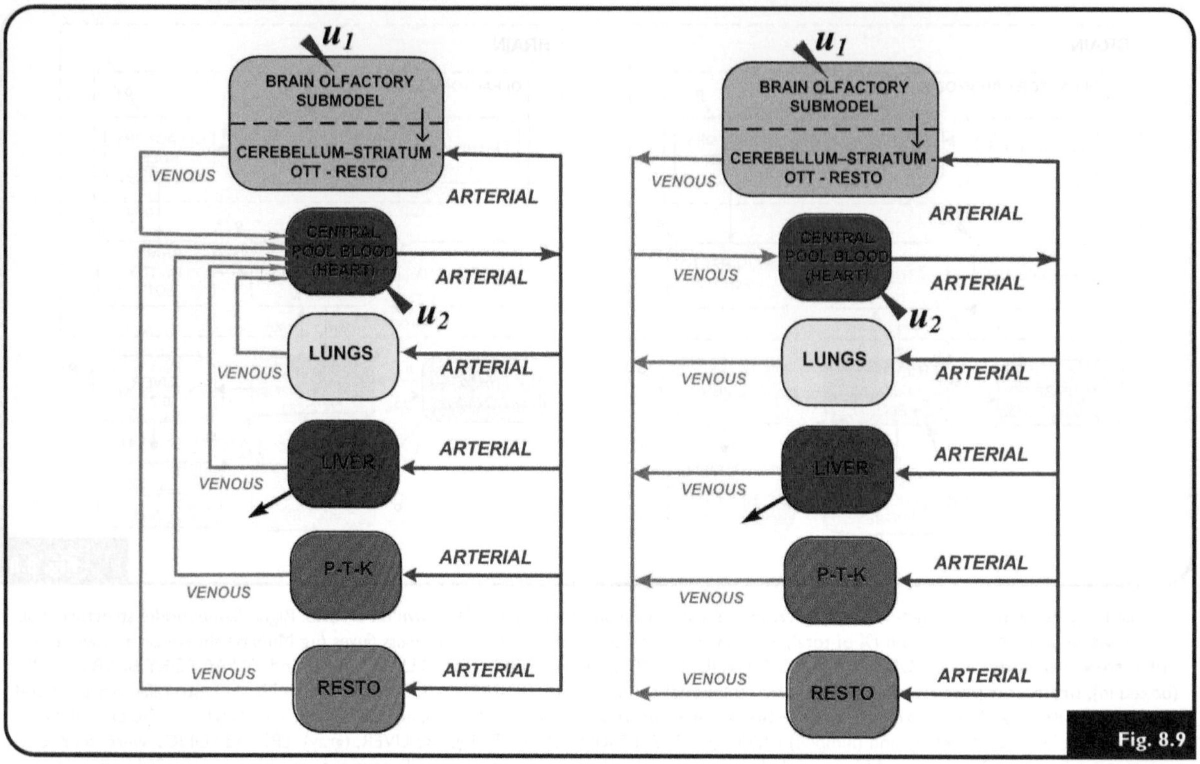

Fig. 8.9

Modern PB version equivalents of the Rat MC Mn distribution model.

quantify this model, test-input 1 (olfactory input) is into olfactory BRAIN and test-input 2 (lung input) is into the CENTRAL pool, as shown.

This PBTK model is primarily linear, with only a single nonlinear component – representing saturated influx of Mn into LIVER, based on data from earlier experimental studies. State variables are expressed in amount (mass) terms. The figure legend provides most of the physiological details. Equations, data and quantification details are deferred to Chapter 15, where we illustrate use of this model in testing the main hypothesis prevalent in the literature at the time: inhaled manganese is transported to brain cell primarily transneuronally, rather than via the cardiovascular and pulmonary systems.

Allometric Scaling in PB Models

Scaling is a way of quantifying relationships among variables or parameters in physical systems, or models of physical systems, in a relative rather than absolute way. We introduced scaling in Chapter 3 when normalizing ODEs, to render their dynamics less dependent on specific conditions (e.g. initial conditions), but readily scalable to any others. The same idea can be used to take advantage of relationships among similar properties of systems in nature, to model biosystem dynamics across species, of specific interest here. Numerous physiological processes (e.g. blood flow rates F) and anatomical entities (e.g. organ weights M) are related to the size (volume, surface area, body weight BW) across animal species. **Scaling laws** are the primarily empirical mathematical equations that relate common biophysical characteristics of structure and function across species. These are also common characteristics in evolution, maturation and seasonal variation.

In PB modeling, scaling up from small experimental animal models (for mice or rats) to models for humans is common. However, scaling across any species also can be important, for example if missing data in one species (e.g. rat) is available in another (e.g. mouse or human), or even within the same species, e.g. males vs. females, normal vs. abnormal conditions, etc.

The most obvious difference across species is size – weight and volume – most readily measured as body weight (*BW*). Adult humans weigh roughly 50–100 kg, mice about 20–30 g, rats about 150–450 g, ~2,000-fold and ~200-fold differences in relative *BW*s. The main question is: how do specific system-model parameters of interest relate to *BW*? The **allometric scaling law** quantifies this relationship for physio-anatomic rate (e.g. basal metabolic rate – *BMR*) and size (e.g. blood volume V_B) parameters, as summarized in the general model of (Brown et al. 1997; West et al. 1997):

$$Y_i = k_i(BW_i)^{\lambda_i} \tag{8.18}$$

where k_i and λ_i are constants and Y_i is either a rate or size parameter. As an example, the classic scaling law for energy metabolism yields an exponent of ¾ (West et al. 1997; Brown et al. 1995). For many physiological parameters, λ is close to ¾. For example, a default value for cardiac output (L/min) in rats is 0.235 $BW^{0.75}$, with *BW* in kg; this yields 83 ml/min for a 250 g rat.

The usual assumption, when comparing the same characteristic index *Y* from one species to another, is that both *k* and λ are the same for both species. Thus, to scale Y_i between two species with body weights BW_1 and BW_2, we have:

$$Y_2 = Y_1 \frac{k_1}{k_2}\left(\frac{BW_2}{BW_1}\right)^{\lambda} = Y_1\left(\frac{BW_2}{BW_1}\right)^{\lambda} \tag{8.19}$$

However, not all parameters change with *BW*, including red cell size, hematocrit (*Hct*), most plasma protein concentrations, mean blood pressure, body core temperature, and others. In vertebrates, heart and lung weights, lung tidal volumes, vital capacities and compliances, and blood volume, muscle and skeletal masses are ~linearly proportional to *BW*, meaning λ is closer to 1. For others, like heart and breathing rates, it is inversely (negatively) related to λ. Brown et al. (1997) provide a comprehensive survey and review of physiological parameters and allometric relationships for PB modeling across several species, including mouse, rat, dog, human and others. Exponent λ values for commonly used and measured mass, volume and other physiological parameters can be found in various references (Brown et al. 1997; Hu and Hayton 2001; Brown et al. 2005; Duncan et al. 2007; Savage et al. 2008; Makarieva et al. 2008; and others).

Example 8.3: ***Allometric Scaling Between Mouse, Rat & Human Parameters***

A mix of mouse, human and rat data were used to build a dynamic system model of *angiogenesis*[15] in mammalian cells under treatment with two potentially useful (at the time) anti-angiogenesis drugs (Tee and DiStefano III 2004). The dynamical structure

15 *Angiogenesis* means growth of new blood vessels. The context here is that cancer cell growth is supported by new blood vessels stimulated by endogenous cellular growth factors enhanced by the mutated cancer cell. Anti-angiogenesis drugs inhibit production or proliferation of these growth factors.

of the model was created primarily from mouse data, mice weighing an average of 20 g. To introduce exogenous drug input pharmacokinetics (PK) into the model, the (volume) clearance rates *CL* (vol/time) of each were required.[16] But only 200 g rat and 70 kg human clearance data were available, which had to be converted to mouse clearance rates, in *L/s*, the units for the mouse dynamics model. The clearance rate *CL* of the anti-angiogenesis drug *angiostatin* was reported for a 200 g *BW* rat as 0.128 ml/min. The exponent used for the conversion was reported be to be 0.74. Thus, *CL* of angiostatin in a 20 g *BW* mouse from Eq. (8.19) was estimated to be:

$$CL^{mouse} = CL^{rat}\left(\frac{BW_{mouse}}{BW_{rat}}\right)^{0.74} = 0.128\left(\frac{20}{200}\right)^{0.74} = 0.0233\frac{\text{ml}}{\text{min}}$$

$$= 0.0057\left(\frac{\text{ml}}{\text{min}}\right)\times\frac{1}{60}\left(\frac{\text{min}}{\text{s}}\right)\times\frac{1}{1000}\left(\frac{\text{L}}{\text{ml}}\right) = 3.88\times10^{-7}\text{L/s}$$

Similarly, the clearance rate *CL* of the potential anti-angiogenesis drug *endostatin* was reported for a 70 kg *BW* human as 0.107 L/s. In this case, *CL* in a 20 g *BW* mouse from Eq. (8.19) was estimated to be:

$$CL^{mouse} = CL^{human}\left(\frac{BW_{mouse}}{BW_{human}}\right)^{0.74} = 0.107\left(\frac{20}{70000}\right)^{0.74} = 2.55\times10^{-4}\text{L/s}$$

Allometry has been a relatively macroscale data-inspired topic – yielding models less mechanistic than one might like. It has long been a popular and somewhat controversial topic for both theorists and practitioners, with the basic theory apparently continually being retested and refined. Savage and coworkers (Savage et al. 2008) revisited the basis of allometric scaling laws, particularly as presented by West et al. (1997), concluding that λ is weakly dependent[17] on actual size (*BW*) and suggesting that the basic law is in need of some reevaluation. Additional extensions of the theory also have been developed from meso- and quantum size scale models in cellular membranes of various species (Demetrius and Tuszynski 2010), and more are to be expected in the future. If used cautiously and judiciously, the basic scaling laws – while incomplete – can be reasonable approximations in practice.

Experiment Design Issues in Kinetic Analysis (Caveats)

We've emphasized throughout this book the importance of modeling the input–output experiment together with modeling biosystem dynamics. This is particularly important in clinically oriented studies, as in pharmacology, toxicology and other areas involving human health and safety – where model quantification needs extra care. We discuss two of the theoretical underpinnings of the kinetic analysis methods described here (and for some methods in other chapters), circumscribing two of their limitations, or caveats – if you will – on their use in

16 Clearance rates *CL* define simple single-compartment models for the PK in this example, with $\dot{c} = -(CL)c + u/V$, where *c* is the plasma drug concentration in plasma volume *V*. Clearance concepts are developed in the next section.

17 Eq. (8.19) is based on the assumption that λ is independent of *BW*.

practice. *System/experiment linearity* and *stationarity* are needed for most of the methods described. These conditions govern *how* to do an experiment and *when* the "state" of the system should be observed for data collection. They are presented here to raise awareness of conditions for their potential misapplication. Adherence and validation of these conditions might be difficult in practice, especially if experimentation or field-testing is needed, as in drug development studies. Simulation of alternative experiments can be quite helpful – which of course requires a model, or different models – one for each experiment.

Linear vs. Nonlinear (NL)

Strictly speaking, kinetic analysis methods described in the next sections are in large part based on model linearity, or model-experiment linearity, if the biosystem model is nonlinear. **Kinetic analysis** here means matching experimental output data with an appropriate system model for the purpose of computing biosystem attributes, usually in the form of **kinetic indices** (parameters) reflecting physiological or pathophysiological states of an organism. The issue here is *how* the data is collected, from a possibly inherently nonlinear (NL) system.

Most NL biosystems behave linearly over a limited range of operation.[18] For this reason, experiments are often *linearizing*, meaning they are done in a way that keeps system model state variables in this linear range. Tracer experiments, for example, are usually small-perturbation linearizing experiments.[19] Thus, linearity can be achieved if the kinetic experiment for quantifying kinetic indices is linearizing, i.e. small enough to keep the outputs or state variables in an approximately linear range, so the resulting model of the biosystem *and* experiment is linear. For example, achieving dose–therapeutic-response (input–output) linearity for a commercial drug is a desirable characteristic. For reasons like these, linear or linearized multicompartmental[20] or noncompartmental models are typically chosen for kinetic studies.

The requirement for (at least approximate) biosystem or experiment linearity in modern kinetic analysis needs emphasis. For example, virtually all whole-organism parameter results determined from pulse-dose or constant infusion studies (presented below) are in the form of kinetic indices, formally dependent on linearity of system or experiment. Whether study results that don't meet these conditions are acceptable in a particular problem depends on the context and (as always) the degree to which linearity conditions are not met. Pharmacokineticists deal with this issues often and have developed numerous workarounds for modeling nonlinear biosystems and extracting useful information from them (Rowland and Tozer 2011).

Time-varying vs. Time-Invariant/Stationary[21]

Theory: We recall from Chapter 2 that a linear or nonlinear model M is time-invariant (or stationary) if its inherent characteristics are independent of *when* inputs are applied. If model

18 Without starting a debate about this issue, NL systems with rapidly switching modes of operation may be exceptions, but they too are likely to behave approximately linearly in *some* range.

19 Chapter 5, section on linearization of ODEs.

20 Traditional (modern) PB or less anatomical MC models.

21 The term stationary is used synonymously with time-invariant in many fields because of the similarity of this property to time-invariance in stochastic models. Stationary stochastic processes have random variables with time-invariant probability distributions.

M is represented by the general input–output relationship: $\boldsymbol{y}(t) = \boldsymbol{H}[\boldsymbol{u}(t)]$, where $\boldsymbol{u}$ is the input, $\boldsymbol{y}$ is the output and $\boldsymbol{H}[\cdot]$ embodies the inherent system dynamical properties, then M is a **time-invariant model** if $\boldsymbol{y}(t+\tau) = \boldsymbol{H}[\boldsymbol{u}(t+\tau)]$ for *any* value of τ. (Appendix C has more on time-invariance, but this equation summarizes it).

Practice: For an input–output (stimulus–response) experiment, with data collected over an arbitrary finite time interval $t_0 \leq t \leq T$, the system is **stationary** if its *internal* properties are independent of when the study is done, which means two things. First, internal system properties – model parameters – remain constant during the experiment. *It also means that parameter values constant over one time interval are not different over another time interval during which the experiment may be done.*[22] In many kinetic studies, e.g. in biochemical or metabolic systems, system stationarity can be essential, because an experiment design or data analysis approach dependent on stationary models – with constant parameters, is normally used, and these parameters should remain (at least approximately) constant during the study.

Experimentalists normally control for the time-varying nature of biological systems. For example, relatively short-term studies in animals are usually performed at the same time of day in biosystems exhibiting mild circadian (or other) rhythms, during a period when the system is **quasi-stationary**, i.e. approximately time-invariant. But the results of such a study are also likely to be different than those obtained during another part of the cycle, at a different time of day. At a more microscopic level, the dynamics of individual proteins involved in cell division are likely changing in the various stages of the cell-cycle – another example. A related problem is that an inherently stationary and nonlinear system exhibiting a time-varying *output*, e.g. a periodic output, has a nonstationary (time-varying parameter) *linearized* model (see Chapter 5, Eq. (5.37)) and thus shares all the problems of this class of models.

Example 8.4: ***Clinical Measurements in Humans***

Accurate measurement of blood-borne compounds associated with disease processes is essential in medical diagnosis. Many of these, e.g. hormones, have normal circadian cycles and some can vary by orders of magnitude over 24 hours. Cortisol, a primary steroidal secretion of the adrenal gland, and thereby an indicator of adrenal function, has a nadir around 4 am, reaching a maximum value about 10 times higher at 8 am. Thyroid stimulating hormone (*TSH*), a secretion of the anterior pituitary gland and a very sensitive measure of thyroid function, varies ~60% over 24 hours, with a peak around 2 am, reaching minimum range and still varying values late morning into the afternoon (Sarapura et al. 2002). It's not practical to measure these hormones more than once a day in a typical clinical situation, so they are (ideally) measured at particular times, for consistency – cortisol at about 8 am, when it is maximum, and *TSH* later in the day, when it's lowest and less variable. Core body temperature, in contrast, is normally regulated within narrow limits, varying by only a few percent at most throughout the day. It usually can be measured anytime, with the expectation that it will be higher during exercise and minimal during sleep or

22 In Chapter 4, we saw that even time-invariant (stationary) NL models can have time-varying linearized models.

upon awakening, before physical activity has begun, but – again – the range should be small. Clearly, greater care would likely be needed in choosing times of the day to do short-term kinetic studies for developing dynamic system models of cortisol dynamics, or other molecules that depend on cortisol.

In the spirit of matching model to experiment, Rescigno published a particularly insightful essay on the relative validity of similar modeling assumptions and some of their nuances in the face of real data collected under real experimental conditions (Rescigno 1996).

Whole-Organism Parameters: Kinetic Indices of Overall Production, Distribution & Elimination

A variety of indices are used for quantifying production, distribution or elimination of drugs, hormones, metabolites and other substances at the whole organism level. These include: compartment sizes, as distribution volumes (V) or masses (q), defined and discussed in Chapter 4, and also volume clearance rates – as in the example above – volume clearance rates specifically from plasma (PCR), urine (UCR) or other spaces (vol/time units); fractional clearance (elimination) rates (FCR) – (time^{-1} units); half-lives ($t_{1/2}$); and mean residence times (MRT), in time units.

These V, q, CL, PCR, FCR and MRT indices are important because they reflect physically meaningful aspects of interest about the dynamic or steady state behavior of the substance at the level of the whole organism and, in practice, they are often computable from data collected in feasible kinetic experiments. This is typically done using structured multicompartmental (MC) models or, under certain conditions, from basic biophysical principles, a *noncompartmental* (NC) approach, or from classical engineering approaches based on flow systems or tank systems. These indices or variants of them are also important in PB modeling.

Measures of Elimination & Production & their Data-Driven Models

Plasma and/or urinary clearance rates are commonly measured in whole-body kinetic studies, usually for assessing the rate of elimination of an endogenous (e.g. hormone) or exogenous substance (e.g. drug), i.e. for quantifying *sinks* in compartments, pools or the whole organism. Under certain steady state conditions, such measures can be used to quantify sources, i.e. production rates (mass/time units).

Plasma Clearance Rate (PCR)

The **plasma clearance rate**[23] PCR is the steady state rate of elimination of X from plasma per unit steady state concentration of the substance in plasma: $\dot{q}_{elim\infty}/c_{p\infty} \equiv PCR$ (mass/t/(mass/vol) = vol/t). Physically, dividing numerator and denominator by the mass of X in plasma, PCR can be said to be the equivalent volume of plasma cleared of a substance X per unit time (vol/t units) in steady state.

23 *PCR* is defined and measured in plasma, the component of whole blood excluding the red cells, because most substances separate from red cells when centrifuged and remain in the plasma portion.

Remark: *PCR* is sometimes called the "metabolic" clearance rate, or *MCR*, but *PCR* is more appropriate. Clearance of a substance from plasma is the result of more than metabolic transformations occurring anywhere in the organism, possibly including plasma. Excretory processes are also generally included, like urinary or fecal excretion, either of which can be a dominant path for clearance.

The *PCR* often serves as a unique measure of *overall elimination rate* of substance *X* in steady state. Thus, when multiplied by the plasma concentration of *X* in steady state, $c_{p\infty}$, it *might* provide the mass flux production rate *PR*, a common application of *PCR* measurements. The caveat is that this works only if all *sources of X* enter the plasma directly, as discussed further below.

PCR Experiment Designs

To quantify the *PCR* of substance *X* in an experiment, one of two inputs are typically used, *constant infusion* (step) or *pulse dose* (impulse). For a *constant infusion rate input* experiment, exogenous *X* is infused into plasma at rate *IR* (mass/time) until a constant steady state is achieved, i.e. when efflux = influx: $\dot{q}_{elim\infty} = IR$. Then, from the definition:

$$PCR \equiv \frac{\dot{q}_{elim\infty}}{c_{p\infty}} = \frac{IR}{c_{p\infty}} \tag{8.20}$$

$c_{p\infty}$ here is the constant steady state concentration of *exogenous X* in plasma, distinguishable from any *endogenous X* that may (or may not) also be in plasma.

Alternatively, we can introduce *X* as an *impulse-dose (pulse-dose)*[24] input of magnitude D_{ose}. For this transient response experiment, we write *PCR* as $\dot{q}_{elim}(t) = PCR \cdot c_p(t)$. Then, integrating both sides from 0 to ∞, we get back the total dose D_{ose}, i.e.

$$\int_0^\infty \dot{q}_{elim}(t)dt = \int_0^\infty PCR \cdot c_p(t)dt = D_{ose} \tag{8.21}$$

Then, *if PCR is constant*, it can be moved from under the integral, yielding:

$$PCR = \frac{D_{ose}}{\int_0^\infty c_p(t)dt} \equiv \frac{D_{ose}}{AUC_p} \tag{8.22}$$

AUC_p in the denominator is the **area under the curve**, the integral of plasma concentration of *X*.

Remark: This derivation of *PCR* for an impulse dose input is a roundabout way of showing that *PCR* is a constant steady state concept. *PCR* has no readily interpretable meaning when the biosystem is not in steady state (DiStefano III 1976). Eq. (8.22) is actually valid for any finite nonnegative input of any shape (e.g. a finite pulse, $|\sin \omega t|$, etc.), with only two requirements: first, $c_p(t)$ must be measured over a time period sufficient to characterize its shape (or the *AUC*) over all time, extrapolated out to time ∞, so the integral can be computed accurately. This is a potential limitation of this approach when extrapolation

24 The term pulse-dose is often used instead of impulse-dose. In modeling, we usually approximate the pulse of a rapid injection or introduction of test input material as an impulse. But, in fact, it is a short pulse.

(prediction) is weakly or insufficiently supported by data or knowledge of longer-term system or process dynamics. Second, the biosystem must be linear or the experiment linearizing. This constraint is demonstrated next.

Example 8.5: ***AUC & PCR Computations from Pulse-Dose Transient Response Data***

We illustrate – by simulation – the effects of model nonlinearities in AUC and PCR calculations for an exogenous material X (e.g. a drug) introduced as a brief pulse-dose (approximate impulse), e.g. into blood. The experimenter measures the output concentration $y(t) = q(t)/V \equiv c_p(t)$ in blood plasma until it becomes negligible, and AUC is calculated using some form of integration of this data, $AUC = \int_0^\infty c_p(t)dt$. Then $PCR = dose/AUC$ (Eq. (8.22)).

We compare results for the simple one-compartment NL model of **Fig. 8.10**, which includes a fairly common first-order NL Hill function elimination rate: $k_{el}(q) \equiv 50q/(q+25)$, versus its linear (linearized) counterpart – with elimination rate of material X simply proportional to the amount $q(t)$ of X in the compartment, i.e. $k_{el}(q) = \text{constant}$ ($=1.88$ in this example). This k_{el} value has been chosen to approximately linearize the NL model for "low-dose" inputs. It is anticipated that different pulse-dose input magnitudes (simulated as ICs here) will yield different AUC and PCR values, with differences dependent on how far the dose magnitudes are from the linear low-dose range of the Hill function.

$$\dot{q} = -\left[\frac{50}{q(t)+25}\right]q(t)$$

$$q(0) = D_{ose}, \quad y(t) = q(t)/V \equiv c_p(t)$$

The linear (top left), NL-linearized (top right) and NL models (bottom right) are shown implemented in *Simulink* in **Fig. 8.11**. Results for a 10-fold difference in dose magnitude, $D_{ose} = 3$ and $D_{ose} = 30$ are given in the boxes after the divider.

For doses between 3 and 30 (not shown), the linear model generates exactly the same AUC and PCR values, the NL model operating in the linear range gives almost the same result (1.887), whereas the NL model produces larger and larger (incorrect) AUC and smaller and smaller (incorrect) PCR values. $D_{ose} = 30$ results

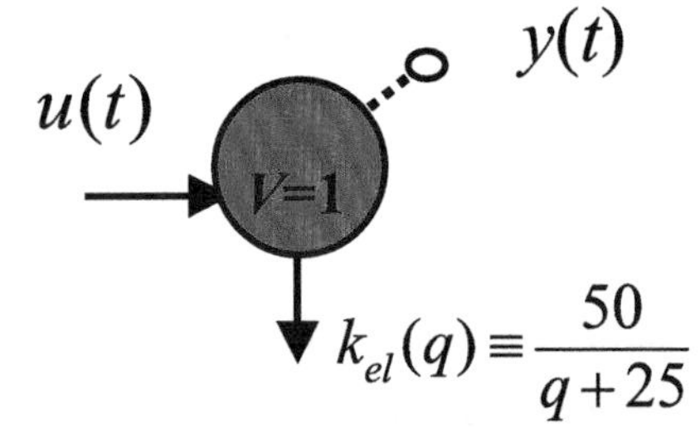

Fig. 8.10

PCR impulse–response study with a NL model.

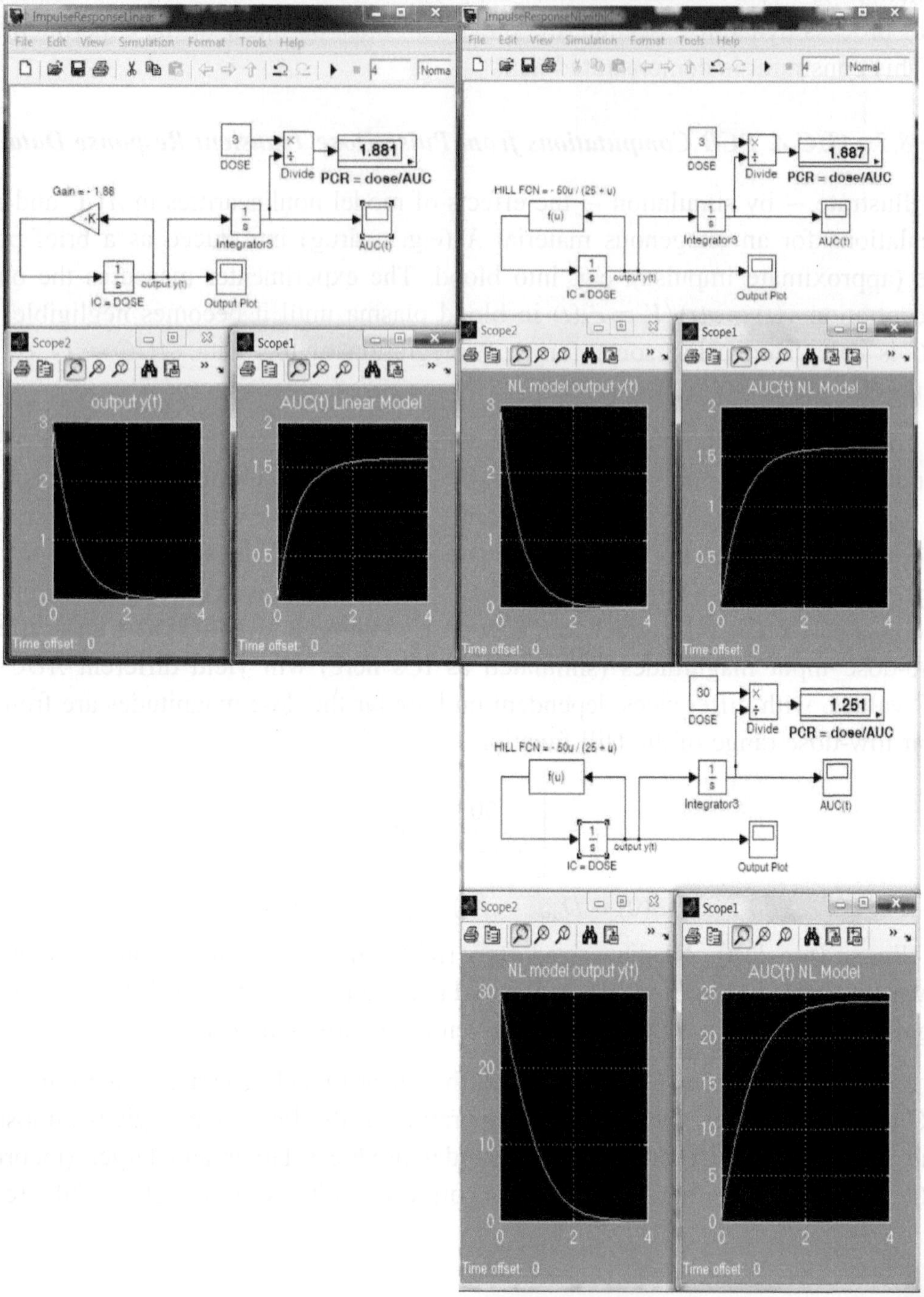

Fig. 8.11

($PCR = 1.251$) are $\sim$ 2/3 of $D_{ose} = 3$ ($PCR = 1.887$) results. For larger doses than 30, the Hill function eventually saturates and $PCR \rightarrow \sim 0.25$ for very high doses. (Why?)

These differences are very difficult to detect by comparing the plots. The transient output responses, which reflect what would be predictively obtained from data collected on a system under the same conditions, nearly superimpose on the same axes – scaled by the dose difference factor of 10. Differences in the $AUC(t)$ values are evident as time increases to its constant value.

Example 8.6: ***AUC and PCR Computations from Constant Infusion Response Data***

Again consider the same one-compartment NL and linearized models of the previous example, this time with two different step-input magnitudes, $IR = 3$ and 30. The experimenter gives a constant infusion (step) input $u(t) = A\mathbf{1}(t)$ into blood plasma (for example) for a period long enough for the output to be approximately constant (approximate steady state), and then measures the single output concentration $c_{p\infty}$. Then, from Eq. (8.20), $PCR = IR/c_{p\infty}$. The *Simulink* program in **Fig. 8.12** illustrates the simulated experiments with the two different inputs applied to the linear and NL models.

For the *linear* model, only the simulation for $IR = 3$ is shown (left side). For $IR = 30$ (not shown), $PCR = 1.881$ is exactly the same as for $IR = 3$. For the *NL model* (right side), PCR is 1.882 for $IR = 3$ (not shown). Only the program solution for $IR = 30$ is shown, with PCR reported as 0.9625 in the figure. The correct value should be 0.8, obtained by running the solution for much longer times. The reason it's wrong can be seen in the output plot, which has not yet reached a constant value (approximate steady state), whereas the linear model did by $t = 4$.

Remark: The previous two examples numerically illustrate a number of issues with PCR computations from experimental data. First, if the biosystem is linear, and the experiment is run long enough to achieve negligible output $y(t)$ or constant steady state, no problem. If it's NL, the

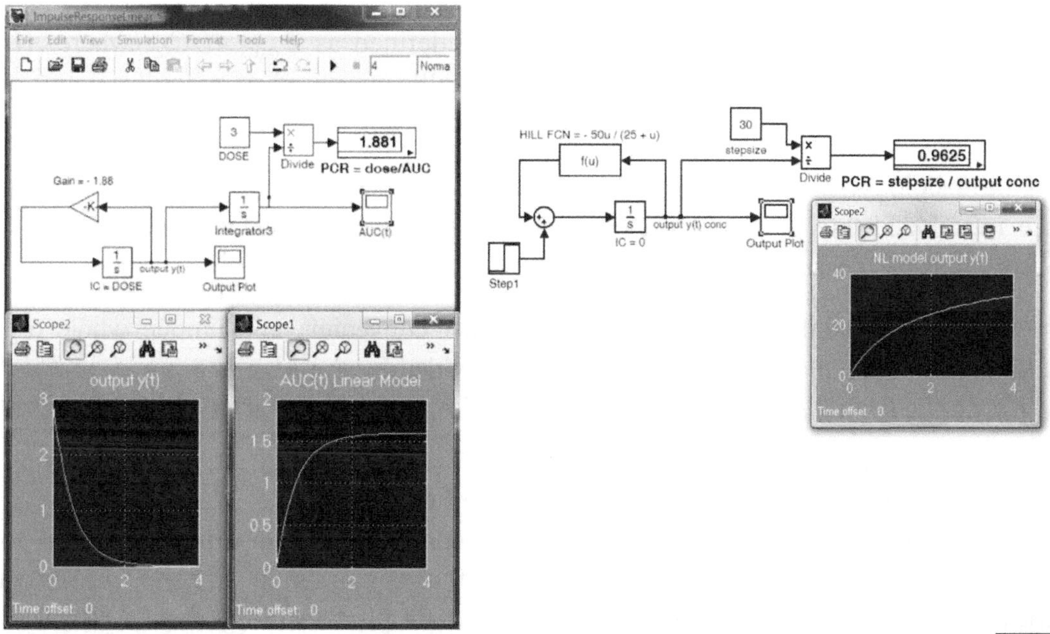

Fig. 8.12

experiment must be (approximately) linearizing to get accurate results; and it must be run long enough to achieve negligible output $y(t)$ for pulse-dose inputs, or approximate steady state for step inputs IR. In all NL cases the input magnitude must be a "small-enough" perturbation to generate a transient response approximately equal to that of a linearized model. Also note that the experiment can be done with a pulse-dose D_{ose} (approximate impulse) or a constant step-input IR, but the magnitude of the step that generates a linearized response is not the same as that of an impulse that does the same, as illustrated in the second example. The reason is that the single pulse-dose terminates shortly after it begins – and delivers a fixed perturbation amount of material to the system, whereas the step input continues to deliver material until the experiment ceases, so the perturbation itself increases with time.

Remark: If the biosystem had been assumed to be linear at the outset, perhaps because it was mistakenly believed that a receptor mechanism represented by the Hill function was nonsaturable – and operating in its linear range, for small-enough inputs, two AUC computations with sufficiently different input magnitudes would certainly negate this hypothesis. What if only one experiment were done, with the larger input?

Remark: Molecules introduced exogenously, e.g. pharmaceuticals (drugs), often act on saturable cellular receptors and are typically modeled using Hill functions in PK or PD equations – like in the examples above. Indeed, in testing efficacy and safety of new drugs,[25] PK and PD modeling plays a central role. Pulse-dose studies for PK and PD parameters like AUC are standard practice for comparing efficacy and safety of different doses, or products made by different manufacturers (**bioequivalence** studies). Given that these models are nonlinear, extra care is called for in interpreting results obtained from them, particularly in comparative studies like those reported in (Eisenberg et al. 2006; Eisenberg and DiStefano III 2009).

Other Clearance Rates: UCR & FCR

The **urinary clearance rate (*UCR*)** is similarly defined as D_{ose}/AUC_U, where AUC_U is the area under the curve for urinary excretion, typically measured as a *total amount of X* in total accumulated urine (letting the system do the integration for us!).

In pharmacology, we have a slightly different clearance measure of the total amount of the dose introduced into plasma excreted in urine unchanged: renal clearance $\equiv D_{ose}/AUC_p$ (AUC from plasma measurements).

The ***FCR* (Fractional Clearance Rate)** is another convenient and unique whole organism parameter, equal to the plasma clearance rate PCR divided by the *plasma equivalent distribution volume* V_D, i.e. $FCR = PCR/V_D$. In pharmacology V_D is called the **apparent distribution volume**.

Distribution Volume, Partition Coefficient & Pool Size Relationships

Steady state equivalent distribution volume concepts were developed in Chapter 4 for MC models. They provide a means for expressing real pool or compartment *sizes*, in true *mass* terms, in *equivalent volume* terms (same as *apparent volumes* in pharmacology lingo). Because organ

25 For example, in applications to pharmaceutical assessment agencies like the US Food and Drug Administration (FDA).

pool/compartment *partition coefficients* P are effectively a pseudo-steady state (equilibrium) concept, they are readily factored into steady state distribution volume equations. The theory here is extended for linear models, but is applicable to NL models – with the usual caveats about degree of approximation and experiment linearity in quantification studies.

We review the concept first. Consider blood *plasma* as the measured *reference compartment*, with steady state X concentration $c_{ref} \equiv c_{p\infty}$. The **equivalent distribution volume V_i of single compartment i** is the equivalent volume substance X *would occupy* in steady state if it were distributed uniformly and with the same concentration in compartment i as in plasma in steady state, i.e. $c_{i\infty} \equiv c_{j\infty} \equiv c_{p\infty}$. (Presumaby, compartment $i \neq p$ is not measured.) The total system **plasma equivalent distribution volume V_D** then is the equivalent volume the substance *would occupy* in steady state if it were distributed uniformly and with the same concentration in *every* (equivalent) compartment in steady state, i.e. $c_{1\infty} \equiv c_{2\infty} \equiv \ldots \equiv c_{n\infty} \equiv c_{p\infty}$. Therefore, for an n-compartment model, $V_D = \sum_{i=1}^{n} V_i$.

This *equivalence* equation interpretation of compartment volumes (actually the reason for its being) is a ***surrogate equation for total mass*** of X, q_{tot}, in the system in steady state, i.e. $q_{tot} = q_{1\infty} + q_{2\infty} + \cdots + q_{n\infty}$, where all $q_{i\infty}$s are *real* masses (e.g. kg, mole). To see this, recall Eq. (4.64) of Chapter 4, for the two-compartment model: $\frac{q_{1\infty}}{q_{2\infty}} = \frac{V_1}{V_2}$, where the V_is here are *equivalent* (not real) distribution volumes. This model is reproduced in **Fig. 8.13**, for "tissue" compartment 2 exchanging with "plasma" compartment 1 and both real and equivalent distribution volumes shown. A constant input to 1 is assumed, so all variables are assumed in constant steady state. Compartment 1 is the reference (measurement) compartment here, so concentrations and volumes are both real and equivalent in compartment 1.

We now show how equilibrium (steady state) partition coefficients P, defined as a ratio of *real* concentrations, are related to real and equivalent distribution volumes. With *real* concentrations $c_{1\infty} \equiv c_{p\infty}$ and $c_{2\infty}^{real} \neq c_{p\infty}$, $P \equiv c_{2\infty}^{real}/c_{1\infty}$, as illustrated in compartment 2 of the figure. Since real concentrations are real masses divided by real volumes, the real volumes $V_1^{real} \equiv V_p \neq V_2^{real}$ are related to P and the equivalent distribution volumes V_1 and V_2 as:

$$P \equiv c_{2\infty}^{real}/c_{1\infty} \equiv \frac{q_{2\infty}/V_2^{real}}{q_{1\infty}/V_1} = \frac{q_{2\infty}}{q_{1\infty}}\frac{V_1}{V_2^{real}} \quad \text{or} \quad P\frac{V_2^{real}}{V_1} = \frac{q_{2\infty}}{q_{1\infty}} \equiv \frac{V_2}{V_1} \tag{8.23}$$

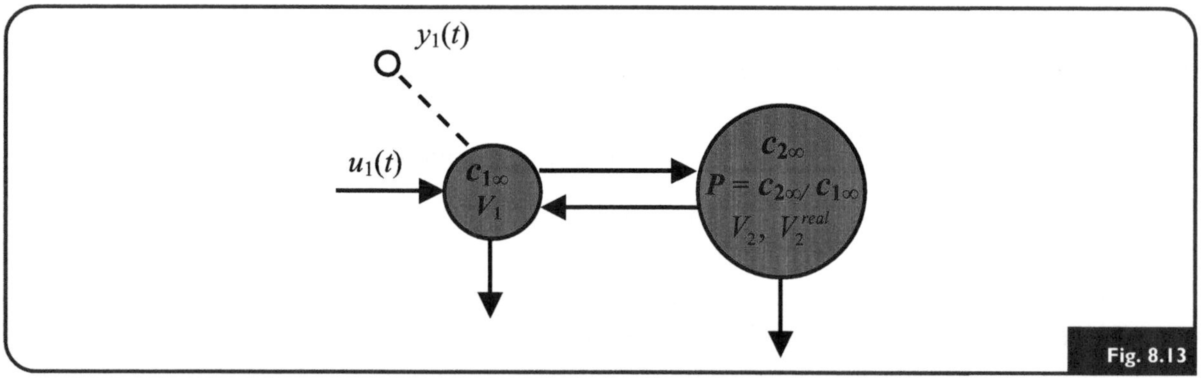

Fig. 8.13

This means:

$$\text{Tissue equivalent distribution volume } V_2 = PV_2^{real} \tag{8.24}$$

and the total equivalent (apparent) distribution volume V_D and total pool size of X are:

$$V_D \equiv V_1 + V_2 = V_p + PV_2^{real}$$

$$\text{and} \tag{8.25}$$

$$q_{tot} \equiv q_{1\infty} + q_{2\infty} = V_p c_{p\infty} + V_2^{real} c_2^{real} = V_p c_{p\infty} + V_2^{real} P c_{p\infty} = (V_p + V_2) c_{p\infty} \equiv V_D c_{p\infty}$$

These practical relationships involving Ps, Vs and qs are readily extended to any number of tissue compartments in steady state, each by simply adding $P_i V_i^{real}$ terms for each tissue compartment with real volume V_i^{real}.

Residence Times

The whole-body **mean residence time** of a substance, ***MRT***, is the ***average time it spends inside the entire organism prior to irreversible removal.*** This temporal parameter of residence, unlike the more historically pervasive half-lives ($t_{1/2}$), is a global whole organism measure applicable to systems of any complexity. It is global in the sense that *MRT* has only a single value for any system, whereas complex systems have multiple "half-lives."

Residence times $\bar{t}$ are also defined for individual compartments in a multicompartmental model, in terms of the compartmental matrix K. Specifically,

$$R \equiv [r_{ij}] = -K^{-1} = \begin{bmatrix} r_{11} & \cdots & r_{1n} \\ \vdots & \ddots & \vdots \\ r_{m1} & \cdots & r_{mn} \end{bmatrix} \tag{8.26}$$

is the **matrix of mean residence times for each compartment** i, i.e. the mean total time a molecule X spends in compartment i before leaving the system (model) irreversibly, given that it started in compartment j (Perl et al. 1957). Numerous combinations of mean times that X spends in various combinations of compartments in the model can be computed from these matrix entries. Column j values r_{ij} are mean residence times r_{ij} in compartment i for X entering compartment j. The mean residence time in the entire system for material entering compartment j is the *sum* of the jth column entries $\sum_i r_{ij}$.

Example 8.7: ***Residence Times from the Compartmental Matrix***

Matrix $R = \begin{bmatrix} 14 & 12 \\ 37 & 40 \end{bmatrix}$ is the negative inverse of K for a two-compartment model. Each of the entries r_{ij} represents the mean residence time in the single ij^{th} compartment for X starting from compartment j, i.e. mean residence in compartment 2 for X entering compartment 1 is 37 time units, etc. Mean residence in the whole system for X entering compartment 2 is 12 + 37 = 49 time units.

Other residence time relationships obtainable from R, including variances for these residence times, can be found in several references, some based in probabilistic notions (Eisenfeld 1980; Covell et al. 1984; Garcıa-Meseguer et al. 2003). Whole system mean residence time (*MRT*) computations and formulas for estimating them by noncompartmental (NC) methods are presented in the next major section.

Whole-Organism Parameter Relationships

The kinetic parameters listed above are related by the following equations, where $c_{p\infty}$ denotes steady state concentration of substance in blood plasma. These equations reflect whole-body kinetics in steady state.[26] They are particularly useful for characterizing and thus differentiating normal physiological from pathophysiological steady states, e.g. well vs. sick states.

$$\text{Total Body Pool Size} = q_{tot} = V_D c_{p\infty} \quad \text{(mass units)} \tag{8.27}$$

$$\text{Fractional Clearance Rate} \equiv FCR = \frac{PCR}{V_D} = \frac{1}{MRT} \quad (\text{time}^{-1}) \tag{8.28}$$

$$\text{Total Body Production Rate} \equiv PR = \frac{q_{tot}}{MRT} = PCRc_{p\infty} \quad \text{(mass/time)} \tag{8.29}$$

In steady state, PR is equal to total body elimination rate, ER, due to all irreversible elimination processes, degradative and excretory. *Under most circumstances, whole-organism parameters must be determined under endogenous steady state conditions in living systems.* Strictly speaking, they only have meaning when the system is in steady state. The same caveats regarding linearity of the system or experiment are applicable here as well.

Half-Lives

The parameter for *half-life* of a substance, typically denoted $t_{1/2}$, is conspicuously absent from Eqs. (8.20) through (8.29). "Half-life" is an accurate kinetics index only for strictly monoexponential responses. Consequently, when one needs a measure of removal or elimination rate in more complex systems, a single half-life measure can be a poor or misleading approximation. Mean residence time (*MRT*, above), for example, is a much better choice.

The half-life concept is based on extension of the half-life property of the single exponential decay. In words, it means the time $t_{1/2}$ at which half of the substance under study remains, typically after injection of a pulse-dose at time $t = 0$. It's computed from: $0.5 = e^{-kt_{1/2}}$ or $\ln 0.5 = -kt_{1/2}$. Then $t_{1/2}$ is the **half-life** of an exponential with exponent k (t^{-1}), i.e. $t_{1/2} = \ln 2/k \equiv T \ln 2$ where $T \equiv 1/k$ is the **time constant**. Whereas $t_{1/2}$ is a perfectly appropriate parameter to describe the rate of decay of e^{-kt}, as are k and T as well, there is no single $t_{1/2}$ that can be accurately associated with more complex data. The line[27] fitted to the final data points in **Fig. 8.14** captures only part of the kinetics. A biexponential response function can be fitted to the entire data set in this figure, with "two half-lives," one for each exponential. But it has no easily interpretable *single* equivalent half-life. How well a single exponential approximates all the data depends on the relative magnitudes (coefficients) of the two exponential components.

26 PCRs obtained from Eq. (8.20) are applicable in real experimental situations only if all sources of substance enter plasma directly, as discussed more fully below.

27 $\ln(\exp(kt)) = kt$.

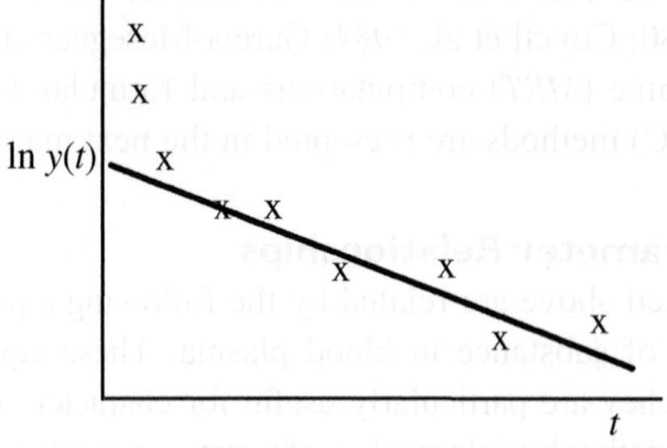

Fig. 8.14

Importantly, inappropriate single half-life comparisons can lead to incorrect conclusions when comparing kinetics in different clinical states, or between different substances, possibly with deleterious consequences. Fortunately, there are alternatives. *Mean residence time MRT* (for example) is a global measure of "disappearance or decay," applicable to (linear) impulse response data of any complexity, and is an appropriate substitute for the half-life on both practical and theoretical grounds. And *MRT* and fractional clearance rates *FCR* work for *linearizing* steady state experiments done on nonlinear systems *of any complexity*.

Remark: In some fields using kinetic modeling, e.g. pharmacology, response data characterized by two exponentials has special interpretations. The first "fast" exponential is said to represent the "mixing phase" or **mixing half-life** of the falling (drug) response. The second "slow" (final) exponential is called the terminal phase or **terminal half-life**. Neither by itself is meant to represent global elimination rate in steady state.

Example 8.8: ***Half-Lives vs. MRTs***

The mean residence time (*MRT*) of thyroxine (T_4) in rat plasma is 18 hours, only a third larger than that (12 hours) for triiodothyronine (T_3), the active thyroid hormone (DiStefano III 1982). This is in contrast with reported half-life values 3–6 times greater for T_4 than for T_3 (*www.thyroidmanager.org*), which suggest that T_3 "disappears" 3 to 6 times faster *from the whole body*, which is far from the truth.

The reason for the large discrepancy is that, whereas T_3 disappears much more rapidly than T_4 from plasma *at very early times* after brief injection, the disappearance (impulse response) curve at later times, when most of the T_3 is already distributed at tissue sites, is more comparable to that of T_4. Corresponding *terminal half-lives* calculated from the slope of the final portion of the same tracer data used to compute the *MRTs* above are 7 h and 12 h, a nearly 100% difference – erroneously suggesting a *much* higher fractional clearance rate ($FCR = 1/MRT$) for T_3 than for T_4. This is a significant overstatement, because *MRT* comparisons, as noted above, indicate that T_3 and T_4 have more comparable times of residence in the organism (DiStefano III 1982).

Noncompartmental (NC) Biomodeling & Analysis (NCA)

The basic noncompartmental (NC) model is illustrated in **Fig. 8.15**. This diagram represents the simplified (lumped) biophysics of the system implicit in noncompartmental analysis, for the most common case of test-input followed by temporal measurements of a single substance in a single pool (single space, not single compartment). The *sources* arrow (mass flux) represents the sum of all *de novo* (new, for the first-time) entry of substance X under study into the central pool, i.e. previously uncirculated material entering the system for the first time, either endogenously or exogenously. Indirect *de novo* entry, following production or infusion in a noncentral pool, eventually arriving (for the first time) in the central pool, renders this an ***equivalent source***. Thus, it is important to note that the NC model makes no assumption about the number of sources that may be distributed throughout the organism (system). Any number may exist, anywhere, and the substance produced by these endogenous sources may or may not appear or reach the central measurement pool.

The *sinks* arrow similarly represents the equivalent sum of all irreversible removal or elimination processes (mass fluxes) of X throughout the system, whether or not they take place in the central pool. All recirculations and exchanges of X with noncentral pools also are lumped together and are represented by the single "loop" shown in **Fig. 8.15**. This simple structure is usually replicated for each additional measurement pool in more complex experimental situations, supplemented by interconnections among the measurement pools (Carson et al. 1983), as illustrated in **Fig. 8.17** (see page 391) of the next section for two measurement pools.

The major conceptual advantage of a noncompartmental model is that any number of recirculations or exchanges of X can occur, with any number of noncentral pools, none of which have to be identified with any physical structures, and *none have to be homogeneous*. Mathematical analyses of time-varying kinetic data measured in the central pool are usually accomplished using equations involving *integrals*, as given in equations (8.30) to (8.35) below, rather than differential equations. This is in contrast to a multicompartmental structure, for which compartmental connectivity must be rendered more specifically, whether or not the physiological or anatomical identity of the several compartments is known, and all compartments must be homogeneous.

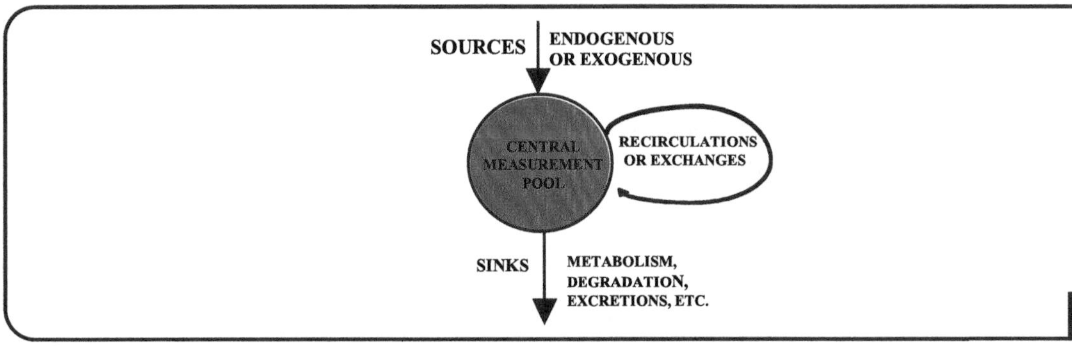

Fig. 8.15

The basic noncompartmental model.

Noncompartmental Analysis Formulas

These are derived directly from the definitions and equations for *PCR*, V_D, *MRT* and *FCR* above and are given here for impulse response (single-injection; pulse-dose) studies, in terms of the integral-area under the concentration (of tracer or substance in plasma) curve, denoted *AUC*, and **area under the moment curve**, *AUMC*. For step (infusion) response experiments carried to (approximate) steady state $c_{p\infty}$, $PCR = IR/c_{p\infty}$ was given by Eq. (8.20), and the others are readily computed[28] (Exercise E8.8).

The superscript NC designates each parameter as its noncompartmental value or approximation. The absence of superscripts denotes true values, $c(t) \equiv$ instantaneous concentration (mass/volume), and $D_{ose} \equiv$ dose (mass).

$$AUC = \int_0^\infty c(t)dt \quad (\text{mass} \cdot \text{time/vol}) \tag{8.30}$$

$$AUMC = \int_0^\infty tc(t)dt \quad (\text{mass} \cdot \text{time}^2/\text{vol}) \tag{8.31}$$

$$PCR^{NC} = \frac{D_{ose}}{AUC} \leq PCR \quad (\text{vol/time}) \tag{8.32}$$

$$V_D^{NC} = \frac{D_{ose} \cdot AUMC}{(AUC)^2} \leq V_D \quad (\text{vol}) \tag{8.33}$$

$$MRT^{NC} = \frac{AUMC}{AUC} \leq MRT \quad (\text{time}) \tag{8.34}$$

$$= \frac{1}{FCR^{NC}} \leq \frac{1}{FCR} \quad (\text{time}) \tag{8.35}$$

The rightmost relations in Eqs. (8.33) to (8.35) are equalities if and only if the source and/or sink requirements for applicability of NC analysis are met, as described in the next sections. Otherwise they are strict inequalities, also as explained in the next sections.

Remark: *AUC* and *AUMC*, as such, are ***models of data***. It is only when NC values in Eqs. (8.32) to (8.35) are given a physiological interpretation implied by the rightmost equalities that the ***NC system model*** (**Fig. 8.15**) is invoked.

Remark: Experiment linearity and stationarity are also important assumptions practically constraining the applicability of the NC analysis approach. This will become more evident as we explore it further.

Noncompartmental Parameters from Multiexponential Models

Suppose the multiexponential function:

$$y(t) = \sum_{i=1}^{n} A_i e^{\lambda_i t} \quad (\text{mass/vol}) \tag{8.36}$$

28 For linear or linearizing biosystem experiments.

is fitted to the output data $c(t)$. Then both AUC and $AUMC$ are easily evaluated analytically using Eq. (8.36). In this case, the NC formulas from Eqs. (8.32) to (8.34) become:

$$PCR^{NC} = \frac{-D_{ose}}{\sum_{i=1}^{n}(A_i/\lambda_i)} \quad \text{(vol/time)} \tag{8.37}$$

$$V_D^{NC} = \frac{D_{ose} \cdot \sum_{i=1}^{n}(A_i/\lambda_i^2)}{\left[\sum_{i=1}^{n}(A_i/\lambda_i)\right]^2} \quad \text{(vol)} \tag{8.38}$$

$$MRT^{NC} = \frac{-\sum_{i=1}^{n}(A_i/\lambda_i^2)}{\sum_{i=1}^{n}(A_i/\lambda_i)} \quad \text{(time)} \tag{8.39}$$

Example 8.9: ***Multiexponential→NC Model***

Suppose a set of (noisy) data $z(t_1), z(t_2), \ldots, z(t_N)$ collected from a system S_a is fitted to a biexponential function (Eq. (8.36), $n = 2$), using a weighted least squares (*WLS*) procedure (Chapter 12), yielding: $y_a(t) = 0.5e^{-1.707t} + 0.5e^{-0.2929t}$. The test-input in this study is a rapidly applied unit dose $D_{ose} = 1$ (approximate impulse). For simplicity, we ignore the effects of the noise (data errors) on computation of whole organism parameters. NC values of whole-organism parameters are easily evaluated from Eqs. (8.37) to (8.39) and the exponential parameters A_1, A_2, λ_1, λ_2, of $y_a(t)$ as $PCR_a^{NC} = 0.5$ (vol/time), $V_{D,a}^{NC} = 1.5$ (vol), and $MRT_a^{NC} = 3$ (time). For a second system, S_b, and a new set of noisy kinetic data, the best fit function is: $y_b(t) = 0.9038e^{-3.179t} + 0.0962e^{-1.32t}$, again for a unit dose test-input. In this case, $PCR_b^{NC} = 2.8$, $V_{D,b}^{NC} = 1.133$, and $MRT_b^{NC} = 0.4048$. This example is extended below, where these numbers eventually have a profound interpretation.

Bioavailability & Bioequivalence in Pharmacology

Bioavailability is a measure (figure-of-merit) of the extent (fraction or percent) to which an administered drug dose reaches the systemic blood circulation from any non-intravenous (non-IV) port, i.e. oral, rectal, transdermal, subcutaneous or sublingual administration, all of which involve drug losses before reaching the circulation. (Intravenous (IV) dosing, by definition, is 100% bioavailable.) The most common – oral dosing of a drug – requires absorption from the gastrointestinal tract to get into blood, which usually entails drug losses via gut excretory and gut metabolic processes or incomplete dissolution, rendering some unavailable for absorption – and thus not bioavailable. Bioavailability is one of the principal pharmacokinetic

(PK) properties of drugs. Its assessment requires two experiments: measurement of AUC values measured in blood, with drug administered both orally and as a single dose IV. **Absolute bioavailability** ($\%AB$) is computed using the NC formula:

$$\%absorbed \equiv \%AB = 100\frac{AUC_{ORAL}/Dose_{ORAL}}{AUC_{IV}/Dose_{IV}} \tag{8.40}$$

Other important PK parameters, based on a single oral dose experiment, are AUC – which quantifies **total drug absorbed**, the **maximum drug concentration** c_{MAX} achieved in plasma following an oral dose, and the **time to reach maximum concentration** t_{MAX}. All are illustrated in **Fig. 8.16**.

Relative bioavailability ***(RB)*** of two different preparations (1 and 2) of the same drug, introduced via the same route, is defined as:

$$relative\ bioavailability\ (RB) = \frac{AUC_1/Dose_1}{AUC_2/Dose_2} \tag{8.41}$$

Bioequivalence BE is a measure of how closely different preparations of the same drug match in terms of their efficacy and safety. Specific statistical tolerances on various PK parameters, defined in particular ways in different countries or regions, are used to establish BE. For example, RB should be close to 1 for BE, with "close" defined by statistical tolerances. Establishing bioequivalence is essential in the approval process for generic drugs by the US Food and Drug Administration (FDA) and other such regulatory agencies throughout the world.[29] FDA requirements for approval are many and usually include most or all of the PK parameters described above.

Noncompartment Model Structure Problems

On the surface, the noncompartment (NC) model appears to be less "constrained" than a compartmental one. Sources, sinks, and measurements are all explicitly associated with one

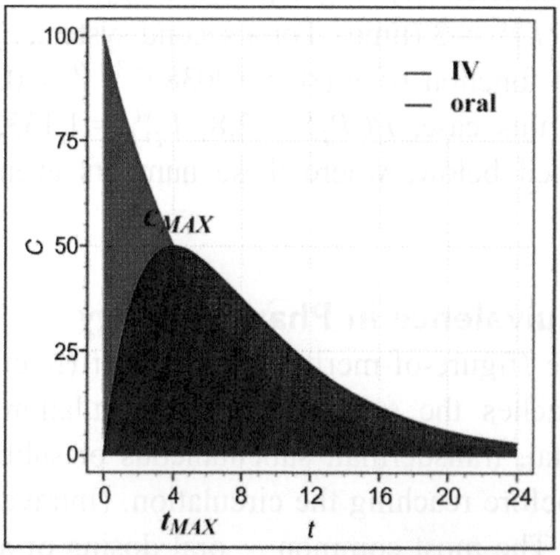

Fig. 8.16

Example dynamic responses (concentrations, C) to equal doses given IV and orally in blood or blood plasma. Absolute bioavailability (AB) is the ratio of areas under these curves. The maximum concentration c_{MAX} and the time the max is reached t_{MAX} are also illustrated.

29 See *http://en.wikipedia.org/wiki/Bioequivalence*

(central) pool in the noncompartmental model (n pools if there are n measurement sites), and the sources and sinks are *equivalent*, reflecting sources that reach the central pool indirectly as well as direct sources and sinks that may appear elsewhere in the system in noncentral pools, e.g. substances synthesized in peripheral tissues (indirect sources) and catabolism in tissues (indirect sinks) (Rescigno and Segre 1966). These seemingly simplifying properties can, however, severely restrict applicability of NC models *in practice* – which means obtaining desired results about a biosystem from available data. Hidden beneath the surface, a *constrained structure* is implied in **Fig. 8.15**, a structure that can limit its usefulness in applications. We distinguish between two types of fundamental structural problems (DiStefano III 1982).

The Equivalent Sink Problem: NC Model Structural Constraint #1

In principle, to accurately compute whole organism parameters using conventional kinetic data collected in feasible kinetic experiments, the NC model requires that the substance under study be *directly* eliminated from *only* the central pool or region in which hormone, drug, or other indicator is introduced and measured, e.g. via intravenous injection, blood sampling and assay. These include distribution volumes (V_D), pool masses (q_{tot}), mean residence times (MRT), fractional clearance rates (FCR) and other size or kinetic indices depending on these parameters.

Plasma clearance rate (PCR) computations are not affected by *this* limitation of the NC model. But, it is the only parameter unaffected. This subtle but very important issue means that relatively little or no elimination can occur in or from extravascular pools in order to evaluate V_D, q_{tot}, MRT or FCR accurately from blood-borne data by NC analysis (NCA). Strictly speaking, this condition is not met for a wide variety of substances, e.g. drugs, hormones, and metabolites deactivated intracellularly or otherwise eliminated directly from noncentral pools. In practice, however, this condition *might* be met approximately, e.g. for conditions where vascular–extravascular exchange rates are very rapid relative to overall elimination rates (see Example 4.4), or if noncentral pools are very small relative to central pools, and/or minimally "leaky." However, the degree of approximation, when approximation is appropriate, is dependent on specific quantitative as well as structural aspects of a system, and this dependency is often not easy to quantify (Feng and DiStefano III 1991). These points are clarified below in the illustrative examples.

The Equivalent Source Problem: NC Model Structural Constraint #2

In principle, the NC model of **Fig. 8.15** applies to systems with any number of sources, distributed anywhere in the system. In practice, when exogenous probes (e.g. tracers) are used as test-inputs to evaluate organ system parameters, use of the NC model requires that all sources of the substance under study be *direct* and *unique* to the measured pool, the same pool into which the tracer or other indicator test-input probe is introduced. This means that noncompartmental analysis (NCA), strictly speaking, is not applicable in analysis of (say) blood-borne data collected from systems in which there is more than one endogenous source, unless the product of all sources enters blood *directly*. This problem affects PCR as well as V_D, q_{tot}, MRT, and FCR computations. On the other hand, noncompartmental analysis can provide *upper or lower bounds* on these indices, or more or less accurate approximations in some situations; but they always will be biased. The important question is "by how much?" This is difficult to answer without additional independent data, or analysis using an independent, modeling methodology, as clarified further below in the examples.

Exchangeable or Circulating Masses & Volumes as NC Parameter Bounds

We now address an important consequence of – and workaround solution for – the equivalent sink and source problems. Consider the common experiment in which an indicator substance or tracer is introduced into a system to glean information about a mother substance (tracee) produced or eliminated all or in part outside the central pool. We assume a multicompartmental (MC) model for this system, with homogeneous compartments, to illustrate the theory here. Assume the test input is infused at a constant rate and measurements are made when the systems achieves (approximate) steady state, with all state variables (approximately) constant. In general, not all compartments of the system are fully revealed by the indicator or tracer. Often, the amount determined from such data will be less than the amount of mother substance actually contained in the system. What can NCA tell us about these?

The concept of *exchangeable mass* (or volume) has been used in the medical physiology literature to denote the actual mass (or volume) measured in this manner (Bergner 1967; Bergner and Lushbaugh 1967; Brownell and Ashare 1967). Exploration of the relationships between the measurable exchangeable masses or volumes and immeasurable true masses or distribution volumes has particular relevance in comparative studies, where differences in species, or system state in the same species, might be masked by comparing only "exchangeable" space parameters – which is what NCA would reveal.

Oppenheimer and Gurpide (1979) introduced an interesting quantitative and appealing variation on the theme of exchangeable quantities, for the purpose of NCA of this problem. In the context of hormones in blood, incompletely "exchanging" with extravascular tissues where they are normally eliminated, they coined the term **recirculating volume** (V_R) for the equivalent exchangeable volume, and the term **nonrecirculating volume** (V_{NR}) for the portion of the total equivalent distribution volume (V_D) not available for exchange after passing from blood:

$$V_D \equiv V_R + V_{NR} \tag{8.42}$$

This is perhaps easier to appreciate intuitively in mass terms: q_{NR} is the net amount of substance that never returns after passing from blood to other tissues, due to extravascular "leaks."

With reference to **Fig. 8.17**, and again for the input PR_1 to the central compartment only ($PR_2 \equiv 0$), we let $q_R = q_1 +$ fraction of q_2 that recirculates to compartment 1, and $q_{tot} = q_1 + q_2 \equiv q_R + q_{NR}$. Direct algebraic analysis[30] of **Fig. 8.17** in steady state then gives:

30 Numbered rate constant subscripts in the compartmental matrix K here are consistent with usage in Chapter 4 and beyond. The standard proposed by (Rescigno et al. 1990) for subscripts does not comply with vector-matrix algebra standards used by most physiologists, engineers and applied mathematicians. (See **Fig. 4.8** of Chapter 4, and accompanying explanation.)

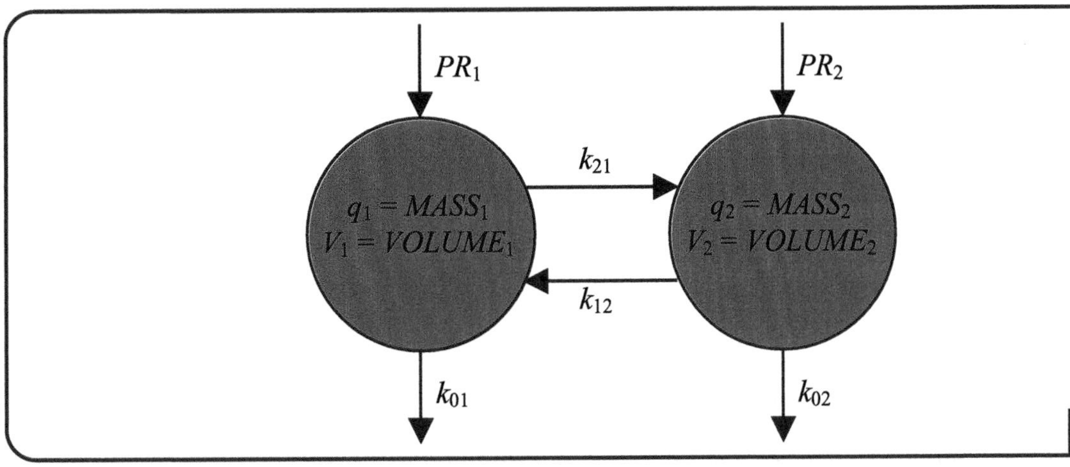

Fig. 8.17

Most general open two-compartment model with two independent sources PR_1 and PR_2 (mass/time), two fractional exchange rates k_{21} and k_{12} (time^{-1}), and two sinks represented by fractional removal rates k_{01} and k_{02} (time^{-1}) from compartments 1 and 2, respectively. q_1 and q_2 are sizes, in mass terms, and V_1 and V_2 are sizes, in equivalent volume terms, of compartments 1 and 2 **in steady state**.

$$q_R = q_1 + \frac{k_{12}q_2}{k_{02} + k_{12}} = \left(1 + \frac{k_{12}k_{21}}{k_{22}^2}\right)q_1 \tag{8.43}$$

$$q_{NR} = q_{tot} - q_R = \left(\frac{k_{02}k_{21}}{k_{22}^2}\right)q_1 \tag{8.44}$$

Equations for V_R and V_{NR} are obtained from Eqs. (8.43) and (8.44) by simply exchanging all qs for Vs. Comparing Eq. (8.44) with Eq. (8.27), we see that $V_R = V_D - V_{NR} = V_D^{NC}$, the non-compartmentally determined (and biased) "distribution volume." Extension to the n-compartment mammillary structure yields:

$$V_{NR} = \left(\frac{k_{02}k_{21}}{k_{22}^2} + \frac{k_{03}k_{31}}{k_{33}^2} + \cdots + \frac{k_{0n}k_{n1}}{k_{nn}^2}\right)V_1 \tag{8.45}$$

and again,[31] $V_R = V_D^{NC}$ (also see Eq. (8.33)).

Let's explore the *Equivalent Source Problem* again, when there is more than one endogenous source, as above. If PR_2 is nonzero in **Fig. 8.17**, it is not difficult to show – using cut-set analysis (Feng and DiStefano III 1991),[32] that Eq. (8.43) becomes:

$$q_R = \left(1 + \frac{k_{12}k_{21}}{k_{22}^2}\right)q_1 + \frac{k_{12}PR_2}{k_{22}^2} \tag{8.46}$$

(see Exercise E8.9). Then, for example, if an appropriate indicator or tracer is injected into compartment 1, the most common case, only the first term in Eq. (8.46) is measurable and the NC formula for V_D^{NC}, Eq. (8.33), does not provide V_R, i.e. $q_{tot} \neq q_R$. The contribution to q_R due to the second term would require a second study based on injection and sampling directly in compartment 2, if it is accessible.

31 Extension of this concept to "circulating" and "recirculating" *FCR* and *MRT* parameters is left for Exercise E8.5.

32 See Chapters 4 and 16 for the basics of cut-set-analysis.

Summarizing these results:

V_D = true steady-state equivalent distribution volume

V_R = true steady-state equivalent recirculating volume

V_{NR} = true steady-state equivalent nonrecirculating volume

PCR = true plasma clearance rate (volume/time)

FCR = true fractional clearance (elimination) rate (time^{-1})

MRT = true mean residence time (time)

x^{NC} = any of the above, but determined by noncompartmental analysis

Then, *for single source systems*, the products of which enter the central measurement pool for the first time directly, but *with sinks or leaks (e.g. catabolism and excretion) in noncentral pools*:

$$PCR^{NC} = PCR \tag{8.47}$$

$$V_D^{NC} = V_R < V_D \tag{8.48}$$

$$MRT^{NC} < MRT \tag{8.49}$$

$$FCR^{NC} > FCR = \frac{1}{MRT} \tag{8.50}$$

When there are *multiple sources* or systemic inputs, not all the products of which enter the central pool directly, and *with sinks in noncentral pools*:

$$PCR^{NC} > PCR \tag{8.51}$$

This means that the product of the measurable PCR^{NC} and the steady-state plasma concentration, $c_{p\infty}$, is an underestimate of the whole organism production rate (DiStefano III 1976; DiStefano III 1982). If we define $PAR \equiv PCR^{NC}c_{p\infty}$, where PAR is called the **Plasma Appearance Rate**, then for multisource systems:

$$PAR \equiv PCR^{NC}c_{p\infty} \equiv PR^{\min} < PR \tag{8.52}$$

That is, PAR is a minimum (lower bound) estimate of the total system production rate PR. Also,

$$V_D^{NC} < V_R < V_D \tag{8.53}$$

and finally, $FCR^{NC} > FCR$ and $MRT^{NC} < MRT$, as above. Measures of bioavailability (%AB) and bioequivalence (BE), discussed in an earlier section, also are affected by the presence of leaks in noncentral pools, i.e. by the equivalent sink problem.

Summarizing Noncompartmental Analysis (NCA) Applicability

The preceding sections elucidated the basis for understanding structural limitations of NC modeling methods. In principle, NCA can provide unbiased values for V_D, q_{tot}, MRT, or FCR

for dynamic biosystems in which *all sources and all sinks* for the substance under study are directly associated with, and only with, the probed, central measurement pool(s). Accurate evaluation of *PCR* or *UCR* requires only one source, directly into the central pool, and no restrictions on sink locations. If these conditions are not met, lower bounds, i.e. minimum values, can be obtained for *PCR, MRT,* V_D and q_{tot}, and maximum values for *FCR*, as indicated next.

Bounds for Whole-Organ System Parameters from NCA

In Eqs. (8.32) to (8.34), we saw that noncompartmental estimates are lower bounds for the parameters *MRT*, V_D and *PCR*. We can also derive upper bounds for these whole organ system parameters, illustrated here for the whole-body mean residence time, *MRT*. From Eqs. (8.33) and (8.39), the lower bound is:

$$MRT^{NC} = \frac{AUMC}{AUC} = \frac{-\sum_{i=1}^{n}(A_i/\lambda_i^2)}{\sum_{i=1}^{n}(A_i/\lambda_i)} \leq MRT \tag{8.54}$$

To obtain an upper bound, we use Eq. (8.26) the matrix of mean residence times: $R \equiv [r_{ij}] = -K^{-1}$. *Assuming X enters compartment 1, MRT for the whole model is the sum of the first column elements of R*, i.e. $\sum_i r_{i1}$. But, $r_{ij} \leq r_{ii}$ (Landaw 1966; DiStefano III et al. 1987). Thus,

$$MRT = \sum_{i=1}^{n} r_{i1} \leq \sum r_{ii} = trace\, R = -\sum_{I=1}^{N}\frac{1}{\lambda_i} \equiv MRT^{\max} \tag{8.55}$$

Note that $MRT = MRT^{max}$ when $r_{i1} = r_{ii}$ for all *i*, e.g. a catenary model with a leak only from compartment *n*.

NC Model Structure Errors & Some Consequences

We explore here some important consequences of not meeting the biological system structure requirements of the noncompartmental approach. **Fig. 8.17** depicts the most general two-compartment model, sufficient to illustrate the primary issues. To keep the analysis as simple as possible, we again assume constant steady state conditions for the unknown substance (as if it were infused or secreted at constant rates PR_1 and PR_2), eliminating the need to deal with differential equations.

The steady state masses in compartments 1 and 2 (individual compartment sizes) are q_1 and q_2; $q_{tot} = q_1 + q_2$ is the whole-body mass or pool size; V_1 is the real distribution volume of compartment 1; V_2 is the equivalent distribution volume of compartment 2, assuming its (steady state) concentration $c_{2\infty}$ in compartment 2 is the same as that in the reference compartment, i.e. $c_{2\infty} \equiv c_{1\infty}$. Thus the steady state equivalent distribution volume for the whole system is $V_D = V_1 + V_2$, and we can immediately also write $q_{tot} = c_{1\infty}V_D$. The two possible sources in this model are PR_1 and PR_2 (mass fluxes), representing independent input production or infusion rates into compartments 1 and 2 individually. The k_{12} and k_{21} are fractional exchange rates (time^{-1}) between compartments, and k_{01} and k_{02} are nonzero fractional leaks (time^{-1}). The compartment turnover rates are $-k_{11} \equiv k_{01} + k_{21}$ and $-k_{22} \equiv k_{02} + k_{12}$.

The *true* $V_D = V_1 + V_2$ is computed from **Fig. 8.15** under the condition that $PR_2 \equiv 0$, which satisfies the single central compartment source but not the single-sink condition for NCA. For compartment 2 in the steady state, $k_{21}q_1 = -k_{22}q_2$. Then, using $c_{2\infty} \equiv c_{1\infty}$, a bit more algebra gives $V_2 = \frac{-k_{21}}{k_{22}} V_1$. Therefore, with $-k_{22} \equiv k_{02} + k_{12}$,

$$V_D = V_1 + V_2 = \left(1 + \frac{k_{21}}{k_{01} + k_{12}}\right) V_1 \tag{8.56}$$

The true total body pool size $q_{tot} = V_D c_{1\infty}$ is then easily obtained from V_D if $c_{1\infty}$ is measurable, usually the case in the context of this problem. Assume the experiment is an indicator or tracer pulse dose study, where $c_1(t)$ (concentration units) is the temporal response in compartment 1. Then the corresponding equivalent total distribution volume V_D^{NC} determined from the NC model of **Fig. 8.15** for the two-pool system (with $PR_2 = 0$) can be computed from Eq. (8.33):

$$V_D^{NC} = \frac{D_{ose} \int_0^\infty t c_1(t) dt}{\left[\int_0^\infty c_1(t) dt\right]^2} \tag{8.57}$$

With some lengthy algebra, it can be shown that:

$$V_D^{NC} = V_D - \left(\frac{k_{02}k_{21}}{k_{22}^2}\right) V_1 \geq V_D \tag{8.58}$$

Therefore $V_D^{NC} = V_D$ if and only if $k_{02} \equiv 0$, i.e. if and only if there is no leak from noncentral compartment 2 (homogeneous pool 2 in the NC model).

This result is readily generalized, by mathematical induction, to n-compartment mammillary structures with no direct exchanges among noncentral compartments:

$$V_D^{NC} = V_D - \left(\frac{k_{02}k_{21}}{k_{22}^2} + \cdots + \frac{k_{0n}k_{n1}}{k_{nn}^2}\right) V_1 \geq V_D \tag{8.59}$$

To summarize, the NC estimate of V_D is the true, unbiased value of total equivalent distribution volume if and only if there are no noncentral pool leaks. The term in parentheses in Eqs. (8.58) or (8.59) is the bias or estimation error using V_D^{NC} (NCA) to estimate V_D for the models under consideration. Clearly, the magnitude of this error cannot be easily assessed without some knowledge of the magnitude of the complex quotients appearing in Eqs. (8.58) or (8.59).

What about the plasma clearance rate (PCR) for this model (with $PR_2 \equiv 0$)? Fortunately, *noncompartmental formulas*, e.g. for brief pulse dose experiments: $PCR^{NC} = D_{ose}/AUC$, *always provide the true PCR, even in the presence of leaks in noncentral compartments.* We forego the algebra to show this result here, leaving this for an exercise instead (E8.10). However, the *mean residence time MRT (with $PR_2 \equiv 0$) is underestimated using NC analysis,* and therefore $FCR = \frac{1}{MRT}$ is overestimated, because $MRT = \frac{V_D}{PCR}$, and V_D is underestimated when $k_{02} \neq 0$. In this case:

$$MRT^{NC} = MRT - \frac{k_{02}k_{21}}{(k_{02} + k_{12})(k_{01}k_{02} + k_{01}k_{12} + k_{02}k_{21})} \leq MRT \tag{8.60}$$

The second term (quotient expression) is the *bias* if MRT^{NC} (from noncompartmental analysis) is used to estimate the true *MRT*, assuming $k_{02} \neq 0$.

Very Limited Compartmental Equivalents of Noncompartmental Models

It is easy to see that the four-compartment model of **Fig. 8.18** is precisely equivalent to, or can be reduced to, the NC model of **Fig. 8.15**. Any multicompartmental model with input only into compartment *j*, "leaks" only from compartment *j* and measurements only in compartment *j*, is equivalent to a NC model. Furthermore, with only one leak, most models of this type are completely quantifiable (identifiable), i.e. all (or most) of the k_{ij} can be determined from either a sum of exponentials quantification (A_i and λ_i) from data, or directly from the differential equations model fitted to the same data. *This is probably why the "central compartment elimination" model is so prevalent in the literature: it usually can be completely quantified.* Unfortunately, the structural analogy of such models with real biological processes is less likely, e.g. for substances metabolized or otherwise eliminated extravascularly. And, when the analogy is inappropriate (no matter how convenient), many kinetic parameters determined from data for such models can have little or no meaning or can be misleading. Comparing results determined from NC and MC analyses illustrates the point in the next example.

Example 8.10: ***Compartmental Model Parameters in Normal & Pathological States***

We again consider systems S_a and S_b of Example 8.9. Suppose S_a is a drug or hormone distribution system and it has been established that this drug or hormone is completely eliminated in a single extravascular organ compartment, e.g. by metabolism in liver. Also, this substance otherwise distributes and exchanges only with the plasma, and it is not eliminated directly in or from plasma (e.g. no metabolism, no urinary excretion, no plasma "leak"). If S_a is linear and stationary, or if the experiment described in Example 8.9 was linearizing, a likely candidate model for S_a would be the two-compartment

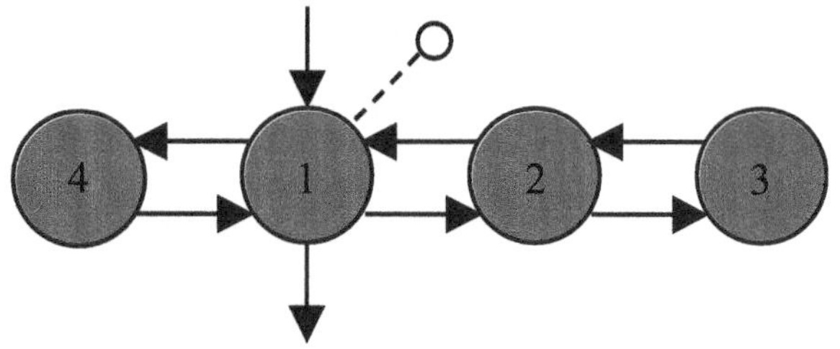

Fig. 8.18

open model with $k_{01} = 0$ (**Fig. 8.17**), designated M_a. It is shown in Chapter 10 that all parameters of this model, i.e. V_1 (and the equivalent V_2), k_{02}, k_{12} and k_{21} are uniquely identifiable from the output measured in compartment 1, $y_1(t)$. The solutions[33] in terms of the biexponential parameters in Example 8.10 are:

$$\begin{aligned} V_1 &= \frac{1}{A_1 + A_2} & k_{02} &= \frac{(A_1 + A_2)\lambda_1\lambda_2 V_1}{k_{21}} \\ V_2 &= \frac{k_{21}V_1}{k_{02} + k_{12}} & k_{12} &= \frac{A_1 A_2 V_1(\lambda_2 - \lambda_1)^2}{k_{21}(A_1 + A_2)} \\ & & k_{21} &= -V_1(A_1\lambda_1 + A_2\lambda_2) \end{aligned} \tag{8.61}$$

Thus, for M_a, $V_1^a = 1$, $V_2^a = 1$, $k_{02}^a = 0.5$, $k_{12}^a = 0.5$ and $k_{21}^a = 1$. Similarly, for system model M_b of Example 8.13, if it too is the same drug (or hormone) distribution system with the same structure but under different (e.g. pathological) conditions, we have: $V_1^b = 1$, $V_2^b = 2$, $k_{02}^b = 1.4$, $k_{12}^b = 0.1$ and $k_{21}^b = 3$.

We are interested now in the true whole-organism parameters for models M_a and M_b. The *plasma equivalent distribution volume* for each model is $V_D = V_1 + V_2$. Thus, $V_D^a = 1 + 1 = 2$ and $V_D^b = 1 + 2 = 3$. Directly from the definition of *plasma clearance rate*, we have $PCR = \frac{k_{02}q_{2\infty}}{c_{1\infty}} = \frac{k_{02}V_2c_{1\infty}}{c_{1\infty}} = k_{02}V_2$. Thus, $PCR_a = 0.5$ and $PCR_b = 2.8$. These numbers are interpreted in the next example.

Example 8.11: ***NC & MC Models Compared: Some Contradictory Results***

Summarizing the results of Examples 8.9 and 8.10, we have for M_a and M_b:

Model M_a	***Model M_b***
$PCR_a^{NC} = 0.5 = PCR_a$	$PCR_b^{NC} = 2.8 = PCR_b$
$V_{D,a}^{NC} = 1.5 < 2 = V_D^a$	$V_{D,b}^{NC} = 1.133 < 3 = V_D^b$
$MRT_a^{NC} = 3 < 4 = MRT_a$	$MRT_b^{NC} = 0.4048 < 1.071 = MRT_b$

We note first that *PCR* values are the same, with the same data analyzed by either NC or MC analysis. This is because there is only one input, directed into the plasma only, and the tracer test-input was introduced into plasma. Second, NC values of V_D and *MRT* are strictly less than MC values, because extravascular "leaks" are present. Third, whereas the true distribution volume V_D *increases* 50%, from 2 to 3, the value of V_D computed by NC analysis *decreases* by 25%, from 1.5 to 1.133, when the system is compared in the two different states *a* and *b*.

Thus, the two modeling approaches, compartmental vs. NC, applied in two different states, can give contradictory results about the direction of changes in kinetic parameters. Without further analysis, such "modeling errors" conceivably could have serious consequences, e.g. in clinical practice.

33 Exercise E8.6.

NC versus MC Modeling: No Easy Choice

The *Equivalent Sink Problem*[34] has received a fair amount of attention in the pharmacokinetics literature (Benet and Ronfeld 1969; Gibaldi and Perrier 1972; Perrier and Gibaldi 1973; Wagner 1976; Benet and Galeazzi 1979; DiStefano III 1982; Wagner 1993). To circumvent the problem, one or more authors suggested that NC equations will make all modelers homogeneous in their approach in situations when the location(s) of leaks (catabolism or excretion sites) are not known. This is not a "good" solution, because there is ample evidence that it can lead to ambiguous conclusions (e.g. the comparison example above) or misinterpretation of results, as noted in (Cobelli et al. 1982; DiStefano III 1982; Yamada et al. 1996). It is all too easy for the biased nature of such an approach to go unnoticed and therefore unquestioned as to its degree of approximation.

Different terms should be used for such approximations, calling them what they really are whenever possible, such as the use of exchangeable or circulating masses or volumes as above, instead of (approximate) distribution volumes or compartment sizes. Similarly, for the *Equivalent Source Problem*, the use of terms like plasma appearance rate (*PAR*) or other more specific identifiers for the product of *PCR* and plasma concentration, for systems with multiple systemic sources (Perl et al. 1969; DiStefano III 1976), are preferable to easily misunderstood terms like blood (or plasma) production rate (Warren et al. 1981).

These problems are subtle and solutions are nontrivial. Certainly, when little is known about the location of all sinks and sources, and when experimental conditions do not permit injecting into or measuring from more than one compartment, MC analysis may not provide any more information than NC analysis about V_D, q_{tot}, *MRT* or *FCR*. But MC analysis, when applied appropriately, can have an edge, because there is a definite informational advantage in additional structure,[35] as there is with additional temporal data collected experimentally. For example, as developed in Chapter 10, specific structure in the form of additional compartments and k_{ij} permits incorporation of additional physiochemically based information about the system in the form of inequalities, constraining what are called *interval identifiable* parameter values to lie within finite ranges (Berman and Schoenfeld 1956; DiStefano III 1982). This approach can provide bounds on otherwise unidentifiable parameters within narrow limits (Chapter 10) and noncentral compartment, systemic sources within similarly narrow limits. On the other hand, NC analysis has many advantages, attributable chiefly to its simplicity. Unfortunately, it was mislabeled "model-independent" (instead of *particular model structure*-independent) in early literature on this subject (Gurpide and Mann 1970), for which many misunderstandings probably can be attributed.

Recap & Looking Ahead

We developed and analyzed physiologically based, whole-organism and noncompartmental methods in this chapter, three specialized biomodeling methodologies for quantitative clinical,

34 The specific location of the leaks in the model has received the attention noted, not necessarily called the Equivalent Sink Problem, as in this chapter.

35 PB modeling – with physiological structure faithfully maintained – also would have this advantage.

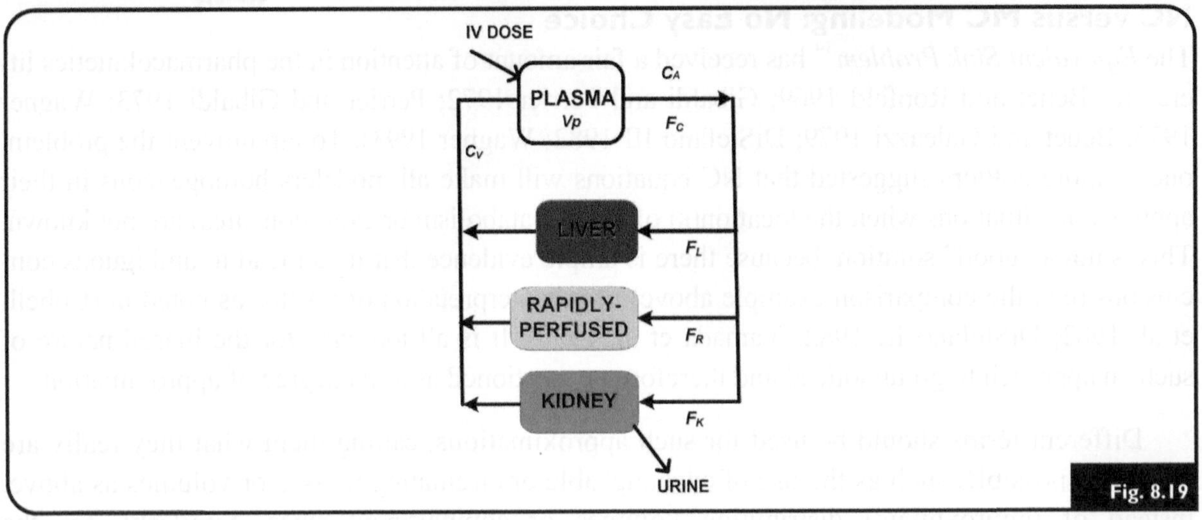

Fig. 8.19

environmental and health-related problems – for the most part at whole-organism and organ-system levels. Lots of new and diverse material here, compared with earlier chapters, and it's the last chapter devoted primarily to developing different modeling methodologies for different domains of life sciences.

Structuring was a main focus in the current and all previous chapters, along with associated quantitative and qualitative analyses (model simulations and solutions) and applicability assessments. We focus on *analysis and quantification* methodology in the next four. Chapter 9 is specifically about stability analysis and oscillations in nonlinear dynamic systems models, where we introduce methods for evaluating qualitative dynamical properties in systems capable of sudden behavioral changes and unusual response profiles. It encapsulates the basics of higher level mathematical approaches to qualitative analysis of dynamical responses to disturbances – primarily for examining these properties in complex nonlinear biosystem models. We turn our attention to model *quantification* in Chapters 10–13, and then to more sophisticated methods and applications in systems biology in the remaining chapters.

EXERCISES

E8.1 Show that the two-compartment PB organ of **Fig. 8.5** satisfies the general mass balance equation:

$$F(c_{in} - c_{out}) = V_1\dot{c}_1 + V_2\dot{c}_2 \tag{8.62}$$

where c_{in} is the inflow concentration and c_{out} is the outflow concentration of substance X, each from the vascular compartment in **Fig. 8.5**. This model works for either flow- or permeability-limited conditions.

E8.2 Using (8.62), show that, for the flow-limited case, the outflow concentration c_{out} is given by:

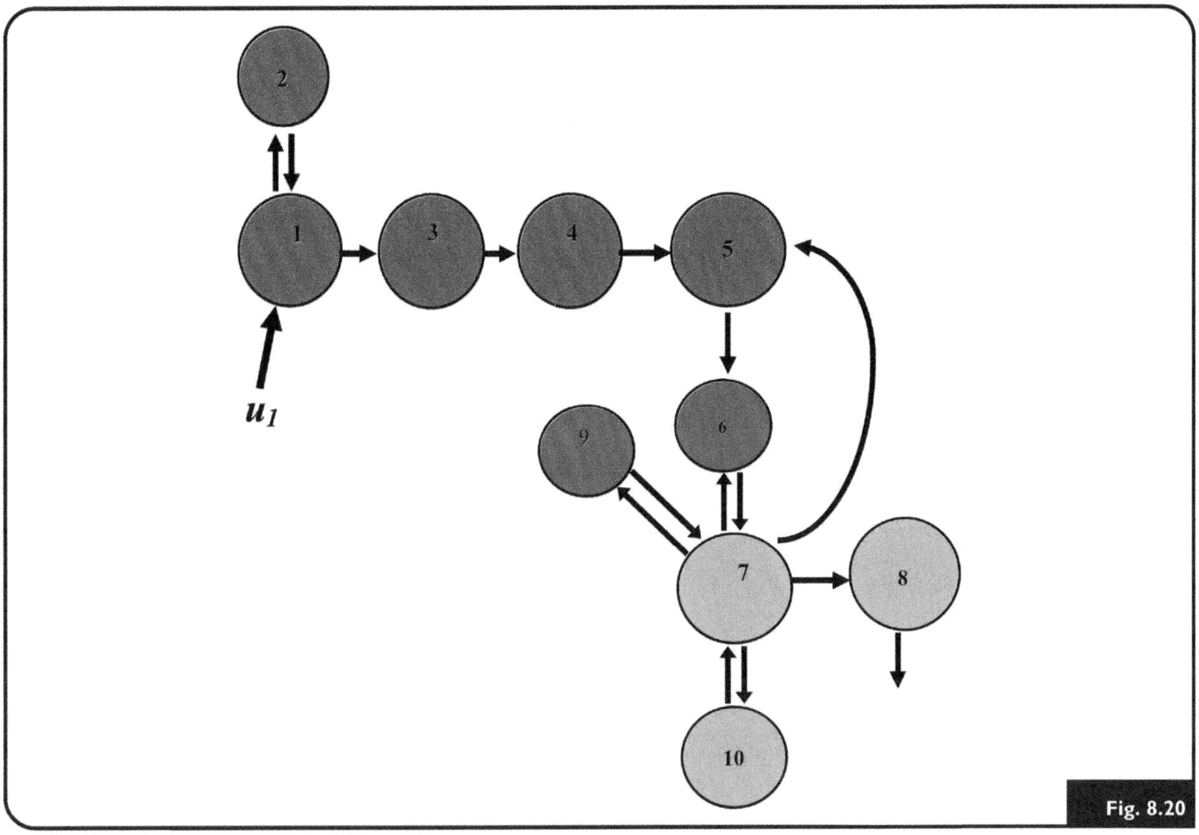

Fig. 8.20

$$c_{out} = (1 + V_1/PV_2)c_2/P - c_{in}V_1/PV_2 \tag{8.63}$$

E8.3 Show that if all of X entering the organ is sequestered, i.e. $c_{out} = 0$, then:

$$Fc_{in} = \left(\frac{V_1}{P} + V_2\right)\dot{c}_2 \quad \text{and} \quad \dot{c}_2 = Fc_{in}/(V_2 + V_1/P) \tag{8.64}$$

E8.4 Suppose R represents source or sink processes for X in an organ. Using (8.62) as an analogy, show that a general mass balance equation for substance X in a three- instead of a two- subcompartment organ has the form:

$$F(c_{in} - c_{out}) + R = V_1\dot{c}_1 + V_2\dot{c}_2 + V_3\dot{c}_3 \tag{8.65}$$

E8.5 Write the ODEs for the four one-compartment flow-limited organs (plasma + three organ compartments) of the PBPK model for compound X dynamics shown in **Fig. 8.19**. The LIVER (L), RAPIDLY-PERFUSED (R) and KIDNEY (K) compartment organs have tissue : plasma partition coefficients P_L, P_R and P_K respectively and plasma gets an impulse dose and urinary excretion is linear. Mass flux of compound X must be balanced for all compartments.

E8.6 Formulate a conventional multicompartmental (MC) structure for the PBPK model in **Fig. 8.7** following the logic in **Fig. 8.9**.

E8.7 The linear multicompartmental model in **Fig. 8.20** is a linearized version of the model of manganese distribution kinetics in brain and other major organs shown in **Fig. 8.8**. Given a trace dose impulse function input into compartment 1, with *all initial conditions (ICs) equal to zero*, how many *modes*, at most, can be measured in *each* of the 10 compartments?

E8.8 Derive equations for *MRT*, V_D and *FCR* for a constant infusion input *IR*.

E8.9 Derive Eq. (8.46) using cut-set analysis (Chapters 4 and 16).

E8.10 Show that $D_{ose}/AUC = PCR$ still works for extravascular pools with leaks.

References

Benet, L., Galeazzi, R., 1979. Noncompartmental determination of the steady-state volume of distribution. J. Pharm. Sci. 68 (8), 1071–1074.

Benet, L., Ronfeld, R., 1969. Volume terms in pharmacokinetics. J. Pharm. Sci. 58 (5), 639–641.

Bergner, P., 1967. The concepts of mass, volume, and concentration. In: Bergner, P., Lushbaugh, C. (Eds.), Compartments, Pools, and Spaces in Medical Physiology. Atomic Energy Commission, Division of Technical Information, Washington, DC, US, pp. 21–37.

Bergner, P., Lushbaugh, C. (Eds.), 1967. Compartments, Pools, and Spaces in Medical Physiology. Atomic Energy Commission, Division of Technical Information, Washington, DC, US.

Berman, M., Schoenfeld, R., 1956. Invariants in experimental data on linear kinetics and the formulation of models. J. Appl. Phys. 27, 1361–1370.

Bischoff, K.B., 1975. Some fundamental considerations of applications of pharmacokinetics to cancer chemotherapy. Cancer Chemother. Rep. 59 (4), 777–793.

Bischoff, K.S., Brown, R.G., 1966. Drug distribution in mammals. Chemical Engineering Progress Symposium Series 62: 33–45.

Brown, J.H., West, G.B., Enquist, B.J., 2005. Yes, West, Brown and Enquist's model of allometric scaling is both mathematically correct and biologically relevant. Funct. Ecol. 19, 735–738.

Brown, R.P., Delp, M.D., Lindstedt, S.L., Rhomberg, L.R., Beliles, R.P., 1997. Physiological parameter values for physiologically based pharmacokinetic models. Toxicol. Ind. Health. 13 (4), 407–484.

Brownell, G., Ashare, A., 1967. Definition of compartment and space. In: Bergner, P., Lushbaugh, C. (Eds.), Compartments, Pools, and Spaces in Medical Physiology. Atomic Energy Commission, Division of Technical Information, Washington, DC, US, pp. 15–19.

Carson, E., Cobelli, C., Finkelstein, L., 1983. The Mathematical Modeling of Metabolic and Endocrine Systems: Model Formulation, Identification, and Validation. J Wiley, New York.

Cobelli, C., Nosadini, R., Toffolo, G., McCulloch, A., Avogaro, A., Tiengo, A., et al., 1982. Model of the kinetics of ketone bodies in humans. Am. J. Physiol. 243 (1), R7–R17.

Covell, D.G., Bernam, M., Delisi, C., 1984. Mean residence time – theoretical development, experimental determination, and practical use in tracer analysis. Math. Biosci. 72 (2), 213–244.

Demetrius, L., Tuszynski, J.A., 2010. Quantum metabolism explains the allometric scaling of metabolic rates. J. R. Soc. Interface. 7 (44), 507–514.

DiStefano III, J., 1976. Concepts, properties, measurements, and computation of clearance rates of hormones and other substances in biological systems. Ann. Biomed. Eng. 4 (3), 302–319.

DiStefano III, J., 1982. Complete parameter bounds and quasiidentifiability conditions for a class of unidentifiable linear systems. Math. Biosci. 65, 51–68.

DiStefano III, J., 1982. Noncompartmental vs. compartmental analysis: some bases for choice. Am. J. Physiol. Regul. Integr. Comp. Physiol. 243 (1), R1–R6.

DiStefano III, J., Jang, M., Malone, T.K., Broutman, M., 1982. Comprehensive kinetics of triiodothyronine (T3) production, distribution, and metabolism in blood and tissue pools of the rat using optimized blood-sampling protocols. Endocrinology. 110 (1), 198–213.

DiStefano III, J., Malone, T.K., Jang, M., 1982. Comprehensive kinetics of thyroxine (T4) distribution and metabolism in blood and tissue pools of the rat from only six blood samples: dominance of large, slowly exchanging tissue pools. Endocrinology. 111 (1), 108–117.

DiStefano III, J., Chen, B.C., Landaw, E.M., 1987. Pool size and mass flux bounds and quasiidentifiability conditions for catenary models. Math. Biosci. 88, 1–14.

DiStefano III, J., Nguyen, T.T., Yen, Y.M., 1992. Sites and patterns of absorption of 3,5,3-triiodothyronine (T3) and thyroxine (T4) along rat small and large intestines. Endocrinology. 131, 275–280.

Douglas, P., Cohen, M.S., DiStefano III, J.J., 2009. Chronic exposure to Mn may have lasting effects: a physiologically based toxicokinetic model in rat. J. Toxicol. Environ. Chem.

Douglas, P., Cohen, M.S., DiStefano III, J.J., 2010. Chronic exposure to Mn may have lasting effects: a physiologically based toxicokinetic model in rat. J. Toxicol. Environ. Chem.(2).

Duncan, R.P., Forsyth, D.M., Hone, J., 2007. Testing the metabolic theory of ecology: allometric scaling components in mammals. Ecology. 88 (2), 324–333.

Eisenberg, M., Distefano III, J.J., 2009. TSH-based protocol, tablet instability, and absorption effects on L-T4 bioequivalence. Thyroid. 19 (2), 103–110.

Eisenberg, M., Samuels, M., DiStefano III, J.J., 2006. L-T4 bioequivalence and hormone replacement studies via feedback control simulations. Thyroid. 16 (12), 1–14.

Eisenfeld, J., 1980. Stochastic parameters in compartmental systems. Math. Biosci. 52 (3–4), 261–275.

Feng, D., DiStefano III, J., 1991. Cut set analysis of compartmental models with applications to experiment design. Am. J. Physiol. Endocrinol. Metab. 261, E269–E284.

García-Meseguer, M.J., Vidal de Labra, J.A., García-Moreno, M., García-Cánovas, F., Havsteen, B.H., Varón, R., 2003. Mean residence times in linear compartmental systems. Symbolic formulae for their direct evaluation. Bull. Math. Biol. 65 (2), 279–308.

Gibaldi, M., Perrier, D., 1972. Drug elimination and apparent volume of distribution in multicompartment systems. J. Pharm. Sci. 61 (6), 952–954.

Gurpide, E., Mann, J., 1970. Interpretation of isotopic data obtained from blood-borne compounds. J. Clin. Endocrino.l Metab. 30 (6), 707–718.

Hu, T.-M., Hayton, W.L., 2001. Allometric scaling of xenobiotic clearance: uncertainty versus universality. AAPS PharmSci. 2000 (3), 1–14.

Innis, R.B., Cunningham, V.J., Delforge, J., Fujita, M., Gjedde, A., Gunn, R.N., et al., 2007. Consensus nomenclature for in vivo imaging of reversibly binding radioligands. J. Cereb. Blood Flow Metab. 27 (9), 1533–1539.

Landaw, E., 1966. Useful parameter bounds for some unidentifiable physiological systems. 13th International Biometrics Conference, Seattle, WA, Biometric Society.

Makarieva, A.M., Gorshkov, V.G., Li, B.L., Chown, S.L., Reich, P.B., Gavrilov, V.M., 2008. Mean mass-specific metabolic rates are strikingly similar across life's major domains: evidence for life's metabolic optimum. Proc. Natl. Acad. Sci. USA. 105 (16), 994–999.

Merrill, E.A., Clewell, R.A., Gearhart, J.M., Robinson, P.J., Sterner, T.R., Yu, K., et al., 2003. PBPK predictions of perchlorate distribution and its effect on thyroid uptake of radioiodide in the male rat. Toxicol. Sci. 73 (2), 256–269.

Merrill, E.A., Clewell, R.A., Robinson, P.J., Jarabek, A.M., Gearhart, J.M., Sterner, T.R., et al., 2005. PBPK model for radioactive iodide and perchlorate kinetics and perchlorate-induced inhibition of iodide uptake in humans. Toxicol. Sci. 83, 25–43.

Modi, S., Stanton, A.W.B., et al., 2007. Clinical assessment of human lymph flow using removal rate constants of interstitial macromolecules: a critical review of lymphoscintigraphy. Lymphat. Res. Biol. 5 (3), 183–202.

Oppenheimer, J., Gurpide, E., 1979. Quantization of the production, distribution, and interconversion of hormones. In: DeGroot, L. (Ed.), Metabolic Basis of Endocrinology, 23. Grune & Stratton, New York, pp. 2029–2036.

Pang, K., Durk, M., 2010. Physiologically based pharmacokinetic modeling for absorption, transport, metabolism and excretion. J. Pharmacokinet. Pharmacodyn. 37 (6), 591–615.

Parham, F.M., Kohn, M.C., Matthews, H.B., Derosa, C., Portier, C.J., 1997. Using structural information to create physiologically based pharmacokinetic models for all polychlorinated biphenyls: i. tissue:blood partition coefficients. Toxicol. Appl. Pharmacol. 144 (2), 340–347.

Patrick, P., Frank-Peter, T., 2000. A priori prediction of tissue:plasma partition coefficients of drugs to facilitate the use of physiologically-based pharmacokinetic models in drug discovery. J. Pharm. Sci. 89 (1), 16–35.

Perl, W., Lassen, N.A., Effros, R.M., 1957. Matrix proof of flow, volume, and mean transit time theorems for regional compartmental systems. Bull. Math. Biol. 37, 573–588.

Perl, W., Effros, R.M., Chinard, F.P., 1969. Indicator equivalence theorem for input rates and regional masses in multi-inlet steady-state systems with partially labeled input. J. Theor. Biol. 25 (2), 297–316.

Perrier, D., Gibaldi, M., 1973. Relationship between plasma or serum drug concentration and amount of drug in the body at steady state upon multiple dosing. J. Pharmacokinet. Biopharm. 1, 17–21.

Poulin, P., Theil, F.-P., 2000. A priori prediction of tissue:plasma partition coefficients of drugs to facilitate the use of physiologically-based pharmacokinetic models in drug discovery. J. Pharm. Sci. 89 (1), 16–35.

Rescigno, A., 1996. Pharmacokinetics, science or fiction? Pharmacol. Res. 33, 227–233.

Rescigno, A., Segre, G., 1966. Drug and Tracer Kinetics. Blaisdell Pub Co., Waltham, MA.

Rescigno, A., Thakur, A., Bertrand, K., Brill, A., Mariani, G., 1990. Tracer kinetics: a proposal for unified symbols and nomenclature. Phys. Med. Biol. 35 (3), 449–465.

Rowland, M., Tozer, T.N., 2011. Clinical Pharmacokinetics and Pharmacodynamics: Concepts and Applications. Lippincott Williams & Wilkins, Baltimore.

Sarapura, V., Samuels, M.H., Ridgway, E.C., 2002. Thyroid stimulating hormone. In: Melmed, S. (Ed.), The Pituitary. Blackwell Science, Malden, MA, pp. 187–229.

Savage, V.M., Deeds, E.J., Fontana, W., 2008. Sizing up allometric scaling theory. PLoS. Comput. Biol. (9), e1000171.

Segel, L.A., Slemrod, M., 1989. The quasi-steady-state assumption: a case study in perturbation. SIAM Rev. 31 (3), 446–477.

Tee, D., DiStefano III, J.J., 2004. Simulation of tumor-induced angiogenesis and its response to anti-angiogenic drug treatment: mode of drug delivery and clearance rate dependencies. Cancer Res. Clin. Oncol. 130 (1), 15–24.

Thompson, M., Beard, D.A., 2011. Development of appropriate equations for physiologically based pharmacokinetic modeling of permeability-limited and flow-limited transport. J. Pharmacokinet. Pharmacodyn. 38 (4), 405–421.

Thompson, M., Beard, D.A., Wu, F., 2012. Use of partition coefficients in flow-limited physiologically-based pharmacokinetic modeling. J. Pharmacokinet. Pharmacodyn. 39 (4), 313–327.

Tofts, P., Brix, G., Buckley, D.L., Evelhoch, J.L., Henderson, E., Knopp, M.V., et al., 1999. Estimating kinetic parameters from dynamic contrast-enhanced T(1)-weighted MRI of a diffusable tracer: standardized quantities and symbols. J. Magn. Reson. Imaging. 10, 223–232.

Wagner, J., 1976. Linear pharmacokinetic equations allowing direct calculation of many needed pharmacokinetic parameters from the coefficients and exponents of polyexponential equations which have been fitted to the data. J. Pharmacokinet. Biopharm. 4 (5), 443–467.

Wagner, J.G., 1993. Pharmacokinetics for the Pharmaceutical Scientist. Technomic Publishing, Lancaster, PA.

Warren, D., LoPresti, J.S., Nicoloff, J.T., 1981. A new method for the measurement of the conversion ratio of thyroxine to triiodothyronine in euthyroid man. J. Clin. Endocrinol. Metab. 53 (6), 1218–1222.

West, G.B., Brown, J.H., Enquist, B.J., 1997. A general model for the origin of allometric scaling laws in biology. Science. 276 (5309), 122–126.

Yamada, H., DiStefano III, J.J., Yen, Y.M., Nguyen, T.T., 1996. Regulation of whole body pool sizes and interconversion rates of thyroid hormones in hypo- and mildly T3-stimulated rats. Endocrinology. 137, 5624–5633.

Chapter 9

Biosystem Stability & Oscillations

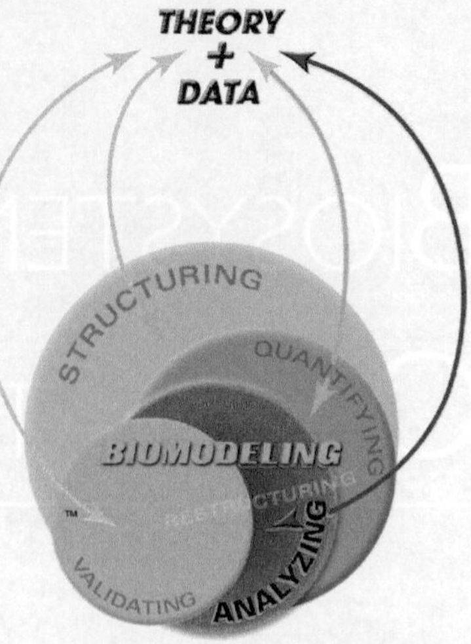

"When young Galileo, then a student at Pisa, noticed one day during divine service a chandelier swinging backwards and forwards, and convinced himself, by counting his pulse, that the duration of the oscillations was independent of the arc through which it moved, who could know that this discovery would eventually put it in our power, by means of the pendulum, to attain an accuracy in the measurement of time till then deemed impossible, and would enable the storm-tossed seaman in the most distant oceans to determine in what degree of longitude he was sailing?"

Hermann von Helmholtz (1877)[1]

1 *On Thought in Medicine.* (1877). Trans. E. Atkinson, Popular Lectures on Scientific Subjects (1881), 234.

Overview/Introduction

Biosystem Stability

Several different notions of the terms *stable* and *unstable* were presented in Chapter 1. As noted, they can mean different things in chemistry and molecular biology than in physics, engineering and dynamic systems biology. All definitions reflect variations of *steadiness* or *unsteadiness*, but dynamic system stability means much more. Stability theory – in particular stability theory for nonlinear dynamic systems – is a fairly complex mathematical subject, reflecting deep and often complex and rich system behavior. We introduce it in this chapter, with illustrative applications in population dynamics and molecular systems biology – an area where it finds much application.

> *"Always remember that the most important thing in a good marriage is not happiness, but stability."*
>
> Gabriel Garcia Marquez, Nobel Laureate in Literature, 1982

Dynamic system stability theory is focused primarily on describing system or subsystem responses to disturbances – usually brief and small disturbances (perturbations) of system steady state operation. They might be disturbances of a potentially deleterious nature, or they might be the kinds of disturbances that are normally expected in the real world. The way a biosystem behaves in response to such perturbations is characterized by its *stability properties*. These can be very predictable and simple to describe, e.g. effects of the disturbance might dissipate asymptotically, or they can be qualitatively different and complex, and surprisingly unpredictable – sometimes even *chaotic*.

Roughly speaking, in the simplest terms, bounded inputs to stable systems should produce bounded output responses – meaning it is **input–output stable**. But knowing only this about system dynamics is hardly enough to characterize it for different operating conditions and different initial conditions. The simple stable, unstable and bistable systems in **Figs. 1.6** and **1.7** illustrate stability notions for disturbances about steady state operating conditions, in terms of disturbances about *equilibrium points* – defined more precisely in the first sections below.

Robust and *fragile* are stability-related terms used to describe *sensitivity* to disturbances, also introduced in Chapter 1. These are addressed briefly here and further in Chapter 11.

Stability notions were developed for linear time-invariant (LTI) models in Chapter 2, in terms of the transfer function (TF). LTI models are (asymptotically) **stable** if their impulse responses approach zero as $t \to \infty$, which translates into the poles of their TF being in the left-half plane (LHP). If any poles are in the right-half plane (RHP), the model is **unstable**; and if any poles are on the imaginary $j\omega$-axis, and all others are in the LHP, the model is **marginally stable/unstable** (and highly fragile/sensitive). The poles of the TF are the same as the eigenvalues of the LTI system matrix (and the roots of the characteristic equation); and we take advantage of these properties in the section below on linearization of nonlinear (NL) models for exploring their stability properties locally, in small regions near *equilibrium points* of the NL system.

In contrast with LTI systems, whose stability properties are completely determined by the system matrix or TF, stability analysis for NL dynamic system models is much more complex.

Quantitative characteristics and solutions to most NL models are obtainable only numerically, and only for specified inputs, initial condition (IC) and parameter values. From a modeler's viewpoint, this leaves too many open questions about what dynamical responses might be if the inputs or ICs or parameters were different than the ones used for numerically solving the equations. We see below that they can be astoundingly different.

Stability analysis of system dynamics is done primarily *qualitatively* – less dependent on particular inputs, ICs and parameters – yielding many (sometimes surprising) insights about dynamical properties. The theory is complex, but accessible – primarily via geometric reasoning, some algebra and by linearization techniques.

Oscillations in Biology

Periodic phenomena abound in nature. Seasonal and **circadian** (24 hour) rhythms are among the most familiar, with circadian behavior synchronized to the daily light–dark cycle – the **zeitgeber** (synchronizer) of this phenomenon. Periodicities come in other shapes and forms as well, including the menstrual cycle – repeating every 28 days – and **ultradian** rhythms – that repeat more often than daily – and these too are synchronized by physical or biological mechanisms or phenomena.

The very word oscillation suggests instability, but oscillations can be stable or unstable, depending on system or model specifics. Several examples of oscillating systems have been presented in previous chapters. These included logistic population growth in Chapters 2 and 3, which can oscillate – either stably or unstably. Chapter 3 examples also illustrated oscillations in models with time delays in the feedback loop – including a multiscale model of protein transcription translation dynamics, models of predator–prey dynamics, and *bifurcating* solutions of a discrete time logistic model – solutions that are asymptotically stable, oscillatory and even chaotic – for different values of model parameters – topics to be amplified on in this chapter. The systems biology literature includes numerous examples of oscillating systems at the cellular level (e.g. Goldbeter 1996; Ciliberto et al. 2007; Wang et al. 2008).

Oscillations are used to convey biological function in subtle ways. Clock-controlled genes facilitate periodic modulation of many physiological processes, such as the cell division and replication cycle, blood pressure, hormone levels, breathing and mental performance. One of the most fascinating questions about such periodic behaviors is their origins: endogenous – from within, or exogenous – by forces external to the system – which can generate oscillations, not just synchronize them. What specific endogenous mechanisms generate biosignal oscillations within organisms or among populations – at molecular, cellular or organ-system levels in a single species, or organisms in a population? Where is the signal generated that elicits the cycling behavior of flapping wings in a bird or fins in the fish? How should these signals be linked and represented by coupled systems of equations exhibiting periodic solutions? Under what circumstances do some model trajectories become *chaotic*?

We address some of these questions in this and subsequent chapters. The main goal here is to introduce some basic methodology for modeling and (mostly) analyzing stability and oscillatory behavior in nonlinear biological systems. More general theory can be found in

(Teschl 2012) and more biochemical and cell biology applications in (Scott 1994; Wang et al. 2008) as well as many other sources.

Finally, in the last section of the chapter, we discuss *nonlinear (NL) modes* – reserved for this chapter, because they are historically associated with oscillations in mechanical systems. We show that the treatment of what are called nonlinear normal modes (NNMs) in mechanics, associated with *two-dimensional manifolds in phase space*, is directly extensible to NL systems biology models – via the quasi-steady state (QSSA) or total quasi-steady state (tQSSA) assumption. Nonlinear modal analysis is about generating ***minimal reduced-order models*** that accurately capture the dynamics of higher-order models, with the help of NNMs in mechanics, and the QSSA or tQSSA in systems biology.

Stability of Nonlinear (NL) Biosystem Models

Phase Space Geometry

Qualitative system behavior can be expressed as *visualizations* of the motions of state variables in state space – also called **phase space** in this context – for any inputs and parameters, or for some feasible set of them. Phase space is n-dimensional and, in phase space geometry, every n can be represented as an axis in a multidimensional Euclidian space, i.e. the n-dimensional extension of orthogonal x-y-z coordinates. Motion of the state of the system evolving over time can be plotted in multidimensional phase space, yielding a **phase space diagram**, or **phase portrait** of this motion. For every possible state of the system, the phase portrait elucidates system properties unlikely to be otherwise apparent. Phase portraits are typically plotted for multiple initial state variable (IC) values, giving a picture of motions of the state starting from different points in state space.

In practice, a phase space is often of dimension $n = 2$, called a **phase plane**. For example, the phase plane of a mechanical system might consist of a plot of momentum or velocity versus position variables. In biochemistry, a phase plane of a reaction might be a plot of reaction rate versus concentration, or the dynamics of a two (molecular) species network, one versus the other.

A two-dimensional NL dynamic system model can be represented classically by either a second-order ODE or, equivalently, by two first-order ODEs:

$$\ddot{x} = f(x;\dot{x}) \quad \Leftrightarrow \quad \dot{x}_1 = x_2 \quad \text{and} \quad \dot{x}_2 = f(x_1, x_2), \quad \text{if} \quad x \equiv x_1 \tag{9.1}$$

The **phase plane** solution for this model is the set of points (x_1, x_2) that defines the motion of the system as it moves thru the plane in time ("velocity" $\dot{x}_1$ versus "position" x_1), from the initial state and time to the final state and time. In other words, $[x_1(t), x_2(t)]$ for $t \geq 0$ is the **path** or **trajectory** in the phase plane, x_2 vs. x_1 parameterized by t. In the Population Interaction Simulation section at the end of Chapter 3, we studied a simple population dynamics example of a predator–prey, shark–tuna ocean life competition system. We solved the two governing NL ODEs equations numerically and plotted a closed curve solution in the phase plane for this model – sharks versus tuna. This is a useful way of characterizing the dynamic response of the model, for both state variables simultaneously, one versus the other.

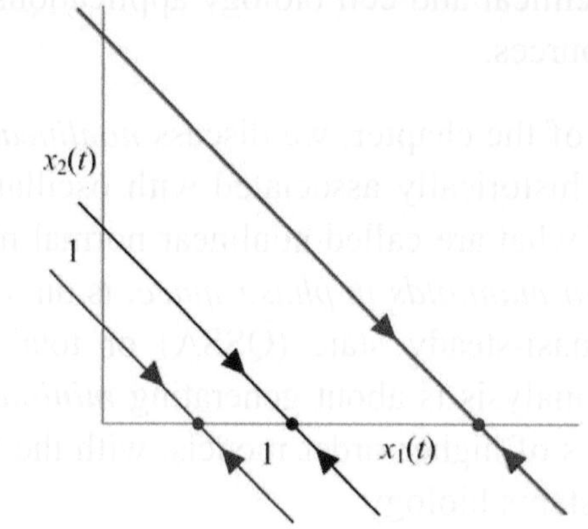

Fig. 9.1

In general, to find the phase plane trajectory, eliminate t from the ODEs by dividing dx_1 by dx_2, i.e. $dx_1/dx_2 = x_2/f(x_1, x_2)$, and solve for $x_2(x_1)$ as a function of x_1.

Remark: This can only be done for time-invariant models. For models with time varying parameters, or time varying terms that do not include state variables, phase plane trajectories cannot be computed, because time t cannot be eliminated. Time-invariant models with constant exogenous inputs, i.e. those that appear to be **autonomous**, are okay. In this case, they appear in the equations like a constant term and time t can be eliminated.

Example 9.1: ***Phase Plane Trajectories for a Simple 2nd-Order Model***

For ODE: $\ddot{x} + \dot{x} = 0$ with ICs: $x(0) = 0$, $\dot{x}(0) = 1$, first convert the second-order ODE to two ODEs: $\dot{x} \equiv \dot{x}_1 \equiv x_2$ and $\dot{x}_2 = \ddot{x}_1 = -x_2$. Then divide dx_1 by dx_2: $dx_1/dx_2 = -x_2/x_2 = -1$, which gives $dx_1 + dx_2 = 0$. Then integrate both sides from the ICs $x(0) = 0$ to $x_1(t)$ and $\dot{x}(0) = 1$ to $x_2(t)$:

$$\int_0^{x_1} dx_1 + \int_1^{x_2} dx_2 = x_1 + x_2 - 1 = 0$$

This line solution $x_1 + x_2 = 1$ is the trajectory of the state of the system in the phase plane for the given ICs and all $t \geq 0$, as shown in black in **Fig. 9.1**.

What about other ICs? All must satisfy the line solution (and have *an equilibrium point* solution – see below). The parallel line solutions for new ICs $x(0) = 0.5$, $\dot{x}(0) = 0.5$ and $x(0) = 0$, $\dot{x}(0) = 2$ are also shown in **Fig. 9.1**, in red and blue, with equilbrium points at (0.5, 0) and (2, 0). *All others will be parallel to these.* The equilibrium points and arrowheads are explained next.

Stability, Equilibrium Points, Steady State Solutions & Nullclines

Stability of a generally nonlinear (NL) dynamic system refers to its behavior *at and near* system *equilibrium points*, representing one or more *stationary* (constant operating, steady state) conditions for the dynamics. This means that, for NL ODE models, stability is about stability of equilibrium points, not the system model as such. For linear systems, stability is about the ODE model (not its equilibrium points). More on this in a later section.

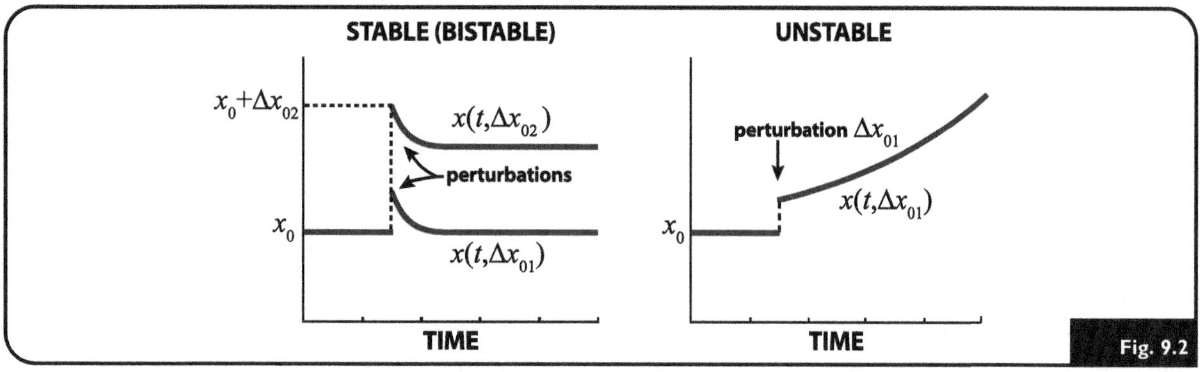

Fig. 9.2

A dynamical system can have zero, one or more equilibria – usually more than one. The question is: what happens if the system is perturbed away from an equilibrium point? Where does it go? Synonyms for equilibrium points are **fixed points, singular points** and **steady state solutions**. The following definitions apply to models in the phase space of any dimension n.

For *continuous-time* NL ODE models of the form: $\dot{x}(t) = \boldsymbol{f}(\boldsymbol{x}(t), \boldsymbol{p}, \boldsymbol{u}(t))$, **equilibrium (fixed, singular, steady state) points** are the $\boldsymbol{x}_e(t)$ in phase space for which $\dot{\boldsymbol{x}}_e(t) = \boldsymbol{f}(\boldsymbol{x}_e(t), \boldsymbol{p}, \boldsymbol{u}(t)) = \boldsymbol{0}$.

For *discrete-time* difference equation (DE) models of the form: $\boldsymbol{x}(k+1) = \boldsymbol{f}(\boldsymbol{x}(k), \ldots)$, the $\boldsymbol{x}_e(k)$ are the solutions of $\boldsymbol{x}_e(k) = \boldsymbol{f}(\boldsymbol{x}_e(k), \ldots)$.

Stability of an equilibrium point $\boldsymbol{x}_e$ defines whether solutions nearby $\boldsymbol{x}_e$ remain close, get closer or farther away from $\boldsymbol{x}_e$. A solution $\boldsymbol{x}(t, \boldsymbol{x}_0)$ with IC $\boldsymbol{x}_0$ is **stable** if other solutions that start near $\boldsymbol{x}_0$ stay close to $\boldsymbol{x}(t, \boldsymbol{x}_0)$. Otherwise it is **unstable**. Solutions don't have to converge to $\boldsymbol{x}(t, \boldsymbol{x}_0)$ to be stable; they only have to stay close.[2]

Each of these notions is illustrated in **Fig. 9.2**. On the left-hand side (LHS), the stable system recovers from the smaller perturbation Δx_{01} by returning to x_0. For the larger perturbation Δx_{02}, the system returns to a different equilibrium point, illustrating a **bistable** system with two stable equilibrium points. The right-hand side (RHS) system is unstable, because the perturbation Δx_{01} at t_0 is amplified for $t > t_0$.

Remark: The dynamic responses of ODE system models depend (of course) on their inputs, ICs and parameters. However, NL ODE models depend in a more complex, nonlinear way on their inputs $\boldsymbol{u}(t)$ than do linear ODE system models. This is one of the reasons stability is about behavior near equilibrium points for NL models, and why their stability is more difficult to pin down. An exception is the stability of NL time-invariant ODE models with *constant* "exogenous" inputs, which can likely be established in the same manner as without inputs. Constant input terms typically appear like any other constant in the equations.[3]

2 **Stability in the sense of Lyapunov** means we have to start *very* close to $\boldsymbol{x}_0$, i.e. the trajectory remains close $\boldsymbol{x}(t, \boldsymbol{x}_0)$ whenever the IC is very close to $\boldsymbol{x}_0$. If it is stable in this sense and different trajectories from different ICs do not converge on each other, it is called **neutrally stable**.

3 In general, fully assessing stability properties of NL models with more time-varying parameters or forcing functions can be considerably more difficult. Some NL models, like the sinusoidally forced van der Pol equation, have been solved numerically and can exhibit complex behavior, including chaos, for particular parameter values (http://en.wikipedia.org/wiki/File:Vanderpol_time_mu%3D8.53_A%3D1.2_T%3D10.svg).

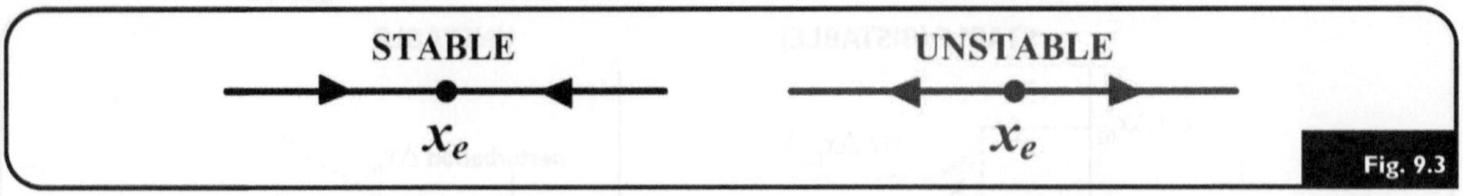

Fig. 9.3

Nullclines are helpful in establishing stability properties. **Nullclines** are the curvilinear functions of time (geometric shapes) for which the time derivatives in each dimension are zero. For example, with $n = 2, \dot{x}_1 \equiv 0$ and $\dot{x}_2 \equiv 0$ define the two nullcline curves in a (2-D) phase *plane*. For $n = 3$, a nullcline is a 2-D *surface*, and so on. The equilibrium points $\boldsymbol{x}_e$ are where all the nullclines intersect. *Local stability* of $\boldsymbol{x}_e$ can be established by the *direction of the nullclines* in the vicinity of $\boldsymbol{x}_e$, determined by the sign of the derivative $\dot{x}_1$ near the equilibrium point for each state variable x_i, i.e. is $\dot{x}_i > 0$, or is $\dot{x}_i < 0$? The ***direction toward (away from) $\boldsymbol{x}_e$ means $\boldsymbol{x}_e$ is stable (unstable)***, as in **Fig. 9.3**.

Nullcline directions are related to *vector fields* in the ODE solution space. At each point in the phase space, the ODEs can be interpreted as a time-dependent n-dimensional state vector defining the direction and distance the dynamical system will go over the next small increment of time dt. This collection of vectors is called a **vector field** in phase space for $\boldsymbol{x}(t)$. Every solution of the ODEs is a curve in the phase space that starts at some initial condition (IC) point and follows the vector field. There is one nullcline for each state variable $x_{1e}(t)$, $x_{2e}(t)$, ..., $x_{ne}(t)$ in the state vector $\boldsymbol{x}_e(t)$ (and for every different IC). Along the x_1-nullcline in the phase plane ($n = 2$), the vector field points North or South. Along the x_2-nullcline, vector field points East or West. Between the nullclines the vector field points in directions NE, SE, SW or NW.[4] *The point of intersection of the nullclines* is the **steady state (equilibrium) solution** of the ODEs, $\dot{\boldsymbol{x}} = \boldsymbol{f} = \boldsymbol{0}$.

Example 9.2: ***Nullclines & Stability for a Simple Second-Order Model***

For the model in Example 9.1, $\dot{x}_2 = -x_2 \equiv 0$ indicates that the *x_2-nullcline* is the (East–West) x_1-axis (abscissa, where $x_2 = 0$) and $\dot{x}_1 = -x \equiv 0$ is also the *x_1-nullcline*. The two nullclines coincide and thus the entire x_1-axis includes infinitely many equilibrium points for this simple model. The first-quadrant trajectory of the black line begins at ICs (0, 1) on the x_2-axis and ends at the point $x_e = (1, 0)$ on the x_1-axis. The parallel line solutions for new ICs $x(0) = 0.5$, $\dot{x}(0) = 0.5$ and $x(0) = 0$, $\dot{x}(0) = 2$, in red and blue, have equilbrium points at (0.5, 0) and (2, 0). All other ICs will similarly have equilibrium points along the x_1-axis.

Stability of any equilibrium point $(a, 0)$ in **Fig. 9.1** is established by examining what happens to the state vector (x_1, x_2) if it is perturbed away from $x_e = (a, 0)$. This is revealed by examining the *sign of the derivative* along the trajectory in the vicinity of any $(a, 0)$ in **Fig. 9.1**. For this simple model, $\dot{x}_2 = -x_2 < 0$, which means it is falling toward x_e, because x_2 is decreasing when it is in the first or second quadrant (down arrow in **Fig. 9.1**), and it is rising when $x_2 < 0$ in the fourth quadrant (up arrow). This means (1, 0)

4 Nullclines and associated vector fields are illustrated in **Fig. 9.8**.

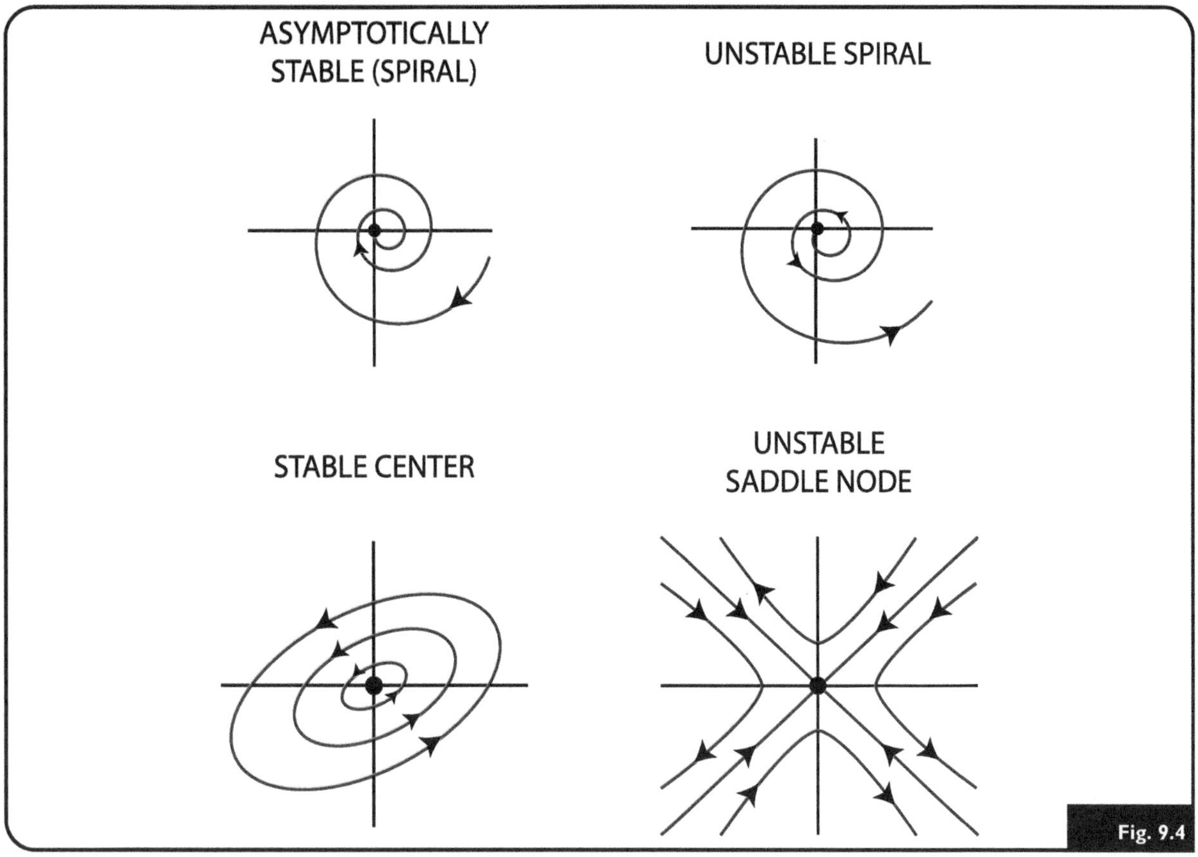

Fig. 9.4

is **a stable equilibrium point (stable node),**[5] as are the other two at (0.5, 0) and (2, 0). The arrows drawn on the solution curves designate stability for all ICs.

Stability Classifications

Nonlinear systems generally have multiple steady state solutions, i.e. multiple equilibria in phase space, with possibly different stability properties in the vicinity of each. In planar geometries, phase plane analysis provides a qualitative understanding of the behavior of model trajectories starting from different ICs in the vicinity of different equilibrium points $\boldsymbol{x}_e$. These different types of equilibria are called *sources, sinks* (or *attractors*), *nodes* and *saddles.*

An equilibrium point $\boldsymbol{x}_e$ is **asymptotically stable**[6] ($\boldsymbol{x}_e$ is a **sink** or **attractor**) **if** $\boldsymbol{x}(t)\rightarrow\boldsymbol{x}_e$ as $t\rightarrow\infty$ (top LHS of **Fig. 9.4**). It's a **stable center (neutrally stable)** if it's stable, but not asymptotically stable, as illustrated by the *orbits* in the lower LHS of the figure.[7]

An **unstable equilibrium point** is called a **source** if all trajectories lead away from it (top RHS of **Fig. 9.4**). It's a **saddle point** if some trajectories lead away and others lead toward it (bottom RHS of **Fig. 9.4**).

5 Interestingly, small perturbations, e.g. to point (1.01, 0) adjacent to (1, 0) on the x_1-axis, would "stay close" but not likely converge back to (1, 0), because all points on the x_1-axis are equilibria. Then, in the sense of Lyapunov stability, these would be better called *neutrally stable* (see Footnote 1).

6 More formally, true asymptotic stability also requires stability in the sense of Lyapunov.

7 Stable limit cycles, defined below, are examples of stable center equilibria.

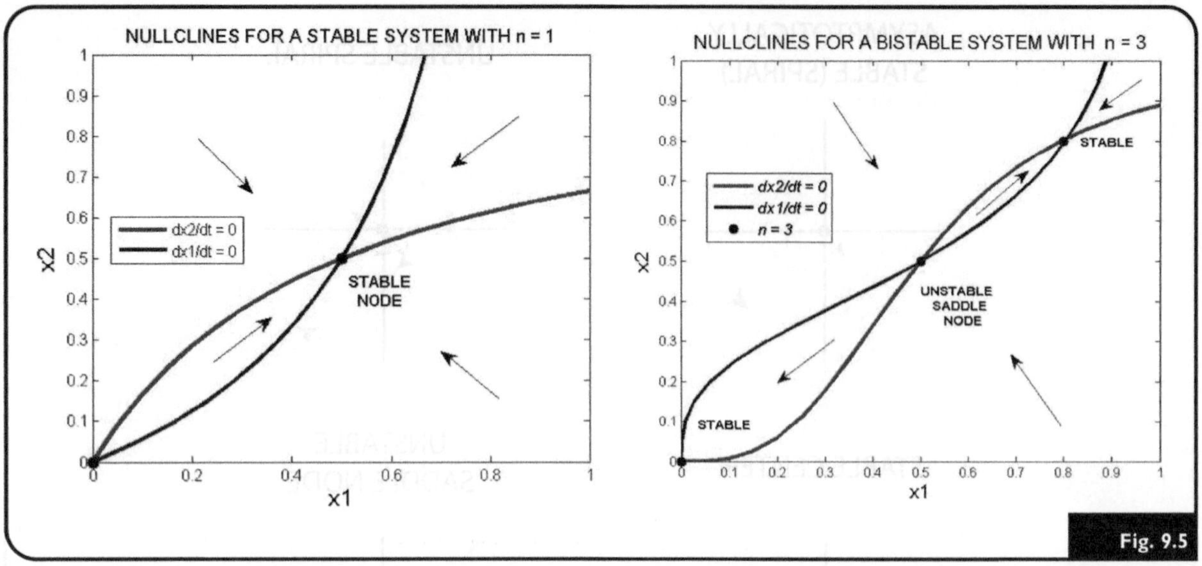

Nullclines in the $x1-x2$ phase plane for Eqs. (9.2). LHS for $n = 1$; RHS for $n = 3$ in the Hill function term. The unstable node for $n = 3$ is called a *saddle node*, in this case with stable equilibria on either side, characteristic of all such *bistable* systems (illustrated in **Fig. 1.6** of Chapter 1). The arrows are representative of the vector field of trajectories in those regions.

Example 9.3: ***Nullclines for a NL Mutual-Activation Model***

Consider the simple 2-D NL model with mutual Hill-function activation and linear degradation of each species (state variable):

$$\begin{array}{c} X_1 \rightleftarrows X_2 \\ \downarrow \quad \downarrow \end{array} \qquad \begin{aligned} \dot{x}_1 &= \frac{x_2^n}{0.5^n + x_2^n} - x_1 \\ \dot{x}_2 &= \frac{x_1^n}{0.5^n + x_1^n} - x_2 \end{aligned} \tag{9.2}$$

The nullclines are the solutions of $\frac{x_2^n}{0.5^n + x_2^n} = x_1$ and $\frac{x_1^n}{0.5^n + x_1^n} = x_2$. *Matlab* solution plots for these two equations are shown in **Fig. 9.5** for $n = 1$ and $n = 3$. For $n = 1$, they intersect in the phase plane at the origin and at $\boldsymbol{x}_e = (0.5, 0.5)$. For $n = 3$, there are 3 equilibrium solutions: the origin, $\boldsymbol{x}_e = (0.5, 0.5)$ and $\boldsymbol{x}_e = (0.8, 0.8)$.

The signs of $\dot{x}_1$ and $\dot{x}_2$ (direction of the nullclines) near $\boldsymbol{x}_e$ determine the local stability of $\boldsymbol{x}_e$. For $n = 1$, the directions can be gleaned by testing several numerical values of x_1 and x_2, yielding downward directions for $x_2 > 0.5$ and upward directions for $x_2 < 0.5$, for all values of x_1. Therefore, the origin is unstable and $\boldsymbol{x}_e = (0.5, 0.5)$ is stable.

For $n = 3$, finding the directions by trial and error is a bit more work, but some numerical trials yields the directions shown in **Fig. 9.5**. Therefore, the origin and $\boldsymbol{x}_e = (0.8, 0.8)$ are stable and $\boldsymbol{x}_e = (0.5, 0.5)$ is unstable. A more formal *linearization* method for establishing stability is developed in the next section.

Stability of Linear System Models

NL models generally have different stability properties at different equilibrium points. That's why we define stability for NL models as stability of their equilibrium points. In contrast, stability for autonomous linear system models $\dot{x} = Ax$ depends only on system matrix A, not particular equilbrium points x_e. For this reason stability of linear systems is considered a property of the *system*, embodied in A. If we include an input in the model, i.e. $\dot{x} = Ax + Bu$, stability analysis is still about A. It doesn't matter what u is.

As noted earlier, $\dot{x} = Ax$ (or $\dot{x} = Ax + Bu$) is **asymptotically stable** if and only if all eigenvalues $\lambda_i = \sigma_i + j\omega_i$ of matrix A have strictly negative real parts, i.e. all $\sigma_i < 0$. It is **unstable** if any $\sigma_i > 0$. If a model output $y = Cx$ is also defined, this condition is equivalent to all of the poles of the system transfer function $H = C(sI - A)^{-1}B$ being in the left half plane (LHP).

The complex part $j\omega_i$ of eigenvalue λ_i is responsible for *oscillations* about x_e, when they exist, and this kind of x_e is called a **focus**. If $\sigma_i < 0$, the oscillations are damped. If $\sigma_i > 0$, the oscillations are amplified. If $\sigma_i = 0$, the oscillations are sustained, a **marginally stable** condition. This is a "fragile" condition, because any small perturbation in system properties that render σ_i slightly greater than or less than 0 will put the pole (eigenvalue) of the system transfer function in the left half (stable region) or right half (unstable region) of the complex plane. Oscillations in nonlinear systems, however, can be sustained and stable, as we see below, when we discuss *limit-cycling* behavior.

Local Nonlinear Stability via Linearization

Local stability[8] of equilibrium points x_e for NL ODE models is often established by Taylor series linearization of the NL ODE $\dot{x}(t) = f[x(t)]$ about x_e. Linearization theory was presented in the last section of Chapter 5. Defining $x(t) \equiv x_e + \Delta x_e(t)$, linearization yields the linear TI ODE:

$$\Delta\dot{x}_e(t) \cong (\partial f/\partial x)|_{x_e} \cdot \Delta x_e \equiv (\partial f/\partial x_e)\Delta x_e \equiv A\Delta x_e \tag{9.3}$$

Analysis of the eigenvalues of the resulting Jacobian matrix $A \equiv \partial f/\partial x_e$ gives the stability properties near x_e: ODE $\dot{x}(t) = f[x(t)]$ is said to be **locally asymptotically stable** (near x_e) if and only if all eigenvalues $\lambda_i = \sigma_i + j\omega_i$ of matrix A have *strictly negative real parts*, i.e. all $\sigma_i < 0$. It is **locally unstable** if any $\sigma_i > 0$.

Example 9.4: ***Stability of x_e by Linearization***

Reconsider the simple mutual activation model of Example 9.3. We examine the eigenvalues $\lambda_i = \sigma_i + j\omega_i$ of $A \equiv \partial f/\partial x_e$ and show by linearization that $x_e = (0.5, 0.5)$ and the origin are stable nodes for $n = 1$.

$$\partial f/\partial x_e = \begin{bmatrix} \partial f_1/\partial x_{1e} & \partial f_1/\partial x_{2e} \\ \partial f_2/\partial x_{1e} & \partial f_2/\partial x_{2e} \end{bmatrix} = \begin{bmatrix} -1 & 0.5/(0.5+x_{2e})^2 \\ 0.5/(0.5+x_{1e})^2 & -1 \end{bmatrix} = \begin{bmatrix} -1 & 0.5 \\ 0.5 & -1 \end{bmatrix} \tag{9.4}$$

The eigenvalues of A are thus $\lambda_{1,2} = -1 \pm \sqrt{0.5}$, both real and negative, so $x_e = (0.5, 0.5)$ is stable. Similarly, the eigenvalues of A for the equilibrium point at the origin

8 Global stability of nonlinear dynamic system equilibria requires more complex mathematical analysis, e.g. Liapunov stability analysis (DiStefano III 1964; Teschl 2012), beyond the scope of this chapter.

x_e are: $\lambda_{1,2} = -1 \pm j\sqrt{2}$, complex conjugates with negative real parts, so the origin is also stable (asymptotically).

Bifurcation Analysis

Bifurcation analysis reveals how the dynamical behavior (stability) of NL ODE models $\dot{x}(t) = \boldsymbol{f}[x(t), \boldsymbol{p}]$ varies with *changes in model parameters* $\boldsymbol{p}$. In general, the location of equilibrium points, their stability and other dynamical characteristics are functions of $\boldsymbol{p}$.

If the behavior of $\dot{x}(t) = \boldsymbol{f}[x(t), \boldsymbol{p}]$ changes *qualitatively* for some $\boldsymbol{p} = \boldsymbol{p}^*$, it has a **bifurcation** at $\boldsymbol{p} = \boldsymbol{p}^*$.

Bifurcation plots, of $\boldsymbol{x}_e(\boldsymbol{p})$ versus $\boldsymbol{p}$, are used to detect qualitative changes in dynamical behavior. Because vector $\boldsymbol{p}$ has P different parameters, P different bifurcation plots can be drawn and not all choices will reveal critical $\boldsymbol{p}^*$. Typically, a parameter p is chosen and the equilibrium point $x_{ei}(p)$ of the ith state variable is plotted versus p, with all other parameters set to their nominal values.

Bifurcation analysis also may employ a **parametric stability diagram**. This maps out the regions in parameter space where model equilibria $\boldsymbol{x}_e(\boldsymbol{p})$ are stable and/or where they are unstable, as functions of the parameters. For two parameters p_i and p_j, for example, the parametric stability diagram delineates stable and unstable regions on the planar surface graph of p_i versus p_j, with all other parameters set to their nominal values. It is usually established numerically.

Equilibrium solutions $\boldsymbol{x}_e(\boldsymbol{p})$ satisfy $\boldsymbol{f}[\boldsymbol{x}_e(\boldsymbol{p}), \boldsymbol{p}] = 0$. As $\boldsymbol{p}$ is varied, $\boldsymbol{x}_e(\boldsymbol{p})$ varies, sometimes in surprising ways – stable $\boldsymbol{x}_e(\boldsymbol{p})$ can become unstable, and *vice versa*. Bifurcations occur if the number of equilibrium solutions changes, or if the stability type changes at $\boldsymbol{p} = \boldsymbol{p}^*$. NL models typically have more than one equilibrium solution, as noted earlier.

Example 9.5: ***Transcritical Bifurcations in Time-Delayed Feedback***

Example 3.5 of Chapter 3 illustrates a bifurcation that separates stable from unstable solutions, called a **transcritical bifurcation**. The simple time-delayed feedback system model in **Fig. 3.8** (lower panel) has the ODE $\dot{y} = 1 - y(t - p)$, where p is a time-delay parameter in the feedback loop. The equilibrium point is $y_e = 1$ for any p. **Fig. 3.9** illustrates asymptotically stable solutions for $p = 0.5$ and $p = 1$, and an unstable solution for $p = 2$. Additional simulation runs using the Matlab code in **Fig. 3.9** indicate that $p = p^* = 1.572$ yields a transcritical bifurcation at $y_e = 1$ that separates stable from unstable solutions.

Other Bifurcation Types

If $x_e(p)$ plotted in the x_e versus p plane is S-shaped, like in **Fig. 9.6**, the points on the curve delimiting the reversals in direction, i.e. $x_e(p_1)$ and $x_e(p_2)$, are called **saddle node bifurcation**

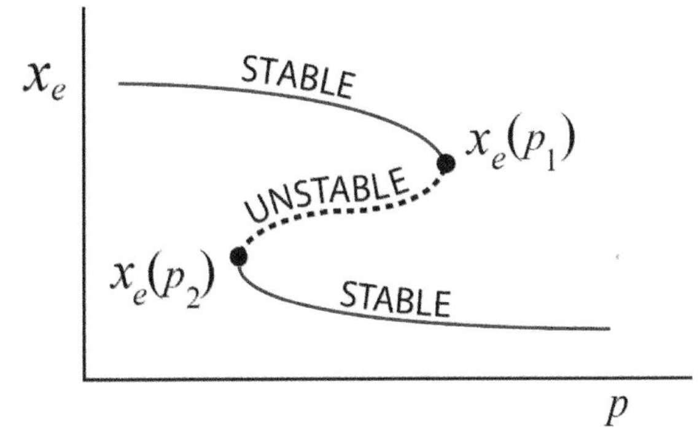

Fig. 9.6

Saddle node bifurcation.

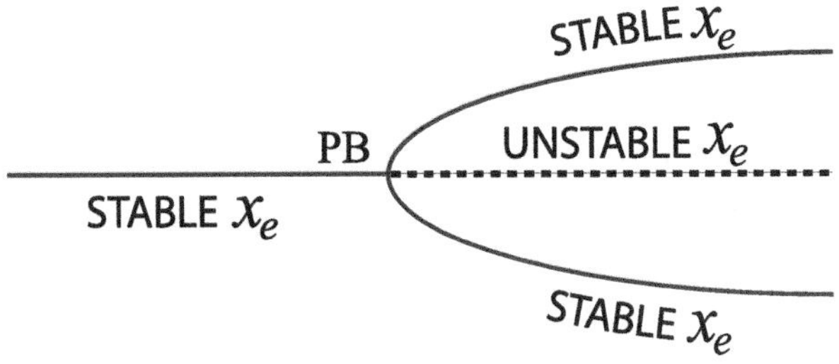

Fig. 9.7

Pitchfork bifurcation (PB).

points. Between these two transcritical values, three steady states are present. Two are stable and one is unstable (the one in the middle), typical of **bistable** systems, e.g. mutual activation models in biochemistry and genetic switches in biology. As p increases, the system *switches* from one stable steady state $x_e(p_1)$ – the high state – to the other stable steady state $x_e(p_2)$ – the low state.

This type of dynamical behavior is called **hysteresis**. In other words, dynamical stability depends on the *direction* of changes in parameter p in bistable or other systems exhibiting saddle node bifurcation points. Normally, two bifurcation points are present, in which case the switching is reversible. In some cases, however, only one point $x_e(p_2)$ is present and the transition is irreversible.

Remark: In biology, *gene-regulatory circuits* displaying multistability and oscillations are found in bacteriophage, Cyanobacteria, circadian oscillators and synthetic bistable gene-regulatory networks in *Escherichia coli* bacteria (Gardner et al. 2000). The simple genetic switch model in this reference is constructed from two repressible promoters arranged in a mutually inhibitory network, very much like the simple NL mutual activation model in Example 9.3 above. **Genetic switches** are bistable and exhibit saddle node bifurcations. A very similar model is described and analyzed in (Santoso and Lanzon 2006).

Coupled subsystems (among others) also can have **pitchfork bifurcation** dynamics, characterized by passage from a stable to an unstable steady state and, at the same time, the appearance of two stable steady states, as illustrated in **Fig. 9.7**. This bifurcation type is

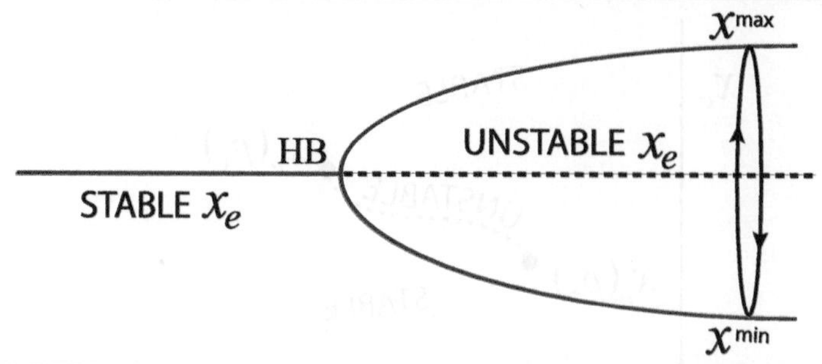

Fig. 9.8

Hopf bifurcation (HB). The closed curve shown in black illustrates a stable limit cycle with an oscillation amplitude $x^{max} - x^{min}$.

fragile: slight asymmetries between the subsystems can break the pitchforking effect at the bifurcation point, generating a saddle node in its place.

Appearance or disappearance of a *periodic orbit* via a local change in the stability properties of an equilibrium point is known as a **Hopf bifurcation**. A Hopf bifurcation occurs when an equilibrium solution $\boldsymbol{x}_e(\boldsymbol{p})$ loses stability as a pair of complex conjugate eigenvalues of the linearized solution around $\boldsymbol{x}_e(\boldsymbol{p})$ cross the imaginary axis of the complex plane. Typically, this generates a small-amplitude **limit cycle** oscillation branching from the bifurcation point – as illustrated in the bifurcation plot in **Fig. 9.8**. Limit cycle oscillations are described further below.

Hopf bifurcations can lead to a variety of complex dynamical responses. They are present in a variety of chemical and biological system models, including the Hodgkin–Huxley model of nerve conduction (Hodgkin and Huxley 1952), the Selkov model of glycolysis and the autocatalytic Brusselator models analyzed in some detail below.

Oscillations in Biology

Harmonic Oscillations

Oscillations of a plucked guitar string or a pendulum have an amplitude that depends on how far it's perturbed from its resting steady state. Larger perturbations generate larger amplitudes, smaller ones smaller amplitudes. In other words, amplitude depends on initial conditions (ICs), characterizing the class of oscillators with **harmonic oscillations**. Examples of harmonic oscillators in population biology include predator–prey models, like the simple shark (S)–tuna (T) predator–prey model: $\dot{S} = k_1ST - k_2S$ and $\dot{T} = k_3T - k_4ST$ developed and simulated in Chapter 3 (**Fig. 3.24**). That model exercise depicted a stable cycling population of sharks and tuna fish, for a particular IC and parameter vector $\boldsymbol{k} = [0.015\ 0.7\ 0.5\ 0.01]^T$. The phase plane for this model was a closed curve, characteristic of all oscillating systems. With different ICs or set of nonzero parameters, the (nonzero) equilibrium solutions $\boldsymbol{x}_e$ of $\dot{S} = 0$ and $\dot{T} = 0$ and the closed curve phase plane would have been different, and the period and amplitude of the oscillations would also have been different.

Limit Cycle Oscillations

Limit cycles are a common type of **sustained oscillations**, also characterized by closed contours in phase space – closed curves in the phase plane ($n = 2$). (See e.g. **Fig. 9.9**, lower left).

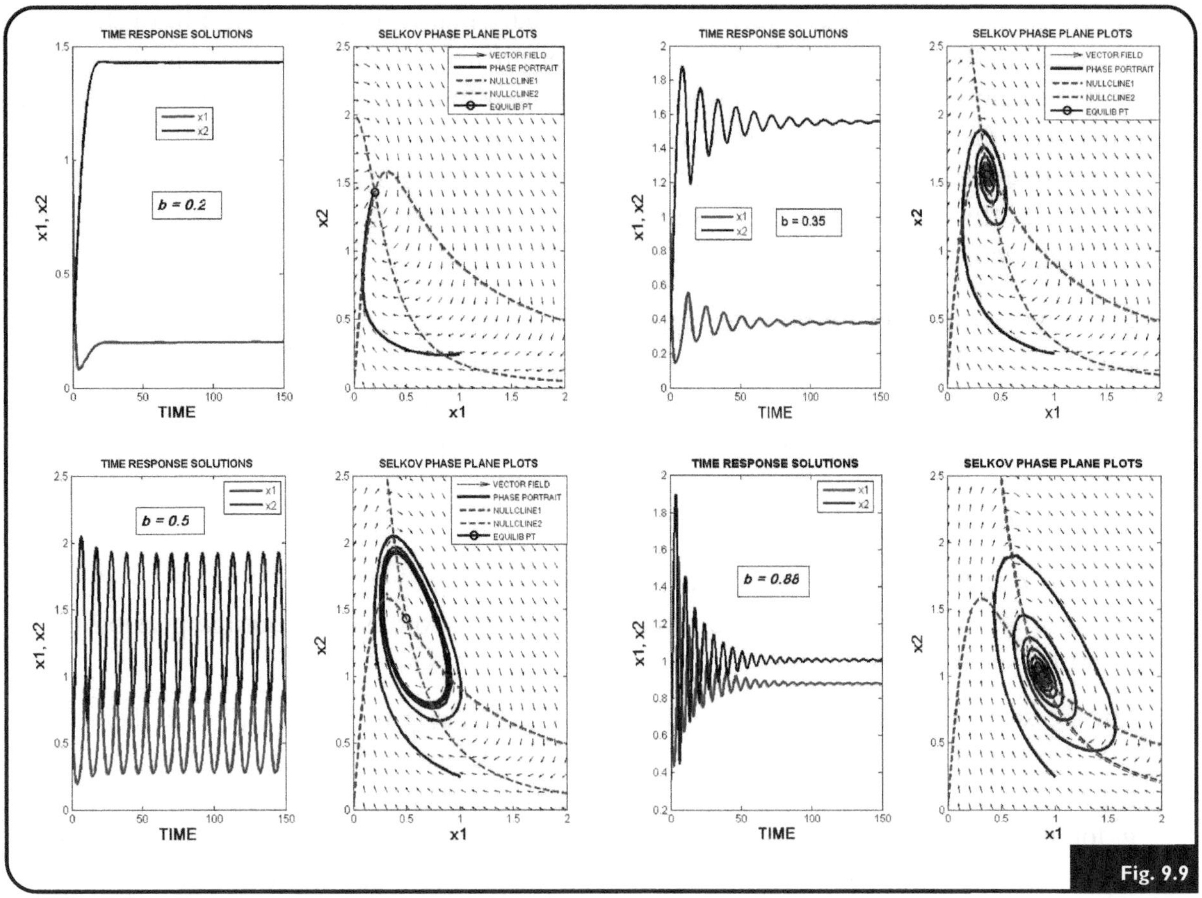

Fig. 9.9

Temporal responses, nullclines and phase portraits in the phase plane for the Selkov model of glycolysis. Solutions are shown for four different values of parameter *b*. Phase portraits pass through two Hopf bifurcations as *b* increases. Trajectories go from being constant stable, to asymptotically stable, to unstable oscillating, to limit cycling and, finally back to a stable constant equilibrium point. Vector fields are also illustrated on the phase planes, as small black arrows surrounding the nullclines.

In contrast to harmonic oscillators, the amplitude of limit cycling systems is not determined by initial conditions. Instead, (stable) limit cycle oscillators automatically return to the same limit cycle – with the same amplitude and period – after perturbation. They have a characteristic amplitude, with an isolated closed trajectory. Neighboring trajectories are not closed but spiral either toward or away from the limit cycle, which means *limit cycles can be stable or unstable*. A stable limit cycle (also called **orbitally stable**) attracts all neighboring trajectories and exhibits self-sustained oscillations. Physically, to maintain this characteristic behavior, limit cycling systems embody a dissipative mechanism to attenuate oscillations that grow too large and an energy source to amplify those that become too small.

Limit cycles play an important role in biology, describing many biological rhythms, not only those that have Hopf bifurcations like that in **Fig. 9.8** above. Circadian clocks, mammalian cell differentiation and the brainwave patterns of an electroencephalogram (EEG) during a siezure, among many others, exhibit limit cycling behaviors.

The **Poincaré–Bendixson theorem**, a central result in nonlinear dynamics (Teschl 2012), says that a *two-dimensional ODE has very regular behavior*, circumscribing the very limited nature of 2-D dynamical system trajectories in the phase plane. In contrast, if a third state variable is added to the model ($n = 3$), the potential for greater complexity takes a profound leap, as we show below, and increases further as the dimensionality goes up. In two dimensions, there are only three possibilities. If a trajectory starting at IC $\boldsymbol{x}_0$ at t_0 remains bounded for $t > t_0$, and does *not* approach an equilibrium point $\boldsymbol{x}_e$ (steady state), then it is either a limit cycle (closed orbit) or it approaches a limit cycle as $t \to \infty$. In two-dimensions, *nothing more complicated can happen!*

The **Bendixson criterion** provides a simple test for existence of limit cycles in the phase plane for a NL ODE model: $\dot{x}_1 = f_1(x_1, x_2)$, $\dot{x}_2 = f_2(x_1, x_2)$. No limit cycle exists for this model if $\nabla \boldsymbol{f}_x \equiv \partial f_1/\partial x_1 + \partial f_2/\partial x_2 \neq 0$ and also does not change sign for all realizable x_1 and x_2. This is a sufficient but not a necessary condition for lack of existence of limit cycles.

Example 9.6: ***Bendixson Criterion Applied***

For the bistable model of Example 9.3 above, $\partial f_1/\partial x_1 + \partial f_2/\partial x_2 = (-1) + (-1) = -2 \neq 0$ for any x_1 and x_2 and -2 cannot change sign, so this rules out the possibility of limit cycles for this model.

For the shark (S)—tuna (T) predator prey model, $\partial f_1/\partial x_1 + \partial f_2/\partial x_2 = k_1 T - k_2 + k_3 - k_4 S \neq 0$ for arbitrary S and T and any $k > 0$ values, so this model meets the first part of the Bendixson criterion. But it can change sign, so it does not satisfy the second part, e.g. for all ks $= 1$, the criterion becomes $T - S \neq 0$ which readily changes sign when $T < S$ or $T > S$. We know this model does not have limit cycles, so the criterion fails to make the distinction, i.e. it is not a *necessary* condition for nonexistence of limit cycles.

Transition to either stable or unstable limit cycling behavior is often found at Hopf bifurcation points. The limit cycle is stable and the Hopf bifurcation is called **supercritical** if a parameter called the *first Lyapunov coefficient* is negative. Otherwise the limit cycle is unstable and the bifurcation is **subcritical** (Hale and Koçak 1991) (Exercise E9.6).

Oscillations & the Selkov Model of Glycolysis

The following example illustrates oscillations and other complex dynamical system behavior in a simple classic NL model of an important metabolic pathway.

Example 9.7: ***Selkov Model of Glycolysis (Selkov 1968)***

This $n = 2$ ODE model: $\dot{x}_1 = x_1 + 0.1x_2 + x_1^2 x_2$ and $\dot{x}_2 = b - 0.1x_2 - x_1^2 x_2$, with one variable parameter b, has trajectories that encounter Hopf bifurcations and a limit cycle for different ranges of b. *Matlab* numerical solutions are illustrated in **Fig. 9.9** for ICs: $\boldsymbol{x}_0 = (1,\ 0.25)$ and four different values of parameter $b = \{0.2,\ 0.38,\ 0.5,\ 0.88\}$, to illustrate the different dynamical regimes of complex qualitative behavior. The nullclines are also shown (red and green dashed lines) in the phase plane plots on the RHS, and equilibrium points are shown as small black circles. Phase portraits are in blue.

For $b = 0.2$, the trajectories approach the constant stable equilibrium point at $\sim$ (0.25, 1.45), as shown in the upper LHS plot. As b increases further to 0.38, the trajectories begin oscillating, but in a dissipative way, and they are asymptotically stable, toward $\sim$ (0.4, 1.55), as shown in the upper RHS plot. Limit cycle oscillations (in blue) appear at the first Hopf bifurcation point at $b = \sim 0.43$, from an unstable equilibrium (oscillations building until they achieve a constant amplitude), at $\sim$ (0.6, 1.45), shown for $b = 0.5$ in the lower LHS plot. The limit cycle then begins to disappear, the time response eventually returning to a stable equilibrium point, at another Hopf bifurcation $b = \sim 0.88$, at $\sim$ (0.88, 1.05), as shown in the lower RHS plot.

Oscillations & the Brusselator Model

The **Brusselator** (*Brussels-oscillator*) is a theoretical model for a type of (cubic) **autocatalytic** chemical reaction that characterizes a number of chemical and biological systems (Turing 1952; Prigogine 1980). Autocatalytic reactions have at least one product that's also a reactant. Familiar ones include glycolysis – with models different than the Selkov one (Chandra et al. 2011); clock reactions (product and reactant concentrations change periodically); ozone formation from atomic oxygen – (via triple collision of oxygen atoms and enzymatic reactions); and the more esoteric but widely quoted BZ-reaction, which generates an autocatalysis from a mixture of potassium bromate, malonic acid, and manganese sulfate in a solution of heated sulfuric acid (Scott 1994). As we show here, the Brusselator model produces stable steady states as well as self-sustaining limit cycle oscillations at a Hopf bifurcation, from a pattern-forming instability.

The full Brusselator system is a *reaction–diffusion system*, common in chemistry and biology. **Reaction–diffusion** models depict how the concentration of one or more substances distributed in space (e.g. in a cellular space) changes under the influence of local chemical reactions in which the substances are transformed into each other, in conjunction with diffusion of the substances as they spread out over the surface. The reaction components of the Brusselator model contains a pair of intermediate state variables X_1 and X_2 with time varying concentrations x_1 and x_2 reacting with product chemicals A, B and C whose concentrations a, b and c are kept constant. The reaction model ODEs are established by straightforward application of mass action to the Brusselator model reactions: $A \xrightarrow{k_1} X_1$, $B + X_1 \xrightarrow{k_2} X_2 + C$, $2X_1 + X_2 \xrightarrow{k_3} 3X_1$, $X_1 \xrightarrow{k_4} D$. The *full Brusselator model* is obtained by adding partial derivative (PD) *diffusion* terms $\nabla^2 x_1$ and $\mu\nabla^2 x_2$, with μ a parameter proportional to the diffusion ratio of the two species x_1 and x_2. The full equations are:

$$\begin{aligned} \dot{x}_1 &= a - bx_1 + x_1^2 x_2 - x_1 + \nabla^2 x_1 \\ \dot{x}_2 &= bx_1 - x_1^2 x_2 + \mu\nabla^2 x_2 \end{aligned} \tag{9.5}$$

For convenience and no loss of generality, the rate constants k_i have been set to 1. Local stability analysis is normally done under the assumption of negligible diffusion, i.e. $\mu \approx 0$, so the

PD terms $\nabla^2 x_1$ and $\mu\nabla^2 x_2$ on the LHS of (9.5) are assumed negligible[9] in the following analysis.

Example 9.8: ***Brusselator Stability Analysis***

Setting the ODE derivatives equal to zero, the a and b concentration-dependent equilibrium steady state solutions of (9.5) are: $\boldsymbol{x}_e = \{a, b/a\}$.

We examine the eigenvalues $\lambda_i = \sigma_i + j\omega_i$ of $A \equiv \partial \boldsymbol{f}/\partial \boldsymbol{x}_e$ in Eq. (9.5) evaluated at $\boldsymbol{x}_e = \{a, b/a\}$.

$$\partial \boldsymbol{f}/\partial \boldsymbol{x}_e = \begin{bmatrix} \partial f_1/\partial x_{1e} & \partial f_1/\partial x_{2e} \\ \partial f_2/\partial x_{1e} & \partial f_2/\partial x_{2e} \end{bmatrix} = \begin{bmatrix} b-1 & a^2 \\ -b & -a^2 \end{bmatrix} \tag{9.6}$$

From matrix $[sI - A]$, the characteristic polynomial is: $s^2 + (a^2 + 1 - b)s + a^2$. The eigenvalues of A are then:

$$\lambda_{1,2} = \left(-(a^2 + 1 - b) \pm \sqrt{(a^2+1-b)^2 - 4a^2}\right)/2 \tag{9.7}$$

The condition for a stable equilibrium is that the real parts of $\lambda_{1,2}$ must be negative, i.e. $a^2 + 1 - b > 0$ or $a^2 + 1 > b$. At the transition point, when $a^2 + 1 \equiv b$, Re $\lambda_{1,2} = 0$ and the eigenvalues are purely imaginary – conditions for a *Hopf bifurcation*. When b becomes $> a^2 + 1$, local stability is lost and further stability properties are determined by the sign of the discriminant under the square root sign, rearranged in easier-to-analyze form as: $D \equiv (a^2+1-b)^2 - 4a^2 \equiv (b-(a+1)^2)(b-(a-1)^2)$. The eigenvalues are thus complex for: $(a-1)^2 < b < (a+1)^2$ and real for: $b < (a-1)^2$ or $(a+1)^2 < b$.

With Re $\lambda_{1,2}$ remaining negative, i.e. $a^2 + 1 > b$, equilibrium points change from a stable node to a stable limit cycle (focus) when D becomes negative, and then it loses stability, becoming an unstable limit cycle when Re $\lambda_{1,2}$ becomes positive. Finally, it becomes an unstable node when both D and Re $\lambda_{1,2}$ become positive.

Fig. 9.10 depicts phase plane trajectories and their corresponding time domain solutions for a = 1 and three values of $b = 0.2$, 2 and 3. A Hopf bifurcation occurs at $b = 2$. Initial conditions in all three are zero. Solutions were computed numerically, using a *Matlab* script.[10] The plots illustrate a stable fixed-point node, an asymptotically stable oscillatory response, and a limit cycling response.

Other Complex Dynamical Behaviors

In addition to periodic oscillations, other complex behaviors are found in biological systems. These include: **bursting:** alternating between oscillatory and non-oscillatory phases, e.g. in

9 Stability analysis is more complicated with diffusion included. Adding diffusion components retains most local stability properties, but greatly broadens the kinds of parameter-dependent and patterned dynamical responses – often termed *Turing-patterned* behaviors (Turing 1952). See e.g. (Pena and Perez-Garcia 2001; Golovin et al. 2008).

10 Available on the Elsevier textbook website.

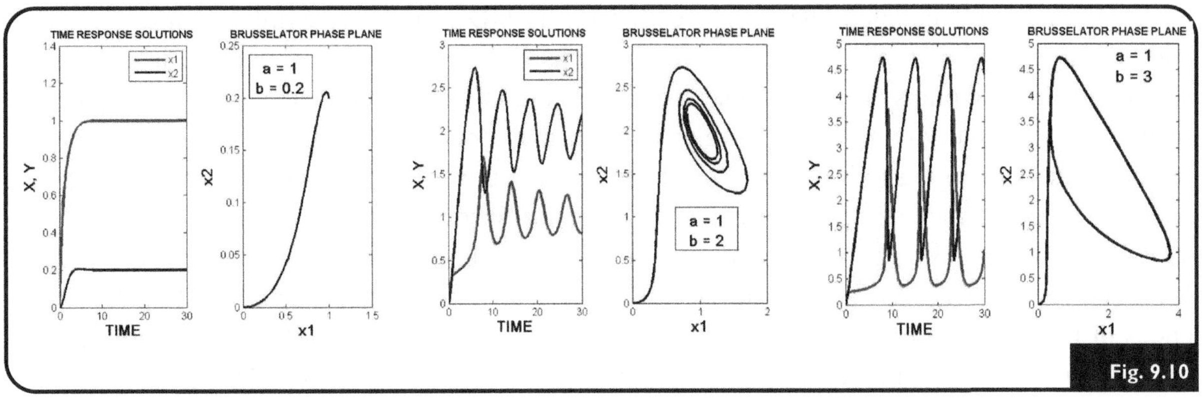

Fig. 9.10

Time response and phase plane solutions for the Brusselator model with $a = 1$ and $b = 0.2$, 2 and 3. Fixed point stable, asymptotically stable and limit cycling are illustrated, the separations occurring at a Hopf bifurcation.

pancreatic beta-cells, modeled and analyzed in (Keener and Sneyd 1998); **quasi-periodic oscillations**, as in the Hodgkin–Huxley and other models of neural conduction (Aihara et al. 1984); **synchronization** of coupled oscillators, e.g. in models of the cardiac pacemaker (Strogatz and Stewart 1993); and **excitability**, e.g. in models of the transmission of impulses between neurons, or in cardiac arrhythmias (Ballarò et al. 2007). Another complex biosystem oscillation – one that remains incompletely classified – is the multi-pulse-like activity of tumor suppressor protein p53 and its coupled regulatory enzyme MDM2 in response to damage-stimulated intracellular kinase (Lev Bar-Or et al. 2000). These oscillations have been described as **digital oscillations**, because their number rather than their amplitudes has been observed to be dependent on the magnitude of the damage signal (Ma et al. 2006). Other mechanisms including input-pulsing, multi-feedback loop, and possibly limit cycling have been suggested for p53–MDM2 oscillatory behavior, with their precise nature, yet to be established (Proctor and Gray 2008; Batchelor et al. 2009; Batchelor et al. 2011). More on this system in Chapter 15.

Chaos

Chaos – in the mathematical systems sense[11] – is a particularly special kind of complex and irregular behavior of deterministic dynamical systems. It is characterized first and foremost by *extreme sensitivity to initial conditions*. Responses to ever so slightly different initial conditions are typically widely divergent from each other. This is quite surprising given that dynamic responses to determinisitic ODEs are completely determined by initial conditions! Indeed, the responses of chaotic systems – like all others – are always precisely the same for the same ICs, but typically very different for slightly different ICs. Solutions of chaotic systems with different ICs never converge to a cycle of any length. And their deterministic nature, oxymoronically, does not make them predictable. Overall, dynamical responses of chaotic systems are quite irregular, nonperiodic and have the appearance of randomness.

11 Chaos theory, sometimes called **deterministic chaos** theory, is highly complex mathematically and we present only an outline of it here, to illustrate the basics and how they have been and can be used in modeling and studying biological systems. More complete treatments and additional biological applications can be found in many references (e.g. Weiss et al. 1994; Alligood et al. 1997; Devaney 2003).

At least three[12] coupled ODEs are needed for an ODE model to exhibit chaotic behavior, and each equation typically has several terms, at least one of which is nonlinear. Topologically, chaotic system response regions also must "mix" with each other as they evolve over time, eventually overlapping – like colored dyes mixing in a liquid. Chaotic responses, while never converging to a periodic orbit, do come arbitrarily close, and for this reason have been given the name **strange attractors**. Unlike equilibrium node attractors and limit cycles, strange attractors are highly detailed and complex, with chaotic behavior exhibited only in localized regions of phase space. Minute differences in initial conditions generate large differences in future points of any chaotic orbital trajectory. Two famous models, the simplest *logistic population growth* model and the *Lorenz weather dynamics model*, described next, both illustrate this chaotic behavior.

Complex Dynamic System Behavior & Chaos for the Discrete-Time Logistic Population Growth Model

We analyze the different stability regimes of the simple discrete-time logistic model of periodically renewing populations – the first-order polynomial recurrence relation of degree 2: $x_{k+1} = ax_k(1 - x_k)$. This is *the* classic example of how complex, chaotic behavior can arise from very simple NL dynamical equations, in this case an innocent-looking *first-order difference equation* (May 1976). Complex dynamic system behavior for this model – from constant and stable to chaotic – was illustrated by simulation in **Fig. 3.22** and Example 3.27 of Chapter 3. We explain these dynamics in some detail here.

The model equation depicts two population phenomena in simplified form. The first term, ax_k, represents a rate of population growth assumed to be proportional to the current population when its size x_k is small. The second term, $-ax_k^2$, approximates density-dependent mortality, i.e. a decrease in growth rate proportional to the square of the population number – the competition factor in the presence of limited carrying capacity. The state variable x_k is normalized, as the ratio of the actual population to the maximum possible population in the kth generation (k^{th}-step or period); x_{k+1} means the same thing in the next generation.

The model: $x_{k+1} = ax_k(1 - x_k)$, with $a > 0$, describes a *parabolic arc* – the shape of the phase plane portrait (x_{k+1} vs. x_k) in the phase plane. (Like plotting the continuum phase plane for $\dot{x}$ vs. x.) Simple analysis indicates that this arc goes through the points (0, 0) and (1, 0) and has a maximum at (0.5, a/4). Equilibrium for a difference equation is defined as $\boldsymbol{x_{k+1} \rightarrow x_k \rightarrow x_e}$ as $k \rightarrow \infty$, so the model has a single equilibrium function: $x_e = ax_e\,(1 - x_e)$, with two distinct solutions, the second dependent on parameter a: $x_{e1} = 0$ and $x_{e2} = (a-1)/a$. To retain $x_{e2} = (a-1)/a > 0$ and the parabolic phase plane in the first quadrant, a must be constrained as $0 < a < 4$. This results in the entire solution bounded within the first quadrant unit square: $0 < x_{k+1} < 1$ and $0 < x_k < 1$. Therefore all x_{k+m} (future generations) remain in the first quadrant for any m. We show below that x_{e1} is stable and constant on the interval $0 < a < 1$ and x_{e2} is asymptotically stable on $1 < a < 3$. Beyond that, the dynamics are much more interesting.

12 For $n = 2$ (or 1), the Poincaré–Bendixson theorem guarantees no chaos!

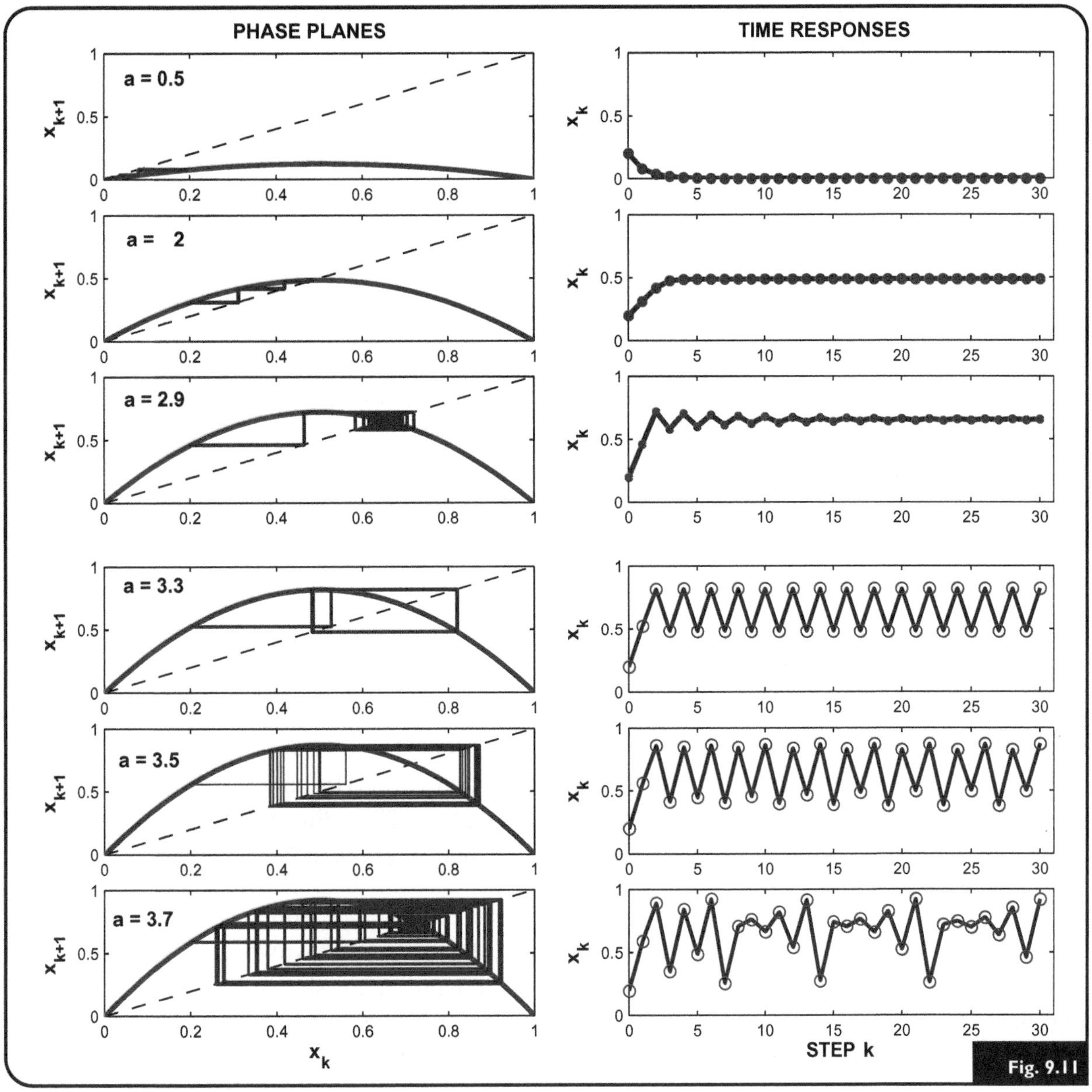

Numerical solutions for the logistic equation – phase plane (x_{k+1} vs. x_k) and time domain solutions (x_k vs. t_k) – computed numerically for six values of a. Each row of figures illustrates different stability regimes. From top to bottom, stable and null, stable and constant, asymptotically stable, limit cycling, cycling with period doubling, and chaos. The phase plane is typically generated automatically at the same time as the equation is solved over some time period. Bifurcations and periodic orbits with frequency doublings begin around 3.45, as shown, until chaos ensues. Also see **Fig. 9.12**.

Fig. 9.11 illustrates graphical solutions of this model for various values of growth rate parameter a, with a different a values for each row of the figure, as shown. The LHS column plots are phase plane solutions. The RHS column plots are the corresponding time domain solutions. Computations and plotting were accomplished with a *Matlab* script, available on the book website. This could have been accomplished manually, however, with pencil and graph paper, and a simple calculator. The 45° *diagonal* of the unit square in the first quadrant is drawn as a *dashed line* in all phase plane illustrations on the LHS of **Fig. 9.11**, along with the parabolic phase plane plots. This line is helpful in describing model solutions – discrete-time-step by discrete-time-step – as a function of parameter a.

To see how this works, start at any initial condition x_k between 0 and 1 and rise vertically to the parabolic arc. All solutions in **Fig. 9.11** start at $x_k = 0.2$, the IC. For $a = 2$ (second row solutions – easiest visualization), we get $x_{k+1} = ax_k(1 - x_k) = 0.32$, the point on the parabola above $x_k = 0.2$, as shown for $a = 2$. To see where the solution goes next, we move horizontally to the 45° diagonal line – in this case to the right from (0.2, 0.32) – as shown, and then up again to the parabolic arc, and so on, like in the figure. Repeating this procedure in this web-like way, we get what's called a **cobweb plot** and eventually arrive at an equilibrium solution. In this example, for $a = 2$, constant steady state $x_e = 0.5$ is reached in three steps. As we see below, this solution can be constant, oscillatory, or chaotic, as a function of a. We are interested specifically in how the phase plane and time domain solutions change qualitatively as a varies over its region of validity: $0 < a < 4$.

First, with $a < 1$, exemplified by $a = 0.5$ in the top row of **Fig. 9.11**, the parabola on the LHS lies completely below the 45° line, so horizontal shifts to the diagonal line are always toward the *left*. This means the equation evolves with a gradual *decrease* of x_k to the *stable equilibrium point* at the origin. This is shown in the time response solution on the RHS, which fall from $x_k = 0.2$ to zero in about 5 time steps. This also is established with little difficulty analytically. For $x_{k+1} = ax_k(1 - x_k)$ and $a < 1$, we have $x_{k+1} < ax_k$ and at the mth-step $x_{k+m} < a^m x_k$, so $x_k \to 0$ as $k \to \infty$.

Things change when a reaches 1. For $a > 1$, the fixed point at the origin is replaced by new equilibrium point $x_{e2} = (a - 1)/a$, moving away from the origin as a increases beyond 1. The parabola rises above the diagonal for all $x_k < x_{e2}$ and the solution gradually converges to the stable and *constant* steady state x_{e2} (as in the second row RHS panel of **Fig. 9.11**), for all values of a up to but still $a < 2$. The solutions labeled "$a = 2$" in **Fig. 9.11** are what the phase plane and dynamic response looks like for a infinitesimally less than 2.

At $a = 2$, $x_{e2} = (a - 1)/a = 0.5$ and for $a > 2$ the steady state solution is no longer constant. x_{k+1} passes over the parabolic peak value, then starts downward, but continues to increase in value, taking on an *oscillating* cobweb pattern, as shown in the third row of **Fig. 9.11**, for $a = 2.9$. To establish stability of this x_{e2}, we check the slope (derivative) of the tangent to this parabolic function (recall the direction of the derivative in the vicinity of x_{e2} criterion for stability). We have $d(ax_{e2}(1 - x_{e2}))/dx_{e2} = a(1 - 2x_{e2}) = 2 - a$. This slope is negative for $a > 2$, so we have *local stability* for $a = 2$, and it must be *asymptotic stability*, because it's approaching equilibrium. Asymptotic stability of x_{e2} continues within $2 < a < 3$.

The solution *bifurcates* at $a = 3$ (also see **Fig. 9.12**) and for a values slightly greater than 3, the trajectory loses stability and transitions into a two-cycle approach toward a stable limit cycle. This is shown in the fourth row of **Fig. 9.11** for $a = 3.3$, where x_k varies up and down between two constants.

Then solutions become much more interesting and complex, with analysis details also becoming increasingly tedious. We outline and abbreviate the results (May 1976). First solve the model for x_{k+2}, the model *mapping done twice*:

$$x_{k+2} = a2x_k(1 - x_k)[1 - ax_k(1 - x_k)] \tag{9.8}$$

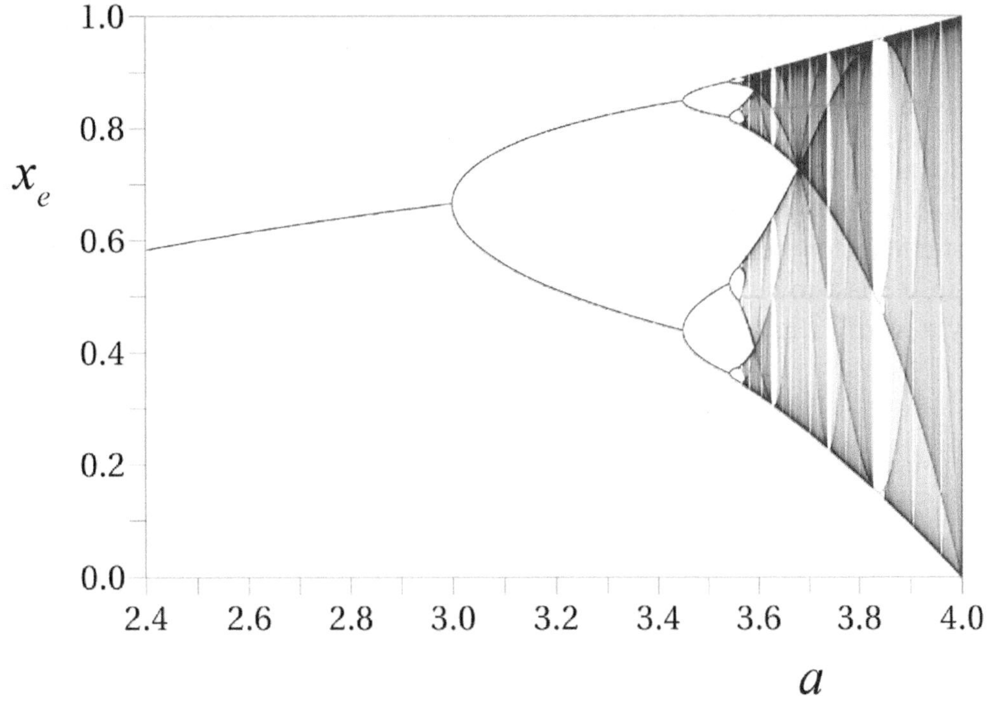

Fig. 9.12

Partial bifurcation diagram, x_e versus a, for the logistic equation, illustrating bifurcations at a ~ {3, 3.45, ...}, with periodic orbits exhibiting frequency doubling beginning around 3.45, and then doubling again and again until chaos ensues. This diagram can be obtained for successive values of a by either computing steady state solutions algebraically, or by computing the full transient responses up to steady state and plotting x_k values for each k after the transients have disappeared. Not shown is x_k versus a for null solutions, i.e. $x_k = 0$ over $0 < a < 1$ and the smooth curve representing asymptotically stable damped-oscillatory solutions, from $x_k = 0$ at $a = 1$ up to the first bifurcation at $a = 3$.

This mapping is fourth-order, so it has four roots. Equilibrium solutions therefore become what is called **doubly periodic** – meaning the trajectory repeats every four steps. For $a > 3$, two of the new roots coincide with x_{e2} – a critical (bifurcation) point where solutions spiral outward. The two others move away as a increases and first become doubly periodic – approaching limit cycling behavior between *two different equilibria* x_{e3} and x_{e4}, as illustrated in the fourth row of **Fig. 9.11**. This means that successive generations limit cycle between x_{e3} and x_{e4}, for parameter a between 3 and $1 + \sqrt{6} \cong 3.45$ (approximately 3.45), from almost all ICs. For a between approximately 3.45 and 3.54, population x_k approaches permanent limit-cycle-like *oscillations among four values*, from almost all ICs. In the fourth row of **Fig. 9.11**, for $a = 3.5$, time responses oscillate between *four* (instead of two) values; and these repeat every *fourth* step.

The ***bifurcation diagram***[13] in **Fig. 9.12** illustrates the ever changing behavior of this model – from one stability regime to another – as a function of the growth rate parameter a, in particular for $a = 2.4$ and above. At each bifurcation point – a point of transition from one stability regime

13 To draw this bifurcation diagram, pick any IC, generate a large number of iterations, discard the first several (transient responses) and plot the rest as a function of a. For a values where the orbit is fixed, the bifurcation diagram is a single line; for periodic values, a series of lines; and for chaotic values, a wash of dots. Not shown is x_k versus a for null solutions, i.e. $x_k = 0$ over $0 < a < 1$ and the smooth curve representing asymptotically stable damped-oscillatory solutions, from $x_k = 0$ at $a = 1$ up to the first bifurcation point shown. Only the more complex ranges for equilibrium solutions as a function of a are shown in **Fig. 9.12**.

to another – stability is determined by the slope of the tangent to the parabolic function at the new x_{ei} values as they appear. For $a > 3.54$, $x_k \rightarrow$new bifurcating equilibria and *multiple periodic orbits* among eight values, then among 16, 32, etc. – a **multiple period-doubling**[14] **cascade** – with corresponding multiple equilibria (**strange attractors**).

For **critical value** $a \sim 3.57$ *chaos* ensues at the end of the period-doubling cascade, with *no oscillations of finite period.* Slight variations in ICs (x_0 values) then yield dramatically different chaotic trajectory results over time, for a in the range: $3.57 < a < 4$. Chaotic behavior is illustrated in the last row of **Fig. 9.11**, for $a = 3.7$ and $x_0 = 0.2$. (Exercise E9.8, provides an opportunity to verify sensitivity to ICs for this model.)

The Lorenz Model & Lorenz Attractor

In 1963, Edward Lorenz developed the following ODE model for atmospheric convection in a weather system (Bergé et al. 1984):

$$\begin{aligned} \dot{x}_1 &= a(x_2 - x_1) \\ \dot{x}_2 &= x_1(b - x_3) - x_2 \\ \dot{x}_3 &= x_1 x_2 - c x_3 \end{aligned} \tag{9.9}$$

The parameters a, b, c are nonnegative. These equations have a linear component, with strong quadratic coupling. In the linear part, x_1 and x_2 are not coupled with x_3. But in the quadratic part x_2 and x_3 are cross-coupled to one another with x_1. These properties give the Lorenz model much of its unusual chaotic character when visualized in 3-D space (Morrison 1991). These properties are shared to some degree with another model, the *Van der Pol equation*, that does not exhibit chaos in its autonomous form, but can if forced with a sinusoidal input.

Stability analysis begins with computing the nullclines (Exercise E9.5) and finding the steady state solutions for $\dot{\mathbf{x}} = 0$. Normally, this can be difficult. But for the Lorenz equations, we recognize from the first ODE that equilibrium solutions exist only if $x_1 = x_2$ (the x_1-nullcline is a 45° line in the (x_1, x_2) plane). Making this substitution into $\dot{x}_2 = 0$ and $\dot{x}_3 = 0$ then yields the following unique, symmetric steady state solutions:

$$\begin{aligned} x_{1e} &= (0, 0, 0) \\ x_{2e} &= \left(+\sqrt{c(b-1)},\ +\sqrt{c(b-1)},\ b-1 \right) \\ x_{3e} &= \left(-\sqrt{c(b-1)},\ -\sqrt{c(b-1)},\ b-1 \right) \end{aligned} \tag{9.10}$$

The dynamical solutions of Eq. (9.9) near these equilibria are starkly different over the three-dimensional parameter space of (a, b, c).

14 Period-doubling occurs when the period of a time series or function doubles so that it passes through twice as many states before repeating. For example, a time series with period two that oscillates between max and min values can undergo period doubling to exhibit a period four oscillation, which has four states before it repeats, as in the fourth row panel of **Fig. 9.11**. In the discrete-time logistic equation, period-doubling occurs over and over again, at all bifurcation points shown in **Fig. 9.12**.

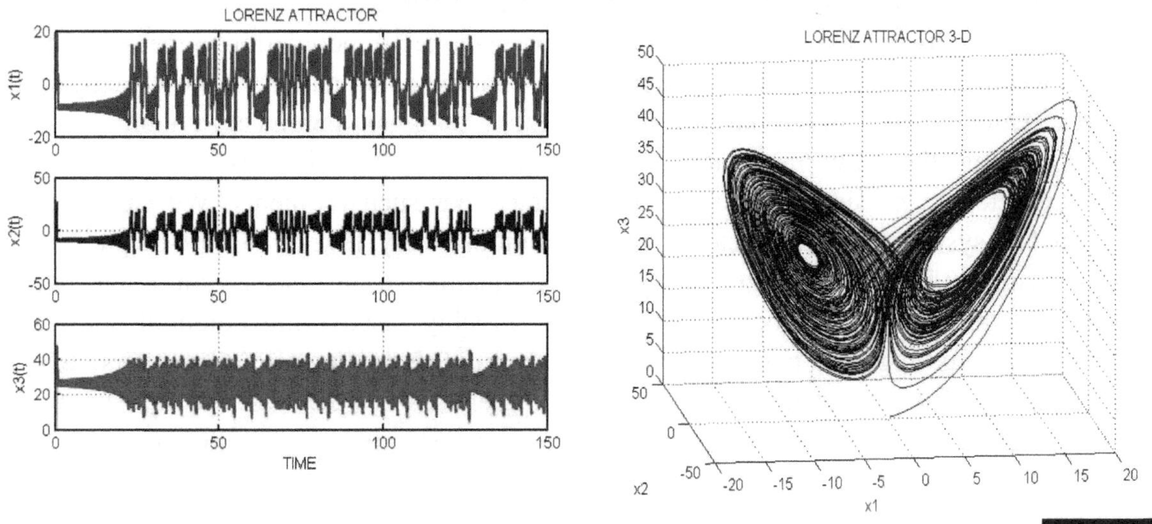

Fig. 9.13

Chaotic Lorenz model temporal solutions and the "butterfly" 3-D phase plane.

For $a,\ c > 0$ and $b < 1$, it is apparent that the only real equilibrium point is the *origin*, and it is *stable*, i.e. all solutions converge to the origin, for any initial conditions. A saddle node bifurcation then occurs at $b = 1$; and, for $b > 1$, two additional (bistable) equilibrium points appear at: $\left(\pm\sqrt{c(b-1)},\ \ \pm\sqrt{c(b-1)},\ \ b-1\right)$, corresponding to *steady* (constant) convection in the weather. Subsequent analysis is complex and involves linearizing about these equilibria and resolving multiple bifurcations and periodic orbits of (9.9), yielding several different complex qualitative behaviors in the dynamics. This pair of equilibrium points is stable only if $b < a(a + c + 3)/(a - c - 1)$ and, for $b > 0$, this holds if $a > c + 1$.

The solutions then become much more interesting at critical point $a = c + 1$, where both equilibrium points (nodes) lose stability via a *Hopf bifurcation* and the solution becomes *chaotic* (Hirsch et al. 2003). This chaotic behavior is characterized by a spiraling in towards one of these two unstable nodes, *never reaching it*, and then suddenly spiraling out and away from that node toward the other. This pattern is repeated over and over between the nodes, the whole geometric figure being the strange attractor for the Lorenz model. With slight changes in ICs the *timing* of the jumps in this pattern is altered, but the double-spiraling attractor remains localized in 3-D phase space.

The Lorenz model was solved numerically using *Matlab*.[15] Results are shown in **Fig. 9.13** for ICs $\boldsymbol{x}_0 = [0.3\ 0.3\ 0.3]^T$ and parameter values $a = 10$, $b = 28$ and $c = 8/3$, the values originally used by Lorenz to exhibit chaotic behavior of his equations. The three temporal solutions shown on the LHS are clearly not periodic. The spiraling 3-D Lorenz attractor solution shown on the RHS resembles the wings a butterfly.[16]

15 *Matlab* code is located on the book website.

16 **The butterfly effect** is attributed to a comment by Lorenz about the possibility of chaotic effects on weather patterns in Brazil made by the flapping of a butterfly's wings on the other side of the world.

Concluding Remarks About Chaos & Complexity

Illustrated in the two classic NL examples above – one of population dynamics and the other of weather dynamics – is the fact that a healthy combination of both analysis and numerical calculations can go far toward analyzing complex nonlinear dynamical system behavior. Along the way, recognition of symmetries in analysis was quite helpful and can be expected to be similarly helpful in solving other such problems. We cannot expect that such NL systems will have closed form solutions. Very few do – not in physics, not in engineering and not in biology. In this regard, qualitative analysis is important, because it helps generalize about a model – as well as possibly discovering potential complexity. Qualitative results augment numerical solutions computed for specific parameters, ICs and inputs. Numerically computed solutions and their graphs, combined with a little knowledge and computation of qualitative measures of dynamical stability, e.g. nullclines, phase portraits and bifurcation plots, can provide a great deal of information about nonlinear system dynamics.

The "butterfly effect" illustrated by the Lorentz model solutions is generally interpreted as reflective of exquisite sensitivity to initial conditions in models of chaos, rather than supporting the idea that minute changes in the weather in China – presumably caused by the flapping of butterfly wings – can be the cause of horrible weather on the other side of the world. In a cause and effect sense, this is unlikely. But it does open our eyes to the possibility of such surprises in our models and some of the systems they represent. Notions like the butterfly effect can provide guidance in understanding how unexpected dynamical responses of complex dynamical systems might be approximated and modeled, and also possibly initiated – given that dynamical models that exhibit chaotic behavior are extremely sensitive to initial conditions. This can be useful in real world modeling. Chaos, or chaotic-like behavior, has been observed in biological systems, for example, in the beating heart (Weiss et al. 1994). Even in situations where actual chaos may not be present, limits to predictability of candidate models of complex dynamical systems can be evaluated for situations where nonlinear dynamic systems are highly sensitive to background noise, for example. And stability analysis can go far toward understanding them. We'll exploit these methods in several NL dynamic system models in biochemistry and biology in subsequent chapters.

Nonlinear Modes

The field of mechanics has a developed theory of nonlinear modes associated with certain oscillatory phenomena in mechanical systems. Nonlinear modes in biological systems – alluded to, but not defined in earlier chapters – share many of the same properties, and are not recognized as such in the field. We make the connection precise here, first reviewing linear modes.

In Chapter 2, linear modes of time-invariant linear ODEs like $\dot{\boldsymbol{x}} = A\boldsymbol{x} + \boldsymbol{u}$ were defined as fundamental solutions, represented in simplified form as n linearly independent exponential components $t^k e^{\lambda_j t}$, with $k = 0, 1, \ldots, n-1$ and $j = 1, 2, \ldots, n$, with n the minimum dynamical dimension of $\dot{x} = Ax + u$. A most important feature of linear modes is that they are a **minimal set of invariants** of the dynamical system and, for any set of ICs or input(s), the modes accurately capture essential system dynamics. Importantly, λ_j can be complex as well as real.

In mechanics, modal analysis is typically concerned with vibrations in mechanical structures. When their models are linear, or NL and linearized about their equilibrium solutions – a common practice in analyzing stability properties – some λ_j are complex. This means oscillatory components in their behavior. In this context, linear modes are called **linear normal modes**, or LNMs, and the governing equations of motion can be decoupled into LNMs, e.g. by diagonalization of their system matrix (see Appendix B). In decoupled form, a linear system vibrates as if it consisted of independent oscillators governed by the (complex) modes, a property not shared with NL systems. Linear modal solutions also obey the superposition principle, i.e. free and forced oscillations are conveniently expressed as linear combinations of individual LNM motions in phase space. Importantly, modal solutions have an *invariance* property, i.e. a motion initiated on one specific LNM – *representable as a two-dimensional planar surface in phase space* – has no effect on the remaining LNMs, because they're decoupled. This property permits the modal concept to be readily extended to NL systems.

By analogy, for NL models of the form $\dot{x} = f(x, u)$, a **nonlinear normal mode** (NNM) is defined similarly, as a *two-dimensional invariant manifold in phase space*. Nonlinear systems, however, can exhibit highly complex behaviors not possible with linear systems. These include (non-separable) modal interactions, saturation, limit cycles, bifurcations, jumps, subharmonic, superharmonic and internal resonances, and chaos – as described above. For this reason, NNM analysis requires more complex mathematical handling (Kerschen et al. 2009).

Nonlinear modal analysis is about generating *minimal reduced-order models* that accurately capture the dynamics of higher-order models, with the help of NNMs. An effective and computationally efficient methodology for accomplishing this is based on the two-dimensional invariant-manifold approach, as developed in (Pierre et al. 2006), for example. Geometrically, (NL) NNMs – like LNMs – are represented by two-dimensional *nonplanar* surfaces in phase space. They are, however, tangent to the planar surfaces of their *linearized* ODE LNMs, at the *equilibrium point* of their NL ODE system.

For mechanical system applications, the nonplanar surface NNM is parameterized by a single pair of state variables (displacement and velocity), chosen as master coordinates, with the remaining variables being functionally related to the chosen pair. In this special form, the original system behaves on the manifold surface like a nonlinear single-degree-of-freedom system. Solutions beginning in the manifold remain in it for all time, which extends the invariance property of LNMs to nonlinear systems.[17]

17 More general theory and applications of NNM analysis are available in (Kerschen et al. 2009).

Nonlinear Modes in Systems Biology

Conceptually, this treatment of NL normal modes in mechanics, as two-dimensional manifolds in phase space, is directly extensible to NL systems biology models. The analogous application is reducing complex biosystem models using conservation relations, combined with dynamical approximations common in biochemical reactions. For example, enzymatic and cellular protein interaction networks (PIN) typically involve complex interactions among numerous cellular components and their models typically involve higher-order ODEs. A dominant characteristic of these systems is the presence of numerous inter-variable conservation relations, which serve to reduce system order, and numerous higher-order interactions that obey the quasi-steady state (QSSA) or total quasi-steady state (tQSSA) assumptions. In each of these latter interactions, the dominant dynamics are second-order, one each for the substrate and product, with all enzymatic reactions substituted by algebraic equations, based on the QSSA or tQSSA. On a microscopic level, each interaction obeying one of these assumptions also has two nonlinear modes, a fast one – on the "fast manifold" – and a slow one – on the "slow manifold." The fast one disappears nearly instantaneously relative to the slow one, which represents the relatively longer-term behavior of the mechanisms that dominate overall biosystem dynamics. A single mode, the slow one, strongly dominates the dynamics and represents the interaction subsystem fairly accurately.[18]

This is a general extension of the QSSAs to higher-order biological systems, and is directly analogous to model reduction of NNM analysis of invariant manifolds on the phase space of mechanical systems. This means *NNM analysis is equivalent to QSSA analysis*, which likely came first (Segel and Slemrod 1989; Shaw and Pierre 1994), with invariant manifolds of the biosystem in the substrate-product phase space of each reaction submodel.

I believe this may be the first published recognition in print that the two theories coincide and individual QSSA approximations for elementary chemical reactions are the "nonlinear modes" of NL systems characterized all or in part by these biomodels.

Recap & Looking Ahead

The several chapters preceding this one developed different dynamic biomodeling methodologies for different domains of life sciences. The main focus was on *structuring* models, with associated *quantitative* analyses (model solutions and simulations), as needed. The present chapter, about stability and oscillations in nonlinear dynamic system models, has been about *analyzing* models. It encapsulated the basics of higher level mathematical approaches to *qualitative* analysis of these models – primarily for examining the dynamical properties of more complex nonlinear biosystem models. These models and methods are especially prominent in molecular and cell systems biology. In the last section, we formally introduced *nonlinear modes* in biochemical and cell biology models – modal solutions on "fast and slow" manifolds in phase space – along with their intimate mathematical relationship to nonlinear normal modes in mechanics. Chapter 14 includes discussion of another *linear* modal concept, called *elementary flux modes*, developed for metabolic control flux analysis, using biochemical network models in (linear) stoichiometric matrix form.

18 This behavior is illustrated in the phase space figures (with nullcline "manifolds") associated with the *Slyke-Cullen reaction* feedback system model discussed in Chapter 14.

We turn our attention in the next three chapters to *model quantification*. Chapter 10, on structural identifiability, is about analytically discovering what's possible and what's not possible to quantify in a dynamic system model – step one in the quantification process. Chapter 11, on sensitivity analysis methods, is more about qualitative aspects of model quantification – primarily discovering how sensitive (robust, fragile) a model output is to variations in different model parameters. Finally, Chapters 12 and 13 are about methodologies for actually accomplishing quantification – which usually means estimating model parameter values (and their statistics) that best fit real output data, collected from real dynamic system experiments. Optimization and search are paramount; and the statistical assessment part of this process is quite important.

Exercises

E9.1 ***Poles in the Complex Plane.*** Which of the following output transforms are stable and which are unstable?

$$Y_1(s) = \frac{(s+k_2)(s-k_1)}{(s-k_3)(s+k_4)}, \quad Y_2(s) = \frac{(s-k_2)(s-k_1)}{(s+k_3)(s+k_4)}, \quad Y_3(s-k_4) = \frac{(s-k_2)(s-k_1)}{(s-k_3)(s-k_4)}, \text{ all } k_i < 0$$

E9.2 ***Lotka–Volterra Model.*** The following Lotka–Volterra equations are a *Kolmogorov-type* model of predator–prey relationships for interacting populations, e.g. hosts and parasites, yeasts and sugars, sharks and surfers.

$$\begin{aligned} \dot{x}_1 &= p_1x_1 - p_2x_1x_2 \qquad x_1(0) = x_{10} \\ \dot{x}_2 &= p_3x_2 + p_1p_4x_1x_2 \quad x_2(0) = x_{20} \end{aligned} \tag{9.11}$$

Parameters p_1 to p_4 are constant death and birth rates; x_1 is the host population, x_2 is the parasite population, and the product x_1x_2 represents the "getting-together" of the two species.

a) Find the equilibrium steady state solutions for these equations, $[x_e\ y_e]^T = \boldsymbol{x_e}$ in terms of the parameters.

b) Derive the small-perturbation equations for this NL model linearized about $\boldsymbol{x_e}$.

Analyze the local stability of the equilibrium points, their phase plane, etc.

E9.3 ***Hares vs Lynxes.*** Consider a somewhat more complex predator–prey model:

$$\begin{aligned} \dot{x}_1 &= rx_1\left(1 - \frac{x_1}{k}\right) - \frac{ax_1x_2}{c+x_1} \\ \dot{x}_2 &= b\frac{ax_1x_2}{c+x_1} - dx_2 \end{aligned} \tag{9.12}$$

This model might represent numbers of hares (x_1, prey) and lynxes (x_2, predators). Nominal parameter values are $a = 3.2$, $b = 0.6$, $c = 50$, $d = 0.56$, $k = 125$, and $r = 1.6$.

a) Find the nonzero equilibrium solutions $\boldsymbol{x}_e$ of the model in terms of the parameters.

b) Do a bifurcation analysis of this model, numerically, varying parameters a and c – keeping all others at their nominal values – and find the stable and unstable regions of the a–c plane.

c) Plot values of x_{1e} and x_{2e} versus parameter a from 0 to 8 (a bifurcation diagram), showing that the unstable equilibrium solutions converge to a stable orbit (simple harmonic motion).

E9.4 ***Saddle Node Bifurcation***. Again consider the simple mutual activation model of Example 9.3. Show by linearization that $\boldsymbol{x}_e$ is an unstable saddle node bifurcation point for all $n > 2$.

E9.5 ***Nullclines for the Lorentz Model.*** Show that the nullclines for Eq. (9.9) are: x_1-nullcline: $x_1 = x_2$; x_2-nullcline: $x_1 = 0$ or $x_3 = b - 1$; and x_3-nullcline: $x_3 = 0$ or $x_1 = \sqrt{c(b-1)}$. Plot the result in 3-D.

E9.6 ***Supercritical–Subcritical Bifurcations***. The canonical (normal) form of a Hopf bifurcation solution is: ($\dot{z} = z\lambda + b|z|^2$), with z and b both complex, λ is a parameter and $b \equiv \alpha + j\beta$. The number α is called the **first Lyapunov coefficient**.

a) Show that if $\alpha < 0$, a *stable limit cycle* exists for $\lambda > 0$ in the solution: $z(t) = \left(\sqrt{-\lambda/\alpha}\right)e^{j\beta r^2}$. This bifurcation is called **supercritical**.

b) If $\alpha > 0$, an *unstable limit cycle* exists for $\lambda < 0$, called **subcritical**.

E9.7 ***Autocatalytic Glycolysis Stability Analysis***. Assess the stability properties of the following NL glycolysis model by linearization about normalized steady state $x_{1ss} = 1$.

$$\begin{aligned} \dot{x}_1 &= -k_1 \frac{Ax_1^{k_1}}{1 + k_2 x_1^{k_2}} + (k_1 + 1)k_3 x_2 - 1 \\ \dot{x}_2 &= \frac{Ax_1^{k_1}}{1 + k_2 x_1^{k_2}} - k_3 x_2 \end{aligned} \qquad (9.13)$$

This model incorporates ATP autocatalysis and inhibitory feedback control and captures essential dynamics observed experimentally, including limit cycle oscillations. System performance is limited by the inherent autocatalytic stoichiometry and higher levels of autocatalysis exacerbate stability and performance (Chandra et al. 2011).

E9.8 ***Chaos and IC Sensitivity***. Exercise the discrete-time logistic equation for chaotic growth rate parameter $a = 3.7$ and three different ICs: $x_0 = (0.1999, 0.2, 0.2001)$. The equation is very easy to code. A *Matlab* script for computing and printing it can be found on the book website.

E9.9 ***Eigenvalues of the Jacobian Matrix***. Establishing local stability properties is facilitated by finding the roots of the characteristic equation: $\det[sI - A] = 0$, where A is the Jacobian matrix $\partial \boldsymbol{f}/\partial \boldsymbol{x}$ of the nonlinear ODEs: $\dot{\boldsymbol{x}} = \boldsymbol{f}(\boldsymbol{x})$. These roots are the eigenvalues of A.

a) For $n = 2$, show that the characteristic polynomial has the form: $s^2 + (\text{trace } A)s + (\det A)$, where trace A is the trace[19] of matrix A and det A is the determinant of matrix A.

b) Show that the eigenvalues of A can be expressed as:

$$\lambda_{1,2} = \frac{\text{trace } A}{2} \pm \sqrt{\frac{(\text{trace } A)^2}{4} - \det A} \tag{9.14}$$

Clearly, local stability of any kind requires trace $A < 0$. If both eigenvalues are also real (have no imaginary parts, because the square root term is zero), then the equilibrium point in question is a *stable constant node* in the phase plane. For the other possible forms of real and complex eigenvalues, show the following:

c) If both eigenvalues are complex, with negative trace $A < 0$, the equilibrium point is *asymptotically stable*, i.e. has a *stable focus* in the phase plane.

d) If both eigenvalues are real and one is positive, the equilibrium solution is an *unstable saddle point* in the phase plane.

e) If the eigenvalues are complex and trace $A > 0$, the equilibrium solution is an *unstable focus*, i.e. it spirals outward in the phase plane.

f) If Trace $A = 0$, the equilibrium solution is a *neutrally stable center*, i.e. it can go either way.

g) Show that (9.14) is valid for matrices A of any dimension n, i.e. linearized ODE models of any[20] dimension n.

REFERENCES

Aihara, K., Matsumoto, G., et al., 1984. Periodic and non-periodic responses of a periodically forced Hodgkin-Huxley oscillator. J. Theor. Biol. 109 (2), 249–269.

Alligood, K.T., Sauer, T., Yorke, J.A., 1997. Chaos: An Introduction to Dynamical Systems. Springer-Verlag, New York.

Ballarò, B., Reas, P.G., Riccardi, R., 2007. Mathematical models for excitable systems in biology and medicine. Riv. Biol. 100, 247–266.

Batchelor, E., Loewer, A., Lahav, G., 2009. The ups and downs of p53: understanding protein dynamics in single cells. Nat. Rev. Cancer 9, 371–377.

Batchelor, E.-L.A., Mock, C., Lahav, G., 2011. Stimulus-dependent dynamics of p53 in single cells. Mol. Syst. Biol. 10, 7.

Bergé, P., Yves, P., Vidal, C., 1984. *Order within Chaos: Towardsa Deterministic Approach to Turbulence.* John Wiley & Sons, New York.

Chandra, F.A., Buzi, G., Doyle, J.C., 2011. Glycolytic oscillations and limits on robust efficiency. Science 187–192.

19 Although not necessary, recognizing symmetries like this can sometimes be helpful in analyzing a model.

20 For $n > 2$, finding the roots of the characteristic equation (eigenvalues of $[sI-A]$) is easy to do numerically, when the elements of A are specified numerically. But finding the eigenvalues *analytically* can be difficult or impossible. Fortunately, methods exist for analytically testing the sign of the real parts of matrix eigenvalues, or – equivalently – the sign of the trace of A. The Routh–Hurwitz criterion (Gantmacher 1959) can be handy for this purpose. The root-locus method for locating poles and zeros in the complex plane also can be helpful in stability analysis. These methods (and others) are available in most books on linear systems or control theory methods, e.g. (DiStefano III et al. 1990), and in specialized software tools in widely available program packages like *Matlab* and *Mathematica.*

Ciliberto, A., Capuani, F., et al., 2007. Modeling networks of coupled enzymatic reactions using the total quasi-steady state approximation. PLoS Comput. Biol. 3 (3), e45.

Devaney, R.L., 2003. An Introduction to Chaotic Dynamical Systems. Westview Press, Boulder CO.

DiStefano III, J., 1964. The choice of state variables for the determination of the stability of nonlinear systems utilizing the second method of Liapunov. IEEE Trans. Automat. Control 9, 279–280.

DiStefano, J.J.I., Stubberud, A.R., Williams, I.J., 1990. Feedback and Control Systems. McGraw-Hill, New York.

Gantmacher, F.R., 1959. Applications of the Theory of Matrices. Interscience, New York.

Gardner, T.S., Cantor, C.R., et al., 2000. Construction of a genetic toggle switch in *Escherichia coli*. Nature. 403 (6767), 339–342.

Goldbeter, A., 1996. Biochemical Oscillations and Cellular Rhythms: The Molecular Bases of Periodic and Chaotic Behaviour. Cambridge University Press, Cambridge, MA.

Golovin, A.A., Matkowsky, B.J., et al., 2008. Turing pattern formation in the brusselator model with superdiffusion. SIAM J. Appl. Math. 69 (1), 251–272.

Hale, J., Koçak, H., 1991. Dynamics and Bifurcations. Springer-Verlag, New York.

Hirsch, M.W., Smale, S., Devaney, R.L., 2003. Differential Equations, Dynamical Systems & An Introduction to Chaos. Academic Press, Boston, MA.

Hodgkin, A., Huxley, A., 1952. A quantitative description of membrane current and its application to conduction and excitation in nerve. J. Physiol. 117, 500–544.

Keener, J., Sneyd, J., 1998. Mathematical Physiology. Springer, New York.

Kerschen, G., Peeters, M., et al., 2009. Nonlinear normal modes, Part I: a useful framework for the structural dynamicist. Mech. Syst. Signal Processing 23 (1), 170–194.

Lev Bar-Or, R., Maya, R., Segel, L.A., Alon, U., Levine, A.J., Oren, M., 2000. Generation of oscillations by the p53–Mdm2 feedback loop: a theoretical and experimental study. Proc. Natl. Acad. Sci. USA (PNAS). 97 (21), 11250–11255.

Ma, L., Wagner, J., Rice, J.J., Hu, W., Levine, A.J., Stolovitzky, G.A., 2006. A plausible model for the digital response of p53 to DNA damage. Proc. Natl. Acad. Sci. USA (PNAS). 102, 14266–14271.

May, R.M., 1976. Simple mathematical models with very complicated dynamics. Nature 261, 459–467.

Morrison, F., 1991. The Art of Modeling Dynamic Systems: Forecasting for Chaos, Randomness and Determinism. John Wiley, New York.

Pena, B., Perez-Garcıa, C., 2001. Stability of Turing patterns in the Brusselator model. Phys. Rev. E. 64, 1–9.

Pierre, C., Jiang, D., Shaw, S.W., 2006. Nonlinear normal modes and their application in structural dynamics. Math. Probl. Eng. 10847, 1–15.

Prigogine, I., 1980. From Being to Becoming: Time and Complexity in the Physical Sciences. W. H. Freeman, San Francisco.

Proctor, C.J., Gray, D.A., 2008. Explaining oscillations and variability in the p53–Mdm2 system. BMC Syst. Biol.(75), 1–20.

Santoso, L., Lanzon, A., 2006. On the modelling of a bistable genetic switch. Proceedings of the 45th IEEE Conference on Decision & Control. San Diego, CA.

Scott, S.K., 1994. Oscillations, Waves, and Chaos in Chemical Kinetics. Oxford University Press, Oxford, New York Toronto.

Segel, L.A., Slemrod, M., 1989. The quasi-steady-state assumption: a case study in perturbation. SIAM Rev. 31 (3), 446–477.

Selkov, E.E., 1968. Self-Oscillations in Glycolysis. Eur. J. Biochem. 44, 79–86.

Shaw, S., Pierre, C., 1994. Normal modes of vibration for non-linear continuous systems. J. Sound Vib. 169 (3), 319–347.

Strogatz, S.H., Stewart, I., 1993. Coupled oscillators and biological synchronization. Sci. Am. 269, 102–109.

Teschl, G., 2012. Ordinary Differential Equations and Dynamical Systems. American Mathematical Society, Providence, RI.

Turing, A.M., 1952. The chemical basis of morphogenesis. Philos. Trans. R. Soc. Lond. B. Biol. Sci. 237, The Chemical Basis of Morphogenesis.

Wang, R., Li, C., et al., 2008. Modeling and analyzing biological oscillations in molecular networks. Proc. IEEE. 96 (8), 1361–1385.

Weiss, J.N., Garfinkel, A., Spano, M.L., Ditto, W.L., 1994. Chaos and chaos control in biology. J. Clin. Invest. 93, 1355–1360.

Chapter 10

Structural Identifiability

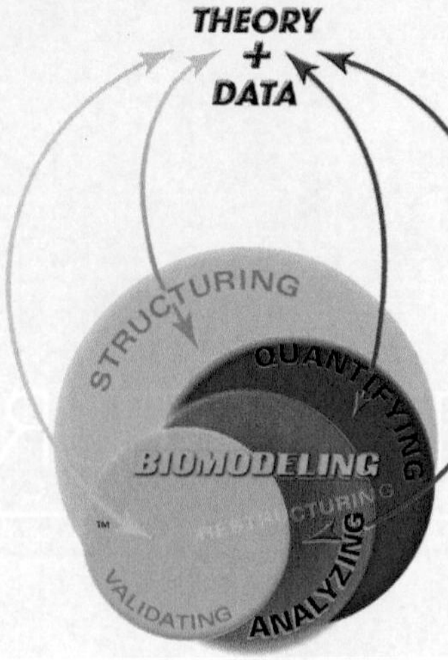

"Once you eliminate the impossible, whatever remains, no matter how improbable, must be the truth."

Arthur Conan Doyle, ~1900

INTRODUCTION

Not enough attention is paid to identifiability (ID) concepts and issues in the study and practice of modeling dynamic biosystems. Yet, ID problems can *plague* biomodelers when they reach the quantification stage of development, even for relatively simple models. Novice modelers often don't even recognize that ID properties of the model are a likely problem when they have difficulties quantifying it. On the other hand, recent recognition of the seriousness of ID problems by the systems biology and other modeling communities is motivating advances toward their solution. Medium to large dimension cell biology network models – with many parameters and few variables to measure – are major computational challenges for existing ID methodologies. And this can be said for ID problems formulated for ideal, noise-free and plentiful output data – the *structural* ID problem – as well as the *numerical* ID problem, with limited and noisy, not-so-ideal output data, as illustrated in **Fig. 10.1**.

Identifiability can be a difficult subject – a large part of the problem. Analysis of ID properties of ODE models involves a spectrum of applied mathematical subjects, including transform and algebraic methods – developed in this chapter, and intermediate to advanced algebra, geometry, probability and statistical methods in chapters that follow. The theory, especially for nonlinear (NL) ordinary differential equation (ODE) models, is deep and nuanced, and has not been easily accessible or applicable, as evidenced in recent literature (Xia and Moog 2003; Chis et al. 2011a,b; Miao et al. 2011). But this is changing, as new algorithms and software tools are becoming available that include various levels of automation and transparency for ID analyses – both structural and numerical (Bellu et al. 2007a,b; Meshkat et al. 2009; Balsa-Canto and Banga 2011; Chis et al. 2011a,b; Meshkat et al. 2014). The goal of this and the next three chapters is to make this

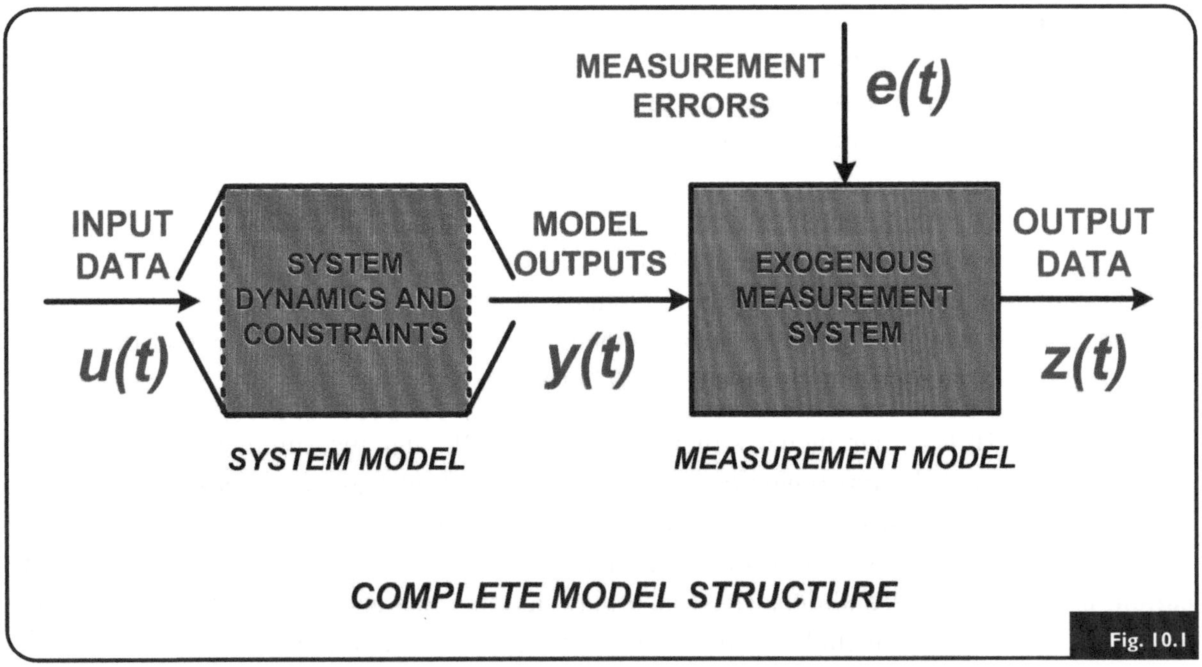

Fig. 10.1

System dynamics and measurement model structures – together forming a complete system-experiment model. "Limited-reach" input and output *ports* of the *system model* are illustrated as small openings (windows) to state variables that can be probed with inputs and measured as outputs. Model constraints are also included. Structural ID is about the model between u and y, numerical ID between u and z.

topic accessible – pedagogy with minimal mathematical detail and plenty of references to make up for the brevity. This chapter is focused only on the structural ID problem. The broader notions and methods for numerical and "practical" ID are developed in Chapters 11–13.

The subject of identifiability occupied a very small niche of the literature in the second half of the 20th century. Math modeling of dynamic systems in the life sciences was practiced on a relatively small scale – primarily in pharmaceutical and macroscale areas of biology – including systems physiology and medicine. The need was there, but the modeling community was small, and the means for producing sufficient data for model development and quantification was much more limited than it became as the century came to a close. Math modeling and awareness of associated ID problems increased substantially in the last two decades, due in large part to the explosion of genomic and other high-throughput data. These new data sources have motivated an ever increasing number of dimensionally complex, network and intracellular signal transduction pathway models. This plethora of candidate network hypotheses has, in turn, greatly motivated interest and developments in molecular systems biology *ODE* models, with a primary goal of quantitatively testing their fidelity. The typically large numbers of nodes in these networks (tens to hundreds of state variables) – with even more interconnections (hundreds of parameters) –have been and remain a formidable, if not overwhelming, challenge to existing modeling methodology. The pharmacology and bioengineering modeling communities also have intensified their interest and contributions toward the solution of model ID problems. They too are benefiting from high-throughput data, as well as awareness and wider distribution of new modeling results via online publishing. We begin exposition of these problems and solutions here and in the next two chapters.

Model/Parameter Quantification & Identifiability (*SI* & *NI*) Defined

Fig. 1.5 in Chapter 1 illustrated the input–output data-driven nature of the model quantification problem in classic system theory terms. This combined model is opened to probes in **Fig. 10.1**, to further emphasize *both* structural components – the dynamics and the input–output ports and data – for building a dynamic system biomodel. The measurement model block on the right realistically represents output data corrupted by measurement noise. The biosystem experiment characterized by this block diagram is modeled by the system ODEs relating inputs $u(t)$ to outputs $y(t)$ – circumscribing the *structural ID problem* – and the input $u(t)$ all the way through to the measurement equations relating measurement data $z(t)$ to $y(t)$ – together circumscribing the *numerical ID problem*.

With **Fig. 10.1** as our model structure, model (ODE and measurement) equations are characterized in part by **parameters:** $p_1, \ldots, p_P$ (vector $\boldsymbol{p}$), at least some unknown in value. Numerical data are needed to quantify the ones that cannot be established from biophysical, biochemical or other biosystem information.[1] This is what is usually meant by **model quantification**, although initial conditions (ICs) also may be unknown and in need of quantification. Modeling goals often include quantifying one or more of these parameters, but any

1 Prior information about parameters can include values established from earlier quantifications, or published reports – often called *literature values*.

quantitative use of the biomodel requires them, or combinations of them,[2] so the model equations can be simulated or otherwise solved uniquely.

Structurally identifiable (*SI*) addresses the question of whether quantification of parameters is possible from a given set of ideal, noise-free input–output (I–O) data ($e(t) = 0$). *A priori*[3] *SI* is thus a *necessary condition* for quantifying any model parameter p from real data. For *SI* analysis, no real data are needed, only the topology, meaning *SI* is a *qualitative* modeling concept. The partly open I–O ports in the LHS block of **Fig. 10.1** are the primary windows for ideal data collection from the biosystem and, together with the ODE model structure, they establish *SI* – the *limits of knowledge* attainable from these data ports. In Chapter 16 we apply structural identifiability methods advantageously for designing alternative experiments, with different I–O ports, to quantify parameters not otherwise quantifiable from a particular I–O experiment.

Numerical identifiability (*NI*) addresses the question of whether useful (practical) quantification of parameters is possible when the I–O data are *real*, with $\mathbf{z}(t)$ noisy. This ID property depends on *a posteriori* data, it is measured by parameter estimation statistics like covariance of $\boldsymbol{p}$ – shown in **Fig. 10.1** as $COV(\boldsymbol{p})$ – and further discussion of it is reserved for Chapter 12 and beyond.

Remark: Model structure development, followed by quantification, is presented pedagogically in this book, but no order is implied. Numerical data often comes first, suggesting some form of structure, as in Examples 5.11 and 5.12 (and others) in Chapter 5. Capturing only the *essential features* of the biosystem structure is often key to ultimate success in model quantification (completion), primarily because the number of model parameters can increase exponentially with each state variable added, i.e. with additional structural complexity: a good reason to strive for parsimony.

Structural Identifiability (*SI*) in the Large

Roughly speaking, the notion of identifiability[4] is akin to the problem of existence and uniqueness of solutions of mathematical relationships. When these relationships represent a model of some characteristics of interest in a biological or other physical system, the identifiability problem usually has meaning in the context of unknown model parameters.[5] As noted above, the question in this case is whether or not it is possible to find one or more sets of solutions for the unknown parameters of the model, from (noise-free) data collected in experiments performed on the real system. Thus the identifiability question is a critical aspect of the *parameter*

2 Parameter combinations (combos) often are sufficient for obtaining unique solutions. For example, computing $x(t)$ from $\dot{x} = (a + b)x$ with $x(0) = 1$, requires knowledge of $a + b$, not a or b separately.

3 *A priori* is redundant in the context of *SI*. It is used by many authors – probably for historical reasons – to emphasize that *SI* is not a property of real data, typically collected *a posteriori*.

4 *Controllability* and *observability* concepts are normally introduced first in studying dynamic system theory, not our purpose here. Identifiability is crucial in model building, so we focus on identifiability. *Controllability, observability, equivalence*, and other qualitative structural properties of dynamic system models are presented in Appendices D and E. As shown there, only the controllable and observable portions of a model can be manipulated and measured via input/output probes. But this does not mean that all *parameters* in the controllable and observable part can be determined from input-output data, nor does it mean that parameters in the unobservable or uncontrollable parts cannot be so determined.

5 The term identifiability also has been used in other ways by some authors, e.g. for the problem of "input identifiability" (Yoshikawa and Bhattacharyya 1975). Our attention is restricted exclusively to parameter identifiability.

estimation problem for state variable models; it is the existence and uniqueness part of this problem.[6]

The question of existence of solutions is easily answered in the affirmative if the overall problem is **well-posed**. By this we mean that the model and the real system must be compatible in some well-defined sense, at least with respect to the data employed to determine the unknown parameters. If certain output variables of the model are made to approximately coincide with their corresponding data variables in the real system, existence is normally guaranteed for other than the simplest models. On the other hand, if the model is not an appropriate representation of the system, i.e. if it is not a "good" one in the context of its application, then compatible relationships among the unknown parameters and the numerical data may not exist. Unfortunately, this aspect of the ID question cannot be so easily resolved by the above reasoning because it is often possible to forcibly, or even brutally, fit a not-so-good model to (good or bad) data, finding at least one set of unknown parameter values that yields a minimum value for, say, the square of the difference between the model outputs and the data (the least-squares output error). For example, it is always possible to fit a straight line to *any* set of data consisting of two or more points in a plane, no matter how wildly (or neatly) scattered they may be. Thus, we assume in the sequel that models under consideration for ID testing are structured well enough so that some form of mathematical compatibility with real outputs is achieved, admitting at least one possible solution to the parameter estimation problem. We begin our exposition of *SI* concepts with some historical perspective.

Historical Perspective[7]

Formal ID notions in biomodeling appeared in the literature more than a half-century ago, probably motivated by early developments in radioactive tracer technology and its application to kinetic analysis of chemistry data, pioneered by Nobel laureate George de Hevesy (1962). In a pioneering work on the problem of quantifying linear kinetic models from experimental biodata, Berman and Schoenfeld (1956) probably ushered in the notion of compartmental model ID, albeit not explicitly. Formal development of the parameter ID concept came more than a decade later.

More often than not, the number of unknown parameters is greater than the number of independent relationships among the parameters derivable from the model structure and the data. These relationships usually have the form of an underdetermined system of *nonlinear* algebraic equations. For models based wholly or in part on physical considerations, such as multicompartmental structures, some equality and/or inequality constraints on and among the parameters are also usually available. However, these additional constraints rarely provide information sufficient to yield unique solutions to the parameter quantification problem for other than the simplest models. On the other hand, when the number of unknown parameters is less than or equal to the number of such independent equations, a finite number of solutions

6 Signals emanating from real dynamic systems are likely never measured exactly; and we usually mean finding *approximations* for parameters (from "noisy" signals) when we refer to parameter *estimation* – a synonym for approximation.

7 This section briefly describes development of identifiability methodology through 2012. It serves to strengthen understanding of the often subtle nature of the problem and how it was encountered and addressed by modelers and mathematicians since the 1970s.

may exist, possibly greater than one; for models *linear* in the parameters, there is only one solution.[8] This can indeed be a problem in applications where one knows that, intrinsically, there can only be one solution, or perhaps worse, when one is not secure on this point and multiple solutions are feasible mathematically, but indistinguishable experimentally.

These aspects of the parameter estimation problem, or parameter *identification* problem, have been treated by numerous authors for multicompartmental models since the mid-1950s. Skinner et al. (1959) were the first to report two feasible solutions for each of two different and open three-compartment structures. They also derived inequality constraints among the parameters of an unidentifiable three-compartment structure. A lengthy treatise on the general subject was reported by Rubinow and Winzer (1973) and later extended by Rubinow (1975). These authors systematically formulated the question of the extent to which the multiplicity of matrices for n-compartment structures compatible with given data can be reduced. They derived minimal matrices (those with the smallest number of parameters) for certain special cases, including specific two-, three-, and four-compartment models and closed or "almost closed" mammillary or catenary structures, some of which were shown to have more than one but less than n solutions.

The question of parameter uniqueness for a more general class of both linear and nonlinear models of dynamic systems evolved in the 1960s into the ID concept in the system theory and applications literature. The notion of *structural ID*, introduced by Bellman and Åström (1970), served to formalize, into a system theoretic framework, the problem of knowing whether it is possible to determine model parameter estimates uniquely from experimental data. They also analytically explored the ID properties of several different linear time-invariant model structures and a few specific classes of compartment models with different input/output configurations. Hájeck (1972) derived the maximum number of determinable rate constants of a multicompartmental model for the case of one test input and one measured compartment, then applied his result to two different three-compartment structures. Milanese and Molino (1975) discussed the utility of the controllability and observability, as well as the structural ID, concepts in interpreting drug kinetics and processing experimental data. These authors applied the above concepts to a six-compartmental model of the hepatic kinetics of the drug bromsulphalein and analyzed four different combinations of input/output measurement configurations to determine which model parameters could be determined uniquely from each experiment. Cobelli and coworkers (Cobelli et al. 1975) applied the structural ID concept to a simple 2-compartment model of bilirubin kinetics, and in two companion papers, Cobelli and Romanin-Jacur (1975, 1976), like Milanese and Molino, applied the controllability, observability, and structural ID concepts to specific classes of multicompartmental models and several simple examples. The relationship between the ID concept and that of controllability and observability is clarified later, when we show that consideration of the latter two structural properties usually is unnecessary in structural ID analysis.

Irvine (1975) algebraically solved the parameter uniqueness problem for a four-compartment model of the metabolism of the prohormone thyroxine (T_4), with one input and three outputs measured. My students and I analyzed a six-compartment model of the combined

8 Appendix B includes a brief section on solving linear equations $y = Ax$.

dynamics of thyroid hormone triiodothyronine (T_3) and T_4 metabolism (DiStefano III et al. 1975), with two inputs and two outputs, and showed it to be only partially identifiable. We also applied the concept directly to the design of tracer experiments for identification of a general class of nonlinear physiological systems, devised a necessary condition for their ID, and presented a formal procedure for kinetic experiment design (DiStefano III 1976). Grewal and Glover (1976) developed a simple ID test for linear time-invariant models in general and derived a sufficient condition for the ID of the same class of nonlinear models that we had considered. We also further developed a formal procedure for tracer experiment design, based on analysis of an *experimental transfer function matrix*, and applied it to identification of the parameters of thyroid hormone metabolism (DiStefano III and Mori 1977). A generalization of these results is included in Chapter 16. Cobelli and I critically reviewed parameter ID concepts and some of the curious ambiguities they often generate (Cobelli and DiStefano III 1980; DiStefano III and Cobelli 1980), developed in some detail in this and subsequent chapters.

In references (DiStefano III 1983a,b) I introduced the notions of *interval ID* and *quasiidentifiability* of parameters, thereby incorporating prior inequality constraint sets on the parameters into the overall ID concept, which provides a means for obtaining finite *ranges* for otherwise unidentifiable parameters of a model. Together with several of my students and faculty collaborators, we subsequently published several extensions of these new concepts to various multicompartmental model configurations, for both noise-free and noisy data cases (Godfrey and DiStefano III 1985; Vicini et al. 2000). These are also presented here and in Chapter 13.

Structural ID analysis for linear ODE models was pretty much resolved by the beginning of the 21st century (Saccomani et al. 2001, 2003; Cobelli et al. 2006), with the exception of computational problems dealing with higher dimensional linear ODE models and sorting out uniqueness issues. For nonlinear (NL) ODE models, ID methods remain under development. Notable methods for addressing this much more difficult problem include the Taylor series (Pohjanpalo 1978; Pohjanpalo and Wahlstrom 1982), generating series (Walter and LeCourtier 1982; Chapman et al. 2003), similarity function (Vajda et al. 1989), tableau (Balsa-Canto et al. 2010), chemical reaction network theory (Davidescu 2008) and differential algebra approaches (Ljung and Glad 1994; Audoly et al. 2001; Saccomani et al. 2003; Margaria et al. 2004; Bellu et al. 2007; Meshkat et al. 2009, 2011, 2012). Application of any can be tedious computationally and symbolic manipulation packages (SMPs) are normally used to facilitate the algebra. Recent advances in this regard by Meshkat and coworkers (Meshkat et al. 2009, 2011, 2012, 2014) are presented in a later section. The Taylor series and differential algebra approaches for NL models are presented in some detail below. A fairly comprehensive review of most methods available through 2011, with analysis and systems biology examples, can be found in Chis et al. (2011a,b).

Basic Concepts

We show first that the overall problem of structural ID (*SI*) is inherently algebraic, for both nonlinear and linear models. We develop the analysis for *linear* dynamic biosystem models here first and introduce nonlinear ID analysis later, each fortified by illustrative examples. As noted earlier, there is no need to consider noise in data, or statistical models of such, because structural ID does not depend on *a posteriori* data, but only on known model structure. It yields *minimal necessary conditions* for obtaining unique estimates from real, limited, noisy data.

Before presenting formal definitions, we explore ID properties of some simple, linear models, thereby illustrating *SI* notions and some of the problems alluded to above.

Example 10.1: ***ODE Model Identifiability at its Simplest***

The following simple first-order model is adapted from Bellman and Åström (1970):

$$\dot{x}(t) = p_1 x(t) + p_2 u(t) \tag{10.1}$$
$$y(t) = p_3 x(t) \tag{10.2}$$

We invent a biological system represented by this model here to enhance the pedagogical result. The state variable $x(t)$ might represent the mass of a drug introduced intravenously (IV) into blood, $u(t)$ its waveform (e.g. a bolus) in mass/time units, and $y(t)$ measurement over time of the drug concentration in blood, say by bioassay.[9] The simplicity of the biomodel here implies that drug distribution in the whole body "appears" uniform (homogeneous), dynamically speaking, at least as far as can be seen via observations in blood.

The model has three unknown parameters: p_1, p_2 and p_3. p_1 is the fractional rate of elimination of the drug from blood; p_2 is another rate constant (t^{-1} units), approximating a reduced unknown fraction entering blood from the intraperitoneal (ip) cavity following direct (ip) injection. The ip injectate is assumed to be taken up in blood vary rapidly, approximating an IV injection, but with losses; and p_3 is the inverse of the drug distribution volume and/or the unknown proportionality constant relating the bioassay result to the true drug concentration in blood, which cannot be measured directly. For any known input u, the explicit solution of Eqs. (10.1) and (10.2) is:

$$y(t) = p_2 p_3 \int_0^t e^{p_1(t-\tau)} u(\tau) d\tau \tag{10.3}$$

If the drug is introduced rapidly as a very brief pulse of unit magnitude, i.e. an approximate impulse $u(t) = \delta(t)$, one obtains the now familiar solution:

$$y(t) = p_2 p_3 e^{p_1 t} \tag{10.4}$$

A semi-logarithmic plot of the data $y(t)$ is a straight line, yielding the coefficient $A \equiv p_2 p_3$ from the intercept and the exponent $\lambda \equiv p_1$ from the slope. Thus, only p_1 and the product $p_2 p_3$ can be determined, and not p_2 or p_3 individually. This means the model is *unidentifiable*. This result is also clear from Eq. (10.3) for any known $u(t)$.

Important: If p_2 or p_3 are known or related in a known way in this example, i.e. if there is additional information about p_2 or p_3, all parameters can be uniquely determined from $y(t)$, and

9 A bioassay is an indirect measure of an unknown quantity, depending on an observable measure of a biological effect of the unknown.

we say the model (or model parameters) is (are) *uniquely identifiable*. This means that such prior information about the parameters must be included in formulating the problem, or the wrong answer to ID questions might result.

Complete Biosystem Model: Constraints as Well as Inputs & Outputs Included

Keeping in mind that independent information about $\boldsymbol{p}$ is part of the problem statement and therefore the model, the result in Example 10.1 generalizes quite easily for linear time-invariant models, but the complete model must be considered, i.e. the dynamics (state equations), outputs, and all *model constraints*, generalized here as $\boldsymbol{h}(\boldsymbol{p}) \geq \boldsymbol{0}$, because constraints can also be inequalities (e.g. nonnegative parameter constraints):

$$\dot{\boldsymbol{x}}(t,\boldsymbol{p}) = A(\boldsymbol{p})\boldsymbol{x} + B(\boldsymbol{p})\boldsymbol{u}(t) \tag{10.5}$$

$$\boldsymbol{y}(t,\boldsymbol{p}) = C(\boldsymbol{p})\boldsymbol{x}(t,\boldsymbol{p}) \tag{10.6}$$

$$\boldsymbol{h}(\boldsymbol{p}) \geq \boldsymbol{0} \tag{10.7}$$

All prior parameter information is embedded in (10.7).

Transfer Functions & *SI*

The transfer function matrix for the model in (10.5) to (10.7) is:

$$H(t,\boldsymbol{p}) \equiv [H_{ij}(s,\boldsymbol{p})] = C(\boldsymbol{p})[sI - A(\boldsymbol{p})]^{-1}B(\boldsymbol{p}) \tag{10.8}$$

with relations (10.7) implicitly included. We recall that each scalar function $H_{ij}(s, \boldsymbol{p})$ represents the Laplace transform of the response in the measurement variable at "port" i, $y_i(t, \boldsymbol{p})$, to a unit impulse test-input at "port" j, $u_j(t) = \delta(t)$. Thus, each element of H, i.e. $H_{ij}(s, \boldsymbol{p})$, reflects an *experiment* performed on the system, between input port j and output port i. The ID properties of such an experimental setup can be therefore determined by analyzing $H_{ij}(s, \boldsymbol{p})$, together with all information available about $\boldsymbol{p}$, i.e. including the constraint relations (10.7) (DiStefano III and Mori 1977).

Preliminary Definitions

In these terms, a parameter p_k is **uniquely identifiable** from *experiment* H_{ij} if it can be evaluated uniquely from the equations for H_{ij}, (10.8), constrained by (10.7). It is **nonuniquely identifiable** from H_{ij}, constrained by (10.7), if it has more than one *distinct* solution. And it is **unidentifiable** from H_{ij} if it has an *uncountably infinite* number of solutions. These definitions are extended below for subsets of the matrix H and subsets of the parameter vector $\boldsymbol{p}$, always constrained by available knowledge about $\boldsymbol{p}$, i.e. relations (10.7). First, we present several examples illustrating other properties and specific problems.

Structural Invariants

A key property of the transfer function (TF) is that *all coefficients* of the numerator and denominator polynomials of transfer function matrix elements $H_{ij}(s, \boldsymbol{p})$ *must be uniquely identifiable*, because H_{ij} cannot be written more simply, a consequence of the fundamental theorem of algebra (Boas 1935). There are at most $2n$ αs and βs, and these coefficients are called **structural invariants** of the model (Vajda 1981). However, they are not the only structural invariants. The following argument demonstrates this. The unit-impulse responses of many linear multicompartmental models have the form

$$\boldsymbol{h}(t,\boldsymbol{p}) = \sum_{k=1}^{n} A_k(\boldsymbol{p})e^{\lambda_k(\boldsymbol{p})t} \tag{10.9}$$

where all the eigenvalues λ_k are distinct (see Chapter 2 and Appendix B).[10] We have made the parameters $\boldsymbol{p}$, as well as time t, explicit in the argument list to emphasize that $\boldsymbol{p}$ is the focus of attention. Now, assuming the model fits the output data perfectly, this yields a unique set of n coefficients A_k and n exponents λ_k. The $2n$ relationships among the A_k, λ_k and p_i then determine the extent to which the p_i are identifiable (or not). It remains to show the equivalence of this approach to that of analyzing the TFs, $H_{ij}(s, \boldsymbol{p})$. For $n = 2$, the Laplace transform of Eq. (10.9) (TF) is:

$$H(s,\boldsymbol{p}) = \frac{A_1}{s-\lambda_1} + \frac{A_2}{s-\lambda_2} = \frac{(A_1+A_2)s - A_1\lambda_2 - A_2\lambda_1}{s^2(\lambda_1+\lambda_2)s + \lambda_1\lambda_2} \equiv \frac{\beta_2 s + \beta_1}{s^2 + \alpha_2 s + \alpha_1} \tag{10.10}$$

Thus, the four parameters A_1, A_2, λ_1, λ_2 map uniquely into the four TF parameters α_1, α_2, β_1, β_2 in Eq. (10.10).[11]

Normally, there is no additional effort involved employing the TF directly. One minor advantage is that some λ_k can be complex numbers, whereas all $2n$ coefficients in the numerator and denominator polynomials of H are real. If there are no common factors in these numerator and denominator polynomials, it is clear that all α and β coefficients of the TF (e.g. Eq. (10.10) for $n = 2$) can be determined from the experiment (data) represented by H, just as all coefficients A and exponents λ of the sum of exponentials solution, Eq. (10.9), can be determined from $y(t)$ data. This may sound easy, or at least straightforward, but unfortunately it's not for anything but the simplest problems, as we see below.

Example 10.2: ***Some Two-Compartment Models***

See **Fig. 10.2**.

$$\dot{q}_1 = -(k_{01} + k_{21})q_1 + k_{12}q_2 + \delta(t) \tag{10.11}$$
$$\dot{q}_2 = k_{21}q_1 - (k_{02} + k_{12})q_2 \tag{10.12}$$
$$y = \frac{q_1}{V_1} \tag{10.13}$$

We assume $k_{ij} > 0$ (a *constraint relation*!), but there are no known relationships among the k_{ij}, for $i \neq j$. This model has five unknown parameters: k_{01}, k_{02}, k_{12}, k_{21}

10 The development and conclusions here are the same even if all λ_k are not distinct, or if there are modes of the form $t^i e^{\lambda_k t}$ in Eq. (10.9).

11 In general, one or more αs or βs may be zero. In such cases, the corresponding combinations of As and λs in equations like (10.10) will be zero and exponential solution parameters will be dependent.

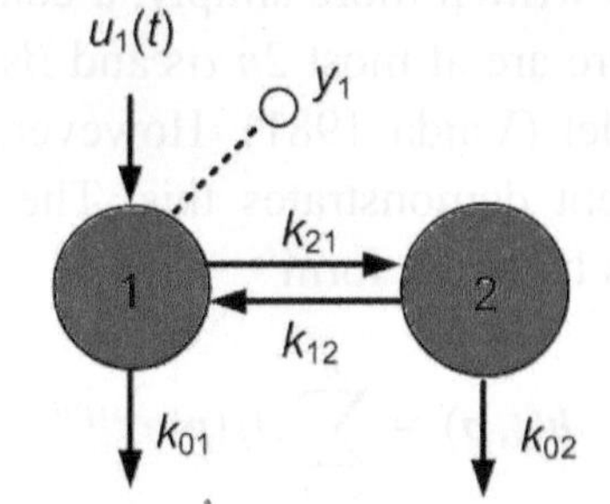

Fig. 10.2

and V_1 (V_2 does not appear in the equations). Eqs. (10.11) to (10.13) have been used to represent the dynamics of very many substances, including distribution and turnover of labeled cholesterol in humans (Nestel et al. 1969), with impulsive injection and concentration measurements of tracer in compartment 1. We assume first that all parameters are nonzero. The TF for this model, $H_{11}(s) = Y(s)/U(s)$, was evaluated in Chapter 4 as:

$$H_{11}(s) = \frac{(s + k_{02} + k_{12})/V_1}{s^2 + (k_{01} + k_{02} + k_{12} + k_{21})s + k_{01}k_{02} + k_{01}k_{12} + k_{02}k_{21}} = \frac{\beta_2 s + \beta_1}{s^2 + \alpha_2 s + \alpha_1} \tag{10.14}$$

where we have defined:

$$\beta_1 = \frac{k_{02} + k_{12}}{V_1} \tag{10.15}$$

$$\beta_2 = \frac{1}{V_1} \tag{10.16}$$

$$\alpha_1 = k_{01}k_{02} + k_{01}k_{12} + k_{02}k_{21} = k_{01}\frac{\beta_1}{\beta_2} + k_{02}k_{21} \tag{10.17}$$

$$\alpha_2 = k_{01} + k_{02} + k_{12} + k_{21} = k_{01} + k_{21} + \frac{\beta_1}{\beta_2} \tag{10.18}$$

We know that all α and β coefficients of the numerator and denominator polynomials of $H_{11}(s)$ can be evaluated from $y(t)$, given by Eq. (10.13), over any time interval. Thus, in principle, we can obtain the four transfer function parameters α_1, α_2, β_1 and β_2 from input/output data. But there are five unknowns in Eqs. (10.15) to (10.18) that cannot be found from only these equations, and hence the model is unidentifiable. Only V_1 can be evaluated uniquely, from Eq. (10.16). Furthermore, even if V_1 is known *a priori*, reducing the number of unknowns by one, still none of the k_{ij} would be identifiable, because there also would be one less equation (try to solve for any k_{ij}!).

Now assume that $k_{01} \equiv 0$ (a *different* two-compartment model). This is a very commonly used kinetic model, e.g. for the disposition of drugs (Cobelli et al. 1975) or hormones (Specker et al. 1984). In this case, there are four unknowns and a *unique* solution is available from Eqs. (10.15) to (10.18), i.e. the model is *uniquely ID*, because:

$$V_1 = \frac{1}{\beta_2} \tag{10.19}$$

$$k_{02} = \frac{\alpha_1}{k_{21}} \tag{10.20}$$

$$k_{12} = \alpha_2 - k_{02} - k_{21} \tag{10.21}$$

$$k_{21} = \alpha_2 - \frac{\beta_1}{\beta_2} \tag{10.22}$$

Similarly, if only $k_{02} \equiv 0$, we obtain a (different) model, which is used even more often to describe drug kinetics (Wagner 1976; Wagner et al. 1981). We again have a unique solution for the remaining four parameters:

$$V_1 = \frac{1}{\beta_2} \tag{10.23}$$

$$k_{01} = \alpha_1 \frac{\beta_2}{\beta_1} \tag{10.24}$$

$$k_{12} = \frac{\beta_1}{\beta_2} \tag{10.25}$$

$$k_{21} = \alpha_2 - \frac{\beta_1}{\beta_2} - k_{01} \tag{10.26}$$

A binary nature for the ID question has been illustrated in Example 10.2. The model is either uniquely identifiable or it is not identifiable at all, depending on *a priori* knowledge of certain parameters. Unfortunately, ID of all parameters does not generally imply a *unique* solution set, as illustrated in the next example.

Example 10.3: ***Some Three-Compartment Models***

The model in **Fig. 10.3** was developed and partially analyzed in Chapter 4, Example 4.11. Leak parameter k_{01} equals zero here. We explore its ID properties.

$$\dot{q}_1 = -(k_{21} + k_{31})q_1 + k_{12}q_2 + k_{13}q_3 + \delta(t) \tag{10.27}$$

$$\dot{q}_2 = k_{21}q_1 - (k_{02} + k_{12})q_2 \tag{10.28}$$

$$\dot{q}_3 = k_{31}q_1 - (k_{03} + k_{13})q_3 \tag{10.29}$$

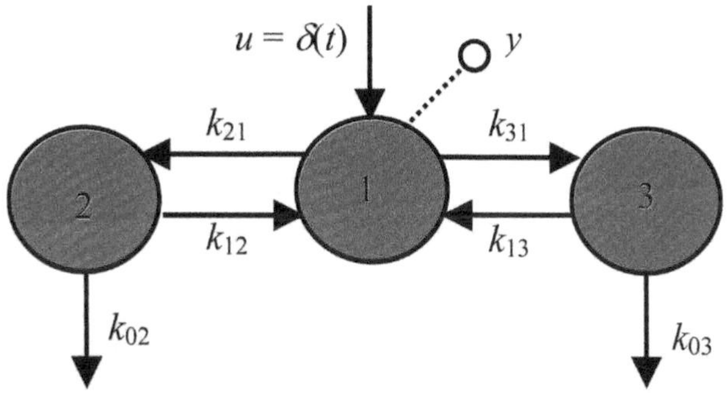

Fig. 10.3

$$y = \frac{q_1}{V_1} \tag{10.30}$$

This model has seven unknown parameters: k_{02}, k_{03}, k_{12}, k_{13}, k_{21}, k_{31} and V_1, all assumed positive (the *constraint relations*!). Note that V_2 and V_3 do not appear in the equations. This model has been used, for example, to represent the intra- and extravascular kinetics of thyroid hormone in response to an intravenous tracer dose (impulse $\delta(t)$) of hormone in the euthyroid human, rat, and sheep (DiStefano III et al. 1975). The measured output $y(t)$ is the concentration of tracer hormone in blood plasma, compartment 1.

The TF for this model, $H_{11}(s) \equiv Y(s)/U(s)$, is again easily evaluated from the Laplace transforms of Eqs. (10.27) to (10.30). To simplify the algebra, we first define:

$$k_{11} = -(k_{21} + k_{31}) \equiv -\text{turnover rate of compartment 1} \tag{10.31}$$

$$k_{22} = -(k_{02} + k_{12}) \equiv -\text{turnover rate of compartment 2} \tag{10.32}$$

$$k_{33} = -(k_{03} + k_{13}) \equiv -\text{turnover rate of compartment 3} \tag{10.33}$$

$$\beta_1 = \frac{k_{22}k_{33}}{V_1} \tag{10.34}$$

$$\beta_2 = \frac{-(k_{22} + k_{33})}{V_1} \tag{10.35}$$

$$\beta_3 = \frac{1}{V_1} \tag{10.36}$$

$$\alpha_1 = k_{12}k_{21}k_{33} + k_{13}k_{22}k_{31} - k_{11}k_{22}k_{33} \tag{10.37}$$

$$\alpha_2 = k_{11}k_{22} + k_{11}k_{33} + k_{22}k_{33} - k_{12}k_{21} - k_{13}k_{31} \tag{10.38}$$

$$\alpha_3 = -(k_{11} + k_{22} + k_{33}) \tag{10.39}$$

Then

$$H_{11}(s) = \frac{\beta_3 s^2 + \beta_2 s + \beta_1}{s^3 + \alpha_3 s^2 + \alpha_2 s + \alpha_1} \tag{10.40}$$

We remark that the sum of k_{11}, k_{22} and k_{33} above is the sum of the exponents (eigenvalues) of the sum of exponentials solution in the time domain, Eq. (10.30), while minus their sum $(-\alpha_3)$ is the sum of all six intrinsic k_{ij} in the model, $i \neq j$. These results are generalized in Chapter 16.

For ID analysis, we again recall that all six α and β coefficients of the numerator and denominator polynomials of $H_{11}(s)$ can be evaluated uniquely from $y(t)$ given by Eq. (10.30), over any time interval, i.e. these TF coefficient parameter combinations are uniquely identifiable. Therefore, we have six equations, (10.34) to (10.39), in the seven unknown parameters and consequently the model, Eqs. (10.27) to (10.30), is *unidentifiable*. As in Example 10.2, the volume V_1 is uniquely identifiable, in this case directly from Eq. (10.36), but none of the six k_{ij} can be determined uniquely from the remaining five equations.

It is again interesting to note that *none* of the k_{ij} would be identifiable, even if V_1 were known *a priori*, or equivalently if the observation variable in Eq. (10.30) were the mass in compartment 1, $q_1(t)$, instead of the concentration $q_1(t)/V_1$. In this case

Eq. (10.36) would be eliminated and V_1 would simply not appear in Eqs. (10.34) and (10.35), leaving five equations with six unknowns.

The final variation of this three-compartment model illustrates a more important result. If $k_{03} \equiv 0$, we have a different model. This one has been used, e.g. to represent bilirubin kinetics (Berk et al. 1969). In this case we obtain six independent equations, (10.34) to (10.39), with six unknowns. This model *is* identifiable, but not uniquely. There are *two* solutions, and both are physiologically feasible if no additional constraints are given *a priori*. This result was first reported in (Skinner et al. 1959) for the same model, in terms of the coefficients and exponents of a three exponential sum solution. The two sets of solutions for the unknown parameters can be determined from the following algorithm, obtained by successive substitution into Eqs. (10.31) to (10.39), with $k_{03} \equiv 0$, i.e.

$$V_1 = \frac{1}{\beta_3} \tag{10.41}$$

$$^*k_{22} = \frac{-\beta_2 V_1 \pm \sqrt{\beta_2^2 V_1^2 - 4\beta_1 V_1}}{2} \tag{10.42}$$

$$^*k_{13} = k_{22} + \beta_2 V_1 \tag{10.43}$$

$$k_{11} = -\alpha_3 + \beta_2 V_1 \tag{10.44}$$

$$\gamma = k_{13} + k_{22} \tag{10.45}$$

$$\Delta = \alpha_1 + k_{11}\beta_1 V_1 \tag{10.46}$$

$$\mu = (\beta_1 + k_{11}\beta_2)V_1 - \alpha_2 \tag{10.47}$$

$$^*k_{31} = \frac{\mu k_{13} + \Delta}{\gamma k_{13}} \tag{10.48}$$

$$^*k_{21} = -(k_{11} + k_{31}) \tag{10.49}$$

$$^*k_{12} = \frac{\mu k_{22} - \Delta}{\gamma k_{21}} \tag{10.50}$$

$$^*k_{02} = -(k_{12} + k_{22}) \tag{10.51}$$

The equations marked with an asterisk each have two solutions, because they all involve k_{22}, which itself has two. Additional information about the model or its parameters is needed to distinguish between these two solutions. Similar results (two solutions) are obtained if only $k_{02} \equiv 0$.

Example 10.4: ***A Simple Glucose–Insulin Regulation Model***

As an example of equations that are not compartmental, we reconsider the classic Bolie model of blood glucose control in mammals (Bolie 1961), developed in

Example 5.14 of Chapter 5. The model is based first on simple mass balance considerations, but also includes hypothesized control signal relationships between glucose and insulin, e.g. insulin secretion controlled by plasma glucose. The model is described by:

$$\dot{q}_1 = p_1 q_1 - p_2 q_2 + \delta(t) \tag{10.52}$$

$$\dot{q}_2 = p_3 q_2 + p_4 q_1 \tag{10.53}$$

$$q_1(0) \equiv q_2(0) \equiv 0 \tag{10.54}$$

where q_1 and q_2 are the deviations of blood plasma glucose and insulin masses, respectively, from their fasting levels, $\delta(t)$ is the impulsive IV injection of glucose, and positive p_1, p_2, p_3 and p_4 (time^{-1} units) are parameters that describe distribution and interactions of glucose and insulin. If plasma glucose concentration is the only measurable variable, the output equation is given by:

$$y_1 = \frac{q_1}{V_p} \tag{10.55}$$

where V_p is the plasma volume. The TF $H(s)$ is then easily calculated:

$$H_{11}(s) = \frac{Y_1(s)}{U(s)} = \frac{(s - p_3)/V_p}{s^2 - (p_1 + p_3)s + p_1 p_3 + p_2 p_4} \tag{10.56}$$

This model is *unidentifiable*, as there are five unknowns (V_p, p_1, p_2, p_3, p_4) and only four independent equations among them. The set of identifiable parameter *combinations* is: V_p, p_1, p_3 and the product $p_2 p_4$. If the plasma concentration of insulin is also measured, i.e.

$$y_1 = \frac{q_1}{V_p} \tag{10.57}$$

$$y_2 = \frac{q_2}{V_p} \tag{10.58}$$

then a second TF is available:

$$H_{21}(s) = \frac{Y_2(s)}{U(s)} = \frac{p_4/V_p}{s^2 - (p_1 + p_3)s + p_1 p_3 + p_2 p_4} \tag{10.59}$$

From both $H_{11}(s)$ and $H_{12}(s)$, *unique ID* is achieved, i.e. V_p, p_1, p_2, p_3 and p_4 can all be uniquely estimated from both plasma glucose and insulin data.

There's a caveat in this example result. In the real world, a "good" fit (qualitative or quantitative) of the Bolie model to *both* real glucose and real insulin data is difficult to achieve for most data. The dynamics of insulin are not adequately described by Eqs. (10.52) and (10.53). This is easy to demonstrate by simulation of both equations and comparisons of the simulated responses with real data. Any reasonable fit to glucose data, for example, generates an insulin dynamic characteristic inconsistent with insulin data generated (for example) in an IV glucose tolerance test (Hagander et al. 1978; Ider and Bergman 1978). In the context of the discussion on model–data compatibility at the beginning of this chapter, this parameter identification problem is *not well posed*. On the other hand, it was not the original goal of the Bolie model to fit both state variables, a point worth emphasizing.

We now examine four example models illustrating several additional properties and nuances of ID issues, using a more general, but still linear, constant matrix model:

$$\dot{\boldsymbol{x}} = A\boldsymbol{x} + B\boldsymbol{u}$$

$$\boldsymbol{x}(0) = \boldsymbol{0}$$

$$\boldsymbol{y} = C\boldsymbol{x}$$

$$A = \text{unknown}, \quad n = 2$$

This model is time-invariant, but otherwise completely general and not necessarily compartmental, with arbitrary A, B and C (no constraint relations).

Suppose we have only one input and one output (SISO), namely u_1 and y_1. Then, since $n = 2$, $B = \begin{bmatrix} 1 & 0 \end{bmatrix}^T = C^T$ and:

$$H_{11} = \frac{s - a_{22}}{s^2 - (a_{11} + a_{22})s + a_{11}a_{22} - a_{12}a_{21}}$$

By inspection, a_{22}, a_{11} and the product $a_{12}a_{21}$ are uniquely identifiable, and a_{12} and a_{21} remain unidentifiable. If, in addition, we measure y_2, then $C = \begin{bmatrix} 1 & 1 \end{bmatrix}$ and we have the additional TF:

$$H_{21} = \frac{a_{21}}{s^2 - (a_{11} + a_{22})s + a_{11}a_{22} - a_{12}a_{21}}$$

The numerator then gives a_{21} uniquely, and a_{12} can then be computed from $a_{12}a_{21}$. So, we see that adding this particular additional measurement gives a fully uniquely identifiable model.

Ambiguities

Now suppose we measure y_2, as above, but we have the additional information that $a_{12} = 0$ (a constraint). Then:

$$H_{21} = \frac{a_{21}}{(s - a_{11})(s - a_{22})}$$

and, again, a_{21} is obviously uniquely identifiable. But, if this were the only experiment done, i.e. if H_{11} were not available, then a_{11} and a_{22} would still be available, from the denominator above. But, it is not clear that these two parameters would be *indistinguishable*, due to the symmetry of H_{21}. There are thus *two solutions* (nonunique ID) for a_{11} and a_{22} from H_{21}. Identifiably clearly depends on parameter constraints, and not necessarily in an obvious way.

Example 10.5: ***Ambiguous Linear Model***

$$\dot{q}_1 = -aq_1 + abu$$
$$\dot{q}_2 = q_1 - nq_2$$
$$q_1(0) = q_2(0) = 0$$
$$y = q_2$$

Let's see if we can find the parameters a, b, and n, and also $q_1(t)$, which is unmeasurable, from $u(t)$ and $y(t)$ data. Solving for $q_1(t)$, we readily get:

$$q_1(t) = ab\int_0^t e^{-a(t-\tau)}u(\tau)d\tau$$

Alternatively, the input-output TF is:

$$H(s) = \frac{ab}{(s+a)(s+n)}$$

From either form we can thus find the product ab uniquely, but a and n have two solutions from the denominator, as in Example 10.3. Therefore, $q_1(t)$ also has two solutions.

Multiple Inputs

Godfrey and DiStefano III (1987) noted that separate analysis of individual transfer functions of multioutput models works well for ID analysis when the model has only a single input. But, when more than one input is applied simultaneously, it's important to examine *output transforms*, rather than individual TFs for each input–output pair. The potential problem is illustrated in the next example.

Example 10.6: ***Ličko-Silvers Model of the IV Glucose Tolerance Test***

In brief, this is a linear model (**Fig. 10.4**) approximating the effects of venous glucose concentration $u(t)$ on pancreatic insulin secretion, and the peripheral distribution and metabolism of insulin (Ličko and Silvers 1975). Note that this model is nearly the

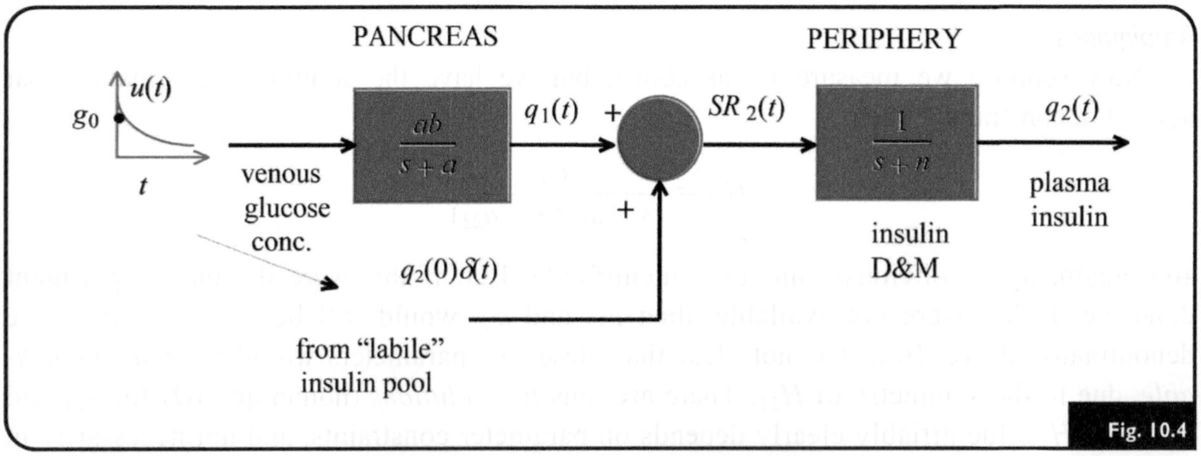

Fig. 10.4

The Ličko–Silvers model of the IV glucose tolerance test.

same as the Bolie model in Example 10.4, with one term added to the second equation due to the initial condition (IC) on the "labile" insulin pool. As we usually do, we interpret the IC as a second, impulsive input, so the model effectively has two inputs and, in this illustrative example, one output $y_2(t)$. The equations are:

$$\begin{aligned} \dot{q}_1 &= -aq_1 + abu(t) \\ \dot{q}_2 &= q_1 - nq_2 + q_2(0)\delta(t) \\ y_2 &= q_2/V \\ q_2(0) &\neq 0 \end{aligned}$$

Taking Laplace transforms for the first equation gives: $H_{11} = \frac{ab}{V(s+a)}$, and for both equations together, the TF for insulin output y_2 with glucose input u is:

$$H_{21} = \frac{ab/V}{(s+a)(s+n)} = \frac{ab/V}{s^2 + (a+n)s + an}$$

Let's "blindly" analyze the two TFs together for identifiability (ID). From H_{11}, a and b/V appear to be uniquely identifiable and then from H_{21}, n additionally appears to be uniquely identifiable. Appearances, however, can be especially deceiving in ID analysis. We made two mistakes here.

First of all, only insulin y_2 is measured in this example, not glucose. So H_{11} is not available. Neither is H_{22}, because only glucose is input exogenously. (This mistake, due to carelessness, is easy to make – the reason it's noted here.) From H_{21} alone, the denominator gives two solutions each for a and n, so the model would be identifiable nonuniquely (go ahead, try to solve for a and n uniquely!). Second, we have so far ignored effects of IC: $q_2(0) \neq 0$. The actual *output transform* for *any* input $u(t)$ is:

$$Y_2(s) = \frac{sq_2(0) + a[bU(s) + q_2(0)]}{V(s+a)(s+n)}$$

If the input $u(t)$ is an impulse at time zero, i.e. $U(s) = 1$, the model is (still) *not* uniquely ID. From the numerator, combinations $q_2(0)/V$ and ab/V are ID, but again, the denominator gives two solutions each for a and n. Therefore, a, n, and b/V each have two solutions and thus *none are uniquely identifiable*. Note that this result does not depend on the value of $q_2(0) \neq 0$.

There's more to learn from this example. If $U(s)$ is *not* an impulse at time zero (in fact anything else will do, including an impulse applied at a later time), then a, n, b/V and $q_2(0)/V$ *are* uniquely ID. Analysis for constant step and exponential inputs is left as an exercise (E10.8). Nevertheless, the most important issue exemplified here is that analysis of *output transforms*, rather than individual TFs separately, is needed to establish ID in multi-input systems.

Reminders: Don't forget to include constraint relationships (10.7), i.e. $\boldsymbol{h}(\boldsymbol{p}) \geq \boldsymbol{0}$, in all ID determinations. And consider analyzing output transforms, instead of just TFs.

Formal Definitions: Constrained Structures, Structural Identifiability & Identifiable Combinations

Consideration of the structural ID (*SI*) properties of a model (always with experiment included explicitly) must take into account the input $\boldsymbol{u}$ and output $\boldsymbol{y}$ and – as we have been emphasizing throughout – all *a priori* information related in any way to the parameters. For this purpose, the *constrained structure*, or its equivalent, must be utilized in ID analysis. The following definitions are based on this paradigm (DiStefano III and Cobelli 1980). Only the noise-free, deterministic case is treated, because the properties under consideration are purely structural. The basic ***system-experiment model*** is:

$$\dot{\boldsymbol{x}}(t,\boldsymbol{p}) = \boldsymbol{f}[\boldsymbol{x}(t,\boldsymbol{p}), \boldsymbol{u}(t), t; \boldsymbol{p}], \quad t \in [t_0, T] \tag{10.60}$$

$$\boldsymbol{y}(t,\boldsymbol{p}) = \boldsymbol{g}[\boldsymbol{x}(t,\boldsymbol{p}); \boldsymbol{p}] \tag{10.61}$$

$$\boldsymbol{x}_0 = \boldsymbol{x}(t_0,\boldsymbol{p}) \tag{10.62}$$

$$\boldsymbol{h}[\boldsymbol{x}(t,\boldsymbol{p}), \boldsymbol{u}(t), \boldsymbol{p}] \geq \boldsymbol{0} \tag{10.63}$$

The four sets of relationships (10.60) to (10.63) are the **constrained structure**, the complete model for a system identification experiment on the interval $0 \leq t \leq T$. Identifiability properties are also generally dependent on the *form* of the input $\boldsymbol{u}(t)$, as well as the period of data observation if the model is time-varying or nonlinear, as indicated in Eq. (10.60).

Following are a set of five definitions encompassing **structural ID (*SI*)** of models and their parameters. We drop the adjective *structurally* when there is no ambiguity, and we often use the abbreviations *un*ID for unidentifiable, as we have done earlier with ID for identifiable.

1. *The single parameter* p_i of the constrained structure (10.60) to (10.63) is structurally **unidentifiable (*un*ID)** on the interval $[t_0, T]$ if there exists an (uncountably) infinite number of solutions for p_i from these relationships. If one or more p_i is (structurally) unidentifiable, then the **model** is **unidentifiable**.

2. The *single parameter* p_i of the constrained structure (10.60) to (10.63) is **nonuniquely structurally identifiable** on the interval $[t_0, T]$ if there exists more than one distinct solution for p_i from these relationships.

3. The *single parameter* p_i of the constrained structure (10.60) to (10.63) is **uniquely structurally identifiable (*SI*)** on the interval $[t_0, T]$ if there exists a unique solution for p_i from these relationships. If all p_i are uniquely structurally identifiable, the **model** is **uniquely structurally identifiable**.

The adjectives *globally* and *locally* are also used in the literature to define ID notions. Globally structurally identifiable (*SI*) simply means uniquely structurally identifiable. Locally structurally identifiable or just identifiable (ID) means structurally identifiable, but not necessarily uniquely. Distinct degrees of ID, or lack thereof, are illustrated in **Fig. 10.5**. In addition

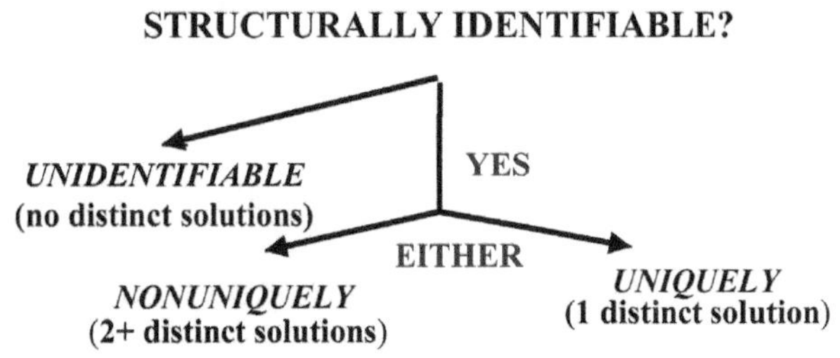

Fig. 10.5

to globally and locally, the adjective *a priori* is associated with or used as a substitute for *structurally* by some authors.

Remark: As we shall see in the next section, unidentifiable parameters – although they have an uncountably infinite number of possible solutions, these solutions are usually implicitly constrained by the model structure. This means they cannot be given any value for, say, model simulation purposes.

If a *model* is structurally *un*ID (i.e. if some p_i are unidentifiable), then there will always exist **identifiable combinations** of parameters (abbrev: **combos**), and each such combination may have one or a finite number of solutions. Combos are important because, even if the original model is unidentifiable, these identifiable combinations can provide a simpler structure for many applications, with some desired input/output properties, e.g. a subsystem model in a feedback control system, or any model in which unidentifiable parameters are not of explicit interest and their identifiable combinations, if known, provide answers to the problem at hand. Combos also can be used to **reparameterize** a model, i.e. fully recast a model in terms of identifiable parameters and parameter combinations, as illustrated later.

Identifiable parameter combinations (**combos**) are also **structural invariants** of the model (Vajda 1981). These definitions, presented in terms of a continuous-time lumped parameter model, are readily extended to distributed parameter, partial differential equation models and to discrete-time dynamic system models, with little more than changes in notation.

Example 10.7: ***Identifiable Parameter Combinations of an Unidentifiable Three-Compartment Model***

It was shown in Example 10.3 that only V_1 for the model shown in **Fig. 10.3** is identifiable from H_{11}, the experiment depicted, where all $k_{ij} > 0$, $i \neq j$. From the diagram, the model equations are easily written as:

$$\begin{aligned}
\dot{q}_1 &= k_{11}q_1 + k_{12}q_2 + k_{13}q_3 + u_1 \\
\dot{q}_2 &= k_{21}q_1 + k_{22}q_2 \\
\dot{q}_3 &= k_{31}q_1 + k_{33}q_3 \\
y &= q_1/V_1
\end{aligned} \tag{10.64}$$

For this model, we now show that the identifiable combinations are:

$$
\begin{aligned}
-k_{11} &\equiv k_{21} + k_{31} > 0 \\
-k_{22} &\equiv k_{02} + k_{12} > 0 \\
-k_{33} &\equiv k_{03} + k_{13} > 0 \\
\gamma_2 &\equiv k_{12}k_{21} > 0 \\
\gamma_3 &\equiv k_{13}k_{31} > 0
\end{aligned}
\tag{10.65}
$$

To see this, we obtain explicit relations for these six identifiable parameter combinations in terms of the $2n = 6$ uniquely identifiable parameters of the *triexponential impulse response*:

$$
y(t) = \sum_{i=1}^{3} A_i e^{\lambda_i t} \tag{10.66}
$$

When $k_{ij} > 0$ and $k_{22} \neq k_{33}$, it is not difficult to show algebraically that the λ_i are real and negative and can be ordered such that $\lambda_1 < \lambda_2 < \lambda_3 < 0$. We *define* compartment 2 as the (relatively) "rapidly exchanging" compartment (exchanging with compartment 1), and compartment 3 as the (relatively) "slowly exchanging" compartment (with compartment 1). This renders $k_{22} < k_{33}$, or $-k_{22} > -k_{33}$, and effectively eliminates one degree of freedom from the unidentifiability problem for establishing the identifiable combos; otherwise, compartments 2 and 3 are indistinguishable in H_{11} data, due to the symmetry of the model. *Note that this designation is additional information (data) about the parameters.* This inequality is part of the model, as required by relationship (10.63) of the constrained structure.

Transfer function analysis is convenient for obtaining expressions for the identifiable parameter combinations of k_{ij} in terms of the impulse response parameters A_i and λ_i of Eq. (10.66). We've already done this for this three-compartment model, in part, in Example 4.11 of Chapter 4 (with $k_{01} \equiv 0$ here) and in Example 10.3 of this chapter. We have:

$$
H_{11} = \frac{(s - k_{22})(s - k_{33})/V_1}{s^3 + \alpha_3 s^2 + \alpha_2 s + \alpha_1} = \sum_{i=1}^{3} \frac{A_i}{s - \lambda_i} \tag{10.67}
$$

where

$$
\begin{aligned}
s^3 + \alpha_3 s^2 + \alpha_2 s + \alpha_1 &\equiv (s - \lambda_1)(s - \lambda_2)(s - \lambda_3) \\
\alpha_1 = k_{12}k_{21}k_{33} - k_{22}(k_{11}k_{33} - k_{13}k_{31}) &= -\lambda_1\lambda_2\lambda_3 > 0 \\
\alpha_2 = k_{11}(k_{22} + k_{33}) + k_{22}k_{33} - k_{12}k_{21} - k_{13}k_{31} &= \lambda_1\lambda_2 + \lambda_1\lambda_3 + \lambda_2\lambda_3 > 0 \\
\alpha_3 = -(k_{11} + k_{22} + k_{33}) &= -(\lambda_1 + \lambda_2 + \lambda_3) > 0
\end{aligned}
\tag{10.68}
$$

For convenience, we define:

$$c_4 \equiv \frac{A_1\lambda_2\lambda_3 + A_2\lambda_1\lambda_3 + A_3\lambda_1\lambda_2}{A_1 + A_2 + A_3} \tag{10.69}$$

Then, after some not-so-trivial algebraic manipulation and inverse Laplace transformation of the summed terms in Eq. (10.67), equated to Eq. (10.66), we obtain the required transformation:

$$k_{11} = \frac{A_1\lambda_1 + A_2\lambda_2 + A_3\lambda_3}{A_1 + A_2 + A_3} = -(k_{21} + k_{31}) \tag{10.70}$$

$$k_{22} = \frac{c_3 - \sqrt{c_3^2 - 4c_4}}{2} = k_{02} - k_{12} \tag{10.71}$$

$$k_{33} = c_3 - k_{22} = -(k_{03} + k_{13}) \tag{10.72}$$

$$k_{12}k_{21} = c_1 = \frac{k_{11}c_4 - k_{22}c_5 + \alpha_1}{k_{33} - k_{22}} \tag{10.73}$$

$$k_{13}k_{31} \equiv c_2 = c_5 - k_{12}k_{21} = c_5 - c_1 \tag{10.74}$$

$$V_1 = \frac{1}{A_1 + A_2 + A_3} \tag{10.75}$$

Only these *six parameter combinations are uniquely identifiable* from H_{11} for $k_{22} < k_{33}$. If $k_{22} > k_{33}$, they are different, but symmetric and thus not difficult to translate notationally from these equations. Without knowing which k_{ii} is bigger, the model has two identifiable combo solutions.

We turn our attention next to the problem of what to do if parameters of interest are unidentifiable (*un*ID) and combos are not good enough for satisfying goals. Fortunately, with some additional effort, *un*ID parameters can often be "localized" in intervals.

Unidentifiable Models

Interval Identifiability & Parameter Interval Analysis

If a model is *un*ID, additional independent experimental data could render it ID, e.g. measurement of an additional state variable of the system. But this means a new model, with a different set of inputs and outputs (different measurement submodel). We'll discuss this remedy in Chapter 16, where we treat experiment design via ID analysis. It may still be possible, however, to obtain useful information about unidentifiable parameters from the original model.

We begin by first formally adding two new definitions to the (structural) ID notions list begun above (DiStefano III 1983a):

4. The *single parameter* p_i of the constrained structure (10.60) to (10.63) is (structurally) **interval identifiable** on $[t_0, T]$ if it is *unidentifiable* (*un*ID) and a finite interval, dependent on these relationships, exists bounding p_i. If all p_i are structurally interval identifiable, the **model** is (structurally) **interval identifiable**.

5. The *single parameter* p_i of the constrained structure (10.60) to (10.63) is (structurally) **quasiidentifiable** on $[t_0, T]$ if it is *interval identifiable* and the interval is small enough to provide a good enough approximate "point" estimate for the *un*ID p_i. If all p_i are quasi-identifiable, the **model** is **quasiidentifiable.**

For many *un*ID models, particularly highly structured ones like compartmental models, it is possible to find the bounds for certain parameters within such finite upper and/or lower limits, a procedure called **interval analysis** (DiStefano III 1983a). No output data (numbers) are required for interval analysis, in the sense that parameter bounds can be determined symbolically as functions of ID parameter *combinations* (*structural invariants*), directly from the model structure. This notion, then, is a structural one, as for the notion of (structural) ID itself. It is thus also examined *a priori* – the reason for this designation by early developers of the concept.

For parameters that are otherwise *un*ID, finite range information about them can be useful in applications. Furthermore, in some cases, ranges can be narrow enough to provide approximate "point" estimates for *un*ID parameters, thus exploiting the additional notion of quasiidentifiability.

These two additional related concepts are developed and illustrated for *un*ID mammillary compartmental models below, first for $n = 3$ and then more generally for any dimension n.

Parameter Bounds & Quasiidentifiability Conditions

As noted earlier, additional data about the k_{ij} in the basic model structure of Example 10.3 (**Fig. 10.3**) might be obtained from additional output kinetic measurements from compartment 2 and/or 3 (e.g. H_{22} or H_{33}), which of course is a different model. Alternatively, we can find *upper and lower bounds* on all of the unidentifiable k_{ij} of the original model, Eqs. (10.64) – with TF H_{11}, using the structural inequality constraints given by relations (10.65).

A greatest lower bound and a least upper bound on k_{21} in **Fig. 10.3** can be derived in terms of the computable (identifiable) parameter combinations in Eq. (10.65). First, we have:

$$k_{21} \equiv \frac{k_{12}k_{21}}{k_{12}} = \frac{-\gamma_2}{k_{02} + k_{22}} > \frac{-\gamma_2}{k_{22}} \tag{10.76}$$

The last inequality was obtained by adding $k_{02} > 0$ to the denominator. Then:

$$k_{11} + k_{21} = -k_{31} = \frac{-k_{13}k_{31}}{k_{13}} = \frac{\gamma_3}{k_{03} + k_{33}} < \frac{\gamma_3}{k_{33}} \tag{10.77}$$

by subtracting $k_{03} > 0$ from the denominator. Thus,

$$k_{21}^{\min} \equiv \frac{-\gamma_2}{k_{22}} < k_{21} < \frac{\gamma_3}{k_{33}} - k_{11} \equiv k_{21}^{\max} \tag{10.78}$$

Similarly,

$$k_{31}^{\min} \equiv \frac{-\gamma_3}{k_{33}} < k_{31} < \frac{\gamma_2}{k_{22}} - k_{11} \equiv k_{31}^{\max} \tag{10.79}$$

A lower bound on k_{12} can be derived in a similar manner as:

$$k_{12} = \frac{k_{12}k_{21}}{k_{21}} = \frac{-\gamma_2}{k_{01} + k_{11} + k_{31}} > \frac{-\gamma_2}{k_{11}} \tag{10.80}$$

But a *greatest* lower bound on k_{12} is obtained using the upper bound on k_{21} in Eq. (10.78) for the denominator, and then $k_{12}^{\min} \equiv \frac{\gamma_2}{k_{21}^{\max}}$, because $k_{12} > \frac{\gamma_2 k_{33}}{\gamma_3 - k_{11}k_{33}}$. The least upper bound is obtained from $k_{12} \equiv -(k_{02} + k_{22}) < -k_{22}$, because $k_{02} > 0$. Thus we have:

$$k_{12}^{\min} \equiv \frac{\gamma_3 k_{33}}{\gamma_3 - k_{11}k_{33}} < k_{12} < -k_{22} \equiv k_{12}^{\max} \tag{10.81}$$

and similarly

$$k_{13}^{\min} \equiv \frac{\gamma_3 k_{22}}{\gamma_2 - k_{11}k_{22}} < k_{13} < -k_{33} \equiv k_{13}^{\max} \tag{10.82}$$

Finally, as $k_{02} \equiv -k_{12} - k_{22}$ and $k_{03} \equiv -k_{13} - k_{33}$, we have:

$$k_{02}^{\min} \equiv 0 < \frac{\gamma_2 k_{02} k_{33}}{k_{11}k_{33} - \gamma_3} - k_{22} \equiv k_{02}^{\max} \tag{10.83}$$

$$k_{03}^{\min} \equiv 0 < \frac{\gamma_3 k_{03} k_{22}}{k_{11}k_{22} - \gamma_2} - k_{33} \equiv k_{03}^{\max} \tag{10.84}$$

Remark: If there also was a leak from pool 1 in **Fig. 10.3**, $k_{01} \neq 0$, then only $k_{11} = -(k_{01} + k_{21} + k_{31})$ would change in all of the above, all relations determined so far would be the same, with k_{11} defined anew, and the bounds on k_{01} would be:

$$k_{01}^{\min} \equiv 0 < k_{01} < \frac{\gamma_2}{k_{22}} + \frac{\gamma_3}{k_{33}} - k_{11} \equiv k_{01}^{\max} \tag{10.85}$$

Bounds for n-Compartment Models

Relations (10.78) and (10.79) are easily generalized, by induction, for the general n-compartment mammillary model, shown in **Fig. 10.6**. Let $\gamma_j \equiv k_{1j}k_{j1}$. For each compartment j, we have $-k_{jj} = k_{0j} + k_{ij} > 0$. Then, for k_{21} and k_{31}, we have:

$$k_{21}^{\min} \equiv \frac{\gamma_2}{k_{22}} < k_{21} < \frac{\gamma_3}{k_{33}} + \cdots + \frac{\gamma_n}{k_{nn}} - k_{11} \equiv k_{21}^{\max} \tag{10.86}$$

$$k_{31}^{\min} \equiv \frac{-\gamma_3}{k_{33}} < k_{31} < \frac{\gamma_2}{k_{22}} + \frac{\gamma_4}{k_{44}} + \frac{\gamma_5}{k_{55}} + \cdots + \frac{\gamma_n}{k_{nn}} - k_{11} \equiv k_{31}^{\max} \tag{10.87}$$

These equations easily generalize for any i:

$$k_{i1}^{\min} \equiv \frac{-\gamma_i}{k_{ii}} < k_{i1} < \sum_{j=2}^{n} \frac{\gamma_j}{k_{jj}} - \left(\frac{\gamma_1}{k_{11}} + \frac{\gamma_i}{k_{ii}}\right) \equiv k_{i1}^{\max} \tag{10.88}$$

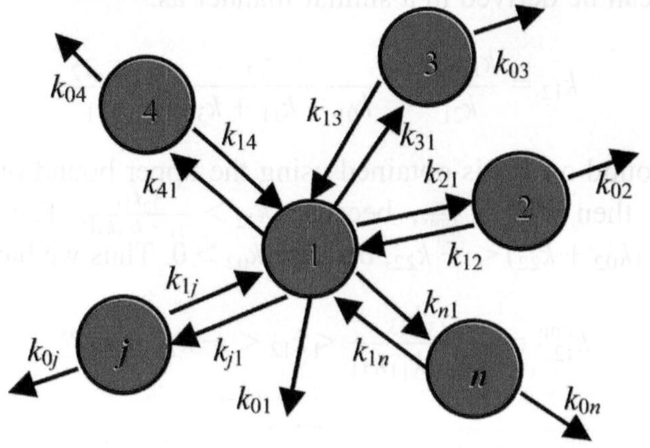

Fig. 10.6

An *n*-compartment mammillary model.

Then, extending Eq. (10.85), we have

$$k_{01}^{\min} \equiv 0 < k_{01} < -\left(k_{11} + \sum_{j=2}^{n} k_{j1}^{\min}\right) \equiv k_{01}^{\max} \tag{10.89}$$

As in Eqs. (10.24) and (10.26), we have $k_{1j} \equiv \frac{k_{1j}k_{j1}}{k_{j1}} > \frac{\gamma_j}{k_{j1}^{\max}}$ and $k_{1j} \equiv -k_{jj} - k_{0j} < -k_{jj}$. Thus, for $j = 2, 3, \ldots, n$:

$$k_{1j}^{\min} \equiv \frac{\gamma_j}{k_{j1}^{\max}} < k_{1j} < -k_{jj} \equiv k_{1j}^{\max} \tag{10.90}$$

Then, as in Eqs. (10.27) and (10.28), with $0 < k_{0j} \equiv -k_{1j} - k_{jj} < -k_{1j}^{\min} - k_{jj}$, $j = 2, 3, \ldots, n$, we have

$$k_{0j}^{\min} \equiv 0 < k_{0j} < -k_{1j}^{\min} - k_{jj} \equiv k_{0j}^{\max} \tag{10.91}$$

Bounds on Equivalent Distribution Volumes

The following is a synopsis of results in (Landaw et al. 1984). Bounds for individual compartment $V_i = -k_{i1}V_1/k_{ij}$ for mammillary compartmental models are available from Eqs. (10.64) in steady state (for $\dot{q}_i = 0$), for $i = 2, 3, \ldots, n$:

$$V_i^{\min} \equiv \frac{V_1\gamma_i}{k_{ii}^2} < V_i < \frac{V_1 k_{i1}^{\max}}{-k_{ii}} \equiv V_i^{\max} \tag{10.92}$$

Then, by direct summation of the min and max values, this inequality easily extends to bounds on the total equivalent distribution volume $V_D \equiv \sum_{i=1}^{n} V_i$ as

$$V_D^{\min} \equiv V_1 + \sum_{i=2}^{n} V_i^{\min} < V_D < V_1 + \sum_{i=2}^{n} V_i^{\max} \equiv V_D^{\max} \tag{10.93}$$

However, these are not the narrowest bounds on V_D, because $\sum_{i=0}^{n} k_{i1} = 0$ constrains the sum of the k_{i1}. The greatest lower bound (GLB) is reached when $k_{02} = \ldots = k_{0n} = 0$ and $k_{01} = k_{01}^{\max}$. The least upper bound (LUB) requires $k_{21}^{\min}$, $k_{31}^{\min}$, $\ldots$, $k_{n-1,1}^{\min}$ and $k_{n1}^{\max}$. The narrowest range for V_D is:

$$V_D^{\min} \equiv V_1 + \sum_{i=2}^{n} V_i^{\min} < V_D < V_1 + \sum_{i=2}^{n-1} V_i^{\min} + V_n^{\max} \equiv V_D^{\max} \tag{10.94}$$

Remark: Noncompartmental analysis (NCA) methods (Chapter 8) are often used for evaluating distribution volumes like V_D from kinetic experiment data. NCA typically yields a lower bound $V_D^{\min}$ for V_D, which may be substantially lower. Relation (10.94) provides the entire range of possible V_D values from identifiable parameter combinations.

Quasiidentifiability Conditions

We derived potentially useful upper and lower bounds on all unidentifiable parameters for the three-compartment mammillary model of **Fig. 10.3** above. We now treat the problem of determining conditions under which any of these parameters are quasiidentifiable, i.e. almost identifiable, for practical purposes (DiStefano III 1982). Loosely speaking, k_{ij} is *quasiidentifiable* if lower bound $k_{ij}^{\min}$ is "sufficiently close" to upper bound $k_{ij}^{\max}$. We return first to the three-compartment example to develop and illustrate the approach.

Consider the bounds on k_{21} derived earlier:

$$k_{21}^{\min} \equiv \frac{-\gamma_2}{k_{22}} < k_{21} < \frac{-\gamma_3}{k_{33}} - k_{11} \equiv k_{21}^{\max} \tag{10.95}$$

As noted above, we want that render $k_{21}^{\min}$ close to $k_{21}^{\max}$ in some sense. From (10.95):

$$\Delta k_{21} \equiv k_{21}^{\max} - k_{21}^{\min} = \frac{\gamma_2}{k_{22}} + \frac{\gamma_3}{k_{33}} - k_{11} = \frac{k_{13}k_{22}k_{31} + k_{12}k_{21}k_{33} - k_{11}k_{22}k_{33}}{k_{22}k_{33}} = \Delta k_{31} \tag{10.96}$$

It is not difficult to show that $\Delta k_{12} = \Delta k_{02}$ and $\Delta k_{13} = \Delta k_{03}$ as well, for the parameter bounds determined for the three-compartment mammillary model of **Fig. 10.3**.

Generalization: For nth-order models, the above results extend as:

$$\begin{aligned} \Delta k_{21} &= \Delta k_{31} = \ldots = \Delta k_{n1} \\ \Delta k_{0j} &= \Delta k_{1j} = \Delta k_{j1}(-k_{jj}/k_{j1}^{\max}) \end{aligned} \tag{10.97}$$

for $j = 2, 3, \ldots, n$. Then, although the incremental ranges Δk_{i1} and Δk_{1i} for the exchange rate constants between compartments are not equal, their *worst-case percentage variations are equal*:

$$\%\Delta k_{j1} = \%\Delta k_{1j} \tag{10.98}$$

for all $j > 1$, because $\%\Delta k_{ij} = 100\Delta k_{ij}/k_{ij}^{\min}$. Given that quasiidentifiability ultimately depends on numbers and is thus problem-dependent, these percentage variations are likely reasonably good measures of quasiidentifiability of individual parameters in applications, as illustrated next.

Example 10.8: ***Interval Analysis & Good & Bad Quasidentifiability Results: Two Numerical Examples Based on Real Experimental Data***

The following two triexponential impulse response functions were obtained by fitting pulse-dose tracer response data collected in two independent rat experiments, A and B, for two different thyroid hormone metabolites, iodide and reverse-triiodothyronine (rT_3):

$$y_A(t) = 5.54e^{-3.75t} + 0.44e^{-0.151t} + 0.47e^{-0.0026t}$$

$$y_B(t) = 5.25e^{-1.25t} + 1.21e^{-0.27t} + 0.41e^{-0.027t}$$

y_A represents iodide and y_B rT_3 tracer kinetics (Jang et al. 1981; Jang and DiStefano 1985). Both models are illustrated in **Fig. 10.7** and both are unidentifiable. Interval ID analysis of both models, using relations (10.78)–(10.85) and (10.92)–(10.94), yields the following bounds and bound percentage ranges for all parameters of each model:

For y_A in Model A (iodide) we have:

$$2.66 < k_{21} < 2.69 \qquad (1\%\ \text{range})$$
$$0.54 < k_{31} < 0.57 \qquad (6\%\ \text{range})$$
$$0.59 < k_{12} < 0.598 \qquad (1\%\ \text{range})$$
$$0.066 < k_{13} < 0.07 \qquad (6\%\ \text{range})$$
$$0 < k_{01} < 0.035 \qquad \text{\_\_\_\_\_\_\_\_}$$
$$0 < k_{02} < 0.008 \qquad \text{\_\_\_\_\_\_\_\_}$$
$$68 < V_2 < 69 \qquad (1\%\ \text{range})$$
$$119 < V_3 < 127 \qquad (6\%\ \text{range})$$
$$204 < V_D < 211 \qquad (3\%\ \text{range})$$

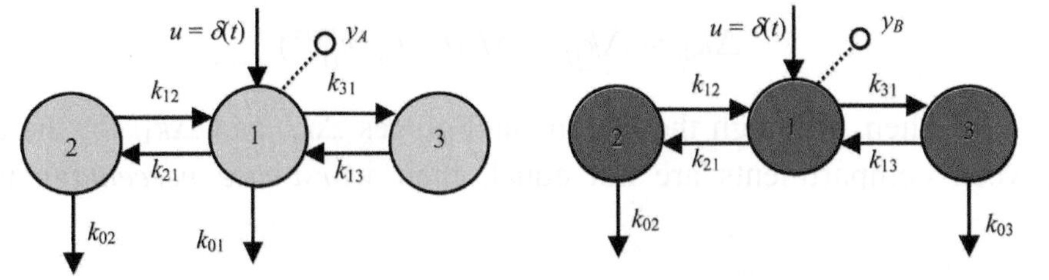

Fig. 10.7

Iodide (LHS) and reverse-T_3 (RHS) linearized tracer kinetics models.

For y_B in Model B (rT_3) we have:

$$0.33 < k_{21} < 0.96 \quad (195\%\ \text{range})$$
$$0.095 < k_{31} < 0.73 \quad (668\%\ \text{range})$$
$$0.15 < k_{12} < 0.46 \quad (195\%\ \text{range})$$
$$0.004 < k_{13} < 0.031 \quad (668\%\ \text{range})$$
$$0 < k_{02} < 0.3 \quad \text{---------}$$
$$0 < k_{03} < 0.02 \quad \text{---------}$$
$$11 < V_2 < 33 \quad (195\%\ \text{range})$$
$$47 < V_3 < 359 \quad (668\%\ \text{range})$$
$$95 < V_D < 385 \quad (305\%\ \text{range})$$

The interval ranges above suggest that it is reasonable to say that Model A *is* and Model B is *not* quasiidentifiable.

It is noteworthy that the equivalent total distribution volume V_D for iodide (y_A) and the total body compartment size (mass) of iodide ($Q_{tot} = V_D \times c_{1\infty}$ iodide concentration) are well estimated from this analysis, because the range is only 3%. However, we also note that V_D, and thus Q_{tot} for rT_3 (y_B), has a very wide feasible range, 305%. The lower bounds in each case are equal to the values that would be obtained by noncompartmental methods (Chapter 8).

Informational Limitations in Unidentifiable Models

We saw above that explicit inclusion of parameter inequality constraint sets to complete a model, Eqs. (10.88) to (10.94) for the n-compartment mammillary model, effectively guaranteed that all structural information was utilized. In other words, all available information about the parameters can be extracted from the input–output data if all such constraints are incorporated in the parameter quantification process. We note, in particular, that the elimination rate constants (*leaks*) k_{0j} for each pool are all bounded below by zero. Thus, while we could consider the minimum and maximum values of the ranges of pool exchange parameters k_{j1} or k_{1j} in a relative sense, e.g. as a maximum percentage variation, $\frac{100\Delta k_{j1}}{k_{j1}^{\min}}$, no such physical interpretation can be given to $\Delta k_{0j} \equiv k_{0j}^{\max}$. Technically, the reason for this difficulty in specifying a level of variability in k_{0j} values is that we implicitly assumed $k_{0j} = 0$ (see Eqs. (10.76)–(10.82)) to obtain ranges on all other parameters, and the k_{0j} received the remainder "left over" from $-k_{jj} = k_{0j} + k_{1j}$ in obtaining their bounds. This was the only choice, because any k_{1j} or k_{j1} set to zero would reduce the effective dimensionality of the model by one for each zero-valued parameter.

This development then explicitly reveals a fundamental limitation on identification of the mammillary structure from data collected under this widely used and practical set of experimental conditions. An experiment conducted in the central compartment only (H_{11}) provides the *least amount of information about unidentifiable elimination rates* from any compartment, relative to information about exchange rates. Of course, if $k_{0j}^{\max} \ll k_{1j}^{\min} = -k_{jj}$, then it may suffice to ignore k_{0j}, or eliminate it from the model for some applications. Nevertheless, if k_{0j}, albeit relatively small, is of specific interest, it cannot be effectively distinguished from zero, based on H_{11} data alone.

From these results, we see why elimination rate constants k_{0j} are difficult to distinguish from zero, whereas exchange rate parameters k_{j1} or k_{1j} may be bounded very tightly in certain circumstances. The notion of quasiidentifiability quantifies these circumstances.

These relative difficulties in quantifying *leaks* indirectly, compared with indirectly quantifying *exchange rates*, are common to most if not all compartmental models, and probably to larger classes of structured models, linear and nonlinear.

SI Under Constraints: Interval Identifiability with Some Parameters Known

In this section, we address interval ID when prior information about one or more model parameters is known beforehand. The major effects of additional parameter knowledge are to narrow the bounds on parameters that remain unidentifiable, as well as to possibly render others identifiable. To develop and illustrate the methodology, we analyze n-compartment (MC) mammillary and catenary models, for the common case of input and output in the first compartment. Closed-form algorithms for bounding the constrained parameter space are then developed algebraically, and also using an independent approach involving the parameters of the uniquely identifiable *submodels* of the original multicompartmental model. Both approaches illustrate general principles of analyzing models for structural properties like ID, useful for model discrimination and experiment design applications as well.

Knowledge of one or more rate constants k_{ij} is common in applications of MC models. For example, we may know (from the biology) that compartment i has no leak: in this case, $k_{0i} = 0$, but this does not necessarily assure structural ID of the model. In general, original (unconstrained) bounds shrink when one or more constraints (like $k_{0i} = 0$) are applied, thereby reflecting the increased available information, consistent with intuition.[12] The reduced range or bounds in this case are termed the **constrained range** or **bounds**.

General Catenary & Mammillary Model Equations

To analyze these models with some fixed parameter equality constraints included, we first develop general equations unified for n-compartment mammillary (**Fig. 10.6**) and catenary (**Fig. 10.8**) models without constraints.

With scalar input $u(t)$ and output $y(t)$ in the first compartment, both have the form:

$$\begin{aligned} \dot{\boldsymbol{q}}(t) &= K\boldsymbol{q}(t) + \boldsymbol{b}^T u(t), \quad \boldsymbol{q}(0) = \boldsymbol{q}_0 \\ y(t) &= \boldsymbol{c}^T \boldsymbol{q}(t) \end{aligned} \tag{10.99}$$

where $\boldsymbol{q}$ is an n-dimensional vector of compartment masses, $\boldsymbol{b} = \begin{bmatrix} 1 & 0 & \cdots & 0 \end{bmatrix}^T$, $\boldsymbol{c} = \begin{bmatrix} 1/V_1 & 0 & \cdots & 0 \end{bmatrix}^T$, $K = [k_{ij}]$, is the matrix of rate constants, with $k_{ii} = \sum_{j=0}^{n} k_{ji}$ (for $i = 1, \ldots, n$), k_{ij} (t^{-1} units) designates transfer to compartment i from compartment j ($j \neq i$), and V_1 is the

12 For the noisy data case, the means of the bounds shrink, but – interestingly – not the variance (Chapter 13, Example 13.11).

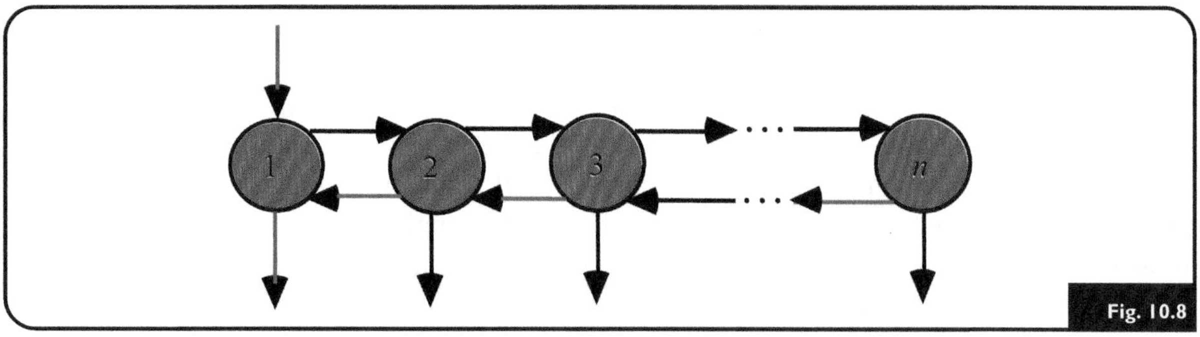

Fig. 10.8

Catenary.

volume of distribution of compartment 1. The *unit-impulse response* for either model is a sum of n distinct exponential *modes*, where $A_1 > 0$ and $\lambda_i < 0$, for all i:

$$y(t) = \sum_{i=1}^{n} A_i e^{\lambda_i t}, \quad t \geq 0 \tag{10.100}$$

when $k_{1i} \neq k_{1j}$, for all $i, j, i \neq j \neq 1$ and $k_{1i}k_{i1} > 0$, for all i for the mammillary model and $k_{i,i+1}k_{i+1,i} > 0$, for all i for the catenary model (Anderson 1983, p. 62–63), and all $k_{0i} \geq 0$ with $k_{0j} > 0$ for at least one j for both models. The input–output TF is the Laplace transform of Eq. (10.100):

$$H(s) = \sum_{i=1}^{n} \frac{A_i}{s - \lambda_i} = \frac{\beta_n s^{n-1} + \beta_{n-1}s^{n-2} + \cdots + \beta_2 s + \beta_1}{s^n + \alpha_n s^{n-1} + \cdots + \alpha_2 s + \alpha_1} \tag{10.101}$$

where each α_i and β_i can be evaluated directly from Eq. (10.99) in terms of the k_{ij}s as *uniquely ID parameter combinations* (abbrev: *combos*), or structural invariants of the model.

For the *catenary* model, these are:

$$\begin{aligned} -k_{11} &= k_{01} + k_{21} \\ -k_{ii} &= k_{0i} + k_{i-1,i} + k_{i+1,i}, \quad 1 < i < n \\ -k_{nn} &= k_{0n} + k_{n-1,n} \\ \lambda_i &= k_{i-1,i}k_{i,i-1}, \quad i = 2, 3, \ldots, n \end{aligned} \tag{10.102}$$

For the *mammillary* model, the combos are:

$$\begin{aligned} -k_{11} &= k_{01} + k_{21} + k_{31} + \cdots + k_{n1} \\ -k_{ii} &= k_{0i} + k_{1i}, \quad 1 < i \leq n \\ \lambda_i &= k_{1i}k_{i1}, \quad i = 2, 3, \ldots, n \end{aligned} \tag{10.103}$$

These combos can be algorithmically derived from output data in three steps: (1) Fit the multi-exponential model output (10.100) to the data $y(t), t \in [0, T]$; (2) algebraically transform the fitted multiexponential parameters $\{A_i, \lambda_i\}$ to TF coefficients $\{\alpha_i, \beta_i\}$; (3) uniquely transform the $\{\alpha_i, \beta_i\}$ to model parameter combos $\{k_{ii}, \gamma_i\}$.

$$y(t),\ t \in [0, T] \underset{(1)}{\Rightarrow} \{A_i, \lambda_i\} \underset{(2)}{\Rightarrow} \{\alpha_i, \beta_i\} \underset{(3)}{\Rightarrow} \{k_{ii}, \gamma_i\}$$

This algorithm was implemented and finite parameter bounds for all $k_{ij} > 0$ and unknown were determined for both model types in (Chen et al. 1985). (Mammillary model bounds also were derived in a previous section of this chapter.) All results are in algorithmic form, easily programmed for automatic computation. They are implemented in web application: *W³MAMCAT* (at www.biocyb.cs.ucla.edu), which handles mammillary and catenary models of any dimension n, with input and output in compartment 1 only, and input accepted in multiexponenential parameter $\{A_i, \lambda_i\}$ form, following Step (1) above.

Now we are ready to take on the main problem in this section: when some k_{ij} are known, (or prior estimates are available) the dimensionality of the space of unknown parameters is reduced and the new (reduced range) bounds must satisfy a modified set of structural relations, as shown below. In addition, prior parameter values or estimates must be *feasible*, i.e. they must be consistent with the model structure and output data $y(t)$, as described next.

Feasible Parameter Ranges (Subspace) for Equality Constraints

Equality constraints on the rate constants of catenary and mammillary models must satisfy the following conditions (Vicini et al. 2000):

1. No additional equality constraints are possible for a one-compartment model, because its single rate parameter is determined uniquely by the available (noise-free) output data $y(t)$, for nonzero input and output, as is $V_1 = q(0)/y(0)$, e.g. $V_1 = 1/y(0)$ for a unit-impulse input.
2. Specified parameter values (or estimates) must fall within their corresponding unconstrained bounds, i.e. bounds for the k_{ij} when none are given *a priori*.
3. Constraints must be *independent*. As a simple example, the following set of equations for the catenary model: $-k_{11} = k_{01} + k_{21}$, $\lambda_2 = k_{12}k_{21}$ provides unique solutions for all three parameters, k_{01}, k_{12}, and k_{21}, when any one of them is given. If there are two conflicting constraints among these three variables, then no solution exists. This condition is called **independence**, which means: *there cannot be more than $n - 1$ independent parameter equality constraints for n compartment catenary or mammillary models*, because there are at most $3n - 2$ unknowns and $2n - 1$ equality relations among the parameters. This leaves at most $n - 1$ degrees of freedom in the model, and thus there cannot be more than $n - 1$ constraints without violating the independence condition (Vicini et al. 2000).

Models with Infeasible or Redundant Constraints

Constraints may be infeasible for a given set of data and thus may yield no solution, e.g. if $k_{12} = 1$ and $k_{21} = 2$ are given, but $k_{12}k_{21} = 0.5$ is estimated from the input–output data, then we have a contradiction. Constraints can also be redundant, e.g. $k_{12} = 1$ and $k_{21} = 0.5$ and $k_{12}k_{21} = 0.5$.

The Constrained Parameter Bounding Algorithms

Catenary Model

The catenary model is illustrated in **Fig. 10.8**. For the unconstrained case, the minimum possible value for any k_{0j} is zero (Chen et al. 1985). However, no more than $n-1$ leaks can be zero at the same time, and this was used in the derivation of the unconstrained algorithm, which used the following recursive relations:

$$k_{i+1,i} = \begin{cases} -k_{11} - k_{01}, & i=1 \\ -k_{ii} - \dfrac{\gamma_i}{k_{i,i-1}} - k_{0i}, & i=2,\ldots,n-1 \end{cases} \tag{10.104}$$

for the k_{ij} elements in the lower diagonal of the system matrix K in Eq. (10.99), and

$$k_{i-1,i} = \begin{cases} -k_{0n} - k_{nn}, & i=n \\ -k_{0i} - k_{ii} - \dfrac{\gamma_{i+1}}{k_{i,i+1}}, & i=n-1,\ldots,2 \end{cases} \tag{10.105}$$

for the k_{ij} elements in the upper diagonal. Then, with the leaks k_{0i} recursively set to zero, the upper bounds on the $k_{i+1,i}$ and $k_{i-1,i}$ were obtained for all i. If one or more rate constants is known, the above equations are still applicable, but algorithmic solution must account for all available information. This is done as follows:

1. Find the largest feasible ranges for the k_{ij}, consistent with the model structure and output data. Then all equality constraints for the k_{ij}s are tested for feasibility. If infeasible, the algorithm is terminated. If any parameter in the first or last compartment is known, the values of the other two are evaluated from

$$\begin{aligned} -k_{11} &= k_{01} + k_{21} \\ \gamma_2 &= k_{12}k_{21} \end{aligned} \tag{10.106}$$

   for the first compartment, and

$$\begin{aligned} -k_{nn} &= k_{0n} + k_{n-1,n} \\ \gamma_n &= k_{n-1,n}k_{n,n-1} \end{aligned} \tag{10.107}$$

   for the nth-compartment.

2. Eqs. (10.104) and (10.105) are solved recursively, substituting the values of known parameters wherever they appear and setting the unconstrained leaks k_{0i} to zero. This gives the new upper bounds on all rate constants but the leaks.

3. New lower bounds on all but the leak parameters are then found in terms of the ID combos, from:

$$k_{i-1,i}^{\min} = \frac{\gamma_i}{k_{i,i-1}^{\max}}, \quad i = 2, \ldots, n$$
$$k_{i+1,i}^{\min} = \frac{\gamma_{i+1}}{k_{i,i+1}^{\max}}, \quad i = 1, \ldots, n-1 \tag{10.108}$$

4. New upper bounds on the leaks are then found from:

$$k_{01}^{\max} = -k_{11} - k_{21}^{\min}$$
$$k_{0i}^{\max} = -k_{ii} - k_{i-1,i}^{\min} - k_{i+1,i}^{\min}, \quad i = 2, \cdots, n-1 \tag{10.109}$$
$$k_{0n}^{\max} = -k_{nn} - k_{n,n-1}^{\min}$$

The new lower bound for each unconstrained leak remains zero, unless the model is uniquely identifiable, in which case the k_{0i}s are uniquely determined from Eq. (10.109).

Remark: If there are n independent constraints, then the model is uniquely identifiable and the algorithm gives $k_{ij}^{\min} \equiv k_{ij}^{\max}$, for all i and j.

The fairly common case of $n-1$ leaks set to zero presents some points of interest, as in **Fig. 10.9**. The remaining unconstrained leak then attains its maximum value and all parameters attain their minimum or maximum values. Conversely, setting any one k_{0j} to its maximum value is equivalent to setting all other k_{0i} to zero; then one equality constraint is enough to make the model uniquely identifiable, with all k_{ij} attaining their minimum or maximum values.

Mammillary Model

The mammillary model is illustrated in **Fig. 10.6**. This algorithm is simpler, because peripheral compartments of the mammillary model are not directly connected with each other. For establishing results, we label the peripheral compartments such that $-k_{22} > -k_{33} > \cdots > -k_{nn}$, where the k_{ii} are generically distinct. We can do this because mammillary models are structurally identical if peripheral compartments are exchanged or renumbered.

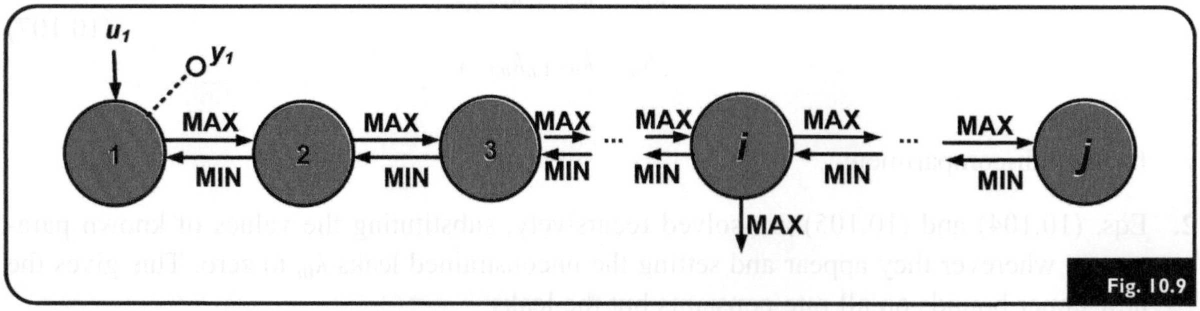

Fig. 10.9

The structural invariant equations associated with peripheral compartment $i > 1$ are:

$$\begin{aligned} -k_{ii} &= k_{0i} + k_{1i} \\ \gamma_i &= k_{1i}k_{i1} \end{aligned} \tag{10.110}$$

By inspection, Eq. (10.110) can be solved for all three variables if any one is known. Furthermore, they do not directly affect the parameters in other peripheral compartments unless there are $n-1$ constraints. These observations motivate the following algorithm:

1. The algorithm for *unconstrained* mammillary models is applied first (see above), to find the largest feasible ranges for the parameters consistent with the model structure and output data. Then all equality constraints for the k_{ij}s are tested for feasibility. If infeasible, the algorithm is terminated. If any parameter in the set $\{k_{0i}, k_{1i}, k_{i1}\}$ is known, the values of the other two are evaluated from Eq. (10.110). More than one equality constraint supplied by a user could be inconsistent with each other and relations (10.110). In this case, the algorithm is terminated.

2. If the parameters associated with any compartment i are all unknown, then $k_{1i}^{\max}$ and $k_{i1}^{\min}$ are calculated from Eq. (10.110) by setting k_{0i} to zero.

3. $k_{i1}^{\max}$, $i > 1$, is calculated from $-k_{11} = k_{01} + k_{21} + \ldots + k_{n1}$ by setting all parameters except k_{i1} to their minimum, i.e.

$$k_{i1}^{\max} = -k_{11} - \sum_{j \neq 1, j \neq i} k_{j1}^{\min} \tag{10.111}$$

for $i > 1$. If any k_{j1} are known $(1 \neq j \neq i)$, then their values are simply substituted into Eq. (10.111) to obtain $k_{i1}^{\max}$.

4. For $i > 1$, $k_{1i}^{\min}$ and $k_{0i}^{\max}$ are calculated from combos (10.110) by setting k_{i1} to its maximum.

5. If k_{01} is unknown, then $k_{01}^{\min} = 0$ and $k_{01}^{\max}$ is found from:

$$k_{01}^{\max} = -k_{11} - \sum_{j > 1} k_{j1}^{\min} \tag{10.112}$$

Remark: If there are $n-1$ independent constraints, the model is uniquely identifiable and the algorithm gives $k_{ij}^{\min} \equiv k_{ij}^{\max}$, for all i and j.

Joint Submodel Parameters: Another Approach to Interval Identifiability

Parameter bounds of unconstrained and unidentifiable compartmental models can be computed using an *identifiable submodels* approach, in which the ranges for each of the parameters of the original unidentifiable model are established from the uniquely identifiable parameters of particular *submodels*. These give either minimum or maximum values of the unconstrained range for a given parameter (Cobelli and Toffolo 1987; Vajda et al. 1989). The ranges for a

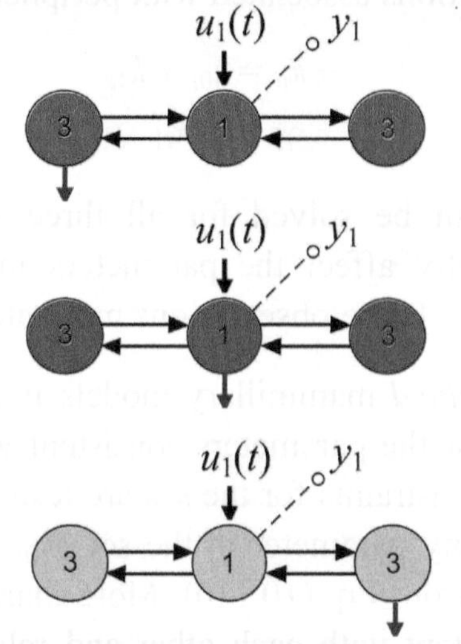

Fig. 10.10

Uniquely identifiable submodels of the *unconstrained* mammillary structure.

constrained model can be determined similarly, by joint estimation of the uniquely identifiable parameters of *submodels of the constrained structure.*

Uniquely identifiable submodels of unidentifiable structures are typically generated by systematically eliminating parameters of the original model until it becomes uniquely identifiable (Vajda 1981). In our case, every unconstrained n-compartment mammillary or catenary model has n uniquely identifiable submodels, each obtained by setting all leaks but one to zero, as described earlier, and illustrated for the catenary model in **Fig. 10.9** above. When parameters are constrained, the number of such submodels is reduced, because leaks then become uniquely identifiable and generally different from zero. For this reason, the joint regions or ranges of values for the constrained model are typically reduced as well (Vicini et al. 2000).

Fig. 10.10 includes all of the *uniquely identifiable submodel structures* of the *unconstrained* three-compartment mammillary model, with input and output in compartment 1 only. We assume $-k_{22} > -k_{33}$ for these configurations, which guarantees unique ID (as discussed in Example 10.7). Each submodel has a single leak from only one of the three compartments, as noted. In the *constrained* model of **Fig. 10.11**, *k_{12} is known* and nonzero, as shown. Then, by fixing k_{12} in Eqs. (10.111) and (10.112), either k_{03}, k_{02} and k_{21} are uniquely identifiable and nonzero, or k_{01}, k_{02} and k_{21} are uniquely identifiable and nonzero, both as shown. These are the two uniquely identifiable submodels of the constrained structure, and parameter bounds of the original constrained structure can therefore be obtained by quantifying these submodels, e.g. by direct estimation using an appropriate parameter search method, which we address in Chapter 12.

The results described in this section provide closure on the deterministic part of the bounding problem for open linear mammillary and catenary compartmental models of any order, with input and output in compartment 1, and for all unknown k_{ij}, with any number of k_{ij} known *a priori.* The overall solutions are closed-form and algorithmic and are implemented in web application: *W*[3]*MAMCAT* (www.biocyb.cs.ucla.edu), which handles mammillary and catenary

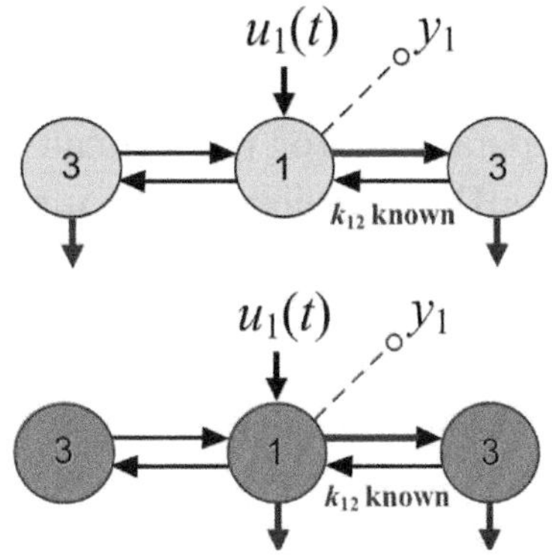

Fig.10.11

Uniquely identifiable submodels of the *constrained* mammillary structure.

models of any dimension n, with input and output in compartment 1 only, and input accepted in multiexponential parameter $\{A_i, \lambda_i\}$ form, following Step (1) above, with and without constraints on the parameters. These algebraic methods are surely applicable to other model types, and other input–output port configurations. However, solutions to these open problems are not likely to be as "neat" as for these two symmetric model classes.

SI Analysis of Nonlinear (NL) Biomodels

Prolog and Overview

To motivate the development, we illustrate – with a simple NL function – a subtle and important difference in analyzing ID of NL compared with linear ODE models. It's a key factor in analyzing NL models and it's best to recognize it from the outset. The issue noted in this example also blurs the distinction between structural and numerical ID for NL models.

In Chapter 6, we briefly addressed the "practical" ID of parameters v_{max} and K_m of the NL Michaelis–Menten (M–M) function, by considering the extremes of $v(t) \equiv \frac{v_{\max}s(t)}{K_m + s(t)}$, i.e. what happens to $v(t)$ for very large and very small substrate concentrations $s(t)$. We found that for "big" $s(t) \gg K_m$, $v(t) \rightarrow v_{max}/K_m$; and for "small" $s(t)$, $v(t) \rightarrow v_{\max}$. In either case, this means only the ratio v_{max}/K_m, or only v_{max} – and *not* K_m, is "practically" identifiable near the extremes of $s(t)$. We deal with the "practicality" – the numerical ID issue for these parameters – later in Chapters 12 and 13. For now – in the limit – *at* the extremes – the function is clearly structurally *un*identifiable (*un*ID). The parameter *combo* v_{max}/K_m, however, is *SI* at one extreme, and the single parameter v_{max} is identifiable at the other extreme.

This example illustrates a major difficulty with establishing ID properties of NL models not present with linear ones. NL model dynamics – structural properties and the temporal evolution of state variables and outputs – depend on the *magnitude*[13] as well as the form of *inputs*

13 In general, most dynamical properties of NL dynamical systems – like stability and identifiability – depend on the form (shape) and location as well as the magnitude of inputs.

$u(t)$, as well as other system properties.[14] In this simple M−M example, $s(t)$ might be either an independent input, or a state variable dependent on input $u(t)$.

In other words, a NL model parameter (or model) can be *SI* in principle, but even with *perfect, noise-free data*, it can be *un*ID, depending on the input.

Paradoxically, we also show below that the M−M function (with two parameters) can be *fully SI* (*v_{max}* **and** *K_m SI*) *in its nonlinear range*, compared with a single parameter linear function! In fact, we conjecture that nonlinearities in ODE models *can improve* ID properties (compared with linear or linearized versions of the same model), even when they increase their parameter number, as illustrated in examples below and in subsequent chapters.

With these interesting features in mind, we turn our attention to examining the *SI* properties of NL ODE models of the fairly general form:

$$\begin{aligned}\dot{x}(t,p) &= f[x(t,p),p] + b[x(t,p),p)]u(t)\\ y(t,p) &= g[x(t,p),p]\\ \dot{x}(0,p) &= x_0(p)\end{aligned} \tag{10.113}$$

with $t \in [0, T]$, and $x(t,p)$, $y(t,p)$, and p are n-, m-, and P-vectors respectively; and f, b and g are NL system, input and output n-, r- and m-vector functions, respectively. NL functions f, b, and g are assumed to be *of rational polynomial* form (the most common), which includes ratios of polynomials, as in M−M, Hill and other ligand−receptor type ODE function terms. We are interested here in structural ID (*SI*)[15] of this class of models. And we have been forewarned: even with perfect input−output data, NL model dependence of results on input magnitudes can make positive *SI* analysis results − which examine only qualitative topological properties − less helpful in practice than *SI* analysis of linear models − but still essential.

A fairly comprehensive review of *SI* analysis methods for NL ODE models was noted in the Historical Perspective section of this chapter (Chis et al. 2011a,b). It included several systems biology examples and compared several methods also mentioned in that section − along with their reference sources − in their analysis of those models. We focus on two of these methods here, the earliest one, the Taylor series (Pohjanpalo 1978; Pohjanpalo and Wahlstrom 1982) approach; and one of the most recent methods based in differential algebra (Ljung and Glad 1994; Audoly et al. 2001); Saccomani et al. 2003; Bellu et al. 2007; Meshkat et al. 2012). We develop them using simpler examples, with analytic or partial analytic solutions, to focus on some of their nuances − shared with other approaches. Application of any is computationally challenging and **symbolic manipulation packages** (SMPs) are normally used to facilitate the computations, e.g. *Maxima, Maple, Reduce, Mathematica, Matlab* (Mueller 2001). Software for doing *SI* analysis of *NL* ODE models by symbolic differential algebra is

14 Example 4.7 of Chapter 2, a two-compartment model with NL transfer flux in M−M form, vividly illustrated this input magnitude dependence. Also, recall that stability analysis of NL models (Chapter 9) similarly suffers from this same difficulty, and linear models do not.

15 *Reminder: SI* of a model is only a necessary condition for model identifiability; it is not sufficient. Numerical identifiability properties must be considered in the presence of real, noisy data (Chapter 12).

available for download (DAISY) or as a web application (COMBOS), both to be discussed later in this section.

SI Analysis by Taylor Series

The basic assumption here is that the output or observation function $\boldsymbol{y}(t, \boldsymbol{p})$ is analytic in a neighborhood of $t = 0$, i.e. $\boldsymbol{y}$ is infinitely differentiable with respect to t on this neighborhood. Under this assumption, $\boldsymbol{y}(t,\boldsymbol{p})$ and its successive time derivatives can be expanded in a Taylor series and evaluated at initial conditions $t = 0^+$ ($t \equiv 0$ for simplicity of notation), in terms of the model parameters $\boldsymbol{p}$:

$$\boldsymbol{y}(t,\boldsymbol{p}) = \boldsymbol{y}(0,\boldsymbol{p}) + \boldsymbol{y}^{(1)}(0,\boldsymbol{p})t + \boldsymbol{y}^{(2)}(0,\boldsymbol{p})\frac{t^2}{2!} + \ldots + \boldsymbol{y}^{(i)}(0,\boldsymbol{p})\frac{t^i}{i!} + \ldots \quad (10.114)$$

where

$$\boldsymbol{y}^{(i)}(0,\boldsymbol{p}) \equiv \frac{d^i\boldsymbol{y}(0,\boldsymbol{p})}{dt^i}, \quad i = 0,\ 1,\ 2,\ \ldots \quad (10.115)$$

Since the coefficients in the Taylor series expansion are unique, the problem reduces to determining the number of solutions for the parameters in a set of algebraic equations generally nonlinear in the parameters. Then, by definition, the parameter set is *un*ID if the set of solutions is uncountable, is nonuniquely (locally) identifiable if it is countable and distinct, and is globally identifiable if there is a unique solution.

For nonlinear models, establishing the upper bound on the number of equations required for this set can be difficult – and therefore demonstrating unidentifiability can be difficult. For linear systems, the Cayley–Hamilton theorem (Appendix B) (Atiyah 1969) tells us there are at most $2n - 1$ independent equations in this set; for bilinear systems, the maximum number of independent equations is $2^{2n} - 1$ (Vajda 1981), while for NL systems of *homogeneous polynomial* form, the corresponding maximum number is $\frac{\eta^{2n} - 1}{\eta - 1}$, where η is the degree of the polynomial (Vajda 1987). For the last two types of NL systems, this upper bound on the number of equations rapidly becomes large as n, the dimension of the state space, increases.

In general, the procedure involves *three steps*, and has provided ID results in many examples of practical interest (Pohjanpalo 1978; Pohjanpalo and Wahlstrom 1982; Godfrey 1983; Godfrey and DiStefano III 1987):

1. Successive differentiation of $y(t, \boldsymbol{p})$.
2. Evaluation of successive derivatives $y^{(i)}(0, \boldsymbol{p})$, by substitution of quantities already known from $y(0,\boldsymbol{p})$ and lower derivatives ($< i$).
3. A check on the independence of the equations in the successive derivatives and on which parameters, if any, can be identified at each stage of the differentiation.

The procedure is terminated at some appropriate stage (discussed further below).

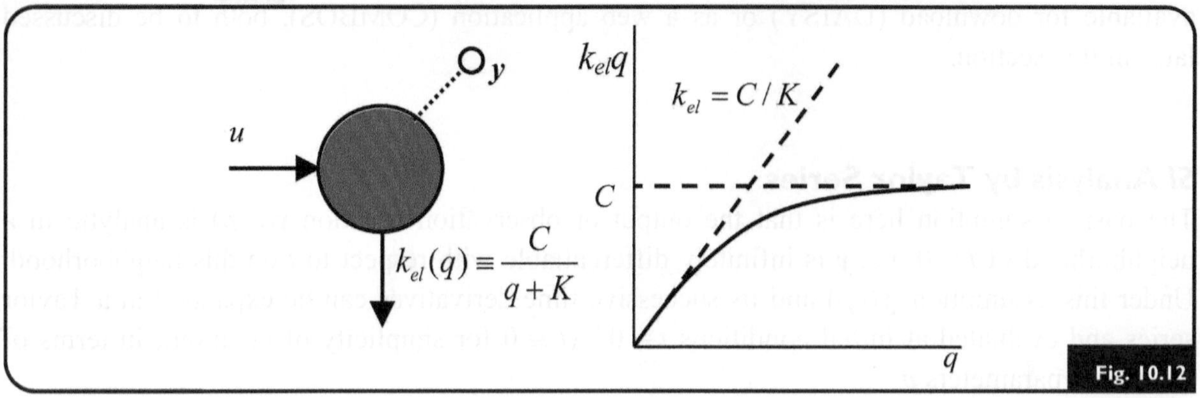

Fig. 10.12

Step 1 of the procedure is usually the easiest of the three, done either manually or by symbolic computation. Step 2 often involves lengthy and tedious calculations, prone to error if carried out by hand, and more easily solved with symbolic manipulations packages. Step 3 can be the toughest, because of the structure of the derivative equations obtained in Step 1, with ID results hard to come by, even using sophisticated solving algorithms available either separately (LeCourtier and Raksanyi 1987) or embedded in certain manipulation packages, for example, the Groebner Bases library routine within REDUCE (Mueller 2001). The following examples clarify the procedure and computational problems.

Example 10.9: ***One-Compartment Model with Nonlinear Elimination***

A one-compartment model with nonlinear elimination is illustrated in **Fig. 10.12**. In this model, $q(t)$ is the amount of substance in the single compartment, and the mass flux (amount/time) irreversibly eliminated $k_{el}q(t)$ depends nonlinearly on $q(t)$, with the fractional loss rate $k_{el}(q)$ shown. We note that $k_{el}q(t)$ is approximately linear for low values of $q(t)$, saturating (zero-order) for high $q(t)$. This model can be appropriate for substances extensively metabolized or eliminated in a single large organ like the liver, such as ethanol, which exhibits saturable inhibition and behaves like a zero-order elimination (constant) process at high concentrations, q in our model (Holford 1987).

We assume a known bolus-dose amount (approximate impulse $u(t) = D\delta(t)$) input into the compartment. The mass balance ODE is simply:

$$\dot{q}(t) = -k_{el}(q)q + D\delta(t), \quad q(0) = 0 \tag{10.116}$$

or, equivalently,

$$\dot{q}(t) = -\frac{Cq(t)}{K + q(t)}, \quad q(0) = D \tag{10.117}$$

With concentration measured in the compartment, we have:

$$y(t) = \frac{q(t)}{V} \tag{10.118}$$

where V is compartment volume.

Question: Is it possible to uniquely estimate the unknown parameters $\boldsymbol{p} = \begin{bmatrix} C & K & V \end{bmatrix}^T$ from this biomodel experiment, focusing on the output y as a function of $\boldsymbol{p}$ as well as t, i.e. $y(t, \boldsymbol{p})$? Assume perfect data.

To apply the Taylor series approach, expand $y(t, \boldsymbol{p})$ as a Taylor series in the neighborhood of $t = 0$:

$$y(t,\boldsymbol{p}) = y(0,\boldsymbol{p}) + t\dot{y}(0,\boldsymbol{p}) + \frac{t^2}{2!}\ddot{y}(0,\boldsymbol{p}) + \ldots \tag{10.119}$$

For the deterministic output function $y(t, \boldsymbol{p})$, we can generate all the derivatives of $y(t, \boldsymbol{p})$, at any time t and for some $\boldsymbol{p}$, in particular at $t = 0$, because, by assumption, $y(t, \boldsymbol{p})$ is an analytic function and can be expanded in a Taylor series. That this is the case is confirmed by the (analytic) ODE being the first-derivative, so subsequent differentiations are guaranteed, i.e. all time derivatives of $y(t, \boldsymbol{p})$ using the original ODE. Solving for $y(0, \boldsymbol{p})$ and the first three of these "known" derivatives in terms of the unknown parameters and the given IC $q(0)$, we get:

$$\begin{aligned}
y(0,\boldsymbol{p}) &= \frac{q(0)}{V} = \frac{D}{V} \\
\dot{y}(0,\boldsymbol{p}) &= \frac{\dot{q}(0)}{V} = -\frac{1}{V}\frac{Cq(0)}{K+q(0)} = -\frac{1}{V}\frac{CD}{K+D} \\
\ddot{y}(0,\boldsymbol{p}) &= \frac{\ddot{q}(0)}{V} = -\frac{1}{V}\frac{C\dot{q}(0)[K+q(0)] - Cq(0)\dot{q}(0)}{[K+q(0)]^2} \\
\dddot{y}(0,\boldsymbol{p}) &= \frac{\dot{q}(0)}{V} = -\frac{1}{V}\frac{C\ddot{q}(0)[K_m+q(0)]^2 - 2CKq(0)[K+\dot{q}(0)]}{[K+q(0)]^2} \\
&= -\frac{CK[\ddot{y}(0,\boldsymbol{p})(K+D) - 2V\dot{y}(0,\boldsymbol{p})^2]}{[K+D]^2}
\end{aligned} \tag{10.120}$$

The assumed known output and its derivatives are related to the basic kinetic parameters K, C, and V by this set of nonlinear algebraic equations (10.120), and the ID problem is reduced to establishing whether any of these parameters can be determined by simultaneously solving these algebraic equations.

We see from the first equation that V is uniquely determined from known D and output $y(0, \boldsymbol{p})$; a little additional algebra yields unique solutions for K and C in terms of output derivatives:

$$\begin{aligned}
V &= D/y(0) \\
K &= \frac{D\ddot{y}(0)^2 - 2V\dot{y}(0)^2\ddot{y}(0) - D\dddot{y}(0)}{\dddot{y}(0) - \ddot{y}(0)^2} \\
C &= \frac{\dddot{y}(0) - \ddot{y}(0)^2}{2D\dot{y}(0)\ddot{y}(0)}
\end{aligned} \tag{10.121}$$

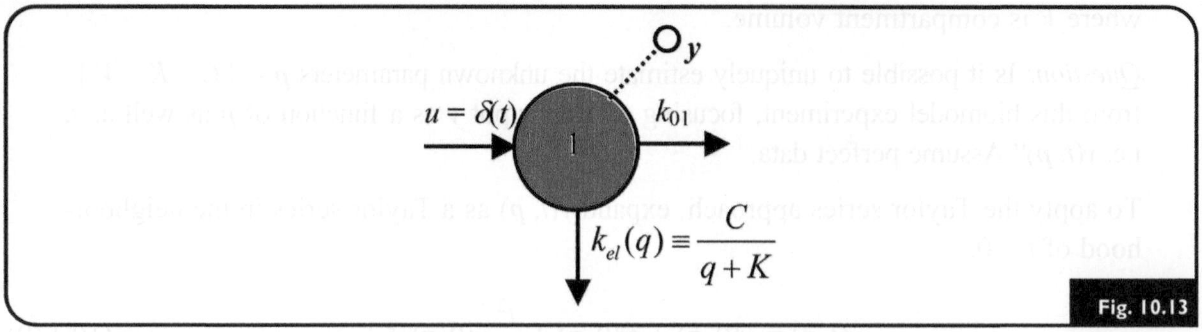

Fig. 10.13

One-compartment model with parallel linear and NL elimination.

We thus see explicitly how all model parameters independently influence the output, and this fact permits their estimation separately and uniquely, i.e. our NL 1-compartment model is uniquely structurally (globally) identifiable.

Example 10.10: ***One-Compartment Model with Both Linear & NL Elimination Pathways***

Now consider the model of **Fig. 10.13**, in which the nonlinear elimination kinetics are again of Hill function form, but we additionally have a linear elimination component. The ODE and output equation for this model are

$$\dot{q}(t) = -\frac{Cq(t)}{K + q(t)} = k_{01}q(t) + u(t)$$
$$y = cq \tag{10.122}$$

where c, C, K and $k_{01} > 0$, and the input is a bolus dose, $u(t) = D\delta(t)$, with D known, as in Example 10.9, i.e. $q(0) = D$. The parameter set to be identified is $\{c, C, K, k_{01}\}$. Note that c in the output equation here may represent any unknown proportionality constant in the measurement system, not just the typical V^{-1} for the measured compartment. We shall see that the single additional unknown parameter k_{01} here renders this problem significantly more difficult to solve without the help of a symbolic manipulation package (SMP).

Taylor Series Analysis: Clearly, $y(0) = cD$ gives c uniquely, and we thus need to check for ID of the set $\{C, K, k_{01}\}$. As above, we need at least three derivatives of $y(t)$ with respect to time. With c established, we can examine derivatives of $q = y/c$ instead of y. The first time derivative of $q(t)$ is the ODE, Eq. (10.122), and the second and third are:

$$q^{(2)}(t) = -\left(\frac{CK}{[K+q(t)]^2} + k_{01}\right)\dot{q}(t) \tag{10.123}$$

$$q^{(3)}(t) = -\left(\frac{CK}{[K+q(t)]^2} + k_{01}\right)q^{(2)}(t) + \frac{2CK}{[K+q(t)]^3}[\dot{q}(t)]^2 \tag{10.124}$$

Successively substituting known values, we get

$$\dot{q}(0) = -\left(\frac{C}{K+D} + k_{01}\right)D = a_1 D \tag{10.125}$$

yielding

$$C = -(a_1 + k_{01})(K + D) \tag{10.126}$$

where $a_1 \equiv \frac{\dot{y}(0)}{y(0)}$ is known. Then

$$q^{(2)}(0) = -\left[\frac{CK}{(K+D)^2} + k_{01}\right]\dot{q}(0) = \left[\frac{(a_1 + k_{01})K}{K+D} - k_{01}\right]\dot{q}(0) = a_2 D \tag{10.127}$$

which gives

$$a_2 = a_1\left(\frac{a_1 K - Dk_{01}}{K+D}\right) \tag{10.128}$$

where $a_2 \equiv \frac{y^{(2)}(0)}{y(0)}$ is known. From this, an expression for k_{01} in terms of K is obtained:

$$k_{01} = a_1\frac{K}{D} - \frac{a_2}{a_1}\left(\frac{K}{D} + 1\right) \tag{10.129}$$

Then

$$q^{(3)}(0) = -\left[\frac{CK}{(K+D)^2} + k_{01}\right]q^{(2)}(0) + \frac{2CK}{(K+D)^3}[\dot{q}(0)]^2 \tag{10.130}$$

yielding

$$a_3 = -\left[\frac{CK}{(K+D)^2} + k_{01}\right]a_2 + \frac{2CK}{(K+D)^3}a_1^2 D \tag{10.131}$$

where $a_3 \equiv \frac{y^{(3)}(0)}{y(0)}$ is known.

The substitutions so far were by hand without too much difficulty. The next step is to substitute for C from Eq. (10.126) and for k_{01} from Eq (10.129) and solve Eq. (10.131) for K. This is lengthy and tedious and hence prone to error if done by hand, whereas the substitutions can be made readily using a symbolic manipulation package (SMP) program. Doing this, we see that terms in K and higher powers of K cancel out, and the resulting equation in K is given by

$$(a_1a_3 - a_2^2 + 2a_1^4 - 2a_1^2a_2)K + (a_1a_3 - a_2^2)D = 0 \tag{10.132}$$

By inspection, this can be solved to give a unique value for K, as long as $(a_1a_3 - a_2^2 + 2a_1^4 - 2a_1^2a_2) \neq 0$. Since K is uniquely valued, a unique value for k_{01} is obtained from Eq. (10.129) and a unique value for C is obtained from Eq. (10.126). Thus the parameter set $\{c, C, K, k_{01}\}$ is uniquely (globally) identifiable, i.e. the entire model is uniquely *SI*.

Remark: **Caveat #1**. For the NL biomodel in Example 10.9, there were three unknowns, and in Example 10.8 there were two, but three derivatives were needed in each to establish *SI*. So, there does not seem to be a general principle that can be gleaned by simply counting. Indeed, when using the Taylor series approach, it is not known *a priori* how many derivatives one has to calculate in order to establish ID. In other words, this method cannot be used to show that a model is *not* ID; after calculating n derivatives, it is always possible that the $(n+1)^{st}$ or higher-order derivative would yield an independent relationship for establishing ID. This is dismaying!

Remark: **Caveat #2.** The model examples above are both NL due to M–M function terms in their equations. We noted earlier that M–M function became either first- or zero-order functions at their limits – either masking or eliminating a parameter at the limits – thereby rendering them "practically" unidentifiable at or near the limits. Again, ID in practice – which we address in Chapter 12 as numerical ID – is all-important. The problem is that these properties depend on the magnitude of inputs $u(t)$, sometimes in a complicated way. Remember, *SI* is only a necessary condition for "practical" ID; it is not sufficient.

SI Analysis by Symbolic Differential Algebra (DA)

This approach is elegant – mathematically speaking – and is showing substantial success, at least with models of low-to-medium dimensionality, with terms in polynomial or quotients-of-polynomials form. This covers a majority if not most ODE biomodels, nonlinear or linear. Furthermore, models with non-polynomial terms can often be approximated by polynomials, so the polynomial constraint is not likely very limiting. The size limitation here is that symbolic algebra programs required for the analysis are limited by computational complexity. Nevertheless, there is progress in reducing this complexity, as we note below.

The differential algebra (DA) method is based on replacing the intrinsic input–output behavior of the system – derived from the ODE model – with a polynomial or rational mapping among the parameters. Differential relationships among the inputs, outputs and parameters are first derived by eliminating *unobservable* state variables, i.e. those that cannot be evaluated or computed from the output variables.[16] The result is an equivalent model expressed exclusively in terms of the input and output variables – plus the parameters – and none of the state variables. It is called the *input–output map* $\Psi(\boldsymbol{y},\boldsymbol{u},\boldsymbol{p}) = \mathbf{0}$. This is usually computed by reducing the ODE model using **Ritt's pseudodivision algorithm** (Bellu et al. 2007a,b) and, as in most developing areas, more efficient algorithms than Ritt's have been recently introduced (Meshkat et al. 2012) and are finding their way into available software (Meshkat et al. 2014). The resulting input–output map is called the **characteristic set**, which is generally non-unique, but can be made unique – an important property – by suitable normalization. This gives a **Gröbner basis** for the model equations plus their derivatives, computed by performing successive substitutions to eliminate the state variables (Shannon and Sweedler 1988). These

16 Model observability properties are developed in Appendix D, but only the basic idea is needed to understand DA algorithms here. In general, not all state variables of a model are connected to the outputs. Parameters in these unconnected paths therefore are not needed in the equivalent input–output model of the system developed here.

eventually provide NL algebraic equations among the parameters, whose solutions would yield all *SI* parameters and parameter combos.

The first m equations of the characteristic set, i.e. those independent of the state variables, are the required input-output map: $\Psi(\boldsymbol{y}, \boldsymbol{u}, \boldsymbol{p}) = \mathbf{0}$. In general, the $\Psi(\boldsymbol{y}, \boldsymbol{u}, \boldsymbol{p}) = \mathbf{0}$ are *polynomial* equations in the variables $\boldsymbol{u}, \dot{\boldsymbol{u}}, \ddot{\boldsymbol{u}}, \ldots, \boldsymbol{y}, \dot{\boldsymbol{y}}, \ddot{\boldsymbol{y}}, \ldots$, with rational coefficients in the parameter vector $\boldsymbol{p}$. That is, we can write $\Psi_i(\boldsymbol{y}, \boldsymbol{u}, \boldsymbol{p}) = \sum_i c_i(\boldsymbol{p})\psi_i(\boldsymbol{u}, \boldsymbol{y})$, where $c_i(\boldsymbol{p})$ is a *rational function* in the parameter vector $\boldsymbol{p}$ and $\psi_i(\boldsymbol{u}, \boldsymbol{y})$ is a *differential function* in the variables $\boldsymbol{u}, \dot{\boldsymbol{u}}, \ddot{\boldsymbol{u}}, \ldots, \boldsymbol{y}, \dot{\boldsymbol{y}}, \ddot{\boldsymbol{y}}, \ldots$, etc.

Example 10.11: ***Input–Output Map for an ODE***

Consider the simple ODE model: $\dot{x} = kx + u$, $y = x/V$, with the *ranking* chosen as $\dot{x} > x > \dot{y} > y > u$. This yields an input–output equation (map) of the form:

$$\Psi(\boldsymbol{y}, \boldsymbol{u}, \boldsymbol{p}) = V\dot{y} - kVy - u = 0$$

In this example, $\psi_1(\boldsymbol{u}, \boldsymbol{y}) = \dot{y}$, $\psi_2(\boldsymbol{u}, \boldsymbol{y}) = y$ and $\psi_3(\boldsymbol{u}, \boldsymbol{y}) = u$.

Mathematically, it is convenient to express *SI* as an **injectivity condition**, as follows. Let $\boldsymbol{y} = \boldsymbol{\Phi}(\boldsymbol{p}, \boldsymbol{u})$ be the input–output map determined by eliminating the state variables $\boldsymbol{x}$. Consider the equation $\boldsymbol{y} = \boldsymbol{\Phi}(\boldsymbol{p}, \boldsymbol{u}) = \boldsymbol{\Phi}(\boldsymbol{p}^*, \boldsymbol{u})$, where $\boldsymbol{p}^*$ is an arbitrary parameter vector and $\boldsymbol{u}$ is the input function. Then: *one* solution $\boldsymbol{p} = \boldsymbol{p}^*$ corresponds to global ID; finitely many distinct solutions for $\boldsymbol{p}$ means local ID; and infinitely many solutions for $\boldsymbol{p}$ means *un*ID.

To form an injectivity condition, we invoke the input–output map $\Psi(\boldsymbol{y}, \boldsymbol{u}, \boldsymbol{p}) = \Psi(\boldsymbol{y}, \boldsymbol{u}, \boldsymbol{p}^*)$. This becomes $\sum[c_i(\boldsymbol{p}) - c_i(\boldsymbol{p}^*)]\psi_i(\boldsymbol{u}, \boldsymbol{y}) = 0$ for each input–output equation. Since $\Psi_i(\boldsymbol{y}, \boldsymbol{u}, \boldsymbol{p}) = 0$ form a basis in the vector space of differential polynomials in $\boldsymbol{u}$ and $\boldsymbol{y}$, they are *linearly independent* and thus $c_i(\boldsymbol{p}) - c_i(\boldsymbol{p}^*) = 0$. Therefore, global *SI* is the same as injectivity of the map $\boldsymbol{c}(\boldsymbol{p})$. That is: the model is *a priori* globally *SI* if and only if $\boldsymbol{c}(\boldsymbol{p}) - \boldsymbol{c}(\boldsymbol{p}^*) = 0$ implies $\boldsymbol{p} = \boldsymbol{p}^*$ for arbitrary $\boldsymbol{p}^*$.

The equations $\boldsymbol{c}(\boldsymbol{p}) = \boldsymbol{c}(\boldsymbol{p}^*)$ are called the **exhaustive summary** (Saccomani et al. 2003). These equations are then solved for the parameter vector $\boldsymbol{p}$ via the **Buchberger algorithm** and successive elimination. The resulting m equations have three possible solutions:

a unique solution (globally/uniquely ID), e.g. $p_i - p_i^* = 0$

a finite number of solutions (nonuniquely ID), e.g. $p_i - p_i^* = 0$ or $p_i - p_j^* = 0$

an infinite number of solutions (*un*ID), e.g. $p_i = F(\boldsymbol{p}, \boldsymbol{p}^*)$

The program *DAISY* (Bellu et al. 2007a,b) singles out which of the three cases it is. However, in the unidentifiable case, *DAISY* does not locate identifiable parameter combos explicitly. Another DA algorithm, an extension of *DAISY* methodology (Meshkat et al. 2009, 2011, 2012 a,b), *does* name all combinations of parameters, as well as single out all identifiable and unidentifiable parameters by name, programmed as web app *COMBOS*.[17] This is

17 Biocyb1.cs.ucla.edu/combos/html/

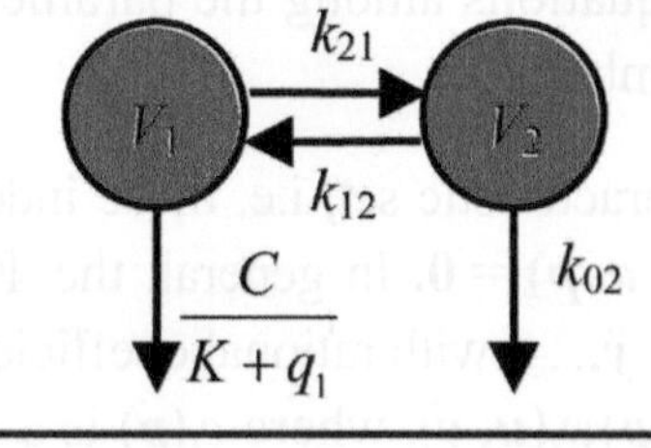

Fig. 10.14

accomplished in *COMBOS* by taking advantage of the fact that there are at least m identifiable coefficients of the input–output relations and, from these, identifiable parameter combos are found by algebraic rearrangement. In the simple Example 10.11 above, the coefficients of the characteristic set are V and combo kV. These are *SI* and are also easily separable, with V and k being identifiable. For most problems, however, combos and such easy separable solutions are not self-evident. Both *DAISY* and *COMBOS* are illustrated in the following examples.

Example 10.12: ***SI Analysis of a NL Two-Compartment Model Using the DAISY Algorithm***

Despite the straightforward nature of the Taylor series approach and its successful application for the one-compartment NL model in Examples 10.9 and 10.10 above, it does not readily extend itself to the NL two-compartment model with NL leak in compartment 1 (**Fig. 10.14**). The algebraic equations obtained from the output and its derivatives evaluated at $t=0$ are extremely complex and have been found to be intractable, even with symbolic manipulation packages, which ran out of memory in the process (Chappell et al. 1990). This problem was alternatively solved using the differential algebra approach, in (Saccomani et al. 2003), as described here. The authors showed the model parameters $\boldsymbol{p} = \begin{bmatrix} k_{02} & k_{12} & k_{21} & C & K \end{bmatrix}^T$ to be globally *SI* for this model.[18] The equations and compartmental diagram are:

$$\dot{q}_1 = -\left[k_{21} + \frac{C}{K + q_1(t)}\right] q_1(t) + k_{12} q_2(t)$$

$$\dot{q}_2 = k_{21} q_1(t) - (k_{02} + k_{12}) q_2(t) \tag{10.133}$$

$$q_1(0) = D, \quad q_2(0) = 0, \quad y(t) = q_1(t)$$

The input is a pulse-dose D, written as an equivalent IC, $q_1(0) = D$ and we set the volume $V = 1$ for convenience. As indicated above, the approach required first ranking of the variables, chosen as $u(t) < y(t) < q_1(t) < q_2(t)$. Instead of the derivatives of the output, as with the Taylor series, the *closed-form normalized characteristic set* is

18 Don't forget Caveat #2 in an earlier *Remark*!

computed, generating the following I/O relationship, containing only the measured (known) variables $y(t)$, $u(t)$, and their derivatives:

$$\begin{aligned}
&\dddot{y}(t)y(t)^2 + k_{21}k_{02}y(t)^3 - (k_{02}+k_{12})y(t)^2u(t) + (k_{02}+k_{12}+k_{21})\dot{y}(t)y(t)^2 \\
&- y(t)^2u(t) + (2k_{02}k_{21}K + k_{02}C + k_{12}C)y(t)^2 + 2Ky(t)\ddot{y}(t) \\
&- 2K(k_{02}+k_{12})u(t)y(t) - 2K\dot{u}(t)y(t) + 2K(k_{02}+k_{12}+k_{21})y(t)\dot{y}(t) \\
&+ K_m(k_{02}k_{21}K + k_{02}C + k_{12}C)y(t) - K^2(k_{02}+k_{12})u(t) + K^2\ddot{y}(t) \\
&+ K(K\,k_{02} + K\,k_{12} + K\,k_{21} + C)\dot{y}(t) - K^2\dot{u}(t) = 0
\end{aligned} \tag{10.134}$$

This expression contains information on model ID in the coefficients of each of its terms. (It is important that the number of coefficients is finite, unlike with Taylor series solutions.) Setting each of these coefficients to presumably known quantities, expressed as ξ_i here, the authors obtain an exhaustive summary of the model parameter relationships, as given in Eq. (10.135) below. They thereby obtain a system of equations in the unknown vector $\boldsymbol{p} = \begin{bmatrix} k_{02} & k_{12} & k_{21} & C & K \end{bmatrix}^T$, which they solve by the Buchberger algorithm, showing that the biomodel (10.133) is globally *SI*, uniquely for each parameter. Actually, this set can be successively solved by hand.[19]

$$\begin{aligned}
&\xi_1 = k_{02}k_{21} \qquad \xi_2 = k_{02}+k_{12} \qquad \xi_3 = k_{02}+k_{12}+k_{21} \\
&\xi_4 = 2k_{02}k_{21}K + k_{02}C + k_{12}C \qquad \xi_5 = 2K \\
&\xi_6 = 2K(k_{02}+k_{12}) \qquad \xi_7 = 2K(k_{02}+k_{12}+k_{21}) \\
&\xi_8 = K(k_{02}k_{21}K + k_{02}C + k_{12}C) \qquad \xi_9 = K^2(k_{02}+k_{12}) \\
&\xi_{10} = K^2 \qquad \xi_{11} = K(Kk_{02}+Kk_{12}+Kk_{21}+C) \qquad \xi_{12} = K^2
\end{aligned} \tag{10.135}$$

Example 10.13: ***SI Analysis of a NL HIV Model Using COMBOS & Other Algorithms***

The following simplified HIV model[20] was analyzed for *SI* using differential algebra algorithms (Xia and Moog 2003) and (Saccomani et al. 2010), differently by each group:

$$\begin{aligned}
&\dot{T} = s - dT - \beta VT && \text{Healthy } T\text{-cells} \\
&\dot{T}_1 = \alpha_1\beta VT - \mu_1 T_1 - k_1 T_1 && \text{Latently-infected } T\text{-cells} \\
&\dot{T}_2 = \alpha_2\beta VT + k_1 T_1 - \mu_2 T_2 && \text{Productively-infected } T\text{-cells} \\
&\dot{V} = k_2 T_2 - cV && \text{Free HIV virus particles} \\
&y_1 = T, \quad y_2 = V && \text{Measured Outputs} \\
&\boldsymbol{p} \equiv \{s, d, \beta, \alpha_1, \alpha_2, \mu_1, \mu_2, k_1, k_2, c\}
\end{aligned} \tag{10.136}$$

19 Exercise E10.9.

20 Parameters in $\boldsymbol{p}$ are all physiologically based. s and d are the constituent T-cell production rate and degradation rate constants, μ_is are cell death rate constants, the α_is and β are infection rate constants, the k_is are infected cell and virus production rate constants and c is the virus death rat constant (Xia and Moog 2003).

Both found the overall model locally *SI*, with three parameters s, d and β uniquely *SI*. The analysis in (Xia and Moog 2003) was mostly analytical – with two levels of analytical identifiability testing (and no automation) – primarily demonstrating new theory. After minimal problem formulation, the (Saccomani et al. 2010) results were completed much more easily and quickly, in one pass using *DAISY*. (*DAISY* algorithms were developed over a decade and are nearly fully automated.)

We analyzed the model for *SI* using *COMBOS* and obtained the same results, even more quickly, facilitated by a simpler model entry interface. And *COMBOS* provided additional identifiability properties. It singled out c and μ_2 as locally identifiable, with three solutions. It also provided explicit *SI* parameter *combinations*, the sum and product $\mu_1 + k_1$ and $a_1k_1k_2$, each designated as having three different solutions; and the product a_2k_2 was found uniquely identifiable. This additional information can be quite useful in the parameter estimation process, as described further in Chapter 12.

What's Next?

In principle, structural ID (*SI*) analysis of linear time-invariant ODE models is a solved problem, at least for models of low to moderate dimension – up to the limits of computational resources. And for nonlinear ODE models, progress has been substantial over the last decade – as illustrated in the last section. In Chapter 13, we apply some of these innovations to problems like reparameterizing an *un*ID model into a *SI* one, in terms of identifiable *combos* – now easily found for many models – with the help of modern software tools.

The cutting-edge for *SI* analysis is currently differential algebra and other symbolic-computation-based methods. For medium to higher-dimensional models, both linear and nonlinear, algorithmic complexity and computational limitations remain challenging. Algorithm expansion and optimization and computer automation (e.g. parallelization) are in order.

Unfortunately, the "ID problem" usually goes much deeper in practice. Knowing that a model or model parameter is *SI* – a necessary condition for quantification – is not enough in the presence of real, noisy or otherwise limited data. The numerical ID (*NI*) problem, what some call *practical ID*, also needs to be addressed. This is a major topic of Chapters 12 and 13, on parameter estimation and numerical ID, and the remaining chapters as well. Parameter sensitivity methods – introduced in the next chapter, actually address both *SI* and *NI* simultaneously, but the *NI* connection and details are deferred to Chapter 12.

We shall see in the next chapters that parameter estimation and parameter sensitivity analysis are strongly interrelated steps in modeling. Estimation is about quantifying a model structure from output measurements. Sensitivity analysis is usually about analyzing intrinsic and emergent properties of the fully developed model; but it's also about exploring *SI* (and *NI*) (estimatable) properties in a different way than done in this chapter. These topics overlap in significant ways, via extended notions of ID – for both ideal data and noisy real data. We shall see that the distinction – ideal vs. noisy data –has quite important consequences in practice.

Exercises

E10.1 Rework Example 10.3 with $k_{03} = 0$, and show that both solutions are feasible by determining the signs of both solutions for all k_{ij}, which must be greater than zero for $i \neq j$ and less than zero for $i = j$.

E10.2 Show that the following two solutions satisfy the ID conditions for the model in E10.1.

$$[k_{02}, k_{12}, k_{13}, k_{21}, k_{31}] = [0.1, 0.2, 0.1, 0.2, 0.2] \text{ and } [0.1, 0.3, 0.1, 0.1, 0.2].$$

E10.3 What ID property does the model in E10.1 have, with $k_{02} = 0$, when $k_{12} > k_{13} - k_{02} > -k_{31}$?

E10.4 Analyze the following two models for ID of *each* k_{ij}, $i \neq j$:

a) $\dot{\boldsymbol{q}} = \begin{bmatrix} k_{11} & 0 & 0 \\ k_{21} & 0 & 0 \\ k_{31} & 0 & 0 \end{bmatrix} q + \begin{bmatrix} 1 \\ 0 \\ 0 \end{bmatrix} \boldsymbol{u}$ and $\boldsymbol{y}_a = \begin{bmatrix} 0 & 1 & 0 \\ 0 & 0 & c_1 \end{bmatrix} \boldsymbol{q}$

b) Same dynamics, but with $k_{01} \equiv 0$ and $\boldsymbol{y}_b = \begin{bmatrix} 0 & 1 & 1 \end{bmatrix} \boldsymbol{q}$

E10.5 **Identifiability Analysis for Two-Compartmental Model (Fig. 10.15).**

a) Analyze all four I/0 configurations individually for ID of all parameters k_{01}, k_{02}, k_{12}, k_{21}, V_1, V_2.

b) Do an *interval ID* analysis for the k_{ij}, i.e. find $k_{ij}^{\min}$ and $k_{ij}^{\max}$ if they exist.

E10.6 Consider the dual-input, single-output three-compartment model. The two experiments depicted are represented by **Fig. 10.16**:

$$H_{11}(s) = \frac{s^2 - (k_{22} + k_{33})s + k_{22}k_{33}}{D} \qquad H_{12}(s) = \frac{k_{12}s + k_{12}k_{33}}{D}$$

$$\begin{aligned} D &\equiv s^3 - (k_{11} + k_{22} + k_{33})s^2 + [k_{11}k_{33} - k_{12}k_{21} - k_{13}k_{31} + k_{22}(k_{11} + k_{33})]s \\ &\quad + k_{12}k_{21}k_{33} - k_{22}(k_{11}k_{33} - k_{13}k_{31}) \\ &\equiv s^3 + \alpha_3 s^2 + \alpha_2 s + \alpha_1 \end{aligned}$$

Analyze this model for the ID of individual parameters: k_{01}, k_{12}, k_{13}, k_{21} and k_{31} when both inputs are unit impulses, $u_1 = u_2 = \delta(t)$, applied independently on different days (or any two different times).

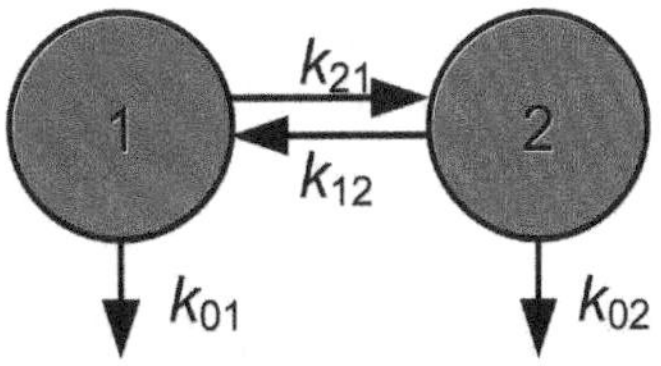

Fig. 10.15

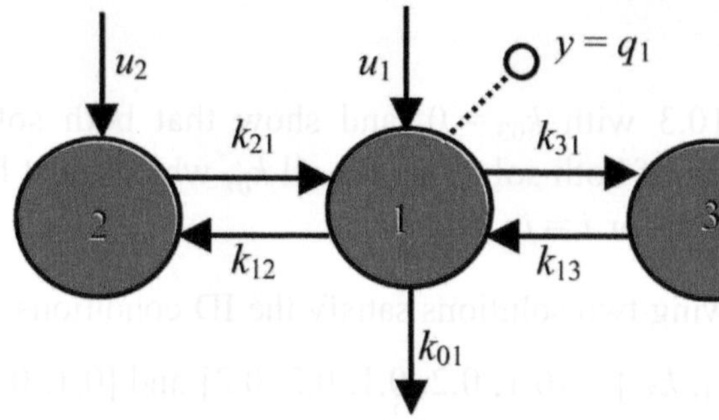

Fig. 10.16

E10.7 Consider the following constant-coefficient linear differential equation model: $\dot{x}_1 = p_1x_1 + p_2x_2 + \delta(t)$ and $\dot{x}_2 = p_3x_1 + p_4p_5x_2$ with initial conditions and outputs: $x_1(0) = x_2(0) = 0$ and $y_i = \frac{x_i}{V}$ for $i = 1, 2$

a) Classify each of the six parameters ($p_1, \ldots, p_5$ and V) as identifiable, uniquely identifiable, or unidentifiable. Use the transfer function approach to solve the problem.

b) Could this model be compartmental? If yes, what conditions are needed on the model parameters?

E10.8 Do ID analyses for the model in Example 10.6 for three different inputs:

a) $U(s) = \frac{1}{s+k}$, an exponential input

b) $U(s) = 1/s$, a step input

c) $U(s) = e^{-sT}$, an impulse delayed by T.

E10.9 Solve Eq. (10.135) for the unique model parameter set.

E10.10 Find the identifiable parameter combinations for the model of Exercise E10.6 using the online program *COMBOS* (at www.biocyb1.cs.ucla.edu/combos/html).

E10.11 The single output transform of a two-input model is given by:

$$Y_2(s) = \frac{k_{12}U_1(s) + (s - k_{22})U_2(s)}{s^2 - (k_{11} + k_{22})s + k_{11}k_{22} - k_{12}k_{21}}$$

The model is: $d\boldsymbol{q}/dt = \boldsymbol{K}\boldsymbol{q} + \boldsymbol{u}$ and $y = [0\ 1]\boldsymbol{q} = q_2$. The unknown parameters are k_{11}, k_{12}, k_{21}, k_{22}. (Forget k_{01}, it's not there!). If input $u_1(t)$ is a unit *impulse* and $u_2(t)$ is a unit *step* function, both applied at time $t = 0$, what parameters or parameter combinations are identifiable? Show your analysis.

E10.12 Show the identifiable combination relationship $k_{12}k_{21} = k_{11}k_{22} - \alpha_1$ for the 2-compartment model of *Example 10.2.* This result shows that identifiable combinations of this model can be written either as uniquely identifiable transfer function coefficients or as SI combos $k_{12}k_{21}, k_{11}, k_{22}$ (and V_1), both sets being structural invariants of the model.

References

Anderson, D., 1983. Compartmental Modeling and Tracer Kinetics. Springer-Verlag, New York.

Atiyah, M.F., MacDonald, I.G., 1969. *Introduction to Commutative Algebra*. Westview Press, Boulder CO.

Audoly, S., Bellu, G., D'Angió, L., Saccomani, M.P., Cobelli, C., 2001. Global identifiability of nonlinear models of biological systems. IEEE Trans. Biomed. Eng. 48, 55–65.

Balsa-Canto, E., A. A., Banga, J., 2010. An iterative identification procedure for dynamic modeling of biochemical networks. BMC Syst. Biol. 4, 11.

Balsa-Canto, E., Banga, J.R., 2011. AMIGO, a toolbox for advanced model identification in systems biology using global optimization. Bioinformatics. 27 (16), 2311–2313.

Bellman, R., Åström, K., 1970. On structural identifiability. Math. Biosci. 7, 329–339.

Bellu, G., Saccomani, M.-P., Audoly, S., D'Angiò, L., 2007. DAISY: a new software tool to test global identifiability of biological and physiological systems. Comput. Meth. Programs Biomed. 88 (1), 52–61.

Berk, P., Hower, R., et al., 1969. Studies of bilirubin kinetics in normal adults. J. Clin. Invest. 48, 2176–2190.

Berman, M., Schoenfeld, R., 1956. Invariants in experimental data on linear kinetics and the formulation of models. J. Appl. Phys. 27, 1361–1370.

Boas, R., 1935. A proof of the fundamental theorem of algebra. Am. Math. Mon. 42, 501–502.

Bolie, V., 1961. Coefficients of normal blood glucose regulation. J. Appl. Physiol. 16, 783–788.

Chapman, M., Godfrey, K.R., Chappell, M.J., Evans, N.D., 2003. Structural identifiability for a class of non-linear compartmental systems using linear/non-linear splitting and symbolic computation. Math. Biosci. 183 (2), 215.

Chappell, M., Godfrey, K.R., Vajda, S., 1990. Global identifiability of the parameters of nonlinear-systems with specified inputs – a comparison of methods. Math. Biosci. 102 (1), 41–73.

Chen, B., Landaw, E., et al., 1985. Algorithms for the identifiable parameter combinations and parameter bounds of unidentifiable catenary compartmental models. Math. Biosci. 76, 59–68.

Chis, O.-T., Banga, J.R., Balsa-Canto, E., 2011a. Structural identifiability of systems biology models: a critical comparison of methods. PLoS ONE. 6 (11), e27755.

Chis, O., B. J., Balsa-Canto, E., 2011b. GenSSI: a software toolbox for structural identifiability analysis of biological models. *Bioinformatics*. 27 (18), 2610–2611.

Cobelli, C., DiStefano III, J.J., 1980. Parameter and structural identifiability concepts and ambiguities: a criticial review and analysis. Am. J. Physiol. Regul. Integr. Comp. Physiol. 239 (1), R7–R24.

Cobelli, C., Frezza, M., Tiribelli, C., 1975. Modeling, identification, and parameter estimation of bilirubin kinetics in normal, hemolytic, and Gilbert's states. Comput. Biomed. Res. 8 (6), 522–537.

Cobelli, C., Romanin-Jacur, G., 1975. Structural identifiability of strongly connected biological compatmental systems. Med. Biol. Eng. 13 (6), 831–838.

Cobelli, C., Romanin-Jacur, G., 1976. Controllability, observability, and structural identifiability of multi-input and multi-output biological compartmental systems. IEEE Trans. Biomed. Eng. 23 (2), 93–100.

Cobelli, C., Sparcino, G., Caumo, A., Saccomani, M.P., Toffolo, G., 2006. Compartmental models of physiological systems. In: Bronzino, J. (Ed.), Biomedical Engineering Fundamentals. CRC Press, Boca Raton, pp. 9–14.

Cobelli, C., Toffolo, G., 1987. Theoretical aspects and practical strategies for the identification of unidentifiable compartmental systems. In: Walter, E. (Ed.), Identifiability of Parametric Models. Pergamon Press, New York, pp. 85–91.

Davidescu, F., J. S., 2008. Structural parameter identifiability analysis for dynamic reaction networks. Chem. Eng. Sci. 63, 4754–4762.

DiStefano III, J., 1976. Tracer experiment design for unique identification of nonlinear physiological systems. Am. J. Physiol. 230 (2), 476–485.

DiStefano III, J., 1982. Complete parameter bounds and quasiidentifiability conditions for a class of unidentifiable linear systems. Math. Biosci. 65, 51–68.

DiStefano III, J., 1983a. Complete parameter bounds and quasiidentifiability conditions for a class of unidentifiable linear systems. Math. Biosci. 65, 51–68.

DiStefano III, J., 1983b. Comprehensive dynamics of distribution and metabolism of thyroid hormones in blood and tissue pools using sequentially optimized experiment designs and a priori inequality analysis. 6th IFAC Symposium on Identification and System Parameter Estimation. Pergamon Press, New York.

DiStefano III, J., Cobelli, C., 1980. On parameter and structural identifiability: nonunique observability/reconstructability for identifiable systems, other ambiguities, and new definitions. IEEE Trans. Automat. Control. 25 (4), 830–833.

DiStefano III, J., Mori, F., 1977. Parameter identifiability and experiment design: thyroid hormone metabolism parameters. Am. J. Physiol. Regul. Integr. Comp. Physiol. 233 (3), R134–R144.

DiStefano III, J., Wilson, K.C., Jang, M., Mak, P.H., 1975. Identification of the dynamics of thyroid hormone metabolism. Automatica. 11, 149–159.

Godfrey, K., 1983. Compartmental Models and their Application. Academic Press, New York.

Godfrey, K., DiStefano III, J., 1985. Identifiability of Model Parameters. 7th IFAC/IFORS Symposium of Identification & Systems Parameter Estimation.

Godfrey, K., DiStefano III, J., 1987. Identifiability of model parameters. In: Walter, E. (Ed.), Identifiability of Parametric Models. Pergamon Press, Oxford, pp. 1–20.

Grewal, M., Glover, K., 1976. Identifiability of linear and nonlinear dynamical systems. IEEE Trans. Automat. Control. 21, 833–837.

Hagander, P., Tranberg, K., et al., 1978. Models for the insulin response to intravenous glucose loads. Math. Biosci. 42, 15–29.

Hajek, M., 1972. A contribution to the parameter estimation of a certain class of dynamical systems. Kybernetika. 8, 165–172.

Hevesy, G., 1962. Adventures in Radioisotope Research: The Collected Papers of George Hevesy in Two Volumes. Pergamon Press, New York.

Holford, N., 1987. Clinical pharmacokinetics of ethanol. Clin. Pharmacokinet. 13, 273–292.

Ider, Y., Bergman, R., 1978. Quantitative estimation of peripheral insulin sensitivity. Diabetes. 27, 408–468.

Irvine, C., 1975. A 4-compartment model of thyroxine metabolism. International Conference on Thyroid Hormone Metabolism. Pergamon Press, London.

Jang, M., Broutman, M., DiStefano III, J.J., 1981. T3, T4, and rT3 distribution and metabolism in obesity and aging. Endocrinology.T44, Supplement, 109.

Jang, M., DiStefano III, J.J., 1985. Some quantitative changes in iodothyronine distribution and metabolism in mild obesity and aging. Endocrinology. 116 (1), 457–468.

Landaw, E., Chen, B.C., DiStefano III, J.J., 1984. An algorithm for the identifiable parameter combinations of the general mammillary compartmental model. Math. Biosci. 72, 199–212.

LeCourtier, Y., Raksanyi, A., 1987. The testing of structural properties through symbolic computation. In: Walter, E. (Ed.), Identifiability of Parameter Models. Pergamon Press, Oxford, pp. 75–84.

Ličko, V., Silvers, A., 1975. Open-loop glucose-insulin control with threshold secretory mechanism: analysis of intrevenous glucose tolerance tests in man. Math. Biosci. 27, 319–332.

Ljung, L., Glad, T., 1994. On global identifiability of arbitrary model parameterizations. Automatica. 30, 265–276.

Margaria, G., Riccomagno, E., White, L.J., 2004. Structural identifiability analysis of some highly structured families of statespace models using differential algebra. J. Math. Biol. 49, 433–454.

Meshkat, N., Eisenberg, M., DiStefano III, J.J., 2009. An algorithm for finding globally identifiable parameter combinations of nonlinear ODE models using Gröbner Bases. Math. Biosci. 222 (2), 61–72.

Meshkat, N., Anderson, C., DiStefano III, J.J., 2011. Finding identifiable parameter combinations in nonlinear ODE models and the rational reparameterization of their input–output equations. Math. Biosci. 233, 19–31.

Meshkat, N., Anderson, C., DiStefano III, J.J., 2012. Alternative to Ritt's pseudodivision for finding the input-output equations of multi-output models. Math. Biosci. 239, 117–123.

Meshkat, N., Kuo, C., DiStefano III. J.J., 2014. COMBOS: web app for finding identifiable parameter combinations in nonlinear ODE models (Submitted to PLOS Computational Biology).

Miao, H., Xia, X., Perelson, A.S., Wu, H., 2011. On identifiability of nonlinear ode models and applications in viral dynamics. *SIAM Rev.* 53, 3–39.

Milanese, M., Molino, G., 1975. Structural identifiability of compartmental models and pathophysiological information from the kinetics of drugs. Math. Biosci. 26, 175–190.

Mueller, F., 2001. High-level parallel programming models and supportive environments. Lecture Notes in Computer Science. Springer Berlin, Heidelberg, 2026.

Nestel, P., White, H., et al., 1969. Distribution and turnover of cholesterol in humans. J. Clin. Invest. 48, 982–991.

Pohjanpalo, H., 1978. System identifiability based on the power series expansion of the solution. Math. Biosci. 41, 21–34.

Pohjanpalo, H., Wahlstrom, B., 1982. Software for solving identification and identifiability problems, e.g. in compartmental systems. Math. Comput. Simul. 24 (6), 490–493.

Rubinow, S., 1975. On closed or almost closed compartment systems. Math. Biosci. 18, 245–253.

Rubinow, S., Winzer, A., 1973. Compartment analysis: an inverse problem. Math. Biosci. 11 (3), 203–247.

Saccomani, M., Audoly, S., D'Angió, L., 2003. Parameter identifiability of nonlinear systems: the role of initial conditions. Automatica. 39 (4), 619–632.

Saccomani, M., D'Angió, L., Audoly, S., Cobelli, C., 2001. A priori identifiability of physiological parametric models. In: Carson, E., Cobelli, C. (Eds.), Modelling Methodology for Physiology and Medicine. Academic Press, San Diego, pp. 77–105.

Saccomani, M.P., Audoly, S., Bellu, G., D'Angio, L., 2010. Examples of testing global identifiability of biological and biomedical models with the DAISY software. Comput. Biol. Med. 40, 402–407.

Shannon, D., Sweedler, M., 1988. Using Groebner bases to determine algebra membership, split surjective algebra homomorphisms and determine birational equivalence. J. Symbolic Comput. 6, 267–273.

Skinner, S., Clark, R., et al., 1959. Complete solution of the three-compartment model in steady state after single injection of radioactive tracer. Am. J. Physiol. 196 (2), 238–244.

Specker, J., DiStefano III, J.J., Grau, E.G., Nishioka, R.S., Bern, H.A., 1984. Development-associated changes in thyroxine kinetics in juvenile salmon. Endocrinology. 115 (1), 399–406.

Vajda, S., 1981. Structural equivalence of linear systems and compartmental models. Math. Biosci. 55, 39–64.

Vajda, S., 1987. Deterministic identifiability and algebraic invariants for polynomial systems. IEEE Trans. Automat. Control. 32 (2), 182–184.

Vajda, S., DiStefano III, J.J., Godfrey, K.R., Fagarasan, J., 1989. Parameter space boundaries for unidentifiable compartmental models. Math. Biosci. 97, 27–60.

Vicini, P., Su, H.T., DiStefano III, J.J., 2000. Identifiability and interval identifiability of mammillary and catenary compartmental models with rate parameter constraints. Math. Biosci. 167, 145–161.

Wagner, J., 1976. Linear pharmacokinetic equations allowing direct calculation of many needed pharmacokinetic parameters from the coefficients and exponents of polyexponential equations which have been fitted to the data. J. Pharmacokinet. Biopharm. 4 (5), 443–467.

Wagner, J., DiSanto, A., et al., 1981. Reversible metabolism and pharmacokinetics: application to prednisone-prednisolone. Res. Commun. Chem. Pathol. Pharmacol. 32, 387–405.

Walter, E., LeCourtier, Y., 1982. Global approaches to identificability testing for linear and non-linear state-space models. Math. Comput. Simul. 24 (6), 472–482.

Xia, X., Moog, H., 2003. Identifiability of nonlinear systems with application to HIV/AIDS models. IEEE Trans. Automat. Control. 48 (2), 330–336.

Yoshikawa, T., Bhattacharyya, S., 1975. Partial uniqueness: observability and input identifiability. IEEE Trans. Automat. Control. 21, 713–714.

Chapter 11

Parameter Sensitivity Methods

* All or parts of the sections denoted with asterisks are more advanced topics – mathematically speaking – and might be skipped over on first pass, or in a more introductory course of study. This does *not* mean they are unimportant, or less important, for understanding the practical modeling topics of this chapter.

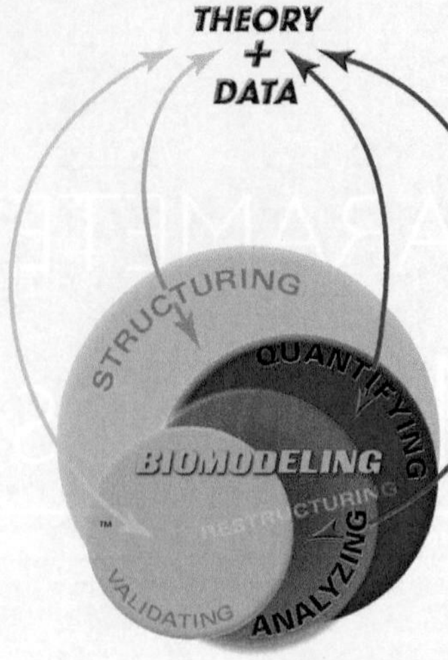

"The sensitivity of men to small matters, and their indifference to great ones, indicates a strange inversion."

Blaise Pascal ~1650

INTRODUCTION

Parameter sensitivity analysis is about quantifying the extent of variations in particular model *variables* – outputs, or functions of outputs, etc. – in response to variations in parameters. Model state variables and parameters are intrinsically coupled via the model equations, and the extent of particular couplings generally varies over a wide range – thereby affecting dynamical (which includes steady state) characteristics of the model in similarly differential ways. If model variables change very little or not at all when a particular parameter value varies, they are said to be **robust** (insensitive) to variations in that parameter. At the opposite extreme, they can change dramatically (be exquisitely sensitive) to parameter variations – and be called (speaking loosely) "unstable," or **fragile**. Everything in between is also possible.

These are important issues. For example, looking at parameter sensitivity from a model-building perspective, less sensitive (more robust) parameters have less influence on output variables, making them more difficult to quantify. If a measured model output is *insensitive* to a parameter, then that parameter cannot be readily quantified from data associated with that output – bringing identifiability issues to mind. By the same reasoning, if the model output is very sensitive to a parameter, that parameter is "more identifiable." We develop these notions below, showing that parameter sensitivities and structural identifiability (*SI*) notions are directly related via output sensitivities, for models with or without uncertainties (noise) in the data. In the presence of noisy data, sensitivity analysis involves additional extensions of the identifiability notions of the last chapter. We develop these in the next chapter, on parameter *estimation* methodology from real (noisy) data. Parameter estimation *quality* and identifiability notions and measures merge in the presence of imperfect data.

Metabolic control analysis (MCA), which we study in Chapter 14, is an important application of sensitivity analysis. MCA addresses the sensitivity of steady state concentrations and fluxes in metabolic systems to variations in system parameters. More generally, these are important tools for synthetic biology/biotechnology. An example is the engineering of biological systems to produce energy or food, often by manipulating metabolic system parameters.

An important application of sensitivity analysis is in *model reduction*, i.e. finding simpler models – often far simpler – that fit the data. We addressed model reduction in different contexts in Chapters 5–7, for reducing/simplifying model structures using graph theory and other algebraic and geometric properties of MC and enzyme and PIN models. Model reduction based on sensitivity theory is a different approach, particularly useful for complex molecular systems biology models of large-scale cellular signal transduction pathways and other biochemical networks. Identifying their essential (dominating) dynamical features – characterized by parameters with the largest relative parameter sensitivities – is a key to reducing their complexity and thereby exposing key biological mechanisms (for example). We introduce the tools for model reduction by sensitivity analysis in this chapter and develop them for application to large systems biology models in later chapters, in Chapter 17 especially.

The math here is intermediate level, presented in a tutorial, expository manner, primarily using vector-matrix notation, to simplify exposition visuals as well as rendering results applicable to problems and models of any order. Simple first-order scalar models or developments are used to introduce concepts, as in the next sections, and extended to higher-order models later. *Open-loop* and *closed-loop feedback system* examples are featured, which also illustrate – as an aside – a major advantage of feedback loops in biological and other physical systems. Additional complexity is included in the Exercises section, and these topics are extended further in subsequent chapters as we develop their applications to current problems in the life sciences.

Sensitivity to Parameter Variations: The Basics

In many modeling applications – in biology, engineering, econometrics, policy making, etc. – it is of interest (or necessary) to know how much a model output y (or internal state variable x) might change, say by Δy, if a model parameter p changes by Δp. For the present, we assume only one output and one parameter and formalize the notion of sensitivity in simple scalar terms, using differentials Δp and $\Delta y(t)$. We consider deterministic, noise-free models first, with extensions to noisy data in the next chapter. Open- and closed-loop control systems illustrate the meaning of sensitivity well.

Example 11.1: ***Open-Loop Parameter Sensitivity***

Suppose $p = k$, a constant, and the model is simply $y(t) = ku(t)$. Then, with u assumed fixed, if k changes by Δk, we have $y(t) + \Delta y(t) = (k + \Delta k)u(t)$, $\Delta y(t) = \Delta k u(t)$, and:

$$\frac{\Delta y(t)}{y(t)} = \frac{\Delta k}{k}$$

We interpret this equation as indicating that, for this simple **open-loop** (*no feedback*) model, the fractional change in the output, $\Delta y(t)/y(t)$, is equal to the fractional change in the parameter k, $\Delta k/k$, i.e.

$$\frac{\Delta y(t)/y(t)}{\Delta k/k} = 1$$

This is called the **relative sensitivity** *of* y w.r.t k, which in this special case linear model is equal to one. In general, it can be greater than or less than 1.

Remark: Relative sensitivity is more intuitive than (absolute) sensitivity, for the same reason it is easier to understand – for example – relative amounts, or increases, or decreases in anything.

Relative Sensitivity of Outputs & Functions of Outputs to Parameter Variations

More formally, first consider output function $y(t,p)$ dependent on p, as above, and use infinitesimal differentials $\partial y(t)$ and ∂p instead of $\Delta y(t)$ and Δp. Then, for any output $y(t,p)$ and any constant parameter p, the **relative sensitivity of** $y(t,p)$ w.r.t. p is:

$$S_p^{y(t,p)} \equiv \frac{\partial y(t,p)/y(t)}{\partial p/p} = \frac{\partial y(t,p)}{\partial p}\frac{p}{y(t,p)} \tag{11.1}$$

To generalize, instead of $y(t,p)$, we define the relative sensitivity of any model function $F(t, \mathbf{p})$ generally depending on (vector of parameters) $\mathbf{p}$ as:

$$S_{\mathbf{p}}^{F(t,\mathbf{p})} \equiv \frac{\partial F(t,\mathbf{p})}{\partial \mathbf{p}} \frac{\mathbf{p}}{F(t,\mathbf{p})} \tag{11.2}$$

$F(t, \mathbf{p})$ can be a function of any differentiable function of model variables. Examples include the model output – as above – the *output integrated over time* (called "area under the curve" (AUC)), any state variables, the model s-domain *transfer function* – illustrated below – its time-domain *impulse response* function, a *"cost" function* of parameter estimation errors to be optimized, etc.

Eqs. (11.1) and (11.2) and, in fact, all parameter sensitivity functions – in any form, are explicitly dependent on the value of parameter p. In other words, they are *context-dependent* – the context being the value of p (in parameter space) about which the output variation $\partial y(t,p)$ or $\Delta y(t)$ is of interest when varying p by ∂p or Δp. This means parameter sensitivities have different values for different (usually constant) values of the parameters and, in this sense, are not unique and, as we shall see, relatively difficult to pin down in practice.

Example 11.2: ***Closed-Loop Relative Parameter Sensitivity: Feedback Control Characteristics***

Consider the simple **feedback** (f.b.) model in **Fig. 11.1**, with scalar parameters $G > 0$ and $F > 0$, which means the model is an algebraic one (the G and F are constants rather than functions of s, as in more general transfer functions). The input–output relation is easily shown to be $y(t) = \frac{Gu(t)}{1 \pm GF} = Hu(t)$, where the plus (+) signifies *negative* and the minus (−) *positive* f.b. (DiStefano III et al. 1990). If only G changes by $\Delta G > 0$:

$$y(t, G) + \Delta y(t, G) \equiv \frac{(G + \Delta G)u(t)}{1 \pm (G + \Delta G)F}$$

Subtracting y from both sides and then dividing by y gives (after a little algebra):

$$\frac{\Delta y(t, G)}{y(t, G)} = \frac{\Delta G}{G}\left(\frac{1}{1 \pm F(G + \Delta G)}\right)$$

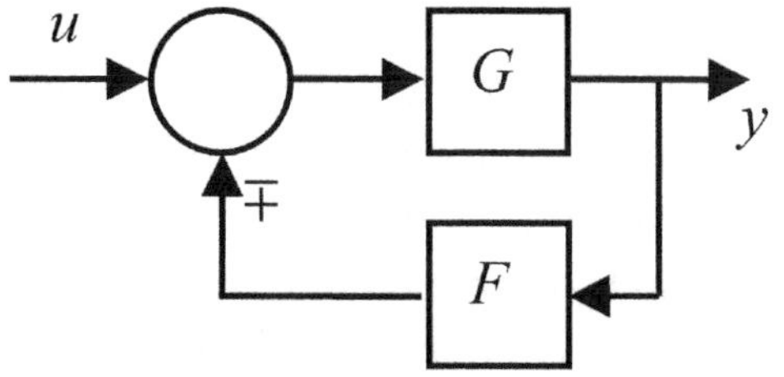

Fig. 11.1

Then the closed-loop relative sensitivity is approximately:

$$S_G^{y(t,G)} \simeq \frac{\Delta y(t,G)/y(t,G)}{\Delta G/G} = \frac{1}{1 \pm F(G+\Delta G)} = \begin{cases} <1 & \text{for negative f . b.}(+\,sign) \\ >1 & \text{for positive f . b.}(-\,sign) \end{cases} \tag{11.3}$$

This example demonstrates a **primary attribute of negative feedback (f.b.) systems**, i.e. negative feedback generally reduces output sensitivity to (forward loop) parameter variations, whereas the opposite is true in positive feedback systems. Using Eq. (11.1), we get the corresponding result for any differential change ∂G:

$$S_G^y = \frac{\partial y}{\partial G}\frac{G}{y} = \left(\frac{\partial}{\partial G}\left(\frac{Gu}{1 \pm GF}\right)\right)\frac{G(1 \pm GF)}{Gu} = \frac{1}{1 \pm GF} = \begin{cases} <1 & \text{for neg. f.b.} \\ >1 & \text{for pos.f.b.} \end{cases} \tag{11.4}$$

for any $GF > 0$. This result is consistent with (11.3), as $\Delta G \to 0$: $\lim_{\Delta G \to 0} S_G^{y(t,G)} \simeq \frac{\partial y/y}{\partial G/G} = \frac{1}{1 \pm FG}$.

We used simplistic algebraic constants for G and F in this example, and thus algebraic equations, to illustrate the point as simply as possible. The same concepts and definitions apply to differential equation models, where G and F are not simply constants, but become Laplace domain transfer functions and $Y = Y(s)$ and $U = U(s)$ are Laplace transformed functions of s (Appendix A).

Model outputs $y(t)$ are generally functions of time t and, therefore, so are their sensitivities to each parameter p_i. As we see later, this dependence of sensitivities on time can complicate matters. For the present, to simplify the notation, we often omit the time arguments on variables like $y(t)$, $u(t)$, $x(t)$ below, without losing sight of the fact that sensitivity functions are generally time-varying.

Example 11.3: ***Parameter Sensitivity for a Simple ODE***

We compute the sensitivity of the output $y(t) = x(t)$ to the single parameter $a > 0$ of the scalar ODE $\dot{x} = -ax + u \equiv f(x,a)$. We first obtain a new equation for $\partial x/\partial a \equiv v$, where v is called a **parameter sensitivity function.** This entails changing the order of differentiation:

$$\dot{v} \equiv \frac{d}{dt}\left(\frac{\partial x}{\partial a}\right) = \frac{\partial}{\partial a}\left(\frac{\partial x}{\partial t}\right) = \frac{\partial}{\partial a}f(x,a) = \frac{\partial f}{\partial x}\frac{\partial x}{\partial a} + \frac{\partial f}{\partial a}\frac{\partial a}{\partial a}$$

$$= \frac{\partial(-ax+u)}{\partial x}\frac{\partial x}{\partial a} + \frac{\partial(-ax+u)}{\partial a}\cdot 1 = -a\frac{\partial x}{\partial a} - x = -av - x$$

The $\dot{v}$ and $\dot{x}$ equation are clearly coupled and therefore we must solve $\dot{x} = -ax + u$ for $x(t)$ and $\dot{v} = -av - x$ for $v(t)$, *simultaneously*. The simulation program and sensitivity solution is illustrated in **Fig. 11.2** for a unit-step input and parameter value $a = 1$, over the interval $0 < t < 8$. The model output, in red, rises to 1 asymptotically, as expected for the unit step input. The sensitivity function $v(t) \equiv \partial x/\partial a$, in blue, falls from 0 to -1, indicating that the sensitivity to parameter a increases in magnitude with time.

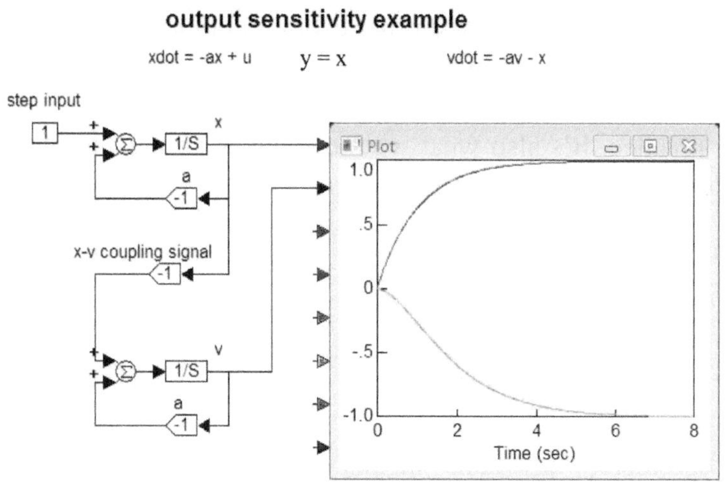

Fig. 11.2

Multiple Outputs & Parameters

In general, since models can have more than one output and more than one parameter, we can ask how much each output y_1, y_2, ..., y_m (or m-vector $\boldsymbol{y}$ of outputs) of a multi-output (MO) model might change in response to variations Δp_1, Δp_2, ..., Δp_P in each parameter p_1, p_2, ..., p_P (variations $\Delta \boldsymbol{p}$ in P-vector $\boldsymbol{p}$). As one might expect, the *sizes* of variations, their possible dependence on each other, and the values of p_i about which they vary all matter, to varying degrees. We develop the linearized (*local*) variational theory of sensitivity analysis first below. We assume first that parameter variations Δp_1, Δp_2, ... are *independent* of each other. This means that simultaneous variations in more than one parameter at a time, if they matter, are neglected. Intuition suggests that, depending on the circumstances – and the model – such "higher-order" parameter variations can matter. The mathematical and computational complexity of this issue (called *global* sensitivities) is introduced in this chapter as the theory unfolds.

State Variable Sensitivities to Parameter Variations

For ODE models, model outputs $\boldsymbol{y}$ are (of course) generated by model state variables $\boldsymbol{x}$. It follows that output sensitivities to parameters $\boldsymbol{p}$ are generated by state variable sensitivities to parameters $\boldsymbol{p}$. To see this, consider nonlinear ODE models of the form:

$$\dot{\boldsymbol{x}}(t,\boldsymbol{p}) \equiv \dot{\boldsymbol{x}} = \boldsymbol{f}[\boldsymbol{x}(t,\boldsymbol{p}),\boldsymbol{u}(t),t;\boldsymbol{p}] \tag{11.5}$$

$$\boldsymbol{x}(t_0,\boldsymbol{p}) \equiv \boldsymbol{x}_0(\boldsymbol{p}) \tag{11.6}$$

defined on the time interval $t_0 \leq t \leq T$, or in equivalent math language: $t \in [t_0, T]$. The time-varying (exogenous) input r-vector $\boldsymbol{u} \equiv \boldsymbol{u}(t)$ in Eq. (11.5) is independent of $\boldsymbol{p}$, but Eq. (11.6) includes the occasional situations where the initial condition (IC) n-vector $\boldsymbol{x}_0$ are a function of unknown parameters $\boldsymbol{p}$, rather than the constant ICs we usually encounter. We develop the theory next.

Total (Global) & First-Order (Local, Linear) Variations & Sensitivities

We use vector-matrix algebra[2] here to keep the notation compact (simple). References to vectors in the singular, as practiced below, i.e. small *variation* $\Delta\boldsymbol{p}$ in $\boldsymbol{p}$, means variations Δp_1, $\Delta p_2, \ldots, \Delta p_P$ in all components of vector variation $\Delta\boldsymbol{p}$ in parameter vector $\boldsymbol{p}$. This is conventional mathematical practice. It's also what we mean when we use the plural *parameters* in $\boldsymbol{p}$. We begin by considering a small variation $\Delta\boldsymbol{p}$ in $\boldsymbol{p}$.

The Taylor series (TS) expansion of $x_i(t,\boldsymbol{p}+\Delta\boldsymbol{p})$, i.e. the ith-component of state vector $x_i(t,\boldsymbol{p}+\Delta\boldsymbol{p})$, about $\boldsymbol{p}$ is:

$$x_i(t,\boldsymbol{p}+\Delta\boldsymbol{p}) = x_i(t,\boldsymbol{p}) + \frac{\partial x_i}{\partial \boldsymbol{p}}\Delta\boldsymbol{p} + \frac{1}{2}\Delta\boldsymbol{p}^T\left(\frac{\partial^2 x_i}{\partial \boldsymbol{p}^2}\right)\Delta\boldsymbol{p} + \text{higher-order terms} \tag{11.7}$$

Note that time t is considered "invariant" in this expansion, because we expand $x_i(t,\boldsymbol{p}+\Delta\boldsymbol{p})$ only about $\boldsymbol{p}$ due to variations $\Delta\boldsymbol{p}$. Defining $\Delta x_i(t,\boldsymbol{p}) \equiv x_i(t,\boldsymbol{p}+\Delta\boldsymbol{p}) - x_i(t,\boldsymbol{p})$, (11.7) yields the **total (global) variation** $\Delta x_i(t,\boldsymbol{p})$ of the ith-component of state vector $\boldsymbol{x}(t,\boldsymbol{p})$ for all $t \geq t_0$:

$$\Delta x_i(t,\boldsymbol{p}) = \frac{\partial x_i}{\partial \boldsymbol{p}}\Delta\boldsymbol{p} + \frac{1}{2}\Delta\boldsymbol{p}^T\left(\frac{\partial^2 x_i}{\partial \boldsymbol{p}^2}\right)\Delta\boldsymbol{p} + \text{higher-order terms} \tag{11.8}$$

Remark: In Eqs. (11.7) and (11.8) the first partial derivative of x_i w.r.t. vector $\boldsymbol{p}$ is (defined as) a row vector: $\partial x_i/\partial\boldsymbol{p} \equiv [\partial x_i/p_1 \quad \partial x_i/p_2 \quad \cdots \quad \partial x_i/p_P]$; and the second partial derivative $\partial^2 x_i/\partial\boldsymbol{p}^2 \equiv [\partial^2 x_i/\partial p_j \partial p_k] \equiv H$ is a P by P **Hessian matrix** of x_i w.r.t. $\boldsymbol{p}$, with components nonzero in general (Appendix B). Third-order and terms higher-order than three similarly include yet higher-order derivatives of x_i with respect to three or more parameters at a time. All this means that the total (global) variation $\Delta x_i(t,\boldsymbol{p})$ depends on these higher-order, nonlinear derivative terms as well as the linear first-order ones. First-order linear (local) variations $\Delta x_i(t,\boldsymbol{p})$ depend only first-order derivatives $\partial x_i/\partial\boldsymbol{p}$.

First-Order (Local, Linear) Variations & Sensitivities

Effects of higher-order terms on $\Delta x_i(t,\boldsymbol{p})$ are discussed later. In the remainder of this section we only need terms to first-order, so we lump the second-order (nonlinear) term in Eq. (11.8) with higher-order (nonlinear) terms (abbrev: h.o.t.). Preserving only first-order terms, and multiplying row vector $\partial x_i/\partial\boldsymbol{p}$ by column vector $\Delta\boldsymbol{p}$, Eq. (11.8) becomes:

$$\Delta x_i(t,\boldsymbol{p}) \equiv \Delta x_i = \frac{\partial x_i}{\partial p_1}\Delta p_1 + \frac{\partial x_i}{\partial p_2}\Delta p_2 + \ldots + \frac{\partial x_i}{\partial p_p}\Delta p_p + \text{h.o.t.} = \sum_{j=1}^{P}\frac{\partial x_i}{\partial p_j}\Delta p_j + O(\Delta\boldsymbol{p}^2) \tag{11.9}$$

Taylor's theorem was applied to obtain the rightmost equality, i.e. all h.o.t. in Eq. (11.9) can be represented by second-order terms $O(\Delta\boldsymbol{p}^2)$, with the second-order derivatives $\partial^2 x_i/\partial p_j \partial p_k$ evaluated somewhere on the interval of variations $\Delta\boldsymbol{p}$ (Kline 1998).

2 See Appendix B for aspects of vector-matrix algebra used in this and subsequent sections.

It follows then that the **first-order (local) variation** in the ith-component of state vector $\boldsymbol{x}(\boldsymbol{t}, \boldsymbol{p})$ is:

$$\Delta x_i(t,\boldsymbol{p}) \equiv \Delta x_i \cong \sum_{j=1}^{P} \frac{\partial x_i}{\partial p_j} \Delta p_j \quad (11.10)$$

This first-order approximation of Δx_i is the basis of the theory for computing local sensitivity functions, defined next. **Global sensitivity analysis** (GSA), which depends also on the h.o.t. terms in Eq. (11.9), is discussed later in the chapter and developed more fully in Chapter 17, along with methods for doing GSA.

Local State Variable Sensitivities

The (first-order, local) **sensitivity function,** i.e. **the local sensitivity of state variable x_i to parameter p_j,** is the time-varying partial derivative function:

$$v_{ij}(t,\boldsymbol{p}) \equiv v_{ij} \equiv \frac{\partial x_i(t,\boldsymbol{p})}{\partial p_j} \equiv \frac{\partial x_i}{\partial p_j} \quad (11.11)$$

For models with n state variables x_i and P parameters p_j, the $n \times P$ **matrix of sensitivity functions** $V(t, \boldsymbol{p}) \equiv V$ is then:

$$V(t,\boldsymbol{p}) \equiv V \equiv [v_{ij}] \equiv \left[\frac{\partial x_i}{\partial p_j}\right] \equiv \frac{\partial \boldsymbol{x}}{\partial \boldsymbol{p}} \quad (11.12)$$

Eq. (11.10) can then be written more simply as:

$$\Delta x_i \cong \sum_{j=1}^{P} v_{ij} \Delta p_j \quad (11.13)$$

and, in vector-matrix form, for all state variables, even more simply:

$$\Delta \boldsymbol{x}(t,\boldsymbol{p}) \cong V(t,\boldsymbol{p}) \Delta \boldsymbol{p} \quad (11.14)$$

For ODE models, state variable sensitivity functions are not usually computed by straightforward partial differentiation. Supplementary ODEs, dependent upon and therefore coupled with the state $\boldsymbol{x}(t)$, are formulated and solved. To determine differential equations for v_{ij} (and V), assume the state variable functions are suitably well behaved so that we can exchange the order of total and partial differentiation in the following equations:

$$\dot{v}_{ij} \equiv \frac{dv_{ij}}{dt} = \frac{d}{dt}\left(\frac{\partial x_i}{\partial p_j}\right) = \frac{\partial \dot{x}_i}{\partial p_j} \quad (11.15)$$

Thus, from Eq. (11.5),

$$\dot{v}_{ij} = \frac{\partial}{\partial p_j} f_i[\boldsymbol{x}(t,\boldsymbol{p}), \boldsymbol{u}(t), t; \boldsymbol{p}] = \left(\frac{\partial f_i}{\partial \boldsymbol{x}}\right)\left(\frac{\partial x}{\partial p_j}\right) + \left(\frac{\partial f_i}{\partial \boldsymbol{p}}\right)\left(\frac{\partial \boldsymbol{p}}{\partial p_j}\right) \quad (11.16)$$

Differentiating the parameter vector $\boldsymbol{p}$ term by term yields:

$$\frac{\partial \boldsymbol{p}}{\partial p_j} = [0\ldots0 \quad 1 \quad 0\ldots0]^T$$

$$\uparrow$$

$$j^{th} \text{ position}$$

Consequently, Eq. (11.16) reduces to

$$\dot{v}_{ij} = \left(\frac{\partial f_i}{\partial \boldsymbol{x}}\right)\left(\frac{\partial \boldsymbol{x}}{\partial p_j}\right) + \frac{\partial f_i}{\partial p_j} \tag{11.17}$$

for $i = 1, 2, \ldots, n$ and $j = 1, 2, \ldots, P$. Expanding (11.17) yields

$$\begin{aligned}\dot{v}_{ij} &= \begin{bmatrix} \frac{\partial f_i}{\partial x_1} & \frac{\partial f_i}{\partial x_2} & \cdots & \frac{\partial f_i}{\partial x_n} \end{bmatrix} \begin{bmatrix} \frac{\partial x_1}{\partial p_j} \\ \vdots \\ \frac{\partial x_n}{\partial p_j} \end{bmatrix} + \frac{\partial f_i}{\partial p_j} = \sum_{k=1}^{n} \frac{\partial f_i}{\partial x_k}\frac{\partial x_k}{\partial p_j} + \frac{\partial f_i}{\partial p_j} \\ &= \sum_{k=1}^{n} \frac{\partial f_i}{\partial x_k} v_{kj} + \frac{\partial f_i}{\partial p_j}\end{aligned} \tag{11.18}$$

This key equation is readily written in matrix (simplest, compact) form as:

$$\dot{V} = \frac{\partial \boldsymbol{f}}{\partial \boldsymbol{x}} V + \frac{\partial \boldsymbol{f}}{\partial \boldsymbol{p}} \tag{11.19}$$

The initial conditions (ICs) for Eqs. (11.18) or (11.19) are:

$$\begin{aligned} v_{ij}(t_0, \boldsymbol{p}) &\equiv \frac{\partial x_i(t_0)}{\partial p_j} \quad \text{for all } i \text{ and } j \\ V(t_0, \boldsymbol{p}) &= \frac{\partial \boldsymbol{x}_0}{\partial \boldsymbol{p}}, \quad n \text{ by } P \text{ matrix} \end{aligned} \tag{11.20}$$

Remark: Note that Eq. (11.19) is an $n \times P$ *matrix* ODE,[3] and it is *linear*, despite the fact that the original system ODE Eq. (11.5) (i.e. $\dot{\boldsymbol{x}} = \boldsymbol{f}$) is NL.

Remark: If the ICs are given as fixed constants in Eq. (11.6), the $v_{ij}(t,\boldsymbol{p})$ are all zero, i.e. the $n \times P$ matrix is null: $V(t_0,\boldsymbol{p}) \equiv [0]$, in Eq. (11.20). However, it's not uncommon in parameter estimation problems to have ICs as functions of the parameters. For example, when estimating parameters of a biomodel perturbed exogenously while in (nonzero) endogenous steady-state, the ICs are the endogeous steady-state values of the model state vector (at perturbation time t_0) and these are indeed functions of the unknown parameters of the not-yet-quantified model.

3 Solution of *matrix* ODEs is straightforward, as described in Appendix B.

Example 11.4: ***State Sensitivity for a Simple ODE with $x(t_0, p) = x_0(p)$ Not Constant***

Consider the same system model as in the previous example: $\dot{x} = -ax + u$, $a > 0$ and $u = 1$ for $t > 0$. The unit step input has been "on" for sufficiently long, so the model response is in constant steady state $x_\infty = 1/a$ (at $\sim t = 10$), obtained as the solution of $\dot{x} = -ax + 1 \equiv 0$. If we do a parameter estimation experiment with the system in this steady state, say at time $t_0 = 10$, then $x_\infty \to x(10)$ and the IC for the model with unknown parameter a is: $x(0) = 1/a$. **Fig. 11.3** illustrates this scenario for impulse perturbation (e.g. brief dose) = 200 applied at $t = 10$, and for $a = 5$ ($x_\infty \equiv x_0 = 1/5 = 0.2$). v jumps (further) down at $t = 10$, as it did beginning at $t = 0$, because of the $-x$ term in the $\dot{v} = -av - x$ equation.

We remind that the negative sign of the sensitivity function does not mean "less sensitive." It's the magnitude that usually counts.

Some Notable Features of State Variable Sensitivity Functions

1. Sensitivity functions $v_{ij}(t, \boldsymbol{p})$, and differentials $\Delta x_i(t, \boldsymbol{p})$, are functions of *time* as well as $\boldsymbol{p}$.

2. $\frac{\partial f}{\partial x}$ and $\frac{\partial f}{\partial p}$ in Eq. (11.19) are the (local) *Jacobian* matrices of the original ODE ($\dot{\boldsymbol{x}} = \boldsymbol{f}(\boldsymbol{x}, \boldsymbol{p}, t)$), Eq. (11.5). Each is generally dependent on $\boldsymbol{x}(t)$ on the interval $[t_0, T]$. Thus, linear ODE (11.19) for $V(t, \boldsymbol{p}) = \partial \boldsymbol{x}/\partial \boldsymbol{p}$ is generally time-varying, even if $\dot{\boldsymbol{x}} = \boldsymbol{f}(\boldsymbol{x}, \boldsymbol{p}, t)$ is time-invariant. One consequence of this is that methods for analysis of constant coefficient (time-invariant) linear dynamic system ODE models, e.g. Laplace transforms, are not generally applicable for computing sensitivity functions of these linear ODE models. They are, however, if the matrices in (11.19) have only constant values, as illustrated later.

3. If parameters $\boldsymbol{p}$ of $\dot{x} = \boldsymbol{f}(x, \boldsymbol{p}, t)$ are unknown – as in a parameter estimation problem – parameter values, called **nominal values**, denoted $\boldsymbol{p}^0$ – are nevertheless needed to solve for parameter sensitivities, because Eqs. (11.19) are functions of $\boldsymbol{p}$. Nominal values are first-approximations of $\boldsymbol{p}$, i.e. the best you can come up with (e.g. literature values). It is thus important to note that sensitivities v_{ij} are *local* sensitivities of the states x_i to

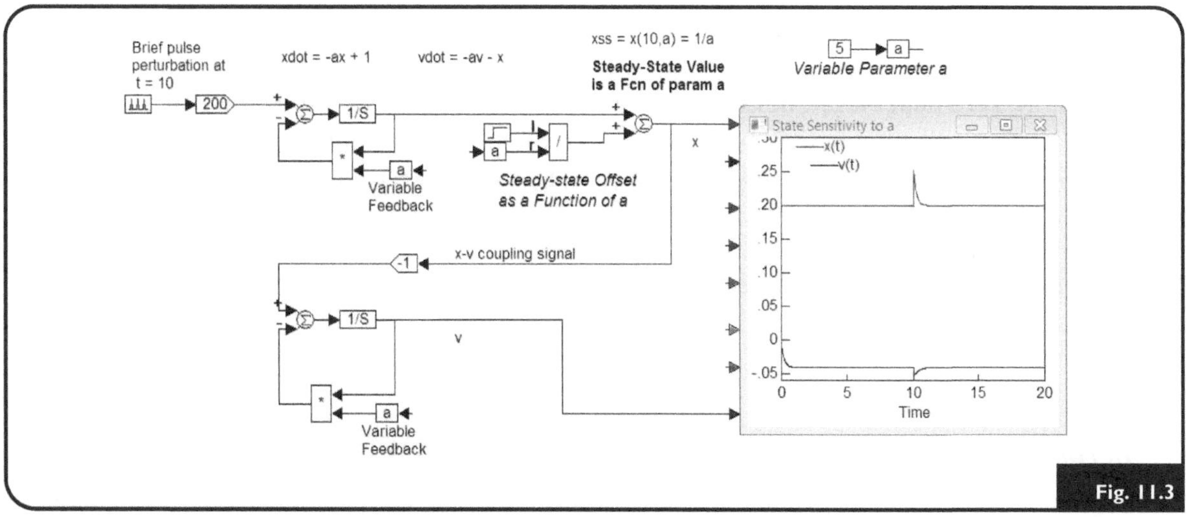

Fig. 11.3

parameters p_j, which means *in a neighborhood of this nominal* $\boldsymbol{p}^0$. Some numerical experimentation with different nominal parameter values may be needed.

4. The first-order sensitivity functions $v_{ij}(t, \boldsymbol{p}^0)$, are local variations of the system state $x_i(t, \boldsymbol{p}^0)$ caused by perturbations Δp_j of parameter p_j about its nominal p_i^0. The total sum of these variations, given by Eq. (11.10), then yields the total $\Delta x_i(t,\boldsymbol{p})$, but only under the assumption of independence of these variations in their effects on $\Delta x_i(t,\boldsymbol{p})$, i.e. assuming higher-order derivatives, e.g. second-order sensitivities $\partial x_i^2/\partial p_j \partial p_k$ and above, are zero or can be neglected. So, the $v_{ij}(t, \boldsymbol{p}^0)$ don't tell the whole sensitivity story. *They do not indicate what happens if two or more* p_j *vary simultaneously, or for different nominal values.* The assumption here depends on the particulars of the system, model and experiment under investigation and should be tested. We address this later in the context of *global sensitivity* analysis.

5. The original model $\dot{\boldsymbol{x}} = \boldsymbol{f}(\boldsymbol{x},\boldsymbol{p},t)$ is n-dimensional. The dimensionality of the parameter sensitivity equations is typically much larger: nP-dimensional ($P > 1$).

6. If the original model $\dot{\boldsymbol{x}} = \boldsymbol{f}(\boldsymbol{x},\boldsymbol{p},t)$ is *linear*, i.e. $\dot{\boldsymbol{x}}(t,\boldsymbol{p}) = \mathrm{A}(\boldsymbol{p})\boldsymbol{x}(t,\boldsymbol{p}) + B(\boldsymbol{p})\boldsymbol{u}(t)$, or more concisely, $\dot{\boldsymbol{x}} = A\boldsymbol{x} + B\boldsymbol{u}$, Eq. (11.19) becomes:

$$\dot{V} = AV + \frac{\partial(A\boldsymbol{x})}{\partial \boldsymbol{p}} + \frac{\partial(B\boldsymbol{u})}{\partial \boldsymbol{p}} \tag{11.21}$$

where $A\boldsymbol{x}$ and $B\boldsymbol{u}$ are vectors, and differentiation of a vector with respect to a vector is a matrix (Appendix B). Derivation is requested in Exercise E11.3.

7. As noted in the first section of this chapter, sensitivity functions at each time t, i.e. $\partial x_i(t,\boldsymbol{p}^0)/\partial p_j$, are easier to interpret if given a relative interpretation. Generalizing Eq. (11.1) or (11.2), the **relative sensitivity function** $S_{p_j}^{x_i}$ of x_i with respect to p_j, and the *matrix* S of these s_{ij}, are:

$$\begin{aligned} s_{p_j}^{x_i}(t,\boldsymbol{p}^0) &\equiv \frac{\partial x_i(t,\boldsymbol{p}^0)/\partial p_j}{x_i(t,\boldsymbol{p}^0)/p_j^0} \\ S(t,\boldsymbol{p}^0) &= \left[s_{p_j}^{x_i}(t,\boldsymbol{p}^0)\right] \equiv [s_{ij}] \end{aligned} \tag{11.22}$$

where the zero superscript means that matrix elements $s_{p_j}^{x_i}(t,\boldsymbol{p}^0)$ are evaluated at the nominal $\boldsymbol{p}^0$, also sometimes written as $\boldsymbol{p}_{NOM}$. If there is no ambiguity, we can also write $s_{p_j}^{x_i}(t,\boldsymbol{p}^0) \equiv s_{ij}$, as in the last equality above. Then the **matrix of relative sensitivity functions** s_{ij} is denoted S.

8. For small variations Δp_j in p_j, each element of S in (11.22) can be approximated as:

$$s_{ij} = \frac{\partial x_i/x_i}{\partial p_j/p_J^0} \simeq \frac{\Delta x_i/x_i}{\Delta p_j/p_j^0} \tag{11.23}$$

In Other Words: s_{ij} is a first-order measure of the fractional change Δx_i, in x_i due to a "small" fractional change Δp_j, in p_j about p_j^0. Strictly speaking, Δp_j does not have to be small, and this formula can be used to approximate sensitivities to larger perturbations, e.g. 5%, 10% etc., a common practice in many model quantitative analyses, (e.g. in Ferl et al. 2006; Douglas et al. 2010). But, again, these are linearized (local) sensitivity estimates (approximations). Total (global) Δx_i, which account for nonlinear effects of parameter variations (and different $\boldsymbol{p}^0$), could be significantly different. (More on this later.)

9. Relative sensitivity is mathematically equivalent to **logarithmic sensitivity** (DiStefano III et al. 1990), i.e.

$$s_{ij} = \frac{\partial \ln x_i}{\partial \ln p_j} \tag{11.24}$$

*Example 11.5: **Relative Sensitivities of Biomodel Parameters***

We extend the model in the previous example, but with $a = 2$ instead of 5, and relative sensitivity $s(t)$ of x w.r.t. $a > 0$ evaluated as well as $v(t)$. All is shown in **Fig. 11.4**.

The Canonical Sensitivity System

The sensitivity equations (11.18) or (11.19) can be formally combined with the state equations (11.5) for direct solution, because they are coupled and must be solved together. This single system of equations is called the **canonical sensitivity system**. Let's do this first for linear time-invariant models: $\dot{\boldsymbol{x}} = A\boldsymbol{x} + B\boldsymbol{u}$. We write V (first step below) as a matrix of column vectors $\boldsymbol{v}_j$ (second step), i.e.

$$V \equiv \begin{bmatrix} v_{11} & v_{12} & \cdots & v_{1j} & \cdots & v_{1,P-1} & v_{1P} \\ v_{21} & v_{22} & \ddots & v_{2j} & \ddots & v_{2,P-1} & v_{2P} \\ \vdots & \ddots & \ddots & \ddots & \ddots & \ddots & \vdots \\ v_{n-1,1} & v_{n-1,2} & \ddots & \ddots & \ddots & v_{n-1,P-1} & v_{n-1,P} \\ v_{n1} & v_{n2} & \cdots & v_{nj} & \cdots & v_{n,P-1} & v_{nP} \end{bmatrix} \tag{11.25}$$

$$\equiv \begin{bmatrix} \boldsymbol{v}_1 & \boldsymbol{v}_2 & \cdots & \boldsymbol{v}_j & \cdots & \boldsymbol{v}_P \end{bmatrix} = \begin{bmatrix} \frac{\partial \boldsymbol{x}}{\partial p_1} & \frac{\partial \boldsymbol{x}}{\partial p_2} & \cdots & \frac{\partial \boldsymbol{x}}{\partial p_j} & \cdots & \frac{\partial \boldsymbol{x}}{\partial p_P} \end{bmatrix}$$

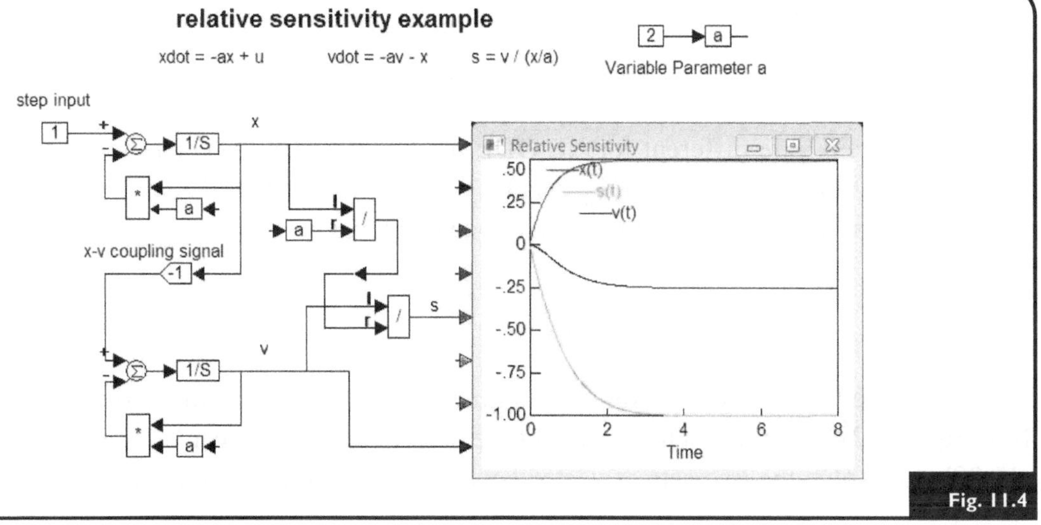

Fig. 11.4

Then Eq. (11.21) yields

$$\dot{v}_{1j} = \sum_{k=1}^{n} a_{1k} v_{kj} + \sum_{k=1}^{n} \frac{\partial a_{nk}}{\partial p_j} x_k + \sum_{k=1}^{n} \frac{\partial b_{nk}}{\partial p_j} u_k$$

$$\vdots$$

$$\dot{v}_{nj} = \sum_{k=1}^{n} a_{nk} v_{kj} + \sum_{k=1}^{n} \frac{\partial a_{nk}}{\partial p_j} x_k + \sum_{k=1}^{n} \frac{\partial b_{nk}}{\partial p_j} u_k$$

Therefore, the jth column of V is:

$$\dot{\boldsymbol{v}}_j = A\boldsymbol{v}_j + \frac{\partial A}{\partial p_j}\boldsymbol{x} + \frac{\partial B}{\partial p_j}u, \quad j = 1, \ldots, P \tag{11.26}$$

Next, define an **augmented state vector** $\boldsymbol{X}$ of dimension $n(P + 1)$:

$$\boldsymbol{X} \equiv [\boldsymbol{x}^T \boldsymbol{v}_1^T \ldots \boldsymbol{v}_j^T \ldots \boldsymbol{v}_P^T]^T = \left[\boldsymbol{x}^T \frac{\partial \boldsymbol{x}^T}{\partial p_1} \cdots \frac{\partial \boldsymbol{x}^T}{\partial p_j} \cdots \frac{\partial \boldsymbol{x}^T}{\partial p_P}\right]^T \tag{11.27}$$

Combining $\dot{\boldsymbol{x}} = A\boldsymbol{x} + B\boldsymbol{u}$ and (11.26) with (11.27) then gives the **canonical sensitivity system** for the *linear time-invariant* case:

$$\dot{X} = \underbrace{\begin{bmatrix} A & 0 & \cdots & 0 & 0 \\ \frac{\partial A}{\partial p_1} & \ddots & \ddots & \ddots & 0 \\ \vdots & 0 & A & \ddots & \vdots \\ \frac{\partial A}{\partial p_{P-1}} & \vdots & \ddots & \ddots & 0 \\ \frac{\partial A}{\partial p_P} & 0 & \cdots & 0 & A \end{bmatrix}}_{n(P+1) \times n(P+1)} X + \underbrace{\begin{bmatrix} B \\ \frac{\partial B}{\partial p_1} \\ \vdots \\ \frac{\partial B}{\partial p_P} \end{bmatrix}}_{n(P+1) \times r} u \tag{11.28}$$

The initial conditions for (11.28) are:

$$\boldsymbol{X}(t_0, \boldsymbol{p}) = \left[\boldsymbol{x}_0^T \frac{\partial \boldsymbol{x}_0^T}{\partial p_1} \frac{\partial_0^T}{\partial p_2} \cdots \frac{\partial \boldsymbol{x}_0^T}{\partial p_P}\right]^T \tag{11.29}$$

The details are a little different for NL ODE models $\dot{\boldsymbol{x}} = \boldsymbol{f}(\boldsymbol{x}, \boldsymbol{p}, \boldsymbol{u})$. The sensitivity matrix ODE (from Eq. (11.19)) is $\dot{V} = \frac{\partial f}{\partial x} V + \frac{\partial f}{\partial p}$, where $\frac{\partial f}{\partial x}$ and $\frac{\partial f}{\partial p}$ are both $nP \times nP$ matrices. In this case, the P columns of V are then the sensitivity vector ODEs:

$$\dot{\boldsymbol{v}}_j = \frac{\partial \boldsymbol{f}}{\partial \boldsymbol{x}} \boldsymbol{v}_j + \frac{\partial \boldsymbol{f}}{\partial p_j}, \quad j = 1, \ldots, P \tag{11.30}$$

The augmented vector $\boldsymbol{X}(t, \boldsymbol{p})$ for the nonlinear ODE would be the same (Eq. (11.27)), with ICs Eq. (11.29), but the ODEs cannot be expressed in "linear" vector-matrix form, like (11.28). Just solve the two sets of vector equations $\dot{\boldsymbol{x}} = \boldsymbol{f}(\boldsymbol{x}, \boldsymbol{p}, \boldsymbol{u})$ and (11.30) simultaneously.

Solution of (11.28), or $\dot{\boldsymbol{x}} = \boldsymbol{f}(\boldsymbol{x}, \boldsymbol{p}, \boldsymbol{u})$, with (11.30), requires solving $n(P + 1)$ coupled linear ODEs, a possibly formidable task for large n and P. Approaches do exist for reducing the number of ODEs to be solved, but these can be more trouble than they are worth (Markussen and DiStefano III 1982; Landaw and DiStefano III 1984). Simulating the full canonical sensitivity ODEs is often the best approach, with suitable derivative approximations (Leis and Kramer 1988; Corliss et al. 2001) – to simplify computations. These approximations are typically based on a variety of backward difference formulas (BDFs) for partial derivatives, like those described in Chapter 3 on Taylor series methods for numerical solution of ODEs. *Matlab* based program packages *AMIGO* (Balsa-Canto and Banga 2011) and *SensSB* (Rodriguez-Fernandez and Banga 2010) implement the ODESSA approach described in (Leis and Kramer 1985).

Output Sensitivities to Parameter Variations

Our interest here is how parameter perturbations affect *outputs* $\boldsymbol{y}$. Output equations have the general NL form:

$$\boldsymbol{y}(t, \boldsymbol{p}) = \boldsymbol{g}[\boldsymbol{x}(t, \boldsymbol{p}), \boldsymbol{p}] \tag{11.31}$$

This says that output $\boldsymbol{y}$ is an m-vector function of state vector $\boldsymbol{x}(t, \boldsymbol{p})$ and parameters $\boldsymbol{p}$, with $\boldsymbol{x}$ also considered a function of the vector $\boldsymbol{p}$ of P constant parameters p_j. We seek the m by P **matrix of output sensitivities** $\frac{\partial y}{\partial p}$ (the mP functions of time inside the matrix):

$$\frac{\partial \boldsymbol{y}}{\partial \boldsymbol{p}} \equiv \left[\frac{\partial y_i}{\partial p_j}\right] \equiv \begin{bmatrix} \frac{\partial y_1(t)}{\partial p_1} & \cdots & \frac{\partial y_1(t)}{\partial p_P} \\ \frac{\partial y_2(t)}{\partial p_1} & \cdots & \frac{\partial y_2(t)}{\partial p_P} \\ \vdots & & \vdots \\ \frac{\partial y_m(t)}{\partial p_1} & \cdots & \frac{\partial y_m(t)}{\partial p_P} \end{bmatrix} \tag{11.32}$$

Each term in this seemingly simple matrix is the partial derivative of the ith output $y_i(t, \boldsymbol{p}) = g_i[\boldsymbol{x}(t, \boldsymbol{p}), \boldsymbol{p}]$ w.r.t. p_j. We compute this $\partial y_i / \partial p_j$ using the chain-rule:

$$\begin{aligned} \frac{\partial y_i}{\partial p_j} &= \frac{\partial g_i}{\partial \boldsymbol{x}} \frac{\partial \boldsymbol{x}}{\partial p_j} + \frac{\partial g_i}{\partial \boldsymbol{p}} \frac{\partial \boldsymbol{p}}{\partial p_j} = \frac{\partial g_i}{\partial \boldsymbol{x}} \frac{\partial \boldsymbol{x}}{\partial p_j} + \frac{\partial g_i}{\partial p_j} \\ &= \sum_{k=1}^{n} \frac{\partial g_i}{\partial x_k} \frac{\partial x_k}{\partial p_j} + \frac{\partial g_i}{\partial p_j} = \sum_{k=1}^{n} \frac{\partial g_i}{\partial x_k} v_{kj} + \frac{\partial g_i}{\partial p_j} \end{aligned} \tag{11.33}$$

More compactly, the matrix of output sensitivities $\partial \boldsymbol{y} / \partial \boldsymbol{p}$ is:

$$\frac{\partial \boldsymbol{y}}{\partial \boldsymbol{p}} = \frac{\partial \boldsymbol{g}}{\partial \boldsymbol{x}} V + \frac{\partial \boldsymbol{g}}{\partial \boldsymbol{p}} \tag{11.34}$$

The matrix S_p^y of relative output sensitivities is denoted:

$$S_p^y \equiv \frac{\partial \boldsymbol{y}}{\partial \boldsymbol{p}} \bigg/ \frac{\boldsymbol{y}(t)}{\boldsymbol{p}} = \left[\frac{\partial y_i(t)}{\partial p_j} \bigg/ \frac{y_i(t)}{p_j} \right] \tag{11.35}$$

Some Notable Features of Output Sensitivity Functions

1. Output sensitivity functions $\partial y_i/\partial p_j$ depend on state sensitivities v_{ij}. State sensitivities must be computed to get output sensitivities.

2. As with state variable sensitivities, output sensitivities $\partial y_i/\partial p_j$ are time-varying functions, because the y_i are time-varying.

3. As with state sensitivities, the *Jacobian* matrices $\partial \boldsymbol{g}/\partial \boldsymbol{x}$ and $\partial \boldsymbol{g}/\partial \boldsymbol{p}$ are needed to solve for $\partial y_i/\partial p_j$ and these generally require a *nominal* $\boldsymbol{p}^0$ and complete knowledge of $\boldsymbol{x}(t)$ over the time interval of solution $[t_0,T]$.

4. If the state variable model is linear, i.e. $\dot{\boldsymbol{x}} = \boldsymbol{f}(\boldsymbol{x},\boldsymbol{p},\boldsymbol{u}) \equiv \dot{\boldsymbol{x}} = A\boldsymbol{x} + B\boldsymbol{u}$ then $\partial \boldsymbol{y}/\partial \boldsymbol{p}$ is:

$$\frac{\partial \boldsymbol{y}}{\partial \boldsymbol{p}} = CV + \frac{\partial (C\boldsymbol{x})}{\partial \boldsymbol{p}} \tag{11.36}$$

   where $C\boldsymbol{x}$ is a vector, and differentiation of a vector with respect to a vector is a matrix, as noted earlier.

5. Output sensitivity functions $\partial y_i/\partial p_j$ (or their relative versions $S_{p_j}^{y_i} = \frac{\partial y_i/y_i}{\partial p_j/p_j^0}$) usually are the sensitivities of interest in parameter estimation problems, because $y_i(t)$ are the *measured* variables, i.e. the *data* for "fitting" or quantifying model parameters. As we see in the next chapter, many parameter **search methods** depend on outputs being sufficiently sensitive to parameter variations as parameters are varied to improve the fit of model outputs to data.

6. If the outputs are *sampled*, and expressed at N discrete-times $y_i(t_1)$, $y_i(t_2)$, etc., $\partial \boldsymbol{y}/\partial \boldsymbol{p}$ is usually expressed as an *augmented mN by P matrix of numerical values* rather than functions, i.e. has mN rows – N rows of the y_i at each time instant, for each $i = 1, \ldots, m$. In this form, numerical operations can be readily performed on $\partial \boldsymbol{y}/\partial \boldsymbol{p}$, as illustrated in the next section.

Example 11.6: ***Output Sensitivities for Second- and First-Order ODEs***

We illustrate only the dimensionality issues and problem setup here. Numerical examples follow in subsequent sections.

For a second-order model, i.e. two ODEs, with a single output measurement of, say, state x_1, and three parameters, p_1, p_2 and p_3, the time-varying *output sensitivities* are: $v_{11} \equiv \partial y(t)/\partial p_1, v_{12} = \partial y(t)/\partial p_2, v_{13} = \partial y(t)/\partial p_3$, obtained by solving Eq. (11.18) for v_{11}, v_{12} and v_{13}, together with the two ODEs, which they depend on. The relative output sensitivities are then obtained from these solutions and Eq. (11.35) as:

$v_{11}/(x_1(t)/p_1), v_{12}/(x_1(t)/p_2)$ and $v_{13}/(x_1(t)/p_3)$. The output $y(t) = x(t)$ in the previous three simple first-order model examples, so the output sensitivities are equal to the state sensitivities here, but they are not in general.

Time-Averaged Relative Output Sensitivity Functions

Sensitivity functions are time-varying and therefore generally have different values at different times. Averaging over time can often be a meaningful metric for expressing the magnitude of relative sensitivity functions, particularly for establishing relative importance of model outputs, state variables or parameters in governing model dynamics. The **Euclidian** or $\boldsymbol{L_2}$**-norm**[4] of the sensitivity function of time is convenient for this purpose. This is the *root mean square* RMS of $s_{p_j}^{y_i}(t,\boldsymbol{p})$ over the time interval $(0,T)$ (or other interval). We consider output functions in both continuous-time and discrete-time form. Both are useful in applications.

If s is a continuous function of time, $s(t)$, the **continuous function root mean square (*RMS*)** of relative sensitivity function $s_{ij} = (\partial y_i/\partial p_j)/(y_i/p_j)$, is:

$$RMS\left[s_{ij}(t,\boldsymbol{p})\right] = \sqrt{\frac{1}{T}\int_0^T s_{ij}^2(t,\boldsymbol{p})dt} = \sqrt{\frac{1}{T}\int_0^T \left(\frac{\partial y_i(t)}{\partial p_j}\right)^2 \left(\frac{p_j}{y_i(t)}\right)^2 dt} \equiv \boldsymbol{length}\left[s_{ij}(t)\right] \tag{11.37}$$

If the data are a discrete-time sequence of N measurements, the **discrete-time function *RMS*** of relative sensitivities $s_{ij}(t_k) = (\partial y_i(t_k)/\partial p_j)/(y_i(t_k)/p_j)$ is:

$$RMS\left[s_{ij}(t_1,t_2,\ldots,t_N)\right] = \sqrt{\frac{1}{N}\sum_{k=1}^{N}\left(\frac{\partial y_i(t_k)}{\partial p_j}\right)^2\left(\frac{p_j}{y_i(t_k)}\right)^2} \equiv \boldsymbol{length}\left[s_{ij}(t_1,t_2,\ldots,t_N)\right] \tag{11.38}$$

Remark: Instead of $s_{ij}{}^2(t,p)$ in Eq. (11.37), the absolute value $\left|s_{ij}(t,p)\right|$ can be used under the integral sign (or in the summation in Eq. (11.38) – if the data are discrete-time), an integrand favored by some authors (Bentele et al. 2004). This is called the $\boldsymbol{L_1}$***-norm***.

Remark: In *network inference* and *model reduction* problems (Chapter 17), Equations (11.37) and (11.38) can be used to explore different state variables (network nodes) $x_1, x_2, \ldots, x_j$ (or functions of state variables), as *candidate outputs* for measurement – by substituting different x_i for y_i in these equations. The *relative importance* of different state variables (network nodes) to overall model dynamics can be established by comparing these metrics. Insensitive state variables (candidate outputs), or the parameters closely associated with them (e.g. edges joining these nodes), can possibly be eliminated from the model, or otherwise neglected. In some cases, insensitive parameters – if not eliminated – can be fixed to nominal values without affecting model outputs. Methods for model **parameter ranking** (ordering) – for their relative importance in affecting the dynamic response – are developed in Chapter 17.

4 The Euclidean norm of an n-vector $\boldsymbol{x}$ is given by: $\|x\|_2 = \sqrt{\sum_{i=1}^{n} x_i^2}$ (see Appendix B). This norm is the **length** of the vector.

Example 11.7: ***Average Magnitude (Length) of Relative Sensitivity Functions***

We take the simulated results of the previous example above one step further in **Fig. 11.5**, computing the *RMS* of the relative sensitivity of $y = x$ w.r.t. a. This is an accumulated average value (number) over the time interval of observation [0, 8] of x(t) (equivalently, the Euclidean length of $\partial x(t)/\partial a$ on [0, 8]). The number 0.3259 in the box on the bottom right is the *RMS* over [0, 8].

Sensitivity Measures of Model Quantification Results (Robustness)

Biomodels quantified from data are used for a variety of purposes. A typical application might be to use it to predict outcomes of different input–output (I–O) experiments that might be performed on the real biosystem represented by the model. To quantify such results, we usually define specific **objective functions**, or **figures-of-merit**, that take on numerical values for comparative analysis. For example, we might want to quantify effects of different drug dosage inputs on system output characteristics – like peak drug concentrations in blood – a number, or total drug absorbed over time – another number. Another application might be to use the model as a subsystem component in a larger model of a closed-loop optimal biocontrol system, with optimization criterion – **objective (loss)** – $L(y,p,u)$. For example, find a temporal pattern of drug input $u(t)$ over a given time interval that minimizes the total population of "microbial invaders" – represented for example by $y_1(t)$ in the model. Let's call all such different kinds of quantifiable goal points **objectives.**

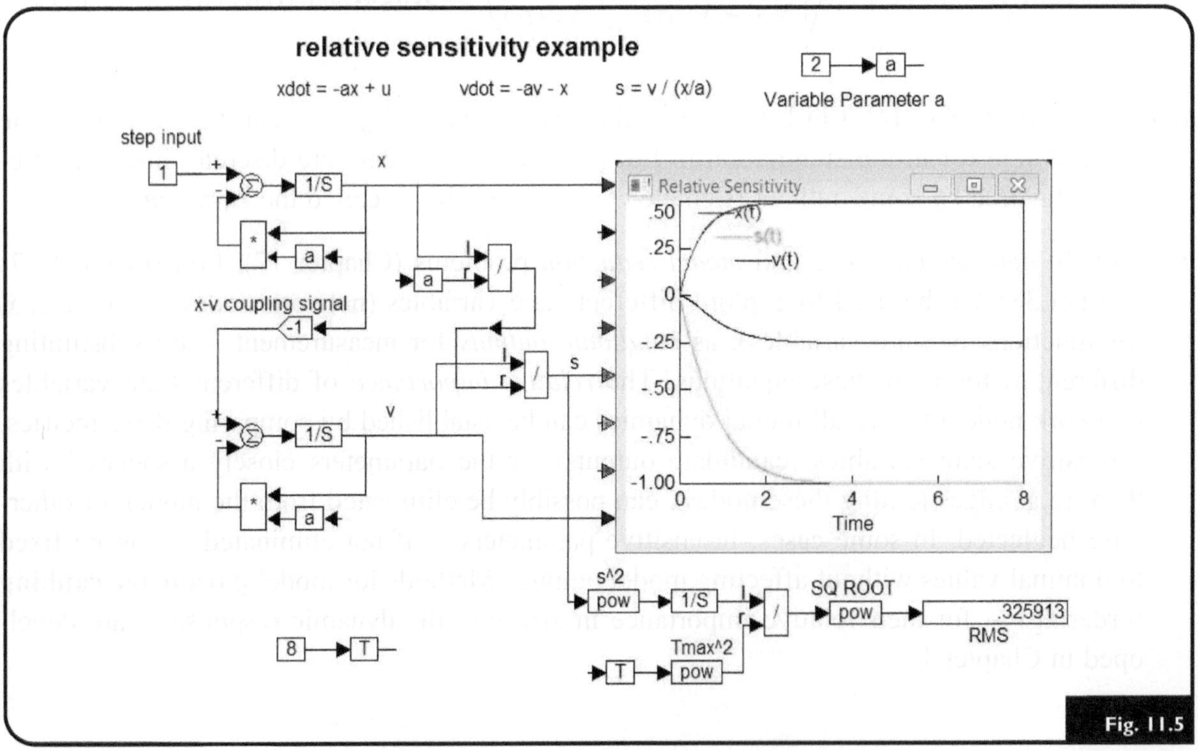

Fig. 11.5

Most such applications require **robustness analysis** and this means quantifying how objectives vary with parameter variations.[5] This involves computation of $\partial y/\partial \boldsymbol{p}$, or $\partial f(\boldsymbol{y})/\partial \boldsymbol{p}$, a (usually) scalar function of the model output. Examples are $f(\boldsymbol{y},\boldsymbol{p}) = \int_0^T y_1(t,\boldsymbol{p})dt \equiv AUC(\boldsymbol{p})$, or an optimization criterion like $L(\boldsymbol{y},\boldsymbol{p},\boldsymbol{u})$ noted above.

Following are *two different ways to evaluate robustness (sensitivity) measures*, in decreasing order of difficulty. We focus on relative rather than absolute sensitivities, because they are easiest to interpret.

1. Implement all state variable and sensitivity equations of the model (the canonical sensitivity system), as described in earlier sections, with a suitable program. Compute $\partial y/\partial \boldsymbol{p}$ and, from this, $\partial f(\boldsymbol{y})/\partial \boldsymbol{p}$, which usually is constant because it is computed over an interval, as illustrated for AUC directly above. This is the most direct approach. Several packages include computationally efficient algorithms (Leis and Kramer 1988), or variants of them, for evaluating sensitivities and ODEs simultaneously. *AMIGO*[6] (Balsa-Canto and Banga 2011) and *SensSB*[7] (Rodriguez-Fernandez and Banga 2010) are example program packages for systems biology models.
2. Solve the approximate relative sensitivity equation:

$$S_{p_j}^{f(y)} \cong \frac{\Delta f(\boldsymbol{y},\boldsymbol{p}_{NOM})/f(\boldsymbol{y},\boldsymbol{p}_{NOM})}{\Delta p_j/p_j} \tag{11.39}$$

   In (11.39), f is the scalar function of model output $\boldsymbol{y}$, $\boldsymbol{p}_{NOM}$ is the quantified (nominal) parameter vector, and p_j is the particular parameter of interest. The $\Delta p_i/\Delta p_j$ is the **fractional parameter variation** and $\Delta f/f$ is the **fractional response variation** to Δp_j. S is computed for several values of $\Delta p_j/p_j$, typically in the range 0.01 to 0.1 (1 to 10%), and for each p_j of interest.

Method 2 is popular, because it's easy, and $\Delta p_j/p_j$ doesn't have to be "small." It's not even unreasonable to ask how S varies with say 20% or even 50% variations in p_j. Results are approximate, of course, but typically useful. Application of Eq. (11.39) in a moderately complex model of manganese accumulation in rats and its relative sensitivity to variations of all model parameters quantified from real rat data is illustrated in Chapter 15.

An even easier and qualitative way to accomplish robustness (sensitivity) analysis for all model parameters $p_1, \ldots, p_P$ is provided by the *parameter covariance matrix* available following full model quantification. This is presented in the next chapter, in the section on parameter estimate quality assessments.

We now revisit structural identifiability (*SI*) and its important relationships to parameter sensitivities.

5 Objectives can be treated as function of other model entities, e.g. initial conditions or inputs instead of parameters. These are also of interest. For example, if a biosystem is operating with its output in saturation, small variations in the input likely will have no effect on the output. This would be important to know!

6 www.iim.csic.es/~amigo

7 www.iim.csic.es/~gingproc/SensSB.html

Output Parameter Sensitivity Matrix & Structural Identifiability[8]

The notions of structural identifiability (*SI*) of the last chapter are expressible in terms of parameter sensitivities. The relationship is quite direct and important in practice. Following is a useful way to view the meaning of parameter identifiability (ID), in the context of sensitivity, sometimes called **sensitivity identifiability**.

From this viewpoint, parameter ID is assessed quantitatively by knowing how changes Δp in each parameter p affect or change outputs $y(t)$, say by Δy. That is, p is **ID** if Δp *does* affect $y(t)$, i.e. $\Delta y(t)/\Delta p \neq 0$, for some t, and it is *un***ID** if it does not, i.e. if $\Delta y(t)/\Delta p = 0$ for all t.

We present only the basic relationship and its meaning here, to be developed further in the next chapter. *Both* structural and numerical ID properties of the model are inherent in the properties of the output parameter sensitivity matrix $\partial \mathbf{y}/\partial \mathbf{p}$, given by Eq. (11.32), and repeated here for convenience:

$$\frac{\partial \mathbf{y}}{\partial \mathbf{p}} \equiv \left[\frac{\partial y_i}{\partial y_j}\right] \equiv \begin{bmatrix} \frac{\partial y_1(t)}{\partial p_1} & \cdots & \frac{\partial y_1(t)}{\partial p_P} \\ \frac{\partial y_2(t)}{\partial p_1} & \cdots & \frac{\partial y_2(t)}{\partial p_P} \\ \vdots & & \vdots \\ \frac{\partial y_m(t)}{\partial p_1} & \cdots & \frac{\partial y_m(t)}{\partial p_P} \end{bmatrix}$$

Key Output Sensitivity Matrix–SI Relationships

1. *The (entire) parameter vector p is structurally identifiable (SI) if and only if the m (row) by P (column) matrix* $\partial \mathbf{y}/\partial \mathbf{p}$ *of time-functions* $\partial y_i(t)/\partial p_j$ *has constant rank P (full column rank)* (Bard 1974). In linear algebra terms, *this means all of the columns of* $\partial \mathbf{y}/\partial \mathbf{p}$ *are linearly independent, for all*[9] *t.*

Remark: For matrices of *constants*, rather than time-functions, the row rank is the same as the column rank. For time-varying sensitivity functions $\partial y_i(t)/\partial p_j$ populating this matrix, *SI* requires only that the columns of $\partial \mathbf{y}/\partial \mathbf{p}$ be linearly independent, not the rows. For example, if a model has only one measured output $y(t)$, then the number of rows in $\partial \mathbf{y}/\partial \mathbf{p}$ is $m = 1$, and the row rank cannot be > 1. The number of model parameters P, however, is usually > 1, so there's more than one column. In this case, linear independence[10] of the P sensitivity functions $\partial y(t)/\partial p_1$, $\partial y(t)/\partial p_2, \ldots, \partial y(t)/\partial p_P$ needs to be tested for *SI*.

8 This section is more mathematically advanced, but should be read first time through, so the important concepts and sensitivity-identifiability relationships are not overlooked. Even if you don't get the math, do read the words!

9 *Minor point*: because sensitivities are time-varying, it is possible the rank could change over time. If it does, then at least one parameter is not *SI* for some *t*.

10 For example, the polynomials $t, t^2, \ldots, t^P$ are linearly independent, because only $c_1 = c_2 = \ldots c_P = 0$ satisfies the equation: $c_1 t + c_2 t^2 + \ldots + c_P t^P = 0$ (Appendix B).

2. If $\partial \boldsymbol{y}/\partial \boldsymbol{p}$ has full column rank, i.e. if the columns are linearly independent, then $\boldsymbol{p}$ is *SI*, but not necessarily uniquely *SI*. The condition is local rather than global, and therefore one or more other (distinct) solutions for $\boldsymbol{p}$ may exist.

3. **Rank-deficient** sensitivity matrices, i.e. those with rank less than the max P, have some linearly dependent columns (sensitivity vectors). For example, if the rank of $\partial \boldsymbol{y}/\partial \boldsymbol{p}$ is $P-2$, two parameters are not locally *SI*. In this case, at least one parameter *combination* pair is *SI*. In vector space, these (sensitivity vector) columns are geometrically parallel.

4. If the m outputs $\boldsymbol{y}$ are measured or expressed at N discrete-time instants, i.e. $y(t_1), y(t_2), \ldots, y(t_N)$, the mN by P **discrete-time sensitivity matrix** $\partial \boldsymbol{y}(t_k)/\partial \boldsymbol{p}$ is:

$$\frac{\partial \boldsymbol{y}}{\partial \boldsymbol{p}} \equiv \left[\frac{\partial y_i(t_k)}{\partial p_j(t_k)}\right] = \underbrace{\begin{bmatrix} \frac{\partial y_1(t_1)}{\partial p_1} & \frac{\partial y_1(t_1)}{\partial p_2} & \cdots & \frac{\partial y_1(t_1)}{\partial p_P} \\ \frac{\partial y_1(t_2)}{\partial p_1} & \frac{\partial y_1(t_2)}{\partial p_2} & \cdots & \frac{\partial y_1(t_2)}{\partial p_P} \\ \vdots & \vdots & \cdots & \vdots \\ \frac{\partial y_1(t_N)}{\partial p_1} & \frac{\partial y_1(t_N)}{\partial p_2} & \cdots & \frac{\partial y_1(t_N)}{\partial p_P} \\ \frac{\partial y_2(t_1)}{\partial p_1} & \frac{\partial y_2(t_1)}{\partial p_2} & \cdots & \frac{\partial y_2(t_1)}{\partial p_P} \\ \vdots & \vdots & \cdots & \vdots \\ \frac{\partial y_m(t_N)}{\partial p_1} & \frac{\partial y_m(t_N)}{\partial p_2} & \cdots & \frac{\partial y_m(t_N)}{\partial p_P} \end{bmatrix}}_{P\text{-columns}} \quad m \ x \ N \ \text{Rows} \tag{11.40}$$

The rank of this matrix is readily computed by **singular value decomposition (SVD)**[11] (Appendix B). If the matrix has no **zero singular values**, the model is *SI*.

Remark: Eq. (11.40) is a matrix of constants, so now we can apply the notion that row and column ranks must be equal. *This means the output must be sampled at least as many times as the number of identifiable parameters P*, which is the row or column rank of this matrix.

5. Column sensitivity vectors corresponding to an **identifiable combination** pair are parallel. These correspond to *singular values equal to zero* (at least two singular values equal to zero).

Remark: The subject of **numerical identifiability** (*NI*) is developed systematically in the next chapter. Roughly speaking, *NI* means *practical* identifiability in the presence of noisy or otherwise limited data. Somewhat prematurely, we note here that, for noisy data, SVD analysis of $\partial \boldsymbol{y}/\partial \boldsymbol{p}$ provides qualitative information about *NI* as well as about *SI*. If matrix

11 Many mathematical software packages, e.g. *Matlab* and *Mathematica*, have commands for SVD analysis. The singular values of the matrix of constants $\partial \boldsymbol{y}/\partial \boldsymbol{p}$ (Eq. (11.40)) are also equal to the magnitude of its eigenvalues.

$\partial y/\partial \boldsymbol{p}$ has full column rank but is *ill-conditioned*,[12] then noise in data $z(t)$ measured at outputs $\boldsymbol{y}(t)$ will result in large variability (variance) in estimated parameter values, i.e. the model is not identifiable for practical purposes (not *NI*) – despite being *SI*. It will be more or less difficult to fully quantify. Geometrically, the reason is that (at least) two matrix columns representing two otherwise *SI* parameters are *almost* parallel, the angle between them being "small." In this case, all singular values are positive, but at least two are close to zero.

6. An alternative sensitivity criterion for *SI* is based on the *squared P* by *P* dimensional *output sensitivity matrix* $F' \equiv \left(\frac{\partial y}{\partial \boldsymbol{p}}\right)^T \left(\frac{\partial y}{\partial \boldsymbol{p}}\right)$. If F' is nonsingular[13] (invertible), $\boldsymbol{p}$ is *SI* (Cobelli and DiStefano 1980). In practice, this is a powerful result, especially when the data weighting matrix W, developed from the data variance model is incorporated, giving the **Fisher information matrix**[14] $F \equiv \left(\frac{\partial y}{\partial \boldsymbol{p}}\right)^T W \left(\frac{\partial y}{\partial \boldsymbol{p}}\right)$, from which parameter variances and correlations can be estimated. We develop this application in the next chapter.

Example 11.8: ***Output Sensitivity Equations and SI for a One-Compartment NL Model***

The NL one-compartment model in Example 10.11 of Chapter 10 was shown to be *SI* using an analytical Taylor series approach. This means all three parameters *C, K* and *V* are uniquely ID from noise-free output concentration measurements $y(t) = x(t)/V$. Taking a different approach here, we reassess identifiability properties of this model by analyzing its sensitivity matrix, beginning with computation of matrix $\partial y/\partial \boldsymbol{p}$. We modify the model a bit first, and use a unit step instead of an impulse input, to see if this makes a difference. We also assume $V = 1$ is given, to reduce the complexity of the example somewhat, without losing its didactic value. So $y(t) = q(t)/V = q(t)$. The unknown and nonzero parameter vector is thus $\boldsymbol{p} = [C \quad K]^T$, the unknown parameters of the nonlinear term. The model (see **Fig. 11.6**) is:

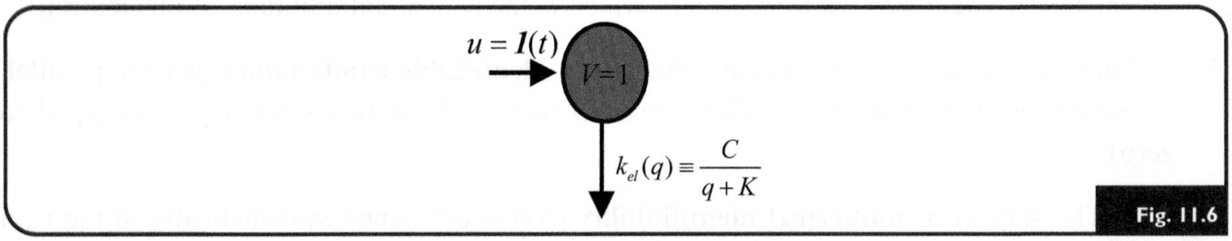

Fig. 11.6

12 Technically, $\partial y/\partial \boldsymbol{p}$ is **ill-conditioned** if the ratio of its largest to smallest singular values, called its **condition number** γ, is relatively large. Intuitively, a matrix A is ill-conditioned if the solution $\boldsymbol{x}$ of $A\boldsymbol{x} = \boldsymbol{y}$ is very sensitive to A or $\boldsymbol{y}$ values. This happens when at least two of the columns of A (or $\partial y/\partial \boldsymbol{p}$) are almost linearly dependent (almost collinear), but not quite. We develop this important concept further when we discuss numerical identifiability further, in Chapter 12.

13 If F' is singular (det $F' = 0$), the *condition number* γ of F' is a measure of how ill-conditioned it is, larger γ yielding greater sensitivity to numerical errors and hence poorer estimates of some parameters. We discuss this further in Chapter 12.

14 Abbreviated ***FIM*** by some authors.

$$\dot{q} = -\left[\frac{C}{q(t)+K}\right]q(t) + 1 \equiv f(q,u,\boldsymbol{p})$$
$$q(0) = 0, \quad y(t) = q(t) \tag{11.41}$$

The sensitivity matrix is $\partial y/\partial \boldsymbol{p} \equiv [\partial q/\partial C \quad \partial q/\partial K]$. Unfortunately, the sensitivity functions $v_1 \equiv \partial q/\partial C$ and $v_2 \equiv \partial q/\partial K$ are not available analytically (by partial differentiation) because the solution $q(t)$ of this NL ODE cannot be determined analytically in closed form.[15] Instead, we must simultaneously and numerically solve the model ODE (11.41), along with the following sensitivity ODEs from Eq. (11.15):

$$\dot{v}_1 \equiv \dot{v}_{11} = \frac{\partial f}{\partial q}\frac{\partial q}{\partial C} + \frac{\partial f}{\partial C} = \frac{\partial f}{\partial q}v_1 + \frac{\partial f}{\partial C} = \frac{-CK}{(q+K)^2}v_1 - \frac{q}{q+K}$$
$$\dot{v}_2 \equiv \dot{v}_{12} = \frac{\partial f}{\partial q}\frac{\partial q}{\partial K} + \frac{\partial f}{\partial K} = \frac{\partial f}{\partial q}v_2 + \frac{\partial f}{\partial K} = \frac{C(q-Kv_2)}{(q+K)^2} \tag{11.42}$$

Note that – unlike the case for the solution $q(t)$ – we *are* able to analytically evaluate the partials $\partial f/\partial q, \partial f/\partial C$ and $\partial f/\partial K$ in (11.42), as shown, because f is available from the ODE. Even better, these partials of f w.r.t. q, C and K are also obtainable automatically, without manual differentiation, using a *symbolic differentiation program command* – called *diff* in *Matlab* (or other program), which we use here to illustrate how this is done (see **Fig. 11.7**). This functionality is especially useful when the partials are not so easy to evaluate, as with more complex models.

We solve and plot the three ODEs (11.41) and (11.42), with nominal parameter values $C = 3$ and $K = 1$, over the observation interval $[0,T] = [0, 5]$, using the *Matlab* program code in **Fig. 11.8**, which includes two separate *Matlab* files, one for ODE functions and the main control script that calls the function file.

To check *SI*: Recall that the model is *SI* if the sensitivity matrix $\partial y/\partial \boldsymbol{p}$ has full column rank. We cannot readily assess the rank, or other key properties of this matrix for our problem, because we do not have analytical expressions for these time-varying solutions to work with. Instead, we have numerical solutions for the two functional elements of the matrix, which represent output sensitivities computed using the *Matlab* solver ODE45 at numerous discrete-time instants[16] over the interval [0, 5], and these are interpolated and illustrated as smoothed functions in **Fig. 11.8**, at resolutions appropriate for expressing their nature as continuous functions of time. To complete the *SI* analysis, we must first systematically discretize the solutions for sensitivity functions v_1 and v_2 at particular time-points of interest,[17] as in Eq. (11.40), and then analyze the resulting discretized sensitivity matrix for identifiability properties, as exemplified next.

15 This NL ODE is actually solvable analytically, using a computer algebra system, but not in explicit form, a result I owe to my friend, Professor Chris Anderson at UCLA.

16 Recall from discussion of solver ODE45 and other *variable-step* solvers in Chapter 3 that the number of time steps taken, and the location of these steps, is model- and model-parameter dependent.

17 These may be the times at which data are collected in a real experiment.

Matlab code for finding & returning the sensitivity functions symbolically

setting up equations for the sensitivity matrix elements

```
% declaration of parameter symbols (syms)
syms C K q v1 v2;

% state equation for the single compartment: qdot = f
f =  @(t, q) (-C*q / (K + q)) + 1;
```

symbolic computation of partial derivative functions

```
dfdq = diff(f, sym('q'))
dfdC = diff(f, sym('C'))
dfdK = diff(f, sym('K'))

dfdq =
(C*q)/(K + q)^2 - C/(K + q)

dfdC =
-q/(K + q)

dfdK =
(C*q)/(K + q)^2
```

sensitivity ODEs in above terms

```
v1dot = v1* dfdq + dfdC
v2dot = v2* dfdq + dfdK

v1dot =
- q/(K + q) - v1*(C/(K + q) - (C*q)/(K + q)^2)

v2dot =
(C*q)/(K + q)^2 - v2*(C/(K + q) - (C*q)/(K + q)^2)
```

substitute parameter values for symbols in the ODEs

```
f = subs(f, {C K}, {3 1})
v1dot = subs(v1dot, {C K}, {3 1})
v2dot = subs(v2dot, {C K}, {3 1})

f = 1 - (3*q)/(q + 1)

v1dot = v1*((3*q)/(q + 1)^2 - 3/(q + 1)) - q/(q + 1)

v2dot = (3*q)/(q + 1)^2 + v2*((3*q)/(q + 1)^2 - 3/(q + 1))
```

Fig. 11.7

Example 11.9: ***SI by Singular Value Decomposition (SVD) of the Discretized*** $\partial y/\partial p$

Besides the rank, SVD analysis also gives the singular values and condition number of $\partial y/\partial \boldsymbol{p}$. These yield information about numerical identifiability (next example) as well as SI. We assume first that the output $y(t) = q(t)$ in the above example is measured at five equally spaced time instants: $y(1), y(2), \ldots, y(5)$. Whereas the continuous-time sensitivity matrix for this model is 1-by-2, the discrete-time sensitivity matrix is a 5-by-2 matrix of numbers:

$$\partial y/\partial \boldsymbol{p} = \begin{bmatrix} -0.1128 & 0.2546 \\ -0.1943 & 0.4060 \\ -0.2298 & 0.4669 \\ -0.2432 & 0.4890 \\ -0.2478 & 0.4965 \end{bmatrix}$$

Solving NL 1-comp sensit model, using *Fig. 11.7* results, & plotting results

```
global C K;
C = 3;
K = 1;
[t, y] = ode45(@NL1comp, [0,5], [0,0,0]);  % NL1comp defined below
plot(t, y)
```

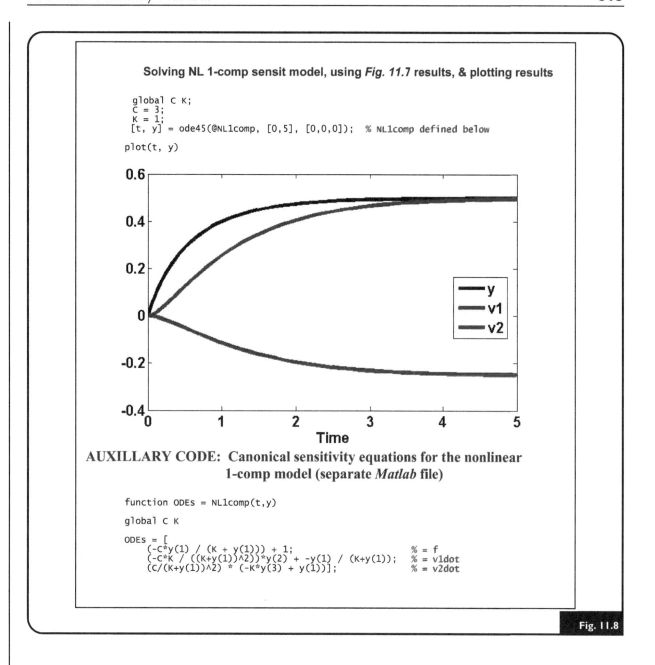

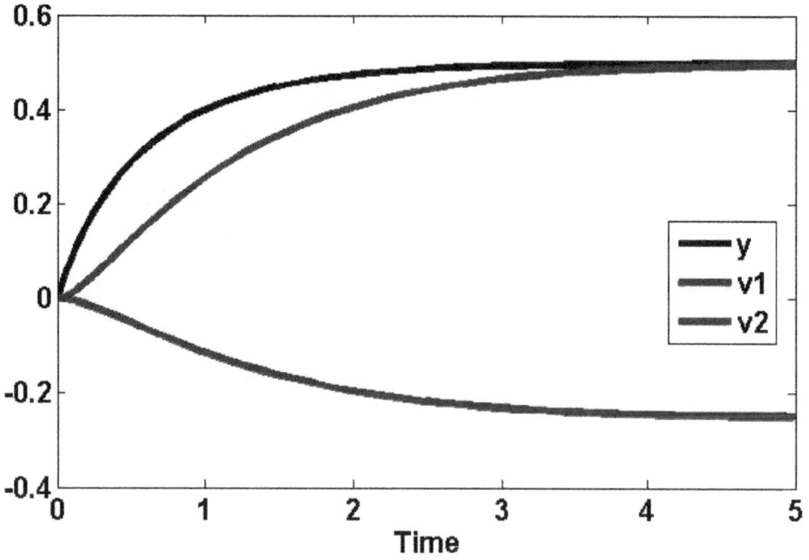

AUXILLARY CODE: Canonical sensitivity equations for the nonlinear 1-comp model (separate *Matlab* file)

```
function ODEs = NL1comp(t,y)
global C K
ODEs = [
    (-C*y(1) / (K + y(1))) + 1;                          % = f
    (-C*K / ((K+y(1))^2))*y(2) + -y(1) / (K+y(1));       % = v1dot
    (C/(K+y(1))^2) * (-K*y(3) + y(1))];                  % = v2dot
```

Fig. 11.8

Matlab program code for computing this $\partial y/\partial \boldsymbol{p} \equiv$ dydp is given in **Fig. 11.9**.

Using additional *Matlab* commands, which follow in **Fig. 11.10**, we get the *rank* of $\partial y/\partial \boldsymbol{p}$ as: rank(dydp) = 2; the *singular values* as: svds(dydp) = 1.0756, 0.0126; and *condition number* as: cond(dydp) = 85.36 (largest over smallest singular value). That there are no zero singular values indicates SI, consistent with the rank of 2 (two independent columns, full column rank) – thereby satisfying the SI conditions of properties 1 and 4 above. And the finite condition number triples the affirmation.

Example 11.10: ***SI from*** $F' \equiv (\partial \boldsymbol{y}/\partial \boldsymbol{p})^T(\partial \boldsymbol{y}/\partial \boldsymbol{p})$

To test for *SI*, we can check the invertibility of F' using property 6 above, as in **Fig. 11.11**. The determinant is $|F| = 1.84 \times 10^{-4} \neq 0$ and thus F' is invertible, confirming the model is *SI*.

Building the discretized sensitivity matrix (DSM) for analysis

```
function [DSM] = DSM_builder()

% specify model parameters

global C K;
C = 3;
K = 1;

% solve the cannonical sensitivity system

[t, X]=ode45(@NL1comp, [0,1,2,3,4,5], [0,0,0]);

% build the discretized sensitivity matrix

sens_matrix = [];   % initialize the matrix

for i = 2:1:6    % need v1 and v2 at times 1 to 5. Thus, skip row 1 (t = 0)

    dydp_iminus1 = [X(i, 2), X(i, 3)];    % where X(i, 2) = v1(i-1)
                                          % and   X(i, 3) = v2(i-1)

% enter newly calculated values into sensitivity matrix

    sens_matrix = [sens_matrix; dydp_iminus1];

end

% print the DSM entries
DSM = sens_matrix

DSM =

   -0.1128    0.2546
   -0.1943    0.4060
   -0.2298    0.4669
   -0.2432    0.4890
   -0.2478    0.4965
```

Fig. 11.9

Example 11.11: ***Qualitative Information from the Discretized*** $\partial y/\partial p$***:***
Effects of Sampling Rate

It is of interest to repeat the matrix computations for higher sampling rates than used in the two examples above, to get some qualitative information[18] about the potential information content of the output data sampled at different frequencies. We test twice as many, and then ten times as many output samples, over the same interval.

For 2 times the number of output samples, the matrix $\partial y/\partial \mathbf{p}$ (not shown) has 10 rows of data. Computations of singular values, rank and condition number yields: *svds* = 1.468, 0.0180; rank = 2 (as expected); and *cond* = 81.42 (rounded). This condition number reduction – from 85.36 to 81.42 – is not substantial, indicating that doubling the sampling rate would not improve the precision of estimates of parameters C and K, at least for the nominal values of 3 and 1 chosen for this illustrative example.

18 Qualitative here and in the next example means *numerical identifiability* measures, as described in Chapter 12.

Analysis of $\partial y / \partial p$

% Rank

```
rank(DSM_Builder)

ans =
    2
```

% Singular Values

```
svds(DSM_Builder)

ans =
   1.0756
   0.0126
```

% Condition Number

```
cond(DSM_Builder)

ans =
  85.2337
```

Fig. 11.10

Is $F' \equiv (\partial \mathbf{y} / \partial \mathbf{p})^T (\partial \mathbf{y} / \partial \mathbf{p})$ Invertible?

```
% call discretized matrix dydp

dydp

dydp =

   -0.1128    0.2546
   -0.1943    0.4060
   -0.2298    0.4669
   -0.2432    0.4890
   -0.2478    0.4965

% evaluate F':

F' = dydp'*dydp

F' =

    0.2238   -0.4568
   -0.4568    0.9334

% To check if it's invertible, compute the
determinant:

det(F')

ans =

    1.8426e-004
```

Fig. 11.11

For ten times the sampling rate – fifty samples over [0, 5] – the condition number is again reduced only slightly, so – the greater sampling rate is again not very helpful. It appears that the original five samples capture the slowly changing solution over the interval 0 to 5 in **Fig. 11.8** quite well.

Example 11.12: ***Qualitative Information from the Discretized $\partial y/\partial p$: Effects of Nomimal Parameter Values & Sampling Rate Doubling***

Example 11.11 is repeated here, using different nominal parameter values, namely $C = 3$ and $K = 0.1$, to illustrate how the extent of the nonlinearities in this simple model qualitatively affect results qualitatively. This is an important result, with implications in experiment design for model development and quantification. This topic is developed in Chapter 16.

The new computations and plots in **Fig. 11.12** below were obtained by changing the nominal parameters to $C = 3$ and $K = 0.1$ in the *Matlab* code in **Fig. 11.7** and **Fig. 11.8**. The transient dynamic response to approximately constant steady state values is much quicker.

$$\partial y/\partial \mathbf{p} = \begin{bmatrix} -0.024997 & 0.49995 \\ -0.02499 & 0.49985 \\ -0.024997 & 0.49996 \\ -0.025002 & 0.50003 \\ -0.025003 & 0.50004 \end{bmatrix}$$

The *singular values* of $\partial y/\partial \mathbf{p}$, computed with the earlier *Matlab* program, are: 1.11 and 2.7×10^{-6}; the *rank* of $\partial y/\partial \mathbf{p}$ is 2; and the *condition number* is 4.2×10^5. The singular value 2.7×10^{-6}, which is quite close to zero, still guarantees *SI*.

When we double the sampling rate and thus the number of output samples, SVD computations yield: *singular values* $= 1.58$ and 4.32×10^{-5}; *rank* $= 2$; *condition number* $= 0.37 \times 10^4$. The smallest singular value is about 5 times larger, and the condition number is >10 times smaller – both indicating that doubling the sample rate significantly improves the precision of estimates of the parameters for C and K for the nominal values of 3 and 0.1, respectively. This result is different than in the previous Example 11.11, where increasing the sampling 10-fold had little effect on qualitative measures.

These results make intuitive sense, supported by visualization of the curves in **Fig. 11.12**. Twice as many samples during early times captures the fast transient, whereas the first sample at $t = 1$ misses it completely. Most of the dynamical change is gone by about 0.5 time units. That means only one sample is registered during that period. The increased sampling rate provides more "information" about the evolution of the model dynamic response for the new set of nominal parameter values. Nevertheless, the still very large condition number predicts relatively poor parameter estimates from real data. The reason is that reducing the K value by a factor of 10 has

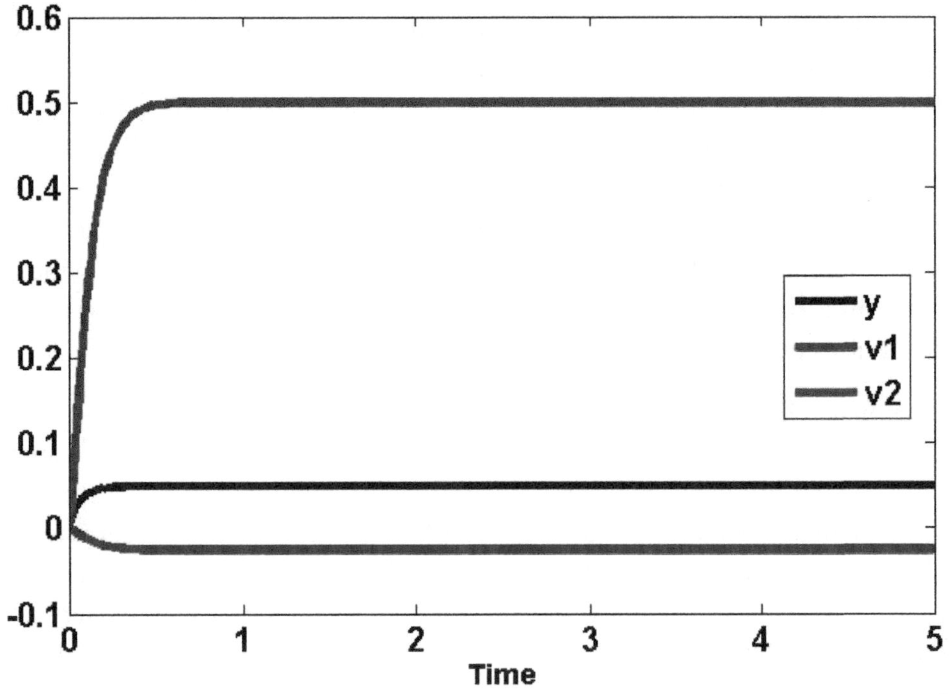

Fig. 11.12

the M−M leak function quickly operating near saturation, so only C/K, and not both C and K, would be numerically ID.[19] This is an example of what is called **overparameterization** (OPM), i.e. the complexity of the model structure is not a good match for the data. On the other hand, if a sufficient number of data points were collected during the linearly rising phase of the response (see the graph), say in the first 0.1 time units, then both C and K would become distinguishable (numerically ID, *NI*) and the model would not then be OPM.

This is an important result, with implications in experiment design for model development and quantification. This problem is extended further, with additional computations of quantitative measures of parameter estimation precision in the next chapter (Example 12.6). The overall topic is developed further in Chapters 16 and 17.

Global Parameter Sensitivities

In earlier sections, we developed and utilized *first-order (linear, local)* parameter sensitivities in model analyses − sensitivities of outputs $y_i(t)$ (or state variables $x_i(t)$) w.r.t. uninteracting single-parameters-p_j-at-a-time, at a *single* nominal point $\boldsymbol{p}^0$ (~best guesstimate) in parameter space. This is only part of the output-parameter sensitivity story, because different model parameters are definitively linked by the ODEs that define them. The effects of these couplings on

19 See the discussion of this issue at the beginning of the section on NL *SI* in Chapter 10.

parameter sensitivities is given by the complete Taylor series expansion of state variable variations $\Delta x_i(t,\boldsymbol{p})$, Eq. (11.9), to simultaneous parameter variations $\Delta p_1, \Delta p_2, \ldots, \Delta p_P$ *in toto*. This includes higher-order terms nonlinear in $\boldsymbol{\Delta p}$, beginning with second-order terms like $(\partial^2 x_i/\partial p_j \partial p_k)\Delta p_j \Delta p_k + \ldots$. In the theoretical development that follows Eq. (11.9), we used only the first-order term approximation for $\Delta x_i(t,\boldsymbol{p})$, namely Eq. (11.10), implicitly assuming no effects of one upon the other. Furthermore, we computed these at a *single* nominal point $\boldsymbol{p}^0$. **Global sensitivity analysis** (GSA) permits investigation of all parameter variation effects – both linear and nonlinear – on model state variables and outputs, and relaxes the dependence of sensitivities on a single $\boldsymbol{p}^0$ ("guesstimate") in parameter space. It generally includes assessment of sensitivities over the entire parameter space of possible values of $\boldsymbol{p}$.

To accomplish GSA, the variational theory of the first sections of this chapter can be developed further, to include two-parameters-at-a-time, and so on. The complexity and dimensionality of the problem, however, can grow substantially. For P parameters, state and output sensitivity matrices can grow by a factor of P for each additional parameter sensitivity variation computed, e.g. from $n(P+1)$ for one to $nP(P+1)$ for two parameters. Fortunately, more efficient methods have been developed[20] for addressing the problem. They are nevertheless fairly involved computationally and we reserve their development and discussion until Chapter 17.

Recap & Looking Ahead

In this chapter, we developed the basics and much of the detailed methodology for analyzing local output, state and other system function sensitivities to parameters. We illustrated these notions and their applications with relatively simple and completely worked out examples – to fortify the pedagogy. Robustness and fragility were defined as system characteristics reflected by relatively low and relatively high sensitivities of model variables and properties. Positive and negative feedback were shown to be enhancers and reducers of system sensitivity to parameters.

We also introduced the topic of global parameter sensitivities, to be developed for applications in Chapter 17. An important issue throughout was the intimate relationship between output sensitivity and identifiability of parameters, with important consequences in biomodel reduction applications – especially for reducing the dimensionality of systems biology network (and other) models. This topic also will be continued in Chapter 17.

We'll develop an important systems biology topic based on sensitivity analysis – metabolic control analysis (MCA) – in biocontrol system Chapter 14. MCA is about characterizing effects of small perturbations in variables and parameters in metabolic pathways operating in steady state. These effects, expressed as properties called control coefficients and elasticities, are local relative sensitivity functions in metabolic pathways. This is a key application of

20 And continue to be developed. This is an open research problem.

sensitivity functions in understanding and controlling metabolic regulation systems – a basis for biofuel production and synthetic biology.

The next two chapters address parameter estimation and numerical identifiability methodologies, the basics of model quantification and model quality measures – all strongly dependent – full circle – on parameter sensitivities. The identifiability thread continues with simultaneous development of numerical identifiability methodology along with parameter estimation reliability assessment measures.

Exercises

E11.1 ***Sensitivity With & Without Feedback***. The open-loop and closed-loop (feedback) models below have the *same* plant and *same* overall TF for gain $G = 2$, i.e. $H_1(s) \equiv H_2(s)$.

$$H_1 \equiv \left[\frac{Y}{U}\right]_1 = \frac{G}{s^2 + 4s + 5}$$

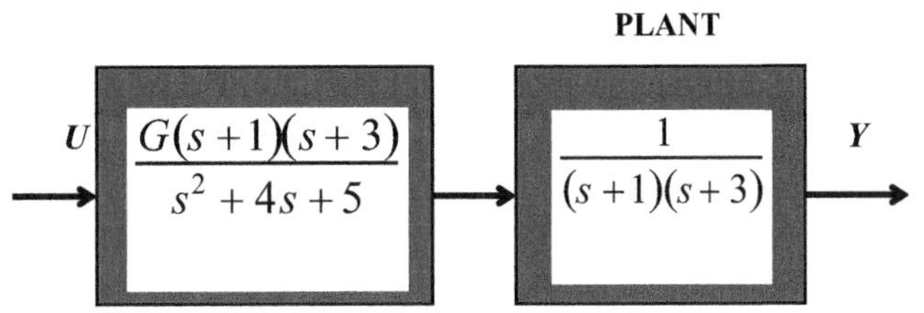

$$H_2 \equiv \left[\frac{Y}{U}\right]_2 = \frac{G}{s^2 + 4s + 3 + G}$$

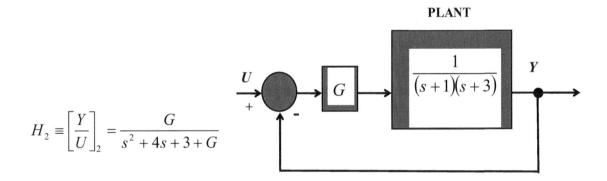

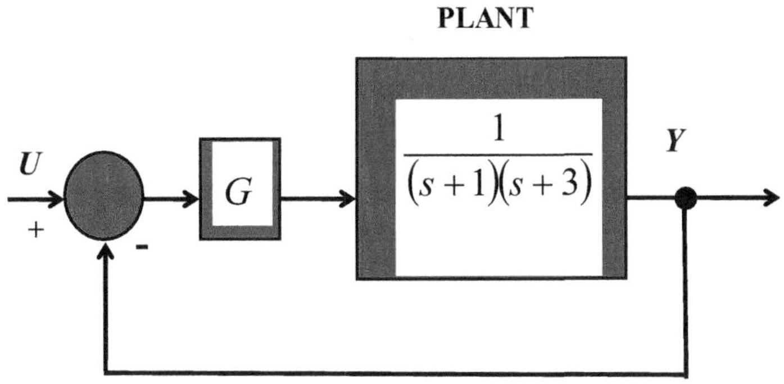

Show that, relative sensitivity $S_G^{H_1} = 1$ for all G, and $S_k^{H_2} < 1$, for all frequencies ω rad/sec of the sinusoidal transfer functions $H_1(j\omega)$ and $H_2(j\omega)$. This illustrates a major advantage of negative feedback (less relative sensitivity to parameter G).

E11.2 ***Higher-Order Sensitivity Functions***. Derive variational equations and ODEs for computing second-order state variable sensitivity functions, i.e. measures $v_{kij}^{(2)}$ of time-varying perturbations in x_k due to small variations Δp_i and Δp_j in *both* p_i and p_j. Assume a linear ODE model: $\dot{\boldsymbol{x}} = A\boldsymbol{x} + B\boldsymbol{u}$ with output $\boldsymbol{y} = C\boldsymbol{x}$. Also derive second-order *output* sensitivity functions.

E11.3 Derive the parameter sensitivity equations for constant-coefficient linear models.

E11.4 ***Sensitivities to initial states.*** In some biomodeling problems, one needs to find sensitivities of time-varying state variables $x_i(t)$ to perturbations or uncertainties in the initial state vector $\mathbf{x_0}$, e.g. in optimal therapy control problems, where initial states are typically not fixed, but vary with conditions requiring therapy. Derive state sensitivity equations for perturbations in initial states $\mathbf{x_0}$. Instead of redefining the sensitivity matrix V, do this by appending new variables $x_{j0} \equiv x_j(t_0)$ to the ODEs and separately derive the sensitivity equations they satisfy.

E11.5 ***Sensitivities to input perturbations.*** Derive sensitivities $\partial \mathbf{x}/\partial \mathbf{u}$ of state variables $x_i(t)$ to perturbations in inputs $u(t_j)$ at *fixed* times $t = \tau$, about constant nominal $\mathbf{u}_N$.

REFERENCES

Balsa-Canto, E., Banga, J.R., 2011. AMIGO, a toolbox for advanced model identification in systems biology using global optimization. Bioinformatics. 27 (16), 2311–2313.

Bard, Y., 1974. Nonlinear Parameter Estimation. Academic Press, New York.

Bentele, M., Lavrik, I., et al., 2004. Mathematical modeling reveals threshold mechanism in CD95-induced apoptosis. J. Cell Biol. 166 (6), 839–851.

Cobelli, C., DiStefano III, J.J., 1980. Parameter and structural identifiability concepts and ambiguities: a criticial review and analysis. Am. J. Physiol. Regul. Integr. Comp. Physiol. 239 (1), R7–R24.

Corliss, G., Faure, C., Griewank, A., Hascoet, L., Naumann, U. (Eds.), 2001. Automatic Differentiation: From Simulation to Optimization, Springer, New York.

DiStefano III, J., Stubberud, A.R., Williams, I.J., 1990. Feedback and Control Systems, 2nd Edition: Continuous (Analog) and Discrete (Digital). McGraw-Hill, New York.

Douglas, P., Cohen, M.S., DiStefano III, J.J., 2010. Chronic exposure to Mn may have lasting effects: a physiologically based toxicokinetic model in rat. J. Toxicol. Environ. Chem. 92 (2), 279–299.

Ferl, G., Kenanova, V., Wu, A.M., DiStefano III, J.J., 2006. A two-tiered physiologically-based model for dually labeled single-chain Fv-Fc antibody fragments. Mol. Cancer Ther. 5 (6), 1550–1558.

Kline, M., 1998. Calculus: An Intuitive and Physical Approach. Dover, London.

Landaw, E., DiStefano III, J., 1984. Multiexponential, multicompartmental, and noncompartmental modeling: physiological data analysis and statistical considerations, part II: data analysis and statistical considerations. Am. J. Physiol. Regul. Integr. Comp. Physiol. 246, R665–R677.

Leis, J.R., Kramer, M.A., 1985. Sensitivity analysis of systems differential and algebraic equations. Comput. Chem. Eng. 93–96.

Leis, J.R., Kramer, M.A., 1988. The simultaneous solution and sensitivity analysis of systems described by ordinary differential equations. ACM Trans. Math. Softw. 14, 45–60.

Markussen, C., DiStefano III, J., 1982. Evaluation of four methods for computing parameter sensitivities in dynamic system models. Math. Biosci. 61, 135–148.

Rodriguez-Fernandez, M., Banga, J.R., 2010. SensSB: a software toolbox for the development and sensitivity analysis of systems biology models. Bioinformatics. 26 (13), 1675–1676.

Chapter 12

PARAMETER ESTIMATION & NUMERICAL IDENTIFIABILITY

* All or parts of the sections denoted with a single asterisk are more advanced topics – mathematically speaking – and might be skipped over on first pass, or in a more introductory course of study. This does *not* mean they are unimportant, or less important, for understanding the practical modeling topics of this chapter.

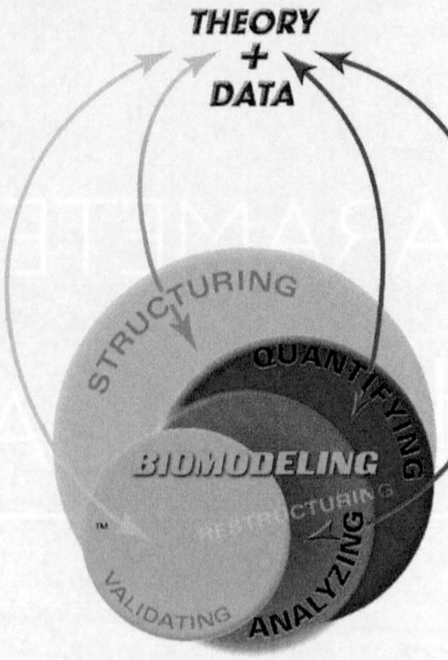

"The best we can do is size up the chances,
calculate the risks involved,
estimate our ability to deal with them,
and then make our plans with confidence."

Henry Ford ~1900

Biomodel Parameter Estimation (Identification)

Model Quantification by any Other Name

Model quantification is called the **inverse modeling problem**[2] in some fields, model or parameter **identification**[3] in systems engineering and related areas, and **model calibration**, or **tuning** or **training** or **fitting** in others. It usually means quantification of model parameters – the more direct connection with parameter identifiability notions. **Parameter estimation** extends the connection in the context of real data, which means quantification in the presence of uncertainties (noise) in data, or *approximation* of parameter "values." This is usually couched in a statistical context: a **parameter estimate**, e.g. a mean or median value, with a statistical variability measure associated with it, e.g. a **standard deviation** (*SD*) or **standard error** (*SE*), or a **confidence interval,** to quantify how good or bad the estimate may be.

Practically speaking, parameter estimation is one of the toughest problems in biodynamic system model building, with many books and research and review articles devoted to the subject (e.g. Bard 1974; Beck and Arnold 1977; Ljung 1987; Moles et al. 2003; Aster et al. 2005; Rodriguez-Fernandez et al. 2006; Chou and Voit 2009). It behooves us to know as much as possible about the process, so our efforts in practice will bear fruit rather than despair. Parameter estimation methods – methods shared with the area of *statistical inference*,[4] are based in numerical analysis, probability and statistical theory. In the current exposition, accessibility to the basic concepts and utility of the methodology are emphasized and theory is minimized.[5] Before getting into deeper verbiage about the math and other technical details of parameter estimation methodology, let's be clear about what we mean by **fitting a model to data**.

Model Fitting

We illustrate with two examples, a simple, familiar one, and a little more complex one involving a dynamic biosystem ODE model fitting to data. In the first simple example, we quickly realize that parameter estimation is many-faceted, with no small amount of terminology. We develop some of the common nomenclature as we go along.

Example 12.1: ***Fitting a Line to Data: Linear Regression***

The graph in **Fig. 12.1** is a plot of a straight line $y = mx + b$ fitted to several y versus x data points. We assume the data were collected in a steady state (nondynamic) biological experiment. **Error bars**, representing two-sided standard deviation (*SD*) or standard error (*SE*) bars, are shown for each data point, all assumed to represent equal data variability here. The assumed goal of the experiment is to see if response y and

2 The inverse problem means generating parameter values from data generated from a physical system (or model). It's "inverse" in the sense that modeling from theory (for example) is often based on knowing parameter values from the physical principles that specify the model of the physical system.

3 Model identification more generally involves establishing quantitative other characteristics of a model, such as its dimensionality, or unknown input, as well as its parameters.

4 Using statistics and random data sampling to make inferences about unknown parameters of a population model (Casella and Berger 2001).

5 The methodology is intermediate level engineering math. For theory, see the references cited.

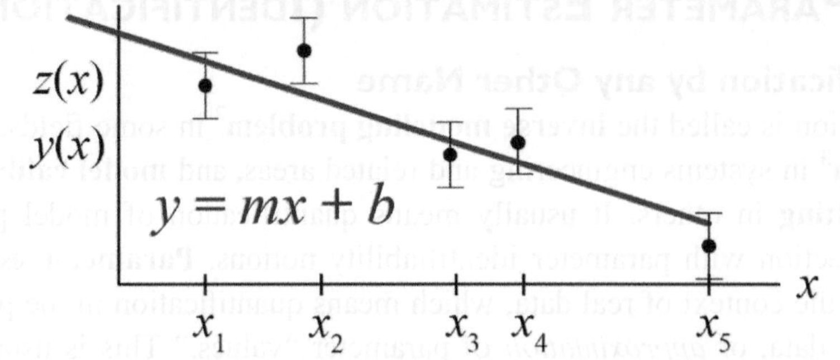

Fig. 12.1

stimulus x are linearly related and, if so, how closely related – a common scientific question. To answer it, we first have to decide how we're going to "fit" the model to the data and then how we're going to assess the quality of the result. The fitting task requires a fitting **(optimization) criterion** as well as an **algorithm** (procedure) to find or **search** for the "best" fit line, the optimized **regression line**. A **least squares** (*LS*) fitting criterion is the most common – invented by Gauss in the 18th century to estimate distances to the planets from data collected through his telescope (Stigler 1986). The line fitted to the data in the figure was presumed computed by **linear least squares**. (*LS* details deferred.) However, as we see below, linear *LS* problems are solvable analytically – no sophisticated numerical search algorithm needed. Either way, the fitted line is the "best" result obtainable using the *LS* method. It's important to appreciate here that this means "best" in a *LS* sense; an optimization criterion other than *LS* would produce a *different* "best" line!

How do we qualitatively assess such results? Most methods and programs provide a common measure of the quality of the fit, typically the **correlation coefficient**, e.g. $r = 0.88$, where $r = 1$ would mean a perfect fit.

Remark: If the output data z are hypothesized to be a "linear" function of two variables, say x and y, a *planar surface* can be similarly fitted by three-dimensional linear least squares, e.g. using the ***regress*** function in *Matlab*. Fitting *nonlinear* functions to data (nonlinear regression) is much more complicated, as discussed below.

Example 12.2: ***Fitting an ODE Model to Data***

Fig. 12.2 illustrates initial and final results of fitting a biodynamical ODE model of the form $\dot{x} = A(\boldsymbol{p})x + Bu,\ \ y = Cx$ to the discrete-time model output data points shown. The data are from blood sampling in experimental animals in an HIV vaccine development study (all detailed in Chapter 13). The criterion of optimality is *LS*, as above, and a sophisticated **gradient search** algorithm was used for numerically and iteratively finding the optimum (best) parameters $\boldsymbol{p}$. Fitting was accomplished using the *SAAMII*[6] package, which has a gradient-based search algorithm built into it (details

6 tegvirginia.com/solutions/saam-ii/

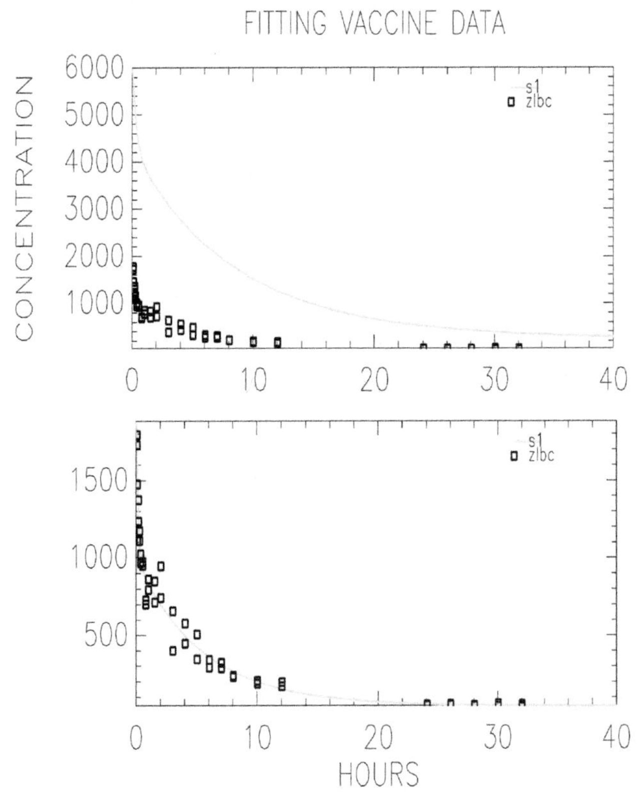

Fig. 12.2

later), and provides statistical measures of the quality of the fit after a successful convergence. Model output is first simulated with a **best guestimate** (guessed estimate) for the parameters, as illustrated in the solid line of the upper panel. The model output initially does not visually fit the data very well, but it has the right shape. When the search eventually converges, the best fit of the output to the data is obtained – shown in the bottom panel passing through the data – and the optimum parameter estimates are reported along with their variability estimates (reserved for Chapter 13).

That's it. Looks easy enough, especially with a program package with parameter estimation as well as simulation capabilities, and a fairly simple ODE model designed to demonstrate success. This elucidates the process. But its apparent ease is deceiving.

Model quantification in the real world is not so readily resolved. Existing program packages for parameter estimation are improving, but few (if any) provide all that is needed in a single easy to understand package. And not for want of trying; the problem is complicated and highly dependent on model specifics and user knowledge and experience. For a wide enough class of dynamic systems biology models, a variety of computational tools are needed. And users need to know how to choose the most effective ones. Quantification usually requires: a choice of optimization criteria – to accommodate different modeling goals, measurement models and data statistics; a choice of search algorithms – because a single "all-purpose" one may not solve even moderately complex models; and provision for computing appropriate and usually fairly sophisticated statistics for resulting parameter estimates. This is a lot to ask for in a single package,[7] as we see as the chapter unfolds – and continues to unfold into the next chapter as well.

Parameter Estimation, Robustness and Sensitivity

What/Which Parameters are Estimated?

Unless otherwise noted, model parameters $\boldsymbol{p}$ subject to estimation in the remainder of this chapter are assumed to be structurally identifiable (*SI*). It doesn't make any sense to attempt to estimate *un*ID parameters, which have an uncountably infinite number of solutions. Models *reparameterized* in terms of *SI* combinations (combos) might then be the object of parameter estimation. Development of reparameterization procedures is deferred to Chapter 13.

We continue to designate parameter vectors as $\boldsymbol{p}$ in the developments that follow, so as to not introduce additional notational complexity, with the understanding that the original model parameters may be different, especially if some are *un*ID. Context should keep things unambiguous.

Robustness, Sensitivity & Parameter Estimation

Parameter sensitivity is integral to parameter estimation. When a model output representing measurement data is insensitive (robust) to a parameter, that parameter estimate is unreliable. It has a very large variability, as measured by *standard deviation SD* or *standard error SE,* when fitted (*optimally estimated*) to that data. So parameter "robustness" is not an advantage if that parameter is to be estimated. When it is highly sensitive (fragile), it more likely can be estimated reliably, with relatively high precision[8] – with a small *SE.*

The following simple example illustrates the connection between parameter sensitivities and estimation – in particular their interdependencies imbedded in the most common class of parameter estimation algorithms: *gradient* formulas. We also codify the notions of *optimal* estimation, *search*, and algorithm *convergence* here. The example ties together these key elements of model quantification and numerical analysis.

Example 12.3: ***Parameter Sensitivities in Model Parameter Estimation***

Consider the simple one-parameter ODE model with continuous-time[9] measurement model:

$$\begin{aligned} \dot{x}(t,a) &= -ax(t,a) + u(t), \quad a > 0, \quad x(0) = x_0 \text{ given} \\ z(t) &\equiv y(t) + e(t) = x(t,a) + e(t), \quad 0 < t \leq T \end{aligned} \tag{12.1}$$

7 Large general-purpose packages like *Matlab* and *Mathematica* have the tools but they're not easily gathered and implemented. Existing toolboxes – like the *SysID* in *Matlab* proper, or *SBtoolbox* – based in *Matlab,* or *Amigo,* and others like it, are for the highly skilled.

8 This is usually true in the context of fitting all parameters simultaneously.

9 If measurements are discrete-time instead of continuous, the integrals in Eqs. (12.2) and (12.3) are replaced by summations over the measured time-series: $z(t_1), z(t_2), \ldots, z(t_N)$ and states $x(t_1,a), x(t_2,a), \ldots, x(t_N,a)$.

The noisy continuous-time data $z(t)$ collected over interval $[0, T]$ has time-varying **output measurement errors** $e(t) = z(t) - y(t,a) = z(t) - x(t,a)$. Note that state $x(t,a)$ is written as a function of a as well as t, because a is the unknown in this problem.

An **optimal** estimate a^* of a is required, with any estimate of a denoted $\hat{a}$. The **optimal estimate** here is defined as that which minimizes the *least-squares (LS) output error*. This means the output errors (called **residuals** below) are integrated ($\sim$averaged[10]) over the entire time interval:

$$LS(a) \equiv \int_0^T e^2(t)dt \equiv \int_0^T [z(t) - x(t,a)]^2 dt \tag{12.2}$$

We need a search algorithm to solve the optimization problem:

$$LS(\hat{a}) \equiv \min_{\hat{a}} \int_0^T [z(t) - x(t,\hat{a})]^2 dt \tag{12.3}$$

One of the simplest is the **first-order gradient**, or **steepest-descent**, method:

$$\hat{a}^{(i+1)} = \hat{a}^{(i)} - K \frac{\partial LS(\hat{a})^{(i)}}{\partial \hat{a}} \tag{12.4}$$

In this recursive algorithm formula, the **stepsize** is $K > 0$, $\hat{a}^{(i)}$ is the estimate of a at the ith iteration, and $\partial LS^{(i)}/\partial \hat{a}$ is the gradient (directional derivative) of LS with respect to $\hat{a}$ at the ith iteration. We surround the iteration number indices in parentheses in superscripts here, to distinguish them – first – from discrete-time steps in integration algorithms; and from subscripts that otherwise represent discrete-time indices, as in $z_k \equiv z(t_k)$, etc., below. Evaluating the gradient analytically we get:

$$\frac{\partial LS}{\partial \hat{a}} = \int_0^T 2e \frac{\partial e}{\partial \hat{a}} dt = \int_0^T 2e \frac{\partial x}{\partial \hat{a}} dt = 2 \int_0^T ev \ dt \tag{12.5}$$

We recognize that $v \equiv \partial x/\partial \hat{a}$ is the parameter sensitivity function for this model and therefore v satisfies the sensitivity ODE (Eq. (11.30) of Chapter 11) $\dot{v} \equiv (\partial f/\partial x)v + \partial f + \partial p$:

$$\dot{v} = -\hat{a}v + \frac{\partial(-\hat{a}x)}{\partial \hat{a}} = -\hat{a}v - x \tag{12.6}$$

To run the algorithm, we begin with an initial guess (**guesstimate**) $\hat{a}^{(0)}$. Then coupled Eqs. (12.1) and (12.6) are solved over interval $[0, T]$. The resulting $v^{(0)}(t)$ and $x^{(0)}(t)$ are substituted into (12.5) yielding $\partial L^{(0)}/\partial \hat{a}$, which in turn is substituted into (12.4). This gives the first improved estimate $\hat{a}^{(1)}$ in the search for the optimum $\hat{a}$. The process is repeated until

$$100\left(\frac{\Delta \hat{a}^{(i+1)}}{\hat{a}^i}\right) = 100\left(\frac{\hat{a}^{(i+1)} - \hat{a}^{(i)}}{\hat{a}^{(i)}}\right) < \varepsilon\% \tag{12.7}$$

10 Normally, the average value would include multiplication of the integral by $1/T$. This is unnecessary because the minimum value of a is the same for any constant multiple of the integral.

where $\varepsilon\%$ is a small positive number chosen to indicate when the algorithm has **converged** with *sufficient precision.*[11] Note that the objective is to drive the gradient $\frac{\partial LS}{\partial a} \rightarrow 0$. This is of course what one would do to obtain a (relative) minimum analytically. Sensitivity function computations, one way or another, are necessary in gradient type algorithms.

Remark: In general, if the measured output *y*(*t*) is not sensitive to variations or changes in a model parameter p, then $v \equiv \partial y/\partial p = 0$, $\partial LS/\partial p = 0$ in Eq. (12.5) and the gradient algorithm (12.4) won't work, at least not for estimating p. Why? *Because p is not SI*!

Remark: The converse logic is also important here. If the output *y*(*t*) *is* sensitive to parameter variations, then $v \equiv \partial y/\partial p \neq 0$ and p can be "estimated" by minimizing the cost function (12.3). The term estimated suggests statistical issues and, as we see below, the *quality* of the estimate is dependent on $\partial y/\partial p$, the bigger the better the estimate.

We defer development of direct parameter estimation methods further until later in the chapter, because indirect methods are easier to develop and comprehend first. We then turn our attention to some widely applicable algorithms useful for both direct and indirect parameter estimation.

Different Quantification Approaches

One way of classifying methods for quantifying parameters of dynamic system ODE models is *direct* and *indirect* (or a *hybrid* of direct and indirect) which addresses the problem in the context of model *forms* most effective for quantification. Another classification is *local* or *global*, which relates to parameter *search schemes*, not whether a *SI* parameter is nonuniquely ("locally") or uniquely ("globally") *SI*. Local searches search in limited regions of parameter space, typically close to their starting point. Global searches span the entire parameter space. Either one might find a locally (nonuniquely) ID or globally (uniquely) ID parameter. Different concept. More on this later.

Direct vs. Indirect Parameter Estimation

Direct methods use the ODE form of the model: $\dot{\boldsymbol{x}} = \boldsymbol{f}(\boldsymbol{x}, \boldsymbol{u}, \boldsymbol{p})$, $\boldsymbol{y} = \boldsymbol{g}(\boldsymbol{x}, \boldsymbol{p})$ in formulating and solving the parameter estimation problem. The model ODEs are solved and the simulated ODE model outputs $\boldsymbol{y} = \boldsymbol{g}(\boldsymbol{x}, \boldsymbol{p})$ are optimally fitted to the output data $\boldsymbol{z} = \boldsymbol{y} + \boldsymbol{e}$ to obtain the optimal estimate $\boldsymbol{p}^*$ (like in the example above). *Indirect methods* circumvent the ODE model, instead fitting a suitable input–output (I–O) form of the model: $\boldsymbol{y} = \boldsymbol{g}'(\boldsymbol{u}, \boldsymbol{p})$ to the same data (illustrated below). As noted earlier, in an appropriately formulated estimation problem, only the *SI* parameters or parameter combinations (combos) of the model are "optimized," using a suitable search algorithm. Hybrids of direct and indirect methods are also useful, as in the *forcing function* method presented in the next chapter.

11 For example, stop when the current estimate is within 1%, 0.1% etc. of the last estimate.

Indirect Parameter Estimation

Indirect methods focus on estimation of identifiable p_i in two steps, first by fitting a suitable *output function* $\boldsymbol{y}(t, \boldsymbol{p})$ to measurement data, instead of solving the ODEs. This works best if some form of structural invariants $\phi(\boldsymbol{p})$ of the model can be estimated from *output function* $\boldsymbol{y}[t, \phi(\boldsymbol{p})]$ measurements, and then computing identifiable p_i (or combos, for unidentifiable p_i) from an algebraic transformation of $\phi(\boldsymbol{p})$.

A popular indirect and didactively useful approach is *multiexponential response modeling*, particularly applicable to linear ODE models with multiexponential outputs, e.g. multicompartmental models.[12] We develop this approach here for single-input–single-output (SISO) models, but it's applicable to multiinput–multioutput (MIMO) models as well, with a bit more effort. Indirect here means first estimating multiexponential model parameters $\phi = \{A_i s$ and $\lambda_i s\}$ and then transforming them to multicompartmental ODE model parameters, e.g. $\boldsymbol{p} = \{k_{ij} s$ and $V_i s\}$. We present this approach first – focusing on the output model only – because it's instructive, intuitive and parameter identifiability is more readily apparent in multiexponential output responses than in ODE models.[13]

Linear Constant-Coefficient Models: Multiexponential Response Approach

This is particularly convenient for impulse responses (or responses to initial conditions) in asymptotically stable models with only one state variable measured – in continuous-time or as a discrete-time sequence. Model linearity guarantees that it also works for step inputs, or any input that is a higher-order integral of impulses (ramp, etc.). We introduce more terminology, optimization criteria and statistical figures of merit for parameter estimation as we develop this approach. First, a simple example, for the same ODE model optimization problem formulated for direct solution in the previous example.

Example 12.4: ***Least Squares Optimization for a Multiexponential Output Function***

If the input is an impulse $u(t) = \delta(t)$ in ODE model $\dot{x}(t, a) = -ax(t, a) + \delta(t)$, $z(t) = x(t, a) + e(t)$, we readily evaluate the output analytically as $x(t, a) = x_0 e^{-at}$, and the sensitivity function for this output: $v \equiv \frac{\partial x}{\partial \hat{a}} = \hat{a} x_0 e^{-\hat{a}t}$. In this case, we can formulate the optimization criterion directly in terms of this output, with no ODEs to solve numerically:[14]

$$LS(\hat{a}) \equiv \min_{\hat{a}} \int_0^T [x_0 e^{-\hat{a}t} - z(t)]^2 dt \tag{12.8}$$

12 Polynomial and other well-known functions are commonly used as parameterized (output) response models in various scientific areas. Multiexponential functions are particularly well suited for linear dynamic system response modeling, because they are their natural homogeneous solutions (modes), and are transformable homeomorphically into compartmental models. Multiexponential, polynomial or other functional form for fitting to output data are also useful in the *forcing function approach* to nonlinear (or linear) biomodel quantification, as presented in the next chapter.

13 ODE model parameter identifiability still must be established, and the required transformation from one model form to another usually reveals these identifiability properties.

14 This kind of optimization problem is also called *nonlinear regression* in the statistics and engineering literature.

Then, from Eq. (12.5), the directional derivative for the gradient search algorithm then becomes

$$\frac{\partial LS}{\partial \hat{a}} = 2\int_0^T [x_0e^{-\hat{a}t} - z(t)](-\hat{a}x_0e^{-\hat{a}t})\ dt = x_0^2(e^{-2\hat{a}T} - 1) - 2\hat{a}x_0\int_0^T z(t)e^{-\hat{a}t}dt \qquad (12.9)$$

The integral in the last term is dependent on the known data $z(t)$. For the first-order gradient algorithm in Eq. (12.4), computing the gradient in (12.9) over the measurement interval [0, T] and iteratively solving the gradient search (12.4) effectively quantifies the optimum $\hat{a}$, i.e. $a*$, when the search has converged.

Remark: With only one parameter to estimate, this example does not require any further indirect algebraic solution for any parameters, whereas higher-order models, considered next, do.

We generalize the above example for nth-order models, with discrete-time rather than continuous-time measurements. Consider first the homogeneous nth-order single-output impulse response model with ICs: $\boldsymbol{x}(0,\boldsymbol{p}) \equiv \boldsymbol{x}_0(\boldsymbol{p}) \neq 0$ and known exponential form of output:

$$\begin{aligned} &\dot{\boldsymbol{x}}(t,\boldsymbol{p}) = F\boldsymbol{x}(t,\boldsymbol{p}) \equiv [f_{ij}]\boldsymbol{x}(t,\boldsymbol{p}) \\ &y(t,\boldsymbol{p}) = y_1(t,\boldsymbol{p}) = \left[\frac{1}{V_1}\ 0\cdots0\right]\boldsymbol{x} = \frac{x_1}{V_1} = \frac{e^{Kt}\boldsymbol{x}_0(\boldsymbol{p})}{V_1} \end{aligned} \qquad (12.10)$$

The f_{ij} are the constant coefficients of matrix F. To simplify the notation, we suppress the dependence on $\boldsymbol{p}$ of $\boldsymbol{x}(\boldsymbol{p})$, the IC: $\boldsymbol{x}_0(\boldsymbol{p})$ and output $y(t,\boldsymbol{p})$ here. We restore it later, as needed. If all eigenvalues λ_i of F are real and distinct,

$$y(t) = \sum_{i=1}^{n} A_ie^{\lambda_it} \qquad (12.11)$$

and a unique solution for (12.11) is determined by the initial condition: $y(0) = x_1(0)/V_1 = \sum_{i=1}^n A_i$. If the data function $z(t) = y(t) + e(t) = \sum_{i=1}^n A_ie^{\lambda_it} + e(t)$ consist of $N+1$ discrete-time measurements,[15] with $z(t_k) \equiv z_k$, $e(t_k) \equiv e_k$, etc, then

$$z_k = \sum_{i=1}^{n} A_ie^{\lambda_it_k} + e_k \qquad k = 0, 1, 2, \ldots, N \qquad (12.12)$$

The **primary computational problem** here is to find the A_i and λ_i, $i = 1,\ldots, n$, and possibly also n, that fit the data $z_0, z_1, z_2, \ldots, z_N$ "best" in some sense. All of the coefficients and exponents (*structural invariants*) are *SI* for this optimization problem. Least-squares (*LS*) analysis, for example, can be used to solve this problem – readily formulated in algebraic terms, instead of ODEs, as in Eq. (12.8) above. In this case, we would be finding the best-fit model, *in terms of the A_i and λ_i*, in a *LS* sense for this data set.

15 The zero-time measurement is the IC, assumed unknown to the extent that, in general, it is dependent on $\boldsymbol{p}$. Then, as is usually the case in practice, N measurements are taken at times $t > 0$.

The **secondary problem** is to compute numerical values for the unknown parameters of the original ODE model from the optimal parameters A_i^* and λ_i^*. Finding the ODE parameters is usually the primary problem, but it's secondary when using the multiexponential approach as an indirect method. It requires algebraic evaluation of the matrix constants f_{ij} in terms of the the As and λs, which typically are *ID combos* of the f_{ij}. To illustrate, see Example 10.7 of Chapter 10, where the $f_{ij} \equiv k_{ij}$ were evaluated for a particular three-compartment linear model in terms of the A_i and λ_i.

Linear & Nonlinear in the Parameters Models

Ideally, we would like computational solutions for parameter estimates in analytic, closed-form – readily evaluated in a single or just a few steps, as in Example 12.1. Unfortunately, for most dynamic system models, only open-ended, iterative search numerical solutions are available. The deciding factor between closed-form and iterative solutions is the form of $y(t,\mathbf{p})$ to be optimized w.r.t. $\mathbf{p}$. The following two definitions encompass the distinction.

If $\mathbf{p}$ is the parameter vector, a **linear-in-the-parameters model** has the form:

$$y(t,\mathbf{p}) = p_1 \mathbf{f}_1(t,\mathbf{x}) + p_2 \mathbf{f}_2(t,\mathbf{x}) + \ldots + p_j \mathbf{f}_j(t,\mathbf{x}) \tag{12.13}$$

where the $\mathbf{f}_i$ are nonlinear functions of $\mathbf{x}$ and t in general, but all are *independent* of $\mathbf{p}$.

A **nonlinear-in-the-parameters** model has the form:

$$y(t,\mathbf{p}) = p_1 \mathbf{f}_1(t,\mathbf{x},\mathbf{p}) + p_2 \mathbf{f}_2(t,\mathbf{x},\mathbf{p}) + \ldots + p_j \mathbf{f}_j(t,\mathbf{x},\mathbf{p}) \tag{12.14}$$

with at least one $\mathbf{f}_i$ a function of $\mathbf{p}$.

Remark: The sum of exponentials function y in Eq. (12.11) above is linear in the A_i parameters *but* nonlinear in parameters λ_i, so it cannot be solved analytically. The reason is that when we take derivatives of y w.r.t. p_i – the usual way we find optima – linear in the parameters models yield explicit solutions for the p_i, whereas derivatives remain functions of the p_i when y is nonlinear in the parameters. In other words, to estimate the optimum exponent of a multiexponential function, iterative search is required because every term $e^{\lambda t}$ is nonlinear in λ.

Residual Errors & Parameter Optimization Criteria

This is the estimation problem formulation section. Generally speaking, different optimization criteria can be used to compute "best" or "optimal" parameter estimates for a particular model – each yielding more or less different results. Modeling goals and type, precision required, computational resources – available software and hardware – and time available for solution all affect choice of criteria. We first precisely define what we mean by estimation or optimization *errors* (*residual errors*) and then formally present two popular criteria for parameter estimation (optimization).

Error functions expressed in optimization criteria for dynamic systems parameter estimation are either: *residual output errors* (*ROE*) or *residual equation errors* (*REE*). *ROE* errors are more prevalent for continuous-time dynamic system (ODE) model formulations and we use only this formulation in this book.[16] For ODE models, the outputs (measurements) can be either continuous-time: $y(t)$ on an interval[17] $[t_0, T]$, or discrete-time, as a time series (sequence) $y(t_1),\ldots,y(t_N)$. In some circumstances, the data can include an initial datum at t_0, $y(t_0)$, if it is known, i.e. the measurement interval is $[t_0, T]$.

Residual Output Errors for Discrete-Time Measurements

Consider the general nonlinear ODE model, which includes ICs dependent on $\boldsymbol{p}$:

$$\begin{aligned} \dot{\boldsymbol{x}} &= \boldsymbol{f}(\boldsymbol{x},\boldsymbol{u};\boldsymbol{p}) \\ \boldsymbol{y} &= \boldsymbol{g}(\boldsymbol{x},\boldsymbol{p}) \end{aligned} \tag{12.15}$$

$$\boldsymbol{x}_0 = \boldsymbol{x}(t_0,\boldsymbol{p}) \tag{12.16}$$

The experiment time interval is $t_0 \le t \le T$, $\boldsymbol{p}$ is the P by 1 vector of unknown parameters, and we assume any constraint equations (e.g. $\boldsymbol{h}(\boldsymbol{x}, \boldsymbol{p}, \boldsymbol{u}, t) \ge \boldsymbol{0}$) have already been incorporated into the model equations above, so the model is complete[18] and the $\boldsymbol{p}$ are *SI*. The $N+1$ discrete-time measurements $z(t_k) \equiv z_k$ ($N+1$ measurements if initial conditions are known) are assumed to include additive noise $\boldsymbol{e}(t_k) \equiv \boldsymbol{e}_k$:

$$\begin{aligned} z_k &= \boldsymbol{y}_k(\boldsymbol{x},\boldsymbol{p}) + \boldsymbol{e}_k \\ k &= 0,1,2,\ldots,N \end{aligned} \tag{12.17}$$

Residual output error $\equiv$ *ROE* is defined as:

$$ROE \equiv \boldsymbol{e}_k = \boldsymbol{y}_k(\boldsymbol{x},\boldsymbol{p}) - z_k \tag{12.18}$$

Minimization of some criterion function of ROE is the most common parameter estimation problem formulation for ODE models, in this case with noisy discrete-time measurements of the output. ROE is called **residual(s)**, **output error**, or just **error**, when the context is clear.

Least-Squares (LS) Criteria

For single output models, residual output error at t_k is $e_k = z_k - y_k$. In these terms, we write the **least-squares output error** (***LS***) as the quadratic form:

$$LS \equiv \sum_{k=0}^{N} e_k^2 = \sum_{k=0}^{N} (z_k - y_k)^2 = \boldsymbol{e}^T\boldsymbol{e} \equiv RSS \tag{12.19}$$

16 **Residual equation errors** $\equiv$ *REE* can be defined in different ways. For discrete-time dynamic system (difference equation, DE) models of the form: $\boldsymbol{x}(t_{k+1}) = \boldsymbol{f}[\boldsymbol{x}(t_k),\boldsymbol{u}(t_k),\boldsymbol{p}]$, $\boldsymbol{y}(t_k) = \boldsymbol{g}[\boldsymbol{x}(t_k),\boldsymbol{u}(t_k),\boldsymbol{p}]$, one form is: $REE \equiv e(t_k,\boldsymbol{p}) \equiv x(t_{k+1}) - f[x(t_k),u(t_k),\boldsymbol{p}]$ (Ljung 1987).

17 The notation $(t_0,T]$ means $t_0 < t \le T$, i.e. in practice, output data are collected *after* t_0, not exactly at t_0.

18 Model constraints must be included, because identifiability properties of the model depend on them (see Chapter 10).

where $\boldsymbol{e} = [e_0\ e_1\ e_2\ \cdots\ e_N]^T$. *LS* is also called **residual sum of squares (*RSS*)** (sum of squared residuals). Let's call the P parameters to be optimized $\boldsymbol{p}$. Then, noting that the data z_k are independent of $\boldsymbol{p}$ and the model is not, we have the following **minimization** problem to solve:

$$\min_{\boldsymbol{p}}\{LS(\boldsymbol{p})\} = \min_{\boldsymbol{p}}\left\{\sum_{k=0}^{N}[z_k - y_k(\boldsymbol{p})]^2\right\} \tag{12.20}$$

For example, from Eq. (12.19), we write the least-squares error function for a sum of exponentials model (for example) (12.12) as:

$$LS(A_i, \lambda_i) = \sum_{k=0}^{N}\left[z_k - \sum_{i=1}^{N}A_i e^{\lambda_i t_k}\right]^2 \quad i = 1, 2, \ldots, n \tag{12.21}$$

Remark: In Eq. (12.8), the measurements were continuous-time and the *LS* criterion function was an integral instead of a summation. The summation also effectively "integrates" the sum of squared residuals, so they are conceptually the same.

Weighted Least-Squares (*WLS*) or Residual Sum of Squares (*WRSS*) Criteria

Real output data are typically not all equally reliable. If there is justification for assigning different weights $w_k > 0$ to different data points, then we can/should minimize:

$$WLS(\boldsymbol{p}) \equiv WRSS(\boldsymbol{p}) \equiv \sum_{k=0}^{N} w_k[z_k - y_k(\boldsymbol{p})]^2 = \sum_{k=0}^{N} w_k e_k^2 \tag{12.22}$$

To be precise in fitting models to real data, evidence of unequal variability (errors) in data means unequal weighting should be used in optimization. Unequal variability is probably more common than not, e.g. when early data, or larger magnitude data, are more reliable than late data, or smaller magnitude data – or *vice versa*. In such cases, use of this information and minimizing *WLS* (rather than unweighted *LS*) maximizes the quality (correctness) of the resulting quantified model.

A *good choice* for data weights – with a solid theoretical basis (Bard 1974; Ljung 1987) – is the *inverse of the data variance*: $w_i \equiv 1/VAR_i = 1/\sigma_i^2$ at each time t_i. Data variance estimates are often estimated from the data, or are part of the output data model. It is common for VAR_i (variance at time t_i) to be approximated by the **sample variance** of M replicate data at time t_i: $\overline{\sigma}_i^2 = \frac{1}{M-1}\sum_{j=1}^{M}(z_i^j - \overline{z}_i)^2$, with $\overline{z}_i$ the **sample mean** of the M replicates: $\overline{z}_i = \frac{1}{M}\sum_{j=1}^{M} z_i^j$. If the data have approximately equal variance over the observation interval of the data, all of the weights can be chosen as 1. If the **percent relative error** – also called the **percent coefficient of variation** – is approximately constant over the observation interval of the data, i.e. $\%\text{CV}_i = 100\overline{\sigma}_i/\overline{z}_i \cong$ constant for all i, then inverse-variance weighting can be approximated using the inverse of the data point as the weight for each datum. More about data weighting later in this and the next chapter.

Writing Eq. (12.22) in vector-matrix form simplifies the look of the equation and makes generalizations to higher dimension easier, even though we are still dealing with scalar (single) outputs and data. The usual procedure is to define **augmented** measurements, outputs and errors as vectors:

$$\boldsymbol{Z} \equiv [z(t_0)\, z(t_1)\cdots z(t_N)]^T \equiv [z_0\;\, z_1\cdots z_N]^T \tag{12.23}$$

$$\boldsymbol{Y} \equiv [y(t_0)\cdots y(t_N)]^T \equiv [y_0\;\, y_1\cdots y_N]^T \tag{12.24}$$

$$\boldsymbol{E} \equiv [(e_0)\cdots e(t_N)]^T \equiv [e_0\;\, e_1\cdots e_N]^T \tag{12.25}$$

If the measurement data are uncorrelated with each other,[19] the matrix of data weights is diagonal, the most common case. Then Eq. (12.22) becomes the quadratic form:

$$WLS(\boldsymbol{p}) = \boldsymbol{E}^T(\boldsymbol{x};\boldsymbol{p})\boldsymbol{W}\boldsymbol{E}(\boldsymbol{x};\boldsymbol{p}) \equiv \boldsymbol{E}^T\boldsymbol{W}\boldsymbol{E} \tag{12.26}$$

where

$$W = \begin{bmatrix} w_0 & & 0 \\ & w_i & \\ 0 & & w_N \end{bmatrix} > 0 \tag{12.27}$$

Remark: Lots of notation, but it is instructive to see it all explicitly, even for this scalar, one-output case.

*WLS for More Than One Output: $y(t) = [y_l(t)\ \cdots\ y_m(t)]^T$

We generalize the scalar *WLS* criterion Eq. (12.22) – and also (12.26) – for *m* outputs, spelling out the arguments under the summation explicitly, to render the optimization problem clearer:

$$WLS(\boldsymbol{p}) = \sum_{k=0}^{N}\sum_{l=1}^{m} w_{lk}[y_l(t_k,\boldsymbol{x};\boldsymbol{p}) - z_l(t_k)]^2 \tag{12.28}$$

Now define $e_{lk} \equiv y_l(t_k, x;\ \boldsymbol{p}) - z_l(t_k) \equiv y_{lk} - z_{lk}$, and the new $mN \times 1$ dimensional augmented vector:

$$\boldsymbol{Z} \equiv [z(t_1)\cdots z(t_N)]^T = [z_1(t_1)\, z_2(t_1)\cdots z_m(t_1)|\; z_1(t_2)\;\, z_2(t_2)\cdots|\cdots z_m(t_N)]^T \tag{12.29}$$

The $mN \times 1$ dimensional augmented output vector $\boldsymbol{Y}$ and augmented error vector $\boldsymbol{E}$ are defined similarly. In these terms, Eq. (12.28) becomes:

$$WLS(\boldsymbol{p}) = \boldsymbol{E}^T(\boldsymbol{x};\boldsymbol{p})\boldsymbol{W}\boldsymbol{E}(\boldsymbol{x};\boldsymbol{p}) = \boldsymbol{E}^T\boldsymbol{W}\boldsymbol{E} \tag{12.30}$$

19 This assumption can be relaxed if the data are correlated in a known way. In that case, the off-diagonal terms are nonzero. However, if individual data are processed independently, in a statistical sense – so they are not correlated with each other – then W remains diagonal.

where W is the $mN \times mN$ diagonal matrix given by:

$$W = \begin{bmatrix} w_{11} & & & & & & & & 0 \\ & w_{12} & & & & & & & \\ & & \ddots & & & & & & \\ & & & w_{1N} & & & & & \\ & & & & w_{21} & & & & \\ & & & & & \ddots & & & \\ & & & & & & w_{2N} & & \\ & 0 & & & & & & \ddots & \\ & & & & & & & & w_{mN} \end{bmatrix} \tag{12.31}$$

Eqs. (12.26) and (12.30) have exactly the same form, but different dimensionality. Therefore we can formally state and solve the *same optimization problem for weighted least squares parameter estimation,* for both scalar and vector output cases. Only the magnitude (dimensionality) problems are different. In each case, we desire the estimate $\boldsymbol{p}^*$ of $\boldsymbol{p}$ that minimizes (12.30).

Extended Weighted Least Squares (EWLS)

To be precise in fitting models to real data, evidence of unequal weighting should be used in optimization. However, it is often difficult to specify data weights precisely. No information, or only partial information about data variability may be available, and a poor choice for the error model can lead to imprecise estimates and models. Extended *WLS* (*EWLS*) provides one possible solution to this problem, by incorporating a data variance model, with unknown parameters, into the search criterion. The form of the variance versus data equation $VAR(z, a, b, c)$ must be specified, and more data are typically needed, because there are more parameters to be estimated, *a, b* and *c* as well as $\boldsymbol{p}$.

Parameters *a, b, c* of the **data variance model** $VAR(z, a, b, c)$ augment the vector of parameters $\boldsymbol{p}$ to be optimized, i.e. $\boldsymbol{p}' \equiv [p_1 \ldots p_P \; a \; b \; c]^T$. The *EWLS* criterion has the form:

$$\begin{aligned} EWLS(\boldsymbol{p}') &= WLS + \ln VAR(z, a, b, c) \\ VAR(z, a, b, c) &= \text{variance model} \end{aligned} \tag{12.32}$$

The equation for VAR_i, the variance of each data point, can take a number of forms, e.g.

$$VAR_i = az_i^b + c \tag{12.33}$$

often with $c = 0$ and $b = 1$ to represent variances that might be proportional to the data, or $c = 0$ and $b = 2$ if variances are proportional to the square of the data.

Use of *EWLS* in high data noise situations can give more precise parameter estimates than *WLS*, assuming the underlying model is correct. And, with more parameters to estimate, more data might be needed to satisfy basic identifiability criteria. Many math modeling programs (e.g. *SAAMII, ADAPT, AMIGO* and others) have provision for including unknown data errors in their parameter search algorithms.

Maximum Likelihood (ML)

ML estimation, introduced by R.A. Fisher a century ago (Aldrich 1997), is likely the most widely quoted scheme for parameter estimation. The computed **sample mean** $\bar{z} \equiv \sum_{i=1}^{N} z_i/N$ of the data $z_1, \ldots, z_N$, for example, is a good approximation of the *ML* estimator of the (theoretical) population mean μ of the data (the parameter). Similarly, the computed **sample variance** of the data is a good approximation of the *ML* estimator of the population variance (another parameter). This criterion function was ingeniously designed to select parameter estimates that make the *data z* "most likely," i.e. *ML* parameter estimates maximize the likelihood of observed results (data). The *ML* estimator is thus expressed in terms of the *joint probability density function*[20] *of the data z conditioned on the parameters* $L \equiv f(z_1, \ldots, z_N|\boldsymbol{p})$, abbreviated $f(z|\boldsymbol{p})$ and aptly termed the **likelihood function** $\boldsymbol{L}$. The usual assumption is that the data are mutually (statistically) independent, and thus $L = f(z_1, \ldots, z_N|\boldsymbol{p}) = f(z_1|\boldsymbol{p})\, f(z_2|\boldsymbol{p}) \ldots f(z_N|\boldsymbol{p})$, i.e. the product of the **marginal probability densities**. To make the data "most likely," the *ML estimator* $\boldsymbol{p}^*_{ML}$ of $\boldsymbol{p}$ maximizes the product of the marginal joint densities:

$$\boldsymbol{p}^*_{ML}(z) = \underset{p}{\arg}\ L = \underset{p}{\arg\ \max}\ f(z_1, \ldots, z_N|\boldsymbol{p}) = \underset{p}{\arg\ \max} \prod_{i=1}^{N} f(z_i|\boldsymbol{p}) \tag{12.34}$$

It is helpful to also define the *logarithm* of the likelihood function, or the **log likelihood function** $\equiv LL$, which has the same max as L, and *sums* the marginal densities, because the log of the product is the sum:

$$\begin{aligned} \boldsymbol{p}^*_{ML}(z) &= \arg\ LL_p = \arg\ \max_p\ \log f(z_1, \ldots, z_N|\boldsymbol{p}) = \\ &\arg\ \max_p\ \log \prod_{i=1}^{N} f(z_i|\boldsymbol{p}) = \underset{p}{\arg\ \max} \sum_{i=1}^{N} f(z_i|\boldsymbol{p}) \end{aligned} \tag{12.35}$$

To apply *ML* estimation in a particular problem, the form of the probability density function f must be defined. This is often a Gaussian or "normal" probability density function, because the data are often approximately Gaussian distributed.

What about optimization? From the calculus *WLS*($\boldsymbol{p}$) – or any criterion function defined above – has a minimum at $\boldsymbol{p}$ if and only if the (first) derivative matrix **Jacobian** $\partial \boldsymbol{WLS}/\partial \boldsymbol{p} = 0^T$, and second-derivative matrix (**Hessian**) $\partial^2 \boldsymbol{WLS}/\partial \boldsymbol{p}^2 > 0$ at $\boldsymbol{p}$. The latter inequality means that the Hessian of *WLS* with respect to $\boldsymbol{p}$ must be positive definite at $\boldsymbol{p}$. This minimization problem is generally *nonlinear in the parameters* $\boldsymbol{p}$, with no analytic solution in general. Iterative numerical solution is usually required, developed next.

Parameter Optimization Methods 101: Analytical and Numerical

We begin by developing and solving two quite simple parameter optimization problems, both of which are intuitive, pedagogically elucidating, and solvable analytically – without the need

20 Some probability theory is needed here, because ML is a probabilistic concept.

for iterative numerical computation. The first is for linear least squares parameter estimation – fitting a simple line to data, as illustrated in **Fig. 12.1**. The second is for estimating parameters of multiexponential models by the *curve-peeling* method – using a log-linear transformation, thereby avoiding the fact that exponentials are nonlinear in the exponent parameters. With this approach, curve-peeling becomes linear least squares parameter estimation, repeated for each exponential mode of the output response function.

We then formulate nonlinear dynamic biosystem model parameter estimation problems more formally and explicitly as iterative search optimization problems – illustrated using *WLS* criteria, and we present iterative algorithms for solving and assessing their validity or usefulness in applications. Again, we assume $\boldsymbol{p}$ is structurally identifiable (*SI*), or that we are attempting to estimate only the identifiable combinations of p_i, or structural invariants of the model.

The Calculus of Linear Least Squares

The idea and the math are simple. We set up the problem for fitting a line $y \equiv mt + b$ to data,[21] the same problem as in **Fig. 12.3**, but we develop the math here. The raw data are shown as circles in **Fig. 12.3**. By definition,[22] we have

$$LS(m,b) = \sum_{k=1}^{N} [y(t_k, m, b) - z_k]^2 = \sum_{k=1}^{N} [mt_k + b - z_k]^2 \tag{12.36}$$

LS is a linear function of *m* and *b*, so we can find the minimum least squares error by setting the derivatives of *LS* equal to zero and solving for *m* and *b*. Differentiating:

$$\frac{\partial(LS)}{\partial m} = \sum 2[\cdots]\frac{\partial(\cdots)}{\partial m} = 2\sum[mt_k^2 + bt_k - t_k z_k] = 2\left[m\sum t_k^2 + b\sum t_k - \sum t_k z_k\right] \equiv 0$$

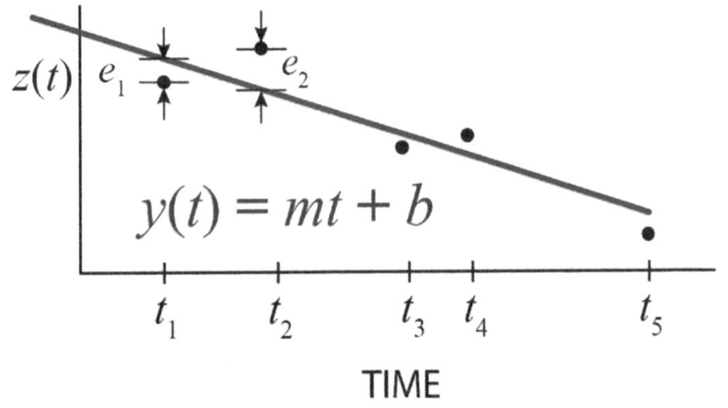

Fig. 12.3

21 *Reminder*: Fitting a line is not *too* simple here, because NL-in-the-parameters exponential functions have a linear natural log transform (for example).

22 From Eq. (12.19), with $e^2 \equiv (-e)^2$.

$$\frac{\partial(LS)}{\partial b} = \sum 2[\cdots]\frac{\partial(\cdots)}{\partial b} = 2\sum[mt_k + b - z_k]\cdot 1 = 2\left[m\sum t_k + b\sum 1 - \sum z_k\right] \equiv 0$$

We have two equations in two unknowns m and b: $(\sum t_k)m + (\sum 1)b = \sum z_k$ and $(\sum t_k^2)m + (\sum t_k)b = \sum t_k z_k$. Set up these equations as: $T\boldsymbol{a} \equiv \boldsymbol{d}$, with $\boldsymbol{a} = [m\ b]^T$ and T a matrix of the data $\sum$ terms, and solve for m and b by **Kramer's rule**:

$$m = \frac{\det\begin{bmatrix} \sum z_k & \sum 1 \\ \sum t_k z_k & \sum t_k \end{bmatrix}}{\det\begin{bmatrix} \sum t_k & \sum 1 \\ \sum t_k^2 & \sum t_k \end{bmatrix}} = \frac{\sum t_k \sum z_k - N\sum t_k z_k}{(\sum t_k)^2 - N\sum t_k^2} \qquad b = \left[\sum z_k - m\sum t_k\right]/N$$

Remark: The polynomial in the denominator above is NL in t, but it is *linear-in-the-parameters* m and b, and so we are able to solve for m and b analytically, as shown.

Remark: For higher dimension linear models of the form $\boldsymbol{z} = T\boldsymbol{p}$, if T is invertible, $\boldsymbol{p} = T^{-1}\boldsymbol{z}$ is the obvious solution for the parameter vector. If T is not square, and therefore not invertible, least squares solutions for $\boldsymbol{p}$ still can be found analytically, as shown in Appendix B.

Remark: Analytic solutions of least-squares problems are not generally available for nonlinear-in-the-parameters models. However, if the NL function can be linearized, linear optimization methods can be applied effectively, as we show next.

The Curve-Peeling (or Stripping) Method for Exponential Curve Fitting

This is perhaps the oldest approach to fitting sums of exponentials as "direct" models-of-data (Riggs 1963), or as indirect models in structural compartmental analysis (Berman et al. 1962). We fit Eq. (12.10) to data as follows. First, order the exponential components so that $\lambda_1| > |\lambda_2| > \ldots > |\lambda_n|$. Then solve for $A_n e^{\lambda_n t}$:

$$A_n e^{\lambda_n t} = y(t) - [A_1 e^{\lambda_1 t} + \cdots + A_{n-1} e^{\lambda_{n-1} t}] \tag{12.37}$$

Now, as $t \to \infty, y(t) \to A_n e^{\lambda_n t}$. Therefore, for large enough t,

$$\ln[y(t)] - \ln(A_n) \simeq \lambda_n t \tag{12.38}$$

This is a straight line on a semi-log plot, with slope λ_n and intercept $ln(A_n)$, similar to **Fig. 12.3**. Thus, a (linear) least-squares line fitted to the "final" portion of the data $z(t)$ will provide estimates of λ_n and A_n, if we substitute $z(t)$ for $y(t)$ in (12.38). The fitting can be done analytically, as for Eq. (12.36). Or, it can be done numerically, using a *LS* optimization program found in most math packages, e.g. in *Mathematica* or *Matlab*.

Next, we subtract $A_n e^{\lambda_n t}$ from $y(t)$ ($\equiv z(t)$), and solve for $A_{n-1} e^{\lambda_{n-1} t}$:

$$A_{n-1} e^{\lambda_{n-1} t} = (y(t) - a_n e^{\lambda_n t}) - (A_1 e^{\lambda_{n-1} t} + \cdots + a_{n-2} e^{\lambda_{n-2} t}) \tag{12.39}$$

Now, as $t \to \infty, (y(t) - A_n e^{\lambda_n t}) \to A_{n-1} e^{\lambda_{n-1} t}$. Therefore, as before, a linear least-squares line fitted to the *transformed* data, $z(t) - A_n e^{\lambda_n t}$, provides estimates of A_{n-1} and λ_{n-1}. This line-fitting procedure is repeated until the sequentially transformed data are exhausted, which usually doesn't take very long.

Remark: A linearized model, like this one, fitted to real noisy data typically needs some statistical adjustments during or after parameter optimization. Residual errors known for the nonlinear model are not the same when the linearized version is fitted (Bard 1974).

This well-known technique is good for sequentially obtaining *rough* estimates of the parameters of $\sum A_i e^{\lambda_i t}$, when $A_i \geq 0$, $\lambda_i \leq 0$, because – besides the statistical distortion issue noted above – it can be too sensitive to computational errors. Exponentials are not *orthogonal* (independent) functions and thus errors made in estimating each parameter propagate into the estimates of parameters subsequently estimated. Douglas Riggs (1963) presents a worthwhile critique of this approach and its pitfalls.

W^3DIMSUM, an interactive expert system program for fitting log-linear transformed multi-exponential functions to discrete-time data, incorporates a *WLS* algorithm to generate initial estimates of the A_i and λ_i of the linearized model (Harless and DiStefano III 2005). It uses them as starting values for a more sophisticated and precise algorithm for obtaining optimal estimates using the original, untransformed nonlinear model, thereby eliminating the distorted statistics problem as well. Web application *W^3DIMSUM* is available for interactive use at www.biocyb.cs.ucla.edu.

Iterative Search Algorithms, Local & Global

Optimization and search belong to fairly large and complex applied numerical and statistical math fields. We circumvent theory here, presenting the basics of several widely used parameter search methods, hopefully with enough technical and motivational detail for understanding results provided by programming packages that include functionality for parameter estimation in ODE models.

We distinguish between local and global optima with a simple one-parameter example. Suppose we seek an optimal estimate p^* of p that minimizes the weighted least-square residuals: $WLS(p) \equiv \sum_{k=0}^{N} w_k[z_k - y_k(p)]^2$. A plot of $WLS(p)$ vs. p might look like the curve in **Fig. 12.4**. It's easy to locate all the minima on the graph and the global min is clearly achieved at $p^{(3)}$. Unfortunately, we don't usually have the luxury of knowing the shape of $WLS(p)$. In a typical parameter estimation problem, it must be computed iteratively for different values of p.

A local iterative search scheme would likely locate any of the three minima shown, if it were initiated near enough to any of them, i.e. if was close enough to $p^{(1)}$, $p^{(2)}$ or $p^{(3)}$. But, even then, we would not know that $p^{(3)}$ is global without more effort. To find the global minimum, we need to somehow scan the entire parameter space, in discrete steps, using an appropriately fine grid of sample points, without "missing" the optimum.

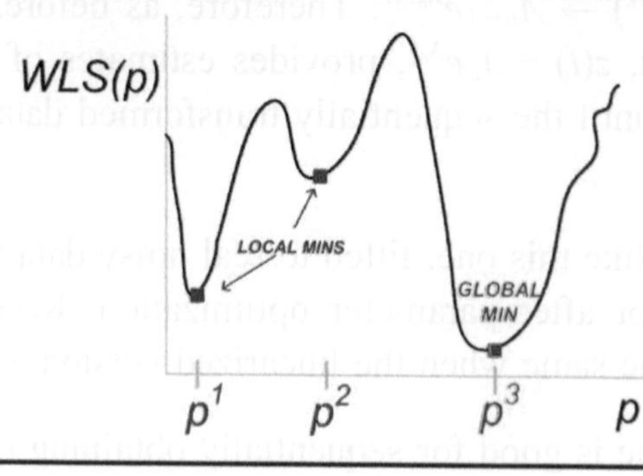

Fig. 12.4

As one can imagine, this can be a formidable computational task. For P different parameters tested for R different values each, the multiparameter $WLS(\boldsymbol{p})$ function must be evaluated at R^P times. This means $100^5 = 10^{10} = 10$ billion evaluations for a five-parameter model tested at 100 different points, or >300 million evaluations of $WLS(\boldsymbol{p})$ if only 50 points are tested. This also means the ODE model must be solved for every $\hat{\boldsymbol{p}}$ tested to get $\boldsymbol{y}(\hat{\boldsymbol{p}})$ and then $WLS(\boldsymbol{p})$, contributing greatly to computational complexity.

The gradient search methods, described first below, first appear to have the disadvantage of being *local,* in that they seek local optima in a usually small region of parameter space. However, local searches are typically initiated at a point $\boldsymbol{p}^{\mathbf{0}}$ in a region of the expected optimum, so – in this sense – they are practical and thus widely used. This "close-to-the-optimum" region may, in fact, have been achieved using a global search method first (together a **hybrid search**), to hone in on the region to find the best local minimum.

Local Search

The **first-order (steepest-descent) gradient** algorithm in Example 12.3 generalizes as:

$$\hat{\boldsymbol{p}}^{(i+1)} = \hat{\boldsymbol{p}}^{(i)} - K\left(\frac{\partial WLS^{(i)}}{\partial \hat{\boldsymbol{p}}}\right)^T \tag{12.40}$$

$\hat{\boldsymbol{p}}^{(i)}$ is the **estimate** of $\boldsymbol{p}$ at the *ith* iteration, $i = 0, 1, 2$, etc.; L is the optimization **criterion** or **loss function**; K is the iteration **stepsize**; and $\frac{\partial WLS^{(i)}}{\partial \hat{p}}$ is the **directional derivative (Jacobian) of *WLS*** in the search direction (Eq. (12.30)) as:

$$\frac{\partial (WLS)^{(i)}}{\partial \hat{\boldsymbol{p}}} = 2(\boldsymbol{E}^{(i)})^T W \frac{\partial \boldsymbol{Y}^{(i)}}{\partial \hat{\boldsymbol{p}}} \tag{12.41}$$

This gradient changes each iteration *i,* along with the (changing) residual E as in Eqs. (12.28) and (12.29). W is the diagonal **data-weighting matrix**. And we should recognize the last term as the (large, augmented) mN by P first-order **output sensitivity matrix**:

$$\frac{\partial \boldsymbol{Y}}{\partial \hat{\boldsymbol{p}}} = \frac{\partial}{\partial \hat{\boldsymbol{p}}}[y_1(t_1)y_2(t_1)\cdots y_m(t_N)]^T$$

$$= \begin{bmatrix} \frac{\partial y_1(t_1)}{\partial \hat{p}_1} & \cdots & \frac{\partial y_1(t_1)}{\partial \hat{p}_P} \\ \frac{\partial y_2(t_1)}{\partial \hat{p}_1} & \cdots & \frac{\partial y_2(t_1)}{\partial \hat{p}_P} \\ \vdots & & \vdots \\ \frac{\partial y_m(t_N)}{\partial \hat{p}_1} & \cdots & \frac{\partial y_m(t_N)}{\partial \hat{p}_P} \end{bmatrix} \tag{12.42}$$

Remark: The algorithm is run recursively, initiated with estimate $\hat{\boldsymbol{p}}^{(0)}$, then with $\hat{\boldsymbol{p}}^{(1)}$, $\hat{\boldsymbol{p}}^{(2)}$, etc., computed from (12.40), until convergence, i.e. when a condition similar to (12.7) is satisfied.

Remark: Eq. (12.40) is used often for minimizing *WLS* (Eq. (12.30)), either by itself, or as part of a more sophisticated hybrid scheme.

The second-order Gauss–Newton algorithm is:

$$\hat{\boldsymbol{p}}^{(i+1)} = \hat{\boldsymbol{p}}^{(i)} - (\hat{H}^{(i)})^{-1}\left(\frac{\partial (WLS)^{(i)}}{\partial \hat{\boldsymbol{p}}}\right)^T \tag{12.43}$$

The 1 by P dimensional directional derivate (Jacobian) $\frac{\partial (WLS)^{(i)}}{\partial \hat{\boldsymbol{p}}}$ for *WLS* here is as in Eq. (12.41). $\hat{\boldsymbol{H}}$ in Eq. (12.43) is an approximation of the symmetric P by P **second-order sensitivity function (Hessian) matrix** $H \equiv \partial^2 \boldsymbol{Y}/\partial \boldsymbol{p}^2$. It is difficult to solve for H exactly and is typically approximated as the ("squared") *quadratic form* of sensitivities[23] weighted by the data-weighting matrix W:

$$\hat{H}^{(i)} \simeq \left(\left(\frac{\partial \boldsymbol{Y}}{\partial \hat{\boldsymbol{p}}}\right)^T W \left(\frac{\partial \boldsymbol{Y}}{\partial \hat{\boldsymbol{p}}}\right)\right)^{(i)} \tag{12.44}$$

The superscript (i) **iterate index** in (12.44) refers to the whole set of matrices inside the big brackets.

Remark: First-order gradient algorithms are more robust to starting values (initial guesstimates) than higher-order gradient methods, i.e. they have a relatively larger region of convergence, but they are slower to converge. Second-order algorithms – e.g. Gauss–Newton,[24] and also the Newton–Raphson[25] algorithm, are more sensitive to starting values (smaller region of convergence), but converge faster (Bard 1974). This suggests combining them in algorithms that yield their best properties.

The **Marquardt–Levenberg (*M–L*) algorithm** is a hybrid iterative scheme designed to retain convergence properties of second-order algorithms like (12.43) in a larger search region

23 To be derived in Exercise E12.3.

24 Exercise E12.4.

25 Exercise E12.5.

than usually feasible with a second-order scheme. It behaves like a first-order steepest descent (*S-D*) scheme at values of $\hat{\boldsymbol{p}}$ far from the optimum (first-order functions better at longer distances), and then like a Gauss–Newton (*G–N*) scheme (converges faster) as it nears the optimum $\hat{\boldsymbol{p}}$. The basic *M–L* algorithm is:

$$\hat{\boldsymbol{p}}^{(i+1)} = \hat{\boldsymbol{p}}^{(i)} - [\hat{H}^{(i)} + \mu^{(i)} I]^{-1} \left(\frac{\partial WLS^{(i)}}{\partial \hat{\boldsymbol{p}}} \right)^T \tag{12.45}$$

where $\mu^{(i)} > 0$ is a number chosen to maintain the above described behavior. As $\mu^{(i)} \to 0$, *M–L* behaves like *G–N*; and as $\mu^{(i)} \to$ large, *MQ* behave like *S-D*, with decreasingly small iterative step sizes – but (ideally) always in a converging direction.

Remark: Computationally intensive sensitivity function computations for $\frac{\partial WLS^{(i)}}{\partial \hat{p}}$ and $\hat{H}^{(i)}$, Eqs. (12.41) and (12.44), dominate the *N–R* and *M–L* algorithms and their variants.

**Global Search*

This remains a fertile area of research. Popular search methods for exploring the entire parameter space include **evolutionary, machine learning** and **directed random search** algorithms. Statistical sampling theory and algorithms underlie efficient choices of sample points in most global search schemes. Two popular methods are **simulated annealing**, inspired by statistical thermodynamics, and **differential evolution**, inspired by genetics (Moles et al. 2003; Rodriguez-Fernandez et al. 2006). These and/or other global search algorithms are available in popular math modeling programs, including *Matlab, Mathematica, COPASI* and *Simbiology*.[26]

Global algorithms typically direct the search to regions near the global minimum, but often not close enough. Hybrid methods are effective for addressing this problem. They utilize global search to get close to and local search to hone in on the minimum. Global search is discussed further in Chapter 17.

AMIGO (*Advanced Model Identification using Global Optimization*) is one of the most comprehensive software packages for accomplishing many dynamic biosystem modeling quantification tasks (Balsa-Canto and Banga 2011). It's a freely downloadable[27] multi-platform (*Windows, MAC* and *Linux*) toolbox, with provision for local and global sensitivity analysis, numerical identifiability analysis, parameter estimation, ranking of parameters (Chapter 17), and some optimal experiment design tasks (Chapter 16). For parameter estimation, it provides a variety of local, global and hybrid algorithms, including global stochastic and sequential and parallel hybrid optimization methods (Balsa-Canto et al. 2010). *AMIGO* functionality is illustrated in examples of Chapters 15–17.

26 Unless you are familiar with using global search, it might be useful to try local methods first.

27 This package is well worth checking out early in the modeling quantification process (http://www.iim.csic.es/~amigo/index.html). A new version, *AMIGO2* should be available in Fall 2013.

Parameter Estimation Quality Assessments

Numerical (*a Posteriori*, Practical) Identifiability (*NI*): Distinguishing the Practical from the Possible

We studied *a priori* structural identifiability (*SI*) concepts and methods for deterministic models in Chapter 10, i.e. ODE models with no noise or error in the (input or) output measurements. The main question addressed in that context is binary: is a parameter p identifiable or not?[28] The answer requires no real data for solution; and it can be addressed beforehand, by analysis of the system-experiment model (system and measurement model structures together). In the presence of measurement noise, the *SI* question remains the same, because it is a necessary (*a priori*) condition for identifiability or (quantifiability). The question now added is: *how* identifiable or unidentifiable is p? This is often the BIG, difficult question, and it depends on all aspects of the output data – as well as the model structure.

We presented the basic relationship of *SI* to output parameter sensitivity in the last chapter, where we introduced the notion that the output sensitivity matrix $[\partial y_i(t)/\partial p_j]$ holds information about (*a posteriori*) numerical identifiability as well, in the relative geometry of its column vectors. We develop this methodology further here, with an eye toward anticipating quantitative numerical identifiability properties when we design experiments as well.

Numerical (*a posteriori*, practical) identifiability (*NI*) codifies the practical question of *how well* a parameter can be quantified (estimated) from limited, noisy data, and with what precision – expressed in statistical terms by variances and covariances, or on quantified measures of output parameter sensitivities. It is thus strongly dependent on actual input–output data. Most approaches for assessing numerical identifiability of parameters $\boldsymbol{p}$ are based on computing the **parameter covariance matrix** $COV(\boldsymbol{p})$, or the **Fisher information matrix** F, from Eqs. (12.50) and (12.52) below, whose inverse is a lower bound on $COV(\boldsymbol{p})$ (Cramér–Rao theorem),[29] or more directly on model **output sensitivity matrix** $[\partial y_i/\partial p_j]$ computations. These approaches are intimately related to each other and all provide local (not necessarily unique) solutions – a property of all methods dependent on local Jacobian (output sensitivity) computations.[30] We develop these methods and illustrate them with several examples, here and in Chapters 13–17. As we shall see, numerical ID problems[31] can be the Achilles heel of real-world modeling efforts.

Remark: Structural identifiability (*SI*) – usually evaluated *a priori* – also can be assessed *after* fitting a model to real data (*a posteriori*), by analysis of the estimated $COV(\hat{\boldsymbol{p}})$ matrix evaluated at the optimum $\hat{\boldsymbol{p}} \equiv \boldsymbol{p}^*$ estimate following a successful and converged search. If $\boldsymbol{p}$ is not *SI*, its variance in a numerical setting will be very large – theoretically infinity. *SI* also can be assessed in this manner using **simulated data**, as we show in the next chapter.

28 We sidestep uniqueness issues for the present.

29 Presented in the next section.

30 Like the gradient search algorithms presented in the last section.

31 Numerical identifiability remains an "open" problem, ripe for fresh approaches – not just in quest of more robust or efficient algorithms, but additional emphasis on effective design of experimental input and output probes and data sampling protocols (Chapter 16).

It is common for at least some *SI* parameters to be badly estimated (*NI*) from real data, i.e. have very large variances and thus be effectively *un*ID for practical purposes. Importantly, this can prevent convergence of parameter search algorithms that depend on derivative computations involving parameters – and thus yield no results! With some caveats, *NI* also can be addressed beforehand, e. g. by **predictive simulation** – along with *SI* – also as shown in the next chapter – an approach that can also improve convergence properties in an otherwise unsuccessful parameter optimization. This approach also can help develop more effective experiment designs (Chapters 16 and 17).

Output Sensitivity Matrix & Numerical Identifiability

For dynamic system models, output sensitivities $[\partial y_i(t)/\partial p_j]$ are time-varying functions, because the $y_i = y_i(t)$ are time-varying outputs. In practice, biodynamic system outputs are more likely sampled at N discrete-times t_1, t_2, etc. and expressed as sequences of data. The output sensitivity matrix then can be expressed as an augmented mN by P matrix of discrete values rather than functions:

$$\frac{\partial \boldsymbol{Y}}{\partial \hat{\boldsymbol{p}}} = \left[\partial y_i(t_k)/\partial \hat{p}_j\right] = \begin{bmatrix} \dfrac{\partial y_1(t_1)}{\partial \hat{p}_1} & \cdots & \dfrac{\partial y_1(t_1)}{\partial \hat{p}_P} \\ \dfrac{\partial y_2(t_1)}{\partial \hat{p}_1} & \cdots & \dfrac{\partial y_2(t_1)}{\partial \hat{p}_P} \\ \vdots & & \vdots \\ \dfrac{\partial y_m(t_N)}{\partial \hat{p}_1} & \cdots & \dfrac{\partial y_m(t_N)}{\partial \hat{p}_P} \end{bmatrix} \equiv \left[\frac{\partial \boldsymbol{y}}{\partial \hat{p}_1}\frac{\partial \boldsymbol{y}}{\partial \hat{p}_2}\cdots\frac{\partial \boldsymbol{y}}{\partial \hat{p}_P}\right] \tag{12.46}$$

We noted earlier that if this **discrete-time output sensitivity matrix** $[\partial y_i(t_k)/\partial \hat{p}_j]$ has full (column or row) rank = P (the number of unknown parameters) the model and thus $\boldsymbol{p}$ are *SI*. However, if $[\partial y_i/\partial p_j]$ is **ill-conditioned,** $\boldsymbol{p}$ is *not* numerically identifiable (*NI*), even if it is *SI*. This means data noise will result in relatively unreliable parameter estimates. Examples 11.11 and 11.12 of Chapter 11 illustrated this numerically with a simple model. Geometrically, the reason is that (at least) two matrix columns representing two otherwise *SI* parameters are *almost* parallel, the angle between them being small. The same properties apply to the relative output sensitivity matrix $[(\partial y_i(t_k)/\partial \hat{p}_j)(\hat{p}_j/y_i)] \equiv [\partial \boldsymbol{s}/\partial \hat{p}_1 \cdots \partial \boldsymbol{s}/\partial \hat{p}_P]$, which is often the one computed.

Note that $[\partial y_i(t_k)/\partial \hat{p}_j]$ is written as a matrix of *columns* in the last equality of Eq. (12.46), thus making it easier to geometrically evaluate the angles between sensitivity matrix columns. **Independent columns** of $\left[\dfrac{\partial \boldsymbol{y}}{\partial \hat{p}_1}\dfrac{\partial \boldsymbol{y}}{\partial \hat{p}_2}\cdots\dfrac{\partial \boldsymbol{y}}{\partial \hat{p}_P}\right]$ are *perpendicular,* or 90° apart. Fully **dependent columns** are parallel. **Partially dependent** columns are between 0° and 90° apart. A simple measure of **the degree of dependence** between two sensitivity column vectors $\dfrac{\partial \boldsymbol{y}}{\partial \hat{p}_i}$ and $\dfrac{\partial \boldsymbol{y}}{\partial \hat{p}_j}$, or *relative* sensitivity vectors $\dfrac{\partial \boldsymbol{s}}{\partial \hat{p}_j}$ and $\dfrac{\partial \boldsymbol{s}}{\partial \hat{p}_j}$, is given by the cosine of the angle between them:[32]

32 The expression $\|\partial y/\partial p_i\|_2 \equiv \sqrt{(\partial y/\partial p_i)^T(\partial y/\partial p_i)}$ is the L_2 norm or *length* of the vector $\partial y/\partial p_i$ (see Chapter 11, Eq. (11.37) or (11.38), or Appendix B).

$$\cos\phi_{ij} = \frac{|(\partial \boldsymbol{y}/\partial p_i)^T(\partial \boldsymbol{y}/\partial p_j)|}{\|\partial \boldsymbol{y}/\partial p_i\|_2\|\partial \boldsymbol{y}/\partial p_j\|_2} = \frac{|(\partial \boldsymbol{s}/\partial p_i)^T(\partial \boldsymbol{s}/\partial p_j)|}{\|\partial \boldsymbol{s}/\partial p_i\|_2\|\partial \boldsymbol{s}/\partial p_j\|_2} \tag{12.47}$$

Values close to 1 means maximal dependence, or *not NI*. Values close to 0 means close to independence and maximally discernable from each other in data (fragile, *NI*). This measure can be used *to prioritize relative importance of parameters* in their quantitative dynamical effects on outputs $\boldsymbol{y}(t)$, a subject of Chapter 17.

Analysis of $\left[\frac{\partial \boldsymbol{y}}{\partial \hat{p}_1} \cdots \frac{\partial \boldsymbol{y}}{\partial \hat{p}_P}\right]$ or $\left[\frac{\partial \boldsymbol{s}}{\partial \hat{p}_1} \cdots \frac{\partial \boldsymbol{s}}{\partial \hat{p}_P}\right]$ can thus provide information about both *SI* and *NI* of a dynamic biosystem model. Alternatively, covariance (*COV*) and correlation (*CORR*) matrices, discussed next, can provide additional information – in statistical terms – some of it equivalent to, but in a different form – than that provided by local sensitivity analysis, and without much more effort. *COV* and *CORR* also are a measures of second-order error effects, among pairs of parameters.

Remark: *COV* and *CORR* are provided by some parameter estimation software programs (e.g. *SAAMII*) by default. In these (lucky) circumstances local sensitivity analysis may not be needed, as discussed further later in the chapter.

Parameter Variances, Covariance/Correlation Matrices & Identifiability Relationships

Like with most approximations, parameter estimates $\hat{p}$ have little quantitative value without some knowledge of their reliability (variability, quality). Variability is typically expressed statistically, as **parameter estimate variance** $\sigma^2(\hat{p}_i)$ for each parameter p_i, and **parameter estimate covariance** $\equiv COV(\hat{p}_i\,\hat{p}_j)$ (or $CORR(\hat{p}_i\,\hat{p}_j)$) between each parameter pair.

Parameter estimates of structured dynamic system models are generally correlated with each other, which effects the quality of their estimates; so computation and reporting of such variability measures, along with their parameter estimates, is essential. Without the full matrix of variances and covariances,[33] $COV(\hat{\boldsymbol{p}})$, $\hat{p}$ is at best a speculation; at worst it is *un*ID and specific values obtained have little or no meaning with regard to the real system the model represents. We present several common schemas for estimating $COV(\hat{\boldsymbol{p}})$ below, called the **mean-square error matrix** in some circles, defined and written out as:

$$COV(\hat{\boldsymbol{p}}) \equiv E[(\hat{\boldsymbol{p}} - \bar{\boldsymbol{p}})(\hat{\boldsymbol{p}} - \bar{\boldsymbol{p}})^T] = \begin{pmatrix} \sigma^2(\hat{p}_1) & \cdots & \mathrm{cov}(\hat{p}_1\,\hat{p}_P) \\ \vdots & \ddots & \vdots \\ \mathrm{cov}(\hat{p}_P\,\hat{p}_1) & \cdots & \sigma^2(\hat{p}_P) \end{pmatrix} \tag{12.48}$$

where $\bar{\boldsymbol{p}}$ is the mean of $\boldsymbol{p}$ and E is the *expected value* (averaging) operation in probability and statistics, i.e. $E(\boldsymbol{p}) \equiv \bar{\boldsymbol{p}}$. The $\sigma^2(\hat{p}_i)$ are the **variance** (*VAR*) estimates, $COV(\hat{p}_i\,\hat{p}_j)$ are the estimates of covariance between $\hat{p}_i$ and $\hat{p}_j$, and these may be expressed by the more familiar –**normalized covariances**, i.e. the *correlation coefficients* between each parameter pair:

33 Or equivalent measures of quality, such as parameter sensitivities.

$CORR(\hat{p}_i \hat{p}_j) = COV(\hat{p}_i \hat{p}_j)/\sqrt{VAR(\hat{p}_i)VAR(\hat{p}_j)}$. **Correlation coefficients** are a normalized measure of the angle between parameters in vector space, ranging in magnitude from 0 to 1. When they are near 0, the vectors are nearly perpendicular – or orthogonal – and thus can be estimated best. When they are near 1, they are nearly parallel – or **collinear** – and thus cannot be estimated (distinguished) well.

Parameter Standard Errors (SEs) & Standard Deviations (SDs)

Parameter estimation variances $VAR(\hat{p}_i) \equiv \sigma^2(\hat{p}_i)$ are primary estimates for assessing parameter estimate variability, and they are expressed in different ways in computed results. The most common are: *standard error* $SE(\hat{p}_i)$ or *standard deviation* $SD(\hat{p}_i)$, either likely denoted $= \sigma(\hat{p}_i)$. *SE* and *SD* are often loosely used interchangeably. They are the square root of the variance estimates located on the diagonal of $COV(\hat{\mathbf{p}})$. Strictly speaking, the **parameter standard error** $SE(\hat{p}_i)$ is an estimate of the true standard deviation *SD* of the estimation error for parameter $\hat{p}_i$. Other common measures include: **fractional standard error (or deviation)** $FSE \sim FSD(\hat{p}_i) = \sigma(\hat{p}_i)/\hat{p}_i$, or **percent** $\boldsymbol{FSD \equiv \%CV(\hat{p}_i)}$ $\equiv$ **percent coefficient of variation**[34] $= 100\sigma(\hat{p}_i)/\hat{p}_i$ of the parameter estimate $\hat{p}_i$. Another is **parameter confidence interval**, an estimate of an interval that includes the parameter estimate with a certain probability (Bard 1974).

*SE*s or *SD*s for the $\hat{p}_i$s (or %*CV*s, or other measure of parameter estimate variability) are needed for assessing overall goodness of fit. They should be computed and reported in some form at the end of a successful parameter optimization (estimation) run.[35] As noted earlier, when they are small (equivalently: fragile, with large parameter sensitivity), the estimate is "good"; when they are small[36] other values more or less near the reported optimum estimate are equally probable, which means the estimate is unreliable, insensitive, i.e. poorly identifiable numerically.

Remark: Parameter reliability measures are certainly important in assessing model reliability, and even poor estimates are okay in some applications, if poor ones and good ones can be distinguished accurately. For example, if the model was fitted to data representing outputs for optimum closed-loop therapy, with the model in the closed-loop, parameters with large %*CV*s are relatively unimportant, meaning they have little or no influence on controlled outputs, so can be fixed to some nominal values. Parameters that affect the outputs strongly need to be relatively reliable.

34 It is convenient – and more intuitive – to specify variability in relative terms, e.g. the *SE* or *SD* relative to the mean of an estimate, rather than absolute *SE* or variance, which can take on any value. For parameter estimates, we like to use percent coefficient of variation. %*CV* is akin to what's usually meant by "percent error."

35 *SAAMII*, for example, normally provides parameter estimate values and their *SE* and %*CV* (or equivalent) estimates; *Copasi* provides relative *SE*s, because, at the time of this writing, it had no provision for individual data weighting.

36 Alas, it is difficult to say in general what "small" and "large" parameter estimation errors might be. This is surely problem-dependent. In my experience – for small dimension biomodels ($n = 2$ to 6 state variables) with data errors in the range of 10 to 25%, parameter %CVs roughly between 10 and 100% can be expected from biodata numbering up to 3 times the number of identifiable parameters or parameter combinations.

$VAR(\hat{p}_i)$ & $CORR(\hat{p}_i\,\hat{p}_j)$ for Identifiability Testing

True variances $\sigma^2(p_i)$ should be finite if the model is *SI*, and infinite if they are not *SI*. Estimates $\sigma^2(\hat{p}_i)$ are difficult to interpret if they are very large. If the variance estimate $\sigma^2(\hat{p}_i)$ of $\hat{p}_i$ is very large, $\hat{p}_i$ is likely *not NI*. Parameter correlations $CORR(\hat{p}_i\,\hat{p}_j)$ also are key in assessing *NI* because, if they are too large (close to 1), parameters are likely *not NI*.

Example 12.5: ***Numerical Identifiability & Correlation Coefficients***

Following is a correlation matrix of a three-parameter *SI* model:

$$\begin{matrix} & p_1 & p_2 & p_3 \\ p_1 \\ p_2 \\ p_3 \end{matrix}\begin{pmatrix} 1 & 0.1 & -0.98 \\ 0.1 & 1 & 0.5 \\ -0.98 & 0.5 & 1 \end{pmatrix} = \textit{correlation}\ \text{matrix} \equiv CORR(\boldsymbol{p})$$

Low correlation $(=0.1)$ pair p_1 and p_2 are likely candidates for being *NI*, because they are far from being collinear. However, p_1 and p_3 are NOT separately *NI*, because they are quite close to being collinear ($corr = -0.98$). They are likely *NI* in some combination (e.g. product or sum). With a correlation coefficient of 0.5, the *NI* of p_2 or p_3 are not clear. So this matrix doesn't tell the whole story, because there remains ambiguity about p_1 or p_2 with p_3. If p_1 or p_2 has a small parameter estimation variance, then the other of the two might be separately quantifiable from the way p_1 and p_3 are quantifiable in combination. This then may or may not yield full *NI* for this *SI* model.

$COV(\hat{\boldsymbol{p}}) \cong \hat{H}(\hat{\boldsymbol{p}})$ from the Hessian Matrix

Importantly, H is asymptotically equal to $COV(\boldsymbol{p})$ (Bard 1974), so $COV(\hat{\boldsymbol{p}})$ can be estimated from approximations of H. Second-order and similar search algorithms typically include terms requiring explicit computation of the Hessian matrix ("squared" output sensitivities) $\partial^2 Y/\partial \hat{\boldsymbol{p}}^2 \equiv (\partial^2 Y/\partial \boldsymbol{p}^2)_{p=\hat{p}}$, or approximations of it,[37] $\hat{H} \simeq ((\partial \boldsymbol{Y}/\partial \hat{\boldsymbol{p}})^T W(\partial \boldsymbol{Y}/\partial \hat{\boldsymbol{p}}))$, with data weighting matrix W included. For this reason, $\hat{H}$ for the final (converged) value of $\hat{H}$ is often used to estimate $COV(\hat{\boldsymbol{p}})$. The next section ties this all together practically and theoretically via the Cramér–Rao theorem.

Remark: To correctly compute $COV(\hat{\boldsymbol{p}})$ from $\hat{H}$, data weighting matrix W should quantitatively reflect output data ($z = y + e$) error (e) variability, as described earlier in the section on weighted least squares (*WLS, WRSS*) criteria. W is typically a diagonal matrix with inverse data variances on the diagonal. Any known data correlations also can be included in the off-diagonal terms; otherwise they are all zero. This is how the final estimate of $COV(\hat{\boldsymbol{p}})$ includes data error variability propagated through the model parameter estimation process into variability of parameter estimates.

37 See the previous section and Exercise E12.3.

$COV(p) \geq F^{-1}(p)$ from the Fisher Information Matrix

Following are some simplified, key results of this interesting and useful information theoretic result.[38] To motivate the presentation, we begin with the simplest linear *SISO* scalar ODE model: $\dot{x} = -ax + u$, with $a > 0$ unknown and N discrete-time noisy measurements: $z_i = x_i + e_i \equiv y_i + e_i$. We assume the residuals e_i (noise) are Gaussian (Normally) distributed, uncorrelated, with mean zero and variance σ_i^2. In this case, the **Fisher information matrix**[39] F (a scalar here) is simply:

$$F = \sum_{i=1}^{N} \left(\frac{\partial y(t_i)}{\partial a} \right)^2 \frac{1}{\sigma_i^2} \tag{12.49}$$

In Other Words: F is the sum of the squared output sensitivity functions weighted by the inverse of the variances of the data. The $\partial y(t_i)/\partial a$ indicate that quantitative information about the parameter is directly dependent on how sensitive each y_i output is to parameter variations, over all t_i in the observation interval. The square $(\cdot)^2$ indicates that both positive and negative sensitivities contribute. And the $w_i \equiv 1/\sigma_i^2$ indicate that data points z_i contribute information in inverse proportion to their variability (variance) – less variable data giving more quantitative info; and more data providing less of such information. A formidable and intuitive result, derived from information theory by brilliant statistician R.A. Fisher in 1925 (Fisher 1950).

To generalize, for multi-input–multi-output (*MIMO*) time-invariant nonlinear models, F is not that much more complicated, if vector-matrix notation is used: $\dot{\boldsymbol{x}} = \boldsymbol{f}(\boldsymbol{x},\boldsymbol{u},\boldsymbol{p})$ with outputs $\boldsymbol{y} = \boldsymbol{g}(\boldsymbol{x},\boldsymbol{u},\boldsymbol{p})$ and N discrete-time noisy measurements: $z_i = y_i + e_i$, and $\boldsymbol{p} = [p_1 p_2 \cdots p_P]^T$. Then F is:

$$F(\boldsymbol{p}) = \sum_{k=1}^{N} \left(\frac{\partial \boldsymbol{y}(t_k, \boldsymbol{p})}{\partial p_i} \right) W \left(\frac{\partial \boldsymbol{y}(t_k, \boldsymbol{p})}{\partial p_j} \right)^T = \left(\frac{\partial \boldsymbol{Y}}{\partial \boldsymbol{p}} \right)^T W \frac{\partial \boldsymbol{Y}}{\partial \boldsymbol{p}} \tag{12.50}$$

As above, the residuals $\boldsymbol{e}_i$ are assumed (approximately) Gaussian distributed, uncorrelated, with mean zero and known data covariance matrix $COV(\boldsymbol{z}) = W$, diagonal in this case – with the inverse variances of the residuals (data) on the diagonal. $\partial \boldsymbol{Y}/\partial \boldsymbol{p} \equiv [\partial y_i(t_k)/\partial p_j]$ is the (Jacobian) matrix of output sensitivity functions.

The Cramér–Rao Theorem

The following result (Cramér 1999) provides the basis for using F^{-1} as an unbiased, albeit lower bound, estimate for $COV(\hat{\boldsymbol{p}})$:

$$COV(\hat{\boldsymbol{p}}) \geq F^{-1}(\hat{\boldsymbol{p}}) \tag{12.51}$$

38 A more general development of the Fisher information matrix F is presented in Appendix F in its original stochastic (statistical) context.

39 The Fisher information Matrix F is designated ***FIM*** by some authors.

where $\hat{\boldsymbol{p}}$ is an unbiased estimate of $\boldsymbol{p}$. **Unbiased** means the expected value of $\boldsymbol{p}$ equals the mean $\bar{\boldsymbol{p}}$ of $\boldsymbol{p}$, i.e. $E(\hat{\boldsymbol{p}}) = E(\boldsymbol{p}) = \bar{\boldsymbol{p}}$. Thus, F^{-1} is a lower bound estimate for the covariance matrix of the parameters. It is, in fact, what is called in statistics the **asymptotic covariance matrix** of $\boldsymbol{p}$ (Bard 1974).

Remark: This strong and very useful result illustrates the important *duality* of parameter sensitivities and covariances. And it is consistent with results obtained from other statistical approaches. The $COV(\hat{\boldsymbol{p}})$ matrix estimate obtained from Eqs. (12.50) and (12.51) coincides with that given by the Hessian matrix approximation $\hat{\boldsymbol{H}} \simeq \left(\frac{\partial \boldsymbol{Y}}{\partial \hat{\boldsymbol{p}}}\right)^T W \frac{\partial \boldsymbol{Y}}{\partial \hat{\boldsymbol{p}}}$ in Eq. (12.44) for the same data error statistics.

In Other Words: This "weighted and squared sensitivity function" is a lower bound estimate of the COV matrix of the parameters, which means the relative magnitudes of local output sensitivities to parameters can be assessed directly from the COV matrix. Roughly speaking, large relative parameter variances (on the diagonal of $\hat{\boldsymbol{H}}$) and small relative sensitivities are one and the same, as are small variances and large sensitivities.

Remark: Again, it is essential to incorporate the weighting matrix when computing $COV(\hat{\boldsymbol{p}})$ from F using Eqs. (12.50) and (12.51). Data weights equal to the inverse of the variance of the data are common, i.e. $w_i = 1/VAR_i$, as illustrated in the next examples.

To see more explicitly how the Eq. (12.51) lower bound estimate of $COV(\hat{\boldsymbol{p}})$ depends directly on output sensitivities (and data variances) for systems with more than one output and more than one parameter, we compute F^{-1} for 2-by-2 systems. In this case, Eq. (12.51) spells out as:

$$\begin{bmatrix} \sigma^2(\hat{p}_1) & COV(\hat{p}_1\hat{p}_2) \\ COV(\hat{p}_1\hat{p}_2) & \sigma^2(\hat{p}_2) \end{bmatrix} \geq \left[\begin{pmatrix} \partial y_1/\partial p_1 & \partial y_1/\partial p_2 \\ \partial y_2/\partial p_1 & \partial y_2/\partial p_2 \end{pmatrix}^T \begin{pmatrix} \frac{1}{\sigma_1^2} & 0 \\ 0 & \frac{1}{\sigma_2^2} \end{pmatrix} \begin{pmatrix} \partial y_1/\partial p_1 & \partial y_1/\partial p_2 \\ \partial y_2/\partial p_1 & \partial y_2/\partial p_2 \end{pmatrix} \right]^{-1}$$

$$= \left[\begin{pmatrix} \frac{(\partial y_1/\partial p_1)^2}{\sigma_1^2} + \frac{(\partial y_1/\partial p_2)^2}{\sigma_2^2} & \frac{(\partial y_1/\partial p_1)\partial y_1/\partial p_2}{\sigma_1^2} + \frac{(\partial y_2/\partial p_1)\partial y_2/\partial p_2}{\sigma_2^2} \\ \frac{(\partial y_1/\partial p_1)\partial y_1/\partial p_2}{\sigma_1^2} + \frac{(\partial y_2/\partial p_1)\partial y_2/\partial p_2}{\sigma_2^2} & \frac{(\partial y_1/\partial p_2)^2}{\sigma_1^2} + \frac{(\partial y_2/\partial p_2)^2}{\sigma_2^2} \end{pmatrix} \right]^{-1} \quad (12.52)$$

Even without completing the inversion, Eq. (12.52) makes explicit how $COV(\hat{\boldsymbol{p}})$ depends on all the (local) output sensitivity functions $\partial y_i/\partial p_j$ and the variances of all of the data.

Example 12.6: ***COV(p), CORR(p) & Structural & Numerical Identifiability Properties from F***

We extend Example 11.11 of Chapter 11 here by computing approximate $COV(\boldsymbol{p})$ and $CORR(\boldsymbol{p})$ matrices from the Fisher Information matrix F, using the the Cramér-Rao lower bound to estimate these matrices. We assume data errors are approximately constant standard deviations $\sigma = 0.1$. Then $\sigma^2 = 0.01$. If the data were continuous-time – the weighting matrix would be 2×2, one dimension for each sensitivity function $\partial y/\partial C$ and $\partial y/\partial K$. But our data and sensitivity matrix are discrete-time, with five timed data points for each of two variables. The W is 5×5:

$$W = \begin{bmatrix} 100 & 0 & 0 & 0 & 0 \\ 0 & 100 & 0 & 0 & 0 \\ 0 & 0 & 100 & 0 & 0 \\ 0 & 0 & 0 & 100 & 0 \\ 0 & 0 & 0 & 0 & 100 \end{bmatrix} \equiv 100I$$

Matlab code leading to results in Example 11.11 of Chapter 11 was extended to include five additional lines at the end, to incorporate W into the computations and approximate $COV(\boldsymbol{p})$ and $CORR(\boldsymbol{p})$:

$$\begin{aligned} &\mathrm{W = 100 * eye(5)} \\ &\mathrm{F = dydp' * W * dydp} = \begin{pmatrix} 22.38 & -45.7 \\ -45.7 & 93.3 \end{pmatrix} \\ &\mathrm{format\ short\ g;} \\ &\mathrm{COVmin = inv\,(F)} \\ &\mathrm{CORmat = corrcov(COVmin)} \end{aligned} \tag{12.53}$$

The Cramér-Rao lower bound estimate F^{-1} of $COV(\boldsymbol{p})$ computed with the *Matlab* program (**Figs. 11.6–11.10** plus the above code) is:

$$COV_{min}(\boldsymbol{p}) = \begin{bmatrix} 50.7 & 24.8 \\ 24.8 & 12.1 \end{bmatrix} \begin{matrix} C \\ K \end{matrix}$$
$$\qquad\qquad C \quad K$$

The asymptotic (predicted best) variances of the parameters are on the diagonal. The smallest is $\sigma^2(K) = 12.1$. It has a % coefficient of variation $\%CV = 100\sqrt{12.1}/K(=1) \cong 348\%$, indicating a very poor estimate of K (a clearer measure of error than variance alone). The correlation matrix $CORR(\boldsymbol{p})$, reproduced below, describes *interparameter* variability in the estimation process – with nearly complete correlations ($\approx \pm 1$) among C and K. All this underscores that the model, despite being *SI*, is hardly *NI*.

$$CORR_{\min}(\boldsymbol{p}) = \begin{bmatrix} 1 & 0.9996 \\ 0.9996 & 1 \end{bmatrix} \begin{matrix} C \\ K \end{matrix}$$
$$\qquad\qquad C \quad K$$

This goes far toward explaining the poor accuracy. Doubling the sampling rate (not shown here) didn't help much either, because the additional data points are still insufficient for capturing enough of the linear rising portion of the M–M leak function.

Residual Mean Square (*RMS*) Error & *COV*(p^*) for Unweighted Regression

We noted earlier that optimum weighting for $WLS \equiv WRSS$ is the inverse variance of the data at each data point collected. A typical iterative search scheme for minimizing the $WRSS$

(Eq. (12.22)) is terminated after the specified convergence criterion is met. The parameter values $\boldsymbol{p}^*$ corresponding to the minimum *WRSS* (usually last iteration) are the **optimum parameter estimates**. We need to know $COV(\boldsymbol{p}^*)$ and $VAR(\boldsymbol{p}^*)$ for this $\boldsymbol{p}^*$, to assess the quality/reliability of these best parameter estimates. The **residual mean square (*RMS*)** can be helpful, because it is a measure of the fitting error variance (σ_i^2) for every point z_i on the regression curve (Ljung 1987). RMS is defined as:

$$RMS = \frac{min\ WRSS(\hat{\boldsymbol{p}})}{df} = \frac{WRSS(\boldsymbol{p}^*)}{N-P} \tag{12.54}$$

N is the number of (discrete) data points fitted, P is the number of structurally identifiable parameters and $df \equiv$ degrees of freedom $\equiv N - P > 0$.

When little or no information is available about data variances, a common practice is to run the estimation procedure with all data weighted equally, which then makes all variances equal as well, otherwise called **unweighted regression** in statistics. RMS can be particularly useful for unweighted regression estimation, where *RMS* provides an unbiased estimate of the (constant) variance σ^2 of each datum, with $COV(\boldsymbol{z}) = \sigma^2 I$. Then $W = I/\sigma^2$ and from (12.50) the parameter *COV* matrix can be estimated from:

$$COV(\boldsymbol{p}^*) \geq F^{-1}(\boldsymbol{p}^*) = \left(\left(\frac{\partial \boldsymbol{Y}}{\partial \boldsymbol{p}^*}\right)^T \frac{I}{\sigma^2}\frac{\partial \boldsymbol{Y}}{\partial \boldsymbol{p}^*}\right)^{-1} = \sigma^2\left(\left(\frac{\partial \boldsymbol{Y}}{\partial \boldsymbol{p}^*}\right)^T \frac{\partial \boldsymbol{Y}}{\partial \boldsymbol{p}^*}\right)^{-1} \tag{12.55}$$

The unweighted Fisher information matrix – last term in parentheses above – is supplied without provision for weighting by W in some packages, e.g. in *Copasi*. Eq. (12.55) can be used to obtain the weighted $COV(\boldsymbol{p}^*)$ matrix in these circumstances. However, we emphasize that, if prior knowledge indicates the error variance is not constant (previous section), then weighting proportional to the inverse of the *actual* error variance for each data point is needed for computing $COV(\boldsymbol{p}^*)$. If the error variances are known beforehand, the best approach is to incorporate them directly into the parameter optimization problem, e.g. into the *WRSS* as $W = [1/\sigma_i^2]$ and evaluate $COV(\hat{\boldsymbol{p}})$ (for example) as $\left(\left(\frac{\partial \boldsymbol{Y}}{\partial \hat{\boldsymbol{p}}}\right)^T W \frac{\partial \boldsymbol{Y}}{\partial \hat{\boldsymbol{p}}}\right)^{-1}$ if the software accommodates.

Covariance & Fisher Information Matrices for Functions *b*(*p*) of *p*

Estimated variability of functions $\boldsymbol{b} = \boldsymbol{b}(\boldsymbol{p})$ of estimated parameters $\boldsymbol{p}$ is sometimes needed. For example, $\boldsymbol{b} = \boldsymbol{b}(\boldsymbol{p})$ might be a set of *structurally identifiable* (*SI*) *parameter combinations* in an otherwise unidentifiable model. In this case, $\boldsymbol{b}$ is *SI* and the model, in principle, can be reparameterized and quantified in terms of $\boldsymbol{b}$ (the next chapter has an example of reparameterization). Asymptotic theory in statistics (Bard 1974) provides a convenient way of computing this as

$$COV(\boldsymbol{b}) = \frac{\partial \boldsymbol{b}}{\partial \boldsymbol{p}} COV(\boldsymbol{p}) \left(\frac{\partial \boldsymbol{b}}{\partial \boldsymbol{p}}\right)^T \tag{12.56}$$

Example 12.7: ***Covariance for the Area Under the Curve (AUC) Function***

In pharmacology, a pharmacokinetic (*PK*) parameter of interest in the study of drug disposition is the area under the curve (*AUC*) of drug concentration measured after an oral or intravenous dose is given.[40] If the drug concentration curve has been fitted, say, by a single exponential function $\hat{A}e^{-\hat{k}t}$, over an interval [0, *T*], then the *AUC* is simply the integral of this function over [0, *T*], readily evaluated as $\hat{A}(1 - \hat{A}e^{-\hat{k}T})$. The *COV* matrix for *AUC* in this case is a scalar, the variance *VAR*(*AUC*), and is computed from (12.56) as:

$$V\hat{A}R(AUC) \equiv \sigma^2_{AUC} = \begin{bmatrix} \dfrac{\partial AUC}{\partial \hat{A}} & \dfrac{\partial AUC}{\partial \hat{k}} \end{bmatrix} \begin{pmatrix} \sigma^2_{\hat{A}} & \operatorname{cov}(\hat{A},\hat{k}) \\ \operatorname{cov}(\hat{A},\hat{k}) & \sigma^2_{\hat{k}} \end{pmatrix} \begin{bmatrix} \dfrac{\partial AUC}{\partial \hat{A}} \\ \dfrac{\partial AUC}{\partial \hat{k}} \end{bmatrix}$$

$$= \sigma^2_{\hat{A}} \left(\frac{\partial AUC}{\partial \hat{A}}\right)^2 + \sigma^2_{\hat{k}} \left(\frac{\partial AUC}{\partial \hat{k}}\right)^2 + 2\operatorname{cov}(\hat{A},\hat{k}) \left(\frac{\partial AUC}{\partial \hat{A}} \frac{\partial AUC}{\partial \hat{k}}\right) \tag{12.57}$$

Substitution of $\dfrac{\partial AUC}{\partial \hat{A}} = 1 - 2\hat{A}e^{-\hat{k}T}$ and $\dfrac{\partial AUC}{\partial \hat{k}} = -\hat{A}\hat{k}e^{-\hat{k}T}$ into (12.57), along with the estimates $\hat{A}$, $\hat{k}$, $\sigma^2_{\hat{A}}$, $\sigma^2_{\hat{k}}$ and $\operatorname{cov}(\hat{A},\hat{k})$ from *COV*(**p**), then yields the variance estimate of *AUC*. Note that this variance result is dependent on the *covariance* of the two parameters $\hat{A}$, $\hat{k}$, as well as their variances $\sigma^2_{\hat{A}}$, $\sigma^2_{\hat{k}}$, because $\hat{A}$ and $\hat{k}$ are related by their exponential "parent" $\hat{A}e^{\hat{k}t}$.

This example result, Eq. (12.57), clearly underscores the relationship between parameter estimation precision (reliability) – in this case estimation of a function of model output (AUC) – the sensitivity of this function to estimated parameters and their variances, and the correlations among these sensitivity functions.

*COV(p) by Stochastic Monte Carlo Simulation

Monte Carlo (M-C) simulation is a way of generating simulated outcomes of an experiment on a computer. For us, this would be a simulation of the experiment generating output data for (re)fitting the ODE biomodel (a seemingly circular problem), with useful results. The sample statistics for the parameter estimates computed for these simulated outcomes (simulated data) then readily provide *sample means, variances* and *covariances* of the parameter estimates. This approach is developed further and illustrated in the first section of the next chapter.

In brief, the M-C procedure entails generating *Q* sets of times-series model output data, each set consisting of *N* stochastic (random) output data points – *N* times *Q* data points in all.

40 *PK* and *AUC*s are developed in Chapter 8. *AUC* is important is establishing bioequivalence and bioavailability of drugs, also developed in Chapter 8.

These are chosen from a random process based on the assumed or known statistics of the output data, superimposed on simulated deterministic model outputs, based on the best estimate of $\boldsymbol{p}$ available – call it $\hat{\boldsymbol{p}}^0$. This simulated stochastic data is then used to obtain a sequence of Q new model parameter estimates $\hat{\boldsymbol{p}}^1, \hat{\boldsymbol{p}}^2, \ldots, \hat{\boldsymbol{p}}^Q$ by fitting the model output(s) to the Q sets of stochastic data. The algorithmic steps are as follows:

1. Generate Q sets of time-series data, each consisting of N data points, by simulating the noisy outputs $z_k = \boldsymbol{y}_k(\hat{\boldsymbol{p}}^0) + e_k, \ \ k = 1,\ldots,N$, where the means $\bar{\boldsymbol{e}}_i$ and $COV(\boldsymbol{e}_i)$ of the data noise are given. The $\boldsymbol{y}_k(\hat{\boldsymbol{p}}^0)$ are the simulated outputs of the deterministic ODE model, using the best estimate $\hat{\boldsymbol{p}}^0$ available, and assuming initial conditions are known.

2. Run the parameter search algorithm for each of the Q sets of simulated data, thereby determining a sequence of Q new parameter estimates $\hat{\boldsymbol{p}}^1, \hat{\boldsymbol{p}}^2, \ldots, \hat{\boldsymbol{p}}^Q$.

3. Determine the sample statistics for the estimated sequence of parameters, i.e. the sample means $\hat{\bar{\boldsymbol{p}}}$, sample variances $\bar{\sigma}^2$ and sample covariances $C\bar{O}V(\hat{\boldsymbol{p}}^j)$ for the parameter estimates. The **sample mean** of the parameter vector is

$$\hat{\bar{\boldsymbol{p}}} = \frac{1}{Q}\sum_{j=1}^{Q} \hat{\boldsymbol{p}}^j \tag{12.58}$$

And the **sample covariance matrix** of the $\hat{\boldsymbol{p}}^j$ is obtained from the first equality in (12.48) by averaging the square of the matrix of differences of the $\hat{\boldsymbol{p}}^j$ from their sample mean vector estimates, i.e.

$$C\bar{O}V(\hat{p}^j) \equiv \frac{1}{Q-1}\sum_{j=1}^{Q} \left(\hat{\boldsymbol{p}}^j - \hat{\bar{\boldsymbol{p}}}\right)\left(\hat{\boldsymbol{p}}^j - \hat{\bar{\boldsymbol{p}}}\right)^T \tag{12.59}$$

The **sample variances** $\bar{\sigma}^2$ of the parameters are on the diagonal of $C\bar{O}V(\hat{\boldsymbol{p}}^j)$.

Remark: The $Q-1$ in the denominator (12.59), instead of Q, eliminates bias in sample estimation of covariances (Cramér 1946).

Robustness and Sensitivities from COV($\boldsymbol{p}$)

Many applications require some kind of robustness analysis and this means quantifying how objective functions vary with parameter estimate variations. As discussed in the previous chapter, this involves computation of $\partial y/\partial \boldsymbol{p}$, or $\partial f(\boldsymbol{y})/\partial \boldsymbol{p}$ a (usually) scalar function of the model output. Examples are, $f(\boldsymbol{y},\boldsymbol{p}) = \int_0^T y_1(t,\boldsymbol{p})dt \equiv AUC(\boldsymbol{p})$, or any optimization criterion $L(\boldsymbol{y},\boldsymbol{p},\boldsymbol{u})$, e.g. $WLS(\boldsymbol{p})$.

The easiest way to evaluate robustness of f to p_j is to estimate the variance of f from $COV(\boldsymbol{p})$, when it is available, using Eq. (12.56), as in Example 12.7 above. To justify this approach we invoke the duality of sensitivity and covariance analyses noted in the *Remarks* associated with the Cramér-Rao lower bound, Eq. (12.51). If f is the objective for optimally estimating $\boldsymbol{p}$, e.g. $WLS(\boldsymbol{p})$, variance measures for all p_j are provided directly from $COV(\boldsymbol{p})$

without additional sensitivity analysis, and relative variances (e.g. fractional standard deviations *FSD*s or coefficients of variation *CV*s), are perfectly good surrogates for relative sensitivities. With complete parameter estimation or modeling software, when the optimization criterion like $WLS(\mathbf{p})$ is satisfied at the end of a parameter estimation run, $COV(\mathbf{p})$ is often given via sensitivity function approximations like Eqs. (12.50) and (12.51) and we can interpret the $WLS(\mathbf{p})$ as $f(\mathbf{y},\mathbf{p})$ in Eq. (11.39) of Chapter 11. This also is the easiest way to accomplish *sensitivity analysis* of the optimally final fitted model w.r.t. local variations in all model parameters[41], $p_1, \ldots, p_P$.

The problem addressed here using local sensitivity analyses also can be addressed more completely – but with considerably greater computational effort – using *global sensitivity* methods.[42] This means quantifying the effects on *objective f* due to variations in all parameters simultaneously about their nominal values, over a large but finite subset of points in the entire range of feasible nominals (parameter space).[43] This is addressed further in Chapter 17.

Other Biomodel Quality Assessments

Goodness-of-Fit Criteria & Figures of Merit – Subjective & Statistical

The following criteria are applicable for fitting (estimating the parameters of) any type of model to sequential data.

The "Eyeball" Test: Visual Inspection

Does the fitted function look like it fits the data as well as expected? This test is certainly not objective, but it usually should be a first consideration, as it can be important, especially if the fit is visibly poor by inspection, which can suggest what's wrong with the estimation problem formulation, or the model.

Analysis of Residuals

These tests are objective, because the residuals $e(t_k)$ represent the (± signed) distance between the data point at time t_k and the fitted function at each point and are thus a direct measure of the fit quality at each point. Following are several statistical procedures for analyzing residuals. The "eyeball-test" is also useful in assessing residuals, as illustrated in **Fig. 12.5**.

Appropriate Weighting of the Data

As noted above, the correctness as well as the quality of the model fitted to measured data depends in part on the correctness of the assumed or measured data variance $\sigma^2(t_k)$. This can be a major weakness in the entire modeling process. A plot of the unweighted residuals $e(t_k)$ vs. time t_k will reveal how the error variance $\sigma^2(t_k)$ varies along a fitted (regression) curve. For

41 This duality property is not widely recognized.

42 See Chapter 17.

43 Some modeling software packages (e.g. *COPASI*) include functionality for *parameter scanning*, which accomplishes a very small part of global sensitivity analysis.

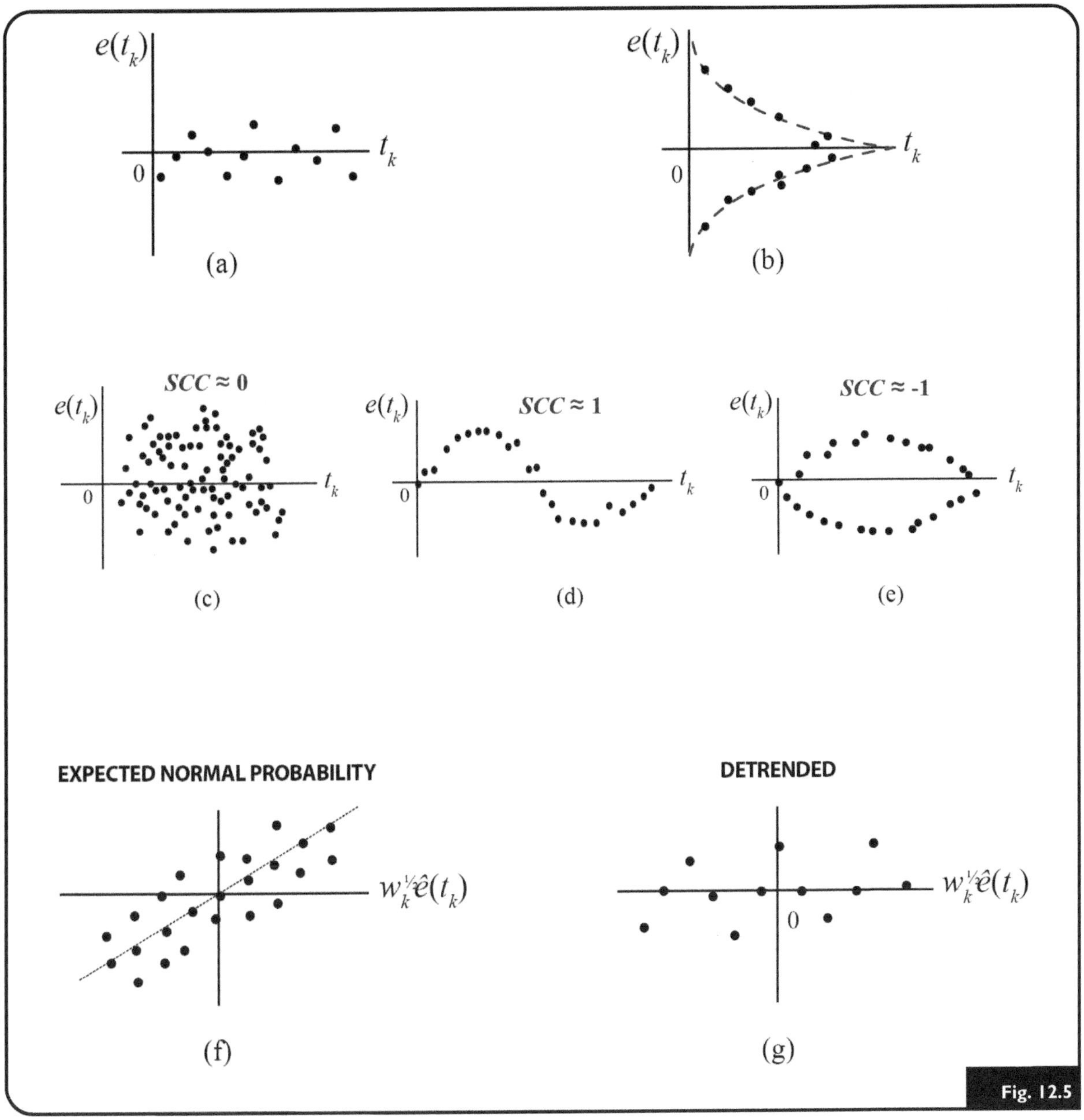

Fig. 12.5

Analysis of residuals $e(t_k)$. Common patterns: (a) uniform $e(t_k)$; (b) $e(t_k)$ for constant CV data errors, (specifically for decaying exponential $y(t_k)$ data); (c) uniform and uncorrelated residuals have a serial correlation coefficient SCC ≈ 0; (d–e) correlated residuals have SCC $\approx \pm 1$; (f) approximately Gaussian sampled probability distribution as a function of weighted residuals $\sqrt{w(t_k)}e(t_k)$; (g) detrended version of (f).

unweighted (effectively, equal weighted) regression, one expects a constant $\sigma^2(t_k)$ and this should correspond to *uniform scattering with time* of the residuals about zero (**Fig. 12.5a**). If the scattering is not uniform (for example, largest scattering at small times), this can suggest a more appropriate weighting. If the data variability has constant coefficient of variation (CV), the standard error (SE) $\sigma(t_k)$ should be proportional to the magnitude of the mean response $y(t)$. For example, *for a monotonically decreasing response $y(t)$*, the CV is also monotonically decreasing,[44] as illustrated in **Fig. 12.5b** for data exponentially decaying with time.

44 For a more general response $y(t)$ – not just decreasing – the $e(t_k)$ will *not* look like the pattern in **Fig. 12.5b**. It will be more varied, larger for larger $y(t)$, etc.

Independence of Residual Errors

An important assumption if the model fits well is that no two measurement errors for different data points are correlated. A visual check of the plot of residuals $e(t_k)$ vs. time t_k can often detect violations of this assumption if there is a discernible pattern in the sequential scattering of the points. The **serial correlation coefficient** (SCC) is one basis for a statistical test on independence of *sequential pairs* of residuals (Landaw and DiStefano III 1984). It is calculated as follows:

$$SCC = \frac{\sum_{k=2}^{N} e(t_k)e(t_{k-1})}{\sqrt{\sum_{k=1}^{N} e^2(t_k) \sum_{k=2}^{N} e^2(t_{k-1})}} \tag{12.60}$$

where $e(t_{k-1})$ is the residual error for the $(k-1)$st data point and N is the total number of data points. If the residual errors are sequentially *uncorrelated* (with the sum of the residuals equal to zero), SCC will be close to zero – the most desirable result. If they are sequentially highly correlated, SCC will be close to ± 1 as illustrated in **Fig. 12.5c–e**. Tables for testing if SCC is significantly different from zero are available (Cramér 1946).

Besides problems in the actual measurements, a common cause of nonrandom behavior in sequential residuals $e(t_k)$ is failure to include enough terms or parameters in the model (fitting equation), e.g. too-few exponentials in a multiexponential model or compartments (state variables) in a compartmental model. This means the model is too simple, not a good enough model, **underparameterized**. In this case, $SCC \cong 1$.

Gaussian Distribution & Independence of Residuals

A desirable property is that the residual errors be (approximately) Gaussian (normally) distributed and statistically independent (Landaw and DiStefano III 1984). Indeed, independence of residuals yields optimally fitted models for most model classes of practical importance (Moran 1950). As a rule of thumb, if residual *independence* is present, SCC is distributed asymptotically *normally* with mean $-1/(N-1)$ and variance $1/(N-1)$ for large N. *For these conditions, least square estimators are also maximum likelihood estimators.* If the residuals are all Gaussian with constant variance $\sigma^2(t_k)$, **normal probability** and **detrended normal probability** plots of the **weighted residuals** $(w_k^{1/2}e_k)$ should be straight lines, where w_k are data weights at time t_k. Detrended plots are horizontal through the zero intercept – see **Fig. 12.5f–g**. Note that this is *not* a test of (statistical) *independence* of the residuals.

Recap and Looking Ahead

Basic methodology for model parameter estimation was provided and illustrated in this chapter, including problem setup, optimization criteria, algorithms for parameter search and methods for assessing parameter estimate reliability and other measures of model quality.

Interrelationships among sensitivity, identifiability and covariance measures were emphasized in analyzing quantified model reliability.

Mathematically, this material is fairly advanced and likely needs time and restudy for mastery. A full reading of it first time around (even without deep understanding), to become familiar at least with the language of parameter estimation methodology, is important for developing meaningful biomodels. Increasing availability of software tools with comprehendible interfaces and reliable numerics will go far toward making the theory and methodology accessible and useful.

The next chapter extends these notions with practical methods and extensions for simplifying and facilitating the quantification problem using simulation tools, parameter space constraints, data-driven forcing function inputs, reparameterizations and realistic noisy data illustrations.

"Young man, in mathematics you don't understand things.
You just get used to them."

John von Neumann

Exercises

E12.1 ***Linear-in-the-parameters model.*** Find the least-squares solution for the following data pairs: (0,1), (2,6), (3, 9), (5,5) fitted to the parabola: $y(t) = a_1 + a_2 t + a_3 t^2$. Set up your three equations and three unknowns in the form: $\boldsymbol{d} = T\boldsymbol{a}$, with $\boldsymbol{a} = [a_1 a_2 a_3]^T$.

E12.2 ***Linear-in-the-parameters embedded in a nonlinear-in-the-parameters model.*** Repeat E12.1 for $y(t) = Ae^{at} + Be^{bt}$, assuming a and b are known already. What if they are unknown?

E12.3** ***Jacobian and Hessian of WLS(p). Find the Jacobian and Hessian of quadratic loss function $L(\boldsymbol{p}) \equiv WLS(\boldsymbol{p}) = \boldsymbol{E}^T(\boldsymbol{x},\boldsymbol{p})W\boldsymbol{E}(\boldsymbol{x},\boldsymbol{p})$. $\boldsymbol{E} \equiv \boldsymbol{Y}(\boldsymbol{x},\boldsymbol{p}) - \boldsymbol{Z}(t)$ and W is a symmetric weighting matrix ($W = W^T$).

E12.4** ***Gauss–Newton Algorithm. Using the result in the solution of Exercise E12.3 above, derive the iterative Gauss–Newton (Gauss) second-order gradient search algorithm.

E12.5** ***Newton–Raphson Algorithm. Follow the same steps used in deriving the first-order gradient search algorithm, i.e. calculus and the Taylor series, and derive the second-order **Newton–Raphson** iterative search algorithm.

E12.6** ***Maximum Likelihood Estimation. Maximum-likelihood and weighted least squares optimization methods yield the same parameter estimation results when residual output errors are zero mean, independent and Gaussian distributed (common in applications like ours). Demonstrate this result for the multiexponential output model $y(t) = \sum_{k=1}^{n} A_k e^{\lambda_k t}$ with unknown A_k and λ_k.

References

Aldrich, J., 1997. R. A. Fisher and the Making of Maximum Likelihood 1912–1922. Statist. Sci. 12 (3), 162–176.

Aster, R.C., Borchers, B., Thurber, C.H., 2005. Parameter Estimation and Inverse Problems. Elsevier. Academic Press, Amsterdam.

Balsa-Canto, E., A. A., Banga, J., 2010. An iterative identification procedure for dynamic modeling of biochemical networks. BMC Syst. Biol. 4, 11.

Balsa-Canto, E., Banga, J.R., 2011. AMIGO, a toolbox for advanced model identification in systems biology using global optimization. Bioinformatics. 27 (16), 2311–2313.

Bard, Y., 1974. Nonlinear Parameter Estimation. Academic Press, New York.

Beck, J., Arnold, K., 1977. Parameter Estimation in Engineering and Science. J Wiley, New York.

Berman, M., Shahn, E., Weiss, M.F., 1962. The routine fitting of kinetic data to models. Biophys. J. 2, 275–287.

Casella, G., Berger, R.L., 2001. Statistical Inference. Duxbury Press.

Chou, I.C., Voit, E.O., 2009. Recent developments in parameter estimation and structure identification of biochemical and genomic systems. Math. Biosci. 219 (2), 57–83.

Cramér, H., 1946. Mathematical Methods of Statistics. Princeton University Press, Princeton, NJ.

Cramér, H., 1999. Mathematical methods of statistics. Princeton University Press, Princeton.

Fisher, R.A., 1950. Contributions to Mathematical Statistics. John Wiley, New York.

Harless, C., DiStefano III, J., 2005. Automated expert multiexponential biomodeling interactively over the internet. Comput. Meth. Programs Biomed. 79 (2), 169–178.

Landaw, E., DiStefano III, J., 1984. Multiexponential, multicompartmental, and noncompartmental modeling: physiological data analysis and statistical considerations, part II: data analysis and statistical considerations. Am. J. Physiol. Regul. Integr. Comp. Physiol. 246, R665–R677.

Ljung, L., 1987. System Identification: Theory for the User. Prentice-Hall, Englewood Cliffs, NJ.

Moles, C.G., Mendes, P., Banga, J.R., 2003. Parameter estimation in biochemical pathways: a comparison of global optimization methods. Genome Res. 13 (11), 2467–2474.

Moran, P.A.P., 1950. A test for the serial independence of residuals. Biometrika. 37 (1–2), 178–181.

Riggs, D., 1963. The Mathematical Approach to Physiological Problems. Williams & Wilkins, Baltimore, MD.

Rodriguez-Fernandez, M., Mendes, P., Banga, J.R., 2006. A hybrid approach for efficient and robust parameter estimation in biochemical pathways. Biosystems. 83 (2–3), 248–265.

Stigler, S.M., 1986. The History of Statistics: The Measurement of Uncertainty Before 1900. Belknap Press of Harvard University Press, Cambridge, MA.

Chapter 13

Parameter Estimation Methods II: Facilitating, Simplifying & Working With Data

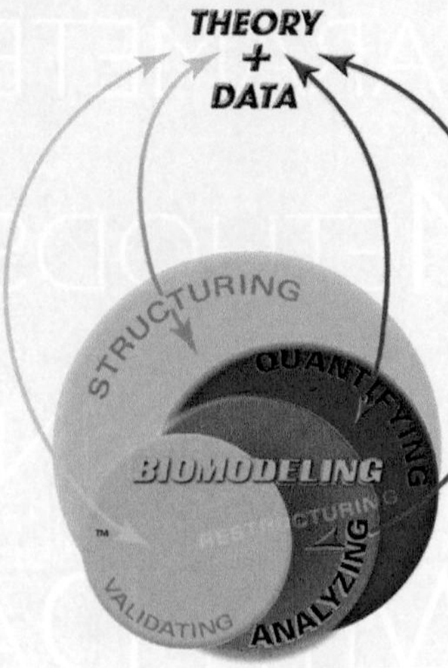

"With four parameters I can fit an elephant,
and with five I can make him wiggle his trunk."

John von Neuemann

Overview

This chapter extends exposition of parameter estimation methodology, building on concepts, models and examples developed in the last and earlier chapters. Model quantification is technically difficult, and we need all the help we can get from practice as well as theory. Several early examples underscore some key issues and illustrate some challenging side-effects of the parameter estimation process itself. Some time-honored ingenuity and some new methodology and insights are presented in each section, all focused on facilitating the process.

Let's begin by restating the parameter estimation problem succinctly. The state equations of biodynamic system models have the general form $\dot{\boldsymbol{x}} = \boldsymbol{f}(\boldsymbol{x}, \boldsymbol{u}, t, \boldsymbol{p})$, where the vector of P unknown and constant parameters $\boldsymbol{p}$ is expressed explicitly in the arguments of $\boldsymbol{f}$. With little disagreement – model quantification usually means assigning or *finding appropriate values* for $\boldsymbol{p}$ from data. "Finding" and "appropriate values," however, can mean different things, depending on modeling goals, approaches, styles and tools available for model quantification. **Finding** usually means using optimization methods and (usually noisy) output measurement data, e.g. $z(t)$ over some finite time interval, to estimate parameter values. *Appropriate values* is not as easy to pin down, because it depends on modeling goals, and therefore qualitative issues, so we have to be specific about goals. For one, maybe we don't need all the parameters to satisfy goals; often we need only a subset of $\boldsymbol{p}$.

Suppose model parameters, as such, are associated with biosystem characteristics that reflect modeling goals, rendering at least some parameter "values" important to know with some degree of precision – with reliability measures made explicit. For example, such goals might be understanding a protein production mechanism, or quantitatively evaluating absorption of a drug. Then **appropriate** means **with statistics**, i.e. parameter values that include measures of parameter estimate uncertainty. **Complete parameter estimation** then means quantifying not only **central-value** estimates for parameters of interest, e.g. mean values, but also their variances, covariances, correlations or other measures of parameter uncertainty. **Complete model quantification** means doing this (at least) for all parameters of interest. Incomplete quantifications are those that do not provide both.[1] Model form or optimality criterion may vary – depending on modeling "styles" or particular circumstances – but complete parameter estimation results include the needed qualitative measures.[2]

We know from developments in Chapters 10–12 that identifiability issues, often hidden or otherwise disguised or unrecognized, can be major encumbrances in model quantification. Success depends very much on the structural identifiability (*SI*) and numerical identifiability (*NI*) properties of the combined structural and measurement models and data. We illustrate this for *SI* properties quite simply and numerically in the first section examples, while formally introducing a new modeling tool. *SI* properties are the easier of the two to establish, particularly with the current and growing availability of differential algebra (and other approaches)

1 Some modeling packages that also do model parameter "fitting," e.g. *Copasi*, *SAAMII*, *AMIGO*, the *Matlab-Parameter Identification Toolbox*, etc., do provide some form of statistical evaluation of results. Others, as of the time this writing, do not provide any (e.g. *Simulink*). For those that don't, other software tools are needed to complete the quantification process.

2 Quoting my wise statistics professor of yesteryear: "An estimate is worthless without a measure of its variability."

and software for *SI* analysis (Chapter 10). *NI* properties, what some call "practical identifiability," remain the most difficult hurdle and are usually established numerically *a posteriori*. Numerical identifiability (*NI*) analysis embodies the needed reliability measures. If a model is structurally identifiable (*SI*), its *NI* properties are revealed by the estimated parameter covariance matrix $COV(\hat{\boldsymbol{p}})$ – with estimated parameter variances on the diagonal and pairwise parameter covariances (unnormalized correlations) on the off-diagonal (Chapter 12).

Interestingly, the overall identifiability problem can be addressed *a priori*, predictively, by *prospective simulation*, and both *SI* and *NI* might be established together using this approach. We develop this methodology and exemplify it in the first section. The "might be" caveat here is because the methodology follows a circular path, with the correctness or effectiveness of the solution depending on the answer. It's nevertheless a useful approach, with paths for circumventing its limitations. The several simple examples in the first section also illustrate numerically what to expect when attempting to fit structurally identifiable (*SI*) versus *un*identifiable (*not-SI*) models to the same "perfect" data. The visual fit in the examples is perfect enough – but deceivingly so in the *not-SI* case, because parameter estimate statistics are awful, or nonexistent, yet another demonstration of the hopelessness of trying to estimate unidentifiable parameters. Optimization side-effects are in focus as well in these examples.

The second section addresses another way – a classical one – to deal with identifiability issues in parameter estimation. It is quite reasonable to look for additional ways to "reshape" the problem, to render it more identifiable, with enhanced reliability. Structural simplifications like quasi-steady state assumptions (QSSA or tQSSA – Chapter 6), where appropriate, should have high priority. We focus on simplifications that do not require model structure approximations. Model *parameter constraints*, when they can be found, are ideal for this purpose, because they reduce identification problem dimensionality, usually by reducing the number of parameters to be estimated, or by confining them to feasible ranges – as illustrated in Chapter 10. Quantifying parameters of any constraint-simplified biomodels from measurement data is usually easier with than without such constraints. All "extra" information about model parameters and variables – independent information extraneous to the basic ordinary differential equations (ODEs) and measurement model – is useful. Constraint-based and parameter-bounding methods are developed further, beginning in the second section, for measurement models with noisy data included.

Constraint-simplified models often still have some p_i of $\boldsymbol{p}$ unidentifiable. A next logical step then is to check for all other things *SI*, namely the *SI parameter combinations* (combos, $\boldsymbol{c}$) are usually apparent or can be derived. Combos are structural invariants of the model and are thus *SI* – by definition. The third section is about developing *reparameterized models* in terms of these $\boldsymbol{c}$. **Reparameterized models** are *input–output (I–O)-equivalent* models that conserve as much as possible of the original model ODE structure. This section includes quantification of parameter ranges (intervals) for unidentifiable but *interval identifiable* models, along with their statistics. Upper and lower bounds on these intervals are functions of *SI* parameter combinations $\boldsymbol{c}$.

The next to last section is about how to quantify *COMBOS* $\boldsymbol{c}$ (or $\boldsymbol{p}$) by simplifying the estimation problem using an *equivalent-input transformation* called the *forcing function* approach.

This is followed by a section on parameter estimation for *multiexponential* (*ME*) input–output (I–O) models, a class of models particularly useful with the forcing function approach. They are also algebraically transformable into reparameterized and structured (e.g. compartmental) ODE models, as described in Chapters 4, 5 and 10.

The last section details the parameter estimation process and results for a pharmacokinetic (PK) model used in a real human immunodeficiency virus (HIV) vaccine development study. The model is pedagogically developed in stages in earlier sections. Additional testing of an alternative PK model for this data is also presented. Overall this example illustrates several topics developed in this and other chapters concerning the fitting and refitting of alternative models to data.

Finally, a point sometimes overlooked in quantifying a model is specifically how data quality (as well as quantity) affects ODE model reliability, and how it should be introduced into the parameter estimation process. Numerically speaking, more data (bigger N) is clearly better than less, but successful model quantification also depends strongly on output measurement data variability, e.g. $VAR(\mathbf{z})$ or $COV(\mathbf{z})$. During an estimation procedure, data variability essentially *propagates* through the model, like inputs, numerically affecting $COV(\hat{\mathbf{p}})$. Normally, it is included in the optimization criterion used for parameter estimation, e.g. in the weighting matrix W in the weighted least squares criterion, e.g. diagonal elements $w_{ij} = 1/VAR(z_i)$ (Chapter 12). It's included in all our examples. This property is also fundamental to prospective design of experiments for minimizing parameter estimate variability (Chapter 16).

Prospective Simulation Approach to Model Reliability Measures

We can leverage the intimate relationships among parameter estimation covariance, identifiability and sensitivity measures to predictively evaluate all of these model reliability measures prospectively, using *simulation* as an analytical tool.

The idea is this, with an initial focus on identifiability. If we don't know a model is *SI*, we might discover whether it is (or not) following parameter search – using *simulated data* – by evaluating the resulting parameter estimation statistics. In other words, if we are able to successfully run a parameter search algorithm to convergence – using simulated (or real) data – and also successfully estimate the $COV(\hat{\mathbf{p}})$ matrix,[3] we can evaluate $COV(\hat{\mathbf{p}})$ for *SI, NI* and other parameter estimation reliability measures. It is sufficient to have finite diagonal terms, i.e finite variances, to ascertain the model is *SI*. Very large but still finite diagonal terms[4] are associated with possibly *SI* but *numerically unidentifiable* (*not-NI)* parameters, and so on. If no convergence is achieved after many attempts, *SI* (and *NI*) remains an open question and must

3 Supplied with some modeling and parameter estimation packages (e.g. *SAAMII, ADAPT, NONLIN, AMIGO*).

4 Large relative to parameter values.

be dealt with by a more formal identifiability analysis, e.g. using a differential algebra or other approach. The examples below illustrate the approach.

Remark: A limitation of the prospective simulation approach is that nominal parameter values must be chosen to simulate the model to generate the data. The method is thus not fool-proof, because it depends on nominal parameter values chosen to run the analysis. But effects of this limitation can be minimized, e.g. by repeating it for different parameter ranges – a functionality supplied by some modeling packages, e.g. *Copasi*.

Structurally Identifiable or Not

We begin with two numerical examples, using the two simple biomodels in **Fig. 13.1** and **Fig. 13.2**, both implemented in *SAAMII.*[5] These linear two-compartment models were shown by transfer function analysis in Chapter 10 to be structurally identifiable (*SI*) and *not-SI* respectively, but we assume here – for pedagogical purposes – that we do not know this.

We also assume, in a rather circular – but useful – fashion, that we know the values of the model parameters perfectly beforehand and we use these to generate "perfect" data – by simulation – for fitting the model. Our goal is to run our model-fitting program with these parameter values and this data and observe the process along the way. We might anticipate that we should be able to recover these parameter values, if we initiate our parameter search algorithm with values close enough to them, which means within the **region of convergence** of the optimization procedure.[6] We do this to illustrate several issues, first what to expect during and at the end of the numerical fitting procedure when attempting to fit *SI* and *not-SI*

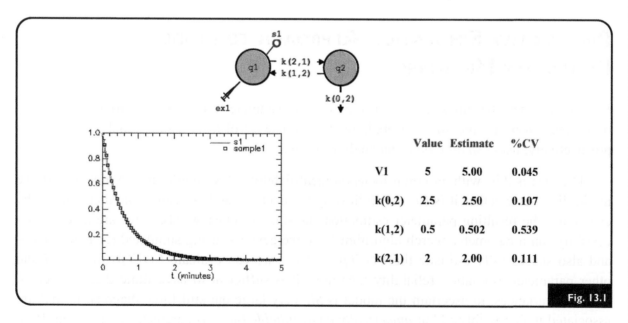

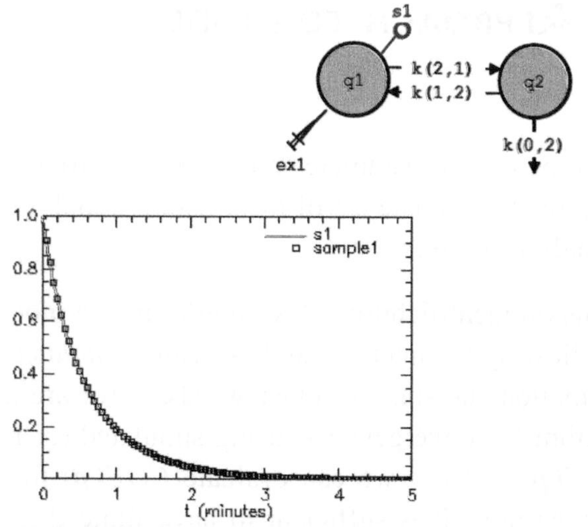

	Value	Estimate	%CV
V1	5	5.00	0.045
k(0,2)	2.5	2.50	0.107
k(1,2)	0.5	0.502	0.539
k(2,1)	2	2.00	0.111

Fig. 13.1

5 Model and data files are available on the book website.

6 This is a difficult technical subject (Bard 1974), which we circumvent here. The important point to be made is that we should *not* anticipate recovering *any* unidentifiable parameters, no matter how close enough starting values!

models, under ideal circumstances. And in the process we illustrate the main steps in setting up and running a parameter estimation problem and interpreting the results – in these examples assuming we do *not* know their *SI* status. This is provided at the end, from the covariance matrix $COV(\boldsymbol{p}^*)$. Second, we begin to demonstrate specifically how simulation can be a very useful tool in biomodel analysis, before or after collecting experimental data for quantifying it, and – by implication – for designing or redesigning experiments to collect that data. We show how to simulate "perfect" data first. **Perfect** here means noise-free and capable of being fitted perfectly – without residual errors – to the chosen ODE structural model. More complex examples to follow, and in subsequent sections and chapters, will reinforce this concept.

Example 13.1: ***Parameter Estimation for a Simple SI Biomodel***

The **Fig. 13.1** model is set up in *SAAMII*, which uses an extended *WLS* (*EWLS*) optimization criterion and a Gauss–Newton search routine to estimate parameters (Chapter 12). The input is programmed as a *bolus*, i.e. impulse $u_1(t) = 5\delta(t)$. To simulate 100 "perfect" data points, over a time interval of 5 here, we assume we know the parameter values perfectly, run the model with these parameter values, generating the solution shown in **Fig. 13.1**. The 100 samples are shown superimposed on the continuous solution shown for $y_1(t) = q_1(t)/V_1$ and are obtainable in tabular form from the program as well. The tabular data are then used for fitting the model by transferring them to the data (*.dat*) file associated with the program. Absolute data weighting is used and, to keep the data "near-perfect" but still realistic, the data are given a standard deviation $SD = 0.0001$, to roughly represent roundoff error in computing the data, which cannot be avoided on finite word-length digital computers.

When the *SAAMII* fitting functionality is exercised for this problem, the model output is fitted perfectly, as expected – no different than that shown at the start and therefore not repeated. The important results are given in the table summary of parameter estimation results in **Fig. 13.1**. Estimated values are near-perfect and – more importantly – parameter estimation variability is very small.

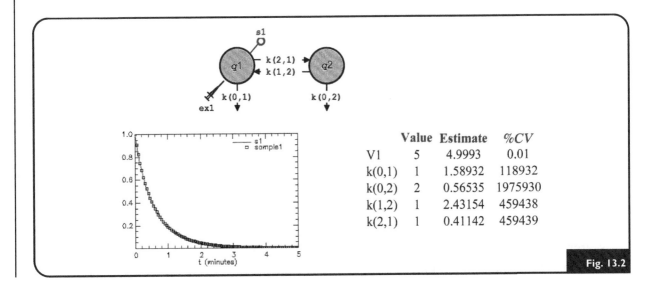

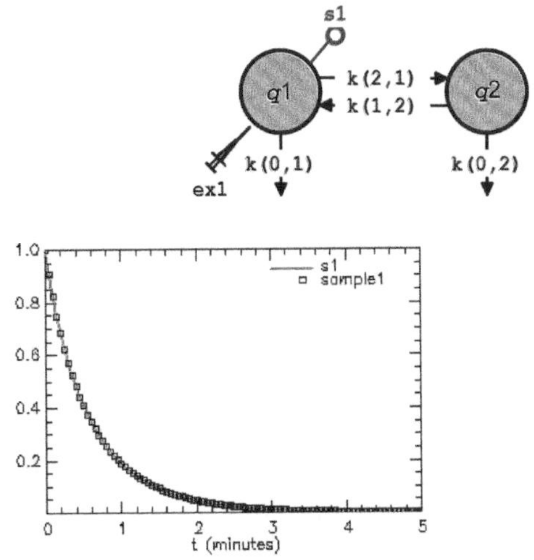

	Value	Estimate	%CV
V1	5	4.9993	0.01
k(0,1)	1	1.58932	118932
k(0,2)	2	0.56535	1975930
k(1,2)	1	2.43154	459438
k(2,1)	1	0.41142	459439

Fig. 13.2

To test the robustness of this result, starting parameter values for the search were perturbed sequentially, from 3 to 50% of their actual values. The search converged and model fitted nearly perfectly again in every case. The visual ("eyeball") fit was again near-perfect, and the resulting parameter estimates and %*CV*s were about the same.

Overall, this numerical exercise confirms what we already know – the model is *SI*. But if we did not know this beforehand, say for a model we could not or did not analyze for identifiability by other means – it would *strongly suggest* the model is *SI*. And, as we demonstrate later, this simulation approach to identifiability analysis works for models with noisy (Monte Carlo simulated) data, and more sparse data as well, also providing predictive results about *NI* via a simulated $COV(\hat{p})$ matrix or other statistical measures of parameter uncertainty.

Example 13.2: ***Parameter Estimation for a Simple NOT-SI Biomodel***

The **Fig. 13.2** model is also set up in *SAAMII*, as shown. Everything else is the same as in the previous example, except the parameter values are a little different (as in the table) and there is one additional parameter, k(0,1), which makes this model *not-SI*. None of the k(*i,j*) are *SI*, but V1 is (Chapter 10).

The 100 samples are again shown superimposed on the continuous impulse response solution shown in the figure for $y_1(t) = q_1(t)/V_1$. When the *SAAMII* fitting functionality was exercised for this problem, with starting values the same as those that generated the simulated data, the model output again fitted the data perfectly – no different than that shown at the start.

However, the search algorithm in this case was less friendly. A convergent solution was eventually (and surprisingly) found, after many attempts and adjustments, primarily by extending the number of search iterations – by making the convergence criterion more stringent, or by adjusting parameter search space constraints. Most runs of this problem stopped prematurely with messages like: *WARNING: The following parameter limit(s) constrain further optimization:* k(0,1) *hit lower limit.* When this happens, the covariance matrix and parameter statistics usually are not available. In our convergent solution, which did not stop on a boundary, statistics were available, as shown. In the other cases, they were not.

Results are given in the table in **Fig. 13.2**. V1 was estimated very well, with $\%CV = 0.01$, because it's *SI*. The k(*i,j*) were estimated poorly – especially considering they started at their exact values; they drifted away instead of toward them. The key and most telling and important results are the %*CV*s for the k(*i,j*) estimates: all *ridiculously large* – confirming, or indicating, that this model is *not-SI*.

Remark: The *SAAMII* optimizer utilizes inverse second-derivative (Hessian $\hat{H}^{-1}$) approximations in the search algorithm and, in principle, H is not invertible for unidentifiable models. So one would think the method not workable in these cases. But, like many other local search methods that rely on derivative computations, the code includes robustness "workarounds" to avoid computational problems that can prevent even identifiable model runs from converging on some solution. As shown in Chapter 12, $\hat{H}(\boldsymbol{p}^*) \cong COV(\boldsymbol{p}^*)$ at convergence on the optimum $\boldsymbol{p}^*$ also provides the statistics for evaluating the reliability of $\boldsymbol{p}^*$ estimates, i.e. the entries in the table of **Fig. 13.2** taken from *SAAMII* results.

Remark: Search methods that do not utilize derivatives, e.g. random search algorithms, also can provide perfect fits to the data, although likely not as quickly as local search methods for the two examples above. Parameter estimate reliability measures, however, must be estimated independently, by methods that typically involve derivative (sensitivity function) computations (see Chapter 12). $COV(\boldsymbol{p}^*)$ estimates derived by any means would in principle yield the same statistical results summarized in tables of **Fig. 13.1** and **Fig. 13.2**.

Structurally Identifiable NL & Numerically Identifiable (NI) or Not NI

The next example utilizes a nonlinear version of this unidentifiable two-compartment model, to see how a simple nonlinearity can alter the effectiveness of the prospective simulation approach for analyzing model reliability. We recall that – unlike linear model dynamics – NL model dynamics are dependent on the magnitude of input stimuli. They also can have better *SI* properties, as we noted in Chapter 10 and illustrate here. We try two magnitudes in this example and evaluate the simulation results.

Example 13.3: ***Parameter Estimation for a NL Biomodel for Two Different Inputs***

The two-compartment NL model shown in **Fig. 13.3** was previously analyzed and found to be *SI* in Chapter 10, Example 10.14, using a differential algebra approach. We will test this numerically here, using near-perfect simulated and sampled data, for two different input magnitudes: $u_{1a}(t) = 5\delta(t)$ and $u_{1b}(t) = 0.05\delta(t)$.

Everything else is the same as in the previous two examples, with corresponding parameter values the same here as in the last example. Anticipating that parameter estimation is likely to be more difficult for the smaller input, results were computed over a time interval of 1 instead of 5, so as to generate a finer grid for the 100 sample points.

As expected, the fitted outputs are near-perfect for both cases. All statistical results are given in the tables in **Fig. 13.3**. For the larger input (upper graph), the model is confirmed as *SI*, with parameters estimated nearly perfectly and parameter $\%CV$s being very small. In contrast, for the small input (lower graph), convergence was never achieved, despite many repeated attempts, with numerous initial values and adjustments in convergence criteria and parameter ranges. The model with the small input is *not NI* because the NL Hill function elimination term is operating in its linear range at this input level. This means the model is effectively linear and the previous example demonstrated numerically what we already knew – that the linear two-compartment model with both leaks present is *not SI* and therefore also *not NI.*

Variances, Covariances & Correlations

In the next example, we use a more comprehensive numerical simulation approach to analyze output parameter sensitivities, both structural and predictive numerical identifiability, and

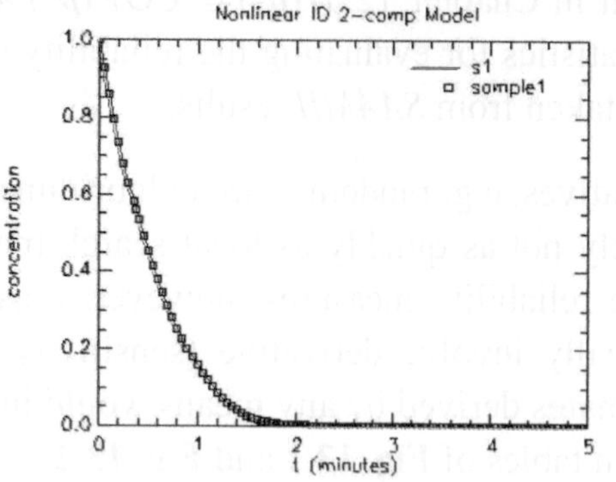

Parameters	True Value	Estimate	%CV
Km	0.5	0.50952	0.74
V	5	4.99989	0.01
Vmax	3	3.04163	0.49
k(0,2)	2	2.01748	1.25
k(1,2)	1	1.04286	1.23
k(2,1)	1	0.99707	0.21

Number of data points used: 100
Input: 5

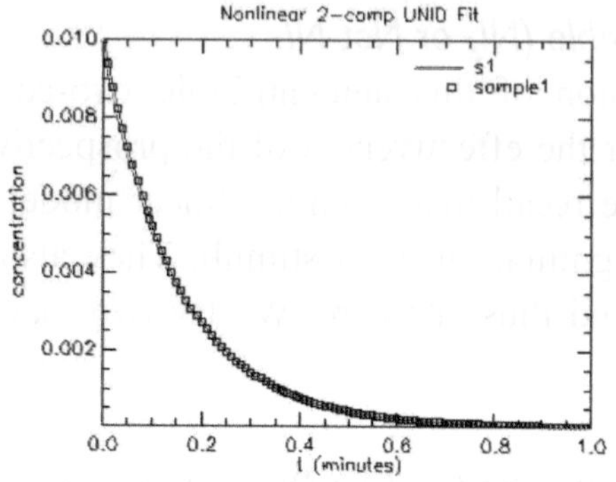

Parameters	True Value	Estimate	%CV
Km	0.5	3.263	**
V	5	4.9909	**
Vmax	3	16.0858	**
k(0,2)	2	0.5206	**
k(1,2)	1	0.1635	**
k(2,1)	1	1.6859	**

Number of data points used: 100
Input: 0.05

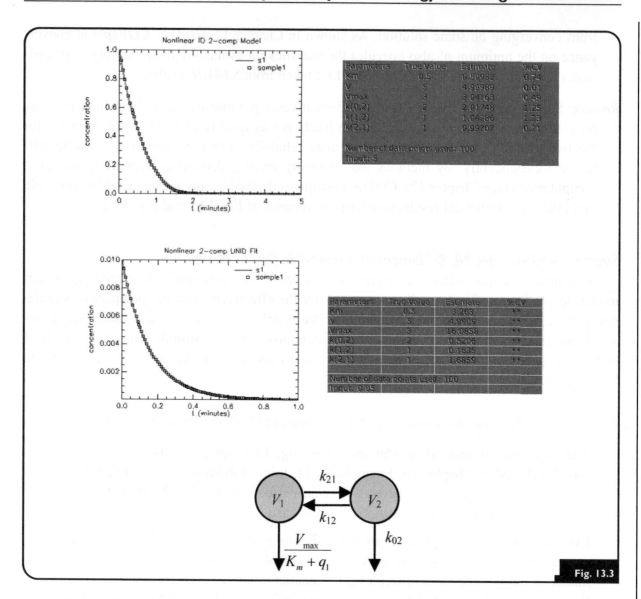

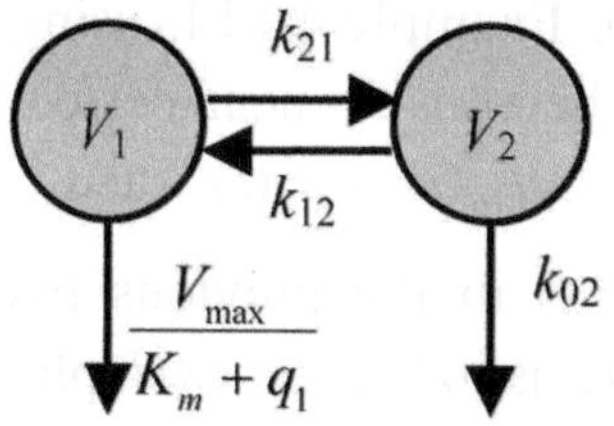

Fig. 13.3

parameter estimation variances and inter-parameter correlations, under ideal (noise-free) data conditions. The simulated data is sampled at different rates. The problem is computationally more complex and, again, we see how variability measures can be disappointing. It's not easy to anticipate such results. But making them explicit here generates some insights.

Example 13.4: ***Sensitivity, Identifiability & Parameter Estimation Reliability for a NL Model with Noise-Free Simulated Data***

The two-compartment NL model of the previous example (**Fig. 13.3**) is analyzed more comprehensively here, using *MATLAB* tools. We again use near-perfect simulated and sampled data, and evaluate *SI*, *NI*, $COV(\hat{p})$ and $CORR(\hat{p})$ measures predictively, but for fewer sampled data points (5, 10 and 50). Input is a unit step instead of an impulse. Everything else is the same.

MATLAB Script –Canonical Sensitivities, Discrete-Sensitivity Matrix & COV(p) and CORR(p) Matrices for both q_1 & q_2 of the NL 2-Comp model

```
function [DSM] = runDSM2comp()
global k12 k21 k02 C K; % model parameters
k12=1; k21=1; k02=2; C = 3; K = 1;
% Canonical sensit system run params
IC = zeros(1,12); % 2 state vars + 10 sensitiv fcns
tstart = 0; tend = 5; deltat = 1; % k = 1, 2, 3 ,4, 5

% solve & plot continuous canonical sensit system
[t, ycont]=ode45(@NL2compDSM, [tstart tend], IC);
plot(t, ycont,'linewidth', 2);
title('STATES & RELATIVE SENSITIVITY FCNS')
xlabel('TIME'); ylabel('RELATIVE SENSITIVITY')
legend('q1(t)','q2(t)','(dq1/q1)/(dC/C)',...
  '(dq1/dK)/(q1/K)','ETC',1)

  % discrete-time solution
[~, y]=ode45(@NL2compDSM, tstart:deltat:tend, IC);

  % discretized sensitivity matrix, both states
param = [C K k12 k21 k02];
datasize = size(y); % returns [numrows, numcolumns]
sens_matrix = [ ]; % matrix declaration
currentrow = [ ]; % evaluating current row
   % rowcounter,skip Xo; 5 params in 2-6
for r = 2:1:datasize(1)
  currentx1 = y(r,1); %x1 at r-th time pt
  currentx2 = y(r,2); %x2 at r-th time pt
  for col = 3:1:7 % cols 3-7 for dx1/dp at r-th time pt
  % current row, with RELATIVE sensit fcns
  currentrow =[currentrow,(y(r,col)/...
    (currentx1/param(col-2)))];
 end;
  sens_matrix = [sens_matrix; currentrow];
  currentrow = [];
   % repeat loop for the x2s and RELATIVE dx2/dps
   % columns 8 to 12 at r-th time pt
 for col = 8:1:12
    currentrow =[currentrow, (y(r,col)/...
    (currentx2/param(col-7)))];
 end;

  sens_matrix = [sens_matrix; currentrow];
  currentrow = []; %loop and do next row;
end;
DSM = sens_matrix;% print the DSM entries
[U,S,V] = svd(DSM) % do SVD
svd(DSM); cond(DSM)
F = DSM.'*DSM; format short g;
COVmin = inv (F)
CORmat= corrcov(COVmin)

%.......................

function dXdt = NL2compDSM(~,X)
global k12 k21 k02 C K;
% canonical sensit system vars (2 states, 10 sensits)
% X(1) = q1 (compartment 1 state variable)
% X(2) = q2 (compartment 2 state variable)
% X(3) = dq1dC (sensitivity of q1 to C)
% X(4) = dq1dK (sensitivity of q1 to K)
% X(5) = dq1dk12 (sensitivity of q1 to k12)
% X(6) = dq1dk21 (sensitivity of q1 to k21)
% X(7) = dq1dk02 (sensitivity of q1 to k02)
% X(8) = dq2dC (sensitivity of q2 to C)
% X(9) = dq2dK (sensitivity of q2 to K)
% X(10) = dq2dk12 (sensitivity of q2 to k12)
% X(11) = dq2dk21 (sensitivity of q2 to k21)
% X(12) = dq2dk02 (sensitivity of q2 to k02)
%dXdt = zeros(12, 1); initialize matrix

% ODEs
dXdt(1) = k12*X(2)-k21*X(1)-(C*X(1)/(K+X(1)))+1;
dXdt(2) = k21*X(1)-(k12+k02)*X(2);
dXdt(3) = k21*X(8) - X(1)/(K + X(1))-X(3)*(k21+...
          C/(K+X(1))-(C*X(1))/(K + X(1))^2);
dXdt(4) = k21*X(9)-X(4)*(k21+C/(K+X(1)) -...
     (C*X(1))/(K + X(1))^2)+(C*X(1))/(K + X(1))^2;
dXdt(5) = X(2) + k21*X(10) - X(5)*(k21 +...
          C/(K + X(1))-(C*X(1))/(K + X(1))^2);
dXdt(6) = k21*X(11)-X(1)-X(6)*(k21 + ...
          C/(K + X(1))-(C*X(1))/(K + X(1))^2);
dXdt(7) = k21*X(12)-X(7)*(k21 + ...
          C/(K + X(1))-(C*X(1))/(K + X(1))^2);
dXdt(8) = k12*X(3)-X(8)*(k02 + k12);
dXdt(9) = k12*X(4) - X(9)*(k02 + k12);
dXdt(10) = k12*X(5)-X(10)*(k02+k12)-X(2);
dXdt(11) = X(1) - X(11)*(k02 + k12) + k12*X(6);
 dXdt(12) = k12*X(7) - X(12)*(k02 + k12) - X(2);
```

Fig. 13.4

The *Matlab* scripts for making all the computations are given in **Fig. 13.4**. The *function script* on the right-hand side (RHS) is for computing the canonical sensitivity system (state + sensitivity) ODEs, i.e. solutions for the two state variables, q_1 and q_2; and the ten *relative sensitivity functions*, $s_{ij} = (dy_i/dp_j)/(y_i/p_j)$. The left-hand side (LHS) run-script first calls the ODE function file for solution and then again for computing the discrete-time relative sensitivity matrix S, row by row, as in Eq. (11.40) – but normalized. This is followed by singular value decomposition (SVD) analysis,

NL 2-comp DSM RESULTS

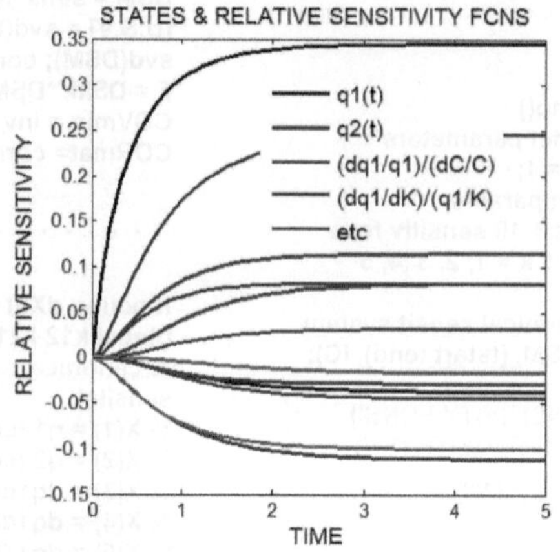

$N = 5$ data points:

$$S = \frac{\partial y / y}{\partial p / p} = \begin{bmatrix} -0.7131 & 0.5606 & 0.0567 & -0.2318 & -0.0284 \\ -0.6077 & 0.4890 & -0.1959 & 0.7986 & -0.4935 \\ -0.8978 & 0.6763 & 0.0868 & -0.2748 & -0.0715 \\ -0.8529 & 0.6484 & -0.2341 & 0.7353 & -0.6871 \\ -0.9454 & 0.7050 & 0.0938 & -0.2842 & -0.0893 \\ -0.9326 & 0.6972 & -0.2379 & 0.7183 & -0.7442 \\ -0.9565 & 0.7115 & 0.0953 & -0.2863 & -0.0942 \\ -0.9534 & 0.7097 & -0.2379 & 0.7143 & -0.7583 \\ -0.9588 & 0.7129 & 0.0955 & -0.2867 & -0.0953 \\ -0.9582 & 0.7125 & -0.2378 & 0.7134 & -0.7615 \end{bmatrix}$$

Singular values = 3.8456, 1.8173, 0.169, 0.0325, 0.0010
Rank = 5, Condition number = 3835

For $N = 10$ data points (matrix not shown):

Singular values = 5.297, 2.554, 0.3577, 0.0653, 0.0046
Rank = 5, Condition number = 1145

Fig. 13.5

computation of the Fisher information matrix F and – from this – the covariance and correlation matrices for the parameters, using the Cramér–Rao theorem (Chapter 12). Results are compared for different sampling rates.

The 12 plots of the ODE solutions are given in **Fig. 13.5** (top), along with the S matrix. SVD results also are given, for both 5 and for 10 data points (program rerun with *deltat* = 0.5), which also doubles the size of S, not shown.

Rank is maximum 5 in all runs, confirming structural identifiability (*SI*). The very large condition number ($CN = SV_{max}/SV_{min} = 3835$), however, suggest difficulties with numerical identifiability (*NI*) for either 5 or 10 data points, but significantly better *NI* properties, CN = 1146, with twice as many equally-spaced data samples.

Running the program for 50 equally spaced samples instead of 5 or 10 yields the lower bound (symmetric) $COV(\hat{\boldsymbol{p}})$ matrix:

$$COV_{min}(\hat{\boldsymbol{p}}) = \begin{bmatrix} 3.8999 & 1.4754 & 1.3660 & -0.0007 & -1.3679 \\ ---- & 0.5585 & 0.5164 & 0.0001 & -0.5160 \\ ---- & --- & 0.4913 & 0.0032 & -0.4812 \\ ---- & ---- & ----- & 0.0017 & 0.0021 \\ ---- & ---- & ---- & ----- & 0.4879 \end{bmatrix} \text{x } 10^5 \begin{matrix} C \\ K \\ k_{12} \\ k_{21} \\ k_{02} \end{matrix}$$

$$\begin{matrix} C & K & k_{12} & k_{21} & k_{02} \end{matrix}$$

Eliminating symmetric entries below the diagonal simplifies visual analysis. The asymptotic (predicted best) variances of the parameters are on the diagonal – all very large (all entries are x 10^5). The smallest is $\sigma^2(k_{21}) = 170$, meaning a % coefficient of variation $\%CV = 100\sqrt{170}/k_{21}(=1) \cong 1300\%$. Pairwise covariances are above the diagonal. We would expect even worse results if the data were noisy.

The correlation matrix $CORR(\hat{\boldsymbol{p}})$, reproduced below, describes interparameter variability – with nearly complete correlations ($\approx \pm 1$) among C and K and k_{12} and k_{02}. Only k_{21} (the forward linear pathway) has low correlations with all other parameters, consistent with it having the smallest parameter variance. All this underscores that the model, despite being *SI*, is hardly *NI*. Five to ten times more simulated near-perfect data samples didn't help much either![7]

$$CORR_{min}(\hat{\boldsymbol{p}}) = \begin{bmatrix} 1 & 0.999 & 0.987 & -0.008 & -0.99 \\ --- & 1 & 0.986 & 0.003 & -0.99 \\ --- & ---- & 1 & 0.11 & -0.98 \\ --- & ---- & ------ & 1 & 0.074 \\ --- & ---- & --- & ---- & 1 \end{bmatrix} \begin{matrix} C \\ K \\ k_{12} \\ k_{21} \\ k_{02} \end{matrix}$$

$$\begin{matrix} C & K & k_{12} & k_{21} & k_{02} \end{matrix}$$

Perhaps the most important lesson here is that we would likely not have had a clue of this predicament without having computed these or other qualitative measures.

Input–Output Models

Parameter precision is just as important for black-box, input–output dynamic system quantification. This model type is designed to match only input–output (I–O) dynamics, and model

7 Entertainer Jimmy Durante would have said: "What a revoltin' development this is!"

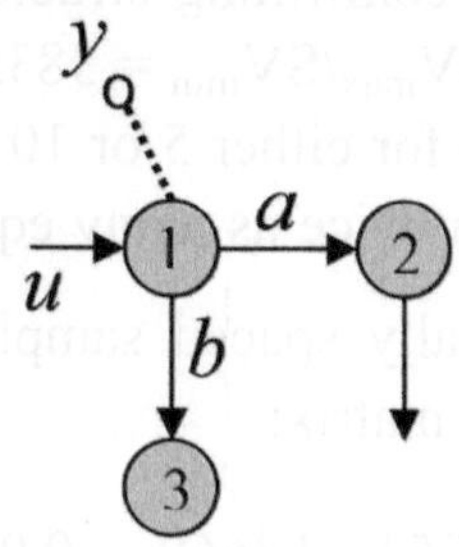

Fig. 13.6

parameters *per se* are not meant to have biological meaning. "Appropriate parameter values" means only that they satisfy the I–O matching goal. As with any model, some I–O model parameters may be *un*ID – and therefore have no effect on outputs, so their specific values may be unimportant, at least to the degree that they satisfy certain physical constraints (e.g. feasible intervals – Chapter 10). But I–O model parameters that determine the transformation dynamics from input to output are important to know reliably, as illustrated in the next example.

Example 13.5: ***Important and Unimportant Parameters in an I–O Transfer Function (TF) Model***

The three-compartment model in **Fig. 13.6** has the I–O TF $H_{11} = 1/(s + a + b)$. The two parameters a and b appear in combination in the denominator, so they are individually *un*ID. Only the combination $a + b = c$ is identifiable, so the combo sum c is important to know precisely if the model is to reproduce the I–O data precisely. a and b are individually unimportant in this regard, but they must nevertheless satisfy $a + b = c$. All parameters are nonnegative, so the values of a and b are restricted to the interval $0 \le a,b \le c$. They can have any values in these constraining intervals that together satisfy $a + b = c$. Questions about the variability of a or b are circumvented by estimating only c and its variability, and choosing a and b to satisfy this constraint.

As an aside, model constraints should be used for directing parameter search, as illustrated in the next section.

Constraint-Simplified Model Quantification

Problems with many feasible solutions are usually easier to solve if the space of possible solutions is reduced. Following this logic, one of the first things to do in formulating a parameter optimization problem is to seek constraints on the problem solution space. This means that any and all independent prior information about the biosystem or its parameters, i.e. information not used in building the model, can (and should) help reduce the dimensionality of the quantification problem. This usually means prior info about parameter values – or ranges of values, or relationships among the parameters, or among the parameters and state variables, or just the state variables. For example, in Example 13.5 above, the parameter space was reduced from 2 to 1.

Model constraints or the need for them are not always anticipated, but if you can find them and apply them before tackling the numerical data fitting issue, you are more likely to get a solution; and it will likely be a "better" one, because it incorporates more system information. If one "blindly" searches for a or b in Example 13.5 – without imposition of constraints – and manages to get a result,[8] it's likely wrong, because neither a nor b is *SI*. This is obvious in this simple example. For more complex ones, it is usually not so readily evident. This, of course, is the reason for defining our formal NL dynamic biosystem model as the *constrained structure* in Chapter 10, Eqs. (10.60) to (10.63), with general constraint Eq. (10.63) as $\boldsymbol{h}(\boldsymbol{x}, \boldsymbol{u}, \boldsymbol{p}) \geq \mathbf{0}$.

Using model constraints to render models easier to solve is commonplace in physics and chemistry and likely as old as mathematics itself. In Chapter 6, for example, physical mass conservation equation constraints simplify classical enzyme kinetic models, replacing an ODE with an algebraic constraint; and QSSA conditions effectively replaces another ODE with a simpler algebraic constraint equation, the famous M–M saturation function. More recently, so-called **constraint-based modeling** has been formalized for solving quantification problems in biochemical reaction networks (Schilling and Palsson 1998), molecular systems biology modeling (Price et al. 2003) and tracer-based metabolomics modeling (Lee et al. 2010). These are introduced in Chapter 14.

Constraint relationships and conditions come in several forms, not just as explicit equations for eliminating variables or parameters by combining them. The distinctions can be subtle, but the principle is always the same. Independently derived information, historical or new, is appended to the model and the augmented model is partially solved in a manner that reduces the number of unknowns. Often, the parameter space is reduced, e.g. from $P = 2$ to $P = 1$ in Example 13.5 above, or the parameter space domain is reduced in other ways. *Interval analysis* (Chapter 10), for example, can be used to restrict the ranges of feasible parameter values. Or the state space is reduced, e.g. from $n = 4$ to $n = 3$ in the derivation of the M–M saturation function, as noted above. Fewer states usually means fewer parameters as well.

Constraints also can be in the form of parameter values (estimates) obtained from the experimental literature, so-called "literature values." These are very common in published systems biology models, often out of necessity – because the parameter space usually is large and – for the most part – *un*ID. Even when model parameters are *SI*, literature values can provide *starting values* for parameter estimation, or parameter estimate *validation*, following quantification.

Often, independent information is obtained from experiments done on the same system under different experimental conditions, which typically yield different measurement and /or modified system models. Examples of these include *wild-type vs. gene-knockout experiments* in molecular biology[9] and experimental media substitution experiments in biochemistry, both of which alter either the system and/or the measurement model.

8 Search algorithms that do not depend on derivatives, e.g. random search, can find the minimum of an optimization criterion function.

9 A **gene-knockout** is a genetic experiment in which a gene is made inoperative, usually for the purpose of drawing inferences from the difference between the knockout organism and normal individuals (called **wild-type**).

Constraints are implemented by combining results and equating things that are the same in all. Typically, some of the parameters in these models represent the same properties in the biosystem, probed under different conditions. They usually are considered equal to each other across the models for the different experiments. This is exemplified below.

Following is an example of constraints delineated for the same system under different experimental conditions and used for quantifying the model. This model-experiment example also serves as *part I* of a case study continued in examples to follow, fully resolving model quantification for this test of an early candidate vaccine for HIV.

Example 13.6: ***HIV Vaccine Development: Part I – Nested Model Setup & Constraints***

This example is based on early kinetic studies of an experimental recombinant DNA molecule – designated *gp120* – under consideration at the Genentech Corporation in the early 1990s for potential use as a vaccine against the HIV (Weissburg et al. 1995). The goal was to examine the pharmacokinetics (PK) of *gp120* in rabbits, an animal-model of choice in vaccine development, under different sample preparation and input conditions. Three separate experiments were done in three different groups of rabbits, at different times, with the *gp120* administered in different ways in each group: (1) dissolved in physiological saline[10] and injected intravenously (IV); (2) dissolved in physiological saline and injected intramuscularly (IM); and (3) mixed with the vaccine adjuvant[11] alum, to which it is reversibly bound, and also injected IM. The input drug dose for all three studies was the same; and discrete-time output data $z_1, z_2, \ldots, z_N$ were collected as concentration measurements in blood plasma only.

Three different models are needed for this study, each reflecting the conditions of the different experimental conditions. The simplest possible ones are shown in **Fig. 13.7** top to bottom, representing experiments 1, 2 and 3 nested within each other, but assumed to have different parameters (in general). At the top (Experiment 1), it is assumed that (free) *gp120* is distributed instantaneously and uniformly throughout a "central" compartment (#1) associated primarily with the blood plasma, i.e. a one-compartment model is assumed. For Experiment 2, it is assumed that free *gp120* is distributed uniformly in a single muscle compartment (#2), with unidirectional uptake into plasma via rate constant k_{12}, elimination from plasma at fractional leak rate k'_{01}, and possible degradation or other path of elimination of *gp120* in muscle via fractional leak rate k_{02}. In Experiment 3, the assumption is that alum-bound *gp120* occupies a third compartment (3) uniformly, with possible elimination of *gp120* via fractional leak rate k_{03}; and it is also assumed the *gp120* ligand is desorbed from the alum irreversibly, i.e. it is not resorbed. This is modeled most simply as an

10 Physiological saline is an *isosmotic* salt solution with a concentration that does not cause loss of fluid (due to osmotic pressure) from the circulation when it is injected.

11 Vaccines are normally made up of antigens, like *gpl20*, combined with an **adjuvant** – an effect enhancer, and/or adsorbent material, to draw immune cells to the site and/or retain the vaccine for long periods *in vivo* following intramuscular (IM) or subcutaneous (SC) administration. *Alum* (aluminum hydroxide) is commonly used for this purpose, as it was in the *gpl20* studies. *Keyhole limpet hemocyanin* (KLH) is another, found in experimental vaccines produced to stimulate a patient's immune system against his or her specific cancer cells (Lamm et al. 1993).

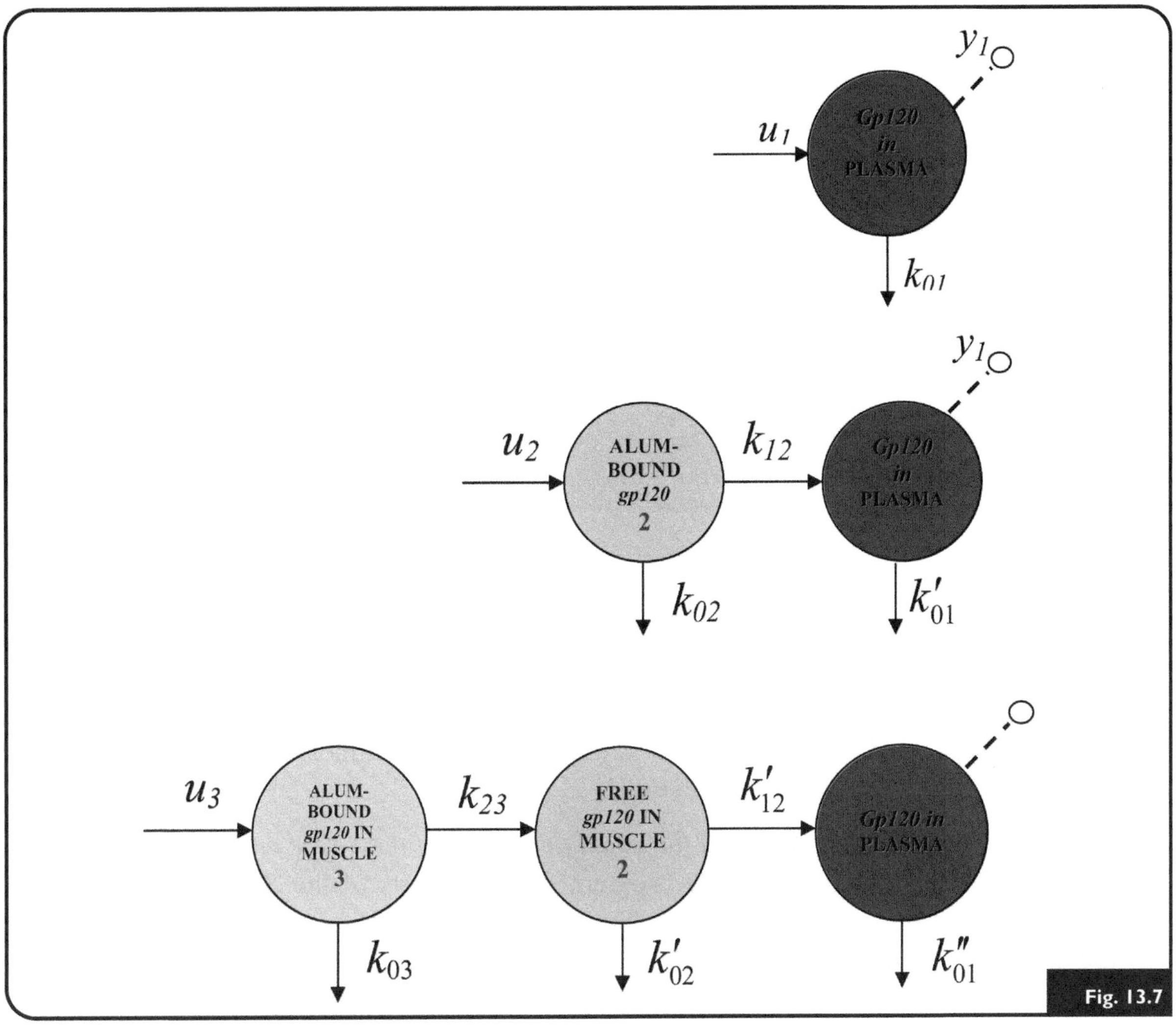

Fig. 13.7

irreversible interaction between two muscle compartments, open to absorption of the *gp120* freed from the complex into muscle at rate k_{23}, as shown in the bottom figure. The free ligand in muscle is then absorbed from muscle into plasma at rate k'_{12}, and further lost irreversibly from plasma at rate k''_{01}. These assumptions are tested and discussed in a subsequent example, when we fit the models to the real data from these studies.

The Experiment 1 model, with one parameter, is evidently *SI.* Some algebra indicates the Experiment 2 model is also *SI*, but the Experiment 3 model, by itself, is not.

To quantify this set of models most effectively, the study group made sure the three experiments were very well matched. This means at least that the rabbit groups were sufficiently similar; the experiments were done the same way, under the same environmental conditions; and all the data were stored after each study and processed simultaneously, all to minimize experimental errors. Under these conditions, models and data can be aggregated into a single parameter estimation problem. All three

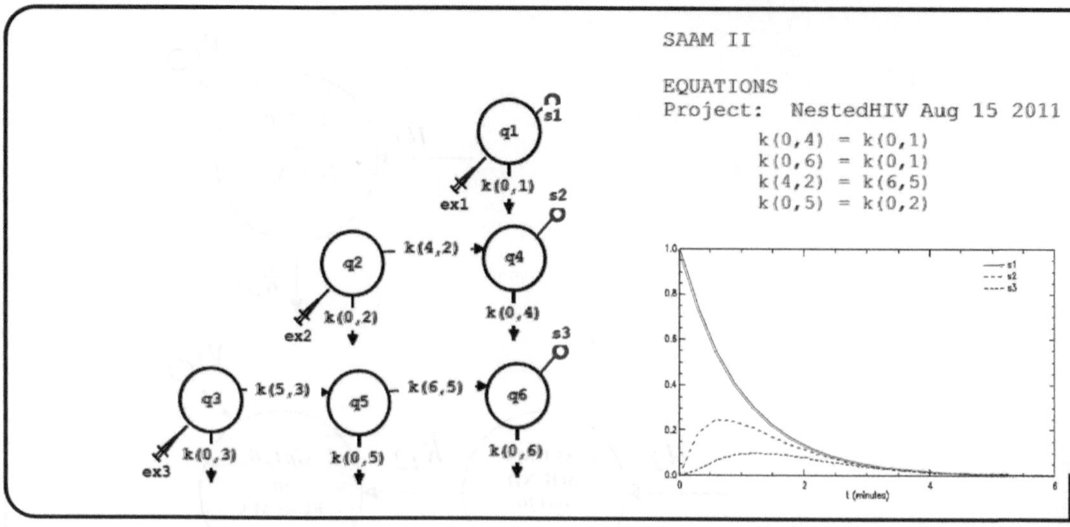

Fig. 13.8

models are programmed for parameter estimation simultaneously, from the aggregated set of data from the three experiments. Importantly here, *the three models are assumed to have common system parameters*, i.e. they are constrained by: $k_{01} = k'_{01} = k''_{01}$ and $k_{12} = k'_{12}$ and $k_{02} = k'_{02}$. With three output measurements and these constraints, the overall model is surely *SI*.

The three models are implemented for quantification as if they were one, in this example using compartmental modeling program *SAAMII*. Parts of the *SAAMII* graphical user interface (GUI) are illustrated in **Fig. 13.8**. The parameter constraint equations are programmed in the EQUATIONS window shown. This guarantees that these constraints will be continually met as the algorithm in the *SAAMII* program searches through parameter space. This model is simulated with all parameters arbitrarily set equal to 1 in the PARAMETERS window, not shown. Simulated results are shown in the graphs. We continue with quantification from real data in Example 13.14 and Example 13.15 a few sections below.

Model Reparameterization & Quantifying the Identifiable Parameter Combinations

A key step in setting up a parameter estimation problem for solution is designating the parameters to search for. We'd like them all, but only the *SI* parameters have unique or nonunique distinct solutions. Reparameterization in terms of structurally identifiable combinations $\boldsymbol{c}$ (*abbrev:* combos) of the p_is is a way to go. Models reparameterized in this way are input–output-equivalent and *SI*. The focus here is finding combos that retain as much of the original model structure as possible, which we call a **structure-preserving (or homomorphic) reparameterization** (**SPR**) and designate it as $\boldsymbol{b}$, to distinguish it from any other combination $\boldsymbol{c}$ – of which there are many. This is neither easy nor always possible (Evans and Chappell 2000). But it is feasible in at least some practical situations, as illustrated in the examples below.

For linear dynamic system models, equivalence theory and practice evolved in the mid-20th century as *equivalence classes of canonical models*, most with inherently identifiable parameters, based on the seminal work of Kalman (1960). Indeed, these **canonical forms** (CFs) are typically derived for the purpose of converting structured and generally *un*ID linear dynamics system models into one having only structural invariants as parameters. To accomplish this, the generally *un*ID parameter set is transformed into a smaller set of ID parameter combinations (combos), i.e. a set of structural invariants. Linear-system CFs are typically derived as sparse matrix canonical transformations, different ones for different purposes – and *not* our purpose in this section. Common ones include *Jordan* CFs – with all eigenvalues on the diagonal of the system matrix (Appendix B), and *controllable* and *observable* CFs – for having certain control system characteristics. Appendices C, D and especially E address these canonical equivalence classes in some detail. These CFs are indeed input–ouput-equivalent to, but are typically structured quite differently than the original model. For this reason, they are not reparameterized models of the type we are likely to be interested in, although we may have to settle for this sparsest type. We want to retain as much original structure of the state variable as possible.

Reparameterization theory and associated computational methodology is still in its infancy, but has advanced substantially in the last decade, in association with advances in automated *SI* analysis of nonlinear (NL) dynamic system models using differential algebra methods. Contributions have come from several groups worldwide (Chappell et al. 1990; Evans and Chappell 2000; Denis-Vidal et al. 2001; Chapman et al. 2003; Boulier 2007; Meshkat et al. 2009, 2011 2012a,b; Kuo et al. 2014).

The structure-preserving reparameterization (SPR) procedure has two steps: first find the combos; then organize them so they are represented in the original model structure, ideally as an SPR. The most accessible and automated tool is the interactive software system available online for evaluating identifiable parameter combos (Kuo et al. 2014) – *Web App: COMBOS.*[12]

The models in the first three of the four examples below are systematically transformed, without use of machine computation, to retain their original structures after reparameterization. The fourth example utilizes *COMBOS* to find the combos and otherwise assist with finding a reparameterization with the same structure.

Example 13.7: ***Unidentifiable Two-Compartment model Reparameterized***

The two-compartment model with both leaks present and input and output in compartment 1 was shown in Chapter 10 to be unidentifiable. None of the k_{ij} are identifiable, but V_1 is. The original equations are:

$$\begin{aligned} \dot{q}_1 &= -(k_{01} + k_{21})q_1 + k_{12}q_2 + u \equiv k_{11}q_1 + k_{12}q_2 + u \\ \dot{q}_2 &= k_{21}q_1 - (k_{02} + k_{12})q_2 \equiv k_{21}q_1 + k_{22}q_2 \\ y &= q_1/V_1 \end{aligned} \tag{13.1}$$

12 www.biocyb1.cs.ucla.edu/combos/html/

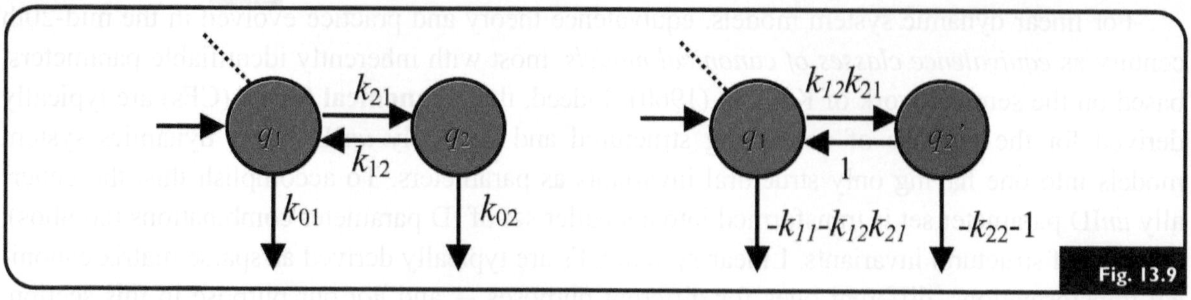

Fig. 13.9

Redefining (*scaling*) the unmeasured state variable $q_2' \equiv k_{12}q_2$ reparameterizes the model with all coefficients having only uniquely identifiable parameters k_{11}, k_{22}, V_1 and $k_{12}k_{21}$ (i.e. the combinations), as in the first sets of equalities below. However, this is not enough, because it does not yet yield a compartmental structure, with balanced exchange fluxes between the compartments and nonzero fluxes out of both.

To preserve the original compartmental structure, we have to add and subtract $q_2' = k_{12}q_2$ from the $\dot{q}_2'$ ODE and also add and subtract $k_{12}k_{21}q_1$ from the $\dot{q}_1$ ODE, as in the second set of equalities for both:

$$\begin{aligned} \dot{q}_2' &\equiv k_{12}\dot{q}_2 = k_{12}k_{21}q_1 + k_{22}q_2' = k_{12}k_{21}q_1 - (-k_{22} - 1)q_2' - q_2' \\ \dot{q}_1 &= k_{11}q_1 + q_2' + u = -(-k_{11} - k_{12}k_{21})q_1 - k_{12}k_{21}q_1 + q_2' + u \end{aligned} \tag{13.2}$$

The model on the RHS of **Fig. 13.9** has parameters $\boldsymbol{b}$ made up only of ID combos. It does, however, have a restricted region of validity, because the leaks in the transformed model must retain their nonnegativity, i.e. $k_{02}' \equiv -k_{22} - 1 \geq 0$ or $k_{22} \leq -1$ and $k_{11} \leq -k_{12}k_{21}$. This additional (add–subtract) affine transformation of the equations to generate an identical structure seems to generalize well, as we see again in Example 13.10 below.

Example 13.8: ***Unidentifiable Three-Compartment Model in Need of Reparameterizaton***

Suppose the following compartmental model is chosen for parameter estimation using measurement data from output y:

$$\dot{\boldsymbol{q}} = \begin{bmatrix} k_{11} & k_{12} & k_{13} \\ k_{21} & k_{22} & 0 \\ k_{31} & 0 & k_{33} \end{bmatrix} \boldsymbol{q} + \begin{bmatrix} 1 \\ 0 \\ 0 \end{bmatrix} u, \qquad y = \begin{bmatrix} c & 0 & 0 \end{bmatrix} \boldsymbol{q} = bq_1 \tag{13.3}$$

Should one attempt to directly search for all eight, using an appropriate parameter estimation algorithm? NO, because ID analysis for this model in Chapter 10, Example 10.7, revealed that only c is *SI*.

If we do not know this model is *un*ID, and "blindly" attempt to quantify all eight parameters directly, a formal search might provide values for all eight, e.g. if we use a search method that does not use derivatives (e.g. a random search or genetic algorithm, Chapter 12). But all k_{ij} estimates will be nonsense values, which can only be detected by examining the estimated parameter covariance matrix, or parameter sensitivities. The variances of *un*ID parameters are infinity and output sensitivities are zero. If, on the other hand, the search algorithm incorporates the inverse of the Hessian matrix or Fisher Information matrix F, it will "blow-up" in some way, because the Hessian and F are singular for unidentifiable models (Chapter 12). We need to reparameterize to quantify this unidentifiable model.

Example 13.9: ***Reparameterization generates a SI model with the Same Structure***

Example 10.7 revealed that only c, k_{11}, k_{22}, k_{33} and the product combinations $k_{12}k_{21}$, $k_{13}k_{31}$ (six total) are identifiable for this model. We can formulate the estimation problem so that only the six identifiable combinations are estimated: c, $k_{12}k_{21}$, $k_{13}k_{31}$, k_{11}, k_{22} and k_{33}. To accomplish this, we redefine (scale) the unmeasured state variables: $q_2' = k_{12}q_2$ and $q_3' = k_{13}q_3$ and substitute these into the ODEs, yielding:

$$\dot{\boldsymbol{q}}' = \begin{bmatrix} k_{11} & 1 & 1 \\ k_{12}k_{21} & k_{22} & 0 \\ k_{13}k_{31} & 0 & k_{33} \end{bmatrix} \boldsymbol{q}' + \begin{bmatrix} 1 \\ 0 \\ 0 \end{bmatrix} u, \qquad y = cq_1 \tag{13.4}$$

All matrix elements of (13.4) are identifiable and, in principle, can be quantified using a suitable search algorithm.

Remark: (13.4) can represent (an unidentifiable) three-compartment mammillary model, with input and output in the first compartment. With this interpretation, $-k_{11} = k_{01} + k_{21} + k_{31}$, $-k_{22} = k_{02} + k_{12}$, $-k_{33} = k_{03} + k_{13}$. Interestingly, it also can represent a three-compartment catenary model, with the k_{ii} defined slightly differently (with a new $\boldsymbol{b}$, see Chapter 4) but with the same reparameterization as in (13.4). Thus, reparameterized models can represent more than one structural model, i.e. they are not necessarily unique. **Fig. 13.10** illustrates the pair of similarly structured mammillary models matching (13.4), with only the SPR one on the right being *SI*. Like the two-compartment transformation in the earlier example, the leak parameters must be nonzero, so this SPR model also has restricted validity.

Remark: The model of (13.4) is also I–O-equivalent to an unstructured (not SPR) three-exponential (I–O) model. The three A_i and three λ_i coefficients and exponents (structural invariants $\boldsymbol{b}$) fitted to the output data can be algebraically transformed into six identifiable combinations (combos $\boldsymbol{c}$) of k_{ij}s, as in Example 10.7 of Chapter 10.

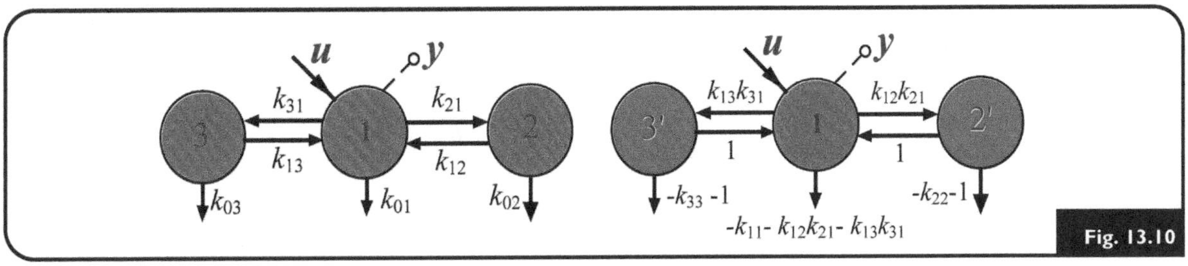

Fig. 13.10

Example 13.10: ***Reparameterization of a Minimal Viral Disease Dynamics Model***

Equations for a classic viral disease model (Perelson 2002) are rewritten in original form on the left in Fig. 13.10 (Target cells, Infected cells, Virions) and in simpler form for analysis on the right below.[13] This model has eight unknown parameters: p_1 to p_8. The (constant) input $p_1u(t) \equiv p_1\mathbf{1}(t)$ is a step function of unknown magnitude p_1.

$$\begin{aligned} \dot{T} &= s - dT - (1-\eta)\beta VT & \dot{x}_1 &= p_1u - p_2x_1 - p_3p_4x_1x_3 \\ \dot{I} &= (1-\eta)\beta VT - \delta I & \dot{x}_2 &= p_3p_4x_1x_3 - p_5x_2 \\ \dot{V} &= (1-\varepsilon)pI - cV & \dot{x}_3 &= p_6p_7x_2 - p_8x_3 \\ y &= V & y &= x_3 \end{aligned} \tag{13.5}$$

Symbolic identifiability analysis using the program *COMBOS* online indicates (in less than 3 seconds computation time) that the model is *not SI*. p_2 and the combos p_3p_4 and $p_1p_6p_7$ are uniquely identifiable, and p_5 and p_8 are both locally identifiable, each with *two solutions*. These make up the combos $\boldsymbol{c}$. The input–output equivalent reparameterized *SI* model (with parameters $\boldsymbol{b}$) is readily written by scaling x_1 by p_6p_7 and x_2 by $1/p_1$. i.e. $x_1' = p_6p_7x_1, \quad x_2' = x_2/p_1$. The resulting ODE model is:

$$\begin{aligned} \dot{T}' &= s(1-\varepsilon)p - dT' - (1-\eta)\beta VT' & \dot{x}_1' &= p_1p_6p_7u - p_2x_1' - p_3p_4x_1'x_3 \\ \dot{I}' &= (1-\eta)\beta VT' - s\delta I' & \dot{x}_2' &= (p_3p_4/p_1p_6p_7)x_1'x_3 - p_1p_5x_2' \\ \dot{V} &= (1-\varepsilon)pI'/s - cV & \dot{x}_3 &= p_6p_7x_2'/p_1 - p_8x_3 \\ y &= V & y &= x_3 \end{aligned} \tag{13.6}$$

The structure is identical to the original, but with new and equivalent set of parameters $\boldsymbol{b}$.

Reparameterized Model Statistics with Noisy Data

Covariance $COV(\hat{\boldsymbol{b}})$ in Terms of Identifiable Parameter Combinations $\hat{\boldsymbol{c}}$

Consider a model with parameters $\boldsymbol{p}$, transformed into one in terms of ID combos $\boldsymbol{c}$, e.g. as in the examples above, so it can be quantified. Thus $\boldsymbol{c}$ can be estimated as $\hat{\boldsymbol{c}}$ (in lieu of unID $\boldsymbol{p}$), and parameter estimation variances and covariances of $\hat{\boldsymbol{c}}$ would be needed for a complete quantification. However, vector $\boldsymbol{c}$ is not unique, because a model can be reparameterized in many ways. Let $\boldsymbol{b}$ be the *SI* parameters of the chosen reparameterization and thus some transformation of $\boldsymbol{c}$. We therefore seek $COV(\hat{\boldsymbol{b}})$, which always exists, because $COV(\hat{\boldsymbol{c}})$ always exists – because combos $\boldsymbol{c}$ are structural invariants and, in principle, thus transformable into $COV(\hat{\boldsymbol{b}})$ (Meshkat et al. 2011).

The $COV(\hat{\boldsymbol{b}})$ matrix for an arbitrary function $\hat{\boldsymbol{b}} = \boldsymbol{b}(\hat{\boldsymbol{p}})$ of *SI* parameters $\hat{\boldsymbol{p}}$ was given in Eq. (12.56) of Chapter 12. An example of computing the variance of scalar $\hat{b} \equiv AUC(\hat{\boldsymbol{p}})$ accompanied that equation. Substituting *SI* combos $\hat{\boldsymbol{c}}$ for $\hat{\boldsymbol{p}}$ in that equation gives the asymptotic covariance matrix for our reparameterization $\hat{\boldsymbol{b}}$:

$$COV(\hat{\boldsymbol{b}}) = \frac{\partial \hat{\boldsymbol{b}}}{\partial \hat{\boldsymbol{c}}} COV(\hat{\boldsymbol{c}}) \left(\frac{\partial \hat{\boldsymbol{b}}}{\partial \hat{\boldsymbol{c}}}\right)^T \tag{13.7}$$

13 This model is developed further in Chapter 14.

Combos, Noisy Data & Parameter Equality Constraint Effects

What about effects of parameter equality constraints on parameter estimation variability measures in the presence of noisy data? Vector $\hat{\boldsymbol{b}}$ is a function of combos $\hat{c}$ and, if an element b_i is fixed (constant), then $\partial b_i / \partial \hat{c} = 0^T$. Some consequences of reducing the dimensionality of the unknown parameter space in this way are anticipated, namely model *SI* properties should be improved.[14] But others – statistical variabilities – can increase, as illustrated below, a somewhat surprising but explainable result.

The following two examples illustrate the combined effects of noisy output data $z(t_i) = y(t_i) + e(t_i)$ and parameter equality constraints on parameter bound variabilities of *interval identifiable* catenary and mammillary models, i.e. the variabilities on the particular set of combos, the structural invariant bounds: $c \equiv [k_{01}^{\min} \ \ldots \ k_{n,n-1}^{\min} \ k_{01}^{\max} \ \ldots \ k_{n,n-1}^{\max}]^T$. These SI interval bounds are computed from the algorithms given in Chapter 10 for mammillary and catenary compartmental models and are themselves functions of alternative sets of structural invariant sets of parameters, e.g. sums and products of unidentifiable k_{ij}, or the coefficients and exponents of input-output equivalent multiexponential models. The principles illustrated here are applicable to any ODE model.

The vector $\boldsymbol{b}$ in these examples is a reduced size subset of $\boldsymbol{c}$. When a model parameter k_{ij} is fixed, $k_{ij}^{\min} = k_{ij}^{\max} \equiv k_{ij}$. Eq. (13.7) is used to evaluate the asymptotic covariance of the bounds on the remaining rate constants, by reevaluating the elements of the matrix $\partial \hat{\boldsymbol{b}} / \partial \hat{c}$ from these same relationships. $COV(\hat{\boldsymbol{b}})$ is reduced in dimension by two for each known parameter, because: $\partial b_i / \partial \hat{c} = 0^T$ for each known constraint. The interactive online program *W³MAMCAT* does all of these computations, evaluating deterministic bounds and their variabilities (www.biocyb.cs.ucla.edu).

Example 13.11: ***Mammillary & Catenary Models with Equality Constraints & Gaussian Measurement Errors (Vicini et al. 2000)***

Unconstrained and constrained three-compartment mammillary and catenary models[15] are shown comparatively in **Table 13.1** and **Table 13.2**, respectively, each unconstrained (LHS) versus constrained (RHS) by $k_{12} = 0.28$. For both, measurement data errors were Gaussian, with 5% (%*CV*) errors. Interval bounds for the unidentifiable k_{ij} named in the first column are given numerically in the second (fifth) and third (sixth) columns. Parameter bound variabilities expressed as *CV*% values in the fourth and seventh columns were computed from $COV(\boldsymbol{b})$ diagonal entries ($CV\% = 100\sqrt{\overline{\sigma}_i^2 / \overline{b}_i}$) provided by *W³MAMCAT*.

Results: Each model remains unidentifiable overall, but with narrower bound means on all interval identifiable k_{ij} in the constrained models. However, almost all bound variabilities (shown in %) increased with k_{12} fixed, several-fold for some.

---

14 For example, the feasible ranges of unidentifiable parameters are reduced, possibly becoming identifiable, as was shown in Chapter 6 for linear multicompartmental models.

15 This structured model was originally computed from a transformed three-exponential model solution (Landaw et al. 1984)

Table 13.1 Unconstrained and Constrained Three-Compartment Mammillary Models

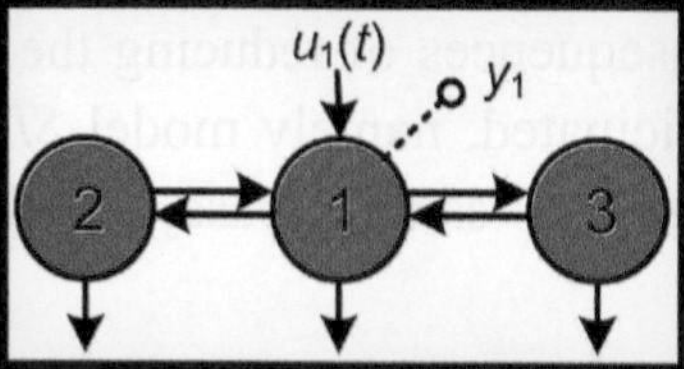

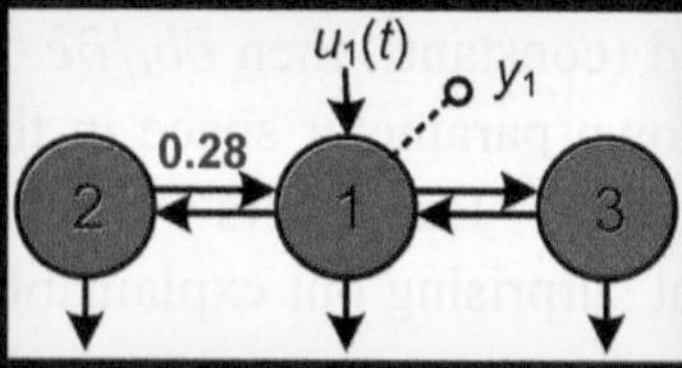

Bound	Unconstrained Case			Constrained Case		
	Value (min^{-1})	SD (min^{-1})	CV (%)	Value (min^{-1})	SD (min^{-1})	CV (%)
k_{01}^{max}	0.60567	0.05190	8.57	0.40014	0.06310	15.8
k_{02}^{max}	0.29592	0.014640	5.55	0.17583	0.03570	20.3
k_{03}^{max}	0.02022	0.00089	4.41	0.01857	0.00095	5.11
k_{12}^{max}	0.45583	0.03570	7.84	0.28000	–	–
k_{12}^{min}	0.15991	0.02210	13.8	0.28000	–	–
k_{13}^{max}	0.02466	0.00114	4.67	0.02446	0.00140	4.7
k_{13}^{min}	0.00424	0.00031	7.19	0.00589	0.00086	14.7
k_{21}^{max}	0.93297	0.09830	10.5	0.053283	0.1140	21.3
k_{21}^{min}	0.32700	0.05210	15.9	0.53283	0.11400	21.3
k_{31}^{max}	0.73300	0.05950	8.11	0.52715	0.06460	12.3
k_{31}^{min}	0.12701	0.00877	6.91	0.12701	0.00877	6.91

Table 13.2 Unconstrained and Constrained Three-Compartment Catenary Models

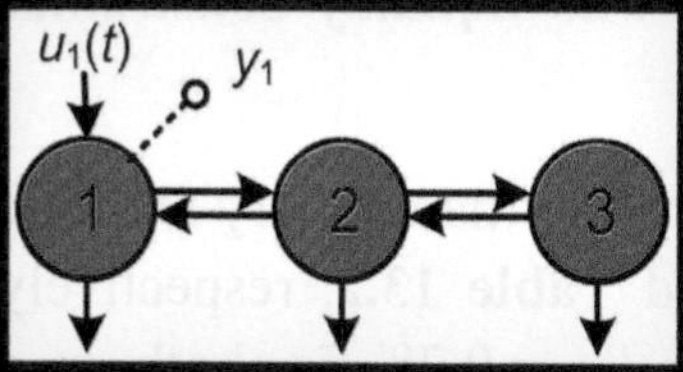

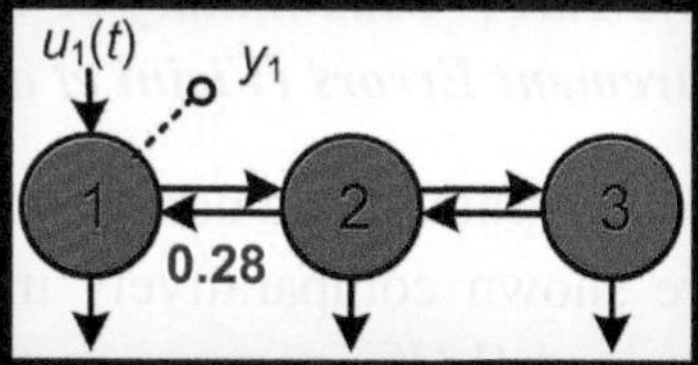

Bound	Unconstrained Case			Constrained Case		
	Value (min^{-1})	SD (min^{-1})	CV (%)	Value (min^{-1})	SD (min^{-1})	CV (%)
k_{01}^{max}	0.60567	0.05191	8.6	0.51605	0.06457	12.5
k_{02}^{max}	0.19154	0.01354	7.1	0.05524	0.03287	59.5
k_{03}^{max}	0.02100	0.00096	4.6	0.01100	0.00411	37.3
k_{12}^{max}	0.33524	0.03287	9.8	0.28000	–	–
k_{12}^{min}	0.14367	0.01990	13.9	0.28000	–	–
k_{21}^{max}	1.05997	0.10581	10	0.54393	0.11407	21
k_{21}^{min}	0.45431	0.05902	13	0.54393	0.11407	21
k_{23}^{max}	0.03326	0.00194	5.8	0.03326	0.00194	5.8
k_{23}^{min}	0.01226	0.00120	9.8	0.02226	0.00479	21.5
k_{32}^{max}	0.30335	0.01843	6.1	0.16703	0.03619	21.7
k_{32}^{min}	0.11180	0.00823	7.4	0.11180	0.00823	7.4

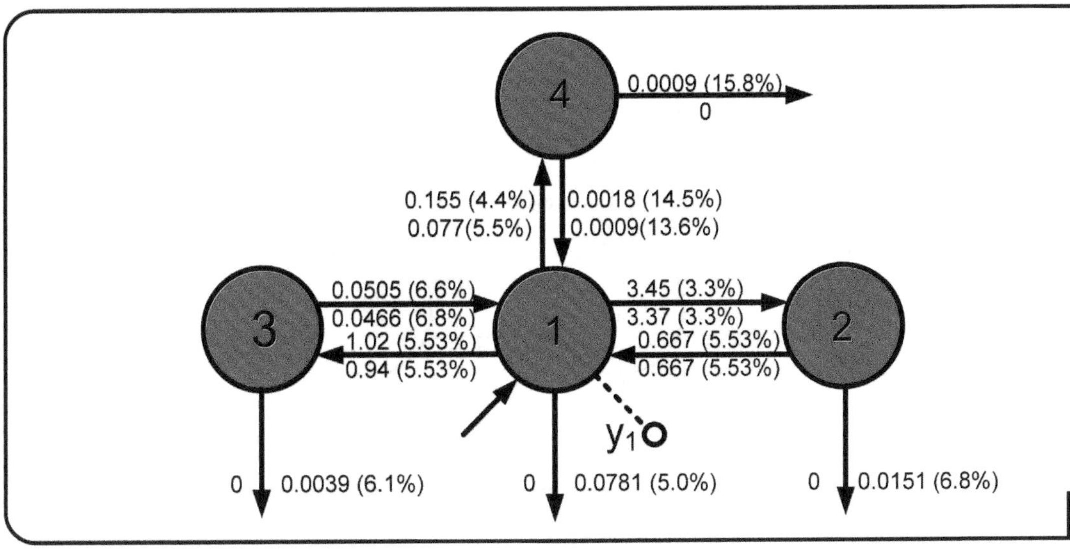

Fig. 13.11

Example 13.12: ***Four-Compartment Mammillary Model with Equality Constraint & Gaussian Measurement Errors***

An unconstrained four-compartment mammillary model[16] is shown in **Fig. 13.11**.

Interval bounds for the unidentifiable k_{ij} – shown in the figure above and below or on either side of the arrows – were computed using *W³MAMCAT*, as in the previous example. Numbers in parentheses are the parameter variabilities expressed as *%CV*. With equality constraint $k_{14} = 0.0014$ given, *W³MAMCAT* yields the constrained model shown in **Fig. 13.12**.

Results: The model remains unidentifiable overall, but with narrower mean bounds on all interval identifiable k_{ij}. In this example, all parameters associated with compartment 4 (k_{21} and k_{02}) are now uniquely identifiable. However, some parameter bound variabilities increase (e.g. all of the k_{0i}).

The Forcing-Function Method

Model quantification from all available data simultaneously usually achieves the best results, because all data statistics (variances and correlations) are considered simultaneously and consistently – in concert with all model structural and other assumptions. This is sensible intuitively, and it has a sound theoretical basis (Ljung 1987). The forcing-function method is an alternative, sequential approach to parameter estimation. It is based on first fitting noisy output data (e.g. $z = y + e$) collected from a single measurable compartment to an arbitrary function, typically a multiexponential, polynomial or other suitable function. The measured compartment is typically a central one, like blood or blood plasma – connected to any number of

16 This model was computed in (Vicini et al. 2000) from a transformed four-exponential model solution in (Landaw et al. 1984).

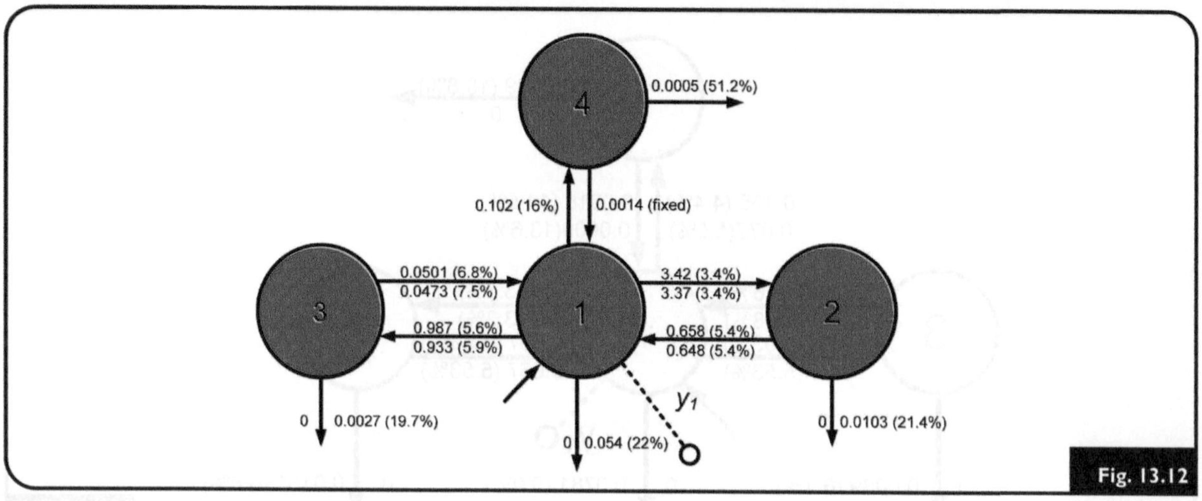

Fig. 13.12

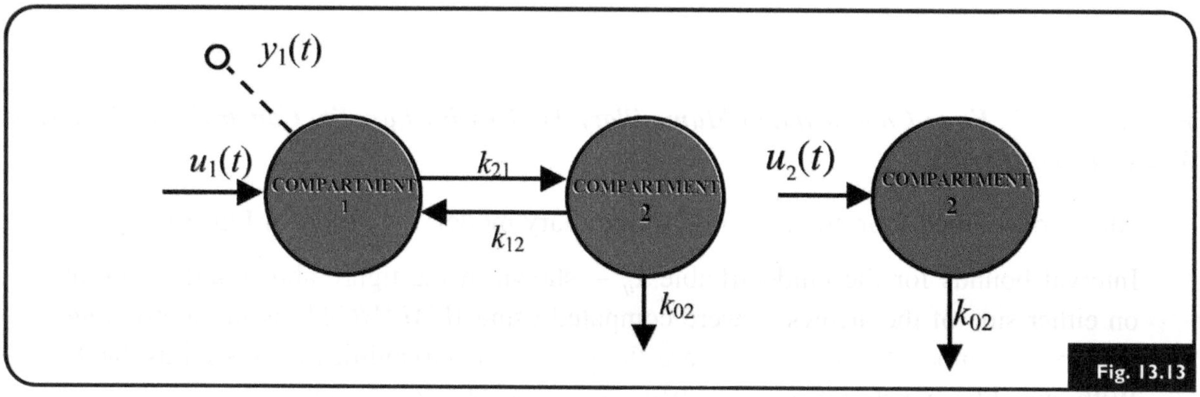

Fig. 13.13

I–O forcing function equivalent models.

other (measured or unmeasured) compartments. The quality of the fit to data from this compartment can be very good, because the measured data are usually relatively reliable (e.g. compared with data from tissue samples) and few compromises are needed in the choice of fitting function. The fitted function – which in effect represents the dynamics (includes all the modes) of the entire model as reflected in the measured compartment – then can be used as a unidirectional **equivalent input forcing function** to any other compartments receiving influx of material from the measured compartment. The chosen other compartment is in effect isolated from the remainder of the model, with the forcing function unilaterally representing the rest of the model. The idea, then, is that (*SI*) parameters of interest associated with other compartments may be easier to estimate, or be more reliably estimated.

We assume not all state variables are measurable. If they were, this approach would be unnecessary. A simple example illustrates the idea and the procedure.

Example 13.13: ***One-Compartment Equivalent of a Two-Compartment Model***

The two-compartment model on the LHS in **Fig. 13.13** was shown in Example 10.2 of Chapter 10 to be uniquely ID. Suppose the input $u_1(t) = \delta(t)$ and $V_1 = 1$. The question

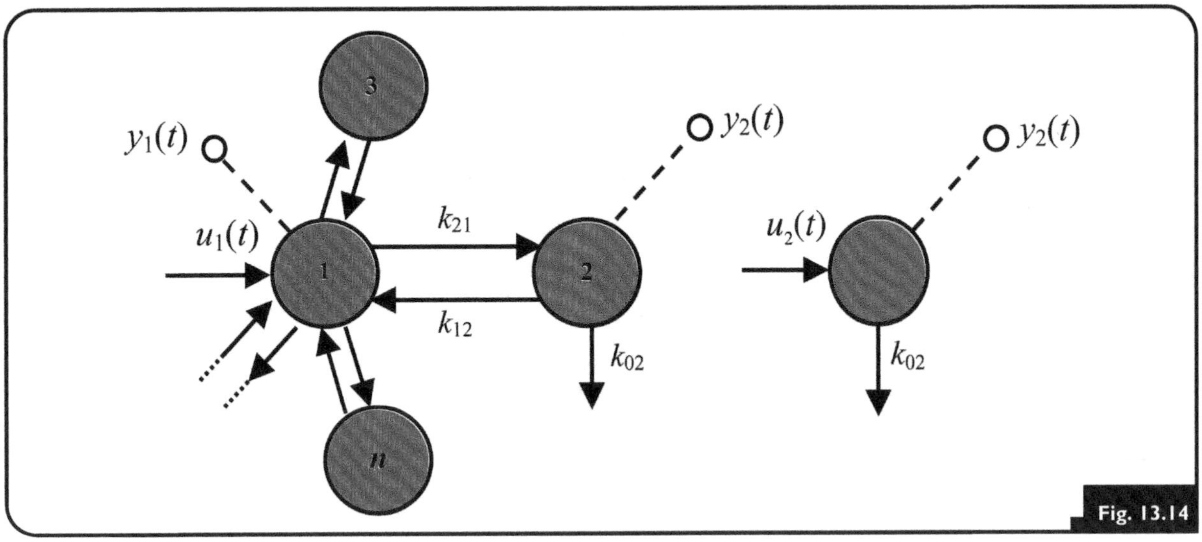

I–O forcing function equivalent models: n-compartments.

is: what equivalent *unidirectional* input $u_2(t)$ into compartment 2 – as in the RHS single compartment 2 – will give identical responses $q_2(t)$ in compartment 2? By inspection, this should equal $u_2 \equiv k_{21}q_1 - k_{12}q_2$, the difference between the mass fluxes between 1 and 2. The ODEs for compartment 2 of the RHS and LHS models are $\dot{q}_2 = u_2 - k_{02}q_2$ and $\dot{q}_2 = k_{21}q_1 - (k_{02} + k_{12})q_2 = k_{21}q_1 + k_{22}q_2$. Substituting $u_2 \equiv k_{21}q_1 - k_{12}q_2$ into the first ODE gives:

$$\dot{q}_2 = u_2 - k_{02}q_2 = k_{21}q_1 - k_{12}q_2 - k_{02}q_2 = k_{21}q_1 + k_{22}q_2 \tag{13.8}$$

This is exactly equal to $\dot{q}_2 = k_{21}q_1 + k_{22}q_2$ for the two-compartment model on the LHS, proving the equality of the forcing function equivalence for this model. This approach generalizes readily to models with any number of compartments.

As implied above, a typical (and popular) application of the forcing function method is in modeling multicompartmental systems where a central exchanging compartment is accessible for measurement, e.g. the blood or plasma compartment in a mammillary model, with *one or more peripheral compartments also measurable*. The dynamic response data from blood (plasma) is fitted first by a mathematical function, e.g. a multiexponential (next section) or polynomial function, and this becomes the forcing function for all compartments connected to the "central" blood (plasma) compartment. The parameters associated with a measured non-central compartment are then separately estimated using this forcing function. A simple example follows.

Example 13.14: ***Single Peripheral Compartment Model Quantification***

The arbitrary n-compartment model on the LHS of **Fig. 13.14** is not generally fully identifiable. Suppose the leak k_{02} is of particular interest and, although it is not *SI* from measurements $y_1(t)$ in compartment 1 only, it is *SI* from measurements in both $y_1(t)$ and $y_2(t)$. The forcing function approach can be used to quantify k_{02} from output $y_2(t)$ data, following fitting of $y_1(t)$ first by a suitable mathematical

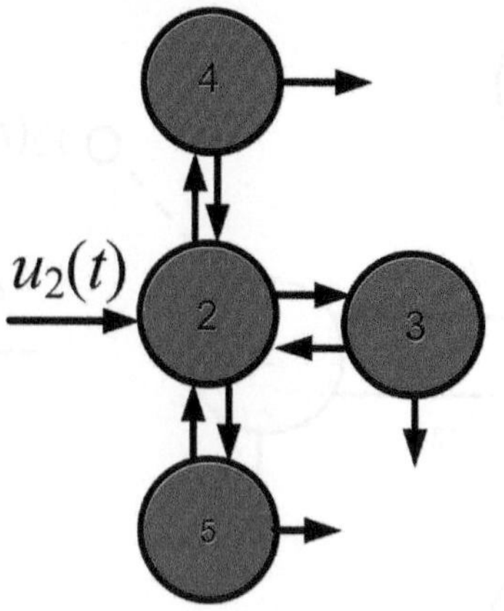

Fig. 13.15

function, e.g. a multiexponential or polynomial. The resulting $u_2(t)$ is the equivalent *unidirectional* input into compartment 2 – as in the RHS of **Fig. 13.14** single compartment 2.

This result is even more extensive. If the equivalent input $u_2(t)$ drives a more complex submodel – e.g. **Fig. 13.15** – instead of single compartment 2, the *SI* parameters of the submodel can be quantified using this forcing function.

We emphasize that the complete system model dynamics are reflected in the forcing function, no matter how many compartments are involved in the main system or the subsystem. As always, only the SI parameters are quantifiable and this, of course, depends on model complexity.

The statistical reliability of models fitted using the forcing function method can be degraded by lack of statistical consistency, due to the sequential nature of the procedure. To remedy this potential disadvantage, the forcing function method can be used effectively to obtain preliminary estimates of all model parameters, say $\mathbf{p}_{FORCING}$, followed by a simultaneous parameter estimation procedure, starting from $\mathbf{p}_{FORCING}$ as initial values of the search (Eisenberg and Distefano 2009).

Multiexponential (ME) Models & Use as Forcing Functions

This class of linear system input–output (I–O) models was discussed and illustrated in several sections of Chapters 5, 10 and 12. In Chapter 5, they helped in establishing the dominant modes and thus minimal dimensionality of a model. In Chapter 10, ME models were useful for analyzing and quantifying a structured compartmental model – as in Example 10.7. ME models fitted to output data are particularly useful as forcing functions, e.g. for fitting the model in **Fig. 13.14**.

The form used most to fit single-input–single-output data is: $y(t) = \sum_{i=1}^{n} A_i e^{\lambda_i t}$, under the assumption that the exponents (eigenvalues) λ_i are distinct – a reasonable one for many applications. The $2n$-dimensional parameter set $\boldsymbol{c} \equiv [A_1\ A_2\ \ldots\ A_n\ \lambda_1\ \lambda_2\ \ldots\ \lambda_n]^T$ is usually a structural invariant, or close to it, with some elements of $\boldsymbol{c}$ possible known from physical or mathematical constraints. For example, if data are collected from compartment 2 of a two-compartment model, with test-input introduced into compartment 1 and otherwise zero initial conditions, then $y_2(0) = A_1 + A_2 = 0$ is a constraint, and $\boldsymbol{c} \equiv [A_1 - A_1\ \lambda_1\ \lambda_2]^T$.

Fitting multiexponential models to data, however, is not without its encumbrances, both structural and numerical. Structurally, exponential sums are not orthogonal,[17] but "skewed" in function space, and they become less skewed (more orthogonal) as their exponents become more separated. When two exponents are close together, which means they differ by less than around a factor of 3 or 4, the two exponential terms are more or less difficult to distinguish from each other in data – because they are so skewed. When exponents are far apart, they are easier to distinguish, because they are more orthogonal.

When $\boldsymbol{c}$ is estimated using a numerical procedure, ill-conditioning of $COV(\boldsymbol{c})$ occurs often enough to be troublesome. As noted in Chapter 12, strong parameter estimate correlations and large standard deviations of the estimates are usual signs of ill-conditioning (*not* NI). When two exponents λ_i and λ_{i+1} are close, they will be highly positively correlated, and the corresponding parameters A_i and A_{i+1} will be strongly negatively correlated. This follows in part because, if the exponential terms are nearly indistinguishable, only the sum $A_i + A_{i+1}$ can be estimated with good precision.

Ill-conditioning is often due to *overparameterization* (OPM), or fitting a model of order greater than n to a system that is really of order n. This mimics the problem of closely spaced exponents because a weighted least-squares (*WLS*) fitting procedure, faced with at least one extra exponential term, attempts to drive its exponent value to the value of another exponent. Another problem can be *poor sampling design*, which fails to distinguish between at least two of the exponential terms. Poor sampling design usually means not carrying sampling far enough out in time to capture the full dynamic range of the model, or not sampling early enough to capture fast transients (Chapter 4). In principle, such a design would cause the same problems as overparameterization because at least two of the terms having the smallest (or largest) magnitude exponents would not be distinguishable. Uniform spacing of sample times or other clearly nonoptimal spacing would be another cause. The solution here would be to use an *optimal sampling design* (see Chapter 16).

Prior information about some model parameters, e.g. from independent experiments or a search of the literature, can be used to overcome structural *un*ID or poor *NI*. Prior

17 Orthogonality means each term in the series is "independent" of all others, in the sense that it is "perpendicular" to the other terms in the approximation subspace (see Appendix B). The function, call it $y(t)$, is approximated by an orthogonal series by projecting it onto this function space, each of whose coordinates are the functional components of the series, and additional components (terms) imply a better approximation. A major advantage of orthogonal series is that, as terms are added to improve the fit, none of the parameters (coefficients) of terms already used in the fit change. Among other benefits, this means that not everything has to be refitted each time a term is added to improve the approximation. Only the parameters of the additional term(s) need be computed. Fourier series terms, for example, are orthogonal.

information is commonly expressed as means, standard deviations or confidence intervals (Box and Tiao 1973).

Multiexponential function fitting to experimental data can be particularly useful in early stages of developing structured models – for establishing their dimensionality, complexity and other properties. Numerical difficulties occurring during the fitting process have a high likelihood of revealing problems in model formulation, assumptions about data errors, or the experiment design. Much of this is revealed in *COV*(*c*). Web application *W*[3]*DIMSUM*[18] (abbrev: *DIMension of a SUM of Exponentials*) can be quite useful for processing data in this way. It provides multiple qualitative statistical measures of several fitted "candidate" models, based on residuals tests as well as *COV*(*c*), and uses an "expert system" based on these measures to assess results and choose a "best" ME model that fits the data. The illustrative examples accessible from within the software GUI illustrate most of the features of multiexponential modeling discussed in this section (Harless and DiStefano III 2005).

Model Fitting & Refitting With Real Data

The HIV vaccine development model begun earlier in Example 13.6 is continued here, in two stages. The first illustrates the typical problem setup and parameter estimation details involved in a real-world problem. The second stage illustrates how better results might be achieved, recognizing that the quality of the fit in the first stage can be readily improved – without much additional effort. The statistics of typical model comparisons are included.

Example 13.15: ***Nested HIV Vaccine Development Part II – Complete Model Quantification***

The three-output concentration data (in ng/ml) for fitting the original nested *gp120* model shown in **Fig. 13.7** is reproduced in **Table 13.4** on the book website www.elsevier.com/distefano2013. It is included in the *data* portion of the *SAAMII* study file for this example.[19] The data are shown with symbols on the LHS plot of **Fig. 13.16**, along with the fitted model curves for each of the three experiment plasma responses. Note that blood sampling times were concentrated over different ranges of the 48 hour study for each of the three rabbit groups, common practice in *in vivo* studies.[20] The "n" entries in **Table 13.4** represent "no data" for this variable at that time. This standard format retains the fidelity of the multioutput data file.

The input dose given to all rabbits was $Q = 3 \times 10^5$ ng. With the data measured in ng/ml concentrations in compartment 1, this introduces equivalent initial distribution volume V_1 as an unknown parameter of the model. IM inputs $Q\delta(t)$ into the muscle compartments 2 and 3 also introduces unknowns V_2 and V_3. We anticipate these equivalent distribution volumes in muscle to be larger than V_1, because the bolus into

18 www.biocyb.cs.ucla.edu/dimsum

19 Model study file included on the book website.

20 Blood sample assay volume requirements limit the number of samples that can be drawn before generating abnormal responses from the subject.

muscle remains effectively concentrated in part, due a "depot" effect, i.e. muscle releases the dose relatively slowly, especially the alum-bound dose – also suggesting $V_3 > V_2$, which was found. The data error model was given as a constant standard deviation $SD = 5$ ng/ml for all data. As discussed in Chapter 12, this data variability propagates through the model during the fitting procedure and is a major factor in generating parameter estimation variabilities.

The original model was assumed to have nonzero leaks in Example 13.6 in all three input compartments, 1, 2 and 3. This is an appropriate starting assumption, because if the leak parameters are *SI* (which they are), their estimated values should answer the validity question. It was not known whether the ligand – bound or unbound – would measurably degrade or be eliminated at the muscle depot site over the 48 hours of the study, other than by absorption into plasma. Elimination from blood, once absorbed, is anticipated ($k_{01} > 0$). However, numerous runs (>20) of the fitting program under these conditions strongly suggested that leaks from muscle, represented by k_{02} and k_{03} in the model, were negligible over the 48 hours of the study, so they were eventually set to zero. Values close to zero gave the same fitting results, or no statistics at convergence, suggesting numerical *un*ID with nonzero values. This does *not* mean these leaks are zero, just that the data does not justify nonzero estimates.

The model outputs fitted to the data are shown in **Fig. 13.16** graphs in the middle section. All statistical results of parameter estimation are summarized in the top section.[21]

The main goal of the study was to quantify three *PK* parameters of the candidate vaccine. From results: $k_{elimination} = k(0,1) = 0.231$ per hour ($\pm$ 10.8% *CV*), $k_{absorption} = k(6,5) = 0.0316\ h^{-1}$ ($\pm$ 21.3% *CV*) and $k_{\mathrm{desorption}} = k(5,3) = 0.15\ h^{-1}$ ($\pm$ 23% *CV*). The magnitude of this last value (0.15), 5 times greater than its absorption rate into plasma ($\sim$0.03), suggested that alum did not provide a good "depot effect" for this ligand.

Discussion of other parameter estimates and comparison of results with another model follows in the next example.

Example 13.16: ***Nested HIV Vaccine Development Part III – Better Fit to the Data***

Visually, the original model appears to fit the plasma data fairly well in the **Fig. 13.16** LHS plot, best in the earliest hours post-injection (pi), with some residual discrepancies dominating the fits of the group 1 data (squares) around 10 hours. Good modeling practice suggests trying for a better fit – if identifiability requirements can still be satisfied – with a slightly extended and physiologically feasible set of experiment models. The original model may have satisfied the goals of the study (Weissburg et al. 1995), but in this case better results were available with a better-fitted model.

21 The two plots are taken directly from *SAAMII* program output results. The statistical tables are slightly reduced versions of direct program output, cleaned of unessentials (numbers rounded from 12 decimal places).

SUMMARY STATISTICS - gp120 vaccine test model, August16,2011

Parameter/Variable	Value	Std Dev	%Coef of Var	95% Confidence Interval	
Q	300000.				
V1	240.75	9.29	3.86	222.	259.1
V2	450.79	93.3	20.7	266.1	635.4
V3	531.36	102.2	19.2	329.0	733.7
k(0,1)	0.231	0.0025	10.8	0.182	0.281
k(0,2)	0				
k(0,3)	0				
k(5,3)	0.15042	0.00346	23.0	0.08	0.218
k(6,5)	0.03160	0.00673	21.3	0.01	0.045

ObjectiveValues

	Objective	ScaledDataVariance
s3 : z1ghi	1.77	5.3
s2 : z1def	2.29	12.7
s1 : z1bc	3.98	1112.0

Total objective = 8.04, AIC = 5.0, BIC = 5.1

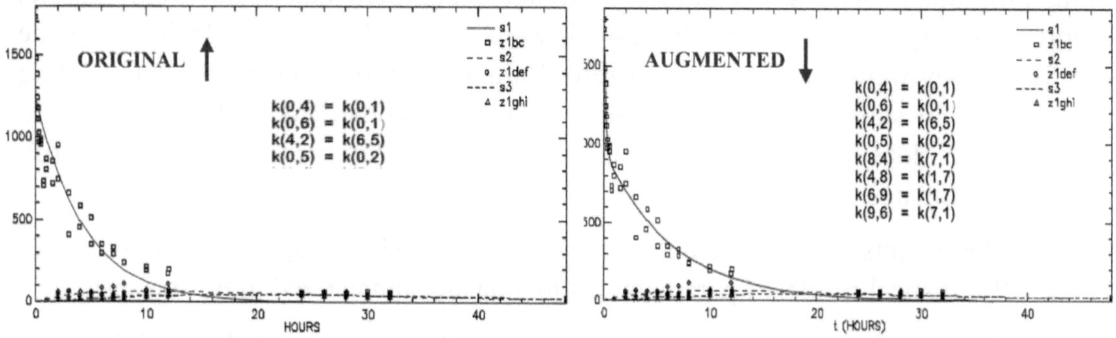

SUMMARY STATISTICS - gp120 Augmented vaccine test model, August 16, 2011

Parameter/Variable	Value	Std.Dev.	%Coef. of Var.	95% Confidence Interval	
Q	300000.				
V1	160.8	8.62	5.36	143.7	177.8
k(0,1)	0.298	0.00273	9.15	0.244	0.352
k(0,2)	0				
k(0,3)	0				
k(1,7)	3.51	0.786	22.4	1.95	5.065
k(5,3)	0.161	0.00255	22.9	0.088	0.234
k(6,5)	0.0438	0.00151	18.6	0.027	0.0599
k(7,1)	3.0	0.765	25.5	1.48	4.51
V2	392.7	63	16.2	266.6	518.9
V3	475.4	70.2	14.8	336.3	614.4

Objective Values

	Objective	Scaled Data Variance
s3 : z1ghi	1.76	5.25
s2 : z1def	2.29	13.0
s1 : z1b	3.57	352.7

Total objective = 7.62, AIC = 4.81, BIC = 4.93

Fig. 13.16

Simplest model results (top & left figure) and augmented model results (bottom & right figure) for the *gp120* vaccine study.

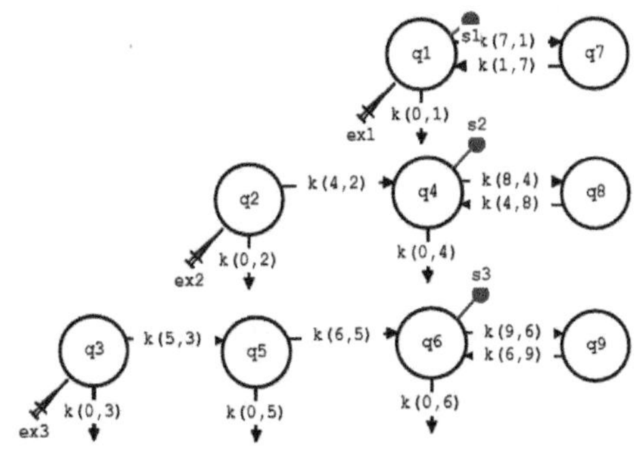

Fig. 13.17

Vaccine study model with one extravascular compartment.

The easiest tack is to augment the model for each experiment with one or two additional "extravascular tissue" compartments, generating one or two additional modes in the plasma responses. These are justified, for example, if non-negligible amounts of *gp120* distribute beyond blood at rates much slower than mixing in plasma. This seemed to be a reasonable assumption, based on preliminary analysis of the IV plasma *gp120* response data, which does not look like a single exponential decay. Adding just one beyond plasma is illustrated in the augmented and *SI SAAMII* model structure of **Fig. 13.17**.

The same fitting issues arose about the muscle compartment leaks, k_{02} and k_{03}. Any values greater than zero for these parameters yielded either poorer or no better fits to the data, or no statistics at convergence, suggesting numerical *un*ID with nonzero values. They were set to zero in the final run with statistics (BOTTOM) and plot (LHS) summarized in **Fig. 13.16**. Again, these leaks may not be zero, but the data does not justify nonzero values.

Model Comparisons: The *total objective* (*EWLS* loss function) value (7.62) and *AIC* value (4.81) for the augmented model are both smaller than *total objective* (8.04) and *AIC* (5.0) values for the original model. These objective measures indicate the augmented model fits better. From augmented model results we have: $k_{elimination} = k(0,1) =$ 0.298 per hour ($\pm$ 9.15% *CV*), $k_{absorption} = k(6,5) = 0.0438\ \mathrm{h}^{-1}$ ($\pm$ 18.6% *CV*) and $k_{\mathrm{desorption}} = k(5,3) = 0.161\ \mathrm{h}^{-1}$ ($\pm$ 22.9% *CV*). These values are *quite similar* to those of the original model, with slightly better variabilities (%*CV*s). Comparison of the two graphs in **Fig. 13.16** admittedly subjective, visually substantiates the better fit of the augmented model. Finally, we compare estimated values of the three compartment *equivalent distribution volumes* (Chapter 4): ~161, 373 and 475 ml. The first (V_1 ~161 ml) is close to the real rabbit plasma volume. The volumes V_2 and V_3 of the muscle depots compartments 2 and 3 clearly indicate concentration of the doses injected, to roughly 2 and 3 times that of the plasma concentration of *gp120*, as anticipated.

Recap and Looking Ahead

This chapter extended the repertoire of parameter estimation and quantitative modeling tools, intended to be directly useful in the process and practice of biomodeling from data. Several time-honored and some new techniques for facilitating the process were developed. Detailed examples, with and without real noisy data, illustrated these innovations and underscore some key issues and illustrate some challenging side-effects of the parameter estimation process itself. As noted in the introduction, model quantification is technically difficult, and we need all the help we can get from practice as well as theory.

Chapter 14 returns attention to structural modeling and analysis methods, formalizing and extending feedback control theory and biocontrol systems methods covered in part in previous chapters. Several illustrative examples include quantification details – the subject of the last two chapters – justifying the "late" placement of these subjects in the book. Developments and examples include more comprehensive assessments of several physiological, biochemical and cellular control system models, including discussion and approaches to solutions of challenges encountered in quantifying these models.

Exercises

E13.1 Draw the three-compartment catenary model with all parameter arrows identifiable, in terms of the identifiable combos given in Eq. (13.4).

E13.2 Repeat E13.1 for the most general three-compartment model.

E13.3 Repeat Example 13.4 using two different sets of nominal parameter values, namely $C = 3$ and $K = 10$, and then $C = 3$ and $K = 100$. Repeat for $u(t)$ = step inputs of magnitude 0.1 and 10. How do these parameter and input magnitude differences effect the quality of numerical identifiability properties for this NL model?

E13.4 In Examples 13.11 and 13.12, constraints consistently reduced mean ranges for unidentifiable k_{ij}, as expected, but some parameter bound variabilities increased, especially for the k_{0j} (leaks). This might be anticipated with reduced mean ranges, for variabilities expressed as $100\ SD/mean$, for k_{ij}. But some *SD*s also increased. Why did this happen? This seemingly paradoxical result can be explained using the properties of variances of sums of random variables.

E13.5 *Hormone Kinetics in the Fish Tilapia – Alternative Quantifications via Indirect Multiexponential and Direct ODE Modeling.* The averages of radioactively-labeled hormone data (percent of dose per volume units) in **Table 13.3** were collected from caudal vein blood of fishes following IV bolus doses of 100% each (DiStefano III et al. 1998). Data sample *SD*s and sample *%CV*s based 100% on residual data error measurements in a larger group of Tilapia collected over a month are also given in the

Table 13.3 Averages of Radioactively-Labeled Hormone (% of Dose per mL Units) Collected from Caudal Vein Blood of Fishes Following IV Bolus Doses of 100% Units Each

t (Hours)	$z(t)$	VAR	SD	$\%CV$
0.0001	46.7	51.3	7.16	15.3
0.0333	16.2	16.4	4.05	25.0
0.0833	10.2	14.8	3.85	37.7
0.167	7.34	9.26	3.04	41.5
0.333	6.42	5.40	2.32	36.2
0.5	5.82	4.04	2.01	34.5
0.75	4.55	3.33	1.82	40.1
1	3.67	1.61	1.27	34.6
1.5	3.22	1.17	1.08	33.6
2	2.67	0.864	0.930	34.8
4	1.77	0.469	0.685	38.7
5	1.55	0.321	0.567	36.6
9	1.42	0.243	0.492	34.7
12	0.834	0.208	0.456	54.7
18	0.479	0.188	0.434	90.5
24	0.262	0.0312	0.177	67.4
36	0.114	0.00543	0.0737	64.6
48	0.0572	0.00105	0.0324	56.6

table. These data errors are considered representative of measurement errors in the actual kinetic studies done separately.

a) Fit a three-exponential model to this data using the web application W^3*DIMSUM* on the website: *www.biocyb.CS.UCLA.edu.*

b) A three-compartment mammillary model with leaks only in peripheral compartments 2 and 3 is the best physiological solution for this data. Exchange occurs between compartments 1 and 2, and 1 and 3. Use the program W^3*MAMCAT* at the same website to generate a formal three-compartment mammillary model with this structure. All parameter values or range values, plus their variabilities, should be given by the program solution.

c) Fit the same three-compartment model using *SAAMII* or other ODE model fitting program.

d) Compare parameter estimates and estimation variability statistics resulting from these two approaches to model quantification, the indirect and direct methods.

References

Bard, Y., 1974. Nonlinear Parameter Estimation. Academic Press, New York.

Boulier, F., 2007. Differential elimination and biological modelling. J. Comput. Appl. Math. 2, 111.

Box, G., Tiao, G., 1973. Bayesian Inference in Statistical Analysis. Addison-Wesley, Reading, MA.

Chapman, M., Godfrey, K.R., Chappell, M.J., Evans, N.D., 2003. Structural identifiability for a class of non-linear compartmental systems using linear/non-linear splitting and symbolic computation. Math. Biosci. 183 (2), 215.

Chappell, M., Godfrey, K.R., Vajda, S., 1990. Global identifiability of the parameters of nonlinear-systems with specified inputs – a comparison of methods. Math. Biosci. 102 (1), 41–73.

Denis-Vidal, L., Joly-Blanchard, G., Noiret, C., 2001. Some effective approaches to check the identifiability of uncontrolled nonlinear systems. Math. Comput. Simul. 57, 35.

DiStefano III, J., Ron, B., Nguyen, T.T., Weber, G.M., Grau, E.G., 1998. 3,5,3′-triiodothyronine (T3) clearance and T3-glucuronide appearance kinetics in plasma of freshwater-reared male tilapia, oreochromis mossambicus. Gen. Comp. Endocrinol. 111, 123–140.

Eisenberg, M., Distefano III, J.J., 2009. TSH-based protocol, tablet instability, and absorption effects on L-T4 bioequivalence. Thyroid. 19 (2), 103–110.

Evans, N.D., Chappell, M.J., 2000. Extensions to a procedure for generating locally identifiable reparameterisations of unidentifiable systems. Math. Biosci. 168 (2), 137–159.

Harless, C., DiStefano III, J., 2005. Automated expert multiexponential biomodeling interactively over the internet. Comput. Methods Programs Biomed. 79 (2), 169–178.

Kalman, R., 1960. A new approach to linear filtering and prediction problems. Trans. ASME – J. Basic Eng. 82, 35–45.

Lamm, D.L., D. J., Riggs, D.R., Delgra, C., Burrell, R., 1993. Keyhole limpet hemocyanin immunotherapy of murine bladder cancer. Urol. Res. 21, 33–37.

Landaw, E., Chen, B.C., DiStefano III, J.J., 1984. An algorithm for the identifiable parameter combinations of the general mammillary compartmental model. Math. Biosci. 72, 199–212.

Lee, W., Wahjudi, P.N., Xu, J., Go, V.L., 2010. Tracer-based metabolomics: concepts and practices. Clin. Biochem. 43, 1269–1277.

Ljung, L., 1987. System Identification: Theory for the User. Prentice-Hall, Englewood Cliffs, NJ.

Meshkat, N., Anderson, C., DiStefano III, J.J., 2011. Finding identifiable parameter combinations in nonlinear ODE models and the rational reparameterization of their input–output equations. Math. Biosci. 233, 19–31.

Meshkat, N., Anderson, C., DiStefano III, J.J., 2012a. Alternative to Ritt's pseudodivision for finding the input–output equations of multi-output models. Math. Biosci. 239, 117–123.

Meshkat, N., Eisenberg, M., DiStefano III, J.J., 2009. An algorithm for finding globally identifiable parameter combinations of nonlinear ODE models using Gröbner Bases. Math. Biosci. 222 (2), 61–72.

Meshkat, N., Kuo, C., Brown, N., DiStefano III J.J., 2012b. Online software for finding identifiable parameter combinations in nonlinear ODE models.

Meshkat, N., Kuo, C., DiStefano III J.J., 2014. COMBOS: web app for finding identifiable parameter combinations in nonlinear ODE models. PLOS Comp. Biol., submitted for publication.

Perelson, A.S., 2002. Modelling viral and immune system dynamics. Nat. Rev. Immunol. 2 (1), 28–36.

Price, N.D., Papin, J.A., Schilling, C.H., Palsson, B.O., 2003. Genome-scale microbial in silico models: the constraints-based approach. Trends in Biotechnol. 21, 162–169.

Schilling, C.H., Palsson, B.O., 1998. The underlying pathway structure of biochemical reaction networks. Proc. Natl. Acad. Sci. U S A. 95 (8), 4193–4198.

Vicini, P., Su, H.T., DiStefano III, J.J., 2000. Identifiability and interval identifiability of mammillary and catenary compartmental models with rate parameter constraints. Math. Biosci. 167, 145–161.

Weissburg, R.P., B. P., Cleland, J.L., Eastman, D., Farina, F., Frie, S., et al., 1995. Characterization of the MN gp120 HIV-1 vaccine: antigen binding to alum. Pharm. Res. 12, 1439–1446.

Chapter 14

Biocontrol System Modeling, Simulation, and Analysis

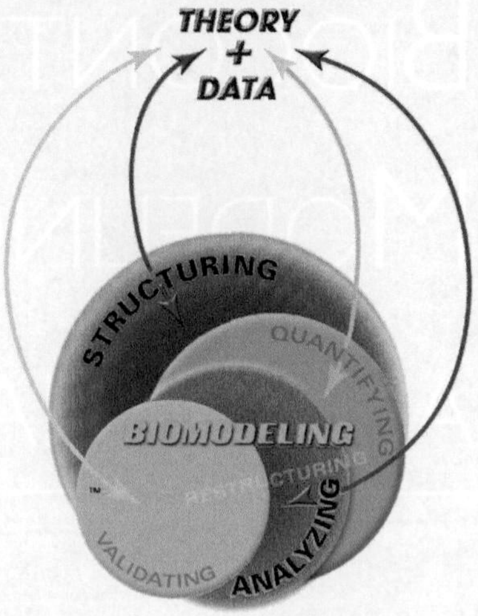

"There is no other company that could make a MacBook Air. And the reason is that not only do we control the hardware, but we control the operating system. And it is the intimate interaction between the operating system and the hardware that allows us to do that. There is no intimate interaction between Windows and a Dell notebook."

Steve Jobs 2008

Overview

Regulation and control processes in biology have been major foci of investigation since the genesis of biochemistry and biology. We have been addressing biocontrol processes and their modeling in all previous chapters, albeit not as a broader topic for special consideration, as we do here. Our goal here is to explore several developments of the last half century specifically for modeling physiological and biomolecular control systems.

Following World War II, engineering feedback control system (fbcs) concepts and methods developed in electrical and mechanical engineering departments and motivated several classes of biocontrol system models, with a following of biocontrol system modelers for each. Macroscopic, physiologically based fbcs models, primarily made up of *blocks* representing organs or organelles interconnected via communication or control *signal pathways*, became a focus of bioengineering, biochemical engineering, and systems physiology modeling literature beginning in the 1960s. Fred Grodins' and John Milsum's contributions were among the earliest books on the subject (Grodins 1963; Milsum 1966). We begin below by examining the architecture and motivating principles of these classical physiological control system models, using the control system block diagram language developed in Chapter 3 for programming models for simulation. A generic multifeedback multifeedforward physiological control system block diagram paradigm is developed and illustrated. This is followed by a detailed example of systematically modeling a neuroendocrine control system from human data to illustrate the methodology. This includes a simulation model that can be exercised via a web application.

The remainder of the chapter delves more deeply than in previous chapters about mechanistically based modeling and analysis of biochemical and cellular control systems – in transient and steady states. The emphasis is on structural and dynamical properties of network biology and metabolism models. Some notable stability properties of feedback in a simple enzyme reaction are developed to illustrate some of the nuances in cellular dynamics. The last sections are focused on metabolic system structures and development of steady-state flux balance analysis (FBA) and associated techniques, for optimizing pathway reactions, and metabolic control analysis (MCA) for analyzing their sensitivity properties. These are particularly useful for applications in synthetic biology, pharmacology, and medicine. ***Appendices C–F*** include additional control theory, focusing on controllability, observability, and equivalence concepts and methods useful for biodynamic system modeling.

Physiological Control System Modeling

The engineering *fbcs* paradigm is natural for modeling physiological control systems, all of which involve feedback biosignaling – neural, endocrine, metabolic, etc. – most with the goal of maintaining **homeostasis**[1] – steady rhythmic or other controlled pattern of behavior in

1 Homeostasis is usually taken to mean *constancy*, or *unchanging*. But, just as steady state in systems terms does not necessarily mean constant steady state (steady oscillations are fine, for example), homeostasis refers more broadly to the maintenance (stabilization, regulation) of internal (not necessarily constant) conditions in the presence of external conditions or disturbances. This requires feedback. If conditions or disturbances persist, systems often *adapt* to them, i.e. change their characteristics to preserve their functionality, but some form of feedback usually persists in the "new" feedback system.

the presence of a changing environment. Core body temperature, for example, is maintained constant (37°C) in a manner not unlike the room temperature control fbcs model in **Fig. 1.3** – but with more complexity.

Fig. 14.1 illustrates the topology and some nomenclature for control system models in biology developed as engineering fbcs. The blocks typically include mathematical descriptions of biological subsystems (specific organs, cells, pathways, networks, etc.) – submodels – interconnected by signals to other submodels. The math is typically a transfer function – say $G(s)$ – if the model is linear and time-invariant, or (for example) – a Hill function – if it is nonlinear. Feedback can be positive or negative. Positive feedback pathways are usually associated with nonlinear saturating functions, which might represent switch-like behavior, or threshold phenomena; otherwise it will be unstable (Chapter 9). Combinations of positive and negative feedback pathways are prevalent in some biological systems and their models (Brandman et al. 2005), representing complex biomolecular network regulation, for example see Tyson (2006). Each type is illustrated later in the chapter.

Remark: As in earlier chapters, the feedforward and feedback pathway *arrows* in **Fig. 14.1** are signal pathways, reflecting signal *direction – not sign*. The arrowheads at the end of a pathway, pointing into a circular summing junction, are designated with a plus or minus sign, to distinguish them as stimulatory or inhibitory signals, as shown. This is a long established convention in control system engineering. In contrast, the arrowhead in biochemical pathway diagrams usually means stimulation, with a + sign implied. Throughout this book, to resolve this ambiguity, plus signs have been included at the ends of stimulatory signal pathways in biochemical diagrams. Negative signs also have been added at the ends of biochemistry inhibition signal pathways, with different end symbol: ⊥. This is done for completeness, even though this symbol is less likely to be misunderstood.

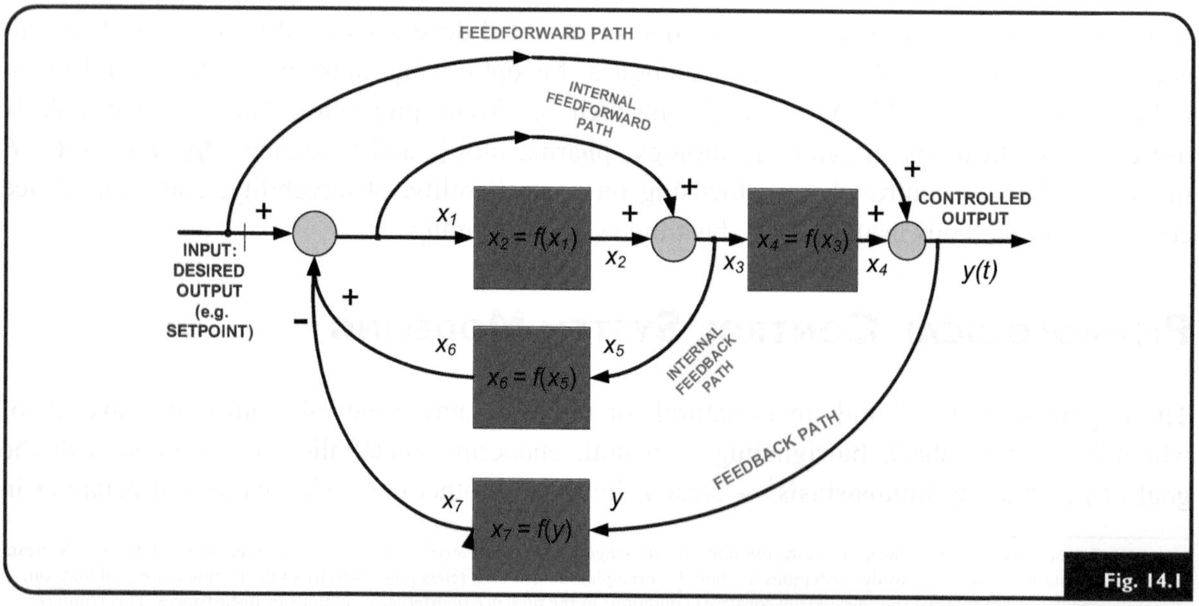

Fig. 14.1

Generic control system block diagram with feedback and feedforward pathways, summing junctions and input-output function subsystem blocks.

Much of the early modeling in this domain was of human and other mammalian physiological control systems. Notable examples – past and present – include fbcs models of the cardiovascular (Guyton et al. 1972), pulmonary (Grodins 1963), adrenocortical (Yates et al. 1969; Perrin et al. 1978; Smiljana 2005; Jelić et al. 2005), thyroid (DiStefano III et al. 1966; DiStefano III and Stear 1968; Eisenberg et al. 2006), pituitary (Eisenberg and DiStefano III 2010), sex hormone (Schwartz and McCobmack 1972), renal (Xingchen et al. 2006), and immunological (Perelson 2002) "body" systems. Individual models represent aspects of cardiovascular functionality such as blood flow or pulse rate or blood volume; or pulmonary gas exchange or other lung function; or production, distribution, action, or elimination of hormones, proteins, drugs, or other biochemicals; or neuroelectric regulatory physiological phenomena. The overall feedback system structures are typically organ-system based, with separate blocks representing individual organs or suborgan components. With reference to **Fig. 14.1**, submodel details inside the organ blocks typically represent their biophysically and/or biochemically based *input–output behavior*, as in the submodel paradigms illustrated in **Fig. 1.7** (Chapter 1). The details typically include mechanistic multicompartmental submodels – with compartments representing physiological or biochemical species; or unstructured (phenomenological) models of input–output data – when biological details are unknown. Most models of this type are *multiscale*, because they typically integrate intracellular or intraorgan phenomena within individual blocks – at one or more scales; and integration and signaling among the blocks – usually involves another time domain (e.g. milliseconds to minutes). All of these submodel types are found in the neuroendocrine fbcs model developed by Eisenberg and coworkers (Eisenberg et al. 2006) described in detail below.

In many such models, the *controlled output signals* are vectors $\boldsymbol{y}(t)$ of controlled or regulated physiological variable(s), thereby requiring more complexity than that in the illustrative block diagram of **Fig. 14.1**. The input signal might be a desired value (setpoint) of a tightly regulated output variable like core body temperature. Or it might be a periodic or other time-varying signal under physiological control, for example, cellular calcium or pituitary hormones in blood. Or it might be the output of another submodel of overall physiological regulation, as in the Guyton group model of overall blood circulation control (Guyton et al. 1972).

The Guyton model is included as an *analog computer simulation diagram* foldout in the original publication and is among the largest. We show it in miniature in **Fig. 14.2**. It is quite similar to a *Simulink* or *VisSim* diagram, because it is a control system model.

Neuroendocrine Physiological System Models

All multicellular organisms, plants as well as animals, produce *hormones* – molecules that function as physiological regulatory *signals*. Hormones are produced in *endocrine* glands, organs, or cells and are usually secreted and transported in the bloodstream. Cells that respond to a hormone carried in blood typically express a specific receptor for that hormone. The hormone binds to the receptor, resulting in activation of a signal transduction mechanism that ultimately leads to cell-specific responses.

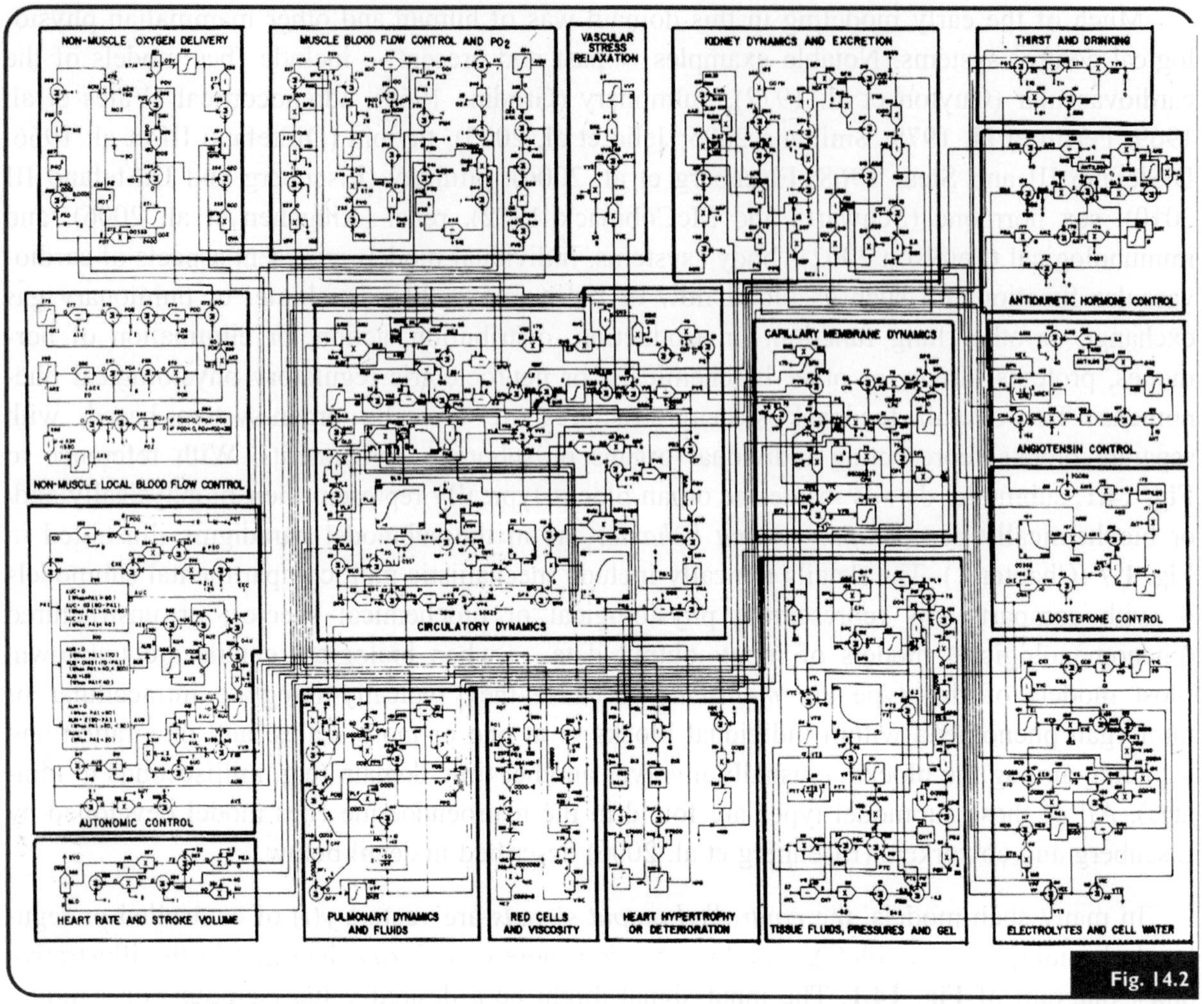

Fig. 14.2

The Guyton group model of the overall cardiovascular system, represented in its originally published analog computer program form in 1972 (Guyton et al. 1972). This was the first large-scale computer model integrating the many factors influencing the heart, peripheral blood circulation, autonomic nervous system, endocrine systems, kidneys and body fluids. (Reproduced with permission, from the American Physiological Society – Copyright Clearance Center © 1972, Vol. 34, *Annual Review of Physiology*).

Neuroendocrine systems are those that include functional interactions among the nervous system and endocrine systems in regulating physiological processes. Their primary interactions occur in the *hypothalamus* in the brain and in the *pituitary* gland. Major ones include the systems regulating thyroid, steroid, sex hormone, growth hormone and cytokine hormones of the immune system.

The following simple example illustrates the basic components of a neuroendocrine fbcs model, patterned after the organ *source–sink paradigm* of **Fig. 1.8**. Only negative feedback is illustrated, but these systems might also include positive feedback loops, and nearly all exhibit oscillatory behavior at some level – in particular, circadian (24 hour) periodicities.

Example 14.1: ***Source–Sink Feedback Control System Biomodel***

> **Fig. 14.3** is representative – in an oversimplified way – of how neuroendocrine fbcs such as thyroid, adrenal and sex hormone regulation systems are regulated by

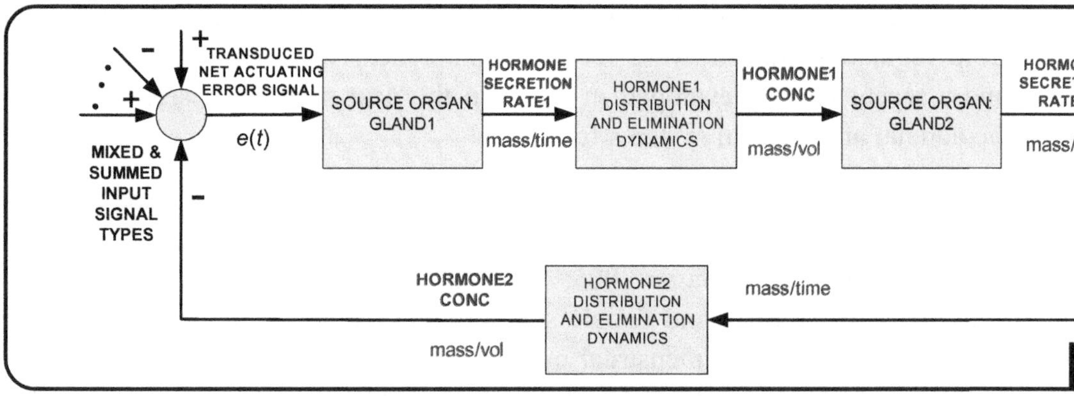

Fig. 14.3

Generic neuroendocrine feedback control system diagram.

negative feedback *at* the organ-system level. We assume only two source organs and two all-encompassing sinks in this generic example. They can be split into more granular submodels, as needed by the application.

The *source* blocks embody the dynamical mechanisms of organ *secretion* into blood (mass flux rates) at their outputs, in response to effector signals stimulating receptors at their inputs, which in turn typically stimulate hormone production as well as secretion rates. The *simplest* math model (rule) for a source block might be: secretion rate $SR_i(t)$ proportional to the concentration of the input signal stimulus $x_j(t)$ in blood, i.e. $SR_i(t) = k\, x_j(t)$. Alternatively, the secretion rate rule might be more specifically receptor mediated. Stimulated and saturable secretion might be modeled as: $SR_i(t) = A_i x_j^n / (B_i^n + x_j^n)$, an nth-order Hill function (often with $n = 1$), as illustrated in Chapter 7. Or, if the model is linear, source components might be represented as a Laplace domain transfer function. These are the simplest source models.

Hormone distribution and elimination (D&E) dynamic sinks can be represented as simply as: $dx_i/dt = -k_{ij}x_i + aSR_i$, a **first-order lag model** (with equivalent transfer function $G_i(s) = a/(k_{ij} + s)$), or as a complex multicompartment model – as in **Fig. 14.6** below, or other paradigm. Note that the D&E sink here is driven by the secretion rate and has concentration x_i as its output.

One or both hormone concentration signals in blood (or the plasma hormone component in blood) usually are the *controlled variables* in this model type, and the *inputs* include an inhibitory (negative feedback) signal – as shown, which effectively closes the feedback loop and stabilizes the hormone levels in blood. Other inhibitory and stimulatory signals that effect hormone levels (e.g. environmental temperature, stress) are included at the input summing junction and some of these signals typically emanate from the nervous system, via the hypothalamus. Together they produce – in classical control systems engineering terms – the *actuating error* $e(t)$ signal of the fbcs. The example in the following section illustrate these choices in greater detail.

Models in the Literature

The neuroendocrine modeling literature of the past several decades includes models based at least in part on the paradigm illustrated in *Example 14.1* and **Fig. 14.3**. It is dominated by models of normal and abnormal control of adrenal steroid, gonadal sex hormones and thyroid hormones. Thyroid regulation models were recently comprehensively reviewed in Dietrich et al. (2012)[2] and a recent one is developed in detail in the next section. A selected list of references for the adrenal and gonadal systems follows.

Adrenal system models are mostly mammalian – primarily human – and include explorations of their periodic behavior and stability properties – including chaos (Yates et al. 1969; Kyrylov et al. 2005; Lenbury 2005; Savic 2005; Smiljana 2005; Gupta et al. 2007; Bairagi et al. 2008; Sriram et al. 2012). Male and female sex hormone regulation models also exhibit oscillatory behavior, for example, the menstrual cycle, and have been developed for humans (Rasgon et al. 2003; Tornøe et al. 2007; Churilov et al. 2009) as well as several fishes (Kim et al. 2006; Watanabe et al. 2009) – motivated by recent interest in hormonal disruptors, particularly prevalent in marine environments.

Anatomy of a Neuroendocrine Model

This section takes us on a more extensive and ambitious journey. It provides a tutorial overview of the process of data-driven, multiscale modeling of the primary neuroendocrine system responsible for regulation of early development in animals and maintenance of metabolic regulation in adults. We explore the biosystem regulating cellular and blood levels of thyroid hormones – in particular in humans. It is probably the simplest of the neuroendocrine systems, dynamically speaking, because – although it exhibits periodic behavior – it is normally quite stable. We show stepwise how this important physiological fbcs has been modeled – how one modeling group selected, organized and classified available data for the purpose of structuring and, eventually, quantifying and validating the model from available human data. This model has evolved and been exercised in a long series of publications over four decades. Motivation, summary diagrams and full equations are included – to fortify the pedagogy – with additional details available in the cited references, or the references cited therein.[3]

Neuroendocrine Regulation of Thyroid Hormones (TH) in the Human: A Feedback Control System (fbcs) Simulation Model

This dynamic biosystem model has a long developmental history. The earliest versions date back to the late 1960s (DiStefano III 1968; DiStefano III 1969a; DiStefano III 1969b; DiStefano III 1969c; DiStefano III 1973; DiStefano III and Fisher 1979; DiStefano III 1985; DiStefano III 1991). The current one, described here, was motivated primarily by several timely clinical applications involving the potency of commercial preparations of hormone for hormone replacement – following thyroidectomy (gland removal) for thyroid cancer – or treatment in hypothyroidism in human adults and children (Eisenberg et al. 2006; Eisenberg et al. 2008; Eisenberg and Distefano III 2009; Eisenberg et al. 2010; Ben-Shachar et al. 2012). Availability of high quality critical human pharmacokinetic (PK) data a decade ago

2 The website: ***tfc.medizinische-kybernetik.de/models.html*** also includes a concise summary of the mathematical complexity of about 20 publications on thyroid system modeling from the 1950s through 2010, and a *simulator* for studying TH regulation.

3 The larger literature on thyroid hormone regulation system modeling was reviewed in (Dietrich et al. 2012). The model described in this section has benefited from other published models, as indicated in the references.

(a fortuitous event) also motivated human modeling and facilitated its development. Until then, experimental laboratory animal data was more readily available in the scientific community and earlier versions of this model – with essentially the same structure – were partially[4] quantified for the rat, sheep, and two fishes – trout and tilapia – all over several decades (DiStefano III et al. 1973; DiStefano III et al. 1975; Eales et al. 1982; DiStefano III et al. 1982a; DiStefano III et al. 1982b; DiStefano III et al. 1993a,b; Sefkow et al. 1996; DiStefano III et al. 1998). We focus only on the more recent human model here, emphasizing that it has been not only developed *and* quantified from data, but it also has been well validated from data not used to quantify it – for a variety of normal and pathological conditions of thyroid hormone regulation and dysregulation in adults and children. The simulation model is also available as a web application (*THYROSIM*),[5] for learning and teaching purposes.

Overall Block Diagram

The closed loop fbcs block diagram is illustrated in **Fig. 1.1**. It is reproduced here, slightly modified, in **Fig. 14.5**, changed little from the earliest versions 40 years ago. It has two major component groups: the *brain*[6] *submodels* and the *TH secretion* and *D&E submodels*, and three major *hormonal signals*: thyroxine (T_4), triiodothyronine (T_3), and thyroid stimulating hormone (*TSH*). The blocks include three isolated *source organs (BLUE)* that synthesize and secrete these hormones and three *distributed sinks (ORANGE)* – located throughout the body – that metabolize or excrete them.

Data

The *primary PK data* for modeling consisted of temporal concentrations in plasma of T_4, T_3, and *TSH*, depicted as block inputs and outputs in **Fig. 14.5**. (The data is plotted in **Figs. 14.7 and 14.8** described in the *Quantification* section.) These data were collected over four days in 33 normal human volunteers given three different oral doses of T_4 in three different studies (Blakesley et al. 2004). Additional literature data was used for modeling aspects of various components. Pertinent ones are noted.

Equations

The complete set of model equations, with *units included,*[7] is given in **Table 14.1**, grouped and numbered in accordance with the submodel descriptions and accompanying figures below. Model parameter estimates, along with their variability (%CV) estimates – too numerous to include here – can be found in Ben-Shachar et al. (2012).

4 Partially here refers to quantification in laboratory animals only for the model block called the *THD&E subsystem* of the overall fbcs model in **Fig. 14.5**. The model presented here for the human is the first full quantification for the entire closed-loop *TH* regulation model for any species.

5 *http://biocyb1.cs.ucla.edu/thyrosim/*

6 "Brain" is not an entirely correct designation for the submodel that includes the anterior pituitary gland as well as the hypothalamus. Anatomically, the hypothalamus is–but the pituitary is not–part of the brain. The pituitary is located in the head just below the brain. It has been loosely associated with brain in much of the literature, the reason the authors adopted that designation, which we retain here for historical purposes.

7 Carrying units through all equations minimizes errors in matching equations to data. Units must balance on both sides of an equation.

Table 14.1 Human Thyroid Hormone Feedback Control Dynamic System Model Under Oral Hormone Dosing Treatment

TH* fbcs: *Brain-Pituitary Submodel Equations	***Eq. #***	
$\dot{q}_7(t) \equiv T\dot{S}H_P(t) = SR_{TSH}(t) - f_{\deg}^{TSH}\,TSH_P(t)$	1	ODE for Plasma *TSH* (μmol/h)
$\dot{q}_8(t) \equiv \dot{T}_{3B}(t) = \frac{f_4}{T_{4P}^{EU}} T_{4P}(t) + \frac{k_3}{T_{3P}^{EU}} T_{3P}(t) - k_{\deg}^{T_{3B}} T_{3B}(t)$	2	ODE for T_3 in the brain (μmol/h)
$\dot{q}_9(t) \equiv \dot{T}_{3B}^{LAG}(t) = f_{LAG}\left(T_{3B}(t) - T_{3B}^{LAG}(t)\right)$	3	ODE for lagged T_3 in brain (μmol/h)
$SR_{TSH}(t) = \left(B_0 + A_0 f_{CIRC}\sin\left(\frac{2\pi}{24}t - \phi\right)\right)e^{-T_{3B}^{LAG}(t)}$	4	TSH secretion rate (μmol/h) $f_{circ} \equiv 1$ for eu- & hyperthyroid model
$f_{def}^{TSH} = k_{degTSH}^{HYPO} + \frac{v_{max}^{TSH}}{K_{50}^{TSH} + TSH_P(t)}$	5	Nonlinear *TSH* degradation rate function (h^{-1})
$f_{LAG} \equiv k_{LAG}^{HYPO} + \frac{2T_{3B}^{11}(t)}{K_{LAG}^{11} + T_{3B}^{11}(t)}$	6	Nonlinear lag-time function for T_{3B} in brain (h^{-1})
$f_4 \equiv k_3 + \frac{5k_3}{1 + e^{2T_{3B}(t)-7}}$	7	Nonlinear rate function for T_4 transport into & conversion of T_4 to T_3 in brain (h^{-1})
$f_{CIRC} \equiv 1 + \left(\frac{A_{\max}}{A_0\ e^{-T_{3B}^{LAG}(t)}} - 1\right)\left(\frac{1}{1 + e^{10T_{3B}^{LAG}(t)-55}}\right)$	8	Circadian rhythm saturation function (h^{-1}) $f_{circ} \approx 1$ for eu- & mild hyper model
TH* fbcs: *Thyroid Secretion & D&E Submodel Equations #	***Eq. #***	
$\dot{q}_1(t) \equiv \dot{T}_{4P}(t) = SR_4(t) + k_{12}q_2(t) + k_{13}q_3(t) - (k_{31}^{free} + k_{21}^{free})FT_{4P}(t) + k_4^{absorb}T_4^{GUT}(t)$	9	ODE for Plasma T_4 (T_{4P}) (μmol/h)
$\dot{q}_2(t) = k_{21}^{free}FT_{4P}(t) - \left(k_{12} + k_{02} + \frac{v_{\max}^{D1fast}}{K_m^{D1fast} + q_2(t)}\right)q_2(t)$	10	ODE for Fast tissue T_4 (μmol/h)
$\dot{q}_3(t) = k_{31}^{free}FT_{4P}(t) - \left(k_{13} + k_{03} + \frac{v_{\max}^{D1slow}}{K_m^{D1slow} + q_3(t)} + \frac{v_{\max}^{D2slow}}{K_m^{D2slow} + q_3(t)}\right)q_3(t)$	11	ODE for Slow tissue T_4 (μmol/h)
$\dot{q}_4(t) \equiv \dot{T}_{3P}(t) = SR_3(t) + k_{45}q_5(t) + k_{46}q_6(t) - (k_{64}^{free} + k_{54}^{free})FT_{3P}(t) + k_3^{absorb}T_3^{GUT}(t)$	12	ODE for Plasma T_3 (μmol/h)
$\dot{q}_5(t) = k_{54}^{free}FT_{3P}(t) + \frac{v_{\max}^{D1fast}q_2(t)}{K_m^{D1fast} + q_2(t)} - (k_{45} + k_{05})q_5(t)$	13	ODE for Fast tissue T_3 (μmol/h)
$\dot{q}_6(t) = k_{64}^{free}FT_{3P}(t) + \frac{v_{\max}^{D1slow}q_3(t)}{K_m^{D1slow} + q_3(t)} + \frac{v_{\max}^{D2slow}q_3(t)}{K_m^{D2slow} + q_3(t)} - (k_{46} + k_{06})q_6(t)$	14	ODE for Slow tissue T_3 (μmol/h)
$FT_{3P}(t) = (a + bT_{4P}(t) + cT_{4P}^2(t) + dT_{4P}^3(t))T_{3P}(t)$	15	free T_3 in plasma (μmol)
$FT_{4P}(t) = (A + BT_{4P}(t) + CT_{4P}^2(t) + DT_{4P}^3(t))T_{4P}(t)$	16	free T_4 in plasma (μmol)
$SR_3(t) = S_3TSH_P(t-\tau) \quad SR_4(t) = S_4TSH_P(t-\tau)$	17	TH secretion rates (μmol/h), τ is hrs time-delay (Mpy by fractions to adjust secretion rates)
2-Compartment Gut Input Submodels	***#***	
$\dot{q}_{10}(t) \equiv \dot{T}_4^{PILL}(t) = -k_4^{dissolve}T_4^{PILL}(t), \quad T_4^{PILL}(0) \equiv T_4Dose$	18	ODE for L-T_4 pill dissolution in gut (μmol/h)
$\dot{q}_{11}(t) \equiv \dot{T}_4^{GUT}(t) = k_4^{dissolve}T_4^{PILL}(t) - (k_4^{excrete} + k_4^{absorb})T_4^{GUT}(t)$	19	ODE for absorbable L-T_4 in gut (μmol/h)
$\dot{q}_{12}(t) \equiv \dot{T}_3^{PILL}(t) = -k_3^{dissolve}T_3^{PILL}(t), \quad T_3^{PILL}(0) \equiv T_3Dose$	20	ODE for L-T_3 pill dissolution in gut (μmol/h)
$\dot{q}_{13}(t) \equiv \dot{T}_3^{GUT}(t) = k_3^{dissolve}T_3^{PILL}(t) - (k_3^{excrete} + k_3^{absorb})T_3^{GUT}(t)$	21	ODE for absorbable L-T_3 in gut (μmol/h)

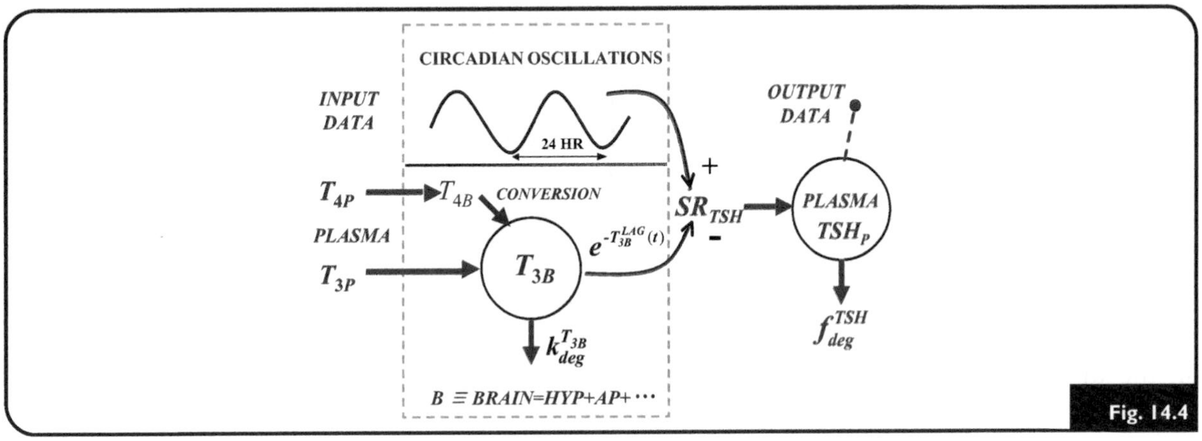

Brain Submodels.

Brain Submodels

The hormonal secretion of the hypothalamus that stimulates pituitary TSH production and secretion – thyrotropin releasing hormone (*TRH*) – is inaccessible to experimental probes and is not explicitly represented in the model, nor is its distribution or degradation. For lack of quantitative data for modeling hypothalamic components explicitly, the hypothalamus and anterior pituitary gland were *functionally integrated* in the brain submodel grouping, as illustrated in **Fig. 14.4.** The primary PK data noted above was used to quantify these combined components as a *single aggregated source* in the model – with plasma T_3 and T_4 as input and plasma *TSH* concentrations as output, as described below. T_4 that enters brain from plasma is converted to T_3 in brain cells and this T_3, along with that entering from plasma, are designated as a single T_3 variable T_{3B} in brain. The well-established circadian *TSH* secretion mechanism is incorporated into this combined source submodel, as shown in Fig. 14.4 and in Eq. (14.4) of **Table 14.1**. The nonlinear inhibitory effect (NL negative feedback) of T_{3B} on pituitary *TSH* secretion is implemented in the SR_{TSH} function as a decaying exponential factor in Eq. (14.4): $e^{-T_{3B}^{LAG}(t)}$. The effect of T_{3B} on secretion is delayed in $e^{-T_{3B}^{LAG}(t)}$ and is represented as a first-order lag function T_{3B}^{LAG} in Eq. (14.3). T_{3B} dynamics – which include enzymatic T_4 to T_3 conversion in these organs (via enzymes D1 or D2) – are expressed in Eq. (14.2). For sinks in **Fig. 14.4**, T_{3B} is distributed and eliminated in a nonlinear (NL) fashion within various brain compartments and pituitary spaces. A NL 1-compartment degradation model was established for *TSH* from a variety of data, with linear plus Michaelis–Menten (M–M) function elimination[8] (Eqs. 14.1 and 14.5). Eqs. (14.1–14.8) in **Table 14.1** completely describe the brain submodel. These are complex dynamics, represented minimally in these equations. Full quantification utilized additional published data on circadian *TSH* patterns and hormone ratios in pituitary as well as the primary PK data (see Eisenberg et al. (2010) for details).

TH Submodels

These include the thyroid gland submodel and the 6-compartment *TH* (T_4 and T_3) D&E submodel.

8 As in **Fig. 10.13** of Chapter 10.

A *very simple* 3-parameter time-delay submodel for the *thyroid gland* (Eq. (14.17) in **Table 14.1**): $SR_3(t) = S_3 TSH(t - \tau)$ and $SR_4(t) = S_4 TSH(t - \tau)$, captured the dynamics of the closed-loop feedback system well, when quantified together with the closed-loop database and nonlinear *TH* D&E submodel. In this model, S_3 and S_4 are constant coefficients representing linear secretory responses to plasma *TSH* concentrations, manifested through the *TSH receptor* (*TSHR*) and its initiation of *G*-coupled second messenger responses, stimulating T_3 and T_4 secretion down the line. Nonlinearities in these equations, for example, a M−M representation of *TSH−TSHR* coupling, instead of a linear one, offered no quantitative advantage in data fitting, nor did inclusion of specific pathways for known intracellular T_4 to T_3 conversion in the thyroid. Parameter τ approximates an apparent 6−8 h delay of thyroidal secretion in response to *TSH* stimulation in the primary PK data.

T_3 & T_4 (= TH) Distribution and Elimination (D&E) Submodel

This multiscale, NL 6-compartment submodel (**Fig. 14.6** and Eqs. (9−16) in **Table 14.1**) essentially represents aggregate *TH* D&E dynamics for the entire organism. Submodel (source) *inputs* are T_3 and T_4 thyroidal secretion rates, as shown; and *outputs* are the plasma total T_3 and T_4 concentrations. Hormone in nonthyroidal tissues exchanging with plasma compartments is classified dynamically as *slow* (slowly exchanging) and *fast* (rapidly exchanging) T_3 and T_4, as shown. To render it useful (robust) over a wide range of hormonal concentrations in applications, two kinds of known structural nonlinearities are included.

M−M kinetics for *cellular enzymatic interconversion* (T_4 to T_3 conversion via enzymes D1 or D2) are reasonably well-established. They are characterized in this submodel by the two pathways from slow and fast T_3 and T_4 compartments in **Fig. 14.6**. Literature data also indicated that fast compartment conversion is only via D1, and ~80% of slow compartment conversion occurs via D2 and ~20% via D1. The K_ms for these enzymes catalyzing these reactions are known,[9] so only v_{max}s need to be quantified. Eqs. (14.10, 14.11, 14.13 and 14.14) in **Table 14.1** includes these NL terms.

Nonlinear plasma protein binding to *TH* also needed to be represented in the model, because free (unbound) rather than protein-bound hormone enters cells. The algebraic (QSSA) plasma protein binding model developed in Mak and DiStefano III (1979) was incorporated in **Fig. 14.6**. This expresses plasma free hormone as polynomial functions of plasma total hormone concentrations, Eqs. (14.15 and 14.16) in **Table 14.1**. This polynomial model represents the reversible and competitive binding of both T_3 and T_4 to three plasma proteins − thyroid binding globulin (TBG), albumin (HSA), and transthyretin (TTR). Binding and debinding reactions are considered instantaneous relative to other hormone dynamics. Binding model parameters were originally estimated by fitting Eqs. (14.15 and 14.16) to solutions of the equilibrium M−M relationships for these reactions, using binding constants and total binding protein concentrations available at the time. For the current model, this binding model was

9 Compartment volume estimates of 1.2 L for the fast (~liver + kidney) and 50 L for the slow compartments were needed to convert the units of the known K_m values to μmoles, the units of the model. K_ms for most enzymes are measured and reported in Molar (mole/L) concentration units. In these units K_ms are the same in all cells for reactions between particular substrate, a particular enzyme. One way or the other, compartment volumes need to be considered in model development.

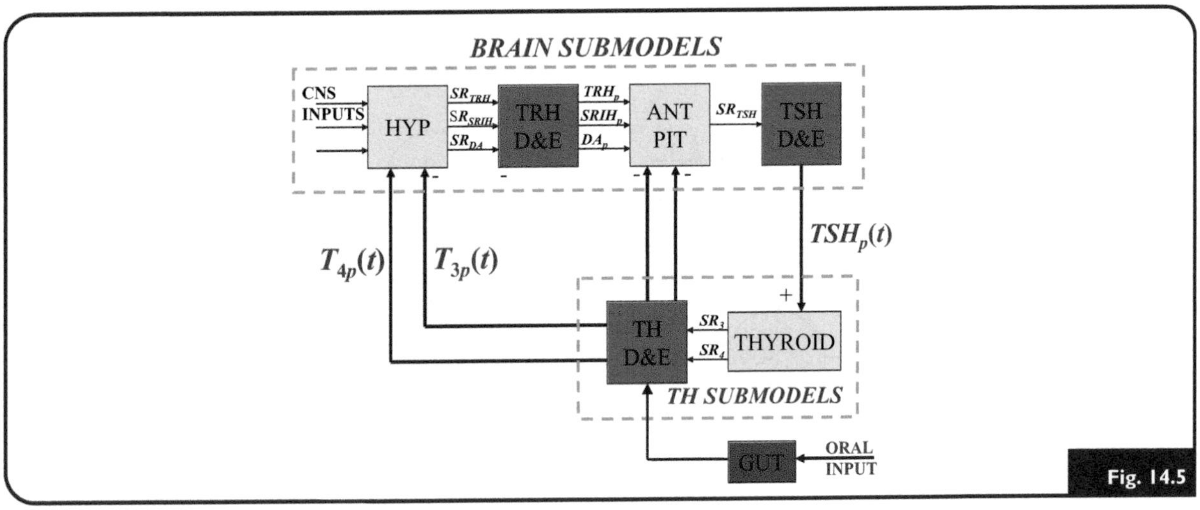

Fig. 14.5

TH fbcs block diagram model.

refitted using updated binding constants and protein concentrations. The binding submodel was then used to convert total hormone uptake rate constants k_{21}, k_{31}, k_{54}, and k_{64} in **Fig. 14.6** into *free hormone uptake rate constants*. This required published normal steady state free fraction values f_3 and f_4, yielding: $k_{54}^{free} \equiv k_{54}/f_3$, where $f_3 \equiv FT_3/T_3 = a + bT_4 + cT_4^2 + dT_4^3$ from Eq. (14.15), and $k_{54}T_3 \equiv k_{54}^{free}FT_3$; and similarly for the three T_4 uptake constants shown in the figure, using Eq. (14.16).

Nonlinearities in the *TH* submodels provide the necessary robustness for handling a wide spectrum of data, both in and out of the physiological range.

Gut Absorption Submodel — Needed for Representing Oral Dosing

Simulated PK studies using oral dose inputs were used for primary model quantification from the primary PK data. This required a gut absorption submodel augmenting the block diagram submodels in **Fig. 14.5**. The gut submodel represents pathway dynamics from the exogenous oral T_4 input, passage through the gut and eventually into the T_4 plasma compartment (see Chapter 4 for absorption modeling basics). The *TH* absorption model structure reported in DiStefano III and Fisher (1979) was used. The T_4 absorption rate into blood was assumed unknown in the current model and simultaneously estimated along with all other model parameters for the PK volunteer subjects, from their own data sets.

Preliminary Quantification

Block diagram models are typically quantified by first quantifying as much as possible about each submodel and then interconnecting them as in **Fig. 14.5**. D&E sink submodels are usually easiest to quantify from input–output kinetics studies and the early literature provided a good linear model qualification for the *TSH* D&E *sink* in **Fig. 14.4**, subsequently augmented with an NL M–M function to represent *TSH* values over a much wider range. The *TH* D&E *submodel* was quantified in *two steps*. Linearized tracer kinetic models are usually a good starting point, even for NL models. Fortuitously, the literature provided human tracer kinetic

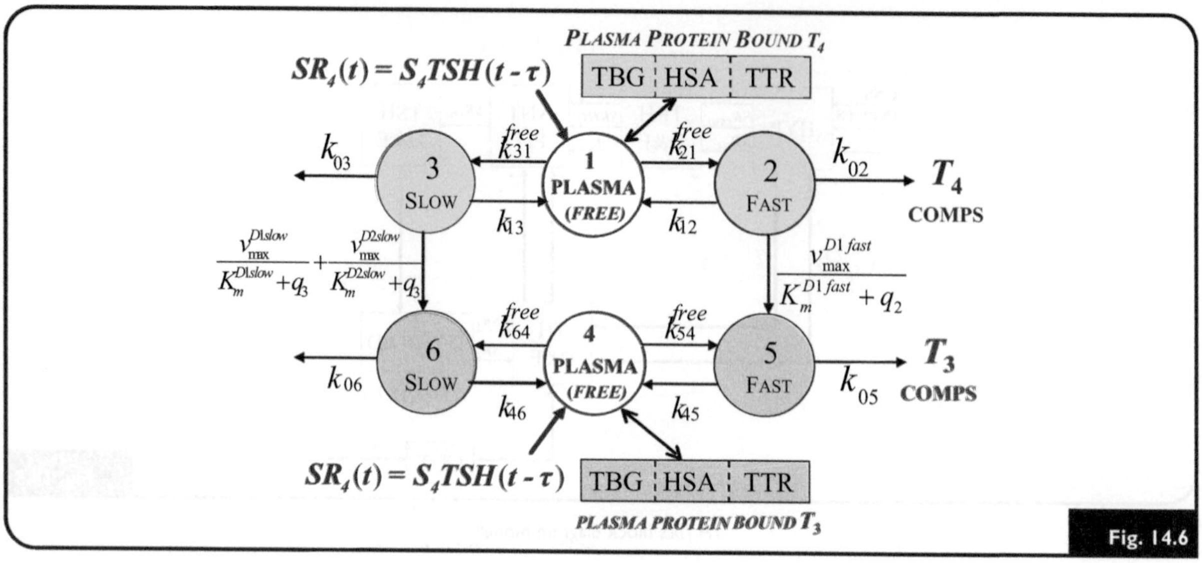

Fig. 14.6

TH Submodels.

data that was resampled and fitted by 3- and 6-dimensional multiexponential functions, providing estimates of the multiexponential A_is and λ_is.

Recall from Chapter 13 that multiexponential models are transformable into structured compartmental models, reparameterized in terms of *SI* combos. None of the k_{ij} rate constants in **Fig. 14.6** are individually SI from the input–output ports available but several *combinations* of them are (**see Table 16.2** and associated text in Chapter 16). This was the approach taken to quantify this submodel first, yielding an *SI* linearized form of the 6-compartment model in **Fig. 14.6**, quantified algebraically then from the multiexponential fitted A_is and λ_is (Eisenberg et al. 2006). This provided good starting values for subsequent closed-loop quantification of the full NL model, as described below.

Source organ submodels are a different matter, because they cannot be sufficiently isolated for quantification; neither feasible nor ethical *in vivo*. The alternative is closed-loop quantification, accomplished in a novel and successful way by this group.

Closed-Loop Quantification

This was done by combining several blocks of **Fig. 14.5** in particular ways, as illustrated in **Figs. 14.7 and 14.8**. The *input forcing function method* described in Chapter 13 played an important role by permitting simultaneous quantification of all submodels and variables circumscribed by the dashed boxes in **Fig. 14.5**, using closed-loop data. The four brain submodel blocks in the dashed box (HYP, TRH D&E, ANT PIT, and TSH D&E) were treated as one big block, *in toto*. Then, recognizing that plasma *TSH data embodies the I–O dynamics for all of the blocks working together*, the *TSH* data was used (as an input forcing function) to help quantify *TH* submodels that respond to *TSH* down the line. The sequence of PK response data points for *TSH* was used as one of the two input forcing functions, the other was the T_4 bolus dose input. T_{3P} and T_{4P} data were the *output variables* for quantifying the model (optimally

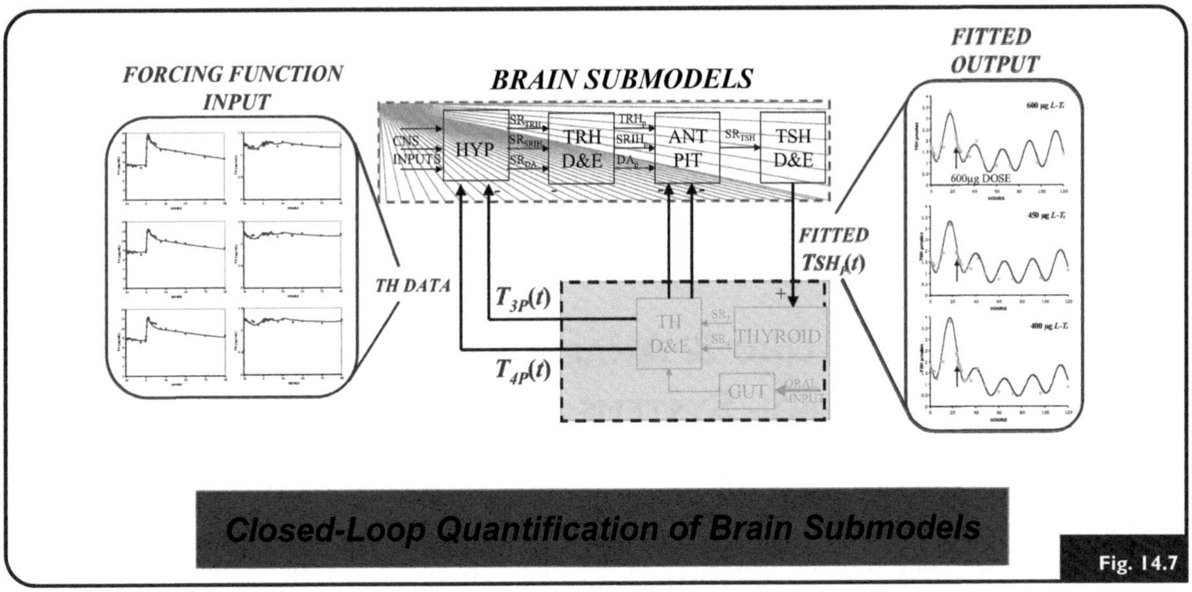

Closed-loop quantification – part 1.

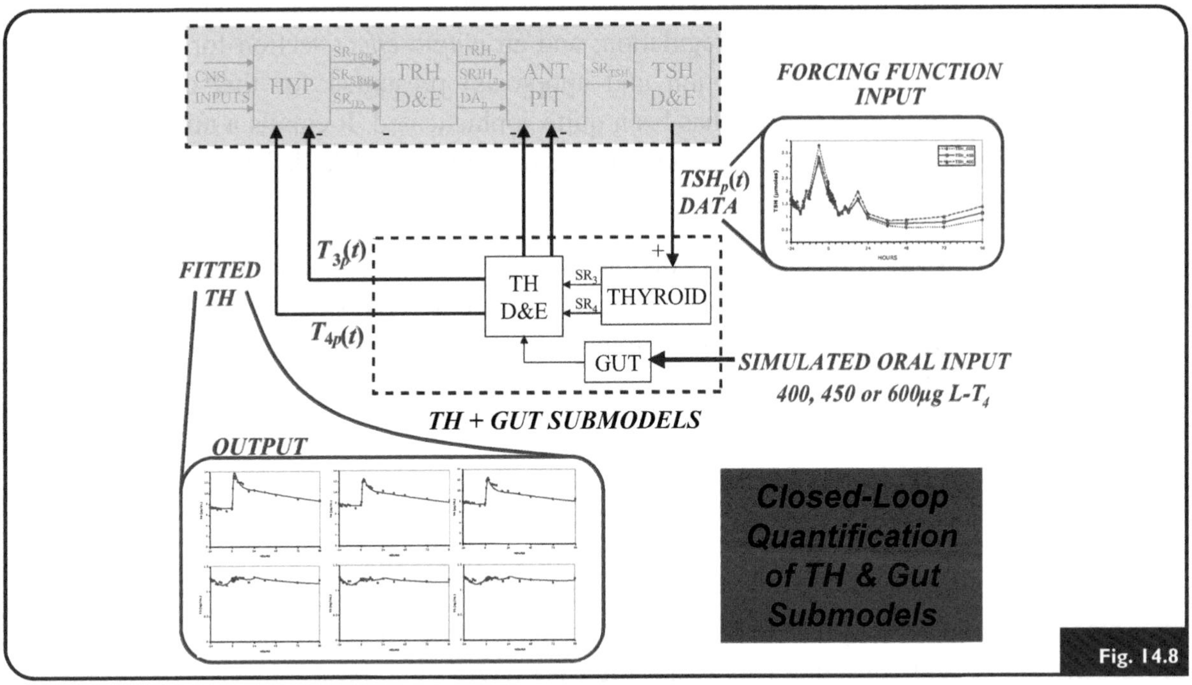

Closed-loop quantification – part 2.

fitting model parameters), as illustrated in **Fig. 14.7**. Conversely, plasma T_3 *and* T_4 *data were used as forcing function inputs* to the brain submodels, for quantifying the brain submodel parameters, as illustrated in **Fig. 14.8**.

Following the closed-loop estimation procedure outlined above, the overall model was fitted simultaneously to PK data sets for all three L-T_4 dose inputs (400, 450, and 600 μg), to capture the nonlinearities appropriately and with fidelity and to provide sufficient data to achieve reasonable statistical precision in model parameter estimates. This rendered the overall FBCS model robust over a wide range of hormone levels. Parameter estimation was accomplished using *SAAMII* by extended weighted least squares optimization with data errors estimated as 10%. Resulting parameter estimates and their variabilities are given in a large table in Ben-Shachar et al. (2012) for normal adults and children and hypothyroid and mildly hyperthyroid adults.[10]

Structural Modeling & Analysis of Biochemical & Cellular Control Systems

The multiloop feedback system (fbcs) and signaling network controlling the dynamical response of the tumor-suppressor protein *p53* to DNA damage has received a great deal of attention from systems biology modelers in the last two decades. This cellular fbcs circumscribes and includes the major dynamic regulatory pathway subsystems operational at cellular levels[11]: an *input* subsystem for detecting exogenous and endogenous stressors; a *signal transduction* subsystem for transforming inputs into *p53* activation signals; several hypothesized *feedback-interaction signals* for self-regulation; and an *output-effect* section for expressing the intended actions (cell-cycle interruption and apoptosis) presumed for this control system. Modeling of this complex biosystem has been quite sophisticated. It entails a number of different structural hypotheses and remains under active investigation. We defer detailed discussion of these efforts to Chapter 15.

We begin this section by analyzing the open- and closed-loop properties of a very simple biochemical reaction mechanism – one that exhibits not so very simple stability characteristics, as well as difficulties with QSSA assumptions. Then we move on to several methodologies practiced in systems biology areas involving metabolism, biofuels, nutrition, and synthetic biology.

Open- vs. Closed-Loop Biochemical Reaction Dynamics

The simple M–M enzyme substrate reaction was developed in Chapter 6 in its classic form, as a closed, autonomous system. Transient responses of this model to initial conditions (ICs) are always stable and constant. This is not the case if the same basic model is *open* and also includes *negative feedback* of reaction product – common in many biochemical pathways. The modified M–M system presented here can exhibit several varieties of complex dynamics. Our primary goal is to expose these interesting properties and, at the same time, further illustrate additional complexities in the modeling process. We utilize and extend several methodologies developed in earlier chapters in model development and analysis here, including magnitude and time scaling – to simplify analysis; multidimensional stability analysis – to expose complex behavior when feedback is introduced; and the QSSA – to simplify the

10 This model is available as an interactive web application simulator, *THYROSIM* (*biocyb1.cs.ucla.edu/thyrosim*).

11 **Figure 15.9** depicts the major functional pathways and biomolecules involved in *p53* regulation.

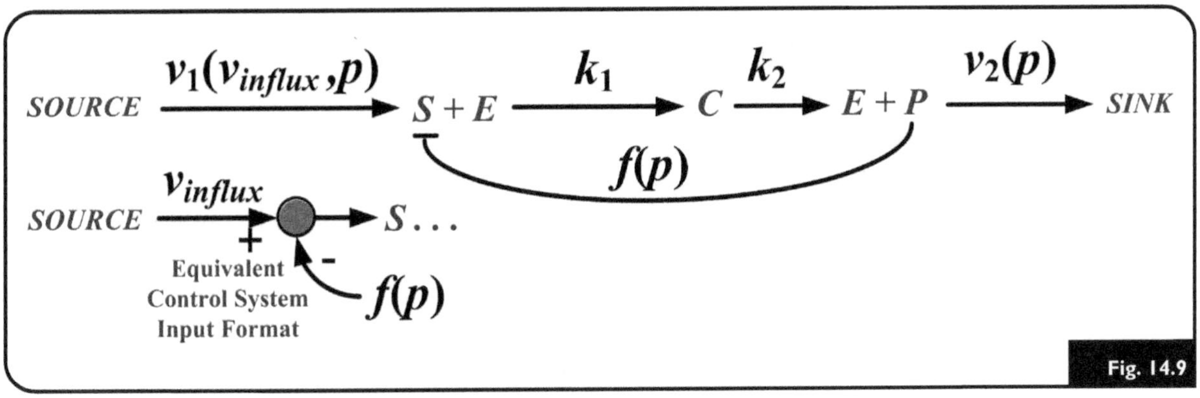

Van Slyke Cullen Model with influx (substrate source) & efflux (product sink). Biochemical reaction with negative feedback notation (TOP). Below it, the same influx and negative feedback interacting via a simple summer in feedback control system notation.

model, which leads to incorrect results when the QSSA approximation is similarly submitted to stability analysis.

In the *Van Slyke–Cullen (VS-C) mechanism* (Van Slyke and Cullen 1914), enzyme E and substrate S bind to form an intermediate enzyme–substrate complex C, at rate k_1, as in the M-M reaction. But in the VS-C model, the binding is, first of all, *irreversible*. The complex then breaks down irreversibly at rate k_2 to form product P. By itself, this unidirectional reaction, like the M–M one, is (so far) stable and constant. However, a negative feedback pathway from product to substrate (*product-feedback*), is also included in the VS-C mechanism, with the feedback signal a function $f(p)$ of instantaneous concentration $p=p(t)$. As might be anticipated, this changes stability properties substantially. To render it more realistic (and interesting), we open it up to an influx (source) v_{influx} of exogenous substrate, and efflux v_2 (sink) of product. Feedback $f(p)$ has the effect of inhibiting (reducing) the substrate source in the reaction. The net influx of S is then equal to the difference: $v_1(v_{influx},p)=v_{influx}-f(p)$. The complete reaction pathway is shown in **Fig. 14.9**.

From the diagram, the governing ODEs for this model are:

$$\begin{aligned} \dot{s} &= v_1(v_{influx},p) - k_1 es \\ \dot{e} &= -k_1 es + k_2 c \\ \dot{c} &= k_1 es - k_2 c \\ \dot{p} &= k_2 c - v_2(p) \end{aligned} \tag{14.1}$$

Applying enzyme conservation constraint $e(t)+c(t)=e_0$ then yields a reduced set of ODEs, with the enzyme ODE eliminated:

$$\begin{aligned} \dot{s} &= v_1(v_{influx},p) - k_1 s(e_0 - c) \\ \dot{c} &= k_1 s(e_0 - c) - k_2 c \\ \dot{p} &= k_2 c - v_2(p) \end{aligned} \tag{14.2}$$

We can proceed to numerically study the dynamical properties of this model directly, as we did in Chapters 6 and 9. Alternatively, we can simplify the equations further by first *magnitude-scaling* the variables, reducing the number of parameters by one – as illustrated in Chapter 3 – and then reducing them further by *time-scaling*, i.e. $\bar{s} \equiv (k_1/k_2)s$, $\bar{c} \equiv (k_1/k_2)c$, $\bar{p} \equiv (k_1/k_2)p$ and then $\bar{t} \equiv k_2 t$, so $d\bar{t} = k_2 dt$. Substituting these new variables into (14.2) gives ODEs easier to work with, without k_1 or k_2, and only initial condition $\bar{e}_0$ as the single "parameter" upon which these equations depend (Flach and Schnell 2006). The simplified core equations are:

$$\begin{aligned} \dot{\bar{s}} &= \bar{v}_1(\bar{p}) - \bar{s}(\bar{e}_0 - \bar{c}) \\ \dot{\bar{c}} &= \bar{s}(\bar{e}_0 - \bar{c}) - \bar{c} \\ \dot{\bar{p}} &= \bar{c} - \bar{v}_2(\bar{p}) \end{aligned} \tag{14.3}$$

The scaled influx, efflux and IC are: $\bar{v}_1(v_{influx}, \bar{p}) \equiv k_1 v_1(v_{influx}, p)/k_2^2$, $\bar{v}_2(\bar{p}) \equiv k_1 v_2(p)/k_2^2$ and $\bar{e}_0 \equiv k_1 e_0/k_2$. Although e has been eliminated from these equations, it will be needed again in stability analysis, so we also scale e: $\bar{e} \equiv k_1 e/k_2$.

Using the simplified equations, the goal here is to compare the qualitative dynamics of this model in *open-loop* versus *closed-loop* configurations. We do this by computing the behavior of these two models in phase space using the methods of Chapter 9. For the closed-loop model, we also study the qualitative dynamics of ODEs (14.3) further simplified using QSSA conditions. The results are anomalous and underscore the need to use the QSSA approximation with great care (Flach and Schnell 2006).

Open-Loop Model

Nullclines are helpful for establishing stability properties. These are the curvilinear functions of time (surfaces) for which the time derivatives in each dimension are zero:

$$\begin{aligned} \dot{\bar{s}} \equiv 0 &\Rightarrow \bar{s} = \frac{\bar{v}_1(\bar{p})}{\bar{e}_0 - \bar{c}} \\ \dot{\bar{c}} \equiv 0 &\Rightarrow \bar{c} = \frac{\bar{e}_0 \bar{s}}{\bar{s} + 1} \\ \dot{\bar{p}} \equiv 0 &\Rightarrow \bar{c} = \bar{v}_2(\bar{p}) \end{aligned} \tag{14.4}$$

Intersection of these parameter-dependent null surfaces – the simultaneous solutions of algebraic equations (14.4) – gives the steady state solutions.

For open-loop analysis, no feedback means $f(p) \equiv 0$ and the nullcline equations simplify. Product is dependent on substrate and complex, but not vice versa. This means $\bar{p}$ is decoupled (free) from the nullclines for s and c. The simultaneous steady state solutions of (14.4) for s and c are then: $\bar{s}_e = \frac{\bar{v}_1}{\bar{e}_0 - \bar{v}_1}$ and $\bar{c}_e = \bar{v}_1$. Physically, $\bar{s}_e$ also establishes a boundary for the solutions, which cannot be negative: $0 < \bar{v}_1 < \bar{e}_0$. Our interest is stability of this equilibrium point for the open-loop model.

We saw in Chapter 9 that local stability of the dynamical system in the vicinity of the steady state solutions can be established by the direction of the phase space nullclines in the vicinity of equilibrium (steady state) points $\boldsymbol{x_e}$. Nullcline directions pointing toward $\boldsymbol{x_e}$ imply stability, and vice versa. We therefore linearize Eqs. (14.3) ($\dot{\boldsymbol{x}} = \boldsymbol{f}(\boldsymbol{x})$) about the steady state established above and examine the eigenvalues of the resulting linearized system matrix – the Jacobian $\partial \boldsymbol{f}/\partial \boldsymbol{x}|_{\boldsymbol{x}_e}$ at $\boldsymbol{x_e}$. Eigenvalues of the linearized equations completely characterize local stability.[12] Local stability also can be established by finding the slope of the nullclines in the vicinity of $\boldsymbol{x_e}$, as illustrated in Chapter 9 as well, but finding eigenvalues can be easier. To simplify derivative computations, we use the original conservation relationship $e = e_0 - c$ and the scaled ones $\overline{e} = \overline{e}_0 - \overline{c}$, plus the scaled one at steady state: $\overline{e}_{eq} = \overline{e}_0 - \overline{c}_{eq} = \overline{e}_0 - \overline{v}_{1eq}$ (subscript *eq* for equilibrium value). This gives:

$$\partial \boldsymbol{f}/\partial \boldsymbol{x}\Big|_{\boldsymbol{x}_e} = \left[\begin{array}{cc} -\overline{e} & \overline{s} \\ \overline{e} & -(\overline{s}+1) \end{array}\right]\Bigg|_{\boldsymbol{x}_e} = \left[\begin{array}{cc} -\overline{e}_{eq} & v_{1eq}/\overline{e}_{eq} \\ \overline{e}_{eq} & -\overline{e}_0/\overline{e}_{eq} \end{array}\right] \tag{14.5}$$

The eigenvalues of this matrix determine local stability. These are the roots of the characteristic equation $\det[s\boldsymbol{I} - \partial \boldsymbol{f}/\partial \boldsymbol{x}_{eq}]$ for this matrix, i.e. the roots of:

$$\lambda^2 + \left(\frac{\overline{e}_0}{\overline{e}_{eq}} + \overline{e}_{eq}\right)\lambda + \overline{e}_{eq} = 0 \tag{14.6}$$

It is easy to show that the roots of Eq. (14.14.6) are real and negative (*Exercise* E14.2).

We choose a constant input/influx: $v_1 = v_{influx} = 1$ and constant efflux: $v_2 = 1$. Together with physical constraint $0 < \overline{v}_1 < \overline{e}_0$, these guarantee satisfaction of mass conservation (net material flow equals zero in steady state), fixed roots of Eq. (14.14.6), and time-invariant phase plane solutions. The negative real roots of Eq. (14.6) guarantee that this open-loop system is stable and constant for any initial conditions that satisfy the constraints. We demonstrate this numerically via *Matlab* solutions.[13]

A plot of the null surface for the complex: $\overline{c} = \frac{\overline{e}_0 \overline{s}}{\overline{s}+1}$ (where $\dot{\overline{c}} = 0$) is shown in **Fig. 14.10** as a curvilinear mesh-grid in three dimensions – c versus s and p in the bottom plane. Product simply accumulates in response to the reaction between c and s, as it does in the simpler M-M reaction. Trajectories are shown for six different IC values, all of which find their way to a stable node (dependent on $\overline{e}_0$) after initial "fast" transients. For all ICs, trajectory flow is initially and directly toward the null surface, then remains close to it, attracted by a unique trajectory on the surface, the *"slow" invariant-manifold* (*nonlinear mode*) solution. Substrate continually diminishes while product accumulates, before arriving at a stable *chemical equilibrium* node as $t \to \infty$, i.e. the constant steady state at $\overline{s}_{eq} = \frac{1}{\overline{e}_0 - 1}$, $\overline{c}_{eq} = 1$ and $\overline{p}_{eq} = \overline{s}_0$. Note that all trajectories move toward this steady state, along the complex-null-surface near the end of their travel.

Remark: The simpler open-loop M–M reaction mechanism (without feedback or influx) has very similar dynamics. After the initial "fast" transient M–M solution (see Chapter 6),

12 Only state variable equations for s and c need be analyzed, because they do not depend on p.

13 *Matlab* code for this problem is located on the book website.

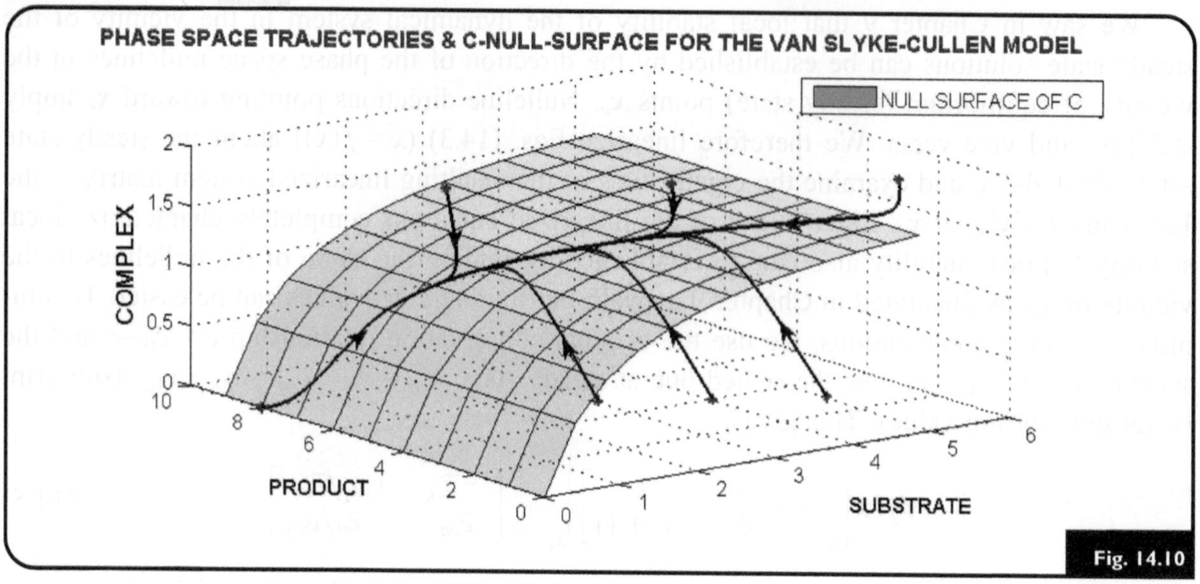

Fig. 14.10

Constant and stable solutions for seven ICs for the open-loop Van Slyke-Cullen reaction open to influx and efflux. All trajectories go quickly to a single steady state via a "slow" dynamics manifold on the C null surface.

trajectories move toward and stay close to the slow invariant-manifold of the "slow" solution at later times, all confined to a region bounded between the null surfaces for complex and substrate (not shown).

Closed-Loop Model

For the closed-loop feedback system, $f(p) \neq 0$. We again choose parameter values that permit qualitative stability analysis of Eq. (14.3) in a time-invariant phase space. As above, constant efflux: $v_2 = 1$ is an easy choice. But $v_1(v_{influx}, p) = v_{influx} - f(p)$ is more of a challenge, because v_1 cannot become negative. Also, we know that nonlinear models can have very different qualitative dynamics with different inputs. We analyze the closed-loop system here for two different inputs: $v_1(v_{influx}, p) = v_{influx} - f(p) = 2 - p$ and $2 - 3.1p$, as in (Flach and Schnell 2006). Both satisfy mass conservation and flow considerations.

The c-nullcline for the complex C remains the same for this model. The p-nullcline surface is simply $p = v_2 = 1$ and the s-nullcline is a little more complex, because $v_1 = v_{influx} - f(p)$ in Eq. (14.4). We only need the c-nullcline, however, to complete our analysis. The steady states $\boldsymbol{x_e}$ for this model are: $\overline{s}_{eq} = 1/(\overline{e}_0 - 1)$, $\overline{c}_{eq} = 1$ and $\overline{p}_{eq} = \overline{s}_0$.

As above, we linearize Eq. (14.3) ($\dot{\boldsymbol{x}} = \boldsymbol{f}(\boldsymbol{x})$), this time for all three state variables, and examine the eigenvalues of the resulting *linearized* system matrix – the Jacobian $\partial \boldsymbol{f}/\partial \mathbf{x}|_{\boldsymbol{x_e}}$ at $\boldsymbol{x_e}$. Eigenvalues of the linearized equations completely characterize *local* stability. We again use the conservation relationships $e = e_0 - c$, $\overline{e} = \overline{e}_0 - \overline{c}$ and $\overline{e}_{eq} = \overline{e}_0 - \overline{c}_{eq} = \overline{e}_0 - \overline{v}_{1eq}(\overline{p})$. This gives:

$$\partial \boldsymbol{f}/\partial \mathbf{x}|_{\boldsymbol{x_e}} = \begin{bmatrix} -\overline{e}_{eq} & v_{1eq}/\overline{e}_{eq} & \partial \overline{v}_1(p)/\partial p \\ \overline{e}_{eq} & -\overline{e}_0/\overline{e}_{eq} & 0 \\ 0 & 1 & 0 \end{bmatrix} \tag{14.7}$$

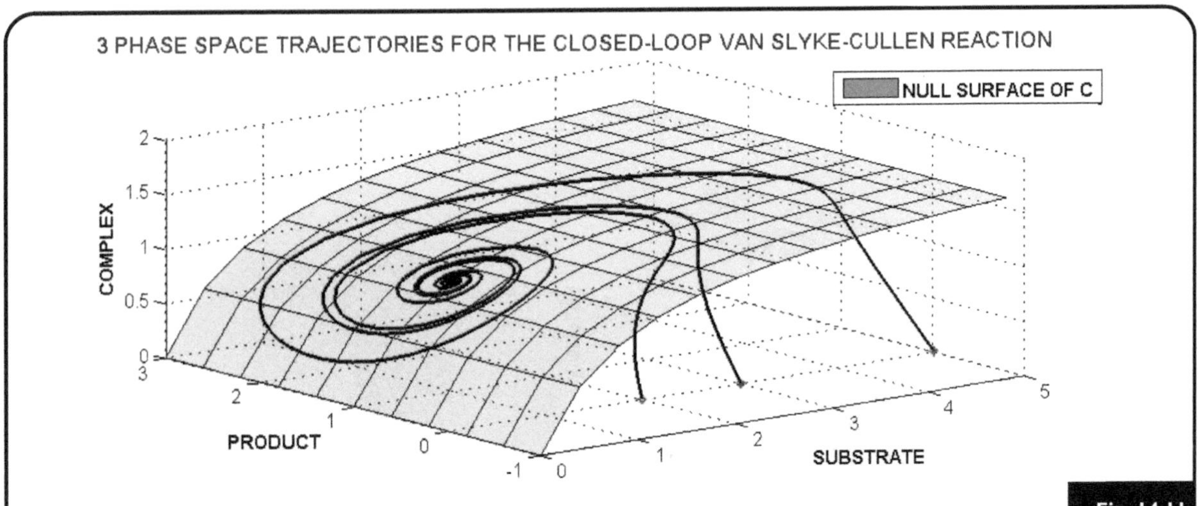

Fig. 14.11

Asymptotically stable oscillatory solutions (blue) of the closed-loop Van Slyke–Cullen reaction, open to influx and efflux, for three different ICs (light blue). Input function is: $v_1 = 2 - p$, i.e. with unity product feedback.

For $\overline{v}_1(\overline{p}) = A - B\overline{p}$, the form of our two influxes, $\partial\overline{v}_1(p)/\partial p = -B$. The eigenvalues of this matrix are thus the roots of:

$$\lambda^3 + \left(\frac{\overline{e}_0}{\overline{e}_{eq}} + \overline{e}_{eq}\right)\lambda^2 + \overline{e}_{eq}\lambda + \overline{e}_{eq}B = 0 \tag{14.8}$$

With $\overline{e}_0$, $\overline{e}_{eq}$, and $B > 0$, all the coefficients of Eq. (14.8) are positive, so we have no positive real roots, and with $n = 3$ we must therefore have at least one negative real root. Complex roots are possible and it can be shown that one of these can have positive real parts (*Exercise* E14.3). This means a local unstable spiral in a plane about the steady state is a possible solution. Other possible solutions include stable spirals and limit cycling behavior. We illustrate two numerical simulations together illustrating two of these possibilities.

Fig. 14.11 is an example numerical phase space solution for the closed-loop model with the first of two inputs: $\overline{v}_1(\overline{p}) = 2 - \overline{p}$, with efflux $v_2 = 1$ and $\overline{e}_0 = 1$ (as before). It includes the same curvilinear mesh-grid c-null-surface in three dimensions as for the open-loop model in **Fig. 14.10**: c versus s and product p in the bottom plane. Trajectories are shown for three different IC values, all of which spiral in to the stable node after initial "fast" transients directed toward the null surface. And, again, trajectories hug the surface, attracted by a unique trajectory, the *slow invariant-manifold* (*nonlinear mode*) solution. Substrate continually diminishes while product accumulates, before arriving at the asymptotically stable node as $t \to \infty$, i.e. the steady state at $\overline{s}_{eq} = \frac{1}{\overline{e}_0 - 1}$, $\overline{c}_{eq} = 1$ and $\overline{p}_{eq} = 1$.

Change the Feedback Parameters: The LHS of **Fig. 14.12** illustrates solution of the same problem, but with a slightly different $(-3.1p)$ feedback function, $\overline{v}_1(\overline{p}) = 2 - 3.1\overline{p}$. In this case, the *qualitative response is completely different than that with unity feedback*: the steady state is now an *unstable limit cycle*, shown as forming three different limit-cycling states from three different ICs.

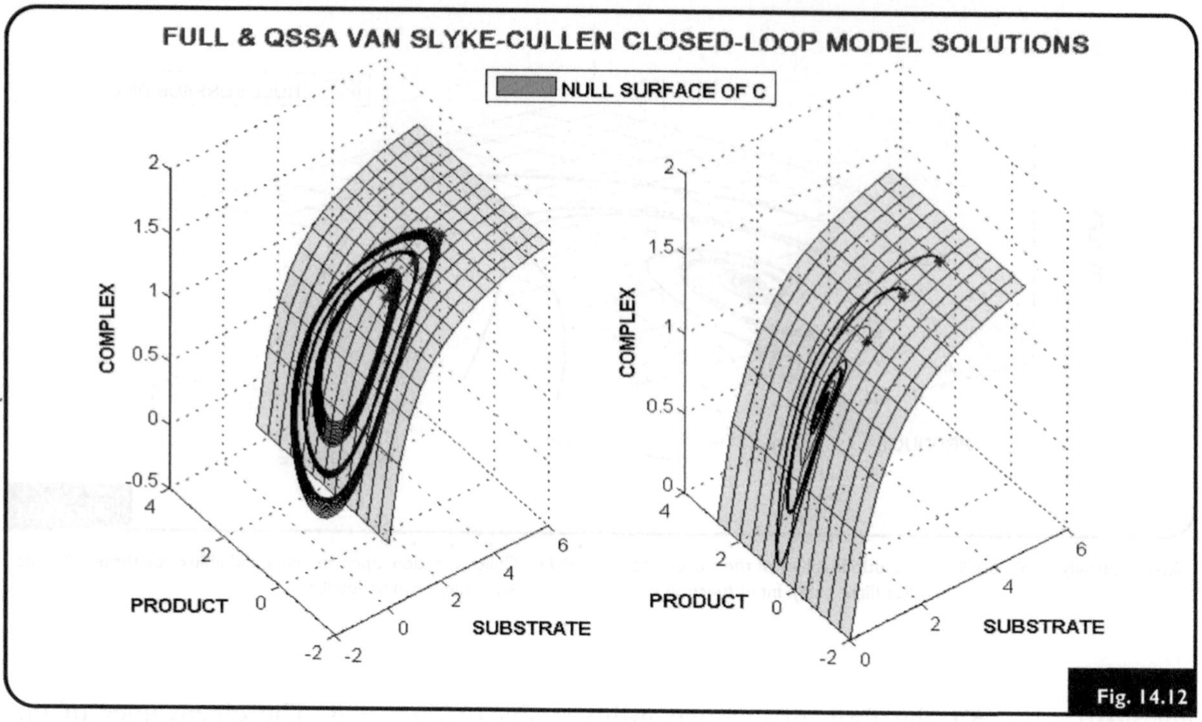

Fig. 14.12

Full model (LHS) vs. QSSA model (RHS) solutions to the Van Slyke–Cullen reaction with same parameters, constant input and efflux and three different ICs. Input function is: $v_1 = 2 - 3.1p$, i.e. with feedback $-3.1p$. Solutions are qualitatively as well as quantitatively different. The QSSA is a bad approximation here.

Approximate QSSA Model Solution is Qualitatively Different (Wrong)

It is common practice to invoke the quasi-steady state assumption (QSSA) in enzyme-substrate problems like this one. This effectively reduces the dimensionality of the model in Eq. (14.3) by one, as follows. Under the assumption that complex reaches approximate equilibrium instantaneously, $\dot{\overline{c}} \cong 0$ yields the approximation: $\overline{c} \cong \overline{e}_0 s/(s+1)$. Substituting this into the remaining equations then gives the QSSA-reduced model:

$$\begin{aligned} \dot{\overline{s}} &\cong \overline{v}_1(\overline{p}) - \frac{\overline{e}_0 s}{s+1} \\ \dot{\overline{p}} &\cong \frac{\overline{e}_0 s}{s+1} - \overline{v}_2(\overline{p}) \end{aligned} \tag{14.9}$$

We again seek the eigenvalues of the linearized model, in this case of Eq. (14.9). The Jacobian is:

$$\partial f/\partial \mathbf{x}|_{x_e} = \begin{bmatrix} -a & \partial \overline{v}_{1eq}(\overline{p})/\partial \overline{p} \\ a & -\partial \overline{v}_{2eq}(\overline{p})/\partial \overline{p} \end{bmatrix} \tag{14.10}$$

where $a \equiv (\overline{e}_0 - \overline{v}_2)^2/\overline{e}_0 > 0$. The characteristic equation for this matrix is:

$$\lambda^2 + (a + \partial \overline{v}_{2eq}(\overline{p})/\partial \overline{p})\lambda + a[\partial \overline{v}_{2eq}(\overline{p})/\partial \overline{p} - \partial \overline{v}_{1eq}(\overline{p})/\partial \overline{p}] = 0 \tag{14.11}$$

With $\overline{v}_{2eq}(\overline{p}) = 1$ and $\overline{v}_{1eq}(\overline{p}) = 2 - 3.1\overline{p}$, this reduces to: $\lambda^2 + a\lambda + 3.1a = 0$, which has solutions: $\lambda_{1,2} = \left[-a \pm \sqrt{a(a - 12.4)}\right]/2$. Then $\text{Re}(\lambda) = -a/2 < 0$, so the steady state is always stable. If $a < 12.4$, the eigenvalues are complex and the trajectories are therefore *asymptotically stable,* spiraling in to the fixed point at $(1, 1, p_{eq})$, as illustrated on the RHS of **Fig. 14.12**. This is quite different than the exact and *unstable limit cycling* solutions for the same parameter values and ICs shown on the LHS of the figure.

Remark: On the one hand, this extensive analysis illustrates how explicit feedback can impart complex dynamical behavior in a simple nonlinear dynamical system. It also illustrates (as in Chapters 6 and 7) why approximations like the QSSA need careful evaluation, because they can lead to results that are qualitatively as well as quantitatively incorrect.

Transient and Steady-State Biomolecular Network Modeling

Modeling Complete Dynamics

Fig. 14.13 illustrates overall signaling and core process dynamics in cellular biomolecular pathways. Interactions among several hypothetical core biomolecules encircled in the middle of the figure are represented by inhibitory or activating "arrows" connecting the various

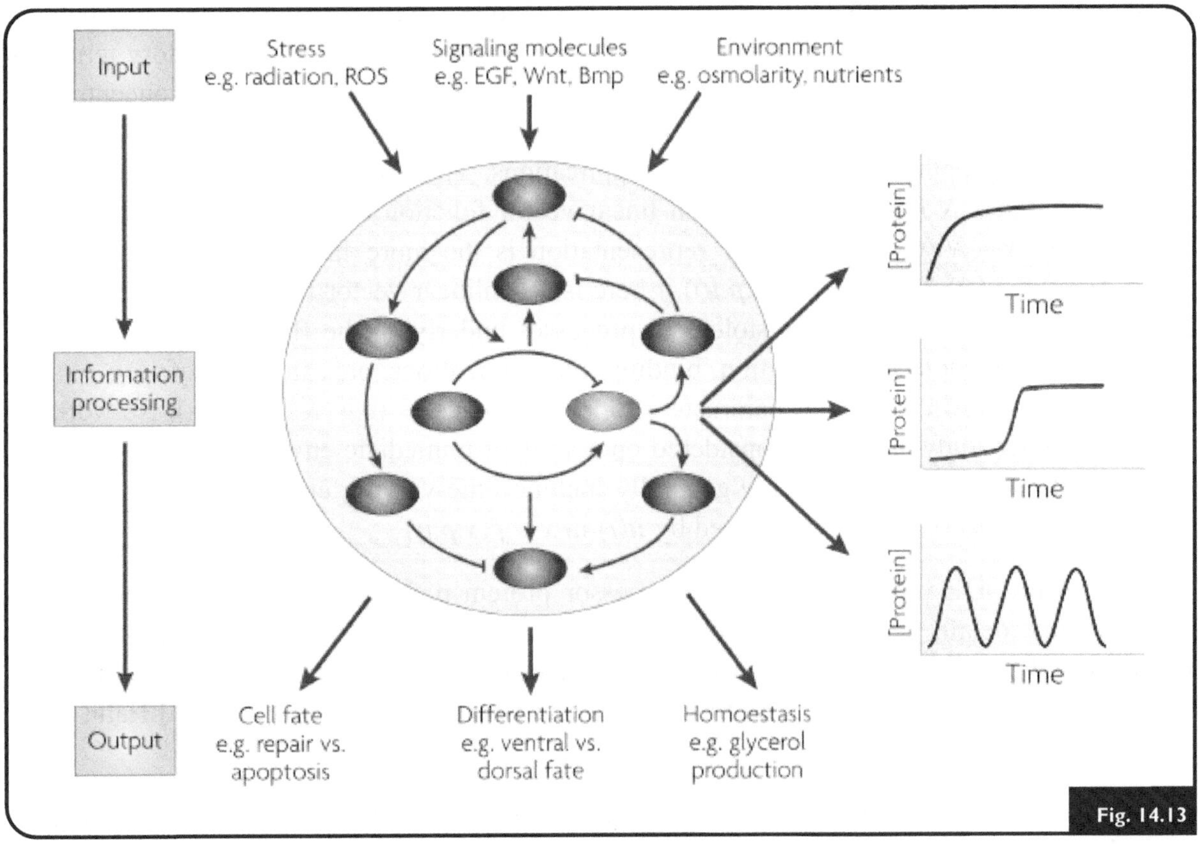

Fig. 14.13

Overall signaling and core process dynamics in cellular biomolecular pathways. (Figure reprinted and adapted by permission from Macmillan Publishers Ltd: *Nature Reviews Cancer* – **Fig. 1** in: The ups and downs of p53: understanding protein dynamics in single cells, copyright 2009).

components of the pathway. In metabolic networks, for example, core interactions typically govern *substance flow* via multienzyme multisubstrate reaction kinetics, often operating in quasi-steady state. ODEs are typically used to represent these coupled reactions. The dynamics of the system would be incomplete if modeling stopped at the borders of the circle. Missing from this conventional representation are the *in vivo* regulatory adjustments made transiently from *transient inputs*, entering via **signal transduction pathways** external to the core network. The focus in these pathways is signal flow, typically regulated by protein phosphorylations and dephosphorylations, generating activated or inactivated molecular forms for signaling. These are illustrated in the upper part of the figure.

Input signaling pathways entering from endogenous and exogenous sources transiently control and influence cell decisions, manifested in network actions – or *output* – on the bottom of **Fig. 14.13**. Importantly, they generally do this differently, and with different outcomes, as illustrated for responses elicited by the *purple* biomolecule in the circle. Three dynamically different outcomes are shown on the RHS of **Fig. 14.13** – illustrating different temporal protein expression outputs, due to three different signal transduction pathway inputs. To capture *all* of these essential dynamics, complete ODE models must include transient input and signal transduction pathways.

Different ODE Model Forms

As shown in Chapter 7, ODEs for systems biology models are normally written as a balance of constitutive and signal-regulated production rate, consumption rate and regulatory terms for the reactions involved, all together in terms of a state vector of n time-varying concentrations or masses $\boldsymbol{x}(t)$ of the interacting species. These models can be written in equivalent linear or nonlinear forms, both of which have their applications. The linear representation is the *stoichiometric matrix* N form: $\dot{\boldsymbol{x}}(t) = N\boldsymbol{v}(t)$, a linear vector function of $m \geq n$ time-varying fluxes $\boldsymbol{v}(t)$ (abbrev: $\dot{\boldsymbol{x}} = N\boldsymbol{v}$). The nonlinear representation is the more familiar state vector ODE: $\dot{\boldsymbol{x}}(t) = \boldsymbol{f}(\boldsymbol{x}(t),\boldsymbol{p}, \boldsymbol{u}(t))$ (abrev: $\dot{\boldsymbol{x}} = \boldsymbol{f}(\boldsymbol{x},\boldsymbol{p},\boldsymbol{u})$). $\boldsymbol{f}$ here is a nonlinear vector function representing the detailed physiocochemical or physiological processes underlying the fluxes $\boldsymbol{v}(t)$, for example, enzyme kinetics, molecular activation, binding, polymerizations, etc., in terms of $\boldsymbol{x}(t)$; $\boldsymbol{p}$ is the vector of P unknown (constant) parameters for these processes (e.g. K_ms, v_{max}s). Networks or pathways under study *in situ* are considered open to their immediate environment and therefore the m reaction fluxes $\boldsymbol{v}(t)$ in $\dot{\boldsymbol{x}} = N\boldsymbol{v}$ generally include time-varying endogenous or exogenous (experimental) influx terms, represented by $\boldsymbol{u}(t)$ in $\dot{\boldsymbol{x}} = \boldsymbol{f}(\boldsymbol{x},\boldsymbol{p},\boldsymbol{u})$.

Complete ODE models for tumor suppressor protein p53 signaling dynamics are developed in some detail, in $\dot{\boldsymbol{x}} = \boldsymbol{f}(\boldsymbol{x},\boldsymbol{p},\boldsymbol{u})$ form, in Chapters 15 and 17, which also include several other examples of cellular signal transduction systems (e.g. NFkB and IL-6), modeled and analyzed in transient and steady states (Yue et al. 2006; Cheong et al. 2008; Chu and Hahn 2008; Kearns and Hoffmann 2009).

Practical circumstances do exist for circumventing – but not ignoring – transient regulatory phenomena in modeling, for example, for modeling metabolism in *steady state*, a topic receiving a great deal of attention in the systems biology community.

Metabolism and Steady State Flux Balance Analysis (FBA)

Metabolism is often treated as a steady state phenomena because it behaves quite homeostatically most of the time *in vivo*, or because it is maintained constant under artificially engineered conditions, as in synthetic biology. Metabolic steady state solutions of $\dot{\boldsymbol{x}} = N\boldsymbol{v}$ are the algebraic equations $\boldsymbol{N}\boldsymbol{v} = \boldsymbol{0}$, the basis for widely used steady state FBA methodology. This methodology – augmented by conservation-based and other constraints to overcome identifiability (parameter ambiguity) issues – has yielded numerous insights about regulation of cellular metabolism. We introduced this subject with a simple plant metabolism model in the *Stoichiometric Matrix* section of Chapter 7, **Fig. 7.6**. We recall that, in a typical network, the rows of stoichiometric matrix N correspond to n metabolites x_i, the columns to m reactions R_j with m metabolic fluxes, each denoted $v_i \geq 0$ (possibly augmented by exogenous inputs or influxes).

A common application for FBA is to optimize some property or parameter of a metabolic network, for example, to discover which metabolic fluxes $v_i \geq 0$ maximize the *growth rate* (*biomass production*) of an organism or to maximize or minimize another aspect of flux, for example, minimize nutrient uptake when the cell is performing its metabolic functions. For the latter, it is important to know the nutrients available to the organism, along with any constraints on typically limited nutrient sources. Another example is minimizing cellular ATP production to determine conditions of optimal metabolic energy efficiency. There are many, and the growing field of synthetic biology is evolving with numerous such applications.

Optimizing Steady-State Fluxes

Linear programming (LP) is usually used to find optimal solutions, because $\boldsymbol{N}\boldsymbol{v} = \boldsymbol{0}$ is linear. A linear optimization criterion L can be written: $L(\boldsymbol{v}) \equiv \boldsymbol{c}^T\boldsymbol{v}$, where $\boldsymbol{c}^T = \begin{bmatrix} c_1 & \cdots & c_m \end{bmatrix}$. If a particular steady-state flux is to be maximized, the optimization problem can be written as:

$$\max \boldsymbol{c}^T\boldsymbol{v} \text{ subject to constraints } \boldsymbol{N}\boldsymbol{v} = \boldsymbol{0} \text{ and } \boldsymbol{v}_{\max} \geq \boldsymbol{v} \geq \boldsymbol{0} \tag{14.12}$$

For example, to maximize the two steady state flux pathways v_1 and v_m, we could choose $\boldsymbol{c}^T = \begin{bmatrix} 1 & 0 & \cdots & 1 \end{bmatrix}$.

The following simple example, adapted from (Lee et al. 2006), illustrates biomass production constrained by no flux exceeding 10. The couplings among the state variables – represented by $\boldsymbol{N}\boldsymbol{v} = \boldsymbol{0}$ in steady state – are the bounded solution space "constraints" more subtly governing the optimization solution.

Remark: The steady state fluxes constrained by $\boldsymbol{N}\boldsymbol{v} = \boldsymbol{0}$ have a notable geometric interpretation. They form a hyperplane in m-dimensional flux space. With individual fluxes constrained by upper and lower bounds, as in Eq. (1.1), the resulting region of *allowed flux distributions* is a convex polyhedron in this flux space. Then, by linearity of the transformation, any linear combination of individual flux distributions is an allowed flux distribution. FBA then maximizes a linear function within this polyhedron, with an optimal solution that lies somewhere on this allowable surface.

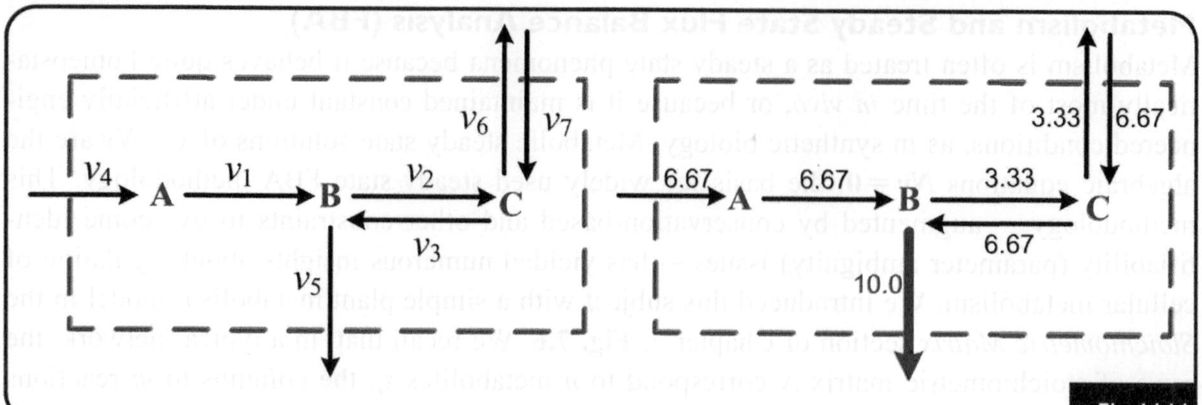

Fig. 14.14

FBA model (LHS) and LP-solution (RHS) that maximizes flux v_5 to its maximum possible value = 10, constrained by $\mathbf{Nv} = \mathbf{0}$ and $10 \geq v_i \geq 0$ for all i.

Example 14.2: ***Reaction Network and Optimized Flux Balance Analysis***

The reaction network shown on the LHS of **Fig. 14.14** has seven reactions, three state variables, and seven fluxes (example from (Lee et al. 2006)).

The stoichiometric matrix is 3×7 and the steady state mass balance equation $\boldsymbol{Nv} = \mathbf{0}$ that linearly constrains the fluxes is given by:

$$\begin{bmatrix} -1 & 0 & 0 & 1 & 0 & 0 & 0 \\ 1 & -1 & 1 & 0 & -1 & 0 & 0 \\ 0 & 1 & -1 & 0 & 0 & -1 & 1 \end{bmatrix} \begin{bmatrix} v_1 \\ v_2 \\ v_3 \\ v_4 \\ v_5 \\ v_6 \\ v_7 \end{bmatrix} = \begin{bmatrix} 0 \\ 0 \\ 0 \end{bmatrix}$$

Fluxes are additionally constrained by the practical flux bounds $0 \leq v_i \leq 10$. The optimization goal is to maximize biomass efflux from the network via v_5 under these conditions. The optimization problem is readily solved using widely available computational tools in *Matlab* and other programs, in particular *Matlab* toolbox *COBRA* (Schellenberger et al. 2011). The optimized flux distribution in steady state is shown on the RHS of **Fig. 14.14**, with $v_{5\max} = 10$.

Experiment Design for FBA by Cutset Analysis

Cutset analysis is a directed graph theoretic method for facilitating design of model-based kinetic experiments for quantifying particular model parameters (Feng and Distefano III 1991). The methodology – developed in Chapter 16 – is particularly well-suited for experiment design in FBA studies, in particular to establish experimental input perturbations needed for quantifying designated fluxes in metabolic pathways, possibly for the purpose of redesigning the pathway – as in synthetic biology (see below). Cutsets are typically drawn and redrawn until experiment design goals are achieved. The procedure effectively isolates parameters (edges) of interest by iteratively grouping different state variables (nodes) into different submodel configurations. Explicit solutions are given in terms of measurable variables and independent inputs, as illustrated by Nguyen et al. (2003), who derived (and did) three input perturbation studies for quantifying thyroid hormone production rates in rat liver and intestine *in vivo*. Variants of the methodology also have been applied to other problems in metabolism (for example by Alves and Savageau 2000), in network thermodynamics (Fidelman and Mikulecky 1986) and in model reduction problems (Jin-Zhi and Chang-Yi 2008), the subject of Chapter 17.

Elementary Mode and Extreme Pathway Analysis

The common objective of these two metabolic network analysis methods is to extract *functionally independent units* of a whole metabolic network, each accomplishing it somewhat differently. Their motivation is that metabolic networks are highly interconnected between exogenous inputs and outputs (influxes and effluxes) and the purpose of all connections is not entirely clear. Some pathways appear to be redundant, others may have no apparent purpose. Given that the network fulfills a biological purpose in the cell, it is important in analysis to delineate the key pathways from input to output needed to satisfy this goal. This is a variation on the question of what is the minimum input−output equivalent model. On the other hand, given that some pathways may appear to be redundant, or have no purpose, a larger question might address what all of the metabolic capabilities of the network and its individual pathways might be – in principle. The set of linear pathways in a particular metabolic network does not always capture the full range of possible behavior of the network and a judgment of "purposeless" or redundancy for some pathways may be premature or incorrect.

The set of **flux modes** (FM) of a network comprehensively describes all possible metabolic routes between input and output that are both stoichiometrically and thermodynamically feasible. The proportions of fluxes in the steady state flux distribution are fixed in FM, and FM establishes the direct routes that possibly lead from one external metabolite to another external metabolite. **Elementary flux modes** (EFMs) are the *minimum* number of reactions/flux paths needed to describe all such metabolic routes, meaning it cannot be decomposed into fewer fluxes. They allow for reversible reactions by breaking them up into forward and backward components. Every network has a unique set of elementary flux modes (Schuster et al. 1999).

Mathematically, elementary flux modes are unlike the modes associated with basis vectors in linear algebra. Instead, they form a basis for *N*, the linear transformation stoichiometric matrix, on a *hyperplane* in *m*-dimensional **flux space**. Elementary mode vectors $\boldsymbol{e}^i$ are uniquely determined up to a multiplicative nonnegative constant, which means any real flux distribution vector $\boldsymbol{v}$ can be represented as a superposition of these elementary flux mode vectors with coefficients $\lambda_i \geq 0$, that is, $\boldsymbol{v} = \sum_{i=1}^{m} \lambda^i \boldsymbol{e}^i$, called a **convex combination** of elementary flux modes.

The *basic algorithm* for elementary mode analysis involves analyzing all reactions and their directionality in the network, to establish the unique set of elementary mode pathways.

The premise of the methodology is that cells can use any combination of these elementary mode pathways. To satisfy particular goals, for example, maximizing production of a certain metabolite, only those that connect the desired input with the desired output can be retained in a new design. In addition, other pathways might be deleted to increase the efficiency of network operation.

Extreme pathways are another unique and minimal set of *convex basis vectors* (pathways) of the stoichiometric matrix *N*. They also completely characterize the steady-state capabilities of the metabolic network. They are analogous but not equivalent to elementary flux modes. In extreme pathways, all reactions are constrained by flux direction, whereas elementary flux modes allow for reversible reactions. In this sense, the set of extreme pathways is a subset of the set of elementary flux modes.

Both elementary flux modes and extreme pathways are powerful ways for exploring the range of metabolic pathways in a network – motivated greatly by problems in synthetic biology and biofuel production – because both are capable of extracting minimal sets of the most relevant pathways. They are used, for example, to screen sets of enzymes for production of a particular product or to screen gene-knockouts – to allow for metabolite redirection, or to detect redundant and nonredundant pathways. In pharmacology, they are used effectively to identify drug targets and, in all application areas, help detail the connectivity of a network from annotated genome sequences.

Elementary flux modes are typically established computationally by convex analysis and other linear algebraic transformations of the linear network structure, often involving *singular value decomposition* of the stoichiometric matrix *N* (Palsson 2006). A number of computational tools have been developed for both of these to accomplish these specialized tasks, including *COBRA* (Schellenberger et al. 2011) already mentioned and others. Following is an example – adapted from (Papin et al. 2003) – illustrating both elementary mode and extreme pathway analysis for a simple reaction network.

Example 14.3: ***Elementary Modes and Extreme Pathways in a Reaction Network***

Three extreme pathways (ExPa) and four elementary modes (ElMo) are shown for the simple three state variable, seven reaction flux network illustrated in **Fig. 14.15**. Both

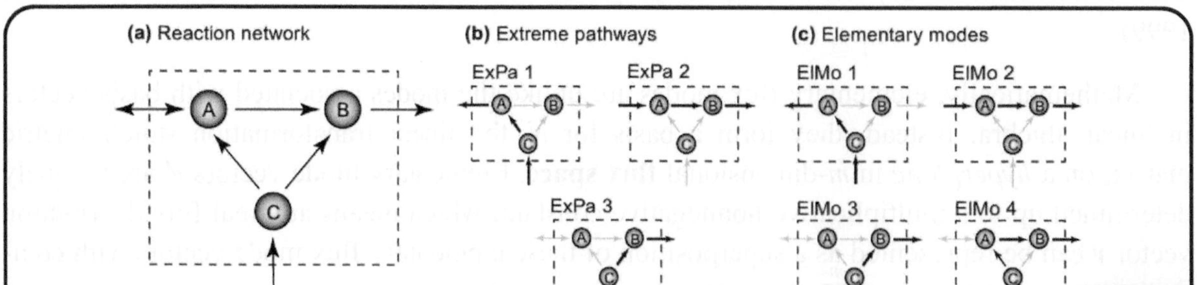

Fig. 14.15

Extreme pathways and elementary modes of a simple reaction network
(Adapted from Papin JA, Price ND, Wiback SJ, Fell DA, Palsson BO. 2003. Metabolic pathways in the post-genome era. *Trends in Biochemical Sciences* **28**:250–258, 2003. Reproduced by permission of Oxford University Press.)

sets satisfy the two shared requirements: each is nondecomposable and each is unique. The extreme pathways are also systemically independent, and ElMo 1 (ExPa 1) and ElMo 2 (ExPa 2) can be combined to give ElMo 4. ElMo 3 and ElMo 4 illustrate two possible modes remaining if reversible exchange of A is prevented, for example, by mutation or inhibition. These two elementary modes are extreme pathways for the reaction network without exchange of A. This analysis is simple because the network is simple. More complex chemical reactions represent exponentially increasing complexity and are significantly more difficult to analyze.

Metabolomics

The **metabolome** is the entire collection of *metabolites* in a cell, tissue, organ or organism comprising the end products (outputs) of cellular metabolic processes. **Metabolomics** is the systematic study of the unique metabolite profiles resulting from these cellular metabolic processes. It is supported by a growing set of genome scale high-throughput data acquisition bioinformatics tools, providing dynamic profiles of metabolites, in conjunction with FBA modeling. In this context, FBA is applied to gene knockout studies and drug target discovery (Lee et al. 2006). For example, the importance of particular enzymes is evaluated by comparing the flux distribution of a mutant metabolic pathway, with an enzyme removed, with that of the wild type. A major software tool in this domain of discovery is a comprehensive FBA *Matlab* toolbox with an *SBML* interface for *COnstraint-Based Reconstruction and Analysis* (*COBRA*) of metabolic networks. The *COBRA* toolbox includes functionality for simulation, optimization, analysis, and prediction of a variety of metabolic phenotypes using FBA and genome-scale models (Schellenberger et al. 2011).

Metabolomics in Synthetic and Mammalian Biology

Metabolic network models analyzed using FBA and bioinformatics tools have been broadly developed and applied quite successfully in synthetic biology studies with simple cellular organisms (Kitney and Freemont 2012). These models and methods have not been exploited as much for mammals – where the biology is much more complex, where modeling that includes analysis of the transient state is at least as important, and modeling goals are distinctly different, e.g. understanding complex relationships among nutrition, carbohydrate metabolism and cancer.

In plants and bacteria, reaction fluxes in a synthetic pathway are usually analyzed to suggest a genetic modification of enzymes to increase production of a compound or produce a new compound. This can be done relatively easily because simple organisms make most organic material from simple nutrients, like glucose and ammonia. The situation is quite different in mammalian cells. Mammalian cells need a rich nutrient environment of amino acids, fats, carbohydrates, vitamins, minerals, etc. Environmental constraints are just as important as enzyme and other intracellular constraints in mammalian cells and many of these biological processes are transient in nature in response to environmental perturbations or signaling changes, for example, cellular state transitions, like cell division, differentiation and apoptosis. In simple bioengineered organisms, "health" of the organism is of little concern. In contrast, in mammalian cells, dynamic system models are needed to understand disease phenotype, or drug treatment effects, for which there are few experimental examples. This is a wide-open research area. An open-source package called *ISODYN* (Selivanov et al. 2006a,b) has been used for stable ^{13}C-labeled-tracer metabolomics FBA studies of metabolism in mammalian cells (Lee et al. 1998; Boros et al. 2002; Selivanov et al. 2004; Selivanov et al. 2005; Selivanov et al. 2006a,b; Maguire et al. 2007).

Metabolic Control Analysis (MCA)

MCA is an important modeling framework, developed originally for studying how perturbations within individual reactions in metabolic pathways affect overall pathway dynamics under steady-state conditions. It is also being used in other systems biology areas, for example, in studying parameter perturbation effects on signal transduction pathway and ecosystem model dynamics (Kholodendo 2000; Westerhoff 2002). In metabolism, it is an important methodology in biotechnology, for example, to optimize biofuel production by manipulating parameters or enzymes in plant metabolism models, and in medicine – for understanding how to up-regulate or down-regulate particular metabolites involved in a clinical disorder – extending the methodology of the last sections. In all applications, it is about characterizing effects of small perturbations in variables and parameters in network models operating in steady state, using models implicitly or explicitly linearized about a system equilibrium or steady state. In MCA, perturbation effects are expressed as *control coefficients* and *elasticities*, characterized as local (first-order) and relative sensitivity functions. This is a key application of local sensitivity functions for understanding and controlling biological regulation systems.

Sensitivity Functions of Metabolic Control Analysis

Control coefficients are a measure of small relative variations in steady state metabolite concentrations or activities, or reaction velocities or other fluxes in a metabolic network, in response to relative variations in key parameters, concentrations or activities. To motivate the math, consider the simple two-metabolite, two-enzyme, one product metabolic pathway illustrated in **Fig. 14.13**. The input u is constant, there is no feedback and the v_i are linear or nonlinear reaction velocities. With the system in steady state, $\dot{m}_1 = \dot{m}_2 = 0$ and therefore $u = v_2 m_{2\infty} = v_1 m_{1\infty}$, which is also the overall flux J of material through the network.

A **flux control coefficient** quantifies a small relative variation $\partial J/J$ in steady-state flux J to a small relative variation $\partial v_i/v_i$ in a v_i:

$$C_{v_i}^{J} \equiv \frac{\partial J}{J} \Big/ \frac{\partial v_i}{v_i} = \frac{\partial \ln J}{\partial \ln v_i} \tag{14.13}$$

A **concentration control coefficient** quantifies the relative change in steady state concentration of metabolite or enzyme to a small relative change in a v_i:

SIMPLE METABOLIC PATHWAY
VARIABLES & PARAMETERS

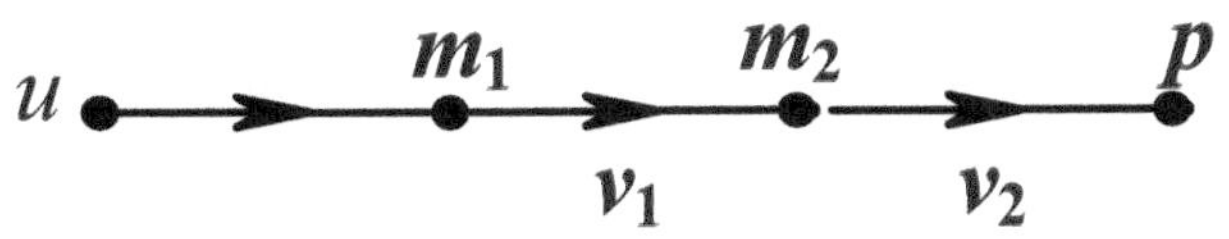

Fig. 14.16

$$C_{v_i}^{m_j} \equiv \frac{\partial m_j}{m_j} \bigg/ \frac{\partial v_i}{v_i} = \frac{\partial \ln m_j}{\partial \ln v_i} \quad \text{or} \quad C_{v_i}^{e_j} \equiv \frac{\partial e_j}{e_j} \bigg/ \frac{\partial v_i}{v_i} = \frac{\partial \ln e_j}{\partial \ln v_i} \tag{14.14}$$

A **response coefficient** quantifies the relative change in steady-state metabolite or enzyme concentration to a small relative change in any endogenous or exogenous parameter p_i:

$$R_{p_i}^{m_j} \equiv \frac{\partial m_j}{m_j} \bigg/ \frac{\partial p_i}{p_i} = \frac{\partial \ln m_j}{\partial \ln p_i} \quad \text{or} \quad R_{p_i}^{e_j} \equiv \frac{\partial e_j}{e_j} \bigg/ \frac{\partial p_i}{p_i} = \frac{\partial \ln e_j}{\partial \ln p_i} \tag{14.15}$$

Elasticity coefficients quantify relative sensitivities for the *inverse* of the variables and parameters defining control coefficients, e.g. they quantify the relative change in a reaction rate v_i in response to a small perturbation in a metabolite or other concentration, or parameter p_i:

$$\varepsilon_{m_i}^{v_j} \equiv \frac{\partial v_j}{\partial m_i} \bigg/ \frac{m_i}{v_j} = \frac{\partial \ln v_j}{\partial \ln m_i} \quad \text{or} \quad \pi_{M_i}^{v_j} \equiv \frac{\partial v_j}{\partial p_i} \bigg/ \frac{p_i}{v_j} = \frac{\partial \ln v_j}{\partial \ln p_i} \tag{14.16}$$

Example 14.4: ***Elasticity Coefficient for Linear and NL Reaction Rates***

Suppose a reaction velocity v in **Fig. 14.16** (or other model) follows a linear reaction rate law: $v = km$. Then

$$\varepsilon_m^v \equiv \frac{\partial v}{\partial m} \bigg/ \frac{m}{v} = \frac{\partial}{\partial m}(km)\frac{m}{km} = 1 \tag{14.17}$$

This is true for any linear law, similar to output parameter sensitivities in open-loop control systems. For a NL M–M law $v = v_{\max} m/(K_m + m)$, perturbations in the metabolite concentration generate an elasticity of less than 1:

$$\varepsilon_m^v \equiv \frac{\partial v}{\partial m} \bigg/ \frac{m}{v} = \frac{\partial}{\partial m}\left(\frac{v_{\max} m}{K_m + m}\right)\frac{m}{v} = \frac{m}{K_m + m} < 1 \tag{14.18}$$

If the Michaelis constant K_m is the perturbed parameter:

$$\varepsilon_{K_m}^v \equiv \frac{\partial v}{\partial K_m} \bigg/ \frac{K_m}{v} = \frac{\partial}{\partial K_m}\left(\frac{v_{\max} m}{K_m + m}\right)\frac{K_m}{v} = \frac{-m}{K_m + m} = -\varepsilon_m^v \tag{14.19}$$

These are like output parameter sensitivities in a closed-loop negative fbcs (Chapter 11).

Properties of Metabolic Control and Elasticity Coefficients

Control Coefficient Constraints

Two interesting and useful *constraint relationships* among the control coefficients provide important insights into network dynamics, in particular about the contributions of individual parts to overall control of steady state fluxes in the network. For a network or pathway with n reactions, these are:

$$\sum_{i=1}^{n} C_{v_i}^{J} = 1 \quad \text{and} \quad \sum_{i=1}^{n} C_{v_i}^{m} = 0 \tag{14.20}$$

The first summation says that the overall change in a flux J through the network, due to fractional perturbations in all of the enzymatically controlled reactions, sums to 1. If all of the flux control coefficients are nonzero, this means that control of metabolic flux is shared by all reactions, that is, balanced. If some go up, others must go down and overall regulation is maintained in the presence of perturbations in any reaction. In general, there is no single one that dominates. The second "conservation" summation says that all concentration control coefficients for given metabolite m sum to 0, that is, there can be *no net change in concentration* (or mass) due to all perturbations acting together.

Control and Elasticity Connectivity Properties

Flux control and elasticity coefficients satisfy the following *constraints*:

$$\sum_{j=1}^{n} C_{v_j}^{J} \varepsilon_{m_i}^{v_j} = 0 \quad \text{and} \quad \sum_{j=1}^{n} C_{v_j}^{m_k} \varepsilon_{m_i}^{v_j} = -\delta_{ki} \tag{14.21}$$

In Eq. (14.21), Kronecker delta $\delta_{ki} = 0$ if $k \neq i$ and $\delta_{ki} = 1$ if $k = i$. (i.e. if the two metabolites are different or the same). These two constraints characterize the interrelationships among the kinetic features of individual reactions and the overall response of the network to perturbations. If only flux control coefficients or elasticity coefficients are known about a reaction pathway, these relationships can be used to evaluate the other coefficients. Qualitatively, they serve to reinforce the concept that flux control in enzymatically regulated networks is distributed among multiple enzymes. A comprehensive review of MCA can be found in Moreno-Sanchez et al. (2008).

Recap and Looking Ahead

In this chapter, we delved deeper than in earlier ones into modeling physiological fbcs and biomolecular and cellular control systems.

One or more positive and/or negative feedback control signaling pathways are present in many dynamic biosystems – adding significantly to the complexity of their dynamics; e.g. sustained oscillations in neuroendocrine signals resulting from multifeedback pathways of somatic hormones stimulating or inhibiting brain or pituitary hormonal secretions. The thyroid hormone regulation model developed here illustrated this behavior. In some biosystems, feedforward signaling pathways are also present. Examples include kinase-dependent signal transduction pathways in cellular systems, also participating in feedback loops and oscillatory behavior in some, such as in tumor suppressor protein p53 regulation (Chapter 15).

Stability analyses of simple structural variations in a M–M kinetics model – with input and product feedback included – generated some surprising results, as well as a demonstration of poor application of the QSSA.

The last sections were focused on metabolic system networks and development of steady state FBA and associated techniques for optimizing pathway reactions, and MCA for analyzing their sensitivity properties. These are major activities in systems biology, with a large and growing body of knowledge and literature, much of it published over the last decade. Applications in synthetic biology, pharmacology and medicine are widespread. The near future should produce further development, consolidation and additional applications, given their commercial motivation (e.g. biofuels) and substantial overlap with whole-genome bioinformatics domains. The combined approaches are already showing a modicum of success beyond synthetic biology – in drug targeting studies and enhancement of basic knowledge of disease processes.

Chapter 15 is about more formally and explicitly testing alternative biosystem hypotheses using alternative, data-driven mechanistic models. The emphasis is on using idealized and real input–output data more actively in the model formulation stage, as well as in quantification and validation. Transfer function analysis, directed-graph modeling methods and cutset analysis and optimization methods for experiment design are the foci of Chapter 16.

Exercises

E14.1 Repeat the closed-loop analysis of the Van Slyke–Cullen model using the total quasi-steady state assumption, tQSSA, for the same parameters and ICs as given in those illustrative examples. Compare the results (three ways) with both the QSSA and the full model stability analysis. Does the tQSSA help?

E14.2 Show that the roots of Eq. (14.6) are real and negative.

E14.3 Show that the roots of Eq. (14.8) can have positive real parts.

E14.4 The phase plots in the Van Slyke–Cullen model examples include c-null surfaces only, that is, surfaces represented by $\dot{c} = 0$. Redo these examples, including the null surface for the substrate, $\dot{s} = 0$, which should clearly delineate equilibrium solutions at their intersections.

E14.5 ***Plant metabolism modeling and analysis.*** Consider the simple plant metabolism model in **Fig. 7.6**.

- **a)** Implement the model in *Matlab, COBRA,* or equivalent software.
- **b)** Simulate its metabolite dynamics with parameter values.
- **c)** Explore the model, by simulation, varying the parameter values, to see how they affect metabolite concentrations.

d) Do a metabolic control analysis (MCA) of this model. Compute the sensitivity functions of MCA that you can compute, using the model augmented with sensitivity equations, or similar functionality in the software package you use. Also compute the **length** of the sensitivity vectors over the time course 0–12 min. Plot all results on the same graph, along with your simulated metabolite concentrations. Discuss your results.

e) Discuss how you might increase the production of metabolite B or C. The idea here is that B or C might be biofuel.

E14.6 ***Elasticity coefficients.*** Find the elasticity coefficients for a NL Hill function law $v = Am^4/(K^4 + m^4)$, for perturbations in the metabolite concentration m and the parameters A and K. What do these sensitivity functions mean?

References

Alves, R., Savageau, M.A., 2000. Systemic properties of ensembles of metabolic networks: application of graphical and statistical methods to simple unbranched pathways. Bioinformatics. 16 (6), 534–547.

Bairagi, N., Chatterjee, S., Chattopadhyay, J., 2008. Variability in the secretion of corticotropin-releasing hormone, adrenocorticotropic hormone and cortisol and understandability of the hypothalamic-pituitary-adrenal axis dynamics–a mathematical study based on clinical evidence. Math. Med. Biol. 25 (1), 37–63, 10.1093/imammb/dqn1003. Epub 2008 March 1014.

Ben-Shachar, R.R., Eisenberg, M.M., Huang, S.A., DiStefano III, J.J., 2012. Simulation of post-thyroidectomy treatment alternatives for triiodothyronine or thyroxine replacement in pediatric thyroid cancer patients. Thyroid. 22, 595–603.

Blakesley, V.A., A. W., Locke, C., Ludden, T.M., Granneman, G.R., Braverman, L.E., 2004. Are bioequivalence studies of levothyroxine sodium formulations in euthyroid volunteers reliable? Thyroid. 14, 191–200.

Boros, L., Cascante, M., Go, V.L.W., Heber, D., Hidvégi, M., Lee, W.N.P., 2002. Metabolic profiling of cell growth and death in cancer: applications in drug discovery. Drug Discovery Today. 7, 366–374.

Brandman, O., Ferrell Jr., J.E., Li, R., Meyer, T., 2005. Interlinked fast and slow positive feedback loops drive reliable cell decisions. Science. 310 (5747), 496–498.

Cheong, R., Hoffmann, A., Levchenko, A., 2008. Understanding NF-kappaB signaling via mathematical modeling. Mol. Syst. Biol. 4 (192), 6.

Chu, Y., Hahn, J., 2008. Parameter set selection via clustering of parameters into pairwise indistinguishable groups of parameters. Ind. Eng. Chem. Res. 48 (13), 6000–6009.

Churilov, A., Medvedev, A., Shepeljavyi, A., 2009. Mathematical model of non-basal testosterone regulation in the male by pulse modulated feedback. Automatica. 45 (1), 78–85.

Dietrich, J.W., Landgrafe, G., Fotiadou, E.H., 2012. TSH and thyrotropic agonists: key actors in thyroid homeostasis. J. Thyroid Res. 2012, 29.

DiStefano III, J., 1968. Endocrine regulation: models and experimental design for system identification. Seminar on Large Scale Systems in the Matematical Biosciences, Dubrovnik, Yugoslavia.

DiStefano III, J., 1969a. Hypothalamic and rate feedback in the thyroid hormone regulation system: a hypothesis. Bull. Math. Biophys. 31, 233–246.

DiStefano III, J., 1969b. A model of the normal thyroid hormone glandular secretion mechanism. J. Theor. Biol. 22, 412–417.

DiStefano III, J., 1969c. A model of the regulator of circulating thyroxin unbound and bound to plasma proteins and its response to pregnancy, drugs, long-acting thyroid stimulator, and temperature stress. Math. Biosci. 4, 137–152.

DiStefano III, J., 1973. The thyroid hormone feedback control system. In: Brown, J., Gann, D. (Eds.), Applications of Engineering Principles in Physiology. Academic Press, New York, pp. 261–289.

DiStefano III, J., 1985. Modeling approaches and models of the distribution and disposal of thyroid hormones. In: Hennemann, G. (Ed.), Thyroid Hormone Metabolism. Marcel Dekker, New York, pp. 39–76.

DiStefano III, J., 1991. Kinetic modeling methods in theory and practice. In: Wu, J. (Ed.), Thyroid Hormone Metabolism - Regulation and Clinical Implications. Blackwell Scientific Publications, Oxford, pp. 65–89.

DiStefano III, J., Fisher, D.A., 1979. Peripheral distribution and metabolism of the thyroid hormones: a primarily quantitative assessment. In: Hershman, J., Bray, G. (Eds.), International Encyclopedia of Pharmacology and Therapeutics, Section 101, The Thyroid: Physiology and Treatment of Disease. Pergamon Press, Oxford, pp. 47–82.

DiStefano III, J., Mak, P.H., 1979. On model and data requirements for determining the bioavailability of oral therapeutic agents: application to gut absorption of thyroid hormones. Am. J. Physiol. Regul. Integr. Comp. Physiol. 236, R137–R141.

DiStefano III, J., Stear, E.B., Moore, G.P., 1966. A model of the thyroid hormone regulator and a proposal for its experimental validation. Proceedings of the 19th Annual Conference on Engineering in Medicine and Biology 8(3).

DiStefano III, J., Stear, E., 1968. Neuroendocrine control of thyroid secretion in living systems: a feedback control system model. Bull. Math. Biophys. 30, 3–26.

DiStefano III, J., Durando, A.R., Jang, M., Jenkins, D., Johnson, D.J., Mak, P.H., et al., 1973. Estimates and estimation errors of hormone secretion, transport, and disposal rates in the maternal-fetal system. Endocrinology. 93 (2), 324–342.

DiStefano III, J., Wilson, K.C., Jang, M., Mak, P.H., 1975. Identification of the dynamics of thyroid hormone metabolism. Automatica. 11, 149–159.

DiStefano III, J., Jang, M., Malone, T.K., Broutman, M., 1982a. Comprehensive kinetics of triiodothyronine (T3) production, distribution, and metabolism in blood and tissue pools of the rat using optimized blood-sampling protocols. Endocrinology. 110 (1), 198–213.

DiStefano III, J., Malone, T.K., Jang, M., 1982b. Comprehensive kinetics of thyroxine (T4) distribution and metabolism in blood and tissue pools of the rat from only six blood samples: dominance of large, slowly exchanging tissue pools. Endocrinology. 111 (1), 108–117.

DiStefano III, J., Morris, W.L., Nguyen, T.T., Van Herle, A.J., Florsheim, W., 1993a. Enterohepatic regulation and metabolism of triiodothyronine (T3) in hypothyroid rats. Endocrinology. 132, 1665–1670.

DiStefano III, J., Nguyen, T.T., Yen, Y.M., 1993b. Transfer of 3,5,3-triiodothyronine (T3) and thyroxine (T4) from rat blood to small and large intestines, liver, and kidneys in vivo. Endocrinology. 132, 1735–1744.

DiStefano III, J., Ron, B., Nguyen, T.T., Weber, G.M., Grau, E.G., 1998. 3,5,3′-triiodothyronine (T3) clearance and T3-glucuronide appearance kinetics in plasma of freshwater-reared male tilapia, oreochromis mossambicus. Gen. Comp. Endocrinol. 111, 123–140.

Eales, J., Chang, J.P., Van Der Kraak, G., Omeljaniuk, R.J., Vin, L., 1982. Effects of temperature on plasma thyroxine and iodide kinetics in rainbow trout. Gen. Comp. Endocrinol. 47, 295.

Eisenberg, M., Distefano III, J.J., 2009. TSH-based protocol, tablet instability, and absorption effects on L-T4 bioequivalence. Thyroid. 19 (2), 103–110.

Eisenberg, M.C., DiStefano III, J.J., 2010. TSH regulation dynamics in central and extreme primary hypothyroidism. Thyroid.

Eisenberg, M., Samuels, M., DiStefano III, J.J., 2006. L-T4 bioequivalence and hormone replacement studies via feedback control simulations. Thyroid. 16 (12), 1–14.

Eisenberg, M., Samuels, M., DiStefano III, J.J., 2008. Extensions, validation, and clinical applications of a feedback control system simulator of the hypothalamo-pituitary-thyroid axis. Thyroid.1071–1085.

Eisenberg, M.C., Santini, F., Marsili, A., Pinchera, A., DiStefano, J.J., 2010. TSH regulation dynamics in central and extreme primary hypothyroidism. Thyroid. 20 (11), 1215–1228.

Feng, D., Distefano, J.J., 1991. Cut set analysis of compartmental models with applications to experiment design. Am. J. Physiol. Endocrinol. Metab. 261 (2), E269–E284.

Fidelman, M.L., Mikulecky, D.C., 1986. Network thermodynamic modeling of hormone regulation of active Na + transport in cultured renal epithelium (A6). Am. J. Physiol. Cell Physiol. 250 (6), C978–991.

Flach, E.H., Schnell, S., 2006. Use and abuse of the quasi-steady-state approximation. IEE Proc. Syst. Biol. 153 (4), 187–191.

Grodins, F.S., 1963. Control Theory and Biological Systems. Columbia University Press, New York.

Gupta, S., A. E., Gurbaxani, B.M., Vernon, S.D., 2007. Inclusion of the glucocorticoid receptor in a hypothalamic pituitary adrenal axis model reveals bistability. Theor. Biol. Med. Model. 4, 8.

Guyton, A.C., Coleman, T.G., Granger, H.J., 1972. Circulation: overall regulation. Annu. Rev. Physiol. 34, 13–46.

Jelić, S., Cupić, Z., Kolar-Anić, L., 2005. Mathematical modeling of the hypothalamic-pituitary-adrenal system activity. Math. Biosci. 197 (2), 173–187.

Jin-Zhi, L., Chang-Yi, W., 2008. On the reducibility of compartmental matrices. Comput. Biol. Med. 38 (8), 881–885.

Kearns, J.D., Hoffmann, A., 2009. Integrating computational and biochemical studies to explore mechanisms in NF-κB signaling. J. Biol. Chem. 284 (9), 5439–5443.

Kholodendo, B., 2000. Diffusion control of proteins phosphorylated in signal transduction pathways. Biochem. J. 350, 901–907.

Kim, J., Hayton, W.L., Schultz, I.R., 2006. Modeling the brain-pituitary-gonad axis in salmon. Mar. Environ. Res. 62, S426–432.

Kitney, R., Freemont, P., 2012. Synthetic biology - the state of play. FEBS Lett. 586 (15), 2029–2036, 2010.1016/j.febslet.2012.2006.2002. Epub 2012 June 2013.

Kyrylov, V., Severyanova, L.A., Vieira, A., 2005. Modeling robust oscillatory behavior of the hypothalamic-pituitary-adrenal axis. IEEE Trans. Biomed. Eng. 52 (12), 1977–1983.

Lee, J.M., Gianchandani, E.P., Papin, J.A., 2006. Flux balance analysis in the era of metabolomics. Brief. Bioinform. 7 (2), 140–150, Epub 2006 April 2026.

Lee, W.N.P., B. L., Puigjaner, J., Bassilian, S., Lim, S., Cascante, M., 1998. Investigation of the pentose cycle using[1, 2-13C2]-glucose and mass isotopomer analysis: estimation of transketolase and transaldolase activities. Am. J. Physiol. 274, E843–E851.

Lenbury, Y., P. P., 2005. A delay-differential equation model of the feedback-controlled hypothalamus-pituitary-adrenal axis in humans. Math. Med. Biol. 22, 15–33.

Maguire, G., B. L., L. W., 2007. Development of tracer-based metabolomics and its implications for the pharmaceutical industry. Int. J. Pharm. Med. 21 (3), 217–224.

Mak, P., DiStefano III, J., 1979. Optimal control policies for the prescription of clinical drugs: dynamics of hormone replacement for endocrine deficiency disorders. In: Leondes, C. (Ed.), Control and Dynamic Systems: Advances in Theory and Application. Academic Press, New York, pp. 2–41.

Milsum, J.H., 1966. Biological Control Systems Analysis. McGraw-Hill, New York.

Moreno-Sanchez, R., Saavedra, E., Rodriguez-Enrquez, S., Olin-Sandoval, V., 2008. Metabolic control analysis: a tool for designing strategies to manipulate metabolic pathways. J. Biomed. Biotechnol.1–30.

Nguyen, T.T., Mol, K.A., DiStefano, J.J., 2003. Thyroid hormone production by rat liver and intestine in vivo: a novel graph theoretic/experimental solution. Am. J. Physiol. Endocrinol. Metab. 285, E171–E181.

Palsson, B.O., 2006. Systems Biology: Properties of Reconstructed Networks. Cambridge University Press, Cambridge.

Papin, J.A., Price, N.D., Wiback, S.J., Fell, D.A., Palsson, B.O., 2003. Metabolic pathways in the post-genome era. Trends Biochem. Sci. 28 (5), 250–258.

Perelson, A.S., 2002. Modelling viral and immune system dynamics. Nat. Rev. Immunol. 2 (1), 28–36.

Perrin, F., Normand, M., Fortier, C., 1978. Analysis through mathematical models of the dynamics of the adrenal cortical response to ACTH in the rat. Ann. Biomed. Eng. 6 (1), 1–15.

Rasgon, N.L., Prolo, P., Elman, S., Negrao, A.B., Licinio, J., Garfinkel, A., 2003. Emergent oscillations in a mathematical model of the human menstrual cycle. CNS Spectr. 8 (11), 805–814.

Savic, D., J. S., 2005. A mathematical model of the hypothalamo-pituitary-adrenocortical system and its stability analysis. Chaos Solitons Fractals. 26, 427–436.

Schellenberger, J., Que, R., Fleming, R.M.T., Thiele, I., Orth, J.D., Feist, A.M., et al., 2011. Quantitative prediction of cellular metabolism with constraint-based models: the COBRA Toolbox v2.0. Nat. Protoc. 6 (9), 1290–1307.

Schuster, S., Dandekar, T., Fell, D.A., 1999. Detection of elementary flux modes in biochemical networks: a promising tool for pathway analysis and metabolic engineering. Focus. 17, 1–8, TIBTECH, Elsevier.

Schwartz, N.B., McCobmack, C.E., 1972. Reproduction: gonadal function and its regulation. Annu. Rev. Physiol. 34 (1), 425–472.

Sefkow, A., DiStefano III, J.J., Eales, J.G., Himick, B.A., Brown, S.B., 1996. Thyroid hormone secretion and interconversion in 5-day fasted rainbow trout, oncorhynchus mykiss. Gen. Comp. Endocrinol. 101, 123–138.

Selivanov, V.A., P. J., Sillero, A., Centelles, J.J., Ramos-Montoya, A., Lee, P.W., et al., 2004. An optimized algorithm for flux estimation from isotopomer distribution in glucose metabolites. Bioinformatics. 20, 3387–3397.

Selivanov, V.A., M. L., Solovjeva, O.N., Kuchel, P.W., Ramos-Montoya, A., Kochetov, G.A., et al., 2005. Rapid simulation and analysis of isotopomer distributions using constraints based on enzyme mechanisms: an example from HT29 cancer cells. Bioinformatics. 21 (17), 3558–3564.

Selivanov, V.A., Marin, S., Lee, P.W.N., Cascante, M., 2006a. Software for dynamic analysis of tracer-based metabolomic data: estimation of metabolic fluxes and their statistical analysis. Bioinformatics. 22 (22), 2806–2812.

Selivanov, V.A., Sukhomlin, T., Centelles, J.J., Lee, Paul W.N., Cascante Serratosa, M., 2006b. Integration of enzyme kinetic models and isotopomer distribution analysis for studies of in situ cell operation. BMC Neurosci. 7 (Suppl 1), S1–S7.

Smiljana, J., Zeljko, C., Kolar-Anic, L., 2005. Mathematical modeling of the hypothalamic-pituitary-adrenal system activity. Mathematical Biosciences. 197, 173–187.

Sriram, K., R.-F. M., Doyle III, F.J., 2012. Modeling cortisol dynamics in the neuro-endocrine axis distinguishes normal, depression, and post-traumatic stress disorder (PTSD) in humans. PLoS Comput. Biol. 8, 2.

Tornøe, C., Agersø, H., Senderovitz, T., Nielsen, H.A., Madsen, H., Karlsson, M.O., et al., 2007. Population pharmacokinetic/pharmacodynamic (PK/PD) modelling of the hypothalamic–pituitary–gonadal axis following treatment with GnRH analogues. Br. J. Clin. Pharmacol. 63, 648–664.

Tyson, J.J., 2006. Another turn for p53. Mol. Syst. Biol. 2 (2006), 13.

Van Slyke, D.D., Cullen, G.E., 1914. The mode of action of urease and of enzymes in general. J. Biol. Chem. 19 (2), 141–180.

Watanabe, K.H., Li, Z., Kroll, K.J., Villeneuve, D.L., Garcia-Reyero, N., Orlando, E.F., et al., 2009. A computational model of the hypothalamic-pituitary-gonadal axis in male fathead minnows exposed to 17α-Ethinylestradiol and 17β-Estradiol. Toxicol. Sci. 109 (2), 180–192.

Westerhoff, H., et al., 2002. ECA: control in ecosystems. Mol. Biol. Reprorts. 29, 113–117.

Xingchen, C., Lonie, A., Harris, P.S., Randall, T., Rajkumar, B., 2006. KidneyGrid: a grid platform for integration of distributed kidney models and resources. Proceedings of the Fourth International Workshop on Middleware for Grid Computing. ACM, Melbourne, Australia.

Yates, F.E., Brennan, R.D., Urquhart, J., 1969. Application of control systems theory to physiology. Adrenal glucocorticoid control system. Fed. Proc. 28, 71–83.

Yue, H., Brown, M., Knowles, J., Wang, H., Broomhead, D.S., Kell, D.B., 2006. Insights into the behaviour of systems biology models from dynamic sensitivity and identifiability analysis: a case study of an NF-kB signalling pathway. Mol. Biosyst. 2.

Data-Driven Modeling and Alternative Hypothesis Testing

Chapter 15

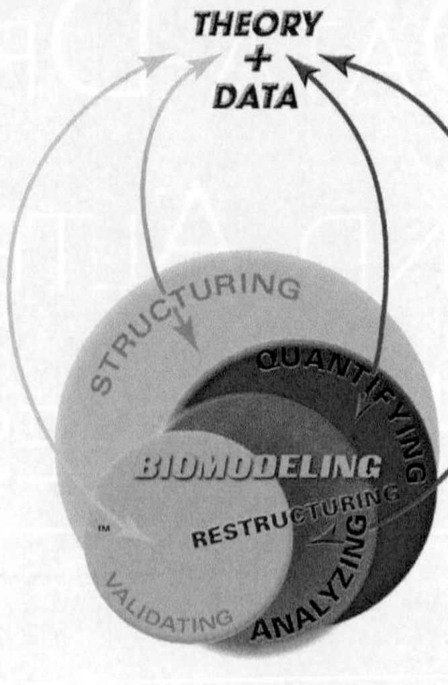

The purpose of models is not to fit the data but to sharpen the questions.

Samuel Karlin, ~1990

Overview

Let the Data Speak First

Among the oldest dictums in science – perhaps most of all in biology – is that while theory is essential to progress, it can only suggest what's possible; data hold all the truths. Good data, of course, tell the truth better than bad data, but we're not going there. Instead, as biomodelers, we are interested first in what *the data we have for modeling* can tell us about the system. The data are surely limited – that's a given; but it has something imminently truthful, useful, and essential to say. And my point here is that we should listen to it first – or at least early on in the model development process. Otherwise, we are likely to waste a lot of time and effort and, at worst, fail. A simple example, to make the point.

Example 15.1: ***One Exponential Is One Exponential Is One Exponential...***

> Suppose you are interested in better understanding a dynamic biosystem for which little is known, and you are able to test it using input–output (I–O) probes. If single-output data collected from it are fitted well by a single exponential function, and not well by anything else, only additional *data* can support a model more complex than a single ODE. This is *not* subtle. Only simplified, to make the point. Extrapolate this notion to a complex biological system, with little data available to build a model of it, and "thinking twice" about taking it on (without more data) might be the way to go.

Formalizing the Modeling Process Based on the Data

Explicit models provide a means for comparing experimental results (data) with expected results (hypotheses) – the whole premise of this book. Models can be our best friends when testing the validity of assumptions about a system; but the data are nevertheless in control of the extent of what we can discover. In this chapter, we develop one formal approach to testing alternative hypotheses or assumptions about a dynamic biosystem, using explicit models of these alternatives, using data collected in experiments designed to distinguish among them.

Testing *alternative hypotheses* expressible in explicit mathematical terms is among the most important applications for biomodeling. Assessing competing mechanistic hypotheses about how a system or process (e.g. an enzyme regulatory pathway) is specifically structured, or how it works is an important example. Procedurally, the different hypotheses are formulated explicitly, each as a distinct mathematical model. Each of them is then tested against the *same set of data*, analyzing and statistically comparing how well each fits this data.

Hypothesis testing is likely what we do implicitly if not purposefully when we model. We don't usually think of it in these terms, because we do it piecemeal (sequentially, over some time) – often from *different data sets*, rather than simultaneously from one set of data.

So much for the easy part (conceptualizing). Let's see how it can be done effectively, using mathematical, engineering design and statistical tools, keeping in mind that the data are limited and may not be good enough to distinguish our hypotheses. Nevertheless, let's be optimistic. The methodologies available to us are certainly useful for solving some such problems – when the data are good enough and we "listen."

Statistical Criteria for Discriminating Among Alternative Models

Fitting *single* models to data invariably involves optimization, and thus *optimization criteria*. These criteria, presented in Chapter 12, formally express how to choose the "best" set of model parameters $\boldsymbol{p}$ that minimize some function of the residual output errors, i.e. $\boldsymbol{r}(t_k, \boldsymbol{p}) = \boldsymbol{y}(t_k, \boldsymbol{p}) - \boldsymbol{z}(t_k)$ for $k = 1, 2, \ldots, N$. The most common are least-squares (*LS*) and weighted least-square (*WLS*) ≡ residual-sum-of-square (*WRSS*) criteria (e.g. Eq. 12.22). Following a successful parameter estimation run, the final *WRSS*, for example, is used as a statistical measure of the quality of the fit of the *single* model to the single set of data, and this measure is assessed using various residuals tests – along with parameter variance and covariance statistics for assessing results of parameter optimization runs, all as described in Chapter 12. Discriminating among several *alternative model hypotheses* is done with *different statistical criteria*, usually after each has been fitted individually to the *same* data set, or possibly simultaneously to the same data set. The basic idea is to compute statistics for each model and compare these statistics using various *measures of hypothesis discrimination*. The four criteria described below are generally applicable to most model types, with the exceptions noted (Landaw and DiStefano III 1984).

Multiexponential/Multicompartmental Model Discrimination: How Many Modes?

Model discrimination is readily illustrated for I–O multiexponential models – a class of ***nested-models***[1] – with *unknown* numbers of exponential terms n. The unknown n might also be the dimensionality or number of compartments (modes) in a structured multicompartmental model input–output-equivalent to the multiexponential model.

If results are available for a sequence of attempts to fit the data to the sum of n exponential terms, for different n, several statistical discrimination methods are available for helping to decide which is the "best" model. The four presented below are imbedded in the multiexponential "expert" model discrimination programs *DIMSUM* (Marino et al. 1992) and *W³DIMSUM* (Harless and DiStefano III 2005). This program is used to independently and automatically fit sums of 1-, 2-, 3- and 4-exponential term models to the data prior to the model discrimination phase. The web application *W³DIMSUM* can be accessed and run online at http://biocyb.cs.ucla.edu/biocybmodeling.html.

F-test of Significance

For multiexponential models, if the residuals are approximately normally distributed, the F-test is a reasonably robust measure of how much better a k-exponential model is than a $(k-1)$-exponential model. Each successive addition of an exponential term will lower the *WRSS*, but one can define as "best" the *highest order* model which *significantly lowers the*

1 Monoexponential ⊂ biexponential ⊂ triexponential ⊂ etc.

WRSS over the models below it. Thus, an order k-model can sequentially be tested against an order $k-1$ model with the F-test statistic:

$$F = \left(\frac{WRSS_{k-1} - WRSS_k}{WRSS_k}\right)\frac{df}{2} \tag{15.1}$$

where $df \equiv$ degrees of freedom $= N - 2k$, N is the number of data points, and $WRSS_k$ is the weighted residual sum of squares for the fit with the sum of k-exponential terms. Under the *null hypothesis* of no significant reduction in *WRSS*, this statistic is distributed asymptotically as $F_{(2,\mathrm{df})}$. As a rough rule of thumb, F values greater than 3 indicate a significant reduction in the *WRSS* at the 99% confidence level ($p < 0.01$). However, because of the nonrobustness of the F statistic, departures from a Gaussian distribution of the error residuals may strongly affect the validity of the F-test and we should look further.

Testing Estimates of Exponential Coefficients A_i

This test can be used in conjunction with the F-test. Here one defines the "best" fit as the highest order model which has all of its A_i estimates significantly different from zero. The *Student t-test* can be used with each of the A_i estimates and their respective standard deviation estimates SD_i at the appropriate number of degrees of freedom (df). As a rough rule, if $SD_i/|A_i| > 0.6$ (i.e. a coefficient of variation of 60%) the estimate is not significantly different from zero ($p > 0.05$).

Akaike Information Criterion (AIC)

The *AIC* is a model selection (discrimination) criterion based in information theory.[2] It is not restricted to nested models or to models with Gaussian residuals (Akaike 1974, 1978), both of which limit the validity of the F-test. If data variances $\sigma^2(t_i) \equiv \sigma_i^2$ (or estimates of them) are known, the *AIC* is:

$$AIC(\boldsymbol{p}^*) = WRSS\,(\boldsymbol{p}^*) + 2P \tag{15.2}$$

where the * notation indicates optimum estimate $\boldsymbol{p}^*$ and P is the number of identifiable parameters of the model. If data variances are known only within an unknown proportionality constant k, i.e. $\sigma_i^2 \equiv k s_i^2$, the *AIC* is:

$$AIC(\boldsymbol{p}^*) = N \log_e WRSS(\boldsymbol{p}^*) + 2P \tag{15.3}$$

These results make intuitive sense. The $WRSS(p^*)$ term favors the model with the smallest weighted residual sum of squares; and the second term favors the candidate model with the least number of parameters. In this manner *AIC* provides a "balance" between accuracy and parsimony. For model discrimination, *AIC* is computed for each model fitted to the same data and the model with the smallest *AIC* is considered best by this criterion.

2 The theory of the *AIC* and some pertinent applications are given in *Appendix F*.

The Schwarz Criterion (SC or BIC)

This minor modification of the *AIC* (Schwarz 1978) is also (humorously) called the *BIC*. If data variances σ_i^2 are known beforehand (Case 1):

$$SC^* \equiv BIC^* = WRSS^* + P^*\log_e N \tag{15.4}$$

When σ_i^2 are known only up to a proportionality constant k (Case 2):

$$BIC = N\log_e WRSS^* + P^*\log_e N \tag{15.5}$$

Remark: If $\log_e N > 2$, the BIC^* penalizes the model with P^* free (identifiable) parameters more so than does the *AIC*.

We apply these criteria in the next example below, where we find the best n value (number of exponential terms or modes in the model output) for some specific data.

Example 15.2: ***How Many Exponential Terms n Fit the Data Best?***

The data shown on the RHS of the upper part of **Fig. 15.1** were obtained in a hormone kinetics study in a dog, adapted from Example 1 of online application program *W3DIMSUM* described in detail in (Harless and DiStefano III 2005). The data are expressed as percent of the initial dose injected and the error model is based on constant coefficient of variation data. The data, best fits for 1-, 2-, 3- and 4-exponential sum models, and weighted residuals for each are shown in the graphs. Both 3- and 4-exponential models (blue and purple) fit equally well visually, best seen in the superimposed residuals plots. But the 4-exponential model bests the 3- when compared using the objective statistical tests summarized (in part) in the lower half of the figure. The most striking differences are in the probability of the *F*-statistic (different at the ~0% vs. 48% level) and the Akaike (13.48 vs. 17.31), and Schwartz criteria (19.45 vs. 25.27) comparisons: smaller values → better model.

The expert system for choosing the best model based on the above statistics is described in the *W3DIMSUM* program online, with additional details in (Marino et al. 1992). The expert score for the 3-exponential model is 100 vs. 40 for the 4-exponential model.

Macroscale and Mesoscale Models for Elucidating Biomechanisms

Minimal Macroscale Disease Dynamics Models: Treatment of Viral Infections

Models that are necessarily or purposefully macroscopic have components that aggregate lower level (microscopic) details. Usually, these microscopic details are either not well-understood or not of particular interest at the level being studied. Macroscopic cellular-organ[3] models capture essential features of intercellular or interorgan dynamics, rather than minute or

3 The term *cellular-organ model* is used here to represent intercellular dynamics of infection of cells in specific organs, e.g. liver and blood for HCV in this example, or HIV infection of T-cells of the immune system, etc.

W3 DIMSUM

DATA MODELS EXPERT SYSTEM RESULTS HELP

Show Data and Best Fits | Show Parameter Estimates | Show Model Statistics

y and z versus t

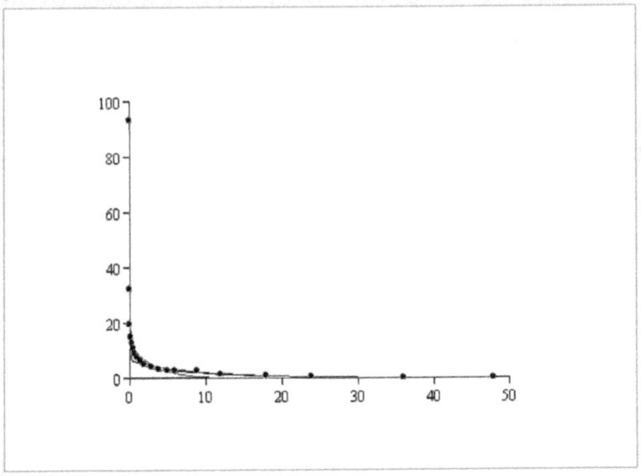

Data

Time	Observation	Variance
0.000E-1	93.63	96.65
0.03330	32.37	33.80
0.08330	19.84	28.42
0.1670	15.37	18.23
0.3300	12.92	10.73
0.5000	11.36	8.086
0.7500	9.315	6.648
1.000	7.733	3.230
1.500	6.485	2.346
2.000	5.330	1.737
3.000	4.130	0.9300
4.000	3.455	0.9379
5.000	3.020	0.6405
6.000	3.044	1.389
9.000	2.848	0.4857
12.00	1.642	0.4067
18.00	0.9695	0.3753
24.00	0.5135	0.06333
36.00	0.2244	0.01081
48.00	0.1043	0.002048

Weighted Residuals

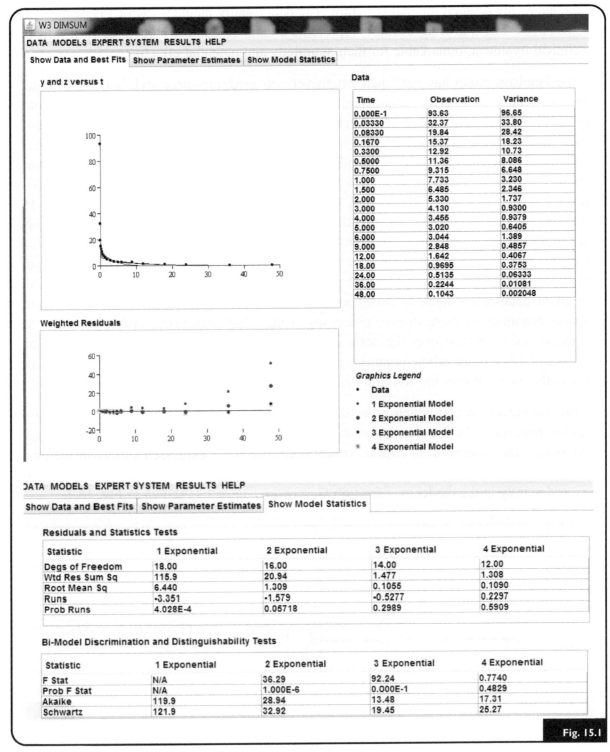

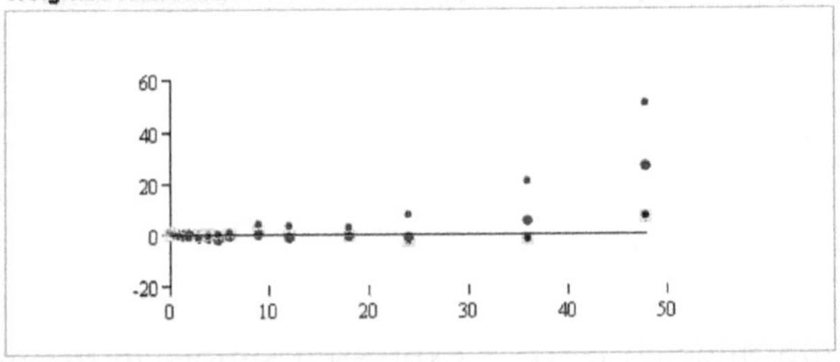

Graphics Legend

- Data
- 1 Exponential Model
- 2 Exponential Model
- 3 Exponential Model
- 4 Exponential Model

DATA MODELS EXPERT SYSTEM RESULTS HELP

Show Data and Best Fits | Show Parameter Estimates | Show Model Statistics

Residuals and Statistics Tests

Statistic	1 Exponential	2 Exponential	3 Exponential	4 Exponential
Degs of Freedom	18.00	16.00	14.00	12.00
Wtd Res Sum Sq	115.9	20.94	1.477	1.308
Root Mean Sq	6.440	1.309	0.1055	0.1090
Runs	-3.351	-1.579	-0.5277	0.2297
Prob Runs	4.028E-4	0.05718	0.2989	0.5909

Bi-Model Discrimination and Distinguishability Tests

Statistic	1 Exponential	2 Exponential	3 Exponential	4 Exponential
F Stat	N/A	36.29	92.24	0.7740
Prob F Stat	N/A	1.000E-6	0.000E-1	0.4829
Akaike	119.9	28.94	13.48	17.31
Schwartz	121.9	32.92	19.45	25.27

Fig. 15.1

WLS fitting of a 3-exponential model to data using *W*^{3}*DIMSUM* (biocyb.cs.ucla.edu/dimsum). Three-exponential model (blue) fits best on nearly all objective statistical counts.

microscopically detailed behavior and, in this sense, they are usually more data-driven than theory-driven. This approach is particularly useful in clinical, especially, therapy applications, where the goal is to characterize or predict dominant dynamical behavior of the biosystem under pathological as well as normal conditions, over extended periods of time, with minimal knowledge of the detailed internal structure. The data usually includes PK or PD data in these circumstances. The following example represents a class of clinical models of viral infection and clinical treatment, remarkable for their simplicity and potential effectiveness.

The Perelson group has published extensively on *predator–prey* style models of human immunodeficiency virus (HIV) and hepatitis-C virus (HCV) dynamics and their application in drug therapy (Neumann et al. 1998; Perelson 1999, 2002; Perelson et al. 2005; Dahari et al. 2008). Their basic structure for HCV is a standard viral infection model, where V represents the number (#) of virus (***virions***) in blood, T the number of normal liver cells targeted, and I the number of infected liver cells, all at time t. The model equations are on the right (RHS) in **Fig. 15.2**. These ODEs are transformed into compartmental (Chapter 4) form[4] in **Fig. 15.2** (LHS). *Fluxes* (v_{ij}s, # of cells per unit time) instead of rate constants are shown associated with the arrows in the figure, to accommodate the nonlinearities. For consistency, the linear cell death rates (leaks) are also represented as fluxes. This model and its HIV counterparts (not shown) are necessarily macroscopic and simple, primarily because of incomplete knowledge about detailed mechanisms involved when they were built.[5] Nevertheless, they capture pertinent dynamics of these disease processes in the whole organism, under treatment with antiviral drugs – at least over the scale of weeks. The model shown captured the pertinent dynamics of HCV in the whole organism under treatment with antiviral drugs and was used to test hypotheses about how the antiviral drug interferon-α (IFN) affects HCV action *in vivo*.

The simple manner in which potential drug effects are represented in the model are its dominant features – the features that *explicitly* represent the *alternative* hypotheses: IFN could either *reduce the production of virions* within infected cells – expressed simply in the model by a fraction $(1 - \varepsilon)$; or it could *reduce the de novo rate*[6]*of infection of liver cells* – also expressed simply in the model by a fraction $(1 - \eta)$, *or both*, all as shown in **Fig. 15.2**.

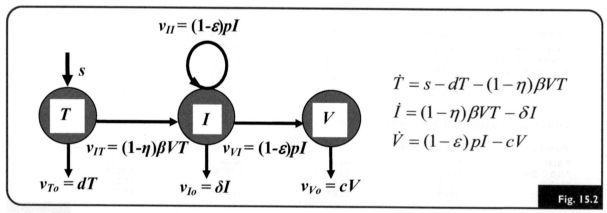

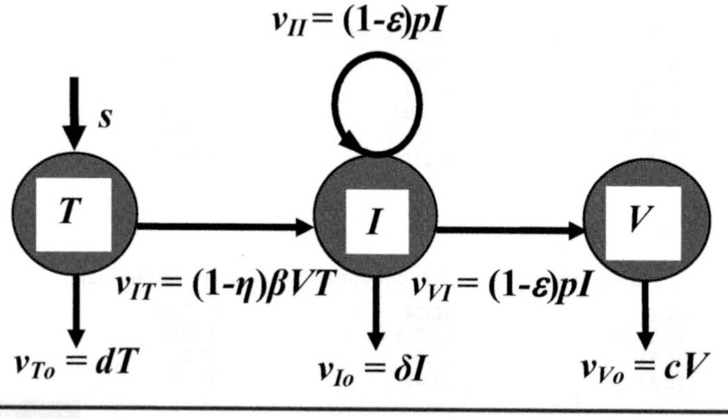

$$\dot{T} = s - dT - (1-\eta)\beta VT$$
$$\dot{I} = (1-\eta)\beta VT - \delta I$$
$$\dot{V} = (1-\varepsilon)pI - cV$$

Fig. 15.2

Early Perelson-group model of HCV disease dynamics (Neumann 1998), transformed here into compartmental form (see **Fig. 4.18**). The v_{ij}s are NL or linear *fluxes* (# cells/time). All arrows have flux units. Its *predator–prey standard viral infection* model equations are shown on the right.

4 This representation is similar to the compartmental disease dynamics models shown in **Fig. 4.18**. The self-loop on compartment I is used to "cancel" the effect of the apparent flux $(1 - \varepsilon)pI$ between compartments I and V in the diagram, which is not in the ODE for I in the model. NL compartmental models are not as easy to graph as linear ones. This is one way to do it, and is consistent with stoichiometric ODE modeling methods in biochemistry.

5 Maximum parsimony in model structure also was likely a practical goal of the modeling strategy, given the paucity of clinical data available.

6 Rate cells are newly infected.

The model incorporates and demonstrates *parametric forcing* – one way to represent parameter adjustments under control of exogenous forces (multiplicative "inputs"). Infection of liver cell targets T by the HCV virus is effected in the model via *adjustment of the rate parameter* β in the pathway from T to I, by attenuating β by $1-\eta$ coming from the (dashed) pathway from V. The mass flux term $(1-\eta)\beta VT$ appears in the first two equations and the model structure thus retains its compartmental nature. Effects on production appear only in the third, $\dot{V}$ equation, so they are easier to implement, as shown, by simply attenuating the fractional rate p by $(1-\varepsilon)$ – a second parametric forcing. Before drug therapy, both ε and η are zero (no attenuation). After therapy is begun, either $0<\varepsilon<1$ or $0<\eta<1$, or both. Estimation of ε and η from the data then provides quantitative information about the relative effects of IFN treatment on each mechanism.

The authors estimated the model parameters from clinical data, with results that did not fully clarify the mechanisms of action of IFN. They did, however, suggest that the *major initial effect* of IFN is to block virion production or release – with percent blocking efficacies of $100(1-\eta)=81$ to 96% (Neumann et al. 1998).

Mesoscale Mechanistic Models of Biochemical/Cellular Control Systems

The HCV modeling study described above, with model in **Fig. 15.2**, was simple, macroscopic, and effective for its limited clinical goals – modeling at its parsimonious-best (Neumann et al. 1998). It required no actual details about the mechanism of action of the virus – unavailable at the time – only explicit macroscopic "parameter adjustment" terms in the equations where viral sources and infection rates were located. This was a step forward, but it could not clearly identify the mechanism(s) of IFN action, because the simple parametric forcing implementation of drug effects was too simple. A more microscopic approach would have been needed if the goal were to more clearly identify and understand the mechanisms of action of the virus.

Models for elucidating molecular and cellular mechanisms usually are developed based on experimental data from studies targeting the molecular or cellular processes involved. Models at this level of detail typically involve individual molecules and their interactions, based on known biochemistry, molecular and cell biology, and – unfortunately – their completion usually also requires incomplete or unavailable information – from experiments often yet to be done.

This is where dynamic systems biology modeling can come to the rescue, with the added power to systematically elucidate mechanisms through an iterative process of alternative model construction, comparison with data, subsequent model adjustment, and new experiment design to reveal the missing links. A mechanistic model of a system, functioning at its best, is a hypothesis to be supported or refuted (and adjusted) by experimental observations. It can help discover how a system is or is not put together, morphologically and dynamically.

There's a caveat to be considered. Mechanistic modeling, as the name suggests, means mathematizing actual physical mechanisms, as isomorphically as possible, with equations reflecting functional biochemical or biophysical details. The problem is that there's a definite danger in trying to codify too much in the model. When there's no need to pin down all the details, or – more likely – there's not enough data, one can resort to using more *homeomorphic* modeling paradigms that might fit the data well, but don't necessarily capture all the details. NL Hill functions are commonly used successfully for this purpose, as illustrated in the integrated modeling and experimental studies described in the following sections.

We illustrate the power of this approach – first – with the help of two published molecular systems physiology models that delve into increasingly more microscopic, subcellular detail. We examine here how two different groups of investigators handled data-driven modeling problems associated with discriminating between competing mechanistic hypotheses. The first is typical of *switch-like responses in cell signaling subsystems* – developed from *molecular biochemistry data*, using several different-order and macroscopic Hill functions in their otherwise microscopic, molecular-level models (Ferrell and Machleder 1998). The second is of *gene network regulation of steroid metabolism* – from *pharmacogenomics and other molecular biology data* (Jin et al. 2003), which also employs Hill functions in modeling. Both mechanistic modeling examples are developed in greater detail in Javier and DiStefano III (2009) and complete details are available in the original publications (Ferrell and Machleder 1998; Jin et al. 2003).

Then we return to the manganese toxicokinetics model developed in part in Chapter 8. This PBTK model was developed specifically to test alternative hypotheses about possible routes of environmental manganese into the brain, which in excess causes ***manganism***, a Parkinson-like disease. Eleven ODE equations are developed here, based on data collected from as many organs in the rat. The manner in which these data "speak for themselves" – several merging into single compartment output dynamics – draws on several of the data-inspired analytical methods described in earlier chapters, and is key to successful quantification. The resulting model is quantified from this data, and the hypotheses are tested by analyzing the results. As a last example in this chapter, we continue exploring dynamical models of the tumor-suppressor protein p53, introduced piecemeal in several earlier chapters. We present several structural model hypotheses proposed for this system over the last decade in this chapter and carry one model further into Chapter 17, where we deal with complex numerical issues involving quantification of this *overparameterized* model (OPM models).

What Signals Control Frog Egg Maturation (Cellular Decision-Making)?

Ferrell and Machleder used systematic math modeling, experimentation and simulations in a series of studies designed to help elucidate the mechanisms that facilitate cell maturation in Xenopus oocytes (frog eggs) (Ferrell and Machleder 1998). It was known that a reproductive system time-varying hormonal signal, progesterone, produces an "all-or-none" switch-like response, resulting in maturation, as illustrated in **Fig. 15.3**. It was also known that a cascade of three kinase enzymes – phosphorylated Mos (Mos-P), Mek, and MAPK – were involved in an intrinsic positive feedback loop, also as shown. Positive feedback, when it does not lead to system instabilities, can generate a switch-like response. It was not known, however, whether the MAPK cascade was specifically responsible for converting the time-varying progesterone input signal into an all-or-none output and, if so, which component within this pathway actually

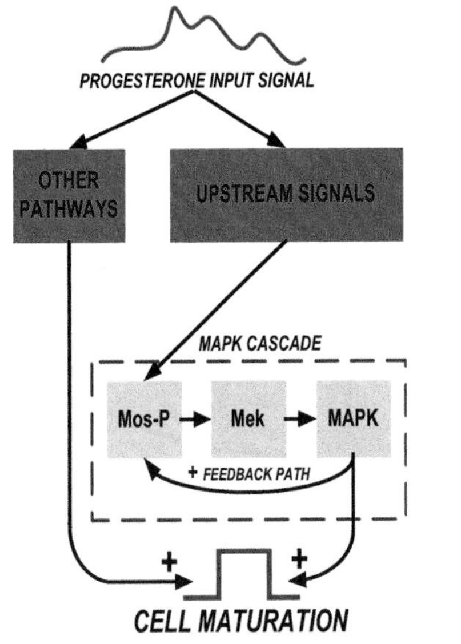

Fig. 15.3

Pathways leading to *Xenopus* oocyte cell maturation. A time-varying hormonal signal, progesterone, is converted into an "all-or-none" output (cell maturation). Each block is a potential generator of this switch-like response. Macroscopic use of Hill functions helped identify the pathway components actually responsible.

generated the switch-like behavior. Two critical steps in answering these questions involved fitting different-order Hill functions to data from several different experiments.

First, the authors recognized that the MAPK cascade, shown boxed in **Fig. 15.3**, can exhibit ultrasensitivity – resembling **positive cooperativity** (Chapter 7), a property well-characterized by a higher-order Hill function at its output: $y(u) = \gamma u^n/(u^n + K^n)$, where y is the response (MAPK activation/phosphorylation) to an input $u(t)$, an upstream signal – in this case Mos or progesterone (Ferrell and Machleder 1998). They hypothesized that the MAPK cascade was the key component generating the "all-or-none" output. They fitted this n^{th}-order Hill function to *phosphorylated/active* MAPK output response data under various progesterone input u conditions, and estimated the Hill coefficient of this response as $n = 42$ in individual cells. We have seen in Chapter 7 and elsewhere that the Hill function becomes more and more switch-like as n increases. The high n-value clearly established the output of the MAPK cascade in response to progesterone as indeed "all-or-none," resolving the first part of the problem.

Second, to test whether the MAPK cascade likely *generated* the ultrasensitive switch-like behavior, or simply propagated it from upstream signals, as in **Fig. 15.3**, they used a more direct experimental input into the MAPK subsystem block. Various doses of the enzyme malE-Mos were administered, which functions like the phosphorylated (activated) kinase Mos-P at the start of the cascade. The MAPK phosphorylation output response was once again switch-like, with $n = 35$, suggesting that the known ultrasensitivity of the MAPK cascade *itself* was responsible for generating the switch-like behavior.

The authors addressed and answered two additional questions in these studies. First, they found more rapid switching in Hill function-fitted MAPK cascade outputs from *intact cell data* ($n = 35$) than with *cell extract data* ($n \approx 3$). They postulated a critical role for **positive feedback** involving *other signals downstream* from *MAPK* that may be present only in intact cells,

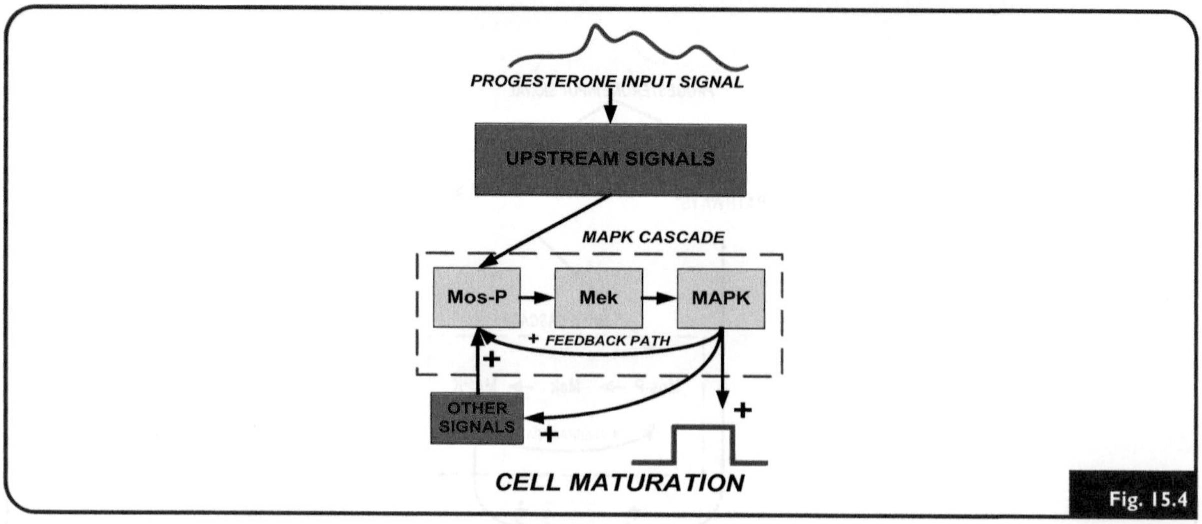

Fig. 15.4

Resulting structure for *Xenopus* oocyte cell maturation, after hypothesis discrimination experiments. Similar to **Fig. 15.3** except that other pathways have been excluded as potential generators of the step-like output, and a second positive feedback pathway present only in intact cells (via the OTHER SIGNALS block) is included.

and hypothesized that these signals involved new protein synthesis (**Fig. 15.4**). This was tested by measuring the MAPK response to malE-MOS input (as before), *with and without* an inhibitor of protein synthesis (independent experimental conditions to match these alternative hypotheses with alternate models). Again fitting a Hill function to the data, they found that protein synthesis inhibition slowed the response in intact cells to the degree found in cell extracts. They concluded that a positive feedback loop involving *protein synthesis* (in the OTHER SIGNALS block in **Fig. 15.4**) is likely to be important for switch-like behavior.

Finally, *simulations* combining both mechanisms – MAPK ultrasensitivity and protein synthesis-dependent positive feedback – produced a more desirable output than either mechanism alone, in two ways. First, the combined system showed *two stable steady states* (see Chapter 9), corresponding to "on" and "off" switch-like behavior, unlike the unstable "off" state with only positive feedback. Second, the combined mechanisms produced a faster response than ultrasensitivity alone, with ultrasensitivity establishing the switching threshold. They concluded that these two mechanisms together give rise to the all-or-none maturation signals of *Xenopus* oocytes in response to a continuously variable progesterone input (**Fig. 15.4**).

It is to be noted that systematic math modeling using Hill functions and simulation in these studies were instrumental in revealing important information about the mechanisms controlling frog egg maturation. Importantly, however, the subsystems analyzed do *not* fit the real biological meaning of the Hill function, i.e. *n* ligands binding to a single receptor. The Hill function, in a more macroscopic way, did however successfully capture the essential dynamics. And, as discussed in Javier and DiStefano (2009), the authors moved the solution to a higher level of detail using this relatively macroscopic approach, one that better elucidated specific physical properties of the biosystem structure.[7]

7 The Ferrel group also used the same cell decision-making approach in other molecular systems biology studies, notably (Brandman et al. 2005).

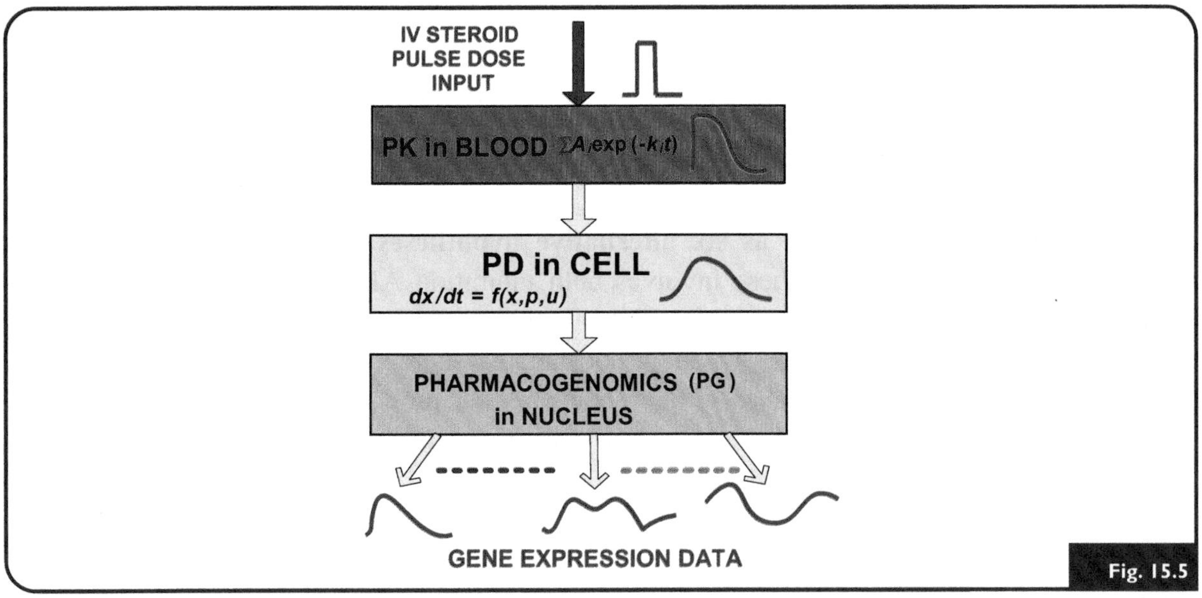

Fig. 15.5

Integrated data-driven model of steroid action. Sketches of dynamic responses at each level are also illustrated. A pulse-dose input of steroid, its PK response in blood, and steroid levels at the PD level site of action. A variety of potential mechanisms at the pharmacogenomic (PG) level produce multioutput gene expression dynamic data responses of different genes involved in steroid action.

Testing Six Alternative Steroid Dynamics Hypotheses Using Biochemistry, PK, PD, and Pharmacogenomic Data

Jin and coworkers used a novel ***pharmacogenomic*** (PG) methodology to explore several hypotheses about mechanisms of cellular corticosteroid metabolism, combining new genomics data with existing biochemistry, pharmacokinetic (PK), and pharmacodynamic (PD) data and modeling results (Jin et al. 2003). *Gene expression data* obtained from *microarrays* were the *multivariable outputs*, measured in liver cells in response to intravenous (IV) injection of this steroid hormone into blood, as illustrated in **Fig. 15.5**, based on the Jin group model. When a cell extract is added to the microarray preparation, any mRNA transcripts present bind to their respective probes in the microarray.[8] They were therefore able to measure multiple outputs (expression of many genes) in response to a single input (IV-injected steroid). Their primary goal was to examine the validity of several alternative cellular mechanism candidate models that might fit this data.

Their model integrated three known dynamical features of the system: drug PK in blood after injection; the PD dynamics of the drug entering cells and forming a complex with a receptor, the presumed drug effect target; and the dynamics of the drug–receptor complex affecting gene transcription – the pharmacogenomics (PG), all as shown in **Fig. 15.5**. Steroid dynamics in blood were represented by a simple multiexponential *forcing function* stimulating downstream components (Chapter 13). The dynamics of the drug and its effects within cells – elucidation of which was the goal of the studies – were described in extensive mechanistic detail. Formation of a

8 **Microarrays** use a matrix-shaped grid containing a different "probe" in each element space that binds to a specific gene transcript (mRNA). By labelling these transcripts with fluorescent dyes, *expression* (quantification) of individual genes is measured fluorometrically. Microarrays are especially useful for measuring expression of many genes, and thus very many outputs – simultaneously – because each *chip* (grid) can contain many thousands of probes (Mandy 2005; Stoughton 2005).

drug–receptor complex and its entry into the cell nucleus was modeled with mass action kinetics at the level of single molecules of drug and receptor. Dynamics of drug–receptor complexing in the nucleus (the PG level) were modeled with equations based primarily on mass action and macroscopic *stimulation/inhibition functions* (Chapter 7), to reflect the multiple known intercouplings between several proteins and processes involved in gene expression.

Six models were created – as **six alternative hypotheses** – representing six possible mechanisms of action. One of these involves both inhibition AND stimulation of gene transcription, by the drug–receptor complex DR_N in the nucleus, and also by a (hypothetical) *intermediate* regulatory biosignal, which we designate I here (Javier and DiStefano 2009). Both are represented in the first term of the model (mRNA production rate), as NL parametric forcing (adjustment/regulation) of the presumed constitutive mRNA production rate p_{mRNA}:

$$\frac{dmRNA}{dt} = p_{mRNA}\left(1 - \frac{DR_N}{K_{50} + DR_N} + k_S I\right) - d_{mRNA}mRNA \tag{15.6}$$

Inhibition by the drug–receptor complex is represented by a special Hill-type inhibition function, $1 - \frac{DR_N}{K_{50} + DR_N}$ (see Eq. 7.16), which captures the saturation nonlinearity and ensures that the whole term will not go negative. Since negative value is not an issue with stimulation, the stimulatory effects of the intermediate biosignal I were assumed linear,[9] with constant coefficient k_S. The second term in Eq. (15.6) represents mRNA degradation (assumed linear). The single hypothetical intermediate regulator I is a reasonable hypothesis, as most signaling pathways have multiple intermediates, with one dominating its dynamics, in this case expressed by the authors as:

$$\frac{dI}{dt} = k(DR_N - I) \tag{15.7}$$

The forward signaling and elimination rate constants of I in Eq. (15.7) were assumed equal (k), presumably to render their model more identifiable. The other five models of the Jin group were similarly described, all reflecting physiologically feasible mechanisms, and all different in particular details.

The Jin group measured dynamic expression of 143 genes in liver cells over time and this information was processed using **cluster analysis**, a statistical method that groups together data sets with similar temporal dynamics. Six clusters were found, one shown in **Fig. 15.6**. For each gene, all six models were fitted to the data and best fits were determined and statistically compared. Incompatible results (bad fits) rejected implausible mechanisms. Good fits revealed mechanisms most likely (among the six tested) controlling expression of individual genes. The dynamics of most probes in **Fig. 15.6** were best captured by the model in Eqs. (15.6) and (15.7) above. Data from one of these genes are also shown in **Fig. 15.6**, with the fitted simulation using this model. In some cases, more than one model fit the data well, suggesting multiple mechanisms as plausible candidates. In summary, the authors of this work were able to deduce which **putative mechanisms** (mechanistic hypotheses) controlled each of many observed behaviors.

9 This simplification also decreased the parameter space dimension, in principle rendering the model more identifiable.

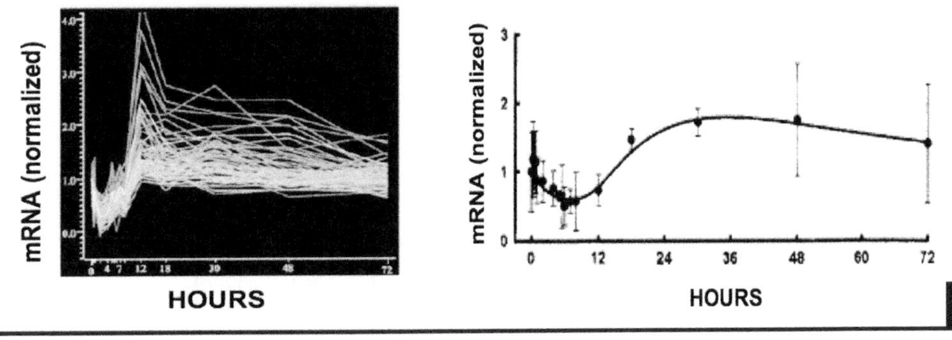

Left: Data from 66 probes grouped together by cluster analysis (Jin et al. 2003). Right: Model fit to the data for one gene in the same cluster. Reproduced with permission from figures in Jin et al. (2003).

The Jin group model is another good example of *multiscale modeling*, spanning multiple levels of detail over different time scales. The PK model of steroid in blood was macroscopic – indeed – without morphological structure (a model of the data); but the PD and PG models were microscopic, involving equations reflecting specific mechanistic details, with dynamics over shorter and shorter time scales. Many elements of these models, however, retained a relatively macroscopic character, such as the assumed linear stimulation of gene transcription by a hypothetical intermediate biosignal and the more generalized use of Hill functions, similar to the egg-cell maturation example above. This modeling approach was effective in part because the biophysics of several possible mechanisms were sufficiently well-known microscopically, and microarrays provided much more than the usual amount of multioutput data, from outputs emanating from coupled molecular (genomic) signals.

This pioneering work was among the first to integrate complex pharmacogenomics (PG), PK, and PD data and intermediate results quite effectively, over several scales, clearly demonstrating that complex hypotheses about biology at molecular and cellular levels can be systematically tested using dynamic systems biology methods.

Pedagogic Perspective on Structural Assumptions

As with all modeling endeavors, Jin and coworkers made a number of simplifying assumptions about biosystem structure (e.g. one dominating intermediate biosignal), kinetic parameters (e.g. some are bilaterally equal), dynamical characteristics (e.g. linear instead of NL degradation), and others, in developing their several models. Most follow conventional modeling practice in this domain, and some are likely acceptable without further evidence. Results, of course, depend more or less on the validity of these assumptions – "more or less" here referring to the extent model details and assumptions approximate reality. Whether or not they are ultimately justified, a few, e.g. bilaterally equal rate constants, were made to reduce the number of unknown parameters needed to achieve identifiability and thus get *any* quantitative results. This is often the practice when the data is insufficiently discriminating, but a step forward – cautiously taken – until better data become available. The most important factor in this regard is that modeling assumptions be out in the open (as they are here), to be questioned by others – as well as the modelers themselves. History will judge their validity and relevance of their modeling results.

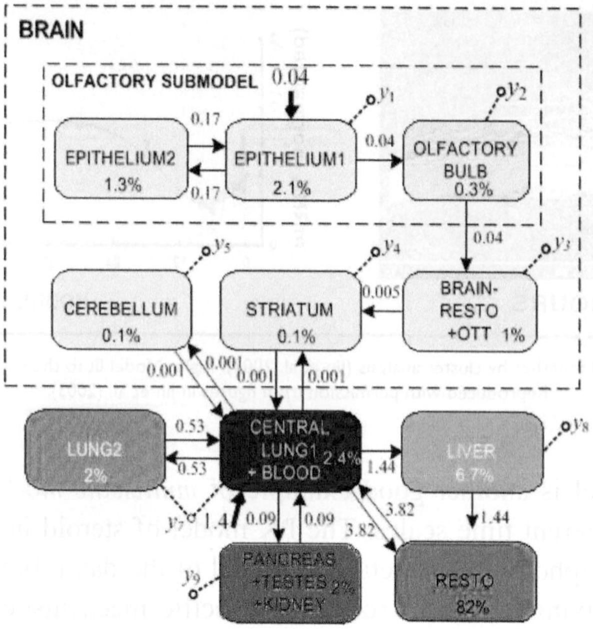

Fig. 15.7

Whole body *PBTK* model structure with whole-body Mn distribution (% *of total*) and Mn compartment interchange rates as *mass fluxes* (*μg Mn/day*) shown on arrows.
Figure adapted from Douglas et al. (2010) and reprinted by permission of Taylor & Francis (www.tandfonline.com).

Testing Manganese Brain Entry Pathway Hypotheses from Quantified Distribution Dynamics of Multiorgan Rat Data

The biology, clinical motivation, physiological details and structure for a hybrid PBTK model for manganese (Mn) distribution dynamics in the rat were developed in *Example 8.2*, based on the work of Douglas et al. (2010). The primary goal of the modeling was to test the hypothesis: *Environmental manganese is transported to brain cells in the rat primarily transneuronally, rather than via the cardiovascular and pulmonary systems after inhalation.* Both pathways are included in the model and their quantification was needed to resolve the issue. Equation development and quantification are described here.

Model Structure

This was described briefly in *Example 8.2* and illustrated in **Fig. 8.8**. **Figure 15.7** reproduces the RHS of **Fig. 8.8**, which includes quantitative results. Dominant BRAIN Mn compartments are divided into the OLFACTORY subsystem and the CEREBELLUM-STRIATUM-OTT-RESTO, each containing 3 Mn subcompartments each. There is local unidirectional intertransfer of Mn between OLFACTORY Mn and the Mn other subcompartments. In the experimental studies used to quantify this model, test-input 1 (olfactory input of inhaled Mn) is into olfactory BRAIN and test-input 2 (lung input of inhaled Mn) is into the CENTRAL compartment. The CENTRAL and 4 other nonbrain Mn organ compartments were defined in part by their physioanatomic interconnectivity and by the dynamic Mn response data in 11 organ compartments, as described below. There is saturated influx of Mn into LIVER and elimination of Mn from LIVER, presumably via the gut. The whole-body RESTO compartment represents Mn in all other unmeasured tissues that exchange with blood and concentrate Mn. State variables are expressed in amount (mass) terms, and the arrows between compartments represent mass fluxes (μg Mn/day). The legend of **Fig. 8.8** provides most of the physiological details.

Data, Data-Driven Structural Assumptions and Equations

The multiorgan rat data is summarized in **Fig. 15.8**. The experiment input was a 90-min exposure to an aerosol of radiolabeled Mn ($^{54}MnHPO_4$), modeled as an impulse function because the kinetic response data for 11 organs in **Fig. 15.7** were collected over 21 days.

Aggregation of several Mn organ dynamic responses in preliminary modeling steps rendered the model more tractable – the most important steps leading to successful quantification (*Let the data speak first*!). *Visual inspection of the data sets* indicated that data from pancreas, testes, and kidneys (P-T-K) had very similar kinetics, and were combined (*data was summed*), and represented as an aggregate P-T-K compartment for Mn (**Fig. 15.7**). Data sets from measured residual brain (B-RESTO) and olfactory tract and tubercle (OTT) also had qualitatively similar Mn kinetics. These *also* were combined (*data summed*), first because of their kinetic similarity and, second, because Mn exchange between these two compartments was much faster than between exchange parameters in the rest of the model, rendering them readily mergible. It was also found that *additional dynamics were needed for the lungs*; two compartments were postulated for lungs, which ultimately fitted the overall data well, as shown in **Fig. 15.8**.

Based on rat brain circuitry, it was postulated that Mn transport from olfactory bulb to B-RESTO/OTT was unidirectional, making the 3-compartment olfactory submodel *separable*, with two compartments for the olfactory epithelium and one for the olfactory bulb. This means *it can be separately quantified* (first below). The olfactory submodel equations describing masses of Mn in compartments 1, 2, and 3 (only) are:

$$\begin{aligned}\dot{q}_1 &= k_{12}q_2 - (k_{21} + k_{31})q_1 + u_1\\ \dot{q}_2 &= k_{21}q_1 - k_{12}q_2\\ \dot{q}_3 &= k_{31}q_1 - k_{43}q_3\end{aligned} \tag{15.8}$$

Equation (15.9), with fractional rate constants k_{ij}, are the ODEs for the remaining compartments. The only model nonlinearity (NL), fractional flux of Mn into LIVER, was modeled as a first-order Hill function in the equation for $\dot{q}_{10}$. Mn is transported actively into LIVER via the carrier protein transferrin and assumed to saturate at toxic levels.

$$\begin{aligned}\dot{q}_4 &= k_{43}q_3 - (k_{54} + k_{64})q_4 + k_{46}q_6\\ \dot{q}_5 &= k_{54}q_4 + k_{56}q_6 - k_{65}q_5\\ \dot{q}_6 &= u_2 + k_{64}q_4 + k_{65}q_5 + k_{67}q_7 + k_{68}q_8 + k_{69}q_9 + k_{6,11}q_{11}\\ &\quad - (k_{46} + k_{56} + k_{76} + k_{86} + k_{96} + k_{10,6} + k_{11,6})q_6\\ \dot{q}_7 &= k_{76}q_6 - k_{67}q_7\\ \dot{q}_8 &= k_{86}q_6 - k_{68}q_8\\ \dot{q}_9 &= k_{96}q_6 - k_{69}q_9\\ \dot{q}_{10} &= \gamma q_6/(K + q_6) - k_{0,10}q_{10}\\ \dot{q}_{11} &= k_{11,6}q_6 - k_{6,11}q_{11}\end{aligned} \tag{15.9}$$

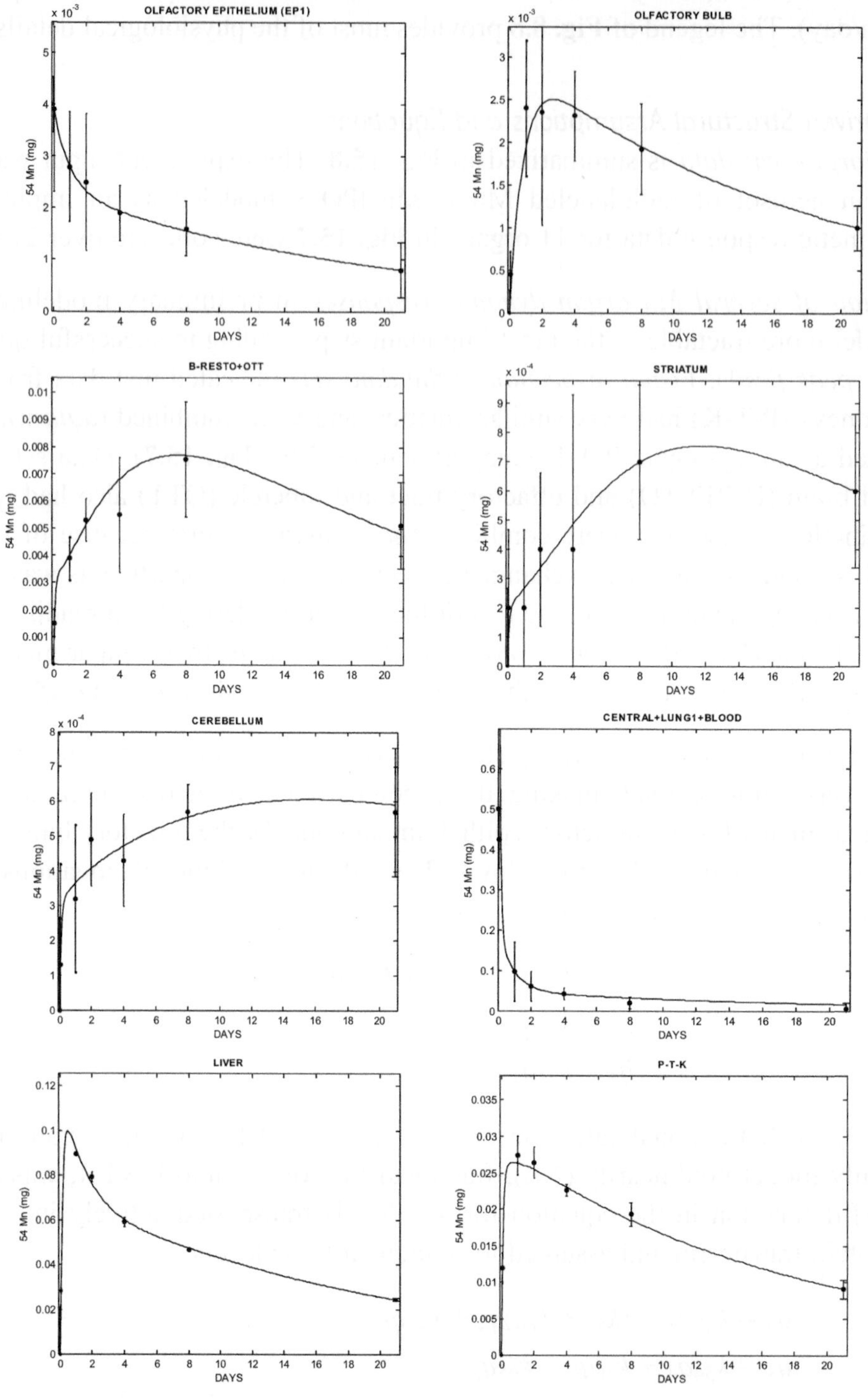

Fig. 15.8

Experimental data with standard deviations (SDs) and optimally fitted model outputs plotted for the body compartments shown in **Fig. 15.7**. The model was implemented and parameters were optimally fitted using *SAAMII* software (Extended *WLS* criterion function). The data and compartments representing the BRAIN-RESTO/OTT (B-RESTO/OTT) were lumped, as were the CENTRAL LUNG + BLOOD and the PANCREAS, TESTES and KIDNEYS — because the data and corresponding lumped compartments had nearly identical dynamic responses, differing only in magnitude. The model was fitted to the *summed outputs* from these compartments, after conversion to mass measurements (mass = conc. × organ wet wt). *Aggregations were important for successful model quantification.* Parameter CV estimates averaged 11%, with a maximum of 31%. Note that the model fitted the last two striatum data points very closely; important because Mn mass in striatum needed to be predictively simulated out to ≈ 100 + days to compute the integrated relative rates of Mn (*AUC* ratios) entering via the two input pathways shown. Relative sensitivities of the *AUC* for striatum ($AUC_{STRIATUM}$) to most estimated parameters were quite low (all but one <11%), with the key sensitivity being ~30% for a 10% perturbation in k_{46}, the fractional rate of Mn transfer to STRIATUM from the CENTRAL COMPARTMENT. 30% is not small, but it's the best that can be gleaned from this data, based on its inherent variability. Figure adapted from Douglas et al. (2010) and reprinted by permission of Taylor & Francis, www.tandfonline.com.

Model Quantification

Model parameters were optimally and simultaneously fitted to the multiorgan rat data graphed in **Fig. 15.8**, in two steps. The experiment input was a 90-min exposure to an aerosol of radiolabeled Mn ($^{54}MnHPO_4$), modeled as an impulse function because the kinetic response data for 11 organs were collected over 21 days. Outputs $y_1 \ldots y_8$, shown in the figure, comprise the single and combined-compartment measurement model.

The olfactory submodel was implemented and fitted to brain data first, using *SAAMII* software. Next, optimized parameter values from this submodel were *fixed* and the whole model was then implemented and quantified. *Structural identifiability* of all model parameters was established *numerically* as part of the numerical optimization procedure, as described in Chapter 12. This yielded *finite variance estimates* (average 11% CVs) for all optimized parameters (see **Fig. 15.8** legend for parameter estimation details).

Integrated **relative parameter sensitivities** were computed to assess how robust model results were to changes in optimized parameter estimates. The *area under the curve* (*AUC*) was selected as the test objective of sensitivity, because it *represents accumulated Mn over time*. The relative parameter sensitivities of the integral of striatum response $AUC_{STRIATUM}$ were computed for 10% variations ($\Delta p = 0.1$) in parameters p_j (Eq. 11.23) from:

$$S_{p_j}^{AUC} = (\Delta AUC_i(t,\mathbf{p})/\Delta p_j)(AUC_i(t,\mathbf{p})/p_j)^{-1}$$

The **relative steady state distribution of whole-body Mn** for each organ compartment were computed by simulating the quantified model out to 200 days and calculating: $\%q_{i200} = 100q_{i200}/\sum_{i}^{n} q_{i200}$, where q_{i200} is the simulated steady state mass of Mn in compartment i at 200 days. All are shown in **Fig. 15.7**. The LIVER, which actively concentrates Mn, contains the greatest percentage ($\sim$6.7%) of Mn among the measured organs. Both the STRIATUM and CEREBELLUM had ($\sim$0.1%) of total Mn, which is notable given that striatum volume is approximately 100-fold less than cerebellum. Most Mn (82%) is in other large and small organs (RESTO). Steady state pool size distributions of toxins or drugs are not usually reported for *PBTK* or *PBTK* models. They are easy to compute for compartments where concentrations are measured: mass $\approx$ toxin or drug concentration times organ wet weight. They are quite useful for where the toxins or drugs accumulate in body compartments.

The quantified model provides *a rejection of the original hypothesis* projected for this model. Instead of being one-sided, the relative contributions of olfactory ($\sim$52%) and pulmonary ($\sim$48%) pathways were estimated to be about equal.

Candidate Models for p53 Regulation

We extend development of the tumor-suppressor-protein p53 regulation modeling here, begun in *Examples 7.1*, *7.7*, and *7.8*. The p53 signaling system is one of the most important and studied control systems in cellular biology and its detailed structure is still very much in discovery mode.[10] It's

10 As of 2012, approximately 65,000 papers have been published about p53 since its discovery in 1979.

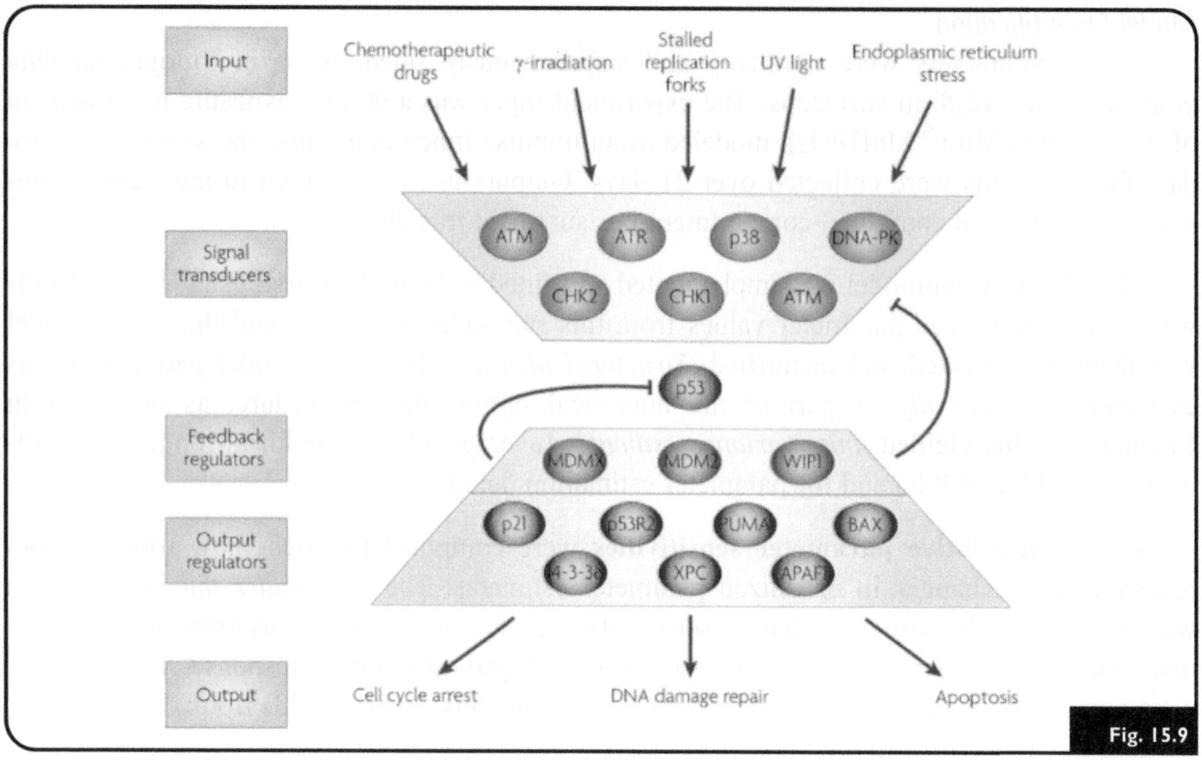

Fig. 15.9

***p53* Dynamic Signaling — From Inputs to Actions.** Input stress signals are detected by signal-transducer *kinases*, which *activate* p53 by phosphorylation. Some p53 targets act as *feedback regulatory signals*, altering kinase activity or p53 degradation rate. Other p53 targets trigger specific cellular *actions*, including cell-cycle interruption or apoptosis.
Figure reprinted by permission from Macmillan Publishers Ltd; Batchelor et al. (2009), copyright 2009.

known to be dysfunctional in most forms of cancer and this has been a major factor motivating numerous modeling efforts – generating a variety of different candidate structures as alternative mechanistic hypotheses – all directed toward better understanding of this system. Together, this variety of paradigms exemplify bottom-up modeling and remodeling well, for discovering what's included, excluded and how components postulated to be included are interconnected. The main biological features of the p53 signaling system and highlights of recent modeling efforts are developed below.

What Does p53 Do? – Modeler's Perspective

The overall p53 signaling process – from input to output – is illustrated in **Fig. 15.9**, adapted from (Batchelor, 2009). In response to DNA damage as *input* – our interest here – e.g. double-stranded breaks (DSBs) in DNA caused by external radiation, this specialized control system, normally resting quiescently, comes into action – acting as "guardian of the genome." Protein kinases at the *signal transduction* level sense DNA damage and activate p53. This initiates transcription of genes – at the final *output regulator* level – whose protein products suspend cell replication long enough for repair processes to occur, or lead to death of the cell (apoptosis) when there is irreparable damage. In this manner, p53 ensures that cells with damaged DNA do not divide to create more cells with errant DNA – namely, tumors.[11]

11 Beyond tumor suppression, p53 is also involved in reproduction, metabolism, aging and many other abnormal stress conditions.

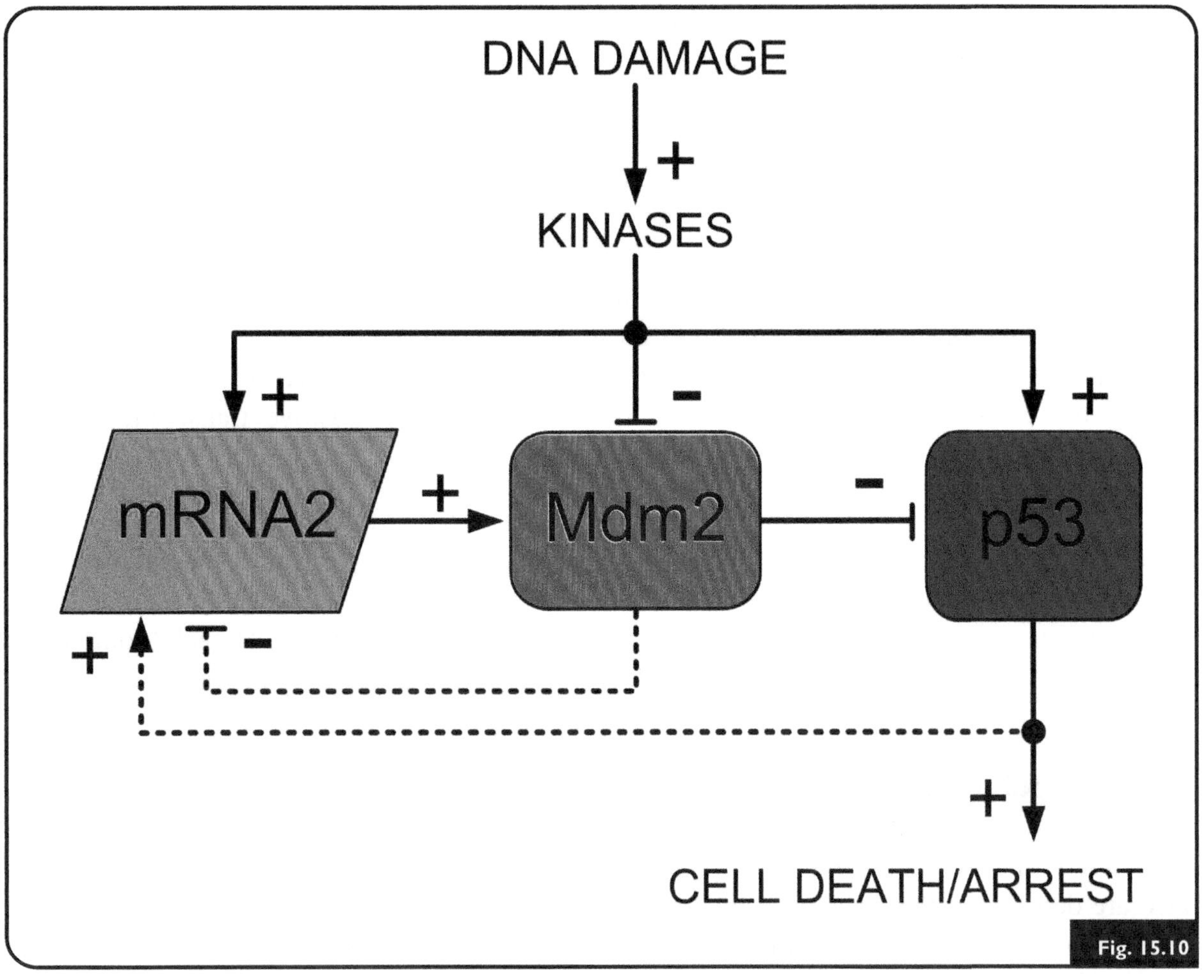

Fig. 15.10

Core subsystem block diagram model of p53 feedback regulation, with mRNA2 included explicitly in the feedback pathway. The outer negative feedback loop between p53 and Mdm2 represents inhibition of p53 degradation by Mdm2. The inner loop represents interruption of p53 transactivation of Mdm2. Also shown are the input to the system (what sets it into action) and the output (what it does). Both engineering (+ , −) and biochemistry sign designations are shown in the diagram. Dashed lines represent transcriptional effects. Only regulatory components and interactions are shown. Constitutive production and degradation of each biomolecule are excluded from the diagram.

p53 Multifeedback Regulation

Cell-cycle arrest or apoptosis may be desirable when DNA is damaged, but these specialized p53 actions must be tightly regulated, so they don't interfere with normal proliferation of healthy, undamaged cells. Elucidation of the mechanisms controlling these actions has been (and remains) a primary goal of modelers and other investigators of this system.

All current models include involvement of ligase Mdm2, widely acknowledged as a *negative feedback regulator* of p53, functioning as a bidirectional feedback subsystem called **core regulation** of p53 (**Fig. 15.10**). Here's how it works.

ATM[12] and other kinases sense DNA damage and activate p53 by phosphorylation. The positive transcription factor action of activated p53 stimulates production of Mdm2

12 ATM (*ataxia telangiectasia mutated*) is one of the protein kinases recruited and activated by double-strand DNA breaks (DSBs). It *phosphorylates* several key proteins that initiate activation of *p53* and other DNA damage repair system components.

(transactivation) by promoting transcription of Mdm2 messenger RNA, designated mRNA2. We modeled this in *Example 7.8*, as a *source* stimulation represented by NL mass action kinetics. Mdm2 negatively regulates p53, first of all, by promoting its ubiquitination and degradation by the proteosome, i.e. it enhances degradation of p53 – stimulation of a *sink*. This effectively decreases p53 abundance levels, thereby maintaining p53 levels low in the absence of DNA damage. This is represented by an arrow with minus sign (−) directed into the p53 block in **Fig. 15.10**. In addition, Mdm2 also binds to and blocks a critical transactivation location on the p53 protein, thus effectively and indirectly blocking *action* of p53 as a transcription factor for its target genes – the gene for mRNA2 as well as those that act to interrupt cell division. This second negative feedback regulator is illustrated in **Fig. 15.10** by the dashed inner loop arrow entering the mRNA2 block, alongside the dashed positive transcription factor action arrow entering the same block.[13]

These interactions occur at the gene level in the nucleus. Most published models of p53 regulation include these basic negative feedback relationships between p53 and Mdm2, some without explicit inclusion of mRNA2. Only regulatory components and signaling interactions are shown in the block diagram. Constitutive production and degradation of each biomolecule are excluded from the diagram, to maintain focus on regulatory signaling interactions.

Other Regulatory Factors and Variables

In addition to core regulation components, there is compelling evidence for modulatory effects of a variety of other biomolecules on p53 regulation (Wahl et al. 2005; Batchelor et al. 2009). Two are depicted at the *feedback regulator stage* in **Fig. 15.9**, protein MdmX and phosphatase Wip1. One cartoon/schematic model, with Wip1 and two kinases included – ATM and Checkpoint kinase 2 (Chk2) – is shown[14] in the LHS of **Fig. 15.11**, adapted from Batchelor et al. (2009). DNA damage activates ATM, which then activates Chk2. Both in turn then upregulate p53 by disrupting its interaction with (inhibitor) Mdm2. In this model p53 also upregulates transcription of Wip1, which has inhibitory (−) or stimulatory (+) effects on all components in the loops, by dephosphorylating ATM, Chk2, p53, and Mdm2. Dephosphorylation of Mdm2 then effectively generates another negative feedback pathway to p53 (**Fig. 15.11**) – indirectly with Mdm2 via Wip1 – in addition to those shown in **Fig. 15.10**.

Different p53 Models are OK

Evidence is accumulating for different mechanisms in different tissue types (Wade et al. 2010). This means potentially different molecular signaling components dominating the dynamics in different tissues – alternative hypotheses representing multiple putative interactions between different molecules considered relevant in different tissues. Another candidate model, with MdmX instead of Wip1 in the feedback regulation subsystem – possibly functioning in some tissues and not others – is illustrated in block diagram **Fig. 15.12**. Negative feedback between p53 and Mdm2 is represented by the outer feedback pathway, an extension of the core model

13 Whether the net effect of feedback in a loop is positive or negative is established by following the signs of the signals around the loop. Both feedback paths in **Fig. 15.10** have net negative effects.

14 Biochemistry notation is used by the authors of this figure, → for " + " and ⊥ for " − " signaling effects on the target molecule.

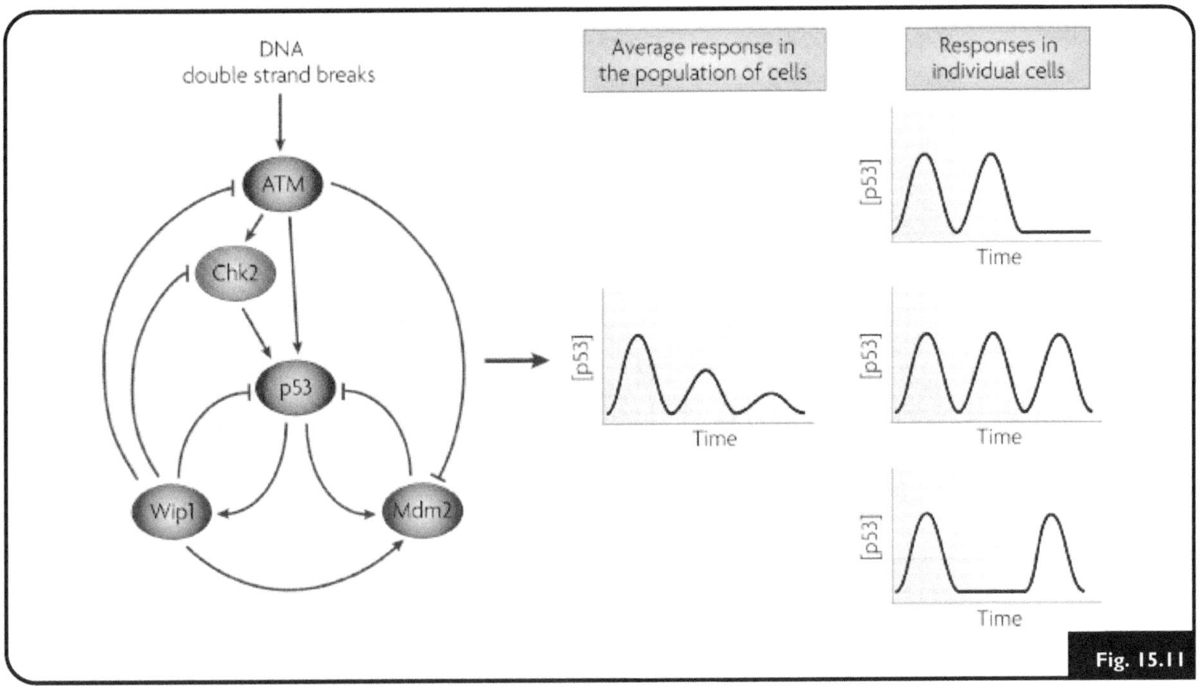

LHS — p53 feedback regulation model schematic, with two kinase and two enzyme components. DNA damage is detected by kinase ATM which then activates kinase Chk2. Both then upregulate p53 by disrupting its interaction with Mdm2. p53 also upregulates transcription of phosphatase Wip1, which then forms other feedback pathways to p53 in this model — by dephosphorylating ATM, Chk2, p53, and Mdm2. RHS — Typical pulsatile dynamic responses to DNA damage in cell population averages and in individual cells. The number of constant amplitude output pulses is proportional to the extent of DNA damage in this model, supported by data in individual and in populations of cells. Single cell measurements depict a series of undamped pulses, with different cells showing different numbers of pulses. Overall, they should appear damped when averaged, as in the left-most graph.

Figure reprinted by permission from Macmillan Publishers Ltd; Batchelor et al. (2009), copyright 2009.

of **Fig. 15.10**. The negative feedback arrow from Mdm2 to MdmX in the diagram represents degradation of MdmX via enzymatic ubiquitination by Mdm2-recruited E2 (the ubiquitin-conjugating enzyme). Association between the deubiquitinating enzyme HAUSP and MdmX is decreased when MdmX is phosphorylated, which occurs in the presence of DNA damage – hence the negative inhibitory arrow in the MdmX block. The negative arrow from MdmX into mRNA2 represents the path in which MdmX dimerized with Mdm2 inhibits p53 transcriptional activity in the forward path (Wahl et al. 2005). The strength of this effect on p53 is highly dependent on MdmX levels and may be different in different cell types (Gu et al. 2002; Wang et al. 2007; Wade et al. 2010). As a first approximation, the net effect of the inner feedback loop between MdmX and Mdm2 is "cooperative," or positive, i.e. *positive feedback* (Gu et al. 2002). The two inhibitory forward paths, from kinases to MdmX to mRNA2, also generate a net positive feed-forward effect of kinases on Mdm2.

p53 Oscillations and Models of How They Arise

It is well-established that the p53 system responds in a periodic or pulsatile fashion when stimulated by double-stranded DNA breaks (DSBs). Data supporting digital-like p53 output pulsing, with pulse number proportional to the extent of DNA damage, have driven much of the modeling in this field (Lev Bar-Or et al. 2000; Lahav et al. 2004; Ma et al. 2005; Geva-Zatorsky et al. 2006; Proctor and Gray 2008; Sun et al. 2011; Purvis and Lahav 2012).

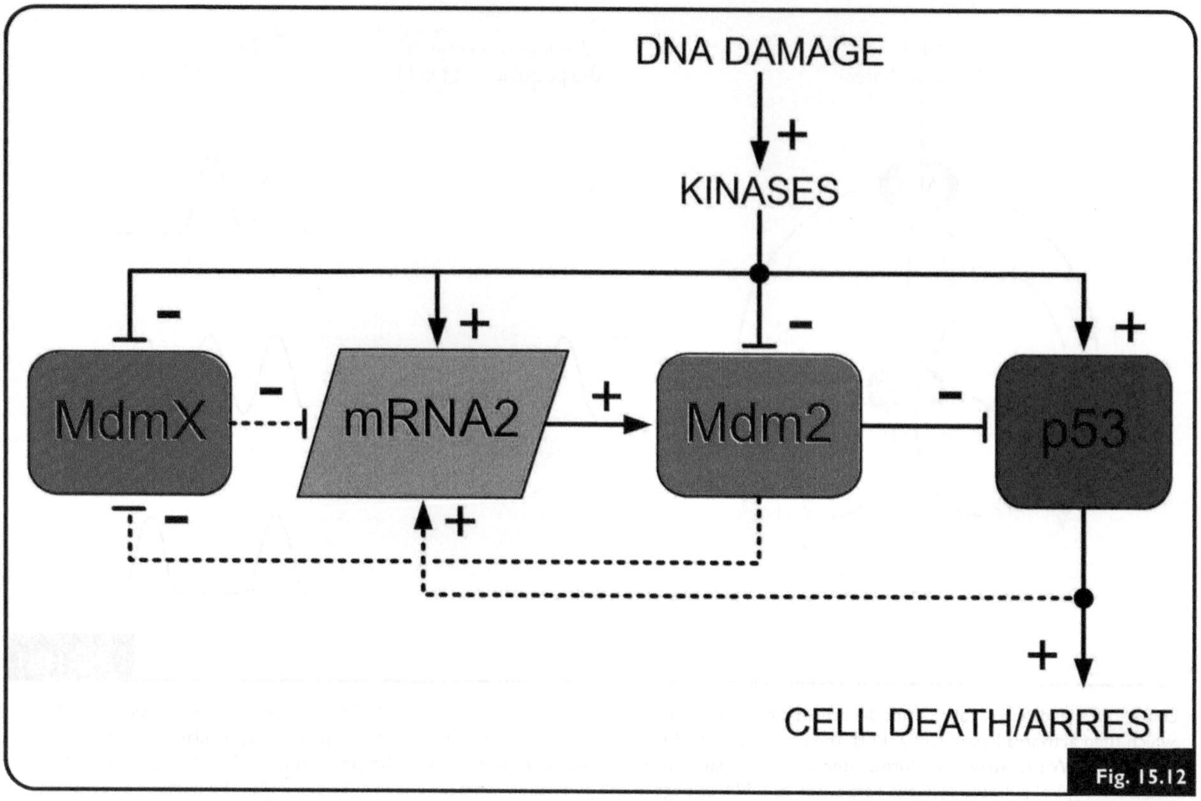

Fig. 15.12

p53 regulation with MdmX included with the core regulator. Constitutive production and degradations excluded. Two signaling feedback loops are present: the (outer) negative one between p53 and Mdm2; and the (inner) positive one between MdmX and Mdm2.

Elucidating a functional role for and identifying the mechanism(s) for cycling behavior in p53 signaling have together been major foci.

Cycling behavior in dynamical systems is typically due to parameter-value-dependent feedback loops inherent in the basic regulatory mechanism dynamics. One or more of the closed loops in **Fig. 15.11**, for example, can have parameter values that yield oscillatory outputs, whereas other parameter values may generate nonoscillatory responses.[15] But there are other sources for periodic behavior. We saw in Chapter 3 that signal time delays in a feedback loop also can contribute to oscillations at the output, and this hypothesis is prominent in some models (Rice et al. 2004; Wagner et al. 2005). Another possibility is that some "exogenous" pulse-generator, or periodic or switching signal at the input, or within the signal transduction subsystem, generates (or contributes to) oscillations at the output, as in the Sun group model (Sun et al. 2011). Oscillations also can be due to or sustained by inherent persistently exciting biological noise in the system, as reported for the models in (Geva-Zatorsky et al. 2010), or from simulated stochastic noise (Cai and Yuan 2009). More on this in the following paragraphs.

We saw in Chapter 9 that an NL ODE model with only two coupled state variables, e.g. p53 and Mdm2, can exhibit inherent limit cycling behavior. More state variables in the loop can elicit more complex dynamics and make it easier to generate observed oscillations.

15 We saw many examples of this in earlier chapters, e.g. a system transfer function representation with poles that go from purely negative to complex in the complex (σ vs. $j\omega$) plane as parameters are varied. Oscillations are damped if $\sigma < 0$ and undamped if $\sigma = 0$.

Most models typically include at least one more "intermediate" variable in the feedback loop. "Intermediate" messenger mRNA2 is commonly included, as in **Fig. 15.10**. For example, one of two Proctor-Gray models (Proctor and Gray 2008) includes mRNA2 as the third state variable, in the core loop; a second model in this report includes only p53 and Mdm2 as state variables in the core loop, driven by a third state variable – nucleolar protein activator ARF – that senses DNA damage at the input level, binds to Mdm2, and enhances its proteosomal degradation. Both models elicit limit cycle oscillations.

ODEs in most p53 models are multiscale and describe lumped local interactions in and among different cellular spaces. Reactions in the nucleus (e.g. between p53 and mRNA2) occur on the scale of milliseconds, and translocation of biomolecules interacting with p53 to and from the cytoplasm (e.g. Mdm2, MdmX, and Wip1) – and their production/degradation processes in cytoplasm – occur on the scale of minutes. Many include pure time delays, lumping molecular translocation, production, and degradation dynamics into delayed mRNA effects (extreme time-scaling), as in *Example 3.6*, and in similar examples throughout this book.

The Rice–Wagner models (Rice et al. 2004; Wagner et al. 2005) include ODEs for mRNA2, Mdm2, and both active and inactive p53 coupled as separate state variables, as well as time delays in two interaction pathways. With delayed feedback, Mdm2 molecules induced by p53 at one time mark the pool of p53 molecules for degradation at a later time – after the Mdm2 have been transcribed, exported from the nucleus, translated, and returned to the nucleus. In their model, this time lag combines with the ATM-controlled feedback strength, effectively damping the negative feedback signal enough to produce limit-cycling behavior after undergoing a supercritical Hopf bifurcation (Chapter 9) (see Exercise E15.2).

The Sun group (Sun et al. 2011) takes a substantially different approach, postulating a damage-dependent ATM activation *switch*, with bistable behavior between two saddle node limit points and dependent on two positive feedback loops – one due to ATM *autophosphorylation* (Chapter 9). Switching between *on* (damage still present) and *off* (damage repaired) then controls downstream p53 actions. Their model for p53–Mdm2 dynamics involves only Mdm2 and activated ATM as other variables, and its highly nonlinear and limit cycles at one of two Hopf bifurcation points.

Whereas most p53 models are deterministic (including Batchelor et al. 2009; Sun et al. 2011; Purvis and Lahav 2012; Alam et al. 2013), some are stochastic or include stochastic components (Ma et al. 2006; Cai and Yuan 2009; Geva-Zatorsky et al. 2010; Leenders and Tuszynski 2013). Stochastic models have been used to characterize p53 and Mdm2 dynamics in part because individual cells behave differently from each other in response to DNA damage. In particular, oscillations in single-cultured cells are constant amplitude, whereas oscillations are damped in measurements of averaged whole cell dynamic responses (Proctor and Gray 2008; Batchelor et al. 2011). This distinction is illustrated in the RHS of **Fig. 15.11**.

Stochastic p53 Models

Two atypical stochastic modeling studies are discussed further here. The **Cai–Yuan group** was the first to include the MdmX protein as an intermediate variable, in an unusually complex stochastic simulation model. It involves the p53, MdmX and Mdm2 interactions postulated in Wahl et al. (2005). Reaction kinetics in the Cai–Yuan model are characterized by 48 biochemical reactions, 21 molecular species and several time delays in the various pathways. Forty three model rate parameters were (heroically) quantified from literature data and biochemical/biophysical reasoning; the remaining five were calibrated by trial and error comparisons with published data on p53, Mdm2 and MdmX levels and kinetics (Wang et al. 2007). The model was implemented in an exact stochastic simulation algorithm, similar to the Gillespie algorithm described in Chapter 7 (Cai and Yuan 2009). Model simulations were consistent with published experimental interventions and oscillatory responses reported in experimental studies after DNA damage in the presence of noise. Oscillation amplitudes were much higher than oscillation periods, and simulation predictions suggest that intrinsic biological noise contributes 60–70% of the total variation in oscillation amplitudes and periods.

The ***Geva-Zatorsky group*** took an entirely different approach, which began with using two very simple alternative ODE models of core p53 dynamics, each driven by stochastic noise. Each was fully quantified (a first in the p53 modeling domain) from oscillating time course data on p53 and Mdm2 protein levels collected from several hundred human breast cancer (MCF7) cells in dynamical stimulus–response experiments. Large variability in oscillation amplitudes between individual cells made analysis in the time domain challenging, so they used Fourier (frequency domain) analysis to transform the varying waveforms from the time to the frequency domain.[16] The Fourier transform converts time domain data into equivalent frequency domain data. This is done by breaking up the time-based waveform into a series of sinusoidal terms, each with a unique magnitude, frequency and phase. This Fourier series of sinusoids, when added together, reproduces the original waveform exactly (Boashash 2003). This is a first step toward using a well-established engineering system identification approach for quantification – fitting the transformed model to the transformed data.

The Fourier ***power spectra*** (amplitude of the Fourier-transformed signal vs. frequency ω) of the pooled data were computed by Fourier transforming the average data over all cells, resulting in the root mean square (RMS) Fourier power spectra of p53 and Mdm2 data – the transformed data for fitting the model. The purpose of estimating these power spectra (also called *spectral densities*) is, first, to detect any *periodicities in the data*, by observing peaks at the frequencies corresponding to these periodicities. Appropriately structured models fitted to this data then yield useful estimates of structurally identifiable model parameters or combos of parameters.

Both stochastic ODE models studied by the Geva-Zatorsky group are structurally simple and *linear* (linearized): a two-dimensional (p53 and Mdm2 only) and a three-dimensional (p53, ATM and Mdm2 only) model. We illustrate this biomodeling approach for their 3-D model:

$$\begin{aligned}\dot{z} &= a_{zx}x - a_{zz}z + N_1\\ \dot{x} &= a_{xz}z + a_{xy}y - a_{xx}x + N_2\\ \dot{y} &= a_{yx}x - a_{yy}y + N_3\end{aligned} \tag{15.10}$$

16 This is equivalent to *steady state frequency response analysis* (see Chapter 6 in DiStefano III et al. 1990).

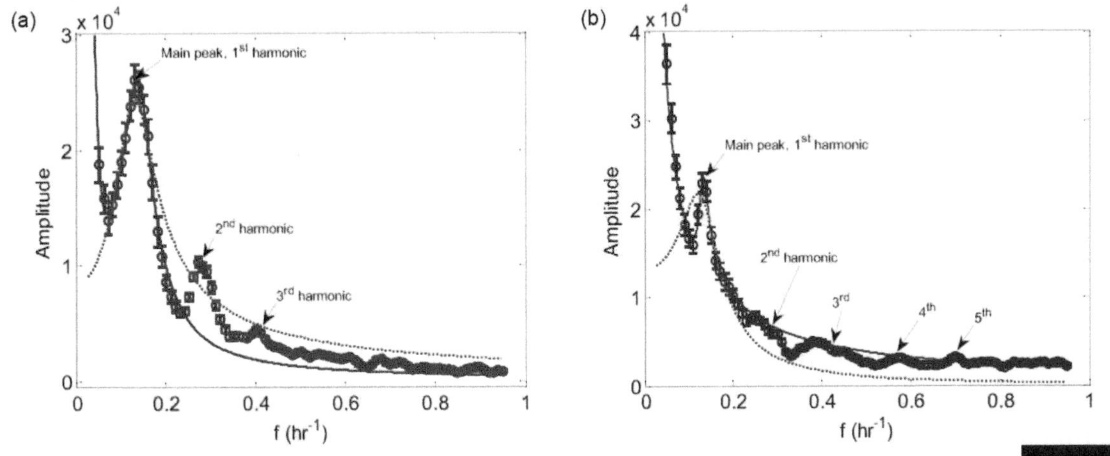

Fig. 15.13

Power spectra for p53 (LHS) and Mdm2 (RHS) in response to DSBs. Frequency response amplitude data are shown as circles with error bars. Peaks represent fundamental and harmonic oscillatory components in the data. Solid lines are fitted responses of the third-order model, Eq. (15.11), with optimized parameter estimates. Dashed lines are fitted responses of the second-order model, which doesn't fit as well. Reproduced with permission from Geva-Zatorsky et al. (2010).

The N_i are white-noise inputs representing stochasticity in the reactions, x is p53, y is Mdm2, z is ATM, and a_{ij} are rate constants (edges) representing the state variable interactions. x and y but not z were measured. The Fourier transform of an ODE is readily obtained from the Laplace transform of the ODE, by setting $s = j\omega$ and ICs = 0 (Chapter 2 and Appendix A). This gives:

$$
\begin{aligned}
j\omega Z &= a_{zx}X - a_{zz}Z + n_1 \\
j\omega X &= a_{xz}Z + a_{xy}Y - a_{xx}X + n_2 \\
j\omega Y &= a_{yx}X - a_{yy}Y + n_3
\end{aligned}
\tag{15.11}
$$

where ω is radian frequency, $j = \sqrt{-1}$, the state variables are complex frequency functions $X \equiv X(j\omega)$, $Y \equiv Y(j\omega)$ and $Z \equiv Z(j\omega)$, and n_i are the constant Fourier transforms of the white noise inputs.

Quantifying this transformed model means optimally fitting[17] the frequency-dependent *power spectra* of transformed state variables $X(j\omega)$ and $Y(j\omega)$ to the power spectra of the data $\overline{X}(j\omega)$ and $\overline{Y}(j\omega)$, functions of ω instead of t. Power spectra are computed as amplitudes[18] (magnitudes) of complex frequency functions. This results in estimates of the 10 free parameters of the model[19] in Eq. (15.11), the eight a_{ij} and two n_i.

Results: The Fourier spectrum of the data (**Fig. 15.13**) has a main peak centered at a frequency corresponding to a 7 ± 1 h period oscillation. Several additional secondary peaks – representing harmonics of the main frequency – occur at higher frequencies, which suggests some nonlinear effects. The computed spectra for the linear third-order model, which represents basic linearized interactions among p53, ATM and Mdm2 following DSB DNA damage,

17 The fitting method, e.g. least-squares minimization (Chapter 12), was unspecified in the reference.

18 Amplitude $= |F(j\omega)| = \sqrt{(Re\, F(j\omega))^2 + (\mathrm{Im}\, F(j\omega))^2}$

19 *SI* analysis of this model, using COMBOS (Chapter 10), indicates that all 10 are structurally uniquely identifiable, if the masses or molecule number of x, y, z are measured. If their concentrations are measured in a constant volume vessel, the answer is the same. But in whole cell preparations – depending on the degree of precision needed – volume corrections may be necessary to obtain accurate parameter estimates (see *Exercise E15.1*).

fitted these data quite well, as shown in **Fig. 15.13**. The model exhibits sustained oscillations when simulated with stochastic white noise inputs, but without the simulated noise, oscillatory outputs are damped (not sustained).

New Hypothesis: To explain this interesting observation, we could speculate that the third-order quantified model undergoes a bifurcation. The damped oscillations in the absence of input noise might be interpreted as the system operating in an asymptotically stable steady state region, where small perturbations are damped out. Then, with sufficient additive noise (signal strength), the bifurcation point is exceeded, resulting in a limit cycle.

The group also fitted the second-order model (ATM feedback loop inactive – red feedback path in **Fig. 15.13** missing) to a separate set of transformed experimental data from experiments with no DNA damage present. The 2-D model also fitted the power spectra of p53 dynamics without damage (not shown), exhibiting sustained oscillations with a longer period (~ 12 h). The power spectra exhibited no additional harmonic oscillations, consistent with the model being simpler – with no ATM feedback present.

From their quantified model, they estimate the rate constant representing activation of Mdm2 by p53 as $a_{yx} = 0.29 \pm 0.03\ \text{h}^{-1}$; and inhibition of p53 by Mdm2: $a_{xy} = -0.55 \pm 0.05\ \text{h}^{-1}$. This model also provides parameter values from which the p53 *half-life* in the presence of Mdm2 can be estimated, i.e. from a_{xx}, as ~ 52 min; and similarly the Mdm2 *half-life*[20] in the presence of p53 is estimated as ~ 100 min – from a_{yy}.

Remark: John Tyson presents a clear and insightful discussion of p53 oscillations and the various ways they might arise, given the varieties of published data with periodic responses that differ in particular aspects (Tyson 2006). Literature data support a variety of different models under different conditions, e.g. models for responses over short or long time periods, and models for small or large damage signals. He discusses two of the several kinds of bifurcations that can produce limit cycles, with both evident in different published data, e.g. p53 signaling over short time versus long time period data. ***Hopf*** bifurcations (Chapter 9) have fixed period and increasing amplitude of response with increasing input signal strength – common in negative feedback systems, particularly in studies with long time periods. Several published models demonstrably produce Hopf bifurcations (Ma et al. 2005; Wagner et al. 2005). The second type noted, ***homoclinic*** bifurcations, would involve both negative and positive feedback and have fixed amplitude and increasing frequency of response with increasing input signal strength – observed particularly over short time periods (Ciliberto et al. 2005). Either of these bifurcations could be present in the Geva-Zatorsky model, as suggested in the ***New Hypothesis*** noted above. Tyson also underscores the distinction between model and real system, by reminding us that limit cycles are a name given to recurrent constant amplitude-constant period solutions of systems of *NL ODEs*, meaning they are simulated responses of *models* with these properties. The very good fit of oscillating models to the Geva-Zatorsky group data does not preclude equivalent (or better, or more accurate) models that exhibit limit cycling, harmonic, or another kind of periodic behavior, possibly

20 These half-life estimates – estimated from their linearized-range tracer-level data – are 2–4 times greater than values reported by other groups, using larger endogenous p53 or Mdm2 perturbations to study these kinetic responses.

yet-unnamed. This story is still unfolding, and it seems more and more likely that no unique model exists for all periodically varying responses to DSBs in all cells.

Different Structures for Different Inputs

Another layer of complexity has recently been demonstrated in p53 signaling. It responds with qualitatively different dynamics to different input stresses. For example, UV radiation exposure in single cells generates a *noncyclic* single dose-dependent pulse, i.e. single output pulse magnitude roughly proportional to dose (Batchelor et al. 2011), as illustrated in the RHS of **Fig. 15.14**, instead of the periodic constant-magnitude pulses, shown in the LHS. Instead of ATM, the model structure that generated this output is replaced by kinase ATR, and Wip1 does not dephosphorylate ATR in this model. Thus, outputs are completely different in response to single-strand DNA breaks (SSBs), in this case UV radiation.

A 4-State Variable Alternative Model for p53–Mdm2–MdmX Signaling

The new model described in some detail here is primarily data-driven, following the block diagram structure in **Fig. 15.12**. The data consist of simultaneous dynamic signal response-level measurements of three protein concentrations (fmol/10^6 cells) and one mRNA – chosen as the four model *state variables*. The model *input* variable was chosen as a phosphor-signal-intensity "surrogate" indicator of kinase activity. All data come from a single experiment done in cultured whole MCF7 (human breast cancer) cells following DSB-DNA damage (Wang et al. 2007).

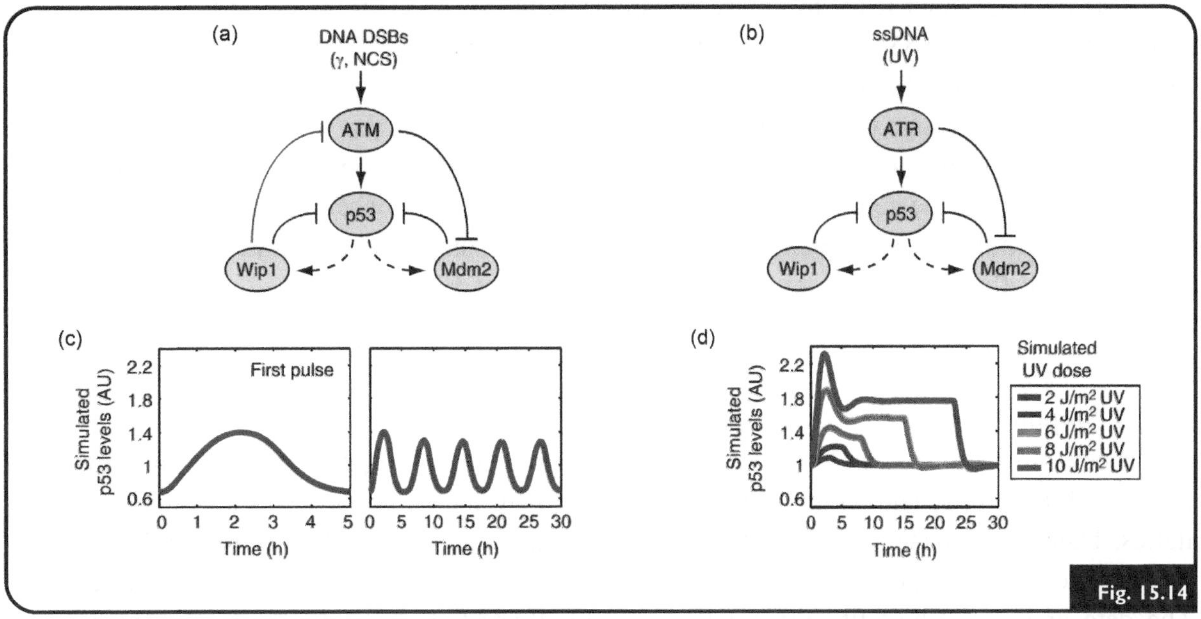

Fig. 15.14

Two different p53 feedback regulation schematics for different damage signals, eliciting different outputs. Dashed lines indicate transcription; solid lines protein–protein interactions. LHS — Key p53 network species, pathways and simulated p53 levels in response to DSBs, e.g. gamma radiation. The amplitude duration and frequency of individual p53 pulses are fixed and do not depend on the damage dose. RHS — Different p53 network species, pathways, and simulated p53 levels in response to SSBs, e.g. UV radiation. Kinase ATR replaces ATM, and Wip1 does not dephosphorylate ATR, as it does in the LHS model. The output of this model is completely different — only a single pulse with amplitude dependent on the magnitude of the damage input signal.

Figure reprinted by permission from Macmillan Publishers Ltd; Batchelor et al. (2011), copyright 2011.

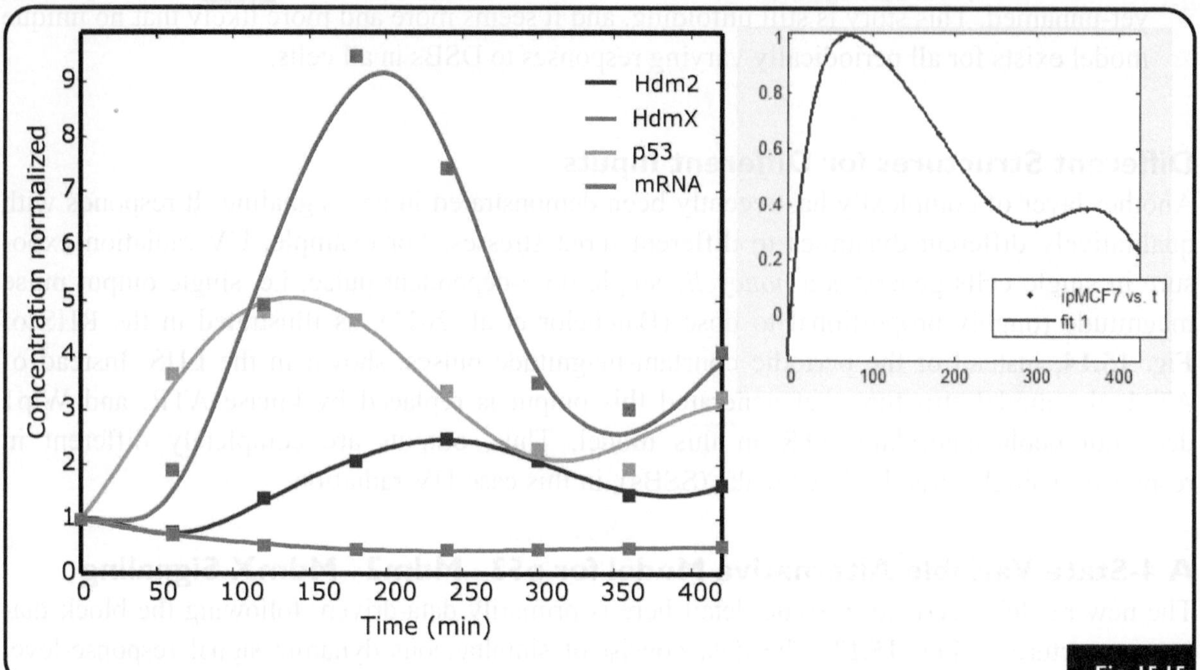

Fig. 15.15

Normalized surrogate kinase input data fitted by a 5th-order polynomial (RHS) and (LHS) output data fitted by p53 signaling model, Eqs. (15.12)–(15.17).

Model Input

DSB damage input leads to kinase phosphorylation of p53, occurring and measured at the serine 15 epitope of p53 in this data set. Normalized serine-15-p53 data are the *input stimulus* in this model, denoted *kinase*(t). This signal acts as a surrogate for the real ATM and other participating kinase signal dynamics actually unfolding during the 7-h experiment. It acts in continuous synchrony with the four other downstream signals, thereby likely accounting for inhomogeneities in the cell population DSB response dynamics observed in the data. To facilitate model simulation and quantification, the *kinase*(t) data were separately fitted by least-squares to a fifth-order polynomial, shown on the RHS of **Fig. 15.15**.

Model Output Measurement Equations

These are $y_1(t) \equiv Mdm2$, $y_2(t) \equiv MdmX$, $y_3(t) \equiv p53$, and $y_4(t) \equiv mRNA2$, the levels of Mdm2, MdmX, p53, and Mdm2-mRNA *normalized* by their initial values and represented in italics. Data errors $e_i(k)$ (%CV = 20% for all data) are included in the discrete-time measurement model: $z_j(k) = y_j(k) + e_j(k)$, for $j = 1, 2, 3, 4$; $k = 0,1,\ldots,7$, the eight hourly sample times. The data and initial best-fit model responses are illustrated in the LHS of **Fig. 15.15** (error bars not shown).[21]

21 This model will be quantified, optimized and reduced in Chapter 17.

Model Equations

The four ODEs of the model are based on the block diagram connectivity in **Fig. 15.12** and molecular biology information and assumptions noted. Normalized[22] whole-cell Mdm2 dynamics are represented by:

$$\begin{aligned}\frac{dMdm2}{dt} &= p_{Mdm2}mRNA2(t-\tau) - d_{Mdm2}Mdm2 \\ &\quad - d'_{Mdm2}Mdm2\left(\frac{Mdm2}{Mdm2 + K_{Mdm2-Ub}}\right)\left(1 + \frac{V \cdot kinase}{kinase + K_{Mdm2-p}}\right)\end{aligned} \tag{15.12}$$

As in Chapter 7, the first term in Eq. (15.12) represents production of *Mdm2* protein via translation from *mRNA2*. Time-delay τ is included to approximate otherwise unmodeled translation and translocation processes (extreme time-scaling). The second term reflects constitutive, basal degradation of *Mdm2*. The third regulatory term represents "auto-degradation" of *Mdm2*, due to enzymatic ubiquitination of itself, and also includes a kinase-dependent stimulation function (Chapter 7), because phosphorylation of Hdm2 blocks its association with the deubiquitinating enzyme HAUSP, thereby increasing auto-degradation.

Normalized whole-cell *MdmX* dynamics are represented by:

$$\begin{aligned}\frac{dMdmX}{dt} &= p_{MdmX} - d_{HdmX}MdmX \\ &\quad - d'_{MdmX}Mdm2\left(\frac{MdmX}{MdmX + K_{HdmX-Ub}}\right)\left(1 + \frac{V' \cdot kinase}{kinase + K_{MdmX-p}}\right)\end{aligned} \tag{15.13}$$

As above, the first two terms reflect constitutive production and degradation of *MdmX*, with rates for both production and fractional degradation assumed relatively constant and independent of the other signals over the time scale of DNA damage repair. The third term represents regulatory functions. Degradation of the substrate *MdmX* via enzymatic ubiquitination by *Mdm2*-recruited E2 (the ubiquitin-conjugating enzyme) is described by the first Michaelis–Menten (M–M) function. Association between the deubiquitinating enzyme HAUSP and *MdmX* is decreased when *MdmX* is phosphorylated, which occurs in the presence of DNA damage. As in Eq. (15.12), this *disinhibition* of *MdmX* ubiquitination and subsequent degradation is represented by the M–M kinase-dependent *stimulation* function. The effects of E2 and HAUSP with *MdmX* are included in d'_{HdmX} of the regulatory term, assuming E2 and HAUSP are approximately constant in this reaction. HAUSP (but not E2) were measured in this study and no change in HAUSP was apparent (Wang et al. 2007).

For p53, it is assumed that the majority of p53 activity is due to stable, phosphorylated p53, with unphosphorylated p53 also active. The data do not discriminate between phosphorylated and unphosphorylated forms, so a single equation for the sum of the two is assumed (denoted *P53*):

$$\frac{dP53}{dt} = p_{P53} - d_{P53}P53 - d'_{P53}Mdm2\left(\frac{P53}{P53 + K_{P53-Ub}}\right)(1 - k_1\, kinase) \tag{15.14}$$

22 All state variables are normalized by their ICs – so they all begin at 1.

The first two terms represent constitutive *P53* production and degradation. The first part of the third, regulatory term represents ubiquitination of *P53* by *Mdm2*, as in Eq. (15.12). Interaction between *P53* and *Mdm2* is primarily regulated by phosphorylation of *Mdm2* (and *MdmX*). When DNA damage is present p53 phosphorylation also inhibits this process, represented in the third term by the linear kinase-dependent *inhibition function*: $(1 - k_1\, kinase)$.

Normalized *Mdm2 mRNA2* dynamics are represented by:

$$\frac{dmRNA2}{dt} = p_{mRNA2} - d'_{mRNA2} mRNA2 + p'_{mRNA2}\left(\frac{P53^4_{total}}{P53^4_{total} + K^4_{transcription}}\right)(1 + k_2\, kinase) I_{m2} I_{mX}$$

$$I_{m2} \equiv 1 - k_5 Hdm2, \quad I_{mX} \equiv 1 - k_6 HdmX \tag{15.15}$$

The first two terms in Eq. (15.15) reflect constitutive production and degradation of mRNA2. The third (regulatory) term includes a fourth-order Hill function for depicting P53-dependent mRNA transcription, via multiple monomers of P53, primarily as a tetramer, and with transcription limited by the availability of p53-specific DNA-binding sites and transcription machinery.[23] Direct effects of *kinase* are included explicitly as the linearized *stimulatory* term: $(1 + k_2\, kinase)$. The two *inhibitory* terms I_{m2} and I_{mX} represent inhibition of p53 transcription factor action due to *Mdm2* and *MdmX* binding, again with different strengths and therefore different parameter values for *Mdm2* and *MdmX*.

Parameter Relationships from Steady State Constraints

Exact relations among constitutive production, degradation and other model parameters can be written in terms of half-lives ($\ln 2/d$) with the system in basal steady state (ODEs = 0). Such relations can reduce the parameter space size if half-life values can be measured independently. The half-life (T_{mRNA2}) of *mRNA2* is the only state variable in this model that might be measured without interference from regulatory interactions with other system biomolecules. With the derivative = 0 in Eq. (15.15) at time 0^- (basal state, just before damage), we have $P53(0^-) = 1$, $MdmX(0^-) = 1$ and $Mdm2(0^-) = 1$ at the initial time. Then, with no damage present, the active kinase signal is absent, so $kinase = 0$. This yields the constraint relationship:

$$p_{mRNA2} = \frac{\ln 2}{T_{mRNA2}} - p'_{mRNA2}\left(\frac{1}{1 + K^4_{transcription}}\right)(1 - k_5)(1 - k_6) \tag{15.16}$$

If T_{mRNA2} can be measured independently, substitution of Eq. (15.16) into Eq. (15.15) then gives an ODE for mRNA2 with one less unknown parameter to be estimated from kinetic data:

$$\frac{dmRNA2}{dt} = \frac{\ln 2}{T_{mRNA2}} - p'_{mRNA2}(1 - k_5)(1 - k_6)\left(\frac{1}{1 + K^4_{transcription}}\right) - d_{mRNA2} mRNA2 + p'_{mRNA2}\left(\frac{P53^4_{total}}{P53^4_{total} + K^4_{transcription}}\right)(1 + k_2\, kinase)(1 - k_5 Hdm2)(1 - k_6 HdmX) \tag{15.17}$$

23 Transcription due to the effects p53 has on the transcription machinery and chromatin, such as the binding of histone acetyl transferases, or recruitment of transcription elongation factors.

If T_{mRNA2} is not available, it can be estimated along with the other unknowns.

Structural Parameter Identifiability

Structural identifiability (*SI*) of this model was analyzed using *COMBOS* (Kuo et al. 2013) with the time delay in the Mdm2 equation set equal to zero. Surprisingly, all 23 unknown parameters are likely globally uniquely *SI*, which means they would be uniquely quantifiable perfectly from noise-free data. The differential algebra algorithm upon which *COMBOS* is based does not handle time delays – the reason for the "likely" qualifier.

Initial Parameter Estimation and Numerical Identifiability

This model is ideally suited for initiating parameter estimation using the *forcing function method* of Chapter 13, because all state variables were measured. Three of the four data sets were fixed as output variables – "fitted" as straight lines connecting the data points – with the parameters of the remaining equation estimated from the data for that single variable. This was repeated for all four state variables. At the end, all 24 model parameters were estimated simultaneously, resulting in a model fitting the data reasonably well (**Fig. 15.15**). This was accomplished first using the limited facilities of *SAAMII* parameter estimation functionality, later confirmed using both *COPASI* and *AMIGO* package tools. Time-delay τ in Eq. (15.12) was fixed to various values from 0 to 30 min. The fitted model did not change so the delay was set to zero. Numerical identifiability results, however – as indicated by parameter estimation statistics from *SAAMII* and *AMIGO* – were quite poor. Not surprisingly, only subsets of model parameters were quantified with reasonable precision from the real, noisy and sparse data. Numerous trial-and-error solutions generated multiple local solutions, typically finding clusters of parameters that appeared to be more identifiable than individual ones, all of which rendered these preliminary results inconclusive.

This lengthy learning experience strongly suggested that this model – structured with many approximations – remains over-parameterized (OPM)[24] and in need of simplification. The estimated parameter correlation matrix, shown in **Fig. 15.16**, illustrates the OPM problem semiquantitatively. This matrix was evaluated from the Cramér Rao lower bound (Chapter 12) for the preliminarily quantified model, computed using the *AMIGO* package as an early step in model reduction.

Solution of the OPM problem for this model is deferred to Chapter 17, where sensitivity analysis, model reduction and global parameter estimation methods are applied. The particular parameter estimation results obtained in the preliminary studies also suggested that *COMBOS* might be helpful first, for qualitatively analyzing *SI* properties of individual nonlinear Hill function terms in the model. This did facilitate use of *AMIGO* for computationally finalizing a reduced-order quantified model solution – as described in Chapter 17.

24 OPM usually means there's not enough or not good enough data to quantify the model adequately. Alternatively, the model structure may be defective. This option – the modeler's dilemma – must always be on the table. The task here was to fit the given data – excellent and self-consistent data – but, as usual, sparse data.

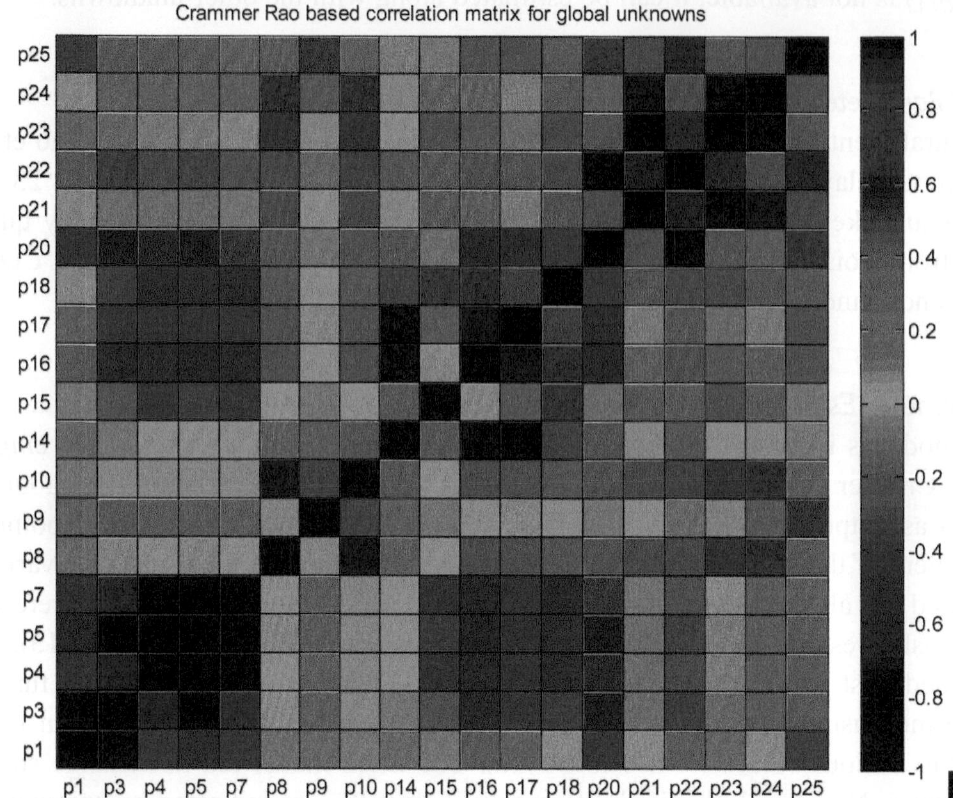

Fig. 15.16

Final parameter correlation matrix for a preliminarily quantified p53 signaling model. Parameters in Eqs. (15.12) through (15.15) were renamed consecutively to simplify the notation for parameter estimation and analysis. Darkest colors mean strongest correlations and poorest numerical identifiability properties for the parameters in the pair. For example, p3 and p5, p4 and p5, p5 and p7, p8 and p10, p21 and p23, and p21 and p24 are highly correlated.

Preliminary p53 Signaling Model Stability Analysis

In lieu of a formal stability analysis (pending at completion), the dynamical responses of a quantified model to nonperiodic (e.g. single pulse) inputs can be explored for possible periodic behavior by simulation.

Figure 15.17 illustrates the simulated responses of the preliminarily quantified and delay-free full p53 signaling model described above to the simple nonperiodic input pulse depicted in red.[25] At the end-time of the pulse, 420 min, it falls to a low constant level (<0.18) representing nonzero basal kinase levels; and it does not repeat. Damped oscillations are evident in all four signal responses following the step input. This implies that oscillatory responses in this model of p53 signaling are due to inherent multifeedback characteristics of the model structure and its parameter values. This model, quantified from real data, oscillates without the need for repeated input pulsing or time delays in the feedback.

Recap and Looking Ahead

This chapter is focused on how dynamic biosystem modeling methodologies are used to formulate alternative mechanistic hypotheses about biosystems, with the purpose of effectively

25 The responses of the model to the real single pulse described by the data in **Fig. 15.15** were similar, as shown in Chapter 17.

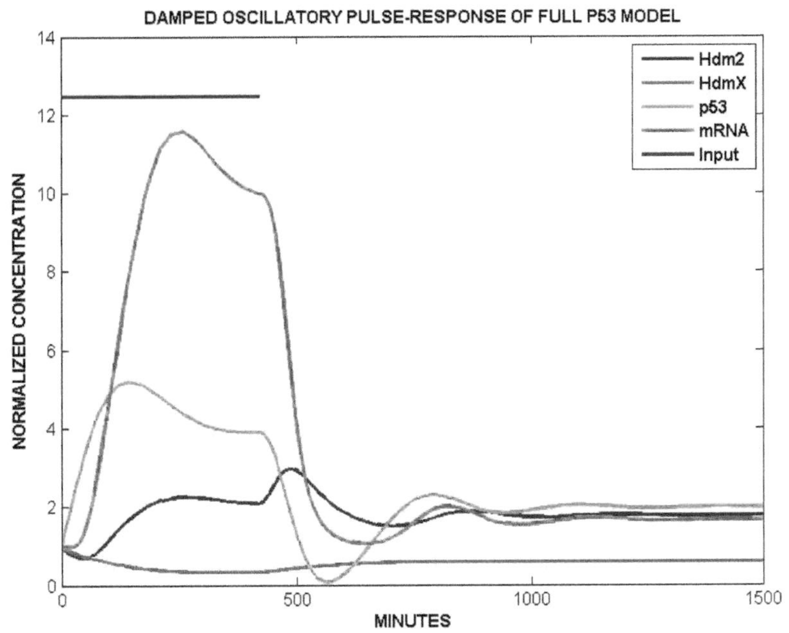

Fig. 15.17

Simulated p53 signaling responses to a 420-min pulse of kinase phosphorylation of p53. This model, quantified from real data, oscillates without the need for repeated input pulsing or time delays in the feedback.

distinguishing among alternative models. And it includes examples – at micro-, meso- and macroscopic scales – to illustrate the process.

Explicit models provide a means for comparing experimental results (data) with expected results (hypotheses) – the tools for testing the validity of assumptions about a system. Procedurally, alternative models are tested against the *same set of data*, analyzing and statistically comparing how well each fits this data. Procedurally, data comes first, i.e. *data rules the modeling kingdom*. Alternative mechanistic hypotheses are both driven and tested by data – existing and new data. The data are in control of the extent of what we can discover. Modeling specifics follow.

An early section covers statistical criteria for discriminating among alternative models, illustrated with multiexponential model discrimination, a class of nested and increasingly complex models capable of revealing the dimensionality of many dynamic system models – a starting point for more structured/mechanistic modeling.

Detailed modeling of the biodynamic systems described or developed in the bulk of the chapter illustrates particulars of the data-driven modeling and discrimination process, based on published and some new results.

The next chapter extends these notions to experiment design – modeling the particular experiments needed for quantifying biodynamic system models. The modeling can be for testing alternative hypotheses (models), or for satisfying other quantitative goals, like quantifying *particular parameters* of interest – not necessarily the whole model.

Exercises

E15.1 ***SI Properties and Compartment Volumes***. Analyze the structural identifiability (*SI*) properties of the p53 model in Eq. (15.11). Do this with three different assumptions about measurement volumes: (a) assume x and y are measured as concentrations in the same volume, i.e. $V_1 = V_2$; (b) assume x and y are measured as concentrations in different volumes $V_1 \neq V_2$; (c) assume x and y are measured as masses, or molecule number – not as concentrations.

E15.2 ***Rice–Wagner Model of p53 Signaling***. The 4 ODE, 11 parameter model is developed and detailed in the reference Wagner et al. (2005). In the appendix of the paper, the authors transformed the model variables and parameters of their model into dimensionless units, as in our Chapter 3. For simulation and analysis purposes, they quantified the scaled model parameters from literature data – all given in the paper.

a) Implement the scaled model with suitable software, e.g. *Simulink* or *VisSim*, using the dimensionless parameters given in Table 1 of the paper. Run the simulation for several cycles, illustrating oscillating responses with approximately 7 h periods, thereby demonstrating constant amplitude cycling behavior.

b) Confirm limit cycling behavior by plotting the phase plane for Mdm2 vs. p53 for the part (a) solution. How do you know this periodic behavior is limit cycling?

c) Sustained oscillations in this model are dependent on the sum of the time lags τ in two loops. The bifurcation diagram for p53 fixed points versus τ in **Fig. 3** of the reference indicates a Hopf bifurcation at $\tau = 0.88$ and sustained oscillations occur for $\tau > 0.88$. For $\tau < 0.88$, the model is stable and does not oscillate. Demonstrate these results in the time domain for several values of τ.

References

Akaike, H., 1974. A new look at the statistical model identification. IEEE Trans. Automat. Control. 19 (6), 716–723.

Akaike, H., 1978. A Bayesian analysis of the minimum AIC procedure. Ann. Inst. Stat. Math. 30A, 9–14.

Alam, M.J., Devi, G.R., et al., 2013. Switching p53 states by calcium: dynamics and interaction of stress systems. Mol. Biosyst. 9 (3), 508–521.

Batchelor, E., Loewer, A., Lahav, G., 2009. The ups and downs of p53: understanding protein dynamics in single cells. Nat. Rev. Cancer. 9, 371–377.

Batchelor, E., L. A., Mock, C., Lahav, G., 2011. Stimulus-dependent dynamics of p53 in single cells. Mol. Syst. Biol. 10 (7).

Boashash, B. (Ed.), 2003. Time-Frequency Signal Analysis and Processing: A Comprehensive Reference. Elsevier Science, Oxford.

Brandman, O., Ferrell Jr., J.E., Li, R., Meyer, T., 2005. Interlinked fast and slow positive feedback loops drive reliable cell decisions. Science. 310 (5747), 496–498.

Cai, X., Yuan, Z.M., 2009. Stochastic modeling and simulation of the p53-MDM2/MDMX loop. J. Comput. Biol. 16 (7), 917–933, doi: 910.1089/cmb.2008.0231.

Ciliberto, A., Novak, B., Tyson, J.J., 2005. Steady states and Oscillations in the p53/Mdm2 network. Cell Cycle. 4 (3), 488–493.

Dahari, H., Shudo, E., Ribeiro, R.M., Perelson, A.S., 2008. Mathematical modeling of HCV infection and treatment. 510, 439–453.

DiStefano III, J., Stubberud, A.R., Williams, I.J., 1990. Feedback and Control Systems Continuous (Analog) and Discrete (Digital), second ed. McGraw-Hill, New York.

Douglas, P., Cohen, M.S., DiStefano III, J.J., 2010. Chronic exposure to Mn may have lasting effects: a physiologically based toxicokinetic model in rat. J. Toxicol. Environ. Chem. 92 (2), 279–299.

Ferrell Jr., J.E., Machleder, E.M., 1998. The biochemical basis of an all-or-none cell fate switch in xenopus oocytes. Science. 280 (5365), 895–898.

Geva-Zatorsky, N., Rosenfeld, N., Itzkovitz, S., Milo, R., Sigal, A., Dekel, E., et al., 2006. Oscillations and variability in the p53 system. Mol. Syst. Biol. 2 (2006), 13.

Geva-Zatorsky, N., Dekel, E., Batchelor, E., Lahav, G., Alon, U., 2010. Fourier analysis and systems identification of the p53 feedback loop. Proc. Natl. Acad. Sci. U S A. 107 (30), 13550–13555.

Gu, J., Kawai, H., Nie, L., Kitao, H., Wiederschain, D., Jochemsen, A.G., et al., 2002. Mutual dependence of MDM2 and MDMX in their functional inactivation of p53. J. Biol. Chem. 277 (22), 19251–19254.

Harless, C., DiStefano III, J., 2005. Automated expert multiexponential biomodeling interactively over the internet. Comput. Methods Programs Biomed. 79 (2), 169–178.

Javier, R.M., DiStefano, J.J., 2009. Dynamical biocontrol systems: insights through mechanistic modelling. Math. Comput. Model. Dynamical Syst. Methods, Tools Appl. Eng. Relat. Sci. 15 (1), 1–16.

Jin, J., Almon, R.R., DuBois, D.C., Jusko, W.J., 2003. Modeling of corticosteroid pharmacogenomics in rat liver using gene microarrays. J. Pharmacol. Exp. Ther. 207, 93–109.

Lahav, G., Rosenfeld, N., Sigal, A., Geva-Zatorsky, N., Levine, A.J., Elowitz, M.B., et al., 2004. Dynamics of the p53-Mdm2 feedback loop in individual cells. Nat. Genet. 36 (2), 147–150, Epub 2004 January 2018.

Landaw, E., DiStefano III, J., 1984. Multiexponential, multicompartmental, and noncompartmental modeling: physiological data analysis and statistical considerations, part II: data analysis and statistical considerations. Am. J. Physiol. Regul. Integr. Comp. Physiol. 246, R665–R677.

Leenders, G.B., Tuszynski, J.A., 2013. Stochastic and deterministic models of cellular p53 regulation. Front. Oncol. 3.

Lev Bar-Or, R., Maya, R., Segel, L., Alon, U., Levine, J., Oren, M., 2000. Generation of oscillations by the p53-Mdm2 feedback loop: a theoretical and experimental study. Proc. Natl. Acad. Sci. U S A (PNAS). 97 (21), 11250–11255.

Ma, L., Wagner, J., Rice, J.J., Hu, W., Levine, A.J., Stolovitzky, G.A., 2005. A plausible model for the digital response of p53 to DNA damage. PNAS. 102, 14266–14271.

Mandy, R., 2005. Microarray: a technique review. J. Young Investig. 5 (4).

Marino, A., DiStefano III, J.J., Landaw, E.M., 1992. DIMSUM: an expert system for multiexponential model discrimination. Am. J. Physiol. Endocrinol. Metab. 262, E546–E556.

Meshkat, N., Kuo, C., Brown, N., DiStefano III J.J., 2014. COMBOS: web app for finding identifiable parameter combinations in nonlinear ODE models. Submitted to PLOS Comp. Biol.

Neumann, A.U., Lam, N.P., Dahari, H., Gretch, D.R., Wiley, T.E., Layden, T.J., et al., 1998. Hepatitis C viral dynamics in vivo and the antiviral efficacy of interferon- therapy. Science. 282 (5386), 103–107.

Perelson, A.S., 2002. Modelling viral and immune system dynamics. Nat. Rev. Immunol. 2 (1), 28–36.

Perelson, A.S., N. P., 1999. Mathematical analysis of HIV-1 dynamics in vivo. SIAM Rev. Soc. Ind. Appl. Math. 41, 3–44.

Perelson, A.S., Herrmann, A., Micol, F., Zeuzem, S., 2005. New kinetic models for the hepatitis C virus. Hepatology. 42 (4), 749–754.

Proctor, C.J., Gray, D.A., 2008. Explaining oscillations and variability in the p53-Mdm2 system. BMC. Syst. Biol. 75, 1–20.

Purvis, J., Lahav, G., 2012. p53 dynamics control cell fate. Science. 336, 1440–1444.

Rice, J.M., Lan, Wagner, J., Stolovitzky, G., 2004. The p53-Mdm2 system modeled as a digital oscillator with a time delay. Proc. Physiol. Soc, Oxford J. Physiol.

Schwarz, G., 1978. Estimating the dimension of a model. Ann. Stat. 6 (2), 461–464.

Stoughton, R.B., 2005. Applications of DNA microarrays in biology. Annu. Rev. Biochem. 74 (1), 53–82.

Sun, T., Y. W., Liu, J., Shen, P., 2011. Modeling the basal dynamics of P53 system. PLoS ONE. 6.

Tyson, J.J., 2006. Another turn for p53. Mol. Syst. Biol. 2 (2006), 13.

Wade, M., Wang, Y.V., Wahl, G.M., 2010. The p53 orchestra: Mdm2 and Mdmx set the tone. Trends Cell Biol. 20 (5), 299–309, doi: 210.1016/j.tcb.2010.1001.1009. Epub 2010 February 1019.

Wagner, J., Ma, L., Rice, J., Hu, W., Levine, A.J., Stolovitzky, G., 2005. p53−Mdm2 loop controlled by a balance of its feedback strength and effective dampening using ATM and delayed feedback. IEE Proc. Syst. Biol. 152, 109−118.

Wahl, G.M., Stommel, J.M., Krummel, K.A., Wade, M., 2005. Gatekeepers of the guardian: p53 regulation by post-translational modification, Mdm2 and Mdmx. 25 Years of p53 Research. K. W. a. P. Hainaut, Springer.

Wang, Y.V., Wade, M., Wong, E., Li.,Y.-C, Rodewald., Luo, W., Wahl, G.M., et al., 2007. Quantitative analyses reveal the importance of regulated HdmX degradation for p53 activation. Proc. Natl. Acad. Sci. USA. 104 (30), 12365−12370.

Experiment Design and Optimization

Chapter 16

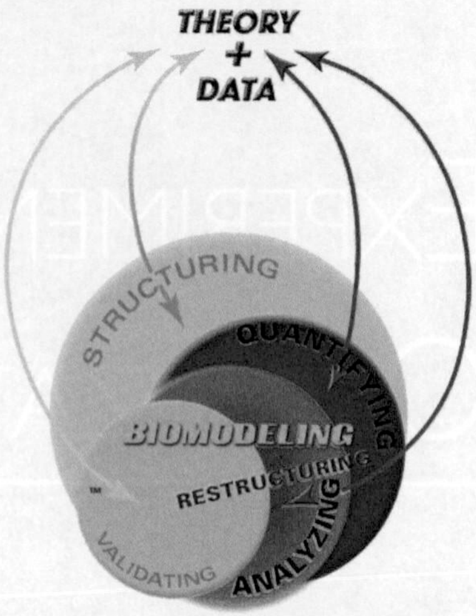

Experiment Design and Optimization

"There is nothing impossible to him who will try."

Alexander the Great ~350BC

Overview

This chapter is about using a variety of identifiability, algebraic, graph theory and optimization methods for systematically designing experiments for quantifying dynamic biosystems. The overall experiment design problem is introduced in the first section. Then methods for choosing experiment designs for systematically maximizing overall structural identifiability (*SI*) properties of a model are developed. The following section is about designing experiments for quantifying particular parameters of interest, using an approach called cutset analysis of the model graph. The last sections address experiment design methodology for optimizing numerical identifiability (*NI*) properties of ordinary differential equation (ODE) models, i.e. maximizing the precision of quantified parameter estimates.

The sections on maximizing overall *SI* properties are based on the proposition that systematic *SI* screening of all feasible and independent input–output (I–O) experiments that might be done on a system should yield all candidate I–O experiment sets for potentially successful quantification. This brute-force approach should work for nonlinear or linear models. We illustrate the method for multi-input—multi-output (MIMO) time-invariant linear (or linearized NL) biomodels in which all possible system input–output port locations are "candidates" for experimentation – via the transfer function (TF) matrix.

Cutset analysis is a graph theory-based approach used here for isolating candidate experiment designs for quantifying particular parameters of interest in a biosystem model. It is applicable to any model representable in graph form, i.e. structured compartmental, pathway, network, etc. models. The primary focus is quantifying parameters rendered *SI* by appropriate choice of input and output ports or measurement models. The method also provides the solution equation(s) for parameters of interest.

The optimal design sections show how *NI* properties of the model can be enhanced by optimally designing quantification experiments in particular ways. This usually means choosing experiment design-degrees-of-freedom that *minimize parameter estimate variability* (maximize parameter estimate reliability). Sequential optimal experiment design and optimum data sampling schedule design are the focus.

A Formal Model for Experiment Design

Designing experiments for quantifying a model usually begins with knowing what's feasible and practical to probe and measure. This means delineating the feasible input and output system "ports" (locations). This might entail augmentation of the system model as well as the measurement model structure, illustrated here for whole-organism studies.

Direct intravenous (IV) or intraarterial (IA) *inputs* of material into blood usually require only a direct arrow into blood, with no intervening compartment dynamics, as in **Figs. 4.1** and **4.3**. No submodel input compartment augmentation is needed. But getting test inputs into and measuring outputs from compartments in blood, if they are not so direct, requires additional structural components. For example, *indirect inputs* into blood, e.g. oral, subcutaneous (SC), intramuscular (IM), and intraperitoneal (IP) inputs require intervening compartments (structural model augmentations) between the input site and blood (or other) compartment. Models for these input modalities were developed in some detail in Chapter 4. Model augmentations also may be needed at the output. Examples encountered in Chapter 4 included urinary and fecal *output* accumulation signals modeled using nonleaky "dead-end" compartments augmenting the others in **Figs. 4.1** and **4.3**.

Before discussing other design-degrees-of-freedom available to experiment designers, let's first formalize the overall experiment design problem. We begin by slightly relaxing specification of the general nonlinear discrete-time noisy-measurement ODE model: $\dot{\boldsymbol{x}} = \boldsymbol{f}(\boldsymbol{x},\boldsymbol{p},\boldsymbol{u})$, $\boldsymbol{y} = \boldsymbol{g}(\boldsymbol{x},\boldsymbol{p})$, $\boldsymbol{z} = \boldsymbol{y} + \boldsymbol{e}$, $k = 1, 2, \ldots, N$. The basic internal structure of the system model $\boldsymbol{f}$ is assumed known – augmented as needed to accommodate additional input and output port modalities – as described above. The parameter vector p is unknown – as in a parameter estimation experiment. Unknown in this model are the input (signal) vector $\boldsymbol{u}$; the way $\boldsymbol{u}$ enters the system – the feasible input "ports"; the structure $\boldsymbol{g}$ of feasible output measurements – the output "ports"; statistical specification of the output measurement noise $\boldsymbol{e}$; and specification of the observation time interval $[t_0, t_N]$ and number of discrete-time samples within that interval. All are open to experiment design.

The **goal of experiment design** is to establish one or more I–O experiments such that all or some of the P parameters of interest in vector $\boldsymbol{p}$ can be estimated from the data, taking into account all practical constraints. A model codifying this goal is given in **Fig. 16.1** – a

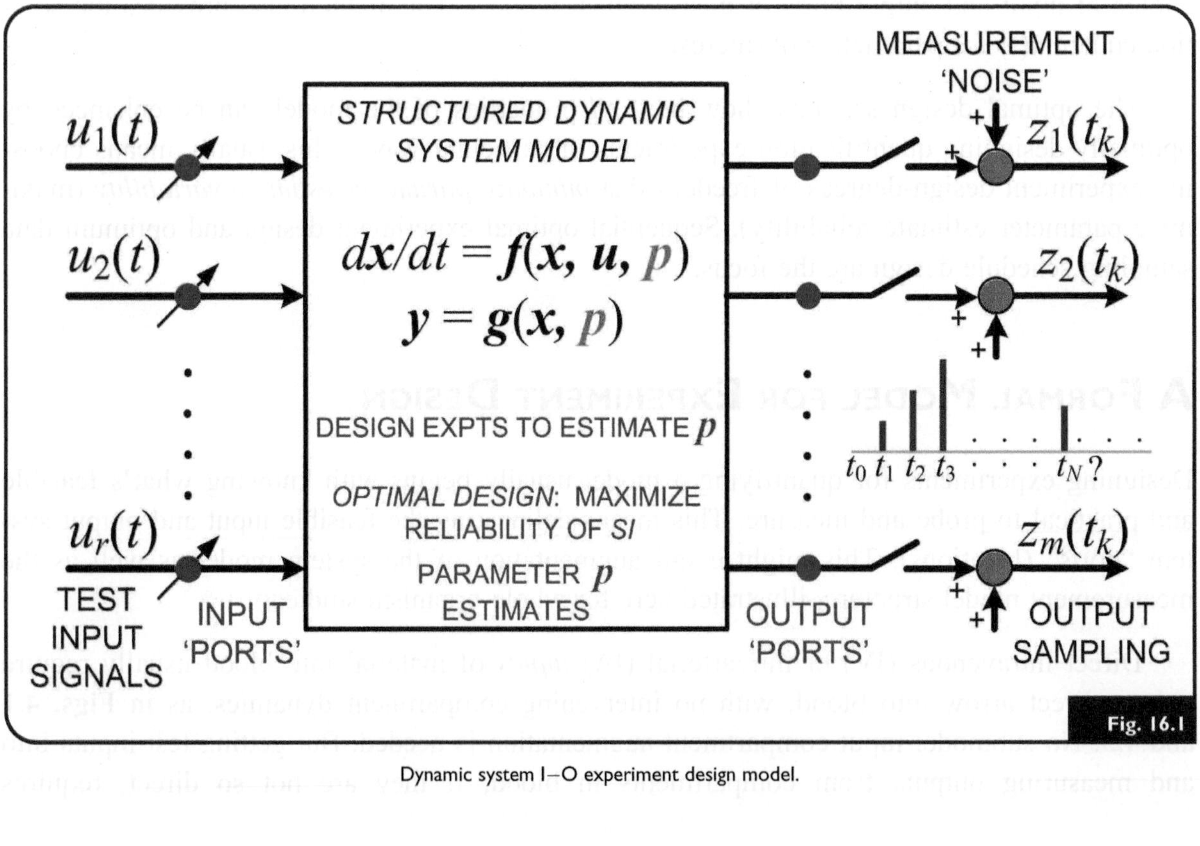

Dynamic system I–O experiment design model.

stimulus–response schematic of the experimental situation, illustrating typical design variables that can be controlled or chosen by an experimenter to satisfy parameter estimation goals. Experiment design-degrees-of-freedom are spelled out in the list below the figure.

Design-Degrees-of-Freedom

TEST-INPUT STIMULI

- PORTS (Where? How Many?)
- SIGNALS (Pulse, Infusion, Other?)

MEASURED OUTPUTS

- PORTS (Where? How Many?)
- OBSERVATION INTERVAL (How Long, t_N?)
- SAMPLE NUMBER (How Many, N?)
- SAMPLING SCHEDULE (When? $t_1, t_2, \ldots t_N$?)
- DATA SAMPLE RELIABILITY (How Precise? $VAR_1, VAR_2, \ldots, VAR_N$?)

SYSTEM OPERATING POINT

- Normal, Hyper-, Hypo- ?

Experiment designers usually have one or more of these design-degrees-of-freedom available in particular situations. They can choose from *r feasible input ports* into which as many *as r test-input signals* can be introduced, possibly simultaneously; and *m feasible output ports* from which as many as *m different output signals* can be measured, also possibly simultaneously. In the absence of economic constraints, more input and output ports should improve quantitative results. But cost/effectiveness considerations might prevail in making these decisions. The types of test-input signals to be employed in the experiment are also likely to be constrained by practical considerations, but they are nevertheless useful design-degrees-of-freedom, because – usually – more than one is available.

Observations of continuous time output signals usually can be made only at discrete instants of time in biological systems with mass-based measurements. This feature is indicated by output sampling switches in **Fig. 16.1**. Output data consists of at most m discrete and different time-series of noisy data signals, each collected at at most N sample times – with N also a design variable. The experiment observation interval $t_0 \leq t \leq t_N$ must be sufficient to capture the essential system/model dynamics, but the number of samples N is typically limited by practical constraints as well – a problem for optimization discussed later in the chapter.

How about data errors? Surely, there must be some maximum tolerable errors in data that can provide useful parameter estimates (sufficient parameter estimate precision). If a particular measurement methodology cannot provide sufficient precision, another (and likely more expensive) one might work, and an optimization goal in this regard might be to find the least expensive method for successful model quantification. Errors in the measurement sampling process – sampling errors, laboratory measurement, and any other errors in the output data – are lumped as measurement noise in **Fig. 16.1**. The statistics, e.g. variances, of these measurement errors also can be expressed as an experiment design variable to be optimized.

The last design variable listed above (*System Operating Point*) might be important in parameter estimation for nonlinear systems, in particular, because they typically express different qualitative as well as quantitative dynamics in different states or regimes of operation. The question is: would an experimental study of the system in another (e.g. supranormal) state provide additional or more precise parameter information?

We've noted several design variables above that might be optimized. The goal of **optimal experiment design problems** is to design one or more I–O experiments such that all or some of the P parameters of interest in vector $\boldsymbol{p}$ can be estimated from the data as precisely as possible,[1] taking into account all practical constraints as part of the optimization procedure. This is primarily a numerical identifiability (*NI*) optimization problem. Importantly, practical constraints can be implemented as part of the experiment design optimization criterion function.

Input–Output Experiment Design from the TF Matrix

Structural identifiability (*SI*) analysis was introduced in Chapter 10. Different structural models with different experiment input-output configurations and parameters were classified as structurally identifiable (*SI*) or not from the different experiments. We generalize the approach here, for all possible input/output experiments that might be performed on a given system. The methods essentially consist of exhaustive structural identifiability (*SI*) analysis of an *experimental TF matrix* (DiStefano III and Mori 1977). We make the following preliminary definitions and observations.

The real system under study can be linear or nonlinear, with constant unknown parameters. The **class of feasible experiments** is defined as the set of linearizing, small-perturbation, impulse response experiments (for NL models) that can be performed on the biosystem, or any impulse response experiments if the system is inherently linear. The usual model equations apply:

$$\dot{\boldsymbol{x}} = A(\boldsymbol{p})\boldsymbol{x} + B(\boldsymbol{p})\boldsymbol{u} \qquad \boldsymbol{y} = C\boldsymbol{x} \tag{16.1}$$

with the assumption that any parameter constraint relations have been included in the matrices, so Eq. (16.1) is a *complete* model. In principle, for single-input deterministic linear models, the response to an impulse input $\delta(t)$, and hence the TF, is universally applicable and analytically convenient. For more than one input, the situation is more complicated.

We distinguish among three different sets of model parameters, first the P unknown **intrinsic model parameters** $p_1, p_2, \ldots, p_P$ in (16.1); second, the **TF parameters**, i.e. the coefficients $\beta_1, \beta_2, \ldots, \alpha_1, \alpha_2, \ldots$ of the numerator and denominator polynomials of all elements H_{ij} of the TF matrix H. These are *structural invariants* of the model. In principle, all α_j and β_j

1 Relative weights can be readily assigned to the different parameters in vector $\boldsymbol{p}$, depending on their relative importance in a given application. In other words, total statistical "power" in an estimation problem can be adjusted so that it is focused on parameters of greatest interest, by weighting them more relative to unimportant ones.

are uniquely structurally identifiable (*SI*) from the I–O experiment(s) defined by H; and the total number of α_j and β_j is the maximum number of intrinsic model parameters p_j that might be structurally identifiable (*SI*). It is convenient to introduce the proposed analysis in terms of a third parameter set, the **matrix parameters** (elements) a_{ij}, b_{ij}, c_{ij} of the n by n A-matrix, n by r B-matrix and m by n C-matrix in (16.1). These matrix elements are algebraic functions of the p_j and it is usually easier to first establish the structural identifiability (*SI*) properties of these matrix parameters, with the p_j dealt with algebraically afterward.

For experiment design, the **test-input matrix** B and **measurement matrix** C each have maximum dimension, n by n, i.e. the largest number of independent test-input variables and measurement variables. It is sufficient to consider B and C diagonal, due to the linearity of the model. These are denoted the **experiment design matrices** $B_n = [b_{ij}]$ and $C_n = [c_{ij}]$ and, for many models, one of these can be defined as the identify matrix I. If the model state variables $x_1, x_2, \ldots, x_n$ represent compartment concentrations, then $C_n = I$ if only concentration measurements are being considered. On the other hand, if the x_i represent amounts or masses, $B_n = I$ is often sufficient. These considerations simplify the model TF and resulting analysis.

The **experimental TF matrix** is the n by n matrix:

$$H_n(s,\boldsymbol{p}) = C_n(\boldsymbol{p})[sI - A(\boldsymbol{p})]^{-1}B_n(\boldsymbol{p}) = \begin{pmatrix} c_{11} & & 0 \\ & \ddots & \\ 0 & & c_{nn} \end{pmatrix}[sI - A(\boldsymbol{p})]^{-1}\begin{pmatrix} b_{11} & & 0 \\ & \ddots & \\ 0 & & b_{nn} \end{pmatrix} \tag{16.2}$$

where, as noted above, it is often the case that either the $c_{ii} = 1$ or the $b_{ii} = 1$ for all i. The word "experimental" designates C_n and/or B_n as **candidate design matrices**.

Algebraic SI analysis of $H_n(s,\boldsymbol{p})$ yields, first, the experiment input-output configurations for which it is impossible to determine parameters of interest, i.e. the globally *unID* parameters. Second, it yields all input-output configurations, singly and in combinations, for which identification of parameters of interest is feasible. If combination test inputs, i.e. test inputs applied to more than one compartment simultaneously, are the only ones possible, they can then be analyzed further by linear transformation of the diagonal test-input matrix B_n. Similarly, linear combinations of state variables in the measurements can be studied by suitable linear transformation of the diagonal measurement matrix C_n.

Remark: In experiments with more than one input, the *form of the various inputs* has a definite effect on identifiability results. This is in contrast to single-input linear models, where identifiability results are independent of the input, either its form or magnitude. This important property can be useful in structural identifiability (*SI*) analysis of the experimental TF matrix and is illustrated in Example 16.6.

The TF matrices of several multicompartment linear models were analyzed in Chapter 10 examples for structural identifiability (*SI*) of parameters in particular model I–O configurations. Comprehensive analysis of more general models in the next sections further clarifies the approach.

Experiment Design by Exhaustive *SI* Analysis of the General 2-Compartment Model

The experimental TF matrix (all four I–O TFs) for this structure is readily written from the TF matrix derived in Eq. (4.34). We assume both outputs are measured and both inputs are present and have the *same form*, i.e. both are steps, impulses, etc. The TF is:

$$H_n(s) = \frac{1}{D(s)} \begin{bmatrix} (s + k_{12} + k_{02})/V_1^{true} & k_{12}/V_1^{true} \\ k_{21}/V_2^{true} & (s + k_{21} + k_{01})/V_2^{true} \end{bmatrix} \tag{16.3}$$

$$D(s) = s^2 + \alpha_2 s + \alpha_1 = s^2 + (k_{12} + k_{21} + k_{01} + k_{02})s + (k_{21}k_{02} + k_{12}k_{01} + k_{01}k_{02}) \tag{16.4}$$

where the feasible outputs were given as concentrations, and thus the volumes are real. H_{11} was analyzed in Example 10.2 and only V_1 was found to be uniquely identifiable. What if y_2 also can be measured? Consider H_{21} in (16.3), the TF for input into 1 and output from 2. As always, we can identify the structural invariants:

$$\beta_1' = k_{21}/V_2^{\text{true}} \tag{16.5}$$

$$\alpha_2 = k_{21} + k_{12} + k_{01} + k_{02} \tag{16.6}$$

$$\alpha_1 = k_{21}k_{02} + k_{12}k_{01} + k_{01}k_{02} \tag{16.7}$$

With three equations and five unknowns, k_{ij} remain *un*ID from H_{21}. Now consider H_{11} and H_{21} in combination, with single input $u_1 = \delta(t)$ and both responses $y_1(t)$ and $y_2(t)$ measured. Only Eq. (16.5) is independent of the others and this equation provides no new information about the k_{ij}. Thus H_{11} and H_{21} together provide no additional parameter information for establishing uniqueness of p_i than does H_{11} alone.

Results of *exhaustive analysis* of all four TFs, singly and in combination, are provided in **Table 16.1**. The last column in the table indicates that:

(a) H_{11} or H_{22}, or H_{11} and H_{22}, provide only compartment volumes.

(b) H_{12} or H_{21}, or H_{12} and H_{21}, provide nothing uniquely.

(c) The single-input/two-output experiments (input into either compartment, measure both) provide only compartment volumes, i.e. H_{11} and H_{21} (Experiment 5) or H_{22} and H_{12} (Experiment 10).

(d) The two-input/single-output experiments (same form of input into compartments, sample only one), i.e. H_{11} and H_{12} (Experiment 6) or H_{22} and H_{21} (Experiment 9), provide all k's and one true V.

(e) All six parameters are provided by any experiment in which both compartments receive inputs of the same form and both are measured (two inputs/two outputs).

Table 16.1 Exhaustive Analysis of the Experimental TF Matrix for the Most General 2-Compartment Model Structure with Concentration Outputs

Expt #	*H_{ij} and TF Combinations*	*Identifiable Parameter Combinations*	*Uniquely SI Parameters*
1	H_{11}	$\beta_2 = 1/V_1^{true}, \beta_1/\beta_2 = k_{12} + k_{02}$ $\alpha_2 = k_{12} + k_{21} + k_{01} + k_{02} = \beta_1/\beta_2 + k_{21} + k_{01}$ $\alpha_1 = k_{21}k_{02} + k_{12}k_{01} + k_{01}k_{02} = k_{21}k_{02} + k_{01}\beta_1/\beta_2$	V_1^{true}
2	H_{21}	$\beta_1' = k_{21}/V_2^{true}, \alpha_1, \alpha_2$	None
3	H_{12}	$\beta_1'' = k_{12}/V_1^{true}, \alpha_1, \alpha_2$	None
4	H_{22}	$\beta_2' = 1/V_2^{true}, \quad \beta_1''/\beta_2' = k_{21} + k_{01}, \alpha_1, \alpha_2$	V_2^{true}
5	H_{11}, H_{21}	$\beta_2 = 1/V_1^{true}, \quad \beta_1' = k_{21}/V_2^{true}, \beta_1, \beta_2, \alpha_1, \alpha_2$	V_1^{true}
6	H_{11}, H_{12}	$\left.\begin{array}{r}\beta_2 = 1/V_1^{true}\\ \beta_1'' = k_{12}/V_1^{true}\\ \beta_1/\beta_2 = k_{12} + k_{02}\end{array}\right\} \Rightarrow V_1^{true}, k_{12}, k_{02}$ $\left.\begin{array}{r}a_2 - \beta_1\beta_2 = k_{21} + k_{01}\\ a_1/k_{02}\beta_1\beta_2 = k_{21}/\beta_1\beta_2 + k_{01}/k_{02}\end{array}\right\} \Rightarrow k_{21}, k_{01}$	V_1^{true} k_{21}, k_{12} k_{01}, k_{02}
7	H_{11}, H_{22}	$\beta_2 = 1/V_1^{true}, \quad \beta_2' = 1/V_2^{true}, \quad \beta_1/\beta_2 = k_{12} + k_{02}$ $\beta_1''/\beta_2' = k_{21} + k_{01}, \quad a_1 - (\beta_1/\beta_2)(\beta_1''/\beta_2') = k_{12}k_{21}$	V_1^{true}, V_2^{true}
8	H_{12}, H_{21}	$\beta_1' = k_{21}/V_2^{true}, \quad \beta_1'' = k_{12}/V_1^{true}, \quad \alpha_1, \alpha_2$	None
9	H_{21}, H_{22}	$\left.\begin{array}{r}\beta_1' = k_{21}/V_2^{true}, \quad \beta_2' = k_{11}/V_2^{true}\\ \beta_1''/\beta_2' = k_{21} + k_{01}\end{array}\right\} \Rightarrow V_2^{true}, k_{21}, k_{01}$ $\left.\begin{array}{r}\alpha_2 - \beta_1''/\beta_2' = \beta_1/\beta_2 = k_{12} + k_{02}\\ (\alpha_1 - k_{01}\beta_1/\beta_2)/k_{21} = k_{02}\end{array}\right\} \Rightarrow k_{02}, k_{12}$	V_2^{true} $k_{12}, k_{21},$ k_{01}, k_{02}
10	H_{12}, H_{22}	$\beta_1'' = k_{12}/V_1^{true}, \quad \beta_2' = 1/V_2^{true}, \quad \beta_1''/\beta_2', \alpha_1, \alpha_2$	V_2^{true}
11	H_{11}, H_{12}, H_{22}	$\beta_2' = 1/V_2^{true}$ plus all parameter available fromH_{11}, H_{12}	ALL
12	H_{11}, H_{21}, H_{22}	$\beta_2' = 1/V_1^{true}$ plus all parameter available fromH_{22}, H_{21}	ALL
13	H_{11}, H_{12}, H_{21}	$V_1^{true}, k_{12}, k_{21}, k_{01}, k_{02}$ from H_{11}, H_{12} and $\beta_1' = k_{21}/V_2^{true}$ gives V_2^{true} from H_{21}	ALL
14	H_{22}, H_{12}, H_{21}	$V_2^{true}, k_{12}, k_{21}, k_{01}, k_{02}$ from H_{22}, H_{21}; and $\beta_1' = k_{12}/V_1^{true}$ gives V_1^{true} from H_{12}	ALL
15	$H_{11}, H_{12},$ H_{21}, H_{22}	Everything	ALL

Remark: The results summarized in **Table 16.1** are based on all inputs having the same form, e.g. $u_1(t) \equiv u_2(t) = \delta(t)$, or delayed impulses $u_1(t) \equiv u_2(t) = \delta(t-\tau)$, or $u_1(t) \equiv u_2(t) = t$, etc. If the two inputs are *different*, e.g. $u_1(t) = \delta(t)$, and $u_2(t) \equiv \mathbf{1}(t)$, a unit step function, the results are different. We illustrate this with another example.

Example 16.1: ***Two-Compartment Model with Different Input Forms Becomes SI***

The output transform:

$$Y_2(s) = \frac{k_{12}U_1(s) + (s - k_{22})U_2(s)}{s^2 - (k_{11} + k_{22})s + k_{11}k_{22} - k_{12}k_{21}}$$

is for the model: $d\boldsymbol{q}/dt = K\boldsymbol{q} + \boldsymbol{u}$ and $y_2 = [0\ 1]\boldsymbol{q} = q_2$. (Mass instead of concentration is measured, to simplify the problem.) This corresponds to the TFs H_{11} and H_{21}, and for Experiment number 5 in **Table 16.1** (compartment volumes not needed). The unknown TF parameters are k_{11}, k_{12}, k_{21}, k_{22}. (k_{01} and k_{02} are not of interest in this example.) If both inputs are the same – the premise of the table – no parameters are structurally identifiable. Let's see if this is the case if input $u_1(t)$ is a unit *impulse* ($U_1(s) = 1$) and $u_2(t)$ is a unit *step* ($U_2(s) = 1/s$), with both applied at time $t = 0$. Substitution of these input transforms gives:

$$Y_2(s) = \frac{k_{12} + (s - k_{22})/s}{s^2 - (k_{11} + k_{22})s + k_{11}k_{22} - k_{12}k_{21}} = \frac{s(k_{12} + 1) - k_{22}}{s^3 - (k_{11} + k_{22})s^2 + (k_{11}k_{22} - k_{12}k_{21})s}$$

By inspection, the numerator provides k_{12} and k_{22} uniquely. The denominator then gives k_{11} from the coefficient of s^2; and then k_{21} is obtained from the last term. All parameters are thus uniquely *SI*.

Any combination of *different* inputs will give the same uniquely *SI* result. This includes two inputs of the same form, but applied at different times.

Experiment Design by Exhaustive *SI* Analysis of a 6-Compartment Biomodel

We address a more complex physiological system model here, the NL submodel representing thyroid hormone (*TH*) distribution, metabolism, and excretion dynamics developed in Chapter 14 and depicted in **Fig. 14.6**. The overall goal here is to find *feasible I–O experiments* that might be done to maximally quantify all the unknown parameters of this model. To use the exhaustive TF analysis approach, we limit the designs to linearizing experiments. This problem and model have been addressed, solved, and generalized in Feng and DiStefano III (1992) and Feng and DiStefano III (1995), using decomposition-based experiment design algorithms applied to linear compartmental models with unidirectional interconnectivity among

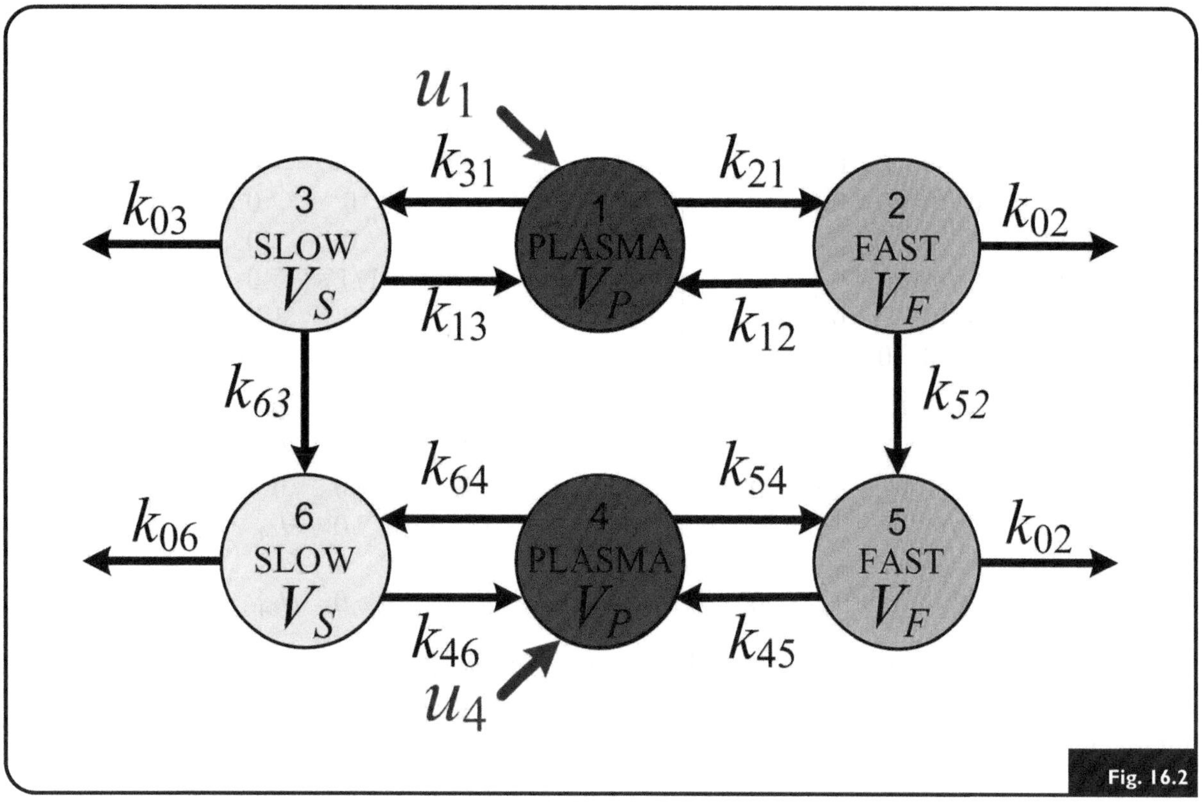

Fig. 16.2

Linearized TH D&E submodel for *SI* analysis.

submodels. The linearized model addressed here – illustrated in **Fig. 16.2** – is in this model class. It has the form $\dot{q} = Kq + Bu, y = Cq$.

Feasible Experiments

This physiological system has been studied comprehensively in several mammalian and nonmammalian species, using a variety of one or two test-input and one to six or more output port configurations, e.g. Hershman et al. (1986), DiStefano III et al. (1993), Nguyen et al. (1993); Sefkow et al. (1996), Nguyen et al. (2003). Test-inputs have been introduced relatively noninvasively into plasma T_4 and T_3 compartments in blood in both transient response and constant steady state studies. In steady state, it has also been possible to introduce inputs into nonvascular compartments in organ tissues, as illustrated in Nguyen et al. (2003), with test-inputs introduced directly into intestinal compartments. We allow for all six possible input ports later in the chapter, when we use cutset analysis to design constant steady state experiments. For purposes of analysis here, we design for *transient* response studies – with only two input ports – as illustrated in **Fig. 16.2**; and all six compartmental ports are available for output measurements.

Transfer Function Matrix

The input, output, system, and experimental TF matrices for this experiment design model are:

$$B_n = \begin{bmatrix} 1 & 0 & 0 & 0 & 0 & 0 \\ 0 & 0 & 0 & 0 & 0 & 0 \\ 0 & 0 & 0 & 0 & 0 & 0 \\ 0 & 0 & 0 & 1 & 0 & 0 \\ 0 & 0 & 0 & 0 & 0 & 0 \\ 0 & 0 & 0 & 0 & 0 & 0 \end{bmatrix}, \quad C_n = \begin{bmatrix} 1/V_P & 0 & 0 & 0 & 0 & 0 \\ 0 & 1/V_F & 0 & 0 & 0 & 0 \\ 0 & 0 & 1/V_S & 0 & 0 & 0 \\ 0 & 0 & 0 & 1/V_P & 0 & 0 \\ 0 & 0 & 0 & 0 & 1/V_F & 0 \\ 0 & 0 & 0 & 0 & 0 & 1/V_S \end{bmatrix} \tag{16.8}$$

$$K = \begin{bmatrix} k_{11} & k_{12} & k_{13} & 0 & 0 & 0 \\ k_{21} & k_{22} & 0 & 0 & 0 & 0 \\ k_{31} & 0 & k_{33} & 0 & 0 & 0 \\ 0 & 0 & 0 & k_{44} & k_{45} & k_{46} \\ 0 & k_{52} & 0 & k_{54} & k_{55} & 0 \\ 0 & 0 & k_{63} & k_{64} & 0 & k_{66} \end{bmatrix}, \quad H_n(s) = \begin{bmatrix} H_{11} & 0 & 0 & 0 & 0 & 0 \\ H_{21} & 0 & 0 & 0 & 0 & 0 \\ H_{31} & 0 & 0 & 0 & 0 & 0 \\ H_{41} & 0 & 0 & H_{44} & 0 & 0 \\ H_{51} & 0 & 0 & H_{54} & 0 & 0 \\ H_{61} & 0 & 0 & H_{64} & 0 & 0 \end{bmatrix} \tag{16.9}$$

Parameters V_P, V_F and V_S in the C_n matrix are the unknown plasma, fast and slow tissue pool volumes (same for T_4 and T_3) for each of T_4 and T_3 concentrations measured in each. Interconnected submodel experiments and parameters are shown in red in K and H_n.

Identifiability Analysis for Each H_{ij}

H_{11} and H_{44} are the same TFs for the three-compartment model in Example 10.3 analyzed for *SI* in Chapter 10. Those results indicated that V_P and the 10 parameter combos: k_{11}, k_{22}, k_{33}, k_{44}, k_{55}, k_{66}, $k_{13}k_{31}$, $k_{12}k_{21}$, $k_{46}k_{64}$ and $k_{45}k_{54}$ are uniquely *SI* from H_{11} and H_{44}.

These two three-compartment submodels are interconnected unidirectionally in **Fig. 16.2** by the red parameters k_{52} and k_{63} and the three additional experiments, designated by red TFs H_{41}, H_{51} and H_{61}. H_{41} is also feasible from measurements only in blood. *SI* analyses and experiment design results developed in Feng and DiStefano III (1995) and Sefkow et al. (1996) show that H_{41} provides unique combos: $k_{21}k_{52}k_{45}$ and $k_{31}k_{63}k_{46}$, in addition to the combos also given by H_{11} and H_{44}. No k_{ij} alone are *SI* from H_{41}. The six remaining components of H_n are analyzed algebraically in the same fashion. Each represents an experiment in which extravascular slow and fast compartment T_3 and T_4 concentrations are measured and each of these H_{ij} provides unique k_{ij}; and H_{54} and H_{64} also provide k_{45} and k_{46} uniquely.

Identifiability Analysis for Combinations of H_{ij}: SI from Multiple Experiments

The results for each H_{ij} were analyzed *combinatorially*, to determined minimal subsets of H_n from which maximum numbers of parameters and/or specific parameters of interest may be obtained. As illustrated in experiments (rows) 7–11 of **Table 16.2** , if test inputs u_1 and u_4 are applied and all six state variables (compartments) are measured, in any one of several possible ways, complete *SI* is achieved. In principle then, all of the physiological parameters (14 rate constants and 3 compartment volumes) and – further – all of the state variables (6 plasma

Table 16.2 Parameters from Combos of TFs (Experiments)

Experiment (#) TF	*Uniquely SI Parameters and Combos*
(1) H_{11}, H_{44}	V_P, k_{11}, k_{22}, k_{33}, k_{44}, k_{55}, k_{66}
(2) H_{11}, H_{44} H_{54}, H_{64}	§: k_{54}, k_{45}, k_{64}, k_{46}
(3) H_{11}, H_{44} H_{21}, H_{31}	§: k_{21}, k_{12}, k_{31}, k_{13}
(4) H_{11}, H_{44} H_{51}, H_{61}	§
(5) H_{11}, H_{44}, H_{54}, H_{64}, H_{51}, H_{21}	§: k_{54}, k_{45}, k_{64}, k_{46} k_{21}, k_{12}, k_{52}
(6) H_{11}, H_{44}, H_{54}, H_{64}, H_{61}, H_{31}	§: k_{54}, k_{45}, k_{64},k_{46} k_{31}, k_{13}, k_{63}
(7) H_{11}, H_{44}, H_{54}, H_{64}, H_{51}, H_{61}	ALL
(8) H_{11}, H_{44}, H_{54}, H_{64}, H_{21}, H_{31}, H_{41}	ALL
(9) H_{11}, H_{44}, H_{54}, H_{64}, H_{21}, H_{31}, H_{51}	ALL
(10) H_{11}, H_{44}, H_{54}, H_{64}, H_{21}, H_{31}, H_{41}	ALL
(11) H_{11}, H_{44}, $H_{54} + H_{51}$, $H_{64} + H_{61}$, H_{21}, H_{31}	ALL

§These are in addition to those obtainable from H_{11} and H_{44}.

and extravascular hormone concentrations) and endogenous hormone secretion and production rates of the model-system can be determined uniquely *from the experiments defined by any of these sets of TFs* (Feng and DiStefano III 1995; Sefkow et al. 1996).

Alternative Combinations of Physiological Experiments

Each of these several groups have four TFs in common: H_{11}, H_{44}, H_{64} and H_{54}. *Physiologically speaking*, this means that for complete quantification, one must minimally measure the T_3 response in plasma to a T_3 injection in plasma, experiment H_{11}; and the T_4 responses in plasma, slow and fast tissue compartments (chiefly skeletal muscle and liver) in response to a T_4 input to blood (experiments H_{44}, H_{64} and H_{54}). One must also measure either the T_3 slow and fast tissue responses to IV T_4 input (experiments H_{51} and H_{61}), or the T_3 slow and fast tissue responses to T_3 input to blood (experiments H_{54} and H_{64}) as well as the plasma T_3 response (Experiment 8), or fast (or slow) tissue responses (Experiments 9 or 10) to IV T_4 input (Experiments H_{41} or H_{51} or H_{61}).

Experiments 5 and 6 indicate that if only one measurement is missing, i.e., the T_3 slow or the T_3 fast tissue compartment is *not* measured, only four of the individual parameters are *SI* – even from six TFs . However, two important parameters of interest, the combo fractional T_4 to T_3 conversion rate constants k_{52} and k_{63} (time^{-1} units) in fast tissue or slow tissue, can be determined from Experiments 5 and 6.

The combinations of four TFs defined by experiments (rows) 2 and 3 in **Table 16.2** thus yield two additional k_{ij} more than can H_{11} or H_{44} alone, whereas Experiment 4 yields no new parameters. Other possible combinations of 2, 3 and 4 TFs in $H_n(s)$ yield little or no new *SI* k_{ij}. In particular H_{41}, by itself or in combination with H_{11} and H_{44}, yields no new parameters. This is unfortunate because the experiment represented by this TF, i.e. measurement of the labeled T_3 response in plasma to an injection of labeled T_4 into plasma, does not require the usually more difficult extravascular measurements.

Aggregated Measurements

Considering the design matrix C_n diagonal exhausts all possible output configurations taken one at a time. If a linear combination of certain measurements is convenient or required or unavoidable, then the *SI* properties of the TFs (experiments) represented by *combined* measurements are generally different than the individual TFs determined by each measurement alone, as illustrated in an earlier example for the two-compartment model. For example, from Experiment 5 of **Table 16.2**, it may be easier to measure the sum of the labeled T_3 and labeled T_4 concentrations in fast tissue, plus the labeled T_4 concentration in fast tissue, following simultaneous administration of trace doses of T_3 and T_4 in plasma (measure $H_{51} + H_{21}$ and H_{51}), rather than both concentrations separately (measure $H_{51} + H_{21}$). The parameters obtainable from the combined experiment would be the same. In fact, in this special case, it is sufficient to measure only the sum of two concentrations, without T_4 separately. Similarly, for Experiment 6 of **Table 16.2,** $H_{61} + H_{31}$ can be determined instead of each separately.

Practical Alternatives via External Monitoring

Noninvasive external monitoring of substances (like tracer T_3 and T_4) over various organs is a feasible measurement scheme for observing *in vivo* extravascular dynamics, e.g. by functional *MRI*, *SPECT* or *PET* imaging following tracer test-input injections. One of the difficulties with this approach is that metabolic products, like *in vivo* metabolites of substances introduced exogenously, typically cannot be distinguished from the tracer or other indicator introduced and thus cannot be individually tracked. For example, tracer T_3 generated from tracer-labeled T_4 in the current analysis cannot be distinguished from the latter by external counting of the isotopic tracer. Aggregated measurements can help solve the *SI* problem under these conditions. For the *TH* subsystem under investigation here, the sum of the labeled $T_3 + T_4$ can be observed in tissues if labeled T_4 is injected IV. $H_{54} + H_{64}$, H_{51} and H_{61} are not separately quantifiable, the aggregated measurements in $H_{54} + H_{51}$ and $H_{64} + H_{61}$ can provide *SI* if labeled T_3 is also injected IV: the kinetics of this label could be observed as T_3 radioactivity alone, H_{44}, because no conversion of T_3 to T_4 takes place. Algebraic analysis of the six TFs: H_{11}, H_{44}, $H_{54} + H_{51}$, $H_{64} + H_{61}$, H_{21} and H_{31} indicates that all parameters are uniquely *SI* from these aggregated experiments. This experiment is listed as number 11 in **Table 16.2**.

Remark: Cutset analysis discussed in the next section is an alternative algebraic approach to experiment design, also capable of exhaustive analysis of experimental I–O configurations like that above. It can provide some of these same results for the model of **Fig. 16.2** somewhat more easily and selectively, from feasible steady state experiment combinations. This is illustrated in **Fig. 16.7** and Example 16.5.

GRAPHS AND CUTSET ANALYSIS FOR EXPERIMENT DESIGN

ODE representations $\dot{x} = f(x, p, u)$ (or algebraic steady state equations $f(x, p, u) = 0$) of metabolic or other network or compartmental models typically associate a single state variable with a single compartment or node of a graph. Model parameters are then typically associated with edges of the graph, or with inter-compartmental or network fluxes or rate constants. **Cutset analysis** is a method for configuring the model graph differently, usually for the purpose of quantifying model parameters – our purpose here. Each part of the reconfiguration usually is a *sum* of state variables, instead of a single one. This facilitates model transformations that isolate model parameters of interest, possibly rendering them quantifiable. Importantly, reconfigurations of the model are done graphically rather than by complex algebraic or other manipulations.

In the context of experiment design, cutset analysis combines aspects of the exhaustive *SI* analyses of earlier sections of this chapter with basic *SI* methods described in Chapter 10 and some additional graph theory notions.[2] However, cutset analysis goes one step further than *SI* analysis by generating relationships for actually computing particular model parameters of interest. Model parameters of interest may or may not be structurally identifiable (*SI*) for a given I–O experiment, and we assume here that one or more I–O experiments can be done that render parameters of interest *SI*.

Operationally, we isolate desired parameters of interest by drawing particular closed curves (called cutsets) that surround and effectively sum one or more nodes (compartments) of the graph – curves that specifically isolate them. If a desired parameter cannot be isolated and computed on a particular graph with designated inputs and outputs, cutset analysis can – in many cases – help discover additional experimental I–O configurations on the same graph that can solve the problem.

Abbreviated theory and methods are developed in the following sections for addressing applications to **steady state inhomogeneous pool models** – a particularly broad class of feasible experiment models. Additional details can be found in the references (Feng and DiStefano 1991; Feng and DiStefano III, 1992; Feng and DiStefano III, 1995; Nguyen et al. 2003). We illustrate the approach first with a relatively simple example.

2 Cutset analysis can also be used to perform *SI* analysis.

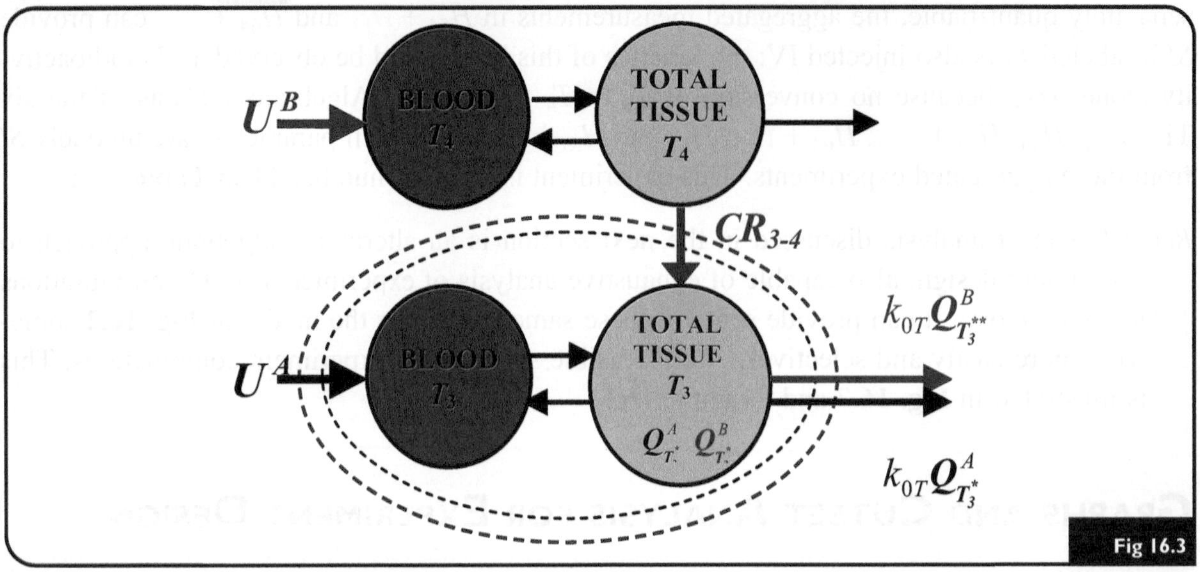

Fig 16.3

Remark: We depart in the current section from our usual notational consistency, adopting the notation for **steady state mass measurements** used in the references. Mass of material measured in pool i in steady state is designated Q_i instead of $q_{i\infty}$, as in earlier chapters.

Example 16.2: ***Whole-Body Hormone Interconversion***

We demonstrate use of cutset analysis here to design a pair of feasible experiments to solve for the *whole-body* hormone INTERCONVERSION RATE $\equiv CR_{3-4}$ (mass/time) in the steady state inhomogeneous pool model given in **Fig. 4.32** (Example 4.14). The conversion rate of T_4 into T_3 is different in different tissues, but CR_{3-4} represents the summed total conversion flux in whole-body measurements,[3] an important measure of overall TH economy – and thus for distinguishing normal from abnormal TH production. Pool models work well for addressing questions like this. The pool model in **Fig. 4.32** – reproduced in **Fig. 16.3** – is an inhomogeneous contraction of the extravascular compartments of the homogeneous compartmental model in **Fig. 16.2**.

We noted in earlier chapters that the enzymes for converting cellular T_4 into T_3 exist only in nonvascular TISSUE pools and no such conversion occurs in blood pools. It follows that the experiment design model in **Fig. 16.3** is the simplest structure accurately reflecting the physiology – biochemistry described that also isolates the desired total conversion flux CR_{3-4}. This pool model also includes two practical and potentially useful exogenous steady state experiments, designated $\boldsymbol{U^A}$ and $\boldsymbol{U^B}$: infusion of tracer T_3^* at constant rate $\boldsymbol{U^A}$ and tracer T_4^{**} at constant rate $\boldsymbol{U^B}$ into blood, as shown (Yen et al. 1994). Use of * and ** superscripts indicates that the two experiments are done simultaneously, with two different isotopic labels, e.g. $^{125}\text{I-}T_3 = T_3^*$ and $^{131}\text{I-}T_4 = T_4^{**}$. The goal is to compute CR_{3-4} from $\boldsymbol{U^A}$ and $\boldsymbol{U^B}$ and the steady state tracer hormone measurements in the 4 pools.

3 Total tissues T_3* and T_4** amount in this experiment are obtained by subtracting the amounts in blood pools from whole-body pools measured in steady state (Yen et al. 1994; Yamada et al. 1996; Nguyen et al. 1998).

Conventional Algebraic Solution: This entails writing mass balance equations for each of the 4 pools in terms of rate constants k_{ij} associated with each of the arrows (edges) and the steady state masses Q_j of tracer hormone in each pool at steady state for each experiment A or B. Each mass flux equals $k_{ij}Q_j$. Experiment A involves two pools and two state variables, whereas experiment B involves all four, for a total of six algebraic equations to solve symbolically for CR_{3-4} in terms of measurable steady state variables $Q^A_{T^*_{3blood}}$, $Q^A_{T^*_3}$, $Q^B_{T^{**}_{3blood}}$, $Q^B_{T^{**}_3}$, $Q^B_{T^{**}_{4blood}}$ and $Q^B_{T^*_4}$ and test-inputs U^A and U^B. Not easy, even for this simple model. Doing the algebra (Exercise **E4.17**), we find that it does generate a unique solution for CR_{3-4} from the six measurables in these experiments.

SI Analysis: That CR_{3-4} is uniquely identifiable from experiments A and B and is readily confirmed using any of the *SI* analysis methods in Chapter 10. *SI* analysis[4] also demonstrates that completing only one of these experiments is insufficient. What's still missing is that *SI* analysis alone does not provide a formula for computing CR_{3-4}.

Cutset Analysis Solution: Much simpler, and more complete. The goal is to split the model in two for each experiment in a way that minimizes unknowns. The black dashed closed-curve for experiment A has only one unknown (cutset k_{0T}), evaluated by equating black arrows into and out of the curve, because the system is in steady state, and the net flux in and out must be zero: $U^A = k_{0T}Q^A_{T^*_3}$. This yields unknown parameter: $k_{0T} = U^A/Q^A_{T^*_3}$. The red dashed closed-curve similarly encloses two fluxes (a second cutset). In steady state, we equate them, and this yields $CR_{4-3} = k_{0T}Q^B_{T^{**}_3}$. Combining the results of both experiments gives the desired result: $CR_{4-3} = k_{0T}Q^B_{T^{**}_3} = U^A Q^B_{T^{**}_3}/Q^A_{T^*_3}$.

Note that both closed cutset curves divided the model in two parts, equivalent to combining state variables in a way that short-circuits the solution graphically, equivalent to solving the six algebraic equations for CR_{3-4}.

More complex versions of this model and its analysis are given in references (Feng and DiStefano 1991; Feng and DiStefano III 1992; Nguyen et al. 2003).

Some Cutset Theory and Applications

First, a few definitions (Feng and DiStefano 1991). **Graph models** M generally include a **zero node**, which for our purposes means where leaks go and inputs come from, i.e. the "environment." A graph model is **connected** if there exists a path from every node to every other node, including the zero node. **Submodel** T of graph model M is a **tree** of M if T is a connected submodel, i.e. if it contains all nodes of M (including node 0), with no **loops** – no path that returns to the same node, ignoring directions. In the 2-pool model in **Fig. 16.4**, edges a and b together with nodes 0, 1, and 2 form a tree. Edges $\{a,b\} \equiv \{k_{01},k_{21}\}$ belongs to that tree, whereas $\{c,d\} \equiv \{k_{12},k_{02}\}$ and $\{c,d,e\}$ do not. Trees in a graph are not unique, and they are important in cutset analysis, as described below.

A **cutset** of the model is a *set of edges* with the following characteristics: removal of all the edges of the cutset leaves the remaining model disconnected; and removal of all but one of the edges of the cutset leaves the remaining model connected.

4 *SI* of the model for each experiment was readily evaluated using web application *COMBOS* (Meshkat et al. 2014).

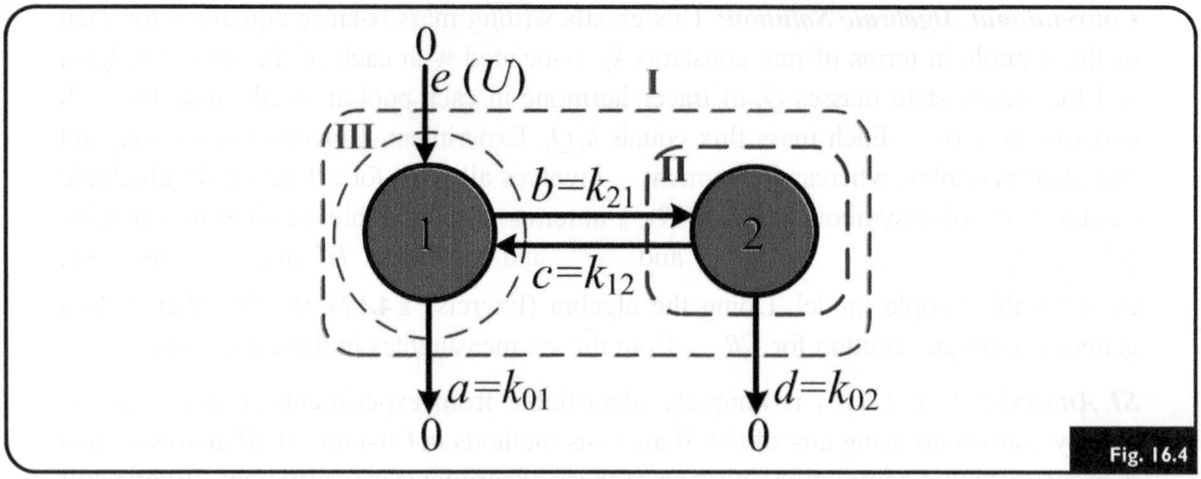

2-pool, 3-node (0, 1, 2) 5-edge (*a*, *b*, *c*, *d*, *e*) graph model. The three cutsets are enclosed by the green (*a*,*d*,*e*), red (*b*,*c*,*d*), and blue (*a*, *b*, *c*, *e*) dashed curves.

The 2-pool (3-node) model in **Fig. 16.4** has three cutsets, (*a*,*d*,*e*), (*b*,*c*,*d*), and (*a*,*b*,*c*,*e*). The dashed closed-curves called **Gaussian curves** operationally serve to define the cutsets – the green (*a*,*d*,*e*), red (*b*,*c*,*d*), and blue (*a*,*b*,*c*,*e*) ones, as shown. Each cutset divides the graph into two parts: graph components inside and outside the Gaussian curves.

Basics of Cutset Analysis

Let's examine each of the three cutsets in **Fig. 16.4**, each enclosed by a Gaussian curve. We are interested in the variety of different equations that equivalently describe this model. Curves II and III contain one node each and therefore generate two independent node equations – the same state equations that would be ordinarily obtained by single node or compartment or pool analysis and flux balance considerations. In contrast, curve I encloses *both* nodes 1 and 2. Flux balance across curve I treats nodes 1 and 2 simultaneously, delineating net flux across Gaussian curve I, i.e. a linear combination of individual node fluxes across I. This equation for net flux, combined with that for single node 1 or 2, is nothing more than a linear combination of the node equations that independently describe model dynamics; and they are equivalent to that which would be obtained for the pair of single node equations for nodes 1 and 2. All the dynamics of the model are preserved, but two equations have been effectively combined – a potential step in successive substitution of variables in individual node equations buried within the Gaussian curves.

In this manner, cutset analysis provides different transformations of the state equations of a model M, I–O equivalent to the original model. It provides linear combinations of the same state variables via an essentially graphic equivalence transformation and, importantly, it retains their physical interpretation – but as sums of these state variables. Among other things, this can simplify solution of equation sets for individual parameters.

Procedural Rules for Experiment Design

From an experiment design viewpoint, a particular set of I–O configurations for graph model M generally does not provide sufficient information to quantify desired parameters.

For example, experiment A in **Fig. 16.4** above, by itself, did not provide CR_{4-3}. Also, none of the limited experiments analyzed for the six-compartment model in **Fig. 16.2**, individually or in combination, provided *SI* solutions for individual T_4 to T_3 interconversion parameters in either of the two extravascular compartments. Thinking prospectively, cutset analysis should be capable of resolving these individual interconversion parameters, by suggesting different or additional experiments not yet suggested and analyzed, as illustrated in examples below.

Let's see what additional methodologic machinery we need for this to happen. We know from Chapter 10 that individual parameters p_i of an ODE model may not be structurally identifiable (*SI*), but there always exists a set of *SI* combinations (structural invariants) that are. Another way of phrasing this is that, in principle, we can always find the *SI* parameters and *SI* combinations (combos) of parameters of a model, so there must be formal relationships among them, relationships that also depend on the I−O measurement components of the model. Generalizing this, we'd like explicit expressions among a subset of the p_i called the **dependent parameters**, in terms of the remaining so-called **independent parameters** and the input and output variables. In a sense, this is what earlier sections on exhaustive TF analysis for linear systems was all about, for all possible I−O configurations. The *dependent parameters here are our parameters of interest.*

Keeping the focus on *steady state experiments* and desired parameters obtainable from such experiments, not having enough information to solve for a particular (dependent) parameter means there aren't enough state equations to solve for it. If there were, dependent parameters could be computed in terms of the independent parameters, inputs and outputs. The way to increase the total number of independent state equations, and thus solve this parameter estimation problem, is to do more independent experiments. Cutset analysis helps find the right ones to do quite efficiently – more efficiently than exhaustive TF analysis, for example.

Basic Property 1: Model parameters can be selected as dependent $\boldsymbol{p^D}$ (desired) ones and expressed in terms of the remaining independent parameters $\boldsymbol{p^I}$ if and only if these $\boldsymbol{p^D}$ correspond to a **tree** of the **connected** graph or compartmental model. (Tree was defined earlier.) This is a property of the graph independent of input and output locations.

Basic Property 2: For any given I−O experimental configuration, ODEs (or steady state equations) obtained from the graph of the model by cutset analysis are *independent* equations if one additional compartment is enclosed by each additional Gaussian curve.

These properties direct the following ***procedure***:

(1) First find a tree in the model that includes a desired parameter p^D. (2) Then draw all the Gaussian curves that include it. (3) Each such curve provides an independent equation that includes p^D and other parameters entering and leaving that curve. Write these out in their simplest form. (4) Then augment the graph with feasible I−O configurations that can provide independent equations capable of evaluating p^D explicitly and uniquely from the equations in step (3). (4) Solve for p^D in terms of measured quantities and known inputs.

Example 16.3: ***Four Steps for the Example 16.2 Model***

Step 1: One of the trees in **Fig. 16.3** that includes the desired edge parameter (flux CR_{3-4}) is defined by the 5 nodes (4 pools plus environment), either of the two edges connecting both blood and tissue pools, and the two red edges – one of which is CR_{3-4}. ***Step 2***: the Gaussian curves that include CR_{3-4} are either of the ones drawn in the figure, another (not shown) enclosing the two T_4 compartments together, and another enclosing the single tissue T_4 or T_3 compartment (not shown). ***Step 3***: Mass balance equations for the Gaussian curve shown in the figure are $U^A = k_{0T}Q^A_{T_3^*}$ and $CR_{4-3} = U^A Q^B_{T_3^{**}}/Q^A_{T_3^*}$. The remaining ones, for the other Gaussian curves, are left as an exercise (Exercise **E16.6**). ***Step 4***: This was completed in Example 16.2 for the Gaussian curves shown in the figure, resulting in $CR_{4-3} = k_{0T}Q^B_{T_3^{**}}$. The solution also could have been obtained from two of the other independent Gaussian curves.

Remark: In most experimental situations, more than one set of I–O ports are available – often many more than one. Cutset analysis permits discovery and screening of the experiments associated with each of these quite effectively. In principle, at least $\prod_{i=0}^{n-1}(2^n - 2^i)$ different transformations of the ODEs can be obtained by cutset analysis – thereby providing a variety of ways to configure experiment designs. For example, for the two-compartment (2-pool) model in **Fig. 16.3**, there are $\prod_{i=0}^{n-1}(2^n - 2^i) = (2^2 - 2^0)(2^2 - 2^1) = 3(2) = 6$ different ways of writing the state equations. With the test-input as shown on edge e, the ODEs can be written three ways, as node 1 and node 2; the sum node 1 + node 2 and node 1; and the sum node 1 + node 2 and node 2. If the test-input were instead introduced into node 2, three different ODEs would result from this new configuration – a total of six.

Parameter k_{ji} from Two or Three Steady State Experiments

The following results address a fairly common experiment design situation (Feng and DiStefano 1991). Suppose we are interested in estimating unknown rate constant k_{ji} between two different pools i and j in an open linear n-pool model, first for the one illustrated in **Fig. 16.5**. The assumption here is that we are dealing with inhomogeneous pool models in which total masses or concentrations in all pools (or key accessible pools) of the model are measurable in steady state.[5] Examples include studies in experimental laboratory animals – as illustrated in the references, e.g. Nguyen et al. (2003).

Parameter k_{ji} from 2 Steady State Experiments (**Fig. 16.5**): If a Gaussian curve G can be found that crosses k_{ji} and this k_{ji} is the *only* parameter pointing into G (from pool i to j), then *two* steady state experiments are sufficient to determine k_{ji}. With k_{ji} directed from pool i *outside* G to pool j *inside* G, the first experiment A is defined by an input U^A to pool j. The second experiment B is defined by an input to *any* pool outside G for which there is some path from the input pool to pool j. In this case, the desired parameter k_{ji} is given by the formula in **Fig. 16.5** for $i, j = 1, 2, \ldots, n$ (Feng and DiStefano 1991). Note that k_{ji} is expressed only in terms of known and measured quantities.

5 Biosystems can usually be probed more easily and extensively under constant steady state conditions. Absence of homogeneity requirements for pool models also make these experiments easier to do than transient response studies.

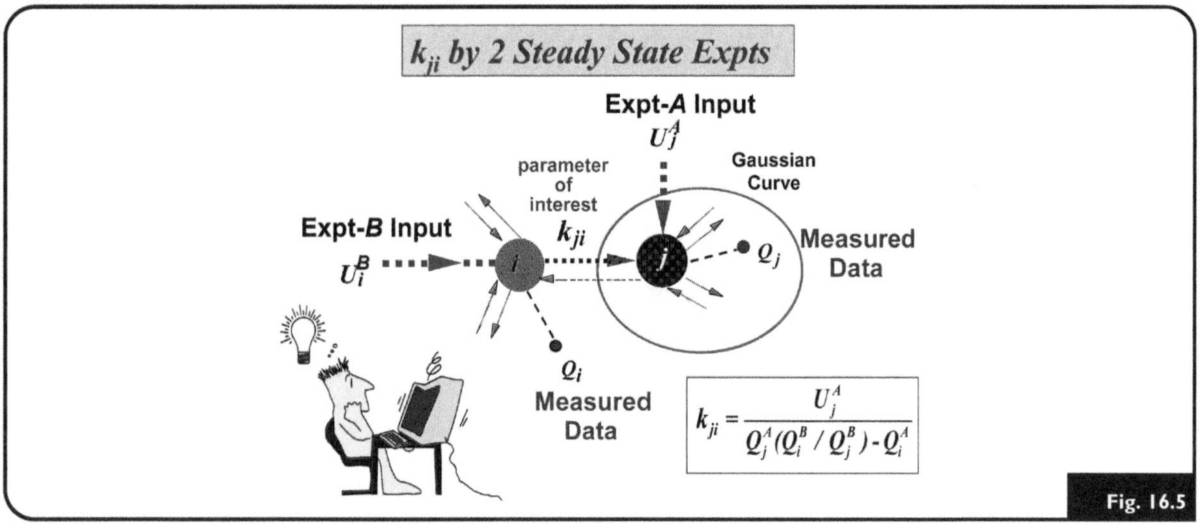

Fig. 16.5

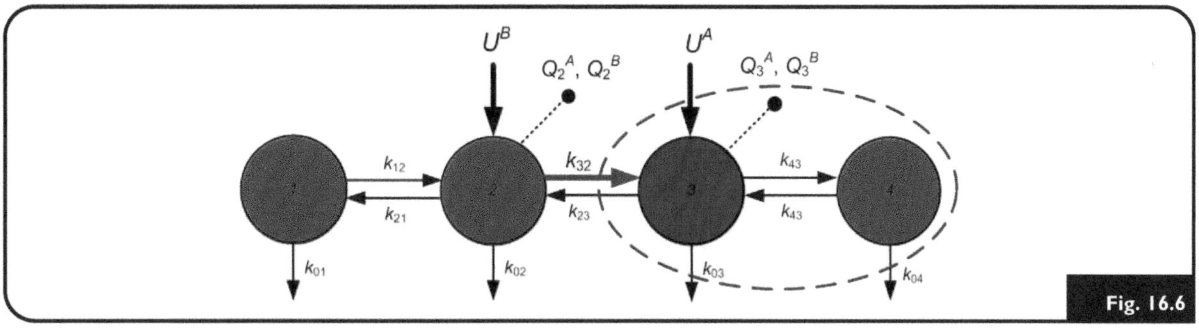

Fig. 16.6

Example 16.4: ***Parameter k_{32} from Two Constant Input Steady State Experiments***

The conditions of **Fig. 16.5** apply to the 4-pool model of **Fig. 16.6** because pool 3 is an accessible input port; k_{32} is the only rate constant pointing into Gaussian curve G; and pools 1 and 2 are accessible input ports that can be used to deliver test-input material to pool 3 via the edge arrows shown. We therefore can apply the formula in **Fig. 16.5** to compute k_{32}.

The two experimental inputs are constant infusion U^A into pool 3 – experiment A; and constant infusion U^B into pool 2 (or pool 1) – experiment B. The output measurements are the steady state masses of test-input material B in pools 2 and 3, i.e. Q_2^B and Q_3^B; and the steady state masses of test-input material A in pools 2 and 3, i.e. Q_2^A and Q_3^A. In these terms:

$$k_{32} = \frac{U^A}{Q_3^A(Q_2^B/Q_3^B) - Q_2^A}$$

Parameter k_{ji} from Three Steady State Experiments: The formula above is not applicable if another rate constant parameter in addition to k_{ji} points into the Gaussian curve G. This situation is illustrated in **Fig. 16.7**, with two parameters – the red and black ones – pointing into

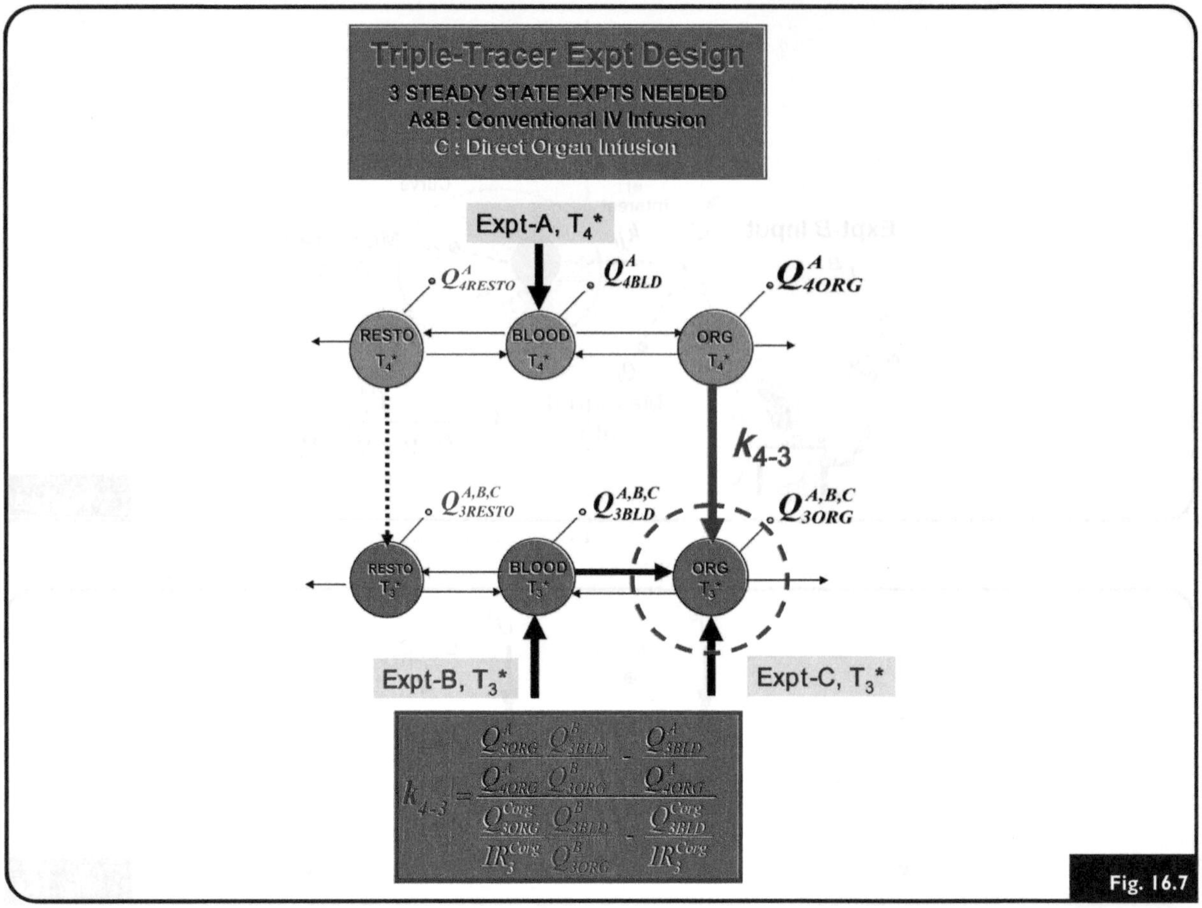

Fig. 16.7

the red dashed Gaussian curve. However, in this case, k_{ji} can be determined from *three* independent experiments.

Example 16.5: ***Conversion Parameter k_{4-3} from Three Constant Input Steady State Experiments***

The triple-tracer experiment design in **Fig. 16.7** has one tracer hormone input directed into an extravascular organ in **Expt-C**. The other two tracer experiments, **A** and **B**, have tracer inputs directly into blood. The equation for solving for parameter k_{4-3} in the **Fig. 16.7** model is given in the bottom of the figure, determined by cutset analysis in Nguyen et al. (2003) using the red Gaussian curve shown in the figure. Note that output measurements are needed in steady state only in blood and the organ ORG under study.

In experimental laboratory animal studies, this experiment design is applicable to conversion of one compound into another in any organ (ORG) separated from the blood pool and the rest of the body (RESTO). RESTO is the whole-body tissue pool – residual carcass after the blood pool and the organ pool of interest has been subtracted/dissected out. The references include results and formulas for other, more complex pool-model situations (Feng and DiStefano 1991; Feng and DiStefano III 1992; Feng and DiStefano III 1995; Nguyen et al. 2003).

Algorithms for Optimal Experiment Design

Quantifying a structured ODE model means finding estimates for the unknown *SI* parameters (or *SI* parameter combinations) of the model, usually by fitting the model to experimental data. We address this problem here as optimal experiment designers – with the goal of obtaining the *most reliable SI* parameter estimates we can, given the practical considerations imposed by experimental resources and other constraints. These "practical considerations" are in part circumscribed by the explicitly named and limited design-degrees-of-freedom illustrated earlier in **Fig. 16.1**.

The major measure of parameter estimate reliability is the parameter covariance matrix (also called parameter variance–covariance matrix) $COV(\hat{\boldsymbol{p}})$, typically computed along with or at the end of a parameter estimation run (Chapter 12). Standard measures of dispersion – parameter variances – appear on the diagonal of $COV(\hat{\boldsymbol{p}})$; and the off-diagonal terms provide the cross correlations (statistical interactions) among the parameters – also quite important in assessing parameter estimation reliability, particularly in the context of experiment design. For experiment design here, the goal is to find a feasible set of dynamical I–O experiments that *minimize a function of the variances* of *SI* model parameter estimates. As we see below, covariances are also included in the optimization process, albeit indirectly.

The Fisher information matrix ($F \equiv FIM$) and Cramér–Rao theorem ($COV(\hat{\boldsymbol{p}}) \geq F^{-1}$) provide a convenient way to develop optimization criteria based on $COV(\hat{\boldsymbol{p}})$: the inverse of F is a computable measure of the smallest possible variance–covariance matrix of the parameter estimates. We do this by formally incorporating adjustable design-degrees-of-freedom in an optimization criterion function for minimizing parameter estimation variances. To help put this notion in perspective, we formulate the optimization problem for a simple example.

Consider the model $\dot{x} = -px + u$, with $p > 0$ unknown and N discrete-time noisy measurements: $z_i = x_i + e_i \equiv y_i + e_i$, $i = 1,\ldots, N$. Assume the residuals e_i (noise) are each Gaussian (Normally) distributed, uncorrelated, with mean zero and variance σ_i^2. In this case, the **Fisher information matrix**[6] F (a scalar here) is simply: $F = \sum_{i=1}^{N} \left(\frac{\partial y(t_i)}{\partial p}\right)^2 \frac{1}{\sigma_i^2}$ (derived in ***Appendix F*** with these statistics). This F is readily rewritten as a function of adjustable design-degrees-of-freedom. The input u (a design variable) is unspecified in the model, so we can write output y as a function of u. The sampling times: $t_1, t_2, \ldots, t_N$ (abbreviation *SS*) are also unspecified, as are the data variances $\sigma_1^2, \sigma_2^2, \ldots, \sigma_N^2$ (vector $\boldsymbol{\sigma}^2$). Clearly, more precise data should yield more precise results, so these too can be design variables.[7] This gives $F(u, SS, p, \boldsymbol{\sigma}^2) = \sum_{i=1}^{N} \left(\frac{\partial y(u,p,t_i)}{\partial p}\right)^2 \frac{1}{\sigma_i^2}$. We almost have an optimization problem.

In the context of optimization, the Cramér Rao theorem suggests "minimizing" some function of the inverse of F, which intuitively, at least, implies maximizing F in some sense. F in our example above is a scalar, so we can attempt to find the best design variables for

6 The Fisher Information Matrix F, designated *FIM* by some authors, is derived in ***Appendix F.***

7 Numerical examples of the influence of data quality on parameter estimation precision, along with an offer of *Matlab* software for the specifics, can be found in Guisasola et al. (2006).

maximizing F for our simple example. More generally, for models of dimension > 1, we need to maximize some *scalar function* of the full information matrix. Again assuming Gaussian zero mean measurement noise, matrix F takes the form:

$$F(\boldsymbol{u}, SS, \sigma^2, \hat{\boldsymbol{p}}) = \sum_{k=1}^{N} \left(\frac{\partial y(\boldsymbol{u}, \hat{\boldsymbol{p}}, t_k)}{\partial \hat{p}_i} \right) R^{-1} \left(\frac{\partial y(\boldsymbol{u}, \hat{\boldsymbol{p}}, t_k)}{\partial \hat{p}_j} \right)^T \equiv Y^T R^{-1} Y \tag{16.10}$$

where $Y \equiv Y(\boldsymbol{u}, \hat{\boldsymbol{p}}, \sigma^2, t_1, \ldots, t_N) = [\partial y_i(\boldsymbol{u}, \hat{\boldsymbol{p}}, t_k)/\partial \hat{p}_j]$ is the **matrix of output sensitivity functions** and R is the (usually) diagonal covariance matrix of the data $R \equiv R(\sigma_1^2, \ldots, \sigma_N^2)$. This is the FIM often computed and used to (asymptotically) estimate the parameter covariance matrix as $COV(\hat{\boldsymbol{p}}) \cong F^{-1}$. In the current application, we are using Eq. (16.10) to design optimal experiments for minimizing variances of $\boldsymbol{p}$ estimates, by adjusting the design variables $\boldsymbol{u}, SS, \boldsymbol{\sigma}^2$ in a manner that maximizes a scalar function of F and, at the same time, accounts for all constraints on the design variables and boundary conditions of the model. Common constraints include practical limitations on the observation interval and the number of sample times, upper and lower bounds and forms of input signal functions, and bounds on sampled data variances.

Scalar Optimization Functions of F

D-optimal and A-optimal criteria are two common scalar functions of F used in practice for finding design variables that give optimal $\boldsymbol{p}$ estimates (DiStefano III 1980a; DiStefano III et al. 1980b; DiStefano III 1981; DiStefano III 1982; Landaw 1982; DiStefano III et al. 1982a; DiStefano III et al. 1982b; DiStefano III 1984; DiStefano III and Landaw 1984; DiStefano III et al. 1985). Others, e.g. E-optimal, are available as well (Balsa-Canto et al. 2007). **D-optimal** means minimizing the *determinant* of matrix F^{-1}, equivalent to maximizing the determinant of F, i.e. max(*det F*). **A-optimal** means minimizing the *trace* of matrix F^{-1}. Directly weighting variances of some parameters differently than others is an advantage of the A-optimal criterion. Weights can be assigned as elements of a diagonal weighting matrix W and then *trace* $(WM)^{-1}$ can be minimized. **E-optimal** means maximizing the minimum eigenvalue of F. Maximization of *det F*, however, is computationally simpler and has other advantages.

The most important advantage of the D-optimality criterion is that optimal designs are invariant under nondegenerate transformations of the model. For linear models, this means by similarity transformation to Jordan or other canonical form. Optimization is then carried out on the model in the most convenient form for analysis, with all results valid for the original model. For many linear models, sums of exponential functions are convenient and they have the added advantage of being analytical functions, so their sensitivity functions $Y \equiv [\partial y_i/\partial \hat{p}_j]$ can be obtained simply by differentiation (DiStefano III and Landaw 1984). We use them below for designing optimal output sampling schedules.

Balsa-Canto and coworkers summarize this same theory in Balsa-Canto et al. (2007) and apply it to a problem in food engineering – computing optimal control input designs using the D-optimal criterion for minimizing the size of the parameter confidence regions and thus parameter estimation variances in a model describing nutrient degradation in the process of sterilizing canned tuna. They use the sophisticated modeling, parameter estimation and optimal

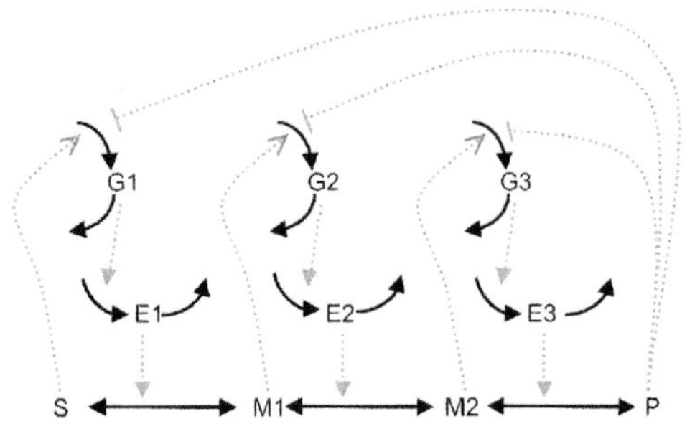

Fig. 16.8

experiment design tools in their *AMIGO*[8] software package for their computations. These tools are comprehensive in *AMIGO* and are particularly useful for optimal design of practical dynamic system experiments, as they describe and illustrate in their User Guide,[9] available at the same web site, and in Balsa-Canto and Banga (2011). The User Guide includes fully worked-out illustrative examples and code for a three-step enzymatically controlled product feedback inhibition pathway (**Fig. 16.8**). The examples involve model formulation, parameter *SI* analysis, estimation, and sensitivity analysis, followed by solution of an optimal experiment design problem for limited (pulse and step) test-input types and magnitudes. They demonstrate an approximately 50% improvement in parameter estimate variance reduction using optimal input and sampling schedule designs.

Optimal Sampling Schedule Design

The times at which output samples are collected are among the easiest to adjust in an experiment; and the payoff in improving parameter estimate reliability can be great when sampling times are chosen optimally. We address sampling schedule design based on the D-optimal criterion here. We maximize *det F*(SS) for a transformed model of *TH* metabolism – demonstrating a very practical result for multiexponential models using the D-optimality criterion.

The following result can be highly efficient in resource terms, satisfying the typical need for minimizing as well as localizing sampling effort in the collection of output data. Optimal sampling schedules (OSS) for **multiexponential models** $M: y = \sum_{i=1}^{n} A_i e^{\lambda_i t}$ of order n – and thus any linear model **input–output equivalent** to M – consist of independent replicates at no more than $2n$ distinct times, equal to the number of *SI* parameters. When some parameters are constrained, e.g. $\sum A_i = 0$, optimal schedules consist of $2n-v$ distinct times for v constraints. In either case, the theory says that any additional sampling effort available should be directed to independent replicates, at the **same** optimal times, rather than in between (Landaw and DiStefano III, 1984). This works very well in practice, as illustrated in the next section for a set of real physiological experiments.

8 http://www.iim.csic.es/ ~ amigo.

9 The *AMIGO* User Guide also includes several illustrative examples and code for analysis and quantification of the Hodgkin–Huxley model, a circadian clock model, and an NFkB kinase pathway model.

Example 16.6: ***Optimal Sampling for One and Two Unknown Parameters***

Consider the simple model of the previous section: $\dot{x} = -px + u$, with one parameter $p > 0$ unknown and the possibility of an unknown number N of discrete-time noisy measurements: $z_i = x_i + e_i \equiv y_i + e_i$, $i = 1,\ldots, N$, the Gaussian distributed residuals e_i have mean zero and are uncorrelated. Scalar F is: $F = \sum_{i=1}^{N} \left(\frac{\partial y(t_i)}{\partial p}\right)^2 \frac{1}{\sigma_i^2}$. We are interested here only in computing optimal sampling times. The input u is fixed as a unit-impulse and all data variances are $\sigma^2 = 1$ (constant variance error model). Then the output is: $y = e^{-pt}$, $y(0) = 1$, $\partial y/\partial p = -t\,e^{-pt}$ and F is a function of only the sampling schedule SS to be optimized: $F(t_1, t_2, \ldots, t_N, p) = \sum_{i=1}^{N} (\partial y(t_i)/\partial p)^2 = t_1^2\, e^{-2pt_1} + t_2^2\, e^{-2pt_2} + \cdots + t_N^2\, e^{-2pt_N}$.

Considering the sequence of discrete times as a vector: $\boldsymbol{t} \equiv [t_1 \quad \cdots \quad t_N]^T$, F has a maximum over $t > 0$ when its derivative $\partial F/\partial \boldsymbol{t} = 0$, i.e. when $\partial F/\partial t_1 = 0, \ldots, \partial F/\partial t_N = 0$. Each of these is of the form $(t\,e^{-pt})^2$, which has the same maximum as $t\,e^{-pt}$, i.e. a maximum at $t^* = 1/p$. Thus, collecting response data at the single optimal time $t^* = 1/p$ is enough in this example to generate a minimum variance estimate of p. This result also means that any additional sampling effort should be at the same time $t^* = 1/p$, with replication generating and even better statistical result, i.e. a smaller variance on the estimate of p. As a corollary, additional samples taken at any other times will improve the statistics of this estimate, but not as well as all sampling effort applied at the single time $t^* = 1/p$.

If the output were a concentration variable, with volume of the single-compartment V unknown as well, i.e. $y = x/V$, then the unknown parameter vector is $\{p, V\}$, and the OSS for *det F* optimization – for the same data error model – consists of *two* distinct time points, $t_0{}^* = 0$ and $t_1{}^* = 1/p$ (Landaw and DiStefano III 1984). Collecting response data at both of these time points is enough to generate minimum variance estimates of p and V. As above, any additional sampling effort would be optimally placed at these same two optimal times.

Given that sampling precisely at $t_0{}^* = 0$ is likely impossible, the earliest possible sampling time consistent with practical considerations would be near-optimal, e.g. the earliest time the input sample is well-mixed in the output compartment – $\sim$1 min in rat blood or $\sim$2 min in human blood.

In general, the choice of times that result in D-optimality depends on four factors, all of which effect optimal sampling times: the order and structure of the model; approximate values or nominals of model parameters; the observation time interval $[t_0, t_f]$ for feasible sampling; and the statistics of the error variance model. For example, with a different data error model, i.e. constant coefficient of variation (CV) error variance model, and feasible observation time interval – from t_0 to t_f, the D-optimal sampling design for the two parameter monoexponential model consists of two times, $t_1{}^* = t_0$ and $t_2{}^*$ is always chosen as the maximum time t_f. Extension of this rule of thumb to higher-order multiexponential models and other error models can be found in Landaw and DiStefano III (1984) and Landaw (1985).

Sequential Optimal Experiment Design

Optimal experiment design methodology requires a preliminary (nominal, initial) parameter estimate, because it is typically nonlinear in the parameters – meaning optimization depends on parameter values, making the problem somewhat circular. This makes it difficult to achieve optimality in a single experiment. A rational approach is then to split the optimization problem into an ordered sequence of quasi-optimized experiments done in several stages. This approach, novel at the time, was applied successfully in a number of hormone and protein distribution and metabolism studies in my lab during the 1980s, e.g. DiStefano III (1980a), DiStefano III et al. (1980b), DiStefano III (1981), DiStefano III (1982), DiStefano III et al. (1982a), DiStefano III et al. (1982b), DiStefano III (1984), DiStefano III and Landaw (1984), Specker et al. (1984), DiStefano III et al. (1985), Hershman et al. (1986), DiStefano III and Sapin (1987a), DiStefano III (1987b), and DiStefano III (1989). The methodology is described here in the context of several of these integrated modeling-experiment studies.[10]

We assume the model structure $\dot{\boldsymbol{x}} = \boldsymbol{f}(\boldsymbol{x}, \boldsymbol{p}, \boldsymbol{u})$, $\boldsymbol{y} = \boldsymbol{g}(\boldsymbol{x}, \boldsymbol{p})$ is fully established and only the parameters $\boldsymbol{p}$ are unknown. The overall goal is to minimize the variability (variances) in *SI* model parameter estimates obtained in dynamical I–O experiments, exemplified specifically in this section by optimally selecting data *sampling times* (OSS designs) to achieve this goal.

A logical flow chart of the steps involved in sequential design of optimal dynamic system experiments is presented in **Fig. 16.9**. The overall algorithm here is expanded to include the more complete modeling problem, beginning with the definition of explicit modeling goals in block 1. Block 2 then provides for recursively considering alternative model structures. Models are tested for structural integrity and adjusted as needed at the halfway point in block 6 and – if they don't pass – on to blocks 12 and 3; and, at the end of the optimization procedure, they are retested in block 11 and returned to 12 and 3 if they don't pass. In the following paragraphs, the focus is on steps 4–10.

OSS Design Software

Interactive program software[11] *OSS1v6* was developed for computing D-OSS specifically for single impulse response experiments with single output multiexponential responses (DiStefano III 1981; Landaw and DiStefano III 1984). It accommodates a variety of output data error models and allows information obtained from pilot experimentation and preliminary *WLS* data fitting to be entered interactively.[12] OSS are then computed for subsequent experiments. The interactive features permit predictive (*in numero*, *in silico*) explorations of optimal and suboptimal designs and their sensitivity to misspecification of error variance structure, parameter nominals and other design degrees of freedom.

10 Also see the review in Walter and Pronzato (1990).

11 OSS1v6 is available on the book website.

12 A batch version called *OSSM* also was developed for *multi-output* sampling designs using user supplied routines for model and sensitivity function computations (Cobelli et al.1985).

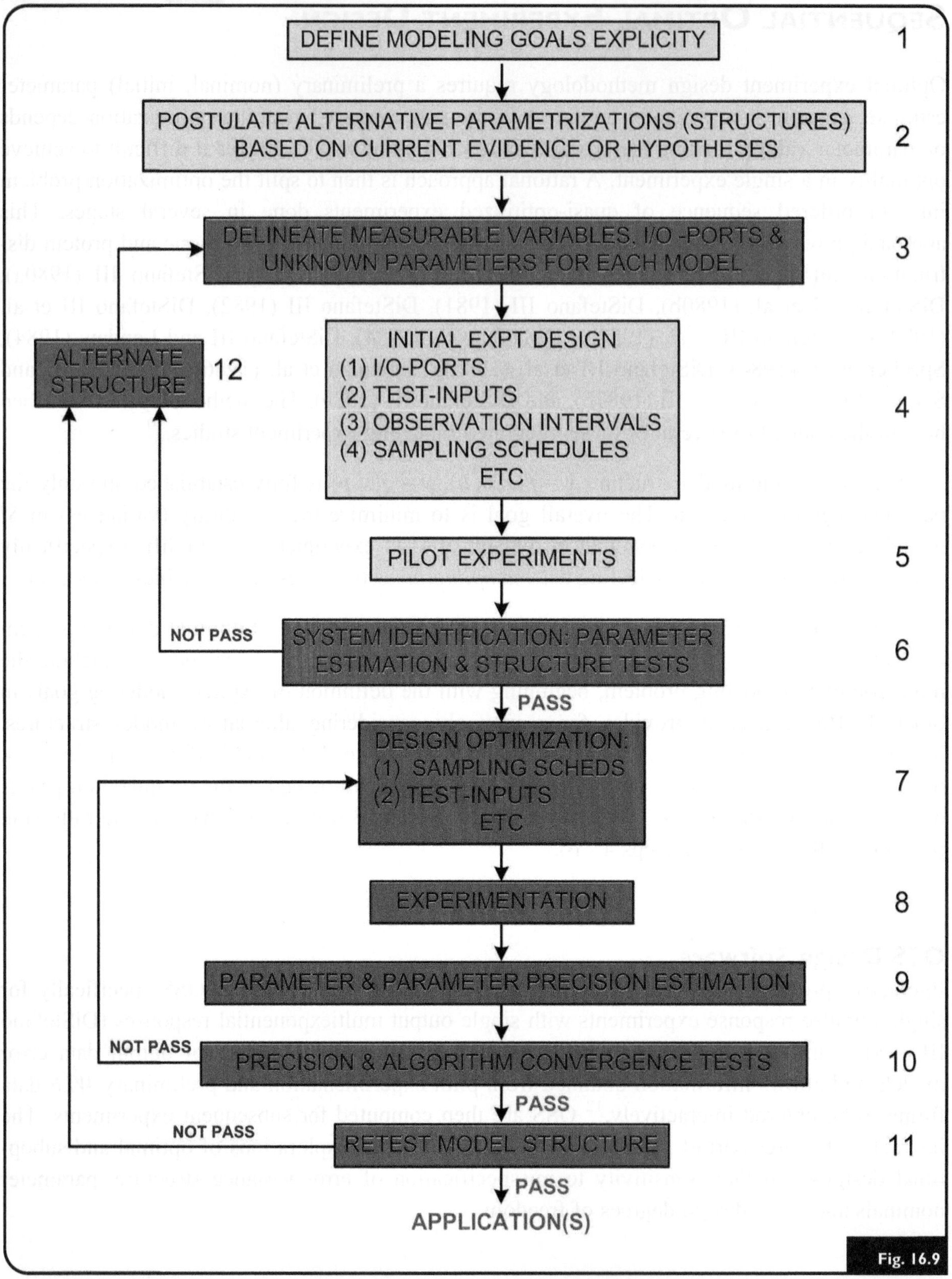

Fig. 16.9

Algorithm for **sequential optimal experiment design** (blocks 7–10) imbedded within experiment design (blocks 3–10) and further embedded within recursive consideration of designs for testing alternate model structures (blocks 1–12) for achieving overall modeling goals, as well as optimum (minimum variance) parameter estimates.

OSS Design Applied in Practice

Models of the dynamics of distribution and metabolism of THs triiodothyronine and thyroxine (T_3 and T_4) were developed and quantified based on this sequential optimal experiment design approach in a sequence of studies in the rat (DiStefano III et al. 1982a; DiStefano III et al. 1982b). *OSS1v6* played an important role in these studies. Pilot studies indicated that the impulse responses for both *THs* consisted of ***three distinct modes, fitted well by three-exponential functions***. The parameter estimation problem was thus formulated to fit this indirect model first, by estimating the *SI* coefficients and exponents of tri-exponential output functions from the data, by weighted least squares optimization, using web application *W³DIMSUM* (www.biocyb.cs.ucla.edu). The parameters of the three-compartment mammillary model[13] shown in **Fig. 10.3** were then computed algebraically from the tri-exponential parameters – as described in Chapters 12 and 13 – from which six parameters, the plasma volume V_P and the five *SI* parameter combinations in Eq. (10.65) were estimated.

Sequential Designs

With reference to **Fig. 16.10**, sampling schedules for 13 data samples in group 1 rats (panel a) were quasi-randomized-to identify possible clustering near five optimal sample times beyond $t > 0$. In group 2 rats (panel b), sample times were optimized on the basis of the group 1 mean rat model, also resulting in clustering around five individual sample times beyond $t > 0$. In group 3 rats (panel c), schedules were optimized on the basis of the group 2 mean rat model – again clustering quite close to the five same sample times and similarly for group 4 rats (panel d). All optimal schedules[14] consisted of $N = 6$ sample times, replicated in different rat subjects, resulting in approximately the same six times in groups 2 through 4: $t = 0$, 1, 4, 44, 202 and 600 min.

Optimally Quantified Model

Complete model quantification results for the T_3 dynamics model are summarized in **Fig. 16.11**. *SI* parameter precision varied from 1% to 26% coefficients of variation, averaging ~9%, reduced from initial values using the nonoptimal randomized schedules in the pilot studies by ~30%. OSS for the optimized groups were fairly robust, with little deterioration in parameter estimate precision for ~20% variations in sampling times.[15]

Recap and Looking Ahead

The overall experiment design problem was introduced formally in the first section, using a schematic block diagram of the problem in the context of systematically choosing practically

13 The six-compartment model in Fig. 16.2 illustrates T_4 and T_3 submodel connectivity, representing the interactions of thyroid hormones in mammalian blood and tissues.

14 In each of the optimized studies, only five samplings of blood were needed, at five (instead of six) different sample times. The sixth data point was provided as a prior estimate of the zero-time normalized tracer concentration, normalized to 100% of the dose. This was established in an independent experiment with radiolabeled albumin – a precise marker of the plasma volume V_P. In this case, $y(0) = 100/V_P$, the same V_P for both albumin and T_3 tracers, and thus the same $y(0)$: the sixth datum in all studies.

15 Please see the references for further details (DiStefano III 1982a; DiStefano III 1982b).

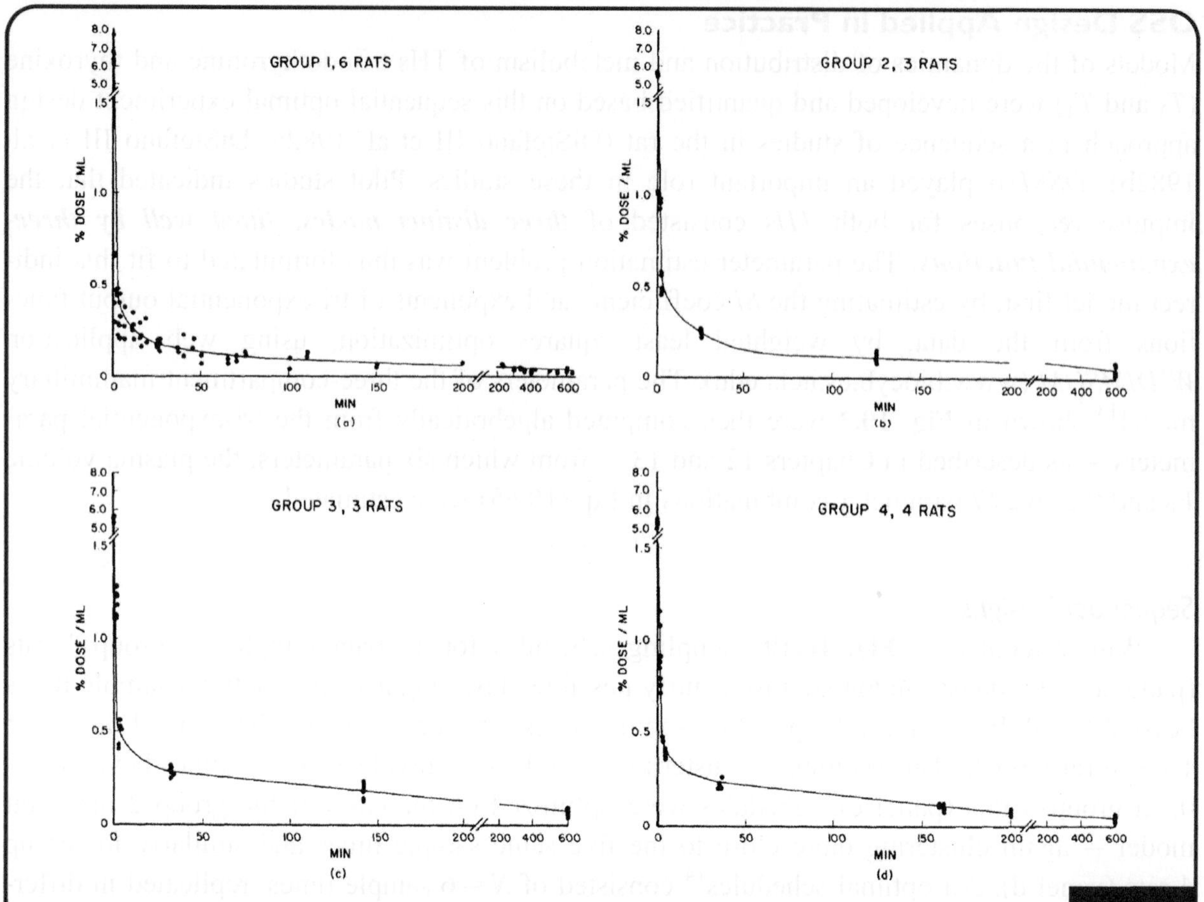

Fig. 16.10

Sequential optimal design results in 16 rat studies. All radioactively labeled T_3 tracer impulse response data (•) and best-fit functions (lines). Sampling schedules in group 1 rats (a) were quasi-randomized. In group 2 rats (b), they were D-optimized on the basis of the group 1 mean rat model. In group 3 rats (c), schedules were D-optimized on the basis of the group 2 mean rat model and similarly for group 4 rats (d). In groups 2 and 4, one rat in each was sampled on a slightly different schedule, with no apparent loss in parameter estimate precisions. At least 12 points, times the number of rats in each group, are actually shown; many data are superimposed on this scale, thus fewer appear. ***Fig. 4*** in DiStefano et al. "Comprehensive kinetics of triiodothyronine (T_3) production, distribution, and metabolism in blood and tissue pools of the rat using optimized blood-sampling protocols." *Endocrinology* 110: 198–213, 1982. Reproduced with permission from the Endocrine Society.

constrained experiment design-degrees-of-freedom for effectively quantifying model parameters.

SI-based methodology, described second, is about screening all feasible candidate parameter estimation experiments *a priori*, to establish the possibility of their success – under ideal conditions. It accomplishes this by exhaustively analyzing the *SI* properties of the model for all feasible I–O experiments that might be done on a system, by exhaustive analysis of all model I–O TF (experiments). It's directly applicable to any constant coefficient linear or linearized model-system. The same approach is applicable to nonlinear ODE models. Computational tools like *DAISY* (Saccomani et al. 2010) and *COMBOS* (Meshkat et al. 2014) – for analyzing NL ODE models for their *SI* properties – can be used to exhaustively screen all feasible I–O combinations for models that fall within the limited domains of applicability of these symbolic algebra methodologies.

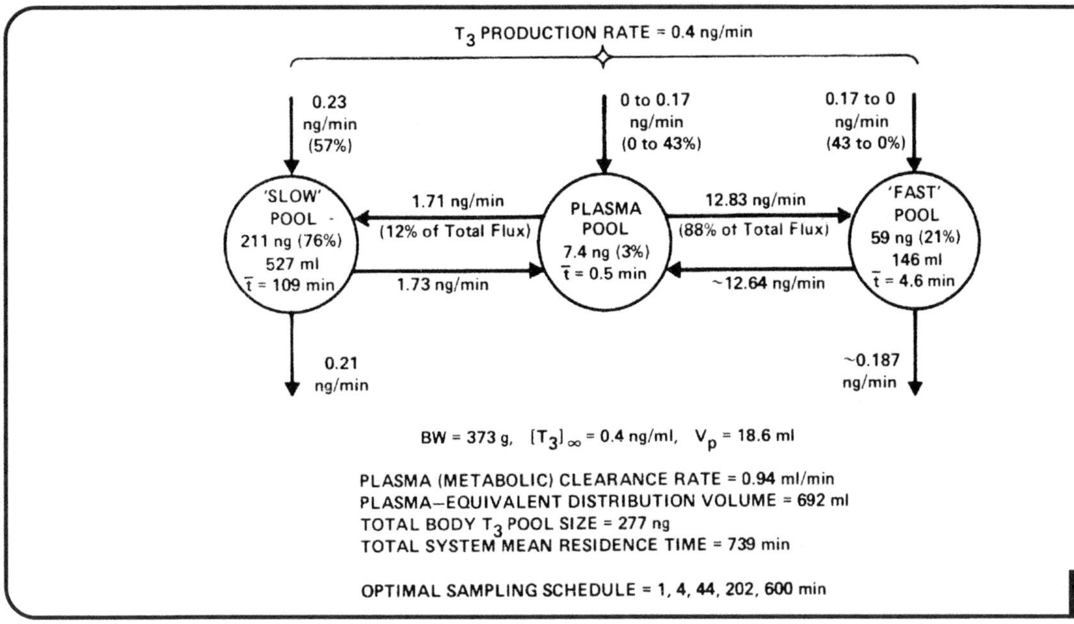

Fig. 16.11

Comprehensive mean model, kinetic parameters and optimal sampling schedule used for quantifying impulse response dynamics in male rats. ***Fig. 5*** in DiStefano et al. "Comprehensive kinetics of triiodothyronine (T3) production, distribution, and metabolism in blood and tissue pools of the rat using optimized blood-sampling protocols." *Endocrinology* 110: 198–213, 1982.
Reproduced with permission from the Endocrine Society.

The methodology for experiment design by cutset analysis is developed and illustrated for application to biological systems under commonly encountered experimental conditions – the constant steady state. This includes quite a broad array of physiological, biochemical and metabolic systems – studied under varieties of *in vivo*, *in vitro* or other conditions in steady state. It has been illustrated here with published example applications to hormone metabolism and physiology studies – completed in experimental animals – complex and difficult experiments – facilitated substantially using cutset analysis in the design process.

The optimal design sections show how *NI* properties of ODE models can be enhanced by optimally choosing experiment design-degrees-of-freedom that maximize parameter estimate reliability measures. Optimization criteria are developed based on optimizing particular scalar functions of the Fisher information matrix. Sequential optimal experiment design and optimum data sampling schedule design are the focus in this chapter and the methodology is illustrated with real and successful experimental laboratory results. OSS are particularly frugal, usually providing improved results with output sampling at fewer selectively placed sample times.

Recent literature indicates increased efforts within the systems biology modeling community in adapting and developing new algorithms and automation tools for addressing identifiability, distinguishability and parameter estimation problems – with some focus on the experiment design problem, e.g. Balsa-Canto et al. (2010) and Balsa-Canto and Banga (2011). Increasingly larger network models, both nonlinear and linear, are at least part of the motivation. We look forward to implementation and widespread adoption of these new tools, probably via the Internet. The last chapter of this book addresses model simplification (reduction) from real data, using all the tools developed in earlier chapters.

EXERCISES

E16.1 Redo Example 16.6, showing that any combination of *different* inputs will give the same uniquely *SI* result.

E16.2 Repeat the previous example for two inputs of the same form, but applied at different times.

E16.3 Consider the *experimental TF matrix* for the most general three-compartment model of **Fig. 4.20** computed in Eq. (4.79). Analyze each I–O TF of this model for its *SI* properties, using results developed in part in the six-compartment model analyzed in this chapter. Only the remaining I–O combinations in **Fig. 4.20** not appearing in **Fig. 16.1** need further analysis. Which individual parameters and parameter combos are *SI* in each experiment?

E16.4 Following up on Exercise 16.2, what *pairs* of experiments are capable of generating improved *SI* properties?

E16.5 Local *SI* properties of $\dot{\boldsymbol{q}} = A\boldsymbol{q} + B\boldsymbol{u}$, $\boldsymbol{y} = C\boldsymbol{q}$ can be explored algebraically by testing the rank of the sensitivity of the **Markov parameter matrix** (G) to $\boldsymbol{p}$:

$$G(\boldsymbol{p}) = \begin{bmatrix} C(\boldsymbol{p})B(\boldsymbol{p}) \\ C(\boldsymbol{p})A(\boldsymbol{p})B(\boldsymbol{p}) \\ C(\boldsymbol{p})A^2(\boldsymbol{p})B(\boldsymbol{p}) \\ \vdots \\ C(\boldsymbol{p})A^{2n-1}(\boldsymbol{p})B(\boldsymbol{p}) \end{bmatrix} \quad (16.11)$$

If the rank of $\partial G(\boldsymbol{p})/\partial \boldsymbol{p} = P$, the number of parameters, the model is *SI*, but not necessarily uniquely (Cobelli and DiStefano 1980). Find $G(\boldsymbol{p})$ and explore the *SI* properties of the three-compartment submodel in **Fig. 16.2** designated by TF H_{44}.

E16.6 Write the conservation equations and solve *them* for CR_{4-3} for each of the combinations of possible Gaussian curves that can be drawn for the model in **Fig. 16.3**.

E16.7 Use cutset analysis to find an expression for k_{31} for the model in **Fig. 16.2** by designing two different two-input constant steady state experiments.

E16.8 For the model in **Fig. 16.7**, find a formula for computing the T_4 to T_3 conversion rate in the whole-body minus the single organ depicted in the figure, i.e. for the *RESTO* compartment.

REFERENCES

Balsa-Canto, E., A. A., Banga, J., 2010. An iterative identification procedure for dynamic modeling of biochemical networks. BMC. Syst. Biol. 4, 11.

Balsa-Canto, E., Banga, J.R., 2011. AMIGO, a toolbox for advanced model identification in systems biology using global optimization. Bioinformatics. 27 (16), 2311–2313.

Balsa-Canto, E., Rodriguez-Fernandez, M., Banga, J.R., 2007. Optimal design of dynamic experiments for improved estimation of kinetic parameters of thermal degradation. J. Food Eng. 82 (2), 178–188.

Cobelli, C., DiStefano III, J.J., 1980. Parameter and structural identifiability concepts and ambiguities: a criticial review and analysis. Am. J. Physiol. Regul. Integr. Comp. Physiol. 239 (1), R7–R24.

Cobelli, C., Ruggeri, A., DiStefano III, J.J., Landaw, E.M., 1985. Optimal design of multioutput sampling schedules – software and applications to endocrine metabolic and pharmacokinetic models. IEEE Trans. Biomed. Eng. BME-32, 249–256.

DiStefano III, J., 1980a. Design and optimization of tracer experiments in physiology and medicine. Fed. Proc. 39, 84–90.

DiStefano III, J., 1981. Optimized blood sampling protocols and sequential design of kinetic experiments. Am. J. Physiol. Regul. Integr. Comp. Physiol. 240 (5), R259–R265.

DiStefano III, J., 1982. Algorithms, software, and sequential optimal sampling schedule designs for pharmacokinetic and physiologic experiments. Math. Comput. Simul. 24, 531–534.

DiStefano III, J., 1984. Optimal experiment design in plasma protein metabolism studies: sequential optimal sampling schedules for quantifying kinetics. In: Mariani, G. (Ed.), Pathophysiology of Plasma Protein Metabolism. McMillan, London, pp. 15–34.

DiStefano III, J., 1987b. Optimizing sampling schedule designs for kinetic experiments: conceptual framework, software, and practical experiences. Pharmacokinetics and Pharmacodynamics: A Meeting of the USC Biomedical Simulation Resource.

DiStefano III, J., 1989. Biomodel formulation and identification using optimal and other effective experiment designs. In: Cobelli, C., Mariani, L. (Eds.), Modelling and Control in Biomedical Systems. Pergamon Press, Oxford, pp. 1–12.

DiStefano III, J., Landaw, E., 1984. Multiexponential, multicompartmental, and noncompartmental modeling: physiological data analysis and statistical considerations, part I: methodological and physiological interpretations. Am. J. Physiol. Regul. Integr. Comp. Physiol. 246, R651–R664.

DiStefano III, J., Mori, F., 1977. Parameter identifiability and experiment design: thyroid hormone metabolism parameters. Am. J. Physiol. Regul. Integr. Comp. Physiol. 233 (3), R134–R144.

DiStefano III, J., Sapin, V., 1987a. Fecal and urinary excretion of six iodothyronines in the rat. Endocrinology. 121, 1742.

DiStefano III, J., Malone, T.K., Jang, M., Broutman, M., 1980b. Maximally accurate parameters of iodothyronine metabolism from only 5 or 6 blood samples: a new and generally applicable approach to kinetic studies and its demonstration in rats. Endocrinology. 106 (Suppl.), 98.

DiStefano III, J., Jang, M., Malone, T.K., Broutman, M., 1982a. Comprehensive kinetics of triiodothyronine (T3) production, distribution, and metabolism in blood and tissue pools of the rat using optimized blood-sampling protocols. Endocrinology. 110 (1), 198–213.

DiStefano III, J., Malone, T.K., Jang, M., 1982b. Comprehensive kinetics of thyroxine (T4) distribution and metabolism in blood and tissue pools of the rat from only six blood samples: dominance of large, slowly exchanging tissue pools. Endocrinology. 111 (1), 108–117.

DiStefano III, J., Jang, M., Kaplan, M., 1985. Optimized kinetics of reverse-triiodothyronine (rT3) distribution and metabolism in the rat: dominance of large, slowly exchanging tissue pools for iodothyronines. Endocrinology. 116 (1), 446–456.

DiStefano III, J., Nguyen, T.T., Yen, Y.M., 1993. Transfer of 3,5,3-triiodothyronine (T3) and thyroxine (T4) from rat blood to small and large intestines, liver, and kidneys in vivo. Endocrinology. 132, 1735–1744.

Feng, D., DiStefano III, J.J., 1991. Cut set analysis of compartmental models with applications to experiment design. Am. J. Physiol. Endocrinol. Metab. 261 (2), E269–E284.

Feng, D., DiStefano III, J.J., 1992. Decomposition-based qualitative experiment design algorithms for compartmental models. Math. Biosci. 110, 27–43.

Feng, D., DiStefano III, J., 1995. An algorithm for identifiable parameters and parameter bounds for a class of cascaded mammillary models. Math. Biosci. 129 (1), 67–93.

Guisasola, A., Baeza, J.A., et al., 2006. The influence of experimental data quality and quantity on parameter estimation accuracy: andrews inhibition model as a case study. Educ. Chem. Eng. 1 (1), 139–145.

Hershman, J.M., Nademanee, K., Sugawara, M., Pekary, A.E., Ross, R., Singh, B.N., et al., 1986. Thyroxine and thriiodothyronine kinetics in cardiac patients taking amiodarone. Acta. Endocrinol. (Copenh). 111 (2), 193–199.

Landaw, E., 1982. Optimal multicompartmental sampling designs for parameter estimation: practical aspects of the identification problem. Math. Comput. Simul. 24, 525–530.

Landaw, E., 1985. Optimal design for individual parameter estimation in pharmacokinetics. In: Rowland, M. (Ed.), Variability in Drug Therapy: Description, Estimation, and Control. Raven Press, New York, pp. 187–200.

Landaw, E., DiStefano III, J., 1984. Multiexponential, multicompartmental, and noncompartmental modeling: physiological data analysis and statistical considerations, part II: data analysis and statistical considerations. Am. J. Physiol. Regul. Integr. Comp. Physiol. 246, R665–R677.

Meshkat, N., Kuo, C., DiStefano III, J.J., 2014. COMBOS: web app for finding identifiable parameter combinations in nonlinear ODE models.

Nguyen, T., DiStefano III, J.J., Yamada, H., Yen, Y.M., 1993. Steady state organ distribution and metabolism of thyroxine (T4) and 3,4,3′-triiodothyronine (T3) in intestines, liver, kidneys, blood, and residual carcass of the rat in vivo. Endocrinology. 133, 2973–2983.

Nguyen, T.T., Chapa, F., DiStefano, J.J., 1998. Direct measurement of the contributions of Type I and Type II 5′-deiodinases to whole body steady state 3,5,3′-triiodothyronine production from thyroxine in the rat*. Endocrinology. 139 (11), 4626–4633.

Nguyen, T.T., Mol, K.A., DiStefano III, J.J., 2003. Thyroid hormone production rates in rat liver and intestine in vivo: a novel graph theory and experimental solution. Am. J. Physiol. Endocrinol. Metab. 285 (1), E171–E181.

Saccomani, M.P., Audoly, S., Bellu, G., D'Angio, L., 2010. Examples of testing global identifiability of biological and biomedical models with the DAISY software. Comput. Biol. Med. 40, 402–407.

Sefkow, A.J., DiStefano III, J.J., Himick, B.A., Brown, S.B., Eales, J.G., 1996. Kinetic analysis of thyroid hormone secretion and interconversion in the 5-day-fasted rainbow trout, oncorhynchus mykiss. Gen. Comp. Endocrinol. 101 (2), 123–138.

Specker, J., DiStefano III, J.J., Grau, E.G., Nishioka, R.S., Bern, H.A., 1984. Development-associated changes in thyroxine kinetics in juvenile salmon. Endocrinology. 115 (1), 399–406.

Walter, E., Pronzato, L., 1990. Qualitative and quantitative experiment design for phenomenological models – a survey. Automatica. 26, 195–213.

Yamada, H., Distefano, J.J., Yen, Y.M., Nguyen, T.T., 1996. Steady-state regulation of whole-body thyroid hormone pool sizes and interconversion rates in hypothyroid and moderately T3-stimulated rats. Endocrinology. 137 (12), 5624–5633.

Yen, Y., Yamada, H., DiStefano III, J.J., Nguyen, T.T., 1994. Direct measurement of whole-body thyroid hormone pools sizes and interconversion rates in fasted rats: hormone regulation implications. Endocrinology. 134, 1700–1709.

Chapter 17

Model Reduction and Network Inference in Dynamic Systems Biology

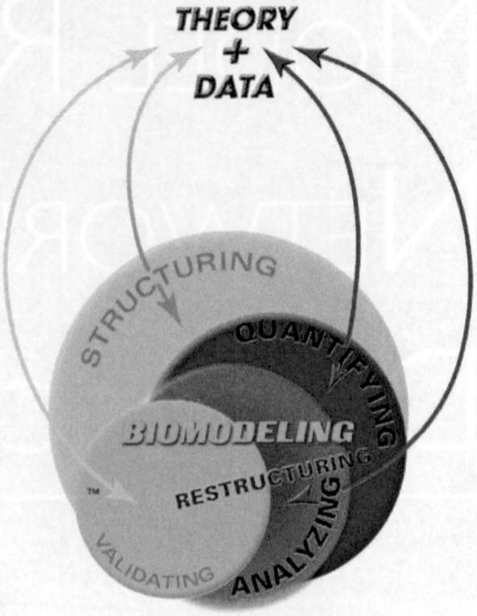

"The architect should strive continually to simplify; the ensemble of the rooms should then be carefully considered that comfort and utility may go hand-in-hand with beauty."

Frank Lloyd Wright

Overview

Models are often oversized, overspecified or overparameterized (OPM) – particularly if the modeling has been accomplished bottom-up. The major symptom of being OPM is failure in model quantification stages, most often manifested in lack of structural identifiability (*SI*) and/or numerical identifiability (*NI*) of one or more parameters. One solution to this problem is to collect more data – from independent input–output ports – as described in Chapter 16 and others. Another often more practical one is to simplify the model, which we address in this chapter using the **model reduction** formalism of system and control theory, currently being actively developed and extended as a systems biology tool. *Network reconstruction* and *network inference* are terms used synonymously in this context.

Candidate models for complex cellular pathways and networks developed from mining of high-throughput -omics and other data are typically overspecified and in need of pruning of unessentials. This means delineating and quantifying the mechanistic dynamics of the dominant molecular interactions imbedded in these systems – important for understanding the biology and developing tractable models for use in biomedical applications. Importantly, the dominant dynamics are often representable by much simpler models and the key to finding them is embedded in their structural and numerical parameter identifiability (*NI*) properties (Chapters 10, 12 and 13), which inherently include their *output sensitivity* characteristics (Chapters 11).

We developed and illustrated first-order (linear, local) sensitivity analysis (abbreviation LSA) measures in Chapter 11. LSA is the basis for most model reduction methods, as illustrated in this chapter. LSA measures are, however, only part of the output-parameter sensitivity story. Global sensitivity analysis (GSA) can also be important. Model parameters are definitively linked by the ordinary differential equations (ODEs) that define them, over the entire parameter space – not just at single best nominal "guestimates" $\boldsymbol{p}^0$. LSA is often extended to the entire parameter space, but GSA additionally addresses nonlinear as well as linear interactions, over the entire parameter space of the model, so GSA should yield better solutions. The first section below summarizes these distinctions in the usual context of sensitivities of model outputs to model parameters, both local and global. Evaluating local sensitivities (LSA) is generally computationally intensive. Evaluating global sensitivities (GSA) is much more so. Further discussion of GSA and several GSA algorithms are presented at the end of the chapter.

For added value, the model simplification methods developed in this chapter can be used for experiment design, i.e. for developing studies to learn more about the real biosystem – and learn it better. Model reduction usually addresses particular models, with given input–output configurations. Instead of including only the given outputs, other measurable state variables can be explored as candidate outputs for measurement. Comparisons of all measurable state variable sensitivities to parameters, in particular ways, then delineate the ones that can provide better parameter estimate precision, and possibly additional fine detail about the real system.

Local and Global Parameter Sensitivities

Local parameter sensitivity analysis (LSA) in Chapter 11 yielded sensitivities $\partial x_i(t)/\partial p_j$ and $\partial y_i(t)/\partial p_j$ of state variables $x_i(t)$ and outputs $y_i(t)$ w.r.t. linearized variations in $\Delta \boldsymbol{p}$. These local parameter sensitivities were based on the Taylor series expansion of state variable variations $\Delta x_i(t,\boldsymbol{p})$ up to first-order linear terms in $\Delta \boldsymbol{p}$:

$$\Delta x_i(t,\boldsymbol{p}) \equiv \Delta x_i \cong \sum_{j=1}^{P} \frac{\partial x_i}{\partial p_j} \Delta p_j \tag{17.1}$$

The effects of nonlinear couplings on parameter sensitivities – not revealed in Eq. (17.1) – are given by the complete Taylor series expansion of state variable variations $\Delta x_i(t,\boldsymbol{p})$, which includes higher-order interacting terms nonlinear in $\Delta \boldsymbol{p}$. The Taylor series (TS) expansion of $x_i(t,\boldsymbol{p}+\Delta\boldsymbol{p})$, i.e. the ith-component of state vector $\boldsymbol{x}(t,\boldsymbol{p}+\Delta\boldsymbol{p})$ about $\boldsymbol{p}$ is:

$$x_i(t,\boldsymbol{p}+\Delta\boldsymbol{p}) = x_i(t,\boldsymbol{p}) + \frac{\partial x_i}{\partial \boldsymbol{p}}\Delta\boldsymbol{p} + \frac{1}{2}\Delta\boldsymbol{p}^T\left(\frac{\partial^2 x_i}{\partial \boldsymbol{p}^2}\right)\Delta\boldsymbol{p} + \text{higher-order terms} \tag{17.2}$$

The **total (global) variation** $\Delta x_i(t,\boldsymbol{p})$ of the ith-component of state vector $\boldsymbol{x}(t,\boldsymbol{p})$ for all $t \geq t_0$ is defined as $\Delta x_i(t,\boldsymbol{p}) \equiv x_i(t,\boldsymbol{p}+\Delta\boldsymbol{p}) - x_i(t,\boldsymbol{p})$:

$$\Delta x_i(t,\boldsymbol{p}) = \frac{\partial x_i}{\partial \boldsymbol{p}}\Delta\boldsymbol{p} + \frac{1}{2}\Delta\boldsymbol{p}^T\left(\frac{\partial^2 x_i}{\partial \boldsymbol{p}^2}\right)\Delta\boldsymbol{p} + \text{higher-order terms} \tag{17.3}$$

This equation expresses total (state variable) variations in response to simultaneous parameter variations $\Delta p_1, \Delta p_2, \ldots, \Delta p_P$ *in toto*, linear effects in the first term and nonlinear effects in the second- and higher-order terms. GSA codifies all parameter variation effects, linear and nonlinear, on model state variables and outputs. GSA also relaxes the dependence of sensitivities on a single nominal $\boldsymbol{p}^0$ point ("guestimate") in parameter space. It generally includes assessment of sensitivities over the entire parameter space of possible values of $\boldsymbol{p}$. Further discussion of and specific algorithms for GSA are presented in the last section.

Model Reduction Methodology

Structural unidentifiability (*un*ID) problems can be the Achilles heel of real-world modeling efforts. They are often difficult to recognize, so delineating *un*ID parameters, to the extent this can be done, can be a very useful first step in reducing big models to their essential features. Sensitivity analysis can help with this. Parameters with high relative sensitivity values $((\partial y/y)/(\partial p/p))$ are relatively important (essential), because they are both structurally identifiable (*SI*) and numerically identifiable (*NI*). In contrast, parameters with low relative sensitivity are relatively less important, and they are *un*ID if their sensitivities are zero. Less important here means they are potential candidates for elimination, fixing or combining with others. For example, *un*ID parameters have zero output sensitivity and if they do not appear in combination with any others in the model, they are unimportant in fitting the model to data – because they are not observable in outputs (***Appendix D***). This means they have no effect on outputs (data) and usually can be fixed to any value in their feasible range[1] – leaving fewer parameters to quantify from data. Sensitivity analysis, as we see below, is central in model reduction.

Remark: Any kind of *SI* analysis – to the extent it can be accomplished – is a good starting point for model reduction; and it should also uncover *SI* parameter combinations (combos) $h(\boldsymbol{p}) \equiv \boldsymbol{b}$. With nominal $\boldsymbol{p}$ values specified, low output sensitivity to structurally identifiable (*SI*) combinations – as with individual *SI* parameters – similarly means poor *NI* of $\boldsymbol{b}$. During a parameter estimation run, this can be mistaken for apparent *clustering* of independently *SI* parameters – *approximately* identifiable parameter combinations – with poor *NI*. These parameter clusters – symptomatic of OPMs – should be recognized and combined, if possible, thereby simplifying the model. This typically renders overall model quantification more successful, because the model is simpler.

Parameter Ranking

The basic idea involves ranking the relative importance of each model parameter p_i in contributing quantitatively to output dynamics and thus the ultimate fit of the model to the output data. This procedure typically reduces the numbers of parameters substantially – by allowing elimination or fixing of those with low-enough ranking – and this effectively reduces the number of state variables as well – sometimes ***greatly*** simplifying the model.

Ranking parameter importance using ***sensitivity-based metrics*** is the most common thread to model reduction of biochemical/cellular pathway models currently developing in the literature – using somewhat different approaches and algorithms, e.g. Cho et al. (2003), Yue et al. (2006), Zi et al. (2008), Cintron-Arias and Banks (2009), Huang and Chu (2010), and Quaiser et al. (2011). New software tools based on these approaches and algorithms are also becoming available (Rodriguez-Fernandez and Banga 2010; Balsa-Canto and Banga 2011). Most approaches reported are based on the ranking of relative sensitivity metrics, based on local sensitivity computations (Yue et al. 2006; Chu and Hahn 2009). Some include singular value decomposition (SVD) of the output or state variable sensitivity matrix with respect to the parameters $\partial\boldsymbol{x}/\partial\boldsymbol{p}$ (Chu and Hahn 2009; Cintron-Arias and Banks 2009). Parameters associated with singular values (metrics) close to zero or equal zero have little or no importance. Some methods are also augmented with covariance matrices estimated from the Fisher information matrix computed from local sensitivities (Cintron-Arias and Banks 2009). Others are based on global and local sensitivity measures (Cho et al. 2003; Zi et al. 2008).

Parameters with very small sensitivity metrics are typically either parallel or nearly parallel (highly correlated) with others in parameter space, suggesting use of the parameter correlation matrix for ranking, as well as other geometrically inspired metrics for assisting with the

1 Feasible range is defined as a property of interval identifiability in Chapter 10.

parameter ranking task. The dominant model dynamics are then represented by the parameters with the largest averaged relative sensitivities and/or parameter estimation variances.

Simple *RMS* Metric for Local Sensitivity-Based Parameter Ranking

Root mean square (*RMS*) sensitivity metrics, developed in Chapter 11, are part of most algorithms. For output measurements at N discrete times, the *RMS* metric is:

$$RMS\left[s_{ij}(t_1, t_2, ..., t_N)\right] = \frac{1}{N}\sqrt{\sum_{k=1}^{N}\left(\frac{\partial y_i(t_k)}{\partial p_j}\right)^2\left(\frac{p_j}{y_i(t_k)}\right)^2} \equiv \boldsymbol{length}\left[s_{ij}(t_1, t_2, ..., t_N)\right] = \|\boldsymbol{s}\|_2 \quad (17.4)$$

This represents the average value over time of the square root of the squared relative sensitivity functions of the outputs with respect to the parameters. It's also the length $\|\boldsymbol{s}\|_2$ (L^2-norm) of the relative sensitivity vector $\boldsymbol{s}$ over the time period – a good average measure of overall relative sensitivity values.

Geometric Metric for Parameter Ranking

Highly correlated parameters are pairwise relatively indistinguishable. A geometric interpretation of this is that the angle between the sensitivity vectors of highly correlated parameters is small. We can use this idea to define a computable metric (similarity measure) of the effect of two parameters on the output as follows:

$$\cos\theta_{\alpha\beta} = \frac{|\boldsymbol{s}_\alpha^T\boldsymbol{s}_\beta|}{\|\boldsymbol{s}_\alpha\|_2\|\boldsymbol{s}_\beta\|_2} = \frac{|\boldsymbol{s}_\alpha^T\boldsymbol{s}_\beta|}{RMS_\alpha \cdot RMS_\beta} \quad (17.5)$$

where $\theta_{\alpha\beta}$ is the is the angle (0–90°) between relative sensitivity vectors $\boldsymbol{s}_\alpha$ and $\boldsymbol{s}_\beta$. The value of this metric ranges from 0 to 1. Unity means the two vectors are parallel ($\theta_{\alpha\beta} = 0$) and the two parameters cannot be distinguished. A zero value ($\theta_{\alpha\beta} = 90°$) means the parameter sensitivity vectors are orthogonal and have a distinct effect on the output.

This metric was used effectively in Chu and Hahn (2009) to cluster parameters into groups based on an **agglomerative hierarchy.**[2] The goal is to have a small number of clusters of parameters with similar angles between them so that a single representative one can be chosen from each.

Added Benefits: State Variables to Measure and Parameters to Estimate

Like *a priori SI* analysis, sensitivity analysis is doubly useful in model quantification tasks, before *and* after data collection. Application of parameter ranking methodology typically begins with a preliminary quantified model based on limited data, enough to provide initial

2 Each parameter starts in its own cluster and pairs of parameters are merged as they move up the hierarchy.

parameter estimates – nominal parameter values for computing relative parameter sensitivities. For follow-up experiment design, the idea is to first quantify the relative sensitivities to parameter variations of all the *potentially measurable* state variables (all *candidate nodes*) in the network, and rank the parameters from most to least important (most to least *fragile*), in accordance with the maximum to minimum values of these relative sensitivities. Then, in principle, the most fragile parameters are the most important ones to be estimated (quantified) from data that might be collected *from the candidate nodes*.

Reducing a Model of NFκB Signaling Dynamics

NF-kappaB (NFκB) is a principal transcription factor in mammalian immune system signaling. This highly regulated cellular signaling pathway is responsible for a variety of different yet specific responses to different somatic stresses that produce inflammatory responses. Multiple feedback loops, crosstalk between multiple NFκB-activating pathways and NFκB oscillations control temporal dynamics and gene expression. Dynamic biosystem models of this system are well developed and moderately complex (Hoffmann et al. 2002; Cheong et al. 2008; Kearns and Hoffmann 2009; Longo et al. 2013).

A highly simplified cartoon of a major feedback subsystem module of the Hoffman group model is illustrated in **Fig. 17.1**. IκB and κBs in the figure are a family of regulatory protein inhibitors of κB proteins, which include NFκB. Subcellular (nuclear and cytoplasmic) molecular interactions governing κB synthesis, degradation, localization, feedback and association/dissociation with NFκB are illustrated highly abbreviated in the figure and multistep events are combined. Proteins are degraded via ubiquitin-dependent proteasome pathways following IKK-mediated phosphorylation at the top. IKK-mediated phosphorylation is rate-limiting, so the simplified model is dominated by a single IKK-dependent protein degradation reaction. IκB-isoform α, β, ε reaction pathways are shown at bottom left control synthesis and

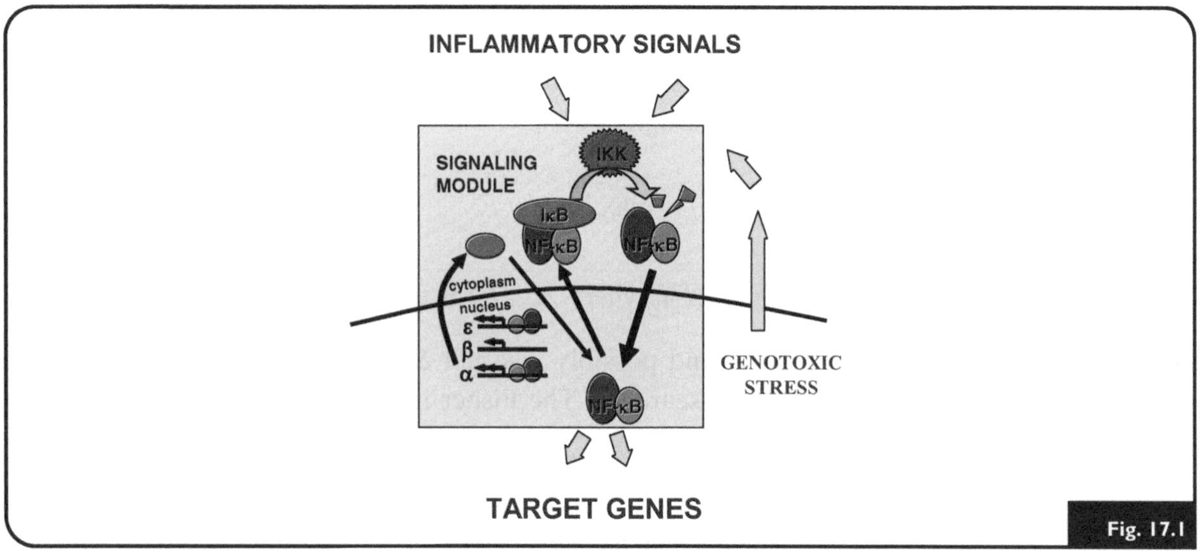

Fig. 17.1

Cartoon schematic of the ***IκB–NFκB signaling module*** with a single delayed negative feedback loop.

degradation of IκB proteins, association and dissociation of IκB proteins NFκB and IKK and cellular localizations. NFκB is a product or reactant in multiple reactions in modules like the ones shown in **Fig. 17.1** and its dynamic responses are oscillatory.

Yue and coworkers used an early version of an ODE model of this NFκB−IκB module (Hoffmann et al. 2002) to illustrate their algorithm for model reduction (Yue et al. 2006). They used the parameter ranking procedure described above, based in large part on computation of *RMS* performance metrics and selection of the p_i with the largest *RMS* values. They also considered one or more outputs measured (Yue et al. 2006). Their results are fairly lengthy and we discuss only two key ones here, the first leading to substantial model simplification and the second indicating which state variables dominate observed oscillatory responses − implying that these might be the only (or primary) ones needing measurement.

Example 17.1: ***Reduced NFκB Signaling Pathway Model***

As described above, the Yue group (Yue et al. 2006) applied their algorithm to ranking the importance of each model parameter p_j of the model illustrated in **Fig. 17.1**. They ranked the full parameter set, with the largest *RMS* contributing most, and showed that the essential dynamics of this 24 ODE model is governed by only 8 of the 64 biomodel parameters, the 8 with the largest *RMS* (**Fig. 17.1**). They also showed that only 3 of the 24 state variables need to be measured to obtain this level of model simplification.

Parameter Ranking Metrics Based on Optimizing the Fisher Information Matrix *F*

The Cramer−Rao theorem, i.e. F^{-1} is a lower bound on $COV(\boldsymbol{p})$, can be used to develop formal criteria for optimization of experiments for selecting the best parameters to estimate or the most informative outputs to measure. These metrics are more useful for model reduction combined with experiment design, as illustrated in *Example* 17.2. In these applications, the *RMS* metric in Eq. (17.4) is augmented by SVD metrics and parameter covariance statistics estimated from $F \equiv FIM$; which is essentially the "square of the sensitivities" (as in Eq. (17.4)) weighted by the covariance of the data (e.g. see Eq. (17.7)). One popular criterion is ***D*-optimality**, also used for optimal sampling schedule design in Chapter 16:

$$L_D = \max \log det\, F \tag{17.6}$$

Another is ***E*-optimality**: $L_E = \max \lambda_{min}(F)$, where $\lambda_{min}(F)$ is the smallest eigenvalue of F.

If some parameters are not *NI* − and possibly also not *SI* − a set suitable for estimation can be found using *D*- or *E*-optimality searches. The Fisher information matrix F_L of a subset of parameters can be written

$$F_L = L^T F L = \left(\frac{\partial \boldsymbol{Y}}{\partial \boldsymbol{p}} L\right)^T W^{-1} \left(\frac{\partial \boldsymbol{Y}}{\partial \boldsymbol{p}} L\right) \tag{17.7}$$

where the **selection matrix** L is given by $L = [e_{i_1} e_{i_2} \cdots e_{i_{(P-r)}}]$. The set $\{i_1, i_2, \ldots, i_{P-k}\}$ are the indices of the (yet unknown) parameters to be selected and e_i is the ith column of the identity matrix. Parameter selection is then reduced to finding the L that maximizes one of the optimality criteria above applied to F_L (Chu and Hahn 2009).

Reducing a Model of Interleukin-6 (IL-6) Signaling Dynamics

Chu and Hahn use the sensitivity-based metrics and SVD analysis described above in combination with a hierarchical **parameter clustering** algorithm to simplify computations for models with large numbers of parameters (Chu and Hahn 2009). In their overall algorithm, they first compute relative sensitivities and eliminate (fix) parameters with less than 5% of the largest relative sensitivities and those with near zero singular values. The remaining ones are hierarchically clustered into several groups based on the geometric metric Eq. (17.5), to quantify how close to parallel they are in parameter space.[3] The parameter with the largest sensitivity vector in each group is chosen as representative of that group. A subset of these representatives is then subjected to optimization, using Eq. (17.6).

Example 17.2: ***Simplified Interleukin-6 (IL-6) Signaling Pathway Model***

> Chu and Hahn applied their parameter clustering and *FIM* optimization methodology to a complex model of a signal transduction pathway in hepatocytes (liver cells) stimulated by IL-6 (Chu and Hahn 2009). IL-6 is a cytokine (immune system hormone) with both pro-inflammatory and anti-inflammatory actions in cells.
>
> The 66 NL ODE model with 115 unknown parameters is illustrated schematically in **Fig. 17.2**. State variables are the protein concentrations in the pathway; the input variable is the concentration of IL-6 outside the cell that initiates signal transduction; and the single measured output variable is the concentration of activated STAT3 dimer in the nucleus.
>
> The 9th to 115th singular values determined by SVD were ~ 0, suggesting a parameter set of size 8. Sensitivity vectors for 70 of the 115 parameters had lengths less than 5% of the highest one, so these were all fixed to their nominal values, leaving 45 to analyze further. The problem was then reduced to finding clusters of size 8 in 45 that maximize the optimality criterion Eq. (17.6). They solved this difficult computational problem using their hierarchical clustering approach, first reducing the set of 45−11. The final results yielded several choices of 8 maximally fragile parameters among 11, and a very highly simplified model capable of reproducing the basic dynamics of the data.

Reducing an Overparameterized (OPM) Model of p53 Signaling Dynamics

In this section, we apply the basic parameter ranking algorithms described above to simplify the new p53 model developed and partially analyzed in Chapter 15. Recall that the model was

3 Recall also from Chapters 11−13 that fully parallel parameter vectors are maximally correlated, and therefore impossible to distinguish. Orthogonal parameter vectors have zero correlation and maximal sensitivity.

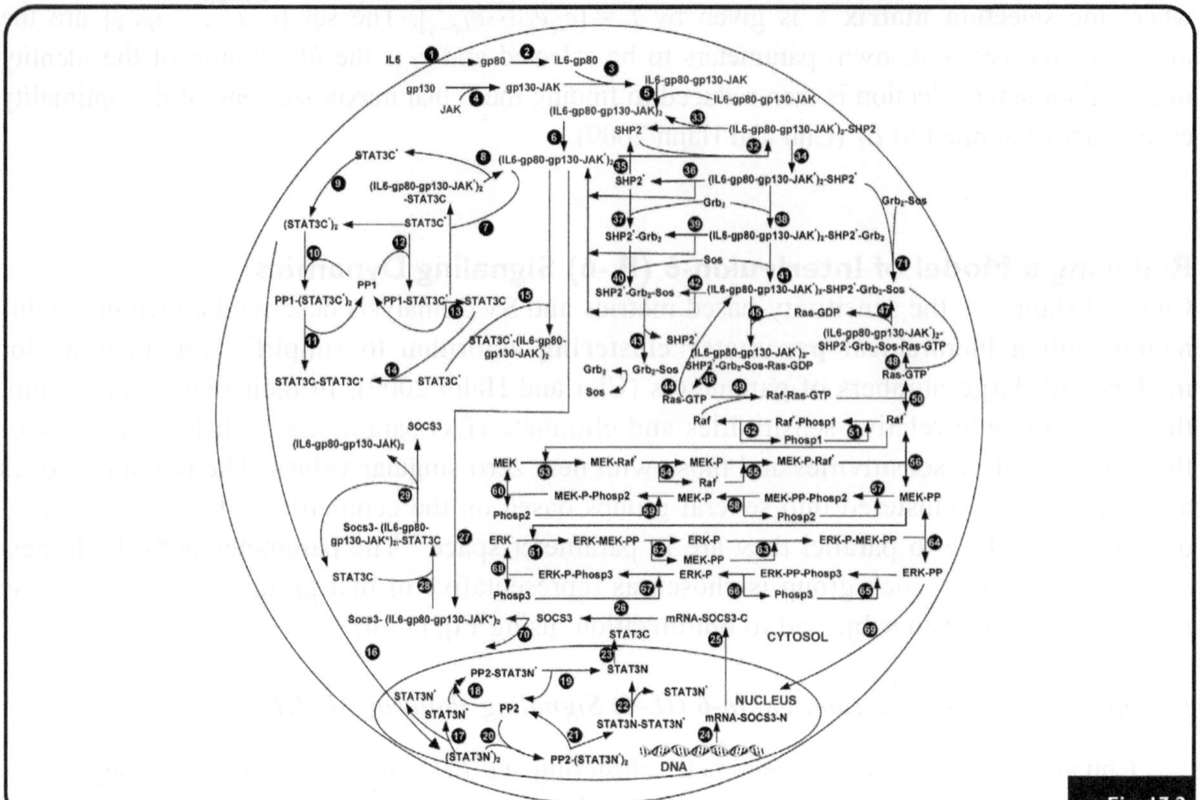

Fig. 17.2

IL-6 signaling pathway.

(Reprinted with permission from Chu, Y, Hahn, J 2009. Parameter Set Selection via Clustering of Parameters into Pairwise Indistinguishable Groups of Parameters. *Industrial & Engineering Chemistry Research* **48**: 6000–6009. Copyright (2009) American Chemical Society.

determined to be uniquely structurally identifiable – in principle – but not *NI*. Most likely it is overparameterized.

In all steps, model parameters were estimated using a maximum likelihood criterion and a variety of search and parameter estimation analysis algorithms in the *AMIGO* software package (Balsa-Canto and Banga 2011). Difficulties in choosing which parameters to fix in the final stages were alleviated using the Cintron-Arias and Banks (2009) algorithm, which systematically compares optimum selection criteria for different parameter sets. Only partial results are presented here, with a focus on the methodology for simplifying the model. Additional details are provided on the book website.

Example 17.3: ***Simplifying an OPM p53 Signaling Model***

Fig. 17.3 includes relative ranking results for both the complete set of 24 parameters of the original model and the reduced set of 14. The blue graph (original model) clearly indicates that p3, p23, p24, and p17 are ***least*** important. Indeed, the values of these parameters estimated for the full model were close to zero. They were set to ***zero*** in building the simplified model. The time-delay parameter τ in the *mRNA2* to *Mdm2* pathway (the first term of Eq. (17.8)) was also set to zero, because nonzero values made no difference in the quality of the resulting fitted model.

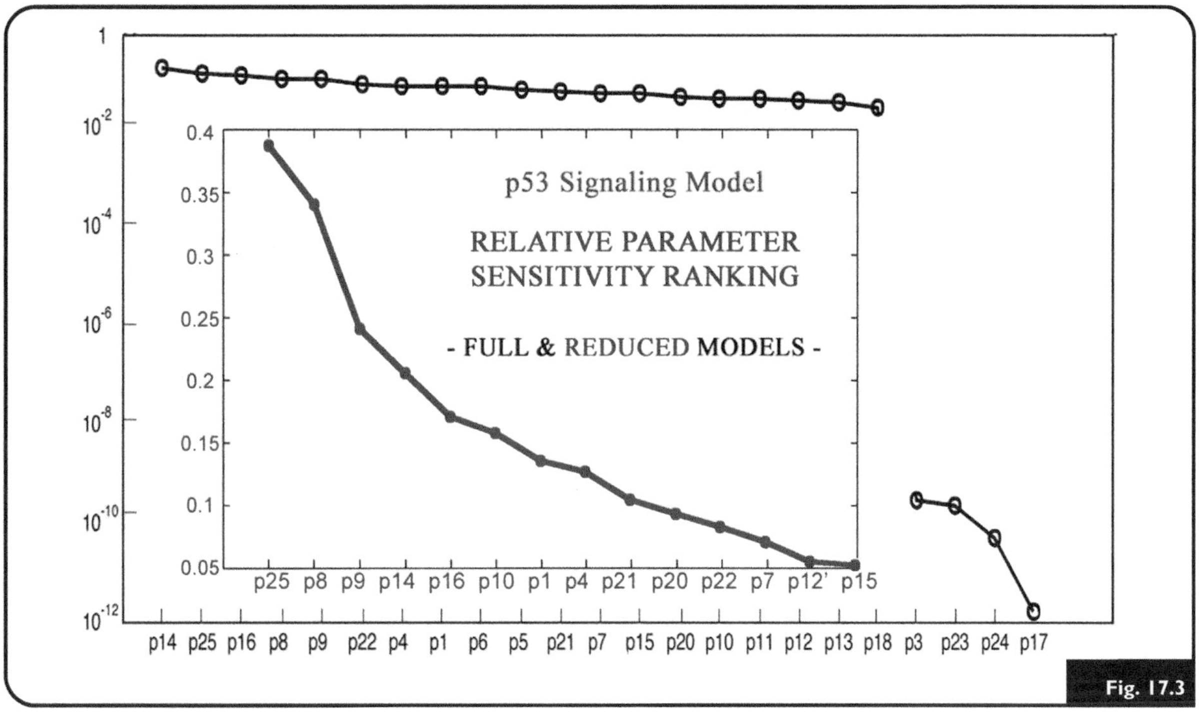

Fig. 17.3

Relative rankings of (1): 23 original p53 signaling model parameters (logarithmic scale – OUTER **BLUE** graph); and (2): the 14 reduced-model parameters to estimate (INNER **RED** graph) in the simplified model. Based on relative sensitivity vector lengths, in decreasing order of importance.

For the remaining parameters, ranking results are less sensitive and therefore less apparent, suggesting trying other metrics. The correlation matrix for the complete set of 24 parameters in Chapter 15, **Fig. 15.16**, was useful in further suggesting model reductions. It indicated that the most highly correlated pairs (darkest colored boxes) were p3–p5, p4–p5, p5–p7, p8–p10, p21–p23. The first three of these pairs appear in the *Mdm2* equation, written in the simplified notation (and with p_3 equal zero) as:

$$\dot{x}_1 \cong p_1 x_4(t-\tau) - p_4\left(\frac{x_1^2}{p_5 + x_1}\left(1 + \frac{p_6 u}{p_7 + u}\right)\right) \tag{17.8}$$

instead of:

$$\frac{dMdm2}{dt} \cong p_{Mdm2} mRNA2(t-\tau) - d'_{Mdm2} Mdm2\left(\frac{Mdm2}{Mdm2 + K_{Mdm2-Ub}}\right)\left(1 + \frac{V \cdot kinase}{kinase + K_{Mdm2-p}}\right)$$

Systematic runs of the web app *COMBOS* indicated that the combo ratio p_4/p_5 is structurally identifiable if $p_5 >> x_1$. The p_5 estimates were of the order 10 and $x_1 < 1$, by definition, so the third regulatory term reduces to $\frac{p_4 x_1^2}{p_5}\left(1 + \frac{p_6 u}{p_7 + u}\right)$. This suggested that either p_4 or p_5 can be fixed.

Highly correlated pairs p5–p7, p8–p10 (and others) were difficult to further unravel without some automation. For this purpose, the Cintron–Arias algorithm was used

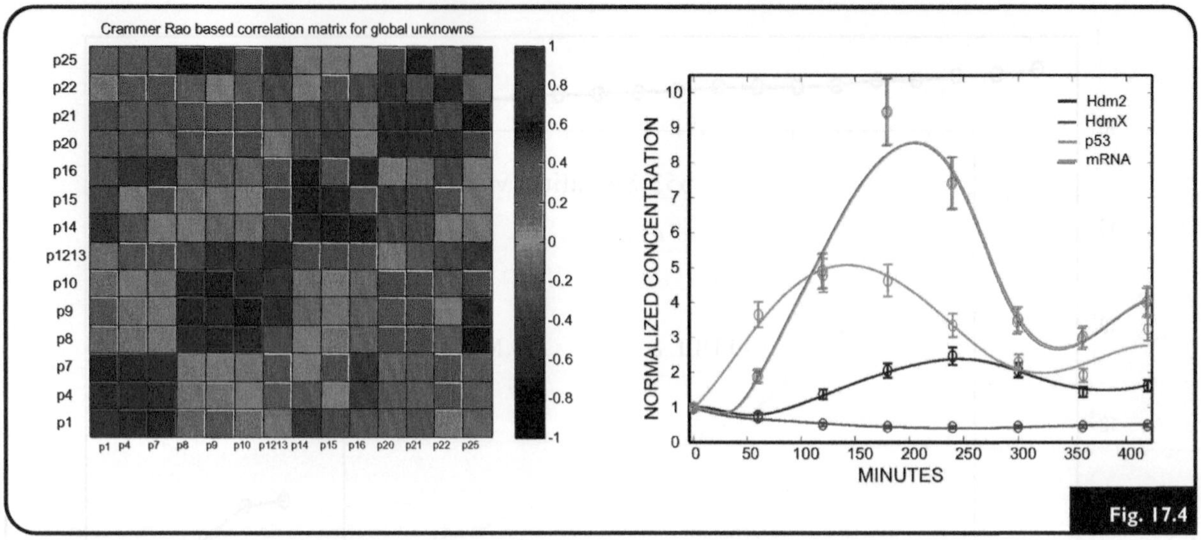

Fig. 17.4

Reduced p53 signaling model results. LHS: parameter correlations for the reduced parameter set – much improved over that in ***Fig. 15.16***. RHS: reduced model fitted to the data.

to compare optimum parameter fittings after iteratively fixing p_6 and three other parameters (p_{11}, p_{12} and p_{13}) from the MdmX equation, because they all appear as potentially approximately constant ratios (*NI* combinations) in NL terms. p_{12}/p_{13} fitted this description well and was redefined as p_{12}' and the others were similarly fixed. The same analytical procedure was applied to the p53 and mRNA2 ODEs. In all, 10 parameters were fixed to either zero or their nominal values, leaving 13 to be optimally estimated. Their relative ranking is shown in **Fig. 17.3**, the red inset graph.

Fig. 17.4 includes two kinds of qualitative results for the simplified p53 signaling model. Parameter correlations for the reduced parameter set indicate much less inter-parameter correlations than for the full model correlation matrix shown in **Fig. 15.16** in Chapter 15. The right-hand side of the figure shows the quality of the fits of the simplified (reduced) model fitted to the data.

Fig. 17.5 demonstrates – by simulation of the quantified full and simplified p53 signaling models – that both exhibit damped oscillatory responses to single pulse inputs. The first input, on top, mimics the real input $u(t)$ in the model and – where there is no further data to be fitted – extends it smoothly downward to a constant basal level (<0.2) after the experimental data run out at 420 min. The lower set of responses has a simpler, unit-valued plain pulse input. Again, all model responses exhibit damped oscillations when the pulses are turned off. Notably, there are no time-delays in the model, so the oscillations are likely due to the multiple internal feedback pathways in the model.

Global Sensitivity Analysis (GSA) Algorithms

Accomplishing GSA directly using higher-order terms in the expansion Eq. (17.3) – in which only terms to second order are shown – would be computationally challenging. For P

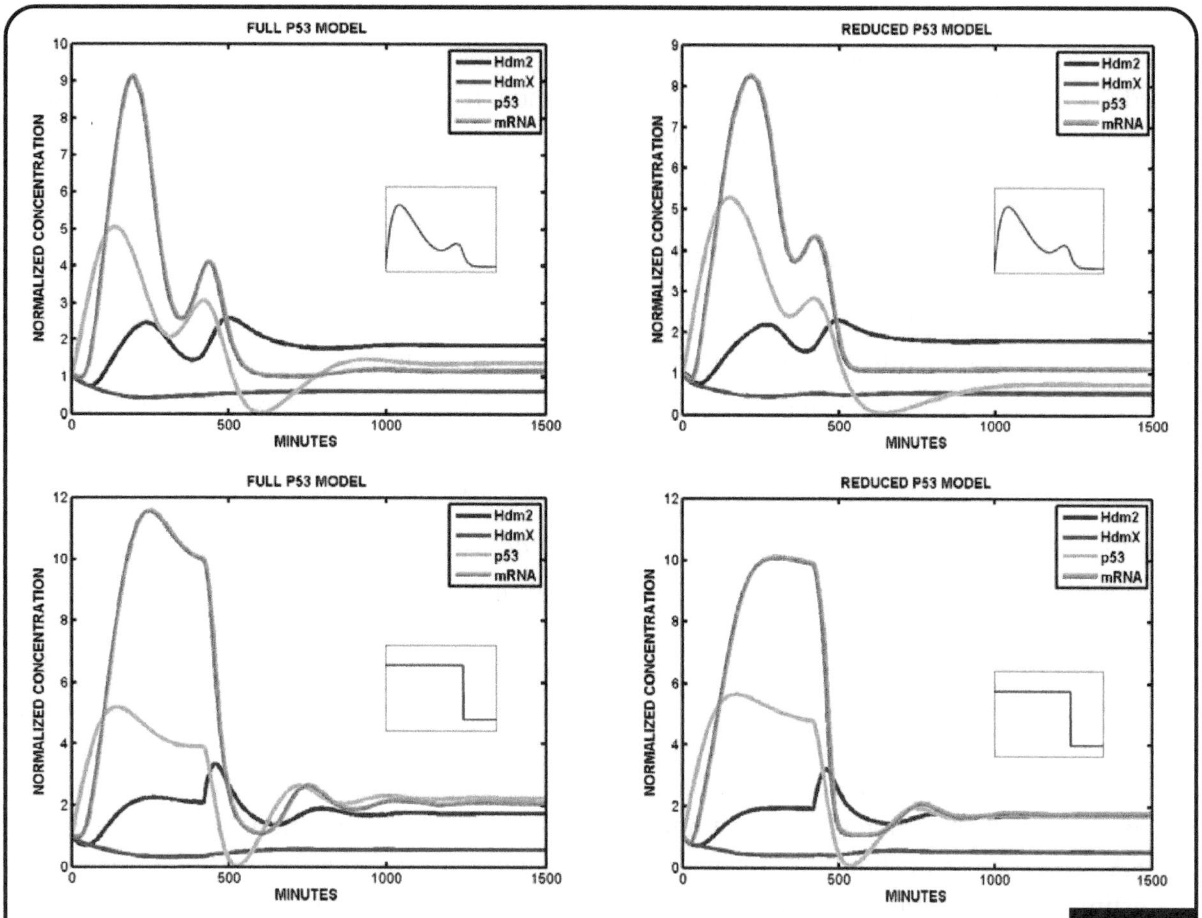

Fig. 17.5

Simulated responses of the full (LHS) and reduced (RHS) p53 signaling models. Responses to two different pulse inputs are shown. The first, on top, mimics the real input $u(t)$ in the model (small red graph), but extends it smoothly downward to a constant basal level (<0.2) after the experimental data run out at 420 min. The lower set is stimulated by a simple unit-valued pulse (lower red graph). All model responses, exhibit damped oscillations when the pulses are turned off.

parameters, state and output sensitivity matrices can grow by a factor of P for each additional parameter sensitivity variation computed, e.g. from $n(P + 1)$ for one to $nP(P + 1)$ for two parameters (second-order sensitivities).

Example 17.4: ***Higher-Order Sensitivity Matrix Computations Can Be Prohibitive***

For an ODE with P parameters, state and output sensitivity matrices based on Eq. (17.3) can grow by a factor of P for each additional parameter sensitivity variation computed, e.g. for only two simultaneous parameter variations, in a *small* model with $n = 3$ and $P = 4$, it would grow from $n(P + 1)$ to $nP(P + 1)$, i.e. 15 to 60. For a *moderate* size model with $n = 6$ and $P = 10$, it would grow from 77 to 770, and that's only for 2 of the 10 parameters that can vary. If all 10 parameters are included, for a global SA, the number of matrices to compute would be $nP^P(P + 1) = 77 \times 10^{10}$!

Fortunately, more efficient ways have been developed for GSA (Saltelli et al. 2008).

Algorithms and software for four approximate GSA methods were presented and compared in Zi et al. (2008), for application to systems biology models. Two of these, *WAR* and *MPSA*, based on similar assumptions and with some common features are described below in some detail, primarily to illustrate the issues involved in solving these kinds of problems. They are algorithms found useful in the solution of several small to medium-size model reduction problems done by the authors. The other two, partial rank correlation coefficient analysis (PRCC) and the Sobol method (SM), are described in the reference.

The Sobol method is an attractive, but computationally expensive variance-based method for GSA. It's a probabilistic approach, allowing for full exploration of the parameter space, and accounts for interactions and nonlinear responses. It typically involves very large numbers of Monte Carlo (M-C) simulations (Archer et al. 1997; Saltelli et al. 2008) – waiting to be optimized and rendered more efficient, perhaps by parallelization of computations.

Remark: To use any of these methods, it is useful to do an identifiability analysis of the model beforehand, if this is feasible, to exclude structurally unidentifiable parameters from the unknown parameter vector $\boldsymbol{p}$, because they will have zero output sensitivities. Usually this results in failure of parameter estimation search algorithms dependent on derivative calculations, or ambiguous results if no derivatives are involved. Identifiability analysis should also define identifiable parameter combinations (combos) $h(\boldsymbol{p}) \equiv \boldsymbol{b}$. Independent clusters of parameters – approximately identifiable parameter combinations – also generate poor *NI*, as illustrated in Chapter 11, and finding them is often the motivation for local or global SA. These clusters should be recognized and combined – if possible – before parameter estimation, so overall model quantification can be more successful.

Weighted-Average of Local Relative Sensitivities (WAR) $S_{\boldsymbol{p}}^{y} = (\partial y/y)/(\partial \boldsymbol{p}/\boldsymbol{p})$

Parameters with smaller *relative* output sensitivities have less effect on outputs (are more robust to changes) than those with higher sensitivities, as emphasized in Chapter 11. The WAR heuristic approach assumes further that this is true for parameters varying together, not just model outputs. It uses simpler local SA results to get a (rough) approximation of global sensitivities, using a Boltzman-distribution-inspired and statistically weighted comparison of averaged local sensitivities. The weighting function is chosen to substitute for lack of independence of coupled effects of different parameters (Zi et al. 2008). The algorithm implements Monte Carlo (M-C) simulation of the model, with systematically varied parameters *over the feasible boundaries of the global parameter space,*[4] in a manner similar to the one presented for estimating the covariance matrix of model parameter estimates in Chapter 12. Relative sensitivities computed for these randomized parameter estimates are weighted by an exponential function of optimization criteria to obtain approximate global sensitivity results.

4 A computationally intensive step, begging for optimization.

The WAR Approximate GSA Algorithm

1. Select the vector of parameters $\boldsymbol{p}$ or parameter combinations (or clusters) $h(\boldsymbol{p}) \equiv \boldsymbol{b}$ to be tested.

2. Establish finite parameter space boundaries or (e.g. uniform) distributions for the search, usually from available literature, separate experimental data, or user knowledge and experience, e.g. from Michaelis–Menten constants, reaction rates, initial concentrations, and half-lives. The $\hat{\boldsymbol{p}}$ (or $\hat{\boldsymbol{b}}$) will be sampled randomly within this space. Alternatively, one might more simply define the search space by defining the range (for example) between 1/5 (or 1/3) of the nominal and 5 (or 3) times the nominal value for each component of $\boldsymbol{p}_{\text{nominal}}$.

3. Simulate[5] the model for each candidate $\hat{\boldsymbol{p}}$ (or $\hat{\boldsymbol{b}}$), sample and save the state $\boldsymbol{x}_{M\text{-}C}(t_k, \hat{\boldsymbol{p}})$ and output vectors $z_{M\text{-}C}(t_k, \hat{\boldsymbol{p}})$ at N (e.g. 50–100) time-points t_k. Simulate and do the same (once), generating the sampled nominal solution $y(t_k, \boldsymbol{p}_{\text{nominal}})$ at N time points.

4. At each candidate $\hat{\boldsymbol{p}}$ (or $\hat{\boldsymbol{b}}$), calculate the corresponding objective function $WRSS(\hat{\boldsymbol{p}})$. (We assume the output is scalar here. Use the vector output case in Chapter 11 if it is not.)

$$WRSS(\hat{\boldsymbol{p}}) \equiv \sum_{k=0}^{N} w_k [z_{M\text{-}C}(t_k, \hat{\boldsymbol{p}}) - y(t_k, \boldsymbol{p}_{\text{nominal}})]^2 \tag{17.9}$$

5. For each $\hat{\boldsymbol{p}}$, compute first-order relative sensitivity matrices $S_p^y(t_k, \hat{\boldsymbol{p}})$ over the same N time points.

6. For each $\hat{\boldsymbol{p}}$, compute the *time-average* of each element of $S_p^y(t_k, \hat{\boldsymbol{p}})$ over the time interval of the study, e.g. using an *RMS* (or similar) index (see Chapter 11, section on *Quantifying Relative Output Sensitivity Functions*).

7. Weight these average indices by

$$e^{-WRSS(\hat{\boldsymbol{p}})/\min WRSS(\hat{\boldsymbol{p}})} \tag{17.10}$$

   This Boltzmann-distribution weighting function amplifies the statistical impact of sensitivities for $\hat{\boldsymbol{p}}$, more consistent with experimental data and therefore more probable, resulting in the $\hat{\boldsymbol{p}}$ with lowest *WRSS* contributing most to calculated sensitivity indices.

8. Order the set of weighted sensitivity indices, to establish the least (smallest) to most important (largest). A 3D histogram plot of these indices – the distribution of weighted relative sensitivity integrals for all the s_{ij} evaluated – e.g. the *RMS* (or similar) indices as in step 6 above, is useful for illustrating the ordering.

5 See the section on M-C simulation for *COV* estimation in Chapter 12.

Remark: The most sensitive (largest) output sensitivities designate the potentially most informative experiments.

Multi-Parametric Sensitivity Analysis (MPSA)

This heuristic method shares much with the WAR approach, but it is more global and does not rely on local sensitivities. The basic idea of MPSA – like with WAR – is to first map parameter variability into the model output by randomly generating parameter values from predefined distributions, but for *all parameters simultaneously in* MPSA – again using an M-C simulation approach (Hornberger 1981; Cho et al. 2003; Zi 2005). It uses **latin hypercube sampling**[6] (L-H-S) to obtain a minimal and efficient set of global candidates $\hat{\boldsymbol{p}}$, for all parameters simultaneously over the feasible ranges of parameter space. L-H-S is an efficient way to assure that individual parameter ranges are evenly covered (Mckay et al. 1979; Iman et al. 1981), especially important for models with more than a few parameters.

Steps 1–4 of MPSA are the same as WAR steps 1–4, with the exception that in step 2 random numbers for M-C simulation are chosen from an independent, Gaussian (normal) distribution across the entire established parameter space. This is to satisfy statistical assumptions of L-H-S, and evenly cover the parameter space with a select subset of $\hat{\boldsymbol{p}}$. Following step 4, where $WRSS(\hat{\boldsymbol{p}})$ is computed, each $\hat{\boldsymbol{p}}$ is then classified as either *acceptable* or *unacceptable* by comparing its $WRSS(\hat{\boldsymbol{p}})$ value to the average of all $WRSS(\hat{\boldsymbol{p}})$ values. Smaller than the average $WRSS(\hat{\boldsymbol{p}})$, are acceptable – because this improves the fit to the simulated data; otherwise $\hat{\boldsymbol{p}}$ is unacceptable.

The final step is statistical evaluation of the occurrences of acceptable and unacceptable cases for each $\hat{\boldsymbol{p}}$. This entails the nonparametric two-sample **Kolmogorov–Smirnov (KS)** test for equality of one dimensional continuous probability distributions of two samples. The KS statistic quantifies the distance between the empirical **cumulative distribution functions**[7] (*cumulative frequencies, CF*) of two samples. The *null distribution* of this statistic is calculated under the *null hypothesis* that the samples are drawn from the same distribution. The distributions must be continuous, but are otherwise unrestricted. The KS statistic is sensitive to differences in both location and shape of the empirical cumulative distribution functions of the two samples, CF_{accept} and CF_{unaccept} in our algorithm.[8] Thus, cumulative frequency is calculated for both acceptable and unacceptable cases, with increasing parameter values. The larger the difference between CF_{accept} and CF_{unaccept}, the more significant the $\hat{\boldsymbol{p}}$. By KS statistics, the global MPSA is the largest vertical distance between CF_{accept} and CF_{unaccept} (Zi et al. 2008), with global MPSA values between 0 and 1. Values closer to 1 designate parameters relatively more important to overall model output.

6 A Latin square is a square grid containing sample positions with only one sample in each row and each column. A **Latin hypercube** has an arbitrary number of dimensions with each sample being the only one in each axis-aligned hyperplane that contains it. This is a well-developed and popular method in statistics.

7 *Cumulative frequency*, or *cumulative distribution function*, is the total sum (running total) of a frequency and all frequencies below it in a frequency distribution. "So far, how much?"

8 The *KS* test is one of the most useful and general nonparametric methods for comparing two samples.

MPSA Algorithm

Steps 1–4 are essentially the same as for the *WAR*. Steps 5 and 6 details follow.

5. Determine if $\hat{\boldsymbol{p}}$ is *acceptable* or *unacceptable* by comparing $WRSS(\hat{\boldsymbol{p}})$ to the average $WRSS(\hat{\boldsymbol{p}})$ ($WRSS_{AVG}$) over the entire set of M-C simulations. If $WRSS(\hat{\boldsymbol{p}}) > WRSS_{AVG}$, $\hat{\boldsymbol{p}}$ is unacceptable; otherwise, it is acceptable.

6. Statistically compare the *cumulative frequency distributions* CF_{accept} and CF_{unaccept} of individual parameter values associated with acceptable and unacceptable cases. For each candidate parameter $\hat{\boldsymbol{p}}$, compute the Kolmogorov-Smirnov (*KS*) statistic:

$$KS = \sup_{\hat{\boldsymbol{p}}} |CF_{\text{accept}}(\hat{\boldsymbol{p}}) - CF_{\text{unaccept}}(\hat{\boldsymbol{p}})| \tag{17.11}$$

   The *KS* statistic is the maximum vertical distance between the cumulative frequency distribution curves for n acceptable and m unacceptable cases.

7. Order the set of weighted sensitivity indices, to establish the least (smallest) to most important (largest). Larger values of *KS* indicate greater sensitivities to that parameter. To visualize the ordering, a 3D histogram plot of the distribution of these indices can be useful – the *RMS* (or similar) indices as in Eqs. (11.37) or (11.38), for all the s_{ij} evaluated.

What's Next?

Model reduction in systems biology modeling is a hot topic these days and we should expect to see substantial progress in the near future in computational algorithm development and software tools for pruning the unessentials from large, overspecified network models. We also anticipate improvements in the efficiency of GSA methods. In principle, GSA should improve the precision of model reduction.

For added value, the same methods used for simplifying models in this chapter can be (and are probably already being) used for the seemingly paradoxical goal of increasing model complexity. The idea is simple and systematic application of it can lead to more detailed knowledge of real biosystem networks. Model reduction addresses particular models, with given input–output configurations. However, complex systems can likely be probed via other input and output ports, bringing to mind the systematic experiment design approaches developed in Chapter 16. Instead of including only the given outputs (as in addressing the model reduction problem), include all measurable state variables as candidate outputs. Comparisons of all measurable state variable sensitivities to parameters should then delineate the ones that can provide maximum sensitivity to parameters and therefore better parameter estimate precision. These will likely include nodal measurements capable of expanding knowledge of network interconnectivity.

Exercises

E17.1 Compute the relative sensitivities for the NL model in *Example 13.4*. Then compute their lengths for each of the five parameters in the model, using an *RMS* formula. Now rank the parameters in order of their importance in generating the dynamical response of the model. Compare your results with the values in the correlation matrix for that example and also the singular values reported in that example. What story do they all tell about *NI* of the parameters of this model? Can you find a reasonable simpler model that generates approximately the same dynamical response?

References

Archer, G., Saltelli, A., Sobol, I.M., 1997. Sensitivity measures, ANOVA-like techniques and the use of bootstrap. J. Stat. Comput. Simul. 58 (2), 99–120.

Balsa-Canto, E., Banga, J.R., 2011. AMIGO, a toolbox for advanced model identification in systems biology using global optimization. Bioinformatics. 27 (16), 2311–2313.

Cheong, R., Hoffmann, A., Levchenko, A., 2008. Understanding NF-kappaB signaling via mathematical modeling. Mol. Syst. Biol. 4 (192), 6.

Cho, K.-H., Shin, S.-Y., Kolch, W., Wolkenhauer, O., 2003. Experimental design in systems biology, based on parameter sensitivity analysis using a monte carlo method: a case study for the TNFα-mediated NF-κ B signal transduction pathway. Simulation. 79 (12), 726–739.

Chu, Y., Hahn, J., 2009. Parameter set selection via clustering of parameters into pairwise indistinguishable groups of parameters. Ind. Eng. Chem. Res. 48 (13), 6000–6009.

Cintron-Arias, A., Banks, H.T., 2009. A sensitivity matrix based methodology for inverse problem formulation. J. Inverse Ill-Posed Probl. 17, 545–564.

Hoffmann, A., Levchenko, A., Scott, M.L., Baltimore, D., 2002. The IkappaB-NF-kappaB signaling module: temporal control and selective gene activation. Science. 298 (5596), 1241–1245.

Huang, Z., Chu, Y., 2010. Model simplification procedure for signal transduction pathway models: an application to IL-6 signaling. Chem. Eng. Sci. 65, 1964–1975.

Iman, R.L.H., J.C., Campbell, J.E., 1981. An approach to sensitivity analysis of computer models, Part 1. Introduction, input variable selection and preliminary variable assessment. J. Qual. Technol. 13 (3), 174–183.

Kearns, J.D., Hoffmann, A., 2009. Integrating computational and biochemical studies to explore mechanisms in NF-κB signaling. J. Biol. Chem. 284 (9), 5439–5443.

Longo, D.M., Selimkhanov, J., Kearns, J.D., Hasty, J., Hoffmann, A., Tsimring, L., 2013. Dual delayed feedback provides sensitivity and robustness to the NF-kB signaling module. PLoS. Comput. Biol. **In press**.

Mckay, M.D., Beckman, R.J., Conover, W.J., 1979. Comparison of 3 methods for selecting values of input variables in the analysis of output from a computer code. Technometrics. 21, 239–245.

Quaiser, T., Dittrich, A., Schaper, F., Monnigmann, M., 2011. A simple work flow for biologically inspired model reduction--application to early JAK-STAT signaling. BMC. Syst. Biol. 5 (30), 1752-0509.

Rodriguez-Fernandez, M., Banga, J.R., 2010. SensSB: a software toolbox for the development and sensitivity analysis of systems biology models. Bioinformatics. 26 (13), 1675–1676.

Saltelli, A., C. K., Scott, E.M., 2008. Sensitivity Analysis. Wiley.

Yue, H., Brown, M., Knowles, J., Wang, H., Broomhead, D.S., Kell, D.B., 2006. Insights into the behaviour of systems biology models from dynamic sensitivity and identifiability analysis: a case study of an NF-kB signalling pathway. Mol. Biosyst. 2 (12), 640–649.

Zi, Z., Zheng, Y., Rundell, A., Klipp, E., 2008. SBML-SAT: a systems biology markup language (SBML) based sensitivity analysis tool. BMC Bioinformatics. 9 (1), 342.

Appendices

Appendix A

A Short Course in Laplace Transform Representations & ODE Solutions

Transform Methods

These are techniques for rendering mathematical expressions easier to solve and represent. Use of the differential operator $D = d/dt$ in Chapter 2 to simplify ordinary differential equations (ODEs) and their solution is one example. Other transforms widely used in engineering, physical and biological sciences are based on replacement of functions of a real variable – usually time or distance – by certain frequency-dependent representations. The frequency-dependent **Fourier transform** and the **Fourier series** representation are used in engineering and signal analysis to represent complicated waveforms, e.g. like a current signal response in some part of a linear time-invariant electrical network when the input voltage is a periodic or repeating waveform. For example, a periodic square-wave voltage can be replaced by its Fourier series representation – a function of *frequency* ω (radians) – and the current produced by each term of the series is determined using network equations and Kirchoff's circuit laws. The total current is then the sum of the individual currents – based on the principle of superposition (Chapter 2). This approach can result in substantial savings in computational effort compared with solving the problem in the time-domain, especially for systems in periodic steady state.

The **Laplace transform** (LT)[1] is another technique relating time functions to frequency-dependent functions of the **complex variable** s. It is more broadly applicable than Fourier transforms and series for building and analyzing linear time-invariant system models and signals, primarily because it represents transient as well as steady state responses. *Linearity and time-invariance* are important limitations of Laplace and Fourier transform methods. Nevertheless, these techniques remain of great value and are surprisingly applicable in modeling practice, because so many systems and their models satisfy these properties, at least approximately, or over some limited region of operation – quite important in biosystem experiments. We first define and exemplify the Laplace transform of $f(t)$, an arbitrary real function of a real variable t, usually time, defined for $t > 0$. Then we show how to apply it for solving linear time-invariant models, in particular linear constant-coefficient differential equations. Readers interested in going deeper, or seeing more applications, are referred to Murray (1965) and DiStefano III et al. (1990), among many others.

The **Laplace transform** (LT) of time function $f(t)$ is given by:

$$F(s) \equiv \mathcal{L}[f(t)] \equiv \lim_{\substack{T \to \infty \\ \varepsilon \to 0}} \int_{\varepsilon}^{T} f(t)e^{-st}dt = \int_{0^+}^{\infty} f(t)e^{-st}dt \tag{A.1}$$

In Eq. (A.1), the symbol s is a **complex variable** defined by $s = \sigma + j\omega$, where σ and ω are real variables and $j = \sqrt{-1}$. The real variable t denotes time in our applications. The **real** part σ of the complex variable s is often written as $Re(s)$ (the real part of s) and the **imaginary** part ω as $Im(s)$ (the imaginary part of s). The lower limit on the integral in (A.1) is ε, where $0 < \varepsilon < T$. This is useful in dealing with functions discontinuous at $t = 0$, e.g. switching on a signal at an arbitrary initial time $t = 0$; real signals don't turn on instantaneously, but take ε time units to be "on." When explicit use is made of this limit, as in (A.1), it is abbreviated $t = \lim_{\varepsilon \to 0} \varepsilon \equiv 0^+$. The unit step and impulse functions in Eqs. (2.15) and (2.16) are two examples of such signals.

Example A.1: The LT of e^{-t} is

$$\mathcal{L}\left[e^{-t}\right] = \int_{0^+}^{\infty} e^{-t}e^{-st}dt = \int_{0^+}^{\infty} e^{-(s+1)t}dt = \left.\frac{-e^{-(s+1)t}}{(s+1)}\right|_{0^+}^{\infty} = \frac{1}{s+1} \tag{A.2}$$

Laplace Transform Representations and Solutions

The s-domain representation of the time-domain function, e.g. $\frac{1}{s+1}$ for e^{-t}, is itself useful in many applications, e.g. as a transfer function (TF). We focus here on solving linear time-invariant ODEs and other time-domain models using the LT.

1 Chapter 2 included minimal development and use of the LT. It's more complete here.

Two-Step Solutions

We first apply the integral transformation in Eq. (A.1) to our model. This effectively transforms the time-domain problem into an algebraic one in the s-domain, as the example above. Algebraic equations in s are usually easier to solve than time-domain ODEs. Fortunately, formal integration, to obtain $F(s)$ from $f(t)$ via Eq. (A.1), is rarely needed. Results can usually be obtained more directly by combining tabulated transforms of functions typically encountered in Laplace transformable modeling problems.

To obtain the time-domain solution from the s-domain solution, we use a special inversion function. Assign $F(s)$ as the LT of a function $f(t)$, for $t > 0$. Then the **inverse Laplace transform (ILT)** of $F(s)$ is defined by the **contour integral**:

$$f(t) \equiv \mathcal{L}^{-1}[F(s)] \equiv \frac{1}{2\pi j}\int_{c-j\infty}^{c-j\infty} F(s)e^{st}ds \tag{A.3}$$

where $j = \sqrt{-1}$ and $c = Re(s) = \sigma_0$ (some real number σ_0). Again we are fortunate that it is rarely necessary in practice to perform the contour integration, which involves the theory of complex analysis. For most applications, only a table of transform pairs is needed, as shown below. First, we examine the key properties of this pair of transforms, the ones most useful for obtaining results.

Key Properties of the Laplace Transform (LT) & its Inverse (ILT)

The following properties facilitate solution of linear constant-coefficient ordinary differential equations (ODEs).

The LT is a ***linear transformation*** between functions defined in the t-domain and functions defined in the s-domain. That is, if $F_1(s)$ and $F_2(s)$ are the Laplace transforms of $f_1(t)$ and $f_2(t)$, respectively, then $a_1F_1(s) + a_2F_2(s)$ is the LT of $a_1f_1(t) + a_2f_2(t)$, where a_1 and a_2 are arbitrary constants.

Example A.2: The Laplace transforms of the functions e^{-t} and e^{-2t} are

$$\mathcal{L}[e^{-t}] = \frac{1}{s+1}, \ \mathcal{L}[e^{-2t}] = \frac{1}{s+2}$$

Then, by the property of linearity,

$$\mathcal{L}[3e^{-t} - e^{-2t}] = 3\mathcal{L}[e^{-t}] - \mathcal{L}[e^{-2t}] = \frac{3}{s+1} - \frac{1}{s+2} = \frac{2s+5}{s^2+3s+2}$$

The ILT is also a ***linear transformation*** between functions defined in the s-domain and functions defined in the *time*-domain. That is, if $f_1(t)$ and $f_2(t)$ are the ILTs of $F_1(s)$ and $F_2(s)$, respectively, then $b_1f_1(t) + b_2f_2(t)$ is the ILT of $b_1F_1(s) + b_2F_2(s)$, where b_1 and b_2 are arbitrary constants.

Example A.3: The inverse Laplace transforms of functions $\frac{1}{s+1}$ and $\frac{1}{s+3}$ are

$$\mathcal{L}^{-1}\left[\frac{1}{s+1}\right] = e^{-t}, \ \mathcal{L}^{-1}\left[\frac{1}{s+3}\right] = e^{-3t}$$

Again, because of the property of linearity,

$$\mathcal{L}^{-1}\left[\frac{2}{s+1} - \frac{4}{s+3}\right] = 2\mathcal{L}^{-1}\left[\frac{1}{s+1}\right] - 4\mathcal{L}^{-1}\left[\frac{1}{s+3}\right] = 2e^{-t} - 4e^{-3t}$$

The LT of the **derivative** df/dt of a function $f(t)$ whose Laplace transformation is $F(s)$ is

$$\mathcal{L}\left[\frac{df}{dt}\right] = sF(s) - f(0^+) \tag{A.4}$$

where $f(0^+)$ is the initial value of f(t), evaluated as the one-sided limit of f(t) as t approaches zero from positive values.

Example A.4: The LT of $\frac{d}{dt}(e^{-t})$ is determined by the above property. Since $\mathcal{L}[e^{-t}] = \frac{1}{s+1}$ and $\lim_{t\to 0} e^{-t} = 1$, then $\mathcal{L}\left[\frac{d}{dt}(e^{-t})\right] = s\left(\frac{1}{s+1}\right) - 1 = \frac{-1}{s+1}$.

The LT of the **integral** $\int_0^t f(\tau)d\tau$ of a function $f(t)$ whose transform is $F(s)$ is

$$\mathcal{L}\left[\int_0^t f(\tau)d\tau\right] = \frac{F(s)}{s} \tag{A.5}$$

Example A.5: The LT of $\int_0^t e^{-\tau}d\tau$ can be determined by application of the above property. Since $\mathcal{L}[e^{-t}] = \frac{1}{s+1}$, then $\mathcal{L}\left[\int_0^t e^{-\tau}d\tau\right] = \frac{1}{s}\left(\frac{1}{s+1}\right) = \frac{1}{s(s+1)}$.

Time Delay: The LT of the function $f(t-T)$, where $T > 0$ and $f(t-T) = 0$ for $t \le T$, is

$$\mathcal{L}[f(t-T)] = e^{-Ts}F(s) \tag{A.6}$$

Example A.6: The transform of the function e^{-t} is $\frac{1}{s+1}$. The LT of the function defined as $f(t) = \begin{cases} e^{-(t-2)}, & t > 2 \\ 0, & t \le 2 \end{cases}$ can be determined by the time-delay property, with $T = 2$: $\mathcal{L}[f(t)] = e^{-2s}\mathcal{L}[e^{-t}] = \frac{e^{-2s}}{s+1}$.

Initial Value Theorem: The initial value $f(0^+)$ of the function $f(t)$ whose transform is $F(s)$ is

$$f(0^+) = \lim_{t\to 0} f(t) = \lim_{s\to\infty} sF(s),\ t > 0 \tag{A.7}$$

Example A.7: The LT of e^{-3t} is $\mathcal{L}\left[e^{-3t}\right] = \frac{1}{s+3}$. The initial value of e^{-3t} can be determined by the initial value theorem as $\lim_{t\to 0} e^{-3t} = \lim_{s\to\infty} s\left(\frac{1}{s+3}\right) = 1.$

Final Value Theorem: If $\lim_{t\to\infty} f(t)$ exists, the final value $f(\infty)$ of $f(t)$ with LT $F(s)$ is

$$f(\infty) = \lim_{t\to\infty} f(t) = \lim_{s\to 0} sF(s) \tag{A.8}$$

Example A.8: The LT of the function $(1 - e^{-t})$ is $\frac{1}{s(s+1)}$. The final value of this function is:

$$\lim_{t\to\infty} (1 - e^{-t}) = \lim_{s\to 0} \frac{s}{s(s+1)} = 1$$

Time Scaling: The LT of a function $f(t/a)$ is

$$\mathcal{L}\left[f\left(\frac{t}{a}\right)\right] = aF(as) \tag{A.9}$$

Example A.9: The LT of e^{-t} is $\frac{1}{s+1}$. The LTof e^{-3t} is determined by application of Time Scaling, with $a = 1/3$, as:

$$\mathcal{L}\left[e^{-3t}\right] = \frac{1}{3}\left(\frac{1}{\frac{1}{3}s+1}\right) = \frac{1}{s+3}$$

Frequency Scaling: The ILT of the function $F(s/a)$ is

$$\mathcal{L}^{-1}\left[F\left(\frac{s}{a}\right)\right] = af(at) \tag{A.10}$$

Example A.10: The ILT of $\frac{1}{s+1}$ is e^{-t}. The ILT of $\frac{1}{\frac{1}{3}s+1}$ is determined by application of frequency scaling as: $\mathcal{L}^{-1}\left[\frac{1}{\frac{1}{3}s+1}\right] = 3e^{-3t}$

Complex Translation: With $F(s) = \mathcal{L}[f(t)]$, the LT of the time-function $e^{-at}f(t)$ is:

$$\mathcal{L}[e^{-at}f(t)] = F(s+a) \tag{A.11}$$

Example A.11: The LT of $\cos t$ is $\frac{s}{s^2+1}$. The LT of $e^{-2t}\cos t$ is determined by complex translation, with $a = 2$, as:

$$\mathcal{L}\left[e^{-2t}\cos t\right] = \frac{s+2}{(s+2)^2+1} = \frac{s+2}{s^2+4s+5}.$$

Convolution: The ILT of the **product of two LTs** $F_1(s)$ and $F_2(s)$ is given by the **time convolution integrals**

$$\mathcal{L}^{-1}[F_1(s)F_2(s)] = \int_{0^+}^{t} f_1(\tau)f_2(t-\tau)d\tau = \int_{0^+}^{t} f_2(\tau)f_1(t-\tau)d\tau \tag{A.12}$$

where $\mathcal{L}^{-1}[F_1(s)] = f_1(t)$, $\mathcal{L}^{-1}[F_2(s)] = f_2(t)$.

Example A.12: The ILT of the function $F(s) = \frac{s}{(s+1)(s^2+1)}$ can be determined by application of the property of convolution. Since $\mathcal{L}^{-1}\left[\frac{1}{s+1}\right] = e^{-t}$ and $\mathcal{L}^{-1}\left[\frac{1}{s^2+1}\right] = \cos t$, then:

$$\mathcal{L}^{-1}\left[\left(\frac{1}{s+1}\right)\left(\frac{s}{s^2+1}\right)\right] = \int_{0^+}^{t} e^{-(t-\tau)}\cos\tau d\tau = e^{-t}\int_{0+}^{t} e^{\tau}\cos\tau d\tau = \frac{1}{2}(\cos t + \sin t - e^{-t})$$

Complex Convolution: The LT of the **product of two time-functions** $f_1(t)$ and $f_2(t)$ is given by the **complex contour integral**:

$$\mathcal{L}[f_1(t)f_2(t)] = \frac{1}{2\pi j}\int_{c-j\infty}^{c+j\infty} F_1(\omega)F_2(s-\omega)d\omega \tag{A.13}$$

where $F_1(s) = \mathcal{L}[f_1(t)]$ and $F_2(s) = \mathcal{L}[f_2(t)]$.

Example A.13: Eq. (A.13) can be used to find the LT of $e^{-2t}\cos t$, determined far more easily by Complex Translation above. With $\mathcal{L}\left[e^{-2t}\right] = \frac{1}{s+2} \equiv F_2$ and $\mathcal{L}[\cos t] = \frac{s}{s^2+1} \equiv F_1$:

$$\mathcal{L}\left[e^{-2t}\cos t\right] = \frac{1}{2\pi j}\int_{c-j\infty}^{c+j\infty}\left(\frac{\omega}{\omega^2+1}\right)\left(\frac{1}{s-\omega+2}\right)d\omega = \frac{s+2}{s^2+4s+5}$$

The details of this contour integration are complicated and unnecessary for solving this problem. There are, however, cases in which complex convolution can be used effectively.

Short Table of Laplace Transform Pairs

This table is far from complete, but when used in conjunction with the properties of the LT described above and the partial fraction expansion techniques described below, it is adequate to handle very many problems. To find the LT of a time function represented by a combination of the elementary functions given in the table, the appropriate transforms are chosen and are combined using the properties of the previous section. A comprehensive table of LT pairs can be found in DiStefano III et al. (1990) or Murray (1965), among others.

Time Function		*Laplace Transform*
Unit impulse	$\delta(t)$	1
Unit step	$u(t)$	$\frac{1}{s}$
Unit ramp	t	$\frac{1}{s^2}$
Polynomial	t^n	$\frac{n!}{s^{n+1}}$
Exponential	e^{-at}	$\frac{1}{s+a}$
Sine wave	$\sin \omega t$	$\frac{\omega}{s^2+\omega^2}$
Cosine wave	$\cos \omega t$	$\frac{s}{s^2+\omega^2}$
Damped sine wave	$e^{-at} \sin \omega t$	$\frac{\omega}{(s+a)^2+\omega^2}$
Damped cosine wave	$e^{-at} \cos \omega t$	$\frac{s+a}{(s+a)^2+\omega^2}$

Example A.14: The LT of the function $f(t) = e^{-4t} + \sin(t-2) + t^2e^{-2t}$ is determined as follows.

The Laplace transforms of e^{-4t}, $\sin t$ and t^2 are given in the table as $\mathcal{L}\left[e^{-4t}\right] = \frac{1}{s+4}$, $\mathcal{L}[\sin t] = \frac{1}{s^2+1}$ and $\mathcal{L}\left[t^2\right] = \frac{2}{s^3}$. Application of the time delay property (A.6) yields $\mathcal{L}[\sin(t-2)] = \frac{e^{-2s}}{s^2+1}$, and the property of complex translation (A.11) leads to $\mathcal{L}\left[t^2e^{-2t}\right] = \frac{2}{(s+2)^3}$. Then, because the LT is a linear transformation, $\mathcal{L}[f(t)] = \frac{1}{s+4} + \frac{e^{-2s}}{s^2+1} + \frac{2}{(s+2)^3}$.

In the same manner, to find the inverse of the transform of a combination of those in the table, the corresponding time functions (inverse transforms) are determined from the table and combined using the various properties of the previous section.

Example A.15: The ILT of $F(s) = \left(\frac{s+2}{s^2+4}\right)e^{-s}$ can be determined as follows. $F(s)$ is

first rewritten as $F(s) = \frac{se^{-s}}{s^2+4} + \frac{2e^{-s}}{s^2+4}$. Now $\mathcal{L}^{-1}\left[\frac{s}{s^2+4}\right] = \cos 2t$ and $\mathcal{L}^{-1}\left[\frac{2}{s^2+4}\right] = \sin 2t$. Application of the time delay property (A.6), for $t > 1$, yields

$\mathcal{L}^{-1}\left[\frac{se^{-s}}{s^2+4}\right] = \cos 2(t-1)$ and $\mathcal{L}^{-1}\left[\frac{2e^{-s}}{s^2+4}\right] = \sin 2(t-1)$. Finally, linearity gives $\mathcal{L}^{-1}[F(s)] = \begin{cases} \cos 2(t-1) + \sin 2(t-1), & t > 1 \\ 0, & t \leq 1 \end{cases}$.

Laplace Transform Solution of Ordinary Differential Equations (ODEs)

Consider:

$$\sum_{i=0}^{n} a_i \frac{d^i y}{dt^i} = u \tag{A.14}$$

where y is the output, u is the input, the coefficients a_i, for $i = 0, 1, \ldots, n-1$, are constants, and $a_n = 1$. The initial conditions for this equation are

$$\left.\frac{d^k y}{dt^k}\right|_{t=0^+} \equiv y_0^k \text{ for } k = 0, 1, \ldots n-1$$

where y_0^k are constants. The LT of Eq. (A.14) is

$$\sum_{i=0}^{n} \left[a_i \left(s^i Y(s) - \sum_{k=0}^{i-1} s^{i-1-k} y_0^k \right) \right] = U(s) \tag{A.15}$$

and the transform of the output is

$$Y(s) = \frac{U(s)}{\sum_{i=0}^{n} a_i s^i} + \frac{\sum_{i=0}^{n}\sum_{k=0}^{i-1} a_i s^{i-1-k} y_0^k}{\sum_{i=0}^{n} a_i s^i} \tag{A.16}$$

When $i = 0$ in (A.15), the summation interior to the brackets is, by definition,

$$\left.\sum_{k=0}^{i-1}\right|_{i=0} = \sum_{k=0}^{k=-1} = 0$$

The right side of Eq. (A.16) is the sum of two terms: a term dependent only on the input transform and a term dependent only on the initial conditions. In addition, the denominator of both terms in Eq. (A.16) is the **characteristic polynomial** of Eq. (A.14), i.e.

$$\sum_{i=0}^{n} a_i s^i = s^n + a_{n-1}s^{n-1} + \cdots + a_i s + a_0$$

The time domain solution $y(t)$ of Eq. (A.14) is the ILT of $Y(s)$:

$$y(t) = \mathcal{L}^{-1}\left[\frac{U(s)}{\sum_{i=0}^{n} a_i s^i}\right] + \mathcal{L}^{-1}\left[\frac{\sum_{i=0}^{n}\sum_{k=0}^{i-1} a_i s^{i-1-k} y_0^k}{\sum_{i=0}^{n} a_i s^i}\right] \tag{A.17}$$

The first term on the right is the **forced response** and the second term is the **free response** of the system represented by Eq. (A.14).

Direct substitution into Eqs. (A.15) to (A.17) yields the transform of the differential equation, the solution transform $Y(s)$, or the time solution $y(t)$, respectively. However, it is often easier to apply the LT properties directly to determine these quantities, especially when the order of the differential equation is low.

Example A.16: The LT of the differential equation $\frac{d^2y}{dt^2} + 3\frac{dy}{dt} + 2y = u(t) = \mathbf{1}(t)$, with initial conditions $y(0^+) = -1$ and $\left.\frac{dy}{dt}\right|_{t=0^+} = 2$, can be written directly from Eq. (A.15) by first identifying n, a_i and y_0^k. These values are $n = 2$, $a_0 = 2$, $a_1 = 3$, $a_2 = 1$, $y_0^0 = -1$ and $y_0^1 = 2$. Substitution of these values into Eq. (A.15) yields $2Y + 3(sY + 1) + 1(s^2Y + s - 2) = \frac{1}{s}$ or $(s^2 + 3s + 2)Y = \frac{-(s^2 + s - 1)}{s}$.

The LT of this second-order ODE also can be found directly using the **second derivative transform** of $\frac{d^2y}{dt^2}$:

$$\mathcal{L}\left[\frac{d^2y}{dt^2}\right] = s^2Y(s) - sy(0^+) - \left.\frac{dy}{dt}\right|_{t=0} \tag{A.18}$$

This is a direct consequence of Eq. (A.4). Applying the linearity property:

$$\mathcal{L}\left[\frac{d^2y}{dt^2} + 3\frac{dy}{dt} + 2y\right] = \mathcal{L}\left[\frac{d^2y}{dt^2}\right] + 3\mathcal{L}\left[\frac{dy}{dt}\right] + 2\mathcal{L}[y]$$

$$= (s^2 + 3s + 2)Y + s + 1 = \mathcal{L}[u(t)] = \frac{1}{s}$$

The output transform $Y(s)$ is determined by rearranging the above equation:

$$Y(s) = \frac{-(s^2 + s - 1)}{s(s^2 + 3s + 2)}$$

The output time solution $y(t)$ is the inverse transform of $Y(s)$, addressed in the next section.

Consider the more general constant coefficient ODE:

$$\sum_{i=0}^{n} a_i \frac{d^i y}{dt^i} = \sum_{i=0}^{m} b_i \frac{d^i u}{dt^i} \tag{A.19}$$

where y is the output, u is the input, $a_n = 1$, and $m \le n$. The LT of Eq. (A.19) is

$$\sum_{i=0}^{n} \left[a_i \left(s^i Y(s) - \sum_{k=0}^{i-1} s^{i-1-k} y_0^k \right) \right] = \sum_{i=0}^{m} \left[b_i \left(s^i U(s) - \sum_{k=0}^{i-1} s^{i-1-k} u_0^k \right) \right] \tag{A.20}$$

where $u_0^k = \left. \frac{d^k u}{dt^k} \right|_{t=0^+}$. The output transform Y(s) is

$$Y(s) = \left(\frac{\sum_{i=0}^{m} b_i s^i}{\sum_{i=0}^{n} a_i s^i} \right) U(s) - \left(\frac{\sum_{i=0}^{m} \sum_{k=0}^{i-1} b_i s^{i-1-k} u_0^k}{\sum_{i=0}^{n} a_i s^i} \right) + \left[\frac{\sum_{i=0}^{n} \sum_{k=0}^{i-1} a_i s^{i-1-k} y_0^k}{\sum_{i=0}^{n} a_i s^i} \right] \tag{A.21}$$

The time solution $y(t)$ is the ILT of $Y(s)$:

$$y(t) = \mathcal{L}^{-1} \left[\left(\frac{\sum_{i=0}^{m} b_i s^i}{\sum_{i=0}^{n} a_i s^i} \right) U(s) - \left(\frac{\sum_{i=0}^{m} \sum_{k=0}^{i-1} b_i s^{i-1-k} u_0^k}{\sum_{i=0}^{n} a_i s^i} \right) \right] + \mathcal{L}^{-1} \left[\frac{\sum_{i=0}^{n} \sum_{k=0}^{i-1} a_i s^{i-1-k} y_0^k}{\sum_{i=0}^{n} a_i s^i} \right] \tag{A.22}$$

The first term on the right is the **forced response**, and the second term is the **free response** of a system represented by Eq. (A.19). Note that the LT $Y(s)$ of the output $y(t)$ is a **rational (algebraic) function,** i.e. a ratio of polynomials in the complex variable s.

When the initial conditions (ICs) are not given for $y(t)$, but for some internal variable (such as the initial voltage across a capacitor not appearing at the output), then y_0^k must be derived using the available information (for $k = 0, 1, \ldots, n-1$). For models represented in the form of Eq. (A.19), i.e. including derivative terms in u, computation of y_0^k also depends on u_0^k.

The restriction $n \ge m$ in Eq. (A.19) is a practical constraint based on the fact that most real systems have a **smoothing** effect on their input. This means variations in the input are made less pronounced, or at least no more pronounced, by action of the system on the input. Since a **differentiator** generates the slope of a time function, it accentuates variations of the function. At the other extreme, an integrator sums the area under the curve of a time function over an interval and thus averages, i.e. smooths, the variations of the function. The output y is related to the input u by an operation that includes m differentiations and n integrations of the input. Hence, for a smoothing effect (at least no accentuation of the variations) between the input and the output, there must be at least as many integrations than differentiations; i.e. $n \ge m$.

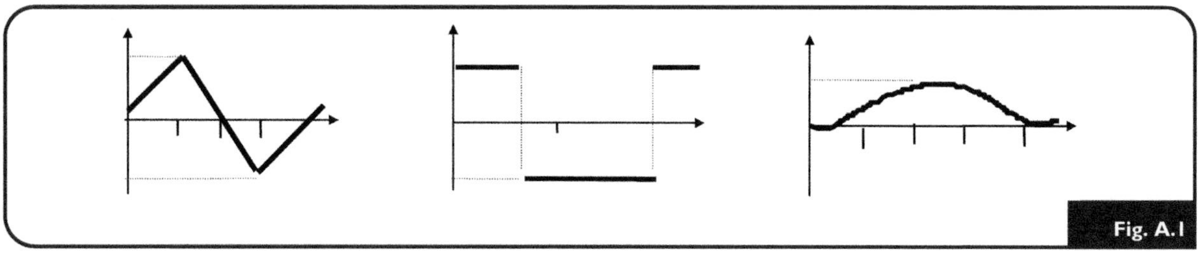

Fig. A.1

Example A.17: A system described by the differential equation

$$\frac{d^2y}{dt^2} = \frac{dx}{dt}, \, y(0^+) = 0, \, \left.\frac{dy}{dt}\right|_{t=0^+} = 0$$

has the input u graphed in Fig. A.1, first plot. The corresponding functions du/dt and $y(t) = \int_{0^+}^{t}\int_{0^+}^{\theta}\frac{du}{d\alpha}d\alpha d\theta = \int_{0^+}^{t}u(\theta)d\theta$ are also shown, as the middle and last plots.

We observe from the graphs that differentiation of u accentuates the variations in u, while integration smooths them.

Example A.18: Consider a system described by

$$\frac{d^2y}{dt^2} + 3\frac{dy}{dt} + 2y = \frac{du}{dt} + 3u$$

with initial conditions $y_0^0 = 1$ and $y_0^1 = 0$. If the input is $u(t) = e^{-4t}$, the LT of the output $y(t)$ can be obtained by direct application of Eq. (A.21) by first identifying m, n, a_i, b_i and u_0^0. We have $m = 1$, $n = 2$, $a_0 = 2$, $a_1 = 1$, $b_0 = 3$, $b_1 = 1$, $x_0^0 = \lim_{t\to 0} e^{-4t} = 1$. Substitution of these values into (A.21) yields $Y(s) = \left(\frac{s+3}{s^2+3s+2}\right)\left(\frac{1}{s+4}\right) + \frac{s+3}{s^2+3s+2} + \frac{1}{s^2+3s+2}$. This transform can also be obtained by direct application of LT properties to the ODE, as in Example A.16.

Partial Fraction Expansions

We saw above that LTs of linear constant-coefficient ODEs generate rational functions of s, i.e. ratios of polynomials of s. The partial fraction expansion is a particularly useful representation of rational functions, greatly simplifying LT inversion. Consider the rational function

$$F(s) = \frac{\sum_{i=0}^{m} b_i s^i}{\sum_{i=0}^{n} a_i s^i} \tag{A.23}$$

where $a_n = 1$ and $n \geq m$. By the fundamental theorem of algebra, the denominator polynomial equation $\sum_{i=0}^{n} a_i s^i = 0$ has n roots, though some may be repeated. For example, the polynomial $s^3 + 5s^2 + 8s + 4$ has three roots: -2, -2, and -1 (-2 is a repeated root). Suppose this denominator polynomial equation has n_1 roots equal to $-p_1$, n_2 roots equal to $-p_2$, ..., and n_r roots equal to $-p_r$, where $\sum_{i=1}^{r} n_i = n$. Then $\sum_{i=0}^{n} a_i s^i = \prod_{i=1}^{r} (s+p_i)^{n_i}$ and the rational function F(s) then becomes:

$$F(s) = \frac{\sum_{i=0}^{m} b_i s^i}{\prod_{i=1}^{r} (s+p_i)^{n_i}}$$

The partial fraction expansion of $F(s)$ is

$$F(s) = b_n + \sum_{i=1}^{r} \sum_{k=1}^{n_i} \frac{c_{ik}}{(s+p_i)^k} \tag{A.24}$$

where $b_n = 0$ unless $m = n$. The coefficient c_{ik} is given by

$$c_{ik} = \frac{1}{(n_i - k)!} \frac{d^{n_i - k}}{ds^{n_i - k}} [(s+p_i)^{n_i} F(s)] \Big|_{s=-p_i} \tag{A.25}$$

The coefficients $c_{i1}, i = 1, 2, \ldots, r$, are called the **residues** of $F(s)$ at $-p_{i1}, i = 1, 2, \ldots, r$. If no roots are repeated, i.e. they are all distinct, then

$$F(s) = b_n + \sum_{i=1}^{n} \frac{c_{i1}}{s + p_i} \tag{A.26}$$

where

$$c_{i1} = (s + p_i) F(s)|_{s=-p_i} \tag{A.27}$$

Example A.19: Consider the rational function $F(s) = \dfrac{s^2 + 2s + 2}{s^2 + 3s + 2} = \dfrac{s^2 + 2s + 2}{(s + 1)(s + 2)}$. Its partial fraction expansion, from (A.26), is $F(s) = b_2 + \dfrac{c_{11}}{s + 1} + \dfrac{c_{21}}{s + 2}$. The numerator coefficient is $b_2 = 1$, and from (A.27), the coefficients c_{11} and c_{21} are

$$c_{11} = (s + 1)F(s)|_{s=-1} = \frac{s^2 + 2s + 2}{s + 2}\Big|_{s=-1} = 1$$

$$c_{21} = (s + 2)F(s)|_{s=-2} = \frac{s^2 + 2s + 2}{s + 1}\Big|_{s=-2} = -2$$

Therefore

$$F(s) = 1 + \frac{1}{s+1} - \frac{2}{s+2}$$

Example A.20: For $F(s) = \frac{1}{(s+1)^2(s+2)}$, the partial fraction expansion is $F(s) = b_3 + \frac{c_{11}}{s+1} + \frac{c_{12}}{(s+1)^2} + \frac{c_{21}}{s+2}$. The coefficients are given by $b_3 = 0$ and

$$c_{11} = \frac{d}{ds}(s+1)^2 F(s)\Big|_{s=-1} = \frac{d}{ds}\left(\frac{1}{s+2}\right)\Big|_{s=-1} = \frac{-1}{(s+2)^2}\Big|_{s=-1} = -1$$

$$c_{12} = (s+1)^2 F(s)\Big|_{s=-1} = \frac{1}{s+2}\Big|_{s=-1} = 1, \quad c_{21} = (s+2)F(s)\Big|_{s=-2} = \frac{1}{(s+1)^2}\Big|_{s=-2} = 1$$

This yields:

$$F(s) = \frac{-1}{s+1} + \frac{1}{(s+1)^2} + \frac{1}{s+2}$$

Inverse Transforms Using Partial Fraction Expansions

We have shown earlier that the solution to a linear, constant-coefficient ODE can be determined as the ILT of a rational algebraic function:

$$\mathcal{L}^{-1}\left[\frac{\sum_{i=0}^{m} b_i s^i}{\sum_{i=0}^{n} a_i s^i}\right] = \mathcal{L}^{-1}\left[b_n + \sum_{i=0}^{r}\sum_{k=0}^{n_i}\frac{c_{ik}}{(s+p_i)^k}\right] = b_n\delta(t) + \sum_{i=0}^{r}\sum_{k=0}^{n_i}\frac{c_{ik}}{(k-1)!}t^{k-1}e^{-p_i t} \tag{A.28}$$

where δ(t) is the unit impulse function and $b_n = 0$ unless $m = n$.

Example A.21: The ILT of $F(s) = \frac{s^2+2s+2}{(s+1)(s+2)}$ is given by

$$\mathcal{L}^{-1}\left[\frac{s^2+2s+2}{(s+1)(s+2)}\right] = \mathcal{L}^{-1}\left[1 + \frac{1}{s+1} - \frac{2}{s+2}\right] = \mathcal{L}^{-1}[1] + \mathcal{L}^{-1}\left[\frac{1}{s+1}\right] - 2\mathcal{L}^{-1}\left[\frac{1}{s+2}\right]$$

$$= \delta(t) + e^{-t} - 2e^{-2t}$$

This is the time domain solution $f(t)$.

References

DiStefano III, J., Stubberud, A.R., Williams, I.J., 1990. Feedback and Control Systems. second ed. McGraw-Hill, New York.

Murray, M., 1965. Laplace Transforms. McGraw Hill, New York.

Appendix B

Linear Algebra for Biosystem Modeling

Overview

Following is a concise summary of numerous basic results about matrices, vector spaces, functions of matrices and other subjects in elementary and intermediate linear algebra. It's presented in list form for easy reference. The topics are selected for their utility in modeling and analyzing nonlinear as well as linear dynamic biosystems represented in state variable vector-matrix form.

Most functions included are easy to evaluate directly using widely available computational math programs like *Mathematica, Matlab, Octave, Mathcad, R*, etc. The last section addresses *singular value decomposition* of a matrix and *principle component analysis*, important tools for computational biology data analysis as well as dynamic systems biology modeling. Basic references for the material in this Appendix include Chen (1985), Friedberg et al. (2003) and Trefethen and Bau (1997), among others cited.

Matrices

Let $A = [a_{ij}]$ be an m (row) by n (column) matrix of real or complex numbers[1] a_{ij}. In expanded form,

$$A = \begin{pmatrix} a_{11} & \cdots & a_{1n} \\ \vdots & \ddots & \vdots \\ a_{m1} & \cdots & a_{mn} \end{pmatrix}$$

Matrix B has the same form, with matrix elements b_{ij}, matrix C with elements c_{ij}, and so on. Matrix manipulations and transformations depend on the number of rows and columns, as noted below. First, for any m and any n:

(1) A is **symmetric** if $A = A^T$

(2) A is **skew-symmetric** if $A = -A^T$

Products: For A $(n \times m)$, B $(m \times r)$

(3) Matrix **product**: $AB = [a_{ik}][b_{kj}] = \sum_{k=1}^{m} a_{ik}b_{kj}$ $(n \times r)$

(4) $AB \neq BA$, i.e. matrix multiplication is **not commutative**

(5) **Transpose**: $A^T = [a_{ji}]$ $(m \times n)$, and $(AB)^T = B^T A^T$ $(r \times n)$

For A $(n \times m)$, B $(m \times r)$, C $(l \times n)$ and D $(r \times \rho)$.

Let $\boldsymbol{a}^i$ be the ith **column** of A and $\boldsymbol{b}^j$ the jth **row** of B. Then:

(6) $$AB = [\boldsymbol{a}^1 \quad \boldsymbol{a}^2 \quad \cdots \quad \boldsymbol{a}^m] \begin{bmatrix} \boldsymbol{b}^1 \\ \boldsymbol{b}^2 \\ \vdots \\ \boldsymbol{b}^n \end{bmatrix} = \boldsymbol{a}^1\boldsymbol{b}^1 + \boldsymbol{a}^2\boldsymbol{b}^2 + \cdots + \boldsymbol{a}^m\boldsymbol{b}^m$$

(7) $$CA = C[\boldsymbol{a}^1 \quad \boldsymbol{a}^2 \quad \cdots \quad \boldsymbol{a}^m] = [C\boldsymbol{a}^1 \quad C\boldsymbol{a}^2 \quad \cdots \quad C\boldsymbol{a}^m]$$

---

1 A **complex number** a has a real (Re) and an imaginary (Im) component, defined by $a \equiv Re\ a + j\ Im\ a$, where $j \equiv j = \sqrt{-1}$. The **complex conjugate** of a is defined by $a^* \equiv Re\ a - j\ Im\ a$.

$$(8)\ BD = \begin{bmatrix} \boldsymbol{b}^1 \\ \boldsymbol{b}^2 \\ \vdots \\ \boldsymbol{b}^m \end{bmatrix} D = \begin{bmatrix} \boldsymbol{b}^1 D \\ \boldsymbol{b}^2 D \\ \vdots \\ \boldsymbol{b}^m D \end{bmatrix}$$

$$(9)\ \sum_{i=1}^{m} \alpha_i \boldsymbol{a}^i = [\,\boldsymbol{a}^1 \quad \boldsymbol{a}^2 \quad \cdots \quad \boldsymbol{a}^m\,] \begin{bmatrix} \alpha_1 \\ \alpha_2 \\ \vdots \\ \alpha_m \end{bmatrix} \equiv A\alpha, \text{ where the } \alpha_i \text{ are scalars.}$$

Square Matrices $m = n$:

(10) **Trace** of $A \equiv \text{tr}\, A = \sum_{i=1}^{n} a_{ii}$ for $(m = n)$.

(11) **Determinant** of $A = \det A \equiv |A| = \sum_{j=1}^{n} A_{i,j}(-1)^{i+j} M_{i,j}$, where $M_{i,j}$ is the **submatrix** of A obtained after removing the ith row and jth column. *Note*: det A is a real number.

For a 2×2 matrix $A = \begin{bmatrix} a & b \\ c & d \end{bmatrix}$, $\det A = ad - bc$.

(12) $\det A^T = \det A$; $\det A^{-1} = \dfrac{1}{\det A}$

(13) $\det(AB) = (\det A)(\det B)$

(14) If A = **triangular**, then $|A| = \Pi a_{ii}$

(15) **Inverse** of $A = A^{-1} = \dfrac{1}{\det A}(\text{adjoint of A})^T = \dfrac{1}{\det A}(-1)^{i+j} {M_{i,j}}^T$, with $M_{i,j}$ defined as for a determinant above. Matrix A is **singular** if it has no inverse. In this case det $A = 0$. **Nonsingular** matrices have an inverse solution and det $A \neq 0$.

(16) Inverse of a 2×2 matrix: $A^{-1} = \dfrac{1}{\det A}\begin{bmatrix} d & -b \\ -c & a \end{bmatrix}$

(17) Inverse of a 3×3 triangular matrix

$$A = \begin{bmatrix} a_{11} & 0 & 0 \\ a_{21} & a_{22} & 0 \\ a_{31} & a_{32} & a_{33} \end{bmatrix} \rightarrow A^{-1} = \begin{bmatrix} \dfrac{1}{a_{11}} & 0 & 0 \\ \dfrac{-a_{21}}{a_{11}a_{22}} & \dfrac{1}{a_{22}} & 0 \\ \dfrac{a_{21}a_{32} - a_{22}a_{31}}{a_{11}a_{22}a_{33}} & \dfrac{-a_{32}}{a_{22}a_{33}} & \dfrac{1}{a_{33}} \end{bmatrix}$$

(18) **Identity** matrix: $AA^{-1} = I = \begin{pmatrix} 1 & \cdots & 0 \\ & \ddots & \\ 0 & \cdots & 1 \end{pmatrix}$

Complex number Matrices: Let a_{ij}^* be the complex conjugate of the number a_{ij}.

(19) The complex conjugate transpose matrix is:

$$A^{\dagger} = [a_{ji}^*] = (A^T)^* = (A^*)^T \text{ and } (AB)^{\dagger} = B^{\dagger}A^{\dagger}$$

(20) A is **Hermitian** if $A = A^{\dagger}$

(21) A is **normal** if $A^{\dagger}A = AA^{\dagger}$

(22) A is **orthogonal** if $A^{\dagger}A = D$ (diagonal) $= \Lambda$

(23) A is **unitary** or **orthonormal** if $A^{\dagger}A = I$

(24) $(A^{-1})^{\dagger} = (A^{\dagger})^{-1}$ and $(AB)^{-1} = B^{-1}A^{-1}$

(25) Kronecker delta $= \delta_{ij} = \begin{cases} 1, & i = j \\ 0, & i \neq j \end{cases}$

Pseudoinverse: When a matrix is not square it has no inverse, but it does have a pseudoinverse. If A is $m \times n$, there exists a unique $n \times m$ dimensional pseudoinverse of A, denoted A^+, with the following properties:

(26) $AA^+A = A$, $A^+AA^+ = A^+$, $A^+A = (A^+A)^T$, and $AA^+ = (AA^+)^T$

(27) If $m = n$ and A is nonsingular, then $A^+ = A^{-1}$

Vector Spaces (V.S.)

(28) ***Short-form definition***: A V.S. is a nonempty set of elements called vectors **closed** under addition and scalar multiplication, i.e. sums of vectors, and vectors multiplied by any number, are also within the V.S. The vector **0** is a member of every V.S.

(29) $\mathbf{v}^1, \mathbf{v}^2, \ldots, \mathbf{v}^n$ are **linearly dependent** if and only if there exist scalars $\beta_1, \beta_2, \ldots, \beta_n$ not all zero such that $\sum_{i=1}^{n} \beta_i \mathbf{v}^i = [\mathbf{v}^1 \quad \mathbf{v}^2 \quad \cdots \quad \mathbf{v}^n] \begin{bmatrix} \beta_1 \\ \beta_2 \\ \vdots \\ \beta_n \end{bmatrix} \equiv V\beta = \mathbf{0}$

(30) $\mathbf{v}^1, \mathbf{v}^2, \ldots, \mathbf{v}^n$ are **linearly independent** if and only if $\sum_{i=1}^{n} \beta_i \mathbf{v}^i = \mathbf{0} \Rightarrow$ (implies) all $\beta_i = 0$.

(31) $\boldsymbol{v}^1, \boldsymbol{v}^2, \ldots, \boldsymbol{v}^n$ form a **basis** in a V.S. if and only if they are linearly independent and every $\boldsymbol{x}$ in the V.S. can be written: $\boldsymbol{x} = \sum_{i=1}^{n} \alpha_i \boldsymbol{v}^i$

(32) The **dimension** of a V.S. is the number of basis vectors in the V.S.

Linear Equation Solutions

(33) The **null space** of A is the V.S. of all solutions of $A\boldsymbol{x} = \boldsymbol{0}$.

(34) $A\boldsymbol{x} = \boldsymbol{0}$ with $\boldsymbol{x} \neq \boldsymbol{0}$ if $|A| = 0$.

(35) Let A be $m \times n$. If A has at most r linearly independent columns (rows), the null space of A has dimension $n-r$; if all columns (rows) are linearly independent, $\boldsymbol{x} = \boldsymbol{0}$ is the unique solution; $n-r$ is the number of nonzero solutions of $A\boldsymbol{x} = \boldsymbol{0}$.

(36) The dimension of the V.S. spanned by the rows of A = the dimension of the V.S. spanned by the columns of A.

(37) The **range space** of A is the V.S. of all $\boldsymbol{y}$ such that $\boldsymbol{y} = A\boldsymbol{x}$, for some $\boldsymbol{x}$.

(38) n = dim(range space) + dim(null space).

(39) **rank** A = dim(range space) = maximum number of linear independent columns (rows) of $A \equiv r$.

(40) rank A = rank A^T = rank(A^TA) = rank(AA^T).

(41) If $A\boldsymbol{x} = \boldsymbol{y}$ has an infinite number of solutions (case $m < n$, an underdetermined system), $\boldsymbol{x} = A^+\boldsymbol{y}$ is the **minimum length (norm) solution**, where A^+ is the **pseudoinverse** of A. In this case, $A^+ = A^T(AA^T)^{-1}$.

(42) If $A\boldsymbol{x} = \boldsymbol{y}$ has no solution ($m > n$, an **overdetermined system**), $\boldsymbol{x} = A^+\boldsymbol{y}$ is the (approximate) **least squares solution**, or the **projection solution**, i.e. the $\boldsymbol{x}$ with minimum (norm). In this case, $A^+ = (A^TA)^{-1}A^T$.

Properties (41) and (42) are discussed more fully next.

Minimum Norm (*MN*) & Least Squares Pseudoinverse Solutions of Linear Equations

The system of linear equations $\boldsymbol{y} = A\boldsymbol{x}$, where A is $m \times n$, has a solution $\boldsymbol{x}$ for any $\boldsymbol{y}$ if and only if rank $A = m$, the dimension of the $\boldsymbol{y}$ vector. If rank $A \neq m$, there are $\boldsymbol{y}$s for which no solution exists. If we consider the additional fact that rank $A \leq \min(m, n)$, then we see that a solution exists only if $m \leq n$, because $m > n$ would mean rank $A < m$.

For the case of multiple solutions $\mathbf{x}^1$, $\mathbf{x}^2$,..., etc. (rank $A = m < n$), we can easily find the one with the smallest **Euclidean (L_2) norm**, i.e. that particular $\mathbf{x}^j$ that minimizes $\|\mathbf{x}\|_2 = \sqrt{\sum_{k=1}^{n} x_k^2}$, where $\|A\mathbf{x}^i - \mathbf{y}\|_2 = \|A\mathbf{x}^j - \mathbf{y}\|_2$, for $i \neq j$. This **minimum norm** solution is $\mathbf{x}_{MN} = A^+_{MN}\mathbf{y}$, where A^+_{MN} is a particular **pseudoinverse** of A, i.e. $A^+_{MN} = A^T(AA^T)^{-1}$. AA^T is nonsingular because rank $A = \mathrm{rank}(AA^T) = m$ and AA^T is $m \times n$.

On the other hand, when no solution exists and rank $A = n < m$, one can find an approximate (**projection**) solution, e.g. the $\mathbf{x} \equiv \hat{\mathbf{x}}$ that minimizes the **least squares** error of the approximation, i.e. $\min\|\mathbf{y} - A\mathbf{x}\|_2^2$. This solution is $\hat{\mathbf{x}} = A^+_{LS}\mathbf{y}$, where $A^+_{LS} = (A^TA)^{-1}A^T$ because rank $(A^TA) = n$ and A^TA is $n \times n$ and thus nonsingular.

Measures & Orthogonality

(43) norm $\mathbf{x} \equiv \|\mathbf{x}\|$ satisfies $\|\mathbf{x}\| \geq 0$, and $\|\mathbf{x}\| = 0$ if and only if $\mathbf{x} = \mathbf{0}$.

(44) $\|\mathbf{x} + \mathbf{y}\| \leq \|\mathbf{x}\| + \|\mathbf{y}\|$

(45) $\|\alpha\mathbf{x}\| = |\alpha| \cdot \|\mathbf{x}\|$

(46) **Euclidean (L_2) norm** of $\mathbf{x} \equiv \|\mathbf{x}\|_2 = \sqrt{\mathbf{x}^\dagger\mathbf{x}} = \sqrt{\sum_{i=1}^{n} x_i^2}$

(47) **Inner Product**, denoted $(\mathbf{x}, \mathbf{y})$ or $<\mathbf{x}, \mathbf{y}>$, satisfies $(\mathbf{x}, \mathbf{y}) \geq 0$, and $(\mathbf{x}, \mathbf{y}) = 0$ if and only if $\mathbf{x}$ and/or $\mathbf{y} = \mathbf{0}$.

(48) $(\mathbf{x}, \mathbf{y}) \equiv \mathbf{x}^T\mathbf{y} = \sum(x_i^* y_i)$

(49) $(\mathbf{x}, \mathbf{x}) \equiv \|\mathbf{x}\|_2^2$

(50) $(\mathbf{x}, \mathbf{y}) = (\mathbf{y}, \mathbf{x})^*$

(51) $(\mathbf{x}, \alpha\mathbf{y}) = \alpha(\mathbf{x}, \mathbf{y})$, including $(\mathbf{x}, \mathbf{0}) = 0$

(52) $(\alpha\mathbf{x} + \beta\mathbf{y}, \mathbf{z}) = \alpha^*(\mathbf{x}, \mathbf{z}) + \beta^*(\mathbf{y}, \mathbf{z})$

(53) Schwartz Inequality: $|(\mathbf{x}, \mathbf{y})|^2 \leq (\mathbf{x}, \mathbf{x})(\mathbf{y}, \mathbf{y}) \equiv \|\mathbf{x}\|^2 \cdot \|\mathbf{y}\|^2$

(54) $\|\mathbf{x} + \mathbf{y}\|_2 \leq \|\mathbf{x}\|_2 \cdot \|\mathbf{y}\|_2$

(55) $\mathbf{x}$ and $\mathbf{y}$ are **orthogonal** if and only if $(\mathbf{x}, \mathbf{y}) = 0$

(56) $\mathbf{x}^1, \mathbf{x}^2, \ldots, \mathbf{x}^n$ are **orthonormal** if and only if $(\mathbf{x}^i, \mathbf{x}^j) = \delta_{ij} = \begin{cases} 1, & i = j \\ 0, & i \neq j \end{cases}$

(57) **Reciprocal basis**: given any basis $\mathbf{v}^1, \mathbf{v}^2, \ldots, \mathbf{v}^n$, a reciprocal basis $\mathbf{r}^1, \mathbf{r}^2, \ldots, \mathbf{r}^n$ satisfies $(\mathbf{r}^i, \mathbf{v}^j) = \delta_{ij}$, i.e. $R \equiv [\mathbf{r}^1 \;\; \mathbf{r}^2 \;\; \cdots \;\; \mathbf{r}^n] = ([\mathbf{v}^1 \;\; \mathbf{v}^2 \;\; \cdots \;\; \mathbf{v}^n]^{-1})^\dagger \equiv (V^{-1})^\dagger$

(58) Any $\boldsymbol{x} \in$ V.S. can be written $\boldsymbol{x} = \sum_{i=1}^{n} \alpha_i \boldsymbol{v}^i = \sum_{i=1}^{n} (\boldsymbol{x}, \boldsymbol{r}_i)\boldsymbol{v}^i$

(59) $G \equiv$ **Grammian** Matrix $\equiv [(\boldsymbol{x}_i, \boldsymbol{x}_j)] = [\boldsymbol{x}^{i\dagger}\boldsymbol{x}^j]$ for a set of vectors $\boldsymbol{x}^1, \boldsymbol{x}^2, \ldots, \boldsymbol{x}^n$. This set of vectors is linearly independent if and only if $|G| \neq 0$.

Matrix Analysis

(60) $A\boldsymbol{x} = \lambda \boldsymbol{x}$, where λ are **eigenvalues** and $\boldsymbol{x}$ are **eigenvectors**.

(61) $|A - \lambda I| = 0$ is the characteristic equation *of A*.

(62) *Distinct* eigenvalues have linearly independent eigenvectors.

(63) **Similarity transform:** A and $\hat{A}$ are similar if $\exists$ a nonsingular matrix $Q \ni \hat{A} = Q^{-1}AQ$

(64) If $|\hat{A} - \lambda I| = |A - \lambda I|$, then all similar matrices have the same λ_i, same traces and same determinants, i.e. $|A| = |Q^{-1}AQ|$, tr $A = \text{tr}(Q^{-1}AQ)$, and $|A| = \prod \lambda_i$.

(65) All **triangular** matrices have λ_i on the diagonal: *tr* $A = \Sigma\lambda_i$.

(66) $Q^{-1}AQ = \Lambda$ (diagonal) if and only if a set of n linearly independent eigenvectors of A (not necessarily all distinct) can be found. In that case, $Q \equiv [\boldsymbol{x}^1 \quad \boldsymbol{x}^2 \quad \cdots \quad \boldsymbol{x}^n]$, $\boldsymbol{x}^i$ are the eigenvectors, and Q is called the **modal matrix**.

(67) The **Jordan form** can always be found, even if Λ cannot. It is almost as simple, in the form of a partitioned block diagonal matrix:

$$\begin{bmatrix} J_1(\lambda_1) & & 0 \\ & \ddots & \\ 0 & & J_p(\lambda_p) \end{bmatrix}$$

where each of the blocks, called a Jordan block, has the form:

$$J_i(\lambda_i) = \begin{bmatrix} \lambda_i & 1 & & 0 \\ & \lambda_i & \ddots & \\ & & \ddots & 1 \\ 0 & & & \lambda_i \end{bmatrix}$$

(68) *One and only one linearly independent eigenvector is associated with each Jordan block.* But one can always find additional **generalized eigenvectors** $\boldsymbol{t}^i$ from

$$\left.\begin{aligned} A\boldsymbol{x}^i &= \lambda_i \boldsymbol{x}^i \\ A\boldsymbol{t}^1 &= \lambda_i \boldsymbol{t}^1 + \boldsymbol{x}^i \\ &\vdots \\ A\boldsymbol{t}^{l-1} &= \lambda_i \boldsymbol{t}^{l-1} + \boldsymbol{t}^{l-2} \end{aligned}\right\} \quad \text{where} \quad \begin{aligned} &l \equiv \text{multiplicity of eigenvalue } \lambda_i \\ &Q \equiv \begin{bmatrix} \boldsymbol{x}^1 & \boldsymbol{t}^1 & \boldsymbol{t}^2 & \cdots & \boldsymbol{t}^{l-1} \end{bmatrix} \\ &J \equiv Q^{-1}AQ \end{aligned}$$

(69) **Adjoint transformation:** a^* of a satisfies $(a\boldsymbol{x}, \boldsymbol{y}) = (\boldsymbol{x}, a^*\boldsymbol{y})$ for all vectors $\boldsymbol{x}$, $\boldsymbol{y}$. For matrices, $A^\dagger \equiv a^*$. If $A = [a_{ij}]$, then $A^\dagger = [a_{ji}^*]$ is its adjoint matrix

(70) $(AB)^\dagger = B^\dagger A^\dagger$; $(A + B)^\dagger = A^\dagger + B^\dagger$; $(\alpha A)^\dagger = \alpha^* A^\dagger$; $(A^\dagger)^\dagger = A$

(71) A is **self-adjoint (Hermitian)** if and only if $A = A^\dagger$, i.e. $a_{ij} = a_{ji}^*$

(72) If $(A = A^\dagger)$, then $(\boldsymbol{x}, A\boldsymbol{x})$ is real.

(73) $f \equiv (\boldsymbol{x}, A\boldsymbol{y}) = \boldsymbol{x}^\dagger A\boldsymbol{y}$ is the **bilinear form**

(74) $q \equiv (\boldsymbol{x}, A\boldsymbol{x}) = (\boldsymbol{x}, A^\dagger\boldsymbol{x}) = \boldsymbol{x}^\dagger A\boldsymbol{x} = \boldsymbol{x}^\dagger A^\dagger \boldsymbol{x}$ is the **quadratic form** $= \sum_{i=1}^{n}\sum_{j=1}^{n} a_{ij}x_i^* x_j$, $a_{ij} = a_{ji}^*$

(75) n orthonormal eigenvectors can be found for an $n \times n$ normal matrix $(N^\dagger N = NN^\dagger)$.

(76) Hermitian matrices are normal. Therefore, eigenvectors belonging to distinct λ_i of Hermitian matrices are orthogonal.

(77) Eigenvalues of an $n \times n$ Hermitian matrix are real.

(78) For any Hermitian matrix A $(=A^\dagger)$, we have $\Lambda = U^{-1}AU = U^{+}AU$ (U is **unitary**). Therefore, $q = (\boldsymbol{x}, A\boldsymbol{x}) = (U\boldsymbol{y}, AU\boldsymbol{y}) = \boldsymbol{y}^\dagger \Lambda \boldsymbol{y} = \sum_{i=1}^{n}(\lambda_i |y_i|^2)$

(79) *Any* matrix A can be expressed as $A = G + jH$, where $G = \dfrac{A + A^\dagger}{2}$, $H = \dfrac{A - A^\dagger}{2j}$ and G and H are Hermitian.

(80) A is **positive definite** $\Leftrightarrow$ $(\boldsymbol{x}, A\boldsymbol{x}) > 0$ for all $\boldsymbol{x}$ $\Leftrightarrow$ all $\lambda_i > 0$

(81) A is **nonnegative definite** $\Leftrightarrow$ $(\boldsymbol{x}, A\boldsymbol{x}) \geq 0$ for all $\boldsymbol{x}$ $\Leftrightarrow$ all $\lambda_i \geq 0$

(82) A is *positive definite* (nonnegative definite) if and only if all $|A_i| > 0$ (≥ 0), where

$$A_1 = a_{11},\ A_2 = \begin{bmatrix} a_{11} & a_{12} \\ a_{21} & a_{22} \end{bmatrix},\ A_3 = \begin{bmatrix} a_{11} & a_{12} & a_{13} \\ a_{21} & a_{22} & a_{23} \\ a_{31} & a_{32} & a_{33} \end{bmatrix}, \ldots,\ A_n = \begin{bmatrix} a_{11} & a_{12} & \cdots & a_{1n} \\ a_{21} & a_{22} & \ddots & \vdots \\ \vdots & \ddots & \ddots & a_{n-1,n} \\ a_{n1} & \cdots & a_{n,n-1} & a_{nn} \end{bmatrix}$$

Matrix Norms

(83) $\|A\| \equiv \max\|A\boldsymbol{x}\|$

(84) $|\lambda_i| \leq \|A\|$ for all λ_i

(85) $\|A\boldsymbol{x}\| \leq \|A\| \cdot \|\boldsymbol{x}\|, \quad \|A+B\| \leq \|A\| + \|B\|$, and $\|AB\| \leq \|A\| \cdot \|B\|$

(86) $\|A\| = 0 \Leftrightarrow A = [0]$

(87) $\|A\|_2 = \sqrt{\lambda_{\max}}$ where $\lambda_{\max}$ is the max eigenvalue of $A^\dagger A$.

Matrix Calculus

(88) $\dfrac{dA(t)}{dt} \equiv \left[\dfrac{da_{ij}}{dt}\right]$ and $\displaystyle\int_0^t A \equiv \left[\int_0^t a_{ij}\right]$

(89) $\dfrac{d(\boldsymbol{x},\boldsymbol{y})}{dt} = \dfrac{d}{dt}(\boldsymbol{x}^\dagger \boldsymbol{y}) = \dot{\boldsymbol{x}}^\dagger \boldsymbol{y} + \boldsymbol{x}^\dagger \dot{\boldsymbol{y}}$

(90) $\dfrac{d(A\boldsymbol{x})}{dt} = \left[\dfrac{dA}{dt}\right]\boldsymbol{x} + A\left[\dfrac{d\boldsymbol{x}}{dt}\right]$

(91) $\dfrac{d(AB)}{dt} = \left[\dfrac{dA}{dt}\right]B + A\left[\dfrac{dB}{dt}\right]$

(92) $\dfrac{d(A^{-1})}{dt} = -A^{-1}\left[\dfrac{dA}{dt}\right]A^{-1}$

(93) Laplace transforms: $\mathcal{L}[A(t)] = [a_{ij}(s)] = A(s)$ and $\mathcal{L}\left[\dfrac{d\boldsymbol{x}}{dt}\right] = s\boldsymbol{X}(s) - \boldsymbol{x}(0)$

(94) Gradient Vector: $\nabla f \equiv \dfrac{\partial f(\boldsymbol{x})}{\partial \boldsymbol{x}} = \begin{bmatrix} \dfrac{\partial f}{\partial x_1} & \dfrac{\partial f}{\partial x_2} & \cdots & \dfrac{\partial f}{\partial x_n} \end{bmatrix}$ and

$$\frac{d(\boldsymbol{c},\boldsymbol{x})}{d\boldsymbol{x}} = \begin{bmatrix} \dfrac{\partial(\boldsymbol{c},\boldsymbol{x})}{\partial x_1} & \dfrac{\partial(\boldsymbol{c},\boldsymbol{x})}{\partial x_2} & \cdots & \dfrac{\partial(\boldsymbol{c},\boldsymbol{x})}{\partial x_n} \end{bmatrix} = \begin{bmatrix} c_1 & c_2 & \cdots & c_n \end{bmatrix} = c^T$$

(95) Jacobian Matrix:

$$J_y(\mathbf{x}) \equiv \frac{d\mathbf{y}}{d\mathbf{x}} \equiv \begin{bmatrix} \frac{\partial y_1}{\partial x_1} & \frac{\partial y_1}{\partial x_2} & \cdots & \frac{\partial y_1}{\partial x_n} \\ \frac{\partial y_2}{\partial x_1} & \frac{\partial y_2}{\partial x_2} & \ddots & \vdots \\ \vdots & \ddots & \ddots & \frac{\partial y_{m-1}}{\partial x_n} \\ \frac{\partial y_m}{\partial x_1} & \cdots & \frac{\partial y_m}{\partial x_{n-1}} & \frac{\partial y_m}{\partial x_n} \end{bmatrix}$$

(96) $\dfrac{d(A\mathbf{x})}{d\mathbf{x}} = A\dfrac{d\mathbf{x}}{d\mathbf{x}} = A$

(97) $\dfrac{df}{d\mathbf{x}} = \dfrac{\partial f}{\partial \mathbf{u}} \cdot \dfrac{d\mathbf{u}}{d\mathbf{x}} \equiv f_u \dfrac{d\mathbf{u}}{d\mathbf{x}}$

(98) $\dfrac{d(\mathbf{x}^\dagger A\mathbf{x})}{d\mathbf{x}} = \left[\dfrac{d(\mathbf{x}^\dagger A\mathbf{x})}{dx_1} \quad \dfrac{d(\mathbf{x}^\dagger A\mathbf{x})}{dx_2} \quad \cdots \quad \dfrac{d(\mathbf{x}^\dagger A\mathbf{x})}{dx_n} \right] = \mathbf{x}^\dagger (A + A^\dagger)$

(99) If $A = A^\dagger$, then $\dfrac{d(\mathbf{x}^\dagger A\mathbf{x})}{d\mathbf{x}} = 2\mathbf{x}^\dagger A$

(100) **Hessian** Matrix:

$$H(f) \equiv \frac{\partial^2 f}{\partial \mathbf{x}^2} = \begin{bmatrix} \frac{\partial^2 f}{\partial x_1^2} & \frac{\partial^2 f}{\partial x_1 \partial x_2} & \cdots & \frac{\partial^2 f}{\partial x_1 \partial x_n} \\ \frac{\partial^2 f}{\partial x_2 \partial x_1} & \frac{\partial^2 f}{\partial x_2^2} & \ddots & \vdots \\ \vdots & \ddots & \ddots & \frac{\partial^2 f}{\partial x_{n-1} \partial x_n} \\ \frac{\partial^2 f}{\partial x_n \partial x_1} & \cdots & \frac{\partial^2 f}{\partial x_n \partial x_{n-1}} & \frac{\partial^2 f}{\partial x_n^2} \end{bmatrix}$$

Computation of *f*(*A*), an Analytic Function of a Matrix

By analogy with the Maclaurin series for $f(x)$, i.e. $f(x) = \sum_{k=0}^{\infty} \left[f^{(k)}(0) \frac{x^k}{k!} \right]$, where $f^{(k)}(0)$ is the kth derivative of f evaluated at $x = 0$, we *define*

$$f(A) = \sum_{k=0}^{\infty} \left[f^{(k)}(0) \frac{A^k}{k!} \right] \tag{1}$$

Then, for example, $\cos A = (\cos 0)I + (-\sin 0)A + (-\cos 0)\frac{A^2}{2!} + \cdots = I - \frac{A^2}{2!} + \frac{A^4}{4!} - \cdots$

Consider the Jordan canonical form of A: $J = Q^{-1}AQ$. We can show from (1) that $f(J) = Q^{-1}f(A)Q$ and $f(A) = Qf(J)Q^{-1}$, where the matrix $Q = \begin{bmatrix} \boldsymbol{x}^1 & \boldsymbol{x}^2 & \cdots & \boldsymbol{x}^n \end{bmatrix}$ is comprised of columns that are the eigenvectors or generalized eigenvectors of A.

Case 1: Suppose J is diagonal, i.e. $J = \Lambda$. Then $A = Q\Lambda Q^{-1}$ and

$$f(A) = \sum_{i=1}^{n} \left[f(\lambda_i) \boldsymbol{x}^i \boldsymbol{r}^{i\dagger} \right] \tag{2}$$

where $R = \begin{bmatrix} \boldsymbol{r}^1 & \boldsymbol{r}^2 & \cdots & \boldsymbol{r}^n \end{bmatrix} = (Q^{-1})^\dagger$ (the reciprocal basis).

Special Case (a): A = Λ

$$f(\Lambda) = \sum \frac{f^{(k)}(0)}{k!} \Lambda^k = \sum \frac{f^{(k)}(0)}{k!} \begin{bmatrix} \lambda_1^k & & 0 \\ & \ddots & \\ 0 & & \lambda_n^k \end{bmatrix}$$

$$= \begin{bmatrix} \sum \frac{f^{(k)}(0)}{k!} \lambda_1^k & & 0 \\ & \ddots & \\ 0 & & \sum \frac{f^{(k)}(0)}{k!} \lambda_n^k \end{bmatrix} = \begin{bmatrix} f(\lambda_1) & & 0 \\ & \ddots & \\ 0 & & f(\lambda_n) \end{bmatrix} \tag{3}$$

Special Case (b): Spectral Representation of A

$$A = \sum_{i=1}^{n} \lambda_i \boldsymbol{x}^i \boldsymbol{r}^{i\dagger}$$

Case 2: $J \neq \Lambda$.

To find $f(A) = Qf(J)Q^{-1}$, we look for $f(J)$:

$$J \equiv \begin{bmatrix} J_{11}(\lambda_1) & & 0 \\ & \ddots & \\ 0 & & J_{mp}(\lambda_p) \end{bmatrix} \quad \text{and} \quad J^k \equiv \begin{bmatrix} J_{11}^k(\lambda_1) & & 0 \\ & \ddots & \\ 0 & & J_{mp}^k(\lambda_p) \end{bmatrix}$$

There exist p distinct eigenvalues, and m is the number of Jordan blocks for each eigenvalue. Then

$$f(J) = \sum \frac{f^{(k)}(0)}{k!} J^k = \sum \frac{f^{(k)}(0)}{k!} \begin{bmatrix} J_{11}^k(\lambda_1) & & 0 \\ & \ddots & \\ 0 & & J_{mp}^k(\lambda_p) \end{bmatrix} \begin{bmatrix} f(J_{11}) & & 0 \\ & \ddots & \\ 0 & & f(J_{mp}) \end{bmatrix} \tag{4}$$

Therefore, we must find $f(J_{ji})$, an $l \times l$ matrix. Since

$$J_{ji}(\lambda_i) \equiv \begin{bmatrix} \lambda_1 & 1 & 0 & \cdots & 0 \\ 0 & \lambda_2 & 1 & \ddots & \vdots \\ \vdots & \ddots & \lambda_3 & \ddots & 0 \\ \vdots & \ddots & \ddots & \ddots & 1 \\ 0 & \cdots & \cdots & 0 & \lambda_l \end{bmatrix}$$

by induction, it is easy to show that

$$J_{ij}^k(\lambda) = \begin{bmatrix} \lambda^k & k\lambda^{k-1} & \frac{k(k-1)}{2!}\lambda^{k-2} & \frac{k(k-1)(k-2)}{3!}\lambda^{k-3} & \cdots \\ 0 & \lambda^k & k\lambda^{k-1} & \frac{k(k-1)}{2!}\lambda^{k-2} & \ddots \\ \vdots & 0 & \lambda^k & \ddots & \ddots \\ \vdots & \ddots & \ddots & \ddots & k\lambda^{k-1} \\ 0 & \cdots & \cdots & 0 & \lambda^k \end{bmatrix}$$

Then, from $f(J) = \sum \frac{f^{(k)}(0)}{k!}(J^k)$, we have, after some manipulation:

$$f\left[J_{jk}(\lambda_i)\right] \equiv f(J_{ji}) = \begin{bmatrix} f(\lambda) & f^{(1)}(\lambda) & \frac{f^{(2)}(\lambda)}{2!} & \frac{f^{(3)}(\lambda)}{3!} & \cdots \\ 0 & f(\lambda) & f^{(1)}(\lambda) & \frac{f^{(2)}(\lambda)}{2!} & \ddots \\ \vdots & 0 & f(\lambda) & \ddots & \ddots \\ \vdots & \ddots & \ddots & \ddots & f^{(1)}(\lambda) \\ 0 & \cdots & \cdots & 0 & f(\lambda) \end{bmatrix} \tag{5}$$

Finally, from Eqs. (4) and (5) we compute $f(J)$ and $f(A) = Qf(J)Q^{-1}$.

I. $$f(J) = e^{Jt}, \text{ where } J = \begin{bmatrix} \lambda & 1 & 0 \\ 0 & \lambda & 1 \\ 0 & 0 & \lambda \end{bmatrix}; \text{ then } e^{Jt} = \begin{bmatrix} e^{t\lambda} & te^{t\lambda} & t^2e^{t\lambda}/2! \\ 0 & e^{t\lambda} & te^{t\lambda} \\ 0 & 0 & e^{t\lambda} \end{bmatrix}$$

II. $$\text{If } J = \left[\begin{array}{ccc:c} \lambda & 1 & 0 & 0 \\ 0 & \lambda & 1 & 0 \\ 0 & 0 & \lambda & 0 \\ \hdashline 0 & 0 & 0 & \lambda \end{array}\right] \text{ then } e^{Jt} = \begin{bmatrix} e^{t\lambda} & te^{t\lambda} & t^2e^{t\lambda}/2! & 0 \\ 0 & e^{t\lambda} & te^{t\lambda} & 0 \\ 0 & 0 & e^{t\lambda} & 0 \\ 0 & 0 & 0 & e^{t\lambda} \end{bmatrix} \text{ with } j_{11}(\lambda) = J$$

(above) and $j_{21}(\lambda) = \lambda$. If J were related to some A by $J = Q^{-1}AQ$, then $e^{At} = Qe^{Jt}Q^{-1}$, where Q is the matrix whose columns are the eigenvectors of A.

III. The solution of $\dot{\boldsymbol{x}} = \begin{bmatrix} 3 & 2 \\ 2 & 3 \end{bmatrix}\boldsymbol{x}$, with $\boldsymbol{x}(0) = \begin{bmatrix} 1 \\ 1 \end{bmatrix}$, has the form $\boldsymbol{x}(t) = \boldsymbol{x}(0)e^{At}$. The characteristic equation is $|A - \lambda I| = (\lambda - 5)(\lambda - 1) = 0$ and the eigenvalues and eigenvectors are $\lambda_1 = 5$, $\lambda_2 = 1, \boldsymbol{x}^1 = \begin{bmatrix} 1 & 1 \end{bmatrix}^T$ and $\boldsymbol{x}^2 = \begin{bmatrix} 1 & -1 \end{bmatrix}^T$. Then $J = \Lambda$, $A = Q\Lambda Q^{-1}$, and

$$\Lambda = \begin{bmatrix} 5 & 0 \\ 0 & 1 \end{bmatrix} T = \begin{bmatrix} 1 & 1 \\ 1 & -1 \end{bmatrix}. \text{ Therefore, } \boldsymbol{x}(t) = \begin{bmatrix} 1 & 1 \\ 1 & -1 \end{bmatrix}\begin{bmatrix} e^{5t} & 0 \\ 0 & e^{t} \end{bmatrix}\begin{bmatrix} 1/2 & -1/2 \\ -1/2 & -1/2 \end{bmatrix}\begin{bmatrix} 1 \\ 1 \end{bmatrix}.$$

IV. The general solution of $\dot{\boldsymbol{x}} = \begin{bmatrix} \lambda & 1 & 0 \\ 0 & \lambda & 0 \\ 0 & 0 & \lambda \end{bmatrix}\boldsymbol{x}$ is found as follows:

$$A = J = \left[\begin{array}{cc:c} \lambda & 1 & 0 \\ 0 & \lambda & 0 \\ \hdashline 0 & 0 & \lambda \end{array}\right] = \begin{bmatrix} A_{11} & 0 \\ 0 & A_{21} \end{bmatrix} \text{ and}$$

$$e^{At} = \begin{bmatrix} e^{A_{11}t} & 0 \\ 0 & e^{A_{21}t} \end{bmatrix}, \quad e^{A_{11}t} = \begin{bmatrix} e^{\lambda t} & te^{\lambda t} \\ 0 & e^{\lambda t} \end{bmatrix}, \quad e^{A_{21}t} = [e^{\lambda t}]$$

$$\text{Therefore, } \boldsymbol{x}(t) = \boldsymbol{x}(0) = \begin{bmatrix} e^{\lambda t} & te^{\lambda t} & 0 \\ 0 & e^{\lambda t} & 0 \\ 0 & 0 & e^{\lambda t} \end{bmatrix}$$

The Cayley–Hamilton Theorem: An Alternative Method for Computing f(A)

An $n \times n$ matrix A satisfies its characteristic polynomial, i.e.

$$\phi(\lambda) = |A - \lambda I| = 0 \text{ and } \phi(A) = [0]$$

Corollary 1: A^k, $\forall$ integers k, can be expressed in the form

$$A^k = \sum_{j=0}^{n-1} \alpha_j A^j \tag{6}$$

V. $\phi(\lambda) = \lambda^2 - 6\lambda + 5 = 0$ and $\phi(A) = A^2 - 6A + 5I = 0$. Therefore, $A^2 = 6A - 5I$ and $A^3 = (6A - 5I)A = 6A^2 - 5A = 6(6A - 5I) - 5A = 31A - 30I$. Also, A^{-1} can be determined as follows: $A^3(A^{-1}) = 31AA^{-1} - 30IA^{-1}$ or $A^2 = 31I - 30A^{-1}$, and $A^{-1} = \frac{31I - A^2}{30} = \frac{31I - 6A + 5I}{30} = \frac{6I - A}{5}$.

Corollary 2:

$$f(A) = \sum_{j=0}^{n-1} \gamma_j A^j \tag{7}$$

where the γ_j are determined from $f(J) = \sum_{j=0}^{n-1} \gamma_j J^j$, with $J = T^{-1}AT$.

VI.

$$A = \begin{bmatrix} 3 & 2 \\ 2 & 3 \end{bmatrix}, \ \lambda_1 = 1, \ \lambda_2 = 5; \ J = \Lambda = \begin{bmatrix} 1 & 0 \\ 0 & 5 \end{bmatrix}, \ e^A = \sum_{j=0}^{1} \gamma_j A^j$$

$$e^J = \begin{bmatrix} e^1 & 0 \\ 0 & e^5 \end{bmatrix} = \sum_{j=0}^{1} \gamma_j J^j = \gamma_0 \begin{bmatrix} 1 & 0 \\ 0 & 1 \end{bmatrix} + \gamma_1 \begin{bmatrix} 1 & 0 \\ 0 & 5 \end{bmatrix}$$

$$\left.\begin{aligned} e^1 &= \gamma_0 + \gamma_1 \\ e^5 &= \gamma_0 + 5\gamma_1 \end{aligned}\right\} \Rightarrow \begin{cases} \gamma_1 = \dfrac{e^5 - e^1}{4} \\ \gamma_0 = \dfrac{5e^{-1} - e^5}{4} \end{cases}$$

$$e^A = \frac{5e^1 - e^5}{4} \begin{bmatrix} 1 & 0 \\ 0 & 1 \end{bmatrix} + \frac{e^5 - e^1}{4} \begin{bmatrix} 3 & 2 \\ 2 & 3 \end{bmatrix}$$

$$e^{Ax} = \frac{5e^x - e^{5x}}{4} \begin{bmatrix} 1 & 0 \\ 0 & 1 \end{bmatrix} + \frac{e^{5x} - e^x}{4} \begin{bmatrix} 3 & 2 \\ 2 & 3 \end{bmatrix}$$

A Combination of the Previous Two Methods for Distinct Eigenvalues

Let A be an $n \times n$ matrix whose (distinct) eigenvalues are arranged in order of increasing absolute value. Then, if $f(\lambda)$ is analytic in a circle about the origin with radius $> |\lambda_n|$,

$$f(A) = \sum_{k=0}^{n-1} \alpha_k A^k \tag{8}$$

where the α_k are determined from

$$f(\lambda_i) = \sum_{k=0}^{n-1} \alpha_k \lambda_i^k, \text{ for } i = 1, 2, \ldots n \quad (9)$$

For Nondistinct (Repeated) Eigenvalues

For each repeated λ_i, i.e. $\lambda_i = \lambda_{i+1} = \ldots = \lambda_{i+m_i-1}$ (λ_i is repeated m_i times), the α_k may be determined from

$$f^{(l)}(\lambda_i) = \frac{d}{d\lambda_i^l}\left(\sum_{k=0}^{n-1} \alpha_k \lambda_i^k\right), \text{ for } l = 0, 1, \ldots, m_i - 1$$

Thus, if λ_i is repeated p times,

$$f(\lambda_i) = \alpha_0 + \alpha_1\lambda_i + \alpha_2\lambda_1^2 + \alpha_3\lambda_1^3 + \cdots$$

$$f^{(1)}(\lambda_i) = \alpha_1 + 2\alpha_2\lambda_i + 3\alpha_3\lambda_1^2 + \cdots$$

$$f^{(2)}(\lambda_i) = 2\alpha_2 + 6\alpha_3\lambda_1 + \cdots$$

$$\vdots$$

$$f^{(p-1)}(\lambda_i) = (p-1)!\alpha_{p-1} + p!\alpha_p\lambda_1 + \cdots$$

yield the appropriate number of equations to determine all of the α_k when these equations are considered with those representing the remaining eigenvalues.

VII. Let $A = \begin{bmatrix} 1 & 1 \\ 0 & 1 \end{bmatrix}$ and $\lambda_2 = \lambda_2 = 1$. Then $e^{At} = \sum_{k=0}^{1} \alpha_k A^k = \alpha_0 \mathrm{I} + \alpha_1 A$. Using $\left\{\begin{array}{l} e^t = \alpha_0 + \alpha_1 \\ te^t = \alpha_1 \end{array}\right\}$ and $\alpha_0 = e^t - te^t$, we determine that $e^{At} = (e^t - te^t)\begin{bmatrix} 1 & 0 \\ 0 & 1 \end{bmatrix} + te^t\begin{bmatrix} 1 & 1 \\ 0 & 1 \end{bmatrix}$

VIII. Consider the difference equations $\begin{array}{l} p_{n+1} = 3p_n + 4q_n \\ q_{n+1} = p_n + 3q_n \end{array}$ with $IC = \begin{bmatrix} p_0 & q_0 \end{bmatrix}^T$. Let $\boldsymbol{x} = \begin{bmatrix} p \\ q \end{bmatrix}$.

Then $\boldsymbol{x}_{n+1} = A\boldsymbol{x}_n$ with $A = \begin{bmatrix} 3 & 4 \\ 1 & 3 \end{bmatrix}$ $(\lambda_1 = 1, \lambda_2 = 5)$. The general solution is $\boldsymbol{x}_n = A^n \boldsymbol{x}_0$.

For $f(A) = A^n$, we have $A^n = \alpha_0 I + \alpha_1 A$ and $\lambda^n = \alpha_0 + \alpha_1\lambda$. This gives $\left\{\begin{array}{l} 1^n = \alpha_0 + \alpha_1 \\ 5^n = \alpha_0 + 5\alpha_1 \end{array}\right\}$;

therefore, $\alpha_0 = \dfrac{5 - 5^n}{4}$ and $\alpha_1 = \dfrac{5 - 5^n}{4}$. $A^n = \dfrac{1}{4}\begin{bmatrix} 2 + 2(5^n) & 4(5^n - 1) \\ 5^n - 1 & 2 + 2(5^n) \end{bmatrix}$ and

$$\begin{array}{l} p_n = 1/2(5^n + 1)p_0 + (5^n - 1)q_0 \\ q_n = 1/4(5^n - 1)p_0 + 1/2(5^n + 1)q_0 \end{array}.$$

Exercise: Show that the solution of $p_{n+1} = 5p_n - 6p_{n-1}$ is $p_{n+1} = (3^{n+1} - 2^{n+1})p_1 - 6(3^n - 2^n)p_0$.

Matrix Differential Equations

The problem here is to solve a *matrix* ODE: $\dot{\Phi} = A\Phi$, where $\Phi(t_0, t_0) = I$ and A and Φ are $n \times n$ matrices.

Solution: Write the partitioned matrix $\Phi = \begin{bmatrix} \Phi_1 & \vdots & \Phi_2 & \vdots & \cdots & \vdots & \Phi_n \end{bmatrix}$, where each $\Phi_i = \begin{bmatrix} \phi_{1i} & \phi_{2i} & \cdots & \phi_{ni} \end{bmatrix}^T$. Also $\Phi_i(t_0, t_0) = [0 \cdots 1 \cdots 0]^T$ where the 1 appears as the *i*th element, e.g. $\Phi_3(t_0, t_0) = [0010 \cdots 0]^T$. Then solve the n vector ODEs ($\Phi_i(t_0, t_0)$ for $i = 1, 2, \ldots, n$) using a package computer simulation program for n first-order equations. Try both stiff and non-stiff solvers if you are not sure about their dynamical properties, as discussed in Chapter 3.

IX. $A \equiv \begin{bmatrix} a_{11} & a_{12} \\ a_{21} & a_{22} \end{bmatrix}$, $\Phi \equiv \begin{bmatrix} \phi_{11} & \phi_{12} \\ \phi_{21} & \phi_{22} \end{bmatrix}$ and $\Phi(t_0, t_0) = \begin{bmatrix} 1 & 0 \\ 0 & 1 \end{bmatrix} = I$. Let $\Phi = \begin{bmatrix} \Phi_1 & \Phi_2 \end{bmatrix} = \begin{bmatrix} \phi_{11} & \phi_{12} \\ \phi_{21} & \phi_{22} \end{bmatrix}$. Then

$$\dot{\Phi} = [\dot{\Phi}_1 \ \vdots \ \dot{\Phi}_2] = A\Phi = \mathrm{A}[\Phi_1 \ \vdots \ \Phi_2] = \begin{bmatrix} a_{11} & a_{12} \\ a_{21} & a_{22} \end{bmatrix} \begin{bmatrix} \phi_{11} & \phi_{12} \\ \phi_{21} & \phi_{22} \end{bmatrix}$$

$$= \begin{bmatrix} a_{11}\phi_{11} + a_{12}\phi_{21} & a_{11}\phi_{12} + a_{12}\phi_{22} \\ a_{21}\phi_{11} + a_{22}\phi_{21} & a_{21}\phi_{12} + a_{22}\phi_{22} \end{bmatrix}$$

$$= \left[\begin{bmatrix} a_{11} & a_{12} \\ a_{21} & a_{22} \end{bmatrix} \begin{bmatrix} \phi_{11} \\ \phi_{21} \end{bmatrix} \vdots \begin{bmatrix} a_{11} & a_{12} \\ a_{21} & a_{22} \end{bmatrix} \begin{bmatrix} \phi_{12} \\ \phi_{22} \end{bmatrix} \right] = \begin{bmatrix} A\Phi_1 & \vdots & A\Phi_2 \end{bmatrix}$$

Therefore $\dot{\Phi}_1 = A\Phi_1$, $\dot{\Phi}_2 = A\Phi_2$, and the ICs for these two vector differential equations are $\Phi_1(t_0, t_0) = \begin{bmatrix} 1 & 0 \end{bmatrix}^T$ and $\Phi_2(t_0, t_0) = \begin{bmatrix} 0 & 1 \end{bmatrix}^T$.

Singular Value Decomposition (SVD) & Principal Component Analysis (PCA)

SVD is a special factorization of a matrix A, decomposing it into a product of three special matrices with many useful applications in applied math, statistics, signal processing, bioinformatics and biomodeling. We use it primarily for sensitivity analysis and model reduction in this book. It's also useful for computing the inverse of a square matrix, the pseudoinverse of a nonsquare matrix, least squares (*LS*) fitting of data, matrix approximation and determining the rank, range and null space of a matrix. Principal component analysis (PCA) is a related method for accomplishing some of the same tasks. A precise relationship exists between SVD and PCA analyses when PCA is computed using the covariance matrix of the data (Wall et al. 2003), as summarized below.

A common application of SVD in biology is in **gene expression analysis**, i.e. for *visualization* of multivariate gene expression data, for expression data *representation* using a smaller number of variables (approximation), and for *detection* of patterns in gene expression data, which are typically noisy. SVD can detect underlying signals from such noisy data. We define and illustrate SVD below in the context of analyzing data from a gene expression experiment – initial computational steps in an approach with broad application in modern medical diagnostics and systems biology. We exemplify it and relate it to parameter sensitivity analysis and model reduction in Chapters 11 and 17.

Singular Value Decomposition (SVD)

Let A be an m-row by n-column matrix of multivariate real-valued data with $m \geq n$. A has rank r and thus $r \leq n$. For gene expression microarray data, our example application, element, a_{ij} is the expression level (signal magnitude) of the ith gene in the jth assay. Matrix A is illustrated in Eq. (10), with different genes represented on each row and different assays represented on each column. Elements of the ith *row* of A form an n-dimensional vector $\boldsymbol{g}_i$ called the **transcriptional response** of the ith gene. Elements of the jth *column* of A form an m-dimensional vector $\boldsymbol{a}_j$ called the **expression profile** of the jth assay. For example, the columns might represent assays sequentially timed expression data points for m genes, e.g at $t_1, t_2, \ldots, t_N$.

$$A = \begin{bmatrix} a_{g1A1} & a_{g1A2} & \cdots & a_{g1An} \\ a_{g2A1} & & & a_{g2An} \\ \vdots & & & \vdots \\ \vdots & & & \vdots \\ a_{gmA1} & \cdots & \cdots & a_{gmAn} \end{bmatrix} \begin{matrix} \text{gene } 1 \\ \text{gene } 2 \\ \vdots \\ \vdots \\ \text{gene } m \end{matrix} \equiv [\boldsymbol{a}_1 \quad \boldsymbol{a}_2 \quad \cdots \quad \boldsymbol{a}_n] \equiv \begin{bmatrix} \boldsymbol{g}_1 \\ \boldsymbol{g}_2 \\ \vdots \\ \boldsymbol{g}_m \end{bmatrix} \tag{10}$$

$$\text{Assay } 1 \quad \cdots \quad \cdots \quad \text{Assay } n$$

The SVD of A is defined as:

$$A = USV^T \tag{11}$$

U is an $m \times n$ **unitary matrix**, S is an $m \times n$ **diagonal matrix** with nonnegative real numbers s_i (**singular-values of** A) on the diagonal, and V is $n \times n$ and also a **unitary matrix**. A has at least one and at most n distinct singular values, with $s_i > 0$ for $1 \leq k \leq r$ (rank is r) and $s_i = 0$ for $(r + 1) \leq k \leq n$.

The m columns of U and n columns of V are called the **left-singular vectors** $\{\boldsymbol{u}_k\}$ (Gantmacher 1960) and **right-singular vectors** $\{\boldsymbol{v}_k\}$ of A, respectively. The $\{\boldsymbol{u}_k\}$ form an **orthonormal basis** of vectors for the *assay expression profiles*, i.e. $<\boldsymbol{u}_i, \boldsymbol{u}_j> = 1$ for $i = j$, and $<\boldsymbol{u}_i, \boldsymbol{u}_j> = 0$ for $i \neq j$ (U is unitary). Similarly, the $\{\boldsymbol{v}_k\}$ form an **orthonormal basis** for the ***gene transcriptional responses***.

Some Properties of the SVD

1. SVD and eigenvalue decomposition are closely related. Namely:
   - The left-singular vectors $\{\boldsymbol{u}_k\}$ of A are eigenvectors of AA^T.
   - The right-singular vectors $\{\boldsymbol{v}_k\}$ of A are eigenvectors of A^TA.

- The non-zero singular values of A are the square roots of the non-zero eigenvalues of both A^TA and AA^T.
- If A is square, SVD is equivalent to diagonalization of A, i.e. a solution of the eigenvalue-eigenvector problem.

2. The **pseudoinverse** A^+ of A with SVD $= USV^T$ is: $A^+ = VS^+U^T$. Matrix S^+ is the pseudoinverse of S, formed by replacing every nonzero diagonal entry by its reciprocal and transposing the result. The pseudoinverse is one way to solve linear least squares problems, as we have seen in an earlier section.
3. The following is an important result for data reduction applications, continued below. Let $A^{(l)}$ be the **closest rank-*l* (truncated-to-*l*) matrix** to A defined as the matrix $A^{(l)}$ that minimizes the sum of square (*LS*) residuals between elements of A and $A^{(l)}$ **(Frobenius norm** of A- $A^{(l)}$), i.e. $\sum_{ij}(a_{ij}-a_{ij}^{(l)})^2$. Then, with singular values and corresponding $\boldsymbol{u}_k$ and $\boldsymbol{v}_k$ vectors ordered with decreasing values, $A^{(l)}$ decomposes as:

$$A^{(l)} = \sum_{k=1}^{l} \boldsymbol{u}_k s_k \boldsymbol{v}_k^T \tag{12}$$

Principal Component Analysis (PCA)

PCA is another orthogonal linear transformation that transforms a set of observations (data) of possibly correlated (random) variables into a set of values of uncorrelated variables called principal components. Assume the data are arranged in a matrix A, as in (10). To accomplish PCA, the data must be centered – which means subtracting the component means from the data (**means centering**). The **centered data matrix** is denoted A_c. PCA can be accomplished by eigenvalue decomposition of the data covariance matrix, or by SVD of the means centered data matrix, as described further below.

The **PCA transformation that preserves dimensionality** is:

$$Y_c = U_c^T A_c = S_c V_c^T \tag{13}$$

Matrices U_c, S_c and V_c are the same as in the SVD section above, but mean centered. Since U_c and V_c are orthonormal, each row of Y_c^T is simply a rotation of the corresponding row of A_c^T. The first column of Y_c^T contains the **scores** of the cases w.r.t. the **first principal component**. The second column contains the scores with respect to the **second principal component**, and so on. In this new coordinate system, the largest variance by any projection of the data lies on the first coordinate – the *first* principal component, i.e. it accounts for as much of the variability of the data as possible. The second largest variance then lies on the second coordinate – the *second* principal component, and so on. PCA results are usually presented in terms of **component scores** – the transformed variable values corresponding to a particular data point – and **loadings** – the weight by which each standardized original variable is multiplied to get the component score. As with all random variables, these uncorrelated principal components are linearly independent only if the data set is jointly normally distributed. The matrix U_c of singular vectors of A_c is the same as the matrix U_c of eigenvectors of the matrix of observed covariances: $A_cA_c^T = U_cS_cS_c^TU_c^T = U_cS_c^2U_c^T$. This is how PCA can be computed as an eigenvalue–eigenvector problem.

PCA from SVD

As noted above, the principle components also can be calculated by SVD from the covariance matrix of the centered data matrix A_c, by means centering each column $\boldsymbol{g}_i$ or each row $\boldsymbol{a}_j$ of A. (i.e. subtracting the means from the data). After centering, the matrix $A_c^TA_c = \sum_i \boldsymbol{g}_{ci}\boldsymbol{g}_{ci}^T$ is proportional to the covariance matrix of the variables of $\boldsymbol{g}_{ci}$, i.e. the covariance matrix of the *assays* in our example application. Similarly, after centering each row $\boldsymbol{a}_j$, $A_cA_c^T = \sum_i \boldsymbol{a}_{cj}\boldsymbol{a}_{cj}^T$ is proportional to the covariance matrix of the variables of $\boldsymbol{a}_j$, i.e. the covariance matrix of the *genes*.

From (11), $A_c^TA_c = V_cS_c^2V_c^T$. This diagonalization then gives the principle components of $\{\boldsymbol{g}_{ci}\}$, i.e. the right singular vectors $\{\boldsymbol{v}_{ci}\}$ *are the same as the principle components* of $\{\boldsymbol{g}_{ci}\}$. The eigenvalues of $A_c^TA_c$ are s_{ck}^2, which are proportional to the **variances of the principal components**. The matrix U_cS_c contains the **principal component scores**, i.e. the coordinates of the genes in the space of principal components.

Similarly, the left singular vectors $\{\boldsymbol{v}_{ci}\}$ are the same as the principal components of $\boldsymbol{a}_j$. The s_{ck}^2 are again proportional to the **variances of the principal components**. Matrix $S_cV_c^T$ again contains the principal component scores, i.e. the coordinates of the assays in the space of principal components.

Data Reduction & Geometric Interpretation

In many applications, PCA also can supply a lower dimensional ($l<n$) picture (**low-rank approximation, shadow**) of the data, by using only the first few principal components – the "most informative" viewpoint – thereby reducing the dimensionality of the transformed data. This was done for SVD in Eq. (12) above. To obtain a reduced-dimensionality representation for PCA, we project A_c downward into the reduced space defined by the first l singular vectors, denoted U_l. The PCA transformation that reduces dimensionality is:

$$Y_c = U_l^TA_c = S_lV_c^T \tag{14}$$

Given a set of multivariate data distributed in Euclidean space, the first principal component corresponds to a line passing through the multidimensional mean, the line that minimizes the least squares distance of the points from the line. The second principle component is computed after all correlation with the first principal component has been subtracted from the points. The singular values are the square roots of the eigenvalues of the covariance matrix $A_cA_c^T$. Each eigenvalue is proportional to the fraction of the least squares distance, or variance, along each eigenvector. The sum of the eigenvalues equals the sum of the squared distances of the data from their multidimensional mean. In effect, then, ***PCA rotates the data about their mean, so as to align them with their principle components, and this moves as much of the variance as possible into the first several dimensions***. Values in the remaining dimensions are relatively much smaller and typically are eliminated with minimal loss of information.

REFERENCES

Chen, C., 1985. Introduction to Linear System Theory. Holt, Rinehart, & Winston, New York.

Friedberg, S., Insel, A., Spence, L.E., 2003. Linear Algebra. Pearson Education, Upper Saddle River, NJ.

Gantmacher, F., 1960. The Theory of Matrices. Chelsea Publishing, New York.

Trefethen, L.N., Bau III, D., 1997. Numerical Linear Algebra. Society for Industrial and Applied Mathematics (SIAM), Philadelphia, PA.

Wall, M.E., Rechtsteiner, A., Rocha, L.M., 2003. Singular Value Decomposition and Principal Component Analysis. Kluwer, Norwell, MA.

INPUT–OUTPUT & STATE VARIABLE BIOSYSTEM MODELING: GOING DEEPER

Appendix C

Inputs & Outputs

Inputs are designated $u_1, \ldots, u_r$ where each $u_i \equiv u_i(t)$; and outputs are designated $y_1, \ldots, y_m$, where each $y_i \equiv y_i(t)$. The system is designated S, the model M (**Fig. C.1**). The column vectors $\boldsymbol{u} \equiv [u_1 u_2 \ldots u_r]^T$ and $\boldsymbol{y} \equiv [y_1 y_2 \ldots y_m]^T$ are the vectors of multiple inputs and outputs, where each is a member of some well-defined *vector space* in a given problem, i.e. $\boldsymbol{u} \in \mathcal{U}^r$, $\boldsymbol{y} \in \mathcal{Y}^m$, called the **admissible *u*** and ***y***, and superscript T is the vector (or matrix) transpose operation. The u_i usually are piecewise-continuous-time functions, and y_i continuous-time functions, each defined on some finite time interval $t_0 < t < \infty$.

Dynamic Systems, Models & Causality

Two classifications of systems and their models are *instantaneous* and *dynamic*. **Instantaneous** (**zero-memory** or **static**) systems and models have outputs $\boldsymbol{y}(t)$ at time t that depend only on inputs $\boldsymbol{u}(t)$ at the *same* time t. Real **dynamic** or **dynamical** systems and models are those for which $\boldsymbol{y}(t)$ depends on *past* as well as present inputs, i.e., $\boldsymbol{u}(\tau)$ for $-\infty < \tau \leq t$. This means physical systems in nature cannot have outputs that anticipate inputs. This property is called **causal**. In this sense, dynamic systems and their models must have a **memory**. Where they can go depends on where they've been.

Input–Output (Black-Box) Models

A system (or model) is said to be **relaxed** if it is "at rest," the same as having no energy stored in it. If M is initially *relaxed* at t_0, then $\boldsymbol{y}(t)$ for $t \geq t_0$ depends only on $\boldsymbol{u}(t)$ for $t \geq t_0$, and we can write

$$\boldsymbol{y} = H\boldsymbol{u} \tag{C.1}$$

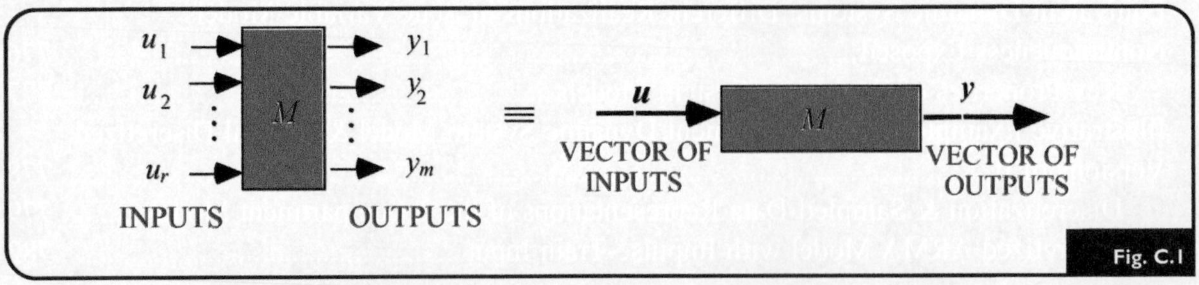

Fig. C.1

where H is some linear or nonlinear operator. In general, M is relaxed at t_0 if and only if $\boldsymbol{u}(t) = \mathbf{0}$ implies $\boldsymbol{y}(t) = \mathbf{0}$ for all t, i.e. if no input is applied, no output is produced.

Eq. (C.1) is the general form of an **input–output model**, also called a **black box model**, linear or nonlinear. A relaxed dynamic system model M is **causal** if its past affects its future, but not conversely, i.e.

$$\boldsymbol{y}(t) = H\boldsymbol{u}(\tau) \text{ only for } -\infty < \tau \leq t \tag{C.2}$$

The expression is somewhat more complicated if M is not relaxed, but the notion of causality is the same. A relaxed system model M is **linear** if and only if the system operator H has the following (linear operator) property:

$$H(\alpha_1\boldsymbol{u}_1 + \alpha_2\boldsymbol{u}_2) = \alpha_1 H\boldsymbol{u}_1 + \alpha_2 H\boldsymbol{u}_2 \tag{C.3}$$

where α_1 and α_2 are arbitrary constants and $\boldsymbol{u}_1$ and $\boldsymbol{u}_2$ are any admissible (vector or scalar) inputs. Otherwise M is **nonlinear**. This is equivalent to the *superposition* principle presented in Chapter 2. If M is not relaxed, (C.3) is not generally true, even if the model is linear.

Time-Invariance (TI)

The **shift operator** Q_α is defined by

$$Q_\alpha \boldsymbol{u}(t) \equiv \boldsymbol{u}(t-\alpha) \tag{C.4}$$

In these terms, a relaxed system (model) is **time-invariant (or fixed, or stationary)** if and only if

$$HQ_\alpha\boldsymbol{u} = Q_\alpha H\boldsymbol{u} = Q_\alpha\boldsymbol{y} \tag{C.5}$$

i.e. the *waveform* of the output does not depend on *when* the input is applied, but it will be shifted in time, by the magnitude of the shift α. The example in **Fig. C.2** illustrates a time-shifted input and the identically time shifted output of a time-invariant model.

Continuous Linear System Input–Output Models

The response of a relaxed, single-input–single-output (SISO) continuous linear dynamic system model to an impulse (Dirac delta function) $\delta(t-\tau)$ input applied at $t=\tau$ is called the **impulse response** $\equiv h(t,\tau)$. In these terms, we can write y as a **superposition, or convolution integral**, in terms of h, with S not necessarily relaxed:

$$y(t) = \int_{-\infty}^{\infty} h(t,\tau)u(\tau)d\tau \tag{C.6}$$

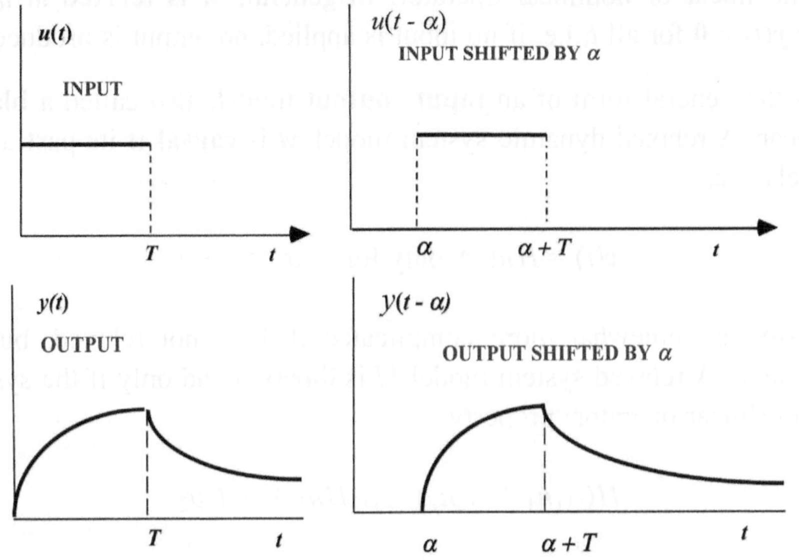

Fig. C.2

Input/output example for a time-invariant *H*.

For *r* inputs and *m* outputs (MIMO models), inputs and outputs are vector-valued:

$$\boldsymbol{y}(t) = \int_{-\infty}^{\infty} H(t,\tau)\boldsymbol{u}(\tau)d\tau \tag{C.7}$$

where *H* is the **impulse response matrix** of the model *M*, defined as: $H(t,\tau) \equiv [h_{ij}(t,\tau)]$, with h_{ij} the scalar elements of the matrix *H*. For a causal, continuous, linear dynamic system model, we always have:

$$\boldsymbol{y}(t) = \int_{-\infty}^{t} H(t,\tau)\boldsymbol{u}(\tau)d\tau \tag{C.8}$$

i.e. the upper limit is finite and equal to *t*, since $H(t,\tau) = [0]$ (the zero matrix) for all $t < \tau$. If *M* is continuous, causal and also *relaxed* at t_0, then

$$\boldsymbol{y}(t) = \int_{t_0}^{t} H(t,\tau)\boldsymbol{u}(\tau)d\tau \tag{C.9}$$

For a continuous linear *time-invariant* model, the impulse response matrix satisfies:

$$H(t,\tau) = H(t-\tau,0) \equiv H(t-\tau) \tag{C.10}$$

and the output $\boldsymbol{y}(t)$ is

$$\boldsymbol{y}(t) = \int_{t_0}^{t} H(t-\tau)\boldsymbol{u}(\tau)d\tau \equiv H(t) * \boldsymbol{u}(t) \tag{C.11}$$

Eq. (C.11) is a **convolution integral** equation, often written using an asterisk (∗), as shown, to represent the convolution operation.

Transfer Function (TF) Matrix for Linear TI Input–Output Models

We begin with the LT of the output vector.

$$\boldsymbol{Y}(s) \equiv \int_0^{\infty} \boldsymbol{y}(t)e^{-st}dt$$

Substituting Eqs. (C.8) and (C.10) we have

$$\boldsymbol{Y}(s) = \int_0^{\infty} \left(\int_0^{t} H(t-\tau)\boldsymbol{u}(\tau)d t \right) e^{-st}dt = \int_0^{\infty} \left(\int_0^{\infty} H(t-\tau)\boldsymbol{u}(\tau)dt \right) e^{-st}dt$$

Then, rearranging the order of integration and changing "dummy" variables of integration:

$$\boldsymbol{Y}(s) = \int_0^{\infty} \left(\int_0^{\infty} H(t-\tau)e^{-s(t-\tau)}dt \right) \boldsymbol{u}(\tau)e^{-st}dt = \int_0^{\infty} H(v)e^{-sv}dv \int_0^{\infty} \boldsymbol{u}(\tau)e^{-st}d\tau \equiv H(s)\boldsymbol{U}(s) \tag{C.12}$$

$Y(s)$ is the product of two transforms, the second being the LT of the input. The first transform, $H(s)$, is the **TF matrix**, which we note is the LT of the impulse response matrix:

$$H(s) = \int_0^{\infty} H(t)e^{-st}dt \tag{C.13}$$

The right-hand side (RHS) of (C.12) is a simple algebraic product in the s-domain, whereas, the RHS of (C.11) is a more complex integral operation in the time-domain.

A **single-input–single-output** model is denoted **SISO**, and a **multi-input–multi-output** model is denoted **MIMO**, also called **multivariable**. In Eq. (C.7), we have a time-domain representation for an input–output description of a multivariable (MIMO) linear model (LM) in the form of the impulse response matrix $H(t,\tau)$; and if the LM is also time-invariant (TI), Eq. (C.13) provides a complex frequency response description in the form of the TF matrix $H(s)$. In either case, nothing need be known about the internal structure of a system this model may represent. These are true **input–output**, or **black-box models**, or **models of data** (Chapter 1).

We emphasize that an input–output description is applicable *only* when S is initially relaxed. If it is not, $y(t) \neq Hu(t)$, $t_0 < \infty$, and a simple input–output (only) description like (C.1) is not applicable.

Structured State Variable Models

State of a System *S* or Model *M*

Given the equations describing the dynamics of a system S (or model M), the **state** of S at t_0 is the quantitative information at t_0 that, together with $\boldsymbol{u}(t)$ for $t \geq t_0$, uniquely determines the behavior of S for all $t \geq t_0$. We show below that:

1. the state of S is not unique; and
2. only dynamic systems (dynamic system models) have a state.

State Variable Models from Input–Output (I–O) Models

Many linear input–output models can be represented as time-domain state variable models. To illustrate, consider the TF of a causal time-invariant linear model:

$$H(s) = \frac{4}{(s+1)(s+3)} = \frac{2}{s+1} - \frac{2}{s+3} \tag{C.14}$$

The response to an impulse at $\tau = 0$ is the inverse LT of $H(s)$: $h(t) = 2e^{-t} - 2e^{-3t}$ (Appendix A). Let's find relationships for the state of this model at $t = t_0$. If M were relaxed at t_0, there could be no output prior to t_0, and the general response to any input $u(t)$, $t \geq t_0$, would be:

$$y(t) = \int_{t_0}^{t} h(t-\tau)u(\tau)d\tau, \quad t_0 \leq t < \infty \tag{C.15}$$

In this case, the state would be zero at $t = t_0$. But if the model were not relaxed at t_0 (the general case), then the output prior to t_0 must also obey the general response relation and we can write:

$$y(t) = \int_{-\infty}^{t} h(t-\tau)u(\tau)d\tau = \int_{-\infty}^{t_0} hu + \int_{-\infty}^{t} hu, \quad -\infty < y < \infty \tag{C.16}$$

The first integral on the RHS embodies the state at t_0, so let's evaluate it:

$$\int_{-\infty}^{t_0} h(t-\tau)u(\tau)d\tau = 2e^{-t}\int_{-\infty}^{t_0} e^{\tau}u(\tau)d\tau - 2e^{-3t}\int_{-\infty}^{t_0} e^{3\tau}u(\tau)d\tau \equiv 2e^{-t}c_1 - 2e^{-3t}c_2 \tag{C.17}$$

where c_1 and c_2 are defined as the definite integrals, which we note are independent of t. Knowledge of c_1 and c_2 would provide $y(t)$ for all t, given any $u(t)$, and *thus* c_1 *and* c_2 *qualify as the state of M*. To obtain c_1 and c_2, we evaluate (C.16) at t_0:

$$y(t_0) \equiv y_0 \equiv \int_{-\infty}^{t_0} hu + \int_{t_0}^{t_0} hu = \int_{\infty}^{t_0} hu = 2e^{-t_0}c_1 - 2e^{-3t_0}c_2 \tag{C.18}$$

This gives one equation in two unknowns, c_1 and c_2. To obtain a second, we differentiate (C.16):

$$\dot{y}(t) = \frac{d}{dt}\int_{-\infty}^{t_0} hu + \frac{d}{dt}\int_{t_0}^{t} hu = \int_{-\infty}^{t_0} \frac{\partial}{\partial t}\left[2e^{-(t-\tau)} - 2e^{-3(t-\tau)}\right]u(\tau)d\tau + \frac{d}{dt}\int_{t_0}^{t} hu$$

$$= \int_{-\infty}^{t_0} (-2e^{\tau}e^{-t} + 6e^{-3\tau}e^{-3t})u(\tau)d\tau + \frac{d}{dt}\int_{t_0}^{t} hu$$

$$= -2e^{-t}\int_{-\infty}^{t_0} e^{\tau}u(\tau)d\tau + 6e^{-3t}\int_{t_0}^{t} e^{3\tau}u(\tau)d\tau + \frac{d}{dt}\int_{t_0}^{t} hu$$

But the two definite integrals are equal to c_1 and c_2, respectively, by (C.17) above. Thus,

$$\dot{y}(t) = -2e^{-t}c_1 + 6e^{-3t}c_2 + \frac{d}{dt}\int_{t0}^{t} hu$$

Then, setting $t = t_0$ gives the required additional (derived) initial condition:

$$\dot{y}(t_0) = -2e^{-t_0}c_1 + 6e^{-3t_0}c_2 + \frac{d}{dt}\int_{t_0}^{t_0} hu = -2c_1e^{-t_0} + 6c_2e^{-3t_0} \tag{C.19}$$

Finally, solving for c_1 and c_2,

$$c_1 = 0.5\, e^{t_0}\left[\frac{y_0}{2} + \dot{y}_0\right], \quad c_2 = 0.5\, e^{3t_0}\left[y_0 + \dot{y}_0\right] \tag{C.20}$$

Thus, if M is not relaxed at t_0, i.e. the general case,

$$y(t) = \left[\frac{y_0}{2} + \dot{y}_0\right]e^{-(t-t_0)} - \left[y_0 + \dot{y}_0\right]e^{-3(t-t_0)} + \int_{t_0}^{t} h(t-\tau)u(\tau)d\tau, \quad t \geq t_0 \tag{C.21}$$

and y_0 *and* $\dot{y}_0$ *qualify as the state of M at* t_0. But, we recall, so do c_1 and c_2. Therefore, among other things, we have illustrated the *nonuniqueness of the state* of M. This means that ***an I–O model can have many, actually infinitely many, state variable models***. With so many possibilities, no wonder finding unique state-variable models that fit input–output data is so hard!

Dynamic State Variable ODE Models for Continuous Systems

The set of equations that describe the unique relations between the input, output and state of S or M are called the **dynamic equations** of S or M.

A fairly general model form for lumped (nondistributed) continuous system models consists of the following two equations, plus additional constraining relationships introduced below:

Process (Plant) Model:

$$\dot{\boldsymbol{x}} = \boldsymbol{f}(\boldsymbol{x}, \boldsymbol{u}, \boldsymbol{w}, t; \boldsymbol{p}) \quad t \geq t_0 \tag{C.22}$$

Output, Observation or Measurement Model:

$$\boldsymbol{z} = \boldsymbol{g}(\boldsymbol{x}, \boldsymbol{v}, t; \boldsymbol{p}) \tag{C.23}$$

In Eqs. (C.22) and (C.23), $\boldsymbol{x} \equiv \boldsymbol{x}(t)$ is the **state vector** at time t, and the n components of $\boldsymbol{x}(t) \equiv [x_1(t)\ x_2(t)\ \ldots\ x_n(t)]^T$, i.e. the $x_i(t)$, are the **state variables** at time t. Also, $\boldsymbol{z}(t)$ are the **measurements** or **output data**, and $\boldsymbol{w}(t)$ and $\boldsymbol{v}(t)$ are **process** and **measurement disturbance** (or error) vectors of appropriate dimension, $\boldsymbol{p}$ is a vector of P unknown and *constant* **parameters** of the process and measurement models, and $\boldsymbol{f}$ and $\boldsymbol{g}$ are nonlinear vector functions, all defined more explicitly below.

If $\boldsymbol{f}$ and $\boldsymbol{g}$ are *linear* vector functions, then we can write:

$$\begin{aligned} \dot{\boldsymbol{x}} &= A(\boldsymbol{p}, t)\boldsymbol{x} + B(\boldsymbol{p}, t)\boldsymbol{u} + E(\boldsymbol{p}, t)\boldsymbol{w} \\ \boldsymbol{z} &= C(\boldsymbol{p}, t)\boldsymbol{x} + F(\boldsymbol{p}, t)\boldsymbol{v} \end{aligned} \tag{C.24}$$

Eqs. (C.24) represent a **stochastic linear dynamic system model**, which has the block diagram shown in **Fig. C.3**.

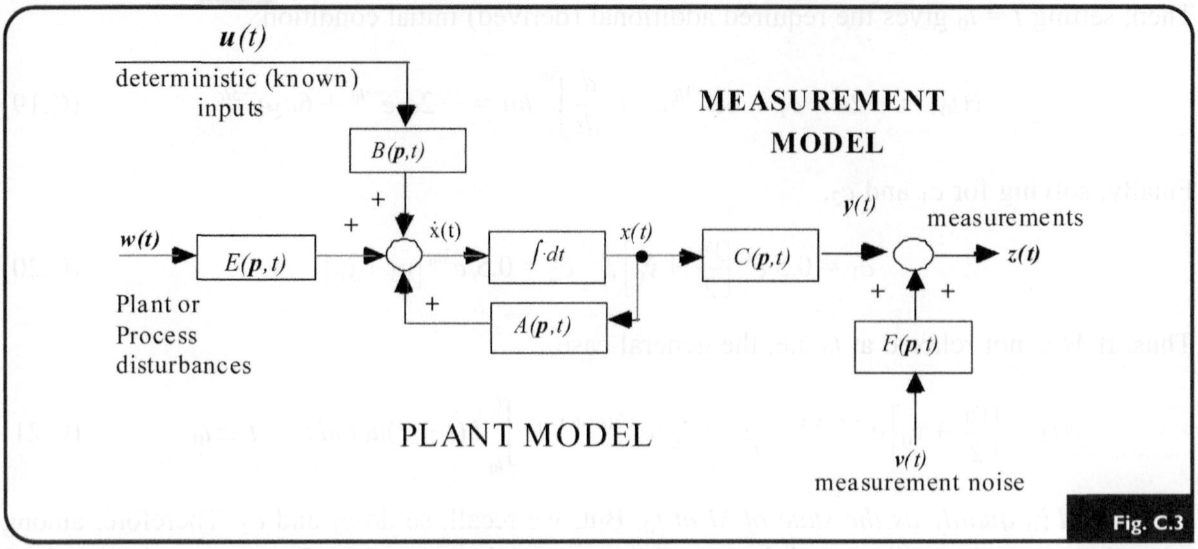

Fig. C.3

General linear dynamic system model block-diagram.

In this formulation: $\boldsymbol{u}(t)$ represents only **deterministic inputs**; $\boldsymbol{w}(t)$ may represent environmental or internally generated **disturbance noise**, or may account for **modeling errors** or **uncertainties**; and the **process (plant)** is **deterministic** if $\boldsymbol{w} \equiv \mathbf{0}$, even if measurement noise is present. $\boldsymbol{y}(t)$ (first term in measurement equation) is the **noise-free output**.

Complete Dynamic System Models: Constrained Structures

The dynamic system models given by Eqs. (C.22) through (C.24) or **Fig. C.3** are classical, but incomplete. To obtain a complete system model description, we must include any additional **constraint equations** *or* **inequality relations** known to exist among the states, inputs and/or parameters $\boldsymbol{p}$, as well as the **boundary conditions** on $\boldsymbol{x}$, usually (but not always) in the form of *initial conditions* (IC), $\boldsymbol{x}(t_0) = \boldsymbol{x}_0$. Algebraic constraint relations can generally be expressed as a nonlinear vector function of $\boldsymbol{x}, \boldsymbol{u}, \boldsymbol{p}$, and possibly t explicitly, i.e. $\boldsymbol{h}(\boldsymbol{x}, \boldsymbol{u}, \boldsymbol{p}, t) \geq \mathbf{0}$. In these terms, we have the following complete deterministic dynamic system description (model M), valid for all $t \geq t_0$:

$$\begin{aligned} \dot{\boldsymbol{x}} &= \boldsymbol{f}[\boldsymbol{x}(t), \boldsymbol{u}(t), t; \boldsymbol{p}], \quad \boldsymbol{x}(t_0) = \boldsymbol{x}_0 \\ \boldsymbol{y}(t) &= \boldsymbol{g}[\boldsymbol{x}(t), t; \boldsymbol{p}] \\ \boldsymbol{h}[\boldsymbol{x}(t), \boldsymbol{u}(t), \boldsymbol{p}, t] &\geq 0 \end{aligned} \tag{C.25}$$

For this model form:

a) $\boldsymbol{f}$ defines the **couplings (interconnections, connectivity)** between the states and inputs, i.e. $\boldsymbol{f}$ defines the **process (plant) structure** (or **topology** or **morphology**).

b) $\boldsymbol{g}$ defines the **couplings** between the outputs (measurements) and states, and any direct couplings between the inputs and outputs.

c) $\boldsymbol{p}$ are the (constant) **parameters** of the model M.

d) $\boldsymbol{h} \geq \mathbf{0}$ defines the algebraic nondynamic **constraints** on or among the parameters, states or inputs.

e) The way $\boldsymbol{f}$, $\boldsymbol{g}$ and $\boldsymbol{h}$ are parameterized by $\boldsymbol{p}$ delineates what is *unknown* about the model.

f) $\boldsymbol{x}_0$ separately represent the *initial conditions* on the state variables, which more generally can be specified as **boundary** conditions at any $t \geq t_0$.

g) In the context of an experimental situation, Eqs. (C.25) represent a ***composite model of a dynamic process and experiment combined***, with $[t_0, t_f]$ denoting the **observation interval** for the experiment. Together, they represent a *class* of systems or models with a specific internal structure and an input–output or measurements relationship that generally reflects, topologically, the (exogenous) stimulus–response conditions of a given experiment design. Alternatively, the same class may represent endogenous (internal) input–output connectivity, or both exogenous and endogenous inputs and outputs, depending on the model application.

(C.25) is called the constrained structure, or **complete model** and, when there is no ambiguity, the **structure**, or **model**, or just M. It is important to keep in mind, however, that *model structure* here does not necessarily imply any isomorphism or homeomorphism with real system structure. A structured model may indeed have quite a different "structure" than the real system. This subtlety and its consequences in interpreting models and modeling results are predominant themes throughout this book. Further explanation begins with our discussion of *equivalence* classes later.

Linear TI State Variable Models

If $\boldsymbol{f}$ and $\boldsymbol{g}$ are linear functions and the system is also time-invariant, then Eqs. (C.25) can written as:

$$\dot{\boldsymbol{x}} = A(\boldsymbol{p})\boldsymbol{x} + B(\boldsymbol{p})\boldsymbol{u}, \quad \boldsymbol{y} = C(\boldsymbol{p})\boldsymbol{x} \tag{C.26}$$

where the unknown parameter vector $\boldsymbol{p}$ is a vector of *unknown* and *constant* parameters of dimension $P \leq (n^2 + nr + mn)$, e.g. the unknown among $a_{11}, a_{12}, \ldots, a_{nn}; b_{11}, \ldots, b_{nr}; c_{11}, \ldots, c_{mn}$. Constraints in (C.25) and (C.26) then form the **linear constrained structure** or **complete linear model** for the time-invariant case.

We purposefully restrict $\boldsymbol{p}$ to include only unknown system parameters. Any known or null (zero) elements of A, B, C, are *not* included in $\boldsymbol{p}$. Any number of elements of $\boldsymbol{p}$, however, may be identical or more generally related by constraint relations, which are generally nonlinear and do not affect the linearity of M in (C.26). If the model has unknown time-varying parameters, these may be expressed as time functions of known form, with unknown *constant* coefficients, e.g. polynomials in t. In this case, the dimension of $\boldsymbol{p}$ may be $P > n^2 + nr + mn$. In some special cases, time-varying parameters represent distinctive system properties, e.g. biological rhythms, and the resulting time-varying model may be treated directly, as in the last section of this appendix. Eqs. (C.26) have a unique solution, which may be determined if $\boldsymbol{p}$, $\boldsymbol{u}$ and the state at any time $t \in [t_0, \infty]$ are known.

Input–Output TF Matrix for a State Variable Model

For Eq. (C.26), the TF matrix is determined by taking LTs (Appendix A):

$$s\boldsymbol{X}(s) - \boldsymbol{x}_0 = A\boldsymbol{X}(s) + B\boldsymbol{U}(s), \quad \boldsymbol{Y}(s) = C\boldsymbol{X}(s) \tag{C.27}$$

where $\boldsymbol{x}_0 \equiv \boldsymbol{x}(t_0)$ and the argument $\boldsymbol{p}$ of each matrix is suppressed, for notational convenience. Therefore,

$$X(s) = (sI - A)^{-1}\boldsymbol{x}_0 + (sI - A)^{-1}BU(s), \quad Y(s) = C(sI - A)^{-1}\boldsymbol{x}_0 + C(sI - A)^{-1}BU(s) \tag{C.28}$$

With $\boldsymbol{x}_0 \equiv 0$ and the argument $\boldsymbol{p}$ reinserted, we have the **TF matrix:**

$$H(s) \equiv [H_{ij}(s)] = C(\boldsymbol{p})[sI - A(\boldsymbol{p})]^{-1}B(\boldsymbol{p}) \tag{C.29}$$

This matrix is generally constrained by relations $\boldsymbol{h} \geq \boldsymbol{0}$ in (C.25). We suppress this important fact in some developments below, for notational convenience, and resurrect constraints as needed.

Discrete-Time Dynamic System Models

If the inputs and outputs of a system S are defined only at discrete instants of time, or if they are piecewise-constant, i.e. if they change only at certain times, say t_k, S may be described by a discrete-time system model, with system variables represented as **sequences** at times t_0, t_1, t_2, ... etc. A continuous-time system model can be modeled as a discrete-time system if its responses are of interest or measurable only at certain instants of time. This is often the case for physiological systems and experiments. This discretization process was illustrated in Chapter 2. The vector input and vector output sequences are denoted:

$$\{\boldsymbol{u}(t_k)\} = \{\boldsymbol{u}_k\} \equiv \{\boldsymbol{u}_k\colon -\infty < k < \infty\}, \quad \{\boldsymbol{y}(t_k)\} = \{\boldsymbol{y}_k\} \equiv \{\boldsymbol{y}_k\colon -\infty < k < \infty\} \tag{C.30}$$

where k is any integer. These are completely general. In practice, k is usually nonnegative, or begins at $k = 0$. Also, the sample times t_k may not be equally spaced, often the case for biological experiments.

Discrete-Time Input–Output Models

If the inputs and outputs of a causal discrete-time system model relaxed at time t_0 satisfy the linearity property, then we can write

$$\boldsymbol{y}_k = \sum_{l=0}^{k} H(k,l)\boldsymbol{u}_l \tag{C.31}$$

where $H(k,l) \equiv [H_{ij}(k,l)]$ is called the **weighting-sequence matrix**, and $H_{ij}(k,l)$ is the response of M at the ith output due to application of the following (Kronecker delta) forcing function at the jth input:

$$u_j(t_k) = \begin{cases} 1, & k = l \\ 0, & k \neq l \end{cases} \tag{C.32}$$

Note that the weighting sequence matrix $H(k,l)$ is analogous to the impulse response matrix $H(t,\tau)$ for continuous systems. Also, $H(k,l) \equiv 0$ for $k < l$, and $H(k,l) = H(k-l,0) \equiv H(k-l)$ for time-invariant discrete-time models.

The Sampled or *z*-Transfer Function

The *z-transform* of a sequence is needed here. The ***z*-transform** of a sequence $\{\boldsymbol{u}_k\} \equiv \{\boldsymbol{u}_0, \boldsymbol{u}_1, \ldots\}$ is the function $\boldsymbol{U}(z)$ of the complex variable z defined as:

$$\boldsymbol{U}(z) \equiv \sum_{k=0}^{\infty} \boldsymbol{u}_k z^{-k} = \boldsymbol{u}_0 + \boldsymbol{u}_1 z^{-1} + \boldsymbol{u}_2 z^{-2} + \cdots \quad \text{(C.33)}$$

a) If $u_k = 1$, for all $k = 0, 1, 2, \ldots$ **(Fig. C.4)**:

$$U(z) = \sum_{k=0}^{\infty} z^{-k} = \frac{1}{1 - z^{-1}} = \frac{z}{z-1} \quad \text{(C.34)}$$

b) For $u_k = e^{-2k}$ **(Fig. C.5)**:

$$U(z) = \sum_{k=0}^{\infty} e^{-2k} z^{-k} = \frac{1}{1 - e^{-2} z^{-1}} = \frac{z}{z - e^{-2}} \quad \text{(C.35)}$$

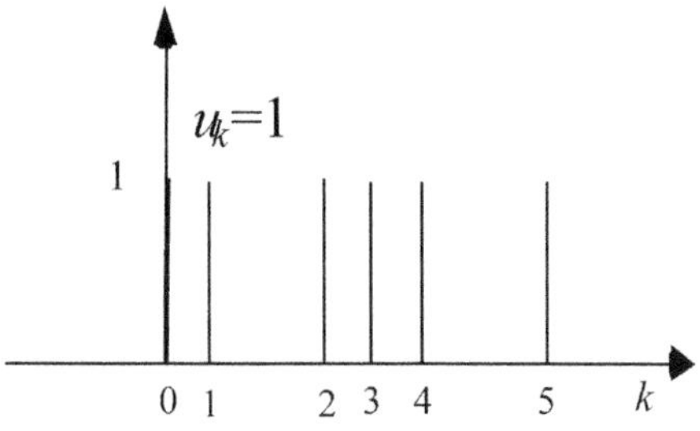

Fig. C.4

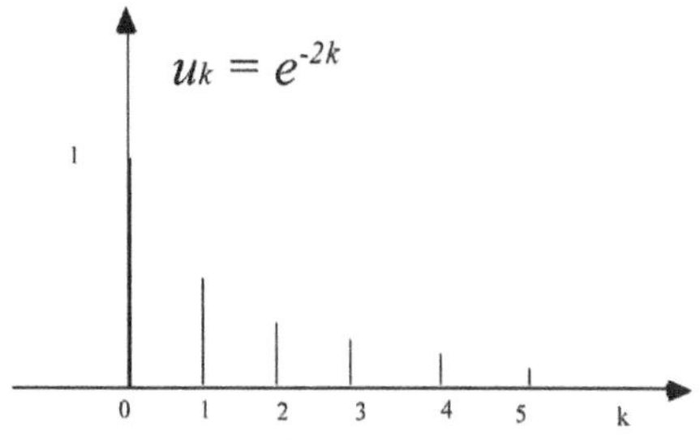

Fig. C.5

The **sampled TF (or *z*-transfer function)**: is the *z*-transform of the weighting sequence $H(0)$, $H(1)$, $H(2)$, ... To justify this definition, we compute $Y(z)$ from Eqs. (C.31) and (C.33):

$$Y(z) \equiv \sum_{k=0}^{\infty} y(k)z^{-k} = \sum_{k=0}^{\infty}\sum_{l=0}^{k} H(k-l)U_l z^{-k} = \sum_{k=0}^{\infty}\sum_{l=0}^{\infty} H(k-l)U_l z^{-(k-l)}z^{-l}$$
$$= \sum_{l=0}^{\infty}\sum_{k=0}^{\infty} H(k-l)z^{-(k-l)}U_l z^{-l} = \sum_{k=0}^{\infty} H(k)z^{-k}\sum_{l=0}^{\infty} U_l z^{-l} \equiv H(z)U(z) \tag{C.36}$$

In Eq. (C.36), we changed the order of summations and used the fact that $H(k-l) = 0$ for $k < l$.

Discrete-Time State Variable Models

The major difference from continuous ODE models here is that we have a set of first-order **difference equations** (DEs) for the process model, with **instantaneous** measurement equations:

$$\boldsymbol{x}_{k+1} = A_k\boldsymbol{x}_k + B_k\boldsymbol{u}_k, \quad \boldsymbol{y}_k = C_k\boldsymbol{x}_k \tag{C.37}$$

Sampled Input–Output Transfer Function Matrix

For the time-invariant case, A, B, and C in (C.37) and are constant matrices. We take the *z*-transform of Eqs. (C.37) and for this case:

$$\boldsymbol{x}_{k+1}(z) = \sum_{k=0}^{\infty} \boldsymbol{x}_{k+1}z^{-k} = z\sum_{k=0}^{\infty} \boldsymbol{x}_{k+1}z^{-(k+1)} = z\left[\sum_{k=-1}^{\infty} \boldsymbol{x}_{k+1}z^{-(k+1)} - \boldsymbol{x}_0\right]$$
$$= z\left[\sum_{k=0}^{\infty} x_k z^{-k} - x_0\right] = z[X_k(z) - \boldsymbol{x}_0] \tag{C.38}$$

where $\boldsymbol{X}_k(z)$ is the ***z*-transform** of $\boldsymbol{x}_k$. Note the similarity of Eq. (C.38) with the LT of dx/dt. To find the *z*-transform of (C.37), we apply (C.38):

$$z\boldsymbol{X}_k(z) - z\boldsymbol{x}_0 = A\boldsymbol{X}_k(z) + B\boldsymbol{U}(z)$$

$$\boldsymbol{Y}(z) = C\boldsymbol{X}_k(z)$$

which yields:

$$\boldsymbol{X}_k(z) = [zI - A]^{-1}z\boldsymbol{x}_0 + [zI - A]^{-1}B\boldsymbol{U}(z), \quad \boldsymbol{Y}(z) = C[zI - A]^{-1}z\boldsymbol{x}_0 + C[zI - A]^{-1}B\boldsymbol{U}(z) \tag{C.39}$$

Setting $\boldsymbol{x}_0 \equiv 0$, the **sampled TF matrix** is:

$$H(z) = C[zI - A]^{-1}B \tag{C.40}$$

Note the similarity of $H(z)$ with $H(s)$ for continuous system models.

Composite Input–Output and State Variable Models

Interconnected models are sometimes called composite models. Equations for two composite model types, the **parallel** and **series** (tandem) connections are developed here. The numbers of inputs and outputs must be compatible, and there must be no "loading" effect in the connection, i.e. impulse responses of each M_i remain unchanged after the connection is made (DiStefano III et al. 1990), otherwise the TFs of individual components are different when they are connected and the methodology of composite models does not work.

Composite Input–Output Models

To illustrate, consider the two-input, two-output linear model described by the superposition integral:

$$y_i(t) = \int_{-\infty}^{t} H_i(t,\tau)\boldsymbol{u}_i(\tau)d\tau, \quad i = 1,2 \tag{C.41}$$

For the **parallel connection (Fig. C.6)**:

$$H(t,\tau) = H_1(t,\tau) + H_2(t,\tau) \tag{C.42}$$

For time-invariant models, TFs also simply add when connected in parallel: $H(s) = H_1(s) + H_2(s)$.

For the **series connection (Fig. C.7)**, the TFs convolve:

$$H(t,\tau) = \int_{\tau}^{t} H_2(t,v)H_1(v,\tau)dv = H_2(t) * H_1(\tau) \tag{C.43}$$

We show (C.43) for the scalar case. First, $h_1(t,\tau) = \int_0^t h_1(t,s)\delta(s-\tau)ds$. This is the input to M_2 in **Fig. C.8**. Then, for the composite model, $h(t,\tau) \equiv y(t) = \int_0^t h_2(t,v)u_2(v)dv =$

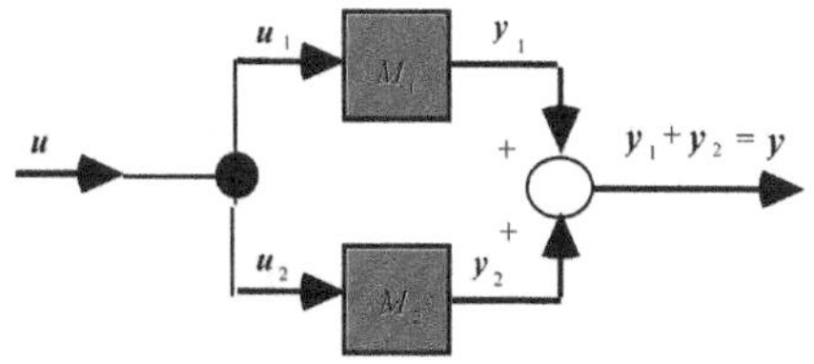

Fig. C.6

Parallel.

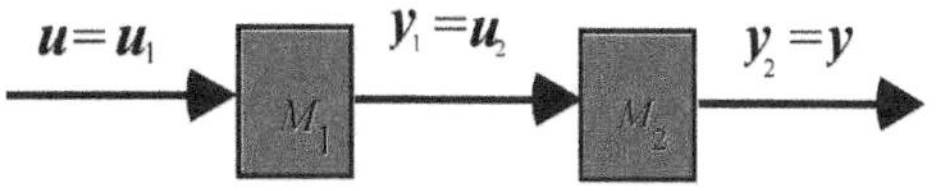

Fig. C.7

Series.

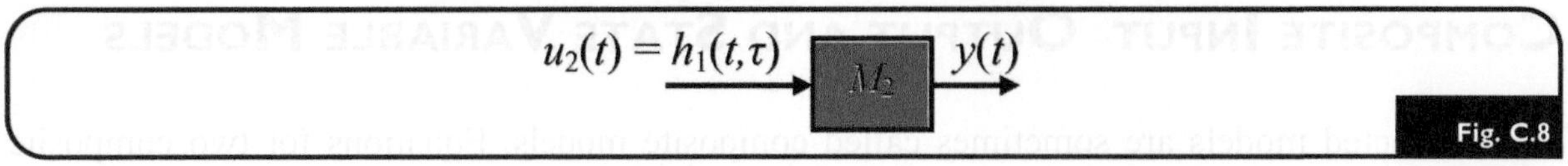

Fig. C.8

$\int_{\tau}^{t} h_2(t,v)h_1(v,\tau)dv$ since $h_1(v,\tau) \equiv 0$, for $v < \tau$. For time-invariant models we of course have a simple product of TFs for the series connection: $H(s) = H_1(s)H_2(s)$.

Composite State Variable Models

We combine two state variable models first, given by:

$$\dot{\boldsymbol{x}}_1 = A_1(t)\boldsymbol{x}_1 + B_1(t)\boldsymbol{u}_1 \text{ and } \dot{\boldsymbol{x}}_2 = A_2(t)\boldsymbol{x}_2 + B_2(t)\boldsymbol{u}_2 \tag{C.44}$$

$$\boldsymbol{y}_1 = C_1(t)\boldsymbol{x}_1 \text{ and } \boldsymbol{y}_2 = C_2(t)\boldsymbol{x}_2 \tag{C.45}$$

It is helpful here to consider the linear vector space $\sum$ as the *direct sum* of two linear spaces $\sum_1$ and $\sum_2$, written as $\sum = \sum_1 \oplus \sum_2$ where every vector in $\sum$ has an *augmented* form:

$$\begin{bmatrix} \boldsymbol{x}_1 \\ \boldsymbol{x}_2 \end{bmatrix}$$

Then, for the **Parallel Connection**: $\boldsymbol{u}_1 = \boldsymbol{u}_2 = \boldsymbol{u}$, $\boldsymbol{y} = \boldsymbol{y}_1 + \boldsymbol{y}_2$, and, we have

$$\begin{bmatrix} \dot{\boldsymbol{x}}_1 \\ \dot{\boldsymbol{x}}_2 \end{bmatrix} = \begin{bmatrix} A_1(t) & 0 \\ 0 & A_2(t) \end{bmatrix} \begin{bmatrix} \boldsymbol{x}_1 \\ \boldsymbol{x}_2 \end{bmatrix} + \begin{bmatrix} B_1(t) \\ B_2(t) \end{bmatrix} \boldsymbol{u}, \quad \boldsymbol{y} = \begin{bmatrix} C_1(t) & C_2(t) \end{bmatrix} \begin{bmatrix} \boldsymbol{x}_1 \\ \boldsymbol{x}_2 \end{bmatrix} \tag{C.46}$$

For the **Series Connection**:

$$\dot{\boldsymbol{x}}_1 = A_1\boldsymbol{x}_1 + B_1\boldsymbol{u}_1 \text{ and } \dot{\boldsymbol{x}}_2 = A_2\boldsymbol{x}_2 + B_2\boldsymbol{u}_2$$

$$\boldsymbol{y}_1 = C_1\boldsymbol{x}_1 \text{ and } \boldsymbol{y}_2 = C_2\boldsymbol{x}_2$$

Therefore,

$$\dot{\boldsymbol{x}}_1 = A_1\boldsymbol{x}_1 + B_1\boldsymbol{u}_1 = \boldsymbol{x}_1 + B_1\boldsymbol{u} \text{ and } \dot{\boldsymbol{x}}_2 = A_2\boldsymbol{x}_2 + B_2\boldsymbol{u}_2 = A_2\boldsymbol{x}_2 + B_2C_1\boldsymbol{x}_1$$

$$\boldsymbol{y}_1 = C_1\boldsymbol{x}_1 = \boldsymbol{u}_2 \text{ and } \boldsymbol{y}_2 = C_2\boldsymbol{x}_2$$

and the composite model becomes:

$$\begin{bmatrix} \dot{x}_1 \\ \dot{x}_2 \end{bmatrix} = \begin{bmatrix} A_1(t) & 0 \\ B_2(t)C_1(t) & A_2(t) \end{bmatrix} \begin{bmatrix} x_1 \\ x_2 \end{bmatrix} + \begin{bmatrix} B_1(t) \\ 0 \end{bmatrix} \boldsymbol{u}, \quad \begin{bmatrix} y_1 \\ y_2 \end{bmatrix} = \begin{bmatrix} C_1(t) & 0 \\ 0 & C_2(t) \end{bmatrix} \begin{bmatrix} x_1 \\ x_2 \end{bmatrix} \tag{C.47}$$

The combination of discrete-time submodels into composite models is performed in the same manner as above and the results for parallel and series connection of discrete-time submodels differ only in form from the continuous case.

State Transition Matrix for Linear Dynamic Systems

Input–Output Model Solutions

If the impulse response matrix $H(t,\tau)$ is known and the continuous linear system model is causal and relaxed at t_0, then for any input vector $\boldsymbol{u}$ the solution for the output vector at any $t \geq t_0$ is:

$$\boldsymbol{y}(t) = \int_{t_0}^{t} H(t,\tau)\boldsymbol{u}(\tau)d\tau \tag{C.48}$$

If the system is also time-invariant and $H(s) \equiv \mathcal{L}[H(t)]$, the output transform is: $\boldsymbol{Y}(s) \equiv \mathcal{L}[\boldsymbol{y}(t)] = H(s)\boldsymbol{U}(s)$. Corresponding results for discrete-time models are: $\boldsymbol{y}_k = \sum_{j=0}^{k-1} H(k,j+1)\boldsymbol{u}_j$ and $\boldsymbol{Y}(z) = H(z)\boldsymbol{U}(z)$. We reemphasize that these solutions are **models of data** (Chapter 1).

State Variable Model Solutions

Continuous Case

Consider

$$\dot{\boldsymbol{x}} = A(t)\boldsymbol{x} + B(t)\boldsymbol{u} \tag{C.49}$$

$$\boldsymbol{y} = C(t)\boldsymbol{x} \tag{C.50}$$

where A, B, C are assumed known and continuous matrix functions of t (possibly constant). The unique solution to the ODE (C.49), for a given $\boldsymbol{u}$ and a given initial state $\boldsymbol{x}_0$ at time t_0 is:

$$\boldsymbol{x}(t) = \Phi(t,t_0)\boldsymbol{x}_0 + \int_{t_0}^{t} \Phi(t,\tau)B(\tau)\boldsymbol{u}(\tau)d\tau \tag{C.51}$$

where $\Phi(t,t_0)$ is called the **state transition matrix** defined below. The solution for the output vector $\boldsymbol{y}(t)$ then follows directly from (C.50).

If the initial state $\boldsymbol{x}_0 \equiv \boldsymbol{0}$, the integral term in (C.51) is called the **zero-state (forced) response** of (C.49):

$$\boldsymbol{x}_{forced}(t) \equiv \int_{t_0}^{t} \Phi(t,\tau)B(\tau)\boldsymbol{u}(\tau)d\tau \tag{C.52}$$

If the input $\boldsymbol{u} \equiv \boldsymbol{0}$, the remaining (first) term in (C.49) is called the **zero-input (free) response**:

$$\boldsymbol{x}_{free}(t) \equiv \Phi(t,t_0)\boldsymbol{x}_0 \tag{C.53}$$

The complete solution is the sum of the two: $\boldsymbol{x}(t) = \boldsymbol{x}_{forced}(t) + \boldsymbol{x}_{free}(t)$. Comparing (C.48) and (C.51), it appears that the state transition matrix $\Phi(t,\tau)$ and the impulse response matrix $H(t,\tau)$ are directly related. To see this, if $\boldsymbol{x}_0 \equiv \boldsymbol{0}$ and if (C.51) is premultiplied by $C(t)$, we get

$$y(t) \equiv \int_{t_0}^{t} C(t)\Phi(t,\tau)B(\tau)\boldsymbol{u}(\tau)d\tau \tag{C.54}$$

We can therefore express the **impulse response matrix** of M in terms of Φ as:

$$H(t,\tau) = \begin{cases} C(t)\Phi(t,\tau)B(t) & \text{for} \quad t \geq \tau \\ 0 & \text{for} \quad t < \tau \end{cases} \tag{C.55}$$

The **state transition matrix** $\Phi(t,\tau)$ is the unique solution to the homogeneous matrix differential equation:

$$\frac{\partial}{\partial t}\Phi(t,\tau) = \mathrm{A}(t)\Phi(t,\tau) \tag{C.56}$$

with initial condition $\Phi(\tau,\tau) = I \equiv$ the identity matrix. A method for solving this equation is given in Appendix B. For convenience, Eq. (C.56) is often written (imprecisely) in the form of a total rather than a partial derivative: $\dot{\Phi}(t,\tau) = A(t)\Phi(t,\tau)$ or more simply as: $\dot{\Phi} \equiv A(t)\Phi \equiv A\Phi$, where the overdot represents differentiation with respect to t (as usual).

The following **fundamental properties of** $\Phi(t,\tau)$ are important. For all t, τ, t_1 and t_2:

a) $\Phi(t,\tau)$ is a nonsingular square matrix

b) $\Phi(t,t) = I$ (C.57)

c) $\Phi^{-1}(t,\tau) = \Phi(\tau,t)$ (C.58)

d) $\Phi(t_2,t_1) = \Phi(t_2,\tau)\Phi(\tau,t_1)$ (C.59)

Property d) is called the **transition property** of Φ.

Φ may also be expressed in terms of the nonsingular **fundamental matrix** $\boldsymbol{F(t)}$, which is a solution of the homogeneous matrix differential equation:

$$\dot{F}(t) = A(t)F(t) \tag{C.60}$$

and the columns of $F(t)$ consists of any n linearly independent solutions of $\dot{\boldsymbol{x}} = A(t)\boldsymbol{x}$. $F(t)$ is nonsingular for all t and is the unique solution of (C.60) for any nonsingular $F(t_0)$. It can be shown that

$$\Phi(t,t_0) = F(t)F^{-1}(t_0) \tag{C.61}$$

In fact, Φ is often defined by this expression and (C.60), and Eqs. (C.52) through (C.59) follow.

Intuitive Explanation of Φ

Physically, or geometrically, the state transition matrix describes the **motion** of $\boldsymbol{x}(t)$ in the linear vector state space, for all t, when the input $\boldsymbol{u}(t) \equiv \mathbf{0}$; i.e. $\Phi(t,t_0)$ is a linear transformation that maps (transfers) $\boldsymbol{x}_0$ at t_0 into $\boldsymbol{x}$ at time t. Multiple transitions, say from $\boldsymbol{x}(t_1)$ to $\boldsymbol{x}(\tau)$ to $\boldsymbol{x}(t)$, are governed by (C.59).

Time-Invariant Case

Because $\frac{d}{dt}(e^{At}) = Ae^{At}$, then e^{At} is a fundamental matrix of $\dot{\boldsymbol{x}} = A\boldsymbol{x}$. Thus, from (C.61),

$$\Phi(t, t_0) = e^{At}[e^{At_0}]^{-1} = e^{A(t-t_0)} = \Phi(t - t_0) \tag{C.62}$$

and the solution of Eq. (C.60) in this case is

$$\boldsymbol{x}(t) = e^{A(t-t_0)}\boldsymbol{x}_0 + \int_{t_0}^{t} e^{A(t-\tau)}B\boldsymbol{u}(\tau)d\tau \tag{C.63}$$

(as in Chapter 2). This very fundamental relationship should be committed to memory.

The **impulse response matrix** is, from Eq. (C.55), $H(t\text{-}\tau) = Ce^{A(t-\tau)}B$, more commonly written:

$$H(t) = Ce^{At}B$$

By LT, the **TF matrix** is

$$H(s) = C[sI - A]^{-1}B \tag{C.64}$$

If $[sI - A]$ has a particularly simple or sparse form, e.g. if it is triangular, or has order less than 4, its inverse is not difficult to compute by hand (Appendix B).

In the time domain, $H(t) = Ce^{At}B$ might be difficult to compute. The computation is simplified if A is represented in its *Jordan form* J_A, with $A = QJAQ^{-1}$, where Q is the *modal matrix* of eigenvectors of A; in this case: $e^{At} = Qe^{J_A t}Q^{-1}$ (Appendix B). Since every element of $e^{J_A t}$ is of the form $t^k e^{\lambda_i t}$, every element of e^{At} is a linear combination of these factors. In many instances, J_A is diagonal, in which case $e^{J_A t}$ is also diagonal. In all cases, the major problem reduces to finding the eigenvalues λ_i and their corresponding eigenvectors for the matrix A. The term $t^k e^{\lambda_i t}$ is (of course) a **mode** of the dynamical ODE equations. Modes are important in modeling, as we have explored in Chapters 5 and 6, and beyond.

The Adjoint Dynamic System

Given

$$\dot{\boldsymbol{x}}(t) = A(t)\boldsymbol{x}(t) \tag{C.65}$$

the equation

$$\dot{\boldsymbol{z}} = -A^T(t)\boldsymbol{z}(t) \tag{C.66}$$

is called the **adjoint** of (C.66). If $\psi(t, t_0)$ is the state transition matrix of (C.66), we have the identities:

$$\Phi^T\Phi = \psi(t, t_0) = \psi^{-1}(t_0, t) \quad \text{(C.67)}$$

$$\Phi^T(t_0, t)\psi(t_0, t) = \psi(t, t_0)\Phi^T(t, t_0) = \psi^T(t, t_0)\Phi(t, t_0) = I \quad \text{(C.68)}$$

Eq. (C.68) means that $\Phi^T(t_0,t)$ is also the transition matrix of the adjoint equation (C.66). To see this, we write $\Phi(t,t_0)\Phi(t_0,t) = I$ and differentiate with respect to t:

$$\frac{d}{dt}[\Phi(t, t_0)\Phi(t, t_0)] = \dot{\Phi}(t, t_0)\Phi(t_0, t) + \Phi(t, t_0)\dot{\Phi}(t_0, t) = 0$$

Then $A(t)\Phi(t, t_0)\Phi(t_0, t) + \Phi(t, t_0)\dot{\Phi}(t_0, t) = 0$ and $\Phi(t, t_0)\dot{\Phi}(t_0, t) = -A(t)$, or $\dot{\Phi}(t_0, t) = -\Phi^{-1}(t, t_0)A(t) = -\Phi(t_0, t)A(t)$. Transposing (C.68) yields the **adjoint equation:**

$$\dot{\Phi}^T(t_0, t) = -A^T(t)\Phi^T(t_0, t) \quad \text{(C.69)}$$

The first equality in Eq. (C.67) then follows.

The adjoint system has a number of important applications. One is that, for any fixed $t > t_0$, Eq. (C.69) can be solved *backwards*, from t down to t_0 (equivalent to $-t$ up to t_0, with a substitution of independent variable $t \equiv -t'$ in the equations), in order to directly obtain $\Phi(t_0,t)$ as a function of the variable argument t_0. This is needed, for example, if Eq. (C.51) or (C.52) is solved directly, i.e. in obtaining the value of the integral for given B, $\boldsymbol{u}$, t_0 and t.

Equivalent Dynamic Systems: Different Realizations of State Variable Models – Nonuniqueness Exposed

This topic is fundamental because it helps reveal what a model does and does not represent. Different forms of models are often called different **realizations**, equivalent in the following sense – if they satisfy several conditions. We present only the time-invariant case.

For the state variable model:

$$\dot{\boldsymbol{x}} = A\boldsymbol{x} + B\boldsymbol{u}, \quad y = C\boldsymbol{x} \quad \text{(C.70)}$$

the following realization (model) is **equivalent** to (C.70), with P n by n and nonsingular,

$$\boldsymbol{x}^* \equiv P\boldsymbol{x}, \text{and:}$$

$$\dot{\boldsymbol{x}} = (PAP^{-1})\boldsymbol{x}^* + (PB)\boldsymbol{u} \equiv A^*\boldsymbol{x}^* + B^*\boldsymbol{u} \quad \text{(C.71)}$$

$$y = (CP^{-1})\boldsymbol{x}^* \equiv C^*\boldsymbol{x}^* \quad \text{(C.72)}$$

The matrix P is called an **equivalence transformation**. If we choose orthonormal vectors as the basis for the state space (Appendix B), then P transforms these basis vectors into the columns of P^{-1}, as $\mathbf{x} = P^{-1}\mathbf{x}^*$.

The matrices A and $A^* = PAP^{-1}$ are said to be **similar**; P^{-1} is called the **modal matrix** (of eigenvectors). When A^* is diagonal, in which case the state variables x_i^* are all uncoupled from each other, the solution $\mathbf{x}^*(t)$ is easily obtained from knowledge of the eigenvalues and eigenvectors of A.

As you might expect, ***equivalent models have the same outputs for the same inputs***, i.e. they are **input–output equivalent**, but not necessarily conversely. Equivalence properties go even deeper.

Key Properties of Equivalent System Models

Two equivalent linear dynamic system models are both **zero-state equivalent** (they have the same impulse response matrix) and **zero-input equivalent** (they have the same zero-input response). But the converse is not necessarily true (see below). Equivalent dynamic equations must have the same dimension, because P must be n by n; and an equivalent model always exists, because every linear, time-invariant dynamic equation has an equivalent Jordan form dynamic equation, with $A^* = PAP^{-1}$, A and A^* similar, and A^* may even be diagonal (Appendix B).

Example C.1

This is a counterexample to the *converse* statement above, adapted from an example in Chen (1970). Electrical circuits 1 and 2 in **Fig. C.9** are **zero-state equivalent** because, with no initial condition (voltage) on the capacitor C in 2, the outputs y across the resistors are identical, because the four resistors combine to form an equivalent resistance of R (ohms). Circuits 1 and 2 are also **zero-input equivalent**, because with any initial condition on C, $y = 0$ if $u = 0$. However, they are **not equivalent** systems because they don't have the same dimension. Circuit 1 has dimension 0, because it's not a dynamic system, and circuit 2 has dimension 1,

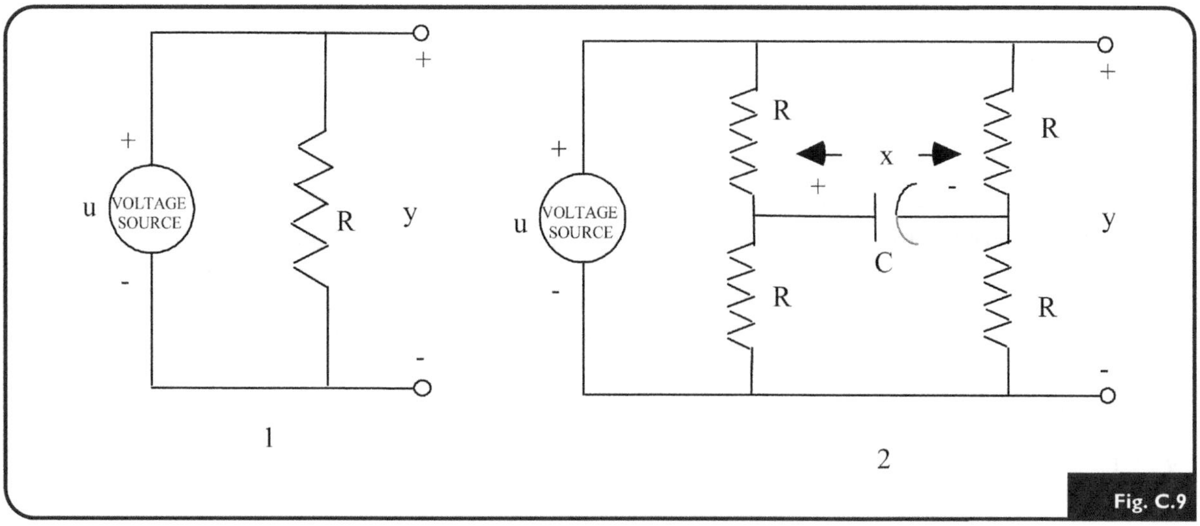

Fig. C.9

because it has one energy-storing element (C). This means, among other things, that the equations describing these two systems generally could not be used interchangeably because, for *arbitrary* inputs and/or initial conditions, not all system signals are identical. However with u the input and y the output, as shown, these nonequivalent models are input-output equivalent, because their TFs $H(s) = Y(s)/U(s) \equiv 1$! Furthermore, equality of the resistances is not needed here, for zero-state or zero-input equivalence. Even if the Rs were all different, $y(t) = u(t)$ for all t and for any initial voltage across C and any $u(t)$.

A lower order model can sometimes be a good approximation of a higher-order model, but the two are not equivalent in the above sense. For example, $\dot{x} + x = u$ may be good approximation for $10^{-6}\ddot{x} + \dot{x} + x = u$, because the second order term is relatively small.

Discrete-Time State Variable Models

The results here are completely analogous to those for the continuous model. The discrete-time linear state variable model has the form:

$$\boldsymbol{x}_{k+1} = A_k\boldsymbol{x}_k + B_k\boldsymbol{u}_k, \quad \boldsymbol{y}_k = C_k\boldsymbol{x}_k \tag{C.73}$$

and the solutions are:

$$\boldsymbol{x}_k = \Phi(k,0)\boldsymbol{x}_0 + \sum_{j=0}^{k-1}\Phi(k,j+1)B(j)\boldsymbol{u}_j, \quad \boldsymbol{y}_k = C_k\Phi(k,0)\boldsymbol{x}_0 + \sum_{j=0}^{k-1}C_k\Phi(k,j+1)B(j)\boldsymbol{u}_j \tag{C.74}$$

Also, the **discrete-time state transition matrix** $\Phi(k,j)$ satisfies the discrete homogeneous equation:

$$\Phi(k+1,j) = A(k)\Phi(k,j) \tag{C.75}$$

$\Phi(k,j)$ can be singular, but otherwise all its properties are the same as for $\Phi(t,\tau)$. For $A_k \equiv A =$ constant (matrix), $\Phi(k,0) = \Phi(k) = A^k$.

Illustrative Example: A 3-Compartment Dynamic System Model & Several Discretized Versions of It

In this and the next section, we illustrate several concepts introduced previously in Chapters 2, 4 and 10, using the three-compartment mammillary model shown in **Fig. C.10**. The q_i represent masses in each compartment and y is a concentration measurement in compartment 1.

The model equations are determined from the diagram by application of mass flux balance to each compartment, as in Chapter 2. For $k_{ij} \geq 0$, $i \neq j$, and $V_i > 0$, these are given by:

$$\dot{\boldsymbol{q}} = \begin{bmatrix} -(k_{01}+k_{21}+k_{31}) & k_{12} & k_{13} \\ k_{21} & -(k_{02}+k_{12}) & 0 \\ k_{31} & 0 & -(k_{03}+k_{13}) \end{bmatrix}\boldsymbol{q} + \begin{bmatrix} 1 \\ 0 \\ 0 \end{bmatrix}u_1, \quad y = [1/V_1\ 0\ 0]\,\boldsymbol{q} \tag{C.76}$$

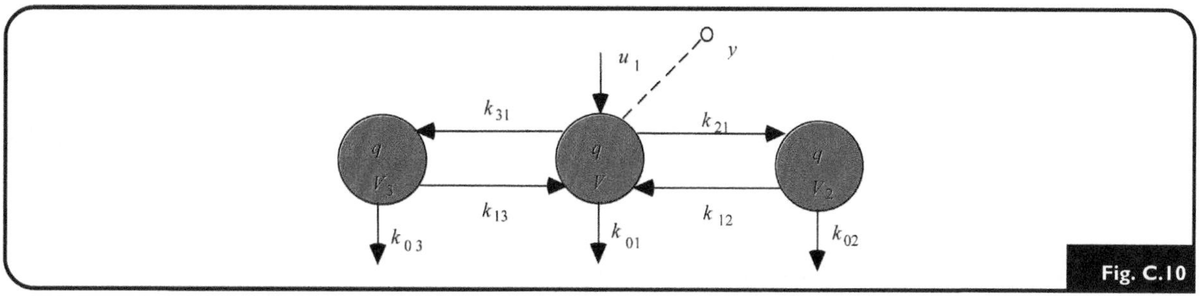

Fig. C.10

More concisely,

$$\dot{\boldsymbol{q}} = \begin{bmatrix} k_{11} & k_{12} & k_{13} \\ k_{21} & k_{22} & 0 \\ k_{31} & 0 & k_{33} \end{bmatrix} \boldsymbol{q} + \begin{bmatrix} 1 \\ 0 \\ 0 \end{bmatrix} u_1 \tag{C.77}$$

where $-k_{11} \equiv k_{01} + k_{21} + k_{31}$, $-k_{22} \equiv k_{02} + k_{12}$ and $-k_{33} \equiv k_{03} + k_{13}$, the **turnover rates** of the three compartments. The unknown parameter vector is: $\boldsymbol{p} \equiv [k_{21}\ k_{12}\ k_{02}\ k_{31}\ k_{13}\ k_{03}\ V_1]^T$. V_2 and V_3 do not appear in the equations (why not?) and therefore are not included in $\boldsymbol{p}$. The **TF matrix** of this model is:

$$H(s) = C[sI - K]^{-1}B = C(\boldsymbol{p})[sI - K(\boldsymbol{p})^{-1}B$$

$$= [c\ 0\ 0] \begin{bmatrix} s-k_{11} & k_{12} & k_{13} \\ k_{21} & s-k_{22} & 0 \\ k_{31} & 0 & s-k_{33} \end{bmatrix}^{-1} \begin{bmatrix} 1 \\ 0 \\ 0 \end{bmatrix} = [c\ 0\ 0] \begin{bmatrix} y_{11} & y_{12} & y_{13} \\ y_{21} & y_{22} & y_{23} \\ y_{31} & y_{32} & y_{33} \end{bmatrix} \begin{bmatrix} 1 \\ 0 \\ 0 \end{bmatrix} \tag{C.78}$$

$$= c y_{11} = \frac{c(s - k_{22})(s - k_{33})}{s^3 + \alpha_3 s^2 + \alpha_2 s + \alpha_1} \equiv h_{11}(s)$$

with

$$\alpha_3 = -(k_{11} + k_{22} + k_{33})$$

$$\alpha_2 = k_{11}(k_{22} + k_{33}) + k_{22}\ k_{33} - k_{12}\ k_{21} - k_{13}k_{31}$$

$$\alpha_1 = k_{12}\ k_{21}\ k_{33} + k_{22}\ (k_{13}\ k_{31} - k_{11}\ k_{33})$$

The TF matrix may be rewritten in terms of the intrinsic parameter vector $\boldsymbol{p}$ by replacing the k_{ii} in (C.78) with the k_{ij} in (C.76), and $V_1 = 1/c$. Either way, the algebra is cumbersome, even for such a simple example.

Discretization & Sampled-Data Representations of the 3-Compartment Model

First, a generalization. Suppose we "instantaneously" (very, very briefly) sample the output of any SISO model M at discrete time instants, $t_0, t_1, \ldots$, as illustrated in **Fig. C.11**. This process is

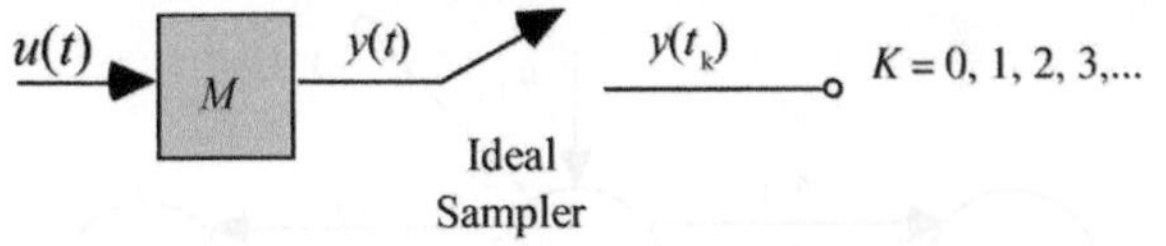

Fig. C.11

called **analog-to-digital (A/D) conversion**. If the model is linear, then as in Chapter 2, we can use any two adjacent time instants k and $k+1$ to describe evolution of the state $\boldsymbol{q}$ explicitly:

$$\boldsymbol{q}(t_{k+1}) = \Phi(t_{k+1}, t_k)\boldsymbol{q}(t_k) + \int_{t_k}^{t_{k+1}} \Phi(t_{k+1}, \tau)B(\tau)\boldsymbol{u}(\tau)d\tau, \quad y(t_{k+1}) = C(t_{k+1})q(t_{k+1}), \quad k = 0, 1, \ldots \tag{C.79}$$

This solution is exact at every time t_k. Returning to the three-compartment model example above,

$$H(s) = H_{11}(s) = \frac{c(s-a_{22})(s-a_{33})}{s^3 + a_3 s^2 + a_2 s + a_1} = \frac{b_1^* c_1^*}{s-\lambda_1} + \frac{b_2^* c_2^*}{s-\lambda_2} + \frac{b_3^* c_3^*}{s-\lambda_3} \tag{C.80}$$

The last equality in (C.80) is in partial fraction form (Appendix A). The λ_i are distinct for this model and b_i^* and c_i^* remain to be determined. In ODE form, we can write the **diagonalized equivalent model** (because the λ_i are distinct):

$$\dot{\boldsymbol{q}}^* \equiv (PAP^{-1})\boldsymbol{q}^* + (PB)u \equiv \begin{bmatrix} \lambda_1 & 0 & 0 \\ 0 & \lambda_2 & 0 \\ 0 & 0 & \lambda_3 \end{bmatrix} \boldsymbol{q}^* + \begin{bmatrix} b_1^* \\ b_2^* \\ b_3^* \end{bmatrix} u, \quad y(t) \equiv [c_1^* c_2^* c_3^*]\boldsymbol{q}^*(t) \equiv (CP^{-1})\boldsymbol{q}^*(t) \tag{C.81}$$

Following (C.79), we discretize (C.81) at two time instants, t_k and t_{k+1}:

$$\boldsymbol{q}^*(t_{k+1}) = e^{A*(t_{k+1}-t_k)}\boldsymbol{q}^*(t_k) + \int_{t_k}^{t_{k+1}} e^{A*(t_{k+1}-\tau)}B^*u(\tau)d\tau, \quad y^*(t_{k+1}) = C^*\boldsymbol{q}^*(t_k+1) \tag{C.82}$$

Equations (C.82) and are exact for any sampling interval, and for all k, in contrast with Eqs. (2.40) – (2.50) in Chapter 2, for uniform sampling only, yielding time-invariant difference equations (1.42). Indeed, Eqs. (C.82) are *time-varying* for all t_k when sampling is nonuniform. Only when $t_{k+1} - t_k \equiv T =$ constant is the model time-invariant. We get a time-varying discrete-time model of a time-invariant continuous model when we sample the latter nonuniformly.

To complete the discretization process, with the objective being to obtain discrete-time state equations of the form of Eqs. (C.73), we must specify the form of the input $u(t)$ and evaluate the integral in (C.82). Two types of inputs are common in applications.

Pulse-Train Inputs

Suppose that sampling is uniform and, also, that $u(\tau) =$ constant between sampling instants for $t_k \le \tau < t_{k+1}$, i.e. u is a **pulse train** over all $k = 0,1,2\ldots$, as shown in **Fig. C.12**. This type

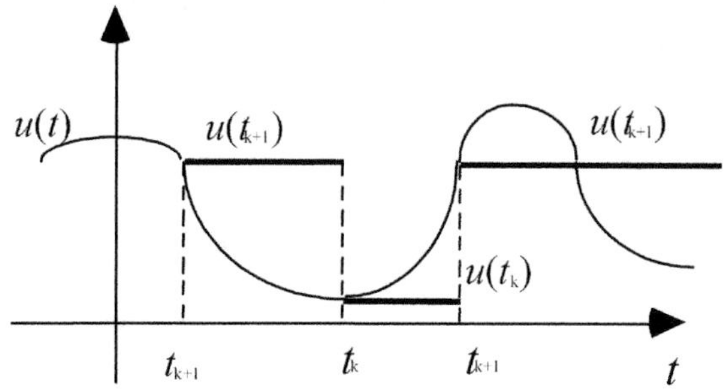

Fig. C.12

Pulse-train representation of $u(t)$.

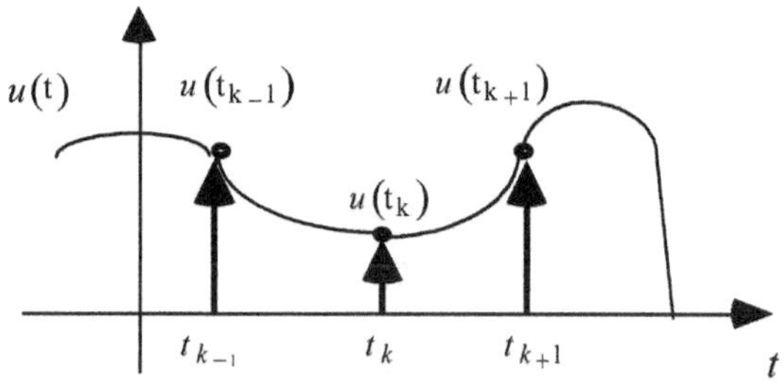

Fig. C.13

Impulse-train representation of $u(t)$.

of input is typical of that following **digital-to-analog (D/A) conversion** of a sampled signal, via a **zero-order-hold** device (DiStefano III et al. 1990). Evolution of the state $\boldsymbol{q}^*$ is then given by the discrete-time solution:

$$\begin{aligned}\boldsymbol{q}^*(t_{k+1}) &= e^{A^*(t_{k+1}-t_k)}\boldsymbol{q}^*(t_k) + \left(\int_{t_k}^{t_{k+1}} e^{A^*(t_{k+1}-\tau)}\boldsymbol{B}^* d\tau\right)u(t_k) \\ &\equiv \Phi^*(t_{k+1}-t_k)\boldsymbol{q}^*(t_k) + \Gamma_p{}^*(t_{k+1}-t_k)u(t_k)\end{aligned} \tag{C.83}$$

Impulse-Train Inputs

Suppose $u(\tau) = u(t_k)\delta(\tau - t_k)$, i.e. impulses of "area" $u(t_k)$ for any $\tau = t_k$ (an **impulse train** for $k = 0,1,\ldots$), e.g. the direct output of an ideally sampled input function $u(t)$), where $u(\tau) = 0$ for all $t_k \leq \tau < t_{k+1}$, i.e. zero in the open interval between sampling instants, as shown in **Fig. C.13**. Then the discrete-time solution is given by:

$$\begin{aligned}\boldsymbol{q}^*(t_{k+1}) &= e^{A^*T}\boldsymbol{q}^*(t_k) + \int_{t_k}^{t_k+T} e^{A^*(t_k+T-\tau)}B^*u(t_k)\delta(\tau - t_k)d\tau \\ &= e^{A^*T}\boldsymbol{q}^*(t_k) + e^{A^*T}B^*u(t_k) \equiv \Phi^*(T)\boldsymbol{q}^*(t_k) + \Gamma_k^*(T)u(t_k)\end{aligned} \tag{C.84}$$

Discretized ARMA Model with Impulse-Train Input

To obtain an equivalent input–output model, first abbreviate $\boldsymbol{q}_k^* \equiv \boldsymbol{q}^*(t_k)$, $\boldsymbol{q}_{k+1}^* \equiv \boldsymbol{q}^*(t_{k+1})$ etc. to simplify notation. Then, for the three-compartment example, the model response to the impulse train, Eq. (C.82), becomes:

$$\boldsymbol{q}_{k+1}^* = e^{A^*T}\boldsymbol{q}_k^* + e^{A^*T}B^*u_k \tag{C.85}$$

and, from (C.81):

$$\Phi = e^{A^*T} = \begin{bmatrix} e^{\lambda_1 T} & 0 & 0 \\ 0 & e^{\lambda_2 T} & 0 \\ 0 & 0 & e^{\lambda_3 T} \end{bmatrix}, \quad \Gamma_1^* \equiv e^{A^*T}B^* = \begin{bmatrix} b_1^* e^{\lambda_1 T} \\ b_2^* e^{\lambda_2 T} \\ b_3^* e^{\lambda_3 T} \end{bmatrix} \tag{C.86}$$

The b_i^* elements are given by Eq. (C.80). We now apply the **time shift operator**[1] Z:

$$Z\boldsymbol{q}_{k+1}^* \equiv \boldsymbol{q}_k^* \tag{C.87}$$

Then (C.82) becomes

$$\begin{aligned} Z\boldsymbol{q}_k^* - \Phi\boldsymbol{q}_k^* &\equiv \Gamma_1^* u_k \\ [ZI - \Phi^*]\boldsymbol{q}_k^* &= \Gamma_1^* u_k \\ \boldsymbol{q}_k^* &= [ZI - \Phi^*]^{-1} e^{A^*T} B * u_k \end{aligned} \tag{C.88}$$

and from (C.81):

$$y_k = C^*\boldsymbol{q}_k^* = C^*[ZI - \Phi^*]^{-1} e^{A*T}B^*u_k = \left[c_1^* c_2^* c_3^*\right] \begin{bmatrix} \dfrac{1}{Z - e^{\lambda_1 T}} & 0 & 0 \\ 0 & \dfrac{1}{Z - e^{\lambda_2 T}} & 0 \\ 0 & 0 & \dfrac{1}{Z - e^{\lambda_3 T}} \end{bmatrix} \begin{bmatrix} b_1^* e^{\lambda_1 T} \\ b_2^* e^{\lambda_2 T} \\ b_3^* e^{\lambda_3 T} \end{bmatrix} u_k$$

$$= \left[\frac{b_1^* c_1^* e^{\lambda_1 T}}{Z - e^{\lambda_1 T}} + \frac{b_2^* c_2^* e^{\lambda_2 T}}{Z - e^{\lambda_2 T}} + \frac{b_3^* c_3^* e^{\lambda_3 T}}{Z - e^{\lambda_3 T}}\right] u_k \equiv \left[\frac{K_1 Z^2 + K_2 Z + K_3}{Z^3 + h_1 Z^2 + h_2 Z + h_3}\right] u_k \tag{C.89}$$

the **sampled TF** between input u_k and output y_k of the sampler. Now multiply numerator and denominator by Z^{-3} and replace $Zy_k \equiv y_{k+1}$, and we have:

$$y_k + h_1 y_{k-1} + h_2 y_{k-2} + h_3 y_{k-3} = K_1 u_{k-1} + K_2 u_{k-2} + K_3 u_{k-3} \tag{C.90}$$

1 Note that the time-shift Z is not the same as the complex variable z in the z-transform. Z and z are related in the same way that the (time-domain) differential operator $\mathscr{D}$ and the complex variable s are related in the Laplace domain.

Remark: As an operational formalism, we can replace Z by z in (C.89), because this equation defines the sampled TF, just like D can be replaced by s in writing the TF from a differential equation model, in each case with zero initial conditions (zero-state).

Eq. (C.90) is called an **autoregressive moving average (ARMA) model**. For this particular example, the coefficients of (C.90) can be mapped 1:1 into the coefficients of (C.80). Eq. (C.90) may also be written:

$$\sum_{i=0}^{n} h_i y_{k-i} = \sum_{i=0}^{n} K_i u_{k-i} \tag{C.91}$$

where $h_0 \equiv 1$, the general form of an **ARMA model**. Clearly, this input–output model is in a form easy to implement on a computer, to obtain a solution y_k in terms of u_k, $k = 0, 1, 2, \ldots$. In general, however, the relationships between the parameters h_i, K_i and the k_{ij} of the compartment model of **Fig. C.10** are complicated and not necessarily unique.

Transforming Input–Output Data Models into State Variable Models: Generalized Model Building

If a state variable description of a linear system is available, the input–output model may be computed from the impulse response matrix:

$$H(t,\tau) = \begin{cases} \boldsymbol{C}(t)\Phi(t,\tau)\boldsymbol{B}(\tau), & t \geq \tau \\ 0, & t < \tau \end{cases} \tag{C.92}$$

or from the TF matrix $H = C[sI - A]^{-1}B$, if the linear model is time-invariant. The problem of finding a state variable model from an input-output model, however, is generally more difficult. In fact, a state variable model may not exist for a given $H(t,\tau)$. In general, the matrices $\{A, B, C\}$ are said to be a **state variable realization** of $H(t,\tau)$ if there exists a model: $\dot{\boldsymbol{x}} = \boldsymbol{A}\boldsymbol{x} + \boldsymbol{B}\boldsymbol{u}$ and $\boldsymbol{y} = \boldsymbol{C}\boldsymbol{x}$ that has $H(t,\tau)$ as its impulse response. The realization gives only the same zero-state response of the system described by $H(t,\tau)$. If it is not in the zero-state, the response of this model may not have any relation to the system. Not every $H(t,\tau)$ is realizable in the sense of the above definition, e.g. $H(t,\tau) = \delta(t - \tau)$, i.e. a τ-unit-time-delay function is not realizable by state variable equations of this form.

Time-Invariant Realizations

We can find time-invariant realizations by exploring the system TF matrix. Consider

$$H(s) = C[sI - A]^{-1}B = \frac{1}{\det[sI - A]} C[Adj(sI - A)]B \tag{C.93}$$

If this is a TF matrix of a causal (real) system, every entry $C[sI\text{-}A]^{-1}B$ is a rational function of s and the degree n of the denominator is at least as high as its numerator. In fact, a TF matrix $H(s)$

with rational function entries is realizable by constant $\{A, B, C\}$ if and only if $H(\infty)$ is a constant matrix (Chen 1970).

$$H(s) = \begin{bmatrix} \dfrac{s^2}{s^3+2} & \dfrac{s+1}{s+2} \\ \dfrac{s-1}{s+1} & s \end{bmatrix} \text{ is unrealizable, because } H(\infty) = \begin{bmatrix} 0 & 1 \\ 1 & \infty \end{bmatrix}.$$

The physical reason is that s represents pure differentiation, which cannot be implemented exactly, because it's anticipatory (not causal).

SISO Models

The most general form of a realizable $H(s)$ for $\dot{\mathbf{x}} = A\mathbf{x} + b\mathbf{u}$ and $\mathbf{y} = \mathbf{c}^T\mathbf{x}$ is

$$H(s) = \frac{\beta_1 s^{n-1} + \cdots + \beta_{n-1}s + \beta_n}{s^n + \alpha_1 s^{n-1} + \cdots + \alpha_{n-1}s + \alpha_n} \tag{C.94}$$

One realization of (C.94) is given by the **companion form** realization:

$$\dot{\mathbf{x}} = \begin{bmatrix} 0 & 1 & 0 & \cdots & 0 \\ \vdots & \ddots & 1 & \ddots & \vdots \\ \vdots & \ddots & \ddots & \ddots & 0 \\ 0 & \cdots & \cdots & 0 & 1 \\ -\alpha_n & -\alpha_{n-1} & \cdots & -\alpha_2 & -\alpha_1 \end{bmatrix} \begin{bmatrix} x_1 \\ x_2 \\ \vdots \\ x_n \end{bmatrix} + \begin{bmatrix} 0 \\ \vdots \\ 0 \\ 1 \end{bmatrix} u, \quad y = [b_n\, b_{n-1}\ldots b_2\, b_1]x \tag{C.95}$$

This representation of (C.94) may be confirmed by showing that (C.94) is the TF of (C.95).

For models with two inputs and two outputs, we seek $rm = 4$ sets of state equations, the maximum number of combinations of 2 inputs and 2 outputs, plus 2 output equations, all of the form:

$$\dot{\mathbf{x}}_{ij} = A_{ij}\mathbf{x}_{ij} + b_{ij}u_j \tag{C.96}$$

$$\mathbf{y}_j = c_{ij}^T\mathbf{x}_{ij} \tag{C.97}$$

for $i, j = 1,2$ and for

$$H(s) = [H_{ij}(s)] = \begin{bmatrix} H_{11}(s) & | & H_{12}(s) \\ \text{---} & | & \text{---} \\ H_{21}(s) & | & H_{22}(s) \end{bmatrix} \tag{C.98}$$

b_{ij} are column vectors and c_{ij}^T are row vectors. One realization is given by the **parallel composite** model:

$$\begin{bmatrix} \dot{x}_{11} \\ \dot{x}_{12} \\ \dot{x}_{21} \\ \dot{x}_{22} \end{bmatrix} = \begin{bmatrix} A_{11} & | & 0 & | & 0 & | & 0 \\ --- & | & --- & | & --- & | & --- \\ 0 & | & A_{12} & | & 0 & | & 0 \\ --- & | & --- & | & --- & | & --- \\ 0 & | & 0 & | & A_{21} & | & 0 \\ --- & | & --- & | & --- & | & --- \\ 0 & | & 0 & | & 0 & | & A_{22} \end{bmatrix} \begin{bmatrix} x_{11} \\ x_{12} \\ x_{21} \\ x_{22} \end{bmatrix} + \begin{bmatrix} b_{11} & | & 0 \\ --- & | & --- \\ 0 & | & b_{12} \\ --- & | & --- \\ b_{21} & | & 0 \\ --- & | & --- \\ 0 & | & b_{22} \end{bmatrix} \begin{bmatrix} u_1 \\ u_2 \end{bmatrix} \tag{C.99}$$

$$\begin{bmatrix} y_1 \\ y_2 \end{bmatrix} = \begin{bmatrix} c_{11}^T & | & c_{12}^T & | & 0 & | & 0 \\ --- & | & --- & | & --- & | & --- \\ 0 & | & 0 & | & c_{21}^T & | & c_{22}^T \end{bmatrix} \begin{bmatrix} x_{11} \\ x_{12} \\ x_{21} \\ x_{22} \end{bmatrix}$$

To confirm the validity of this representation, we compute the TF matrix of the partitioned composite model:

$$H(s) = \begin{bmatrix} c_{11}^T & | & c_{12}^T & | & 0 & | & 0 \\ --- & | & --- & | & --- & | & --- \\ 0 & | & 0 & | & c_{21}^T & | & c_{22}^T \end{bmatrix} \begin{bmatrix} [sI-A_{11}]^{-1} & | & 0 & | & 0 & | & 0 \\ ---- & | & ---- & | & ---- & | & ---- \\ 0 & | & [sI-A_{12}]^{-1} & | & 0 & | & 0 \\ ---- & | & ---- & | & ---- & | & ---- \\ 0 & | & 0 & | & [sI-A_{21}]^{-1} & | & 0 \\ ---- & | & ---- & | & ---- & | & ---- \\ 0 & | & 0 & | & 0 & | & [sI-A_{22}]^{-1} \end{bmatrix}$$

$$\bullet \begin{bmatrix} b_{11} & | & 0 \\ --- & | & --- \\ 0 & | & b_{12} \\ --- & | & --- \\ b_{21} & | & 0 \\ --- & | & --- \\ 0 & | & b_{22} \end{bmatrix} = \begin{bmatrix} c_{11}^T[sI-A_{11}]^{-1}b_{11} & | & c_{12}^T[sI-A_{12}]^{-1}b_{12} \\ -------------------- & | & -------------------- \\ c_{21}^T[sI-A_{21}]^{-1}b_{21} & | & c_{22}^T[sI-A_{22}]^{-1}b_{22} \end{bmatrix} \begin{bmatrix} H_{11}(s) & | & H_{12}(s) \\ ------ & | & ------ \\ H_{21}(s) & | & H_{22}(s) \end{bmatrix}$$

Each $H_{ij}(s)$ defined by the last equality is the TF for each set of equations (C.96) and (C.97), for each $i, j = 1, 2$. Q.E.D.

Several remarks about these realizations are pertinent.

Remark: Each set of matrices A_{ij} in the two-input, two-output model (C.99) may be realized in the SISO companion form Eq. C.95.

Remark: For more than two inputs and/or more than two outputs the procedure is the same as above. Only the dimensionality of the resulting composite system is different.

Remark: The form of the realizations considered above is quite general. But such realizations usually have *larger dimension than the minimum required to realize H(s)*, often much larger. This is usually impractical. *Minimal realizations* are usually more useful (Appendix D).

References

Chen, C., 1970. Introduction to Linear System Theory. Holt, Rinehart, & Winston, New York.

DiStefano III, J., Stubberud, A., Williams, I.J., 1990. Feedback and Control Systems, 2nd Edition: Continuous (Analog) and Discrete (Digital). McGraw-Hill, New York.

Controllability, Observability & Reachability

Basic Concepts and Definitions

Controllability and observability are structural (topological, morphological) properties of dynamic system models. They indicate whether the state or output of a model can be controlled in a prespecified way and whether it is possible to know the state by observing the outputs. These notions have rather obvious utility for control system engineers. In medicine, they can be helpful in applying both state variable and input–output models for clinical therapy and diagnosis. We study them also because they provide a deeper understanding of modeling theory and methodology, in particular connections between model topology and qualitative as well as quantitative dynamical properties. They are the basis of a growing number of computational methodologies for analyzing identifiability, realizability and distinguishability algorithms.

The following definitions are applicable to nonlinear as well as linear and discrete-time as well as continuous system models. We explore the continuous case first. Inputs, states and outputs are denoted, respectively, by the sets of time-dependent vector functions: $\boldsymbol{u}(t) \equiv \boldsymbol{u} \in \boldsymbol{U}^r, \boldsymbol{x}(t) \equiv \boldsymbol{x} \in \boldsymbol{\Sigma}^n$, and $\boldsymbol{y}(t) \equiv \boldsymbol{y} \in \boldsymbol{Y}^m$, where $\boldsymbol{U}^r$, $\boldsymbol{\Sigma}^n$ and $\boldsymbol{Y}^m$ are shorthand notations for *vector spaces* that define the limits of the input, state and output vector domains, respectively. For example, $\boldsymbol{U}^r$ may be the set of all possible r-vectors with *piecewise-continuous* elements $u_i(t)$. The complete model is defined for all $t \geq t_0$ by relations:

$$\begin{aligned} &\dot{\boldsymbol{x}} = \boldsymbol{f}(\boldsymbol{x}, \boldsymbol{u}, t; \boldsymbol{p}) \quad \text{with } \boldsymbol{x}(t_0) \equiv \boldsymbol{x}_0 \\ &\boldsymbol{y} = \boldsymbol{g}(\boldsymbol{x}, \boldsymbol{u}, t; \boldsymbol{p}) \\ &\boldsymbol{h}(\boldsymbol{x}, \boldsymbol{u}, \boldsymbol{p}) \geq \boldsymbol{0} \end{aligned} \tag{D.1}$$

Controllability

Formally, the process model in Eq. (D.1) is **controllable** at time t_0 if and only if (iff) an (unconstrained) control $\boldsymbol{u}_0 \in \boldsymbol{U}^r$ can be found which transfers the state $\boldsymbol{x}$ from any initial state $\boldsymbol{x}_0 \in \boldsymbol{\Sigma}^n$ (at time t_0) to any specified final state $\boldsymbol{x}_1 \in \boldsymbol{\Sigma}^n$ in a *finite* time $t_1 - t_0$ (a practical condition). Otherwise, the process is **uncontrollable**. The process in Eq. (D.1) is **completely controllable** if it is controllable for all initial states $\boldsymbol{x}_0$ and all initial times t_0.

Controllability is a property of the couplings $\boldsymbol{f}$ between the input and the state, and possibly also involves constraint relations in Eq. (D.1) but does not concern the output equation at all. The definition above also implies that the *whole* state vector $[x_1 \quad x_2 \quad \cdots \quad x_n]^T$ is controllable. One could also consider the controllability of any *subset* of state variables x_i, and this would require additional clarification, or the controllability of *specific states* $\boldsymbol{x}(t_0) \equiv \boldsymbol{x}_0$ at a fixed time t_0, or *specific states* at *any* t_0 time. Then some states $\boldsymbol{x}_0$, or some states $\boldsymbol{x}_0$ at some or any time t_0, may or may not be controllable. Controllability is defined somewhat differently by some authors, based on which of these variations is useful in a particular treatment. Little difficulty should arise if the context is kept clear and developments are internally consistent.

In essence, controllability of state variables (or states, for short) means that, under a given set of conditions, inputs can be found that permit *independent manipulation* (control) of the state variables of a model. As we shall see in examples, this is not always possible for a given model, and therefore use of a model for control purposes (e.g. for therapy) depends on this property.

Observability

A model (1) is **observable** if the state $\boldsymbol{x}(\tau) \in \boldsymbol{\Sigma}^n$ can be determined from the input $\boldsymbol{u}(t)$ and the measurements $\boldsymbol{y}(t) \in \boldsymbol{Y}^m$, $t_0 \leq \tau \leq t \leq t_1$, with t_1 *finite* (but unspecified). Otherwise it is **unobservable**. If it is observable for *any* $\boldsymbol{x}(t) \in \boldsymbol{\Sigma}^n$ and *any* $\tau \in [t_0, t_1]$, the process is **completely observable**.

Observability depends on the couplings $\boldsymbol{g}$ between the state vector and the measurements in Eq. (D.1), but it also depends on $\boldsymbol{f}$, which defines the couplings among the state variables and possibly also on the constraints. Thus, all equations in Eq. (D.1) must be examined for observability properties.

If all state variables are available for measurement, the process is obviously always observable. We also could consider the observability of any subset of state variables $x_1, \ldots, x_n$, but this is not what is meant by state observability. Observability of one or more state variables would be required, for example, in a feedback system, where control inputs depend on outputs being controlled, via a closed-loop feedback path connecting one or more state variables to a controller subsystem.

A model is **output controllable** if a $\boldsymbol{u} \in \boldsymbol{U}^r$ can be found that transfers the output $\boldsymbol{y}(t) = \boldsymbol{g}[\boldsymbol{x}(\boldsymbol{u}), t; \boldsymbol{c}]$ from *any* initial $\boldsymbol{y}(t_0) \in \boldsymbol{Y}^m$ to *any* final $\boldsymbol{y}(t_1) \in \boldsymbol{Y}^m$ in a *finite* time $t_1 - t_0$.

Remark: Process or state controllability is defined only for the state variable model, whereas output controllability depends only on the input–output relation. Thus, these two concepts are not necessarily related.

Some authors choose the origin $\boldsymbol{x}_1 \equiv \boldsymbol{0}$ as the *final state* (or initial state, $\boldsymbol{x}_0 \equiv \boldsymbol{0}$) in defining controllability. And the *initial state* $\boldsymbol{x}_0$ is sometimes chosen as the observable one in some definitions of observability. This is possible for linear systems, without loss of generality, for the following reasons. Consider first the solution for the state at time t_1:

$$\boldsymbol{x}(t_1) = \Phi(t_1, t_0)\boldsymbol{x}_0 + \int_{t_0}^{t_1} \Phi B\boldsymbol{u} \equiv \boldsymbol{x}_1 \tag{D.2}$$

or, equivalently, using the transition property of $\Phi(t,\tau)$:

$$\boldsymbol{0} = \Phi(t_1, t_0)\boldsymbol{x}_0 - \Phi(t_1, t_1)\boldsymbol{x}_1 + \int_{t_0}^{t_1} \Phi B\boldsymbol{u} = \Phi(t_1, t_0)\boldsymbol{x}_0 - \Phi(t_1, t_0)\Phi(t_0, t_1)\boldsymbol{x}_1 + \int_{t_0}^{t_1} \Phi B\boldsymbol{u}$$

Then factoring out $\Phi(t_1,t_0)$ gives:

$$\boldsymbol{0} = \Phi(t_1, t_0)[\boldsymbol{x}_0 - \Phi(t_0, t_1)\boldsymbol{x}_1] + \int_{t_0}^{t_1} \Phi B\boldsymbol{u} \tag{D.3}$$

Now, $\boldsymbol{x}^1 \equiv [\boldsymbol{x}_0 - \Phi(t_0, t_1)\boldsymbol{x}_1] \in \boldsymbol{\Sigma}^n$, i.e., the state defined as $\boldsymbol{x}^1$ is in the same state space $\boldsymbol{\Sigma}^n$, because it is a simple linear translation of $\boldsymbol{x}_0$. Thus, (complete) controllability can be defined in terms of any $\boldsymbol{x}^1$ "transferred" to the origin, $\boldsymbol{x}_0 \equiv \boldsymbol{0}$. Also, reexamination of Eq. (D.2) indicates that, if Φ, B and $\boldsymbol{u}$ are known, then any $\boldsymbol{x}(t_1)$ can be determined if $\boldsymbol{x}_0$ is known. Thus, it is sufficient to define observability for $\boldsymbol{x}_0$ only, if linear system models are being considered (and possibly others). Using the same kind of arguments, it is sufficient to consider the zero-input response, i.e. the linear model with $\boldsymbol{u} \equiv \boldsymbol{0}$, when discussing observability, as we see below.

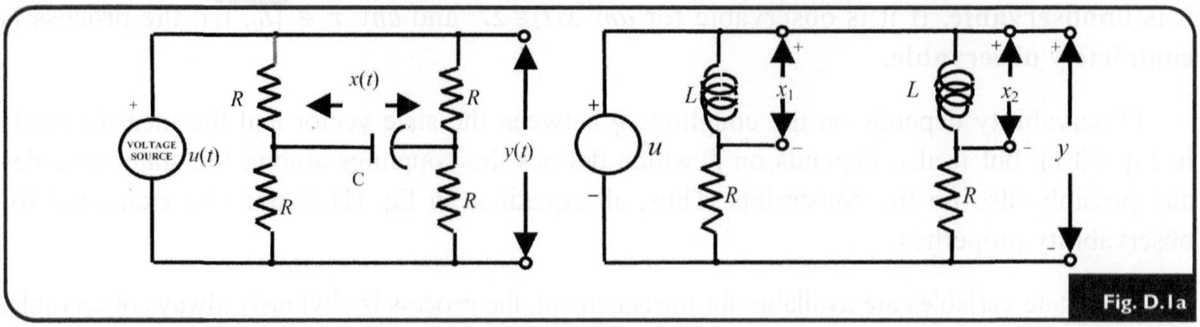

Fig. D.1a

We emphasize that controllability and observability are qualitative properties of *models*. **Two different models of the same system may or may not have the same observability or controllability properties**. Also, these two concepts are defined under the assumption that we have *complete* knowledge of model structure ***f***, ***g*** and ***h***, and sufficient quantitative knowledge of model parameters ***p***. The constraint relations in Eq. (D.1) have not been formally (explicitly) considered part of the overall model in most of the literature, in particular, in the original definitions of these qualitative concepts. Nevertheless, if there are constraints, they must be treated as part of the overall model and should enter into controllability or observability determinations or both (DiStefano III, 1977). Controllability and observability definitions and properties for discrete-time models are virtually identical to those above, with minor and obvious changes in notation.

The following example is adapted from Chen (1970).

Example D.1: ***State and Output Controllability and Observability or Lack Thereof***

The symmetric electrical bridge network in **Fig. D.1a** is neither state controllable nor state observable, but it is output controllable. We argue first from the physics. The circuit is not controllable because if the initial voltage across the capacitor is zero, that is, the initial state $x(t_0) = 0$, then $x(t) = 0$, for all t and *any* $u \neq 0$, as the circuit is completely symmetric and no current is ever produced across C.

It is not observable because, if $x(t_0) \neq 0$ (anything) and $u = 0$, in particular, then $y \equiv 0$ for all t again by symmetry. In other words, the state is impossible to observe (know) from the output, for this input, for any t. It is clearly output controllable, as are all SISO models. Indeed, $y(t) = u(t)$.

The inductive–resistive network in Fig. D.1b is not controllable. Again, arguing from the physics, we see that x_1 or x_2 is controllable, but not x_1 and x_2, because $x_1 \equiv x_2$, due to circuit symmetry. The state variables x_1 and x_2 must be independently controllable for vector $\boldsymbol{x}$ to be controllable. (Is this network output controllable? state observable?)

Observability and Controllability of Linear State Variable Models

The intention of this sections is to summarize the major results in this area pertinent to the study of linear models of biosystems and experiments, as well as other system types modeled by the same equations. Proofs, or partial proofs, are presented only when constructive. Some results for time-varying systems also are included. Periodic phenomena (rhythms) abound in biological systems, and these are sometimes represented by *time-varying* dynamic system models.

Linear Time-Varying Models

Again consider the model:

$$\dot{\boldsymbol{x}} = A(t)\boldsymbol{x} + B(t)\boldsymbol{u}$$
$$y = C(t)\boldsymbol{x} \tag{D.4}$$

possibly constrained by, $\boldsymbol{h}(\boldsymbol{x},\boldsymbol{u},\boldsymbol{p}) \geq \boldsymbol{0}$, and where the dependence of A, B and C on $\boldsymbol{p}$ is suppressed for notational simplicity. The same results apply for discrete-time linear system models, with minor and obvious changes in notation. Most of the results for linear models are based on the following.

In Eq. (D.4), $\dot{\boldsymbol{x}}$ *is* **controllable** at t_0 if and only if there exists a finite time $t_1 > t_0$ such that the *n rows* of the $n \times r$ matrix function $\Phi(t_1,\tau)B(\tau)$, or $\Phi(t_0,\tau)B(\tau)$, are linearly independent over the interval $[t_0, t_1]$. The (complete) model is **observable** at t_0 if and only if there exists a finite $t_1 > t_0$ such that the *n columns* of the $m \times n$ matrix function $C(\tau)\Phi(\tau, t_0)$ are linearly independent over the interval $[t_0, t_1]$, where $t_0 \leq \tau < t_1$ (Chen 1985).

This is technical, and therefore somewhat difficult to understand at first reading, but the idea is simple. It states the *conditions under which the model equations can be solved*: (a) for the input $\boldsymbol{u}$, when the question is whether or not the process state can be transferred from one point to another (controllability); and (b) for the state $\boldsymbol{x}$ from the measured output $\boldsymbol{y}$, when the question is whether or not the state can be observed.

The sufficiency part of the proof is constructive and clarifies the issues further. First we simplify the controllability and observability conditions somewhat. For controllability, we subtract out the zero-input response, so that we need only consider the input/state relation:

$$\boldsymbol{x}^*(t_1) \equiv \boldsymbol{x}(t_1) - \Phi(t_1, t_0)\boldsymbol{x}_0 = \int_{t_0}^{t_1} \Phi B\boldsymbol{u} \tag{D.5}$$

Thus, $x^*(t) \in \Sigma^n$ and we can define the linear operator:

$$\boldsymbol{T}(\boldsymbol{u}) \equiv \int_{t_0}^{t_1} \Phi(t_1, \tau)B(\tau)\boldsymbol{u}(\tau)d\tau \tag{D.6}$$

We can say that the first equation in Eq. (D.4) is *controllable* if and only if and only if there exists a $\mathbf{u}(t) \in \boldsymbol{U}^r$ such that $x^*(t) = \boldsymbol{T}(u)$ has the solution $\boldsymbol{u}(t)$ for arbitrary $\boldsymbol{x}^* \in \Sigma^n$. (Recall: it is controllable if one can find an admissible $\boldsymbol{u}$.) Similarly, for observability, we subtract from the output the term due to the input:

$$\boldsymbol{y}^*(t) \equiv \boldsymbol{y}(t) - C(t)\int_{t_0}^{t} \Phi B\boldsymbol{u} = C(t)\Phi(t, t_0)\boldsymbol{x}_0 \tag{D.7}$$

Thus, $y^*(t_1) \in Y^m$ and we can define the linear operator:

$$S(x_0) \equiv C(t_1)\Phi(t_1, t_0)x_0 \tag{D.8}$$

Now, the model is *observable* if and only if there exists a unique solution $x_0 \in \Sigma^n$ such that $y^*(t_1) = S(x_0)$ for arbitrary $y^* \in Y^m$, with t_1 finite, i.e. if we can solve Eq. (D.8) for x_0 uniquely.

If the rows of $\Phi(t_1,\tau)B(\tau)$ are linearly independent on the interval $[t_0, t_1]$, it can be shown that the $n \times n$ constant matrix defined as:

$$W'(t_0, t_1) \equiv \int_{t_0}^{t_1} \Phi(t_1, \tau)B(\tau)B^T(t)\Phi^T(t_1, \tau)d\tau \tag{D.9}$$

is nonsingular (Casti 1987). What we need now is an input u that satisfies Eq. (D.6). Given any $x^*(t_1) \in \Sigma^n$, let us try the following input function:[1]

$$u'(t) \equiv B^T(t)\Phi^T(t_1, t)[W'(t_0, t_1)]^{-1}x^*(t_1) \tag{D.10}$$

Substituting Eq. (D.10) into Eq. (D.5) gives:

$$\begin{aligned} x^*(t_1) &= \int_{t_0}^{t_1} \Phi(t_1, \tau)B(\tau)\, B^T(\tau)\Phi^T(t_1, \tau)[W'(t_0, t_1)]^{-1}x^*(t_1)d\tau \\ &= [W'(t_0, t_1)][W'(t_0, t_1)]^{-1}x^*(t_1) = x^*(t_1) \end{aligned} \tag{D.11}$$

Thus, we have found such a $u(t) \in U^r$, which serves our purpose here for showing sufficiency, because it is a solution that is valid for arbitrary $x^*(t_1)$. The matrix $W'(t_0, t_1)$ is called a **controllability matrix**.

The sufficiency part of the observability portion of the above theorem also is constructive. We first premultiply both sides of Eq. (D.7) by $\Phi^T(t, t_0)C^T(t)$ and integrate from t_0 and t_1:

$$\int_{t_0}^{t_1} \Phi^T(t, t_0)C^T(t)y^*(t)dt = \int_{t_0}^{t_1} \Phi^T(t, t_0)C^T(t)C(t)\Phi(t, t_0)dt \cdot x_0 \tag{D.12}$$

Now let the term in brackets be defined as the **observability matrix**:

$$V(t_0, t_1) \equiv \int_{t_0}^{t_1} \Phi^T(\tau, t_0)C^T(\tau)C(\tau)\Phi(\tau, t_0)d\tau \tag{D.13}$$

Then, if the n columns of $C(\tau)\Phi(\tau,\ t_0)$ are linearly independent on $[t_0,\ t_1]$, $V(t_0,\ t_1)$ is nonsingular (Casti 1987) and we can solve for x_0:

$$x_0 = V^{-1}(t_0, t_1)\int_{t_0}^{t_1} \Phi^T(\tau, t_0)C^T(\tau)y^*(\tau)d\tau \tag{D.14}$$

Therefore, the system (model) is observable, because a unique solution for x_0 exists, given by Eq. (D.14).

---

1 This input function u' in (D.10) is the solution of a special problem in optimal control theory. u' is a **minimum energy control**.

We thus have the following criteria for controllability and observability for all linear state variable models, stated for pedagogical rather than practical reasons, because they are in terms of $\Phi(t,\tau)$, which generally is difficult to obtain. More practical results are discussed later.

Controllability Criterion

The process in Eq. (D.4) is controllable if and only if $W'(t_0, t_1)$ given by Eq. (D.9) is nonsingular. Corresponding minimum energy controls $\boldsymbol{u}'(t)$ are given by Eq. (D.10).

Observability Criterion

The model in Eq. (D.4) is observable if and only if $V(t_0, t_1)$ in Eq. (D.13) is nonsingular. Then $\boldsymbol{x}_0$ is given by Eq. (D.14) in terms of V.

Linear Time-Invariant Models

All results of the previous section are of course applicable to the time-invariant case. But the situation is much simpler and the tests for these properties are much easier to apply for time-invariant models. Moreover, specification of the times t_0 or t_1 is unnecessary: if the model M is controllable (observable) for any one t, it is controllable (observable) for all t. Thus, when we speak of controllable or observable models here, the results will imply the strongest forms of these properties – completely controllable (observable). That is, controllability of a time-invariant M means we can transfer any state $\boldsymbol{x}_0$, that is, independently control all the state variables, $x_1, \ldots, x_n$, at any t_0, to any other state $\boldsymbol{x}_1$ in as short (or long) a time as we like. And observability similarly means we can determine $\boldsymbol{x}_0$ by observing $\boldsymbol{y}(t)$ for as short (or long) a time as we like. Following are the major practical tests for controllability and observability for such models.

Practical Controllability and Observability Conditions

The process model

$$\dot{\boldsymbol{x}} = A\boldsymbol{x} + B\boldsymbol{u} \tag{D.15}$$

(possibly constrained by $\boldsymbol{h}(\boldsymbol{x}, \boldsymbol{u}, \boldsymbol{p}) \geq 0$) is *controllable* if and only if the $n \times nr$ **controllability matrix** W:

$$W = \left[A^{n-1}B \;\vdots\; A^{n-2}B \;\vdots\; \cdots \;\vdots\; AB \;\vdots\; B\right] \equiv \left[B \;\vdots\; AB \;\vdots\; \cdots \;\vdots\; A^{n-1}B\right] \tag{D.16}$$

has rank n. Moreover, if the rank of B is s, then M is controllable if and only if

$$\operatorname{rank} W_{n-s} = \operatorname{rank}\left[\mathrm{A} \;\vdots\; \mathrm{AB} \;\vdots\; \cdots \;\vdots\; A^{n-s}B\right] = n \tag{D.17}$$

or, equivalently, if the $n \times n$ matrix $W_{n-s}W_{n-s}^T$ is nonsingular. This is a simplification of the rank test for Eq. (D.16). Proof is given in Chen (1985). The overall model is *observable* if and only if the $nm \times n$ **observability matrix** V:

$$V = \left[C \;\vdots\; CA \;\vdots\; \cdots \;\vdots\; CA^{n-1}\right]^T \tag{D.18}$$

has rank n. Moreover, if the rank of C is s, M is observable if and only if the rank of $V_{n-s} = \left[C \ \vdots \ CA \ \vdots \ \cdots \ \vdots \ CA^{n-s}\right]^T = n$, or, equivalently, if the $n \times n$ matrix $V_{n\text{-}s}^T V_{n-s}$ is nonsingular (Chen 1985). We note that both of these conditions, for controllability and observability, are directly in terms of the realization $\{A, B, C\}$ and $\Phi(t-\tau)$ need not be evaluated.

It is instructive to consider a heuristic, albeit incomplete, development of these criteria. Let us look at the discrete time model, for a change of pace:

$$\boldsymbol{x}_{k+1} = A\boldsymbol{x}_k + B\boldsymbol{u}_k, \quad \boldsymbol{y}_k = C\boldsymbol{x}_k \tag{D.19}$$

For a controllability test, we see how we can solve Eq. (D.19) for $\boldsymbol{u}_k$:

$$\begin{aligned}
\boldsymbol{x}_1 &= A\boldsymbol{x}_0 + B\boldsymbol{u}_0 \\
\boldsymbol{x}_2 &= A\boldsymbol{x}_1 + B\boldsymbol{u}_1 = A(A\boldsymbol{x}_0 + B\boldsymbol{u}_0) + B\boldsymbol{u}_1 \\
&= A^2\boldsymbol{x}_0 + AB\boldsymbol{u}_0 + B\boldsymbol{u}_1 \\
&\ \vdots \\
\boldsymbol{x}_n &= A^n\boldsymbol{x}_0 + A^{n-1}\boldsymbol{u}_0 + A^{n-2}B\boldsymbol{u}_1 + \cdots + B\boldsymbol{u}_{n-1}
\end{aligned}$$

Rearranging these equations we get:

$$\boldsymbol{x}_n - A^n\boldsymbol{x}_0 = \left[A^{n-1}B \ \vdots \ A^{n-2}B \ \vdots \ \cdots \ \vdots \ B\right]\begin{bmatrix} \boldsymbol{u}_0 \\ \boldsymbol{u}_1 \\ \vdots \\ \boldsymbol{u}_{n-1} \end{bmatrix} \equiv W\begin{bmatrix} \boldsymbol{u}_0 \\ \boldsymbol{u}_1 \\ \vdots \\ \boldsymbol{u}_{n-1} \end{bmatrix} \tag{D.20}$$

The LHS of Eq. (D.20) is assumed known for this problem and, from linear equation theory (see ***Appendix B***), a unique set (sequence) of $\boldsymbol{u}_k$ can be found only if the matrix W has rank n.

For an observability test, we see how we can solve $\boldsymbol{y}_k = C\boldsymbol{x}_k$ for $\boldsymbol{x}_0$:

$$\begin{aligned}
\boldsymbol{y}_0 &= C\boldsymbol{x}_0 \\
\boldsymbol{y}_1 &= C\boldsymbol{x}_1 = CA\boldsymbol{x}_0 \\
&\ \vdots \\
\boldsymbol{y}_{n-1} &= CA^{n-1}\boldsymbol{x}_0
\end{aligned}$$

Rearranging these equations we get:

$$\begin{bmatrix} \boldsymbol{y}_0 \\ \boldsymbol{y}_1 \\ \vdots \\ \boldsymbol{y}_{n-1} \end{bmatrix} = \begin{bmatrix} C \\ \cdots\cdots \\ CA \\ \cdots\cdots \\ \vdots \\ \cdots\cdots \\ CA^{n-1} \end{bmatrix} \quad \text{and} \quad \boldsymbol{x}_0 = V\boldsymbol{x}_0 \tag{D.21}$$

From linear equation theory (***Appendix B***), a unique solution $\boldsymbol{x}_0$ of Eq. (D.21) can be found only if V has rank n.

Remark: If a linear time-invariant model is controllable, then the input can be made to excite all its modes. Similarly, it can be designed to suppress undesirable, for example, unstable modes. (Shades of the CIA!)

Example D.2: ***Controllable/Uncontrollable Models and Control Input Calculation***

The 3-dimensional model:

$$\dot{\boldsymbol{x}} = \begin{bmatrix} 1 & 1 & 0 \\ 0 & 1 & 0 \\ 0 & 1 & 1 \end{bmatrix} \boldsymbol{x} + \begin{bmatrix} 0 & 1 \\ 1 & 0 \\ 0 & 1 \end{bmatrix} \boldsymbol{u}$$

is not controllable because *rank* $B = s = 2$ and rank $W_{n-1} =$ rank $W_{n-s} =$ rank $W_1 =$

$$\text{rank}\,[\text{B} \ \vdots \ \text{AB}] = \text{rank}\begin{bmatrix} 0 & 1 & 1 & 1 \\ 1 & 0 & 1 & 0 \\ 0 & 1 & 1 & 1 \end{bmatrix} = 2 < 3$$

The following 2-dimensional model is controllable:

$$\begin{bmatrix} \dot{x}_1 \\ \dot{x}_2 \end{bmatrix} = \begin{bmatrix} -1/2 & 0 \\ 0 & -1 \end{bmatrix} \boldsymbol{x} + \begin{bmatrix} 1/2 \\ 1 \end{bmatrix} u$$

because

$$\text{rank}\,[\text{B} \ \vdots \ \text{AB}] = \text{rank}\begin{bmatrix} 1/2 & -1/4 \\ 1 & -1 \end{bmatrix} = 2$$

Let us see now how to construct one possible control $u'(t)$ that will transfer $\boldsymbol{x}(0) = \begin{bmatrix} 10 & 10 \end{bmatrix}^T$ to $\boldsymbol{x}(2) = \begin{bmatrix} 0 & 0 \end{bmatrix}^T$. We employ this result and the more general controllability matrix $W'(t_0, t_1)$ of Eq. (D.9) to find the desired control input:

$$W'(0,2) = \int_0^2 \begin{bmatrix} e^{\tau/2} & 0 \\ 0 & e^{\tau} \end{bmatrix} \begin{bmatrix} 1/2 \\ 1 \end{bmatrix} \begin{bmatrix} 1/2 & 1 \end{bmatrix} \begin{bmatrix} e^{\tau/2} & 0 \\ 0 & e^{\tau} \end{bmatrix} d\tau = \begin{bmatrix} 1.597 & 6.363 \\ 6.363 & 26.8 \end{bmatrix}$$

Therefore, the input function is:

$$u'(t) = [1/2 \quad 1] \begin{bmatrix} e^{\tau/2} & 0 \\ 0 & e^{t} \end{bmatrix} W'^{-1}(0,2) \begin{bmatrix} 10 \\ 10 \end{bmatrix} = -44.2e^{t/2} + 20.6e^{t}$$

OUTPUT CONTROLLABILITY

The general definition was given earlier. A linear model with a continuous impulse response matrix $H(t,\tau)$ is **output controllable** at time t_0 if and only if, for any $\boldsymbol{y}_1$, there exists a finite $t_1 > t_0$ and an input $\boldsymbol{u}(t)$ that transfers the output from a given $\boldsymbol{y}(t_0) = \boldsymbol{y}_0$ to $\boldsymbol{y}(t_1) = \boldsymbol{y}_1$ at t_1. The main result for time-varying models is: a continuous $H(t,\tau)$ is *output controllable* at t_0 if and only if there exists a finite $t_1 > t_0$ such that the *rows of $H(t,\tau)$ are linearly independent* for $\tau \in [t_0, t_1]$ (Chen 1985). Again, for output controllability, it is not necessary to have a state variable model. An input–output model is sufficient.

Time-Invariant (TI) Models

Corresponding to the above result, a TI model with a strictly proper TF matrix (i.e. $H(\infty) = 0$) is output controllable if and only if all the rows of $H(s)$ are linearly independent, i.e. *if and only if rank $H(s) = m$, where m is the number of outputs* (Chen 1985). This result applies to state variable as well as input–output models.

TI State Variable Models

These are output controllable if and only if $H(\infty) = 0$ and

$$\text{rank}\left[CB \;\vdots\; CAB \;\vdots\; \cdots \;\vdots\; CA^{n-1}B\right] = m \tag{D.22}$$

where m is the number of outputs, and $H(s) = C[sI - A]^{-1}B$. The matrices, CA^sB (for $s < n$), are called **Markov matrices** or (sometimes) **Markov parameters** of the model $\dot{x} = Ax + Bu$, $y = Cx$.

We reemphasize that state and output controllabilities are not necessarily related, as illustrated next.

Example D.3: ***Controllable but not Output Controllable***

The model:

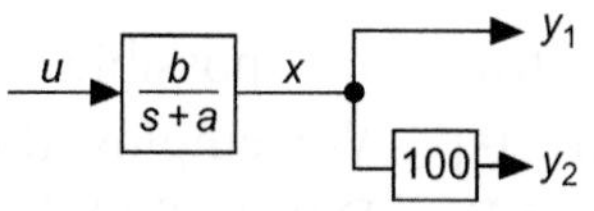

has the TF matrix:

$$H(s) = \begin{bmatrix} \dfrac{b}{s+a} & \dfrac{100b}{s+a} \end{bmatrix}^T$$

The rows are dependent and the model is therefore not output controllable. However, the state equations are:

$$\dot{x} = -ax + bu$$

$$y = \begin{bmatrix} 1 \\ 100 \end{bmatrix} x$$

and we see the model is both (state) controllable and observable.

Output Function Controllability

It is of practical interest to know whether the output $\boldsymbol{y}(t)$ of a model can be made to follow a preassigned curve over an interval $[t_0, t_1]$. If so, the model is **output function controllable**. This question is akin to the *servomechanism* problem in classical control system theory. The following result is for models with proper rational TF matrices. If $H(s)$ is $m \times r$ and $H(\infty) = 0$, then it is output function controllable if and only if rank $H(s) = m$. Note that, for strictly proper rational $H(s)$, the necessary and sufficient conditions for output controllability and output function controllability are equivalent, i.e. one implies the other.

If we wish to follow an output $y(t)$, which has the LT $\boldsymbol{Y}(s)$, and M is initially relaxed, then we can choose:

$$\boldsymbol{U}(s) = H^T(s)[H(s)H^T(s)]^{-1}\boldsymbol{Y}(s) \tag{D.23}$$

if H has rank m. We note that a state variable model is not required here. The matrix $H^T(HH^T)^{-1}$ in Eq. (D.23) is called a **pseudoinverse** of H (see ***Appendix B***).

Consider an $m \times r$ TF matrix $H(s)$, and therefore m outputs and r inputs, i.e. the vector–matrix equation:

$$\boldsymbol{Y}(s) \equiv \boldsymbol{Y} = H(s)\boldsymbol{U}(s) \equiv H\boldsymbol{U} \tag{D.24}$$

If $m = r$, then $\boldsymbol{U} = H^{-1}\boldsymbol{Y}$ if and only if H is nonsingular (rank $H = r = m$). If $m < r$, then we have fewer outputs (m columns of H) than inputs (r rows of H), or fewer equations than unknowns, assuming $\boldsymbol{U}$ is unknown and $\boldsymbol{Y}$ is known. This is the classic underdetermined linear (algebraic) system of equations problem, which has no unique solution (***Appendix B***). But it has an infinite number of solutions for $\boldsymbol{U}$ if rank $H = m$. Then, if we can find one such input solution $\boldsymbol{U}$, the system represented by H is **output function controllable**, as it also would be for $m = r$, if in addition $H(\infty) = 0$.

To find a solution, let rank $H = m$. Then HH^T which is $m \times m$, is nonsingular. Thus, from ***Appendix B***, we have the **minimum norm solution**:

$$\boldsymbol{U}_{MN} = H^T(HH^T)^{-1}\boldsymbol{Y} \tag{D.25}$$

This is the control (in the s-domain) that will transfer the output to any desired $\boldsymbol{y}(t)$, for any or all $t \in [t_0, t_1]$, if and only if $H(\infty) = 0$ and $m \le r$. (For $m = r$, $\boldsymbol{U} = H^{-1}\boldsymbol{Y}$.)

Finally, if $r < m$, i.e. we have fewer inputs than outputs, there is (strictly speaking) no possibility of output controllability, because $\boldsymbol{Y} = H\boldsymbol{U}$ is overdetermined and therefore there is no solution for $\boldsymbol{U}$. However, an **approximate least squares solution** can still be found, if rank $H = r$. In this case, H^TH is $r \times r$ and therefore nonsingular. The $\hat{\boldsymbol{U}}$ that minimizes the Euclidean (L_2) norm $\|\boldsymbol{Y} - H\boldsymbol{U}\|_2^2 = (\boldsymbol{Y} - H\boldsymbol{U})^T(\boldsymbol{Y} - H\boldsymbol{U}) \equiv \boldsymbol{e}^T\boldsymbol{e}$ is the $\hat{\boldsymbol{U}}$ that satisfies $\dfrac{d(\boldsymbol{e}^T\boldsymbol{e})}{d\boldsymbol{U}} = 2\boldsymbol{e}^TI = 2(\boldsymbol{Y} - H\boldsymbol{U})^T \equiv 0^T$, or $\boldsymbol{Y} - H\hat{\boldsymbol{U}} = \boldsymbol{0}$. Premultiplying by $(H^TH)^{-1}H^T$ gives

$$\hat{\boldsymbol{U}}_{\mathrm{LS}} = (H^TH)^{-1}\boldsymbol{U}_{\mathrm{LS}} = (H^TH)^{-1}H^T\boldsymbol{Y} \tag{D.26}$$

as the least-squares approximate solution to this control problem, for $r < m$ and $H(\infty) = 0$. In this case, $(H^T H)^{-1}$ is another **pseudoinverse** of H, for rank $H = r < m$, just as $H^T(HH^T)^{-1}$ in Eq. (D.25) was a pseudoinverse of H for rank $H = m < r$.

Reachability

This concept is complementary to but entirely independent of controllability in general. The differences are important in practice only for time-varying models. Nevertheless, even for time-invariant models, understanding the distinction reinforces comprehension of the concepts. We recall that a state $\boldsymbol{x}_0$ is controllable at t_0 if an admissible $\boldsymbol{u} \in \boldsymbol{U}^r$ can be found which transfers the given system (model) from $\boldsymbol{x}_0 \in \boldsymbol{\Sigma}^n$ (at t_0) to any specified final state $\boldsymbol{x}_1 \in \boldsymbol{\Sigma}^n$ (at $t_1 \geq t_0$) in a finite time. Thus the question of controllability (of $\boldsymbol{x}_0$) refers to whether the particular starting state $\boldsymbol{x}_0$ can be transferred to any given *final* state $\boldsymbol{x}_1$ in a finite time. On the other hand, reachability refers to the set of unspecified final states $\boldsymbol{x}_1$ which can be reached from a specified initial state $\boldsymbol{x}_0$. Formally, a state $\boldsymbol{x}_1 \equiv \boldsymbol{x}(t_1) \in \boldsymbol{\Sigma}^n$ is **reachable** at t_1 if and only if there exists a $t_0 < t_1 < \infty$ and an admissible $\boldsymbol{u} \in \boldsymbol{U}^r$ that can transfer the given system (model) from (a given) $\boldsymbol{x}_0$ at t_0 to $\boldsymbol{x}_1$ at t_1, as shown in **Fig. D.2**. The *process* (model) is **completely reachable** if every $\boldsymbol{x} \in \boldsymbol{\Sigma}^n$ is reachable for all t_1 (Casti 1977).

Reachability (R) and controllability (C) are also illustrated in **Fig. D.3** where the trajectory R is the same as that of C reflected about $t = t_0$, with t_0 and t_1 – and thus the direction of the trajectory – reversed.

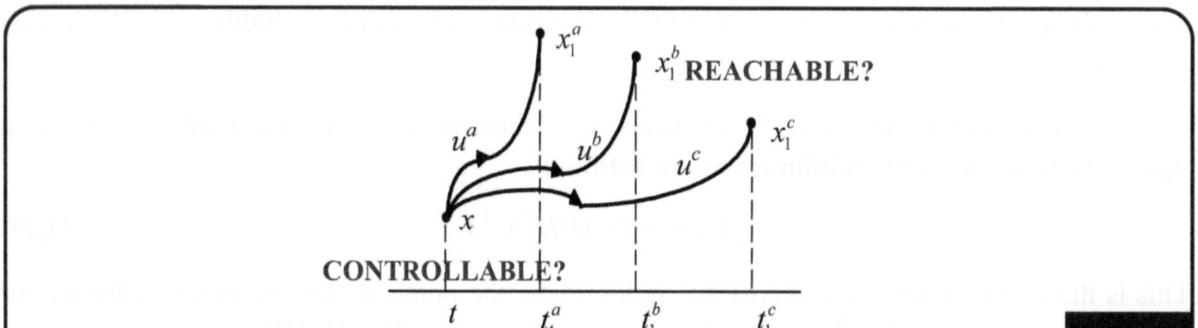

Fig. D.2

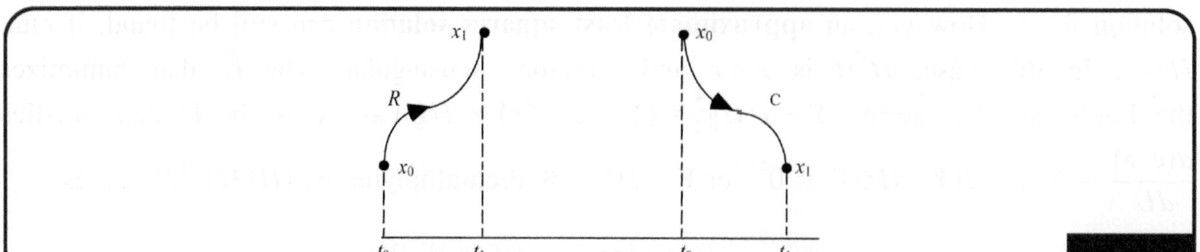

Fig. D.3

Example D.4: ***Controllable but not Reachable***

This very simple example illustrates the duality. Let $a(t)$ be any continuous function such that $a(t) \equiv 0$ for $t < 0$. Clearly, the (directly input-state connected) time-varying model $\dot{x} = a(t)u(t)$ is controllable for any $t_1 > 0$ and any $x_0 = x(0)$. But $x(0) \equiv x' \neq 0$ is not reachable, because $t_1 = 0$ in this case and $a(t) = 0$ for $t_0 < 0$.

Although controllability and reachability are different concepts, they do coincide for linear time-invariant models.

Example D.5: ***Controllable and Reachable States for LTI Models***

Let M be defined by

$$\dot{\boldsymbol{x}} = \begin{bmatrix} 1 & 0 \\ 0 & 1 \end{bmatrix} \boldsymbol{x} + \begin{bmatrix} 0 \\ 1 \end{bmatrix} u$$

where $u(t) \in \boldsymbol{U}^1$, the set of piece wise-continuous inputs, $t > 0$. If we are interested in the states reachable from the origin, $\boldsymbol{x}_0 \equiv \boldsymbol{0}$, these must have the form $\boldsymbol{x} = \begin{bmatrix} 0 & a \end{bmatrix}^T$, where a is arbitrary, because u is *not connected* to x_1 in any way. Note that for this model,

$$\text{rank } W = \text{rank} \begin{bmatrix} 0 & 0 \\ 1 & 1 \end{bmatrix} = 1$$

Thus, it is not controllable, as we have already found, i.e. $\boldsymbol{x}_0 = \begin{bmatrix} b & a \end{bmatrix}^T$, $b \neq 0$, could never be driven to the origin, again because $u(t)$ is never connected to the state variable x_1 and therefore cannot control it.

The following result for time-invariant models can be useful in determining the reachable or controllable states of a linear model.

The **set of reachable states** $\boldsymbol{x}_1 \in \boldsymbol{\Sigma}^n$ (or controllable states $\boldsymbol{x}_0 \in \boldsymbol{\Sigma}^n$) is the subspace of the state space $\boldsymbol{\Sigma}^n$ generated by the columns of the controllability matrix W (Casti 1977). This means that the state space $\boldsymbol{\Sigma}^n$ can be **decomposed** into controllable ($\boldsymbol{\Sigma}_c^{n_1}$) and uncontrollable ($\boldsymbol{\Sigma}_{\bar{c}}^{n_2}$) subspaces:

$$\boldsymbol{\Sigma}^{\rm n} \equiv \boldsymbol{\Sigma}_{\rm c}^{\rm n1} \oplus \boldsymbol{\Sigma}_{\bar{\rm c}}^{\rm n2} \tag{D.27}$$

where $n_1 + n_2 = n$. This construction decomposes the system dynamics into controllable and uncontrollable ($\boldsymbol{x}_{\bar{c}}$) parts:

$$\dot{\boldsymbol{x}}_c = A_{11}\boldsymbol{x}_c + A_{12}\boldsymbol{x}_{\bar{c}} + B_1\boldsymbol{x} \tag{D.28}$$

$$\dot{\boldsymbol{x}}_{\bar{c}} = A_{22}\boldsymbol{x}_{\bar{c}} \tag{D.29}$$

In other words: in problems which intrinsically involve the control input, only the subspace $\boldsymbol{\Sigma}_c^{n_1}$, and hence Eq. (D.28), have meaning; the fact that $n_1 < n$, perhaps much less than ($<<$) n, can be important in applications. (More on this is discussed in ***Appendix E***.)

Constructibility

A state $\boldsymbol{x}(\tau)$ is observable at time τ if it can be determined uniquely from knowledge of **future** inputs $\boldsymbol{u}(t)$ and outputs $\boldsymbol{y}(t)$, $\tau \le t \le T$. Constructibility is a property complementary to observability, but it is generally independent of it. A state $\boldsymbol{x}(t) \in \Sigma^n$ is **constructible** at time t if $\boldsymbol{x}(t)$ can be determined uniquely from *past* inputs $\boldsymbol{u}(\tau)$ and outputs $\boldsymbol{y}(\tau)$, $\tau \le t$ (Casti 1977).

For the observability problem, one attempts to determine the *cause* of observed future *effects* of the current state. In the constructibility problem, *the objective is to reconstruct the present state when there is a lack of complete knowledge of all state transitions, because not all state variables are known.* Constructibility and observability properties are coincident for linear time-invariant systems.

Example D.6: ***Digitoxin Metabolism and Therapy***

This example, adapted from Jeliffe et al. (1970) serves to show how unessential features of a model can be eliminated, as well as exemplifying the constructibility property.

Digitoxin is a powerful drug used to treat certain types of congestive heart failure (CHF). It is metabolized to digoxin and both substances are eliminated in the urine and otherwise degraded elsewhere. The Jeliffe group posed a six-compartment model for this system, the first compartment representing the (lumped) digoxin compartment of the body and the second the corresponding whole-body digoxin compartment. The remaining compartments are the urinary and nonurinary sinks for digitoxin and digoxin, respectively, all as shown in **Fig. D.4**. All state variables are masses, all k_{ij} are fractional transfer rates (time^{-1}), and only urinary concentrations are considered as measurable in this therapy application as y_1 and y_2.

We note that material transported to the nonurinary sinks 5 and 6 is never returned to compartments 1 or 2. Therefore, for purposes of analysis, we can omit compartments

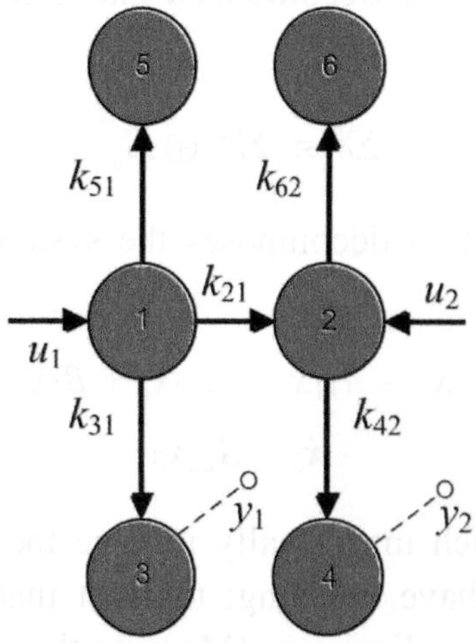

Fig. D.4

5 and 6, retaining the "leaks" to nonurinary sinks – but redefining them as k_{01} and k_{02}, as shown in **Fig. D.5**.

In the reference application, the digitoxin dose D is given as a brief pulse or approximate impulsive injection and a fraction (8%) of this appears virtually instantaneously in compartment 2. This represents 8% contamination of the dose of digitoxin by degradation to digoxin in storage. Writing $k_{11} = -(k_{21} + k_{31} + k_{01})$ and $k_{22} = -(k_{42} + k_{02})$, the resulting model equations are:

$$\dot{\boldsymbol{x}} = \begin{bmatrix} k_{11} & 0 & 0 & 0 \\ k_{21} & k_{22} & 0 & 0 \\ k_{31} & 0 & 0 & 0 \\ 0 & k_{42} & 0 & 0 \end{bmatrix} \boldsymbol{x} + \begin{bmatrix} u_1 \\ u_2 \end{bmatrix}$$
$$\boldsymbol{y} = \begin{bmatrix} 0 & 0 & 1/V_3 & 0 \\ 0 & 0 & 0 & 1/V_4 \end{bmatrix} \boldsymbol{x} \tag{D.30}$$

It is useful to transform the impulse inputs to compartments 1 and 2 into nonzero initial conditions (see Chapter 2). Thus, we have $u_1 \equiv u_2 \equiv 0$ and:

$$\boldsymbol{x}(0) = \begin{bmatrix} 0.92D & 0.8D & 0 & 0 \end{bmatrix}^T \tag{D.31}$$

One of the important clinical problems with digitoxin therapy is the fine line between lethal and therapeutically effective amount of digitoxin. The amount of drug present in the body must be known when contemplating additional doses, and this can only be estimated indirectly from the urinary measurements.

We pose this problem as a **state estimation problem**: can $\boldsymbol{x}(t)$ be determined from $\boldsymbol{y}(t)$, $t \geq 0$? To obtain a solution, the model must be *constructible*. The linear time-invariant

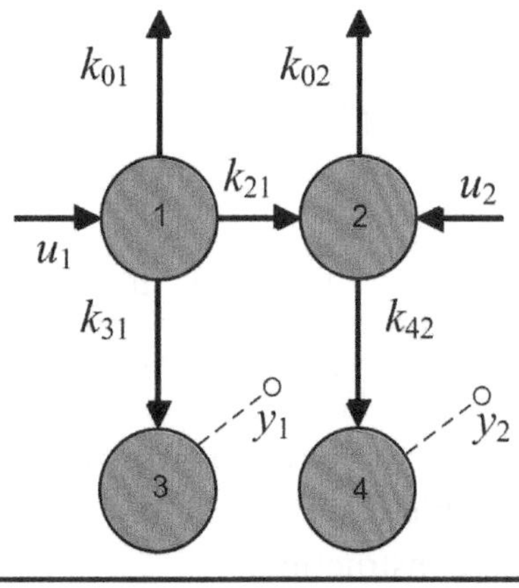

Fig. D.5

properties of this model render the observability and constructibility problems equivalent. Thus, we test the rank of the observability matrix V:

$$V^T = \begin{bmatrix} 0 & 0 & \vdots & \frac{k_{31}}{V_3} & 0 & \vdots & \frac{k_{11}k_{31}}{V_3} & \frac{k_{21}k_{42}}{V_4} & \vdots & \frac{k_{11}^2 k_{31}}{V_3} & \frac{(k_{11}+k_{22})k_{21}^2 k_{42}}{V_4} \\ 0 & 0 & \vdots & 0 & \frac{k_{42}}{V_4} & \vdots & 0 & \frac{k_{22}k_{42}}{V_4} & \vdots & 0 & \frac{k_{22}^2 k_{42}}{V_4} \\ \frac{1}{V_3} & 0 & \vdots & 0 & 0 & \vdots & 0 & 0 & \vdots & 0 & 0 \\ 0 & \frac{1}{V_4} & \vdots & 0 & 0 & \vdots & 0 & 0 & \vdots & 0 & 0 \end{bmatrix} \tag{D.32}$$

The first 4 columns are independent, that is, rank $V = 4 = n$, so the four-compartment model is completely observable/constructible, and $\boldsymbol{x}(t)$ can, in principle, be reconstructed from urinary measurements represented by $\boldsymbol{y}(t)$ over the time interval $t \geq 0$. Some additional algebra would show that the original six-compartment model is not observable, and the two additional states x_5 and x_6 are the unobservable ones. They are not needed to solve the clinical problem, and it is important that we recognized this early in modeling. We also note that only x_1 and x_2 need be observable.

Example D.7: ***Finding Reachable and Constructible States for a Discrete-Time Model***

$$\boldsymbol{x}(k+1) = \begin{bmatrix} a & 0 & 0 \\ 0 & a & 0 \\ 0 & 0 & b \end{bmatrix} \boldsymbol{x}(k) + \begin{bmatrix} 1 \\ 0 \\ 1 \end{bmatrix} u(k)$$

$$\boldsymbol{y}(k) = \begin{bmatrix} 1 & 0 & 0 \\ 0 & 0 & 1 \end{bmatrix} \boldsymbol{x}(k) \quad \text{for } k = 0, 1, 2, \ldots$$

This two-input two-output model is time-invariant, so the (equivalent) controllability and observability conditions apply. We use Eqs. (D.16) and (D.18) and thus check the rank of:

$$W = [B \vdots AB \vdots A^2B] = \begin{bmatrix} 1 & a & a^2 \\ 0 & 0 & 0 \\ 1 & b & b^2 \end{bmatrix}$$

$$V = [C \vdots CA \vdots CA^2] = \begin{bmatrix} 1 & 0 & 0 & a & 0 & 0 & a^2 & 0 & 0 \\ 0 & 0 & 1 & 0 & 0 & b & 0 & 0 & b^2 \end{bmatrix}$$

We have $rankW = 2 = rankV$ if $a \neq b \neq 1$. So two of the three state variables, x_1 and x_3, are reachable and constructible if $a \neq b \neq 1$. If $a = b$, both ranks $= 1$ and only one state variable is reachable and constructible.

Controllability and Observability with Constraints

In studies of the structural properties of dynamic system models we usually consider relatively unrestricted input signals $u(t) \in U^r$ and a virtually unrestricted state space Σ^n. Inputs $\boldsymbol{u}(t)$ are usually only required to be piecewise continuous functions. In many interesting and practical situations, one or both of these conditions may be undesirable, unreasonable, or impossible. For example, models of systems with signals representing masses or concentrations are normally constrained by $\boldsymbol{x}(t) \geq \boldsymbol{0}$ and $\boldsymbol{u}(t) \geq \boldsymbol{0}$ for all t. An exception is the class of models representing the dynamics of deviations, or perturbations, of masses or concentrations from steady state operating points, which may be positive or negative. We briefly consider next several situations in which controllability properties are restricted in one way or another.

Positive Controllability

Suppose the inputs can only be nonnegative (and piecewise-continuous), denoted $\boldsymbol{u}(t) \in \boldsymbol{U}_+^r$, with all such $\boldsymbol{u}(t) \geq \boldsymbol{0}$. The following result is useful for testing for positive controllability of a class of linear time-invariant models with matrices $\{A, B, C\}$.

SIMO (single-input) models are **positively controllable** if and only if: (a) the controllability matrix W has rank n; (b) A has only complex or purely imaginary eigenvalues; and (c) t is sufficiently large (Gabasov and Kirillova 1976).

Note that no single-input model can be positively controllable for *all* t (from (c) above), i.e. over *any* interval. This situation is significantly different than that for controllability of linear time invariant models in general, where complete controllability in some finite time implies complete controllability in an arbitrary short time. And, from (b) above, we note that no single-input model can be positively controllable if it is of odd order n, because complex roots (eigenvalues) always occur in pairs, that is, only real roots can be unpaired.

Relative Controllability (Reachability)

Sometimes it is only necessary to consider whether or not it is possible to transfer an initial state $\boldsymbol{x}_0$ to a given *subspace* of $\boldsymbol{\Sigma}^n$, rather than a *fixed point* $\boldsymbol{x}_1 \in \boldsymbol{\Sigma}^n$ (a *reachability* problem). This situation is typical of many practical control problems, including optimal therapy problems in pharmacology and medicine, where it is only necessary to restore a (state) variable to its **normal range** of values.

We define the given subspace by $\boldsymbol{M} = \{\boldsymbol{x}\colon M\boldsymbol{x} = \boldsymbol{0} \text{ and } \boldsymbol{x} \in \boldsymbol{\Sigma}^n\}$ and also define $\boldsymbol{M}$ as *reachable* if for every $\boldsymbol{x}_0$ there exists a finite time $\tau \geq t_0$ and an admissible control $\boldsymbol{u} \in \boldsymbol{U}^r$ such that $M\boldsymbol{x}(\tau) = \boldsymbol{0}$. We have the following easily applicable result. The linear model $\{A, B, C\}$ is *relatively controllable (reachable)* to the subspace $\boldsymbol{M}$ (or $\boldsymbol{M}$ is reachable) if and only if $rank(MW) = rank M$, where W is the controllability matrix (Gabasov and Kirillova 1976).

Conditional Controllability

Here we have a variation on the constraint on the state space considered above. Suppose a particular subspace $\boldsymbol{N} \equiv \{\boldsymbol{x}_0 : \boldsymbol{x}_0 = N\boldsymbol{x}, \boldsymbol{x} \in \Sigma^n\}$ of Σ^n is given and we want to know whether every state in $\boldsymbol{N}$ is controllable. This situation can occur if we have influence over or prior knowledge of $\boldsymbol{x}_0$, for example, the initial dose of a drug and its metabolites or contaminants. In this case, the linear model $\{A, B, C\}$ is conditionally controllable from a subspace $N = \{\boldsymbol{x}_0 = N\boldsymbol{x}\}$ if and only if $rank\left[N \ \vdots \ W\right] = rank W$ (Gabasov and Kirillova 1976).

Example D.8: ***Conditionally Controllable Subspace***

Consider $\dot{\boldsymbol{x}} = A\boldsymbol{x} + B\boldsymbol{u}$, where

$$A = \begin{bmatrix} 0 & -1 & -5 & -14 \\ 1 & 2 & -1 & 6 \\ 0 & 0 & -1 & -4 \\ 0 & 0 & -1 & 3 \end{bmatrix} \quad \text{and} \quad B = \begin{bmatrix} 0 & 0 & 0 & 0 \\ 3 & 1 & 0 & 0 \\ 4 & 0 & 1 & 0 \\ 5 & 0 & 2 & 1 \end{bmatrix}$$

Suppose $\boldsymbol{x}_0$ lies in the plane defined by $\boldsymbol{x}_0 = \begin{bmatrix} 0 & 0 & x_3 & x_4 \end{bmatrix}^T$. Is every state $\boldsymbol{x}_0$ in the subspace (with $x_1 \equiv x_2 \equiv 0$ and x_3, x_4 arbitrary) controllable? We have, from $\boldsymbol{x}_0 = N\boldsymbol{x}$,

$$N = \begin{bmatrix} 0 & 0 & 0 & 0 \\ 0 & 0 & 0 & 0 \\ 0 & 0 & 1 & 0 \\ 0 & 0 & 0 & 1 \end{bmatrix}$$

Some additional algebra shows that rank $[N \ \vdots \ W] = rank W$ and thus the model is conditionally controllable from the subspace defined by $\boldsymbol{N}$.

Structural Controllability and Observability

The controllability properties of a linear ODE model with matrices $\{A, B, C\}$ generally depend on the specific numerical values the elements of A and B can assume, that is, the rank of W depends on a_{ij} and b_{ij} values. This restriction can be too stringent and a more formal topological concept can be useful, one that depends only on the internal **connectivity** of the model, as with *structural identifiability* (Chapter 10)

Assume the elements of A and B are either fixed zeros, for representing nonconnectedness in the equations, or arbitrary nonzero parameters, which represent connections. Such matrices are called **structured**; a **structured system or model** is an ordered pair of structured matrices. Further, two models M and M' are called **structurally equivalent** if there is a one-to-one correspondence between the locations of their fixed zero and nonzero entries.

A model $\{A, B, -\}$ is **structurally controllable** if there exists a model structurally equivalent to it which is controllable in the usual sense (Casti 1977). Note that only the A and

B matrices need be tested for equivalence, as C takes no part in the controllability properties. Structural observability is defined similarly.

Observability and Identifiability Relationships

The qualitative properties of models, controllability and observability, are related to the identifiability property. However, they are neither necessary nor sufficient conditions for identifiability. In essence, *all that is needed for identifiability of an unknown parameter of interest is that it be included in that part of the model which is both controllable and observable.* But this neither guarantees the identifiability of that parameter nor are the parameters in the uncontrollable or unobservable parts necessarily unidentifiable, a nuance of identifiability concepts not fully appreciated in the early literature (Cobelli and Romanin-Jacur 1976; DiStefano III 1977; Cobelli et al. 1978; DiStefano III 1978; Jacquez 1978). These two points are made by the following two simple counterexamples, first reported in DiStefano III (1977).

Example D.9: ***Controllable and Observable but not Identifiable***

We consider the general two-compartment model shown to be unidentifiable when $k_{ij} > 0$ and $V_i > 0$. This model, however, is controllable and observable; both the controllability matrix W and the observability matrix V have rank 2:

$$W = [B \mid AB] = \begin{bmatrix} 1 & -(k_{21} + k_{01}) \\ 0 & k_{21} \end{bmatrix}$$

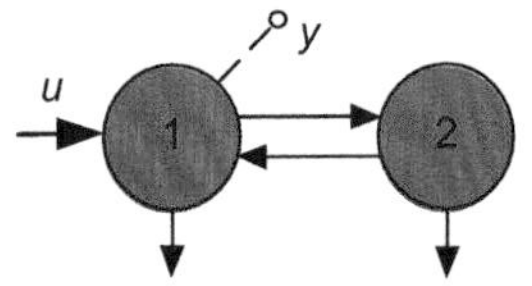

$$V = \begin{bmatrix} C \\ \cdots \\ CA \end{bmatrix} = \begin{bmatrix} 1/V_1 & 0 \\ \dfrac{-(k_{01} + k_{21})}{V_1} & \dfrac{k_{12}}{V_1} \end{bmatrix}$$

Thus, controllability and observability are not sufficient conditions for identifiability.

Example D.10: ***Identifiable but not Controllable or Observable***

Consider the (not necessarily compartmental) model:

$$\dot{x}_1 = -p_1 x_1 + u, \quad \dot{x}_2 = -p_1 x_2, \quad \dot{x}_3 = -p_2 x_3 + u, \quad y_1 = x_1, \quad y_2 = x_3$$

Thus,

$$A = \begin{bmatrix} -p_1 & 0 & 0 \\ 0 & -p_1 & 0 \\ 0 & 0 & -p_2 \end{bmatrix}, \quad B = \begin{bmatrix} 1 \\ 0 \\ 1 \end{bmatrix}, \quad C = \begin{bmatrix} 1 & 0 & 0 \\ 0 & 0 & 1 \end{bmatrix}$$

and the TF matrix is

$$H(s,\boldsymbol{p}) = C[sI - A(\boldsymbol{p})]^{-1}B = \begin{bmatrix} \frac{1}{(s+p_1)} & \frac{1}{(s+p_2)} \end{bmatrix}^T = \begin{bmatrix} H_{11}(s,\boldsymbol{p}) & H_{21}(s,\boldsymbol{p}) \end{bmatrix}^T$$

Thus, the two model parameters p_1 and p_2 are determinable from the two input/output experiments defined by H_{11} and H_{21} and the model is *uniquely SI*. The state variable x_2 and therefore the model is, however, both *uncontrollable* and *unobservable*, which we see by inspection, because the model state variables are completely uncoupled: x_2 is not connected in any way to the input or output. The reason for unique identifiability of p_1 and p_2, of course, is that the a_{22} element of A is equal to a_{11}, and therefore it is not needed via measurement of (unobservable) x_2; it is available via x_1. This is yet another example illustrating the importance of including all parameter knowledge (constraints) in the problem statement.

Thus, it is neither necessary nor sufficient to consider whether or not a model is controllable or observable if only identifiability is of interest. For LTI models, identifiability is imbedded in the properties of the TF (or equivalent for NL models). The TF matrix of the model automatically excludes the unobservable and uncontrollable portions of the model. Such additional analysis may of course yield other useful insights, but it is not required when the identifiability properties of the model are to be determined from its TF.

Controllability and Observability of Stochastic Models

The controllability and observability criteria given by Eqs. (D.9) and (D.13) and the two theorems that follow them extend readily to stochastic system models, with disturbance inputs $\boldsymbol{w}(t)$ and measurement noise $\boldsymbol{v}(t)$, each having probabilistic model descriptions. Consider stochastic models given by

$$\dot{\boldsymbol{x}} = A(t)\boldsymbol{x} + E(t)\boldsymbol{w} \quad \text{(D.33)}$$

$$\boldsymbol{z} = C(t)\boldsymbol{x} + F(t)\boldsymbol{v} \quad \text{(D.34)}$$

The random vectors $\boldsymbol{w}$ and $\boldsymbol{v}$ are members of statistically independent, Gaussian stochastic processes with covariance matrices $Q(t)$ and $R(t)$, respectively. The **stochastic controllability** condition requires that the following matrix be nonsingular for some $t > 0$:

$$W^*(t_0, t_1) \equiv \int_{t_0}^{t} \Phi(t,\tau)E(\tau)Q(\tau)E^T(\tau)\Phi^T(t,\tau)d\tau \quad \text{(D.35)}$$

This condition ensures that all modes are what is termed **persistently excited** by $\boldsymbol{w}$. Similarly, for **stochastic observability**, the following matrix must be nonsingular for some $t > t_0$:

$$V^*(t_0, t_1) \equiv \int_{t_0}^{t} \Phi^T(\tau, t_0) C^T(\tau) F^T(\tau) R^{-1}(\tau) F(\tau) C(\tau) \Phi(\tau, t_0) d\tau \tag{D.36}$$

This matrix is also called the **Fisher information matrix for the state estimation problem** for this model. Note that as the covariance R "increases," this matrix becomes "more singular," i.e. there is "less information" about the state in measurements of $\boldsymbol{z}$. Finally, note the similarity between Eqs. (D.35) and (D.9) and Eqs. (D.36) and (D.13).

References

Casti, J.L., 1977. Dynamical Systems and Their Applications. Academic Press, New York.

Casti, J.L., 1987. Linear Dynamical Systems. Academic Press, New York.

Chen, C., 1970. Introduction to Linear System Theory. Holt, Rinehart, & Winston, New York.

Chen, C., 1985. Introduction to Linear System Theory. Holt, Rinehart, & Winston, New York.

Cobelli, C., Lepschy, A., Romanin-Jacur, G., 1978. Comments on 'On the relationship between structural identifiability and the controllability, observability properties'. IEEE Trans. Automat. Control. 23, 965–966.

Cobelli, C., Romanin-Jacur, G., 1976. Controllability, observability, and structural identifiability of multi-input and multi-output biological compartmental systems. IEEE Trans. Biomed. Eng. BME-23, 93–100.

DiStefano III, J., 1977. On the relationships between structural identifiability and the controllability, observability properties. IEEE Trans. Automat. Control. 22, 652.

DiStefano III, J., 1978. Reply to 'Comments on 'On the relationships betweeen structural identifiability and the controllability, observability properties''. IEEE Trans. Automat. Control. 23, 966.

Gabasov, R., Kirillova, F., 1976. The Qualitative Theory of Optimal Processes. M Dekker, New York.

Jacquez, J., 1978. Further comments on 'On the relationships between structural identifiability and the controllability, observability properties'. IEEE Trans. Automat. Control. 23, 966.

Jeliffe, R., Buell, J., Kalaba, R., Shridar, R., Rockwell, R., 1970. A mathematical study of the metabolic conversion of digitoxin to digoxin in man. Math. Biosci. 6, 387–403.

Decomposition, Equivalence, Minimal & Canonical State Variable Models

Appendix E

Realizations (Modeling Paradigms)

The realization problem is about finding a suitable structural ODE model for matching input–output (I–O) data and at the same time satisfying modeling goals. Some are better than others – even though we assume they all match the I–O data and are therefore dynamically **I–O equivalent**. Realization theory is well developed for linear ODE models, with broad application in engineering and other areas that use dynamic system modeling methodology. For NL models, problems are typically addressed on an ad hoc, case-by-case basis, often using linearization methods. In this appendix, we develop realization and equivalence methodology for linear, time-invariant dynamic system models and classes of biosystems and experiments representable by such models. Linear multicompartmental models based on small-perturbation tracer experiments performed on nonlinear (NL) systems, and any other NL model operating in a linear range, are included.

Modeling goals sometimes can be satisfied with unstructured I–O models. In addition, unstructured I–O models can be a starting point for structured modeling (*Let the Data Speak First!*). If the system is linear and time invariant, or the NL system experiment is linearizing – with linearized model – the simplest and usually quantifiable I–O representation (realization) is the *transfer function* (TF) *H*(*s*), where:

$$Y(s) = H(s)U(s) \tag{E.1}$$

For example, impulse response data might first be fitted with a multiexponential function *y*(*t*), as described in Chapter 13. Inverse Laplace transformation of the result is the TF *H*(*s*). The TF *H*(*s*) also can be derived from I–O data by other methods as well, for example, deconvolution and Bode methods (Chen 1985; Casti 1987; DiStefano III et al. 1990).

The elements of the matrix *H*(*s*) are typically representable as rational algebraic expressions (ratios of polynomials in *s*), with constant coefficients and $H(\infty) =$ constant. Under these conditions, *H*(*s*) can be **realized** in **state variable (ODE) form**, abbreviated {*A, B, C*}, as:

$$\begin{aligned} \dot{\boldsymbol{x}} &= A(\boldsymbol{p})\boldsymbol{x} + B(\boldsymbol{p})\boldsymbol{u} \\ \boldsymbol{y} &= C(\boldsymbol{p})\boldsymbol{x} \end{aligned} \tag{E.2}$$

with constraints $\boldsymbol{h}(\boldsymbol{x}, \boldsymbol{u}, \boldsymbol{p}) \geq \boldsymbol{0}$ included. Suppressing the dependence of *A*(***p***), *B*(***p***), *C*(***p***), and *H*(***p***) on parameters ***p*** and constraints ***h***, the TF then becomes:

$$H(s) = C[sI - A]^{-1}B \tag{E.3}$$

Equation (E.3) defines the precise **equivalence** of **unstructured I–O** representation (E.1) and **structured state variable (ODE)** representation (E.2).

ODE models are our interest. If structural information about the system is available, or if a hypothesized structure is being tested, the matrices *A*, *B* and *C* are more or less well defined topologically, as **structured matrices**. With some exceptions, this is how we have done most modeling in this book. *Realization and equivalence theory expand on these options*. For every realizable *H*(*s*), an *unlimited number* of realizations $\{A, B, C\}$ are still obtainable from *H*(*s*), whether or not structural information is available. *Equivalence transformations* of the ODEs are the main focus here. Many modeling and model analysis algorithms are based on equivalence transformations, including *model discrimination* and *structural identifiability* algorithms presented in Chapters 5 and 10.

State variable (ODE) models are typically capable of reproducing observed I–O behavior to the extent required by modeling goals. However, different structures (realizations or model classes), rather than just fixed structures, can be allowable candidates. The *simplest* models that satisfy these criteria are realizations $\{A, B, C\}$ with **minimum dimension *n***, so-called **minimal models**. Importantly, all realizations of $\{A, B, C\}$ of *H*(*s*) that are both **controllable and observable** are minimal models, as we show below. We also develop realizations for model classes not necessarily of minimal dimension, called **canonical models**.

The Canonical Decomposition Theorem

This important result is due to Kalman (1962). In brief, it says that the dimension of a state variable model can sometimes be reduced without sacrificing any of the I–O (TF) properties of the model. Furthermore, this decomposition results in a model with a more illuminating structure, making it easier to draw conclusions about the reachable (controllable) and observable subspaces of the model (Appendix D). Our purpose here is to illustrate the importance and subtleties of this theorem, which codifies many of the limitations of modeling (at least) linear dynamic systems. It utilizes controllability and observability properties quite explicitly and brings to mind identifiability issues in the process.

Theorem: An n-dimensional realization $\{A, B, C\}$ can be transformed into the following canonical form by **equivalence transformations**:

$$\begin{bmatrix} \dot{x}^*_{c\bar{o}} \\ \cdots \\ \dot{x}^*_{co} \\ \cdots \\ \dot{x}^*_{\bar{c}o} \\ \cdots \\ \dot{x}^*_{\bar{c}\bar{o}} \end{bmatrix} = \left[\begin{array}{c:c:c:c} A^*_{c\bar{o}} & A^*_{12} & A^*_{13} & A^*_{14} \\ \hdashline 0 & A^*_{co} & A^*_{23} & A^*_{24} \\ \hdashline 0 & 0 & A^*_{\bar{c}o} & A^*_{34} \\ \hdashline 0 & 0 & 0 & A^*_{\bar{c}\bar{o}} \end{array}\right] \boldsymbol{x}^* + \begin{bmatrix} B^*_{c\bar{o}} \\ \cdots \\ B^*_{co} \\ \cdots \\ 0 \\ \cdots \\ 0 \end{bmatrix} \boldsymbol{u} \tag{E.4}$$

$$y = [0 \quad C^*_{co} \quad C^*_{\bar{c}o} \quad 0]\boldsymbol{x}^* \tag{E.5}$$

In Eqs. (E.4) and (E.5), the equivalence transformed state vector $\boldsymbol{x}^* \in \Sigma^n$ (n-dimensional linear vector space) is decomposed into four state vectors (**subspaces**) whose dimensions add up to n:

$$\boldsymbol{x}^* \equiv \begin{bmatrix} x^*_{c\bar{o}} \\ x^*_{co} \\ x^*_{\bar{c}o} \\ x^*_{\bar{c}\bar{o}} \end{bmatrix} \equiv \begin{bmatrix} \text{the controllable but unobservable part} \\ \text{the controllable but observable part} \\ \text{the uncontrollable but observable part} \\ \text{the uncontrollable but unobservable part} \end{bmatrix} \tag{E.6}$$

The TF matrix of Eqs. (E.4) and (E.5), and hence of Eq. (E.3), is

$$H(s) = C^*_{co}[sI - A^*_{co}]^{-1} B^*_{co} \tag{E.7}$$

It is clear from Eq. (E.7) that *H(s) depends only on the controllable and observable parts of the model.* This is illustrated in the detailed **top-down** block diagram of Eqs. (E.4) and (E.5) in ***Fig. E.1***, and in the simplified (**bottom-up**) block diagram in ***Fig. E.2***.

In Other Words: *the I–O description (TF, impulse-response-matrix) describes only that part of a system model that is both controllable and observable.* ***Figure E.2*** tells it all.

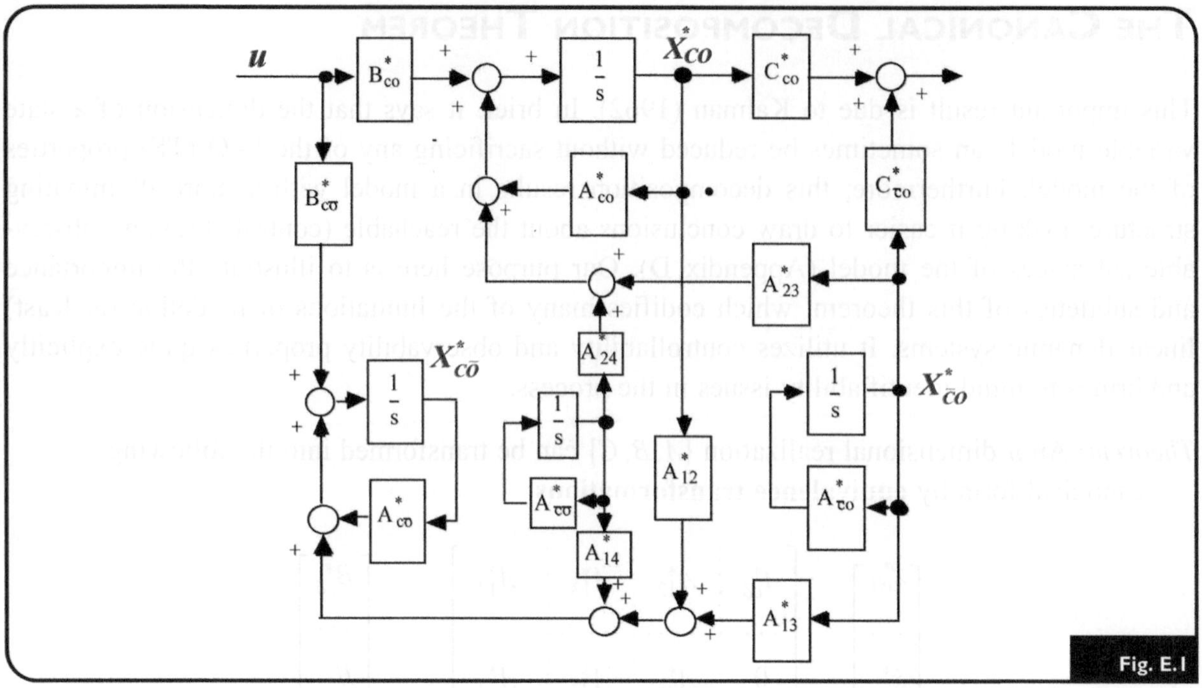

Fig. E.1

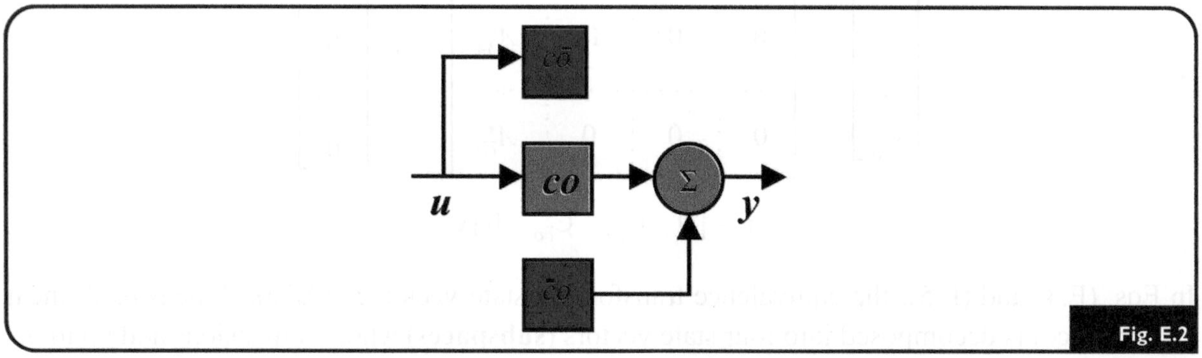

Fig. E.2

Uncontrollable and/or unobservable parts of a model are not and cannot be represented in the TF matrix (i.e. I−O data). In general, $H(s)$ is insufficient to completely describe the full structure of the real system. Furthermore, if M is either uncontrollable or unobservable, it is **reducible**, which means that there exists a state variable model of lesser dimension (and fewer couplings) with the same $H(s)$.

Remark: This profound result is surely one of the main points emphasized and reemphasized over and over throughout this book. ***Dynamical output data tells all about dynamics. But it generally reveals only a fraction of the structural components.***

Also note that M is **irreducible** or **minimal** if and only if it is both controllable and observable. Furthermore, all minimal models are **I−O equivalent** and thus all minimal realizations have the same dimension.

We explore some of the consequences of these important results below. Keep in mind, however, that there's nothing magical or mysterious about systems with uncontrollable or unobservable states in their models. Surely, there is likely some set of input/output experiments with controllable and observable submodels of the same dynamic system, at least in principle. The key here is that *these properties depend completely on input/output experiments as well as system dynamics*.

How to Decompose a Model

Any linear system model can be decomposed as above, but it is generally not easy. The basic idea is that the state space can be divided into four subspaces $(c\bar{o}, co, \bar{c}o, \overline{co})$ by an equivalence transformation. This is easier to see (and do) in several steps. First, $\boldsymbol{\Sigma}^n$ is divided into $\boldsymbol{\Sigma}^{n_1} \oplus \boldsymbol{\Sigma}^{(n-n_1)}$, where n_1 is the rank of the controllability matrix W. The required equivalence transformation is $\boldsymbol{x}^* = P_1\boldsymbol{x}$ and

$$P_1^{-1} \equiv \begin{bmatrix} \boldsymbol{w}_1 & \boldsymbol{w}_2 & \cdots & \boldsymbol{w}_{n_1} & \cdots & \boldsymbol{w}_n \end{bmatrix} \tag{E.8}$$

where $\boldsymbol{w}_1, \ldots, \boldsymbol{w}_{n_1}$ are any n_1 linearly independent *columns* of W and the remaining $(n - n_1)$ columns are *arbitrary* vectors chosen to make P_1 nonsingular. This transformation yields two subspaces: $\boldsymbol{\Sigma}^{n_1}$, consisting of all controllable composite vectors $\begin{bmatrix} x_c^* & 0 \end{bmatrix}^{\mathrm{T}}$, and $\boldsymbol{\Sigma}^{(n-n_1)}$, consisting of all uncontrollable composite vectors $\begin{bmatrix} 0 & x_{\bar{c}}^* \end{bmatrix}^{\mathrm{T}}$. Then, in a similar manner, $\boldsymbol{\Sigma}^{n_1}$ can be divided into two subspaces, $\boldsymbol{\Sigma}^{n_2}$ and $\boldsymbol{\Sigma}(n_1 - n_2)$, where the first is observable and the second is unobservable, by another equivalence transformation:

$$\begin{bmatrix} x_{co}^* \\ x_{c\bar{o}}^* \end{bmatrix} \equiv P_2\boldsymbol{x}_c^* \tag{E.9}$$

In this case,

$$P_2^{-1} \equiv \begin{bmatrix} \boldsymbol{v}_1^{\mathrm{T}} & \boldsymbol{v}_2^{\mathrm{T}} & \cdots & \boldsymbol{v}_{n_2}^{\mathrm{T}} & \cdots & \boldsymbol{v}_{(n_1-n_2)}^{\mathrm{T}} \end{bmatrix}^{\mathrm{T}} \tag{E.10}$$

where $\boldsymbol{v}_1, \ldots, \boldsymbol{v}_{n_2}$ are any n_2 linearly independent *rows* of the observability matrix V, n_2 is the rank of V, and the remaining $(n_1 - n_2)$ rows of P_2^{-1} are arbitrary vectors chosen to make P_2 nonsingular. This process is repeated to further subdivide the state space into its remaining decomposed parts.

Example E.1: ***Equivalent Model***

The model: $\dot{x}_1 = -x_1 + u, \quad \dot{x}_2 = -x_2 + u$ has the block diagram in ***Fig. E.3***. For this model, rank $W =$ rank $\begin{bmatrix} 1 & -1 \\ 1 & -1 \end{bmatrix} = 1$. We can choose $P_1^{-1} \equiv [\boldsymbol{w}_1 \quad \boldsymbol{w}_2] \equiv \begin{bmatrix} 1 & 1 \\ 1 & 0 \end{bmatrix}$, where the second column is linearly independent of the first; any other linearly independent two-vector would also qualify. Then the equivalent model is

$$\dot{\boldsymbol{x}}^* = (P_1AP_1^{-1})\boldsymbol{x}^* + (P_1B)u = \begin{bmatrix} -1 & 0 \\ 0 & -1 \end{bmatrix}\begin{bmatrix} x_1^* \\ x_2^* \end{bmatrix} + \begin{bmatrix} 0 & 1 \\ 1 & -1 \end{bmatrix}\begin{bmatrix} 1 \\ 1 \end{bmatrix}u$$

$$= \begin{bmatrix} -1 & 0 \\ 0 & -1 \end{bmatrix}\begin{bmatrix} x_1^* \\ x_2^* \end{bmatrix} + \begin{bmatrix} 1 \\ 0 \end{bmatrix}u$$

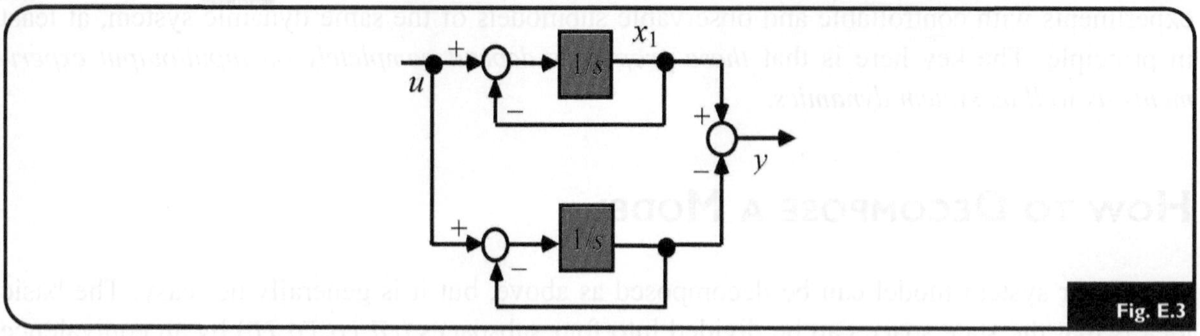

Fig. E.3

By inspection, x_1^* is controllable and x_2^* is not, because x_2^* is completely disconnected from the input (compare the result with Eq. (E.6)). Then, from ***Fig. E.3***, the output equation is

$$y = [1 \quad -1]\mathbf{x} \equiv CP_1^{-1}\mathbf{x}^* = [0 \quad 1]\begin{bmatrix} x_1^* \\ x_2^* \end{bmatrix}$$

which clearly makes x_2^* observable and x_1^* not, by inspection. Thus, the controllable portion of the model is defined by the submodel:

$$\dot{x}_1^* = -x_1^* + u$$
$$y = 0 \cdot x_1^*$$

Thus $C_c^* \equiv 0$ and the observability matrix V for this submodel is

$$\begin{bmatrix} C_c^* \\ \cdots \\ C_c^* \mathrm{A} \end{bmatrix} = \begin{bmatrix} 0 \\ \cdots \\ 0 \end{bmatrix}$$

which renders it unobservable overall. The canonically (partially) decomposed system is thus given by:

$$\begin{bmatrix} \dot{x}_{c\bar{o}}^* \\ \dot{x}_{\bar{c}}^* \end{bmatrix} = \begin{bmatrix} A_{c\bar{o}}^* & A_{12}^* \\ 0 & A_{\bar{c}}^* \end{bmatrix}\begin{bmatrix} x_{c\bar{o}}^* \\ x_{\bar{c}}^* \end{bmatrix} + \begin{bmatrix} B_{c\bar{o}}^* \\ 0 \end{bmatrix} u$$

$$\begin{bmatrix} \dot{x}_1^* \\ \dot{x}_2^* \end{bmatrix} = \begin{bmatrix} -1 & 0 \\ 0 & -1 \end{bmatrix}\begin{bmatrix} x_1^* \\ x_2^* \end{bmatrix} + \begin{bmatrix} 1 \\ 0 \end{bmatrix} u$$

$$y = \begin{bmatrix} 0 & C_{\bar{c}}^* \end{bmatrix}\mathbf{x}^* = \begin{bmatrix} 0 & 1 \end{bmatrix}\begin{bmatrix} x_1^* \\ x_2^* \end{bmatrix}$$

The TF is $H(s) = 0$, and the block diagram of the equivalent model is shown in ***Fig. E.4***.

Remark: The observability and controllability properties are *not obvious in the original block diagram* in ***Fig. E.3***, even though hindsight and deeper understanding and experience with

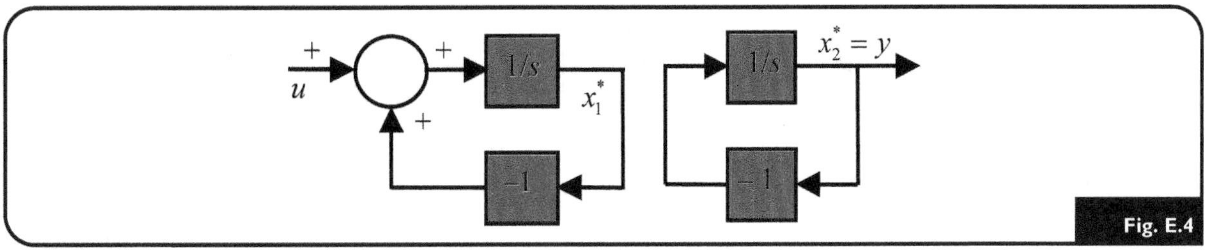

Fig. E.4

such models can generate the same conclusions about observability and/or controllability properties by inspection, *if* the model is simple enough. Even slightly greater complexity can render "inspection" virtually useless, or dangerous, if the conclusion addresses safety issues in some way, for example, as in a biomodel for prescribing drugs.

Example E.2: ***Another Equivalent Model***

We decompose the following two-input/two-output model into canonical form:

$$\dot{\boldsymbol{x}} = \begin{bmatrix} -7 & -2 & 6 \\ 2 & -3 & -2 \\ -2 & -2 & 1 \end{bmatrix} \boldsymbol{x} + \begin{bmatrix} 1 & 1 \\ 1 & -1 \\ 1 & 0 \end{bmatrix} u$$

$$\boldsymbol{y} = \begin{bmatrix} -1 & -1 & 2 \\ 1 & 1 & -1 \end{bmatrix} \boldsymbol{x}$$

The eigenvalues of A are $\lambda_i = -1, \ -3, \ -5$. The eigenvectors are easy to find (Appendix B) and they form the modal matrix and its inverse:

$$P^{-1} = \begin{bmatrix} 1 & 1 & 1 \\ 0 & 1 & -1 \\ 1 & 1 & 0 \end{bmatrix} \qquad P = \begin{bmatrix} -1 & -1 & 2 \\ 1 & 1 & -1 \\ 1 & 0 & -1 \end{bmatrix}$$

Thus, $\boldsymbol{x}^* = P\boldsymbol{x}$, $J = \Lambda = PAP^{-1}$, and the equivalent model is

$$\dot{\boldsymbol{x}}^* = \begin{bmatrix} -1 & 0 & 0 \\ 0 & -3 & 0 \\ 0 & 0 & -5 \end{bmatrix} \boldsymbol{x}^* + \begin{bmatrix} 0 & 0 \\ 1 & 0 \\ 0 & 1 \end{bmatrix} \begin{bmatrix} u_1 \\ u_2 \end{bmatrix}$$

$$\begin{bmatrix} y_1 \\ y_2 \end{bmatrix} = \begin{bmatrix} 1 & 0 & 0 \\ 0 & 1 & 0 \end{bmatrix} \boldsymbol{x}^*$$

with its block diagram shown in ***Fig. E.5***.

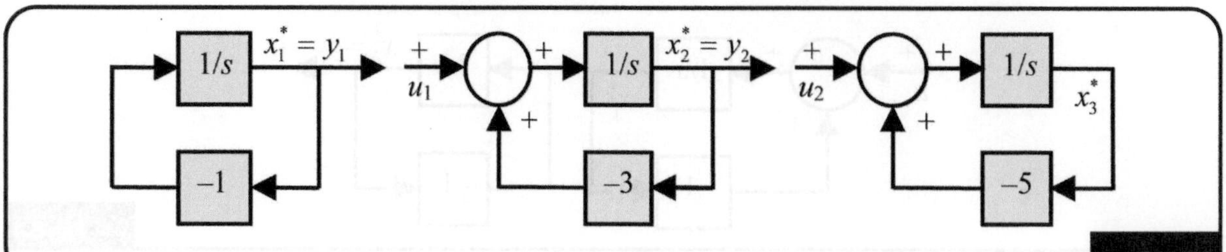

Fig. E.5

It is directly apparent from the equivalent model in ***Fig. E.5*** that $x_1{}^*$ and $x_2{}^*$ are observable and $x_3{}^*$ is not; and $x_2{}^*$ and $x_3{}^*$ are controllable and $x_1{}^*$ is not. Thus, we have the canonical form corresponding to Eqs. (E.4) and (E.5):

$$\begin{bmatrix} \dot{x}_3^* \\ \dot{x}_2^* \\ \dot{x}_1^* \end{bmatrix} = \begin{bmatrix} -5 & 0 & 0 \\ 0 & -3 & 0 \\ 0 & 0 & -1 \end{bmatrix} \begin{bmatrix} x_3^* \\ x_2^* \\ x_1^* \end{bmatrix} + \begin{bmatrix} 0 & 1 \\ 1 & 0 \\ 0 & 0 \end{bmatrix} \begin{bmatrix} u_1 \\ u_2 \end{bmatrix}$$

$$\begin{bmatrix} y_1 \\ y_2 \end{bmatrix} = \begin{bmatrix} 0 & 0 & 1 \\ 0 & 1 & 0 \end{bmatrix} \begin{bmatrix} x_3^* \\ x_2^* \\ x_1^* \end{bmatrix}$$

and the 2×2 TF is

$$H(s) = C_{co}^* \left[sI - A_{co}^* \right]^{-1} B_{co}^* = \begin{bmatrix} 0 \\ 1 \end{bmatrix} \left(\frac{1}{s+3} \right) \begin{bmatrix} 1 & 0 \end{bmatrix} = \frac{1}{s+3} \begin{bmatrix} 0 & 0 \\ 1 & 0 \end{bmatrix}$$

This example was easy because all eigenvalues were distinct and the order n was small. Models with distinct eigenvalues decompose into composite first-order submodels connected in parallel, i.e. they are **parallel composite models**.

Controllability and Observability Tests Using Equivalent Models

Sometimes we can directly test controllability and observability properties by inspection of an equivalent model block diagram, as we did in *Example E.2*. Following are several results, adapted from (Chen, 1985) – further illustrating modeling limitations – and useful in formulation and reduction of canonical models, discussed later.

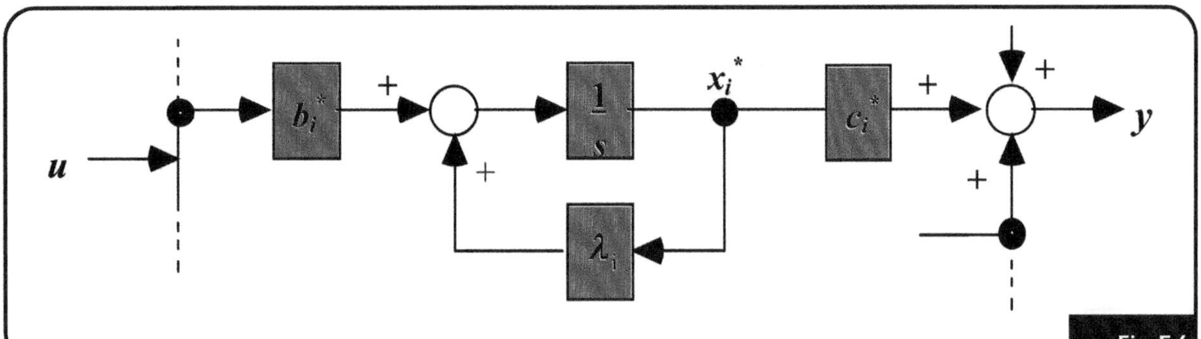

Fig. E.6

SISO Models

Consider

$$\dot{x} = Ax + Bu, \quad y = Cx \tag{E.11}$$

Case 1

If A has *distinct eigenvalues*, M is *controllable* if and only if the vector $B \equiv PB$ has no zero elements, where P^{-1} is the modal matrix, with eigenvectors of A as its columns. Similarly, M is *observable* if and only if $C^* \equiv CP^{-1}$ has no zero elements.

The block diagram in this case (i.e., parallel models for each mode) makes this result obvious, as zero elements in B^* (or C^*) represent **unconnectedness** of the input (or output) to the mode, or equivalent state, where each *mode* is represented by a canonical feedback loop, like the one in the middle of ***Fig. E.6***.

In ***Fig. E.6***, if $b_i^* = 0$ or $c_i^* = 0$, the mode is disconnected from the I–O path.

Case 2 (SISO)

If A is *arbitrary*, then M is *controllable* if and only if:

(i) Each eigenvalue λ_i associated with a Jordan block L_{ji} (λ_i) (Appendix A) is distinct from each eigenvalue associated with every other Jordan block.

(ii) Each element of the column vector $B^* \equiv PB$ associated with the *last row* of each Jordan block is nonzero. Furthermore, M is *observable* if and only if requirement (*i*) above is true.

(iii) Each element of the row vector $C^* \equiv CP^{-1}$ associated with the *first column* of each Jordan block is nonzero.

We illustrate the block diagram for one $l \times l$ Jordan block L_{ji} (λ_i) of

$$\dot{x} = Jx^* + B^*u$$
$$y = C^*x^*$$

as in ***Fig. E.7***:

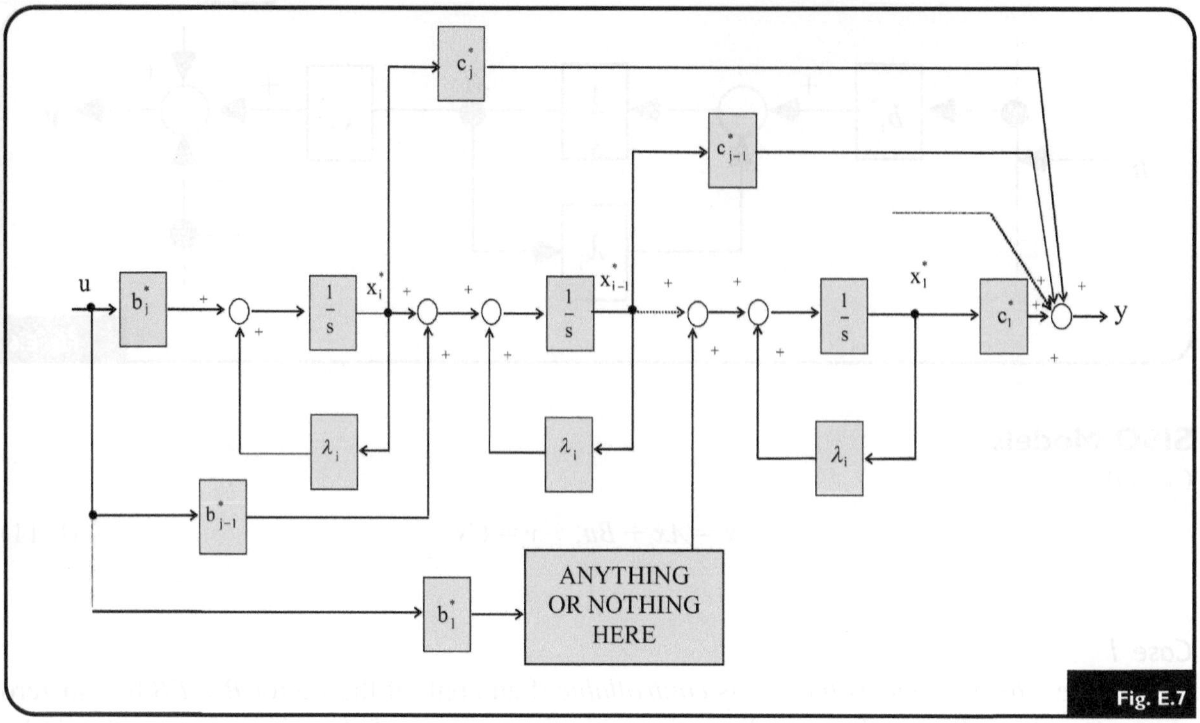

Fig. E.7

It is clear from ***Fig. E.7*** that M is controllable and observable if and only if $b_j^* \neq 0$ and $c_1^* \neq 0$. The remaining b_{j-k}^* and c_{j-k}^*, for $k = 1, 2, \ldots, j-1$, do not matter. Note that *Jordan blocks are series-connected composite submodels with canonical feedback loops.*

MIMO Models

We now consider vector inputs and outputs:

$$\dot{x} = Ax + Bu, \quad y = Cx \tag{E.12}$$

Case 1

If *A has distinct eigenvalues*, then M is *controllable* if and only if $B^* \equiv PB$ has no zero rows; M is *observable* if and only if $C^* = CP^{-1}$ has no zero columns.

Case 2

If A is *arbitrary*, then M is *controllable* (*observable*) if and only if, for each distinct eigenvalue $\lambda_1, \ldots, \lambda_m$, the *last rows* of the matrix $B^* \equiv PB$ (*first columns* of the matrix $C^* \equiv CP^{-1}$) corresponding to each Jordan block are linearly independent.

It is instructive to compare the block diagrams for Case 2, for both SISO and MIMO models. The requirement for SISO models that each λ_i be associated with one and only one Jordan block is relaxed here. It is replaced by the requirement that the first "gain vectors" in the control paths (or observation paths) for each chain associated with the same eigenvalue be linearly independent.

Example E.3: ***Controllability & Observability from Equivalent Models with Distinct Modes***

We showed in *Example E.2* that the MIMO model:

$$\dot{x} = \begin{bmatrix} -7 & -2 & 6 \\ 2 & -3 & -2 \\ -2 & -2 & 1 \end{bmatrix} x + \begin{bmatrix} 1 & 1 \\ 1 & -1 \\ 1 & 0 \end{bmatrix} u$$

$$y = \begin{bmatrix} -1 & -1 & 2 \\ 1 & 1 & -1 \end{bmatrix} x$$

has the (simplest, diagonal) Jordan canonical representation:

$$\dot{x}^* = \left[\begin{array}{c|c|c} -1 & 0 & 0 \\ \hline 0 & -3 & 0 \\ \hline 0 & 0 & -5 \end{array}\right] x^* + \left[\begin{array}{cc} 0 & 0 \\ \hline 1 & 0 \\ \hline 0 & 1 \end{array}\right] \begin{bmatrix} u_1 \\ u_2 \end{bmatrix} \longleftarrow \text{FIRST ROW ZERO}$$

$$y = \left[\begin{array}{c|c|c} 1 & 0 & 0 \\ 0 & 1 & 0 \end{array}\right] x^* \longleftarrow \text{LAST COLUMN ZERO}$$

The eigenvalues are distinct and the presence of the (first) zero row of B^* and (last) zero column of C^* means this system is neither controllable nor observable. We could also apply the last result above (for arbitrary A) and note that the (only) last rows of B^* corresponding to each 1×1 Jordan block are not each linearly independent because of the presence of the zero vector in the first row, which corresponds to eigenvalue $\lambda_1 = -1$.

Example E.4: ***Controllability & Observability from Equivalent SISO Models with Nondistinct Modes***

The following model in Jordan canonical form is representative of Case 2, for SISO systems. The last row of B^* corresponding to the eigenvalue 0 is zero. Thus, the system is not controllable. Each Jordan block has distinct eigenvalues and the first column of C^* corresponding to each block is nonzero. Thus, M is observable.

$$\dot{x} = \left[\begin{array}{ccc|c} 0 & 1 & 1 & 0 \\ 0 & 0 & 1 & 0 \\ 0 & 0 & 0 & 0 \\ \hline 0 & 0 & 0 & 0 \end{array}\right] x + \left[\begin{array}{c} 10 \\ 9 \\ 0 \\ \hline 1 \end{array}\right] u$$

$$y = [\,1\ 0\ 0\ 1\,]x$$

Example E.5: ***Controllability & Observability from Equivalent MIMO Models with Nondistinct Modes***

The following model in Jordan canonical form is representative of Case 2, for MIMO systems.

$$\dot{x} = \left[\begin{array}{cc:c:c:ccc} 1 & 1 & 0 & 0 & 0 & 0 & 0 \\ 0 & 1 & 0 & 0 & 0 & 0 & 0 \\ \hdashline 0 & 0 & 1 & 0 & 0 & 0 & 0 \\ \hdashline 0 & 0 & 0 & 1 & 0 & 0 & 0 \\ \hdashline 0 & 0 & 0 & 0 & 2 & 1 & 0 \\ 0 & 0 & 0 & 0 & 2 & 1 & 0 \\ 0 & 0 & 0 & 0 & 0 & 0 & 2 \end{array}\right] x + \left[\begin{array}{ccc} 0 & 0 & 0 \\ 5 & 0 & 0 \\ \hdashline 0 & 1 & 0 \\ \hdashline 0 & 0 & 2 \\ \hdashline 1 & 0 & 0 \\ 0 & 1 & 2 \\ 0 & 0 & 2 \end{array}\right] u \quad \begin{array}{l} \leftarrow a \\ \\ \leftarrow b \\ \leftarrow c \\ \\ \leftarrow d \\ \\ \end{array}$$

$$y = \left[\begin{array}{c:c:c:ccc} 11 & 5 & 4 & 0 & 2 & 5 \\ 10 & 6 & 2 & 0 & 2 & 1 \\ 11 & 5 & 2 & 0 & 2 & 6 \end{array}\right] x$$

$$\begin{array}{cccc} \uparrow & \uparrow & \uparrow & \uparrow \\ e & f & g & h \end{array}$$

Three Jordan blocks are associated with $\lambda = 1$ and one with $\lambda = 2$. The last rows of B^*, designated *a*, *b* and *c*, and associated with each $\lambda = 1$ block, are linearly independent; and the last row (*d*) of B^* associated with the $\lambda = 2$ block is (separately) linearly independent. Thus, *M* is controllable. Similarly, the *e*, *f*, *g* columns are linearly independent, but the final *h* column is not. Thus, *M* is not observable.

Minimal State Variable (ODE) Models from I–O TFs (Data)

The message in this section, seen many times over in this book in different contexts, is reduced to its simplest and most obvious forms in the context of equivalent models. *Only a* ***minimal (controllable and observable)*** *state variable model (SISO or MIMO) can be obtained from a given I–O record or I–O TF model of a time-invariant linear system.*

Minimal SISO State Variable Models

The state variable model $\dot{x} = Ax + Bu$, $y = Cx$ has the TF matrix:

$$H(s) = C[sI - A]^{-1}B \equiv \frac{N(s)}{D(s)} \tag{E.13}$$

where $D(s)$ is the *characteristic polynomial* of A, that is, $D(s) \equiv \det[sI - A]$. This model is **minimal (controllable and observable)** if and only if the polynomials $N(s)$ and $D(s)$ have no common factors, that is, if and only if the degree of the denominator of the TF equals n, the order of the model. This result also implies that *at least* one pole cancels at least one zero in $H(s)$ if the model is uncontrollable *or* unobservable.

Minimal MIMO State Variable Models

In this case, the state variable model is: $\dot{\mathbf{x}} = A\mathbf{x} + B\mathbf{u}$, $\mathbf{y} = C\mathbf{x}$ and the **characteristic polynomial** of $H(s)$ is the *least* common denominator of all minors[1] of $H(s)$, and the **degree of** $H(s)$ is defined as the degree of the characteristic polynomial of $H(s)$. In these terms, an n-dimensional MIMO model is **minimal** if and only if the characteristic polynomial of its $H(s)$ is equal to $D(s) = \det[sI - A]$, that is, if and only if the degree of $H(s)$ is n.

Example E.6: ***Minimal or Nonminimal State Variable Models from MIMO TFs***

$$H(s) = \begin{bmatrix} \dfrac{2}{s+1} & \dfrac{1}{s+1} \\ \dfrac{1}{s+1} & \dfrac{1}{s+1} \end{bmatrix} \tag{E.14}$$

The minors of order 1 are each of the elements of TF $H(s)$. Each has degree 1. The only minor of order 2 is $1/(s + 1)^2$ (the determinant of A). Thus the characteristic polynomial is $(s + 1)^2$, of degree 2. To test whether this is the TF of a *minimal MIMO* model, we must have the model, in state variable form. So, if the model is second-order ($n = 2$), it is. Otherwise it is not.

Canonical State Variable (ODE) Models from I–O Models (Data)

We begin with the simple nth-order TF with a constant term $\beta \neq 0$ in the numerator, deriving a minimal nth-order state variable model for this $H(s)$, using a transformation introduced in Chapter 2.

Remark: The matrix parameters of canonical ODE models are *structural invariants* (*SI* parameters or parameter combos) of the model.

Companion Canonical Models

$$H(s) = \frac{\beta}{s^n + \alpha_1 s^{n-1} + \ldots + \alpha_{n-1}s + \alpha_n} \tag{E.15}$$

1 The **minors** of an $n \times n$ square matrix A are associated with the element a_{pq} and are each equal to the determinant formed by crossing out the pth row and the qth column. Another minor is det A.

In the time domain, with differential operator $D^i \equiv \frac{d^i}{dt^i}$, we have

$$[D^n + \alpha_1 D^{n-1} + \cdots + \alpha_n]y(t) = \beta u(t) \tag{E.16}$$

Let

$$\begin{aligned} x_1 &= y \\ x_2 &= \dot{y} = Dy = \dot{x}_1 \\ x_3 &= \ddot{y} = D^2 y = \dot{x}_2 \\ &\vdots \\ x_n &= y^{(n-1)} = D^{n-1}y = \dot{x}_{n-1} \end{aligned} \tag{E.17}$$

Then, from Eqs. (E.15) and (E.16), we get

$$\dot{x}_n = y^{(n)} = D^n y = -\alpha_n x_1 - \alpha_{n-1}x_2 - \cdots - \alpha_1 x_n + \beta u \tag{E.18}$$

In vector-matrix form, this transformation yields the so-called **normal-form** representation, or **companion canonical state variable model**, with A in **companion matrix form**:

$$\dot{\boldsymbol{x}} = \begin{bmatrix} 0 & 1 & 0 & \cdots & 0 \\ 0 & 0 & 1 & \ddots & \vdots \\ \vdots & \vdots & \ddots & \ddots & 0 \\ 0 & 0 & \cdots & 0 & 1 \\ -\alpha_n & -\alpha_{n-1} & \cdots & -\alpha_2 & -\alpha_1 \end{bmatrix} \boldsymbol{x} + \begin{bmatrix} 0 \\ \vdots \\ 0 \\ \beta \end{bmatrix} u \tag{E.19}$$

$$\boldsymbol{y} = [1 \quad 0 \quad 0 \quad \cdots \quad 0]\boldsymbol{x} \tag{E.20}$$

The block diagram is given in ***Fig. E.8***. It is easy to show that this model is both observable and controllable (with $\beta \neq 0$), that is, $rankW = rankV = n$, and we see that the state variable model can be written directly from the numerator and the denominator polynomial in $H(s)$.

Another realization of Eq. (E.15) is obtained if we factor the denominator polynomial:

$$H(s) = \frac{\beta}{(s-\lambda_1)(s-\lambda_2)\cdots(s-\lambda_n)} \tag{E.21}$$

The block diagram of Eq. (E.21) may be expressed as a *series-connected composite model* block diagram in ***Fig. E.9***.

Then, with the state variables chosen as shown, we get the following *minimal realization*:

$$\dot{\boldsymbol{x}} = \begin{bmatrix} \lambda_1 & 0 & 0 & \cdots & 0 \\ 1 & \lambda_2 & 0 & \cdots & 0 \\ 0 & 1 & \lambda_3 & \ddots & \vdots \\ \vdots & \ddots & \ddots & \ddots & 0 \\ 0 & \cdots & 0 & 1 & \lambda_n \end{bmatrix} \boldsymbol{x} + \begin{bmatrix} \beta \\ 0 \\ \vdots \\ 0 \end{bmatrix} u \qquad \boldsymbol{y} = [0 \quad 0 \quad 0 \quad \cdots \quad 1]\boldsymbol{x} \tag{E.22}$$

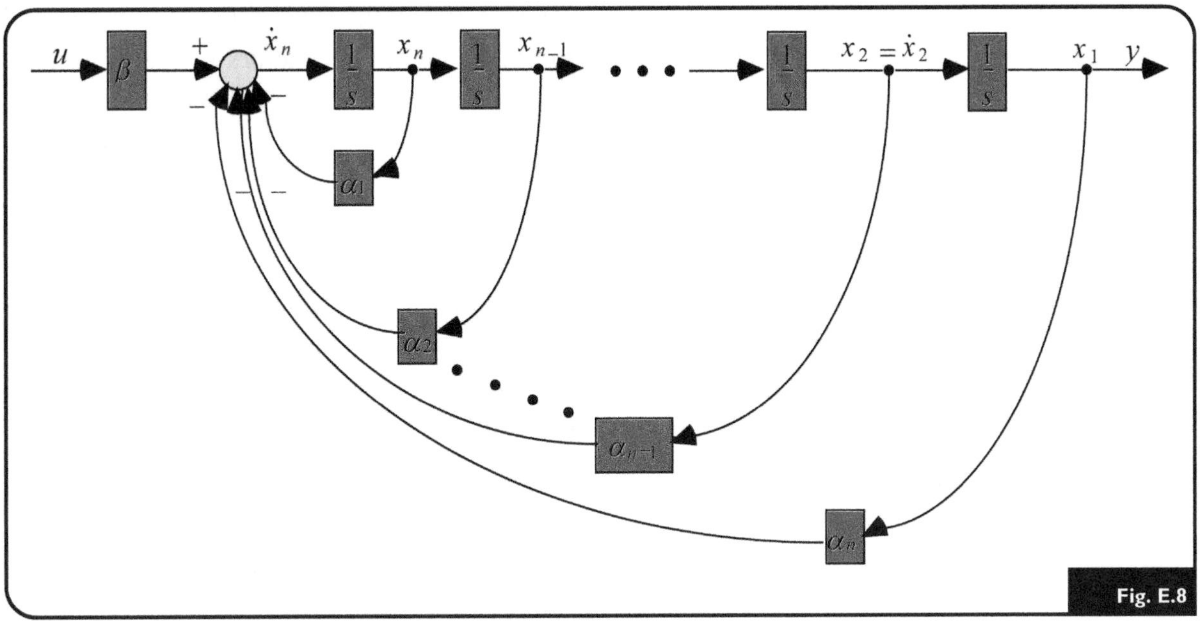

Fig. E.8

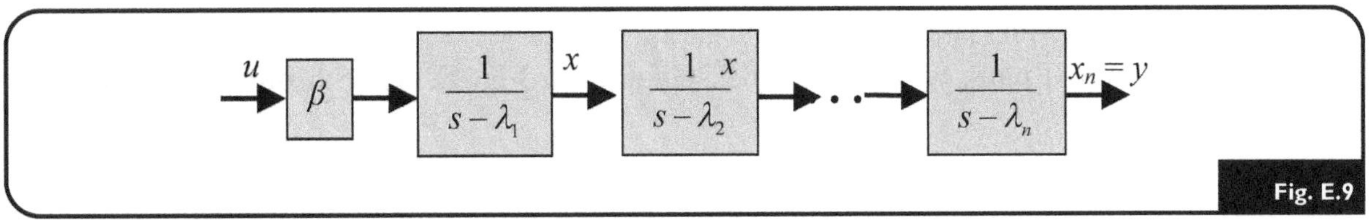

Fig. E.9

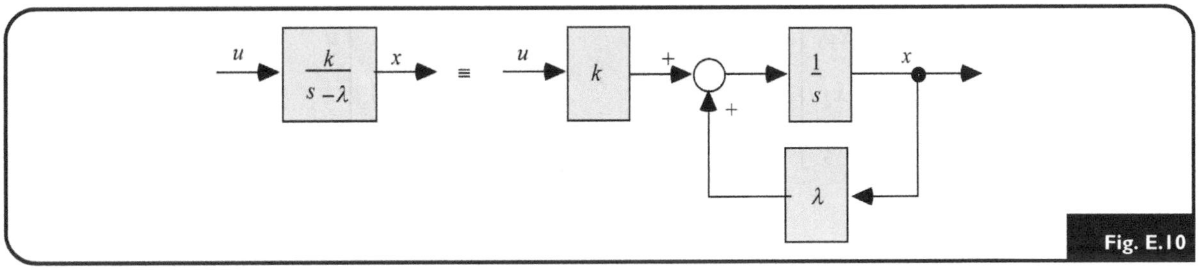

Fig. E.10

Each block in the series (cascade) block diagram of ***Fig. E.9*** can be represented equivalently by the **canonical feedback submodel** shown in ***Fig. E.10***. This block diagram transformation may be used to simplify complicated block diagrams (DiStefano III et al. 1990), as well as providing another representation. Equation (E.22) is a **parallel composite** state variable representation of $H(s)$.

Canonical State Variable Models for More General SISO TFs

$$H(s) = \frac{N(s)}{D(s)} = \frac{\beta_1 s^{n-1} + \beta_2 s^{n-2} + \cdots + \beta_n}{s^n + \alpha_1 s^{n-1} + \cdots + \alpha_n} \tag{E.23}$$

Assume we have already divided out any constant term from the TF (if the numerator and denominator have the same order n) and, also, there are no common factors in $N(s)$ and $D(s)$, i.e. no pole-zero cancellations. If there are any cancellations, then we begin with a smaller n, after canceling those terms.

Jordan-Canonical State Variable Models for N(s)/D(s)

This transformation requires that $H(s)$ be expanded into partial fractions. Thus, the denominator $D(s)$ in Eq. (E.23) must first be factored. We illustrate this realization with an example. The approach may be extended to the general case, recognizing that factoring a symbolic polynomial may not be feasible. For $n > 3$, numerical methods are required (DiStefano III et al. 1990).

Suppose $D(s)$ has three distinct roots λ_1, λ_2, and λ_3, and that λ_1 has multiplicity 3. Then, we can write

$$H(s) = \frac{\rho_{11}}{(s-\lambda_1)^3} + \frac{\rho_{12}}{(s-\lambda_1)^2} + \frac{\rho_{13}}{s-\lambda_1} + \frac{\rho_2}{s-\lambda_2} + \frac{\rho_3}{s-\lambda_3} \tag{E.24}$$

The block diagram model for Eq. (E.24) is not unique, because the *residues* ρ_i and ρ_{ij} can be placed in either the input path or the output path, or both if we let each ρ_{ii} or $\rho_i \equiv b_i c_i$. If we put them in the output path, we get the block diagram in **Fig. E.11**, with **minimal realization** model:

$$\begin{bmatrix} \dot{x}_{11} \\ \dot{x}_{12} \\ \dot{x}_{13} \\ \dot{x}_2 \\ \dot{x}_3 \end{bmatrix} = \left[\begin{array}{ccc:c:c} \lambda_1 & 1 & 0 & 0 & 0 \\ 0 & \lambda_1 & 1 & 0 & 0 \\ 0 & 0 & \lambda_1 & 0 & 0 \\ \hdashline 0 & 0 & 0 & \lambda_2 & 0 \\ \hdashline 0 & 0 & 0 & 0 & \lambda_3 \end{array}\right] \boldsymbol{x} + \begin{bmatrix} 0 \\ 0 \\ 1 \\ 1 \\ 1 \end{bmatrix} u \tag{E.25}$$

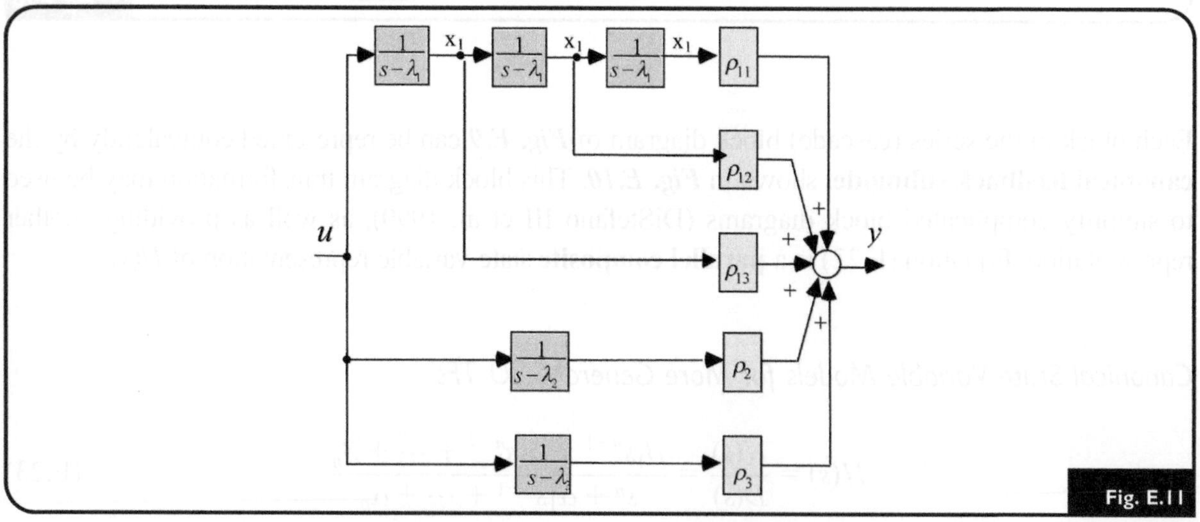

Fig. E.11

$$y = [\rho_{11} \quad \rho_{12} \quad \rho_{13} \quad \rho_2 \quad \rho_3]\boldsymbol{x} \tag{E.26}$$

Alternatively, if we put the ρ_i in the input path (in *front* of the first block in each parallel pathway), the minimal realization is

$$\begin{bmatrix} \dot{x}_{11} \\ \dot{x}_{12} \\ \dot{x}_{13} \\ \dot{x}_2 \\ \dot{x}_3 \end{bmatrix} = \left[\begin{array}{ccc:c:c} \lambda_1 & 0 & 0 & 0 & 0 \\ 1 & \lambda_1 & 0 & 0 & 0 \\ 0 & 1 & \lambda_1 & 0 & 0 \\ \hdashline 0 & 0 & 0 & \lambda_2 & 0 \\ \hdashline 0 & 0 & 0 & 0 & \lambda_3 \end{array}\right] \boldsymbol{x} + \begin{bmatrix} \rho_{11} \\ \rho_{12} \\ \rho_{13} \\ \rho_2 \\ \rho_3 \end{bmatrix} u \tag{E.27}$$

$$y = [0 \quad 0 \quad 1 \quad 1 \quad 1]\boldsymbol{x} \tag{E.28}$$

The choice of whether to use Eqs. (E.25) and (E.26) or Eqs. (E.27) and (E.28), both of which are parallel composite, may depend, for example, on whether the residues ρ_i (perhaps unknown parameters) are better placed in the state equations or in the output equation. This question may be associated with "scaling" the equations for simulation or for parameter estimation. If $H(s)$ has complex poles, the matrices A, B, and C will have complex elements. If simulation of this realization is desired, an equivalence transformation can be found that yields real A^*, B^* and C^*. The Jordan form minimal realization has the desirable property that its parameters – the eigenvalues – are less sensitive to variations in the parameters of $H(s)$ – the coefficients of $N(s)$ and $D(s)$ – than other popular minimal realizations, discussed next. However, the requirement that the denominator $D(s)$ be factored requires more effort. If the order of $D(s)$ is greater than 3, a computational solution is needed, using *Matlab, Mathematica* or other suitable software.

Controllable Canonical State Variable Models for N(s)/D(s)

This minimal model has the form:

$$\dot{\boldsymbol{x}} = \begin{bmatrix} 0 & 1 & 0 & 0 & \cdots & 0 \\ 0 & 0 & 1 & \ddots & \ddots & \vdots \\ 0 & 0 & 0 & \ddots & 0 & 0 \\ \vdots & \vdots & \vdots & \ddots & 1 & 0 \\ 0 & 0 & 0 & \cdots & 0 & 1 \\ -\alpha_n & -\alpha_{n-1} & -\alpha_{n-2} & \cdots & -\alpha_2 & -\alpha_1 \end{bmatrix} \boldsymbol{x} + \begin{bmatrix} 0 \\ 0 \\ 0 \\ \vdots \\ 0 \\ 1 \end{bmatrix} u \tag{E.29}$$

$$y = [\beta_n \quad \beta_{n-1} \quad \cdots \quad \beta_1]\boldsymbol{x} \tag{E.30}$$

with block diagram in ***Fig. E.12***. This realization, also obtained directly from the coefficients of $H(s)$, is always controllable (i.e., rank $W = n$). Note that A^* is in companion matrix form, but there are no simple relationships among the u, y, and x_i. The derivation of this transformation goes as follows. From Eq. (E.23), we can write the differential equations:

$$(\mathscr{D}^n + \alpha_i \mathscr{D}^{n-1} + \cdots + \alpha_n)y(t) = (\beta_i \mathscr{D}^{n-1} + \beta_2 \mathscr{D}^{n-2} + \cdots \beta_n)u(t) \tag{E.31}$$

or $D(\mathscr{D})y(t) = N(\mathscr{D})u(t)$, where $\mathscr{D}$ is the differential operator. Then, we define a new variable $v(t)$ such that:

$$D(\mathscr{D})v(t) \equiv u(t) \tag{E.32}$$

Therefore, $D(\mathscr{D})y(t) = N(\mathscr{D})D(\mathscr{D})v(t) = D(\mathscr{D})N(\mathscr{D})v(t)$ which implies

$$y(t) = N(\mathscr{D})v(t) \tag{E.33}$$

From Eq. (E.32), we get the state equation Eq. (E.29) by defining

$$\begin{aligned} x_1 &\equiv v \\ x_2 &\equiv \dot{v} = \dot{x}_1 \\ &\vdots \\ x_n &\equiv v^{(n-1)} = \dot{x}_{n-1} \end{aligned} \tag{E.34}$$

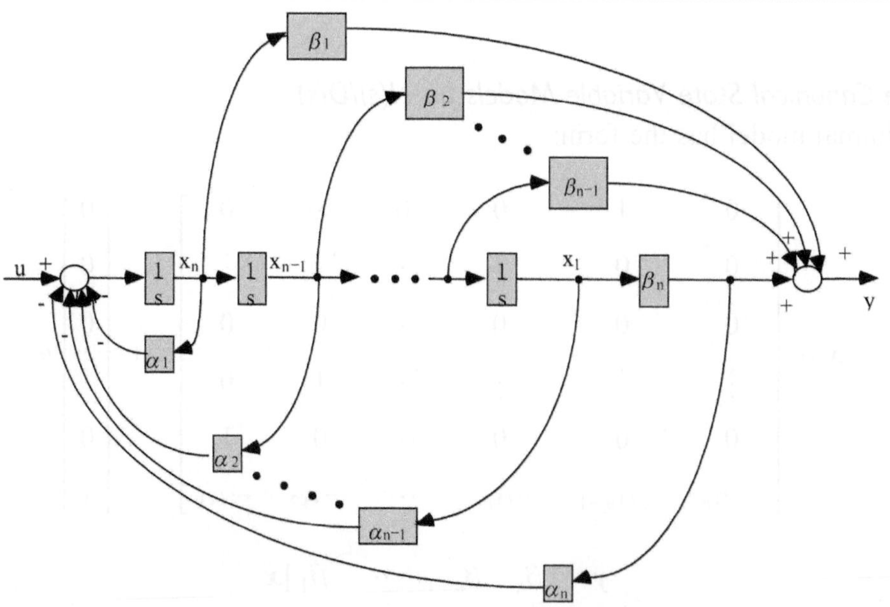

Fig. E.12

just as we did to obtain the companion matrix canonical model earlier. From Eq. (E.33), we get the output equation:

$$\begin{aligned} y(t) &= N(\mathscr{D})v(t) = (\beta_1\mathscr{D}^{n-1} + \cdots + \beta_n)v \\ &= \beta_1 v^{(n-1)} + \cdots + \beta_n v = \beta_1 x_n + \beta_2 x_{n-1} + \cdots + \beta_n x_1 \\ &= [\,\beta_n \quad \beta_{n-1} \quad \cdots \quad \beta_1\,]\boldsymbol{x} \end{aligned} \tag{E.35}$$

If $N(s)$ and $D(s)$ have common factors, i.e. if we have a non-minimal realization to start out with, the controllable canonical form can still be obtained and it will be controllable. But it will not be observable.

Observable Canonical State Variable (ODE) Models for N(s)/D(s)

This minimal model has the form:

$$\dot{\boldsymbol{x}} = \begin{bmatrix} 0 & 0 & \cdots & 0 & \alpha_n \\ 1 & 0 & \cdots & 0 & \alpha_{n-1} \\ 0 & 1 & \ddots & \vdots & \vdots \\ \vdots & \ddots & \ddots & 0 & \alpha_2 \\ 0 & 0 & 0 & 1 & -\alpha_1 \end{bmatrix} \begin{bmatrix} x_1 \\ x_2 \\ \vdots \\ x_{n-1} \\ x_n \end{bmatrix} + \begin{bmatrix} \beta_n \\ \beta_{n-1} \\ \vdots \\ \beta_2 \\ \beta_1 \end{bmatrix} u \tag{E.36}$$

$$y = [\,0 \quad 0 \quad \cdots \quad 0 \quad 1\,]\boldsymbol{x} \tag{E.37}$$

The block diagram is given in ***Fig. E.13***. This realization, which is always observable, (i.e. rank $V = n$), is obtained directly from the coefficients of $H(s)$. Derivation goes as follows.

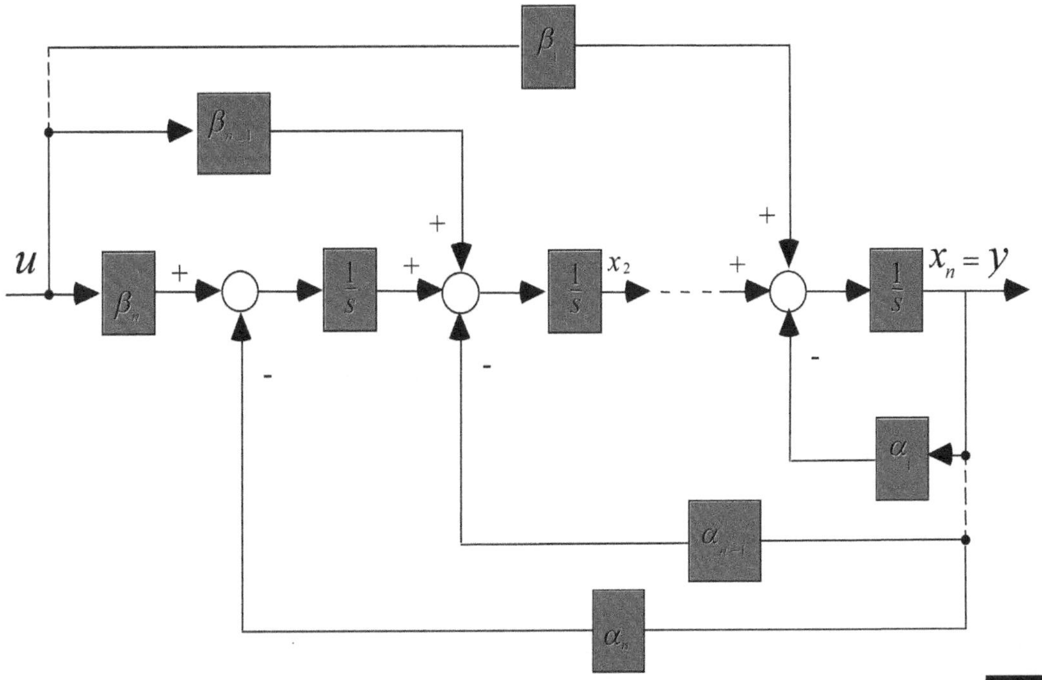

Fig. E.13

First, we take the LT of Eq. (E.31) and group the terms into those due to the input and those due to initial conditions on y and u and their derivatives:

$$Y(s) = \frac{N(s)}{D(s)}U(s) + \frac{1}{D(s)}\Big\{y(0)s^{n-1} + [y^{(1)}(0) + \alpha_1 y(0) - \beta_1 u(0)]s^{n-2} + [y^{(n-1)}(0) + \cdots$$
$$+\alpha_1 y^{(n-2)}(0) - b_1 u^{(n-2)}(0) + a_2 y^{(n-3)}(0) - b_2 u^{(n-3)}(0) + \ldots + a_{n-1}y(0) - b_{n-1}u(0)]\Big\} \tag{E.38}$$

If all coefficients of s^k are known, then the following $\boldsymbol{x}$ qualifies as a state vector:

$$\begin{aligned} x_n &= y \\ x_{n-1} &= y^{(1)} + \alpha_1 y - \beta_1 u = \dot{x}_n + \alpha_1 x_n - \beta_1 u \\ x_{n-2} &= y^{(2)} + \alpha_1 y^{(1)} - \beta_1 u^{(1)} + \alpha_2 y - \beta_2 u \\ &= \dot{x}_{n-1} + \alpha_2 x_n - \beta_2 u \\ &\vdots \\ x_1 &= \dot{x}_2 + \alpha_{n-1} x_n - \beta_{n-1} u \end{aligned} \tag{E.39}$$

and, from Eq. (E.31), we get

$$\dot{x}_1 = -\alpha_n x_n - \beta_n u \tag{E.40}$$

Rearrangement of Eqs. (E.39) and (E.40) yields (E.36). If $N(s)$ and $D(s)$ have common factors, i.e. if the original model is not minimal, the observable canonical form can still be obtained and it will be observable. But it will not be controllable. Also, if $N(s)$ and $D(s)$ have common factors, it is also possible to find a realization that is neither controllable nor observable. Such a realization can be "hidden" from I–O probes.

Observable and Controllable Canonical Forms from Arbitrary State Variable Models Using Equivalence Properties

Some simple algebraic relationships among observable and controllable canonical state variable models are derivable from the controllability and observability matrices, W and V, or the controllable and observable subspaces of the full state space of dimension n (Chen, 1985). We consider only the SISO case here, with model M: $\dot{\boldsymbol{x}} = A\boldsymbol{x} + Bu$ and $y = C\boldsymbol{x}$ and its equivalent model M^*:

$$\dot{\boldsymbol{x}}^* = A^*\boldsymbol{x}^* + B^*\boldsymbol{u} \quad \text{and} \quad y = C^*\boldsymbol{x}^* \tag{E.41}$$

where $\boldsymbol{x}^* \equiv A^*\boldsymbol{x}$ and $A^* = PAP^{-1}$, $B^* = PB$, $C^* = CP^{-1}$. The controllability matrices of M and M^* are

$$W \equiv \left[B \;\vdots\; AB \;\vdots\; \cdots \;\vdots\; A^{n-1}B\right], \quad W^* \equiv \left[B^* \;\vdots\; A^*B^* \;\vdots\; \cdots \;\vdots\; A^{*n-1}B^*\right] = PW \tag{E.42}$$

If M and M^* are controllable, then W and W^* are nonsingular and we can write

$$P = W^*W^{-1} \tag{E.43}$$

Similarly, if V and V^* are the observability matrices of M and M^*, then

$$V \equiv \begin{bmatrix} C \\ \cdots\cdots \\ CA \\ \cdots\cdots \\ \vdots \\ \cdots\cdots \\ CA^{n-1} \end{bmatrix} \quad \text{and} \quad V^* \equiv \begin{bmatrix} C^* \\ \cdots\cdots\cdots \\ C^*A^* \\ \cdots\cdots\cdots \\ \vdots \\ \cdots\cdots\cdots \\ C^*A^{*n-1} \end{bmatrix} = VP^{-1} \tag{E.44}$$

If M and M^* are observable, then V and V^* are nonsingular and $P = (V^*)^{-1}V$. If two models M and M^* are known to be equivalent, and if they are either controllable or observable, then P can be computed from either Eq. (E.43) or Eq. (E.44). Thus, if M^* is to be a controllable or observable canonical form, Eq. (E.43) or (E.44) can be used to get it directly, without first obtaining $H(s)$. Actually, $H(s)$ is also obtained by the procedure.

Now, if M is controllable (observable), then it can be transformed by equivalence transformation into the controllable (observable) canonical form as follows. To obtain $\alpha_1, \ldots, \alpha_n$, evaluate the characteristic polynomial of A, that is,

$$\det[sI - A] = s^n + \alpha_1 s^{n-1} + \cdots + \alpha_n \tag{E.45}$$

To obtain $\beta_1, \ldots, \beta_n$, we note that

$$B^* = PB = (V^*)^{-1}VB = \left[\beta_n \quad \beta_{n-1} \quad \cdots \quad \beta_1\right]^{\mathrm{T}} \tag{E.46}$$

for the observable canonical form and

$$C^* = CP^{-1} = CW(W^*)^{-1} = \left[\beta_n \quad \beta_{n-1} \quad \cdots \quad \beta_1\right] \tag{E.47}$$

for the controllable canonical form. It can be shown (Chen, 1985) that $(V^*)^{-1}$ and $(W^*)^{-1}$ are given by:

$$(V^*)^{-1} = (W^*)^{-1} = \begin{bmatrix} \alpha_{n-1} & \alpha_{n-2} & \cdots & \alpha_1 & 1 \\ \alpha_{n-2} & \alpha_{n-3} & \cdots & 1 & 0 \\ \vdots & \vdots & \iddots & \iddots & \vdots \\ \alpha_1 & 1 & \iddots & 0 & 0 \\ 1 & 0 & \cdots & 0 & 0 \end{bmatrix} \tag{E.48}$$

Consequently, the $\beta_1, \beta_2, \ldots, \beta_n$ can be determined from either Eqs. (E.46) and (E.48) or Eqs. (E.47) and (E.48).

Example E.7: ***Canonical State Variable (ODE) Models by Equivalence Transformations***

We transform the following minimal realization to both controllable and observable canonical models:

$$\dot{\boldsymbol{x}} = \begin{bmatrix} 1 & 2 & 0 \\ 3 & -1 & 1 \\ 0 & 2 & 1 \end{bmatrix} \boldsymbol{x} + \begin{bmatrix} 2 \\ 1 \\ 1 \end{bmatrix} u \text{ and } \boldsymbol{y} = \begin{bmatrix} 0 & 0 & 1 \end{bmatrix} \boldsymbol{x}$$

The characteristic polynomial is $\lambda^3 - 9\lambda + 2 = \lambda^3 + \alpha_1\lambda^2 + \alpha_2\lambda + \alpha_3$. Thus, $\alpha_1 = 0$, $\alpha_2 = -9$, $\alpha_3 = 2$. Then, from Eq. (E.48):

$$(W^*)^{-1} = (V^*)^{-1} = \begin{bmatrix} -9 & 0 & 1 \\ 0 & 1 & 0 \\ 1 & 0 & 0 \end{bmatrix} \text{ and } W = [B \mid AB \mid A^2B] = \begin{bmatrix} 2 & 4 & 16 \\ 1 & 6 & 8 \\ 1 & 2 & 12 \end{bmatrix}$$

Thus, from Eq. (E.47), $\begin{bmatrix} \beta_3 & \beta_2 & \beta_1 \end{bmatrix} = \begin{bmatrix} 3 & 2 & 1 \end{bmatrix}$. The canonical models are then given by the corresponding equations for each model form given earlier, i.e. the controllable model:

$$\dot{\boldsymbol{x}}^* = \begin{bmatrix} 0 & 1 & 0 \\ 0 & 0 & 1 \\ -2 & 9 & 0 \end{bmatrix} \boldsymbol{x}^* + \begin{bmatrix} 0 \\ 0 \\ 1 \end{bmatrix} u$$

$$\boldsymbol{y} = \begin{bmatrix} 3 & 2 & 1 \end{bmatrix} \boldsymbol{x}^*$$

The observable model:

$$\dot{\boldsymbol{x}}^* = \begin{bmatrix} 0 & 0 & -2 \\ 1 & 0 & 9 \\ 0 & 1 & 0 \end{bmatrix} \boldsymbol{x}^* + \begin{bmatrix} 3 \\ 2 \\ 1 \end{bmatrix} u$$

Each of these minimal canonical state variable models has the minimum number of parameters ($2n$ maximum) and each is a set of *structural invariants* of $\boldsymbol{p}$ (combos). The original system had a maximum number of parameters equal to $n^2 + 2n$. These minimal canonical models are thus even simplest in the sense that they have the minimum number of parameters (all combos) as well as state variables.

Example E.8: ***Equivalent Canonical State Variable Models for Human Hormone Metabolism***

Examples 13.8 and *13.9* of Chapter 13 illustrated reparameterization of the unidentifiable compartmental model in ***Fig. C.10*** of ***Appendix C*** into one with identifiable parameters made up of *combinations* (structural invariants) of the original compartmental rate constants. The resulting SISO structure has the form:

$$\dot{\boldsymbol{x}} = \begin{bmatrix} p_1 & 1 & 1 \\ p_2 & p_4 & 0 \\ p_3 & 0 & p_5 \end{bmatrix} \boldsymbol{x} + \begin{bmatrix} 1 \\ 0 \\ 0 \end{bmatrix} u(t) \text{ and } \boldsymbol{y} = [p_6 \; 0 \; 0] \boldsymbol{x} \qquad \text{(E.49)}$$

$$H(s) = C[sI - A]^{-1}B = \frac{p_6(s - p_4)(s - p_5)}{s^3 + \alpha_3 s^2 + \alpha_2 s + \alpha_1} \tag{E.50}$$

$\alpha_1 = p_2p_5 + p_4(p_3 - p_1p_5)$, $\alpha_2 = p_1(p_4 + p_5) + p_4p_5 + p_3 - p_2$, $\alpha_3 = -(p_1 + p_4 + p_5)$. The parameters were estimated from human data in (DiStefano III 1985), resulting in: $p_1 = -9.9$, $p_2 = 20$, $p_3 = 0.04$, $p_4 = -2.2$, $p_5 = -0.067$ and $p_6 = 0.41$. The TF is thus:

$$H(s) = H_{11}(s) = \frac{0.41s^2 + 0.929s + 0.06}{s^3 + 12.17s^2 + 2.55s + 0.031} \tag{E.51}$$

From Eq. (E.48), the **controllable canonical model** is

$$\dot{\boldsymbol{x}}_c = \begin{bmatrix} 0 & 1 & 0 \\ 0 & 0 & 1 \\ -0.031 & -2.55 & -12.17 \end{bmatrix} \boldsymbol{x}_c + \begin{bmatrix} 1 \\ 0 \\ 0 \end{bmatrix} u \text{ and } \boldsymbol{y} = [0.06 \quad 0.929 \quad 0.41]\boldsymbol{x}_c$$

From Eq. (E.47), the **observable canonical model** is

$$\dot{\boldsymbol{x}}_0 = \begin{bmatrix} 0 & 0 & -0.031 \\ 1 & 0 & -2.55 \\ 0 & 1 & -12.17 \end{bmatrix} \boldsymbol{x}_0 + \begin{bmatrix} 0.06 \\ 0.929 \\ 0.41 \end{bmatrix} u \text{ and } \boldsymbol{y} = [0 \quad 0 \quad 1]\boldsymbol{x}_0$$

Factoring the denominator of the TF, the **Jordan canonical model** is diagonal:

$$\dot{\boldsymbol{x}}_J = \begin{bmatrix} -11.95 & 0 & 0 \\ 0 & -0.2 & 0 \\ 0 & 0 & -0.013 \end{bmatrix} \boldsymbol{x}_J + \begin{bmatrix} 1 \\ 1 \\ 1 \end{bmatrix} u \text{ and } \boldsymbol{y} = [0.34 \quad 0.01 \quad 0.061]\boldsymbol{x}_J$$

And the TF has the equivalent partial fraction expanded form:

$$H(s) = \frac{\rho_1}{s - \lambda_1} + \frac{\rho_2}{s - \lambda_2} + \frac{\rho_3}{s - \lambda_3} = \frac{0.34}{s + 11.95} + \frac{0.01}{s + 0.2} + \frac{0.061}{s + 0.013} \tag{E.52}$$

The **multiexponential model** impulse response is written directly from H as:

$$h(t) = \sum_{i=1}^{3} \rho_i e^{\lambda_i t} = 0.34e^{-11.95} + 0.01e^{-0.2t} + 0.061e^{-0.013t} \tag{E.53}$$

Equating Eqs. (E.50) and (E.52) then gives the relationships among the parameters of the Jordan form and the original structure. These may also be interpreted as the relationships among the parameter (coefficients and exponents) of the three-exponential model (ρ_i, λ_i), and those of the original differential equation model ($p_1, p_2, \ldots, p_6$). If $p_4 < p_5$, the unique solution is given by $p_1 = \dfrac{\rho_1\lambda_1 + \rho_2\lambda_2 + \rho_3\lambda_3}{\rho_6}$,

$$p_2 = \frac{p_1K_2 - p_4K_3 - \lambda_1\lambda_2\lambda_3}{p_5 - p_4} \text{ with } K_3 = p_1K_1 + K_2 - \lambda_1\lambda_2 - \lambda_1\lambda_3 - \lambda_2\lambda_3, \; p_3 = K_3 - p_2,$$

$$p_4 = \frac{K_1 - \sqrt{K_1^2 - 4K_2}}{2} \text{ with } K_1 = \lambda_1 + \lambda_2 + \lambda_3 - p_1 \text{ and } K_2 = \frac{\rho_1\lambda_1\lambda_2 + \rho_2\lambda_1\lambda_3 + \rho_3\lambda_2\lambda_3}{p_6}$$

$$p_5 = K_1 - p_4 \text{ and } p_6 = \rho_1 + \rho_2 + \rho_3.$$

Actually, there are two solutions for p_4, because p_4 is the solution of a quadratic equation resulting when Eqs. (E.50) and (E.52) are equated. If $p_4 > p_5$ (the other solution), this is equivalent to the state variables x_2 and x_3 exchanging positions in the state vector. Both are feasible. Additional knowledge about the dynamics is necessary to distinguish between them (see also *Examples 13.8* and *13.9* in Chapter 13).

Extra: Using the controllability matrix W and observability matrix V, it can be shown that a time-invariant model $M = \{A, B, C\}$ is a *minimal realization* of dimension n if the **Hankel matrix** HM of **Markov parameters** CA^iB has rank n:

$$HM \equiv \begin{bmatrix} CB & CAB & \cdots & CA^{n-1}B \\ CAB & CA^2B & \ddots & \vdots \\ \vdots & \ddots & \ddots & CA^{2n-3}B \\ CA^{n-1}B & \cdots & CA^{2n-3}B & CA^{2n-2}B \end{bmatrix}$$

References

Casti, J.L., 1987. Linear Dynamical Systems. Academic Press, New York.

Chen, C., 1985. Introduction to Linear System Theory. Holt, Rinehart, & Winston, New York.

DiStefano III, J., 1985. Modeling approaches and models of the distribution and disposal of thyroid hormones. In: Hennemann, G. (Ed.), Thyroid Hormone Metabolism. Marcel Dekker, New York, pp. 39–76.

DiStefano III, J., Stubberud, A.R., Williams, I.J., 1990. Feedback and Control Systems, 2nd Edition: Continuous (Analog) and Discrete (Digital). McGraw-Hill, New York.

Kalman, R., 1962. Canonical structure of linear dynamical systems. Proc. Natl. Acad. Sci. 48, 596–600.

More on Simulation Algorithms & Model Information Criteria

Appendix F

Additional Predictor-Corrector Algorithms

Understanding more about how these integration algorithms work can be helpful in assessing simulation results and making choices as to which to choose for particular problems.

Modified Euler Second-Order Predictor and Corrector Formulas

This set of formulas involves different modifications of the Euler formula but with errors of the order of Δt^3 instead Δt^2 (Press 1992). E_T is shown explicitly here, using Taylor's theorem, which aggregates all terms from third-order on into one term, evaluated at some point between x^{m-1} and x^{m+1} for this *two-stage* jump. These remainder terms can be used to estimate E_T, for correcting formulas, as shown further below. First, a two-step **predictor formula**:

$$x_P^{m+1} = x^{m-1} + 2\Delta t\,\dot{x}^m + \frac{\Delta t^3}{3}\dddot{x}(\varepsilon) \quad \text{where} \quad x_{m-1} \le \varepsilon \le x_{m+1} \tag{F.1}$$

$$x_P^{m+1} \cong x^{m-1} + 2\Delta t\, f(x^m, t_m) \tag{F.2}$$

The **corrector formula** uses the *mean* of the derivatives as a modified Euler formula:

$$x_C^{m+1} = x^m + \frac{\Delta t}{2}\left[\dot{x}^m + \dot{x}_P^{m+1}\right] - \frac{\Delta t^3}{12}\dddot{x}(\eta) \quad \text{where} \quad t_{m-1} \leq \eta \leq t_{m+1} \tag{F.3}$$

$$x_C^{m+1} \cong x^m + \frac{\Delta t}{2}\left[f(x_P^{m+1}, t_{m+1}) + f(x^m, t_m)\right] = x^m + \frac{\Delta t}{2}\left[f_P^{m+1} + f^m\right] \tag{F.4}$$

The final corrector formula here, an **implicit formula**, is recognized as the estimated slope $f(x^{m+1}, t_{m+1})$ at $t_m + \Delta t$, used as a correction to the slope at t_m, by averaging the two to yield a better estimate of $x(t_m + \Delta t)$.

An Iterative-Implicit Predictor–Corrector Algorithm

Use a self-starting method (e.g. Runge-Kutta) to compute the first two starting values, i.e. at $m = 0$ and 1. Then obtain the first **prediction**, at $m + 1 = 2$, using Eq. (F.2):

$$x_P^{m+1} = x^{m-1} + 2\Delta t x^m \tag{F.5}$$

Then compute the **predicted derivative** from the ODE:

$$\dot{x}_P^{m+1} = f(x_P^{m+1}, t_{m+1}) \tag{F.6}$$

Then obtain a first **correction** from Eq. (F.4):

$$x_C^{m+1} = x^m + \frac{\Delta t}{2}(f^m + f_P^{m+1}) \tag{F.7}$$

Iterate on the predictor–corrector, substituting x_C for x_P in the derivative formula (F.6) and recomputing a predictively better x_C, until

$$100\frac{|(x_C^{m+1})^{i+1} - (x_C^{m+1})^i|}{(x_C^{m+1})^i} < \varepsilon\% \quad i = 1, 2, \ldots \tag{F.8}$$

where ε is a positive number, for example, 1, which means the new value is within 1% of the old. The stopping condition, ε%, can be chosen freely, with smaller values requiring longer computations.

Non-Iterative, Predictor–Modifier–Corrector (P–M–C) Algorithms

This class of algorithms is explicit, avoiding iterations. It yields improved corrections over predicted values by modifying predictions with added estimates of truncation errors E_T at each step. The procedure is

1. *Predict* a value for x^{m+1}.
2. *Modify* the predicted value by adding an estimate of the truncation error.
3. *Correct* the *modified* value of x^{m+1}. The first corrected value of x^{m+1} is accepted.

Truncation Error Estimation

We illustrate this using the remainder terms in Eqs. (F.1) and (F.3):

$$E_T^P = \frac{h^3}{3}\dddot{x}(\varepsilon) \quad E_T^C = -\frac{h^3}{12}\dddot{x}(\eta) \quad t_{m-1} \le \varepsilon, \eta \le t_{m+1} \tag{F.9}$$

We assume the third derivative $\dddot{x}$ remains approximately constant over the time interval (t_{m-1}, t_{m+1}):

$$\dddot{x}(\varepsilon) \cong \dddot{x}(\eta) \equiv \dddot{x} \tag{F.10}$$

By definition, the exact value of x^{m+1} is

$$x^{m+1} \equiv x_P^{m+1} + E_T^P \equiv x_C^{m+1} + E_T^C \tag{F.11}$$

Thus, the difference between the predicted and the corrected values can be computed as:

$$\Delta x^{m+1} \equiv x_P^{m+1} - x_C^{m+1} = E_T^C - E_T^P = -\left[\frac{h^3}{12} + \frac{h^3}{3}\right]\dddot{x} = -\frac{5}{12}\dddot{x} \tag{F.12}$$

Solving for the corrector truncation error, using Eq. (F.9), then yields

$$E_T^C = -\frac{h^3}{12}\dddot{x} = \frac{x_P^{m+1} - x_C^{m+1}}{5} = \frac{\Delta x^{m+1}}{5} \tag{F.13}$$

Equation (F.13) thus provides an approximation of the truncation error E_T^C at each step in the calculation, which also can be used to adjust the step-size Δt during iterations, as illustrated below. Solving Eq. (F.12) again for the corresponding estimate of the predictor truncation error, we get

$$E_P^{m+1} = -\frac{4}{5}(x_P^m - x_C^m) \tag{F.14}$$

A Predictor-Modifier Corrector Algorithm Exemplified

We use the estimate of E_P^{m+1} derived above to *modify* or improve our initial predictor estimate of x, before correcting it with an explicit, higher-order corrector formula, also modified by correcting it further using the truncation error estimate E_T^C. We illustrate this class of noniterative algorithms for the multistep Euler P–C formulas illustrated earlier using an iterative approach. The predictor is the same:

$$x_P^{m+1} = x^{m-1} + 2\Delta t x f^m \tag{F.15}$$

In the second step, we modify this predicted solution by adding the truncation error:

$$m^{m+1} \equiv x_P^{m+1} + E_P^{m+1} = x_P^{m+1} - \frac{4}{5}(x_P^m - x_C^m) \tag{F.16}$$

The predicted derivative is then:

$$\dot{m}^{m+1} = f(m^{m+1}, t_{m+1}) \tag{F.17}$$

And the first corrector is

$$x_C^{m+1} = x^m + \frac{h}{2}(\dot{m}^{m+1} + f^m) \tag{F.18}$$

Finally, we modify this corrected value by adding the truncation error estimate computed in Eq. (F.13):

$$x^{m+1} \equiv x_C^{m+1} + E_T^C = x_C^{m+1} + \frac{x_P^{m+1} - x_C^{m+1}}{5} \tag{F.19}$$

The Backward-Euler Algorithm for Stiff ODEs

The predictor formulas are standard Euler:

$$\begin{aligned} x_P^{m+1} &= x^m + \Delta t f(x^m, t_m) \\ f_P^{m+1} &= f(x_P^{m+1}, t_{m+1}) \end{aligned} \tag{F.20}$$

The corrector formulas are also standard Euler, but iterative (implicit). For the first iteration:

$$\begin{aligned} x_C^{m+1}(0) &= x_m + \Delta t f_P^{m+1} \\ f_C^{m+1}(0) &= f[x_C^{m+1}(0), t_{m+1}] \end{aligned} \tag{F.21}$$

The final corrector formula for remaining iterations is similar:

$$\begin{cases} x_C^{m+1}(i) = x^m + \Delta t f_C^{m+1}(i) \\ f_C^{m+1}(i) = f(x_C^{m+1}, t_{m+1}) \end{cases} \quad i = 1, 2, \ldots \tag{F.22}$$

Iterations stabilize this algorithm, allowing larger step-sizes. ***Figure F.1*** illustrates diverging iterative solutions of $\dot{x} = -kx$ using standard Euler, ***Fig. F.2*** converging, stable solutions using the backward Euler.

As noted in Chapter 3, other popular stiff ODE methods include, Gear's method, the Bulerich-Storer and Rosenbrock methods. Gear's method can be found in *Matlab/Simulink*, Bulerich-Storer and Backward Euler in *VisSim* and Rosenbrock in *SAAMII* and *Matlab/Simulink*.

Derivation of the Akaike Information Criterion (AIC)

For context, we describe the data first, as a random (probabilistic) variable function of the parameters $\boldsymbol{p}$. The *data* time series $z_0, z_1, \cdots, z_N$ is treated as a sequence of *random variables* (r.v.), with a known joint probability density function $f(z_0, z_1, \cdots, z_N)$. The marginal density function of each z_i is $f(z_i)$. The z_i depend on unknown parameter (vector) $\boldsymbol{p}$ and we are

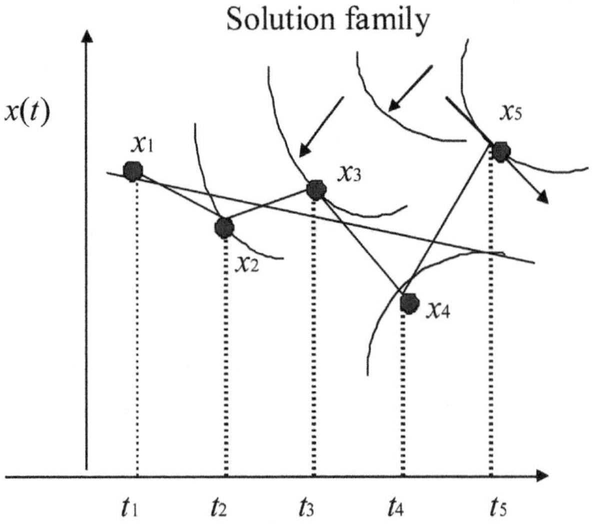

Fig. F.1

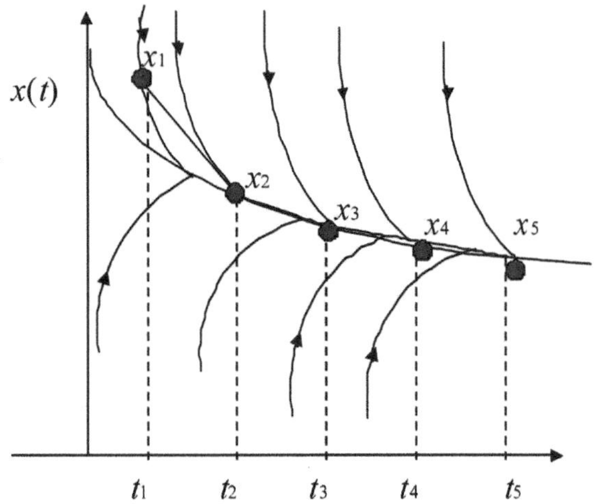

Fig. F.2

interested in the (joint and marginal) conditional probability density functions $f(z_0, z_1, \cdots, z_N|\boldsymbol{p})$, abbreviated $f(\boldsymbol{z}|\boldsymbol{p})$, and $f(z_i|\boldsymbol{p})$. If the r.v. (data) are mutually statistically *independent*, then $f(z_0, \cdots, z_N) = f(z_0) \cdot f(z_1) \cdot \ldots \cdot f(z_N)$ and $f(z_0, \cdots, z_N|\boldsymbol{p}) = f(z_0|\boldsymbol{p}) \cdot f(z_1|\boldsymbol{p}) \cdot \ldots \cdot f(z_N|\boldsymbol{p})$.

The **Akaike Information Criterion (*AIC*)** is a scalar valued function of the conditional densities $f(z_i|\boldsymbol{p})$ based on the **likelihood function** L for the sequence $z_1, z_2, \cdots z_N$ of r.v., each dependent on $\boldsymbol{p}$. This is (classically) defined as $L \equiv \Sigma_{i=1}^{N} f(z_i|\boldsymbol{p})$. The **log likelihood** is then:

$$\text{LL} \equiv \sum_{i=1}^{N} \log_e f(z_i|\boldsymbol{p}). \tag{F.23}$$

The ***AIC*** (Akaike, 1974) is given by:

$$AIC \equiv -2\,\log_e(maximum\ \ likelihood) + 2P^*$$
$$= -2\sum_{i=1}^{N}\log_e f(z_i|\boldsymbol{p}^*) + 2P^* \qquad \text{(F.24)}$$

where $\boldsymbol{p}^*$ is the vector of parameter values that maximizes the (log) likelihood function. Then, given a class of candidate models for the data $z_1, z_2,\ldots,z_N$ each model dependent on a (different) parameter vector $\boldsymbol{p}$, *the "best" model minimizes the AIC.*

Remark: The *AIC* is intuitively appealing, as well as well grounded in information theory. The first term, being (twice) the negative of the log likelihood function of the data dependent on the parameters, should be as "small as possible" if the conditional density functions $f(z_i|\boldsymbol{p})$ are to be as narrow and large as possible (max likely). The second term, two times the number of unknown and identifiable parameters, should also be as small as possible, to satisfy the "principle of parsimony" in modeling – the simplest model that fits the data best. Thus, the *AIC* provides a balance between data fitting precision in the first term and minimum dimensionality in the second.

The AIC can be used for data distributed with any probability density. Most commonly, the z_i are *Gaussian* (normally) distributed,

$$f(z_i) = \frac{1}{\sigma_i\sqrt{2p}}\exp\left\{-\frac{1}{2}\left(\frac{z_i - \mu_i}{s}\right)^2\right\} \qquad \text{(F.25)}$$

where $\mu_i \equiv E(z_i)$ is the mean or expected value of z_i and $\sigma_i^2 \equiv E[(z_i - \mu_i)^2]$ is its variance. If the mean and variance of the data are also unknown, the density of z_i conditioned on μ_i and σ_i^2 has the same form, that is,

$$f(z_i|\mu_i, \sigma_i^2) = \frac{1}{\sigma_i\sqrt{2\pi}}\exp\left\{-\frac{1}{2}\left(\frac{z_i - \mu_i}{\sigma_i}\right)^2\right\} \qquad \text{(F.26)}$$

If all the z_i are *statistically independent*, then

$$f(z|\mu, \sigma^2) \equiv f(z_1, \cdots, z_N|\mu_1, \cdots, \mu_N; \sigma_1^2, \cdots, \sigma_N^2) = \prod_{i=1}^{N}\frac{1}{\sigma_i\sqrt{2\pi}}\exp\left\{-\frac{1}{2}\left(\frac{z_i - \mu_i}{\sigma_i}\right)^2\right\}$$
$$= \left(\frac{1}{2\pi}\right)^{\frac{N}{2}}\frac{1}{\sigma_i\cdots\sigma_N}\exp\left\{-\frac{1}{2}\sum_{i=1}^{N}\left(\frac{z_i - \mu_i}{\sigma_i}\right)^2\right\} \qquad \text{(F.27)}$$
$$= \left(\frac{1}{2\pi}\right)^{\frac{N}{2}}\frac{1}{\sigma_i\ldots\sigma_N}\exp\left\{-\frac{1}{2}(\mathrm{z} - \mu)^{\mathrm{T}}R^{-1}(\mathrm{z} - \mu)\right\}$$

R here is the *covariance matrix* of the data:

$$R \equiv \begin{bmatrix} \sigma_1^2 & \cdots & 0 \\ \vdots & \ddots & \vdots \\ 0 & \cdots & \sigma_N^2 \end{bmatrix} \equiv COV(z) \equiv [E[(z_i - \mu_i)(z_j - \mu_j)]] \equiv [E(e_i e_j)] \tag{F.28}$$

and $e_i \equiv z_i - \mu_i$ are the (r.v.) *residuals.*

Example F.1: ***Conditional Density Function for Multiexponential Models with Normally Distributed Data Residuals***

A sum of exponentials model has the form $y(t) = \Sigma_{k=1}^{n} A_k e^{\lambda_k t}$. A time series of data $z_1, z_2, \ldots, z_N$ can be fitted by this model using nonlinear regression, for measurement model $z_i = y_i + e_i$, assuming the e_i are independent measurement errors (r.v.) with zero mean values, $E(e_i) \equiv 0$. Then $E(z_i) = E(y_i + e_i) = E(y_i) = \Sigma_{k=1}^{n} A_k e^{\lambda_k t_i}$. If we treat the parameters A_k, λ_k as unknowns, we can write $\boldsymbol{p} \equiv [A_1 \ \lambda_1 \ \ldots \ A_n \ \lambda_n]^T$ and $E(z_i|\boldsymbol{p}) = E(y_i|\boldsymbol{p}) = \Sigma_{k=1}^{n} A_k e^{\lambda_k t_i}$. Then the conditional density is

$$f(x_i|\boldsymbol{p}) = \frac{1}{\sigma_i\sqrt{2\pi}} \exp\left[-\frac{1}{2\sigma_i^2}\left[z_i - \sum_{k=1}^{n} A_k e^{\lambda_k t_i} \right]^2 \right] \tag{F.29}$$

Thus, in the conventional nonlinear regression problem, when we seek the best estimates of the A_k and λ_k, we are actually estimating the *mean value* of the function y_i. The variances σ_i^2 also can be treated as unknowns in this problem, i.e. write $\boldsymbol{p} \equiv [A_1 \ \lambda_1 \ A_2 \ \cdots \ A_n \ \lambda_n \ \sigma_1^2 \ \cdots \ \sigma_N^2]^T$ and Eq. (F.29) remains unchanged.

The AIC for Nonlinear Regression

Suppose the z_i are jointly gaussian and $E(z_i) \equiv y_i(\boldsymbol{p})$, with $\boldsymbol{p}$ the unknown parameter vector. We distinguish between two cases here, depending on whether or not the data variances σ_i^2 are known or not.

Case I — σ_i^2 Known

The density function is

$$f(z_i) = \frac{1}{\sigma_i\sqrt{2\pi}} \exp\{ -[z_i - y(\boldsymbol{p})]^2/2\sigma_i^2\} \tag{F.30}$$

From Eq. (F.27), the log likelihood is then:

$$LL(\boldsymbol{p}) = -\frac{1}{2}\left[\sum_{i=1}^{N} \log_e(2\pi\sigma_1^2) + \sum_{i=1}^{N} \frac{1}{\sigma_1^2}(z_i - y_i(\boldsymbol{p}))^2 \right] = -\frac{1}{2}\left[\sum_{i=1}^{N} \log_e(2\pi\sigma_1^2) + WRSS \right] \tag{F.31}$$

where $WRSS$ is the weighted residual sum of squares in regression, with the weights $w_i \equiv 1/\sigma_i^2$. Thus,

$$\max_{p} LL(\boldsymbol{p}) = -\frac{1}{2}\sum_{i=1}^{N} \log_e(2\pi\sigma_i^2) + \frac{1}{2}\min_{p} WRSS \tag{F.32}$$

Finally, from Eq. (F.28), we have

$$AIC = \sum_{k=0}^{N} \log_e(2\pi\sigma_k^2) + WRSS(\boldsymbol{p}^*) + 2P^* \tag{F.33}$$

Now, if different candidate models are being compared with the same set of data via the *AIC*, the first term is the same for all models, because σ_i^2 and N are the same for all. Therefore, it is sufficient to compare only the second two terms, that is, we can consider a slightly modified *AIC*:

$$AIC^*(\boldsymbol{p}^*) = WRSS(\boldsymbol{p}^*) + 2P^* \tag{F.34}$$

Again, this result makes intuitive sense. The first term $WRSS(\boldsymbol{p}^*)$ favors the model with the smallest weighted residual sum of squares; and the second term favors the candidate model with the least number of parameters. In this manner, AIC^* provides a balance between accuracy and parsimony.

Case 2 – The Data Variances $\sigma_i^2 \equiv ks_i^2$ are Known Up to a Proportionality Constant k

In this special case, $LL \equiv LL(\boldsymbol{p},\sigma^2) \equiv LL(\boldsymbol{p},k)$ and the maximum likelihood is taken with respect to k as well as $\boldsymbol{p}$. As before, but for $\sigma_i^2 \equiv ks_i^2$ and s_i^2 known,

$$\begin{aligned} LL &= -\frac{1}{2}\left[\Sigma \log_e 2\pi\sigma_i^2 + \Sigma\frac{e_i^2}{\sigma_i^2}\right] = -\frac{1}{2}\left[\Sigma \log_e 2\pi s_i^2 + \Sigma \log_e k + \Sigma\frac{e_i^2}{ks_i^2}\right] \\ &= -\frac{1}{2}\left[\Sigma \log_e 2\pi s_i^2 + N\log_e k + \frac{1}{k}\Sigma\frac{e_i^2}{s_i^2}\right] \end{aligned} \tag{F.35}$$

Now, define the weights $w_i \equiv 1/s_i^2$ and the last term in Eq. (F.35) becomes $WRSS/k$, or

$$LL = -\frac{1}{2}\left[\Sigma \log_e 2\pi s_i^2 + N\log_e k + \frac{WRSS}{k}\right]$$

Then,

$$\frac{\partial LL}{\partial k} = -\frac{1}{2}\left[\frac{N}{k} - \frac{WRSS}{k^2}\right] \equiv 0$$

has the optimum solution $k^* = WRSS/N$, and

$$\frac{\partial LL}{\partial \boldsymbol{p}} = \frac{\partial(WRSS)}{\partial \boldsymbol{p}} \equiv 0^T$$

provides (by inspection) the same optimum solution WRSS($\boldsymbol{p}^*$) provided by ordinary weighted least squares. Thus,

$$\max_{\boldsymbol{P},k} LL = -\frac{1}{2}\left[\Sigma \log_e 2\pi s_i^2 + N \log_e \frac{WRSS(\boldsymbol{p}^*)}{N} + N\frac{WRSS(\boldsymbol{p}^*)}{WRSS(\boldsymbol{p}^*)}\right]$$

$$= -\frac{1}{2}\left[N \log_e 2\pi s_i^2 + N \log_e WRSS(\boldsymbol{p}^*) + N \log_e\left(\frac{1}{N}\right) + N\right]$$

Therefore, from Eq. (F.28), recognizing that the constant terms are the same for all models, we have for Case 2:

$$AIC = N \log_e WRSS(\boldsymbol{p}^*) + 2P^* \tag{F.36}$$

The Stochastic Fisher Information Matrix (FIM): Definitions & Derivations

The FIM (F) of the data with respect to the parameters (Federov, 1972) is defined as:

$$F \equiv E_{z^*|p}\left\{\left[\frac{\partial \log f(\boldsymbol{z}^*|\boldsymbol{p})}{\partial \boldsymbol{p}}\right]^T \left[\frac{\partial \log f(\boldsymbol{z}^*|\boldsymbol{p})}{\partial \boldsymbol{p}}\right]\right\} \tag{F.37}$$

where the large vector $\boldsymbol{z}^* \equiv [z_1^T z_2^T \ldots z_N^T]^T$ denotes the *composite of data vectors*, $f(\boldsymbol{z}^*|\boldsymbol{p})$ is the probability density function of this data given the parameters, and $E_{z^*|p}$ denotes the expectation over the density $f(\boldsymbol{z}^*|\boldsymbol{p})$. The terms in square brackets are the derivatives of the log likelihood function with respect to the parameters. Note that they are *sensitivity functions*.

We first evaluate F for a simple scalar model, repeated here from Chapter 12, but in more detail. Consider the scalar model:

$$\dot{x} = -ax + u \text{ for } a > 0 \text{ unknown}$$

$$z_i = x_i + e_i \text{ for } i = 0, 1, \ldots, N$$

We show, for e_i gaussian and "white" (white noise), with mean zero and known variance σ_i^2, that:

$$F = \sum_{i=1}^{N}\left(\frac{\partial x_i}{\partial a}\right)^2 \frac{1}{\sigma_i^2} \tag{F.38}$$

The data vectors are jointly gaussian and thus the joint (conditional) density function may be written as:

$$f(\boldsymbol{z}^*|a) = \frac{1}{\sqrt{(2\pi)^N |R^*|}} e^{-\frac{1}{2}(z^*-x^*)^T R^{*-1}(z^*-x^*)} \tag{F.39}$$

where R^* is the covariance matrix of the **white noise** (independent, uncorrelated) data errors and $E(z^*|a) = x^*$. Thus,

$$R^* = \begin{bmatrix} \sigma_1^2 & \cdots & 0 \\ \vdots & \ddots & \vdots \\ 0 & \cdots & \sigma_N^2 \end{bmatrix} \tag{F.40}$$

Evaluate $\log_e f$ for this density function:

$$\begin{aligned} \log_e f &= \log_e[(2\pi)^N |R^*|]^{-\frac{1}{2}} - \frac{1}{2}(z^* - x^*)^{\mathrm{T}} R^{*-1}(z^* - x^*) \\ &= -\frac{N}{2}\log_e(2\pi) - \frac{1}{2}\log_e|R^*| - \frac{1}{2}\sum_{i=1}^{N}\frac{(z_i - x_i)^2}{\sigma_i^2} \\ &= -\frac{N}{2}\log_e(2\pi) - \frac{1}{2}\log_e \prod_{i=1}^{N}\sigma_i^2 - \frac{1}{2}\sum_{i=1}^{N}\frac{[z_i - x_i(a)]^2}{\sigma_i^2} \end{aligned} \tag{F.41}$$

The sensitivity of $\log_e f$ to parameter a is then:

$$\begin{aligned} \frac{\partial \log_e f}{\partial a} &= 0 + 0 - \sum_{i=1}^{N}\frac{1}{2\sigma_i^2}\frac{\partial}{\partial a}[z_i - x_i(a)]^2 \\ &= -\sum_{i=1}^{N}\frac{1}{\sigma_i^2}[z_i - x_i(a)]\left(-\frac{\partial x_i}{\partial a}\right) = \sum_{i=1}^{N}\frac{1}{\sigma_i^2}e_i\frac{\partial x_i}{\partial a} \end{aligned} \tag{F.42}$$

The FIM is then:

$$\begin{aligned} F = E\left\{\left(\frac{\partial \log_e f}{\partial a}\right)^2\right\} &= E\left\{\left(\sum \frac{1}{\sigma_i^2}e_i\frac{\partial x_i}{\partial a}\right)\left(\sum \frac{1}{\sigma_j^2}e_j\frac{\partial x_j}{\partial a}\right)\right\} \\ = E\left\{\sum_i\sum_j \frac{1}{\sigma_i^2\sigma_j^2}e_ie_j\left(\frac{\partial x_i}{\partial a}\right)\left(\frac{\partial x_j}{\partial a}\right)\right\} &= \sum_i\sum_j \frac{1}{\sigma_i^2\sigma_j^2}\left(\frac{\partial x_i}{\partial a}\right)\left(\frac{\partial x_j}{\partial a}\right)E[e_ie_j] \end{aligned} \tag{F.43}$$

Since all e_i are independent,

$$E(e_ie_j) = \begin{Bmatrix} E(e_i^2) = \sigma_i^2 & \text{if } i = j \\ 0 & \text{if } i \neq j \end{Bmatrix}$$

Thus,

$$F = \sum \frac{1}{\sigma_i^4}\left(\frac{\partial x_i}{\partial a}\right)^2\sigma_i^2 = \sum_{i=1}^{N}\frac{1}{\sigma_i^2}\left(\frac{\partial x_i}{\partial a}\right)^2 \tag{F.44}$$

Remark: M is the sum of the squared sensitivity functions weighted by the inverse of the variances of the data (as might have been anticipated).

FIM for Multioutput Models

For a model with vector-valued residuals: $\boldsymbol{e} = \boldsymbol{z} - \boldsymbol{y}$, with r.v. $\boldsymbol{e}$ gaussian distributed and white – with mean zero and covariance R^*, the FIM Eq. (F.37) becomes

$$F = \sum_{k=1}^{N} \left(\frac{\partial y(\boldsymbol{p}, t_k)}{\partial p_i} \right) R^{*-1} \left(\frac{\partial \mathbf{y}(\boldsymbol{p}, t_k)}{\partial p_j} \right)^{\mathrm{T}} = Y^{\mathrm{T}} R^{*-1} Y \tag{F.45}$$

where $Y \equiv [\partial y_i / \partial p_j]$ is the *matrix of output sensitivity functions*. This is the FIM often computed and used to estimate the parameter covariance matrix as: $COV(\boldsymbol{p}) \geq F^{-1}$.

INDEX

Note: Page numbers followed by "*f*" and "*t*" refer to figures and tables, respectively.

E

F

G

Q

R

T

Printed and bound by CPI Group (UK) Ltd, Croydon, CR0 4YY
03/10/2024
01040321-0020

W

Z